The Supramolecular Chemistry of Organic–Inorganic Hybrid Materials

The Supramolecular Chemistry of Organic–Inorganic Hybrid Materials

Edited by

Knut Rurack and Ramón Martínez-Máñez

WILEY

A JOHN WILEY & SONS, INC., PUBLICATION

Published by John Wiley & Sons, Inc., Hoboken, New Jersey
Published simultaneously in Canada

Colored versions of Figures 3.1–3.5, 3.7–3.17, 3.19, 3.20, 3.22, 3.24–3.29, 4.9, 4.12, 5.2, 5.10, 6.1–6.3, 6.5, 6.10, 6.12, 6.14, 7.2–7.5, 7.7–7.11, 8.3, 8.5, 8.9–8.11, 8.13–8.16, 9.3, 9.6, 9.8, 9.10, 10.2, 10.7, 10.10, 11.9–11.11, 11.14–11.16, 11.18, 12.2, 12.4–12.7, 13.1–13.19, 15.2–15.6, 15.9, 16.2, 16.5–16.12, 19.1–19.6, 19.8, 19.12, 19.13, 19.15–19.19, 20.1, 20.4, 20.5, 21.1, 21.2, 21.5–21.10, 21.12, 21.13, 21.15, 21.21, 21.24, 21.25, 22.1, 22.2, 22.5, 23.9, 24.2, 25.1, 25.2, 25.4–25.6 and Schemes 5.1–5.5, 5.8, 5.9, 6.3, 6.4, 20.1 are available from the FTP site.

Library of Congress Cataloging-in-Publication Data:

The supramolecular chemistry of organic-inorganic hybrid materials / edited by Knut Rurack and Ramón Martínez-Máñez
p. cm.
Includes index.
ISBN 978-0-470-37621-8 (cloth)
1. Supramolecular chemistry. 2. Nanochemistry. 3. Nanostructured materials.
I. Rurack, Knut. II. Ramón Martínez-Máñez
QD882.S87 2010
620.1′1—dc22

2009019352

10 9 8 7 6 5 4 3 2 1

Contents

Preface

Supramolecular chemistry, which is basically devoted to the study of the interaction between molecules, and materials chemistry, dealing with the development of solids with specific properties, are two powerful disciplines that have traditionally been poorly interrelated. Only the drive to create ever faster, ever more affordable, and ever more convenient technologies with a myriad of new and advanced features has tempted materials scientists to push the boundaries to ever smaller components and chemical researchers to design ever larger supramolecular structures, both entering into the interfacial zone of nanotechnology and nanochemistry. *Function* is the keyword here, especially when the aim is to design "smart" or "intelligent" materials. Inorganic supports are often inert and do not display many functions. In contrast, organic molecules can have a rich functionality, yet an ensemble of them in a disordered state—whether in solution or randomly adsorbed on a surface—often does not perform as desired. Thus, at the dawn of nanotechnology research in the late 1980s, chemists and materials scientists realized that a combination of their skills might be more successful in approaching the aims than to stay on the beaten tracks. Hence, the rapidly growing world of organic–inorganic hybrid materials emerged, producing nanoscopic matter with a plethora of novel properties and functions. Although the basic idea to *combine inorganic materials with functional organic molecules* might sound straightforward, its realization is connected to several challenges. Of course, "smart" hybrid materials cannot be obtained simply by teaching organic molecules to sit on an inorganic support and solve a Sudoku, play tennis, or sing a number-one hit. Organic functions on inorganic supports have to be organized and have to be orchestrated in their action, which often involves sophisticated chemistry and a structuring and patterning of the inorganic partner at molecular dimensions. At this stage, supramolecular concepts and tools from nanotechnology come into play. Only a clever combination of these strategies and techniques allows the creation of tailor-made "hetero-supramolecular" functionalities, showing new synergisms and unprecedented performance. Compared to the vast amounts of macroscopic devices available in society today and molecular biological processes operative in living organisms, naturally, only considerably few active functions have been realized in the young research field covered here. However, this book shows how a plethora of promising ideas has arisen from the combination of supramolecular chemistry, inorganic solids, and nanotechnology and has already accomplished significant advances in many areas such as sensing, controlled motion, or delivery. The objective here is to provide a compendium that gives an overview of the present state and upcoming challenges in this rapidly growing, highly cross- or interdisciplinary research field.

Flanked by three general chapters, the book is divided into five thematic sections. After a brief introduction to basic terms and concepts in the areas of supramolecular chemistry and hybrid (nano)materials in Chapter 1, Ariga et al. sketch general aspects of supramolecular chemistry related to hybrid materials and structures at the mesoscale in Chapter 2. The chapters collected in the first thematic section on Organic–Inorganic Hybrid Nanomaterials provide an in-depth introduction to synthetic strategies, major properties, characterization techniques, key features, and selected applications of today's most important families of hybrid materials: mesoporous organic–inorganic hybrid silica (Chapter 3 by Hoffmann and Fröba), modified gold nanoparticles and surfaces (Chapter 4 by Pengo and Pasquato), and organically functionalized semiconductor nanocrystals (Chapter 5 by Reiss et al.). Chapter 6 by Gu et al. deals with the functionalization of carbon nanotubes and their bioanalytical and biomedical applications and in Chapter 7, Kitagawa and Noro unfurl the world of porous coordination polymers or MOFs.

The second, third, and fourth sections comprise detailed introductions to design strategies, collective properties, signaling aspects, and/or application-oriented features of a broad variety of hybrid materials in the context of major supramolecular concepts such as assembly, sensing, switching, gating, catalysis, and molecular machinery. Part Two, Improvement of Signaling and Sensing by Organization on Surfaces, basically shows how the organization of molecular entities on surfaces can be used to enhance electrochemical or optical signaling and sensing processes for materials such as gold or silica nanoparticles and quantum dots. In Chapter 8, Polsky et al. report on biomolecular–nanoparticle hybrid systems for electrochemical signaling, followed by Guo et al.'s Chapter 9 on the use of modified nanoparticles for electrocatalysis and as amplifying sensors. The use of gold and silica nanoparticles and quantum dots for optical sensing and imaging applications is demonstrated in Chapters 10, 11, and 12 by Wei and Wei, Mancin et al., and Fernández Argüelles et al., respectively.

The section Control of Supramolecular Nanofabrication, Motion, and Morphology is devoted to state-of-the-art applications of certain supramolecular tools and functions on solid supports. In Chapters 13 and 14, Ling et al. and Rożkiewicz et al. discuss different strategies for the directed self-assembly of nanoparticles on surfaces and give an overview of immobilization and patterning techniques for the attachment of biomolecules on surfaces. The other chapters elaborate on the realization of advanced supramolecular functions on solid scaffoldings related to switchable host–guest chemistry (Chapter 15 by Kong et al.), the control of mass transport by gating in mesoporous hybrid silica (Slowing et al. in Chapter 16), the directed motion of molecular machines on surfaces (Chapter 17 by Credi et al.), and controlled changes in morphology of mesostructured hybrid materials (Dunphy et al. in Chapter 18).

The subsequent section Biomimetic Chemistry presents hybrid solids that were developed according to signaling and recognition processes established in nature (Chapters 19 and 20 by Rurack et al. and Collinson) and concludes with Chapter 21 by Kamperman and Wiesner, who show how nature's strategy of combining biomacromolecules and inorganic skeletons can be transferred to block copolymers and inorganic nanomaterials, yielding hybrid materials with unprecedented properties and functions.

The chapters in the last section have a "wildcard" character, each touching a very particular aspect of the area of nanoscopic hybrid materials in a rather short and concise manner yet each having a background of more general importance. In Chapter 22, Shchukin et al. report on the use of hybrid nanocontainer materials as self-healing anticorrosion coatings. The ways in which adaptive or stimuli-responsive "schizophrenic" materials with a dual character might revolutionize chemo- and bio-sensing systems is discussed by Byrne and Diamond in Chapter 23. After all the chemistry highlighted in the previous chapters, Chapter 24 sheds light on a particular keyword that is often used by scientists in the field themselves as well as by policy makers, interdisciplinarity. Rafols et al. approach an answer to the question of how far hybrid nanomaterials research really is interdisciplinary with scientometric tools, that is, with a bibliometric analysis of the field as presented in the book. Another short chapter written by the editors completes the book by looking ahead on four exemplary research directions that have developed only in the last two to three years, during the making of the book, or that are on the verge of developing in the near future.

Finally, we would like to thank all the authors of this book wholeheartedly for their enthusiastic participation and the effort they made in preparing such interesting and stimulating chapters. To work on this book has been an exciting and pleasurable experience for us, and we are also grateful to Anita Lekhwani and Rebekah Amos of John Wiley & Sons for their belief in the book and for their help in realizing it. Of course, a book like this cannot be complete yet we hope that through this collection of excellent contributions the reader will gain profound insight into this fascinating and emerging research area, will appreciate what has been already achieved by scientists around the globe, will be captivated to keep an eye on the field in the future, and, perhaps, will be inspired to join in and discover future advances in the supramolecular chemistry of organic–inorganic hybrid materials.

KNUT RURACK, BERLIN, D
RAMÓN MARTÍNEZ-MÁÑEZ, VALENCIA, E

Note: Additional color versions of selected figures printed in Chapters 3–13, 15, 16, and 19–25 are available on ftp://ftp.wiley.com/public/sci_tech_med/supramolecular_chemistry.

Editors and Contributors

EDITORS

RAMÓN MARTÍNEZ-MÁÑEZ, Instituto de Reconocimiento Molecular y Desarrollo Tecnológico, Centro Mixto Universidad Politécnica de Valencia – Universidad de Valencia. Departamento de Química, Universidad Politécnica de Valencia, Camino de Vera s/n, E-46022 Valencia, Spain

KNUT RURACK, Div. I.5 Bioanalytics, BAM Bundesanstalt für Materialforschung und –prüfung, Richard-Willstätter-Strasse 11, D-12489 Berlin, Germany

CONTRIBUTORS

DARIA V. ANDREEVA, Max-Planck Institute of Colloids and Interfaces, D-14424 Potsdam, Germany

KATSUHIKO ARIGA, World Premier International (WPI) Research Center for Materials Nanoarchitectonics (MANA), National Institute for Materials Science (NIMS), Japan and Supermolecules Group, National Institute for Materials Science (NIMS), Japan

C. JEFFREY BRINKER, The University of New Mexico/NSF Center for Micro-Engineered Materials, Chemical and Nuclear Engineering Department, Albuquerque, New Mexico, 87131, and Sandia National Laboratories, Advanced Materials Lab, 1001 University Blvd. SE, Albuquerque, New Mexico 87106, USA

SUSAN M. BROZIK, Biosensors & Nanomaterials, Sandia National Laboratories, PO Box 5800, MS-0892, Albuquerque, New Mexico 87185, USA

ROBERT BYRNE, National Centre for Sensor Research, Dublin City University, Dublin 9, Ireland

MARYANNE M. COLLINSON, Department of Chemistry, Virginia Commonwealth University, Richmond, Virginia 23284-2006, USA

JOSÉ M COSTA-FERNÁNDEZ, Department of Physical and Analytical Chemistry, University of Oviedo, c/ Julián Clavería, 8, 33006 Oviedo, Spain

ALBERTO CREDI, Dipartimento di Chimica "G. Ciamician", Università di Bologna, via Selmi 2, 40126 Bologna, Italy

Julia De Girolamo, INAC/SPrAM (UMR 5819 CEA-CNRS-Univ. J. Fourier-Grenoble I), Laboratoire d'Electronique Moléculaire Organique et Hybride, CEA Grenoble, 17 Rue des Martyrs, 38054 Grenoble Cedex 9, France

Ana B. Descalzo, Div. I.5 Bioanalytics, BAM Bundesanstalt für Materialforschung und –prüfung, Richard-Willstätter-Strasse 11, D-12489 Berlin, Germany

Dermot Diamond, National Centre for Sensor Research, Dublin City University, Dublin 9, Ireland

Darren R. Dunphy, The University of New Mexico/NSF Center for Micro-Engineered Materials, Chemical and Nuclear Engineering Department, Albuquerque, New Mexico 87131, USA

María Teresa Fernández Argüelles, Department of Physical and Analytical Chemistry, University of Oviedo, c/ Julián Clavería, 8, 33006 Oviedo, Spain

Michael Fröba, Institute of Inorganic and Applied Chemistry, University of Hamburg, Martin-Luther-King-Platz 6, D-20146 Hamburg, Germany

Lingrong Gu, Department of Chemistry and Laboratory for Emerging Materials and Technology, Clemson University, Clemson, South Carolina 29634-0973

Shaojun Guo, State Key Laboratory of Electroanalytical Chemistry, Changchun Institute of Applied Chemistry, Chinese Academy of Sciences, Changchun 130022, Jilin, China and Graduate School of the Chinese Academy of Sciences, Beijing, 100039, China

Jason C. Harper, Biosensors & Nanomaterials, Sandia National Laboratories, PO Box 5800, MS-0892, Albuquerque, New Mexico 87185, USA

Jonathan P. Hill, World Premier International (WPI) Research Center for Materials Nanoarchitectonics (MANA), National Institute for Materials Science (NIMS), Japan and Supermolecules Group, National Institute for Materials Science (NIMS), Japan

Frank Hoffmann, Institute of Inorganic and Applied Chemistry, University of Hamburg, Martin-Luther-King-Platz 6, D-20146 Hamburg, Germany

Jurriaan Huskens, Molecular Nanofabrication Group, MESA$^+$ Institute for Nanotechnology, University of Twente, P.O. Box 217, 7500 AE, Enschede, The Netherlands

Chunming Jiang, Department of Chemistry and Institutes of Biomedical Sciences, Fudan University, Shanghai, 200433, China

Marleen Kamperman, Department of Materials Science & Engineering, Cornell University, Ithaca, New York 14853

Susumu Kitagawa, Department of Synthetic Chemistry and Biological Chemistry, Graduate School of Engineering, Kyoto University, Katsura, Nishikyo-ku, Kyoto 615-8510, Japan, Kitagawa Integrated Pore Project, Exploratory Research for

Advanced Technology (ERATO), Japan Science and Technology Agency (JST), Kyoto Research Park, 134 Chudoji Minami-machi, Shimogyo-ku, Kyoto 600-8813, Japan and Institute for Integrated Cell-Material Sciences (iCeMS), Kyoto University, 69 Konoe-cho, Yoshida, Sakyo-ku, Kyoto 606-8501, Japan

JILIE KONG, Department of Chemistry and Institutes of Biomedical Sciences, Fudan University, Shanghai, 200433, China

VICTOR S.-Y. LIN, Department of Chemistry, U.S. Department of Energy Ames Laboratory, Iowa State University, Ames, Iowa 50011-3111, USA

XING YI LING, Molecular Nanofabrication Group, MESA$^+$ Institute for Nanotechnology, University of Twente, P.O. Box 217, 7500 AE, Enschede, The Netherlands

FUSHEN LU, Department of Chemistry and Laboratory for Emerging Materials and Technology, Clemson University, Clemson, South Carolina 29634-0973, USA

PENGJU G. LUO, Department of Chemistry and Laboratory for Emerging Materials and Technology, Clemson University, Clemson, South Carolina 29634-0973, USA

FABRIZIO MANCIN, Dipartimento di Scienze Chimiche, Università di Padova, via Marzolo 1, I-35131 Padova, Italy

MARTIN MEYER, SPRU – Science and Technology Policy Research, University of Sussex, Brighton, BN1 9QE, England

MOHAMMED J. MEZIANI, Department of Chemistry and Laboratory for Emerging Materials and Technology, Clemson University, Clemson, South Carolina 29634-0973, USA

HELMUTH MÖHWALD, Max-Planck Institute of Colloids and Interfaces, D-14424 Potsdam, Germany

TOSHIYUKI MORI, Fuel Cell Materials Center, National Institute for Materials Science (NIMS), Japan

LI MU, Department of Chemistry and Institutes of Biomedical Sciences, Fudan University, Shanghai, 200433, China

SHIN-ICHIRO NORO, Research Institute for Electronic Science, Hokkaido University, N20W10, Kita-ku, Sapporo 001-0020, Japan.

JAE-HWAN PARK, SPRU – Science and Technology Policy Research, University of Sussex, Brighton, BN1 9QE, England

LUCIA PASQUATO, Dipartimento di Scienze Chimiche, Università degli Studi di Trieste, via L. Giorgieri 1, 34127 Trieste, Italy

PAOLO PENGO, XEPTAGEN S.p.A., Life Nano-Biotechnology, VEGA Science Park – Building Auriga, Via delle Industrie 9, 30175 Marghera Venezia, Italy

ROSARIO PEREIRO, Department of Physical and Analytical Chemistry, University of Oviedo, c/ Julián Clavería, 8, 33006 Oviedo, Spain

RONEN POLSKY, Biosensors & Nanomaterials, Sandia National Laboratories, PO Box 5800, MS-0892, Albuquerque, New Mexico 87185, USA

ADAM PRON, INAC/SPrAM (UMR 5819 CEA-CNRS-Univ. J. Fourier-Grenoble I), Laboratoire d'Electronique Moléculaire Organique et Hybride, CEA Grenoble, 17 Rue des Martyrs, 38054 Grenoble Cedex 9, France

ISMAEL RAFOLS, SPRU – Science and Technology Policy Research, University of Sussex, Brighton, BN1 9QE, England

BART JAN RAVOO, Organic Chemistry Institute and Center for Nanotechnology (CeNTech), Westfälische Wilhelms-Universität Münster, Corrensstrasse 40, D-48149 Münster, Germany

DAVID N. REINHOUDT, Supramolecular Chemistry and Technology and Molecular Nanofabrication Groups, MESA$^+$ Institute for Nanotechnology, University of Twente, P.O. Box 217, 7500 AE, Enschede, The Netherlands

PETER REISS, INAC/SPrAM (UMR 5819 CEA-CNRS-Univ. J. Fourier-Grenoble I), Laboratoire d'Electronique Moléculaire Organique et Hybride, CEA Grenoble, 17 Rue des Martyrs, 38054 Grenoble Cedex 9, France

GARY J. RICHARDS, Supermolecules Group, National Institute for Materials Science (NIMS), Japan and Fuel Cell Materials Center, National Institute for Materials Science (NIMS), Japan

DOROTA I. ROŻKIEWICZ, Supramolecular Chemistry and Technology, MESA$^+$ Institute for Nanotechnology, University of Twente, P.O. Box 217, 7500 AE Enschede, The Netherlands

FÉLIX SANCENÓN, Instituto de Reconocimiento Molecular y Desarrollo Tecnológico, Centro Mixto Universidad Politécnica de Valencia – Universidad de Valencia. Departamento de Química, Universidad Politécnica de Valencia, Camino de Vera s/n, E-46022 Valencia, Spain

ALFREDO SANZ-MEDEL, Department of Physical and Analytical Chemistry, University of Oviedo, c/ Julián Clavería, 8, 33006 Oviedo, Spain

DMITRY G. SHCHUKIN, Max-Planck Institute of Colloids and Interfaces, D-14424 Potsdam, Germany

SERENA SILVI, Dipartimento di Chimica "G. Ciamician", Università di Bologna, via Selmi 2, 40126 Bologna, Italy

KATJA SKORB, Max-Planck Institute of Colloids and Interfaces, D-14424 Potsdam, Germany

IGOR I. SLOWING, Department of Chemistry, U.S. Department of Energy Ames Laboratory, Iowa State University, Ames, Iowa 50011-3111, USA

BERND SMARSLY, Physikalisch-Chemisches Institut, Justus-Liebig-Universität Gießen, Heinrich-Buff-Ring 58, D-35392 Gießen, Germany

Ya-Ping Sun, Department of Chemistry and Laboratory for Emerging Materials and Technology, Clemson University, Clemson, South Carolina 29634-0973, USA

Paolo Tecilla, Dipartimento di Scienze Chimiche, Università di Trieste, via Giorgeri 1, I-34127 Trieste, Italy

Umberto Tonellato, Dipartimento di Scienze Chimiche, Università di Padova, via Marzolo 1, I-35131 Padova, Italy

Brian G. Trewyn, Department of Chemistry, U.S. Department of Energy Ames Laboratory, Iowa State University, Ames, Iowa 50011-3111, USA

Margherita Venturi, Dipartimento di Chimica "G. Ciamician", Università di Bologna, via Selmi 2, 40126 Bologna, Italy

Ajayan Vinu, World Premier International (WPI) Research Center for Materials Nanoarchitectonics (MANA), National Institute for Materials Science (NIMS), Japan

Erkang Wang, State Key Laboratory of Electroanalytical Chemistry, Changchun Institute of Applied Chemistry, Chinese Academy of Sciences, Changchun 130022, Jilin, China and Graduate School of the Chinese Academy of Sciences, Beijing, 100039, China

Haifang Wang, Department of Chemistry and Laboratory for Emerging Materials and Technology, Clemson University, Clemson, South Carolina 29634-0973, USA

Alexander Wei, Department of Chemistry, Purdue University, 560 Oval Drive, West Lafayette, Indiana 47907, USA

Qingshan Wei, Department of Chemistry, Purdue University, 560 Oval Drive, West Lafayette, Indiana 47907, USA

Ulrich Wiesner, Department of Materials Science & Engineering, Cornell University, Ithaca, New York 14853, USA

Xiurong Yang, State Key Laboratory of Electroanalytical Chemistry, Changchun Institute of Applied Chemistry, Chinese Academy of Sciences, Changchun 130022, Jilin, China and Graduate School of the Chinese Academy of Sciences, Beijing, 100039, China

Abbreviations

2D	two-dimensional
3D	three-dimensional
3PL	three-photon luminescence
4VP	4-vinylpyridine
AA	alginic acid
AA2024	trade name of an aluminum alloy
Ab	antibody
ac	acetate
adip	5,5′-(9,10-anthracenediyl)di-isophthalate
ADP	adenosine diphosphate
AFM	atomic force microscopy
Aib	α-aminoisobutyric acid
Ala	alanine
AM 1.5	air mass 1.5 conditions
AMP	adenosine monophosphate
AMS	α-methylstyrene
ANA	analcime
AOT	sodium bis(2-ethylhexyl)sulfosuccinate
APC	2,4-bis(4-dialkylaminophenyl)-3-hydroxy-4-alkylsulfanylcyclobut-2-enone
apoB-100	ligand of LDL receptor
Apt	aptamer
APTES	3-aminopropyltriethoxysilane (frequently abbreviated as APTS in the literature)
APTMS	3-aminopropyltrimethoxysilane
Asn	asparagine
ASTM	American Society for Testing and Materials
atp	aminoterephthalate
ATP	adenosine triphosphate
ATR	attenuated total reflection (spectroscopy)
ATRP	atom transfer radical polymerization
AuNP	gold nanoparticle
Az	azurin
azpy	trans-4,4′-azopyridine

αCP	affinity contact printing
αG	α-L-guluronic acid (frequently abbreviated as G in the literature)
b	block (in block copolymer nomenclature)
B50-6600	trade name of a block copolymer $EO_{39}BO_{47}EO_{39}$
BaM	barium ferrite nanocrystals
BBDA	*N*,*N*′-bis(4-*tert*-butylphenyl)-*N*,*N*′-bis(4-((*E*)-2-(triethoxysilyl)vinyl)phenyl)biphenyl-4,4′-diamine
bbim	1,3-dibutylimidazolium
bBSA	biotinylated bovine serum albumin
BCAm	benzo[18]crown-6-acrylamide
bdc	benzene dicarboxylate
BDD	boron-doped diamond
BDMS	bis(*tert*-butyldimethylsilyl)
BEA	beta (zeolite beta)
BET	Brunauer–Emmett–Teller
BFc	biferrocene
BHEEEN	1,5-bis[2-(2-(2-hydroxyethoxy)ethoxy)ethoxy]naphthalene
BINAP	2,2′-bis-(diphenylphosphino)-1,1′-binaphthyl (also: binap)
BJH	Barrett–Joyner–Halenda
BO	butyleneoxide
Boc	*tert*-butoxycarbonyl
BP	benzophenone
bpbp	4,4′-bis(4-pyridyl)biphenyl
bpethe	1,2-bis(4-pyridyl)ethene
bpy	bipyridine/yl
Brij 56	trade name of a poly(ethylene glycol hexadecyl ether) detergent and emulsifier
BS^3	bis(sulfosuccinimidyl)suberate
BSA	bovine serum albumin
btapa	1,3,5-benzenetricarboxylic acid tris[*N*-(4-pyridyl)amide]
btb	1,3,5-benzenetribenzoate
btc	benzene tricarboxylate
BTEB	1,4-bis(triethoxysilyl)benzene
BTEBP	4,4′-bis(triethoxysilyl)biphenyl
BTEE	1,2-bis(triethoxysilyl)ethane
BTEMEB	1,4-bis(triethoxysilyl)-2-(1-methoxyethyl)benzene
BTESM	bis-triethoxysilylmethane
BTET	2,5-bis(triethoxysilyl)thiophene
BTEX	benzene, toluene, ethylbenzene, xylenes
BTEY	1,2-bis(triethoxysilyl)ethene
BTME	1,2-bis(trimethoxysilyl)ethane
BTMSEB	1,4-bis(trimethoxysilylethyl)benzene
BTMSPA	bis[3-(trimethoxysilyl)propyl]amine

btt	1,3,5-benzenetristetrazolate
Bu	butyl
bza	benzoate
βM	β-D-mannuronic acid (frequently abbreviated as M in the literature)
CA	cancer antigen
CALNN	Cys-Ala-Leu-Asn-Asn
cAMP	cyclic AMP
CASH	combined assembly by soft and hard (chemistries)
CB	conduction band
CB[6]	cucurbit[6]uril
CBED	convergent beam electron diffraction (patterns)
CBPQT	cyclobis(paraquat-*p*-phenylene)
CCD	charge-coupled device
CCM	Cornell composition of matter (family of materials)
CD	cyclodextrin
CD4	(cluster of differentiation 4) glycoprotein, a co-receptor of the T (thymus) cell receptor
CDA	cell-directed assembly
CE	capillary electrophoresis
CE	cholesterol esterase
CFU	colony forming units
chbt	cyclohexylbutyrate
CHI	chitosan
CHO	Chinese hamster ovary
CMC	critical micelle concentration
CMK-*n*[1]	carbon mesostructured by Korea Advanced Institute of Science and Technology (family of mesoporous carbon materials)
CNT	carbon nanotubes
co	*co*polymer
cod	cyclooctadiene
COx	cholesterol oxidase
CP	conjugated polymer
CP/MAS	cross-polarization MAS (NMR)
CPB	hexadecylpyridinium bromide
CPC	hexadecylpyridinium chloride
CPL	coordination pillared layer
CPO	chloroperoxidase
CPU	central processing unit
CS	core–shell
CT	charge-transfer
CT	computed tomography
CTAB	cetyltrimethylammonium bromide

CTAC	cetyltrimethylammonium chloride
CTES	carboxyethylsilanetriol, sodium salt
CV	cyclic voltammetry
CVD	chemical vapor deposition
CW	continuous wave
CXCR4	CXC chemokine receptor; CXC stands for a C-X-C motif with C = cysteine and X = arbitrary amino acid
Cy3	carbocyanin 3
Cy3.5	carbocyanin 3.5
Cys	cysteine
DA	dodecylamine (frequently abbreviated as DDA in the literature)
dabco	diazabicyclo[2.2.2]octane
DAP	diaminopyrimidine
DAR	diazo resin
DART	direct analysis in real time (ionization technique in MS)
DAT	diaminotriazine
DB24C8	dibenzo[24]crown-8
DBM	dibenzoylmethane
DBS	dibutyl sebacate
DDA	discrete dipole approximation
DDAB	dilauryldimethylammonium bromide
DFT	density functional theory
dhbc	2,5-dihydroxybenzoate
DHLA	dihydrolipoic acid
DIEA	diisopropylethylamine
diglyme	diethylene glycol dimethyl ether
dipn	*N,N*-di(3-aminopropyl)amine
diPyNI	*N,N′*-di-(4-pyridyl)-1,4,5,8-naphthalenetetracarboxydiimide
DMAP	4-dimethylaminopyridine
DMB	(1*R*,2*S*)-(-)-*N*-dodecyl-*N*-methylephedrinium bromide
DMF	*N,N*-dimethylformamide
DMPA	dimyristoylphosphatidyl
DMSO	dimethyl sulfoxide
DNA	deoxyribonucleic acid
DOE	U.S. Department of Energy
DON	1,5-dioxynaphthalene
DOPA	D-/L-3,4-dihydroxyphenylalanine
DOPO-Br	*p*-bromobenzyl-di-*n*-octylphosphine oxide
DOX	doxorubicin
DPAR	4-*n*-dodecyl-6-(2-pyridylazo)phenol
DPC	diphenylcarbazide
DPN	dip-pen nanolithography
dpp	4,4′-diphenylphenanthroline
DPV	differential pulse voltammetry

dpyg	1,2-di(4-pyridyl)glycol
dsDNA	double-stranded DNA
DSG	disuccinimidylglutarate
DsRed	mutant of red fluorescent protein
DT	dodecanethiol
DTA	differential thermal analysis
DTAR	4-*n*-dodecyl-6-(2-thiazolylazo)resorcinol
DTC	dithiocarbamate
dtoa	dithiooxamide
DTPA	diethylenetriaminepentaacetic acid
DTT	dithiothreitol
DVB	divinylbenzene
dx	decylxanthate
DZ	diphenylthiocarbazone
E	energy
e^-	electron
eAuNP	enlarged AuNP
ECL	electrochemiluminescence
EDAC	see EDC
EDC	1-ethyl-3-[(3-dimethylamino)propyl]carbodiimide hydrochloride
EDOT	3,4-ethylenedioxythiophene
EDS	energy-dispersive spectroscopy
EDTA	ethylenediaminetetraacetic acid
EELS	electron energy loss spectroscopy
EFTEM	energy-filtered TEM
EG4	tetra(ethylene glycol)
EG6	hexa(ethylene glycol)
EGDMA	ethylene glycol dimethacrylate
EGFP	enhanced green fluorescent protein
EGFR	epidermal growth factor receptor
EHTES	5,6-epoxyhexyltriethoxysilane
eim	2-ethylimidazolate
EISA	evaporation-induced self-assembly
ELISA	enzyme linked immunosorbent assay
EO	ethyleneoxide
EPR	enhanced permeability and retention
ER	electrorheological
ESR	electron spin resonance
Et	ethyl
EU	European Union
ex	ethylxanthate
F88	trade name for a Pluronic surfactant
F127	trade name for a Pluronic surfactant

FA	folic acid
FAD	flavin adenine dinucleotide
Fc	ferrocene or ferrocenyl
Fc	fragment, crystallizable (of an antibody)
Fc-D	ferrocenyl-tethered dendrimer
FCC	fluid catalytic cracking
FcN	(*R*)-/(*S*)-*N*,*N*′-dimethylferrocenylethylamine
FDA	U.S. Food and Drug Administration
FDM	ferrocene dimethanol
FDMDG	ferrocene dimethanol diethylene glycol
FDTD	finite difference time domain (method)
FDU-*n*	Fudan University (family of mesoporous silicas)
FET	field effect transistor
FGD	flue gas desulfurization
FIA	fluoroimmunoassay
FITC	fluorescein isothiocyanate
Fmoc	fluorenylmethoxycarbonyl
FND	fluorescent ND
FNP	fluorescently doped nanoparticles
FR^+	receptors for FA ligands
FRET	Förster resonance energy transfer (frequently yet inappropriately referred to as fluorescence resonance energy transfer, see IUPAC recommendations)[2]
FSM-*n*	(family of) folded sheet mesoporous (materials)
FTIR	Fourier transform IR
FUM	fumaramide
Fur	ferric uptake regulator
FWHM	full width at half maximum
G	generation (of dendrimer)
GAG	glycosaminoglycan
Gal	galactose
GBP	gold binding polypeptide
GCE	glassy carbon electrode
GCNF	graphitic carbon nanofiber
GEPI	genetically engineered peptides for inorganics
gFND	green FND
GFP	green fluorescent protein
Glu	glutamine
GLYMO	3-glycidyloxypropyltrimethoxysilane
GMP	guanosine monophosphate
GOx	glucose oxidase
GSH	reduced form of glutathione
GSSG	oxidized form of glutathione

h^+	hole
h-h	head-to-head coupling
HDA	hexadecylamine
hdx	hexadecylxanthate
HeLa	Henrietta Lacks (HeLa cells: immortal cell line derived from H.L.'s cervical cancer cells)
HETCOR	heteronuclear correlation (NMR)
HFBI	hydrophobin
HFE	gene responsible for hereditary hemochromatosis
HH	hexadecyl hexadecanoate
HHCC	horse heart cytochrome c
HI	hot-injection method
His	histidine
HIV	human immunodeficiency virus
HMDS	hexamethyldisilazane
HOMO	highest occupied molecular orbital
HOPG	highly ordered pyrolytic graphite
HPLC	high pressure liquid chromatography
HREELS	high-resolution EELS
HRP	horseradish peroxidase
HRS	hyper-Rayleigh scattering
HRTEM	high resolution TEM
HSMA	hydrolyzed poly(styrene-*alt*-maleic anhydride)
HU	heating up method
I	inorganic species
IC_{50}	half-maximum inhibitory concentration
iCCD	intensified charge-coupled device
ICS	ion channel sensor
ICS	isocyanurate
ICTES	(3-isocyanatopropyl)triethoxysilane (frequently abbreviated as ICPES in the literature)
IgE	immunoglobulin E
IgG	immunoglobulin G
IL	ionic liquid
im	imidazolate
IMAC	immobilized metal ion affinity chromatography
IR	infrared
IRMOF-n^3	isoreticular metal–organic framework
ISE	ion-selective electrode
ISI	ISI Web of Knowledge^{SM} (database by Thomson Reuters)
ITO	indium tin oxide
IUPAC	International Union of Pure and Applied Chemistry

KIT-*n*[4]	(family of) large mesopore *Fm3 m* silica(s)
LA	lauric acid
LB	Langmuir–Blodgett
LbL	layer-by-layer
LC	liquid chromatography
LC	liquid crystal/crystalline
LCST	lower critical solution temperature
LD	lethal dose
LDL	low density lipoprotein
LD-SAM	low density SAM
LED	light emitting diode
Leu	leucine
LOD	limit of detection
LPEI	linear poly(ethylene imine)
LSPR	localized surface plasmon resonance
LTA	Linde type A (zeolite A)
LUMO	lowest unoccupied molecular orbital
M41S[5,6]	family of mesoporous molecular sieves
MA	myristic acid
MAA	methacrylic acid
mAb	monoclonal Ab
MagMOON	magnetically-modulated optical nanoprobes
MAL	maleamide
MALDI-TOF	matrix assisted laser desorption/ionization time-of-flight (MS)
MAS	magic angle spinning NMR
MAXSORB	trade name of an activated carbon adsorbent
MBP	maltose binding protein
MBP-zb	maltose binding protein with a positive leucine zipper domain
MC	merocyanine
MCF-7	human breast cancer cells
MCM-*n*	Mobil Composition of Matter (family of mesoporous silicas)
MDMO-PPV	poly[2-methoxy-5-(3′,7′-dimethyloctyloxy)-1,4-phenylenevinylene]
Me	methyl
MEH-PPV	poly(2-methoxy-5-(2′-ethyl-hexyloxy)-1,4-phenylene-vinylene)
MeOTAD	2,2′,7,7′-tetrakis(*N*,*N*-di-*p*-methoxyphenylamine)9,9′-spirobifluorene
MFI	Mobil five (ZSM-5; zeolite)
MHDA	16-mercaptohexadecanoic acid (frequently abbreviated as MHA in the literature)
MHT	1-(6-mercaptohexyl)thymine
MHV68	murine gammaherpesvirus 68

MIL-*n*[7]	Materials of Institut Lavoisier (family of metal–organic frameworks)
mim	2-methylimidazolate
MIMIC	micromolding in capillaries
MIP	molecularly imprinted polymer
MM	methyl myristate
MNP	magnetic nanoparticles
MO	molecular orbital
MOF	metal–organic framework
MOR	methanol oxidation reaction
MPA	mercaptopropionic acid
MPC	monolayer protected cluster
MPL	multiphoton luminescence
MPMD	(mercaptopropyl)methyldimethoxysilane
MPTES	3-mercaptopropyltriethoxysilane
MPTMS	3-mercaptopropyltrimethoxysilane (frequently abbreviated as MPMS in the literature)
MRI	magnetic resonance imaging
MS	mass spectrometry
MSH	α-melanocyte stimulating hormone
MSN	mesoporous silica nanoparticles
MSU-*x*	Michigan State University (family of mesoporous silicas)
mTERT	murine telomerase reverse transcriptase
MTG	methanol-to-gasoline
MTMOS	methyltrimethoxysilane
MTX	methotrexate
MUA	11-mercaptoundecanonic acid
MWNT	multiwalled carbon nanotube
my	myristate
μCP	microcontact printing
μFN	microfluidic network
μTM	microtransfer molding
n	wildcard character for numbers
NBIC	nanotechnology, biotechnology, information technology, and cognitive science
NBTC	Nanobiotechnology Center (Cornell University)
NC	nanocrystal
ND	nanodiamond
Nd:YAG	neodymium-doped yttrium aluminium garnet (laser)
ndc	2,6-naphthalenedicarboxylate
NDR	2-nitro-*N*-methyl-4-diazonium-formaldehyde resin
NEST	New and Emerging Science and Technology
NEXAFS	near-edge x-ray absorption fine structure (spectroscopy)

NHA	carbon nanohorn aggregate
NHS	*N*-hydroxysuccinimide/yl
$NHSC_{11}SH$	11-mercaptoundecanoyl-*N*-hydroxysuccinimide ester
NIL	nanoimprint lithography
NIR	near infrared
NLO	nonlinear optics
NLS	nuclear localization sequence (peptide sequence for nuclear targeting)
NMR	nuclear magnetic resonance
NN	next-neighbor (interactions)
NOM	nano-on-micro
NP	nanoparticle
NR	nanorod
NTA	nitrilotriacetic acid
NTS	nonadecenyltrichlorosilane
NV	nitrogen vacancy (defect sites)
NVOC	nitroveratryloxycarbonyl
NVP	naphthalenylvinylpyridine
OA	oleic acid
OAm	oleylamine
OCT	optical coherence tomography
ODA	octadecylamine
ODE	1-octadecene
ODPA	octadecylphosphonic acid
ODT	octadecanethiol
OFMS	organic-functionalized molecular sieve
ON	oxynaphthalene
OPTA	*o*-phthalic hemithioacetal
OR	alkoxy group
ORMOSIL	organically modified silicate
ORR	oxygen reduction reaction
ORTEP	Oak Ridge thermal ellipsoid plot
OT	octanethiol
OTAB	octadecyltrimethylammonium bromide
OTAC	octadecyltrimethylammonium chloride
OTf	trifluoromethanesulfonate
OTS	octadecyltrichlorosilane
ox	oxalate
P123	trade name for a Pluronic block copolymer
P3HT	poly(3-hexylthiophene)
P3(ODAP)HT	poly(3-(6-oxy-2,4-diaminopyrimidine)hexylthiophene)
P3TOPA	poly(3-(3′-thienyloxy)propyltrimethylammonium)

P4VP	poly(4-vinylpyridine)
PA	polyaniline
PAA	poly(acrylic acid)
PAG	photoacid generator
PAH	polycyclic aromatic hydrocarbon
PAMAM	poly(amido amine)
PANI	polyaniline
pAP	*p*-aminophenol
PAR	4-(2-pyridylazo)-resorcinol
PAT	photoacoustic tomography
PB	Prussian blue
PBS	phosphate buffered saline
PC	personal computer
PC	photovoltaic cell
PCBM	(1-(3-methoxycarbonyl)propyl-1-phenyl[6,6]C_{61})
PCP	porous coordination polymer
PCPM	porous coordination polymer magnets
PDA	personal digital assistant
PDA	poly(diacetylene)
pdc	pyridine-3,5-dicarboxylate
PDDA	poly(diallyldimethylammonium)
PDITC	1,4-phenylenediisothiocyanate
PDMAEMA	poly(dimethylaminoethyl methacrylate)
PDMS	poly(dimethylsiloxane)
PEBBLE	probes encapsulated by biologically located embedding (more recently: photonic explorers for bioanalysis with biologically localized embedding)
PEDOT	poly-(3,4-ethylenedioxythiophene)
PEELS	parallel electron energy loss spectroscopy
PEG	poly(ethylene glycol)
PEG-Si	2-[methoxypoly(ethyleneoxy)propyl]trimethoxysilane
PEI	poly(ethylene imine)
PEO	poly(ethylene oxide)
PEP	poly(ethylene-*alt*-propylene)
PER	photoelectrorheological (effect)
PET	positron emission tomography
PG	adaptor protein G
PG-zb	Fc domain of PG
Ph	phenyl
phim	benzimidazolate
PHMA	poly(*n*-hexyl methacrylate)
PI	polyisoprene
pia	*N*-4-pyridylisonicotinamide
PIXIES	protein imprinted xerogels with integrated emission sites

PL	phospholipid
PL	photoluminescence
PLA	poly(lactic acid)
PLGA	poly(DL-lactic acid-*co*-glycolic acid)
PLO	poly(L-ornithine)
PMAA	poly(methacrylic acid)
PMDs	periodic mesoporous dendrisilicas
PMMA	poly(methyl methacylate)
PMOs	periodic mesoporous organosilicas
PNA	peptide nucleic acids
PNIPAAm	poly(*N*-isopropyl acryl amide)
PNMOF	polymer-nucleated metal–organic framework
pNP	*p*-nitrophenol
PO	propyleneoxide
POP	polyolefin plastomer
ppm	parts per million
PPO	poly(propylene oxide)
PPV	poly(*p*-phenylene vinylene)
ppy	2,2′-phenylpyridyl
PPY	polypyrrole
PQD	polymer-coated QD
Pr	propyl
PR	pyrogallol red
PSA	prostate specific antigen
PSD	pore size distribution
PSf	polysulfone
PSMA	prostate specific membrane antigen
pSPNIPAAm	spirobenzopyran-functionalized poly(*N*-isopropylacrylamide)
PSS	poly(styrene sulfonate)
PSS-MA	PSS containing maleic acid groups
PT	polythiophene
PTAA	poly(3-thiophene acetic acid)
PTMOS	phenyltrimethoxysilane
ptmtc	(4,4′,4″-tricarboxydodecachlorotriphenyl)methyl radical
ptz	5-(3-pyridyl)tetrazolate
PVDF	poly(vinylidene fluoride)
PVMP	poly(vinyl-*N*-methylpyridinium)
PVP	poly(vinylpyrrolidone)
PVSA	poly(vinylsulfonic acid)
PXRD	powder XRD
py	pyridine
pyz	pyrazine
pzdc	pyrazine-2,3-dicarboxylate

QCM	quartz crystal microbalance
QD	quantum dot
QDL	quantized double layer (charging)
QI	*p*-quinone imine
QSY9	trade name for an efficient xanthene-based fluorescence quencher dye
QY	quantum yield
R	alkyl or aryl group
RAFT	reversible addition-fragmentation chain transfer (polymerization)
RAIRS	reflection absorption infrared spectroscopy
RAM	random access memory
RDE	rotating disk electrode
REM	replica molding
RES	reticuloendothelial system
RGD	arginine-glycine-aspartic acid peptide sequence
RHO	zeolite rho
RI	refractive index
RITC	rhodamine isothiocyanate
RIU	refractive index unit
RNA	ribonucleic acid
ROS	reactive oxygen species
RTP	room temperature phosphorescence
S	surfactant
SA	stearic acid
SAED	selected area electron diffraction (patterns)
SAM	self-assembled monolayer
SAMIM	solvent-assistant microcontact molding
SATI	self-assembly, transfer, and integration
SATP	*N*-succinimidyl *S*-acetylthiopropionate
SAv	streptavidin
SAXS	small angle X-ray scattering
SBA-*n*	Santa Barbara Amorphous (family of mesoporous silicas)
SBU	secondary building unit
SC-ISE	solid contact ISE
SCO	spin-crossover
SDA	structure-directing agent (frequently: structure-directing additive)
SE	secondary electron
SEB	staphylococcal enterotoxin B
SECM	scanning electrochemical microscopy
SEF	surface-enhanced fluorescence
SEM	scanning electron microscopy

SERS	surface-enhanced Raman scattering
SFM	scanning force microscopy
SHG	second harmonic generation
SILAR	successive ion layer adsorption and reaction
sip	5-sulfoisophthalate
siRNA	short interfering RNA
SNARF-1	trade name of a seminaphthorhodamine fluorophore
SNOM	scanning near-field optical microscopy
SNP	single nucleotide polymorphism
SNU-*n*	Seoul National University (family of mesoporous carbon materials)
SOD	sodalite
SOMC	surface organometallic chemistry
SP	spiropyran
SP-CP	scanning probe contact printing
SPG	schizophyllan
SPL	scanning probe lithography
SPR	surface plasmon resonance spectroscopy
SQ	squaraine
SQUID	superconducting quantum interference device
SR	mercapto group
ssDNA	single-stranded DNA
SSMCC	sulfosuccinimidyl 4-(*N*-maleimidomethyl)cyclohexane-1-carboxylate
s-SPG	single chain schizophyllan
st	stearate
STEM	scanning transmission electron microscopy
STM	scanning tunneling microscopy
STP	standard temperature and pressure equivalent
STS	scanning tunneling spectroscopy
SVET	scanning vibrating electrode technique
SWNT	single-walled carbon nanotube
T7	bacteriophage T7 promoter, sequencing primer
TA	thioglycolic acid or mercaptoacetic acid
TBP	tributylphosphine
TCFePc	tetra-carboxyl-substituted iron phthalocyanine
TCNQ	7,7,8,8-tetracyano-*p*-quinodimethane
TDDFT	time-dependent DFT
TDPA	tetradecylphosphonic acid
TEAF	tetraethylammonium fluoride
TEG	triethylene glycol
TEGME	triethylene glycol methylether
TEM	transmission electron microscopy
TEMABr	triethylmethylammonium bromide

TEOS	tetraethoxysilane (frequently: tetraethyl orthosilicate)
TFS	tetrafluorosuccinimide
TG	thermogravimetry
THF	tetrahydrofuran
THG	third harmonic generation
THH	tetrahexahedral
thpdc	tetrahydroxypyrenedicarboxylate
TI	trypsin inhibitor
TIR	total-internal reflection
TLCT	true liquid-crystal templating
TMOS	tetramethoxysilane (frequently: tetramethyl orthosilicate)
TMPyp	α,β,γ,δ-tetrakis(1-methylpyridinium-4-yl)porphine/ porphyrinato
TMR	tetramethylrhodamine
TMS	trimethylsilyl
TMSCl	trimethylsilyl chloride
TMV	tobacco mosaic virus
TNF-α	tumor necrosis factor-alpha
TNT	2,4,6-trinitrotoluene
TOA	trioctylamine
TOF SIMS	time-of-flight secondary ion mass spectrometry
TOP	trioctylphosphine
TOPO	trioctylphosphine oxide
TPA	tripropylamine
TPAF	tetrapropylammonium fluoride
TPAOH	tetrapropylammonium hydroxide
TPL	two-photon luminescence
TPPS	tetraphenylporphine tetrasulfonic acid
tpt	2,4,6-tris(4-pyridyl)triazine
tresyl	2,2,2-trifluoroethanesulfonyl
Tris	tris(hydroxymethyl)aminomethane
TRITC	tetramethylrhodamine isothiocyanate
Try	trypsin
t-SPG	triple helix schizophyllan
TT	name of a certain phase (Chapter 21), no particular abbreviation
TTF	tetrathiafulvalene
Tween-80	trade name of a surfactant/emulsifier derived from polyoxyethylene sorbitan monooleate
UDD	ultradispersed diamond
UKON	University of Konstanz (family of PMOs)
UMC	(coordinatively) unsaturated metal center
UPD	underpotential deposition
UV-S1	sunscreen UV absorber
UW-1	SWNT-binding peptide

VB	valence band
VCT	variable contact time (CP/MAS NMR)
VOC	volatile organic compound
VTES	vinyltriethoxysilane
W	wurtzite
WGA-TRITC	wheat germ agglutinin conjugated to tetramethylrhodamine isothiocyanate
x	wildcard character for letters
XPS	X-ray photoelectron spectroscopy
XRD	X-ray diffraction
ZB	zinc blende
ZIF	zeolitic imidazolate framework
ZOL	zeolites with organic group as lattice
%ID	percent injected dose

REFERENCES

1. H. J. Shin, R. Ryoo, M. Kruk, M. Jaroniec, *Chem. Commun.* **2001**, 349–350.
2. S. E. Braslavsky, *Pure Appl. Chem.* **2007**, *79*, 293–465.
3. M. Eddaoudi, J. Kim, N. Rosi, D. Vodak, J. Wachter, M. O'Keeffe, O. M. Yaghi, *Science* **2002**, *295*, 469–472.
4. F. Kleitz, D. Liu, G. M. Anilkumar, I.-S. Park, L. A. Solovyov, A. N. Shmakov, R. Ryoo, *J. Phys. Chem. B* **2003**, *107*, 14296–14300.
5. C. T. Kresge, M. E. Leonowicz, W. J. Roth, J. C. Vartuli, J. S. Beck, *Nature* **1992**, *359*, 710–712.
6. J. S. Beck, J. C. Vartuli, W. J. Roth, M. E. Leonowicz, C. T. Kresge, K. D. Schmitt, C. T. W. Chu, D. H. Olson, E. W. Sheppard, *J. Am. Chem. Soc.* **1992**, *114*, 10834–10843.
7. C. Sassoye, T. Loiseau, F. Taulelle, G. Férey, *Chem. Commun.* **2000**, 943–944.

Chapter 1

Hybrid (Nano)Materials Meet Supramolecular Chemistry: A Brief Introduction to Basic Terms and Concepts

KNUT RURACK AND RAMÓN MARTÍNEZ-MÁÑEZ

1.1 INTRODUCTION

Standing at the crossroads of two broad, diverse, and rapidly expanding research areas such as *supramolecular chemistry* and *hybrid (nano)materials*, the latter being deeply rooted in the even vaster area of *nanotechnology*, the intention of this brief introductory chapter is to focus on and collect some of the basic terms, concepts, and approaches of these nanochemical sciences. During the last decades, *supramolecular chemistry* and *nanomaterials*, with all the "hybridness" and "technology" attached to the latter, became increasingly influential and their impact on various disciplines in the natural sciences grew strongly, spreading beyond the traditional perception of chemistry, physics, and biology. Naturally, such cross- or interdisciplinarity (see Chapter 24 for a discussion) manifested, for instance, in an amalgamation of molecular physics and cellular chemistry to create, say, a novel stimuli-responsive drug delivery

The Supramolecular Chemistry of Organic–Inorganic Hybrid Materials. Edited by Knut Rurack and Ramón Martínez-Máñez

system often involves a phase in which the different researchers participating in such a project have to settle terms and definitions and have to agree on a common "language," because similar terms or even identical abbreviations can have different meanings in different scientific (sub)communities. For illustrative examples, one does not have to embark on lengthy searches but can simply take a look at the List of Abbreviations of this book. Attempting the unequivocal use of abbreviations book-wide, we were rather soon confronted with the impossibility of doing so. Whereas for some abbreviations used in various chapters with different meanings the unification was simple (e.g., MAA stood formerly for *methacrylic acid* and *mercaptoacetic acid*, the latter being now abbreviated as TA for its synonym *thioglycolic acid*),* for others, it was not possible. In several cases, both abbreviations have become so common in different areas of the chemical sciences, the natural sciences, or even society that we had to accept their "schizophrenic nature" and kept them with two different meanings in the book. Examples include

Fc	*ferrocene* (or ferrocenyl)	vs.	*fragment, crystallizable* (of an antibody)
PL	*photoluminescence*	vs.	*phospholipid*
PDA	*poly(diacetylene)*	vs.	*personal digital assistant*
CT	*charge-transfer*	vs.	*computed tomography*
PC	*photovoltaic cell*	vs.	*personal computer*

Therefore, the key features of supramolecular chemistry and nanomaterials important in the present context are introduced in the next two sections.

1.2 TERMS AND CONCEPTS IN SUPRAMOLECULAR CHEMISTRY

Historically, the term *supramolecular chemistry* has been put onto the chemical map by J.-M. Lehn in the middle of the 1980s,[1] and received broad coverage from 1987 on, the year Lehn, D. J. Cram and C. J. Pederson won the Nobel prize in chemistry.[2–4] In short, Lehn defined the term as the "chemistry of molecular assemblies and of the molecular bond," which is often used in its condensed form of the "chemistry beyond the molecule." According to Lehn, the main characteristics of a supramolecular ensemble are a certain degree of order and/or symmetry of the packing as well as an interaction between the subunits and/or specific types of intermolecular interactions. Since Lehn distinguishes between the *supramolecular complex* and the *covalent molecule*, these interactions are supposed to be *noncovalent* in nature. With regard to function or properties, *supramolecular systems* show certain abilities for recognition, catalysis, and/or transport. In contrast, the *covalent molecule* has a unique chemical nature, shape, and polarity (and chirality) as well as redox, optical, and/or magnetic properties. In this sense, the covalent grafting of, for instance, a receptor unit onto a silica nanoparticle does not create a supramolecular ensemble, but a hybrid material. The latter nonetheless may exert a supramolecular function upon binding of a guest (see, e.g., Chapter 11) or may provide a suitable platform to build a supramolecular

*This is actually a more complicated case since the second most obvious abbreviation, TGA, also leads to a conflict with a term widely used in the materials sciences, *thermogravimetric analysis*.

ensemble (see, e.g., Chapter 13). The same holds for molecules attached to a metal surface through bonds that have a pronounced covalent character, such as the gold–thiol bond (see, e.g., Chapters 4 and 8). These hybrids are not supramolecular in nature but *composite* or *hybrid* and can be used in a supramolecular context as discussed for instance in Chapters 15 and 23. In addition, complex molecules comprising various active and/or addressable subunits in a covalent fashion such as for example **10** in Figure 17.9 in Chapter 17 are also not supramolecular ensembles, but the supramolecular function is only generated by using, in this case, the ferrocenyl-appended cyclodextrin that interacts with different parts of switchable **10** through hydrophobic forces, depending on the state the molecular machine is in.

Closely connected to the definition of Lehn is the term *host–guest chemistry* as laid out by Cram.[3] The same forces are at play and the main constraint with respect to the more general concept of supramolecular chemistry pertains to the binding site features that the two partners bring into the complex: whereas the guest can possess *divergent* binding sites (extreme cases: spherical guests such as halide anions or metal cations), the host has to be equipped with *convergent* binding sites (simple cases: a crown ether as host for sodium ions[5] or an oligopyrrole for the binding of chloride ions).[6] Very much related to host–guest complex formation is the term *molecular recognition*, which implies that a certain receptor can discriminate between a number of potential guests, that is, recognizes the designated guest.[7] Historically, all the concepts that are based on molecular specificity in binding are rooted in the biochemical background of the *biological receptor*[8] and the principle of *lock-and-key*[9] (initially coined for enzyme–substrate interactions; however, later modified for these specific interactions to the principle of the *induced fit*).[10] P. Ehrlich and J. N. Langley developed the concept of the *biological receptor* from the observations they made during their studies of how toxins and drugs can influence certain cellular or synaptic functions (cited in Reference 8) and their dose-effect experiments foreshadowed analytical trends such as competitive assays or displacement assays, the latter for instance being discussed here in Chapter 19. Regarding host–guest chemistry and molecular recognition, at the opposite end of the scale of complexity of hosts such as crown ethers and oligopyrroles thus lie antibodies[11,12] or their synthetic analogs, molecularly imprinted polymers.[13,14] The latter are featured in this book in their organic–inorganic hybrid form in Chapter 20.

A typical feature of larger and more complex hosts is that they usually bind the guest through *multiple interactions* in which the binding sites act in a concerted fashion, that is, the overall binding constant between host and designated guest is often the sum or even larger than the sum of the binding constants of the individual binding sites with the guest. Such *cooperative effects* are also found in simple systems, for example, for crown ethers or ethylenediamine tetraacetic acid (EDTA) and metal ions, and are termed here *macrocyclic effect*[15] or *chelate effect*.[†16] To be able to

[†]Note that in analogy to cofactors and substrates in enzyme chemistry, the general definition of *cooperativity* usually concerns two different binding sites for two different guests and includes *positive* as well as *negative cooperativity*: Cooperativity results when occupation of a given binding site leads to a change on the binding features of the other site(s), making binding either easier or more difficult [J. M. Lehn, *Supramolecular Chemistry* (Weinheim: VCH, **1995**), 141 ff].

bind a certain guest selectively, not only does the host have to possess the adequate binding sites, but the latter have to be arranged in an optimal fashion. This means that in supramolecular systems, *preorganization* often plays a critical role. Within the context of this book, *preorganization* plays an outstanding role because the use of inorganic, nanoscopically sized, structured or textured supports to preorganize (bio)organic functionalities in such a way that they can be used in supramolecular recognition or signalling is one of the main driving forces in the field of hybrid materials design (see, e.g., Chapter 14). Ultimately, the preorganization of the binding sites of the host leads to a *complementarity* of host and guest that entails the unique response.

Having introduced the major principles of complexation, association, and organization, it is important to review the physicochemical forces that lead to supramolecular ensemble formation. As mentioned above, supramolecular interactions are by definition noncovalent. In the order of the polarity of the partners involved, they comprise *ionic* or *electrostatic* interactions,[17,18] *ion–dipole* interactions,[19] *dipole–dipole* interactions,[20] (ionic) *hydrogen bonding*,[21–23] *cation-π*[24] and *anion-π* interactions,[25] *π-π stacking*,[26] interactions based on *van der Waals forces*[27] and *hydrophobic effects*,[28] that is, the specific exclusion of polar solvents, in particular water, with specific *packing effects* in the solid state taking a special position.[29] Whereas polar and electrostatic forces are the key players to hold together tightly the porous coordination polymers discussed in Chapter 7, van der Waals forces and solvent exclusion effects are decisive forces in porous hybrid materials that mimic biological receptors like binding pockets of proteins (see, e.g., Chapter 19). Because of the special physical parameters involved, a delicate balance of (very) polar and (very) apolar interactions usually governs the possibilities of self-assembly at interfaces and the behavior of objects such as *self-assembled monolayers* (SAMs)[30] or *Langmuir–Blodgett films* (LB films).[31]

In terms of the physicochemical control of the selectivity in supramolecular systems, both *thermodynamic* and *kinetic gain* are important. Thermodynamic aspects most of all govern a steady-state discrimination between different guests, the best known examples being perhaps the binding of molecular oxygen to hemoglobin in the presence of a number of potentially competing species such as molecular nitrogen, water, and carbon dioxide, partly being present in rather high excess.[32] Kinetic driving forces are more important in catalytically acting systems such as enzymes since here often the binding of the designated target is comparatively weak, yet the kinetics of turnover for the guest of choice are much faster than for other potential guests as well as for the same reaction in liquid solution.[33] The loose binding is important because the educt and the product of the catalytic reaction commonly have different shapes, topologies, and chemical structures and the host has to rearrange during the reaction. If enzymes would be preorganized in a rigid way to achieve stronger binding of the (educt) guest, the reaction would be unfavorable.

Having mentioned self-assembled objects as examples already above in the context of supramolecular forces, this paragraph briefly introduces *self-assembly*[34] and *templating* or the *template effect*,[35] both concepts being deeply rooted in supramolecular chemistry. While the term *self-assembly* relates to a system that performs spontaneously several steps in a single operation to create a supramolecular ensemble, the DNA double helix being perhaps the most prominent biochemical example, *templating* or *template-directed synthesis* requires a certain, temporary or permanent, species that aids the conversion of reactants into a product through the formation of a supramolecular complex, for example, an alkali metal ion that promotes oxacrown synthesis.[36] Besides objects such as molecular rectangles,[37] capsules[38] or rosettes,[39] self-assembly also includes mechanically interlocked architectures such as catenanes and rotaxanes,[40] extremely relevant in the framework of hybrid molecular machines (see Chapter 17), as well as coordination polymers (see Chapter 7). On the other hand, recent advancements in the supramolecular chemistry of interfaces have propelled the typical design of functional monolayers to another level of sophistication, allowing the construction of programmed 3D architectures through *layer-by-layer* (LbL) assembly techniques.[41] Several chapters in this book discuss the many facets related to layered assembly in detail, including Chapter 2. In addition, micelles, liposomes, and vesicles can be considered as dynamic supramolecular assemblies and gained paramount importance in the biological context of the protocell.[42,43] Finally, besides the role of templates in synthesis, templating of course is intrinsically connected to molecular imprinting and its use in chemical sensing.[13,14]

1.3 TERMS AND CONCEPTS RELATING TO NANOMATERIALS

As fuzzy and buzzwordy as the words may sound, *nanomaterials*, *nanochemistry*, and *nanotechnology*, including their nanobiological variations, form the cornerstones of the research map that spans the size range between 1 nm and 1 μm.[44,45] In view of traditional chemistry, 1 nm is considered as being large, whereas the traditional engineer or physicist who was educated in the area of microtechnology regards 1 μm as small;[46,47] the interface is the main playground of the nanosciences. In particular, the term *nanomaterials* encompasses inorganic, (bio)organic, or hybrid objects of nanometric size and bulk material that is structured or patterned at nanometric dimensions. Typical terms of classification include dimensionally rather constrained objects such as *nanoparticles*, *nanocrystals*, and *nanotubes*,[48] dimensionally less defined objects such as *nanolayers* and *nanofibers*, and nanosized voids like *nanopores* or *nanocavities*. Some classes can be subdivided into certain categories such as, for instance, nanoparticles into *nanospheres*, *nanorods*,[49] or *nanocubes*.[50] It is interesting to note that the term *nanoparticle*,[51–53] even in a classic supramolecular sense, that is, as colloidal drug delivery system, was established in the pharmaceutical community even a decade before the term *nanotechnology* saw a first significant increase in published papers in the late 1980s, after the release of Drexler's visionary book in 1986.[‡45] Moreover, the functionalization of nanoparticles with complex hosts such as antibodies was also already realized before all the influential treatises on nanotechnology appeared.[54] Returning once more to the triangle of *nanomaterials*, *nanochemistry*, and

[‡]Colloidal metal nanoparticles, however, have been known for centuries since gold and silver nanoparticles are responsible for the brilliant reds and yellows seen in stained glass windows. Scientific research on metal nanoparticles dates back at least to M. Faraday; see, for example, M. Faraday, *Philos. Trans.* **1857**, *147*, 145–181.

nanotechnology, today, *nanomedicine* most likely constitutes the fourth cornerstone and is perhaps the fastest developing one of the nano-areas, which was foreseen in the late 1980s[55] and closes the loop to the early development of nanoparticles in the pharmaceutical sciences.

Among the nanoscopic objects, terms such as tubes, spheres, rods, and cubes are self-explanatory and, as described in many chapters of the book, have today been obtained from many different materials, including carbon, silica, or metals; see, for example, Chapters 3, 4, and 6. These objects are commonly characterized by sizes in the lower nanometer regime. *Nanofibers*, which can easily cross the border to micrometer-sized objects, can have very diverse features, not only in size but also in composition, ranging from pure carbon materials (e.g., graphitic fibers)[56] via polymer assemblies[57] to metal oxide fibers.[58] Regarding topology, the term *nanowire* is almost synonymous, yet it is basically used in connection with either metallic or semiconductor materials,[59,60] which can, however, be also assembled by supramolecular templating,[61] or for *molecular wires*. Traditionally, molecular wires are conjugated polymers designed for molecular electronics,[62] yet they can be equipped with adequate groups to perform tasks of molecular recognition and act as sensory devices[63] or they can show certain features that can only be obtained through the help of supramolecular forces, for example, in the case of insulated wires.[64] The term *nanocrystal*, on the other hand, is commonly used for typically small (mostly <10 nm) particles of inorganic origin, containing two different types of atoms such as CdSe, PbS, or TiO_2.[65] They possess special properties arising from *confinement effects*, that is, in these nanoscale objects the movement of an electron is confined in all three dimensions. The electrons can thus only populate discrete energy levels. Accordingly, energy bands (e.g., of the bulk semiconductor) converge to discrete, atom-like states, with the oscillator strength being compressed into few transitions, leading to the typically observed high absorption coefficients.[66] Since in semiconductors the electronic properties are strongly governed by the transitions between the edges of valence and conduction band, confinement effects are more dramatic for these materials than for metals (see, e.g., Chapters 5 and 10). When metallic materials are concerned which are typically obtained by the reduction of a metal salt dissolved in an organic liquid, producing also dimensionally small particles (mostly <10 nm), these are mainly referred to as (atomic) *clusters*.[67] Inherent to both types of particles is the fact that they are not very stable or soluble in neat form and usually carry a protective organic shell, tightly adsorbed to the inorganic core. In the case of metal particles, particles stabilized with an organic layer are frequently referred to as *monolayer-protected clusters*.[68] Because *surface defects* (dangling orbitals or dangling bonds) govern especially the luminescence properties of semiconductor nanocrystals, such a capping layer also determines the performance of these particles in desired applications (see, e.g., Chapter 12). Besides organic ligands, the passivation procedure for semiconductor nanocrystals often comprises the growth of a second semiconductor layer onto the core in a *core-shell* type of fashion (e.g., ZnS@CdSe) with subsequent capping with organic ligands.[69] In contrast to nanocrystals, the term *nanocrystalline* is often used with a different connotation. These materials do not have to have nanoscale size, but are usually composed of nanosized crystals or crystallites in a consolidated state, the single crystallites being held together by interatomic forces acting between the atoms in the crystals on either side of the *defect core* or *defect site*.[70] In other words, these materials consist of numerous defect cores that are embedded into (an elastically distorted) crystal lattice(s). Since atoms at these defect sites possess unique properties, the bulk material retains these properties in its entirety and not only at its surface. Such materials are commonly produced by vapor deposition techniques, the most prominent example today being perhaps nanocrystalline silicon, which consists of crystalline grains of approximately 10 to 1000 nm size. *Nanolayer* finally is a very general term and is today used mainly in conjunction with composite materials prepared by deposition techniques. The prefix *nano* relates here basically to the thickness or the structured pattern of the layer.

Talking about fabrication techniques, they fall basically in one of the two concepts of nanotechnology, *top-down* and *bottom-up*.[44] Top-down is the classic engineering approach,[71] while bottom-up relates to (bio)chemical synthesis.[72] With respect to discrete nanomaterial objects, carbon nanotubes isolated from soot on one hand and silica materials synthesized through the sol–gel route or metal clusters through redox reactions on the other hand can be seen as the two antipoles. The chemical bottom-up strategies are intrinsically connected to supramolecular synthetic concepts, discussed above. The top-down approaches important within the present context include various deposition techniques for layered objects such as *physical*[73] or *chemical vapor deposition*[74] and, to a lesser extent, epitaxy techniques such as *molecular beam epitaxy*;[75] the latter are much more frequently used in device construction, for instance in optoelectronics. Particles can be generated either by synthetic *reduction* (e.g., metal particles), *precipitation* (e.g., semiconductor nanocrystals), and *condensation* techniques (e.g., using the sol–gel route), constituting the major routes, as well as by *electrochemical deposition* (e.g., for modified electrodes, see Chapter 9) or by physical methods such as *grinding* or *ball-milling* (e.g., for polymer-nanoparticle hybrids),[76] *laser ablation* (e.g., for Raman-active hybrid nanoparticles)[77] and *thermal decomposition*.[78] In addition, nanostructuring techniques comprise first of all *lithographic methods*. For inorganic or rigid organic (e.g., polymer) supports, photolithography is mostly used,[79] other lithographic techniques such as electron beam lithography[80] playing a minor role. Major advances in the field discussed in this book, however, rely on the development of *soft lithography* for the attachment and patterning of organic entities onto substrates,[81] rendering the preparation of *nanoarrays* possible.[82]

Relatively straightforward is the definition of nanoscopic voids. *Nanopores* and *nanocavities* are elongated voids or voids of any shape, and nanomaterials can incorporate especially nanopores in an ordered or disordered way. The former is of crucial importance for many of the hybrid materials discussed in the book (e.g., in Chapters 16 or 18). *Nanochannel* is also frequently used instead of *nanopore*, often in biological or biochemical contexts. Besides *nanoporous*, the term *mesoporous* is often found in hybrid materials research. Interestingly, the IUPAC has defined the terms *mesoporous* (pores with diameters between 2 and 50 nm), *microporous* (pores with diameters <2 nm) and *macroporous* (pores with diameters >50 nm), yet has not given a definition of *nanoporous* in the IUPAC Recommendations on the "Nomenclature of Structural and Compositional Characteristics of Ordered Microporous and

Mesoporous Materials with Inorganic Hosts."[83] Nanoscopic voids surrounded only by a comparatively thin layer of hybrid material are often referred to as *nanocontainers*. Intensive research on these hybrid analogs of micelles or vesicles has already accomplished rather sophisticated functions as presented in Chapter 22.

The last paragraph in the *nano*-section lists some terms that are frequently used in connection with "nanoscience," that are, however, rather ill- or very broadly defined only. Among these, *nanocomposite*, *nanodevice*, *nanosystem*, and *nanodomain* are perhaps the least specific ones because they concern any kind of composite material, device, system, or domain that is related to nanometric dimensions. *Nanofluidics*, *nanoelectronics*, *nanolithography*, and other such linguistic "nanocomposites" encompass broad definitions, yet are self-explanatory, because they are rooted in the parent, macroscopic technique. With respect to alluding research accomplishments in nanochemistry or nanophysics to applications of the macroscopic world, a whole new list of terms can easily be compiled from browsing (scientific) journals. People publishing in or writing about nanomaterials and nanotechnology often tend to illustrate their findings with more tangible terms such as *nanoshells*, *nanotweezers*, *nanoreactors*, *nanorobots*, or *nanofabrication*.[84]

1.4 HYBRIDS

Finally, the term *hybrid* is not intrinsically connected to supramolecular chemistry or nanomaterials but is commonly used for any combination of materials that can be classified into the various subdisciplines of the chemical sciences. As the title of the book suggests, most important here are *organic–inorganic* hybrids, stretching into the fields of *biomolecular–inorganic*[85] and *polymeric–inorganic*[86] materials such as those discussed, for example, in Chapter 21. Other combinations such as *biomolecular–polymeric*, *organic–biomolecular*, or *inorganic–inorganic* hybrids (e.g., alloy nanoparticles or AuNPs on supports consisting of other metals) are of course also possible but have less relevance in the present context. The noun or adverb *composite* is often used synonymously. In the scientific literature, the term *organic–inorganic hybrid material* was first used by G. L. Wilkes' group in the context of organically modified silica materials prepared by the sol–gel route.[87]

REFERENCES

1. J.-M. Lehn, *Science* **1985**, *227*, 849–856.
2. J.-M. Lehn, *Angew. Chem. Int. Ed. Engl.* **1998**, *27*, 89–112.
3. D. J. Cram, *Angew. Chem. Int. Ed. Engl.* **1998**, *27*, 1009–1020.
4. C. J. Pedersen, *Angew. Chem. Int. Ed. Engl.* **1998**, *27*, 1021–1027.
5. S. Maleknia, J. Brodbelt, *J. Am. Chem. Soc.* **1992**, *114*, 4295–4298.
6. J. L. Sessler, D. An, W.-S. Cho, V. Lynch, M. Marquez, *Chem. Commun.* **2005**, 540–542.
7. J.-M. Lehn, *Struct. Bonding* **1973**, *16*, 1–69.
8. M. R. Bennett, *Neuropharmacology* **2000**, *39*, 523–546.
9. E. Fischer, *Ber. Dtsch. Chem. Ges.* **1894**, *27*, 2985–2993.
10. D. E. Koshland, *Proc. Natl. Acad. Sci. U.S.A.* **1958**, *44*, 98–104.

11. C. Milstein, *Proc. R. Soc. Lond. Ser. B Biol. Sci.* **1990**, *239*, 1–16.
12. P. J. Hudson, *Curr. Opin. Biotechnol.* **1998**, *9*, 395–402.
13. K. Haupt, K. Mosbach, *Chem. Rev.* **2000**, *100*, 2495–2504.
14. G. Wulff, *Chem. Rev.* **2002**, *102*, 1–27.
15. D. K. Cabbines, D. W. Margerum, *J. Am. Chem. Soc.* **1969**, *91*, 6540–6541.
16. G. Schwarzenbach, *Helv. Chim. Acta* **1952**, *35*, 2344–2363.
17. G. V. Oshovsky, D. N. Reinhoudt, W. Verboom, *Angew. Chem. Int. Ed.* **2007**, *46*, 2366–2393.
18. R. C. Ahuja, P. L. Caruso, D. Möbius, G. Wildburg, H. Ringsdorf, D. Philp, J. A. Preece, J. F. Stoddart, *Langmuir* **1993**, *9*, 1534–1544.
19. P. Cintas, *J. Inclusion Phenom. Mol. Recognit. Chem.* **1994**, *17*, 205–220.
20. C. Hilger, M. Draeger, R. Stadler, *Macromolecules* **1992**, *25*, 2498–2501.
21. D. C. Sherrington, K. A. Taskinen, *Chem. Soc. Rev.* **2001**, *30*, 83–93.
22. M. Meot-Ner, *Chem. Rev.* **2005**, *105*, 213–284.
23. T. Steiner, *Angew. Chem. Int. Ed.* **2002**, *41*, 48–76.
24. J. C. Ma, D. A. Dougherty, *Chem. Rev.* **1997**, *97*, 1303–1324.
25. B. L. Schottel, H. T. Chifotides, K. R. Dunbar, *Chem. Soc. Rev.* **2008**, *37*, 68–83.
26. C. A. Hunter, J. K. M. Sanders, *J. Am. Chem. Soc.* **1990**, *112*, 5525–5534.
27. H.-J. Schneider, T. Schiestel, P. Zimmermann, *J. Am. Chem. Soc.* **1992**, *114*, 7698–7703.
28. D. B. Smithrud, E. M. Sanford, I. Chao, S. B. Ferguson, D. R. Carcanague, J. D. Evanseck, K. N. Houk, F. Diederich, *Pure Appl. Chem.* **1990**, *62*, 2227–2236.
29. U. Ziener, E. Breuning, J.-M. Lehn, E. Wegelius, K. Rissanen, G. Baum, D. Fenske, G. Vaughan, *Chem. Eur. J.* **2000**, *6*, 4132–4139.
30. F. Schreiber, *J. Phys. Condens. Matter* **2004**, *16*, R881–R900.
31. J. A. Zasadzinski, R. Viswanathan, L. Madsen, J. Garnaes, D. K. Schwartz, *Science* **1994**, *263*, 1726–1733.
32. M. C. Marden, L. Kiger, C. Poyart, S. J. Edelstein, *Cell. Mol. Life Sci.* **1998**, *54*, 1365–1384.
33. J. J. Perona, C. S. Craik, *Protein Sci.* **1995**, *4*, 337–360.
34. D. Philp, J. F. Stoddart, *Angew. Chem. Int. Ed. Engl.* **1996**, *35*, 1155–1196.
35. S. Anderson, H. L. Anderson, J. K. M. Sanders, *Acc. Chem. Res.* **1993**, *26*, 469–475.
36. B. R. Bowsher, A. J. Rest, *J. Chem. Soc. Dalton Trans.* **1981**, 1157–1161.
37. P. Thanasekaran, R. T. Liao, Y. H. Liu, T. Rajendran, S. Rajagopal, K. L. Lu, *Coord. Chem. Rev.* **2005**, *249*, 1085–1110.
38. M. M. Conn, J. Rebek, Jr., *Chem. Rev.* **1997**, *97*, 1647–1668.
39. G. M. Whitesides, E. E. Simanek, J. P. Mathias, C. T. Seto, D. N. Chin, M. Mammen, D. M. Gordon, *Acc. Chem. Res.* **1995**, *28*, 37–44.
40. F. M. Raymo, J. F. Stoddart, *Chem. Rev.* **1999**, *99*, 1643–1663.
41. K. Ariga, J. P. Hill, Q. Ji, *Phys. Chem. Chem. Phys.* **2007**, *9*, 2319–2340.
42. P. L. Luisi, F. Ferri, P. Stano, *Naturwissenschaften* **2006**, *93*, 1–13.
43. S. Mann, *Angew. Chem. Int. Ed.* **2008**, *47*, 5306–5320.
44. G. A. Ozin, *Adv. Mater.* **1992**, *4*, 612–649.
45. K. E. Drexler, *Engines of Creation*, New York: Anchor Books/Doubleday, **1986**; electronic version: http://e-drexler.com/p/06/00/EOC_Cover.html.
46. G. M. Whitesides, J. P. Mathias, C. T. Seto, *Science* **1991**, *254*, 1312–1319.
47. H. Rohrer, *Microelectron. Eng.* **1996**, *32*, 5–14.
48. S. Iijima, *Nature* **1991**, *354*, 56–58.
49. H. Dai, E. W. Wong, Y.-Z. Lu, S. Fan, C. M. Lieber, *Nature* **1995**, *375*, 769–772.
50. C. J. Murphy, *Science* **2002**, *298*, 2139–2140.
51. J. J. Marty, R. C. Oppenheim, P. Speiser, *Pharm. Acta Helv.* **1978**, *53*, 17–23.
52. J. Kreuter, *Pharm. Acta Helv.* **1978**, *53*, 33–39.
53. R. C. Oppenheim, *Int. J. Pharm.* **1981**, *8*, 217–234.
54. L. Illum, P. D. E. Jones, J. Kreuter, R. W. Baldwin, S. S. Davis, *Int. J. Pharm.* **1983**, *17*, 65–76.
55. P. A. Hansson, *Futures* **1991**, *23*, 849–859.
56. N. M. Rodríguez, A. Chambers, R. T. K. Baker, *Langmuir* **1995**, *11*, 3862–3866.

57. K. Jayaraman, M. Kotaki, Y. Zhang, X. Mo, S. Ramakrishna, *J. Nanosci. Nanotechnol.* **2004**, *4*, 52–65.
58. M. Macias, A. Chacko, J. P. Ferraris, K. J. Balkus, *Microporous Mesoporous Mater.* **2005**, *86*, 1–13.
59. T. R. Kline, M. Tian, J. Wang, A. Sen, M. W. H. Chan, T. E. Mallouk, *Inorg. Chem.* **2006**, *45*, 7555–7565.
60. H. J. Fan, P. Werner, M. Zacharias, *Small* **2006**, *2*, 700–717.
61. E. Gazit, *FEBS J.* **2007**, *274*, 317–322.
62. J. M. Tour, *Acc. Chem. Res.* **2000**, *33*, 791–804.
63. T. M. Swager, *Acc. Chem. Res.* **1998**, *31*, 201–207.
64. M. J. Frampton, H. L. Anderson, *Angew. Chem. Int. Ed.* **2007**, *46*, 1028–1064.
65. Y. Yin, A. P. Alivisatos, *Nature* **2005**, *437*, 664–670.
66. M. G. Bawendi, M. L. Steigerwald, L. E. Brus, *Annu. Rev. Phys. Chem.* **1990**, *41*, 477–496.
67. J. P. Wilcoxon, B. L. Abrams, *Chem. Soc. Rev.* **2006**, *35*, 1162–1194.
68. A. C. Templeton, W. P. Wuelfing, R. W. Murray, *Acc. Chem. Res.* **2000**, *33*, 27–36.
69. S. Pokrant, K. B. Whaley, *Eur. Phys. J. D* **1999**, *6*, 255–267.
70. H. Gleiter, *Adv. Mater.* **1992**, *4*, 474–481.
71. R. P. Feynman, *Eng. Sci.* **1960**, *23*, 22–26, 30, 34, and 36, accessible online at http://www.zyvex.com/nanotech/feynman.html.
72. J.-M. Lehn, *Supramolecular Chemistry: Concepts and Perspectives*, Weinheim: VCH, **1995**.
73. K. Reichelt, X. Jiang, *Thin Solid Films* **1990**, *191*, 91–126.
74. B. D. Fahlman, *Curr. Org. Chem.* **2006**, *10*, 1021–1033.
75. K. Ploog, *Angew. Chem. Int. Ed. Engl.* **1988**, *27*, 593–621.
76. M. Xiong, M. Wu, S. Zhou, B. You, *Polym. Int.* **2002**, *51*, 693–698.
77. K. Siskova, B. Vlckova, P. Y. Turpin, A. Thorel, A. Grosjean, *Vib. Spectrosc.* **2008**, *48*, 44–52.
78. M. Nakade, K. Ichihashi, M. Ogawa, *J. Porous Mat.* **2005**, *12*, 79–85.
79. M. Qhobosheane, P. Zhang, W. H. Tan, *J. Nanosci. Nanotechnol.* **2004**, *4*, 635–640.
80. J. Joo, S. Moon, J. M. Jacobson, *J. Vac. Sci. Technol. B* **2006**, *24*, 3205–3208.
81. Y. Xia, G. M. Whitesides, *Annu. Rev. Mater. Sci.* **1998**, *28*, 153–184.
82. J. Gu, X. Xiaoiao, B. R. Takulapalli, M. E. Morrison, P. Zhang, F. Zenhausern, *J. Vac. Sci. Technol. B* **2008**, *26*, 1860–1865.
83. L. B. McCusker, F. Liebau, G. Engelhardt, *Pure Appl. Chem.* **2001**, *73*, 381–394.
84. S. A. Edwards, *The Nanotech Pioneers*, Weinheim: Wiley-VCH, **2006**, 2–3.
85. E. Katz, I. Willner, *Angew. Chem. Int. Ed.* **2004**, *43*, 6042–6108.
86. H.-T. Chen, Y. Ofir, V. M. Rotello in *Molecular Recognition and Polymers* (Eds. V. M. Rotello, S. Thayumanavan), Hoboken, NY: Wiley, **2008**, 137–157.
87. B. Wang, G. L. Wilkes, J. C. Hedrick, S. C. Liptak, J. E. McGrath, *Macromolecules* **1991**, *24*, 3449–3450.

Chapter 2

Supramolecular Chemistry at the Mesoscale

KATSUHIKO ARIGA, GARY J. RICHARDS, JONATHAN P. HILL, AJAYAN VINU, AND TOSHIYUKI MORI

2.1 INTRODUCTION

Dimensions between the atomic/molecular and the bulk macroscopic scales are sometimes called "mesoscopic." Because the mesoscopic scale corresponds roughly to the electron free path, unusual phenomena such as quantum effects can be observed, some of which could be used in the development of single-electron devices or quantum computers. Top-down-type nanofabrication techniques are now capable of producing structures in this size range, and research on this subject has received significant attention, especially in the field of semiconductor science and technology.

The mesoscopic size regime is also relevant to organic chemistry and biochemistry since many biological mechanisms, including information transmission across cell membranes, ribosomal protein synthesis, mitochondrial energy conversion, and enzymatic reactions, occur within complexes at the mesoscopic level. For construction of functional mesoscopic structures from organic and biological components, bottom-up

The Supramolecular Chemistry of Organic–Inorganic Hybrid Materials. Edited by Knut Rurack and Ramón Martínez-Máñez

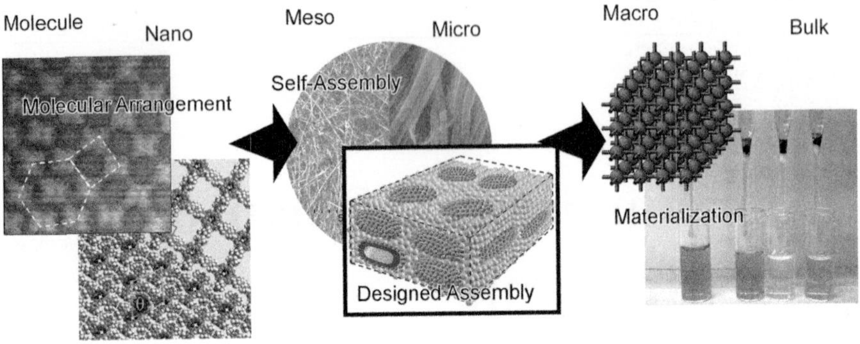

Figure 2.1 Schematic illustration of the bottom-up approach from molecular to bulk scales.

approaches based on supramolecular self-assembly techniques have to be utilized, because such structures require sophisticated arrangement of their components in order to express fine functions. However, although formation of self-assembling structures at the molecular- and nano-scales is well developed, fabrication of well-organized structures in the mesoscale range remains a challenging target.[1]

In Figure 2.1, schematic representations illustrate strategies for creation of bulk materials using molecular scale interactions through bottom-up supramolecular approaches. Control of molecular arrangement,[2–6] self-assembly or designed assembly processes,[7–9] and structural transcription to materials[10,11] all play important roles. Additionally, construction of mesoscopic structures is one of the most important steps. Assembly processes involving only organic materials are sometimes not suitable for obtaining well-structured materials at the mesoscopic level. Combination of organic or biological moieties with inorganic materials is one of the practical solutions to this problem because inorganic materials can improve stability of the resulting hybrids. Fabrication of organic–inorganic hybrid materials is thus an excellent route for the construction of mechanically stable mesoscopic structures. Immobilization of organic and biological components on inorganic supports leads to sophisticated arrangements of these functional components, while transcription of supramolecular self-assembled structures into inorganic materials results in novel functional inorganic structures. These ambitious challenges can now be achieved at the mesoscopic scale.

In this chapter, supramolecular chemistry related to developments in materials fabrication and functionalization at the mesoscale are discussed, with an emphasis on those systems based on organic–inorganic hybrid structures. The contents of this chapter are classified into (1) supramolecular chemistry within mesoscopic media, (2) supramolecular assembly at the mesoscale, and (3) supramolecular materials at the mesoscale. Despite this classification these topics have considerable similarities.

2.2 SUPRAMOLECULAR CHEMISTRY IN MESOSCOPIC MEDIA

Molecular recognition is a key event in supramolecular chemistry since it determines how molecules assemble in a predesignated way. Most molecular recognition events in traditional supramolecular chemistry have been researched in molecularly dispersed systems in the solution phase,[12–17] although most of the important phenomena in natural systems do not occur in such a simple medium. For example, biological recognition occurs at a variety of surfaces, including cell membranes, within protein cavities, and at various points on nucleic acid structures. Thus, these can be seen as mesoscopic interfaces. Therefore, investigation of molecular recognition at mesoscopic interfaces should lead to deeper understanding of biological phenomena.

Some pioneering research has revealed the importance of the size and nature of the medium on molecular recognition. If molecular recognition sites are embedded in a mesoscopic hydrophobic environment dispersed in or located at an aqueous interface, the strength of hydrogen bonding interactions is significantly enhanced. For example, Nowick and coworkers demonstrated binding of adenine and thymine moieties by concealing them in the hydrophobic core of aqueous micelles,[18] and Bonar-Law and Sanders applied a similar approach using a porphyrin receptor.[19] Kitano and Ringsdorf reported pioneering work on base-pair mimicry at the air–water interface evaluated by changes in surface pressure–molecular area (π-A) upon base pairing.[20] Later, Kunitake and coworkers systematically investigated molecular recognition at the air–water interface, as exemplified in Figure 2.2.[21] Efficient molecular recognition at the air–water interface was found for nucleobase recognition in the cases of thymine recognition[22] and adenine recognition.[23] High specificity and efficiency in biological molecular recognition often rely on multipoint interactions. The air–water interface is a medium appropriate for the formation of self-assembled systems for multipoint recognition. Noncovalent self-assembly of functional components to form multifunctional receptor sites with precisely defined spatial disposition can be constructed at the air–water interface. Kunitake and coworkers reported several examples of multipoint recognitions including recognition systems for flavin adenine dinucleotide (FAD) as shown in Figure 2.2c.[24] Such multipoint molecular recognition often induces spontaneous alignment of amphiphilic molecules, resulting in mesoscopic patterning.[25]

Sakurai and coworkers proposed the application of quantum chemical calculations to interfacial molecular recognition in order to clarify the origin of the highly efficient nature of molecular recognition at the air–water interface.[26–28] The calculation was performed based on a multidielectric model for the guanidinium-phosphate system placed at an interface, as illustrated in Figure 2.3. The complexes consisting of guanidinium and phosphate were placed on various positions of the interface, denoted by parameter d (Fig. 2.3a). For a given d value, the free energies of the complex double-layer systems were plotted as a function of the distance R (Fig. 2.3b). When the interface is set on the guanidinium central carbon, the calculated energy corresponds to the experimentally obtained values. This means that complete exclusion of water from the binding sites is not always necessary for efficient guest binding through hydrogen bonding. Even when recognition sites are exposed to the aqueous phase, electronic influences from the adjacent nonpolar phase can strengthen hydrogen bonding between host and guest.

Kunitake and coworkers experimentally investigated the effect of interface size on molecular recognition efficiency with a guanidinium-phosphate interaction (Fig. 2.4). Binding constants of aqueous phosphates such as adenosine monophosphate (AMP) to a monolayer of a guanidinium amphiphile were determined to be in the range of 10^6

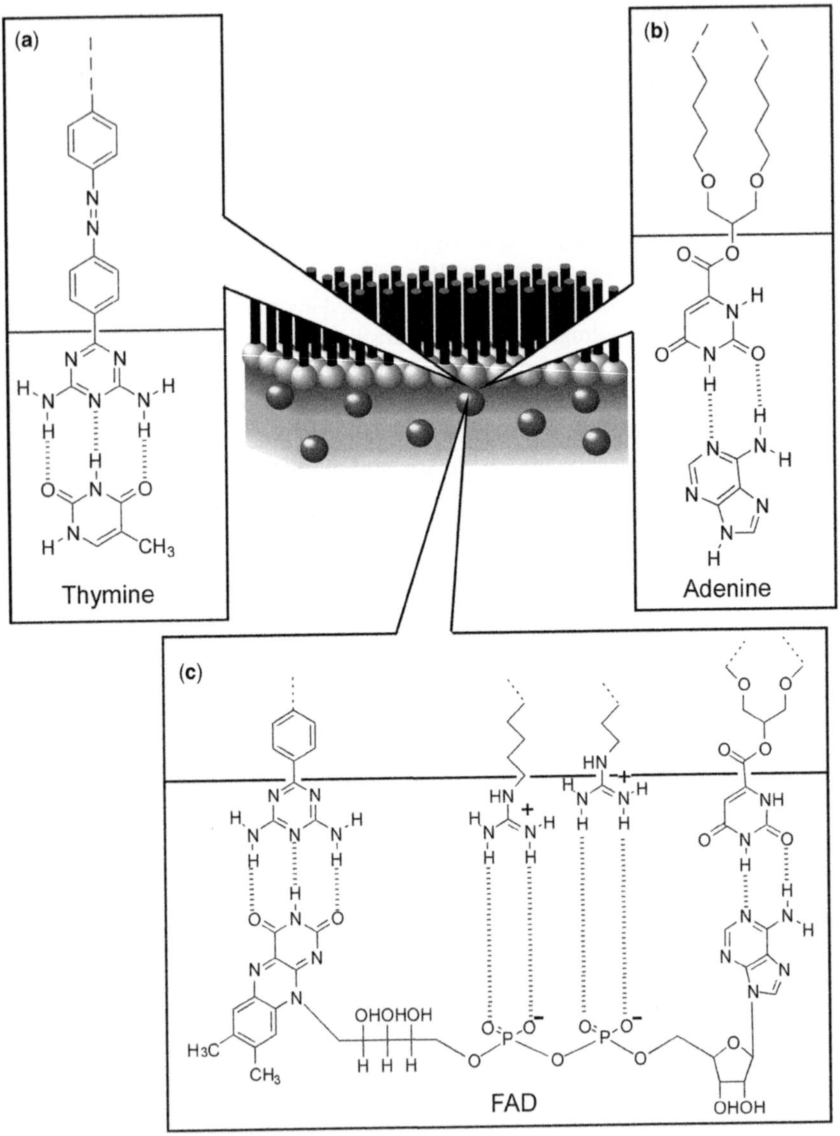

Figure 2.2 Typical examples of hydrogen bonding-based molecular recognition: (a) thymine recognition; (b) adenine recognition; (c) flavin adenine dinucleotide recognition.[21] (Reprinted with permission from K. Ariga and T. Kunitake, *Acc. Chem. Res.* **1998**, *31*, 371–378. Copyright 1998 American Chemical Society.)

to 10^7 M^{-1}.[29] These values are significantly larger than that obtained for molecularly dispersed guanidinium and phosphate (1.4 M^{-1}).[30] They also investigated molecular recognition efficiency between guanidinium and phosphate at the surfaces of micelles and bilayers that correspond to mesoscopic interfaces.[31] A Langmuir analysis with

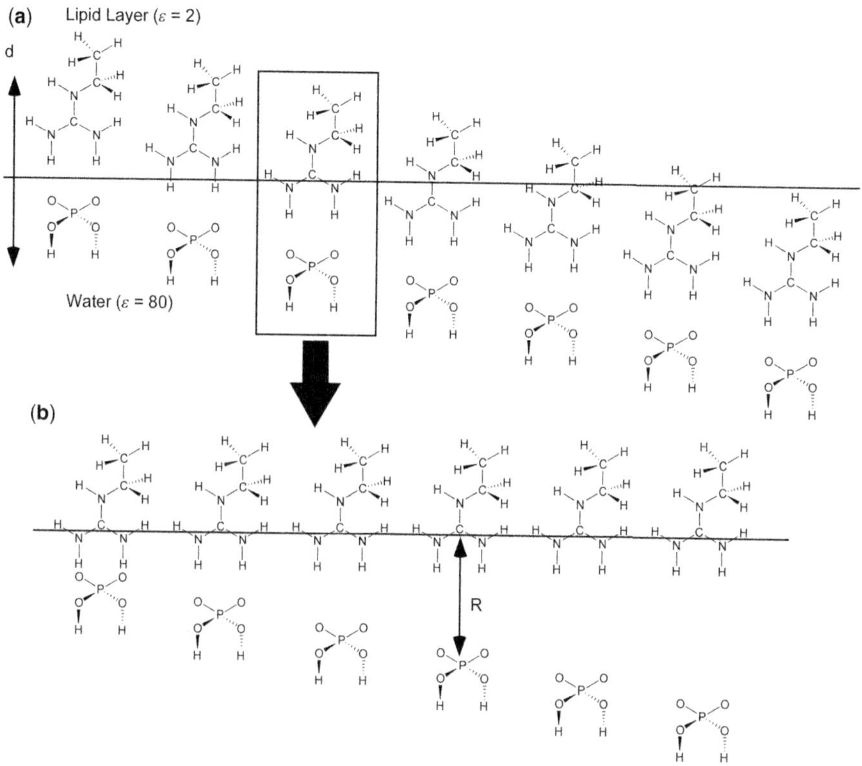

Figure 2.3 Model system (guanidinium-phosphate recognition) for quantum chemical calculation: (a) setting of interfacial position with parameter d; (b) distance between guanidinium and phosphate (R).[27] (Reprinted with permission from M. Sakurai et al., *J. Phys. Chem. B* **1997**, *101*, 4810–4816. Copyright 1997 American Chemical Society.)

variation of guest concentrations gave binding constants in the range 10^2 to 10^4 M^{-1} for the binding of aqueous phosphate to guanidinium groups at mesoscopic interfaces. As indicated by these data, the binding strength between guanidinium and phosphate appears to depend on the size of the interface, that is, there is a measurable "size effect" which is confirmed even in supramolecular systems.

Other unusual phenomena reflecting mesoscopic size effects have also been reported. Mesoporous structures sometimes provide mesoscopic media in which functional supermolecules can be confined. Several researchers reported the behaviors of functional polymers entrapped within mesopores of silicate materials. Aida and Tajima synthesized mesoporous silica containing polydiacetylene in microfibrous form. The silica polymer composite obtained exhibited a red-shifted electronic absorption spectrum.[32] It is possible that confinement of the polydiacetylene resulted in an elongated effective conjugation length within the silica nanochannel. Lu et al. also prepared polydiacetylene-containing mesoporous silica films through casting, spin-coating, or dip-coating methods.[33] The resulting nanocomposites displayed unusual chromatic changes in response to thermal, mechanical, and chemical stimuli.

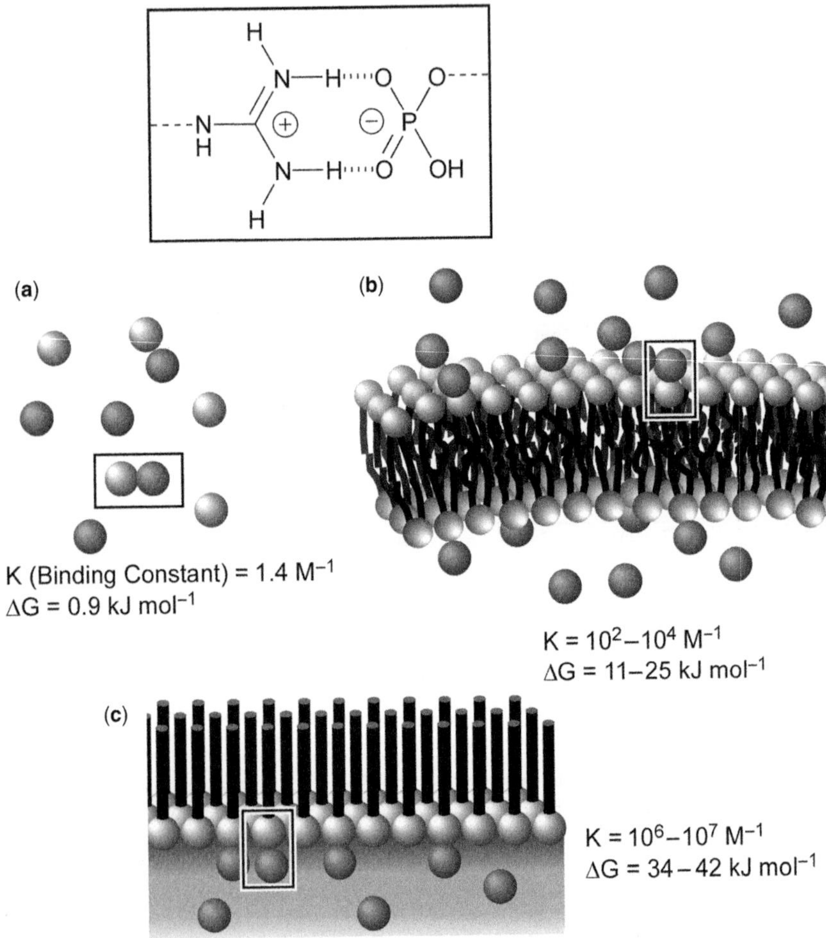

Figure 2.4 Effect of size of interface on molecular recognition efficiency as exemplified by guanidinium-phosphate interaction: (a) molecularly dispersed system; (b) mesoscopic interface (surfaces of micelles and bilayers); (c) macroscopic interface (air–water interface).[31] (Reprinted with permission from M. Onda et al., *J. Am. Chem. Soc.* **1996**, *118*, 8524–8530. Copyright 1996 American Chemical Society.)

Aida and coworkers developed poly(pyrrole)-containing mesoporous silica films in both hexagonal and lamellar phases.[34] The polypyrrole chains are highly constrained and insulated when incorporated within hexagonal mesoscopic channels and the possibility of the polarons recombining into bipolarons is significantly suppressed. In contrast, the two-dimensional lamellar phase affords spatial freedom for electron recombination.

From the viewpoint of supramolecular chemistry, behavior of noncovalent supramolecular assemblies confined in mesopores would be a more interesting target. Aida and coworkers reported the first example of immobilization of one-dimensional columnar charge-transfer (CT) assemblies in a mesoporous silica film through sol–gel

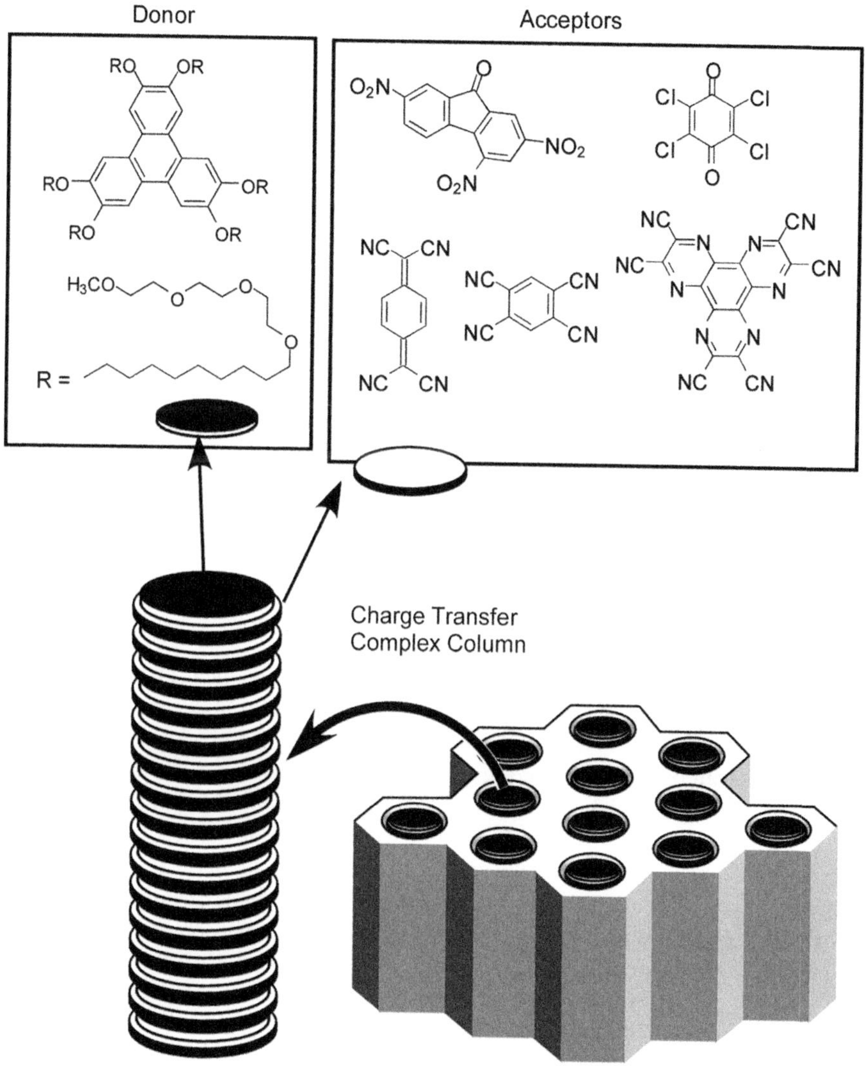

Figure 2.5 Immobilization of one-dimensional columnar charge-transfer (CT) assemblies of an amphiphilic triphenylene donor and various acceptors confined within mesoporous silica nanochannels.

reaction with CT complexes of an amphiphilic triphenylene donor and various acceptors (Fig. 2.5).[35] The films obtained were highly transparent and color-tunable by the intercalating acceptor. In hexagonal mesoporous silica, the silica wall segregates the individual CT columns, which display neither solvatochromism nor guest exchange activity. Interestingly the films exhibit red-shifted absorption bands, which are possibly due to a long-range structural ordering.

Mesoscopic inorganic wires entrapped within an envelope of organic materials were also reported. One class of halogen-bridged one-dimensional mixed valence

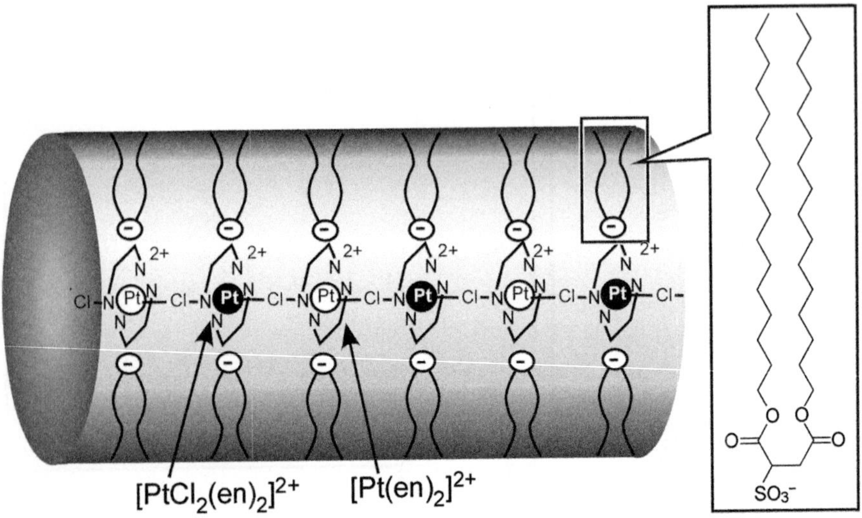

Figure 2.6 Chloro-bridged linear platinum chains dispersed as aggregates of a polyion complex with a sulfonate amphiphile.

complexes is of great interest due to its unique physicochemical properties. These properties include intense intervalence charge-transfer absorption, semiconductivity, and large, third-order, nonlinear optical susceptibilities. Kimizuka and coworkers demonstrated preparation of chloro-bridged linear platinum chains dispersed as aggregates of a polyion complex with a sulfonate amphiphile whose lipophilic alkyl chains are oriented toward the organic media (Fig. 2.6).[36] Dispersion of the low-dimensional structures within the envelope of the organized lipid molecules can stabilize the mixed valence chains in solution. Heating the linear platinum complex caused dissociation into its individual components, which were then reassembled on cooling to lower temperatures. In this example, a reversible structural transition between aggregated mesoscopic and individual molecular metal complexes was thus realized by the formation of a self-assembling amphiphilic supramolecular structure.

2.3 SUPRAMOLECULAR ASSEMBLY AT THE MESOSCALE

Supramolecular self-assembly is one of the most useful methods to form mesoscopic-size structures by low-energy processes. Addition of a secondary interaction such as hydrogen bonding to conventional solvophilic–solvophobic interactions sometimes leads to formation of highly extended assemblies.[37] Typical examples are displayed in Figure 2.7 where a peptide-lipid containing Ala-Ala-Glu residue was used as a component material.[38,39] Self-assembled structures of this peptide-lipid in carbon tetrachloride (CCl_4) solutions can be assigned as well-extended thin fibers (Fig. 2.7a).

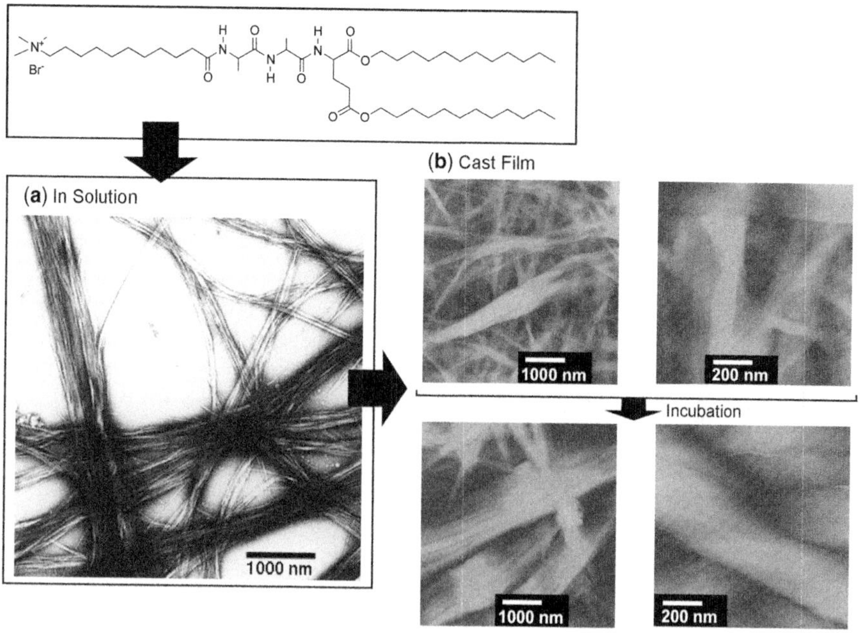

Figure 2.7 Mesoscopic assemblies composed of a peptide-lipid containing the Ala-Ala-Glu residue: (a) TEM images of the assemblies from carbon tetrachloride (CCl_4) solution; (b) AFM images of assemblies in cast films before and after incubation.[39] (Reprinted with permission from K. Ariga et al., *Langmuir* **2000**, *16*, 4929–4939. Copyright 2000 American Chemical Society.)

Similar structures can be seen also in atomic force microscopic (AFM) images of its cast films (Fig. 2.7b). Interestingly, these fibrous structures become thick, rope-like structures on incubation over several days. Infrared (IR) spectra of this peptide-lipid assembly in CCl_4 solution contain amide A, amide I, and amide II peaks at approximately 3300, 1630, and 1540 cm^{-1}, respectively, and no peak at approximately 1690 cm^{-1}. These spectral features indicate that the derivatives form parallel β-sheet structures. The formation of a parallel β-sheet structure and exposure of the hydrophobic tails to the bulk nonpolar solvent result in formation of a fibrous structure having a cross-section of a single or a multiple reversed micelle structure. After an appropriate incubation time, the fibers are aligned, and the aligned fibers probably fuse together through solidification of the alkyl chains, resulting in the thick structure. The solidification of the alkyl chains upon fusion is indicated by the $\nu_{as}(CH_2)$ vibration in their Fourier transform (FT)-IR spectra. Combination of a hydrogen-bonded network and interactions between huge numbers of alkyl chains enable this simple compound to form mesoscopic-level structures.

Hierarchic design of assembled structures should be important for preparation of functional mesoscopic structures. Typical examples can be seen in biological systems where tissues of organisms in living systems consist of assemblies of cells with cell membranes composed of self-assembled lipids, proteins, saccharides, etc. Therefore, tissues and organisms can be regarded as (at least) two-level assemblies: lipid to cell and cell to tissue. Analogous assembly methods were realized using organic–inorganic hybrid structures, as reported by Katagiri et al.[40,41] This research team applied the covalent linkage of a siloxane framework to a lipid bilayer vesicle in order to reinforce the vesicle structures for organization of multicellular systems (Fig. 2.8). The resulting vesicles are called "cerasome," derived from the words "ceramic" + "soma," because they have an inorganic siloxane network covalently attached to the bilayer membrane surface (Fig. 2.8a). In the preparation of the cerasome structure, an alkoxysilane-bearing amphiphile was first dispersed in an acidic aqueous solution using a vortex mixer. The siloxane bonds were formed immediately after hydrolysis. Formation of vesicular structures was confirmed by transmission electron microscopy (TEM) using hexaammonium heptamolybdate tetrahydrate as contrast agent (Fig. 2.8a). Subjecting the cerasome structure to layer-by-layer (LbL) techniques resulted in multicellular mimics in a predesigned way. Of two types of

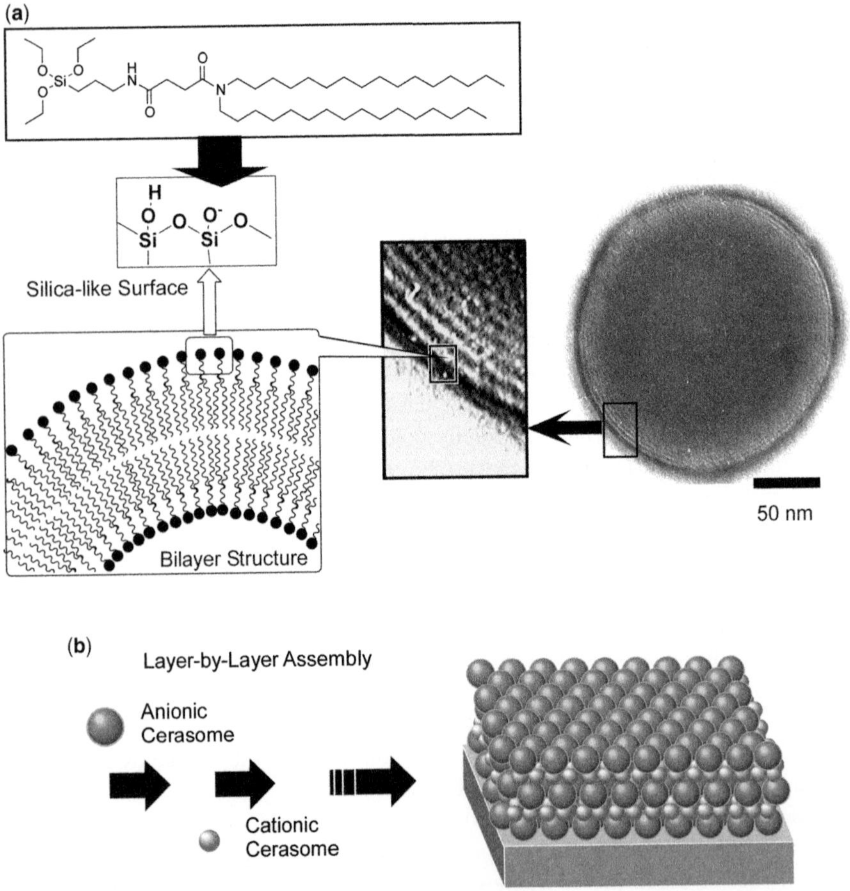

Figure 2.8 (a) Illustration of an organic–inorganic hybrid vesicle cerasome and its TEM image. (b) LbL assembly between an anionic cerasome and a cationic cerasome as a multicellular mimic.

cerasome LbL assemblies reported by Katagiri et al. LbL assembly between the cationic polyelectrolyte, poly(diallyldimethylammonium) chloride (PDDA), and anionic vesicles was the first demonstrated.[42] Subsequently, direct LbL assembly between anionic and cationic cerasomes (Fig. 2.8b) was realized.[43] In the latter case, the presence of closely packed cerasome particles like a stone pavement in both layers was clearly confirmed by AFM observations of the surface of the assembled structures. These assemblies could be subsequently used as bioreactors or biosensors. Further covalent functionalization of the cerasome surface with various biomolecules, such as enzymes and antibodies, suggests a great potential for creating various kinds of biomimetic silica nanohybrids.

Inorganic nanoparticles themselves can be assembled into mesoscopic structures. Dinsmore et al. proposed an approach for the fabrication of solid capsules from colloidal particles with precise control of size, permeability, mechanical strength, and compatibility (Fig. 2.9).[44] This unusual mesoscopic structure is called "colloidosome" and is prepared through emulsion droplets at a water–oil interface. Following the "locking" together of the particles to form elastic shells, the emulsion droplets were transferred to a fresh continuous-phase fluid identical to that contained inside the droplets. The resultant structures are hollow, elastic shells whose permeability and elasticity can be precisely controlled.

Inorganic supports effectively strengthen organic mesoscopic supramolecular assemblies. The example depicted in Figure 2.10 presented the worlds thinnest lipid membrane with permeation controllability based on organic–inorganic hybrid structures in which only a single monolayer could regulate permeation of aqueous probe molecules.[45,46] In order to prepare a mechanically stable Langmuir monolayer, an organosilane monolayer was first prepared and polymerized on an air–water interface, which was then transferred for immobilization onto a solid substrate with small pores (Fig. 2.10a). The Langmuir–Blodgett (LB) films prepared under acidic conditions showed bands in their infrared spectra at 1026 and 1095 cm^{-1} assignable to Si–O–Si, indicating that the presence of an acid catalyst in the subphase stimulated polymerization of the monolayer constituents. The transferred monolayer was covalently immobilized on the porous glass plate by heat treatment. The glass plate with the immobilized monolayer was attached to the bottom of a polyethylene tube and

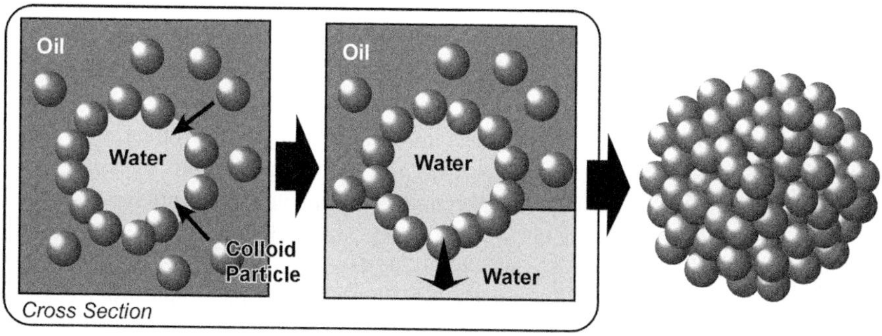

Figure 2.9 Preparation of a colloidosome.

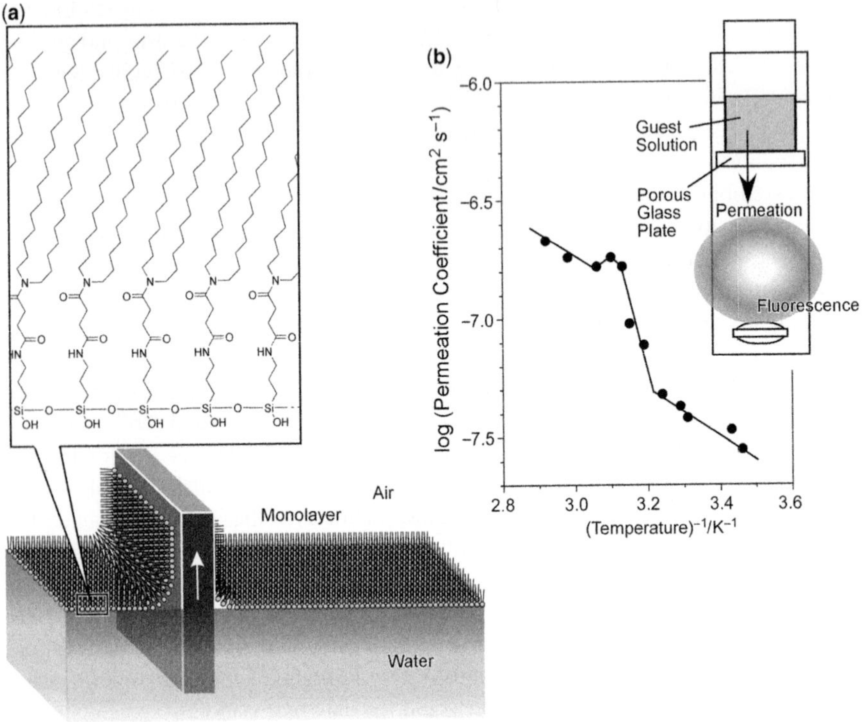

Figure 2.10 (a) Langmuir monolayer with an organosilane network. (b) Permeation coefficient of an aqueous fluorescent probe through a monolayer structure immobilized on a porous glass plate.[46] (Reprinted with permission from K. Ariga and Y. Okahata, *J. Am. Chem. Soc.* **1989**, *111*, 5618–5622. Copyright 1989 American Chemical Society.)

soaked in a quartz cell. Permeation of a water-soluble fluorescent probe was estimated through the increase in fluorescence intensity of the lower solution. As shown in an Arrhenius-type plot (Fig. 2.10b), permeation through the monolayer exhibited a discontinuity at around 45°C, which is very close to the phase transition temperature (crystal state to liquid crystalline state) of the polymerized monolayer. It is the first example of permeation control using a monolayer only 2 nm thick and, so far, it could be the thinnest lipid film operating as a permeation valve. This demonstration of control of organic thin film permeation became possible only through hybridization of the fragile monolayer with the mechanically stable inorganic substrate.

There are some unique examples of organic–inorganic hybrids in natural systems. Some kinds of protein possess inorganic components at their interior. Because such bio-originated hybrid structures have discrete size and composition, they can be used for preparation of fine mesoscopic structures. For example, ferritin, which is composed of 24 self-assembled peptide subunits and capable of including iron oxide, is a good candidate to fulfill this demand. Yamashita demonstrated use of a ferritin array for preparation of a mesoscopic nanodot array (Fig. 2.11).[47] A Langmuir

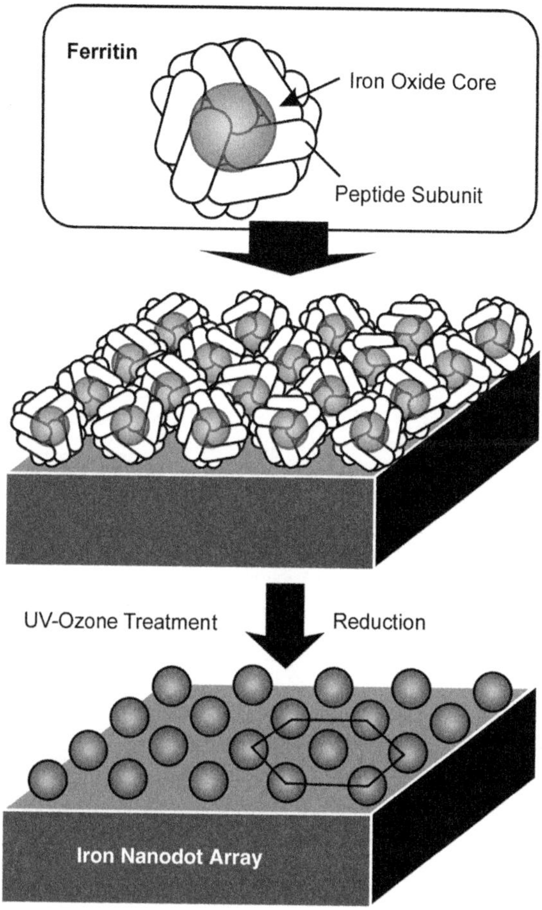

Figure 2.11 Monolayer array of ferritin and conversion into a nanodot array.

monolayer of ferritin was first transferred onto a hydrophobized silicon substrate to form a two-dimensional hexagonal array of ferritin. Destruction of the organic components by UV/ozone treatment and heat treatment under hydrogen flow left iron particles with sub-6 nm diameters arranged in a hexagonal arrangement. Supramolecular arrays prepared from inorganic substances can be used for device preparation. Immobilization of such arrayed structures, for instance, as the gate part of a transistor could lead to the preparation of a floating quantum dot gate transistor for multilevel logic gates capable of room temperature operation.

As briefly described previously in this chapter, layer-by-layer (LbL) assembly is a versatile method for construction of layered structures from various materials.[48,49] Therefore, the LbL technique should be a very useful method for constructing layered mesoscopic structures. The LbL assembly concept was first suggested by Iler,[50] but realized and established by Decher and coworkers.[51] An outline of this technique is shown in Figure 2.12, where assembly between a cationic polyelectrolyte and anionic

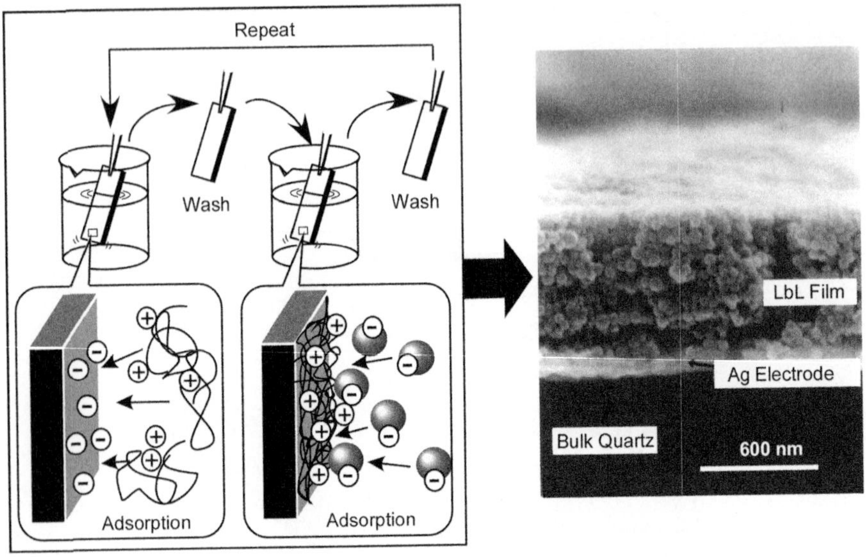

Figure 2.12 Illustration of an LbL assembly between polyelectrolytes and colloidal particles and TEM image of the assembled film (cross-section).[77] (Reprinted with permission from Y. Lvov et al., *Langmuir* **1997**, *13*, 6195–6203. Copyright 1997 American Chemical Society.)

nanoparticles through electrostatic interaction is exemplified. Excess adsorption of the substances to an oppositely charged surface leads to charge neutralization and resaturation, resulting in charge reversal. Continuous assembly between oppositely charged materials can be conducted continuously with great freedom in the number of layers and the layering sequence. In addition, assemblies with other interactions such as hydrogen bonding,[52] metal coordination,[53] and charge transfer interaction[54] have also been demonstrated.

The variability in applicable materials is another of the most pronounced advantages of this method. Applicable materials are not limited to conventional polyelectrolytes[55–57] so that not only biochemical materials, including proteins,[58–64] DNA strands,[65] and virus particles,[66] but also charged inorganic substances such as colloidal nanoparticles,[67] clay,[68] nanosheets,[69] modified zeolite crystals,[70] two-dimensional perovskite,[71] and polyoxometalates[72] have been included. Furthermore, charged supramolecular assemblies such as bolaamphiphile monolayers,[73] lipid bilayers,[74] LB films,[75] and stacked dye molecules[76] have also been applied in LbL approaches. Among them, charged colloidal nanoparticles are typical components. Kunitake and coworkers successfully demonstrated the assembling processes of three sizes of SiO_2 particles (25, 45, and 78 nm in diameter) with the aid of poly(diallyldimethylammonium) chloride (PDDA).[77] In this method, many kinds of assembly sequence between different particles can be realized if we consistently insert an intermediate polycationic layer. Figure 2.12 also displays scanning electron microscopy (SEM) images of the SiO_2-PDDA LbL film. The SiO_2 particles are closely packed in the layer, but there is no long-range ordering.

2.4 SUPRAMOLECULAR MATERIALS AT THE MESOSCALE

Some supramolecular mesoscopic structures remain mere objects of scientific interest because of their low yields. If we prepare materials having mesoscopic structures in useful quantities, such materials would be useful for "real" applications such as highly selective material separation and storage.

Self-assembly processes based on directionally defined interactions such as metal cation coordination or hydrogen bonding can be used for syntheses of materials with precise nanostructures. In particular, rigid porous frameworks have been constructed through a coordination-type self-assembly process. Some of these are known as metal organic frameworks (MOFs) and/or coordination polymers.[78] As illustrated in Figure 2.13a, appropriate selection of organic ligands can lead to regular porous structures with various sizes and geometries. These materials are often prepared in high yield by very simple procedures. Yaghi and coworkers have developed novel types of MOF structures[79] and some of them have even reached commercialization. They recently reported the preparation of three-dimensional covalent organic frameworks.[80]

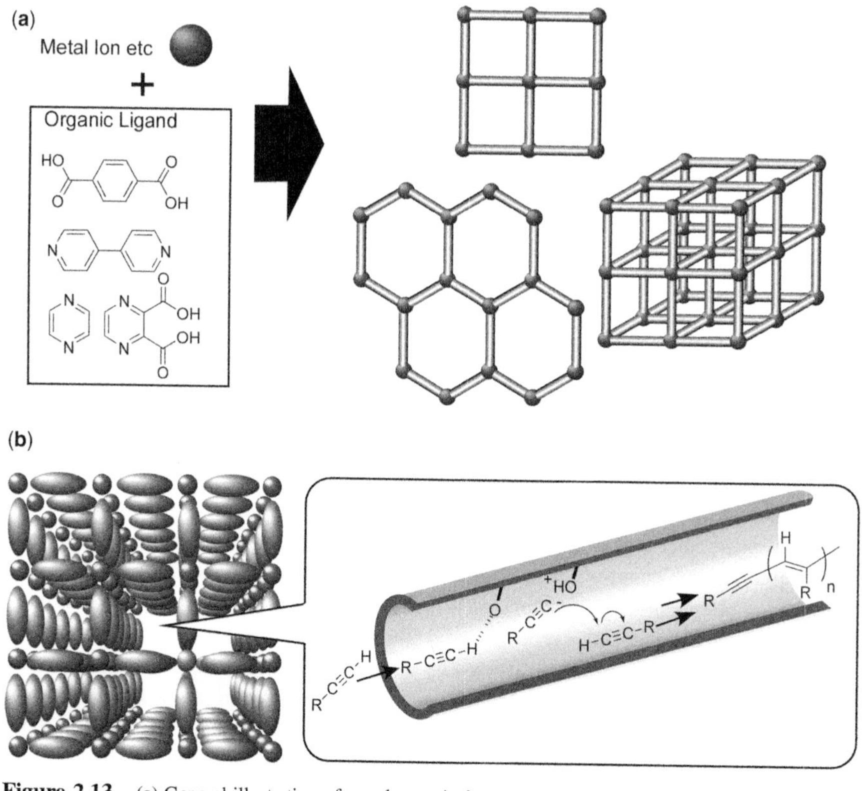

Figure 2.13 (a) General illustration of metal organic frameworks and/or coordination polymers. (b) Acetylene polymerization within designed pores.

These materials are anticipated to have excellent capacities for material storage. Rowsell and Yaghi reported the use of MOFs for storage of gaseous guests such as methane or hydrogen.[81] Yaghi and coworkers also reported that related materials, zeolitic imidazolate frameworks (ZIF), selectively capture carbon dioxide from several different gas mixtures at room temperature.[82] Some kinds of ZIF are capable of storing 28 L of gas per liter of material at standard temperature and pressure. These materials could be promising candidate materials for strategies aimed at ameliorating increasing atmospheric carbon dioxide levels.

Kitagawa and coworkers have also developed various coordination polymers and investigated the unusual properties of trapped guest molecules.[83] For example, they reported a one-dimensional ladder structure of O_2 dimers physisorbed in the nanochannels of a microporous copper coordination polymer.[84] The one-dimensional nanochannels of the coordination polymer result in a confinement effect and restricted geometry leading to a specific molecular assembly. Such mesoscopic arrays cannot exist as a bulk fluid and/or solid. The same research group also reported high levels of selective sorption of acetylene molecules, relative to the similar molecule carbon dioxide, onto the functionalized surface of some coordination polymers.[85] At room temperature, acetylene could be stored at a density 200 times greater than the safe compression limit of free acetylene. As illustrated in Figure 2.13b, the controlled and selective polymerization of substituted acetylenes in one-dimensional specific nanochannels containing basic carboxylate oxygen atoms as catalytic interaction sites on the pore walls was demonstrated.[86]

Structure transcription from supramolecular self-assembled structures to mechanically stable materials has been reported. Several research groups including the Shinkai group[87] and the Shimizu group[88] have prepared inorganic mesoscopic structures from self-assembled objects formed in gels. The class of organic molecules known as gelators can form various supramolecular structures such as helices, fibers, tubes, and ribbons in appropriate solvents. Jung et al. reported the use of crown-ether-based gelators as templates for inorganic nanostructure formation (Fig. 2.14).[89] Sol–gel reaction of tetraethoxysilane (TEOS) on the surface of the supramolecular structures provided silica-coated supramolecular structures. Subsequent removal of the gelator resulted in a hollow tubule structure.

There exist a series of techniques in the syntheses of mesoporous materials, which can be used as powerful methods for preparation of bulk quantities of materials through structure transcription processes.[90,91] In 1990, Yanagisawa et al. first reported the preparation of mesoporous silica with uniform pore size distribution from the layered polysilicate kanemite A.[92] A significant breakthrough in mesoporous materials research came when Mobil scientists disclosed the M41S family of materials.[93] Mesoporous silica materials were prepared through silica formation around template mesoscopic micelle assemblies followed by template removal by appropriate methods such as calcination (Fig. 2.15a). Mesoporous carbon, a carbon material with regular mesopores, was synthesized using mesoporous silica as a hard template (Fig. 2.15b).[94] In this case, the carbon source was introduced into the mesopore channels and then carbonized. Selective removal of the silica framework results in mesoporous carbon materials. This template synthesis is not limited to the synthesis

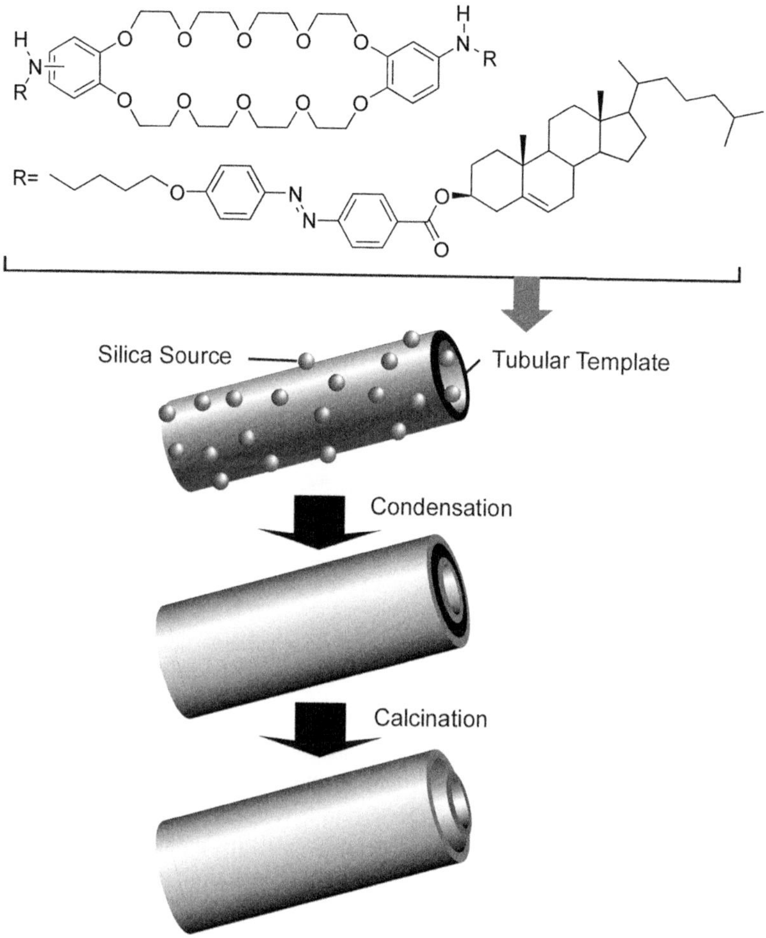

Figure 2.14 Structure transcription from a tubular structure in a gel to silica tubes.[89] (Reprinted with permission from J. H. Jung et al., *Chem. Mater.* **2002**, *14*, 1445–1447. Copyright 2002 American Chemical Society.)

of carbon materials with mesoporous structures. Synthesis of mesoporous carbon nitride was reported by Vinu and coworkers.[95] They also proposed a third method for the synthesis of mesoporous materials, the elemental substitution method.[96] In this method, component elements are substituted by other elements with retention of the mesoporous structure. Vinu and coworkers first applied this concept to the synthesis of mesoporous boron nitride and mesoporous boron carbon nitride.[97]

Because many functional biosystems have mesoscopic dimensions, integration of biocomponents into mesoporous structures should be an interesting research target. Confinement of biomimetic systems in mesoporous media is an important subject when pursued from the view point of biomaterial applications. Incorporation of biomaterials such as proteins,[98,99] amino acids,[100] and vitamins[101] have been investigated

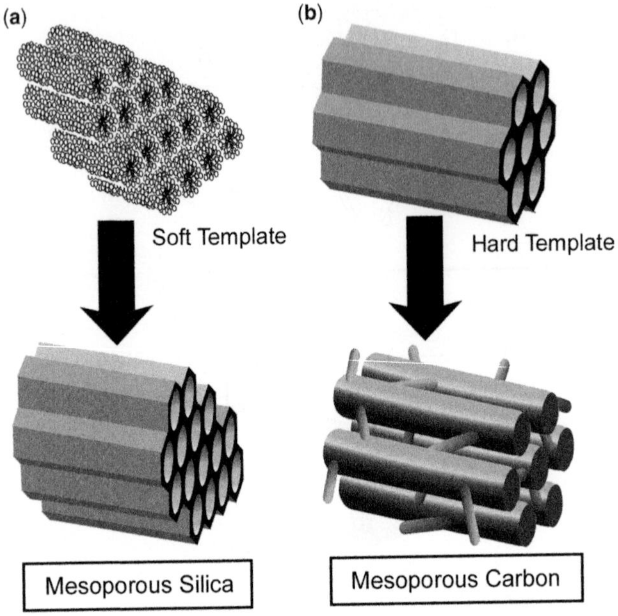

Figure 2.15 General concept of the syntheses of mesoporous materials: (a) mesoporous silica synthesis with soft template; (b) mesoporous carbon synthesis with hard template.

through simple adsorption processes. In contrast to these approaches, Ariga proposed the preparation of mesoscopic hybrid structures of peptide assemblies and mesoporous silica frameworks.[102] The concept of the hybrid is roughly illustrated in Figure 2.16, where mesoporous silica pores include densely packed peptide assemblies. These materials are called proteosilica. For proteosilica preparation, amphiphilic peptides, which have a polar quaternary ammonium salt at the peptide N-terminal and a hydrophobic alkyl chain at the C-terminal, were used as templates for mesoporous silica in both transparent film and powder form. Photochromic dye molecules such as spiropyran and diarylethene can be doped in these chiral environments of the proteosilica films, where asymmetric photoreaction was demonstrated.

Because pores of proteosilica are densely packed with amphiphilic peptides, penetration of materials through the pore channels is obstructed. For further improvement of mesoscopic hybrids between peptides and mesoporous silica, Zhang et al. developed a new synthetic method for mesoporous silica that enables both dense modification of the pore interior with biofunctionality and excellent accessibility for incoming guests (Fig. 2.17).[103] An organosilane compound having a condensable head and a cleavable alkyl tail was used as a structure-directing reagent. The organosilane was covalently attached to the silica framework on sol–gel reaction with TEOS, resulting in mesoporous silica channels filled with the organic group of the template. Cleavage and removal of the alkyl tail by selective hydrolysis of the ester at the C-terminal leaves open pores with a surface covalently functionalized by the alanine residue. This template behaves like a lizard whose head bites the silica wall and whose

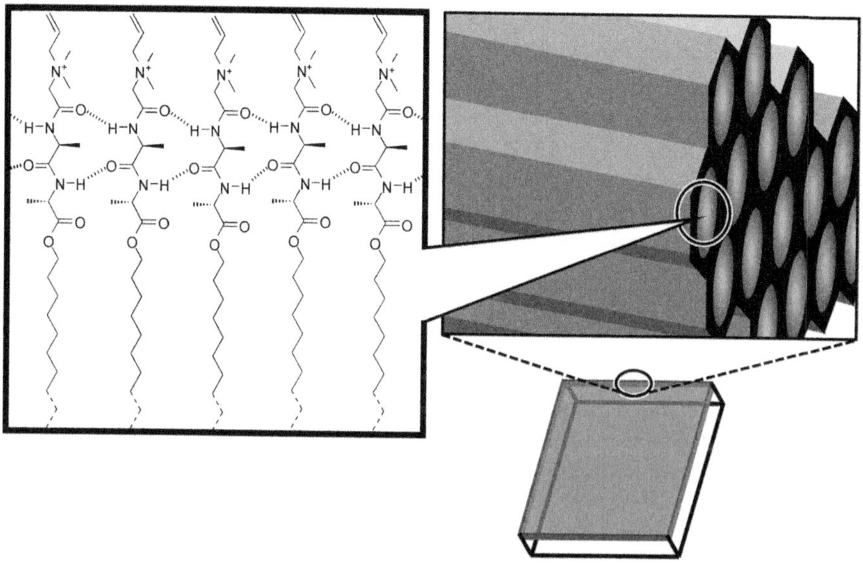

Figure 2.16 Schematic illustration of a proteosilica film.

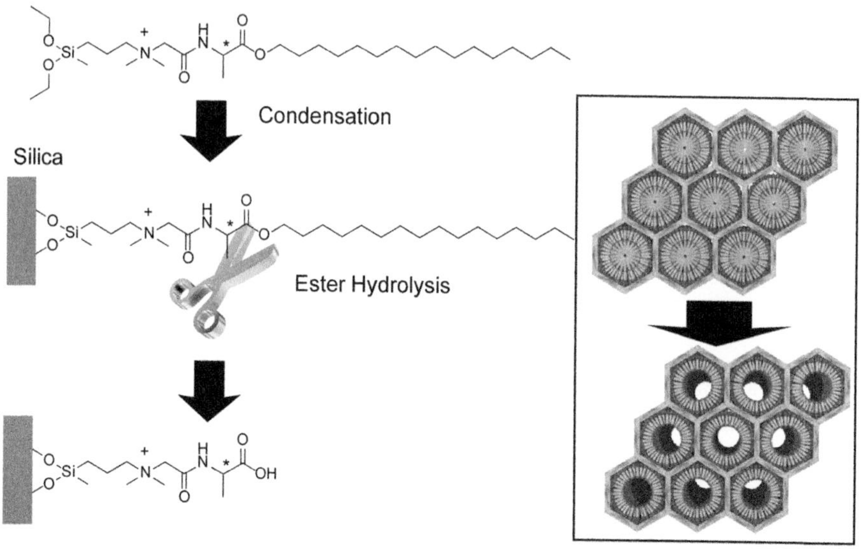

Figure 2.17 Synthetic method for mesoporous silica that enables both dense modification of the pore interior with biofunctions and good accessibility by external guests (Lizard templating method).[103] (Reprinted with permission from Q. Zhang et al., *J. Am. Chem. Soc.* **2004**, *126*, 988–989. Copyright 2004 American Chemical Society.)

tail can be cleaved off. Therefore, the term "lizard template" was applied to describe this method. Aida and coworkers used the hybrid structure (uncleaved material) for reactor applications and illustrated the catalytic capability of unhydrolyzed materials on the acetalization of a ketone, such as cyclohexanone, in ethanol under mild conditions.[104]

Selection of the template structure is one of the most important factors in syntheses of mesoporous materials. Vinu and coworkers developed a novel nanocarbon, carbon nanocage, through template synthesis using three-dimensional, large cage-type, face-centered-cubic mesoporous silica materials (KIT-5) as hard templates (Fig. 2.18a).[105] As shown in TEM images, highly regular structures are preserved even after template removal (Fig. 2.18b). Nitrogen adsorption and desorption

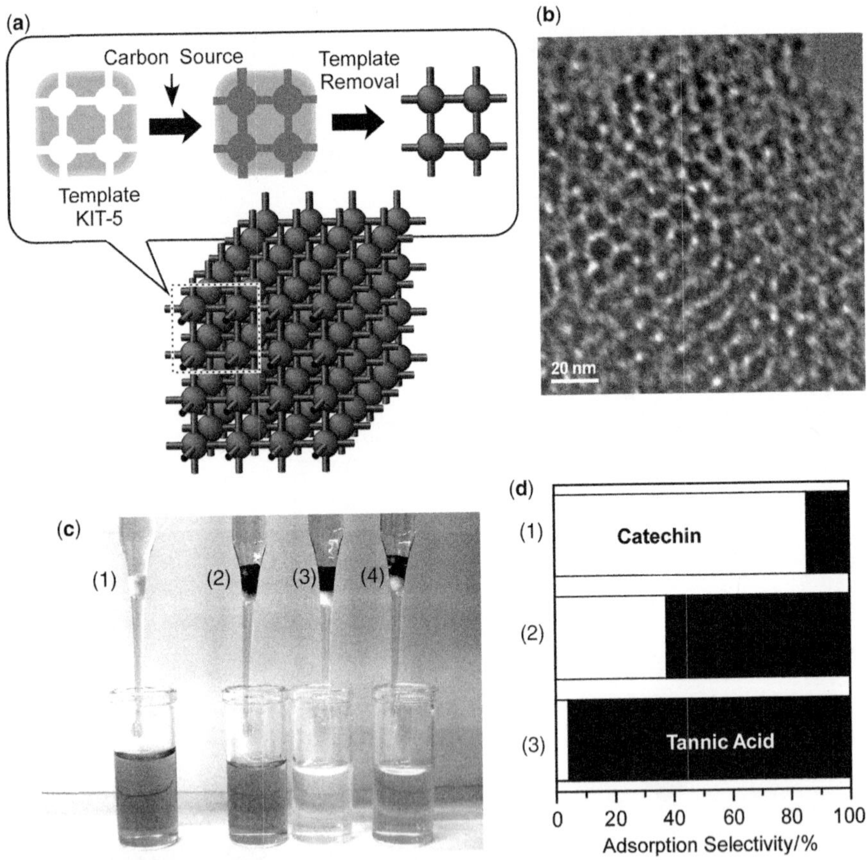

Figure 2.18 (a) Synthetic scheme of carbon nanocage materials. (b) TEM image of carbon nanocage material. (c) Removal experiments of Alizarin Yellow by a simple filtration procedure: (1) no carbon, (2) activated carbon, (3) carbon nanocage, and (4) conventional mesoporous carbon CMK-3. (d) Competitive adsorption of catechin and tannic acid: (1) activated carbon, (2) conventional mesoporous carbon CMK-3, and (3) carbon nanocage material.[106] (Reprinted with permission from K. Ariga et al., *J. Am. Chem. Soc.* **2007**, *129*, 11022–11023. Copyright 2007 American Chemical Society.)

experiments revealed that the specific surface area and specific pore volume reaches 1600 $m^2\ g^{-1}$ and 2.1 $cm^3\ g^{-1}$, respectively, and are much larger than those reported for conventional mesoporous carbon materials. Its huge pore volume results in superior properties in separation applications.[106] Carbon nanocages can completely remove a typical dye (Alizarin Yellow) by a simple filtration procedure [in Fig. 2.18c: (1) no carbon, (2) activated carbon, (3) carbon nanocage, and (4) conventional mesoporous carbon CMK-3]. In addition, simple, economical, and toxic solvent-free

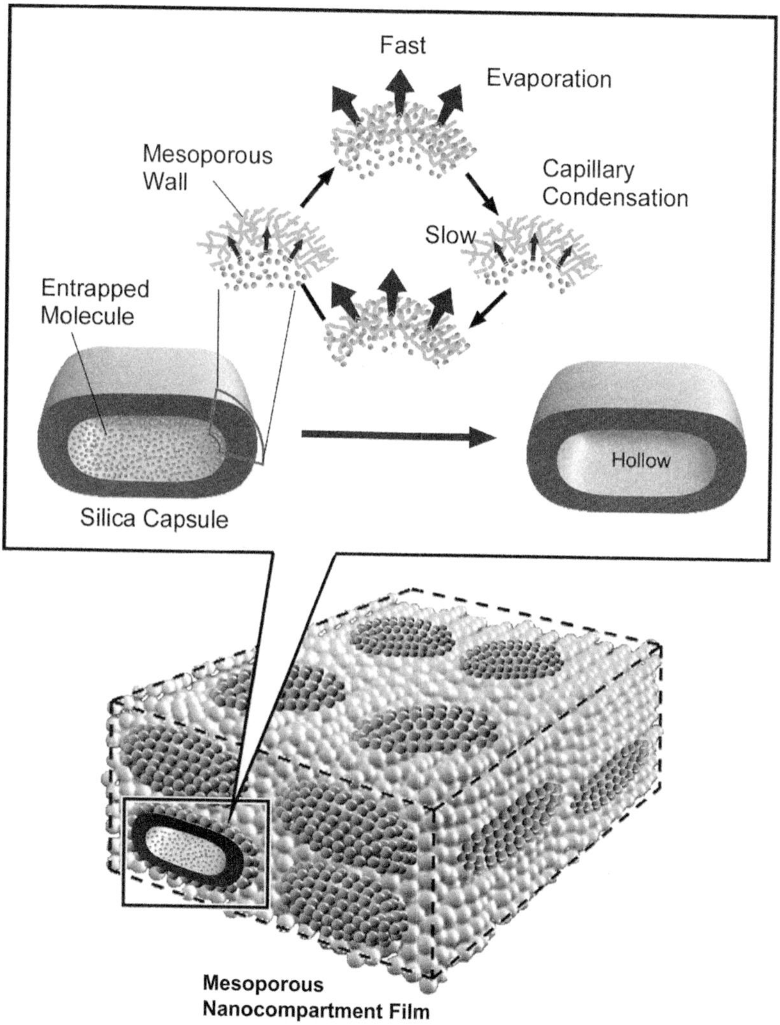

Figure 2.19 Mesoporous nanocompartment films and stimuli-free, automodulated stepwise release of materials based on the nonequilibrated rates between evaporation and capillary penetration.[108] (Reprinted with permission from Q. Ji et al., *J. Am. Chem. Soc.* **2008**, *130*, 2376–2377. Copyright 2008 American Chemical Society.)

one-pot separation of tea components, such as catechin and tannic acids, has also been demonstrated. Competitive adsorption of catechin and tannic acid to the carbon nanocage material results in highly selective adsorption of tannic acid (about 95%) using an easy one-pot process [in Fig. 2.18d: (1) activated carbon, (2) conventional mesoporous carbon CMK-3, and (3) carbon nanocage].

Silica capsule structures with mesoporous walls have been invented.[107] The mesoporous silica capsules were co-assembled with silica particles by the LbL technique with the aid of appropriate polyelectrolytes, resulting in mesoporous nanocompartment films (Fig. 2.19).[108] The mesoporous nanocompartment films possess special molecular encapsulation and release capabilities so that stimuli-free, automodulated, stepwise release of water or drug molecules was achieved through the mesopore channels of robust silica capsule containers embedded in the film. Stepwise release of water was reproducibly observed that originates in the nonequilibrated rates between evaporation of water from the mesopore channels to the exterior and the capillary penetration of water from container interior to the mesopore channels. This process was generalized to the evaporation of other substances such as fragrances. Application was also tested in the controlled release of the sunscreen UV absorber (UV-S1) for circumvention of its rapid dissolution in water and prolongation of its prophylactic effect towards harmful ultraviolet radiation. UV-S1 was entrapped within the mesoporous nanocompartment films and was released in a prolonged stepwise mode. The nanocompartment films developed in this research are promising materials for drug delivery since they allow gradual release of therapeutic agents, with likely related improvements in their efficacy.

2.5 FUTURE PERSPECTIVES

In this chapter, several aspects of supramolecular chemistry at the mesoscopic scale, including basic supramolecular chemistry such as molecular recognition, methods for preparing mesoscopic assemblies and materialization of mesoscopic structures, have been introduced. Although the selected examples cannot cover every feature of supramolecular chemistry at the mesoscale, together they indicate that the science and technology of this field maintains a position at the forefront of recent research and development. However, many challenges remain to be met for supramolecular chemistry in the mesoscopic range.

Bottom-up approaches based on supramolecular chemistry have only recently reached the mesoscopic scale range. Top-down nanofabrication is also available at this level following miniaturization from the macroscopic range. Therefore, these complementary approaches can meet at the mesoscale. Combining these two approaches at the mesoscale should become increasingly important and some recent research has realized this concept, such as that conducted by Lvov and coworkers, shown in Figure 2.20. This example demonstrates cantilever objects using both top-down (photolithography) and bottom-up approaches (LbL assembly).[109] Apart from this example, combination of supramolecular chemistry with top-down nanofabrication can create many possibilities such as ultrafine microsensor arrays and ultrasmall

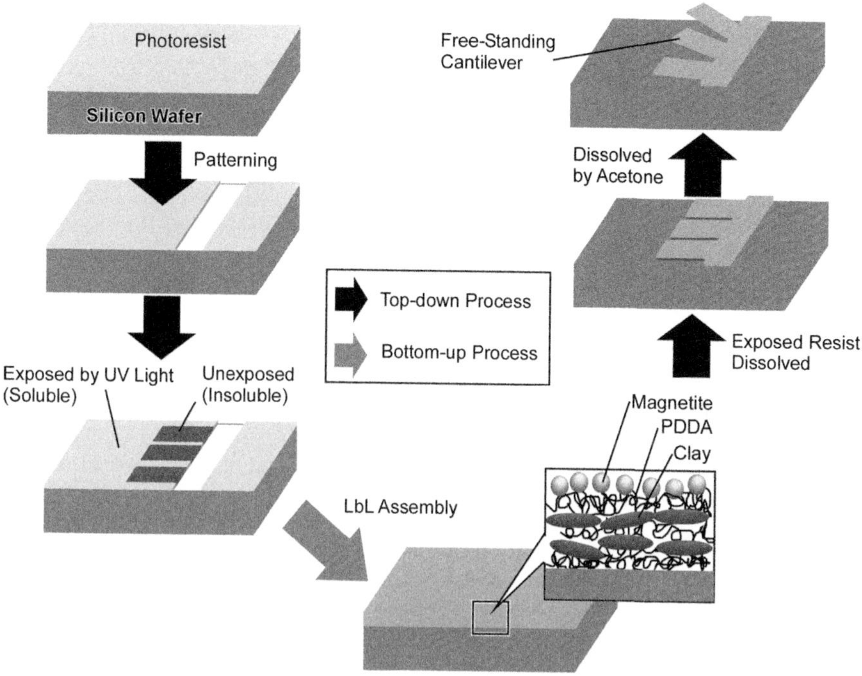

Figure 2.20 Cantilever fabrication based on both top-down fabrication and bottom-up assembly.[109] (Reprinted with permission from F. Hua et al., *Nano Lett.* **2004**, *4*, 823–825. Copyright 2004 American Chemical Society.)

sequential reactors. If we can successfully merge the top-down and the bottom-up approaches at the mesoscale, all dimensions from bulk to molecular or atomic scale will become available to supramolecular science.

ACKNOWLEDGMENT

The research described in this review was partially supported by Grant-in-Aid for Scientific Research on Priority Areas "Chemistry of Coordination Space" and a Grant-in-Aid for Science Research in a Priority Area "Super-Hierarchical Structures" from Ministry of Education, Science, Sports, and Culture, Japan; World Premier International Research Center Initiative (WPI Initiative) on Materials Nanoarchitronics, from Ministry of Education, Science, Sports, and Culture, Japan; and Grants-in-Aid for Scientific Research (B) from Japan Society for the Promotion of Science.

REFERENCES

1. K. Ariga, T. Nakanishi, J. P. Hill, *Curr. Opin. Colloid Interface Sci.* **2007**, *12*, 106–120.
2. J. P. Hill, Y. Wakayama, W. Schmitt, T. Tsuruoka, T. Nakanishi, M. L. Zandler, A. M. McCarty, F. D'Souza, L. R. Milgrom, K. Ariga, *Chem. Commun.* **2006**, 2320–2322.

3. T. Nakanishi, M. Miyashita, T. Michinobu, Y. Wakayama, T. Tsuruoka, K. Ariga, D. G. Kurth, *J. Am. Chem. Soc.* **2006**, *128*, 6328–6329.
4. J. P. Hill, Y. Wakayama, K. Ariga, *Phys. Chem. Chem. Phys.* **2006**, *8*, 5034–5037.
5. Y. Wakayama, J. P. Hill, K. Ariga, *Surf. Sci.* **2007**, *601*, 3984–3987.
6. J. P. Hill, Y. Wakayama, M. Akada, K. Ariga, *J. Phys. Chem. C* **2007**, *111*, 16174–16180.
7. T. Nakanishi, W. Schmitt, T. Michinobu, D. G. Kurth, K. Ariga, *Chem. Commun.* **2005**, 5982–5984.
8. T. Nakanishi, K. Ariga, T. Michinobu, K. Yoshida, H. Takahashi, T. Teranishi, H. Möhwald, D. G. Kurth, *Small* **2007**, *3*, 2019–2023.
9. T. Nakanishi, T. Michinobu, K. Yoshida, N. Shirahata, K. Ariga, H. Möhwald, D. G. Kurth, *Adv. Mater.* **2008**, *20*, 443–446.
10. K. J. B. van Bommel, A. Friggeri, S. Shinkai, *Angew. Chem. Int. Ed.* **2003**, *42*, 980–999.
11. K. Ariga, *J. Nanosci. Nanotechnol.* **2004**, *4*, 23–34.
12. K. Ariga, E. V. Anslyn, *J. Org. Chem.* **1992**, *57*, 417–419.
13. D. M. Kneeland, K. Ariga, V. M. Lynch, C. Y. Huang, E. V. Anslyn, *J. Am. Chem. Soc.* **1993**, *115*, 10042–10055.
14. J. Smith, K. Ariga, E. V. Anslyn, *J. Am. Chem. Soc.* **1993**, *115*, 362–364.
15. P. D. Beer, P. A. Gale, *Angew. Chem. Int. Ed.* **2001**, *40*, 486–516.
16. J. P. Hill, A. L. McCarty, F. D'Souza, J. Labuta, C. Redshaw, M. R. J. Elsegood, M. Aoyagi, T. Nakanishi, K. Ariga, *Inorg. Chem.* **2006**, *45*, 8288–8296.
17. Y. Xie, J. P. Hill, R. Charvet, K. Ariga, *J. Nanosci. Nanotechnol.* **2007**, *7*, 2969–2993.
18. J. S. Nowick, J. S. Chen, G. Noronha, *J. Am. Chem. Soc.* **1993**, *115*, 7636–7644.
19. R. P. Bonar-Law, J. K. M. Sanders, *J. Am. Chem. Soc.* **1995**, *117*, 259–271.
20. H. Kitano, H. Ringsdorf, *Bull. Chem. Soc. Jpn.* **1985**, *58*, 2826–2828.
21. K. Ariga, T. Kunitake, *Acc. Chem. Res.* **1998**, *31*, 371–378.
22. K. Kurihara, K. Ohto, Y. Honda, T. Kunitake, *J. Am. Chem. Soc.* **1991**, *113*, 5077–5079.
23. T. Kawahara, K. Kurihara, T. Kunitake, *Chem. Lett.* **1992**, 1839–1842.
24. K. Taguchi, K. Ariga, T. Kunitake, *Chem. Lett.* **1995**, 701–702.
25. Y. Oishi, Y. Torii, T. Kato, M. Kuramori, K. Suehiro, K. Ariga, K. Taguchi, A. Kamino, H. Koyano, T. Kunitake, *Langmuir* **1997**, *13*, 519–524.
26. M. Sakurai, H. Tamagawa, T. Furuki, Y. Inoue, K. Ariga, T. Kunitake, *Chem. Lett.* **1995**, 1001–1002.
27. M. Sakurai, H. Tamagawa, Y. Inoue, K. Ariga, T. Kunitake, *J. Phys. Chem. B* **1997**, *101*, 4810–4816.
28. H. Tamagawa, M. Sakurai, Y. Inoue, K. Ariga, T. Kunitake, *J. Phys. Chem. B* **1997**, *101*, 4817–4825.
29. D. Y. Sasaki, K. Kurihara, T. Kunitake, *J. Am. Chem. Soc.* **1991**, *113*, 9685–9686.
30. B. Springs, P. Haake, *Bioorg. Chem.* **1977**, *6*, 181–190.
31. M. Onda, K. Yoshihara, H. Koyano, K. Ariga, T. Kunitake, *J. Am. Chem. Soc.* **1996**, *118*, 8524–8530.
32. T. Aida, K. Tajima, *Angew. Chem. Int. Ed.* **2001**, *40*, 3803–3806.
33. Y. Lu, Y. Yang, A. Sellinger, M. Lu, J. Huang, H. Fan, R. Haddad, G. Lopez, A. R. Burns, D. Y. Sasaki, J. Shelnutt, C. J. Brinker, *Nature* **2001**, *410*, 913–917.
34. M. Ikegame, K. Tajima, T. Aida, *Angew. Chem. Int. Ed.* **2003**, *42*, 2154–2157.
35. A. Okabe, T. Fukushima, K. Ariga, T. Aida, *Angew. Chem. Int. Ed.* **2002**, *41*, 3414–3417.
36. N. Kimizuka, N. Oda, T. Kunitake, *Inorg. Chem.* **2000**, *39*, 2684–2689.
37. T. Shimizu, M. Masuda, H. Minamikawa, *Chem. Rev.* **2005**, *105*, 1401–1443.
38. N. Yamada, K. Ariga, M. Naito, K. Matsubara, E. Koyama, *J. Am. Chem. Soc.* **1998**, *120*, 12192–12199.
39. K. Ariga, J. Kikuchi, M. Naito, E. Koyama, N. Yamada, *Langmuir* **2000**, *16*, 4929–4939.
40. K. Katagiri, K. Ariga, J. Kukuchi, *Chem. Lett.* **1999**, 661–662.
41. K. Katagiri, M. Hashizume, K. Ariga, T. Terashima, J. Kikuch, *Chem. Eur. J.* **2007**, *13*, 5272–5281.
42. K. Katagiri, R. Hamasaki, K. Ariga, J. Kikuchi, *Langmuir* **2002**, *18*, 6709–6711.
43. K. Katagiri, R. Hamasaki, K. Ariga, J. Kikuchi, *J. Am. Chem. Soc.* **2002**, *124*, 7892–7893.
44. A. D. Dinsmore, M. F. Hsu, M. G. Nikolaides, M. Marquez, A. R. Bausch, D. A. Weitz, *Science* **2002**, *298*, 1006–1009.

45. Y. Okahata, K. Ariga; H. Nakahara, K. Fukuda, *J. Chem. Soc., Chem. Commun.* **1986**, 1069–1071.
46. K. Ariga, Y. Okahata, *J. Am. Chem. Soc.* **1989**, *111*, 5618–5622.
47. I. Yamashita, *Thin Solid Films* **2001**, *393*, 12–18.
48. G. Decher, *Science* **1997**, *277*, 1232–1237.
49. K. Ariga, J. P. Hill, Q. Ji, *Phys. Chem. Chem. Phys.* **2007**, *9*, 2319–2340.
50. R. K. Iler, *J. Colloid Interface Sci.* **1966**, *21*, 569–594.
51. G. Decher, J. D. Hong, J. Schmitt, *Thin Solid Films* **1992**, *210*, 831–835.
52. W. B. Stockton, M. F. Rubner, *Macromolecules* **1997**, *30*, 2717–2725.
53. H. Lee, L. J. Kepley, H. G. Hong, T. E. Mallouk, *J. Am. Chem. Soc.* **1988**, *110*, 618–620.
54. Y. Shimazaki, M. Mitsuishi, S. Ito, M. Yamamoto, *Langmuir* **1997**, *13*, 1385–1387.
55. Y. Lvov, K. Ariga, M. Onda, I. Ichinose, K. Kunitake, *Colloid Surf. A* **1999**, *146*, 337–346.
56. Y. Lvov, M. Onda, K. Ariga, T. Kunitake, *J. Biomater. Sci. Polym. Ed.* **1998**, *9*, 345–355.
57. S. S. Shiratori, M. F. Rubner, *Macromolecules* **2000**, *33*, 4213–4219.
58. Y. Lvov, K. Ariga, T. Kunitake, *Chem. Lett.* **1994**, 2323–2326.
59. Y. Lvov, K. Ariga, I. Ichinose, T. Kunitake, *J. Am. Chem. Soc.* **1995**, *117*, 6117–6123.
60. Y. Lvov, K. Ariga, I. Ichinose, T. Kunitake, *J. Chem. Soc. Chem. Commun.* **1995**, 2313–2314.
61. M. Onda, Y. Lvov, K. Ariga, T. Kunitake, *J. Ferment. Bioeng.* **1996**, *82*, 502–506.
62. M. Onda, Y. Lvov, K. Ariga, T. Kunitake, *Biotechnol. Bioeng.* **1996**, *51*, 163–167.
63. F. Caruso, D. N. Furlong, K. Ariga, I. Ichinose, T. Kunitake, *Langmuir* **1998**, *14*, 4559–4565.
64. M. Onda, K. Ariga, T. Kunitake, *J. Biosci. Bioeng.* **1999**, *87*, 69–75.
65. Y. M. Lvov, Z. Lu, J. B. Schenkman, X. Zu, J. F. Rusling, *J. Am. Chem. Soc.* **1998**, *120*, 4073–4080.
66. Y. Lvov, H. Haas, G. Decher, H. Möhwald, A. Mikhailov, B. Mtchedlishvily, E. Morgunova, B. Vainshtein, *Langmuir* **1994**, *10*, 4232–4236.
67. K. Ariga, Y. Lvov, M. Onda, I. Ichinose, T. Kunitake, *Chem. Lett.* **1997**, 125–126.
68. Y. Lvov, K. Ariga, I. Ichinose, T. Kunitake, *Langmuir* **1996**, *12*, 3038–3044.
69. T. Sasaki, Y. Ebina, T. Tanaka, M. Harada, M. Watanabe, *Chem. Mater.* **2001**, *13*, 4661–4667.
70. G. S. Lee, Y.-J. Lee, K. B. Yoon, *J. Am. Chem. Soc.* **2001**, *123*, 9769–9779.
71. T. Matsui, A. Yamaguchi, Y. Takeoka, M. Rikukawa, K. Sanui, *Chem. Commun.* **2002**, 1094–1095.
72. J. Liu, L. Cheng, B. Liu, S. Dong, *Langmuir* **2000**, *16*, 7471–7476.
73. G. Decher, J. D. Hong, *Makromol. Chem. Macromol. Symp.* **1991**, *46*, 321–327.
74. I. Ichinose, K. Fujiyoshi, S. Mizuki, Y. Lvov, T. Kunitake, *Chem. Lett.* **1996**, 257–258.
75. Y. Lvov, F. Essler, G. Decher, *J. Phys. Chem.* **1993**, *97*, 13773–13777.
76. K. Ariga, Y. Lvov, T. Kunitake, *J. Am. Chem. Soc.* **1997**, *119*, 2224–2231.
77. Y. Lvov, K. Ariga, M. Onda, I. Ichinose, T. Kunitake, *Langmuir* **1997**, *13*, 6195–6203.
78. C. Sanchez, B. Julián, P. Belleville, M. Pooall, *J. Mater. Chem.* **2005**, *15*, 3559–3592.
79. M. Eddaoudi, D. B. Moler, H. L. Li, B. L. Chen, T. M. Reineke, M. O'Keeffe, O. M. Yaghi, *Acc. Chem. Res.* **2001**, *34*, 319–330.
80. H. M. El-Kaderi, J. R. Hunt, J. L. Mendoza-Cortes, A. P. Cote, R. E. Taylor, M. O'Keeffe, O. M. Yaghi, *Science* **2007**, *316*, 268–272.
81. J. L. C. Rowsell, O. M. Yaghi, *Angew. Chem. Int. Ed.* **2005**, *44*, 4670–4679.
82. B. Wang, A. P. Côté, H. Furukawa, M. O'Keeffe, O. M. Yaghi, *Nature* **2008**, *453*, 207–212.
83. S. Kitagawa, R. Kitaura, S. Noro, *Angew. Chem. Int. Ed.* **2004**, *43*, 2334–2375.
84. R. Kitaura, S. Kitagawa, Y. Kubota, T. C. Kobayashi, K. Kindo, Y. Mita, A. Matsuo, M. Kobayashi, H.-C. Chang, T. C. Ozawa, M. Suzuki, M. Sakata, M. Takata, *Science* **2002**, *298*, 2358–2361.
85. R. Matsuda, R. Kitaura, S. Kitagawa, Y. Kubota, R. V. Belosludov, T. C. Kobayashi, H. Sakamoto, T. Chiba, M. Takata, Y. Kawazoe, Y. Mita, *Nature* **2005**, *436*, 238–241.
86. T. Uemura, R. Kitaura, Y. Ohta, M. Nagaoka, S. Kitagawa, *Angew. Chem. Int. Ed.* **2006**, *45*, 4112–4116.
87. J. H. Jung, Y. Ono, S. Shinkai, *Langmuir* **2000**, *16*, 1643–1649.
88. Q. Ji, R. Iwaura, T. Shimizu, *Chem. Mater.* **2007**, *19*, 1329–1334.
89. J. H. Jung, H. Kobayashi, K. J. C. van Bommel, S. Shinkai, T. Shimizu, *Chem. Mater.* **2002**, *14*, 1445–1447.
90. A. Vinu, K. Z. Hossain, K. Ariga, *J. Nanosci. Nanotechnol.* **2005**, *5*, 347–371.

91. K. Ariga, A. Vinu, J. P. Hill, T. Mori, *Coord. Chem. Rev.* **2007**, *251*, 2562–2591.
92. T. Yanagisawa, T. Shimizu, K. Kuroda, C. Kato, *Bull. Chem. Soc. Jpn.* **1990**, *63*, 988–992.
93. C. T. Kresge, M. E. Leonowicz, W. J. Roth, J. C. Vartuli, J. S. Beck, *Nature* **1992**, *359*, 710–712.
94. R. Ryoo, S. H. Joo, S. Jun, *J. Phys. Chem. B* **1999**, *103*, 7743–7746.
95. A. Vinu, K. Ariga, T. Mori, T. Nakanishi, S. Hishita, D. Golberg, Y. Bando, *Adv. Mater.* **2005**, *17*, 1648–1652.
96. A. Vinu, T. Mori, K. Ariga, *Sci. Technol. Adv. Mater.* **2006**, *7*, 753–771.
97. A. Vinu, M. Terrones, D. Golberg, S. Hishita, K. Ariga, T. Mori, *Chem. Mater.* **2005**, *17*, 5887–5890.
98. A. Vinu, M. Miyahara, K. Ariga, *J. Phys. Chem. B* **2005**, *109*, 6436–6441.
99. A. Vinu, M. Miyahara, K. Ariga, *J. Nanosci. Nanotechnol.* **2006**, *6*, 1510–1532.
100. A. Vinu, K. Z. Hossain, G. Satish Kumar, K. Ariga, *Carbon* **2006**, *44*, 530–536.
101. M. Hartmann, A. Vinu, G. Chandrasekar, *Chem. Mater.* **2005**, *17*, 829–833.
102. K. Ariga, *Chem. Rec.* **2004**, *3*, 297–307.
103. Q. Zhang, K. Ariga, A. Okabe, T. Aida, *J. Am. Chem. Soc.* **2004**, *126*, 988–989.
104. W. Otani, K. Kinbara, Q. Zhang, K. Ariga, T. Aida, *Chem. Eur. J.* **2007**, *13*, 1731–1736.
105. A. Vinu, M. Miyahara, V. Sivamurugan, T. Mori, K. Ariga, *J. Mater. Chem.* **2005**, *15*, 5122–5127.
106. K. Ariga, A. Vinu, M. Miyahara, J. P. Hill, T. Mori, *J. Am. Chem. Soc.* **2007**, *129*, 11022–11023.
107. J.-S. Yu, S. B. Yoon, Y. J. Lee, K. B. Yoon, *J. Phys. Chem. B* **2005**, *109*, 7040–7045.
108. Q. Ji, M. Miyahara, J. P. Hill, S. Acharya, A. Vinu, S. B. Yoon, J.-S. Yu, K. Sakamoto, K. Ariga, *J. Am. Chem. Soc.* **2008**, *130*, 2376–2377.
109. F. Hua, T. Cui, Y. M. Lvov, *Nano Lett.* **2004**, *4*, 823–825.

Part One

Organic–Inorganic Hybrid Nanomaterials

Chapter 3

Silica-Based Mesoporous Organic–Inorganic Hybrid Materials

FRANK HOFFMANN AND MICHAEL FRÖBA

The Supramolecular Chemistry of Organic–Inorganic Hybrid Materials. Edited by Knut Rurack and Ramón Martínez-Máñez

3.1 AIMS, SCOPE, AND ORGANIZATION OF THIS CHAPTER

In this chapter, we summarize and highlight recent developments in the field of mesoporous organic–inorganic hybrid materials that are based on silica. Our aim is *not* to render the "whole story" in all details. For this, we will refer the reader needing more specialized information to the respective comprehensive review articles and to two books.[1–11] Instead, the main focus will be placed on (1) the explanation of the principles and underlying concepts that are relevant to the different methods of functionalization of silica-based mesoporous materials and (2) to present some interesting and illustrative examples of the rich world of hybrid materials. The organization of this chapter is as follows: In the Introduction, some remarks concerning the concept and general aspects of organic–inorganic hybrid (meso)porous materials are made, followed by a section in which the genealogy and emergence of this class of materials is presented. In Section 3.4, the three principal ways to organofunctionalize mesoporous silica phases are elaborated on, their respective assets and drawbacks are expounded, and a number of examples of recent work are compiled. In Section 3.5, finally, a survey of selected highlights pointing to current formative trends is presented.

3.2 INTRODUCTION

The general interest in developing hybrid materials arises from the expectation that the assembly of organic and inorganic building blocks in a single material can combine their particular advantages. That is, first of all the manifold chemical functionality of the organic component combined with the chemical, mechanical, and thermal robustness of the inorganic part, which is itself functionally very limited. Through this symbiosis, materials are formed whose properties differ significantly from those of the single, isolated components; completely new properties of these hybrid materials can emerge. This is valid in particular for porous materials, that is, for those with high surface-to-volume ratios. Although also compact composite and hybrid materials exhibit modified bulk properties, such as density, hardness, elasticity, and so on, the actual advantage of *porous* hybrid materials arises out of the number of and accessibility to localized functional or interacting centers. Of course, this can be realized better the larger the interface and the more functional centers are placed at this interface. To give only two examples: Tuning the polarity of the pore surfaces by integrating organic components into an inorganic matrix is supposed to extend the spectrum of materials that are optimized for a particular chromatographic problem. Furthermore, by functionalizing an inorganic host material with organic functionalities like C—C multiple bonds, alcohols, thiols, carboxylic or sulfonic acids, amines, etc., for instance, chemical or biochemical reactions can be carried out on a stable, solid, stationary inorganic matrix.

Three pathways are available for the synthesis of porous hybrid materials based on (organo-)silica units: (1) the subsequent modification of the pore surface of a purely inorganic silica material (grafting), (2) the simultaneous condensation of corresponding silica and organosilica precursors (co-condensation), and (3) the incorporation of organic groups as bridging components directly and specifically into the pore walls by the use of bis- or multi-silylated single-source organosilica precursors [creation of (periodic) mesoporous organosilicas; PMOs]. The syntheses of the supporting silica phases are carried out usually under so-called *hydrothermal conditions* (in an autoclave), that is, in an aqueous medium over 100°C and 1 bar,[12] whereas syntheses of PMOs are carried out commonly at temperatures slightly below 100°C.

From a higher-ranking viewpoint concerning the topic structuring of matter, all three possibilities, which will be explained in detail below, belong to the so-called *bottom-up approach*, which is the constitution of an *organized unit* composed of (much) smaller building units.* When not only loose unstructured aggregates but ordered porous materials should be formed, then certain aspects have to be considered, which are addressed in the following remarks.

3.2.1 Porosity: Thermodynamic and Kinetic Stability

Concerning the assembly of inanimate matter, nature shows a tendency towards compact forms, or in other words: the creation of interfaces is being avoided due to the energetic cost of its formation. However, this active principle does not exclude the existence of porous matter, for example, materials with a high surface-to-volume ratio, as long as they are sufficiently kinetically stable. This is demonstrated by the existence of diverse porous objects, whether from natural sources or manmade, ranging from porous volcanic rocks or basalt, corks, clay jugs, and styrofoam to skeletal inorganic structures of diatoms and sponges. These general considerations can also be applied to mesostructured silica and organosilica phases. The above-mentioned principle holds for the synthesis of composite materials, too, being composed of micellar aggregates of surfactants on one hand and the surrounding silica or organosilica framework on the other hand. The formation of micelles from surfactants implies minimizing of the interface. In addition, the negative imprint of this liquid-crystalline phase, the (organo-)silica phase as such, is also in a compact and complete space-filling state. The porosity is generated only by removal of the micellar phase. The materials obtained in this way remain in their (X-ray amorphous) state predetermined through the condensation process. These materials are only meta-stable from a thermodynamic

*An impressive example of the opposite approach, the top-down approach, can be found in the chip industry, where structures of only 45 nm length can be realized, as Intel demonstrated with the actual Xeon chip.

point of view with respect to both the pore system and the molecular structure; they exist in their specific (porous) state only due to their kinetic stability.

3.2.2 Periodicity: Spatially and/or Molecularly Ordered versus Amorphous

The result of a template-directed synthesis is not necessarily an *ordered* material. This is only the case when the micelles form a spatially ordered array. Due to the fact that in such synthesis many components are involved [water, acid or base, (organo-)silica precursor, surfactants, etc.], the attractive and repulsive energy terms which, at a given composition, lead to a particular phase (unordered, wormhole-like, lamellar, hexagonal, cubic, etc.) are hard to estimate. In the respective phase diagram of such a multicomponent system, the region in which an ordered phase will actually be formed is often very small. Meanwhile, there are several recipes available for the preparation of high-quality pure silica mesoporous phases. Therefore, concerning the grafting procedure, which leaves the parent phase unaffected, this is not an issue. In contrast, in co-condensation reactions or when dealing with PMO syntheses much less is known about the phase behavior of the involved species, so that predicting which molar ratios of the reaction mixture will lead to which kind of phase is virtually impossible. For the experimentalists this often means that the phase diagram has to be scanned by the trial-and-error method. Other difficulties arising in practice when applying these two methods will be explained in the respective sections. For now, we would like to address only the more general question, whether an ordered phase is advantageous over unordered ones and suggest the following aspects for consideration.

The ordered and uniform porosity of mesoporous organic–inorganic hybrid materials may have the following advantages: (1) the narrow pore size distribution is useful in all size selective applications (sieve effect), for instance chromatography or catalysis; (2) the absence of slit-like or bottle-neck pores guarantees a fast and efficient mass transport of species with appropriate sizes (i.e., those that are smaller than the pore diameter) within these materials, which is advantageous, for instance, for sorption applications like wastewater treatment; and (3) regarding the field of host–guest chemistry the regular arrangement of the pores provides opportunities to design materials with completely new properties, for instance, those in which guest molecules or clusters of semiconductors are adsorbed in a uniform manner and orientation and thus may show interesting collective properties like nonlinear optical behavior or special magnetic coupling effects; this may be of particular interest for PMO thin films and monoliths.

3.3 GENEALOGY OF SILICA-BASED MESOPOROUS ORGANIC–INORGANIC HYBRID MATERIALS

To provide readers with a more systematic view on silica-based mesoporous organic–inorganic hybrid materials, a short review of the genealogy of this and related material classes will be given in the following (see Fig. 3.1).

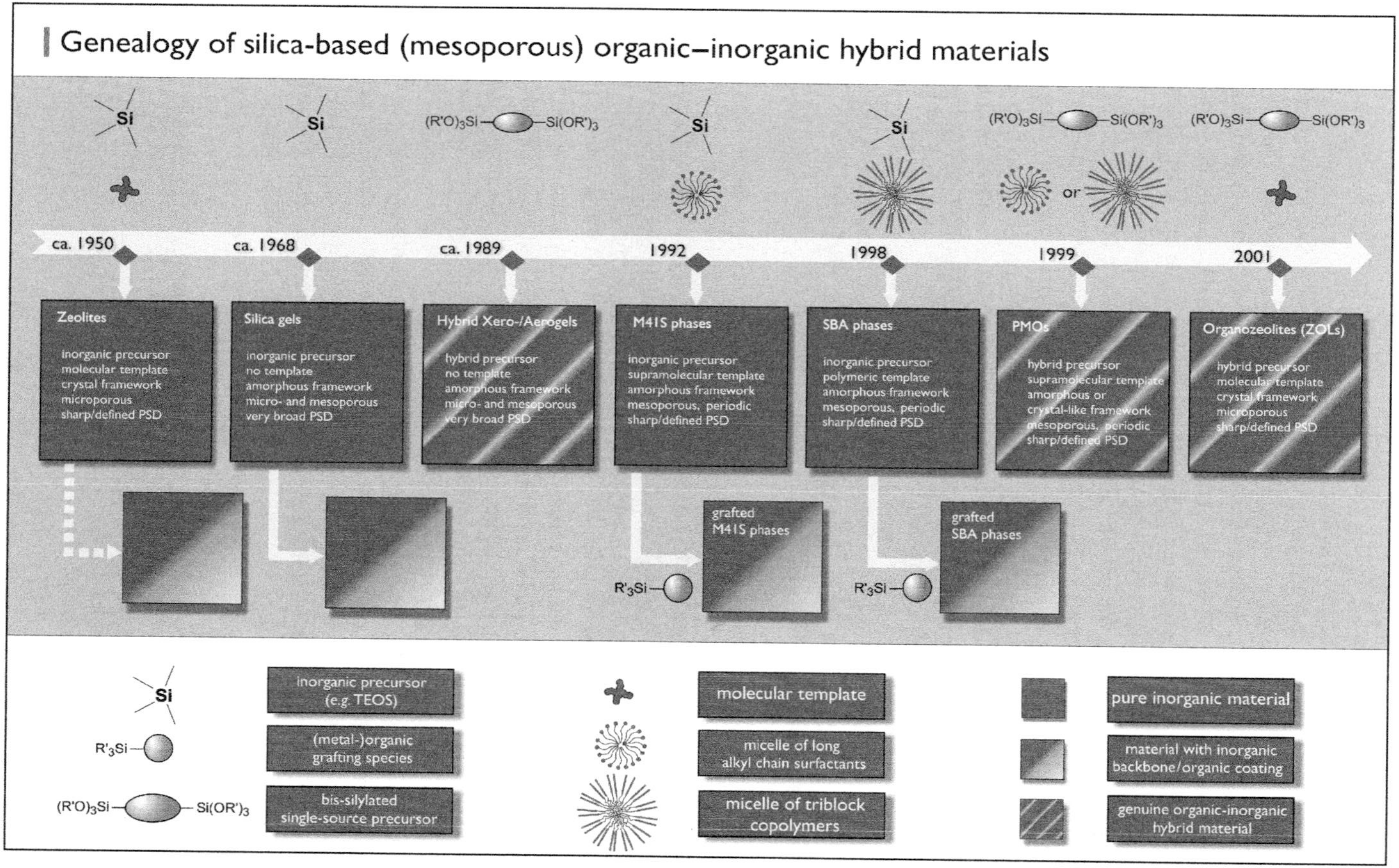

Figure 3.1 Synoptic scheme of the history of the development of silica-based organic–inorganic hybrid and related classes of materials (PSD = pore size distribution, TEOS = tetraethoxysilane).

3.3.1 Zeolites

As is well known, the history of ordered *microporous*† silica-based materials began with the discovery of some natural zeolites in 1756 by the Swedish mineralogist Freiherr Axel Frederic von Cronstedt. These zeolites were considered mineralogical curiosities, which are formed as secondary minerals in cavities and hydrothermal veins of volcanic lava. For the next 200 years, zeolites sank almost into oblivion. Only in the 1950s, with the geological discovery that zeolites are formed in large quantities, intensive investigations of their properties and the exploration of synthesis pathways were initiated. Zeolites became a major industrial raw material. The group of zeolite minerals is one of the largest groups of minerals (48 naturally occurring species are known) and nearly 150 have been synthesized in the laboratory. Nowadays, the annual production rate is approximately 1 million tons. The most important technical applications are (1) as catalysts in the fluid catalytic cracking (FCC) process of medium-weight raw oil fractions and (2) in the methanol-to-gasoline (MTG) process. The spectrum of other application fields is also enormous.‡

3.3.2 Silica Gels and Organic–Inorganic Hybrid Xero- and Aerogels

3.3.2.1 Silica Gels

Silica gel has been known since 1640, but remained a scientific curiosity until its adsorbent properties were found useful during World War I. The synthetic sol–gel route for silica gel—made of sodium silicate—was patented by chemistry professor Walter A. Patrick at Johns Hopkins University, Baltimore, Maryland, in 1919 and was used for the adsorption of vapors and gases in gas mask canisters, as part of his patent.[13] In a sol, colloidal particles with diameters in the range of 1 to 1000 nm, built up from hydrolysis and polycondensation of suitable precursors, are dispersed in a liquid. Gelation of this suspension occurs if these particles agglomerate to build a sponge-like, three-dimensional solid network whose pores are filled with the liquid. The resulting "wet" gels are therefore called aquagels, hydrogels, or alcogels. When the pore liquid is replaced by air without significantly altering the network structure, highly porous aerogels with very low densities (0.004 to 0.5 g cm^{-3}) are obtained. A xerogel (xero = Greek word for "dry") is formed upon conventional

†According to the IUPAC definition, porous materials are divided into three different classes, depending on their pore sizes. Mesoporous materials are described as materials whose pore diameters lie in the range between 2 and 50 nm. Solids with a pore diameter below 2 nm or above 50 nm belong to the class of micro- and macroporous materials, respectively.

‡These include ion-exchange for water softening, isolation and drying in refrigerators and composite windows, filler for paper and cement, separation of gases, for example, oxygen from air, cleaning of radioactive waste, flue gas desulfurization (FGD), as abrasives in toothpaste, for the regeneration of dialysis fluids, freeze-drying of food, as adsorption agent in cigarette filters, for reconditioning of used engine oil, as additives in animal feed, for the disposal of radioactive isotopes, in solar energy storage technology, and as catalyst in organic syntheses.

drying of the wet gels, that is, by increase in temperature or decrease in pressure, with concomitant large shrinkage and collapse of the delicate, highly porous inorganic networks of the aquagels or alcogels. The first synthesis of an aerogel—still made of an aqueous solution of sodium silicate, that is, water glass—was accomplished by Kistler in 1931 by means of supercritical fluid extraction of the water from the pores of the network.[14,15] A decisive simplification and acceleration of the synthesis was achieved by Teichner et al. in the late 1960s.[16,17] They dissolved tetramethoxysilane (TMOS) in methanol and added a defined amount of water to start the hydrolysis and condensation reactions. The chemical reactions during sol–gel processing of alkoxysilanes can be formally described by three equations (shown in Fig. 3.2).

Condensation occurs even if not all of the OR groups are hydrolyzed. Small oligomers (clusters) are initially formed by the condensation reactions. Growing of these clusters will form the sol particles which finally build up the oxidic network through gelation, aging, and ripening. However, all intermediate species still contain Si-OR and/or Si-OH groups. Hydrolysis therefore takes place parallel to condensation during all steps of the sol–gel process. By now, this route constitutes the standard sol–gel process. Apart from their extreme low bulk densities (typically ~0.1 g cm^{-3}), aerogels are structurally characterized by their porosity (80% to 99.8%) accompanied by pore diameters of 20 to 150 nm and specific inner surface areas up to 1600 $m^2\ g^{-1}$. Many publications are available concerning sol–gel chemistry and physics, not only of silica but also of other metal oxides. We would like to draw attention only to two of them, the 900 page book by Brinker and Scherer,[18] and the excellent review of aerogels by Hüsing and Schubert.[19]

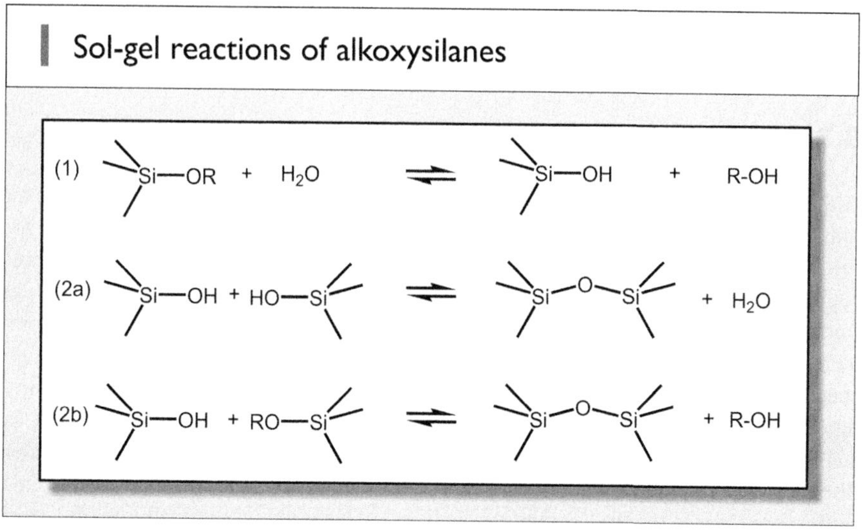

Figure 3.2 The three main reaction steps occurring in the sol–gel process of alkoxysilanes: (1) hydrolysis of an alkoxy group resulting in a silanol group and the respective alcohol; (2a) condensation of two silanol-bearing species; (2b) mixed condensation of a silanol group and an alkoxy group-bearing species.

3.3.2.2 *Organic–Inorganic Hybrid Xero- and Aerogels*

Early considerations to introduce organic components into silica (or titania, zirconia, alumina, etc.) networks or, inversely, to create polymers with new properties by combining them with inorganic network builders, go back to the mid-1980s and efforts in this direction were undertaken particularly by H. Schmidt from the Fraunhofer Institute for Silicate Research, Würzburg, and G. L. Wilkes and coworkers from the Chemical Engineering and Polymer Materials and Interfaces Laboratory in Virginia.[20–23] In these early stages of development mainly organotrialkoxysilanes were employed as co-condensation reaction partners together with TMOS/TEOS.

One drawback of the co-condensation approach is that phase separations are likely to occur, which means that the ratio of the organic and inorganic amounts cannot be completely controlled/tuned. A greater degree of control can be achieved by means of using bridged bis(trialkoxysilyl) compounds, as they can be applied as real hybrid *single-source* precursors, that is, they exhibit their hybrid nature on a molecular scale. These silsesquioxane precursors of the type $(R'O)_3Si{-}R{-}Si(OR')_3$ readily undergo hydrolysis and subsequent polycondensation reactions to form organic–inorganic hybrid xero- or aerogels. First and foremost, this field of research is inevitably connected with the names D. A. Loy and K. J. Shea, who, together with O. W. Webster, introduced these kinds of materials in the late 1980s,[24] and continue to report progress,[25–29] as more and more precursors with varying organic bridges R are accessible. The broad range of precursors, comprising those with alkylene, alkylidene, arylene, and heteroarylene organic bridges, will play a prominent role in this book chapter as they are exactly the same type of precursors that are used for the synthesis of periodic mesoporous organosilicas. Thus, hybrid xero- and aerogels on the one hand and pure PMOs on the other (made of 100% bis-silylated compounds, see below) can be considered as brothers and sisters, the former being fabricated without, the latter with application of structure-directing agents.

3.3.3 M41S, SBA, and Related Phases

Although zeolites are an extremely versatile class of materials, they bear one inherent drawback that their pore sizes are limited to approximately 15 Å. This means that, for instance, catalytic processes within the channels are restrained to relatively small molecules, being particularly important for the oil industry: the higher and heavier raw oil fractions cannot be cracked with microporous molecular sieves. For that reason, efforts were undertaken to overcome this limitation and to develop materials with high specific surfaces and larger pore sizes. In 1992, researchers of the Mobil Oil company reported on the successful synthesis of new, mesoporous silica phases, known as M41S phases,[30,31] which are formed with the help of supramolecular structure-directing agents (SDAs), that is, self-assembled aggregates of ionic surfactants (long chain alkyltrimethylammonium halides). These phases typically exhibit pore diameters from 2 to 5 nm. The most well-known representatives of this class include the silica solids MCM-41 (with a hexagonal arrangement of the mesopores, space group *p6mm*), MCM-48 (with a cubic arrangement of the mesopores, space group

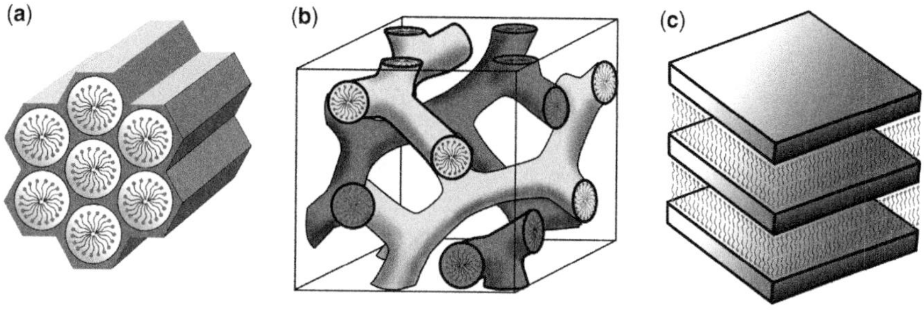

Figure 3.3 Idealized models of mesostructured M41S materials: (a) MCM-41 (2D hexagonal, space group *p6mm*), (b) MCM-48 (cubic, space group *Ia*-3*d*), and (c) MCM-50 (lamellar, space group *p*2).

Ia-3*d*),[§,32] and MCM-50 (with a lamellar structure, space group *p*2), where MCM stands for Mobile Composition of Matter (Fig. 3.3).[¶,33]

What is remarkable about the formation process of this class of materials is that different kinds of forces, directed and nondirected, on different length scales, act in a specific structure-directing way, which leads to an interconnected composite material and not to two separated phases, a surfactant phase on one and a condensed silica phase on the other hand. First, there is the formation of the mesophase, that is, the liquid-crystalline phase of the supramolecular ensemble of the surfactants, determined itself by the concentration of the surfactant, the length of the hydrophobic chain, the steric requirement and/or charge of the headgroup, the temperature and pH value of the solution. Second, there is the hydrolysable and condensable silica precursor, whose corresponding reaction rates are dependent again on the pH value, ionic strength of the solution, and temperature, and both to different degrees. Finally, there is the interaction between the inorganic species (silica precursor) and the headgroups of the surfactant, which is the fundamental prerequisite for this method to ensure inclusion of the structure-directing agent without phase separation taking place, and which can be mediated by different forces. The mesoporous materials are obtained by subsequent removal of the surfactant by extraction or calcination after the condensation/polymerization process is complete.

In-depth investigations into the formation process of these composite materials—which can be rightly called complex—have found that two different mechanisms can be involved: On one hand, in true liquid-crystal templating (TLCT), the concentration of the surfactant is so high that under the prevailing conditions (temperature, pH) a

§The structure of MCM-48 is an example of a so-called gyroid: an infinitely connected triply periodic minimal surface discovered by Alan Schoen in 1970.[32] The channels run through the gyroid labyrinths in the (100) and (111) directions; passages emerge perpendicular to any given channel as it is traversed, the direction at which they do so gyrating down the channel, giving rise to the name "gyroid."

¶Independently of the researchers at Mobil, Yanagisawa et al.[33] discovered somewhat earlier another method to prepare mesoporous silicon dioxide by the intercalation of surfactants into lamellar silicas, the so-called FSM materials (Folded Sheet Mesoporous Materials). However, this is not an actual template mechanism; rather, the preparation involves a "swelling" of lamellar silicas from which the three-dimensional structures were eventually obtained."

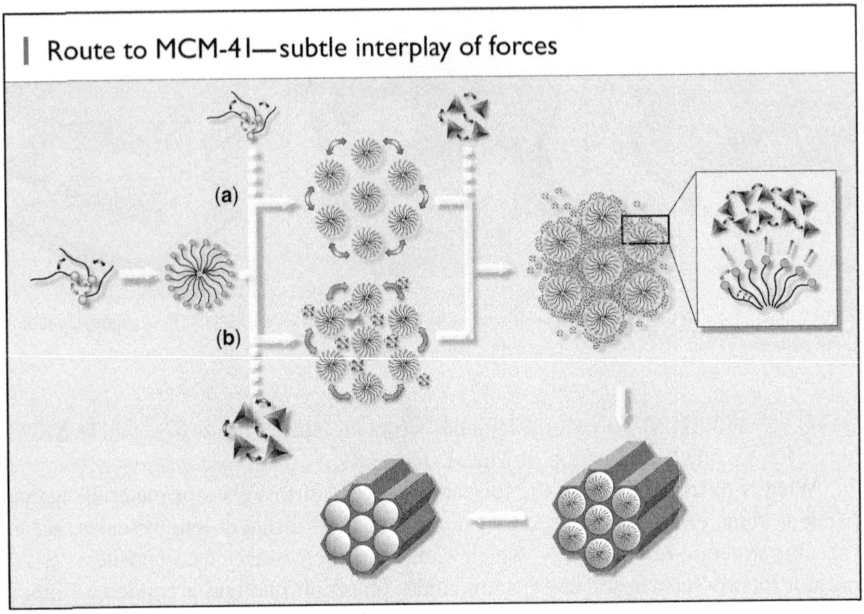

Figure 3.4 Formation of mesoporous materials by structure-directing agents: (a) true liquid-crystal template mechanism, (b) cooperative liquid-crystal template mechanism.

lyotropic liquid-crystalline phase is formed without requiring the presence of the inorganic precursor [normally tetraethoxy-(TEOS) or tetramethoxysilane (TMOS)].[34] On the other hand, it is also possible that this phase forms even below surfactant concentrations that are normally required to build such a mesophase of surfactants only. In this case, the addition of the inorganic species induces a cooperative self-assembly process, in which case a liquid-crystalline composite phase with hexagonal (as depicted exemplarily in Fig. 3.4), cubic, or lamellar arrangement can develop.[35,36]

Meanwhile, the original approach has been extended by a number of variations, for example, by use of other, nonionic triblock copolymer surfactants such as PEO-PPO-PEO (so-called Pluronics®)** (P123, F127), PEO-PLGA-PEO††, and B50-6600 ($EO_{39}BO_{47}EO_{39}$) as SDAs in neutral or acidic media. These triblock copolymers were used in the synthesis of large-pore (5 to 30 nm) mesoporous pure silica phases such as SBA-15 (*p6mm*),[37–39] SBA-16 (*Im-3m*),[38] MSU-H,[40] and FDU-1 (*Fm-3m*) materials.[41] Excellent reviews on this topic can be found in References 42 and 43. What is special about the 2D hexagonal SBA-15 phase is that XRD and physisorption

**Pluronics are the trade name of Poloxamers from the BASF company, which are composed of a central hydrophobic chain of polyoxypropylene [poly(propylene oxide)] flanked by two hydrophilic chains of polyoxyethylene [poly(ethylene oxide)]. For the generic term "poloxamer," these copolymers are commonly named with the letter "P" (for poloxamer) followed by three digits. For the Pluronic trade name, coding of these copolymers starts with a letter to define its physical form at room temperature [L = liquid, P = paste, F = flake (solid)] followed by two or three digits.

††Poly(ethylene oxide)-poly(DL-lactic acid-*co*-glycolic acid)-poly(ethylene oxide).

measurements are not consistent with the existence of ideal arrays of mesopores imbedded in a uniform silica matrix but can be accounted for by models involving a corona (with a density considerably lower than the surrounding solid matrix), which arises from the partial occlusion of the PEO chains into the silica matrix, and which becomes microporous upon calcination.[44] As a consequence, the mesopores in SBA-15 are interconnected by microporous channels, which might be advantageous in terms of mass transport properties regarding possible catalytic applications. The formation of SBA-15 phases with the help of Pluronic SDAs is depicted schematically in Figure 3.5.

Furthermore, aggregates of nonionic surfactants like polyoxyethylene alkyl/aryl ethers of the Tergitol, Triton,[45] and Brij®[38] series were used as templates for the formation of mesoporous silica materials in neutral or acidic media. The pore diameters of the materials that can be obtained with these surfactants are restricted to around 5.5 nm. The advantages of these surfactants over triblock copolymers are that they are cheap, nontoxic, and biodegradable. An overview of the most commonly used SDAs is compiled in Table 3.1.

The employment of polymeric SDA in acidic media generally leads to higher hydrolysis and condensation rates of the precursor, higher product yields, and a

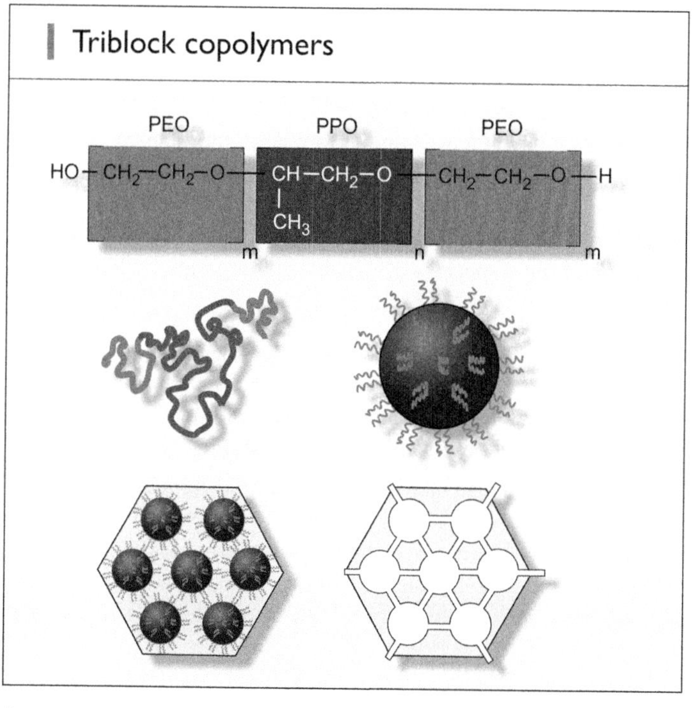

Figure 3.5 General chemical composition of triblock copolymers with hydrophilic PPO and hydrophobic PEO parts (top), giving rise to micelles with a corona of PEO blocks (middle), leading to composites/mesoporous materials in which the mesopores are connected via micropores (bottom).

Table 3.1 Overview of Commonly Used Surfactants as SDAs

	Ionic/ nonionic	Type	Chemical composition[a]	Typical pore diameter (nm)
CTAB/ CTAC	Ionic	Long-chain alkyl trialkyl ammonium salt	$C_{16}H_{33}N(Me)_3^+Br/Cl^-$	2–3.5
OTAB/ OTAC	Ionic	Long-chain alkyl trialkyl ammonium salt	$C_{18}H_{37}N(Me)_3Br/Cl^-$	2.5–4
CPB/ CPC	Ionic	Long-chain alkyl pyridinum salt	$C_{16}H_{33}$-$C_5H_5N^+Br/Cl^-$	2–3
Brij 56	Nonionic	Copolymer[b]	H-$[EO]_{10}$-O-$C_{16}H_{33}$	2–4.5
Brij 76	Nonionic	Copolymer[b]	H-$[EO]_{10}$-O-$C_{18}H_{37}$	2.5–5.5
B50-6600	Nonionic	Triblock copolymer	$EO_{39}BO_{47}EO_{39}$	5–10
P123	Nonionic	Triblock copolymer	$EO_{20}PO_{70}EO_{20}$	6–10
F127	Nonionic	Triblock copolymer	$EO_{100}PO_{65}EO_{100}$	5.5–10

[a]EO = ethyleneoxide, PO = propyleneoxide, BO = butyleneoxide.
[b]To be precise, this class of surfactants is composed of a hydrophilic poly(ethyleneglycol) moiety and a hydrophobic fatty alcohol component.

tendency towards thicker pore walls, which is desirable with respect to thermal stability. In addition, the extraction procedure is easier because of the weaker interaction between the micellar and silica interface.

Many studies that have focused on the formation mechanism of various types of templated mesoporous materials have been summarized in a number of recent reviews,[46–49] and it is still an ongoing topic of research.[50,51] Here, we want to recapitulate only very briefly the useful interaction classification scheme between the surfactants (S) and the inorganic species (I), which was suggested by Huo et al.[52,53]

- If the reaction takes place under basic conditions then the silica species are present as anions, that is, deprotonated silanol groups (Si-O^-); in this case the surfactants have to be charged positively to ensure interactions between both components; commonly cationic quaternary ammonium surfactants are used as the SDA; this synthesis pathway is termed S^+I^- (Fig. 3.6a).
- The preparation can also take place under acidic conditions below the isoelectric point of the Si–OH-bearing inorganic species (pH $\approx$ 2); then the silica species are positively charged, that is, protonated silanol groups (Si-OH_2^+); in order to produce an interaction with cationic surfactants, it is necessary to add a mediator ion X^- (usually a halide), which gives rise to the $S^+X^-I^+$ pathway (Fig. 3.6b).
- Conversely, when negatively charged surfactants (e.g., long-chain alkyl phosphates) are used as SDAs, it is possible to work in basic media, whereby again a mediator ion, in this case a positively charged one (M^+), must be added to

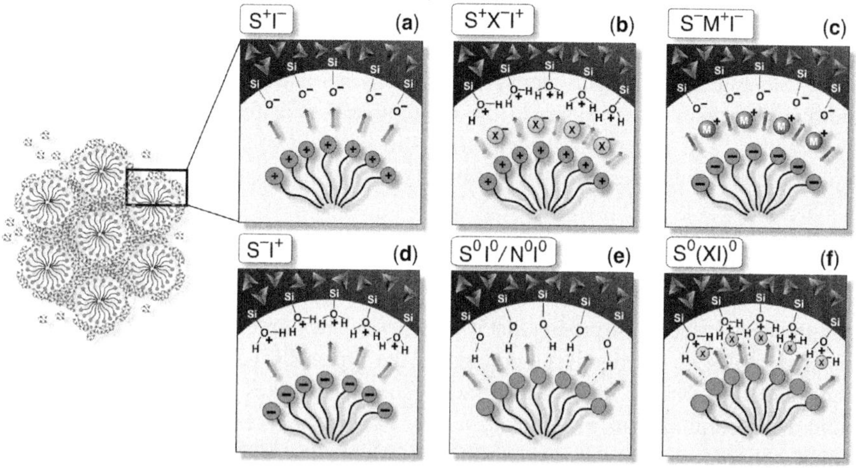

Figure 3.6 Interactions between the inorganic species and the head group of the surfactant with consideration of the possible synthesis pathway in acidic, basic, or neutral media. Electrostatic: S^+I^-, $S^+X^-I^+$, $S^-M^+I^-$, S^-I^+; through hydrogen bonds: S^0I^0/N^0I^0, $S^0(XI)^0$. (See color insert.)

ensure interaction between the equally negatively charged silica species; here the $S^-M^+I^-$ pathway is present (Fig. 3.6c).

- When working in acidic media and using negatively charged surfactants a mediator ion is not required; the S^-I^+ pathway (Fig. 3.6d).
- Thus, the dominating interactions in pathways (a) to (d) in Figure 3.6 are of electrostatic nature; moreover, it is possible that the attractive interactions are mediated through hydrogen bonds; this is the case when nonionic surfactants are used (e.g., S^0: a long-chained amine; N^0: polyethylene oxide), whereby uncharged silica species (S^0I^0 pathway; Fig. 3.6e) or ion pairs [$S^0(XI)^0$ pathway; Fig. 3.6f] can be present.

Meanwhile, SDA-mediated syntheses have also been used successfully in the preparation of nonsilica mesoporous metal oxides,[54–60] metal sulfides,[61] and metal phosphates.[62,63] An account devoted to nonsiliceous mesoporous materials and the host–guest chemistry of such systems can be found in a review by Zhao and coworkers.[64]

The syntheses of ordered mesoporous solids described above are classified as endotemplate methods (soft-matter templating). In exotemplate methods (nanocasting), a porous solid is used as the template in place of the surfactant. This method is also known as hard-matter templating. In this way, the pore system of the rigid template is copied as a negative image. This replication method was used for the first time by Ryoo et al.[65] for the synthesis of mesoporous carbon (CMK-1), using MCM-48 as the rigid structure matrix. A short time later, Hyeon and coworkers[66] independently presented very similar approaches for the preparation of mesoporous carbon materials, known as SNU-X materials. By now, these negative replicas are frequently used themselves again as rigid templates for repeated nanocasting processes. The terms endo- and exotemplate are formally derived from the terms endo- and exoskeleton used in biology. An excellent overview of the concepts and their usage has been presented by Schüth.[67] The conceptual developments in the field of structure replication and repeated templating approaches were summarized by Tiemann.[68]

3.3.4 PMOs

In retrospect, it seems surprising that it took seven years until the sol–gel approach of bis-silylated organobridged compounds of the type $(R'O)_3Si{-}R{-}Si(OR')_3$ on one hand and the concept of using supramolecular aggregates as SDAs on the other hand were combined to give finally a new class of mesoporous hybrid materials, called periodic mesoporous organosilicas (PMOs). In 1999, indeed three research groups reported independently of each other for the first time on the successful synthesis of these materials.[69–71] As PMOs will be discussed in more detail in Section 3.4, here, only some remarks regarding the conceptual placement of PMOs within the class of silica-based organic–inorganic, porous materials will be made: (1) in contrast to completely amorphous aero- and xerogels, PMOs are characterized by a periodically organized pore system and a very narrow pore radius distribution; (2) some PMO precursors, in particular those with rigid, inflexible organic bridges, exhibit the tendency to undergo (at least partially) self-assembling processes on the *molecular scale*, too, which may give rise to PMOs that are not only ordered with respect to their channel system but exhibit additionally a crystal-like arrangement of their organic bridges within the pore walls; (3) with some exceptions it is usually not possible to remove the SDAs from the composite material by calcination due to the limited thermal stability of the organic moieties; (4) with increasing length of the organic bridge the hydrophobicity of the PMO precursor will increase; this can cause an interference with the SDAs, and in the worst case can lead to a massive disturbance of the liquid-crystalline phase, which may finally result in an unordered (or nonperiodic) mesoporous organosilica material with a spaghetti- or worm-like arrangement of the pores.[‡‡,72]

3.3.5 ZOLs

Although micro- and not mesoporous but worth mentioning in any case is a new class of materials in this evolution series of porous silica-based hybrid materials, the so-called ZOLs (zeolites with organic group as lattice). As we will see, many studies in the past focused on the organic functionalization of amorphous silica porous

‡‡Based on the experiences with PMO precursors of varying polarity, M. Cornelius hypothesized in his Ph.D. thesis that precursors with a log K_{OW} value larger than 6.7 (in their ethoxy form) do not allow an easy transformation to a mesoscopically ordered organosilica, at least not with standard procedures (i.e., without the addition of further structure-forming agents or by addition of cosolvents).[72]

materials. In contrast, only very few studies were carried out on crystalline zeolitic material. The organic–inorganic hybridization of zeolites began with the work of Jones et al. who succeeded in modifying BEA-type zeolite with various terminal organic groups by employing organomonosilanes such as phenylethyl-trimethoxy-silane and 3-mercaptopropyl-trimethoxy-silane as a part of the silicon source. They were named OFMSs (organic-functionalized molecular sieves).[73–76]

In ZOL materials, introduced by Tatsumi et al.[77] in 2003, not a monovalent pendant organic group but a divalent methylene group is introduced into the zeolite. This new family of hybrid zeolite materials is synthesized from bis-triethoxysilylmethane (BTESM), an organosilane with a bridging methylene group between two silicon atoms ($Si\text{-}CH_2\text{-}Si$), substituting a siloxane bridge (Si–O–Si). In this approach, ideally the inserted organic moiety will not form a structural defect or spoil the microporosity. Because the Si-C bond (~1.9 Å) is usually longer than the Si–O bond (~1.6 Å), the insertion of the methylene moiety into the zeolite framework looks difficult. However, the Si-C-Si angle (~109°) is smaller than the Si–O–Si angles in the zeolite framework; for example, those in MFI-type zeolite range from 140° to 170°.[78] This smaller bond angle would compensate for the distance of two silicon atoms to enable the insertion of methylene species into the zeolite framework. Employing density functional theory, Astala and Auerbach demonstrated theoretically that methylene fragments can substitute oxygen atoms in LTA- and SOD-type zeolite frameworks in a stable manner.[79] Six different ZOLs are known (for a recent review, see Reference 80) and their characteristic features are summarized in Table 3.2.

Although ZOLs have physical properties similar to those of their inorganic counterparts, they exhibit a higher hydrophobicity because of the presence of the organic framework group. This could be exemplarily shown for ZOL-A by the much higher adsorption capacity for *n*-hexane in comparison to its inorganic counterpart zeolite A.[81]

Until now it was not possible to obtain organozeolites from other bridged organosilanes that have been used in the synthesis of inorganic–organic hybrid amorphous silica materials. Presumably, long bridging organic groups cannot be incorporated into crystalline materials to be substituted for oxygen atoms. However, methylene-bridged three-membered organosilane rings that have been used for the successful synthesis of PMOs by Ozin's group (see below) or other similar subunits could be introduced into the framework. This leads to an expectation that the zeolite framework can be designed

Table 3.2 Characteristic Features of Known ZOL Materials

	Silicon source	SDA	Al/(Si + Al)	Time (d)	Topology
ZOL-1	BTESM	TPAOH	0	5	MFI
ZOL-2	BTESM	TEMABr	0	20	MFI
ZOL-5	BTESM	–	0.034	7	MFI
ZOL-A	BTESM	–	0.450	14	LTA
ZOL-B(F)	BTESM + TEOS	TEAF	0	14	BEA
ZOL-1(F)	BTESM + TEOS	TPAF	0	14	MFI

by building the structure with prebuilt favorable subunits, which could formulate a new strategy for synthesizing hybrid zeolites.

3.4 THREE WAYS TO OBTAIN ORGANICALLY FUNCTIONALIZED MESOPOROUS SILICA PHASES

In the following, a more detailed look at the three principal ways to obtain organofunctionalized, mesoporous silica phases will be offered. These include (1) the subsequent modification of the pore surface of a formerly pure inorganic silica material with organic or organometallic compounds (also known as grafting or post-synthetic functionalization), (2) the simultaneous condensation of corresponding silica and organosilica precursors (also known as co-condensation, direct synthesis, or one-pot synthesis), and (3) the incorporation of organic components as bridging groups directly and specifically into the pore walls by using bis- or multi-silylated *single-source* organosilica precursors [synthesis of (periodic) mesoporous organosilicas].

1. In this context, the grafting procedure refers to the subsequent modification of the outer or inner surface of silica phases with organic or metal–organic groups. This process is carried out primarily by reaction of organosilanes of the type $(R'O)_3SiR$, that is, organotrialkoxysilanes, and less frequently by chlorosilanes $ClSiR_3$ or silazanes $HN(SiR_3)_2$, with the free silanol groups of the pore surfaces of the inorganic silica phase (M41S, SBA-15, and related phases) in terms of a simple condensation reaction (Fig. 3.7). As the reactions of silica with chlorosilanes are hard to control and give rise to unwanted polymerization reactions[82] the utilization of alkoxyorganosilanes or silazanes is preferable.

   In principle, functionalization with a variety of organic groups can be realized in this way by variation of the organic residue R. This method of modification has the advantage that, under the synthesis conditions used, the mesostructure of the starting silica phase is usually retained. One of the disadvantages is that this method is, of course, dependent on the number (which can

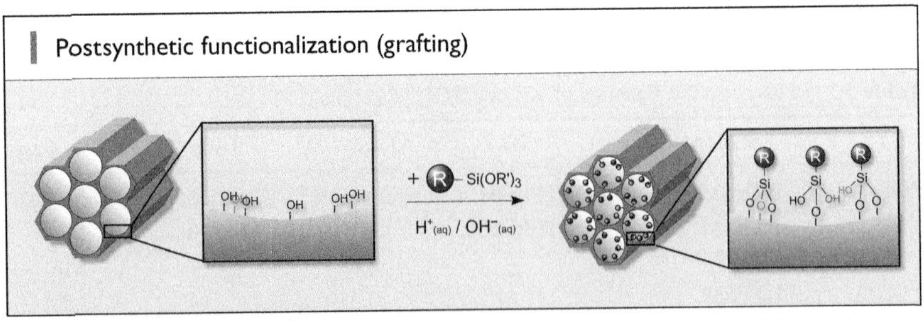

Figure 3.7 Grafting (postsynthetic functionalization) for organic modification of mesoporous pure silica phases with terminal organotrialkoxysilanes of the type $(R'O)_3SiR$. R = organic functional group.

vary considerably) and density or distributions (which are mostly nonuniform) of the surface silanol groups of the parent material; in other words, the degree of control is limited and the result is often a nonuniform coverage with organic residues. If the organosilanes react preferentially at the pore openings during the initial stages of the grafting process, the diffusion of further molecules into the center of the pores can be impaired, which can in turn lead to an even more nonhomogeneous distribution of the organic groups within the pores and a lower degree of occupation. In extreme cases (e.g., with very bulky grafting species), this can lead to a complete closure of the pores (pore-blocking effect; Fig. 3.8). Another inherent drawback of the grafting method is that the lining of the pore walls is accompanied by a reduction in the porosity of the hybrid material (albeit depending on the size of the organic residue and the degree of occupation).

2. An alternative method to synthesize organically functionalized mesoporous silica phases is the co-condensation method (also called one-pot synthesis or direct synthesis). Historically, the classical co-condensation reaction involved tetraalkoxysilane [$(RO)_4Si$ (predominantly TEOS or TMOS)] and a terminal organotrialkoxysilane of the type $(R'O)_3Si\text{-}R$ in the presence of structure-directing agents (Fig. 3.9, top). Meanwhile, not only bis- and multi-silylated precursors are used, that is, precursors of the type $[(R'O)_3Si]_mR$, $m \geq 2$ (Fig. 3.9, bottom), together with TMOS/TEOS, but

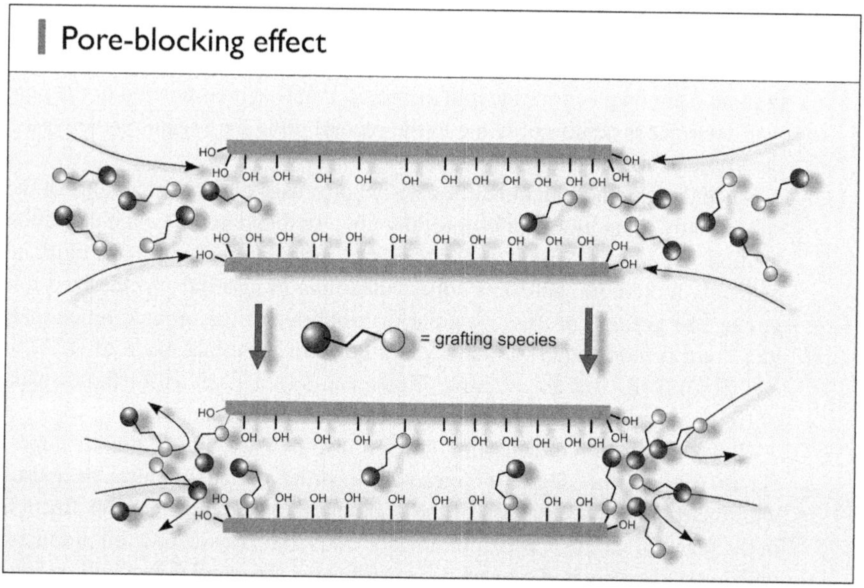

Figure 3.8 Illustration of the pore-blocking effect caused by bulky grafting species reacting preferentially at the pore openings and thus giving rise to a nonhomogeneous distribution of the organic components within the channel system.

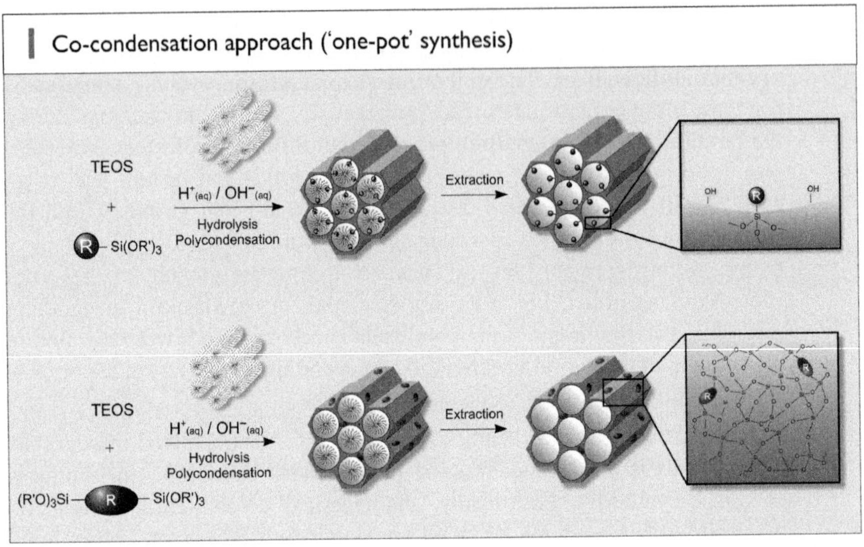

Figure 3.9 Co-condensation method (direct synthesis) employing TEOS and a terminal organotrialkoxysilane (top) or TEOS and an organo-bridged silsesquioxane (bottom) as mixed precursors for the organic modification of mesoporous pure silica phases. R = organic functional group.

also ternary and quaternary reaction compositions are applied, for instance, those composed of TEOS, one mono- and one bis-silylated precursor.

The main difference between the two approaches, as indicated in Figure 3.9, is that the first route leads to hybrid materials in which the main fraction of the organic components protrude into the pores, that is, a modification of the pore wall interface is achieved, while in the second route the organic moieties are incorporated directly into the framework.

Since the organic functionalities are in both cases direct components of the silica matrix, pore blocking is generally not a problem in the co-condensation method, in contrast to the grafting approach. Furthermore, the organic units are generally more homogeneously distributed than in materials synthesized following the grafting process. Regular distributions of the organic functionalities were demonstrated for some cases both for materials made of $(RO)_4Si$ and $(R'O)_3Si$-R (see, for example, Reference 83), as well as for those made of $(RO)_4Si$ and $[(R'O)_3Si]_mR$ (see, for example, Reference 84).

However, the co-condensation method also has two major disadvantages. (1) In general, the degree of mesoscopic order of the products decreases with increasing concentration of the organotrialkoxysilane precursor fraction in the reaction mixture, which ultimately leads to totally disordered products; with every incorporated organosilane precursor a structural defect will be generated. Consequently, the content of organic functionalities in the modified silica phases does not normally exceed 40 mol%; typical values are even lower (5 to 15 mol%). (2) Furthermore, the proportion of organic groups

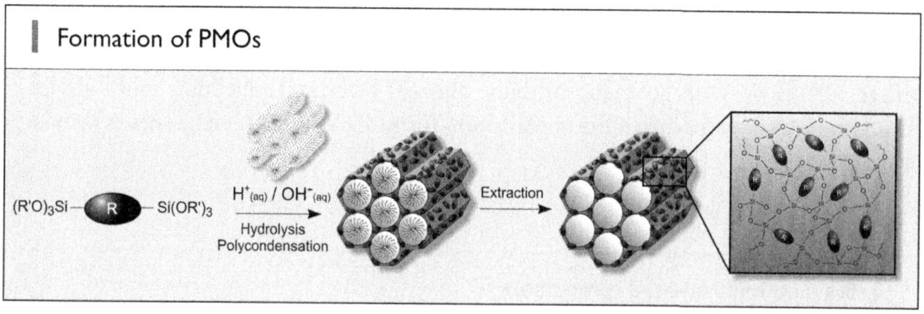

Figure 3.10 General synthesis pathway to PMOs that are constructed from bis-silylated organic bridging units. R = organic bridge.

that are incorporated into the pore wall network is generally lower than would correspond to the starting concentration of the reaction mixture. These observations can be explained by the fact that an increasing proportion of organosilanes in the reaction mixture favors homocondensation reactions, at the cost of cross-linking heterocondensation reactions with the silica precursors. The tendency towards homocondensation reactions, which is caused by the different hydrolysis and condensation rates of the structurally different precursors, is a constant problem in co-condensation. In extreme cases, the situation is comparable to the grafting scenario: The very high hydrolysis and condensation rates of the pure inorganic precursors can lead to a pure inorganic network, while the organosilane precursors are initially almost unaffected and will only slowly hydrolyze and subsequently condense onto the outer and inner surface of the already formed silica matrix.

3. The third concept to obtain mesoporous organic–inorganic hybrid materials is based on a combination of two approaches: (1) utilizing surfactants that act as structure-directing agents as in the synthesis of mesoporous pure silica on one hand, but instead of using inorganic precursors, (2) bridged bis- or multi-silylated precursors as in the synthesis of hybrid xero- or aerogels are applied. The first synthesis of these types of material was conducted with silsesquioxane precursors of the type $(R'O)_3Si{-}R{-}Si(OR')_3$ as *single-source* precursors leading to materials in which the organic components are as bridging groups directly and specifically incorporated into the pore walls and which were named periodic mesoporous organosilicas (PMOs; Fig. 3.10).[§§,69–71,85]

[§§]Since the synthesis of PMO or related organosilica materials is usually conducted in a relatively harsh environment (hydrothermal treatment at very low or very high pH values), the question whether and to what extent the Si—C bonds were cleaved during the synthesis procedure is always a key issue. Sometimes IR or Raman spectroscopy is applied in order to examine the status of the Si—C bonds. However, Hoffmann et al.[85] have recently shown that vibrational spectroscopy is not a reliable method for monitoring the integrity of the Si—C bonds in PMO or related organosilica materials; therefore, for valid statements regarding the extent of Si—C bond cleavage, carrying out solid-state NMR spectroscopy is an absolute necessity.

PMO precursors can be divided roughly into three classes: bis-silylated compounds, which contain (1) short-chained alkylene or alkenylene groups as the organic bridge, (2) those with aromatic bridges, and (3) branched or cyclic multi-silylated compounds. An overview of the constitution formulae of PMO precursors is provided in Figure 3.11.

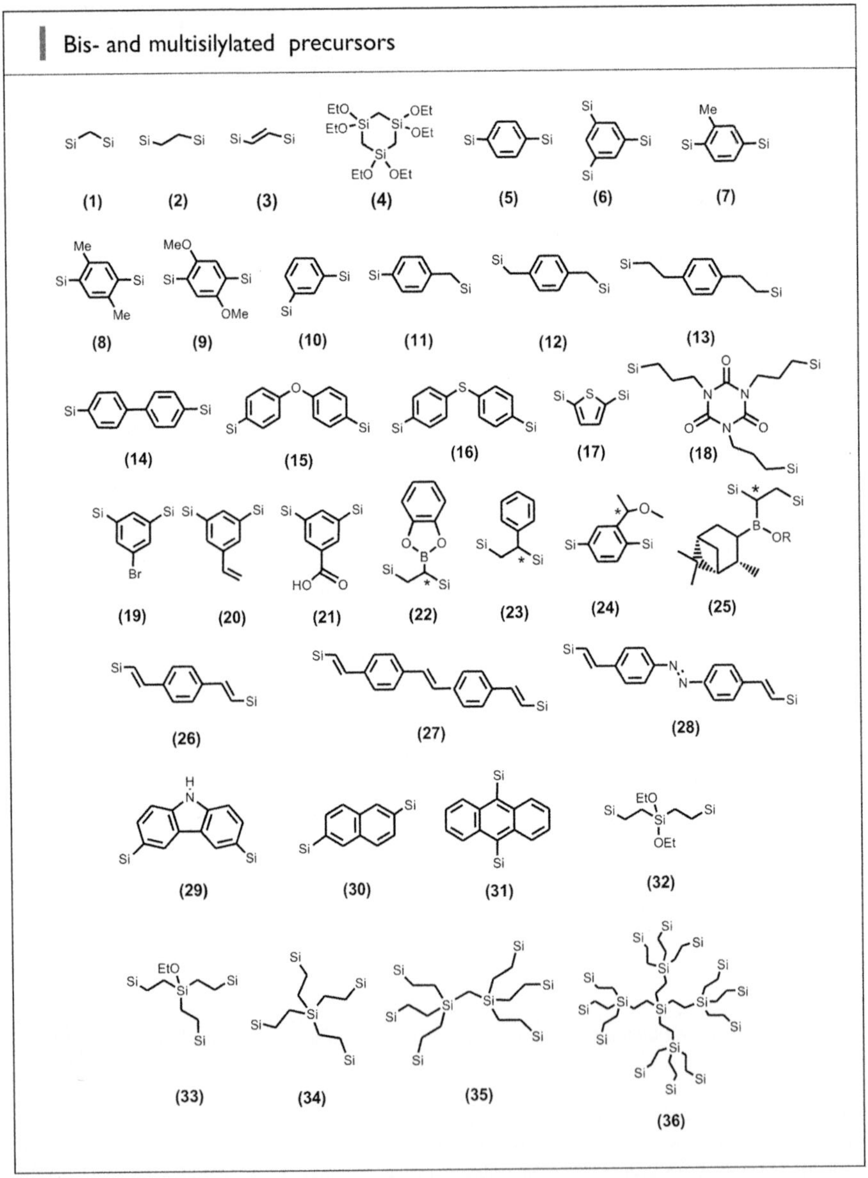

Figure 3.11 Overview of the chemical structures of organosilica precursors that have been converted into PMOs. Terminal Si atoms: Si = $Si(OR)_3$ with R = Me, Et, or *iso*-Prop.

The most frequently used ionic SDAs are again CTAB/CTAC, OTAB/OTAC as well as hexadecylpyridinium bromide/chloride (CPB/CPC), whereby the syntheses are carried out under basic conditions, and most commonly surfactants of the Pluronic and Brij type[¶¶,86] are used as nonionic SDAs under acidic conditions. In rare cases also Gemini surfactants[87–90] or long alkyl chained ionic liquids[91,92] were applied as SDAs, whereby an ethane-bridged PMO with a *Fm*-3*m* symmetry was successfully prepared for the first time.[88]

In the meantime, the variability of PMOs was further extended by the construction of bifunctional PMOs through co-condensation reactions of mixtures of either (1) bridged bis(trialkoxysilyl)-organo-silanes $[(R'O)_3Si{-}R{-}Si(OR')_3]$ and terminal organotrialkoxysilanes $[(R'O)_3SiR'']$ in the presence of an appropriate SDA, in complete analogy to the co-condensation reactions of TEOS/TMOS and terminal organotrialkoxysilanes $[(RO)_3SiR'']$); or (2) by using a mixture of two or more different bridged bis(trialkoxysilyl) precursors which will co-condense in the presence of a SDA. In the first case, the resulting bifunctional PMOs then consist of a combination of bridging organic units and terminal organic groups whose ends point mainly into the pore interiors and are thus accessible for further chemical reactions. In the latter case, the PMOs consist then of two (or more) different organic bridges that are bonded covalently within the framework of the pore walls.

It is evident that a completely consistent classification of organic–inorganic mesoporous hybrid materials based on silica is hard to deduce. Before we proceed, some (hopefully) clarifying remarks on this issue will be made.

1. One classification criterion could be the question of whether only one single source or two (or more) different precursors are utilized. Then, for instance, materials made of two different silsesquioxane precursors have to be classified into the same class as a material that is made of TEOS and another terminal organotrialkoxysilane, because in both cases a *co-condensation* approach is followed; is this truly justified?
2. Another criterion could be whether the organic functions are only incorporated into the framework or only the pore wall interface is functionalized. However, how then should a co-condensation reaction between a silsesquioxane precursor and a terminal organotrialkoxysilane be classified?
3. Since a silsesquioxane precursor is also involved in the second co-condensation route (Fig. 3.9, bottom), that is, a precursor that leads to PMO materials when it is used as the only single-source precursor, many research groups also call the resulting mesoporous organic–inorganic hybrid materials PMOs. Of course, there are two possibilities: These materials can be considered as based on pure mesoporous silica phases in which a certain amount of organic components is integrated, or they can be considered as PMOs that are (in most cases extremely) diluted with the actual, pure inorganic network builder. One has to bear in mind, however, that in co-condensation procedures the risk of

¶¶By the way, it was Markowitz's group that established the Brij-56/76-mediated synthesis protocol for PMOs, not only for monofunctional ones but also for those composed of two and more bis-silylated precursors.[86]

phase separation is always present, which is circumvented inherently by using precursors that are hybrid even on their molecular scale and which are used as single-source precursors.

4. In addition, the clear distinction between the grafting approach on one hand and co-condensation on the other hand becomes more and more blurred, because the products of co-condensation reactions are used frequently as the basis for subsequent grafting reactions or the integrated bridged organic components of PMOs are chemically modified afterwards.

In summary, this situation is somehow unsatisfying. A possible solution could be to adopt the nomenclature that is used in the silicone area: The organosiloxane compounds are classified according to the number of oxygen atoms that are grouped around the silicon core. The silicon atom can form one to four siloxane bridges whereby mono-, di-, tri-, and tetra-substitutions are named with the letters M, D, T, and Q, respectively. M41S and SBA-15 phases would then be named as Q phases, genuine PMOs as T^2 phases, materials achieved by the second co-condensation route as Q/T phases, a phase composed of silsesquioxanes and terminal organotrialkoxysilanes as T^2/T phases, and, to give a last example, a material built-up from the siloxane-disilsesquioxane PMO precursor (**32**) and TEOS as DT^2/Q phases (Fig. 3.12). However, this approach has at least one disadvantage: it is in a way very abstract and not very intuitive for nonspecialists, and it is questionable if this approach will be accepted by the majority of the research community.

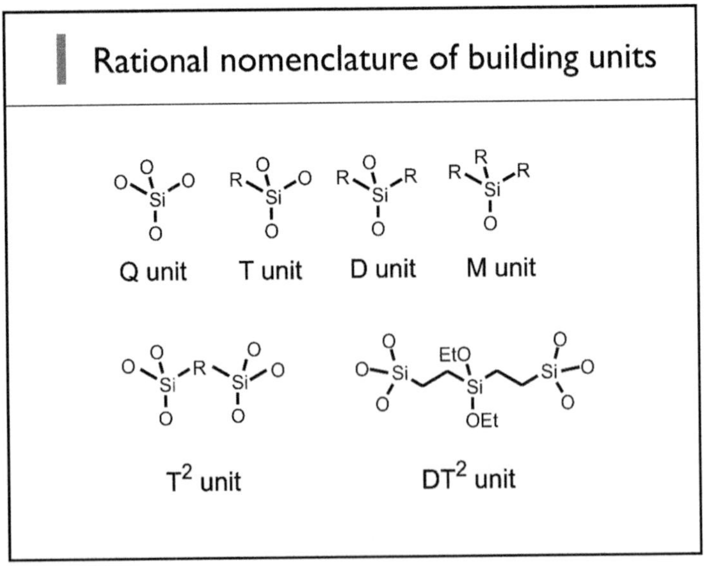

Figure 3.12 Examples of building blocks of organic–inorganic hybrid materials along with their names according to a rational nomenclature system that is already applied in the field of silicone research.

We decided to structure the sections within this chapter according to an increase of organic content in the framework of the resulting phases. Therefore, we will start with grafting, that is, with phases in which the organic component is only postsynthetically anchored on a formerly pure silica matrix. Subsequently, an account of the co-condensation reaction procedure and the respective products will be given, starting with the classical co-condensation reactions between TEOS/TMOS and terminal organotrialkoxysilanes, followed by those reactions between TEOS/TMOS and bis- or multi-silylated organobridged precursors. Finally, in the section on PMOs, materials are presented that are composed of at least one bis-silylated organosilica precursor, whether as the only single-source precursor or in co-condensation reactions with other bis- or multi-silylated compounds (bi- and multifunctional PMOs) or in co-condensation reactions with terminal organotrialkoxysilanes.

A huge number of projects in the context of organic–inorganic mesoporous hybrid materials has dealt with the preparation of heterogeneous catalysts, for which the mesoporous (organo)silica phases built the support. Among these efforts a large fraction is concerned with grafting of metal–organic compounds and the rich world of the respective catalytic applications. This constitutes its own area of research, for which J. M. Basset coined the term surface organometallic chemistry (SOMC).[93,94] The preparation of catalysts will not be covered in detail in this chapter. A good overview of the design and preparation of organic–inorganic hybrid catalysts up to the year 2002 is provided in Reference 95. A more recent summary of organic transformations over silica materials modified by covalently bonded surface functional groups is given in Reference 96. The special aspects of cooperative catalysis by silica-supported organic functional groups are elucidated in Reference 97. Finally, a review focusing solely on the synthesis and catalytic applications of organosulfonic-functionalized meso-structured materials, not limited to grafting but also covering the functionalization route by means of co-condensation, has been published by Melero et al.[98]

Instead, we would like to give in the following a survey of a variety of examples mainly from other fields covering a broad and diverse spectrum of research and development of possible applications.

3.4.1 Grafting of (Mesoporous) Silica Phases (Postsynthetic Functionalization)

The modification of the surface of silica materials is carried out primarily in order to (1) get rid of unwanted superficial silanol groups (passivation), (2) modify the polarity (hydrophobicity) and/or sorption properties, or (3) introduce specific organic functionalities. Early research goes back to the 1950s, when the formation of surface silicic acid esters by the reaction of surface silanol groups with alcohols was reported.[99] For understandable reasons, the organosilane chemistry of surfaces or the modification of silica surfaces, respectively, experienced a renaissance with the invention of the mesoporous materials. The extremely high specific surface area values of $\sim 1000\,m^2\,g^{-1}$ permit remarkably high loading densities with organic groups. To give the reader an idea about the amount of the organic fraction in a grafted silica material this can be exemplified by the following consideration: If the underlying silica material exhibits a surface silanol group density of 4 mmol g^{-1} and if—a complete reaction assumed (see also below)—the grafting species possesses a molar mass of approximately 100 g mol^{-1}, then the organic fraction of the hybrid material amounts to approximately 30 wt.%. Assuming further that the pore volume is approximately 0.75 cm^3 g^{-1}, then the concentration of the organic functionalities inside the pores is approximately 4 mol L^{-1} (for comparison, liquid hexane possesses a molarity of ~9 mol L^{-1}). This example may illustrate that the tethered organic groups not only can provide a high density of potentially functional centers but also can exert an enormous influence on the polarity at the interface of the channels, giving rise to new possibilities for catalyst design in confined reaction spaces.

3.4.1.1 Passivation/Hydrophobizing and Adsorbents

Due to their specific properties, that is, large pore volumes, high specific surface areas, and pore diameters large enough to accommodate relatively sterically demanding molecules, mesoporous silica phases are potentially excellent adsorbents. However, only in rare cases are unmodified phases applied. For the uptake of nonpolar substances, the silica phases are lined with hydrophobic compounds, for instance, with trimethylsilyl compounds or with silazanes; for the adsorption of polar substances or (metal) ions, they are lined with hydrophilic groups, that is, Lewis bases or acids. In order to do so, there are, in principle two possibilities, namely to introduce these moieties by means of co-condensation reactions or by employing the grafting approach; the main disadvantage of the latter might be that the accessible pore volume is already more or less reduced by the grafted species itself.

Anwander and coworkers[100,101] silylated the inner surfaces of MCM-41 samples and phases built by co-condensation of TEOS and BTEE with disilazanes of the type $HN(SiRR'_2)_2$ (R, R′ = H, Me, Ph, vinyl, *n*-butyl, *n*-octyl), whereby the degree of silylation depended naturally on the spatial requirements of the silylating reagent. Complete passivation could be achieved with hexamethyldisilazane, which can be used to determine the number of free silanol groups. The vinyl-functionalized MCM-41 samples proved to be very amenable to subsequent surface modification by hydroboration.

Besides the three silylation reagents already mentioned, the concept of the surface silicic acid esters was taken up again and extended: Via a two-step reaction, the first being the esterification by alcohols to form Si—OR′ from Si—OH, the successive reaction with Grignard reagent to convert Si—OR′ to Si—R being the second, stable Si—C bonds were formed to give finally MCM-41/SBA-15 phases, which exhibited high hydrophobicity and remarkable hydrothermal stability.[102,103]

Furthermore, MCM-41, MCM-48, and SBA-15 silica materials have been lined/functionalized with, for example, alkyl,[104,105] amino or aminopropyl groups,[106–120] diamine,[121,122] triamine,[122,123] ethylenediamine,[124] imidazole,[120,125–127] triazol,[120] *N*-benzylidenebutan-1-amine,[128] malonamide,[129] carboxy(late),[108,112,114] thiol,[106,111,130] 1-allyl,[131] 1-benzoyl-3-propylthiourea[132,133] and dithiocarbamate,[134] as well as saccharides.[135] Not in all of these publications the use as adsorbents, as efficient as possible, was intended, but some dealt also with the characterization of these materials in terms of average distance and homogeneity of the grafted species and related issues.

What kinds of adsorptives were examined in which environments? Besides the well known affinity of thiophilic elements like mercury, and, to a lesser extent copper, towards thiol or thiourea groups, adsorption experiments were reported for:

- Cu^{2+}, Zn^{2+}, Cr^{3+}, and Ni^{2+} adsorption onto amino-functionalized SBA-15 materials.[106]
- Selective adsorption of Pd^{2+} or Pt^{2+} in the presence of Ni^{2+}, Cu^{2+}, and Cd^{2+}, even when these were present in high excess, on imidazole- and thiol-functionalized SBA-15 samples.[126]
- Selective adsorption of Au^{3+} and Pd^{2+} in the presence of competitive Co^{2+}, Ni^{2+}, Cu^{2+}, and Zn^{2+} salts, as well as the separation of Au^{3+} from Pd^{2+} (at $pH = 1$) by amino-functionalized MCM-41.[117]
- Adsorption of the radionuclide ions Am^{3+} (^{241}Am) and Eu^{3+} (^{152}Eu) by malonamide-functionalized MCM-41 materials.[129]
- The removal of toxic oxyanions such as arsenate and chromate from contaminated effluents by means of amino-functionalized mesoporous silicas.[107]
- Efficient borate ion adsorption on saccharide-functionalized MCM-41 materials.[135]
- Selective and efficient adsorption of $Cr_2O_7^{2-}$ on H_2N-MCM-41 samples in the presence of Cu^{2+}, and, inversely, selective but only moderately efficient adsorption of Cu^{2+} in the presence of $Cr_2O_7^{2-}$ on HOOC-MCM-41 phases.[114]
- Adsorption of nitrate and phosphate, occurring in agricultural effluents, by amino-functionalized materials.[118,119]
- Selective and efficient adsorption of anthraquinone blue (Acid Blue 25) in the presence of methylene blue on H_2N-MCM-41, and, inversely, selective adsorption of methylene blue in the presence of anthraquinone blue with carboxyl-functionalized phases.[112]
- Efficient adsorption of alkylanilines as well as selective adsorption of 4-nonylphenol*** in the presence of phenol on octylsilane-lined samples.[104,105]

***Nonylphenol is commonly obtained as a mixture of isomers, whereby the alkyl chain can be linear or branched. These compounds are representatives of the class of so-called endocrine disruptors, that is, a compound that interacts unfavorably with the hormone system of mammals. 4-Nonylphenol interacts with the estrogen receptor and is therefore classified as xenoestrogen. Nonylphenol was used in large quantities in pharmaceuticals and fungicides, as well as plastizers in cellulose esters. The industrial usage is prohibited in the EU since 2003 as nonylphenols are not only endocrine disruptors but are also toxic to a variety of organisms, for instance, coalfish, daphnia magna, green algae, crab, and lobster. The LD_{50} values lie between 0.1 and 1 mg L^{-1}.

In conclusion, functionalized mesoporous materials have proven to be efficient adsorbents for a broad spectrum of adsorptives; however, it is yet unclear if they will find commercial applications as there are already established and highly developed adsorbents for many purposes on the market.

3.4.1.2 Photochemistry, Photoswitches, Stimuli-Response and Drug Delivery Systems, and Sensors

In 2001, MCM-41 was first proposed as a suitable drug-delivery system by Vallet-Regí and coworkers,[136] and since then related mesoporous materials have been discussed intensely as drug carriers and controlled release systems. Controlled drug-delivery systems are one of the most promising applications for human health care. A very well written review of the activities in this field has been published in Reference 137. Particularly noteworthy in this context is the work of Mal et al.[138,139] who designed in a clever way a photochemically controlled system for compound uptake and release by anchoring coumarin to the pore openings of MCM-41 silica phases.

Furthermore, Fu et al.[140] developed a transport system that responds to thermal stimuli. This system is based on chains of poly-*N*-isopropylacrylamide (a known thermosensitive polymer), which exists in a collapsed, hydrophobic state when exposed to heat, but in an expanded, hydrophilic state in the cold. In this way, samples of mesoporous, spherical silica particles (particle diameter 10 μm) that were lined and coated with the thermosensitive polymer by atom transfer radical polymerization

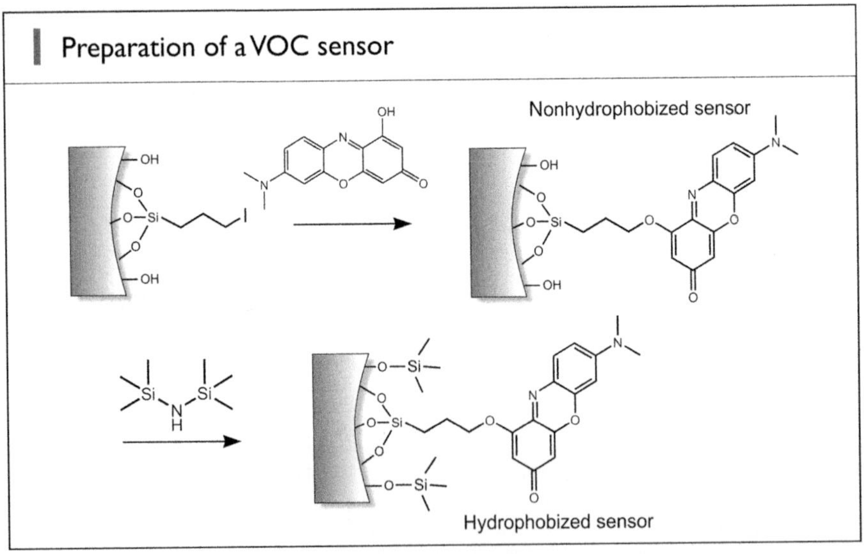

Figure 3.13 Two-step grafting process in order to construct a sensor for the detection of volatile organic compounds based on the well-known solvatochromic fluorescent phenoxazinone dyes. Hydrophobizing of the surface of MCM-41 with silazanes leads to better performance of this sensor.

could take up greater or lesser amounts of the dye fluorescein. This transport process was followed spectroscopically.

In addition, several sensor systems based on MCM-41 or SBA-15 were developed, for instance those that are able to differentiate between neurotransmitters with different amino acid functionalities[141] or which can detect adenosine triphosphate (ATP) in aqueous solution towards concentrations as low as 0.5 ppm.[142,143] More recently, Descalzo et al. developed a chemosensor based on covalently anchored phenoxazinone dyes, a known solvatochromic and fluorescent chromophore, on MCM-41 samples (Fig. 3.13), capable of detecting polar, volatile organic compounds (VOCs) such as methanol or acetone.[144] Passivation of the remaining silanol groups with hexamethyldisilazane after grafting the dye yielded a hydrophobic hybrid sensory material with improved performance in terms of the spectroscopic response, the detectable polarity range of organic vapors, response times and reversibility, and minimized interference from environmental humidity.

Likewise, Fiorilli et al.[145] covalently anchored solvatochromic Reichardt's dye [2,6-diphenyl-4-(2,4,6-triphenyl-*N*-pyridinio)-phenolate] on both aniline- and

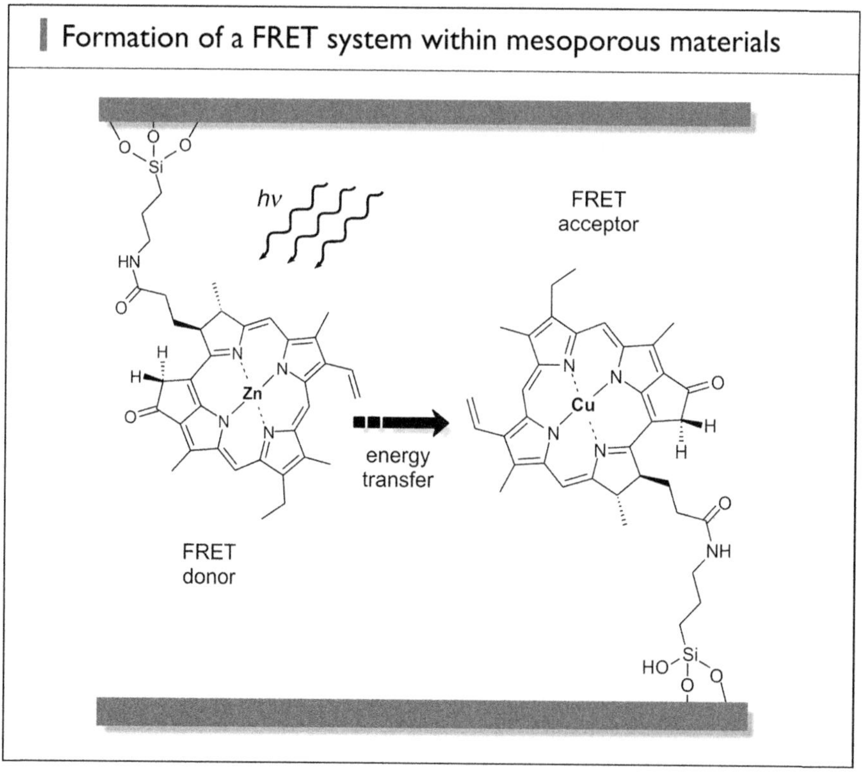

Figure 3.14 FRET system within the pores of an FSM silica phase; the energy transfer is initiated by irradiation with light and takes place from the FRET donor (Zn^{2+} as central ion) to the FRET acceptor (Cu^{2+} as central ion).

propylamine-functionalized SBA-15 mesoporous silicas and examined the influence on the visible absorption spectrum on the interactions with molecules from the gas or vapor phase, such as chloroform, ammonia, and ethanol.

Based on the change of fluorescence intensity, Kim et al.[146] prepared an anthraquinone-based sensor, which was proven to selectively detect fluoride anions (with a detection limit of ~0.50 μM) in the presence of competitive Cl^-, Br^-, I^-, and HSO_4^- anions. Recently, Banet et al.[147] demonstrated that hybrid mesoporous organosilicas functionalized with grafted β-diketones, namely dibenzoylmethane (DBM), exhibit fluorescence that can be enhanced or quenched in the presence of BF_3 or BCl_3 gases, respectively. This selective behavior opens up promising avenues for the design of new hybrid nanosensors for toxic gases commonly used in semiconductor industries. Wang et al.[148] developed a fluorescent chemosensor for Fe^{3+} on the basis of the immobilized bis-Schiff base *N,N′*-(1,4-phenylenedimethylidyne)bis(1,4-benzenediamine) onto previously 3-chloropropyl-functionalized SBA-15 samples. This sensor showed a high selectivity towards Fe^{3+} ions and a good linearity between the fluorescence intensity and Fe^{3+} ion concentration (with a detection limit of 1.98 μM).

A complete FRET (Förster resonance energy transfer) system based on chlorophyll in the pores of FSM materials was accomplished by Kuroda's group.[149,150] They first functionalized the FSM samples with 3-aminopropyl groups to guarantee an ideal position of the macroscopic chlorin units (in the pore center) and prevent their denaturation. Then they ligated chlorophyll derivatives that possess 3-(triethoxysilyl)-*N*-methylpropan-1-amine groups to the pore walls. Zinc (for the FRET donor) and copper (for the FRET acceptor) were chosen as the central ions of the chlorins, which made it possible to initiate and record an efficient FRET process (Fig. 3.14).

3.4.2 Co-Condensation (One-Pot Synthesis, Direct Synthesis)

An alternative method to synthesize organically functionalized mesoporous silica phases is the co-condensation method (also called one-pot synthesis or direct synthesis). It is possible to prepare mesostructured silica phases by the co-condensation of tetraalkoxysilanes [$(RO)_4Si$ (predominantly TEOS or TMOS)], either (1) with terminal organotrialkoxysilanes of the type $(R'O)_3Si$-R or (2) with bis- or multi-silylated precursors of the type $[(R'O)_3Si]_m$, $m \geq 2$ in the presence of structure-directing agents (see Fig. 3.9). As also ternary mixtures were applied, that is, functionalization of both the pore wall interface and the framework were carried out simultaneously, we will not explicitly subdivide this section according to these two routes.

3.4.2.1 Hybrid Materials from Mixtures of Tetra(m)ethoxysilane and Mono-, Bis- as well as Multi-Silylated Organic Precursors

In principle, the integration of any desired organic functional group into mesoporous organic–inorganic hybrid materials is possible by employing the co-condensation approach as long as the respective organosilane precursor is accessible, stable under the employed hydrolysis/polycondensation reaction conditions, and if its amount in the reaction mixture does not exceed a certain threshold. Since the early work in 1996 to 1997 of the groups of Mann (phenyl-triethoxysilane, *n*-octyl-triethoxysilane),[151] Macquarrie (3-aminopropyl-trimethoxysilane (APTMS), 2-cyanoethyl-trimethoxysilane),[152] and Stein (vinyl-triethoxysilane),[153] a large number of other organically modified silica phases have been synthesized by co-condensation. The mesoporous materials obtained by direct synthesis can exhibit interesting catalytic and adsorption properties, or, by subsequent chemical transformation of the organic groups on the pore surfaces, can act as starting compounds for the synthesis of new organically modified silica phases. Here, only a few examples will be presented in more detail.

Further projects dealt with the fabrication of heterogeneous, basic or acidic solid-state catalysts or adsorbents carrying, for instance, amino or sulfonic acid groups. Amino-functionalized silicas were prepared and analyzed for the catalytic activity in Knoevenagel condensation reactions of aldehydes or ketones with ethyl cyanoacetate ions by Macquarrie et al.[154,155] Recently, Zhang et al.[156,157] reported on the successful preparation of amino-functionalized silica thin films by means of the EISA approach.

Thiol-functionalized silicas serve as a basis for the construction of solid-state acids, as the SH groups in the pore channels can be transformed into sulfonic acid groups by suitable oxidizing agents such as HNO_3 or H_2O_2.[158–162] Stucky and coworkers[163] have shown that the oxidation of thiol groups does not necessarily need to be carried out after the synthesis of the mesostructured products, but can take place *in-situ* by the addition of H_2O_2 to the reaction mixture of the co-condensation reactants. The catalytic activities of sulfonic acid-functionalized silicas have been determined, for example, by means of esterifications and ether syntheses.[164–166]

Alauzun et al.[167] have shown that it is possible to prepare bifunctional mesoporous materials containing two antagonist functions, that is, an acidic group in the framework and a basic one in the channel pores. This novel material was synthesized by co-condensation of a ternary mixture of the silsesquioxane α,ω-bistrimethoxysilyl-4,5-dithiooctane $[(MeO)_3Si\text{-}(CH_2)_3\text{-}S\text{-}S\text{-}(CH_2)_3\text{-}Si(OMe)_3]$, the terminal organosilane 3-*tert*-butyloxycarbonyl(aminopropyl)trimethoxysilane $[(MeO)_3Si\text{-}(CH_2)_3\text{-}NHBoc]$ and TEOS in the presence of P123 as template in low acidic medium. The bifunctional material, having an acidic framework and basic functionalities protruding into the pores, was obtained by post-treatments of the resulting material, that is, reduction of the disulfide units followed by their oxidation to sulfonic acid groups and deprotection of the amino groups.

Acidic organo-functionalized silicas could also be prepared by converting the cyano groups of cyanoalkyl-functionalized phases by hydrolysis with H_2SO_4 into the corresponding carboxylic acid functionalities[168,169] and by ester hydrolysis of silica phases functionalized with diethyl phosphonate groups, resulting in the respective phosphoric acid bearing phase.[170] Sulfonic acid functionalization was also achieved by postsynthetic sulfonation of 1,4-diethylbenzene groups within the framework of mesoporous hybrid materials, which were obtained by co-condensation of 1,4-bis(trimethoxysilylethyl)benzene (BTMSEB, **13**) and TMOS under acidic conditions using P123 as SDA.[171] Although attempts by Burleigh et al. [172] to prepare the corresponding genuine PMOs, that is, out of 100% BTMSEB (using CTAB as SDA under basic conditions), resulted only in amorphous materials, here, well-ordered samples up to a content of 70 mol% in the initial reaction mixture could be obtained. Li et al. reported the successful preparation of the respective PMO materials, again by using P123 as SDA under acidic conditions.[173]

Thiol-functionalized mesoporous silicas not only serve as a basis for sulfonic acid-functionalized phases, but can be used for several other applications, for instance as adsorbents. Thiol-functionalized silicas can also be prepared as small nanoparticles with spherical morphologies by employing modified Stöber reaction conditions.[174] The high affinity of these compounds for thiophilic heavy metals, especially toxic Hg^{2+} ions, has been demonstrated by several authors.[175–178]

Jaroniec and coworkers were able to incorporate (up to ~30 mol%) tris[3-(trimethoxysilyl)-propyl]isocyanurate (ICS) (**18**) into mesoporous silica phases by co-condensation with TEOS.[179,180] They also tested multifunctional hybrid materials based on ICS as potential Hg^{2+} adsorbents by direct co-condensation of ternary mixtures comprising TEOS (70 to 88 mol%), ICS, and another organosilane, namely bis[3-(triethoxysilyl)propyl]tetrasulfide, *N*-(3-triethoxysilylpropyl)-4,5-dihydroimidazole, or ureidopropyltrimethoxysilane.[181] Ternary systems comprising TEOS, ICS, and MPTES, which showed a high adsorption capacity for Hg^{2+}, were also reported (for PMOs based on ICS, see Section 3.4.3).[182]

It is possible to functionalize silicas with far more complex organic groups by means of co-condensation reactions, which opens up paths to further materials with interesting chelating or adsorbing properties. Li et al. succeeded in the incorporation of 1,4-diureylenebenzene moieties into SBA-15-like silicas through co-condensation of bis-silylated 1,4-diureylenebenzene and TEOS using P123 as SDA in acidic solutions.[183]

Furthermore, Corriu et al.[184] anchored chelating cyclam molecules by substitution of the chlorine atoms on previously synthesized 3-chloropropyl-functionalized silicas and showed that almost all cyclam units were localized on the pore surface and were thus freely accessible to complexation by Cu^{2+} and Co^{2+} ions (Fig. 3.15a); cyclam = 1,4,8,11-tetraazacyclotetradecane). Additionally, Corriu's group demonstrated that it is possible to introduce cyclam groups into mesoporous hybrid materials directly by co-condensation reactions. The respective cyclam derivatives were used both as terminal organosilanes (in this case the cyclam units are located rather on the pore surfaces) as well as tetra-silylated compounds (in this case the cyclam units are incorporated preferentially in the pore wall). Also ternary mixtures were subjected to co-condensation reactions, whereby the tetra-silylated cyclam derivative was one component and the second organosilane consisted of a mono-silylated cyclam derivate or another mono-silylated functional group; the third component was TEOS. These clever recipes to build up bifunctional mesoporous hybrid materials were further refined by combining them with subsequently conducted grafting procedures.[185–188]

The particular versatility of an integrated iodopropyl group was recently demonstrated by Corriu and coworkers.[189] In comparison with the chloropropyl group, they could show that the iodo group allows much easier and faster postsynthetic chemical

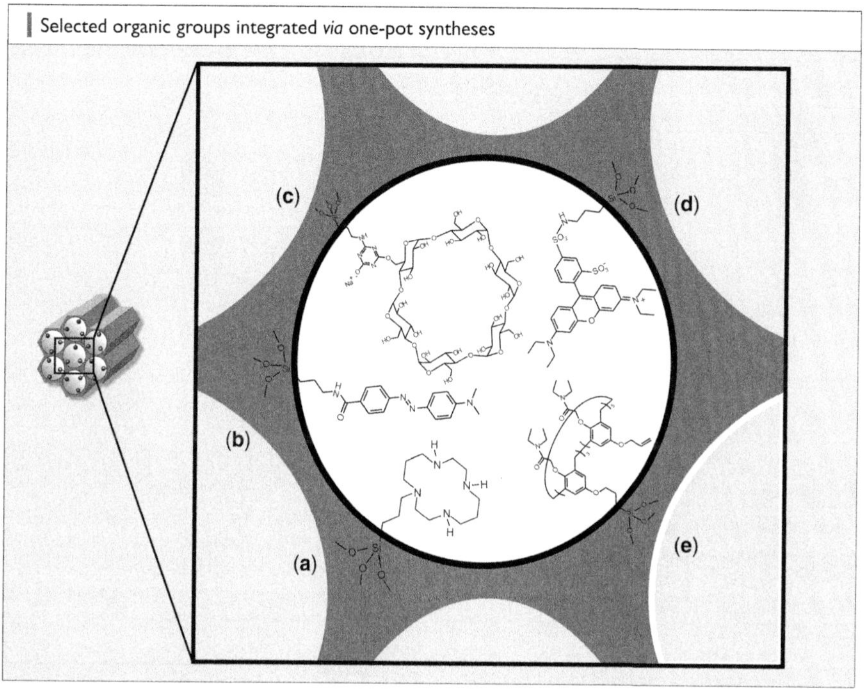

Figure 3.15 Selected organic functionalities anchored by co-condensation on the pore surface of silica phases (see text for details).

transformations, giving rise to a variety of mesoporous materials with large and/or hydrophilic functional groups located within the channel pores that are difficult or impossible to obtain by other routes, among them ethylenediamine, dipropylenetriamine, tris(2-cyanoethyl)phosphine, triethylphosphite, ephedrine, and cyclam groups (see Fig. 3.16).

Recently, Ozin and coworkers[190] reported on the synthesis of a mesoporous hybrid material, containing covalently integrated bucky balls, that is, C_{60} moieties. Interestingly, the synthesis failed under basic reaction conditions (CTAB as SDA)—these gave rise to a phase separation—but was successful under acid-catalyzed conditions (P123 as SDA), which gave well-ordered materials (at least up to a content of 21 wt% of the multi-silylated C_{60} precursor in the reaction mixture with TEOS, corresponding to approximately 11 mol%) with uniformly distributed C_{60} moieties. Furthermore, it was demonstrated that a portion of these C_{60} groups is chemically accessible through selective reaction of C_{60} with osmium tetroxide.

Huq and Mercier[191] synthesized cyclodextrin modified silicas by first coupling the cyclodextrin units to 3-aminopropyltriethoxysilane (APTES) and then subjecting it to a co-condensation reaction with TEOS (Fig. 3.15c). Interestingly, all attempts to anchor cyclodextrins by grafting onto previously thiol-functionalized silicas have been unsuccessful. It has also been demonstrated that the cyclodextrin units on the

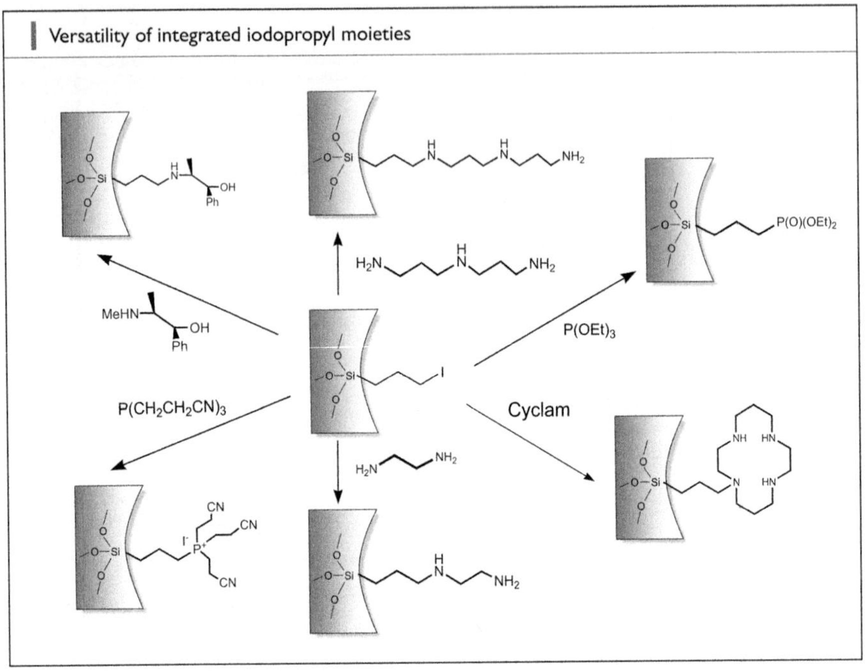

Figure 3.16 Possible transformations of iodopropyl groups integrated into mesoporous silica phases via the co-condensation method.

surface of the silicas obtained by direct synthesis are able to adsorb *p*-nitrophenol from aqueous solutions. Liu et al.[192] synthesized silicas functionalized with calix[8]arene amide (Fig. 3.15e) and showed that they were suitable for the adsorption of humic acid from aqueous solutions. The synthesis of spherical mesostructured particles modified with the dipeptide carnosine (β-alanyl-L-histidine) has been published by Walcarius et al.[193] The peptide units of the ordered mesoporous materials obtained were more accessible than those of analogously functionalized amorphous porous particles, as was demonstrated by complexation reactions with Cu^{2+} ions.

Other projects focus on the modification of silica matrices with different chromophores. Mann and coworkers[194,195] synthesized 3-(2,4-dinitrophenylamino)propyl-functionalized MCM-41 samples and obtained materials in the form of powders, thin films, and monoliths. Ganschow et al.[196] anchored the photochromic azo dye 4-[(4-dimethylaminophenyl)azo]benzoic acid and the fluorescent laser dye sulforhodamine B into 3-aminopropyl-MCM-41 samples (Figure 3.15b, d). The coupling of the dye molecules to the amino group of the organosilane through the carboxylic and sulfonic acid groups, respectively, and the actual co-condensation reaction with the inorganic precursor could be carried out simultaneously in a microwave apparatus. It was possible to significantly reduce the reaction time by microwave-supported synthesis relative to conventional methods, and thus, degradation of the dye molecules during hydrothermal treatment could be considerably reduced.

Besides interesting work on photochemical switches based on MCM-41 phases obtained by means of co-condensation, which is covered in Chapter 18 of this book,[197–199] also the preparation of viologen containing highly ordered 2D hexagonal silica phases by García's group is worth mentioning.[200–202] Interestingly, the viologen units of this material can be almost completely transformed into bipyridinium radical cations upon irradiation or thermal treatment, which showed lifetimes from femtoseconds up to minutes. In further work, they studied the electrochemical behavior and the catalytic potential of the viologen in the electrochemical oxidation of hydroquinone.[203]

Further functionalization of mesoporous films with the pH-sensitive dye fluorescein was accomplished by Wirnsberger et al.[204] The organosilane used for the actual co-condensation reaction was first prepared by reaction of fluorescein isothiocyanate (FITC) with APTES. The possible use of the dye-modified films as pH sensors was investigated by measurement of the fluorescence after excitation with an Argon laser (488 nm); a dramatic change in fluorescence intensity was observed around pH 8 with a response time of a few seconds.

An interesting potential optical sensor for explosives was recently presented by Balkus and coworkers.[205] They synthesized a redox active mesoporous material by co-condensation of TMOS with bis(3-trimethoxysilylpropyl)-diazapyridinium diiodide (up to a content of 2.3 mol%) and observed a pronounced photoluminescence quenching on exposure to several nitrated, volatile, so-called explosive taggants.[†††]

Finally, recently, electrochemically active mesoporous organosilica materials with amino-capped aniline trimer units as bridges have also been reported.[206] This aniline trimer derivative was chosen because it exhibits similar intrinsic physical properties and electroactive behavior to polyaniline due to the same oxidation and protonic doping mechanisms. The silsesquioxane precursor *N,N′*-bis-(4′-(3-triethoxysilylpropylureido)phenyl)-1,4-quinonene-diimine (Fig. 3.17) was used, which underwent hydrolysis and co-condensation with TEOS and CTAB as SDA in basic medium using acetone as the cosolvent. The cyclic voltammograms indicated that the electroactivity from the electroactive aniline trimer moiety is maintained in the respective hybrid materials. However, substantially lower intensities of the redox peaks were observed compared to the precursor under the same conditions, since TEOS or the polycondensation product of TEOS, respectively, is electrochemically inert. At any rate, the introduction of electroactivity with retention of the mesoporous reservoirs makes this kind of hybrid materials promising for designing electrochemically switchable devices.

[†††]Explosives manufacturers add identification markers, called taggants, to their explosives to aid in detection and tracking. There are two types of taggants: detection taggants and identification taggants. Identification taggants are various types of materials that have been placed in the explosive which allow the explosive to be traced, postdetonation, to its point of origin and possibly determine how the explosive was acquired. Because plastic explosives have low vapor pressures and are therefore difficult to detect through chemical means prior to detonation, detection taggants are added. These taggants are high vapor pressure compounds whose volatility enables the explosive to be detected more easily.

Electroactive tri-aniline-based precursor

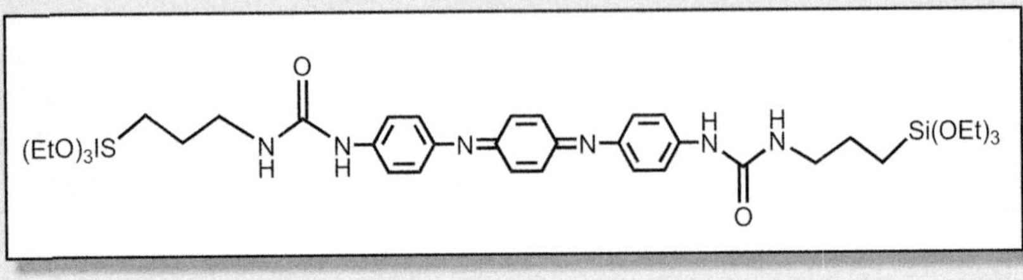

Figure 3.17 Chemical structure of *N,N'*-bis-(4'-(3-triethoxysilylpropylureido)phenyl)-1,4-quinonene-diimine.

3.4.3 Single-Source Precursor Method: Using Bis- and Multi-Silylated Compounds for the Synthesis of (Periodic) Mesoporous Organosilicas (PMOs)

The largest class of PMOs is made up of those that exhibit short-chained alkylene or alkenylene groups as the organic bridge. Apart from the instances already mentioned (one ethane-bridged and two ethene-bridged organosilica materials, all synthesized in basic media, following the $S^+ I^-$ pathway with OTAC or CTAB as SDA) methane-, (**1**)[207–210] ethane-, (**2**)[90,208,211–228] and ethene-bridged (**3**)[229–235] PMOs were prepared with all imaginable SDAs (with and without swelling-agents), leading to a variety of symmetries with a broad coverage of pore diameters [3.1 to 5.0 for (**1**), 2.2 to 20.0 for (**2**), 2.4 to 14.7 for (**3**)].

In the recent past, several projects dealt with the question of the way in which inorganic salts may influence the structural properties when added to the reaction mixture (for instance, with regard to the resulting phase, mesoscopic order, and wall thickness) of the resulting PMO materials. Zhang et al.[227] demonstrated for the first time for Brij-templated ethane-silicas under neutral conditions that the addition of certain amounts of divalent inorganic salts such as $NiCl_2$ can improve the structural (mesoscopic) order considerably. Concerning the triblock copolymer templated synthesis pathway, it was recently shown by Zhai et al.[236] that it is not necessary to add extra acids. It is sufficient to add relatively small amounts of hydrolysable, acidic reacting metal (oxy) chlorides, like $ZrOCl_2$ or $FeCl_3$. In combination with NaCl, which promotes the interaction between the pluronic-composed micelles and the precursors further, very well-ordered SBA-15-like ethane-bridged (and in a subsequent work also benzene-bridged)[237] PMOs were obtained. The effect of trivalent aluminum salts as well as mono-(KCl) and divalent ($CaCl_2$) salts was investigated in the synthesis of large-pore ethane- and benzene-bridged PMOs by Cho et al.[238] They found that sodium aluminate had a pronounced effect on the porosity of 2D hexagonal ethane-silicas (larger S_{BET} and pore volume, thinner pore walls) and that highly ordered benzene-silicas can be prepared with the addition of $AlCl_3$ instead of the mineral acid HCl. In this context, also the work of Shylesh et al.[239] is noteworthy, who prepared, in analogy to the microporous zeolitic and mesoporous M41S- or SBA-15-based acidic alumosilicates, *organo* alumosilicas by using aluminum isopropoxide as an additional precursor. The aluminum-containing ethane-silicas obtained showed considerably improved hydrothermal stability.

In 2008 Yoon et al. reported that it is possible to reduce the synthesis time for ethane-bridged PMOs drastically by using microwave heating.[240]

A very interesting and, at first glance, quite surprising application of Brij-templated ethane-bridged PMOs (pore diameter: 4.8 nm) was presented by Wang et al.[241] who used this kind of material to assist the refolding procedure of denaturated hen egg white lysozyme proteins (molecular dimensions of $1.9 \times 2.5 \times 4.3$ nm) and yielded over 80% refolding at about 0.6 mg mL^{-1}. The refolding protocol is shown schematically in Figure 3.18. For comparison, SBA-15 samples were also tested for this purpose. The trick consists in using a releasing filter which has the correct dimensions and a suitable polarity of the inner surface: (1) the channels were first loaded with the denaturated and reduced (through 8 M urea and dithiothreitol) lysozyme, whereby the SBA-15 samples showed a higher adsorption capacity (280 mg g^{-1}) than the PMO samples (168 mg g^{-1}). Due to the uniform mesoporous channels tailored to accommodate individual proteins, protein aggregation was minimized. (2) At pH = 8.5, the interaction of lysozyme with the SBA-15 surface is much stronger than the interaction between lysozyme and the PMO surface (due to the lower number of silanol groups). (3) Poly(ethylene glycol) (PEG) was selected as a trigger

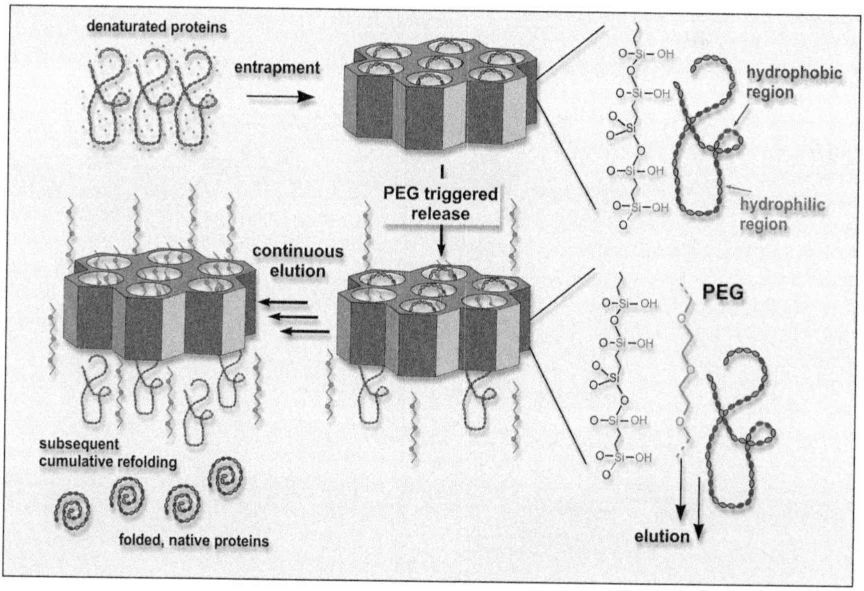

Figure 3.18 Schematic diagram of the hen egg lysozyme refolding procedure with the help of ethane-bridged PMOs.[241] (Reprinted with permission from X. Wang et al., *Langmuir* **2007**, *23*, 5735–5739. Copyright 2007 American Chemical Society.) (See color insert.)

to release encapsulated denaturated lysozyme from the mesopores into the refolding buffer (containing reduced and oxidized glutathione; GSH and GSSG) because PEG can form strong hydrogen bonds with the silanol groups. As a result, the PMO samples performed best in release, whereas the amount of protein released from SBA-15 was only small under the same conditions. Although the formation of inactive aggregates at high PEG trigger concentrations reduces the refolding yield, this approach illustrates how the specific porosity and surface properties of a material can be decisive for a specific application.

While the saturated methane and ethane bridges are almost chemically inert, the unsaturated ethene bridge enables postsynthetic modifications, that is, to carry out addition reactions. That the ethene bridge is accessible for such reactions was demonstrated by Asefa et al.[71] in 1999 by means of bromination reactions, although, interestingly, the degree of bromination was much lower than expected (only 10%). Particularly noteworthy is the work of Nakajima et al.[232,233] who prepared ethene-bridged PMOs in a first step, and successively modified them via a two-step chemical reaction—the first step involves a Diels–Alder reaction with benzocyclobutene, the second step consists of a sulfonation of the aromatic ring system—to produce finally a sulfonic acid-functionalized hybrid mesoporous solid acid catalyst (Fig. 3.19). This material exhibited high activity for various acid-catalyzed reactions and can be used repeatedly without deactivation.

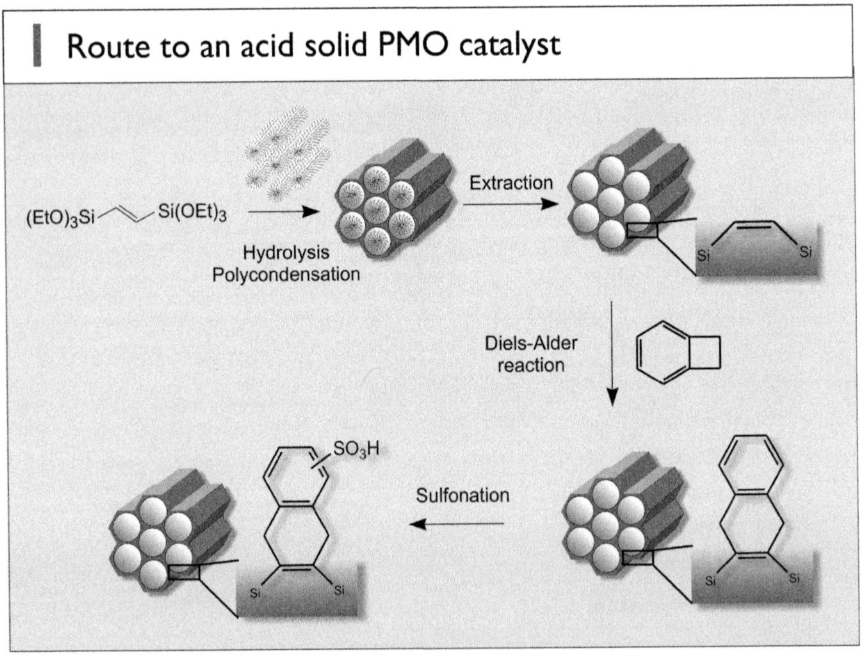

Figure 3.19 Schematic illustration of the synthetic pathway to benzene sulfonic acid PMOs; the first step involves a Diels–Alder reaction of the ethene bridges with benzocyclobutene; in the second step the benzene moieties were sulfonated.

A similar strategy to obtain sulfonic acid-functionalized PMOs was pursued by Dubé et al.[234] Again based on ethene organosilicas, they carried out a two-step chemical modification involving an arylation of the ethene sites with benzene in the presence of $AlCl_3$ [prior, some of the PMO samples were treated with hexamethyldisilazane (HMDS) to eliminate surface silanol groups (capping) in order to avoid any reaction of $AlCl_3$ with surface OH groups] followed by sulfonation in concentrated H_2SO_4. This material exhibited a high catalytic activity for the self-condensation of heptanal, owing to the high density of acid sites and the presence of a hydrophobic bridged framework. This underlines the potential advantages of using PMOs as acid catalysts in all water-generating reactions whenever the acid sites are deactivated by water.

3.4.3.1 PMOs with Crystal-like Pore Walls and Aromatic PMOs

Meanwhile, there are many indications that the *precursor* must obey certain structural requirements for the realization of the *periodicity* at the *mesoscale* of organosilicas:

1. The precursor or, to be more precise, the organic bridge must not be too flexible; interestingly, so far, PMOs with propyl, butyl, or even longer hydrocarbon chains as the organic bridge have not been realized.
2. The precursor must not be too voluminous; empirically it turned out that very bulky silsesquioxanes have to be diluted with another, much smaller, more flexible network building component, whether a pure inorganic silica precursor like TEOS/TMOS (see above) or another silsesquioxane with a very short organic bridge (methylene or ethane, see below), in order to give well-ordered mesostructured products.
3. Precursors with a distinctive tendency to undergo molecular self-assembly processes (i.e., even without the presence of SDAs), for instance driven by π-stacking or hydrogen bonds, also facilitate the onset of mesoscale periodicity in templated synthesis approaches.

It is still not clear how these relationships can be rationalized. However, as (hetero-)aromatic compounds are rigid and are able to form π-stacks, numerous syntheses of PMOs with aromatic bridges have been reported, thereby introducing another sort of functionality into PMOs.

The first synthesis of PMO materials with aromatic bridges was reported by Yoshina-Ishii et al.[242] as early as 1999. They used 1,4-bis(triethoxysilyl)benzene (BTEB, **5**) and 2,5-bis(triethoxysilyl)thiophene (BTET, **17**) as precursors in the presence of CPC as SDA in acidic medium. The benzene-bridged organosilica is, among the aromatic PMOs, by far the best studied system.[231,237,238,243–260] This is also the system for which for the first time a *crystal-like arrangement* of the organic bridges within the pore walls was detected (see also Fig. 3.20). This means that apart from the periodicity at the mesoscale the organic bridges exhibit a long-range order, which is reflected in the corresponding X-ray diffractograms by (dependent on the degree of order) one to several additional reflections in the wide-angle region ($2\theta > 10°$).

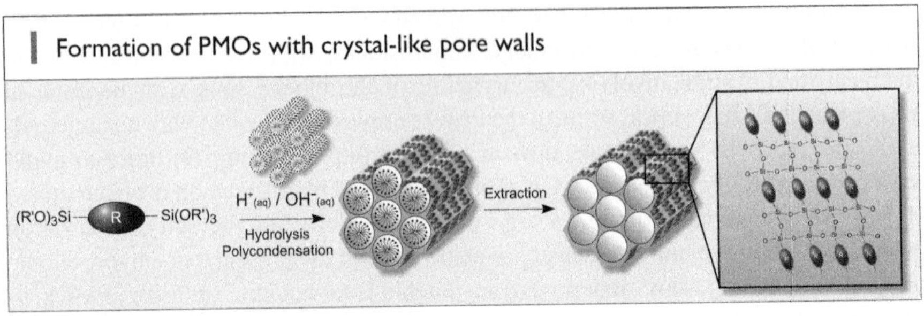

Figure 3.20 Schematic synthesis pathway of PMOs with a crystal-like arrangement of the bridging organic units R in the pore walls. This representation is idealized: the bridges can be slightly tilted or twisted with respect to each other.

This interesting feature was first reported by Inagaki et al. in 2002 for 2D hexagonally mesostructured benzene-bridged PMOs (pore diameter: 3.8 nm).[243] This crystal-like organization of the organic bridges within the pore walls (a model of the pore wall along with the experimental powder X-ray diffractograms is shown in Fig. 3.21) was confirmed by HRTEM images, which showed numerous lattice fringes along the pore axis, also indicating a separation distance of 7.6 Å. Note, the term *crystal-like* instead of crystalline is used here because the organic bridges do not have to

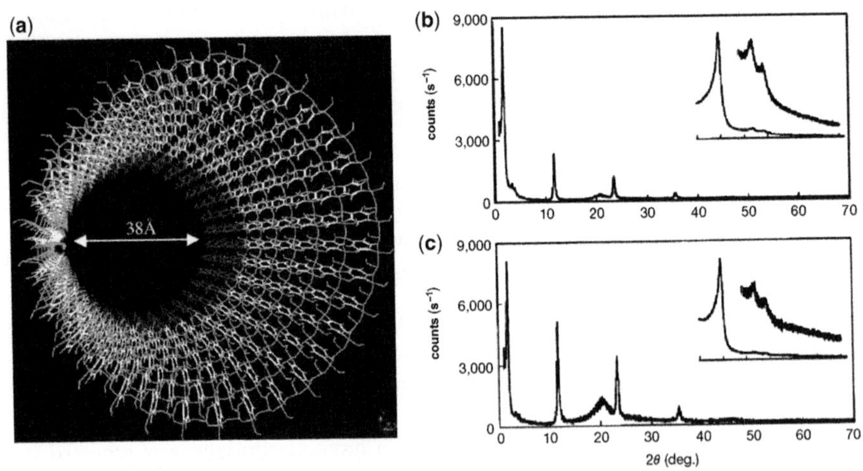

Figure 3.21 (a) Model of a pore wall of benzene-silicas. The benzene rings are aligned in a circle around the pore, fixed at both sides by silicate chains. The silicate is terminated by silanol groups (Si—OH) at the surface. Hydrophobic benzene layers and hydrophilic silicate layers array alternately at an interval of 7.6 Å along the channel direction. Silicon, orange; oxygen, red; carbon, white; hydrogen, yellow. (b) Powder X-ray diffraction pattern of a mesoporous benzene-bridged PMO sample after removal of the surfactant; (c) the pattern of the composite material that still contains the surfactant. Insets: Reflections in the small-angle area ($1 < 2\theta < 7$). This material shows periodicity both on the mesoscopic ($d = 45.5$, 26.0, and 22.9 Å) and on the molecular scale ($d = 7.6$, 3.8, and 2.5 Å).[243] (Adapted with permission from S. Inagaki et al., *Nature* **2002**, *416*, 304–307. Copyright 2002 Macmillan Publishers.) (See color insert.)

possess strict translational symmetry: (1) because of the free rotation along the Si–R–Si longitudinal axis the bridges can possess alternating orientations in relation to the flanking silicate layers; (2) furthermore, the organic units can be slightly tilted with respect to each other.

Meanwhile, Inagaki's group developed an alternative route to benzene-bridged PMOs not by using bridged bis(trialkoxysilyl)organosilanes but bridged bis(tris-allyl)organosilanes as precursors.[253,254] The innovation of this approach consists in the first instance not in the result as such—this is identical: a benzene-bridged PMO with both mesoscopic and molecular order—but because it opens up new possibilities concerning the synthesis of PMO precursors. Note that the range of suitable alkoxysilane precursors is limited because alkoxysilane precursors containing relatively large organic groups are difficult to obtain in high purity due to the limitations of distillative and chromatographic separation for such nonvolatile compounds.

Goto and Inagaki[255] were also able to prepare mesoporous benzene-silicas with cubic symmetry and 3D cage-like pores, both with small pores using CTAC or CPC as SDA as well as with large pores using F88 $[(EO)_{100}-(PO)_{39}-(EO)_{100}]$ as SDA. In contrast to the 2D hexagonally ordered samples, these cubic benzene-bridged PMOs exhibited no order at the molecular level. Syntheses of benzene-bridged PMOs with Brij surfactants under acidic conditions gave only molecularly amorphous products.[251]

Benzene-bridged PMOs are also accessible as thin films.[257] PXRD patterns of these thin films reveal the presence of a hexagonal mesostructure as well as an ordered molecular-scale structure, but, interestingly, the diffraction peaks correspond to d values of $d = 10.1$, 5.0, 3.3, 2.5, and 2.0 Å, which means an additional basal spacing between the silicate layers of 2.5 Å in comparison to the bulk samples of Inagaki et al.[243] A possible explanation of this observed effect is the presence of extensive hydrogen bonding as a result of a considerably low degree of silicate condensation of the organobis-silanetriols, which should effectively elongate the benzene–silica layers. This hypothesis would be consistent with the molecular packing structure of benzene-bridged bis-silanetriol $[(HO)_3Si-C_6H_4-Si(OH)_3]$, solved by Cerveau et al.[261] in which sheets of the bis-silanetriol molecules are stabilized via hydrogen bonding, and the unit-cell parameter a of this solid equals 10.056 Å. However, there is an ongoing and interesting discussion if this is the correct interpretation since it cannot be excluded that the structural features are the result of a phase separation process, that is, that the film obtained is a blend of a mesoporous thin film with amorphous pore walls along with a small fraction of *non-porous* 1,4-bis(benzenesilyl-triol) $[(HO)_3Si-C_6H_4-Si(OH)_3]$ particles having molecular-scale periodicity. Readers are invited to make up their own minds by studying the arguments in the current debate, in References 262 and 263.

Since the discovery of this first benzene-bridged mesoporous hybrid material with crystal-like pore walls, the series of organosilica precursors that give PMOs exhibiting molecular-scale periodicity was considerably extended. Kapoor et al. demonstrated with the formation of PMO products from the precursor 1,3-bis(triethoxysilyl)benzene (**10**) that a crystal-like organization of the organic bridges within the pore walls is not restricted to linear-substituted precursors.[264]

Further aromatic PMO systems that show both periodic ordered mesoporosity and periodicity at the molecular level, include those with a biphenyl (**14**)[265,266] and a divinylbenzene (**26**) bridge.[267,268] The biphenyl- and divinylbenzene-bridged PMOs exhibit approximately the same interlayer distance of 11.6 and 11.9 Å, respectively.

That PMOs with crystal-like pore walls are not limited to aromatic organic bridges at all was shown by the group of Mokaya.[269–272] In a series of publications, his group reported on the preparation and extensive studies of mesoporous, molecularly ordered ethene-bridged organosilicas as well as organoalumosilicas. Here, the basal spacing between the alternating organic/silicate layers amounts to 5.6 Å. Table 3.3 summarizes all precursors that—up to now—gave molecularly ordered PMOs together with their periodicity length along the pore walls.

Although PMOs with crystal-like pore walls have gained considerable attention within the research community, relatively little is known about the specific formation mechanism. In the authors' opinion, this is due to the fact that the elucidation is very hard to tackle even from a technical point of view. For instance, many of the standard exploration techniques look at the wrong length-scale (X-ray, TEM) or are not easily applied *in situ*. Only a few projects[247,248,259,273] have addressed the question of how the building units are organized at the mesoscopic and molecular levels during the course of the hydrothermal reaction and in which way these two processes are correlated explicitly; however, no conclusive results could be inferred with regard to the mechanism.

Further aromatic PMOs—*without* crystal-like pore walls—comprise those with 2-methylbenzene (**7**), 2,5-dimethylbenzene (**8**), and 2,5-dimethoxybenzene (**9**) bridges,[274] as well as those with diphenyl ether (**15**) and diphenyl thioether (**16**)[275]

Table 3.3 Overview of Bis-Silylated Precursors that Show a Crystal-like Arrangement Within the PMO Pore Walls, Together with their Molecular Periodicity Length

Precursor	Periodicity [Å]	Ref.
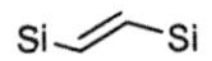	5.6	269–272
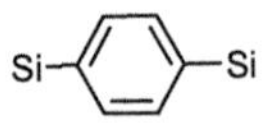	7.6	243
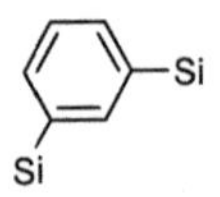	7.6	264
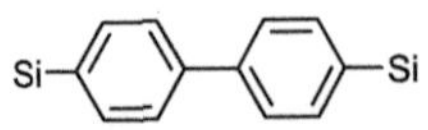	11.5 (acidic medium)	266
	11.6 (basic medium)	265
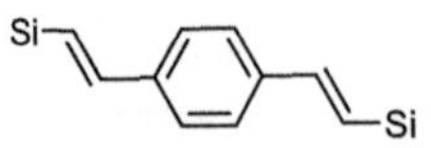	11.9	267,268

and toluene (**11**) and *p*-xylene bridges (**12**).[276] For quite a long time, it seemed that it was not possible to prepare aromatic PMOs in which the aryl moiety is connected to the linking Si units through alkyl chains longer than C_1, because the flexible alkyl chains could counteract on the advantageous structuring conditions provided by the rigid aryl core. However, in 2007 Li et al.[173] were able to obtain PMOs with 1,4-diethylbenzene bridges (**13**).

In the meantime, also two heteroaromatic precursors have been prepared and successfully converted into the respective PMO materials: Apart from a carbazole-based PMO (see below), Morell et al.[277] reported the synthesis of a highly ordered, large-pored thiophene-bridged PMO material [made of (**17**)] with an SBA-15-analogous mesostructure (using P123 as SDA). Interestingly, other monocyclic heteroaromatic-based PMO systems such as furane-, pyrrole- or pyridine-bridged PMOs have not been realized so far.

3.4.3.2 Postmodification of Aromatic PMO Precursors and PMO Materials in the Solid State

Although they are not as reactive as, for instance, unsaturated alkyl bridges, the organic bridges of aromatic PMOs were successfully subjected to postsynthetic chemical modifications in the solid state in some cases. This was demonstrated by Inagaki et al.[243] in 2002 by means of sulfonation (with fuming sulfuric acid) of benzene-bridged PMOs with crystal-like pore walls, while preserving both the meso- and molecular-scale ordered structure. However, the conversion of phenylene groups was less than 10%. Recently, Inagaki and coworkers introduced aminobenzene-functionalized PMOs, which were obtained via a two-step chemical transformation procedure of the parent benzene-silicas (the overall conversion rate was 28%).[278] In addition, it was shown that this material can act as an effective and reusable basic solid-state catalyst for the Knoevenagel reaction.

An interesting approach of developing new PMO materials was presented in 2008 by Kuschel and Polarz,[279] who succeeded in preparing 1,3-bis-(tri-*iso*-propoxysilyl)-5-bromobenzene (**19**) as a new PMO precursor. This approach has the distinction of being versatile because halogen-aryl systems can easily be chemically modified. Not only chemical modifications of the precursor itself are possible (resulting in further new precursors) but also postsynthetic conversions of the solid PMO materials can be realized (Fig. 3.22). The bromobenzene precursor were converted to 1,3-bis-(tri-*iso*-propoxysilyl)benzoic acid (**21**) and 1,3-bis-(tri-*iso*-propoxysilyl)styrene (**20**). These three new precursors were transferred into the respective mesoporous organic–inorganic hybrid materials (although the mesoscopic order was limited and only wormhole-like pore systems were detected), which are called UKON (obviously an abbreviation for University of Konstanz, Germany) materials.[‡‡‡] Furthermore, the solid-state bromobenzene-bridged PMO were converted postsynthetically via the

[‡‡‡]It can be left to anyone's judgment if the trend of universities emulating companies, that is, profit-gaining entities, to create abbreviations of their products mirroring the companies names is an added benefit.

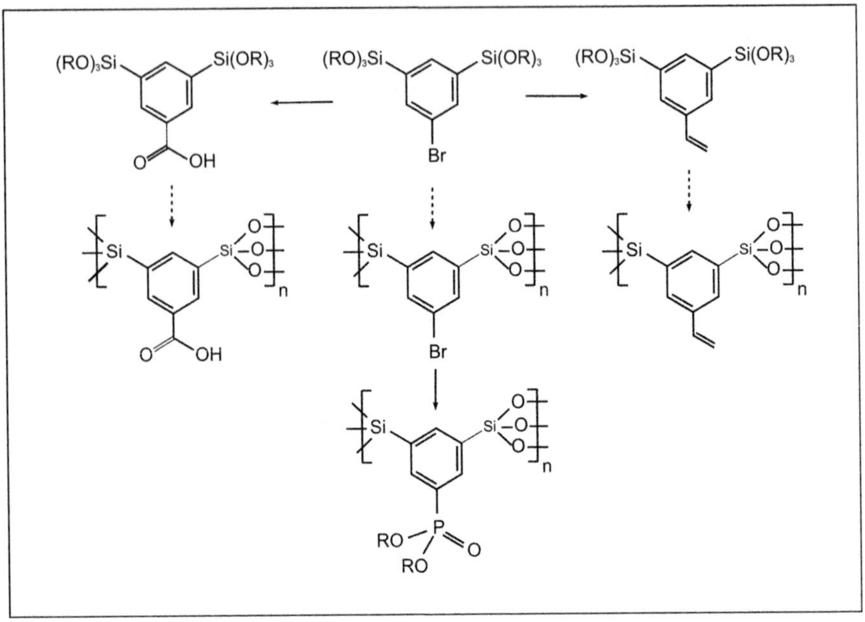

Figure 3.22 Starting from the bis-silylated 5-bromobenzene precursor which can be easily converted to the respective vinyl (right) and benzoic acid derivative (left) a series of four new PMOs, called UKON materials, are accessible. The fourth is prepared via a postsynthetic conversion of 5-bromobenzene-bridged PMO in the solid state.

reaction with $P(OEt)_3$ in the presence of $NiCl_2$ as a catalyst to give a PMO material with phenylphosphonic ethylester as a bridging group.

3.4.3.3 PMOs with Interesting Optical Properties

In the more recent past, intensified efforts were undertaken to design new PMOs that may find application in the optical or optoelectronic field. PMOs are particularly suitable for optical applications due to the large amount of densely packed organic chromophores that can be incorporated within their pore walls. For instance, Maegawa et al.[280] reported on the successful preparation of various bis-silylated carbazole derivatives, one of which (**29**) could be converted into the respective PMO material.

A very interesting thermo-stimuli responsive PMO material on the basis of bridged diacetylene containing silsesquioxanes was prepared by Peng et al.;[281] this precursor was previously subjected to a pure sol–gel process (without SDA).[282] After the surfactant removal of the mesoscopically 2D hexagonal well-ordered composite with molecularly aligned diacetylenic units a topo-polymerization creates the responsive PMO embedded with polydiacetylene (PDA). The reversible chromatic responses to external stimuli were demonstrated by subjecting the PDA-PMO to thermal cycles between 20°C and 103°C: The PDA-PMO rapidly (within approximately

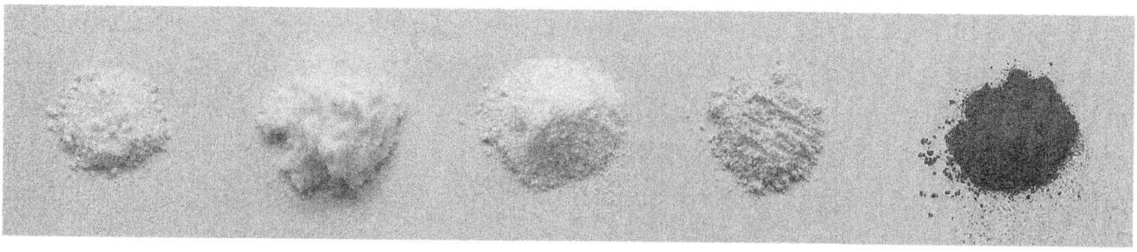

Figure 3.23 Photographs of powders of mesoporous hybrid materials with various conjugated π-electron systems (from left to right): ethene-, benzene-, 1,4-divinylbenzene-, 4,4′-divinylstilbene-, and 4,4′-divinylazobenzene-bridged organosilicas. (See color insert.)

10 seconds) changes its color *reversibly* between blue and red. This fast response may be of great interest for sensing applications.

Cornelius et al.[283] presented recently a project showing that the optical absorption properties of the incorporated organic functionalities are tunable by adjusting the length of the conjugation of the π-electron systems or by integration of heteroatoms therein. For this purpose, two new bis-silylated compounds with 18 π-electron systems [(**27**), (**28**)] and their related mesoporous hybrid materials were synthesized. On the basis of solid-state UV-vis absorption spectra it was demonstrated that for highly conjugated organic bridges the absorption maximum reaches the visible region, resulting in a yellow powder for a pure hydrocarbon backbone and a deep red powder for an azobenzene-containing material (Fig. 3.23).

Finally, Goto et al.[284] carried out a comparative fluorescence study on thin films of benzene-, biphenyl-, naphthalene-, and anthracene-bridged PMOs, thereby introducing new PMOs with 2,6-naphthalene (**30**) and 9,10-anthracene (**31**) organic bridges. The fluorescence spectra of the PMO films were significantly red-shifted compared with those of their precursor solutions, suggesting excimer formation.

3.4.3.4 PMOs from Tris- and Multi-Silylated Precursors

PMO materials that are constructed from bis-silylated precursors may be regarded as MCM-41/SBA-15 phases in which, in the ideal case, a quarter of all Si—O—Si units is replaced by Si—R—Si units, which corresponds to a formal molecular formula of $[R_{0.5}SiO_{1.5}]$. Unlike the bivalent oxygen atom, however, the organic bridges can in principle form bonds to more than two silicon atoms. In this way, the structural motifs already realized in PMOs can be extended considerably. At the same time, the mechanical and thermal stability of PMOs may be increased, since tris- and multi-silylated precursors can act as cross-linkers.

In 2002, Kuroki et al.[285] published the synthesis of an aromatic PMO material that was obtained from the three-point coupling precursor 1,3,5-tris(triethoxysilyl)benzene (**6**). Benzene groups began to separate from the 1,3,5-benzene-bridged PMO above around 600°C (under a nitrogen atmosphere), whereas xerogels made of the 1,3-benzene-bridged and the corresponding mono-silylated compound, which were prepared for comparison, began to decompose at 500°C and 450°C, respectively. Another structurally interesting motif was constructed by Landskron et al.[286] who used the cyclic precursor 1,1,3,3,5,5-hexaethoxy-1,3,5-trisilacyclohexane (**4**), which led finally to PMO samples with linked $\{Si(CH_2)\}_3$ ring structures. These materials could also be obtained as oriented thin films by spin coating onto glass plates that exhibited unusually low dielectric constants *k* (*k*: relative permittivity or relative dielectric constant, also known as ε_r), therewith demonstrating impressively that PMOs may find a new field of applications.[§§§] The *k* value for the film prepared completely from the cyclohexane derivative was as low as 2.0.

In 2004, Landskron and Ozin[287] transferred a structure building concept from organic chemistry, the dendrimer concept, into the world of PMO chemistry. Self-assembly of dendrimer building blocks (**34–36**) with hydrolyzable alkoxysilyl groups at the outer edge through ionic (with OTAC) and nonionic (with triblock copolymers) synthesis pathways gave highly ordered *periodic mesoporous dendrisilicas* (PMDs) with pore diameters and wall strengths that are typical for the respective synthesis routes. One year later, Hunks and Ozin[288] introduced another new class of bifunctional PMOs that were formed from single-source precursors and contained either siloxanedisilsesquioxane (DT^2 type; formed from **32**) or siloxytrisilsesquioxane units (MT^3 type; formed from **33**).

Soon after isocyanurate-(ICS)-based hybrid materials were reported by Olkhovyk and Jaroniec[179] and Jaroniec et al.[289] the synthesis of the respective PMO, made of 100% (**18**), was also accomplished. Zhang et al.[290] synthesized ICS-PMOs by self-assembly in the presence of P123 and additional inorganic additives (NaCl, NH_4F) under acidic conditions.

3.4.3.5 PMOs from Mixtures of Bis- or Multi-Silylated and Monosilylated Precursors

Looking back, it seemed for a while that the number of new PMOs was increasing only very slowly with time, most likely for two main reasons: Not all organosilica precursors can be converted into mesostructured products, since they lack the necessary structural requirements, especially adequate rigidity of the organic bridges. Another reason is that the syntheses of those precursors that are not commercially available are often not trivial. However, with more and more experience in template-steered syntheses of organosilica precursors, by testing new pathways and by collaboration between synthetically working organochemists and experts in the field of

§§§The continually increasing density of electronic components, switching elements, and circuit paths in modern, highly integrated circuits require insulators that have sufficiently small *k* values to prevent transfer of charge and signal losses that can arise through undesirable charge accumulation between the switching elements and circuit paths. These insulating materials must have relative permittivity values significantly lower than that of silicon dioxide ($\varepsilon_r \approx 3.8$), and for components smaller than the typical size of 0.1 μm, even materials with ε_r values less than 2 are necessary (so-called ultralow-*k* materials). Owing to their large pore volumes [ε_r(air) $\approx$ 1] and their comparatively high organic fraction, PMO films, whose atoms are lighter and less extensively polarizable than those of silicon dioxide, are in principle very suitable as low-*k* dielectrics. As the materials must be moisture repellent, the free silanol groups on the surface of the film are generally hydrophobized with HMDS or trimethylsilyl chloride (TMSCl).

soft-matter chemistry, continuous progress can be noticed. Since mid-2006 (see Reference 7), at least 13 new PMO precursors have been reported.

However, the variability of PMOs can be further extended by the construction of bifunctional PMOs through co-condensation reactions of mixtures of bridged bis(trialkoxysilyl)organosilanes [$(RO)_3Si{-}R'{-}Si(OR)_3$] and terminal organotrialkoxysilanes [$(RO)_3SiR''$] in the presence of an appropriate SDA, in complete analogy to the co-condensation reactions of TEOS/TMOS and terminal organotrialkoxysilanes [$(RO)_3SiR''$]). The resulting bifunctional PMOs then consist of a combination of bridging organic units and terminal organic groups whose ends point mainly into the pore interiors and are thus accessible for further chemical reactions.

Reports are found in the literature on syntheses in which the bridging components consist of, for example, ethane (of the majority), ethene, benzene, or biphenyl species, while the terminal functionalities include different alkyl, amino, thiol (predominantly), cyano, isocyanate, urea, vinyl, alkoxy, aromatic, and heteroaromatic groups. A summary of the published syntheses of these bifunctional PMOs is provided in Table 3.4. In principle, the same negative effects with regard to the degree of mesoscopic order and porosity of the products appear in these co-condensation reactions as those already discussed in the main co-condensation section above. In some cases, the terminal organic groups of the bifunctional PMOs were further chemically modified, for example, by selective transformation of terminal vinyl groups into hydroxyl groups by hydroboration and subsequent oxidation,[376] or by oxidative conversion of thiol groups into sulfonic acid groups.[384,385,387,390,391]

Recently, also two chiral terminal organotrialkoxysilanes were subjected to co-condensation reactions with BTME (**2**), namely *N*-[(triethoxysilyl)-propyl]-*trans*-(1*R*,2*R*)-diaminocyclohexane and *N*-[(triethoxysilyl)propyl]-L-prolinamide.[394]

3.4.3.6 PMOs from Mixtures of Two or More Different Bis- or Multi-Silylated Precursors

A further possibility for the synthesis of bifunctional PMOs is to use a mixture of two, three, or more different bridged bis(trialkoxysilyl) precursors which will co-condense in the presence of a SDA. The PMOs obtained in this way consist then of two (or more) different organic bridges that are bonded covalently within the framework of the pore walls. In dependence on the difference of the physicochemical properties of the different precursors in the reaction mixture this can have several consequences, for example, regarding the degree of order, the homogeneity of the distribution of the two (or more) organic bridges, and the outer morphology of the particles as well. As always in co-condensation approaches, the risk that phase separations (homocondensation) will occur is present here, too. Another question is, how the different precursors will assemble on the molecular scale, in particular if aromatic precursors are involved, with their tendency to form ordered stacks: will the two kinds of precursors really be completely homogeneously distributed within the pore walls or will they form separated domains or islands, respectively? This question cannot be answered easily, especially if the precursors are very similar. For now, we will defer consideration of this issue.

Table 3.4 Overview of the Syntheses Based on Bis- or Multi-Silylated and Mono-Silylated Precursors in the Presence of Ionic or Nonionic Surfactants as Structure-Directing Agents

Monosilylated precursor	Bis- or multi-silylated precursor	SDA	Content reaction mixture (mol%)	Content product (mol%)[a]	Ref.
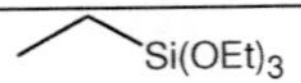	**2**	CTAB/Brij 30	25	n.d.	374
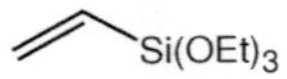	**2**	CTAB/Brij 30	25	n.d.	375
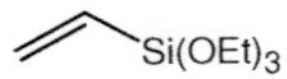	**3**	CTAB	33	n.d.	376
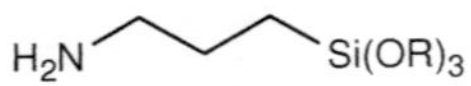	**2**	CTAC	25	~17	377
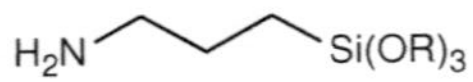	**2**	P123	10–40	20	378
	**5**	OTAC	20–66	n.d.	379
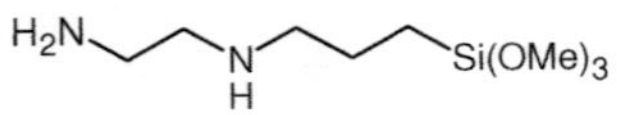	**2**	CTAC	25	13	377, 380, 381
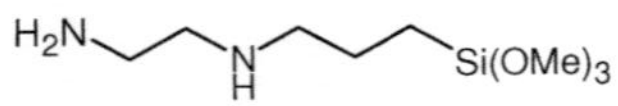	**2**	CTAC	30	16	382
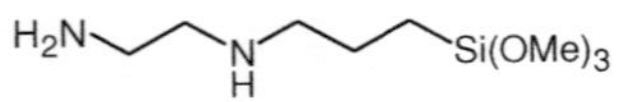	**5**	OTAC	20–66	n.d.	379

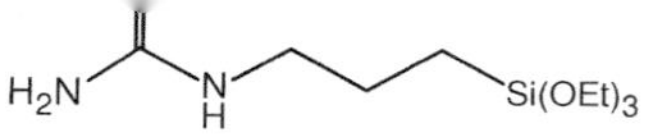	**2**	CTAB/Brij 30	5–40	n.d.	383
	**2**	CTAC	25	21	377
	**2**	CTAC	50	33	384
	**2**	OTAC	25	n.d.	385
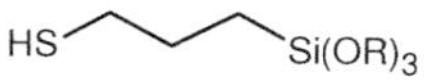	**2**	Brij 56	25	13	386
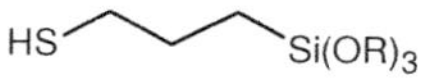	**2**	Brij 76	40	n.d.	387
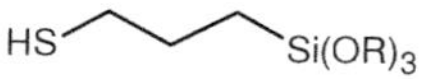	**2**	Brij 76	30	n.d.	388
	**2**	P123	25	16	386
	**2**	P123	30	n.d.	387
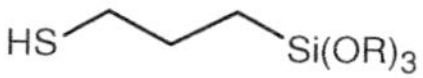	**2**	P123	34	n.d.	389
	**5**	OTAC	67	24	390
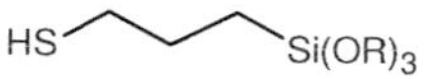	**9**	OTAC	70	n.d.	391

(*Continued*)

TABLE 3.4 *Continued*

Monosilylated precursor	Bis- or multi-silylated precursor	SDA	Content reaction mixture (mol%)	Content product (mol%)[a]	Ref.
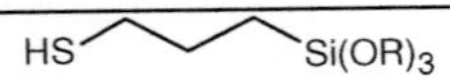	**18**	P123	3–17	3–17	392
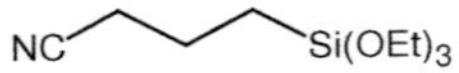	**2**	CTAB	50	n.d.	393
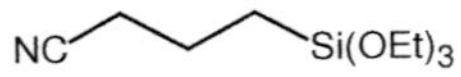	**2**	CTAB/Brij 30	25	n.d.	375
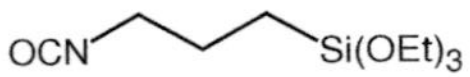	**2**	CTAB/Brij 30	5–40	n.d.	383
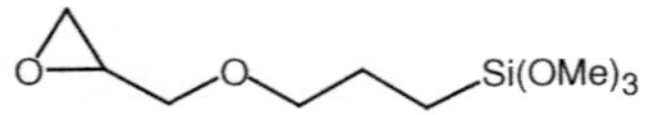	**2**	CTAB/Brij 30	25	n.d.	375
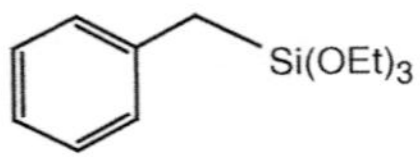	**2**	CTAC	25	23	377
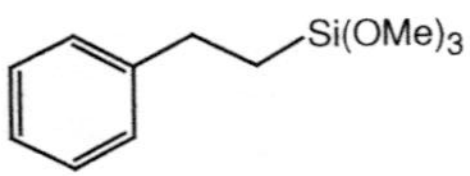	**2**	CTAC	25	22	377
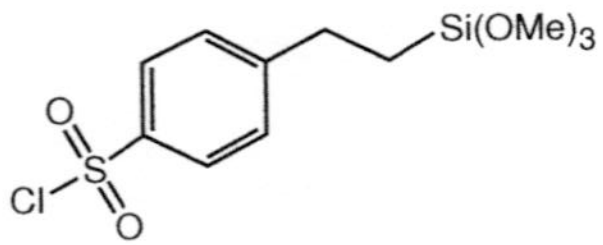	**2**	Brij 56	22	9	386

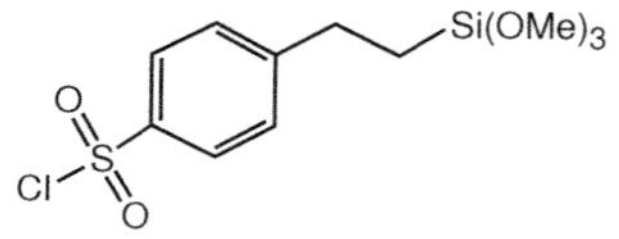	**2**	P123	22	16	386,389
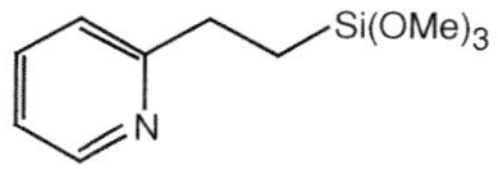	**2**	CTAC	25	16	377
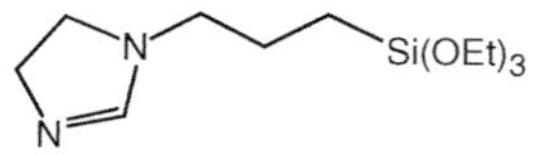	**2**	CTAC	25	10	377
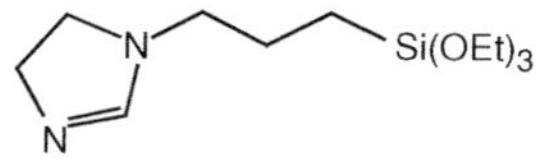	**5**	OTAC	20–66	n.d.	379
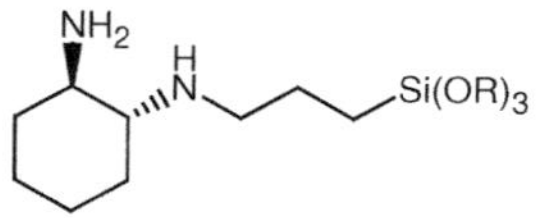	**2**	CTAC	20	n.d.	394
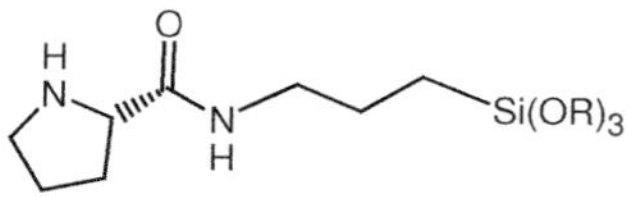	**2**	CTAC	20	n.d.	394

[a] n.d. = not determined.

Several publications report on the synthesis of amine- and ethylenediamine-bridged PMOs, in which a second bis-silylated precursor acts as the main network builder.[291–293]

Morell et al.[294] reported on the first syntheses of bifunctional PMOs containing different amounts of aromatic thiophene and benzene bridging groups. Independent of the molar ratios of the organosilanes in the initial reaction mixtures, highly ordered PMO materials were obtained in all cases (Fig. 3.24).

Zhang et al.[290] prepared not only pure ICS-PMOs and mesoporous hybrid materials from ICS (**18**) and the mono-silylated mercaptopropyl precursor (see also Table 3.4) but also bifunctional PMOs made of varying amounts of ICS (**18**) and BTEE (**2**). Cho et al.[295,296] synthesized bifunctional and even trifunctional PMO particles (some of spherical morphology), containing a combination of phenylene, thiophene, and ethane bridging groups in the presence of a PEO-PLGA-PEO triblock copolymer template under slightly acidic conditions. However, the question as to whether one or more phases were built was not specifically addressed in all these studies. Recently, Treuherz and Khimyak[297] synthesized hexagonal PMO materials using ethane- and ethene-bridged silsesquioxane precursors and used $^{1}H-^{13}C$ and $^{1}H-^{29}Si$ cross-polarization MAS NMR measurements to specifically address the issue as to whether both organic spacers are located within the same "bifunctional" mesophase or in different, highly dispersed mesophases. The cross-polarization kinetics data seem to indicate that both functionalities are incorporated in the same PMO structure, however, with heterogeneous distribution of the organic spacers.

An interesting bifunctional PMO with photoluminescence (PL) properties was recently reported by Wahab et al.[298] By employing the EISA approach, they incorporated (up to 5 wt%) *N,N′*-bis(4-*tert*-butylphenyl)-*N,N′*-bis(4-((*E*)-2-(triethoxysilyl)-vinyl)phenyl)biphenyl-4,4′-diamine (BBDA) (Fig. 3.25) into the matrix of thin films of a well-ordered organosilica matrix based on BTEE (**2**). While films made of the pure BBDA showed a broad featureless PL spectrum with a maximum at

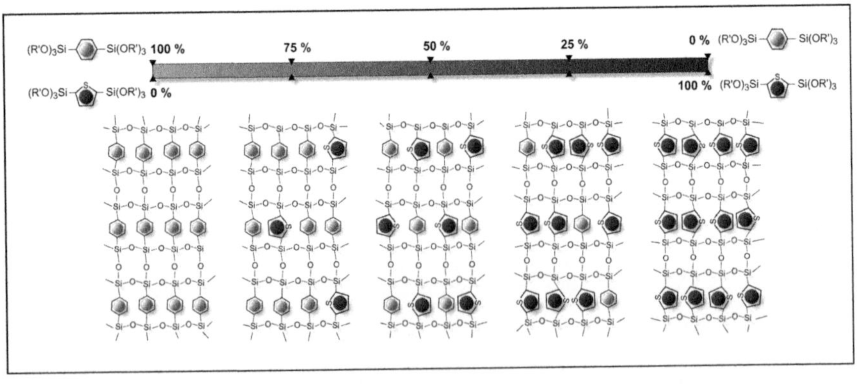

Figure 3.24 A bifunctional PMO material that consists of two different organic bridging groups and whose content is freely adjustable (thiophene, dark grey; benzene, light grey), assuming that they are completely homogeneously distributed within the pore walls. (Adapted from Reference 294.)

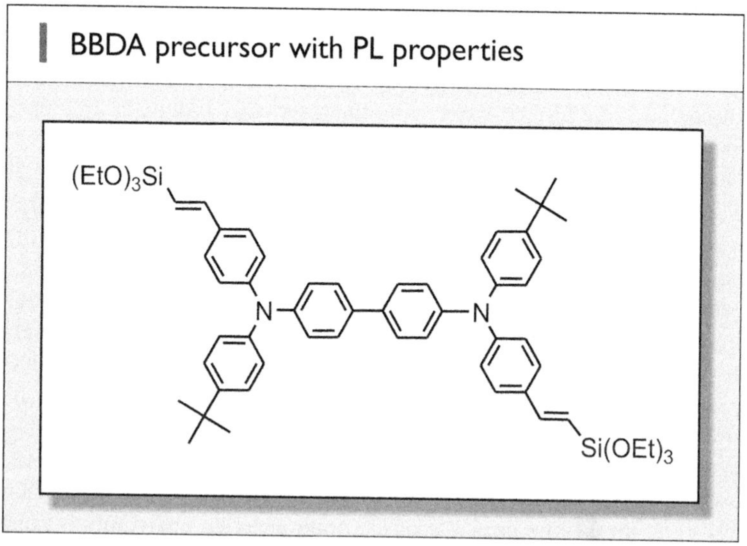

Figure 3.25 Chemical structure of *N,N′*-bis(4-*tert*-butylphenyl)-*N,N′*-bis(4-((*E*)-2-(triethoxysilyl)vinyl)phenyl)biphenyl-4,4′-diamine (BBDA).

455 nm and a FWHM of 77 nm, the bifunctional PMO films showed a 5 nm blue-shifted maximum of the PL band at 450 nm with a significant shoulder at 475 nm. Additionally, these films have a much more narrow FWHM (~55 nm), indicating that BBDA is isolated within the pore walls of the PMO, which suppresses the ability for BBDA to form long-range aggregates that are often responsible for red-shifted and broad PL spectra. Furthermore, these PMO films exhibit increased hardness over their nonordered counterparts.

In an interesting study carried out by Yang and Sayari,[299] mesoporous materials were prepared using mixtures of phenyl- (BTEB, **5**) and biphenyl-silsesquioxanes (BTEBP, **14**) under basic conditions (OTAC as SDA) where, in addition to the long-range order of the pore system, molecular order within the pore walls is expected to occur. Under such conditions, the one- or two-phase issue may be delineated by inspecting the X-ray diffraction pattern in the wide-angle region. In fact, the materials derived from mixed precursors exhibited two distinct molecular-scale periodicities indicative of the occurrence of both phenylsilica and biphenyl silica lamellar structures within the pore walls. With increasing BTEB : BTEBP ratios in the synthesis mixture, the relative intensity of the X-ray diffraction peaks corresponding to the molecular-scale periodicity of both types of bridging groups changed accordingly. All other characterization techniques cannot discriminate between a single-phase PMO and a highly dispersed mixture of two PMOs. Because of the fact that the pore sizes and unit-cell dimension of both phenylene and biphenylene-bridged PMOs are very close, it is difficult to detect for example a bimodal pore size distribution in mixed-precursor materials (by TEM or nitrogen adsorption) or two well resolved d_{100} diffraction peaks. Also, the SEM images did not show particles with different morphologies in mixed-precursor materials that would be associated with separate PMO mesophases. But the XRD data indicate that either phase separation or segregation took place. In the former case, separate yet highly dispersed phenylene and biphenylene PMO mesophases are formed, whereas in the latter case, islands of different PMOs are formed within the same particles.

The issue of the homo- or heterogeneity of the organic bridges in bi- or multifunctional PMOs was taken up in an inventive manner by Jones et al.[300] using a modified synthesis procedure involving separate or joint prehydrolysis conditions for the two different precursors [BTEE (**2**) and BTEY (**3**)] in order to control the distribution of the organic bridges within the final PMO products. In joint prehydrolysis conditions a homogeneous distribution of the two organic groups was observed. However, when separate prehydrolysis conditions were used a domain type structure was obtained containing segregated regions of $-CH_2CH_2-$ and $-CH=CH-$ bridges within the same mesostructural framework. The distribution of the organic functional groups was confirmed by $^1H-^{29}Si$ HETCOR and variable contact time (VCT) CP/MAS NMR spectroscopy. In this way they were able to obtain materials with a sort of domain functionality, so to speak. The principle of this joint/separate prehydrolysis concept is depicted in Figure 3.26.

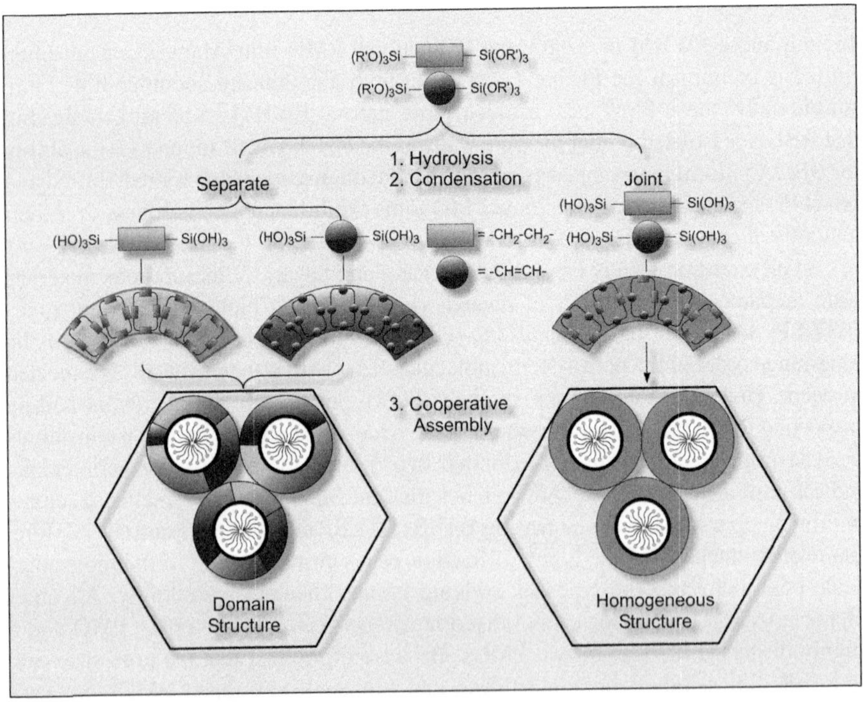

Figure 3.26 Formation of homo- and hetero-organo-functionalized clusters (a sort of secondary building units) and the resulting bifunctional PMOs prepared using joint and separate prehydrolysis protocols. (Adapted from Reference 300.)

3.5 SELECTED TOPICS AND SOME OTHER HIGHLIGHTS

3.5.1 Chirality

Chirality is an interesting topic in the context of mesoporous solids in general and PMOs in particular, both in regard to the fundamental aspects of chirality and in relation to possible applications, such as enantioselective separation and heterogeneous asymmetric catalysis. One of the major problems linked with homogeneous asymmetric catalysis is due to the difficulty of separation and recycling of the chiral catalysts, which is the main reason why many homogeneous asymmetric catalysts have not yet been industrialized. In this context, the development of asymmetric catalysts immobilized on an organic, inorganic or organic–inorganic hybrid support is highly desirable. The two special advantages of using *mesoporous* supports resulted from their large surface (which means that a high density of chiral centers can be achieved) and that relatively voluminous chiral species can be fixed in the nanopores.

In order to introduce chirality into mesoporous hybrid materials, principally two different approaches are conceivable. One possibility is to employ the chirality transfer or chirality induction principle, respectively.[¶¶¶] While this principle is well-known and widely applied in organic chemistry and the physico-chemistry of liquid-crystals, hitherto only very few examples have been reported in the field of inorganic chemistry. Some pioneering work on chiral sol–gel products have been conducted by the group of Shinkai. The polycondensation of TEOS in cholesterol and diaminocyclohexane-based organogel systems resulted in unique right-handed (for cholesterol) and both right- and left-handed (for diaminocyclohexane) spiral silicas with helical motifs.[301,302] The chirality transfer concept was applied to mesoporous MCM-41-like materials for the first time in 2004.[303,304] Chiral anionic surfactants (*N*-acyl-L-alanate), which formed a chiral, hexagonal-twisted mesophase was used as the SDA and the polycondensation of TEOS around this mesophase gave well-ordered chiral mesoporous silica. This process could be called chirality casting. As a result, the mesoporous material is not built-up by chiral components but exhibits an external helical morphology as well as hexagonally arranged channels which themselves are twisted, that is, possess a chiral topology (Fig. 3.27). A study that suggests that the handedness of the chiral surfactants is correlated with the resulting handedness of the helical mesoporous silica particles was conducted by Yokoi et al.[305] although this is in general not straightforward for a supramolecular collective of chiral molecules as was expounded by Harris, Kamien, and Lubensky in the context of chiral cholesteric liquid-crystal

[¶¶¶]The terminus of chirality induction is used for processes, in which the structural information of a chiral molecule is transferred to an initially achiral collective which then will form a superstructural chiral phase. One of the most prominent examples can be found in the field of liquid crystals: The doping of a nematic LC phase with chiral mesogenes (dopants) can lead to a twisted, helical cholesteric phase. Noteworthy is the fact that the length scales of the chiral information that characterizes the involved species can differ by several orders of magnitude: a few Ångstrøms in a single chiral molecule, but the pitch of a helical cholesteric phase amounts typically a few microns.

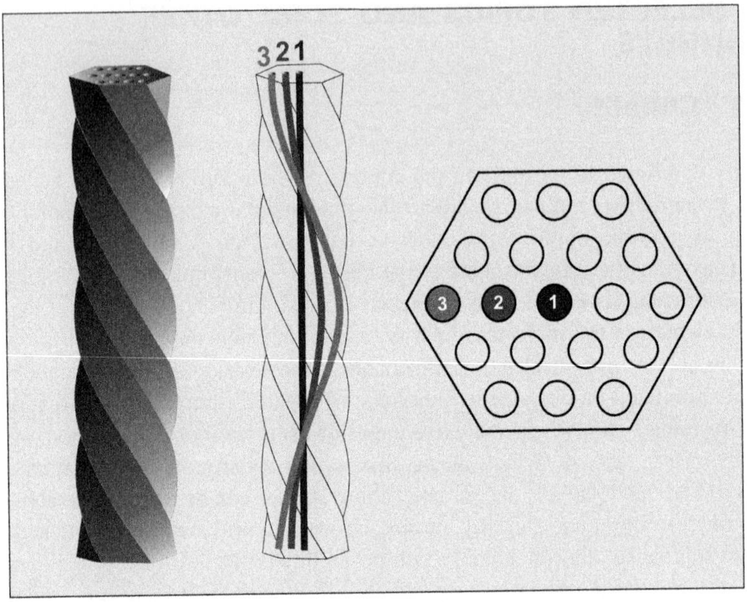

Figure 3.27 Schematic drawing of a structural model of individual particles of chiral mesoporous silicas with an outer helix-like morphology (left), showing three channels inside these particles, which are running along the long axis of the particle and are twisted around each other, and (right) the respective cross-sections. (Adapted from Reference 303.)

phases, for which they demonstrated that the handedness of individual molecules does not determine the handedness of the collective structure; see also Figure 3.28.[306,307]

Moreover, recently, even the synthesis of mesoporous pure silica[308,309] as well as mesoporous organosilica phases[310] with chiral topology in the absence of any chiral transfer additive has been reported. The minimization of the free surface energy has been proposed as the main driving force for the formation of such chiral morphologies.[309]

The second way to introduce chirality into mesoporous hybrid materials is the use of condensable chiral components, which will be integrated into the framework or onto the inner surface, irrespective whether a chiral grafting species with which the inner surface of mesoporous silica phases can be lined, a chiral co-condensing component in a one-pot synthesis approach, or a chiral or pro-chiral bis-silylated PMO precursor (the latter can then *in situ* or postsynthetically be converted into a chiral organobridge) is employed. Concerning the grafting procedure, also two- or multi-step anchoring methods are used, that is, consisting of a first grafting reaction with an achiral organic or metal-organic compound which then acts as a linkage to the actual chiral species introduced in a second or third step. The advantages and disadvantages are mainly the same as already explained in the introductory part.

The multi-step grafting strategy was applied, for instance, to line the surface of MCM-41 samples with chiral tartaric acid,[311] which was linked onto an

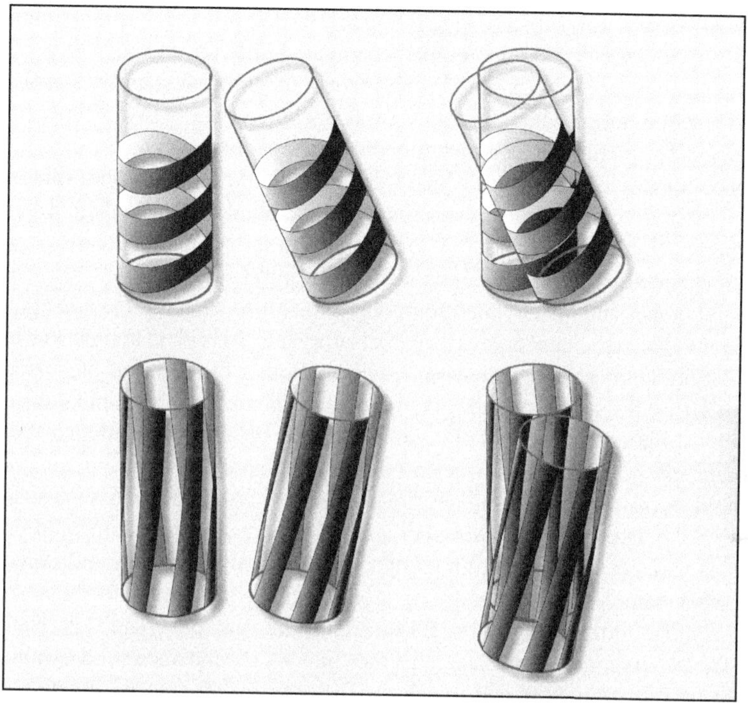

Figure 3.28 Visual demonstration of the fact that the handedness of individual molecules does not determine the handedness of the collective structure, assuming only steric interactions between these chiral entities. The ensemble produces a net relative rotation of their long axes, if the protrusions on one molecule (the shaded regions) are forced to fit "hand in glove" into the grooves on another (the unshaded regions). The result is that the long axes of the two molecules acquire a relative twist determined by the pitch of the stripes: (top) the molecules exhibit a tight right-handed pitch, and the ensemble shows a right-handed twist, too; (bottom) the pitch of an ensemble of two molecules, both exhibiting a weak right-handed pitch, is now left-handed.

APTMS-modified MCM-41 surface. In a last step, titanium tetra-*iso*-propoxide reacted with the chiral organic-inorganic hybrid material, to give the heterogeneous variant of the asymmetric epoxidation catalyst of allylic alcohols, proposed by Katsuki and Sharpless.[312]

Many other projects in the past dealt with the immobilization of the famous chelating salen**** ligand or salen metal complexes, respectively—the Mn^{III} salen Schiff base complex is also known as Jacobsen's catalyst, and it is used, among others, in asymmetric epoxidation of olefins††††—mainly onto the surface of MCM-41 and SBA-15.

****H_2 salen = *N,N′*-bis(salicylidene)ethylenediamine or 2,2′-ethylene-bis(nitrilomethylidene)diphenol; the name salen is a contraction of the two components of this compound: salicylic aldehyde and ethylene diamine.

††††See also the entry "Molecule of the Week" of May 29, 2007 on the ACS website, http://portal.acs.org/portal/PublicWebSite/molecule/archive/WPCP_006862.

A good overview on this subject can be found in an article by Li et al. which summarizes the work up to 2006.[313] Serrano et al.[314] compared different methods to anchor the Jacobsen catalyst on SBA-15 supports and found a significant stability increase (i.e., reduced leaching rates) when the catalyst was anchored on a SBA-15 support which was previously functionalized with aminophenyltrimethoxysilane.

Other chiral grafting species which were linked to mesoporous pure silica phases in order to study chiral catalysis reactions include alkaloids and alkaloid derivates (cinchona[315–317] and ephedrine[318,319]), bis(oxazoline),[320] amino alcohols,[321] proline and benzylpenicillin derivatives[322] as well as binaphtyl species.[323]

Chiral mono- and bis-silylated compounds have also been used to integrate chiral building blocks into mesoporous organic–inorganic hybrid materials via the co-condensation approach together with TEOS or another bis-silylated precursor [e.g., BTME (**2**)] as the actual framework builder. With the inherent limitation of this approach—without a severe loss of mesoscopic order the amount of integrated organic component rarely exceeds 15%—at least the integration of chiral organic moieties like binaphthyl, *trans*-(1*R*,2*R*)-diaminocyclohexane, and tartrate derivatives was accomplished.[324–330] The modified hydrophobicity of the organosilicas compared with the pure silica variants has an interesting impact on the catalytic activity and stereoselectivity: After complexing $[Rh(cod)Cl]_2$ (cod = cyclooctadiene) onto the surface of the co-condensation product of BTME and the mono-silylated derivative of *trans*-(1*R*,2*R*)-diaminocyclohexane, this mesoporous organosilica solid exhibits conversion rates of 82% to 96% and *ee* values of 19% to 23% for the asymmetric transfer hydrogenation of acetophenone, which is much higher than for the co-condensation product with TMOS instead of BTME (48% conversion and an *ee* value of 14%).[325] However, also the reverse effect could be observed in another system: Corma and coworkers[331–333] reported that the co-condensation product of TEOS and a bis-silylated chiral vanadyl Schiff base complex showed only enantioselectivity values of 30% for the asymmetric catalytic cyanosilylation of benzaldehyde with trimethylsilylcyanide, while the *ee* values for the VO(salen) catalyst covalently grafted to a mesoporous pure silica phase was considerably higher (63%). This reveals again that (1) the polarity of the surrounding chiral species as well as (2) the pore diameter (i.e., the confinement effect) can play a crucial role for the enantioselectivity of the asymmetric catalysts.

A very interesting observation pointing in a similar direction was made by Thomas and coworkers[335] with respect to the positive influence of the confinement on the chiral discrimination ability of the chiral catalyst: They reported that a chiral Rh^{I} complex could be confined in nanopores by an ion exchange method for the heterogeneous asymmetric catalytic hydrogenation of methylbenzoylformate, an α-ketone. The chiral catalyst confined in nanopores of 3.8 nm showed an *ee* value of 77% for the asymmetric hydrogenation, while the homogeneous chiral catalyst gave the racemic products. An increase of the pore size from 3.8 to 6.0 nm led to decreased *ee* values of 61%. When the pore size was further increased to 25 nm, the *ee* value was decreased to zero, indicating that the confinement effect plays an important role in the chiral induction.[334] This is currently an active area of research and some highlights have been summarized in an account.[335]

3.5.1.1 Genuine Chiral PMOs

Only very recently have hybrid materials composed of chiral single-source precursors emerged. The difficulties to obtain chiral PMOs or nonperiodic organosilicas composed of single-source precursors, respectively, are at least twofold: (1) chiral precursors are not commercially available and their synthesis is often not trivial, (2) the stereogenic center located at the organic bridge introduces to some extent an asymmetry that may perturb the self-assembly process at least at the molecular level, which could have an unfavorable impact on the order at the mesoscale as well. Of course, the magnitude of the perturbation exerted by the chiral moiety is dependent on its size.

Polarz and Kuschel[336] were the first who reported on the successful synthesis of a PMO composed of chiral building blocks. Based on the achiral BTEY PMO precursor (**3**), first an enantioselective hydroboration in the presence of the chiral (*R*)-BINAP catalyst was carried out. Subsequently, the polycondensation in the presence of CTAB as the SDA and an organic peroxide (in order to convert the hydroborated moiety (**22**) *in situ* into the respective alcohol) finally gave the chiral PMO. However, neither information on the enantiomeric excess of the precursor was given nor a proof of the chirality of the final solid PMO material.

Recently, Inagaki et al.[337] reported on the synthesis of a new PMO material carrying chiral phenylethylene bridging groups (**23**), for which the enantiomeric purity was determined in a destructive manner by eluting the organic groups from the solids. New trails to develop a nondestructive way to proof the chirality of PMOs were blazed by Morell et al.[338] They synthesized the novel chiral organosilica precursor 1,4-bis(triethoxysilyl)-2-(1-methoxyethyl)benzene (BTEMEB) (**24**) carrying benzylic ether bridging groups, which was used as a single-source precursor to prepare a homochiral genuine PMO and demonstrated that it is possible to proof the chirality directly by measuring the optical activity of the solid samples. In fact, the rotation of the plane of linear-polarized light of a PMO suspension was measured, which required the invention of a new experimental set-up.

A different approach to obtain mesoporous organosilicas with chiral bridges was employed by the group of Thomas. They designed a precursor that combines the functions of a network builder, a chiral latent functional group, and a porogen in one molecule. The precursors are formed by a convenient enantioselective hydroboration using (*S*)-monoisopinocampheylborane on an ethene-bridged silica precursor to gave (**25**). These precursors do self-organize when hydrolysis of their inorganic moiety takes place via an aggregation of their organic moiety into hydrophobic domains. After a condensation-ammonolysis sequence, mesoporous organosilicas functionalized with chiral amine groups are obtained, with the complete chiral functionalities located at the pore wall surface, which are therefore accessible to chemical processes. The pore size of the resulting organosilicas can be fine-tuned using different organic moieties attached to the boron group in the first step. While a wormlike arrangement of pores is observed for the pure precursor, common surfactants like F127 can be admixed to further control and tailor the resulting mesoporous system. In this way, it was also possible to obtain chiral periodic mesoporous organosilicas.[339]

This all-in-one approach was employed by the same group to prepare highly amine-functionalized mesoporous organosilicas.[340]

Noteworthy also is the work by the group of Li et al. who were able to obtain chiral PMOs possessing a morphology that is suitable for applications in the field of enantioselective separations. Monodispersed spherical PMO particles (with particle sizes of 6 to 9 μm) were prepared by co-condensation of *N,N′*-bis-[(triethoxysilyl)-propyl]-*trans*-(1*R*,2*R*)-bis-(ureido)-cyclohexane and BTME by a hierarchical double templating method, using CTAC as SDA and ethanol as a cosolvent, and *N,N*-dimethyldecylamine as a postsynthetically applied pore expanding agent, resulting in pore diameters from 1.7 to 10 nm. This hybrid material was employed as a novel kind of chiral stationary phase for HPLC columns, which can efficiently separate the *R*/*S*-1,10-bi-2-naphthol enantiomers even at high sample loadings and high flow rates.[329,330]

3.5.2 Surfactant-Free Synthesis of Siloxane-Based Nanomaterials

Another very interesting approach to prepare (micro- and) mesoporous silica-based materials was developed by Kuroda and coworkers in the last few years,[341–345] for which no extra SDA is needed, because the component that drives the self-assembly process is already integrated into the hydrolysable precursor, meaning that the precursor itself also functions as a surfactant. For that, they designed a new class of special alkyl-alkoxysilane and siloxane-based precursors consisting of a long-chain alkyl-, alkenyl-, or alkynylsilane core (the surfactant-like component) and two or three trimethoxylsilyl or trimethoxysilyloxy groups (the hydrolysable network-forming component) and carried out sol–gel reactions, some of them by adding TMOS as an additionally network-building source. Depending on the chain length of the surfactant-like component, interdigitated lamellar, worm-like, or 2D hexagonally ordered phases of hybrid materials were obtained. The covalently bonded long-chained component could then be removed by calcination. This led to a collapse of the structure in the case of the layered products, but the structure of the 2D hexagonally ordered samples was retained, leading to micro- and mesoporous silica phases (with pore diameters between 1.7 and 3.3 nm); an example of the synthesis approach is depicted schematically in Figure 3.29. These phases could also be successfully prepared as thin films. The interested reader is referred to a recent short review on this topic.[346]

3.5.3 Chromatography

Bearing in mind the famous crystallographic correspondence principle between the outer shape of a crystal and its inner lattice structure, it is conceivable that it is impossible to form structurally *ordered* particles of arbitrary shapes so long as the reactions are carried out at or near the thermodynamic equilibrium. Inversely, special

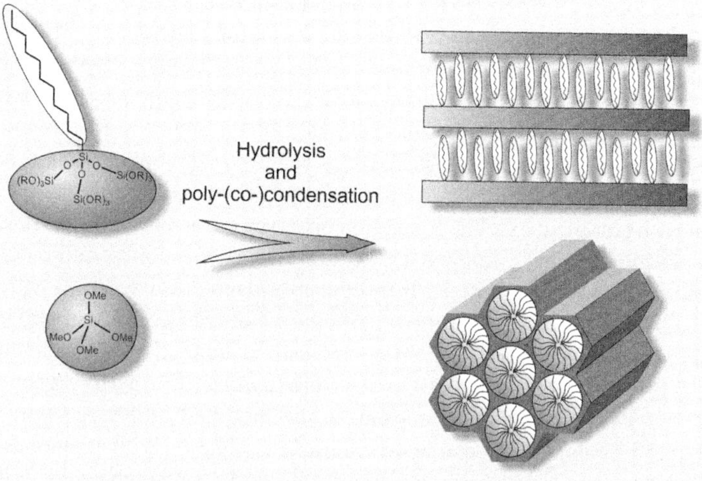

Figure 3.29 Synthesis of a PMO derivative with the use of a siloxane-based oligomer that consists of an alkylsilane nucleus (white) and three branching trimethyoxysilyl groups (light grey) This precursor acts simultaneously both as organosilica source and surfactant. The alkyl chain length (n) determined the phase of the composite: for $n = 10$ a 2D hexagonal phase is formed; for $n = 16$, a lamellar phase is obtained. This approach can also be employed as co-condensation reaction together with TMOS.

conditions have to be applied if particles with a specific morphology are intended. Several studies concerning the morphology of ethane-bridged PMOs[347–349] underline the fact that the symmetry of the micellar arrangement also determines the outer shape of the PMO particles. This circumstance is reflected in the difficulties in preparing PMO particles with other morphologies, especially spherical particles, which are, monodispersity assumed, most suitable for chromatographic applications, in particular HPLC.

Martin et al.[350] developed a method for the preparation of octyl-functionalized, spherical MCM-41 particles, which they used as packing materials in reverse-phase HPLC columns that in some aspects exhibited separation superior to conventional reversed silica phases. Kapoor and Inagaki[251] prepared benzene-bridged PMO particles with spherical morphology and particle diameters between 0.6 and 1.0 μm (average: 0.8 μm). On the basis of electron microscopy investigations at different times during the growth of the particles, the authors deduced that mild basicity and very low reaction and condensation rates are necessary conditions for the formation of spherical morphologies. These conditions are required for the nuclei to form simultaneously and for each individual nucleus to grow uniformly at a constant but slow rate until all particles have achieved simultaneously their final size and adjust their growth on the basis of the consumed condensation material. By means of a microwave oven, Kim et al.[351] synthesized ethane-bridged PMO particles with spherical morphology and narrow size distributions with diameters between 1.5 and 2.5 μm—the lower limit for materials for application in HPLC. The material was tested for its separation capabilities as a packing material in a HPLC microcolumn: The quality of the separation of a mixture of eight substances of medium to high hydrophobicity with this column was compared with that of a column packed with conventionally prepared ethane-bridged PMO particles of octadecahedral morphology and an average particle size of 8 μm. Separation of the eight substances was achieved with both columns, but the particles produced by the microwave method proved to be a significantly better separating medium with complete baseline separation observed for at least three of the eight compounds. Rebbin et al.[352] employed modified Stöber reaction conditions to prepare almost perfectly spherical particles from ethane-bridged PMO materials with diameters between 0.4 and 0.5 μm and a narrow particle-size distribution. Meanwhile, considerably larger benzene-bridged PMO particles with average diameters tunable between 3 and 15 μm and acceptably narrow size distributions have been prepared.[353] This synthesis was carried out not according to the modified Stöber reaction; instead, a mixture of P123 and CTAB in a hydrochloric acid solution with ethanol as cosolvent was used. These particles were successfully tested as a new phase for HPLC applications, which separated more clearly three different mixtures containing up to four components with a broad polarity range than the commercial Nucleosil 50–10 phase. Furthermore, Zhu et al.[354] synthesized phenyl-functionalized ethane-silicas [by co-condensation of phenyltrimethoxysilane and (**2**)] with spherical morphology for use as reversed phase in HPLC applications. It was demonstrated that this packing material is a promising stationary phase by separation of several polycyclic aromatic hydrocarbons.

3.5.4 Enzyme Immobilization and Biocatalysis

The pore systems of mesoporous silica phases have sizes that can accommodate at least small enzymes or proteins; this applies to a lesser extent to MCM-41/48-like phases and to a greater extent to SBA-15-like phases. Therefore, these materials are attractive for a range of applications in biotechnology, including purification, environmental applications, and enzymatic catalysis for fine chemicals. In particular, applications in the separation of proteins and in immobilizing enzymes are the most widely studied. Enzyme immobilization is used commercially, for example, in the food industry for the synthesis of fructose from starch[355] and in the pharmaceutical industry to prepare 6-aminopenicillanic acid from native penicillins.[356] The advantages of using robust, heterogeneous supports include increased ease of separation, together with possibilities for storage and reuse, increased stability and the ability to use them in organic media. The immobilization can involve physisorption or actual chemical bonding of the enzyme or protein to the surface of the pore walls, although in the latter case there is always the risk of partial or total denaturation and hence a considerable decrease in activity. Conversely, it is expected that the stronger bonding will result in a smaller amount of material washed away during recycling. Chemisorption is generally carried out with functionalized phases, either via the grafting process or through co-condensation reactions, or sometimes also PMOs are used.

A few selected examples, mainly in order to give a hint of the broad range of research in this field, are highlighted below; a more comprehensive overview on this topic can be found in excellent and extensive reviews of Hartmann,[357] Yiu and Wright,[358] and Kim et al.[359]

Unmodified MCM-41/48 and SBA-15 phases[360] as well as carboxy-, aminopropyl-, thiol-, cyano-, and phenyl-modified SBA-15 phases[361] have been used by Yiu et al. for the immobilization of trypsin. The activity of the trypsin was measured on the basis of the hydrolysis of *N*-α-benzoyl-DL-arginine-4-nitroanilide, and the thiol-functionalized phases proved to be the most promising. Size-selective protein immobilization on thiol-functionalized phases has also been reported.[362] Lei et al.[363] successfully used amino- and carboxy-functionalized SBA-15 phases to immobilize the enzyme organophosphorus hydrolase, which in this state had twice the activity of that in the free state. Ma et al.[364] carried out studies on the activity of porcine pancreas lipase, which was immobilized onto the surface of MCM-41 samples solely by hydrogen bonds. They noticed a strong decrease in activity following recovery due to leaching of the enzyme from the pores. Functionalization of the substrate with vinyl groups after the immobilization of the enzyme took place, which was anticipated to lead to stronger binding of the enzyme to the surface, but did not lead to a significant improvement. Salis et al.[365] immobilized *Mucor javanicus* lipase in the channel system of SBA-15 materials at different pH values (pH 5 to 8). The loading and hydrolysis activity (in the hydrolysis of tributyrin and triolein) were highest at pH 6. Chemical adsorption was achieved by functionalization of the support medium with glutardialdehyde (pentaldial); however, hydrolysis activity was lost in the case of triolein. Also, the immobilization of conalbumin,[366] cytochrome c,[367] subtilisin,[368] and lysozyme[369] in SBA-15 phases has been reported. Qiao et al. showed that large-pore ethane-silicas do not have higher adsorption capacity than SBA-15 silica phases.[370] Moreover, Li et al.[173] studied pH-dependent lysozyme adsorption properties of bifunctional PMOs composed of different ratios of ethane and 1,4-diethylbenzene bridges. The maximum lysozyme adsorption amount of 360 mg g^{-1} (at pH 11) has been achieved on a PMO sample with a ratio of 70:30, while each of the respective pure monofunctional PMOs exhibited significantly lower adsorption values.

Furthermore, some promising attempts were undertaken to immobilize chloroperoxidase (CPO)—a versatile heme peroxidase that exhibits halogenase and peroxidative activity and has significant potential for use in the fine chemical industry, for example, for stereoselective epoxidations—onto mesoporous materials. However, its stability to heat, pH, and denaturants still has to be enhanced. Aburto et al.[371] reported an increase in the stability of CPO immobilized on pure silica and functionalized SBA-16 compared with free CPO in the presence of the denaturant urea. Hartmann and Streb[372] found that immobilized CPO (on SBA-15) selectively oxidized indole to 2-oxoindole and was active over a wider pH range than the free enzyme. Hudson et al.[373] successfully immobilized CPO on a variety of (sonificated) mesoporous materials obtained via co-condensation reactions of varying molar ratios of TEOS and the silsesquioxane precursor bis[3-(trimethoxysilyl)propyl]amine (BTMSPA) in the presence of P123 under acidic conditions. When the enzyme was immobilized onto materials prepared with 20 mol% of the bis-silylated precursor, it could be reused 20 times with little loss in activity, while the enzyme lost all activity upon reuse in materials made of 100% or 11 mol% BTMSPA, indicating that the CPO molecules did not have access to the inner surface of the mesopores.

The last two examples have demonstrated in particular how subtle the effects can be and that much work is required to fine-tune the properties of a mesoporous material as a suitable host for active enzymes. Besides the stability of the enzyme itself, important factors include the pore size, the robustness of the host, and the distribution of functional groups within the material.

3.5.5 Complex and Hierarchical Structures

Finally, we want to return to the statement in the Introduction concerning the structuring of matter by means of the bottom-up approach and would like to present, as an example, one project that shows which stepladder is meanwhile accomplished. At least four levels of hierarchy are brought together within the formation of magnetic Fe_3O_4 (magnetite) nanocrystals inside hollow spheres of ethane-bridged PMOs, published recently by Zhang et al.[374] Monodisperse Fe_3O_4 nanocrystals were synthesized by the high-temperature decomposition of iron stearic acid, $Fe(SA)_3$, and then dispersed in chloroform. These stearic acid-capped magnetic nanocrystals were transferred into an aqueous solution by using CTAB as surfactant, producing water-dispersible nanoparticles with an outer CTAB shell through interactions between the stearic acid ligands and CTAB. Application of the fluorocarbon surfactant FC4,[‡‡‡‡] which tends to form vesicles in aqueous solution, results in an encapsulation of the CTAB-stabilized Fe_3O_4 nanoparticles into the vesicles. Addition of the organosilica precursor BTME, which hydrolyzes quickly at high temperature under basic conditions, produces composite micelles in the vesicle containing solution with the ethane-bridged PMO shell incorporating the vesicles that contain the magnetic nanoparticles in their core. This elegant self-assembly of the complex hybrid material was accomplished within 2 hours (Fig. 3.30). The modification or coating of magnetic nanoparticles with silica or polymers has been studied extensively for potential applications in biomedical fields, such as targeted drug delivery, magnetic resonance imaging (MRI) contrast agents, protein separations, cell sorting, and immunoassays. Mesoporous silica is a promising, appropriate matrix for making magnetic nanocomposites, because this material has a low cytotoxicity, high surface area, large pore volume, and tunable pore size, in addition to being biocompatible and biodegradable. The authors realized a high saturation magnetization of these trapped magnetite nanoparticles and also carried out drug release experiments (with aspirin and ibuprofen) with the respective amino-functionalized PMO shell systems (synthesized by co-condensation with BTME and APTES), thereby demonstrating the creation of a very efficient protocol to targeted drug delivery assisted by magnetic fields.

‡‡‡‡ $C_3F_7O(CFCF_3CF_2O)_2CFCF_3CONH(CH_2)_3N^+(C_2H_5)_2CH_3I^-$

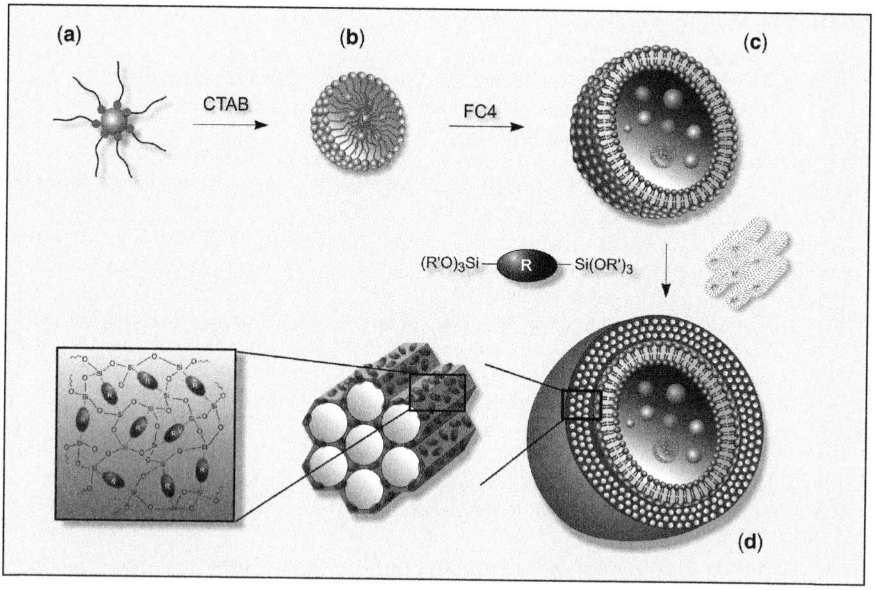

Figure 3.30 Synthesis route of PMO magnetic hollow spheres: (a) hydrophobic, stearic acid-capped Fe_3O_4 nanoparticles in an organic phase are treated with CTAB to produce (b) water-dispersible nanoparticles. (c) Formation of FC4 vesicles leads to encapsulation of the CTAB-stabilized Fe_3O_4 nanoparticles. (d) Addition of BTME/CTAB forms the outer ethane-bridged PMO shell surrounding the vesicles. (See color insert.)

3.6 CONCLUDING REMARKS

From what has been summarized here, it is evident that the spectrum of mesoporous organic–inorganic hybrid materials and their respective fields of applications has expanded progressively in the recent past. Currently, there are no indications that this development will stop in the near future. It still remains to be seen, though, if any of these materials will ever find a *commercial* application.[§§§§]

Looking at the three coarse categories, grafted materials, those originating from co-condensation reactions, and, finally, PMOs, we think that all of them have their particular advantages—it depends mostly on the intended purpose of the resulting materials. From our point of view, particularly all kinds of chiral mesoporous (organo)silicas bear great potential for future developments, especially for catalytic and chromatographic applications.

Finally, the authors would like to thank the research community whose inventive research has made compiling this chapter very exciting and enjoyable.

[§§§§]In a way, it is thought-provoking that even such a famous industrial process as the Haber-Bosch process would not have been invented today in times of the completely globalized capitalism, because no company would have been willing to invest such an amount of time and money anymore in advance (to adapt freely Dr. U. Müller, BASF SE, conference talk in Marburg, Germany, **2004**).

REFERENCES

1. D. M. Antonelli, J. Y. Ying, *Curr. Opin. Colloid Interface Sci.* **1996**, *1*, 523–529.
2. P. Behrens, *Angew. Chem. Int. Ed.* **1996**, *35*, 515–518.
3. A. Sayari, *Chem. Mater.* **1996**, *8*, 1840–1852.
4. J. Y. Ying, C. P. Mehnert, M. S. Wong, *Angew. Chem. Int. Ed.* **1999**, *38*, 56–77.
5. G. J. de A. A. Soler-Illia, C. Sanchez, B. Lebeau, J. Patarin, *Chem. Rev.* **2002**, *102*, 4093–4138.
6. A. Stein, *Adv. Mater.* **2003**, *15*, 763–775.
7. F. Hoffmann, M. Cornelius, J. Morell, M. Fröba, *Angew. Chem. Int. Ed.* **2006**, *45*, 3216–3251.
8. C. Sanchez, C. Boissière, D. Grosso, C. Laberty, L. Nicole, *Chem. Mater.* **2008**, *20*, 682–737.
9. S. Fujita, S. Inagaki, *Chem. Mater.* **2008**, *20*, 891–908.
10. P. Gómez-Romero, C. Sanchez, Eds., *Functional Hybrid Materials*, Weinheim: Wiley-VCH, **2004**.
11. U. Schubert, N. Hüsing, *Synthesis of Inorganic Materials*, 2nd ed., Weinheim: Wiley-VCH, **2005**.
12. A. Rabenau, *Angew. Chem. Int. Ed.* **1985**, *24*, 1026–1040.
13. M. Feldman, P. Desrochers, *Ind. Innov.* **2003**, *10*, 5–24.
14. S. S. Kistler, *Nature* **1931**, *127*, 741–744.
15. S. S. Kistler, *J. Phys. Chem.* **1932**, *36*, 52–64.
16. G. A. Nicolaon, S. J. Teichner, *Bull. Soc. Chim. Fr.* **1968**, 1900, 1906, 3107, 3555.
17. S. J. Teichner, G. A. Nicolaon, M. A. Vicarini, G. E. E. Gardes, *Adv. Colloid Interf. Sci.* **1976**, *5*, 245–273.
18. C. J. Brinker, G. W. Scherer, *Sol-Gel Science: The Physics and Chemistry of Sol-Gel Processing*, New York: Academic Press, **1990**.
19. N. Hüsing, U. Schubert, *Angew. Chem. Int. Ed.* **1998**, *37*, 22–45.
20. H. Schmidt, *J. Non-Cryst. Solids* **1985**, *73*, 681–691.
21. Ì. Yíigöra, A. K. Sha'abana, W. P. Steckle Jr., D. Tyagia, G. L. Wilkes, J. E. McGrath, *Polymer* **1984**, *25*, 1800–1806.
22. G. L. Wilkes, B. Orler, H. Huang, *Polym. Preprints* **1985**, *26*, 300.
23. H. Huang, B. Orler, G. L. Wilkes, *Polym. Bull.* **1985**, *14*, 557–564.
24. K. J. Shea, D. A. Loy, O. W. Webster, *Chem. Mater.* **1989**, *1*, 572–574.
25. D. A. Loy, K. J. Shea, *Chem. Rev.* **1995**, *95*. 1431–1442.
26. K. J. Shea, D. A. Loy, *Acc. Chem. Res.* **2001**, *34*, 707–716.
27. K. J. Shea, D. A. Loy, *MRS Bulletin* **2001**, *26*, 368–376.
28. M. Khiterer, D. A. Loy, C. J. Cornelius, C. H. Fujimoto, J. H. Small, T. M. McIntire, K. J. Shea, *Chem. Mater* **2006**, *18*, 3665–3673.
29. L. Zhao, M. Vaupel, D. A. Loy, K. J. Shea, *Chem. Mater.* **2008**, *20*, 1870–1876.
30. C. T. Kresge, M. E. Leonowicz, W. J. Roth, J. C. Vartuli, J. S. Beck, *Nature* **1992**, *359*, 710–712.
31. J. S. Beck, J. C. Vartuli, W. J. Roth, M. E. Leonowicz, C. T. Kresge, K. D. Schmitt, C. T.-W. Chu, D. H. Olson, E. W. Sheppard, S. B. McCullen, J. B. Higgins, J. L. Schlenker, *J. Am. Chem. Soc.* **1992**, *114*, 10834–10843.
32. A. H. Schoen, NASA Technical Note, TN D-5541 (**1970**).
33. T. Yanagisawa, T. Shimizu, K. Kuroda, C. Kato, *Bull. Chem. Soc. Jpn.* **1990**, *63*, 988–992.
34. G. S. Attard, J. C. Glyde, C. G. Göltner, *Nature* **1995**, *378*, 366–368.
35. A. Monnier, F. Schüth, Q. Huo, D. Kumar, D. Margolese, R. S. Maxwell, G. Stucky, M. Krishnamurty, P. Petroff, A. Firouzi, M. Janicke, B. Chmelka, *Science* **1993**, *261*, 1299–1303.
36. G. D. Stucky, A. Monnier, F. Schüth, Q. Huo, D. Kumar, D. Margolese, M. Krishnamurty, P. Petroff, A. Firouzi, M. Jaenicke, B. F. Chmelka, *Mol. Cryst. Liq. Cryst.* **1994**, *240*, 187–200.
37. D. Zhao, J. Feng, Q. Huo, N. Melosh, G. H. Fredrickson, B. F. Chmelka, G. D. Stucky, *Science* **1998**, *279*, 548–552.
38. D. Zhao, Q. Huo, J. Feng, B. F. Chmelka, G. D. Stucky, *J. Am. Chem. Soc.* **1998**, *120*, 6024–6036.
39. D. Zhao, P. Yang, N. Melosh, J. Feng, B. F. Chmelka, G. D. Stucky, *Adv. Mater.* **1998**, *10*, 1380–1385.
40. S.-S. Kim, T. R. Pauly, T. J. Pinnavaia, *Chem. Commun.* **2000**, 1661–1662.
41. C. Yu, Y. Yu, D. Zhao, *Chem. Commun.* **2000**, 575–576.

42. G. J. de A. A. Soler-Illia, E. L. Crepaldi, D. Grosso, C. Sanchez, *Curr. Opin. Colloid Interface Sci.* **2003**, *8*, 109–125.
43. S. Förster, *Top. Curr. Chem.* **2003**, *226*, 1–28.
44. M. Impéror-Clerc, P. Davidson, A. Davidson, *J. Am. Chem. Soc.* **2000**, *122*, 11925–11933.
45. S. A. Bagshaw, E. Prouzet, T. J. Pinnavaia, *Science* **1995**, *269*, 1242–1244.
46. J. Patarin, B. Lebeau, R. Zana, *Curr. Opin. Colloid Interface Sci.* **2002**, *7*, 107–115.
47. K. J. Edler, *Austr. J. Chem.* **2005**, *58*, 627–643.
48. J. D. Epping, B. F. Chmelka, *Curr. Opin. Colloid Interface Sci.* **2006**, *11*, 81–117.
49. Y. Wan, D. Zhao, *Chem. Rev.* **2007**, *107*, 2821–2860.
50. S. Cadars, N. Mifsud, A. Lesage, J. D. Epping, N. Hedin, B. F. Chmelka, L. Emsley, *J. Phys. Chem. C* **2008**, *112*, 9145–9154.
51. S. Ruthstein, J. Schmidt, E. Kesselman, R. Popovitz-Biro, L. Omer, V. Frydman, Y. Talmon, D. Goldfarb, *Chem. Mater.* **2008**, *20*, 2779–2792.
52. Q. Huo, D. I. Margolese, U. Ciesla, P. Feng, T. E. Gier, P. Sieger, R. Leon, P. M. Petroff, F. Schüth, G. D. Stucky, *Nature* **1994**, *368*, 317–321.
53. Q. Huo, D. I. Margolese, U. Ciesla, D. G. Demuth, P. Feng, T. E. Gier, P. Sieger, A. Firouzi, B. F. Chmelka, F. Schüth, G. D. Stucky, *Chem. Mater.* **1994**, *6*, 1176–1191.
54. D. M. Antonelli, J. Y. Ying, *Angew. Chem. Int. Ed.* **1995**, *34*, 2014–2017.
55. D. M. Antonelli, J. Y. Ying, *Angew. Chem. Int. Ed.* **1996**, *35*, 426–430.
56. S. A. Bagshaw, T. J. Pinnavaia, *Angew. Chem. Int. Ed.* **1996**, *35*, 1102–1105.
57. Z.-R. Tian, W. Tong, J.-Y. Wang, N.-G. Duan, V. V. Krishnan, S. L. Suib, *Science* **1997**, *276*, 926–930.
58. P. Yang, D. Zhao, D. I. Margolese, B. F. Chemlka, G. D. Stucky, *Chem. Mater.* **1999**, *11*, 2813–2826.
59. D. M. Antonelli, *Microporous Mesoporous Mater.* **1999**, *30*, 315–319.
60. K. L. Frindell, J. Tang, J. H. Harreld, G. D. Stucky, *Chem. Mater.* **2004**, *16*, 3524–3532.
61. M. J. MacLachlan, N. Coombs, G. A. Ozin, *Nature* **1999**, *397*, 681–684.
62. U. Ciesla, S. Schacht, G. D. Stucky, K. K. Unger, F. Schüth, *Angew. Chem. Int. Ed.* **1996**, *35*, 541–543.
63. M. Tiemann, M. Fröba, *Chem. Mater.* **2001**, *13*, 3211–3217.
64. Y. Wan, H. F. Yang, D. Y. Zhao, *Acc. Chem. Res.* **2006**, *39*, 423–432.
65. R. Ryoo, S. H. Joo, S. Jun, *J. Phys. Chem. B* **1999**, *103*, 7743–7746.
66. J. Lee, S. Yoon, T. Hyeon, S. M. Oh, K. B. Kim, *Chem. Commun.* **1999**, 2177–2178.
67. F. Schüth, *Angew. Chem. Int. Ed.* **2003**, *42*, 3604–3622.
68. M. Tiemann, *Chem. Mater.* **2008**, *20*, 961–971.
69. S. Inagaki, S. Guan, Y. Fukushima, T. Ohsuna, O. Terasaki, *J. Am. Chem. Soc.* **1999**, *121*, 9611–9614.
70. B. J. Melde, B. T. Holland, C. F. Blanford, A. Stein, *Chem. Mater.* **1999**, *11*, 3302–3308.
71. T. Asefa, M. J. MacLachlan, N. Coombs, G. A. Ozin, *Nature* **1999**, *402*, 867–871.
72. M. Cornelius, Ph.D. thesis, Justus-Liebig-Universität Gießen, Gießen, Germany, **2007**.
73. C. W. Jones, K. Tsuji, M. E. Davis, *Nature* **1998**, *393*, 52–54.
74. K. Tsuji, C. W. Jones, M. E. Davis, *Microporous Mesoporous Mater.* **1999**, *29*, 339–349.
75. C. W. Jones, K. Tsuji, M. E. Davis, *Microporous Mesoporous Mater.* **1999**, *33*, 223–240.
76. C. W. Jones, M. Tsapatsis, T. Okubo, M. E. Davis, *Microporous Mesoporous Mater.* **2001**, *42*, 21–35.
77. K. Yamamoto, Y. Sakata, Y. Nohara, Y. Takahashi, T. Tatsumi, *Science* **2003**, *300*, 470–472.
78. I. Petrovic, A. Navrotsky, M. E. Davis, S. I. Zones, *Chem. Mater.* **1993**, *5*, 1805–1813.
79. R. Astala, S. M. Auerbach, *J. Am. Chem. Soc.* **2004**, *126*, 1843–1848.
80. K. Yamamoto, T. Tatsumi, *Chem. Mater.* **2008**, *20*, 972–980.
81. K. Yamamoto, Y. Nohara, Y. Domon, Y. Takahashi, Y. Sakata, J. Plévert, T. Tatsumi, *Chem. Mater.* **2005**, *17*, 3913–3920.
82. W. J. Herzberg, W. R. Erwin, *J. Colloid Interface Sci.* **1970**, *33*, 172–177.
83. R. Mouawia, A. Mehdi, C. Reyé, R. Corriu, *New J. Chem.* **2006**, *30*, 1077–1082.
84. E. Besson, A. Mehdi, D. A. Lerner, C. Reyé, R. J. P. Corriu, *J. Mater. Chem.* **2005**, *15*, 803–809.

85. F. Hoffmann, M. Güngerich, P. J. Klar, M. Fröba, *J. Phys. Chem. C* **2007**, *111*, 5648–5660.
86. M. C. Burleigh, S. Jayasundera, M. S. Spector, C. W. Thomas, M. A. Markowitz, B. P. Gaber, *Chem. Mater.* **2004**, *16*, 3–5.
87. Y. Liang, R. Anwander, *Microporous Mesoporous Mater.* **2004**, *72*, 153–165.
88. Y. Liang, M. Hanzlik, R. Anwander, *Chem. Commun.* **2005**, 525–527.
89. Y. Liang, M. Hanzlik, R. Anwander, *J. Mater. Chem.* **2005**, *15*, 3919–3928.
90. Y. Liang, E. Sev. Erichsen, M. Hanzlik, R. Anwander, *Chem. Mater.* **2008**, *20*, 1451–1458.
91. B. Lee, H. Luo, C. Y. Yuan, J. S. Linc, S. Dai, *Chem. Commun.* **2004**, 240–241.
92. B. Lee, H.-J. Im, H. Luo, E. W. Hagaman, S. Dai, *Langmuir* **2005**, *21*, 5372–5376.
93. C. Nédez, A. Théolier, F. Lefebvre, A. Choplin, J. M. Basset, J. F. Joly, *J. Am. Chem. Soc.* **1993**, *115*, 722–729.
94. J. Corker, F. Lefebvre, C. Lécuyer, V. Dufaud, F. Quignard, A. Choplin, J. Evans, J. M. Basset, *Science* **1996**, *271*, 966–969.
95. A. P. Wight, M. E. Davis, *Chem. Rev.* **2002**, *102*, 3589–3614.
96. Á. Molnár, B. Rác, *Curr. Org. Chem.* **2006**, *10*, 1697–1726.
97. E. L. Margelefsky, R. K. Zeidan, M. E. Davis, *Chem. Soc. Rev.* **2008**, *37*, 1118–1126.
98. J. A. Melero, R. van Grieken, G. Morales, *Chem. Rev.* **2006**, *106*, 3790–3812.
99. R. K. Iler, U.S. Patent 2,657,149 (DuPont), **1953** [*Chem. Abstr.* 50, 11564].
100. R. Anwander, I. Nagl, M. Widenmeyer, G. Engelhardt, O. Groeger, C. Plam, T. Röser, *J. Phys. Chem. B* **2000**, *104*, 3532–3544.
101. Y.C. Liang, R. Anwander, *J. Mater. Chem.* **2007**, *17*, 2506–2516.
102. K. Yamamoto, T. Tatsumi, *Chem. Lett.* **2000**, *29*, 624–625.
103. K. Yamamoto, T. Tatsumi, *Microporous Mesoporous Mater.* **2001**, *44–45*, 459–464.
104. K. Inumaru, J. Kiyoto, S. Yamanaka, *Chem. Commun.* **2000**, 903–904.
105. K. Inumura, Y. Inoue, S. Kakii, T. Nakano, S. Yamanaka, *Phys. Chem. Chem. Phys.* **2004**, *6*, 3133–3139.
106. A. M. Liu, K. Hidajat, S. Kawi, D. Y. Zhao, *Chem. Commun.* **2000**, 1145–1146.
107. H. Yoshitake, T. Yokoi, T. Tatsumi, *Chem. Mater.* **2002**, *14*, 4603–4610.
108. C. Lei, Y. Shin, J. Liu, E. J. Ackerman, *J. Am. Chem. Soc.* **2002**, *124*, 11242–11243.
109. A. Matsumoto, K. Tsutsumi, K. Schumacher, K. K. Unger, *Langmuir* **2002**, *18*, 4014–4019.
110. A. C. C. Chang, S. S. C. Chuang, M. Gray, Y. Soong, *Energy Fuels* **2003**, *17*, 468–473.
111. A. Walcarius, M. Etienne, B. Lebeau, *Chem. Mater.* **2003**, *15*, 2161–2173.
112. K. Y. Ho, G. McKay, K. L. Yeung, *Langmuir* **2003**, *19*, 3019–3024.
113. H. Y. Huang, R. T. Yang, D. Chinn, C. L. Munson, *Ind. Eng. Chem. Res.* **2003**, *42*, 2427–2433.
114. K. F. Lam, K. L. Yeung, G. Mckay, *Microporous Mesoporous Mater.* **2007**, *100*, 191–201.
115. Y. Sakamoto, K. Nagata, K. Yogo, K. Yamada, *Microporous Mesoporous Mater.* **2007**, *101*, 303–311.
116. S. Oh, T. Kang, H. Kim, J. Moon, S. Hong, J. Yi, *J. Membrane Sci.* **2007**, *301*, 118–125.
117. K. F. Lam, C. M. Fong, K. L. Yeung, *Gold Bull.* **2007**, *40*, 192–198.
118. S. Hamoudi, R. Saad, K. Belkacemi, *Ind. Eng. Chem. Res.* **2007**, *46*, 8806–8812.
119. R. Saad, S. Hamoudi, K. Belkacemi, *J. Porous Mater.* **2008**, *15*, 315–323.
120. J. Li, L. Wang, T. Qi, Y. Zhou, C. Liu, J. Chu, Y. Zhang, *Microporous Mesoporous Mater.* **2008**, *110*, 442–450.
121. R. A. Khatri, S. S. C. Chuang, Y. Soong, M. Gray, *Ind. Eng. Chem. Res.* **2005**, *44*, 3702–3708.
122. T. Yokoi, H. Yoshitake, T. Tatsumi, *J. Mater. Chem.* **2004**, *14*, 951–957.
123. P. J. E. Harlick, A. Sayari, *Ind. Eng. Chem. Res.* **2007**, *46*, 446–458.
124. F. Zheng, D. N. Tran, B. J. Busche, G. E. Fryxell, R. S. Addleman, T. S. Zemanian, C. L. Aardahl, *Ind. Eng. Chem. Res.* **2005**, *44*, 3099–3105.
125. G. S. Armatas, C. E. Salmas, M. Louloudi, G. P. Androutsopoulos, P. J. Pomonis, *Langmuir* **2003**, *19*, 3128–3136.
126. T. Kang, Y. Park, K. Choi, J. S. Lee, J. Yi, *J. Mater. Chem.* **2004**, *14*, 1043–1050.
127. T. Kang, Y. Park, J. Yi, *Ind. Eng. Chem. Res.* **2004**, *43*, 1478–1484.
128. J. C. Hicks, C. W. Jones, *Langmuir* **2006**, *22*, 2676–2681.

129. P. Trens, M. L. Russell, L. Spjuth, M. J. Hudson, J.-O. Liljenzin, *Ind. Eng. Chem. Res.* **2002**, *41*, 5220–5225.
130. L. Mercier, T. J. Pinnavaia, *Environ. Sci. Technol.* **1998**, *32*, 2749–2754.
131. V. Antochshuk, M. Jaroniec, *Chem. Commun.* **2002**, 258–259.
132. V. Antochshuk, O. Olkhovyk, M. Jaroniec, I.-S. Park, R. Ryoo, *Langmuir* **2003**, *19*, 3031–3034.
133. O. Olkhovyk, V. Antochshuk, M. Jaroniec, *Colloid Surf. A* **2004**, *236*, 69–72.
134. K. A. Venkatesan, T. G. Srinivasan, P. R. Vasudeva Rao, *J. Radioanal. Nuclear Chem.* **2003**, *256*, 213–218.
135. G. Rodríguez-López, M. D. Marcos, R. Martínez-Máñez, F. Sancenón, J. Soto, L. A. Villaescusa, D. Beltrán, P. Amorós, *Chem. Commun.* **2004**, 2198–2199.
136. M. Vallet-Regí, A. Rámila, R. P. del Real, J. Pérez-Pariente, *Chem. Mater.* **2001**, *13*, 308–311.
137. M. Vallet-Regí, F. Balas, D. Arcos, *Angew. Chem. Int. Ed.* **2007**, *46*, 7548–7558.
138. N. K. Mal, M. Fujiwara, Y. Tanaka, *Nature* **2003**, *421*, 350–353.
139. N. K. Mal, M. Fujiwara, Y. Tanaka, T. Taguchi, M. Matsukata, *Chem. Mater.* **2003**, *15*, 3385–3394.
140. Q. Fu, G. V. R. Rao, L. K. Ista, Y. Wu, B. P. Andrzejewski, L. A. Sklar, T. L. Ward, G. P. López, *Adv. Mater.* **2003**, *15*, 1262–1266.
141. D. R. Radu, C.-Y. Lai, J. W. Wiench, M. Pruski, V. S.-Y. Lin, *J. Am. Chem. Soc.* **2004**, *126*, 1640–1641.
142. A. B. Descalzo, D. Jiménez, M. D. Marcos, R. Martínez-Máñez, J. Soto, J. El Haskouri, C. Guillém, D. Béltran, P. Amorós, M. V. Borrachero, *Adv. Mater.* **2002**, *14*, 966–969.
143. A. B. Descalzo, M. D. Marcos, R. Martínez-Máñez, J. Soto, D. Beltrán, P. Amorós, *J. Mater. Chem.* **2005**, *15*, 2721–2731.
144. A. B. Descalzo, M. D. Marcos, C. Monte, R. Martínez-Máñez, K. Rurack, *J. Mater. Chem.* **2007**, *17*, 4716–4723.
145. S. Fiorilli, B. Onida, C. Barolo, G. Viscardi, D. Brunel, E. Garrone, *Langmuir* **2007**, *23*, 2261–2268.
146. E. Kim, H. J. Kim, D. R. Bae, S. J. Lee, E. J. Cho, M. R. Seo, J. S. Kim, J. H. Jung, *New J. Chem.* **2008**, *32*, 1003–1007.
147. P. Banet, L. Legagneux, P. Hesemann, J. J. E. Moreau, L. Nicole, A. Quach, C. Sanchez, T.-H. Tran-Thi, *Sens. Actuators B* **2008**, *130*, 1–8.
148. J.-Q. Wang, L. Huang, M. Xue, Y. Wang, L. Gao, J. H. Zhu, Z. Zou, *J. Phys. Chem. C* **2008**, *112*, 5014–5022.
149. S. Murata, H. Hata, T. Kimura, Y. Sugahara, K. Kuroda, *Langmuir* **2000**, *16*, 7106–7108.
150. H. Furukawa, T. Watanabe, K. Kuroda, *Chem. Commun.* **2001**, 2002–2003.
151. S. L. Burkett, S. D. Sims, S. Mann, *Chem. Commun.* **1996**, 1367–1368.
152. D. J. Macquarrie, *Chem. Commun.* **1996**, 1961–1962.
153. M. H. Lim, C. F. Blanford, A. Stein, *J. Am. Chem. Soc.* **1997**, *119*, 4090–4091.
154. D. J. Macquarrie, D. B. Jackson, *Chem. Commun.* **1997**, 1781–1782.
155. D. J. Macquarrie, D. B. Jackson, S. Tailland, K. A. Utting, *J. Mater. Chem.* **2001**, *11*, 1843–1849.
156. X. Zhang, W. Wu, J. Wang, C. Liu, *J. Am. Ceram. Soc.* **2007**, *90*, 965–968.
157. X. Zhang, W. Wu, J. Wang, X. Tian, *Appl. Surf. Sci.* **2008**, *254*, 2893–2899.
158. M. H. Lim, C. F. Blanford, A. Stein, *Chem. Mater.* **1998**, *10*, 467–470.
159. W. M. Van Rhijn, D. E. De Vos, B. F. Sels, W. D. Bossaert, P. A. Jacobs, *Chem. Commun.* **1998**, 317–318.
160. I. Díaz, C. Márquez-Álvarez, F. Mohino, J. Pérez-Pariente, E. Sastre, *J. Catal.* **2000**, *193*, 283–294.
161. V. Ganesan, A. Walcarius, *Langmuir* **2004**, *20*, 3632–3640.
162. S. Saravanamurugan, Sujandi, E. A. Prasetyanto, S.-E. Park, *Microporous Mesoporous Mater.* **2008**, *112*, 97–107.
163. D. Margolese, J. A. Melero, S. C. Christiansen, B. F. Chmelka, G. D. Stucky, *Chem. Mater.* **2000**, *12*, 2448–2459.
164. W. D. Bossaert, D. E. De Vos, W. M. Van Rhijn, J. Bullen, P. J. Grobet, P. A. Jacobs, *J. Catal.* **1999**, *182*, 156–164.

165. I. Díaz, C. Márquez-Álvarez, F. Mohino, J. Pérez-Pariente, E. Sastre, *J. Catal.* **2000**, *193*, 295–302.
166. J. G. C. Shen, R. G. Herman, K. Klier, *J. Phys. Chem. B* **2002**, *106*, 9975–9978.
167. J. Alauzun, A. Mehdi, C. Reyé, R. J. P. Corriu, *J. Am. Chem. Soc.* **2006**, *128*, 8718–8719.
168. C. Yang, B. Zibrowius, F. Schüth, *Chem. Commun.* **2003**, 1772–1773.
169. R. Mouawia, A. Mehdi, C. Reyé, R. Corriu, *J. Mater. Chem.* **2007**, *17*, 616–618.
170. R. J. P. Corriu, L. Datas, Y. Guari, A. Mehdi, C. Reyé, C. Thieuleux, *Chem. Commun.* **2001**, 763–764.
171. C. Li, J. Yang, X. Shi, J. Liu, Q. Yang, *Microporous Mesoporous Mater.* **2007**, *98*, 220–226.
172. M. C. Burleigh, M. A. Markowitz, M. S. Spector, B. P. Gaber, *Environ. Sci. Technol.* **2002**, *36*, 2515–2518.
173. C. Li, J. Liu, X. Shi, J. Yang, Q. Yang, *J. Phys. Chem. C* **2007**, *111*, 10948–10954.
174. M. Nakamura, K. Ishimura, *Langmuir* **2008**, *24*, 5099–5108.
175. J. Brown, R. Richer, L. Mercier, *Microporous Mesoporous Mater.* **2000**, *37*, 41–48.
176. R. I. Nooney, M. Kalyanaraman, G. Kennedy, E. J. Maginn, *Langmuir* **2001**, *17*, 528–533.
177. A. Bibby, L. Mercier, *Chem. Mater.* **2002**, *14*, 1591–1597.
178. H.-M. Kao, T.-Y. Shen, J.-D. Wu, L.-P. Lee, *Microporous Mesoporous Mater.* **2008**, *110*, 461–471.
179. O. Olkhovyk, M. Jaroniec, *J. Am. Chem. Soc.* **2005**, *127*, 60–61.
180. R. M. Grudzien, S. Pikus, M. Jaroniec, *J. Phys. Chem. B* **2006**, *110*, 2972–2975.
181. O. Olkhovyk, M. Jaroniec, *Ind. Eng. Chem. Res.* **2007**, *46*, 1745–1751.
182. O. Olkhovyk, S. Pikus, M. Jaroniec, *J. Mater. Chem.* **2005**, *15*, 1517–1519.
183. J. Li, T. Qi, L. Wang, Y. Zhou, C. Liu, Y. Zhang, *Microporous Mesoporous Mater.* **2007**, *103*, 184–189.
184. R. J. P. Corriu, A. Mehdi, C. Reyé, C. Thieuleux, *Chem. Mater.* **2004**, *16*, 159–166.
185. R. J. P. Corriu, A. Mehdi, C. Reyé, C. Thieuleux, *Chem. Commun.* **2002**, 1382–1383.
186. R. J. P. Corriu, A. Mehdi, C. Reyé, C. Thieuleux, *Chem. Commun.* **2003**, 1564–1565.
187. R. J. P. Corriu, A. Mehdi, C. Reye, C. Thieuleux, *New J. Chem.* **2003**, *27*, 905–908.
188. J. Alauzun, A. Mehdi, C. Reyé, R. J. P. Corriu, *J. Mater. Chem.* **2007**, *17*, 349–356.
189. J. Alauzun, A. Mehdi, C. Reyé, R. Corriu, *New J. Chem.* **2007**, *31*, 911–915.
190. W. Whitnall, L. Cademartiri, G. A. Ozin, *J. Am. Chem. Soc.* **2007**, *129*, 15644–15649.
191. R. Huq, L. Mercier, *Chem. Mater.* **2001**, *13*, 4512–4519.
192. C. Liu, N. Naismith, L. Fu, J. Economy, *Chem. Commun.* **2003**, 2472–2473.
193. A. Walcarius, S. Sayen, C. Gérardin, F. Hamdoune, L. Rodehüser, *Colloid Surf. A* **2004**, *234*, 145–151.
194. C. E. Fowler, B. Lebeau, S. Mann, *Chem. Commun.* **1998**, 1825–1826.
195. B. Lebeau, C. E. Fowler, S. R. Hall, S. Mann, *J. Mater. Chem.* **1999**, *9*, 2279–2281.
196. M. Ganschow, M. Wark, D. Wöhrle, G. Schulz-Ekloff, *Angew. Chem. Int. Ed.* **2000**, *39*, 161–163.
197. N. Liu, Z. Chen, D. R. Dunphy, Y.-B. Jiang, R. A. Assink, C. J. Brinker, *Angew. Chem. Int. Ed.* **2003**, *42*, 1731–1734.
198. M. Álvaro, B. Ferrer, H. García, F. Rey, *Chem. Commun.* **2002**, 2012–2013.
199. M. Álvaro, M. Benítez, D. Das, H. García, E. Peris, *Chem. Mater.* **2005**, *17*, 4958–4964.
200. M. Álvaro, B. Ferrer, V. Fornés, H. García, *Chem. Commun.* **2001**, 2546–2547.
201. M. Álvaro, B. Ferrer, V. Fornés, H. García, *ChemPhysChem* **2003**, *4*, 612–617.
202. M. Álvaro, B. Ferrer, H. García, S. Hashimoto, M. Hiratsuka, T. Asahi, H. Masuhara, *ChemPhysChem* **2004**, *5*, 1058–1062.
203. A. Domenech, M. Álvaro, B. Ferrer, H. García, *J. Phys. Chem. B* **2003**, *107*, 12781–12788.
204. G. Wirnsberger, B. J. Scott, G. D. Stucky, *Chem. Commun.* **2001**, 119–120.
205. K. J. Balkus, Jr., T. J. Pisklak, G. Hundt, J. Sibert, Y. Zhang, *Microporous Mesoporous Mater.* **2008**, *112*, 1–13.
206. Y. Guo, A. Mylonakis, Z. Zhang, G. Yang, P. I. Lelkes, S. Che, Q. Lu, Y. Wie, *Chem. Eur. J.* **2008**, *14*, 2909–2917.
207. S. Guan, S. Inagaki, T. Ohsuna, O. Terasaki, *J. Am. Chem. Soc.* **2000**, *122*, 5660–5661.
208. A. Sayari, S. Hamoudi, Y. Yang, I. L. Moudrakovski, J. R. Ripmeester, *Chem. Mater.* **2000**, *12*, 3857–3863.

209. S. Hamoudi, Y. Yang, I. L. Moudrskovski, S. Lang, A. Sayari, *J. Phys. Chem. B* **2001**, *105*, 9118–9123.
210. T. Asefa, M. J. MacLachlan, H. Grondey, N. Coombs, G. A. Ozin, *Angew. Chem. Int. Ed.* **2000**, *39*, 1808–1811.
211. O. Muth, C. Schellbach, M. Fröba, *Chem. Commun.* **2001**, 2032–2033.
212. M. C. Burleigh, M. A. Markowitz, E. M. Wong, J.-S. Lin, B. P. Gaber, *Chem. Mater.* **2001**, *13*, 4411–4412.
213. H. Zhu, D. J. Jones, J. Zajac, J. Rozière, R. Dutartre, *Chem. Commun.* **2001**, 2568–2569.
214. T. Ren, X. Zhang, J. Suo, *Microporous Mesoporous Mater.* **2002**, *54*, 139–144.
215. J. R. Matos, M. Kruk, L. P. Mercuri, M. Jaroniec, T. Asefa, N. Coombs, G. A. Ozin, T. Kamiyama, O. Terasaki, *Chem. Mater.* **2002**, *14*, 1903–1905.
216. M. C. Burleigh, M. A. Markowitz, M. S. Spector, B. P. Gaber, *J. Phys. Chem. B* **2002**, *106*, 9712–9716.
217. M. P. Kapoor, S. Inagaki, *Chem. Mater.* **2002**, *14*, 3509–3514.
218. S. Hamoudi, S. Kaliaguine, *Chem. Commun.* **2002**, 2118–2119.
219. W. Guo, J.-Y. Park, M.-O. Oh, H.-W. Jeong, W.-J. Cho, I. Kim, C.-S. Ha, *Chem. Mater.* **2003**, *15*, 2295–2298.
220. W. Guo, I. Kim, C.-S. Ha, *Chem. Commun.* **2003**, 2692–2693.
221. M. C. Burleigh, M. A. Markowitz, S. Jayasundera, M. S. Spector, C.W. Thomas, B. P. Gaber, *J. Phys. Chem. B* **2003**, *107*, 12628–12634.
222. E. B. Cho, K. Char, *Chem. Mater.* **2004**, *16*, 270–275.
223. X. Y. Bao, X. S. Zhao, X. Li, P. A. Chia, J. Li, *J. Phys. Chem. B* **2004**, *108*, 4684–4689.
224. X. Y. Bao, X. S. Zhao, S. Z. Qiao, S. K. Bhatia, *J. Phys. Chem. B* **2004**, *108*, 16441–16450.
225. X. Y. Bao, X. S. Zhao, X. Li, J. Li, *Appl. Surf. Sci.* **2004**, *237*, 380–386.
226. L. Zhao, G. Zhu, D. Zhang, Y. Di, Y. Chen, O. Terasaki, S. Qiu, *J. Phys. Chem. B* **2005**, *109*, 765–768.
227. L. Zhang, Q. Yang, W.-H. Zhang, Y. Li, J. Yang, D. Jiang, G. Zhu, C. Li, *J. Mater. Chem.* **2005**, *15*, 2562–2568.
228. R. M. Grudzien, B. E. Grabicka, M. Jaroniec, *Colloid Surf. A* **2007**, *300*, 235–244.
229. K. Nakajima, D. Lu, J. N. Kondo, I. Tomita, S. Inagaki, M. Hara, S. Hayashi, K. Domen, *Chem. Lett.* **2003**, *32*, 950–951.
230. W. Wang, S. Xie, W. Zhou, A. Sayari, *Chem. Mater.* **2004**, *16*, 1756–1762.
231. M. C. Burleigh, S. Jayasundera, C. W. Thomas, M. S. Spector, M. A. Markowitz, B. P. Gaber, *Colloid Polym. Sci.* **2004**, *282*, 728–733.
232. K. Nakajima, I. Tomita, M. Hara, S. Hayashi, K. Domen, J. N. Kondo, *Adv. Mater.* **2005**, *17*, 1839–1842.
233. K. Nakajima, I. Tomita, M. Hara, S. Hayashi, K. Domen, J. N. Kondo, *Catal. Today*, **2006**, *116*, 151–156.
234. X. Zhou, S. Qiao, N. Hao, X. Wang, C. Yu, L. Wang, D. Zhao, G. Q. Lu, *Chem. Mater.* **2007**, *19*, 1870–1876.
235. D. Dubé, M. Rat, F. Béland, S. Kaliaguine, *Microporous Mesoporous Mater.* **2008**, *111*, 596–603.
236. S.-R. Zhai, S. S. Park, M. Park, M. H. Ullah, C.-S. Ha, *Microporous Mesoporous Mater.* **2008**, *113*, 47–55.
237. S.-R. Zhai, I. Kim, C.-S. Ha, *J. Solid State Chem.* **2008**, *181*, 67–74.
238. E.-B. Cho, D. Kim, M. Jaroniec, *Chem. Mater.* **2008**, *20*, 2468–2475.
239. S. Shylesh, Prinson, P. Samuel, A. P. Singh, *Microporous Mesoporous Mater.* **2007**, *100*, 250–258.
240. S.-S. Yoon, W.-J. Son, K. Biswas, W.-S. Ahn, *Bull. Korean Chem. Soc.* **2008**, *29*, 609–614.
241. X. Wang, D. Lu, R. Austin, A. Agarwal, L. J. Mueller, Z. Liu, J. Wu, P. Feng, *Langmuir* **2007**, *23*, 5735–5739.
242. C. Yoshina-Ishii, T. Asefa, N. Coombs, M. J. MacLachlan, G. A. Ozin, *Chem. Commun.* **1999**, 2539–2540.
243. S. Inagaki, S. Guan, T. Ohsuna, O. Terasaki, *Nature* **2002**, *416*, 304–307.
244. Y. Goto, S. Inagaki, *Chem. Commun.* **2002**, 2410–2411.

245. N. Bion, P. Ferreira, A. Valente, I. S. Goncalves, J. Rocha, *J. Mater. Chem.* **2003**, *13*, 1910–1913.
246. W. Wang, W. Zhou, A. Sayari, *Chem. Mater.* **2003**, *15*, 4886–4889.
247. B. Onida, L. Borello, C. Busco, P. Ugliengo, Y. Goto, S. Inagaki, E. Garrone, *J. Phys. Chem. B* **2005**, *109*, 11961–11966.
248. B. Onida, B. Camarota, P. Ugliengo, Y. Goto, S. Inagaki, E. Garrone, *J. Phys. Chem. B* **2005**, *109*, 21732–21736.
249. Y. Goto, K. Okamoto, S. Inagaki, *Bull. Chem. Soc. Jpn.* **2005**, *78*, 932–936.
250. O. Ohtani, Y. Goto, K. Okamoto, S. Inagaki, *Chem. Lett.* **2005**, *34*, 1342–1343.
251. M. P. Kapoor, S. Inagaki, *Chem. Lett.* **2004**, *33*, 88–89.
252. M. P. Kapoor, N. Setoyama, Q. Yang, M. Ohashi, S. Inagaki, *Langmuir* **2005**, *21*, 443–449.
253. M. P. Kapoor, S. Inagaki, S. Ikeda, K. Kakiuchi, M. Suda, T. Shimada, *J. Am. Chem. Soc.* **2005**, *127*, 8174–8178.
254. M. P. Kapoor, M. Yanagi, Y. Kasama, T. Yokoyama, S. Inagaki, T. Shimada, H. Nanbu, L. R. Juneja, *J. Mater. Chem.* **2006**, *16*, 3305–3311.
255. Y. Goto, S. Inagaki, *Microporous Mesoporous Mater.* **2006**, *89*, 103–108.
256. S. S. Park, D. H. Park, C.-S. Ha, *Chem. Mater.* **2007**, *19*, 2709–2711.
257. S.-Y. Wu, H.-S. Hsueh, M. H. Huang, *Chem. Mater.* **2007**, *19*, 5986–5990.
258. B. Camarota, B. Onida, Y. Goto, S. Inagaki, E. Garrone, *Langmuir* **2007**, *23*, 13164–13168.
259. B. Camarota, S. Mann, B. Onida, E. Garrone, *ChemPhysChem* **2007**, *8*, 2363–2366.
260. T. Suzuki, H. Miyata, K. Kuroda, *J. Mater. Chem.* **2008**, *18*, 1239–1244.
261. G. Cerveau, R. J. P. Corriu, B. Dabiens, J. Le Bideau, *Angew. Chem. Int. Ed.* **2000**, *39*, 4533–4537.
262. T. Tani, S. Inagaki, *Chem. Mater.* **2008**, *20*, 4531–4531.
263. S.-Y. Wu, H.-S. Hsueh, M. H. Huang, *Chem. Mater.* **2008**, *20*, 4532–4532.
264. M. P. Kapoor, Q. Yang, S. Inagaki, *Chem. Mater.* **2004**, *16*, 1209–1213.
265. M. P. Kapoor, Q. Yang, S. Inagaki, *J. Am. Chem. Soc.* **2002**, *124*, 15176–15177.
266. Y. Yang, A. Sayari, *Chem. Mater.* **2007**, *19*, 4117–4119.
267. A. Sayari, W. Wang, *J. Am. Chem. Soc.* **2005**, *127*, 12194–12195.
268. M. Cornelius, F. Hoffmann, M. Fröba, *Chem. Mater.* **2005**, *17*, 6674–6678.
269. Y. Xia, W. Wang, R. Mokaya, *J. Am. Chem. Soc.* **2005**, *127*, 790–798.
270. Y. Xia, R. Mokaya, *J. Mater. Chem.* **2006**, *16*, 395–400.
271. Y. Xia, Z. Yang, R. Mokaya, *Chem. Mater.* **2006**, *18*, 1141–1148.
272. Y. Xia, R. Mokaya, *J. Phys. Chem. B* **2006**, *110*, 3889–3894.
273. J. Morell, C. V. Teixeira, M. Cornelius, V. Rebbin, M. Tiemann, H. Amenitsch, M. Fröba, M. Lindén, *Chem. Mater.* **2004**, *16*, 5564–5566.
274. G. Temtsin, T. Asefa, S. Bittner, G. A. Ozin, *J. Mater. Chem.* **2001**, *11*, 3202–3206.
275. W. J. Hunks, G. A. Ozin, *Chem. Commun.* **2004**, 2426–2427.
276. W. J. Hunks, G. A. Ozin, *Chem. Mater.* **2004**, *16*, 5465–5472.
277. J. Morell, G. Wolter, M. Fröba, *Chem. Mater.* **2005**, *17*, 804–808.
278. M. Ohashi, M. P. Kapoor, S. Inagaki, *Chem. Commun.* **2008**, 841–843.
279. A. Kuschel, S. Polarz, *Adv. Funct. Mater.* **2008**, *18*, 1272–1280.
280. Y. Maegawa, Y. Goto, S. Inagaki, T. Shimada, *Tetrahedron Lett.* **2006**, *47*, 6957–6960.
281. H. Peng, J. Tang, L. Yang, J. Pang, H. S. Ashbaugh, C. J. Brinker, Z. Yang, Y. Lu, *J. Am. Chem. Soc.* **2006**, *128*, 5304–5305.
282. H. Peng, J. Tang, J. Pang, D. Chen, L. Yang, H. S. Ashbaugh, C. J. Brinker, Z. Yang, Y. Lu, *J. Am. Chem. Soc.* **2005**, *127*, 12782–12783.
283. M. Cornelius, F. Hoffmann, B. Ufer, P. Behrens, M. Fröba, *J. Mater. Chem.* **2008**, *18*, 2587–2592.
284. Y. Goto, N. Mizoshita, O. Ohtani, T. Okada, T. Shimada, T. Tani, S. Inagaki, *Chem. Mater.* **2008**, *20*, 4495–4498.
285. M. Kuroki, T. Asefa, W. Whitnal, M. Kruk, C. Yoshina-Ishii, M. Jaroniec, G. A. Ozin, *J. Am. Chem. Soc.* **2002**, *124*, 13886–13895.
286. K. Landskron, B. D. Hatton, D. D. Perovic, G. A. Ozin, *Science* **2003**, *302*, 266–269.
287. K. Landskron, G. A. Ozin, *Science* **2004**, *306*, 1529–1532.
288. W. J. Hunks, G. A. Ozin, *Adv. Funct. Mater.* **2005**, *15*, 259–266.

289. R. M. Grudzien, B. E. Grabicka, S. Pikus, M. Jaroniec, *Chem. Mater.* **2006**, *18*, 1722–1725.
290. W.-H. Zhang, X. Zhang, Z. Hua, P. Harish, F. Schroeder, S. Hermes, T. Cadenbach, J. Shi, R. A. Fischer, *Chem. Mater.* **2007**, *19*, 2663–2670.
291. M. A. Wahab, I. Kim, C.-S. Ha, *J. Solid State Chem.* **2004**, *177*, 3439–3447.
292. H. Zhu, D. J. Jones, J. Zajac, R. Dutartre, M. Rhomari, J. Rozière, *Chem. Mater.* **2002**, *14*, 4886–4894.
293. S. Jayasundera, M. C. Burleigh, M. Zeinali, M. S. Spector, J. B. Miller, W. Yan, S. Dai, M. A. Markowitz, *J. Phys. Chem. B* **2005**, *109*, 9198–9201.
294. J. Morell, M. Güngerich, G. Wolter, J. Jiao, M. Hunger, P. J. Klar, M. Fröba, *J. Mater. Chem.* **2006**, *16*, 2809–2818.
295. E.-B. Cho, D. Kim, M. Jaroniec, *Langmuir* **2007**, *23*, 11844–11849.
296. E.-B. Cho, D. Kim, *Microporous Mesoporous Mater.* **2008**, *113*, 530–537.
297. B. A. Treuherz, Y. Z. Khimyak, *Microporous Mesoporous Mater.* **2007**, *106*, 236–245.
298. M. A. Wahab, S. Sudhakar, E. Yeo, A. Sellinger, *Chem. Mater.* **2008**, *20*, 1855–1861.
299. Y. Yang, A. Sayari, *Chem. Mater.* **2008**, *20*, 2980–2984.
300. J. T. A. Jones, C. D. Wood, C. Dickinson, Y. Z. Khimyak, *Chem. Mater.* **2008**, *20*, 3385–3397.
301. J. H. Jung, Y. Ono, S. Shinkai, *Angew. Chem. Int. Ed.* **2000**, *39*, 1862–1865.
302. J. H. Jung, Y. Ono, K. Hanabusa, S. Shinkai, *J. Am. Chem. Soc.* **2000**, *122*, 5008–5009.
303. S. Che, Z. Liu, T. Ohsuna, K. Sakamoto, O. Terasaki, T. Tatsumi, *Nature* **2004**, *429*, 281–284.
304. T. Ohsuna, Z. Liu, S. Che, O. Terasaki, *Small* **2005**, *1*, 233–237.
305. T. Yokoi, Y. Yamataka, Y. Ara, S. Sato, Y. Kubota, T. Tatsumi, *Microporous Mesoporous Mater.* **2007**, *103*, 20–28.
306. A. B. Harris, R. D. Kamien, T. C. Lubensky, *Phys. Rev. Lett.* **1997**, *78*, 1476–1479.
307. A. B. Harris, R. D. Kamien, T. C. Lubensky, *Rev. Mod. Phys.* **1999**, *71*, 1745–1757.
308. B. Wang, C. Chi, W. Shan, Y. Zhang, N. Ren, W. Yang, Y. Tang, *Angew. Chem. Int. Ed.* **2006**, *45*, 2088–2090.
309. S. Yang, L. Zhao, C. Yu, X. Zhou, J. Tang, P. Yuan, D. Chen, D. Zhao, *J. Am. Chem. Soc.* **2006**, *128*, 10460–10466.
310. X. Meng, T. Yokoi, D. Lu, T. Tatsumi, *Angew. Chem. Int. Ed.* **2007**, *46*, 7796–7798.
311. S. Xiang, Y. Zhang, Q. Xin, C. Li, *Angew. Chem. Int. Ed.* **2002**, *41*, 821–824.
312. T. Katsuki, K. B. Sharpless, *J. Am. Chem. Soc.* **1980**, *102*, 5974–5976.
313. C. Li, H. Zhang, D. Jiang, Q. Yang, *Chem. Commun.* **2007**, 547–558.
314. D. P. Serrano, J. Aguado, C. Vargas, *Appl. Catal. A* **2008**, *335*, 172–179.
315. I. Motorina, C. M. Crudden, *Org. Lett.* **2001**, *3*, 2325–2328.
316. H. M. Lee, S.-W. Kim, T. Hyeon, B. M. Kim, *Tetrahedron-Asymmetry* **2001**, *12*, 1537–1541.
317. A. Corma, S. Iborra, I. Rodríguez, M. Iglesias, F. Sánchez, *Catal. Lett.* **2002**, *82*, 237–242.
318. S. Abramson, N. Bellocq, M. Laspéras, *Top. Catal.* **2000**, *13*, 339–345.
319. S. Abramson, M. Laspéras, D. Brunel, *Tetrahedron-Asymmetry* **2002**, *13*, 357–367.
320. A. Lee, W. Kim, J. Lee, T. Hyeon, B. M. Kim, *Tetrahedron-Asymmetry* **2004**, *15*, 2595–2598.
321. M. S. Whang, Y. K. Kwon, G.-J. Kim, *J. Ind. Eng. Chem.* **2002**, *8*, 262–267.
322. D. Dhar, I. Beadham, S. Chandasekaran, *Proc. Indian Acad. Sci. Chem. Sci.* **2003**, *115*, 365–372.
323. X. Wang, P. Han, X. Qiu, X. Ji, L. Gao, *Catal. Lett.* **2008**, *124*, 418–427.
324. M. Álvaro, M. Benítez, D. Das, B. Ferrer, H. García, *Chem. Mater.* **2004**, *16*, 2222–2228.
325. D. M. Jiang, Q. H. Yang, J. Yang, L. Zhang, G. R. Zhu, W. G. Su, C. Li, *Chem. Mater.* **2005**, *17*, 6154–6160.
326. D. Jiang, Q. Yang, H. Wang, G. Zhu, J. Yang, C. Li, *J. Catal.* **2006**, *239*, 65–73.
327. D. Jiang, J. Gao, J. Li, Q. Yang, C. Li, *Microporous Mesoporous Mater.*, **2008**, *113*, 385–392.
328. R. A. García, R. van Grieken, J. Iglesias, V. Morales, D. Gordillo, *Chem. Mater.* **2008**, *20*, 2964–2971.
329. G. Zhu, D. Jiang, Q. Yang, J. Yang, C. Li, *J. Chromatogr. A* **2007**, *1149*, 219–227.
330. G. Zhu, H. Zhong, Q. Yang, C. Li, *Microporous Mesoporous Mater.* **2008**, *116*, 36–43.
331. C. Baleizão, B. Gigante, D. Das, M. Álvaro, H. García, A. Corma, *Chem. Commun.* **2003**, 1860–1861.

332. C. Baleizão, B. Gigante, D. Das, M. Álvaro, H. García, A. Corma, *J. Catal.* **2004**, *223*, 106–113.
333. C. Baleizão, B. Gigante, H. García, A. Corma, *Tetrahedron* **2004**, *60*, 10461–10466.
334. R. Raja, J. M. Thomas, M. D. Jones, B. F. G. Johnson, D. E. W. Vaughan, *J. Am. Chem. Soc.* **2003**, *125*, 14982–14983.
335. J. M. Thomas, R. Raja, *Acc. Chem. Res.* **2008**, *41*, 708–720.
336. S. Polarz, A. Kuschel, *Adv. Mater.* **2006**, *18*, 1206–1209.
337. S. Inagaki, S. Guan, Q. Yang, M. P. Kapoor, T. Shimada, *Chem. Commun.* **2008**, 202–204.
338. J. Morell, S. Chatterjee, P. J. Klar, D. Mauder, I. Shenderovich, F. Hoffmann, M. Fröba, *Chem. Eur. J.* **2008**, *14*, 5935–5940.
339. A. Ide, R. Voss, G. Scholz, G.A. Ozin, M. Antonietti, A. Thomas, *Chem. Mater.* **2007**, *19*, 2649–2657.
340. R. Voss, A. Thomas, M. Antonietti, G.A. Ozin, *J. Mater. Chem.* **2005**, *15*, 4010–4014.
341. A. Shimojima, K. Kuroda, *Angew. Chem. Int. Ed.* **2003**, *42*, 4057–4060.
342. A. Shimojima, Z. Liu, T. Ohsuna, O. Terasaki, K. Kuroda, *J. Am. Chem. Soc.* **2005**, *127*, 14108–14116.
343. Y. Fujimoto, A. Shimojima, K. Kuroda, *J. Mater. Chem.* **2006**, *16*, 986–994.
344. A. Shimojima, C.-W. Wu, K. Kuroda, *J. Mater. Chem.* **2007**, *17*, 658–663.
345. J. Suzuki, A. Shimojima, Y. Fujimoto, K. Kuroda, *Chem. Eur. J.* **2008**, *14*, 973–980.
346. A. Shimojima, K. Kuroda, *J. Sol-Gel Sci. Technol.* **2008**, *46*, 307–311.
347. C. H. Lee, S. S. Park, S. Joon Choe, D. H. Park, *Microporous Mesoporous Mater.* **2001**, *46*, 257–264.
348. S. S. Park, C. H. Lee, J. H. Cheon, S. J. Choe, D. H. Park, *Bull. Korean Chem. Soc.* **2001**, *22*, 948–952.
349. S. S. Park, C. H. Lee, J. H. Cheon, D. H. Park, *J. Mater. Chem.* **2001**, *11*, 3397–3403.
350. T. Martin, A. Galarneau, F. Di Renzo, D. Brunel, F. Fajula, *Chem. Mater.* **2004**, *16*, 1725–1731.
351. D.-J. Kim, J.-S. Chung, W.-S. Ahn, G.-W. Kang, W.-J. Cheongy, *Chem. Lett.* **2004**, *33*, 422–423.
352. V. Rebbin, M. Jakubowski, S. Pötz, M. Fröba, *Microporous Mesoporous Mater.* **2004**, *72*, 99–104.
353. V. Rebbin, R. Schmidt, M. Fröba, *Angew. Chem. Int. Ed.* **2006**, *45*, 5210–5214.
354. G. Zhu, Q. Yang, D. Jiang, J. Yang, L. Zhang, Y. Li, C. Li, *J. Chromatogr. A* **2006**, *1103*, 257–264.
355. M. J. E. C. van der Maarel, B. van der Veen, J. M. C. Uitdehaag, H. Leemhuis, L. Dijkhuizen, *J. Biotechnol.* **2002**, *94*, 137–155.
356. J. Rajendhran, P. Gunasekaran, *J. Biosci. Bioeng.* **2004**, *97*, 1–13.
357. M. Hartmann, *Chem. Mater.* **2005**, *17*, 4577–4593.
358. H. H. P. Yiu, P. A. Wright, *J. Mater. Chem.* **2005**, *15*, 3690–3700.
359. J. Kim, J. W. Grate, P. Wang, *Chem. Eng. Sci.* **2006**, *61*, 1017–1026.
360. H. H. P. Yiu, P. A. Wright, N. P. Botting, *Microporous Mesoporous Mater.* **2001**, *44–45*, 763–768.
361. H. H. P. Yiu, P. A. Wright, N. P. Botting, *J. Mol. Catal. B* **2001**, *15*, 81–92.
362. H. H. P. Yiu, C. H. Botting, N. P. Botting, P. A. Wright, *Phys. Chem. Chem. Phys.* **2001**, *3*, 2983–2985.
363. C. Lei, Y. Shin, J. Liu, E. J. Ackerman, *J. Am. Chem. Soc.* **2002**, *124*, 11242–11243.
364. H. Ma, J. He, D. G. Evans, X. Duan, *J. Mol. Catal. B* **2004**, *30*, 209–217.
365. A. Salis, D. Meloni, S. Ligas, M. F. Casula, M. Monduzzi, V. Solinas, E. Dumitriu, *Langmuir* **2005**, *21*, 5511–5516.
366. Y.-J. Han, G. D. Stucky, A. Butler, *J. Am. Chem. Soc.* **1999**, *121*, 9897–9898.
367. L. Washmon-Kriel, V. L. Jimenez, K. J. Balkus, Jr., *J. Mol. Catal. B* **2000**, *10*, 453–469.
368. H. Takahashi, B. Li, T. Sasaki, C. Miyazaki, T. Kajino, S. Inagaki, *Microporous Mesoporous Mater.* **2001**, *44–45*, 755–762.
369. A. Vinu, V. Murugesan, M. Hartmann, *J. Phys. Chem. B* **2004**, *108*, 7323–7330.
370. S. Z. Qiao, H. Djojoputro, Q. Hu, G. Q. Lu, *Progr. Solid State Chem.* **2006**, *34*, 249–256.
371. J. Aburto, M. Ayala, I. Bustos-Jaimes, C. Montiel, E. Térres, J. M. Domínguez, E. Torres, *Microporous Mesoporous Mater.* **2005**, *83*, 193–200.
372. M. Hartmann, C. Streb, *J. Porous Mater.* **2006**, *13*, 347–352.
373. S. Hudson, J. Cooney, B. K. Hodnett, E. Magner, *Chem. Mater.* **2007**, *19*, 2049–2055.
374. L. Zhang, S. Qiao, Y. Jin, Z. Chen, H. Gu, G. Q. Lu, *Adv. Mater.* **2008**, *20*, 805–809.
375. M. A. Wahab, I. Imae, Y. Kawakami, C.-S. Ha, *Chem. Mater.* **2005**, *17*, 2165–2174.

376. T. Asefa, M. Kruk, M. J. MacLachlan, N. Coombs, H. Grondey, M. Jaroniec, G. A. Ozin, *J. Am. Chem. Soc.* **2001**, *123*, 8520–8530.
377. M. C. Burleigh, M. A. Markowitz, M. S. Spector, B. P. Garber, *J. Phys. Chem. B* **2001**, *105*, 9935–9942.
378. L. Zhang, J. Liu, J. Yang, Q. Yang, C. Li, *Microporous Mesoporous Mater.* **2008**, *109*, 172–183.
379. D. Coutinho, C. Xiong, K. J. Balkus, Jr., *Microporous Mesoporous Mater.* **2008**, *108*, 86–94.
380. M. C. Burleigh, M. A. Markowitz, M. S. Spector, B. P. Garber, *Chem. Mater.* **2001**, *13*, 4760–4766.
381. M. C. Burleigh, S. Dai, E.W. Hagaman, J. S. Lin, *Chem. Mater.* **2001**, *13*, 2537–2546.
382. M. C. Burleigh, M. A. Markowitz, M. S. Spector, B. P. Garber, *Langmuir* **2001**, *17*, 7923–7928.
383. M. A. Wahab, I. Imae, Y. Kawakami, I. Kim, C.-S. Ha, *Microporous Mesoporous Mater.* **2006**, *92*, 201–211.
384. S. Hamoudi, S. Kaliaguine, *Microporous Mesoporous Mater.* **2003**, *59*, 195–204.
385. X. Yuan, H. I. Lee, J. W. Kim, J. E. Yie, J. M. Kim, *Chem. Lett.* **2003**, *32*, 650–651.
386. S. Hamoudi, S. Royer, S. Kaliaguine, *Microporous Mesoporous Mater.* **2004**, *71*, 17–25.
387. Q. Yang, J. Liu, J. Yang, L. Zhang, Z. Feng, J. Zhang, C. Li, *Microporous Mesoporous Mater.* **2005**, *77*, 257–264.
388. Q. Yang, J. Liu, J. Yang, M. P. Kapoor, S. Inagaki, C. Li, *J. Catal.* **2004**, *228*, 265–272.
389. M. Rat, M. H. Zahedi-Niaki, S. Kaliaguine, T. O. Do, *Microporous Mesoporous Mater.* **2008**, *112*, 26–31.
390. Q. Yang, M. P. Kapoor, S. Inagaki, *J. Am. Chem. Soc.* **2002**, *124*, 9694–9695.
391. M. P. Kapoor, Q. Yang, Y. Goto, S. Inagaki, *Chem. Lett.* **2003**, *32*, 914–915.
392. W.-H. Zhang, X. Zhang, L. Zhang, F. Schroeder, P. Harish, S. Hermes, J. Shi, R. A. Fischer, *J. Mater. Chem.* **2007**, *17*, 4320–4326.
393. M. A. Wahab, I. Kim, C.-S. Ha, *Microporous Mesoporous Mater.* **2004**, *69*, 19–27.
394. G. Zhu, Q. Yang, H. Zhong, D. Jiang, C. Li, *J. Phys. Chem. B* **2007**, *111*, 8027–8033.

Chapter 4

Modified Gold Nanoparticles and Surfaces

PAOLO PENGO AND LUCIA PASQUATO

4.1 INTRODUCTION

Surfaces are the boundaries of the "bulk world"; they are the regions of a material where a dramatic discontinuity exists with respect to all its chemical and physical properties. Such a macroscopic discontinuity affects the microscopic physical and chemical properties of the interface between two media. Naively, atoms or molecules at the surface of a material experience different environments with respect to others in the

The Supramolecular Chemistry of Organic–Inorganic Hybrid Materials. Edited by Knut Rurack and Ramón Martínez-Máñez

bulk; they have different chemical potential and therefore different thermodynamic properties. They can be seen as "unsaturated" with respect to intramolecular interactions and therefore they have their own chemical reactivity. All these considerations are at the basis of the peculiar properties of surfaces and interfaces, which have been considered the fourth state of matter.[1] A surface is the *locus* where certain interface properties such as friction, wettability, or certain optical properties become apparent, but these are neither intrinsic properties of the bulk material as such nor properties of its surface, they are just properties of the interface between two different interacting media. From a general point of view, a metal surface can be physically or chemically modified. As far as the second approach is concerned, a very large number of methodologies can be used, exploiting the "unsaturation" of surface localized atoms and molecules. However, there is a remarkable difference between the dynamics of reactions occurring in solution and those occurring on surfaces. A solution-phase reaction usually proceeds under diluted conditions, where reagents and products are free to move. In the case of a reaction at the interface, the reaction product is a surface-bound species occupying a given reaction site. Other reaction sites next to it will be similarly occupied by other molecules of the same reaction product. Thus, the reaction at the interfaces produces a dense system, ideally a two-dimensional lattice, a self-assembled monolayer (SAM) of closely packed and strongly interacting molecules, where a completely new set of phenomena such as cooperative and next-neighbor effects as well as multivalent interactions are expected to come into play. The modification of a surface with self-assembled monolayers introduces selectivity and specificity. In other words, SAMs allow the preparation of a surface capable of performing a function.

The considerations outlined above apply to SAMs on flat as well as on curved surfaces. From a historical point of view, the study of SAMs has been restricted for a long time to flat surfaces because of the lack of reliable methods for the preparation of species with curved surfaces in which curvature has the same size as the molecular object. This limitation has been overcome by the introduction of methods for the synthesis and functionalization of metal nanoparticles. Nowadays, the development of three-dimensional SAMs on metal nanoparticles represents a well-established and almost mature field of research in its own right. In fact, in the last years, a large number of books, reviews, and articles has been published in the literature focusing on gold nanoparticles (AuNPs) and, more specifically, on AuNPs passivated by organic compounds dubbed MPCs (from monolayer protected clusters). Many aspects of the chemistry, properties, and applications of SAMs on either flat or curved metal surfaces have been reviewed extensively.[1–5] The aim of this chapter is twofold: in the first section the properties of self-assembled monolayers on flat gold surfaces, the chemical procedures used for their functionalization, and some applications will be presented in a concise form. In the second section, the most recent synthetic methods for the preparation of hybrid organic-AuNPs as well as the properties of their monolayers and their metal cores will be discussed. In particular the preparation of monovalent and divalent AuNPs will be presented along with the preparation of AuNPs displaying on their surface ligands and biomolecules such as amino acids, peptides, glycosides, and aptamers, that have been used in the fabrication of new materials, smart responsive materials, or to mimic naturally evolved systems. The interaction of these nanoparticles with biomolecules in recognition processes will also be discussed briefly.

4.2 CHEMISTRY OF ORGANOSULFUR COMPOUNDS AT THE GOLD INTERFACE AND CHARACTERIZATION OF SAMs

Self-assembly of organosulfur compounds to form monolayers on metal surfaces is a spontaneous process. A survey of the literature concerning these SAMs on gold highlights that: (1) an oxidation state of -2 or -1 at the sulfur center is mandatory and (2) at least one of the bonds at the sulfur center should be relatively easy to break homolytically. Organosulfur compounds in which sulfur has a high oxidation state do not form SAMs on gold. The most common compounds used for SAM preparation on gold are thiols, disulfides,[1] thioethers, and thioesters.[6] Dithiocarbamates[7] have also been reported, though to a lesser extent. Thiols and disulfides fulfill both criteria outlined above, thioethers and thioesters fulfill only the first one. Dithiocarbamates fulfill the first criterion; moreover, their bidentate nature introduces some differences in the geometry of binding with respect to thiols and disulfides. However, according to a generally agreed rule, the stability of SAMs increases according to the series RSH > RSSR > RSR. A precise ranking is not easy to make because the stability of the SAM is not just related to the strength of the gold–sulfur interaction but also to interchain interactions that are not always trivial to estimate, especially when functionalized thiols are concerned. Though thiols have been the most common reactants used in the preparation of SAMs, the precise nature of the gold–sulfur bond and the topological arrangement of the sulfur head groups on the gold surface have been rather controversial. However, despite the lack of this important information, the synthetic procedures developed for SAM preparation are reproducible and reliable, indicating that the thiol–gold interaction gives robust systems. Thiols are believed to bind to the threefold hollow sites on the Au(111) surface,[8] the binding process entailing a splitting of the S—H bond, Figure 4.1.

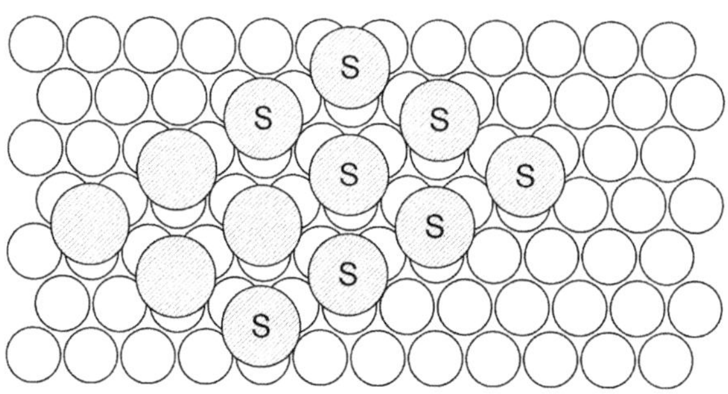

Figure 4.1 Schematic representation of the sulfur binding geometry on Au(111).

Many of the mechanistic details of thiol adsorption on gold surfaces have been obtained from gas–phase studies. The first step in SAM assembly is the formation of a physisorbed phase in which the thiols lay flat on the Au(111) surface (striped phase);[9] at this stage, the S—H bond is believed to remain intact.[10] The striped phase eventually evolves towards the formation of a compact monolayer of upright alkyl chains with a $(\sqrt{3} \times \sqrt{3})R30^\circ$ structure; the driving force for this transition is the establishment of strong interchain hydrophobic interactions. The S—H bond is split during this process. In this arrangement, the distance between the sulfur head-groups is 5 Å and in order to reach the closest packing of the 2D system, maximizing the van der Waals interactions, the chains tilt, forming an angle of 30° from the surface normal. The $(\sqrt{3} \times \sqrt{3})R30^\circ$ structure was recently investigated by Scoles and coworkers for SAMs of methylthiolate on Au(111). The authors reported that a dynamic equilibrium between bridge site adsorption and a novel structure exists where two CH_3S radicals are bound to a gold adatom that has been lifted from the gold substrate.[11] Recent findings by Scoles also suggest that in SAM formation, a competition between SAM ordering, driven by van der Waals interactions, and disordering of interfacial Au atoms, driven by the sulfur–gold interaction, come into play.[12]

Other intra-monolayer interactions such as hydrogen bonding contribute to define its structure.[13] It has been reported that lateral hydrogen bonding due to amide groups present in the thiols confers enhanced thermal stability to self-assembled monolayers.[14,15] The gold–thiol bond is strong, its dissociation energy is 20 to 40 kcal mol^{-1} and represents the highest contribution to the overall stability of the SAMs; additional stabilization derives from the van der Waals interactions between neighboring methylene groups, which amounts to 1.0 kcal mol^{-1} per methylene unit.[16] For SAMs formed by alkanethiols on gold, the typical surface density of molecules (when maximum coverage is obtained) is 4.5×10^{14} molecules cm^{-2} corresponding to about 150 pmol cm^{-2}. The thickness of alkanethiolate SAMs depends of course on the length of the thiol chain and ranges usually from 1 to 3 nm. The adsorption of thiols can be described as the oxidative addition of the S—H bond to the Au(111) surface followed by the reductive elimination of dihydrogen or, in the presence of dissolved oxygen, by the formation of water. The reverse process, that is, the desorption of thiol moieties from the gold surface, occurs with the formation of disulfides, corresponding to a process identical to the reductive elimination of a disulfide from the gold surface. This is a kinetically disfavored process, with an activation energy of about 30 kcal mol^{-1},[17] occurring at high temperatures.[18] As outlined above, disulfides may also be used for the formation of SAMs; in this case, the S—S bond is broken upon adsorption;[19] this process can be seen as the oxidative addition of the S—S bond to the gold surface. After adsorption and splitting of the S—S bond, the structure of the SAMs derived from disulfides should be identical to those formed by thiols. However, the former are reported to have a higher number of defects.[20] Dialkyl disulfides do not form stable self-assembled monolayers on gold because of the fairly weak sulfur–gold interaction observed in these cases. It is likely that the lack of a mechanistic pathway to the formation of a chemisorbed phase in this case limits the gold–sulfur interaction to a simple and unstable physisorption. However, this limitation has been overcome using multidentate thioethers with complex topologies allowing multiple contact interactions.[21]

The characterization of SAMs is generally accomplished exploiting spectroscopic methods and electron or force microscopy. Techniques such as reflection absorption infrared spectroscopy (RAIRS),[22,23] X-ray photoelectron spectroscopy (XPS),[24,25] ellipsometry,[23,26,27] surface plasmon resonance spectroscopy (SPR),[23] Raman spectroscopy,[28] near-edge X-ray absorption fine structure spectroscopy (NEXAFS),[29] high resolution electron energy loss spectroscopy (HREELS),[30] scanning tunneling microscopy (STM) or atomic force microscopy (AFM), and cyclic voltammetry (CV) or contact angle goniometry are commonly used. On the other hand, the composition of mixed monolayers and functionalized SAMs is increasingly being analyzed by mass spectrometric methods such as MALDI-TOF and TOF SIMS[31] and recently the new direct analysis in real time ionization technique (DART).[32]

4.2.1 Preparation of Homogeneous and Mixed SAMs

The self-assembly of organosulfur compounds on gold surfaces occurs spontaneously and rapidly. Thiols are the species of choice for the preparation of SAMs on gold surfaces because of the relatively simple experimental conditions required and the stability of the resulting monolayers. The most common procedure consists of immersing the gold substrates in deoxygenated diluted solutions of thiol. The procedure is robust and has been used to prepare SAMs from thiols of very diverse structure, such as alkanethiols, aromatic thiols, mercaptobiphenyls, PEGylated thiols [PEG = poly(ethylene glycol)], and *p*-terphenylthiols.[1,33–35]

With the advent of soft lithographic techniques such as scanning probe lithograhy and microcontact printing, the possibility to "write" on gold substrates using thiols solutions as "inks" opened the route to the generation of well-defined features and patterns of nanoscopic and microscopic size on gold surfaces.

Another interesting technique to pattern gold surfaces is ink-jet printing, because it allows the preparation of very large surfaces of SAMs.[36] The reaction conditions devised for the solution-phase adsorption of organosulfur compounds are essentially the same regardless of the precise chemical nature of the ligands. SAMs of thiols can be prepared by immersing a clean gold substrate in a millimolar solution (1 to 10 mM) of thiols for 1 to 20 hours at room temperature. The reaction is fast, but to allow the formation of a densely packed monolayer with a reduced number of defects, prolonged contact with the solution is usually recommended. This is because even though the association of thiols with the metal surface is fast, requiring milliseconds to minutes, the reorganization of a large system such as the forming monolayer proceeds slowly and may require days. Electrochemistry,[37] STM,[38] and RAIRS[39] indicate, however, that the structure of the SAM can continue to evolve over immersion times of 7 to 10 days. The main factors affecting the quality of SAMs and their formation rate are ligand concentration, temperature, presence of dissolved oxygen, and purity of the starting material. Of course, the formation of SAMs is an associative process, and therefore it is favored at high concentration of thiols. In these conditions the reaction occurs in a short time, while longer reaction times are required when less concentrated solutions are used. The presence of dissolved oxygen in the thiol solution leads to oxidation, with a decrease in the concentration of the reacting species. Besides this consideration based on the known chemistry of thiols, there are few studies and little rationale on the effect of oxygen on SAM formation. The choice of the solvent is sometimes limited by the solubility of the thiols used for the preparation of SAMs. From a general point of view, the optimal solvent should be able to increase the chemical potential of the unreacted thiol and to stabilize, as much as possible, the adsorbed state and the reaction product (SAM), allowing at the same time a correct packing of the monolayer. This would decrease the activation energy for the reaction and, at the same time, increase the ΔG of reaction. Such a solvent should interact weakly with the thiol, while it should solvate the forming monolayer without hampering its growth. Small polar organic solvents such as ethanol, tetrahydrofuran (THF), and acetonitrile seem to be well suited to solubilize the thiols, while they would be less useful to solvate a rather hydrophobic construct such as the SAM. On the other hand, more hydrophobic solvents will interact strongly with the growing SAM, hampering the addition of new monomers. The experimental results infer that polar organic solvents such as ethanol, THF, acetonitrile, and *N,N*-dimethylformamide (DMF) are well suited for the preparation of stable SAMs, while hydrocarbon solvents such as hexane or heptane give less organized SAMs as confirmed by contact angle measurements and electrochemistry,[40,41] even though the adsorption kinetics are faster in these solvents. The explanation for this behavior is that hydrophobic solvents interact strongly with the monolayer, hampering its reorganization.

A simple procedure for the development of SAMs with embedded functional groups is the coadsorption of mixtures of functional and nonfunctional thiols. In principle, the procedure allows dilution of the functional thiol to obtain surfaces with different reactivity. This procedure is useful when a given molecule unable to form stable or homogeneous SAMs needs to be supported on a surface. The coadsorption with another molecule known to produce stable monolayers gives a stable ensemble, with the first molecule diluted by the second. The thermodynamics of SAM formation from a single chemical species is relatively simple to describe, because the resulting SAM is homogeneous in its composition; but the analysis of the thermodynamics underlying the introduction of two or more different species in a monolayer is rather complex. When more than one thiol is used in the preparation of SAMs, two possible scenarios may be present: in the first case, thiols may be evenly distributed and in the second limiting case, thiols may self-sort in different regions forming patches on the gold surface, giving rise to phase separation. What drives the thermodynamics of the system towards the first or the second equilibrium state is the strength of the next-neighbor interactions in the monolayer. When next-neighbor interactions (NN) are weak or there is no difference between pairs of NN interactions established between the same or different thiols, the molecules will be evenly distributed in the SAM, but if a favorable NN interaction exists, self-sorting will occur to a certain extent. A compelling example of how NN interactions may produce surprisingly well-defined structures was recently reported by Pace et al.[42]

The authors reported the first evidence of completely miscible bicomponent domains formed on metal surfaces, where alternating stripes of two different molecules form well-defined nanodomains.

Since a SAM is a densely packed system where cooperative phenomena may occur, the overall strength of the next-neighbor interactions is difficult to foresee. For this reason, it is not possible to ensure that a feeding solution made of two, or more, thiols in any well-defined molar ratio will produce mixed monolayers with the same composition. The experimental parameters and conditions may be changed to a certain extent to tune the composition of the resulting monolayer. For instance, the solvent may be chosen to favor a polar thiol over an apolar one.[40,43,44]

Other strategies for the preparation of mixed SAMs are the reactions of gold surfaces with mixed disulfides or adsorption of an asymmetric dialkylthioether. Some of the considerations already made for thiols can be applied to disulfides as well. Since the formation of SAMs is a reversible process, leading to the most stable system, the use of a mixed disulfide for the formation of a mixed SAM does not guarantee a 1 : 1 molar ratio of the two moieties in the monolayer, which may indeed contain an excess of one among the two. There are two other problems limiting the use of disulfides in the preparation of SAMs; in the first place, the solubility of disulfides is usually lower than the solubility of the corresponding thiols. Second, disulfides tend to generate SAMs with a higher number of defects compared to thiols.[21] Even though the use of an asymmetric thioether is theoretically possible, the stability of the resulting SAMs is too low for practical applications and is limited to multivalent thioethers.[4]

4.2.2 Postsynthetic Modification of SAMs

The presence of sulfur limits the range of functional groups that can be introduced in SAMs by simple adsorption of thiols. For instance, strongly electrophilic functional groups are not compatible with the presence of the sulfhydryl group. In addition, relatively few functionalized thiols are commercially available and the synthetic routes to more sophisticated ones may not be trivial even for relatively simple molecules. A solid-phase approach to the synthesis of functionalized alkanethiols has been recently proposed by Kiessling and coworkers,[45] providing a great simplification of the purification steps required in the preparation of functionalized thiols. The problem of generating a reasonable diversity in SAMs may be partially overcome by their postsynthetic functionalization.

From a general point of view, there are two reactive centers in a SAM on gold. The first is the gold–sulfur bond and the second may be any other functional group of the thiols amenable to chemical modification and conveniently found on the SAM's outer surface. Reactivity at the gold–sulfur bond allows modification or functionalization of a preformed SAM by place exchange simply immersing the SAM in a solution of a second thiol. Since the formation of SAMs is reversible, the new thiol will start to displace the first from the most reactive sites such as defect sites and grain boundaries (Fig. 4.2). The reaction proceeds rapidly (minutes to hours) at those defect sites, but requires longer times (days) at compact crystalline regions. However, in general, the

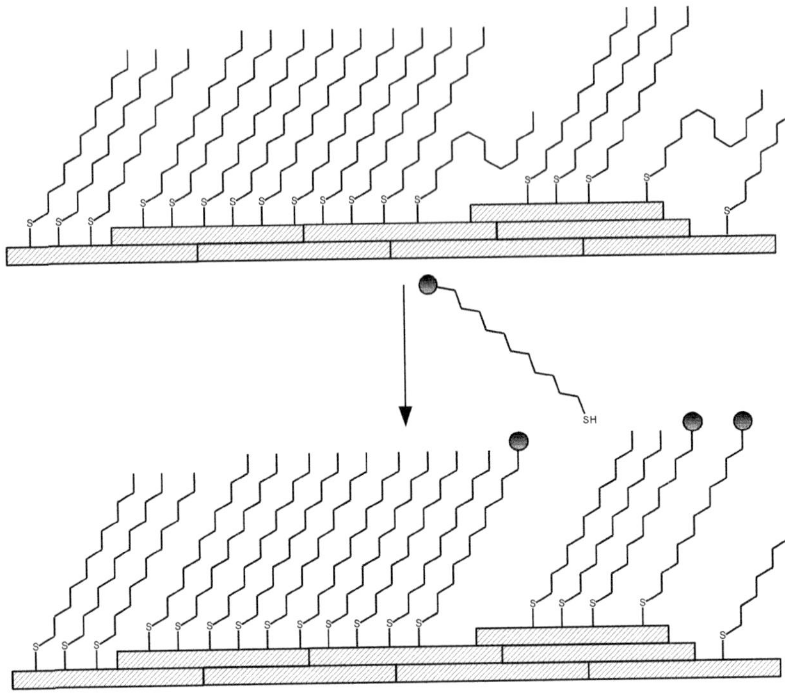

Figure 4.2 Exchange of thiols on preformed SAMs.

process does not provide homogeneous SAMs. The reaction has been studied with the standard techniques used for the characterization of SAMs, such as electrochemistry,[46] contact angle goniometry,[47] and STM.[48] Since short chain thiols are displaced more readily than long chain ones, the exchange reaction may be used to selectively displace a single component from a mixed monolayer or a patterned SAM.

The second route to postsynthetic modification of SAMs is the chemical transformation of functional groups present on their outermost surface; this approach mostly relies on chemistries already established for the functionalization of solid supports (Fig. 4.3). Two important points to bear in mind are: (1) it is essentially impossible to extensively characterize the structure of the reaction products or purify them without destroying the SAM and (2) many solution-phase reactions may be very difficult when carried out on a surface because of the steric hindrance due to the very closely packed end groups.

Clearly, these considerations imply that the preparation of functionalized SAMs via postsynthetic procedures requires reactions proceeding under mild conditions compatible with the structure and the integrity of the self-assembled monolayer. In addition, these reactions should, ideally, proceed quantitatively; these considerations actually limit the reactions commonly used for SAM functionalization to a handful. In fact, the number of reactions used for postsynthetic modifications of SAMs is relatively small, and is essentially limited to acylation reactions (either amidations

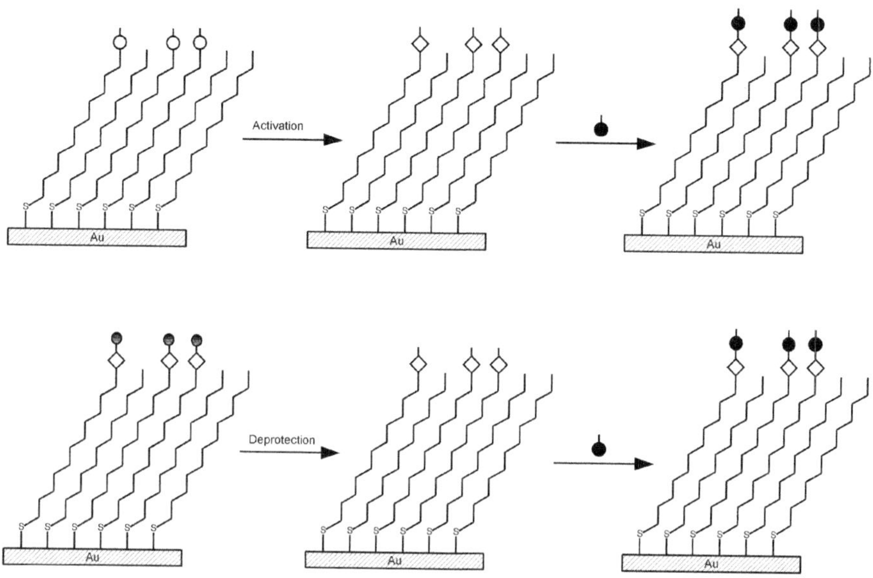

Figure 4.3 Sequential synthesis of functionalized SAMs.

or esterifications), Diels–Alder cycloadditions,[49] 1,3-dipolar cycloadditions (Huisgen reactions), and ruthenium-catalyzed alkene cross metatheses (Table 4.1). Other procedures such as the reaction between maleimide groups with thiols, the formation of imines and their reduction to amines with sodium cyanoborohydride,[50] and the formation of hydrazones[51] have also been reported. Mrksich and coworkers[52] reported a series of 16 chemical reactions for the functionalization of SAMs on gold. The approach takes advantage of a MALDI-TOF method for the rapid characterization of reaction products and for the screening and optimization of reaction conditions. According to Mrksich,[53] mass spectrometry represents a powerful tool for the characterization of reaction products obtained on SAMs.

Another approach to modify the surface of a SAM is the *in situ* formation of a reactive intermediate such as an active ester, an electrophile, or a good leaving group to be further reacted (Table 4.2). A relatively straightforward procedure for SAM functionalization, which is based on the availability of carboxylate-functionalized thiols, is the formation of amides or esters on the SAM surface via activated esters. Other methods such as the generation of acyl chlorides, by action of gaseous thionyl chloride,[54] and the generation of acyl fluorides, by action of cyanuric fluoride and pyridine, on carboxyl-terminated SAMs have been reported.[55] The preparation of acyl fluorides activated SAMs has also been reported by Rotello and coworkers[56] by direct adsorption of acyl fluoride-terminated disulfides on gold.

Carboxylic acids on the outer surface of SAMs may also be activated as symmetric anhydrides by action of trifluoroacetic anhydride as dehydrating agent.[57] The most common active esters are *N*-hydroxysuccinimidyl esters (NHS esters). They are usually prepared by reacting the carboxyl-terminated SAMs with

Table 4.1 Selected Methods for Direct Functionalization of SAMs

Surface bound reactive group	Reagent	Product	Reference
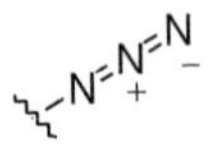	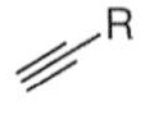	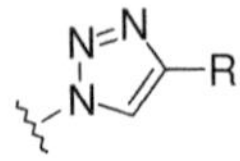	65–68
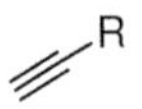	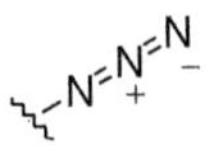	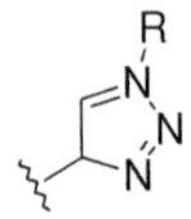	70
	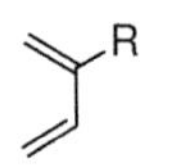	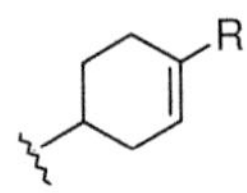	49
	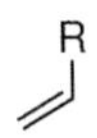	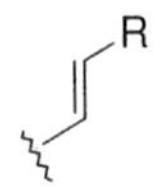	76
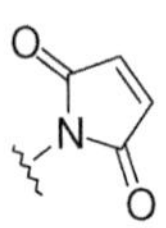	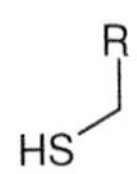	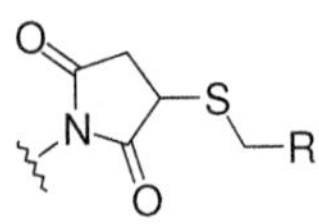	74
	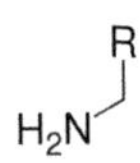	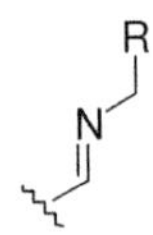	50
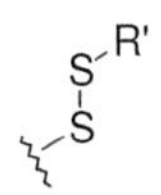	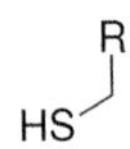	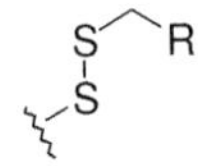	73
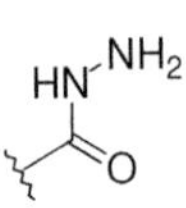	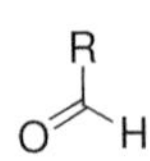	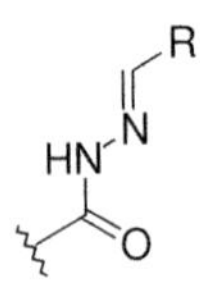	51

N-hydroxysuccinimide in the presence of a soluble coupling agent. Besides NHS esters, other active esters used are pentafluorophenyl esters,[58] and the recently reported tetrafluorophenyl esters.[59] The hydrolytic stability of NHS esters embedded in oligoethylene thin films have been studied using XPS and time-of-flight secondary ion mass spectrometry (TOF SIMS).[60] On standing at room temperature in air for

Table 4.2 Selected Methods for SAMs Functionalization via the Formation of Activated Species

Surface bound reactive group	Activated species	Reagent	Product	Reference
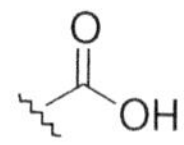	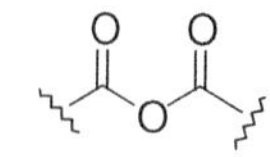	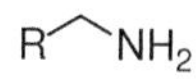	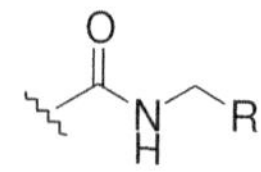	57
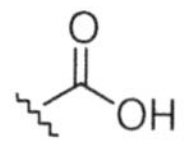	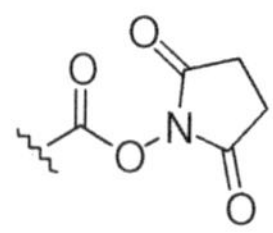	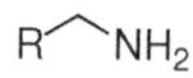	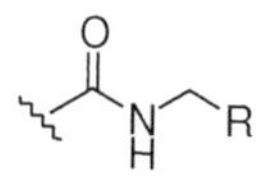	72
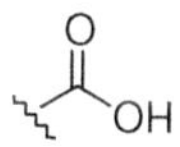	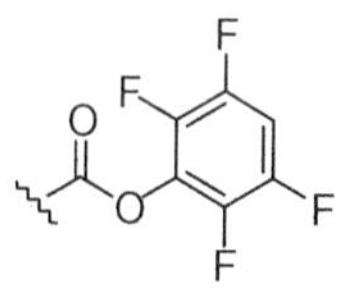	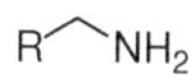	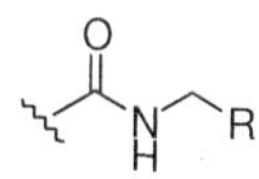	59
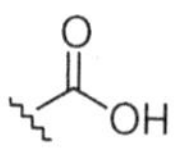	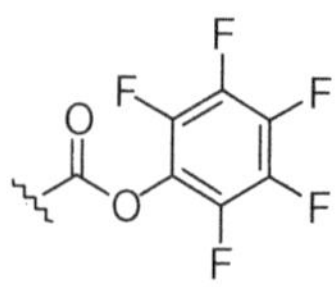	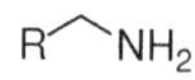	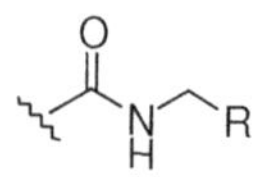	58
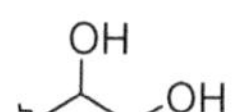	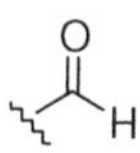	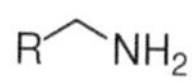	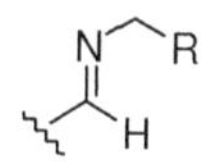	61–63
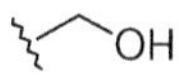	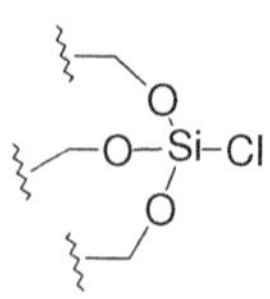	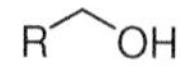	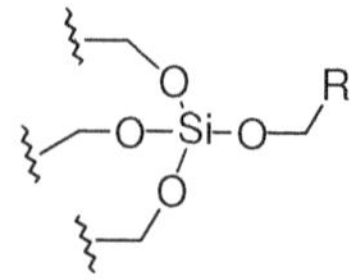	64
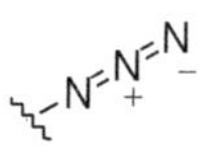	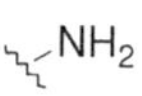	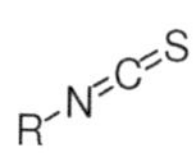	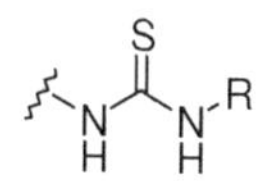	71, 78

7 days, some hydrolysis as well as some thiol oxidation and deposition of adventitious material were observed. The same authors reported overnight hydrolysis of these active esters in presence of bulk water.

The use of tetrafluorophenyl esters[59] allowed the preparation of "activated" monolayers with enhanced hydrolytic stability at basic pH when compared to NHS activation. Moreover, the reaction of the tetrafluorophenylester-functionalized monolayers with DNA molecules resulted in a fivefold greater surface density of DNA molecules compared to that obtained exploiting the NHS chemistry. Ducker and coworkers,[61] taking advantage of SAM surface activation, reported monolayers bearing "masked" aldehyde groups in the form of vicinal diols. The same chemistry has been recently exploited by Hahn et al.[62] and Hölzl et al.[63] for the immobilization of proteins on SAMs.

The preparation of an aldehyde-SAM was achieved by treating the vicinal diol-SAM with sodium periodate. The preparation of an activated tripodal structure on a hydroxyl terminated SAM has been published by Ichimura and coworkers.[64] These authors reported the reaction of SAMs formed by mercapto-alkanols such as 2-mercaptoethanol, 6-mercaptohexanol, and 16-mercaptohexadecanol with $SiCl_4$, with the formation of a tripodal derivative bearing a terminal Si—Cl bond amenable to react with methanol or water.

Functionalization of azide-terminated SAMs exploiting the formation of triazoles (Huisgen reaction) is an increasingly pursued procedure because of the rather simple approach and the mild conditions required.[65] The reaction proceeds in aqueous or aqueous-alcoholic solution in the presence of a freshly produced Cu^I catalyst. The reaction is usually allowed to proceed at room temperature in the dark for some hours. Using time resolved electrochemical techniques, Collman et al.[66] analyzed the kinetics of the reaction between a ferrocene-functionalized alkyne with azido-SAMs on gold electrodes. The authors observed well-behaved second-order reactions, with a second-order rate constant of $1 \times 10^3\ M^{-1}\,s^{-1}$. By exploiting the Huisgen reaction, Devaraj et al.[67] reported also the functionalization of mixed monolayers comprising azidothiols and alkanethiols with alkynyl-functionalized oligonucleotides. The same group described an interesting application of the Cu^I catalysed 1,3-dipolar cycloaddition: by electrochemical switching of the redox state of the copper catalyst from (II) to (I), they were able to control the triazole formation between surface immobilized azides and ethynylferrocene in electrochemically addressable multielectrode arrays.[68] The 1,3-dipolar cycloaddition between azide-functionalized SAMs and alkynes has also been used to immobilize bulky supramolecular structures such as hemoprotein models on gold surfaces.[69] Another approach in the functionalization of SAMs by the Huisgen reaction is the reaction of alkynyl-functionalized SAMs with molecules bearing azido groups. As an example, Zhang et al.[70] reported the functionalization of SAMs with mannose, lactose, and α-Gal to study specific protein–carbohydrate interactions by quartz crystal microbalance (QCM) and SPR.

Besides the reactivity in Huisgen reactions, aryl azide-functionalized SAMs have been photocatalytically reduced to aryl amine-functionalized SAMs using 2 nm CdS quantum dots as photocatalyst.[71] The resulting mixed monolayer has been characterized by X-ray photoelectron spectroscopy (XPS), contact angle measurement, and grazing angle FTIR. The amino groups on the mixed monolayer surface were further reacted with fluorescein isothiocyanate, thus confirming the successful conversion of the azido to amino group.

By exploiting the NHS activation methods, Friedrich et al.[72] reported the functionalization of SAMs on gold with *N*-(5-amino-1-carboxypentyl) iminodiacetic acid, an analog of nitrilotriacetic acid, and used the resulting functionalized surface to immobilize histidine-tagged cytochrome oxidase on gold electrodes, studying the activity of the protein as a function of the packing density.

Disulfide exchange[73] on the outer surface of SAMs has been exploited as a possible route to SAM functionalization provided that the reaction does not involve thiol displacement from the gold surface. This has been found to be the case for bulky thiols such as thiol-modified DNA, peptides, and carbohydrates; it has been reported, in fact, that disulfide exchange occurs more rapidly than thiol displacement from the gold surface. Corn and coworkers[73] reported the introduction of masked sulfhydryl groups on the outer surface of SAMs by reaction of *N*-succinimidyl *S*-acetylthiopropionate (SATP) with a preformed monolayer of 11-mercaptoundecylamine. The sulfhydryl groups could be unmasked by action of diluted alkaline solution. The resulting sulfhydryl surface was reacted with 2,2′-bipyridyl disulfide to form a mixed disulfide amenable to react with thiols in a disulfide exchange reaction. The authors reported this chemistry for the detection of *in situ* hybridization onto DNA arrays. The same group used this chemistry for the immobilization of peptides[73] and thiol-modified carbohydrates to fabricate carbohydrate arrays.[73] Mrksich and coworkers reported the functionalization of SAMs using peptides and carbohydrates bearing a free thiol and maleimide-functionalized SAMs.[74]

A different approach for grafting thiol-modified molecules on SAMs was reported by Kane and coworkers.[75] These authors prepared SAMs of alkanethiols presenting chloroacetylated hexa(ethyleneglycol) groups. The chloroacetyl groups were reacted, under mild conditions, with biologically active molecules such as thiol-functionalized biotin, benzenesulfonamide, and the well-known RGD peptide sequence for the recognition of integrins. The authors reported selective recognition of neutravidin, carbonic anhydrase, and integrin-mediated adhesion of fibroblasts on their functionalized SAMs using the three aforementioned ligands.

The functionalization of SAMs via ruthenium-catalyzed cross metathesis of vinyl-terminated SAMs has been reported by Lee et al.[76] to install a variety of acrylic derivatives on SAMs bearing vinyl groups on their outer surface. The major drawback of this approach is the intra-SAM metathesis which causes the formation of a mixture of surface-bound products, limiting the reproducibility of the method. The formation of urethanes by the reaction of diisocyanates[77] or isothiocyanates[78] with hydroxyl- and amino-terminated SAMs has been reported as well. The reaction of hydroxyl-terminated SAMs with diisocyanates, allowed the preparation of isocyanate SAMs that proved to be reactive towards amines, alcohols, and water, displaying the standard chemistry of the isocyanate groups.[77]

A major advantage of the postsynthetic modification of SAMs, with respect to the exchange reaction, is the very small scale of the reactions. The surface coverage of gold in a typical SAM is in the range of 10^{14} molecules cm^{-2} corresponding to hundreds of pmol cm^{-2}. The amount of reactants or species to be immobilized on the functionalized SAMs is of the same order of magnitude. Among the disadvantages it must be stressed that the organization of the functional groups on the surface remains generally unknown.

4.2.3 Nonfouling Surfaces Based on SAMs and Their Use in Biomolecular Recognition

Hydrophobic SAMs are known to give nonspecific interactions with biomolecules and cells. These properties, along with the possibility to pattern gold surfaces by means of lithographic techniques, allow the addressable immobilization of cells even in the absence of specific recognition moieties. However, for certain purposes, when the development of systems for selective recognition is pursued, the surfaces must display a certain degree of inertness. Among SAMs, those presenting oligo(ethylene oxide) and poly(ethylene oxide) units on the outer surface are increasingly being used to study biomimetic and biological interactions because of their property to minimize nonspecific interactions. From a general point of view, the ability of a surface to resist nonspecific adsorption correlates with its polarity, overall charge, presence of hydrogen bond donors, and conformational flexibility.[79] The importance of oligo(ethylene glycol)-terminated SAMs is witnessed by the large number of studies concerning their structural characterization. Liedberg and coworkers[80] reported a temperature-programmed infrared reflection absorption spectroscopy study on hexa(ethylene glycol) (EG6)- and tetra(ethylene glycol) (EG4)-terminated alkanethiols assembled on polycrystalline gold. At 60°C, the authors observed a phase transition from the helical to the all-trans (zigzag) conformation for the polyether moiety of EG6. EG4 did not show phase transition, and the conformation of the polyether moiety was marginally affected by heating the SAM to 75°C. The observed conformational changes were reversible on returning to 20°C. The conformation of polyether-terminated SAMs enforced by the presence of intramonolayer hydrogen bond donors and acceptors has been studied from the computational point of view by Liedberg and coworkers.[81]

Other nonfouling surfaces different from poly(ethylene oxide)-functionalized alkanethiols have been proposed. Hyperbranched polyglycerols have been reported to have superior nonfouling properties compared to standard poly(ethylene oxide)-functionalized alkanethiols.[82] On the other hand, completely different systems such as zwitterionic SAMs showed promise as nonbiofouling surfaces. Jiang and coworkers[83] reported a mixed self-assembled monolayer comprising an equimolar mixture of positively and negatively charged thiols. The authors infer that the tightly bound water molecules on the outermost part of the resulting SAM are responsible for the resistance to nonspecific protein adsorption. Exploiting a similar approach, the same group reported surfaces based on the polymeric poly(carboxybetaine methacrylate) grafted on gold with excellent nonfouling properties.[84]

Of course, a poly(ethylene oxide)-functionalized SAM may be further functionalized with suitable groups to ensure protein grafting on the SAM surface. The best choice for these purposes are active esters; functionalization of oligo(ethylene oxide) SAMs with biotin[85] has also been reported. The adsorption of poly(ethylene oxide)-functionalized disulfides on gold surfaces has been proven to give SAMs with protein-resistant surfaces comparable to those obtained from oligo(ethylene oxide)-functionalized thiols.[86]

The stability of poly(ethylene oxide)-functionalized SAMs under different storage conditions such as in air, under nitrogen, water, phosphate-buffered saline, and ethanol have been studied as a function of time[87] by cyclic voltammetry, contact angle measurements, grazing angle FTIR and XPS, and by measuring the SPR performance of the resulting biosensor. The authors report the SAMs to be very stable when stored in air or under nitrogen, while storage in ethanol resulted in loss of recognition properties.

The development of recombinant proteins bearing histidine tags gives a facile entry to the introduction of these molecules in SAMs. The chemistry involved is very simple and the well-known ligand for histidine-tagged proteins, the complex of Ni^{2+} with nitrilotriacetic acid, has been embedded into SAMs, allowing straightforward noncovalent protein grafting onto surfaces. The coadsorption of alkylthiols containing multivalent chelators based on mono-, bis-, and tris-nitriloacetic acid and tri(ethylene glycol)-terminated alkanethiols has also been described.[88] The formation of SAMs was studied at different compositions of the feeding solution by contact angle goniometry and ellipsometry. The authors reported the formation of a compact and ordered monolayer when the concentration of the multivalent chelator-functionalized thiols did not exceed 30 mol% in the loading solution. At higher molar fractions, the resulting monolayer displayed a significant degree of disorder.

Castner and coworkers[89] obtained mixed SAMs bearing nitrilotriacetic and oligo(ethylene glycol) moieties by sequential adsorption of the two thiols rather than coadsorption. The monolayer composition was characterized by XPS, NEXAFS, SPR, and TOF SIMS. The nitrilotriacetic acid-functionalized thiol was adsorbed first, forming an incomplete monolayer; the oligo(ethylene glycol)-functionalized thiol was then incorporated until a complete monolayer was formed. The monolayer thus formed was used to bind His-tagged humanized anti-lysozyme variable fragments in the presence of Ni^{2+} ions.

A relevant application of alkanethiols assembled monolayers is the functionalization of optical biosensors based on the excitation of surface plasmons, SPR. Conventional SPR measurements as applied in biosensors are used to monitor the changes in thickness or refractive index of ultrathin organic films at metal surfaces, usually silver or gold. The change in refractive index or thickness occurs as the result of a specific binding interaction between the analyte and a suitable receptor or ligand. Theoretically, the biosensor surface is supposed to be specific for the analyte and to give very low or undetectable background. The first property is easily obtained exploiting the postsynthetic modification of SAMs. In closer details, a receptor or ligand known to interact with the analyte is immobilized on the metal surface, which nonetheless needs to be nonreactive towards undesired species. These issues are rather difficult to address, since the range of analytes assayed by SPR is very large, ranging from the detection of pathogens, toxins, veterinary drug residues present in processed food, and allergens to proteins in dairy products.[90] Since the receptors usually are biomolecules such as antibodies or peptides, the most common approach of conjugation consists in the formation of amide bonds between the protein and an activated carboxylic acid installed at the SAM surface. By exploiting this technique, a large number of sensing devices for food quality and safety analysis has been designed and validated. Another important field of application is medical diagnostics, where modified SAMs are used to immobilize receptors for relevant proteins, including the detection of cancer markers such as the prostate specific antigen (PSA) complexed with antichymotrypsin[91] or the cancer antigen CA 19-9.[92]

4.3 HYBRID ORGANIC-GOLD NANOPARTICLES

The most intensively studied nanosized materials are AuNPs such as colloids and nanocrystals that typically are 1 to 50 nm in diameter. These nanoparticles find useful applications in electronics, energy storage, for the development of tags or staining materials for the analysis of biological samples by electron and optical microscopic techniques, and as carriers of biomolecules or as nanosized drug-delivery systems. Nanoparticles may also act as catalysts, mainly in oxidation processes, or as seeds for the generation of more sophisticated nanostructures such as nanowires of semiconducting materials.[93,94]

These systems display unusual and unique size-dependent chemical and physical properties, and one of the striking differences between nanoparticles and the corresponding bulk materials is the percentage of atoms that are interfacial.[93,95] Nonprotected, "naked" clusters possess surface atoms that are coordinatively highly unsaturated, interacting only with atoms inside the particle and having free valences outside. Their contribution to the electronic structure of the particle is different from that of fully coordinated inner atoms. We need to distinguish between: (1) the total number of metal atoms forming the cluster and (2) those atoms forming its inner part and having only other metal atoms as neighbors. Of course, the percentage of interfacial atoms with respect to the total number of atoms depends on the nanoparticle size. For example, a densely packed cluster of 13 atoms has 12 surface atoms (92.3%), and one single central atom. A two-shell cluster consisting of 55 atoms has a metallic core of 13 atoms, whereas the outer shell consists of 42 atoms (76.4%). A three-shell cluster has 92 interfacial atoms (62.6%) and 147 total atoms. A 2.0 nm nanoparticle has 58% surface atoms and a 5.0 nm cluster presents only 23% of the total atoms on the surface.

The surface of AuNPs may be passivated by organic ligands, forming a self-assembled monolayer on a curved surface.[1,96–98] The structure of these SAMs depends heavily on the curvature and defect rate of a given surface. SAMs on nanoparticles simultaneously stabilize the reactive metal and present organic functional groups at the particle–solvent interface. Tailored organic surfaces on AuNPs are useful for applications in nanotechnology, in different fields of material sciences, chemistry, biology, and medicine.

4.3.1 Technological Relevance

It is important to stress that protection of the metal surface is of paramount importance to enable the facile manipulation of these systems, thus allowing the analysis of individual particles and their assembly in two- and three-dimensional structures without the onset of uncontrolled aggregation.

For instance, functional AuNPs have been implemented into powerful analytical tools[94,99–101] for nucleic acids,[102] proteins,[103] and bacteria,[104] as well as into intracellular probes and gene regulators.[105] Enzyme biosensor devices based on pepetide-AuNP conjugates have been conceived in the last years; a critical discussion can be found in a recent report by Ghadiali and Stevens.[106] Moreover, recent work shows that DNA–AuNP conjugates might serve as basic building blocks to assemble highly ordered macroscopic materials through programmable base-pairing interactions.[107,108]

Schmid was a pioneer in understanding the potential applications of ligand-stabilized colloids and clusters.[109] Together with his collaborators he showed that single particles studies,[110] quantum dot solids,[111] and even nanoelectronic devices were conceivable applications of his Au_{55} clusters.[112] Ligand-stabilized AuNPs have attracted much attention in the quest for new compounds that are potentially useful in future information technology based on single-electron devices.[113]

The extraordinary catalytic properties of gold at the nanometer scale are now well documented.[114–119] In particular AuNPs supported on metal oxides are excellent catalysts for a variety of reactions as, for example, CO oxidation (the reaction most extensively studied), oxidation of propene to propene oxide, NO_x reduction, selective hydrogenation of acetylene, and the water-gas shift reaction. The smaller the size of the NPs the higher the activity of the catalyst to be expected because small AuNPs have a larger percentage of surface atoms and accordingly, a larger percentage of Au atoms is in contact with the support. In the preparation of these catalysts, a small size and an improved monodispersity is achieved by using preformed AuNPs protected by strongly interacting ligands.[120] Therefore, to obtain catalytic activity the ligands must be removed postdeposition.

The optical properties of AuNPs of different size and shape have been studied,[121–123] as well as those of materials based on chromophore-gold nanoparticle conjugates.[124]

The development of simple strategies to produce robust AuNPs that are stable in biological media is an active field of research. An up-to-date discussion of the available bioconjugation strategies involved in interfacing inorganic nanoparticles with biological systems was recently provided by Rotello and coworkers.[125] Interesting examples of this approach are the single-particle coding and tracking experiments *in vitro*,[126] in cells,[127] and *in vivo*.[128] This goal can be achieved, for example, by encapsulation of AuNPs or AuNP bioconjugates within a shell of photo-cross-linked block copolymer surfactant, as reported by Chen et al.[129] This strategy of encapsulation dramatically improves the stability of the particle against heat, etchants, cosolvents, and buffers. Another example is the innovative encapsultation of AuNP conjugates inside a virus capsid with the aim to build new drug delivery vectors.[130]

4.3.2 Synthesis of Spherical Gold Nanoparticles

Hybrid organic-gold nanoparticles may be obtained essentially in two ways, that is, by formation of an organic layer around preformed AuNPs or colloids, and by synthesis of AuNPs in the presence of organic ligands such as thiols, amines, or phosphines.

Spherical gold colloids of different sizes are commercially available in dilute aqueous solutions or may be prepared following standard protocols using different reducing agents such as citric acid,[131–133] ascorbic acid,[123] sodium borohydride,[134] and alcohols.[135–138] The resulting nanoobjects are weakly stabilized by electrostatic interactions and may be further reacted with a large range of ligands.[139–144]

As far as the second approach is concerned, that is, the reduction of a metal precursor (usually tetrachloroauric acid) in the presence of ligands (phosphines, amines, and thiols) or surfactants, a further distinction based on the nature of the specific interaction between gold and the organic additive has to be made: whereas ionic surfactants interact weakly with the gold surface mainly via electrostatic interactions, organic ligands are chemisorbed on the gold surface. The adsorption of surfactants to nucleated nanocrystals lowers the free energy of the surface and, therefore, the reactivity of the particles. The ratio of surfactant to metal precursor can control the size distribution of the nanoparticles. The steric hindrance of the surfactants provides a physical barrier that prevents the metal surfaces from contacting each other directly.

During the formation of nanoparticles, surfactants interact with the surfaces of the particles in a dynamic equilibrium process. On the contrary, ligands that are chemisorbed to the surface of the particle are less prone to desorption compared to physisorbed ones. A first consequence of a low desorption rate is that particles do not grow rapidly after nucleation. As an example, when thiols are chemisorbed to AuNPs particle sizes are typically limited to <5 nm. In contrast, the majority of surfactants used to stabilize nanoparticles associate with the surface of particles through van der Waals or other weak interactions, allowing the preparation of larger particles. The first report of thiol-stabilized AuNPs appeared in 1993 by Giersig and Mulvaney.[145]

A simple method that gave access to large amounts of material and that is still largely used is the synthesis of AuNPs protected by an organic monolayer, reported in 1994 by Brust et al.[134] The authors combined the chemistry already established for the preparation of self-assembled monolayers on flat surfaces with the two-phase approach of Faraday in preparing colloidal gold solutions. In this procedure, the tetrachloroaurate anion is transferred from the aqueous to the organic phase using a phase transfer agent such as tetraoctylammonium bromide. Addition of alkylthiols gives rise to the formation of a polymer with concomitant reduction of Au^{III} to Au^{I}. Subsequent addition of sodium borohydride causes the reduction of Au^{I} to Au^{0}, entailing nucleation of the particle and concomitant organization of the thiolates on its surface, Figure 4.4. The molar ratio between gold and thiol controls the cluster size.[146] The size dispersion seems to be controlled by the rate of the last reduction step; since this process occurs at the interface, the rate of reduction is not linearly correlated to the concentration of borohydride solution.

Nanoparticles can be purified by repeated centrifugation, by size exclusion chromatography, by dialysis or using Soxhlet extraction.[147]

In general, thiols characterized by a hydrocarbon chain length of C5 to C18 impart an extraordinarily high degree of stability to AuNPs, while nanoparticles protected by aromatic thiols, such as 4-mercaptophenol, have been reported to degrade slowly

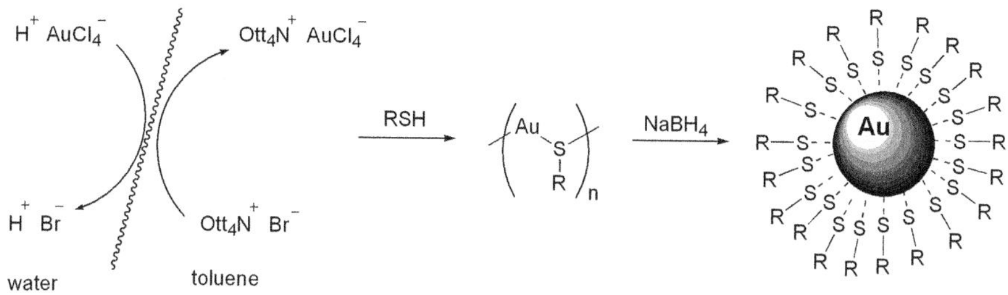

Figure 4.4 Schematic representation of the two phase AuNPs synthesis proposed by Brust and coworkers in 1994.[134]

over several weeks at room temperature.[148] Surprisingly, nanoparticles protected by 4-mercaptobenzoic acid were the first to be characterized by X-ray crystallographic analysis, as will be discussed in Section 4.3.4.[149]

Most synthetic methods produce dispersed samples of nanoparticles and practical limitations of these approaches become apparent only when narrow size distribution or monodispersity is required. This is relevant for instance for electrochemical studies of quantized capacitance charging,[150] single-electron transistor assembly,[151] and in advanced applications of thermal gradient optical imaging.[152] Several methods have been established to separate NPs with different average diameters. The first, proposed by Whetten and coworkers, was the fractional crystallization monitored by mass spectrometric analysis.[153] The quality of separation was excellent but the method suffers from a low yield of purified material. Moreover, gel electrophoresis has been employed for the isolation of 10.4 kDa glutathione-protected AuNPs[154] and size exclusion chromatography has also been used in several cases.[155] In addition, near monodispersity was achieved by reduction of hydrogen tetrachloroaurate with sodium borohydride in the presence of a water-soluble alkyl thioether end-functionalized poly(methacrylic acid) stabilizer. This compound acts as steric stabilizer, binding the gold surface through the thioethers. In addition, the electrostatic repulsion between different particles, due to the negative charges of the polymer backbone, allows the formation of NPs with narrow size distribution. Unfortunately, this polymer is not commercially available but free samples may be obtained from the authors.[156]

Nanoparticles are unstable under strongly acidic conditions. Indeed, NPs protected by a mixed monolayer composed of a 1 : 1 mixture of dodecanethiolate and triethylene glycol thiolates aggregate under strongly acidic conditions. Interestingly, these authors reported that the acid induces the formation of clusters of these MPCs and that the extent and the rate of cluster formation is the result of a subtle balance between hydrophobic and hydrogen bonding interactions between particles and between the particles and the solvent.[157]

Recently, Agasti et al.[158] studied the thermodynamic and kinetic stability of MPCs that differ in the substitution pattern on the first and second carbon next to the sulfur atom (Fig. 4.5). The rate constant of the decomposition of MPCs by potassium cyanide increases in the series: iso-thiolate (a methyl group on C2) $<$ nor-thiolate (two methylene groups next to S) $<$ sec-thiolate (a methyl group on C1). Moreover,

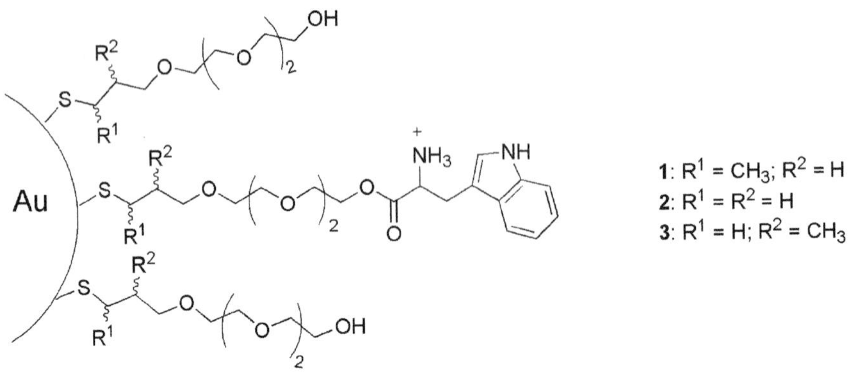

Figure 4.5 Monolayer structure of three structurally-related Au-MPCs.[158]

the rate constants for the *in situ* place-exchange reaction (see Section 4.3.3 for this reaction) of MPCs in the presence of intracellular thiols, such as dithiothreitol (DTT), dihydrolipoic acid (DHLA), and glutathione (GSH), follow the same order. This different reactivity can be exploited to trigger the release of drug molecules from MPC-based drug delivery systems.[158]

The stability of AuNPs in thiol-rich environments is a main concern when NPs are employed in biological matrices. For instance, glutathione, a short peptide containing a thiol group, may reach high concentrations and consequently may displace other functional thiols present in the monolayer. The stability of water-soluble gold nanoparticles has been addressed in a very interesting study carried out using 36 commercially available sulfur ligands for the synthesis of MPCs following different reaction protocols.[159] The key findings were that the majority of the ligands tested did not exhibit the extraordinary stability of alkanethiolate-protected AuNPs; only tiopronin, 4-mercaptobenzoic acid, 1-thio-β-D-glucose and glutathione formed clusters that were stable in solution for months. In particular, AuNPs prepared with the synthetic method reported by Brust using glutathione and 4-mercaptobenzoic acid show a very narrow core size distribution.

Robust protection of gold surfaces has also been achieved using bifunctional ligands such as dithiocarbamates. Gold colloids have been passivated by dithiocarbamate ligands prepared *in situ* from CS_2 and a secondary amine.[160] Finally, AuNPs can also be stabilized by sulfur-containing dendrimers (Fig. 4.6), the size of the nanoparticles obtained depending on the generation and chemical nature of the dendrimer. The comparison of several molecules with multiple sulfur binding sites infers that the dendritic branches play a critical role in successful nanoparticle formation.[161] The surprisingly high degree of stabilization induced by the dendrimers cannot be rationalized by considering only the weak thioether–gold interaction; instead, the electron-rich phenyl rings that lie on the surface of the nanoparticles most probably contribute to the stability. This suggests that one or more dendrimers—depending on the NP size—are wrapped around the gold core, thus forming a very thin and better protective shell in comparison to the monolayer originating from alkanethiols.[161]

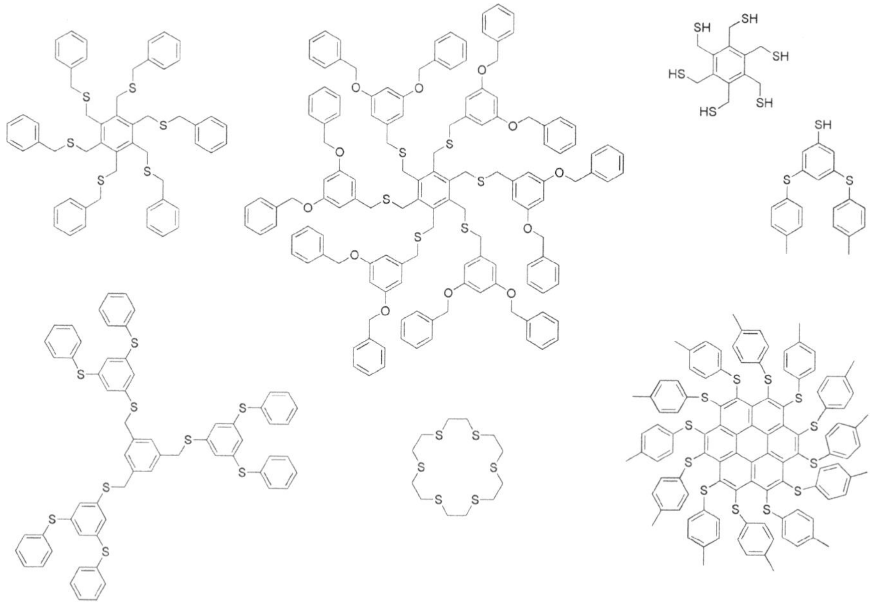

Figure 4.6 Structure of oligothia-dendrimers used for the controlled formation of AuNPs,[161] and of 1,2,3,4,5,6,7,8,9,10,11,12-dodecakis-*p*-tolylsulfanyl-coronene used for the synthesis of 1.0 nm AuNPs.[162]

Very small gold nanoparticles with an average core diameter of 1.0 nm have been obtained by disproportionation of Au^+ ions,[138] in the presence of dodecakis-*p*-tolylsulfanyl-coronene (Fig. 4.6), that favors encounters between Au^+ ions and protects the resulting small nanoparticles from further aggregation.[162]

The need of very diverse synthetic approaches to AuNPs arises either from the desire for a capping thiol layer displaying particular properties or from the intended use of the resulting NPs. It is thus necessary to modify classical methods and to develop new ones specifically for the case in study. For example, this has been accomplished for water-soluble nanoparticles using thiolated derivatives of PEG.[163–167]

Moreover, for specific applications it may be desirable to avoid traces of unbound ligands or capping agents and impurities introduced by the use of surfactants. For this reason, surfactant-free syntheses have been conceived.[168] Hydrogen tetrachloroaurate in diethylene glycol dimethyl ether (diglyme) can be reduced by sodium naphthalenide in diglyme to form AuNPs. The resulting, weakly protected NPs (likely only by the solvent) can be stabilized and functionalized in a straightforward manner by addition of a variety of ligands. The size dispersities of 15% to 20% are narrow and comparable to those obtained with other methodologies.[168]

Electron-poor thiols such as amphiphilic perfluorinated thiols with a perfluorinated chain in α-position to the thiol group are not able to reduce Au^{III} and to form the Au^I-SR polymer. On the other hand, when thiols like HS-F8-PEG (Fig. 4.7), are used in the thiolate form, water-soluble AuNPs protected by a monolayer that

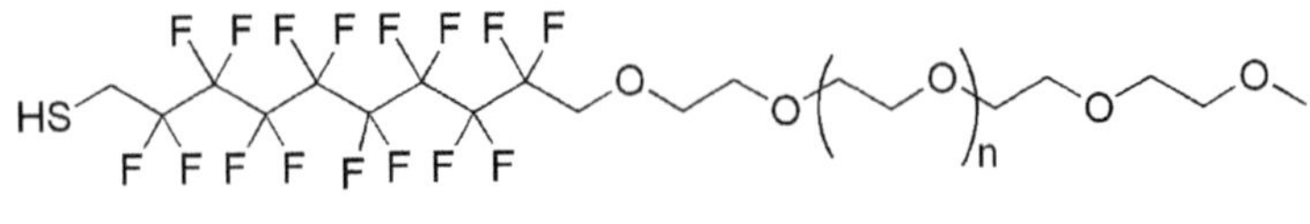

Figure 4.7 Structure of the amphiphilic fluorinated thiol used for the synthesis of water-soluble AuNPs.[169]

presents at the same time a perfluorinated shell close to the gold surface and a short polyethyleneoxide region to impart solubility in water can be obtained.[169]

Synthesis, properties, and applications of polymer-protected AuNPs have been reviewed by Shan and Tenhu and will not be covered in detail here.[139] The basic motivation here is to better control the particle size and dispersity and to obtain robust materials avoiding metal loss during or after utilization of the polymeric material.

4.3.3 Synthesis of Functional Nanoparticles

One of the most attractive features of monolayer-protected AuNPs is the possibility for facile introduction of a wide variety of different thiolated molecules into the ligand shell. This may be achieved using different strategies: (1) direct synthetic methods,[97,124,170] (2) the so-called place-exchange reaction, (3) postsynthetic modifications, or (4) displacement of weak ligands (or protective agents) by functional ligands.

In the *direct synthetic methods* approach, functional thiols are used in the synthesis of AuNPs. The only limitation is that the functional groups should be compatible with the reducing reaction conditions and the basic pH used during NP synthesis. For example, this strategy has been employed for the preparation of highly active catalysts where AuNPs are embedded into a solid support such as silica.[120c] For this purpose, capped AuNPs were prepared in a water-in-oil microemulsion in which a three-phase system composed of water, isooctane, and AOT (sodium bis(2-ethylhexyl)sulfosuccinate) with a molar ratio of 4.5 : 60 : 1 serves as the reaction medium.[171] Under these conditions, ionic gold can be reduced with an excess of hydrazine, in the presence of a mixture of 1-dodecanethiol (DT) and 3-mercaptopropyltrimethoxysilane (MPTMS) in a 13 : 1 molar ratio. Such particles reveal a narrow size distribution centered at approximately 2.0 nm. The use of the two organic functions (DT and MPTMS) is essential since only in the presence of DT, a porous silica network (termed "open shell" by the authors) is formed around the AuNPs instead of a dense coating, still permitting access to the embedded AuNPs for diffusing reactants after sintering of the hybrid. The reduced dimension of the Au particles together with the strong interaction with the support—covalent fixation through MPTMS—is an important factor to convert the inert gold into highly active catalysts.[120c]

Since oligosaccarides are stable in the presence of sodium borohydride, the first synthesis of gold glyconanoparticles has been realized lately with neoglycoconjugate functionalized with a thiol group as ligands in a direct synthetic fashion.[172] The methods for the synthesis of AuNPs presenting carbohydrate groups on the surface have been recently reviewed.[173]

The *ligand place-exchange reaction* was introduced by Murray and coworkers more than 10 years ago.[174] This process occurs under neutral conditions and is an equilibrium process between entering and exiting thiols. The final composition of the monolayer is not easy to predict since it depends on several parameters such as the strength of the S—Au bond, the bulkiness of the entering thiol with respect to the exiting one and the relative length of the two thiols (the longest being favored).

When the ligand exchange is incomplete, that is, when not all the ligands present on the nanoparticles have been displaced, the final product is protected by a mixed monolayer composed of two different ligands. This indeed is one of the most widely used strategies to prepare nanoparticles with mixed monolayers and to tune the composition of their protecting monolayer by changing the ratio of entering and exiting thiols. Examples of this approach are the introduction of amino acids and of short peptides.[175–177] In the latter case, solubility is a major feature of peptide-decorated nanoparticles because, as many proteins do, also AuNPs may aggregate when thiol-containing peptides are used as capping agents.[178] A strategy to overcome this problem is to prepare nanoparticles with a mixed monolayer composed of thiol-peptides and thiols that impart solubility in water.[176,177]

4.3.3.1 Mechanism of the Place-Exchange Process

Theoretical calculations and experimental investigations have been carried out in recent years to understand the mechanism of the exchange process. The self-assembling of the protecting monolayer implies a dynamic process and the exchange of thiolates in solution is macroscopic experimental evidence. Several, though not yet conclusive, kinetic studies of this reaction have been published. The results vary depending on the conditions and the time scale of the experiments used in the assessment.[179–181] However, the exchange reaction can be considered, in a simplified manner, as an equilibrium process involving one entering and one exiting thiol. The experimental results suggest an associative-dissociative mechanism and the concomitant mobility of the thiolates on the gold surface. The thiolates bound to vertexes and edges are more exposed and intrinsically more reactive than those bound to gold atoms on the facets.[180] Disulfides proved to have a reduced reactivity compared to thiols and exchange has been observed with gold NPs protected by amines, sulfides, or short thiolates. In an ESR (electron spin resonance) case study, radical-tagged disulfides did however not exchange with *n*-octadecanethiolate-protected nanoparticles.[182] Exchange of thiolates has been reported to occur also between nanoparticles protected by different thiolates and thiolate hopping between Au nanoparticles has been monitored by ESR spectroscopy using spin-labeled ligands.[183]

Displacement of weak ligands (or protective agents) by functional ligands: thiols can displace other ligands weakly bound to gold (e.g., amines, phosphines, polymers as poly(vinylpyrrolidone) or PVP and citrate anions). This procedure is widely used to functionalize AuNPs (>5 nm) that cannot be obtained by direct synthesis in the presence of thiolates.[184] Contrary to the displacement of thiols, a complete substitution occurs in this case due to the much stronger S-Au interaction. A mixed monolayer may be obtained using a blend of different thiols.[185]

As an example, dipotassium bis(*p*-sulfonatophenyl)phenylphosphane dihydrate was used to protect gold colloids with a diameter of 5 or 10 nm.[186] These nanoparticles did not aggregate in aqueous buffers and could be subjected to repeated cycles of NaCl precipitation for purification from excess of ligand. In addition, triarylphosphines could be displaced by alkylthiol derivatized single-stranded DNA. This surface modification strategy offers different approaches to nanocrystal assembly[144,187] so that the displacement of phosphines by functional thiols became a general protocol to produce functional gold NPs, as demonstrated, for instance, by the preparation of Cys-terminated peptide AuNPs. These systems are very stable and their properties as responsive materials could be studied under different temperature and pH conditions.[188,189]

A different strategy has been used by Mirkin for the synthesis of DNA–AuNP conjugates. The synthesized DNA probe strands were functionalized at either the 3′- or 5′-end with terminal sulfur-containing groups in the form of a monothiol, a cyclic disulfide, or trithiol.[190] It was found that DNA–AuNPs functionalized with probe strands terminated with a trihexylthiol group were stable at higher temperature and in the presence of reducing agents such as dithiothreitol (DTT) and mercaptoethanol for longer periods of time when compared to mono- or cyclic dithiane species.[191] In addition, DNA–AuNPs made with a trithiol linker have higher DNA-probe surface coverages than those made with either the monothiol or cyclic dithiane species.[192] Recently, dithiocarbamate ligands have been also used to prepare AuNP-DNA conjugates.[193]

Under different reaction conditions, both thiol-ssDNA and -dsDNA stabilize gold colloids, but possible nonspecific interactions between the hydrophobic DNA bases and the gold surface have been observed. Moreover, interparticle interactions are promoted by the formation of hydrogen bonds between DNA bases causing aggregation within a rather short period of time.[194]

On the other hand, peptide sequences with a cysteine amino acid at one end have been used to graft peptides on gold colloid surfaces forming robust monolayers. However, a major problem to be solved when preparing a self-assembled peptide monolayer on a 3D substrate is the solubility of the resulting nanoparticles. In some cases, proteins may become insoluble under physiological conditions and this may also happen to peptide-functionalized AuNPs. Brust and Levy described detailed design criteria for peptide capping ligands. A combinatorial approach allowed the authors to screen 3,200,000 peptide sequences, from which 49 peptides were selected (using specific criteria) with the sequence CALNN showing the best performance as capping peptide. The sequence Cys-Ala-Leu-Asn-Asn (CALNN) contains hydrophilic end groups to ensure solubility in water, a thiol group to form a stable bond with the gold surface, and a positively charged ammonium group that may also interact with gold.[195] CALNN-protected AuNPs are robust systems that can be purified by size-exclusion chromatography and ion-exchange chromatography.

Concerning other ligands, zwitterionic disulfides have been recently used to displace citrate anions from the surface of 4 nm gold colloids, producing water-soluble AuNPs.[196] Furthermore, a very recent example of amine displacement has been published by Scrimin. Here, AuNPs were prepared by reduction of tetraoctylammonium tetrachloroaurate with sodium borohydride in the presence of dioctylamine. The molar ratio between gold and amine controlled the nanoparticle size and NPs with an average core diameter between 1.90 and 8.90 nm have been obtained. In a second step, thiols were used to displace the amines, with only a slight excess having been necessary with respect to the theoretical amount needed to fully cover the gold surface. In this way, Scrimin and coworkers prepared mixed monolayers composed of HS-C8-TEGME (TEGME=triethylene glycol methylether) and thiols functionalized with disaccharide, ammonium, and amine groups, as shown in Figure 4.8.[197] This approach to the synthesis of functional AuNPs is compatible with a multitude of functional groups. It allows the use of a precise amount of functional thiolates and a good control of the monolayer composition.

Postsynthetic modifications: another procedure for the preparation of functional self-assembled monolayers on 3D surfaces is the covalent modification of simple functional groups present on the monolayer's surface in analogy to methods extensively used for 2D surfaces, the most useful methods exploiting the "click chemistry" approach.[198–200] Drawbacks of many postsynthetic modification approaches are similar to those known from the synthesis of dendrimers with the complication that NMR and mass analyses are, in these cases, more difficult, or even not practicable at all like in the case of large nanoparticles. A way to overcome this problem is using click chemistry reactions to modify the monolayer.[198] Taking inspiration from nature, this synthetic strategy is based on the use of a set of powerful, highly reliable, and selective reactions for rapid and quantitative synthesis. For this reason, the click chemistry approach has been widely used for the synthesis of AuNP bioconjugates,[199] and to form new materials such as carbon nanotubes decorated with AuNPs.[200]

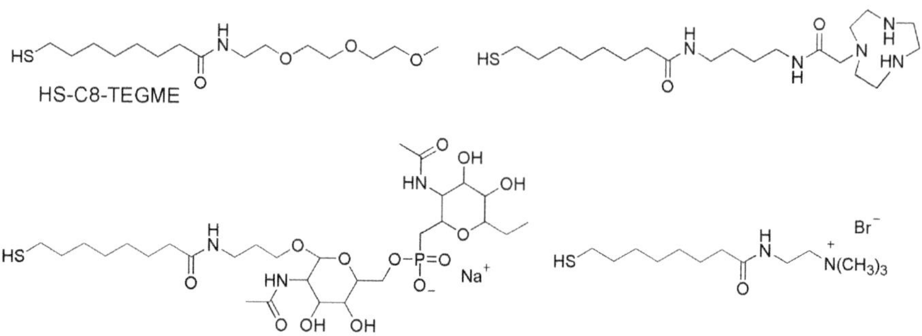

Figure 4.8 Functional thiols used to form mixed monolayers in combination with the amphiphilic thiol HS-C8-TEGME.[197]

4.3.3.2 Monovalent Gold Nanoparticles

Despite the tremendous success in the development of nanomaterials, the level of control of their structure and properties is still very limited. When AuNPs are used as building blocks, monovalent, divalent, or higher valency nanoparticles would be required when aiming at the controlled and well-organized construction of 3D nanonetworks from a simple linear topology.

Different methods have been reported so far for the synthesis of monovalent, that is, monofunctional AuNPs and they have been recently reviewed.[201] Monofunctionalized AuNPs may be produced by using appropriate reaction conditions and then isolated by chromatographic methods.[144,202,203] Alivisatos and coworkers reported the control of the number of oligonucleotides bound to a nanocrystal by electrophoretic isolation of 5 and 10 nm gold nanocrystals bearing discrete numbers of ssDNA.[144,202] A more recent example has been reported by Levy et al.[203] for peptide-capped AuNPs. Peptide-capped NPs were prepared by mixing 6 nm gold NPs with peptide solutions which comprised the peptide CALNN (see above) and a functional peptide composed of the CALNN sequence extended by a His-tag sequence. The His-tag was used for its ability to bind to immobilized metal ions. Working with a low ratio of functional peptide to matrix peptide, the NPs displayed none or only one functional peptide. Using metal ion affinity chromatography, only the NPs containing one functional peptide were thus separated from the other NPs.[203]

Monofunctional AuNPs could also be prepared with a modification of the solid-phase approach previously proposed,[204] using noncovalent interactions with amino-

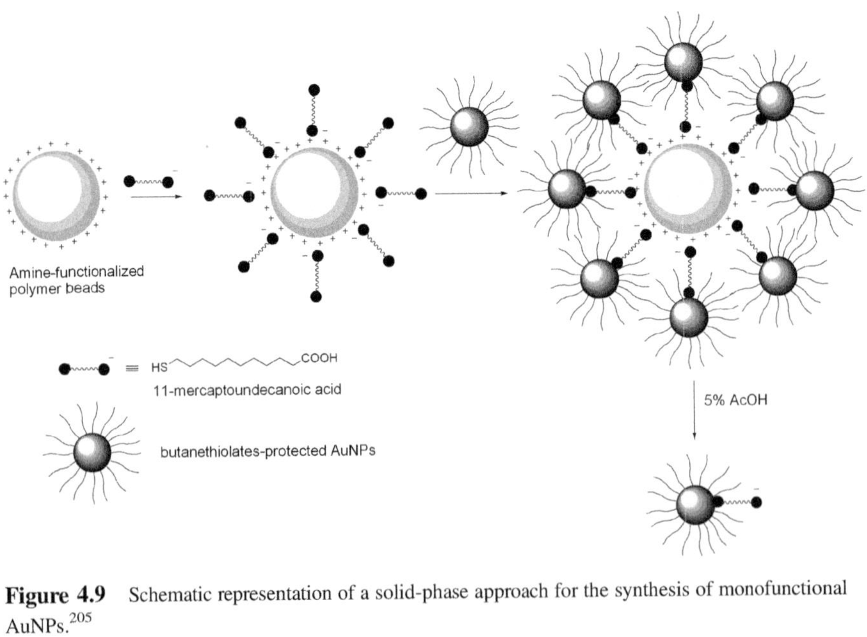

Figure 4.9 Schematic representation of a solid-phase approach for the synthesis of monofunctional AuNPs.[205]

functionalized silica gel as a solid support. The purpose was to reduce the cost and time for the synthesis and to find milder reaction conditions, allowing the modification of a wider range of nanoparticles.[205] The synthetic strategy used in this investigation is shown in Figure 4.9.

4.3.4 Characterization of Passivated Gold Nanoparticles

4.3.4.1 X-Ray Analysis

In 2007, Kornberg's group solved the first solid-state structure of thiol-protected AuNPs at 1.1 Å resolution.[149] These nanoparticles are composed of 102 atoms of gold and are protected by 44 *p*-mercaptobenzoic acid units. The authors were able to obtain suitable crystals because of two important factors: (1) the choice of a rigid thiol, characterized by an aromatic ring and the presence of the carboxylic group that, on the monolayer surface, stabilized the crystal structure by intermolecular hydrogen bonding; (2) the use of a methodology commonly applied for the crystallization of proteins completed the success of this approach. In particular, crystallization occurred at acidic pH (pH value of 2.5) to have the carboxylic group in the acidic form, in a solution of sodium chloride and sodium acetate starting with a 46% of methanol then slowly evaporated forming thin black rods after three days. The structure of the protected-gold cluster has unraveled an unexpected metal core organization and sufur-Au interactions in contrast to the standard model for 2D SAMs. The core architecture could be described as a 49 atom Marks decahedral, with two 20 atom caps of C_5 symmetry on opposite poles, and a 13 atom equatorial band with no apparent symmetry. The geometry of the equatorial band imparts chirality to the cluster. Three types of Au-SR interactions are present in the structure: staple motifs characterized by SR bridges between two gold atoms, while the Au atoms on the outermost external shell binds two SR groups forming a RS-Au-SR motif, and the gold atoms in the next shell are linked to only one SR group.

In such a small cluster, the surface coverage is 70%, much higher than the percentage measured on flat surfaces covered by arylthiolates (33%).[206] The formation of the $Au_{102}(SR)_{44}$ nanoparticle may be explained with the electronic rules of cluster stability. Considering that each gold atom contributes one valence electron, and that each of the 44 SR thiyl radicals formally takes one electron, a 58 electron closed shell is obtained. A deep analysis and discussion on this landmark achievement have been recently published.[207,208]

In agreement with the enhanced stability for metal cores with masses of 5, 8, and 14 kDa,[97,209] crystals of the tetraoctylammonium thiolate gold nanoparticle $N(C_8H_{17})_4^+\ Au_{25}(SCH_2CH_2Ph)_{18}^-$ suitable for X-ray analysis have been isolated. The crystal structure displays also in this case two types of gold atoms bound to thiolates. Twelve gold atoms are involved in six staple motifs (one unit longer than those observed by Kornberg, i.e., SR-Au-SR-Au-SR) and 12 Au atoms, that form the vertices of an icosahedron around the central atom, are bound to a single thiolate group.[210] Density functional theory (DFT) has been used to predict the structure of

$Au_{25}(RS)_{18}$. The preferred structure consists of an icosahedral Au_{13} core protected by 6 RS-Au-RS-Au-RS units as confirmed experimentally. The enhanced stability of the structure as an anion arises from closure of an eight-electron shell for delocalized Au(6s) electrons.[211]

A rationalization of these data and previously reported theoretical calculations and X-ray analyses on $Au_{39}(PR_3)_{14}Cl^{6-}$, $Au_{11}(PR_3)_7X_3$, and $Au_{13}(PR_3)_{10}X_2^{3+}$ (X = halogen or thiolate, R = alkyl or aryl)[212] has been proposed based on the consideration that these compounds have closed shells with a total count of, for example, 8, 34, and 58 electrons, and a large energy HOMO–LUMO gap.[213] The observed exceptional stability of these gold clusters supports the powerful concept of noble gas superatoms. This can also be used to predict the stability of other known gold clusters.

4.3.4.2 Characterization of the Monolayer

The understanding of the organization of functional thiolates in the monolayer protecting AuNPs is a prerequisite to tailor the properties of the NPs. New information concerning the properties and the morphology of self-assembled monolayers and mixed self-assembled monolayers on gold have been disclosed in the last years using direct techniques of investigation such as STM and IR spectroscopy as well as indirect methodologies such as ESR spectroscopy.

4.3.4.3 Packing Arrangement of Ligands around the Core

Despite the experimental results and theoretical calculations reported so far, a conclusive description of the ligand packing for homoligand nanoparticles cannot be drawn. The present idea is that the real structure of the ligand shell results from the interplay between van der Waals interactions and the tendency of the thiolates to reach the optimal tilt angle on each facet, as discussed by Stellacci and coworkers on the basis of STM analyses.[214,215]

Experimental studies reported by Lucarini et al.[216] are in agreement with the hypothesis of the existence of an increasingly available space between thiols upon reduction of the diameter of the NPs. In this investigation, ESR spectroscopy has been used to study the interaction of an amphiphilic nitroxide radical with the monolayer of water-soluble protected gold clusters having a core diameter between 1.6 and 5.3 nm. The solubilization of the nitroxide probe in the more hydrophobic environment of the monolayer strongly depends on the size of the gold core. With decreasing curvature radius of the NPs, the thiolates are more densely packed and the energetic cost for the probe to enter the monolyer also increases, resulting in enhanced binding of the probe to smaller NPs. In particular, the partition equilibrium constant increases as the nanoparticle diameter decreases.

Molecular-dynamic simulations on a cube octahedron composed of 309 Au atoms and protected by 80 hexanethiolates (ensemble diameter approximately 2.3 nm) show that during about 1 ns of simulation time some ligands stay apart from the others, either because they are more separated or because they are more stretched. These ligands are

Figure 4.10 A pictorial illustration of splay vector order parameter configuration on a sphere with formation of two vortices at the north and south poles.

located in the proximity of the edges of the faces of Au_{309}.[217] An important feature that stems from these studies is that thiolates on a gold surface behave as the hair on a ball does; that is, they follow the so-called hairy-ball theorem. In practice, as soon as the thiolate chains tilt to maximize their mutual interactions, the effects of the theorem kick in and there must be (at least) two chains that stands apart from the others (Fig. 4.10).

A very elegant investigation concerning the topological organization of different thiols on self-assembled monolayers on nanoparticles has been reported by the group of Stellacci and coworkers.[214,215,218] They demonstrated that stripe-like patterns are formed by thiols with different tails or thiols of different lengths coadsorbed on the surface of gold (and silver) nanoparticles. The ligand organization has been studied by scanning tunneling microscopy that actually seems to be the only microscopy technique that has the capability of visualizing these molecules. This analysis has been carried out on NPs immobilized on gold surfaces so that they are not free to move during imaging. Dithiol molecules were used to act as linkers, binding the nanoparticles to the underlying substrate. Mixtures of thiols, such as octanethiol (OT) and mercaptopropionic acid (MPA) in a 2:1 molar ratio, phase separate into ordered, ribbon-like domains of alternating composition that encircle the nanoparticle core (Fig. 4.11).

Interestingly, the width of these domains is extremely small, leading to unexpected and unprecedent solubility properties of the nanoparticles[219] and to good resistance towards nonspecific adsorption of protein.[214] Moreover, Verma et al.[220] identified different cell membrane penetration processes of isomeric AuNPs as a function of the organization of the monolayer. AuNPs coated with subnanometer domains of alternating anionic and hydrophobic groups penetrate the plasma membrane without bilayer disruption, whereas AuNPs coated with the same mixture of ligands in a random arrangement are mostly trapped in endosomes.[220] The mechanism of membrane penetration by these strip-like monolayer protected AuNPs has to be unraveled, perhaps leading to novel and more efficient low-toxicity drug-delivery systems.

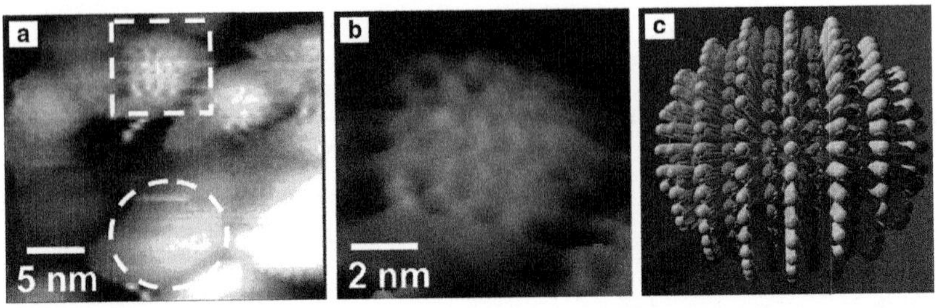

Figure 4.11 (a) STM height image of AuNPs, on Au foil, coated with a 2 : 1 molar ratio of octanethiol and mercaptopropionic acid, showing ribbon-like stripes (ripples) due to the phase separation of the two ligands. One such particle is outlined by the dashed square and shown in an enlarged image in (b). The raised domains that run across the particles are the OT phases which alternate with the MPA phases. A schematic drawing to help the reader visualize this 3D arrangement is shown in (c). The dashed white circle in (a) marks an area in which the underlying gold foil exhibits a curvature comparable to that of the nanoparticles but where, however, no hints of ripples are observed, indicating that ripples are not the result of the STM tip scanning over an area of high curvature.[215] (Reprinted with permission from A. M. Jackson et al., *J. Am. Chem. Soc.* **2006**, *128*, 11135–11149. Copyright 2006 American Chemical Society.) (See color insert.)

The curvature of the metal surface is responsible for this organization of the thiolates in the monolayer. Atomistic and mesoscale simulations show that when the conformational entropy gained via this morphology is sufficient, microphase-separated strip-like patterns form. When the entropic gain is not sufficient, bulk phase-separated Janus particles may be predicted. The arrangement of the stripes depends on the substrate curvature; stripes are not aligned on large spheres or flat surfaces, while they are aligned on medium-sized spheres. These predictions have been confirmed experimentally[221] and nanodomains are formed on nanoparticles with a core diameter ranging from 2.5 to 8.0 nm.[222]

A different situation has been observed using mixtures of thiolates with immiscible chains. AuNPs protected by mixed monolayers of different composition made of HS-C8-TEGME and HS-F8-PEG have been investigated by ESR spectroscopy using a radical probe sensitive to the hydrophobicity of the environment.[223] The ESR spectral parameters of the probe in the monolayer are identical to those of the probe in a completely fluorinated medium when the ratio (R_{CF}) between alkyl and perfluoroalkyl is lower than 2.5. Only at $R_{CF} > 2.5$, the probe starts to experience the environment of alkyl chains. The experimental results support the phase segregation of perfluoroalkyl thiolates in patches on nanoparticles with a core size of 2.5 to 4.0 nm.[223]

DeVries et al.[224] demonstrated experimentally that polar singularities form when a curved surface is coated with ordered monolayers (hairy-ball-theorem). The thiolates bound at the polar positions exchange faster than those in the bulk monolayer, thus allowing introduction of two—and only two—functional groups at the poles of a nanoparticle.[224] This was the first example of divalent metal nanoparticles. This discovery opens up the path to new applications in the field of materials chemistry, and supramolecular chemistry, as stressed by Perepichka and Rosei.[225] Undoubtedly, the regiochemical stability of these disubstituted nanoparticles and the chemical yield of their functionalization should be the subject of future studies.

Functionalization of gold colloids on one face only can be realized anchoring the NPs on a solid surface. Thus, after exposure to a solution of 11-mercapto-1-undecanol a protective monolayer is formed but not on the side anchored. After sonication in the presence of 16-mercaptohexadecanoic acid (or mercaptoethylamine), the nanoparticles are removed from the solid surface and the functional thiol is adsorbed on the free gold surface of the NPs. Reaction of NP-COOH with NP-NH_2 in the presence of coupling reagents produces the formation of gold nanoparticle dimers with a 30% to 65% yield depending on the NP sizes.[226]

4.3.5 Synthesis of Gold Nanoparticles of Different Shapes and Their Functionalization

Gold nanoparticles of different shapes have received considerable attention in recent years because of the strong correlation between the shape and their chemical, physical, electronic, optical, magnetic, and catalytic properties.[123,227–230] Specific synthetic procedures have been developed to fabricate nanorods (NRs), nanowires, and nanotubes of different aspect ratio, as well as nanoprisms, nanocubes, nanocages, and branched multipod gold nanocrystals. Several reviews already cover this field.[122,123,229,231,232]

Besides gold nanocrystals of spherical shape, elongated structures such as nanorods are the second most studied ones. Their first synthesis was reported in 2001[233] and since then this synthetic procedure has been widely studied and improved. Nonetheless, the influence of various parameters such as the presence of seeds, the nature of the solvent, and the presence of surfactants on the reaction yield and the control of shape make the reproducibility of this synthesis still a challenging task.[234–237]

The synthesis is a seed-mediated growth procedure, in which $HAuCl_4$ is reduced initially with a strong reducing agent ($NaBH_4$) in the presence of sodium citrate or CTAB (cetyltrimethylammonium bromide), in water, to produce seed particles of about 4 nm. Subsequent reduction of additional metal salt with a weak reducing agent (ascorbic acid), in the presence of structure-directing agents (SDAs; the most widely used is CTAB) leads to the controlled formation of nanorods (NRs) of specified aspect ratio (ratio between the length and the width of the NR). The aspect ratio is influenced also by the concentration of the gold seeds, CTAB, and ascorbic acid. The presence of silver ions in the growing step allows the control of the aspect ratio and hence, permits control of the optical properies of the nanorods through tuning of the crystal enlargement. However, in these syntheses often a gradual blue-shift in the longitudinal plasmon resonance is observed over a period of hours to days after seed injection. This undesirable optical drift can be greatly reduced by the addition of millimolar solutions of sodium sulfide, with an optimized sulfur-to-metal ratio of 4 : 1, to quench the process and limit nanorod growth.[238]

Nanorods are ideally prisms with ten {111} end faces and five {100} or {110} side faces, or both. It has been demonstrated that CTAB strongly binds to the {100} faces, along the length of the rods, while the interaction with the {111} Au facets is weaker. Likely gold–surfactant complexes are specifically incorporated into the

{100} side edges, whereas noncomplexed ion pairs or Au^0 atoms/clusters are added at the {111} end faces. The selectivity may be due to the increased stability of the close-packed {111} surfaces compared with the edge sites, thus the larger ammonium head-group may be readily accommodated around the edges. This distribution enhances the preferential growth along the common [110] axis rather than others.[239]

As a consequence, functionalization with ligands that bind strongly to gold occurs mainly on the distal ends of the rods. This has been demonstrated by TEM observation of preferential end-to-end self-assembling of gold nanorods using functionalized thiols with biotin–streptavidin,[240] antigen–antibody,[241,242] and groups presenting complementary hydrogen bond partners.[243–245] The end-to-end self-assembly of gold nanorods enables the organization of gold nanorods in long chains. This approach is a first step toward the self-assembling fabrication of new 1D materials that may serve as plasmonic waveguides.[246]

Functionalization of gold nanorods may be obtained using different strategies. A simple methodology takes advantage of the positive charges present on the nanorod surfaces; biomolecules stick on the nanorod surface because of strong electrostatic interactions.[247] A similar strategy uses electrostatic interactions between the positively charged CTAB monolayer covering the nanorods and a negatively charged polymer such as PSS (polystyrene sulfonate). This surface may interact with positive charges on the surface of proteins to enable, for instance, the formation of antibody–nanorod conjugates.[248]

Functional groups present on the polymer surface can be modified with classical chemical reactions. An example has been reported by Gole and Murphy using the click chemistry approach.[249] Gold nanorods coated with a copolymer containing sulfonate and maleic acid groups (PSS-MA) have been conjugated to an amine-PEG-azide. The azide-functionalized nanorods were then reacted with an acetylene-functionalized enzyme, trypsin, by a copper-catalyzed 1,3-dipolar cycloaddition reaction. The enzyme conjugated by click chemistry to the NPs demonstrated improved biological activity compared with other methods of bioconjugation.

A specific peptide sequence for nuclear targeting (NLS) has been conjugated to gold NRs via click chemistry. In particular, a thiol with an azide end group has been reacted via Huisgen 1,3-dipolar cycloaddition to an alkyne-terminated NLS peptide and then anchored to CTAB-protected gold NRs.[250]

Concerning SDAs, however, the presence of CTAB in the shell is not desirable for *in vivo* applications because of its toxicity.[251,252] In fact, the deliberate displacement of CTAB from the AuNR surface with a commercially available mPEG-SH has been demonstrated.[242] However, in comparison to 2D gold surfaces, the assembling of thiolated molecules on the surface of gold nanorods is slower and thus higher excesses of thiolates have to be employed for efficient CTAB removal. For example, 24 h have been used for mPEG-SH and 72 h have been used for thiolated DNA.[253]

Replacement of CTAB on the gold nanorod surface has also been achieved by *in situ* dithiocarbamate (DTC) formation introduced for gold surface protection, see Section 4.3.2.[254] Using this protection procedure, the nonspecific uptake of nanorods by KB cells (a tumor cell line derived from oral epithelium) could be greatly reduced. Moreover, following the DTC strategy and using diamino-PEG, different functional groups may be introduced on the NRs suface. For example, amine-coated NRs were treated with a solution of *N*-hydroxysuccinimidylfolate conjugates to prepare folate-conjugated NRs that were used to study the mechanism of tumor cell death.[255,256]

An innovative method to remove CTAB from gold nanorod surfaces and at the same time to protect them with thiols has been carried out by a place-exchange reaction inside an ionic exchange polymer. The trapping of gold nanorods inside polymer beads allows sufficient time for a stable thiol monolayer to form on the gold nanorod surface.[257]

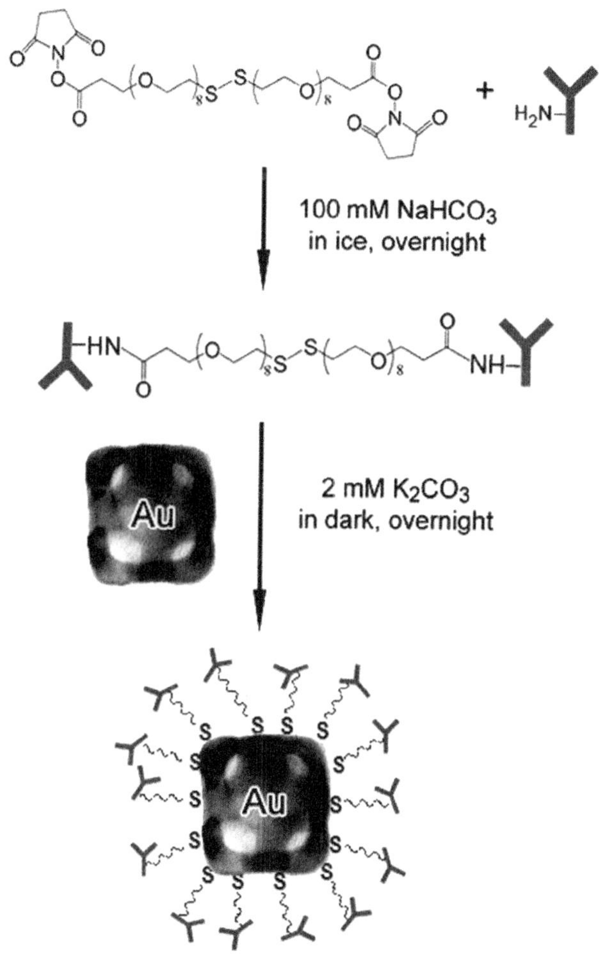

Figure 4.12 A schematic illustration of the two-step protocol used to conjugate antibodies to the surface of a gold nanocage. In the first step, succinimidyl propionyl poly(ethylene glycol) disulfide (NHS-activated PEG, M.W. = 1109) was reacted with the primary amine of an antibody. In the second step, the PEG-antibody complex was bonded to the gold nanocage (as represented by a TEM image) by breaking its internal disulfide bond and forming an Au-S linkage.[262] (Reprinted with permission from J. Chen et al., *NanoLett.* **2005**, *5*, 473–477. Copyright 2005 American Chemical Society.)

Gold nanocages, in spite of their fancy name, are easy to prepare. Nanoboxes with faces smaller than 40 nm have been synthesized using the galvanic replacement reaction between silver nanocubes[135,258] and tetrachlorauric acid in aqueous solution in the presence of poly(vinyl pyrrolidone) (PVP).[259–261] By optimizing the reaction conditions, the dimension and consequently the optical properties of the cages could be readily controlled. The stoichiometry of the redox process requires three silver atoms for the reduction of one gold atom. By exploiting this approach, the plasmon resonance band could be tuned to any specific wavelength in the spectral range from 400 to 1000 nm by titrating a specific amount of aqueous suspension of silver nanocubes with different amounts of a solution of tetrachloroauric acid. Gold nanocages can be easily functionalized since they are protected by PVP that can be easily displaced by sulfur ligands. For example, gold nanocages were conjugated with a secondary antibody, as illustrated in Figure 4.12.[262] The absorption of these gold nanocages is about five orders of magnitude stronger than that of the conventional dye indocyanine green, suggesting their potential use as a new class of absorption contrast agents for optical coherence tomography (OCT) imaging.

The simple route to prepare functional gold nanocages and their tunable surface plasmon resonance bands, which extend into the near-infrared, make these nano-objects extremly interesting for biomedical applications. The synthesis, properties, and applications of gold nanocages has been reviewed.[263]

4.4 CONCLUSIONS AND OUTLOOK

The use of SAMs, supported on flat surfaces or nanosized particles, to prepare interacting interfaces requires the development of methodologies to ensure selectivity and a certain degree of addressability. Selectivity is rather simple to achieve in either 2D or 3D SAMs by the introduction of specific functional groups and a large variety of methods is indeed available to this aim. The addressability of these functional groups is by far less simple and represents a tremendous challenge because it requires control, down to the molecular scale, of the chemical reactions allowing the introduction of specific features in macromolecular objects that are intrinsically heterogeneous and dynamic because of their self-assembled nature. As far as 2D SAMs are concerned, lithographic methods allow the preparation of patterned surfaces, where certain functional groups can be easily positioned and addressed. This approach cannot be applied to SAMs on nanoparticles. The possibility to organize at will functional groups on soluble nanosized scaffolds such as nanoparticles could allow the preparation of intelligent materials able to perform complex programmed tasks involving other species such as biomolecules, or cells. On the other hand, the availability of methods enabling one to precisely define the relative positions of a functional group with respect to others is paramount in developing new materials, catalysts, and devices. Some relevant achievements, such as the development of reliable methods for the preparation of monovalent and divalent nanoparticles, disclosed the route to aggregates with unprecedented topologies. However, the methods for the preparation of patterned monolayers on nanoparticles are still at the early stage, but no doubt, this will be a field of intense research activity involving several aspects of fundamental and applied chemical research.

REFERENCES

1. J. C. Love, L. A. Estroff, J. K. Kriebel, R. G. Nuzzo, G. M. Whitesides, *Chem. Rev.* **2005**, *105*, 1103–1169.
2. (a) A. Ulman, *Chem. Rev.* **1996**, *96*, 1533–1554; (b) G. M. Whitesides, P. E. Laibinis, *Langmuir* **1990**, *6*, 87–96; (c) F. Tao, S. L. Bernasek, *Chem. Rev.* **2007**, *107*, 1408–1453.
3. J. E. Houston, H. I. Kim, *Acc. Chem. Res.* **2002**, *35*, 547–553.
4. X.-M. Li, J. Huskens, D. N. Reinhoudt, *J. Mater. Chem.* **2004**, *14*, 2954–2971.
5. J. Huskens, *Curr. Opin. Chem. Biol.* **2006**, *10*, 537–543.
6. (a) J. M. Tour, L. Jones, D. L. Pearson, J. J. S. Lamba, T. P. Burgin, G. M. Whitesides, D. L. Allara, A. N. Parikh, S. Atre, *J. Am. Chem. Soc.* **1995**, *117*, 9529–9534; (b) L. Cai, Y. Yao, J. Yang, D. W. Price, Jr., J. M. Tour, *Chem. Mater.* **2002**, *14*, 2905–2909.
7. (a) P. Morf, F. Raimondi, H.-G. Nothofer, B. Schnyder, A. Yasuda, J. M. Wessels, T. A. Jung, *Langmuir* **2006**, *22*, 658–663; (b) R. D. Weinstein, J. Richards, S. D. Thai, D. M. Omiatek, C. A. Bessel, C. J. Faulkner, S. Othman, G. K. Jennings, *Langmuir* **2007**, *23*, 2887–2891.
8. P. Fenter, F. Schreiber, L. Berman, G. Scoles, P. Eisenberger, M. J. Bedzyk, *Surf. Sci.* **1998**, *412/413*, 213–235.
9. G. E. Poirier, *Chem. Rev.* **1997**, *97*, 1117–1128.
10. F. Schreiber, *Prog. Surf. Sci.* **2000**, *65*, 151–256.
11. R. Mazzarello, A. Cossaro, A. Verdini, R. Rousseau, L. Casalis, M. F. Danisman, L. Floreano, S. Scandolo, A. Morgante, G. Scoles, *Phys. Rev. Lett.* **2007**, *98*, 016102.
12. A. Cossaro, R. Mazzarello, R. Rousseau, L. Casalis, A. Verdini, A. Kohlmeyer, L. Floreano, S. Scandolo, A. Morgante, M. L. Klein, G. Scoles, *Science* **2008**, *321*, 943–946.
13. (a) E. Barrena, E. Palacios-Lidón, C. Munuera, X. Torrelles, S. Ferrer, U. Jonas, M. Salmeron, C. Ocal, *J. Am. Chem. Soc.* **2004**, *126*, 385–395; (b) M. C. Gurau, G. Kim, S.-M. Lim, F. Albertorio, H. C. Fleisher, P. S. Cremer, *ChemPhysChem* **2003**, *4*, 1231–1233.
14. R. Valiokas, M. Östblom, S. Svedhem, S. C. T. Svensson, B. Liedberg, *J. Phys. Chem. B* **2002**, *106*, 10401–10409.
15. P. A. Lewis, R. K. Smith, K. F. Kelly, L. A. Bumm, S. M. Reed, R. S. Clegg, J. D. Gunderson, J. E. Hutchison, P. S. Weiss, *J. Phys. Chem. B* **2001**, *105*, 10630–10636.
16. L. H. Dubois, R. G. Nuzzo, *Annu. Rev. Phys. Chem.* **1992**, *43*, 437–463.
17. (a) R. G. Nuzzo, B. R. Zegarski, L. H. Dubois, *J. Am. Chem. Soc.* **1987**, *109*, 733–740; (b) D. Fischer, A. Curioni, W. Andreoni, *Langmuir* **2003**, *19*, 3567–3571.
18. (a)H. Kondoh, C. Kodama, H. Sumida, H. Nozoye, *J. Chem. Phys.* **1999**, *111*, 1175–1184; (b) G. J. Kluth, C. Carraro, R. Maboudian, *Phys. Rev. B Condens. Matter* **1999**, *59*, R10449-R10452; (c) J. L. Trevor, K. R. Lykke, M. J. Pellin, L. Hanley, *Langmuir* **1998**, *14*, 1664–1673; (d) N. Nishida, M. Hara, H. Sasabe, W. Knoll, *Jpn. J. Appl. Phys. Part 1* **1997**, *36*, 2379–2385.
19. T. Ishida, S. Yamamoto, W. Mizutani, M. Motomatsu, H. Tokumoto, H. Hokari, H. Azehara, M. Fujihira, *Langmuir* **1997**, *13*, 3261–3265.
20. M. D. Porter, T. B. Bright, D. L. Allara, C. E. D. Chidsey, *J. Am. Chem. Soc.* **1987**, *109*, 3559–3568.
21. (a) M. W. J. Beulen, J. Bugler, B. Lammerink, F. A. J. Geurts, E. M. E. F. Biemond, K. G. C. Van Leerdam, F. C. J. M. Van Veggel, J. F. J. Engbersen, D. N. Reinhoudt, *Langmuir* **1998**, *14*, 6424–6429; (b) A. Friggeri, H. Schoenherr, H.-J. van Manen, B.-H. Huisman, G. J. Vancso, J. Huskens, F. C. J. M. van Veggel, D. N. Reinhoudt, *Langmuir* **2000**, *16*, 7757–7763; (c) J. D. Faull, V. K. Gupta, *Langmuir* **2002**, *18*, 6584–6592.
22. D. L. Allara, R. G. Nuzzo, *Langmuir* **1985**, *1*, 52–66.
23. D. Roy, J. Fendler, *Adv. Mater.* **2004**, *16*, 479–508.
24. C. D. Bain, G. M. Whitesides, *J. Phys. Chem.* **1989**, *93*, 1670–1673.

25. A.-S. Duwez, *J. Electron Spectrosc. Relat. Phenom.* **2004**, *134*, 97–138.
26. M. D. Porter, T. B. Bright, D. L. Allara, C. E. D. Chidsey, *J. Am. Chem. Soc.* **1987**, *109*, 3559–3568.
27. M. Prato, R. Moroni, F. Bisio, R. Rolandi, L. Mattera, O. Cavalleri, M. Canepa, *J. Phys. Chem. C* **2008**, *112*, 3899–3906.
28. M. A. Bryant, J. E. Pemberton, *J. Am. Chem. Soc.* **1991**, *113*, 8284–8293.
29. (a) G. Haehner, M. Kinzler, C. Thuemmler, C. Woell, M. Grunze, *J. Vac. Sci. Technol. A* **1992**, *10*, 2758–2763; (b) H. Rieley, G. K. Kendall, *Langmuir* **1999**, *15*, 8867–8867.
30. H. S. Kato, J. Noh, M. Hara, M. Kawai, *J. Phys. Chem. B* **2002**, *106*, 9655–9658.
31. D. J. Graham, D. D. Price, B. D. Ratner, *Langmuir* **2002**, *18*, 1518–1527.
32. K. Kpegba, T. Spadaro, R. B. Cody, N. Nesnas, J. A. Olson, *Anal. Chem.* **2007**, *79*, 5479–5483.
33. P. E. Laibinis, G. M. Whitesides, D. L. Allara, Y. T. Tao, A. N. Parikh, R. G. Nuzzo, *J. Am. Chem. Soc.* **1991**, *113*, 7152–7167.
34. S. Frey, V. Stadler, K. Heister, W. Eck, M. Zharnikov, M. Grunze, B. Zeysing, A. Terfort, *Langmuir* **2001**, *17*, 2408–2415.
35. A. Ulman, *Acc. Chem. Res.* **2001**, *34*, 855–863.
36. (a) M. J. Tarlov, D. R. F. Burgess, Jr., G. Gillen, *J. Am. Chem. Soc.* **1993**, *115*, 5305–5306; (b) D. Ryan, B. A. Parviz, V. Linder, V. Semetey, S. K. Sia, J. Su, M. Mrksich, G. M. Whitesides, *Langmuir* **2004**, *20*, 9080–9088; (c) J. C. Love, D. B. Wolfe, H. O. Jacobs, G. M. Whitesides, *Langmuir* **2001**, *17*, 6005–6012.
37. C. E. D. Chidsey, *Science* **1991**, *251*, 919–922.
38. L. Sun, R. M. Crooks, *J. Electrochem. Soc.* **1991**, *138*, L23–L25.
39. R.-H. Terrill, T. A. Tanzer, P. W. Bohn, *Langmuir* **1998**, *14*, 845–854.
40. C. D. Bain, E. B. Troughton, Y. T. Tao, J. Evall, G. M. Whitesides, R. G. Nuzzo, *J. Am. Chem. Soc.* **1989**, *111*, 321–335.
41. R. Yamada, H. Sakai, K. Uosaki, *Chem. Lett.* **1999**, 667–668.
42. G. Pace, A. Petitjean, M.-N. Lalloz-Vogel, J. Harrowfield, J.-M. Lehn, P. Samorì, *Angew. Chem. Int. Ed.* **2008**, *47*, 2484–2488.
43. C. D. Bain, H. A. Biebuyck, G. M. Whitesides, *Langmuir* **1989**, *5*, 723–727.
44. J. F. Kang, S. Liao, R. Jordan, A. Ulman, *J. Am. Chem. Soc.* **1998**, *120*, 9662–9667.
45. R. Derda, D. J. Wherritt, L. L. Kiessling, *Langmuir* **2007**, *23*, 11164–11167.
46. (a) D. M. Collard, M. A. Fox, *Langmuir* **1991**, *7*, 1192–1197; (b) T. Felgenhauer, H. T. Rong, M. Buck, *J. Electroanal. Chem.* **2003**, *550-551*, 309–319.
47. (a) H. Schoenherr, G. J. Vancso, B.-H. Huisman, F. C. J. M. Van Veggel, D. N. Reinhoudt, *Langmuir* **1999**, *15*, 5541–5546; (b) C. D. Bain, J. Evall, G. M. Whitesides, *J. Am. Chem. Soc.* **1989**, *111*, 7155–7164.
48. (a) G. Yang, N. A. Amro, Z. B. Starkewolfe, G.-Y. Liu, *Langmuir* **2004**, *20*, 3995–4003; (b) N. Nishida, M. Hara, H. Sasabe, W. Knoll, *Jpn. J. Appl. Phys. Part 1* **1997**, *36*, 2379–2385.
49. Y. Kwon, M. Mrksich, *J. Am. Chem. Soc.* **2002**, *124*, 806–812.
50. D. Peelen, L. M. Smith, *Langmuir* **2005**, *21*, 266–271.
51. Z.-I. Zhi, A. K. Powell, J. E. Turnbull, *Anal. Chem.* **2006**, *78*, 4786–4793.
52. J. Li, P. S. Thiara, M. Mrksich, *Langmuir* **2007**, *23*, 11826–11835.
53. M. Mrksich, *ACS Nano* **2008**, *2*, 7–18.
54. R. V. Duevel, R. M. Corn, *Anal. Chem.* **1992**, *64*, 337–342.
55. Y. S. Chi, I. S. Choi, *Langmuir* **2005**, *21*, 11765–11772.
56. A. Niemz, E. Jeoung, A. K. Boal, R. Deans, V. M. Rotello, *Langmuir* **2000**, *16*, 1460–1462.
57. L. Yan, C. Marzolin, A. Terfort, G. M. Whitesides, *Langmuir* **1997**, *13*, 6704–6712.
58. J. Lahiri, L. Isaacs, J. Tien, G. M. Whitesides, *Anal. Chem.* **1999**, *71*, 777–790.
59. M. R. Lockett, M. F. Phillips, J. L. Jarecki, D. Peelen, L. M. Smith, *Langmuir* **2008**, *24*, 69–75.
60. F. Cheng, L. J. Gamble, D. W. Grainger, D. G. Castner, *Anal. Chem.* **2007**, *79*, 8781–8788.
61. C.-H. Jang, B. D. Stevens, R. Phillips, M. A. Calter, W. A. Ducker, *Nano Lett.* **2003**, *3*, 691–694.
62. C. D. Hahn, C. Leitner, T. Weinbrenner, R. Schlapak, A. Tinazli, R. Tampé, B. Lackner, C. Steindl, P. Hinterdorfer, H. J. Gruber, M. Hölzl, *Bioconjugate Chem.* **2007**, *18*, 247–253.

63. M. Hölzl, A. Tinazli, C. Leitner, C. Hahn, B. Lackner, R. Tampé, H. J. Gruber, *Langmuir* **2007**, *23*, 5571–5577.
64. A. S. Ichimura, W. Lew, D. L. Allara, *Langmuir* **2008**, *24*, 2487–2493.
65. (a) J. P. Collman, N. K. Devaraj, C. E. D. Chidsey, *Langmuir* **2004**, *20*, 1051–1053; (b) J. K. Lee, Y. S. Chi, I. S. Choi, *Langmuir* **2004**, *20*, 3844–3847.
66. J. P. Collman, N. K. Devaraj, T. P. A. Eberspacher, C. E. D. Chidsey, *Langmuir* **2006**, *22*, 2457–2464.
67. N. K. Devaraj, G. P. Miller, W. Ebina, B. Kakaradov, J. P. Collman, E. T. Kool, C. E. D. Chidsey, *J. Am. Chem. Soc.* **2005**, *127*, 8600–8601.
68. N. K. Devaraj, P. H. Dinolfo, C. E. D. Chidsey, J. P. Collman, *J. Am. Chem. Soc.* **2006**, *128*, 1794–1795.
69. R. A. Decréau, J. P. Collman, Y. Yang, Y. Yan, N. K. Devaraj, *J. Org. Chem.* **2007**, *72*, 2794–2802.
70. Y. Zhang, S. Luo, Y. Tang, L. Yu, K.-Y. Hou, J.-P. Cheng, X. Zeng, P. G. Wang, *Anal. Chem.* **2006**, *78*, 2001–2008.
71. C. Radhakrishnan, M. K. F. Lo, M. V. Warrier, M. A. Garcia-Garibay, H.G. Monbouquette, *Langmuir* **2006**, *22*, 5018–5024.
72. M. G. Friedrich, V. U. Kirste, J. Zhu, R. B. Gennis, W. Knoll, R. L. C. Naumann, *J. Phys. Chem. B* **2008**, *112*, 3193–3201.
73. (a) E. A. Smith, M. J. Wanat, Y. Cheng, S. V.P. Barreira, A. G. Frutos, R. M. Corn, *Langmuir* **2001**, *17*, 2502–2507; (b) G. J. Wegner, H. J. Lee, R. M. Corn, *Anal. Chem.* **2002**, *74*, 5161–5168; (c) E. A. Smith, W. D. Thomas, L. L. Kiessling, R. M. Corn, *J. Am. Chem. Soc.* **2003**, *125*, 6140–6148.
74. B. T. Houseman, E. S. Gawalt, M. Mrksich, *Langmuir* **2003**, *19*, 1522–1531.
75. K. V. Gujraty, R. Ashton, S. R. Bethi, S. Kate, C. J. Faulkner, G. K. Jennings, R. S. Kane, *Langmuir* **2006**, *22*, 10157–10162.
76. J. K. Lee, K.-B. Lee, D. J. Kim, I. S. Choi, *Langmuir* **2003**, *19*, 8141–8143.
77. H. H. J. Persson, W. R. Caseri, U. W. Suter, *Langmuir* **2001**, *17*, 3643–3650.
78. E. Delamarche, G. Sundarababu, H. Biebuyck, B. Michel, C. Gerber, H. Sigrist, H. Wolf, H. Ringsdorf, N. Xanthopoulos, H. Mathieu, *Langmuir* **1996**, *12*, 1997–2006.
79. (a) E. Ostuni, R. G. Chapman, R. E. Holmlin, S. Takayama, G. M. Whitesides, *Langmuir* **2001**, *17*, 5605–5620; (b) A. J. Pertsin, M. Grunze, H. Jurgen Kreuzer, R. L. C. Wang, *Phys. Chem. Chem. Phys.* **2000**, *2*, 1729–1733; (c) G. B. Sigal, M. Mrksich, G. M. Whitesides, *J. Am. Chem. Soc.* **1998**, *120*, 3464–3473.
80. R. Valiokas, M. Östblom, S. Svedhem, S. C. T. Svensson, B. Liedberg, *J. Phys. Chem. B* **2000**, *104*, 7565–7569.
81. L. Malysheva, A. Onipko, B. Liedberg, *J. Phys. Chem. A.* **2008**, *112*, 1686–1687.
82. P.-Y. J. Yeh, R. K. Kainthan, Y. Zou, M. Chiao, J. N. Kizhakkedathu, *Langmuir* **2008**, *24*, 4907–4916.
83. S. Chen, F. Yu, Q. Yu, Y. He, S. Jiang, *Langmuir* **2006**, *22*, 8186–8191.
84. Z. Zhang, S. Chen, S. Jiang, *Biomacromolecules* **2006**, *7*, 3311–3315.
85. B. S. Lee, Y. S. Chi, K.-B. Lee, Y.-G. Kim, I. S. Choi, *Biomacromolecules*, **2007**, *8*, 3922–3929.
86. D. J. Vanderah, M. L. Walker, M. A. Rocco, K. A. Rubinson, *Langmuir* **2008**, *24*, 826–829.
87. K. Jans, K. Bonroy, R. De Palma, G. Reekmans, H. Jans, W. Laureyn, M. Smet, G. Borghs, G. Maes, *Langmuir* **2008**, *24*, 3949–3954.
88. R. Valiokas, G. Klenkar, A. Tinazli, A. Reichel, R. Tampé, J. Piehler, B. Liedberg, *Langmuir* **2008**, *24*, 4959–4967.
89. F. Cheng, L. J. Gamble, D. G. Castner, *Anal. Chem.* **2008**, *80*, 2564–2573.
90. J. Homola, *Chem. Rev.* **2008**, *108*, 462–493.
91. C. Cao, J. P. Kim, B. W. Kim, H. Chae, H. C. Yoon, S. S. Yang, S. J. Sim, *Biosens. Bioelectron.* **2006**, *21*, 2106–2113.
92. J. W. Chung, R. Bernhardt, J. C. Pyun, *Sens. Actuators B* **2006**, *118*, 28–32.
93. G. Schmid, Ed., *Nanoparticles. From Theory to Application*, Weinheim: Wiley-VCH, **2004**.
94. G. A. Ozin, A. C. Arsenault, *Nanochemistry. A Chemical Approach to Nanomaterials*, Cambridge, UK: Royal Society of Chemistry, **2005**.

95. G. Schmid, *Chem. Rev.* **1992**, *92*, 1709–1727.
96. M. C. Daniel, D. Astruc, *Chem. Rev.* **2004**, *104*, 293–346.
97. A. C. Templeton, W. P. Wuelfing, R. W. Murray, *Acc. Chem. Res.* **2000**, *33*, 27–36.
98. A. Badia, R. B. Lennox, L. Reven, *Acc. Chem. Res.* **2000**, *33*, 475–481.
99. (a) C. M. Niemeyer, C. A. Mirkin, Eds., *Nanobiotechnology*, Weinheim: Wiley-VCH, **2004**; (b) C. A. Mirkin, C. M. Niemeyer, Eds., *Nanobiotechnology II*, Weinheim: Wiley-VCH, **2007**.
100. V. Rotello, Ed., *Nanoparticles. Building Blocks for Nanotechnology*, New York: Kluwer Academic/Plenum, **2004**.
101. N. L. Rosi, C. A. Mirkin, *Chem. Rev.* **2005**, *105*, 1547–1562.
102. J. J. Storhoff, C. A. Mirkin, *Chem. Rev.* **1999**, *99*, 1849–1862.
103. (a) C.-C. You, A. Chompoosor, V. M. Rotello, *Nano Today* **2007**, *2*, 34–43; (b) C.-C. You, A. Verma, V. M. Rotello, *Soft Matter* **2006**, *2*, 190–204; (c) C.-C. You, O. R. Miranda, B. Gider, P. S. Ghosh, I.-B. Kim, B. Erdogan, S. A. Krovi, U. H. F. Bunz, V. M. Rotello, *Nature Nanotechnol.* **2007**, *2*, 318–323.
104. R. L. Phillips, O. R. Miranda, C.-C. You, V. M. Rotello, U. H. F. Bunz, *Angw. Chem. Int. Ed.* **2008**, *47*, 2590–2594.
105. G. Han, P. Ghosh, M. De, V. M. Rotello, *Nanobiotechnology* **2007**, *3*, 40–45.
106. J. E. Ghadiali, M. M. Stevens, *Adv. Mater.* **2008**, *20*, 4359–4363.
107. D. Nykypanchuk, M. M. Maye, D. van der Lelie, O. Gang, *Nature* **2008**, *451*, 549–552.
108. S. Y. Park, A. K. R. Lytton-Jean, B. Lee, S. Weigand, G. C. Schatz, C. A. Mirkin, *Nature* **2008**, *451*, 553–556.
109. (a) G. Schmid, R. Pfeil, R. Boese, F. Bandermann, S. Meyer, G. H. M. Calis, J. W. A. van der Velden, *Chem. Ber.* **1981**, *114*, 3634–3642; (b) G. Schmid, A. Lehnert, U. Kreibig, Z. Adamczyk, P. Belouschek, *Z. Naturforsch.* **1990**, *45*, 989–994; (c) G. Schmid, Ed., *Clusters and Colloids*, Weinheim: VCH, **1994**; (d) G. Schmid, *J. Chem. Soc., Dalton Trans.* **1998**, 1077–1082.
110. L. F. Chi, M. Hartig, T. Drechsler, Th. Schwaack, S. Seidel, H. Fuchs, G. Schmid, *Appl. Phys. A* **1998**, *66*, S187–S190.
111. U. Simon, G. Schön, G. Schmid, *Angew. Chem. Int. Ed. Engl.* **1993**, *32*, 250–254.
112. G. Schmid, Y.-P. Liu, M. Schumann, T. Raschke, C. Radehaus, *Nano Lett.* **2001**, *1*, 405–407.
113. G. Schmid, U. Simon, *Chem. Commun.* **2005**, 697–710.
114. M. Haruta, N. Yamada, T. Kobayashi, S. Iijima, *J. Catal.* **1989**, *115*, 301–309.
115. G. C. Bond, C. Louis, D. T. Thompson, *Catalysis by Gold*, London: Imperial College Press, **2006**.
116. M. Chen, D. W. Goodman, *Acc. Chem. Res.* **2006**, *39*, 739–746.
117. T. Ishida, M. Haruta, *Angew. Chem. Int. Ed.* **2007**, *46*, 7154–7156.
118. B. K. Min, C. M. Friend, *Chem. Rev.* **2007**, *107*, 2709–2724.
119. A. S. K. Hashmi, *Chem. Rev.* **2007**, *107*, 3180–3211.
120. (a) J. Chou, E. W. McFarland, *Chem. Commun.* **2004**, 1648–1649; (b) N. F. Zheng, G. D. Stucky, *J. Am. Chem. Soc.* **2006**, *128*, 14278–14280; (c) G. Budroni, A. Corma, *Angew. Chem. Int. Ed.* **2006**, *45*, 3328–3331; (d) L. D. Menard, F. Xu, R. G. Nuzzo, J. C. Yang, *J. Catal.* **2006**, *243*, 64–73; (e) N. Hickey, P. A. Larochette, C. Gentilini, L. Sordelli, L. Olivi, S. Polizzi, T. Montini, P. Fornasiero, L. Pasquato, M. Graziani, *Chem. Mater.* **2007**, *19*, 650–651.
121. C. J. Murphy, A. M. Gole, S. E. Hunyadi, J. W. Stone, P. N. Sisco, A. Alkilany, B. E. Kinard, P. Hankins, *Chem. Comm.* **2008**, 544–557.
122. C. Burda, X. Chen, R. Narayanan, M. A. El-Sayed, *Chem. Rev.* **2005**, *105*, 1025–1102.
123. C. J. Murphy, T. K. Sau, A. M. Gole, C. J. Orendorff, J. Gao, L. Gou, S. E. Hunyadi, T. Li, *J. Phys. Chem. B* **2005**, *109*, 13857–13870.
124. K. G. Thomas, P. V. Kamat, *Acc. Chem. Res.* **2003**, *36*, 888–898.
125. M. De, P. S. Ghosh, V. M. Rotello, *Adv. Mater.* **2008**, *20*, 4225–4241.
126. C. Sönnichsen, B. M. Reinhard, J. Liphard, A. P. Alivisatos, *Nature Biotechnol.* **2005**, *23*, 741–745.
127. L. Cognet, C. Tardin, D. Boyer, D. Choquet, P. Tamarat, B. Lounis, *Proc. Natl. Acad. Sci. U.S.A.* **2003**, *100*, 11350–11355.
128. G. F. Paciotti, L. Myer, D. Weinreich, D. Goia, N. Pavel, R. E. McLaughlin, L. Tamarkin, *Drug Deliv.* **2004**, *11*, 169–183.

129. Y. Chen, J. Cho, A. Young, T. A. Taton, *Langmuir* **2007**, *23*, 7491–7497.
130. N. L. Goicochea, M. De, V. M. Rotello, S. Mukhopadhyay, B. Dragnea, *Nano Lett.* **2007**, *7*, 2281–2290.
131. J. Turkevich, P. C. Stevenson, J. Hillier, *Discuss. Faraday Soc.* **1951**, *11*, 55–75.
132. G. Frens, *Nature Phys. Sci.* **1973**, *241*, 20–22.
133. K. C. Grabar, R. G. Freeman, M. B. Hommer, M. J. Natan, *Anal. Chem.* **1995**, *67*, 735–743.
134. M. Brust, M. Walker, D. Bethell, D. J. Schiffrin, R. Whyman, *J. Chem. Soc., Chem. Commun.* **1994**, 801–802.
135. Y. G. Sun, Y. Xia, *Science* **2002**, *298*, 2176–2179.
136. G. Carotenuto, L. Nicolais, *J. Mater. Chem.* **2003**, *13*, 1038–1041.
137. S. Eustis, H.-Y. Hsu, M. A. El-Sayed, *J. Phys. Chem. B* **2005**, *109*, 4811–4815.
138. S. Eustis, M. A. El-Sayed, *J. Phys. Chem. B* **2006**, *110*, 14014–14019.
139. J. Shan, H. Tenhu, *Chem. Commun.* **2007**, 4580–4598.
140. A. Roucoux, J. Schulz, H. Patin, *Chem. Rev.* **2002**, *102*, 3757–3778.
141. M. Zheng, Z. Li, X. Huang, *Langmuir* **2004**, *20*, 4226–4235.
142. T. A. Taton, C. A. Mirkin, R. L. Letsinger, *Science* **2000**, *289*, 1757–1760.
143. C. A. Mirkin, R. L. Letsinger, R. C. Mucic, J. J. Storhoff, *Nature* **1996**, *382*, 607–609.
144. A. P. Alivisatos, K. P. Johnsson, X. Peng, T. E. Wilson, C. J. Loweth, M. P. Bruchez, Jr., P. G. Schultz, *Nature* **1996**, *382*, 609–611.
145. M. Giersig, P. Mulvaney, *Langmuir* **1993**, *9*, 3408–3413.
146. M. J. Hostetler, J. E. Wingate, C.-J. Zhong, J. E. Harris, R. W. Vachet, M. R. Clark, J. D. Londono, S. J. Green, J. J. Stokes, G. D. Wingall, G. L. Glish, M. D. Porter, N. D. Evans, R. W. Murray, *Langmuir* **1998**, *14*, 17–30.
147. C. A. Waters, A. J. Mills, K. A. Johnson, D. J. Schiffrin, *Chem. Commun.* **2003**, 540–541.
148. M. Brust, J. Fink, D. Bethell, D. J. Schiffrin, C. J. Kiely, *J. Chem. Soc., Chem. Commun.* **1995**, 1655–1656.
149. P. D. Jadzinsky, G. Calero, C. J. Ackerson, D. A. Bushnell, R. D. Kornberg, *Science* **2007**, *318*, 430–433.
150. S. Chen, R. S. Ingram, M. J. Hostetler, J. J. Pietron, R. W. Murray, T. G. Schaaff, J. T. Khoury, M. M. Alvarez, R. L. Whetten, *Science* **1998**, *280*, 2098–2101.
151. P. Moriarty, *Rep. Prog. Phys.* **2001**, *64*, 297–381.
152. D. Boyer, P. Tamarat, A. Maali, B. Lounis, M. Orrit, *Science* **2002**, *297*, 1160–1163.
153. (a) R. L. Whetten, J. T. Khoury, M. M. Alvarez, S. Murthy, I. Vezmar, Z. L. Wang, P. W. Stephens, C. L. Cleveland, W. D. Luedtke, U. Landman, *Adv. Mater.* **1996**, *8*, 428–433; (b) R. L. Whetten, M. N. Shafigullin, J. T. Khoury, T. G. Schaaff, I. Vezmar, M. M. Alvarez, A. Wilkinson, *Acc. Chem. Res.* **1999**, *32*, 397–406.
154. T. G. Schaaff, G. Knight, M. N. Shafigullin, R. F. Borkman, R. L. Whetten, *J. Phys. Chem. B* **1998**, *102*, 10643–10646.
155. (a) J. P. Wilcoxon, J. E. Martin, P. Provencio, *Langmuir* **2000**, *16*, 9912–9920; (b) H. Tsunoyama, Y. Negishi, T. Tsukuda, *J. Am. Chem. Soc.* **2006**, *128*, 6036–6037.
156. I. Hussain, S. Graham, Z. Wang, B. Tan, D. C. Sherrington, S. P. Rannard, A. I. Cooper, M. Brust, *J. Am. Chem. Soc.* **2005**, *127*, 16398–16399.
157. P. Pengo, G. Fantuzzi, L. Pasquato, P. Scrimin, *J. Supramol. Chem.* **2002**, *2*, 305–310.
158. S. S. Agasti, C.-C. You, P. Arumugam, V. M. Rotello, *J. Mater. Chem.* **2008**, *18*, 70–73.
159. C. J. Ackerson, P. D. Jadzinsky, R. D. Kornberg, *J. Am. Chem. Soc.* **2005**, *127*, 6550–6551.
160. Y. Zhao, W. Perez-Segarra, Q. Shi, A. Wei, *J. Am. Chem. Soc.* **2005**, *127*, 7328–7329.
161. A. D'Aléo, R. M. Williams, F. Osswald, P. Edamana, U. Hahn, J. van Heyst, F. D. Tichelaar, F. Vögtle, L. De Cola, *Adv. Funct. Mater.* **2004**, *14*, 1167–1177.
162. G. Bergamini, P. Ceroni, V. Balzani, M. Gingras, J.-M. Raimundo, V. Morandi, P. G. Merli, *Chem. Commun.* **2007**, 4167–4169.
163. E. E. Foos, A. W. Snow, M. E. Twigg, M. G. Ancona, *Chem. Mater.* **2002**, *14*, 2401–2408.
164. A. G. Kanaras, F. S. Kamounah, K. Schaumburg, C. J. Kiely, M. Brust, *Chem. Commun.*, **2002**, 2294–2295.

165. P. Pengo, S. Polizzi, M. Battagliarin, L. Pasquato, P. Scrimin, *J. Mater. Chem.* **2003**, *13*, 2471–2478.
166. M. Zheng, F. Davidson, X. Huang, *J. Am. Chem. Soc.* **2003**, *125*, 7790–7791.
167. M. Zheng, Z. Li, X. Huang, *Langmuir* **2004**, *20*, 4226–4235.
168. M. Schulz-Dobrick, K. V. Sarathy, M. Jansen, *J. Am. Chem. Soc.* **2005**, *127*, 12816–12817.
169. C. Gentilini, F. Evangelista, P. Rudolf, P. Franchi, M. Lucarini, L. Pasquato *J. Am. Chem. Soc.* **2008**, *130*, 15678–15682.
170. F. Stellacci, C. A. Bauer, T. Meyer-Friedrichsen, W. Wenseleers, S. R. Marder, J. W. Perry, *J. Am. Chem. Soc.* **2003**, *125*, 328–329.
171. (a) S. Eriksson, U. Nylen, S. Rojas, M. Boutonnet, *Appl. Catal. A* **2004**, *265*, 207–219; (b) D.-H. Chen, C.-J. Chen, *J. Mater. Chem.* **2002**, *12*, 1557–1562.
172. J. M. de la Fuente, A. G. Barrientos, T. C. Rojas, J. Rojo, J. Cañada, A. Fernández, S. Penadés, *Angew. Chem. Int. Ed.* **2001**, *40*, 2257–2261.
173. J. M. de la Fuente, S. Penadés, *Biochim. Biophys. Acta* **2006**, *1760*, 636–651.
174. (a) M. J. Hostetler, S. J. Green, J. J. Stockes, R. W. Murray, *J. Am. Chem. Soc.* **1996**, *118*, 4212–4213; (b) R. S. Ingram, M. J. Hostetler, R. W. Murray, *J. Am. Chem. Soc.* **1997**, *119*, 9175–9178.
175. M. De, C.-C. You, S. Srivastava, V. M. Rotello, *J. Am. Chem. Soc.* **2007**, *129*, 10747–10753.
176. P. Pengo, S. Polizzi, L. Pasquato, P. Scrimin, *J. Am. Chem. Soc.* **2005**, *127*, 1616–1617.
177. P. Pengo, L. Baltzer, L. Pasquato, P. Scrimin, *Angew. Chem. Int. Ed.* **2007**, *45*, 400–404.
178. P. Pengo, Q. B. Broxterman, B. Kaptein, L. Pasquato, P. Scrimin, *Langmuir* **2003**, *19*, 2521–2524.
179. M. J. Hostetler, A. C. Templeton, R. W. Murray, *Langmuir* **1999**, *15*, 3782–3789.
180. Y. Song, R. W. Murray, *J. Am. Chem. Soc.* **2002**, *124*, 7096–7102.
181. M. Montalti, L. Prodi, N. Zaccheroni, R. Baxter, G. Teobaldi, F. Zerbetto, *Langmuir* **2003**, *19*, 5172–5174.
182. (a) P. Ionita, A. Caragheorgheopol, B. C. Gilbert, V. Cechick, *Langmuir* **2004**, *20*, 11536–11544; (b) P. Ionita, A. Caragheorgheopol, B. C. Gilbert, V. Chechik, *J. Am. Chem. Soc.* **2002**, *124*, 9048–9049; (c) P. Ionita, B. C. Gilbert, V. Chechik, *Angew. Chem. Int. Ed.* **2005**, *44*, 3720–3722.
183. M. Zachary, V. Chechik, *Angew. Chem. Int. Ed.* **2007**, *46*, 3304–3307.
184. M. Brust, D. J. Schiffrin, D. Bethell, C. J. Kiely, *Adv. Mater.* **1995**, *7*, 795–797.
185. C. Guarise, L. Pasquato, V. De Filippis, P. Scrimin, *Proc. Natl. Acad. Sci. U.S.A.* **2006**, *103*, 3978–3982.
186. G. Schmid, A. Lehnert, *Angew.Chem. Int. Ed. Engl.* **1989**, *28*, 780–781.
187. C. J. Loweth, W. B. Caldwell, X. Peng, A. P. Alivisatos, P. G. Schultz, *Angew. Chem. Int. Ed.* **1999**, *38*, 1808–1812.
188. M. M. Stevens, N. T. Flynn, C. Wang, D. A. Tirrell, R. Langer, *Adv. Mater.* **2004**, *16*, 915–918.
189. A. Laromaine, L. Koh, M. Murugesan, R. V. Ulijn, M. M. Stevens, *J. Am. Chem. Soc.* **2007**, *129*, 4156–4157.
190. J. J. Storhoff, R. Elghanian, R. C. Mucic, C. A. Mirkin, R. L. Letsinger, *J. Am. Chem. Soc.* **1998**, *120*, 1959–1964.
191. Z. Li, R. Jin, C. A. Mirkin, R. L. Letsinger, *Abstracts Papers Am. Chem. Soc.* **2001**, *222*, U342–U342.
192. C. S. Thaxton, C. A. Mirkin, in *Nanobiotechnology* (Eds. C. M. Niemeyer, C. A. Mirkin), Weinheim: Wiley-VCH, **2004**, 288–307.
193. J. Sharma, R. Chhabra, H. Yan, Y. Liu, *Chem. Commun.* **2008**, 2140–2142.
194. M. Cardenas, J. Barauskas, K. Schillen, J. L. Brennan, M. Brust, T. Nylander, *Langmuir* **2006**, *22*, 3294–3299.
195. R. Levy, N. T. K. Thanh, R. C. Doty, I. Hussain, R. J. Nichols, D. J. Schiffrin, M. Brust, D. G. Fernig, *J. Am. Chem. Soc.* **2004**, *126*, 10076–10084.
196. L. L. Rouhana, J. A. Jaber, J. B. Schlenoff, *Langmuir* **2007**, *23*, 12799–12801.
197. F. Manea, C. Bindoli, S. Polizzi, L. Lay, P. Scrimin, *Langmuir* **2008**, *24*, 4120–4124.
198. H. C. Kolb, M. G. Finn, K. B. Sharpless, *Angew. Chem. Int. Ed.* **2001**, *40*, 2004–2021.

199. (a) H. Nandivada, X. Jiang, J. Lahann, *Adv. Mater.* **2007**, *19*, 2197–2208; (b) J. L. Brennan, N. S. Hatzakis, T. R. Tshikhudo, N. Dirvianskyte, V. Razumas, S. Patkar, J. Vind, A. Svendsen, R. J. M. Nolte, A. E. Rowan, M. Brust, *Bioconjugate Chem.* **2006**, *17*, 1373–1375; (c) M. Fischler, A. Sologubenko, J. Mayer, G. Clever, G. Burley, J. Gierlich, T. Carell, U. Simon, *Chem. Commun.* **2008**, 169–171; (d) W. J. Sommer, M. Weck, *Langmuir* **2007**, *23*, 11991–11995.
200. R. Voggu, P. Suguna, S. Chandrasekaran, C. N. R. Rao, *Chem. Phys. Lett.* **2007**, *443*, 118–121.
201. Q. Huo, J. G. Worden, *J. Nanoparticle Res.* **2007**, *9*, 1013–1025.
202. D. Zanchet, C. M. Micheel, W. J. Parak, D. Gerion, A. P. Alivisatos, *Nano Lett.* **2001**, *1*, 32–35.
203. R. Levy, Z. Wang, L. Duchesne, R. C. Doty, A. I. Cooper, M. Brust, D. G. Fernig, *ChemBioChem* **2006**, *7*, 592–594.
204. (a) J. G. Worden, A.W. Shaffer, Q. Huo, *Chem. Commun.* **2004**, 518–519; (b) K.-M. Sung, D.W. Mosley, B. R. Peelle, S. Zhang, J. M. Jacobson, *J. Am. Chem. Soc.* **2004**, *126*, 5064–5065.
205. X. Liu, J. G. Worden, Q. Dai, J. Zou, J. Wang, Q. Huo, *Small* **2006**, *2*, 1126–1129.
206. L. J. Wan, M. Terashima, H. Noda, M. Osawa, *J. Phys. Chem. B* **2000**, *104*, 3563–3569.
207. R. L. Whetten, R. C. Price, *Science*, **2007**, *318*, 407–408.
208. E. G. Mednikov, L. F. Dahl, *Small* **2008**, *4*, 534–537.
209. R. L. Whetten, M. N. Shafigullin, J. T. Khoury, T. G. Schaaff, I. Vezmar, M. M. Alvarez, A. Wilkinson, *Acc. Chem. Res.* **1999**, *32*, 397–406.
210. M. H. Heaven, A. Dass, P. S. White, K. M. Holt, R. W. Murray, *J. Am. Chem. Soc.* **2008**, *130*, 3754–3755.
211. J. Akola, M. Walter, R. L. Whetten, H. Hakkinen, H. Gronbeck, *J. Am. Chem. Soc.* **2008**, *130*, 3756–3757.
212. (a) D. M. P. Mingos, *J. Chem. Soc. Dalton Trans.* **1976**, 1163–1169; (b) D. M. P. Mingos *J. Chem. Soc., Dalton Trans.* **1996**, 561–566. (c) C. E. Briant, B. R. C. Theobald, J. W. White, L. K. Bell, D. M. P. Mingos, *J. Chem. Soc., Chem. Commun.* **1981**, 201–202. (d) K. Nunokawa, S. Onaka, M. Ito, M. Horibe, T. Yonezawa, H. Nishihara, T. Ozeki, H. Chiba, S. Watase, M. Nakamoto, *J. Organomet. Chem.* **2006**, *691*, 638–642.
213. M. Walter, J. Akola, O. López-Acevedo, P. D. Jadzinsky, G. Calero, C. J. Ackerson, R. L. Whetten, H. Grönbeck, H. Häkkinen, *Proc. Natl. Acad. Sci. U.S.A.* **2008**, *105*, 9157–9162.
214. A. M. Jackson, J. W. Myerson, F. Stellacci, *Nature Mater.* **2004**, *3*, 330–336.
215. A. M. Jackson, Y. Hu, P. J. Silva, F. Stellacci, *J. Am. Chem. Soc.* **2006**, *128*, 11135–11149.
216. M. Lucarini, P. Franchi, G. F. Pedulli, C. Gentilini, S. Polizzi, P. Pengo, P. Scrimin, L. Pasquato, *J. Am. Chem. Soc.* **2005**, *127*, 16384–16385.
217. S. Rapino, F. Zerbetto, *Small* **2007**, *3*, 386–388.
218. A. Centrone, Y. Hu, A. M. Jackson, G. Zerbi, F. Stellacci, *Small* **2007**, *3*, 814–817.
219. O. Uzun, Y. Hu, A. Verma, S. Chen, A. Centrone, F. Stellacci, *Chem. Commun.* **2008**, 196–198.
220. A. Verma, O. Uzun, Y. Hu, Y. Hu, H.-S. Han, N. Watson, S. Chen, D. J. Irvine, F. Stellacci, *Nature Mater.* **2008**, *7*, 588–595.
221. C. Singh, P. K. Ghorai, M. A. Horsch, A. M. Jackson, R. G. Larson, F. Stellacci, S. C. Glotzer, *Phys. Rev. Lett.* **2007**, *99*, 226106.
222. R. P. Carney, G. A. DeVries, C. Dubois, H. Kim, J. Y. Kim, C. Singh, P. K. Ghorai, J. B. Tracy, R. L. Stiles, R. W. Murray, S. C. Glotzer, F. Stellacci, *J. Am. Chem. Soc.* **2008**, *130*, 798–799.
223. C. Gentilini, P. Franchi, E. Mileo, S. Polizzi, M. Lucarini, L. Pasquato, *Angew. Chem. Int. Ed.* **2009**, *48*, 3060–3064.
224. G. A. DeVries, M. Brunnbauer, Y. Hu, A. M. Jackson, B. Long, B. T. Neltner, O. Uzun, B. H. Wuntsch, F. Stellacci, *Science* **2007**, *315*, 358–361.
225. D. F. Perepichka, F. Rosei, *Angew. Chem. Int. Ed.* **2007**, *46*, 6006–6008.
226. R. Sardar, T. B. Heap, J. S. Shumaker-Parry, *J. Am. Chem. Soc.* **2007**, *129*, 5356–5357.
227. M. A. El-Sayed, *Acc. Chem. Res.* **2001**, *34*, 257–264.
228. C. J. Orendorff, T. K. Sau, C. J. Murphy, *Small* **2006**, *2*, 636–639.
229. J. Pérez-Juste, I. Pastoriza-Santos, L. M. Liz-Marzán, P. Mulvaney, *Coord. Chem. Rev.* **2005**, *249*, 1870–1901.

230. H. Wang, T. B. Huff, D. A. Zweifel, W. He, P. S. Low, A. Wei, J.-X. Cheng, *Proc. Natl. Acad. Sci. U.S.A.* **2005**, *102*, 15752–15756.
231. Y. Xia, P. Yang, Y. Sun, Y. Wu, B. Mayers, B. Gates, Y. Yin, F. Kim, H. Yan, *Adv. Mater.* **2003**, *15*, 353–389.
232. L. A. Bauer, N. S. Birenbaum, G. J. Meyer, *J. Mater. Chem.* **2004**, *14*, 517–526.
233. N. R. Jana, L. Gearheart, C. J. Murphy, *J. Phys. Chem. B* **2001**, *105*, 4065–4067.
234. X. C. Jiang, M. P. Pileni, *Colloids Surf. A* **2007**, *295*, 228–232.
235. X. C. Jiang, A. Brioude, M. P. Pileni, *Colloids Surf. A* **2006**, *277*, 201–206.
236. J. Pérez-Juste, L. M. Liz-Marzán, S. Carnie, D. Y. C. Chan, P. Mulvaney, *Adv. Funct. Mater.* **2004**, *14*, 571–579.
237. D. K. Smith, B. A. Korgel, *Langmuir* **2008**, *24*, 644–649.
238. D. A. Zweifel, A. Wei, *Chem. Mater.* **2005**, *17*, 4256–4261.
239. C. J. Johnson, E. Dujardin, S. A. Davis, C. J. Murphy, S. Mann, *J. Mater. Chem.* **2002**, *12*, 1765–1770.
240. (a) K. K. Caswell, J. N. Wilson, U. H. F. Bunz, C. J. Murphy, *J. Am. Chem. Soc.* **2003**, *125*, 13914–13915; (b) A. Gole, C. J. Murphy, *Langmuir* **2005**, *21*, 10756–10762.
241. J. Y. Chang, H. Wu, H. Chen, Y. C. Ling, W. Tan, *Chem. Commun.* **2005**, 1092–1094.
242. H. Liao, J. H. Hafner, *Chem. Mater.* **2005**, *17*, 4636–4641.
243. K. G. Thomas, S. Barazzouk, B. I. Ipe, S. T. S. Joseph, P. V. Kamat, *J. Phys. Chem. B* **2004**, *108*, 13066–13068.
244. P. K. Sudeep, S. T. S. Joseph, K. G. Thomas, *J. Am. Chem. Soc.* **2005**. *127*, 6516–6517.
245. X. Hu, W. Cheng, T. Wang, E. Wang, S. Dong, *Nanotechnology* **2005**, *16*, 2164–2169.
246. M. H. Zareie, X. Xu, M. B. Cortie, *Small* **2007**, *3*, 139–145.
247. R. T. Tom, A. K. Samal, T. S. Sreeprasad, T. Pradeep, *Langmuir* **2007**, *23*, 1320–1325.
248. N. J. Durr, T. Larson, D. K. Smith, B. A. Korgel, K. Sokolov, A. Ben-Yakar, *Nano Lett.* **2007**, *7*, 941–945.
249. A. Gole, C. J. Murphy, *Langmuir* **2008**, *24*, 266–272.
250. A. K. Oyelere, P. C. Chen, X. Huang, I. H. El-Sayed, M. A. El-Sayed, *Bioconjugate Chem.* **2007**, *18*, 1490–1497.
251. R. Cortesi, E. Esposito, E. Menegatti, R. Gambari, C. Nastruzzi, *Int. J. Pharm.* **1996**, *139*, 69–78.
252. D. Mirska, K. Schirmer, S. S. Funari, A. Langner, B. Dobner, G. Brezesinski, *Colloids Surf. B* **2005**, *40*, 51–59.
253. E. Dujardin, S. Mann, L. B. Hsin, C. R. C. Wang, *Chem. Commun.* **2001**, 1264–1265.
254. T. B. Huff, M. N. Hansen, Y. Zhao, J.-X. Cheng, A. Wei, *Langmuir* **2007**, *23*, 1596–1599.
255. L. Tong, Y. Zhao, T. B. Huff, M. N. Hansen, A. Wei, J.-X. Cheng, *Adv. Mater.* **2007**, *19*, 3136–3141.
256. T. B. Huff, L. Tong, Y. Zhao, M. N. Hansen, J.-X. Cheng, A. Wei, *Nanomedicine* **2007**, *2*, 125–132.
257. Q. Dai, J. Coutts, J. Zou, Q. Huo, *Chem. Commun.* **2008**, 2858–2860.
258. B. Wiley, Y. Sun, B. Mayers, Y. Xia, *Chem. Eur. J.* **2005**, *11*, 454–463.
259. Y. Sun, Y. Xia, *J. Am. Chem. Soc.* **2004**, *126*, 3892–3901.
260. Y. Sun, B. Mayers, Y. Xia, *Adv. Mater.* **2003**, *15*, 641–646.
261. Y. Sun, Y. Xia, *Nano Lett.* **2003**, *3*, 1569–1572.
262. J. Chen, F. Saeki, B. J. Wiley, H. Cang, M. J. Cobb, Z.-Y. Li, L. Au, H. Zhang, M. B. Kimmey, X. Li, Y. Xia, *Nano Lett.* **2005**, *5*, 473–477.
263. S. E. Skrabalak, J. Chen, Y. Sun, X. Lu, L. Au, C. M. Cobley, Y. Xia, *Acc. Chem. Res.* **2008**, *41*, 1587–1595.

Chapter 5

Organically Functionalized Semiconductor Nanocrystals: Synthesis, Properties and System Design for Optoelectronic Applications

PETER REISS, JULIA DE GIROLAMO, AND ADAM PRON

The Supramolecular Chemistry of Organic–Inorganic Hybrid Materials. Edited by Knut Rurack and Ramón Martínez-Máñez

5.1 INTRODUCTION

The terms *quantum dots* (QDs) and *semiconductor nanocrystals* (NCs) are used as synonyms in the scientific literature. The term *dots* implies of course that the discussed particles are spherical. Therefore, we will use the term *nanocrystals* throughout this chapter to emphasize the fact that colloidal nanoparticles of any shape (e.g., spherical, rodlike, branched) are being described. Since their discovery in the early 1980s, the research devoted to colloidal semiconductor NCs has grown exponentially, as judged from the increasing number of publications and conferences related to the chemistry, physics, and materials science of these nanoobjects. This makes any attempt to provide an exhaustive review of the domain difficult and therefore, the selection of the data presented in this review reflects to some extent the research interests of the authors. Special emphasis is put on those aspects of NC science that are most closely related to their expected technological and industrial applications in optoelectronic devices. The organization of this paper is as follows.

The first part of the chapter reviews progress in the synthesis of monodisperse semiconductor NCs and gives a basic introduction to their specific physical properties. In conformity with the literature, the term "monodisperse" is used here to describe colloidal samples, in which the standard deviation of the particle diameter does not exceed 5%. Throughout the text we will restrict ourselves to the description of binary II–VI (CdSe, CdS, CdTe, ZnSe, etc.), III–V (InP, InAs), and IV–VI (PbS, PbSe, PbTe) semiconductor NCs. These systems exhibit optical properties that can be varied in the visible part of the spectrum, the near UV or near IR by changing the NC size and/or composition.

The second part of the review is dedicated to organic–inorganic materials fabricated from conjugated polymers and NCs. Their use in optoelectronic devices such as light-emitting diodes or photovoltaic cells is—besides biological labeling and biochemical analysis, see Chapter 12—the most promising application of NCs that has emerged until now.

A number of recent reviews deal with the synthesis and physical properties of NCs,[1–4] whereas less work exists on their organic–inorganic composites.[5,6] This chapter gives an overview of the aforementioned research fields, with special focus on the latest results and developments.

5.2 BASIC PHYSICAL PROPERTIES AND SYNTHESIS OF SEMICONDUCTOR NANOCRYSTALS

5.2.1 Physical Properties of Semiconductor Nanocrystals

5.2.1.1 Quantum Confinement Effect

Colloidal semiconductor NCs are crystalline particles with diameters ranging typically from 1 to 10 nm, comprising some hundreds to a few thousands of atoms. The inorganic core consisting of the semiconductor material is capped by an organic outer layer of surfactant molecules (ligands), which provide sufficient repulsion between the crystals to prevent them from agglomeration. In the nanometer size regime, many physical properties of the semiconductor particles change with respect to the bulk material. Examples of this behavior are melting points and charging energies of NCs, which are, to a first approximation, proportional to the reciprocal value of their radii. At the origin of the great interest in NCs was yet another observation, namely the possibility of changing the semiconductor band gap, that is, the energy difference between the electron-filled valence band and the empty conduction band, by varying the particle size. In a bulk semiconductor, an electron e^- can be excited from the valence to the conduction band by absorption of a photon with an appropriate energy, leaving a hole h^+ in the valence band. Feeling each other's charge, the electron and hole do not move independently from each other because of the Coulomb attraction. The formed $e^- - h^+$ bound pair is called an exciton and has its lowest energy state slightly below the lower edge of the conduction band. At the same time, its wave function is extended over a large region (several lattice spacings), that is, the exciton radius is large, since the effective masses of the charge carriers are small and the dielectric constant is high.[7] To give examples, the Bohr exciton radii in bulk CdS and CdSe are approximately 3 and 5 nm. Reduction of the particle size to a few nanometers produces the unusual situation that the exciton size can exceed the crystal dimensions. To "fit" into the NC, the charge carriers have to assume higher kinetic energies, leading to an increasing band gap and quantization of the energy levels to discrete values. This phenomenon is commonly called the quantum confinement effect,[8] and its theoretical treatment is usually based on the quantum mechanical particle-in-a-box model.[9] With decreasing particle size, the energetic structure of the NCs changes from a band-like one to discrete levels. Therefore, in some cases a description by means of molecular orbital theory may be more appropriate, applying the terms HOMO (highest occupied molecular orbital) and LUMO (lowest unoccupied molecular orbital) instead of conduction band and valence band. This ambiguity in terminology reflects the fact that the properties of semiconductor NCs lie in between those of the corresponding bulk material and molecular compounds. The unique optical properties of semiconductor NCs are exploited in a large variety of applications essentially in the fields of biological labeling and (opto-)electronics.

Initiated by pioneering work in the early 1980s,[10–13] remarkable progress was made in the research on semiconductor NCs after the development of a novel chemical synthesis method in 1993, which allowed for the preparation of samples with a low size dispersion.[14] The physical properties of NCs, and in particular the optical ones, are strongly size dependent. To give an example, the linewidth of the photoluminescence band is directly related to the size dispersion of the NCs, and thus a narrow size distribution is necessary to obtain a pure emission color. The method developed by Murray and coworkers was the first one to allow for the synthesis of monodisperse cadmium chalcogenide NCs in a size range of 2 to 12 nm. It relies on the rapid injection of organometallic precursors into a hot organic solvent. Numerous synthesis methods deriving from the original one have been reported in the literature in the last 15 years. Nowadays, a much better understanding of the influence of the different reaction parameters has been achieved, allowing for the rational design of synthesis protocols.

5.2.1.2 *Optical Properties of Semiconductor Nanocrystals*

As stated in the previous section, absorption of a photon by the NC occurs if its energy exceeds the band gap. Due to quantum confinement, decreasing the particle size results in a hypsochromic (blue-) shift of the absorption onset. A relatively sharp absorption feature near the absorption onset corresponds to the excitonic peak, that is, the lowest excited state exhibiting an important oscillator strength (see Fig. 5.1). While its position depends on the band gap and, consequently, on the particle size, its shape and width are strongly influenced by the distribution in size, as well as the form and stoichiometry of the NCs. Therefore, polydisperse samples typically exhibit only a shoulder in the absorption spectrum at the position of the excitonic transition. Less pronounced absorption features in the lower wavelength range correspond to excited states of higher energy.[15] As a rule of thumb, it can be asserted that the

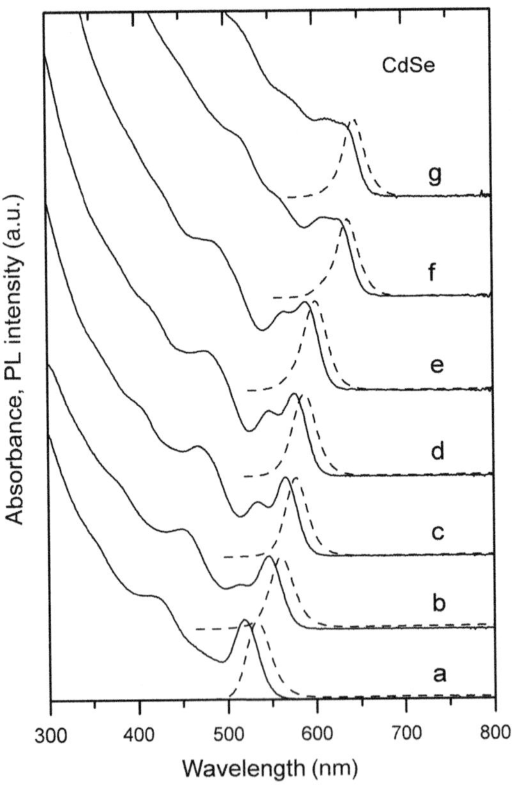

Figure 5.1 Room temperature absorption and photoluminescence spectra of CdSe nanocrystals in the size range of 3 nm (a) to 7 nm (g).[16] (Reprinted with permission from D. V. Talapin et al., *Nano Lett.* **2001**, *1*, 207–211. Copyright 2001 American Chemical Society.)

larger the number of such spectral features and the more distinctly they are resolved in the absorption spectrum, the smaller is the size dispersion of the sample. In Figure 5.1 the absorption and photoluminescence spectra of a series of CdSe NCs differing in size are depicted.

Photoluminescence (PL), that is, the generation of luminescence through excitation by photons, is formally divided into two categories, fluorescence and phosphorescence, depending on the electronic configuration of the excited state and the emission pathway. Fluorescence is the property of a semiconductor to absorb photons with an energy $h\nu_e$ superior to its band gap, and, after charge carrier relaxation via phonons to the lowest excited state, to emit light of a higher wavelength (lower energy $h\nu_f$) after a brief interval, called the fluorescence lifetime (Fig. 5.2). The process of phosphorescence occurs in a similar manner, but with a much longer excited state lifetime, due to the symmetry of the state.

The emitted photons have an energy corresponding to the band gap of the NCs and for this reason the emission color can be tuned by changing the particle size. It should be noted here that efficient room temperature band-edge emission is only observed for NCs with proper surface passivation because otherwise charge carriers are very likely to be trapped in surface states, enhancing nonradiative recombination. Due to spectral diffusion and the size distribution of NCs, the room temperature luminescence linewidths of ensembles lie for the best samples of CdSe NCs in the range of 20 to 25 nm (full width at half maximum, FWHM). As can be seen in Figure 5.1, the maxima of the emission bands are red-shifted by approximately 10 to 20 nm as compared to the excitonic peak in the absorption spectra. This phenomenon is usually referred to as Stokes-shift and has its origin in the particular structure

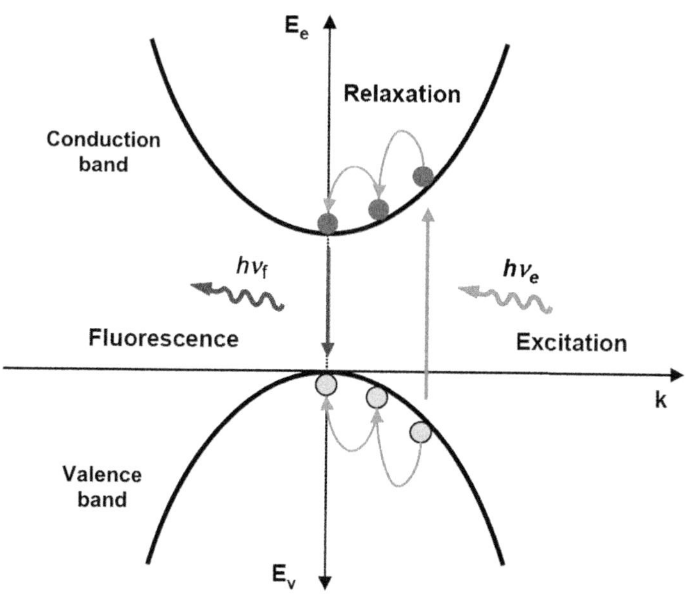

Figure 5.2 Fluorescence in a bulk semiconductor.

of the exciton energy levels inside the NC. Models using the effective mass approximation show that in bulk wurtzite CdSe the exciton state ($1S_{3/2}1S_e$) is eightfold degenerate.[17] In CdSe NCs, this degeneracy is partially lifted and the band-edge state is split into five states, due to the influence of the internal crystal field, effects arising from the nonspherical particle shape and the electron-hole exchange interaction (see Fig. 5.3). The latter term is strongly enhanced by quantum confinement.[18]

Two states, one singlet state and one doublet state, are optically inactive for symmetry reasons. The energetic order of the three remaining states depends on the size and shape of the NC. In the case of weak excitation on a given state, absorption depends exclusively on its oscillator strength. As the oscillator strength of the second and third excited (bright) states is significantly higher than that of the first (dark) one, excitation by photon absorption occurs to the bright states. On the other hand, photoluminescence depends on the product of oscillator strength and population of the concerned state. Relaxation via acoustic phonon emission from bright states to the dark band-edge state causes strong population of the latter and enables radiative recombination (Fig. 5.3). This model is corroborated by the experimental room temperature values of the Stokes-shift, which are consistent with the energy differences between the related bright and dark states.

The emission efficiency of an ensemble of NCs is expressed in terms of the fluorescence quantum yield (QY), that is, the ratio between the number of absorbed photons and the number of emitted photons. The theoretical value of 1 is generally not observed due to a number of reasons. First of all, spectroscopic investigation of single semiconductor NCs revealed that their emission under continuous excitation turns on and off intermittently, that is, at any time a certain number of NCs are in non-emissive "off" states.[19–21] Furthermore, the QY may be additionally reduced as a result of quenching caused by surface trap states. As both of these limiting factors are closely related to the quality of the NCs' surface passivation, they can be considerably diminished by appropriate surface functionalization. The latter can be achieved for example by changing the nature of the organic stabilizing ligands, which are capping the NCs after their synthesis. To give an example, after substitution of the trioctylphosphine oxide (TOPO) cap on CdSe NCs by hexadecylamine (HDA) or allylamine, an increase of the QY from about 10% to values of 40% to 50% has been reported.[16] In this case,

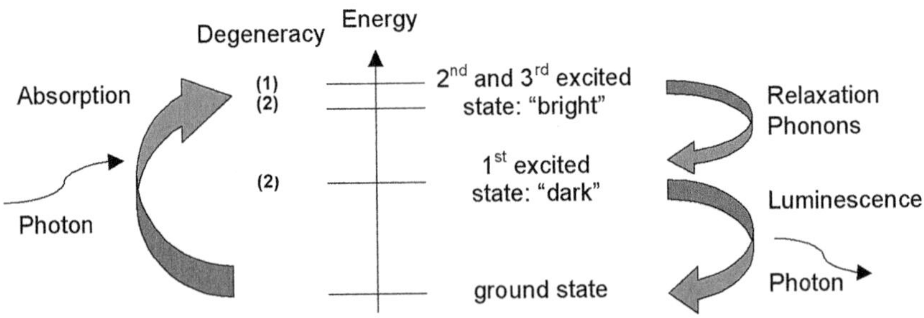

Figure 5.3 Schematic representation of the exciton states of CdSe NCs involved in absorption and emission processes.

better surface passivation probably results from an increased capping density of the sterically less hindered amines as compared to TOPO. The influence of thiol and amine ligands on the PL properties of CdSe-based NCs has been studied in further detail by Munro et al.[22] Jang et al. obtained cadmium chalcogenide NCs with QYs up to 75% after treatment with $NaBH_4$ and explained the better surface passivation by the formation of a cadmium oxide layer.[23] In the case of III–V semiconductors, the groups of Mićić and Talapin reported an enhancement of the PL QY of InP NCs from less than 1% to 25% – 40% on treatment with HF.[24,25] Here, the improved emission properties have been attributed to the removal of phosphorus dangling bonds under PF_3 elimination from the NC surface.

However, in view of further NC functionalization, it is highly desirable to provide a surface passivation, which is insensitive to subsequent ligand exchange. This is obviously not the case with the described procedures for QY enhancement. A suitable and widely applied method consists of the growth of an inorganic shell on the surface of the NCs. The resulting core–shell systems will be described in detail in the following section.

5.2.1.3 Structural Properties of Semiconductor NCs

Most binary octet semiconductors crystallize either in the cubic zinc blende (ZB) or in the hexagonal wurtzite (W) structure, both of which are four-coordinate and vary in the layer stacking along (111), showing an ABCABC or an ABAB sequence, respectively. The room temperature ground state structures of selected II–VI, III–V, and IV–VI semiconductors is given in Table 5.1. In cases of relatively low difference in the total energy between the ZB and the W structure (e.g., CdTe, ZnSe), the materials exhibit the so-called W–ZB polytypism.[26] Depending on the experimental conditions, nucleation and growth of the NCs can take place in either structure and also the coexistence of both structures in the same nanoparticle is possible. Lead chalcogenide NCs crystallize in the six-coordinate rocksalt structure, and it has been shown that also CdSe NCs can exist in this crystal structure at ambient pressure, provided that their diameter exceeds a threshold size of 11 nm, below which they transform back to the four-coordinate structure.[27]

5.2.1.4 Semiconductor NCs of Core–Shell Type

Surface engineering is an important tool to control the properties of the NCs and in particular the optical ones. One important strategy is the overgrowth of NCs with a shell of a second semiconductor, resulting in core–shell systems. This method has been applied to improve the fluorescence QY and the stability against photooxidation but also, by proper choice of the core and shell materials, to tune the emission wavelength in a large spectral window. After pioneering work in the 1980s and the development of powerful chemical synthesis routes at the end of the 1990s,[29–31] a strongly increasing number of articles have been devoted to core–shell NCs in the last five years. Today, almost any type of core NC prepared by a robust chemical synthesis method has been overgrown with shells of other semiconductor materials.

Table 5.1 Structural Parameters of Selected Bulk Semiconductors

Material	Structure (300 K)	Type	E_{gap} (eV)	Lattice parameter (Å)	Density (kg/m^3)
ZnS	Zinc blende	II–VI	3.68	5.4102	4079
ZnSe	Zinc blende	II–VI	2.69	5.668	5266
ZnTe	Zinc blende	II–VI	2.39	6.104	5636
CdS	Wurtzite	II–VI	2.49	4.136/6.714	4820
CdSe	Wurtzite	II–VI	1.74	4.3/7.01	5810
CdTe	Zinc blende	II–VI	1.43	6.482	5870
GaN	Wurtzite	III–V	3.44	3.188/5.185	6095
GaP	Zinc blende	III–V	2.27[a]	5.45	4138
GaAs	Zinc blende	III–V	1.42	5.653	5318
GaSb	Zinc blende	III–V	0.75	6.096	5614
InN	Wurtzite	III–V	0.8	3.545/5.703	6810
InP	Zinc blende	III–V	1.35	5.869	4787
InAs	Zinc blende	III–V	0.35	6.058	5667
InSb	Zinc blende	III–V	0.23	6.479	5774
PbS	Rocksalt	IV–VI	0.41	5.936	7597
PbSe	Rocksalt	IV–VI	0.28	6.117	8260
PbTe	Rocksalt	IV–VI	0.31	6.462	8219

[a]Indirect band gap.
Source: Data from Reference 28.

Depending on the band gaps and the relative position of electronic energy levels of the involved semiconductors, the shell can have different functions in core–shell NCs. Scheme 5.1 gives an overview of the band alignment of the bulk materials, which are mostly used in NC synthesis. Two main cases can be distinguished, denominated type I and type II band alignment, respectively. In the former, the band gap of the shell material is larger than that of the core one, and both electrons and holes are confined in the core. In the latter, either the valence band edge or the conduction band edge of the shell material is located in the band gap of the core. Upon excitation of the NC, the resulting staggered band alignment leads to a spatial separation of the hole and the electron in different regions of the core–shell (CS) structure.

In type I core–shell NCs, the shell is used to passivate the surface of the core with the goal to improve its optical properties. The shell separates physically the surface of the optically active core NC from its surrounding medium. As a consequence, the sensitivity of the optical properties to changes in the local environment of the NCs' surface, induced for example by the presence of oxygen or water molecules, is reduced. With respect to core NCs, core–shell systems exhibit generally enhanced stability against photodegradation. At the same time, shell growth reduces the number of surface dangling bonds, which can act as trap states for charge carriers and reduce the fluorescence QY. The first published prototype system was CdSe/ZnS. The ZnS shell significantly improves the fluorescence QY and stability against photobleaching.

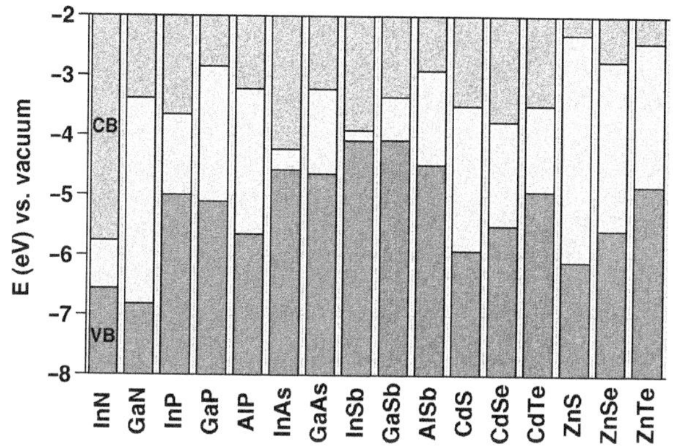

Scheme 5.1 Electronic energy levels of selected III–V and II–VI semiconductors using the valence band offsets from Reference 32 (VB: valence band, CB: conduction band).

Shell growth is accompanied by a *small* red shift (5 to 10 nm) of the excitonic peak in the UV-vis absorption spectrum and the PL wavelength. This observation is attributed to a partial leakage of the exciton into the shell material.

In type II systems, shell growth aims at a *significant* red shift of the emission wavelength of the NCs. The staggered band alignment leads to a smaller effective band gap than each one of the constituting core and shell materials. The interest of these systems is the possibility to tune the emission color with the shell thickness towards spectral ranges, which are difficult to attain with other materials. Type II NCs have been developed in particular for near infrared emission, using for example CdTe/CdSe or CdSe/ZnTe. In contrast to type I systems, the PL decay times are strongly prolonged in type II NCs due to the lower overlap of the electron and hole wavefunctions. As one of the charge carriers (electron or hole) is located in the shell, an overgrowth of type II core–shell NCs with an outer shell of an appropriate material can be used in the same way as in type I systems to improve the fluorescence QY and photostability.

5.2.2 Solution-Phase Synthesis of Semiconductor NCs

5.2.2.1 Synthesis of NCs of II-VI, III-V, and IV-VI Semiconductors

Solution-phase (or wet chemical) methods for the synthesis of semiconductor NCs can be roughly divided into two general categories based on the reaction medium applied:

1. Room temperature procedures consisting of NC precipitation in aqueous media, either in the presence of stabilizers or within microreactors of micellar type.

2. High temperature reactions in high boiling organic solvents of either hydrophobic or amphiphilic nature. These reactions are based on the temporal separation of nucleation and growth of NCs.

Among other reaction media, not discussed in more detail here, ionic liquids[33] and supercritical fluids[34] have been proposed.

Historically, synthesis in aqueous media was the first successful preparation method of colloidal semiconductor NCs. Procedures initially developed comprise NC formation in homogeneous aqueous solutions containing appropriate reagents and surfactant-type or polymer-type stabilizers.[35,36] The latter bind to the NC surface and stabilize the particles by steric hindrance and/or electrostatic repulsion in the case of charged stabilizers. Such stabilizers include in particular thiol derivatives, such as thioglycolic acid, mercaptopropionic acid, or thioglycerol. In parallel to this monophase synthesis, a biphase technique has been developed, which is based on the arrested precipitation of NCs within inverse micelles.[11,12,37,38] Here, nanometer-sized water droplets (dispersed phase) are stabilized in an organic solvent (continuous phase) by an amphiphilic surfactant. They serve as nanoreactors for the NC growth and prevent particle agglomeration at the same time. Both methods provide relatively simple experimental approaches using standard reagents as well as room temperature reactions and were of great importance for the development of NC synthesis. Furthermore, for some materials (e.g., mercury chalcogenides)[39–41] the (monophase) synthesis in aqueous media is the only successful preparation method reported today. On the other hand, the samples prepared by these synthetic routes usually exhibit size dispersions at least on the order of 15% and therefore, fastidious procedures of NC separation into "sharp" fractions have to be applied in order to obtain monodisperse samples. A more recent biphase preparation method is NC synthesis at the oil–water interface. The interested reader is referred to a review on this subject.[42]

The introduction in 1993 of a high temperature preparation method using organic solvents constituted an important step towards the fabrication of monodisperse CdS, CdSe, and CdTe NCs.[14] As demonstrated in classical studies by LaMer and Dinegar,[43] the synthesis of monodisperse colloids via homogeneous nucleation requires a temporal separation of nucleation and growth of the seeds. The LaMer plot (Fig. 5.4) is very useful to illustrate the separation of nucleation and growth by means of a nucleation burst.

Initially the concentration of monomers, that is, the minimum subunits of the crystal, constantly increases by addition from exterior or by *in situ* generation within the reaction medium. It should be noted that in stage I, no nucleation occurs even in supersaturated solution ($S > 1$) due to the extremely high energy barrier for spontaneous homogeneous nucleation. The latter is overcome in stage II for a yet higher degree of supersaturation ($S > S_c$), where nucleation and formation of stable nuclei take place. As the rate of monomer consumption induced by the nucleation and growth processes exceeds the rate of monomer supply, the monomer concentration and hence the supersaturation decreases below S_c, the level at which the nucleation rate becomes zero. In the following stage III, the particle growth continues under further monomer consumption as long as the system is in the supersaturated regime.

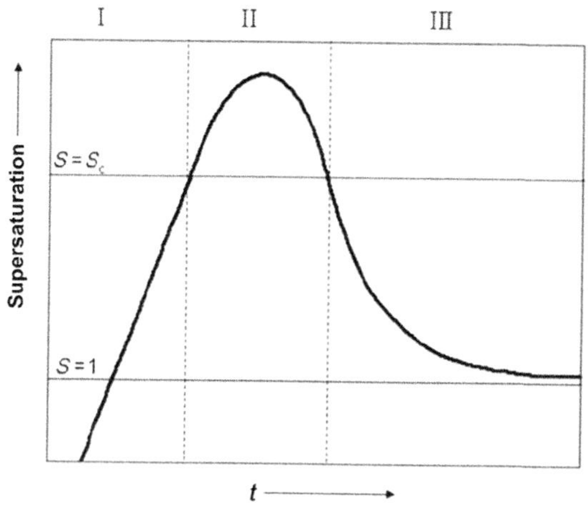

Figure 5.4 LaMer plot depicting the degree of supersaturation as a function of reaction time.[43] (Reprinted with permission from V. K. LaMer and R. H. Dinegar, *J. Am. Chem. Soc.* **1950**, *72*, 4847–4854. Copyright 1950 American Chemical Society.)

Experimentally, the separation of nucleation and growth can be achieved by rapid injection of the reagents into the hot solvent, which raises the precursor concentration in the reaction flask above the nucleation threshold (hot injection method).[44] The hot injection leads to an instantaneous nucleation, which is quickly quenched by the fast cooling of the reaction mixture (the solution to be injected is at room temperature) and by the decreased supersaturation after the nucleation burst. Another possibility relies on attaining the degree of supersaturation necessary for homogeneous nucleation via the *in situ* formation of reactive species upon supply of thermal energy (heating-up method).[2] This method is widely used in the synthesis of metallic nanoparticles, but recently an increasing number of examples of semiconductor NCs prepared by this approach can be found. Table 5.2 gives an overview of the applied method as well as of the combinations of precursors, stabilizers, and solvents mainly used in the synthesis of II–VI, III–V, and IV–VI semiconductor NCs.

In an ideal case, all crystallization nuclei are created at the same time and undergo identical growth. During the growth stage it is possible to carry out subsequent injections of precursors in order to increase the mean particle size without deterioration of the narrow size distribution as long as the concentration corresponding to the critical supersaturation S_c is not exceeded. Crystal growth from solution is in many cases followed by a second distinct growth process, which is referred to as Ostwald ripening.[92] It consists of the dissolution of the smallest particles because of their high surface energy and subsequent redeposition of the dissolved matter onto the bigger ones. Thereby the total number of NCs decreases, whereas their mean size increases. As shown in early studies,[93] Ostwald ripening can lead to reduced size dispersions of micron-sized colloids. In the case of nanometer-sized particles, however, Ostwald ripening generally yields size dispersions of the order of 15% to 20%,[4] and therefore the reaction should be stopped before this stage. Using the hot injection or heating-up

Table 5.2 Synthetic Parameters Used for the Preparation of Various II–VI, III–V, and IV–VI Semiconductor NCs in Organic Solvents

Material	Precursors and stabilizers[a]	Solvent(s)[a]	Method[b]	Reference
CdS, CdSe, CdTe	$CdMe_2$/TOP, $(TMS)_2Se$ or $(TMS)_2S$ or $(BDMS)_2Te$	TOPO	HI	14
CdSe	$CdMe_2$/TOP, TOP-Se	TOPO, HDA	HI	16, 45
CdSe, CdTe	CdO, TDPA, TOP-Se or TOP-Te	TOPO	HI	46
CdSe	CdO or $Cd(ac)_2$ or $CdCO_3$, TOP-Se, TDPA or SA or LA	TOPO	HI	47
CdS, CdSe	CdO, S/ODE or TBP-Se/ODE, OA	ODE	HI	48
CdSe	$Cd(st)_2$, TOP-Se, HH or BP	HDA, octadecane	HI	49
CdSe	$Cd(my)_2$, Se, OA/ODE	ODE	HU	50
CdSe	CdO, Se/ODE, OA	ODE	HI	51
CdSe	CdO, Se, OA	Olive oil	HI	52
CdSe	$Cd(st)_2$, TBP-Se, SA, DA	ODE	HI	53
CdS	$Cd(ac)_2$, S, MA	ODE	HU	54
CdS, ZnS	$Cd(hdx)_2$ or $Cd(ex)_2$ or $Cd(dx)_2$ or $Zn(hdx)_2$	HDA	HU	55, 56
CdS, ZnS	$CdCl_2$/OAm or $ZnCl_2$/OAm/TOPO, S/OAm	OAm	HU	57
CdTe	$CdMe_2$, TOP-Te	DA	HI	58
CdTe	CdO, TBP-Te/ODE or TOP-Te/ODE, OA	ODE	HI	59, 60
CdTe	CdO, TBP-Te, ODPA	ODE	HU	50
ZnS	$Zn(st)_2$, S/ODE	ODE, Tetracosane	HI	61
ZnS	$ZnEt_2$, S	HDA/ODE	HU	62
ZnSe	$ZnEt_2$, TOP-Se	HDA	HI	63
ZnSe	$Zn(st)_2$, TOP-Se	Octadecane	HI	64
ZnTe	Te and $ZnEt_2$ in TOP	ODA, ODE	HI	65
HgTe	$HgBr_2$, TOP-Te	TOPO	HI	66
$Cd_{1-x}Zn_xSe$	$ZnEt_2$/TOP, $CdMe_2$/TOP	TOPO, HDA	HI	67, 68
$Cd_{1-x}Zn_xSe$	$Zn(st)_2$, $Cd(st)_2$, TOP-Se	ODE	HI	69
$Cd_{1-x}Zn_xS$	CdO, ZnO, S/ODE, OA	ODE	HI	70
$CdSe_{1-x}Te_x$	CdO, TOP-Se, TOP-Te	TOPO, HDA	HI	71
InP	$InCl_3$ or $InCl_3/Na_2C_2O_4$, $P(TMS)_3$	TOPO or TOPO/TOP	HU	72, 73
InP, InAs	$In(ac)_3$, $P(TMS)_3$ or $As(TMS)_3$, MA	ODE	HI	74
InP	$InMe_3$, $P(TMS)_3$, MA	MM or DBS	HI	75

(Continued)

TABLE 5.2 *Continued*

Material	Precursors and stabilizers[a]	Solvent(s)[a]	Method[b]	Reference
InP	$In(ac)_3$, Ca_3P_2, MA	ODE	HI	76
InAs	$InCl_3$, $As(TMS)_3$	TOP	HI	77
GaP	$[Cl_2GaP(TMS)_2]_2$	TOPO/TOP	HU	78
GaP	$Ga(P^tBu_2)_3$	TOA, HDA	HI	79
GaP	$GaCl_3$, $P(TMS)_3$	TOPO	HI	80
GaN, AlN, InN	$[M(H_2NCONH_2)_6]Cl_3$ (M = Ga, Al, In)	TOA	HU	81
PbSe	$Pb(ac)_2$, OA, TOP-Se	Et_2O	HI	82–86
PbSe	$Pb(chbt)_2$, TBP-Se	TOPO	HI	87
PbSe	PbO, OA, TOP-Se	ODE	HI	88
PbS	PbO, OA, $(TMS)_2S$	ODE	HI	89
PbS	$PbCl_2$, S/OAm	OAm	HI	57
PbTe	$Pb(ac)_2$, OA, TOP-Te	Et_2O	HI	90
PbTe	PbO, OA, TOP-Te	ODE	HI	91

[a] $CdMe_2$ = dimethylcadmium, $ZnEt_2$ = diethylzinc, TMS = trimethylsilyl, $(BDMS)_2Te$ = bis(*tert*-butyldimethylsilyl) telluride, TDPA = tetradecylphosphonic acid, ODPA = octadecylphosphonic acid, SA = stearic acid, LA = lauric acid, OA = oleic acid, MA = myristic acid, ac = acetate, my = myristate, st = stearate, hdx = hexadecylxanthate, ex = ethylxanthate, dx = decylxanthate, TOPO = trioctylphosphine oxide, HDA = hexadecylamine, DA = dodecylamine, ODA = octadecylamine, TOP = trioctylphosphine, TBP = tributylphosphine, ODE = 1-octadecene, HH = hexadecyl hexadecanoate, BP = benzophenone, MM = methyl myristate, DBS = dibutyl sebacate, TOA = trioctylamine, chbt = cyclohexylbutyrate, OAm = oleylamine, Et_2O = diethylether.
[b] HI = hot injection method; HU = heating-up method.

synthesis methods, it is possible to obtain samples with 5% to 10% standard deviation from the mean size without postpreparative size fractionation.

One of the main disadvantages of the initially reported preparation methods lies in the fact that pyrophoric organometallic precursors were applied. Their use requires special experimental precautions and their extremely high reactivity restricts the batches to laboratory-scale quantities. As a result, in recent years the development of NC high temperature synthesis focused on the replacement of these organometallic precursors by easy to handle standard reagents. To give an example, in the preparation of cadmium chalcogenide NCs, dimethylcadmium has been successfully substituted by cadmium oxide or cadmium salts of weak acids (cadmium acetate, cadmium carbonate) after complexation with long chain phosphonic or carboxylic acids.[46,47] A further modification of the high temperature methods consisted of the appropriate selection of coordinating and noncoordinating solvents, with the goal to determine their influence on the nucleation and growth kinetics of the NCs and to fine-tune the reactivity of the precursors, aiming at narrow size distributions.[48] More recently, mechanistic studies have been undertaken by means of NMR spectroscopy, giving important insight into the chemical reactions occurring during the formation of CdSe NCs using the hot injection method.[94] Concerning other materials, it can be concluded that the synthesis methods initially developed for CdSe NCs have been adapted to the majority of II–VI and a few III–V semiconductors in the last five years (see Table 5.2). The exponentially increasing number of publications proves that this research field is highly active worldwide.

5.2.2.2 Synthesis of Core–Shell Systems

A general requirement for the synthesis of CS NCs with satisfactory optical properties is *epitaxial type shell growth*. Therefore an *appropriate band alignment* is not the sole criterion for choice of materials but, in addition, the core and shell materials should *crystallize in the same structure and exhibit a small lattice mismatch*. In the opposite case, the growth of the shell results in strain and the formation of defect states at the core–shell interface or within the shell. These can act as trap states for photogenerated charge carriers and diminish the fluorescence QY.[95] The structural parameters of selected semiconductor materials are summarized in Table 5.1.

Good precursors for shell growth should fulfill the criteria of high reactivity and selectivity (no side reactions). For practical reasons, and in particular if a scaling up of the reaction or industrialization of the production process is attempted, additional properties of the precursors come into play. Pyrophoric and/or highly toxic compounds require special precautions for their manipulation, especially if used in large quantities. To give an example, for the synthesis of zinc sulfide shells on various core NCs, first diethylzinc (pyrophoric) and hexamethyldisilathiane (toxic) have been proposed. Even though widely used in laboratory-scale syntheses, these compounds are not very suitable for large-scale production of ZnS overcoated NCs. Further criteria that have to be taken into account for the choice of the precursors concern the environmental risks related to the use of these compounds and eventually their degradation products, their price, and their commercial availability. As it is rather difficult to satisfy all of these factors, the development of shell synthesis methods is currently an active area of research.

The control of the shell thickness is a delicate point in the fabrication of CS NCs and deserves special attention. If the shell is too thin, the passivation of the core NCs is inefficient, resulting in reduced photostability. In the opposite case, in turn, the optical properties of the resulting CS NCs deteriorate as, driven by the lattice mismatch of the core and shell materials, defects are created with increasing shell thickness. CS systems are generally fabricated in a two-step procedure, consisting of core NC synthesis, followed by a purification step, and the subsequent shell growth reaction, during which a small number of monolayers (typically one to five) of the shell material are deposited on the cores. The temperature used for the core NC synthesis is generally higher than that used for the shell growth and the shell precursors are slowly added, for example, by means of a syringe pump. The major advantages over a so-called one-pot approach without an intermediate purification step is the fact that unreacted precursors or side products can be eliminated before the shell growth. The core NCs are purified by precipitation and redispersion cycles, and finally they are redispersed in the solvent used for the shell growth. In order to calculate the required amount of shell precursors to obtain the desired shell thickness, the knowledge of the concentration of the core

NCs is indispensable. In the case of cadmium chalcogenide NCs, the concentration of a colloidal solution can be determined in good approximation by means of UV-vis absorption spectroscopy thanks to tabulated relationships between the excitonic peak, the NC size, and the molar absorption coefficient.[96] An advanced approach for shell growth derived from chemical bath deposition techniques and aiming at the precise control of the shell thickness is the so-called SILAR (successive ion layer adsorption and reaction) method.[97] It is based on the formation of one monolayer at a time by alternating the injections of cationic and anionic precursors and has been applied first for the synthesis of CdSe/CdS CS NCs. Monodispersity of the samples was maintained for CdS shell thicknesses of up to five monolayers on 3.5 nm core CdSe NCs, as reflected by the narrow PL linewidths obtained in the range of 23 to 26 nm FWHM.

5.3 SURFACE FUNCTIONALIZATION

NCs possess a very high surface to volume ratio, which increases strongly with decreasing radius. Individual crystals are prevented from agglomeration by means of an organic ligand cap on their surface. At the same time, these ligand molecules complete the coordination sphere of the NC surface atoms and thus reduce the number of dangling bonds, which cause fluorescence quenching. A large number of potential applications of NCs require nonetheless a modification of this organic ligand cap, with the goal to make the colloids dispersible in various solvents and/or to graft them to other molecules of biological or electronic interest, to solid substrates, etc. By consequence, different approaches have been developed in recent years to chemically modify the NC surface. A widely applied method consists of the exchange of the initial surface ligands, that is, the ligands originating form the reaction medium, by bifunctional ligands X–Y–Z.[98–100] Such ligands contain a functional group X, which provides strong interaction with the metal ions on the NC surface and a second chemical moiety Z, which allows the subsequent binding to molecules or substrates. Typical examples of X are thiol or phosphonic acid groups, whereas Z usually comprises carboxyl, amine, aldehyde, or alcohol functions, to name a few. The selection of Z is of crucial importance, not only in view of NCs grafting to given molecules and substrates, but also for their dispersion in polar solvents and in particular in water, indispensable for biological applications. X and Z are usually separated by a spacer, Y, of alkylene or arylene type.

Bifunctional ligands X–Y–Z, in which Z is an ionizable group, can also facilitate the formation of more or less ordered aggregates of NCs with oppositely charged molecules (macromolecules) *via* electrostatic interactions. A typical example of this type is the electrostatic self-assembly of carboxylic acid coated (and thus negatively charged) NCs and a synthetically engineered, positively charged protein.[101] Since this protein enables the conjugation of biological molecules such as antibodies, highly stable and specific conjugates for immunoassays can be created using this method.

The main problem of the described functionalization methods is the fact that, in the majority of cases, the exchange of the original ligands results in a significant decrease of the NCs' fluorescence QY. This problem can be overcome by the use of a so-called NC encapsulation method. Here, the initial optical properties of the NCs can be essentially maintained, since this method does not involve the exchange of the original passivating ligands. As an example, the application of this procedure to hydrophobic NCs with TOPO and/or HDA ligands on their surface will be described here. The encapsulation is carried out in a three-component system consisting of water, oil, and an appropriate surfactant of amphiphilic nature. The surfactant molecules, being at the interface of the water continuous phase and the droplets of the oil, stabilize the formation of a microemulsion. If hydrophobic NCs are introduced into this microemulsion, micellar-type system, they are placed in the hydrophobic environment, that is, within the oil microdroplets. Upon evaporation of both solvents, alkyl chains of the surfactant molecules interdigitate with the alkyl groups of the NC ligands, encapsulating the NC in such a manner, while the polar groups of the surfactant are directed on the outside of the encapsulated particle. An example of the successful application of this technique is the NC encapsulation by phospholipid micelles, enabling *in vivo* imaging experiments.[102] Further examples comprise the use of amphiphilic polymers, which may subsequently be cross-linked.[103] Despite the described appealing features of the encapsulation method, it also shows some drawbacks, among them a significant size increase with respect to the pristine NCs and, in most cases, a rather low stability of the adsorbed micellar type overlayer as compared to chemically bound bifunctional ligands. Anyhow, the present state of the art seems to indicate that the selected functionalization method has to be adjusted to the desired application of the functionalized NCs; no ideal choice exists for all uses.

5.4 ASSOCIATIONS OF QDs AND CONJUGATED OLIGOMERS OR POLYMERS

5.4.1 Preparation of Organic–Inorganic Hybrids

Hybrids of semiconductor NCs and conjugated oligomers or polymers have emerged in the past decade as components of various electronic devices such as light-emitting diodes, photovoltaic cells, and others.[104,105] Before discussing the strategies for the preparation of these new materials, it is instructive to give the broadest definition of a hybrid since different (and sometimes contradictory) definitions of this term can be found in the literature. A hybrid is an association of at least two components of distinctly different chemical nature, linked together via specific interactions, which can involve the formation of covalent, coordination, ionic, or hydrogen bonds. Each hybrid component possesses its chemical identity and can exist independently of the hybrid.

In hybrids discussed here, the electronic and optoelectronic properties of their constituents can be tuned individually and adapted to each other. In particular the positions of the HOMO and LUMO levels in NCs, which determine their electronic, redox, and luminescence properties, can be tuned by changing their chemical composition and additionally by varying their size and shape.[106,107] Similarly, in the oligomeric (polymeric) part of the hybrid the positions of the HOMO and LUMO levels depend not only on the chemical constitution of the molecule (macromolecule) but also on its topology. As a result semiconductor NCs-conjugated oligomers (polymers) hybrids can hardly be matched by other systems as far as the design flexibility is concerned.

Although a large number of electroactive conjugated oligomers and polymers have been synthesized in the past 30 years, the studies on their use as components of organic–inorganic hybrids have been focused, for the most part, on three families of compounds, namely functionalized oligo- and poly(thiophene), oligo- and poly(*p*-phenylene vinylene)s, and oligo- and poly(aniline)s.

5.4.1.1 Simple Blending Methods

The preparation of such composite materials is not trivial. Simple casting from a common solvent leads to phase separation of both components of the hybrid on a micrometer scale.[6] This is caused by two factors: (1) NCs capped with initial ligands, that is, ligands originating from the reaction medium, such as TOPO, stearic acid, and others, have a strong tendency to aggregate on solidification, forming large, densely packed agglomerates and in some cases even highly ordered supercrystals;[1,4] (2) conjugated oligomers easily crystallize, whereas conjugated polymers of regular chain microstructure form semicrystalline structures, which consist of crystallized zones separated by amorphous ones.[108] These two phenomena promote the phase separation of the hybrid components and make control of its morphology at the nanoscale level difficult. Some attempts to overcome this problem, using purely physical approaches, have been reported in the literature. In one of the suggested methods, an electrical field is applied during casting of a composite material consisting of TOPO-capped CdSe NCs and poly(3-hexylthiophene). The phase-separated nanorods align in this case perpendicular to the substrate, leading to an oriented nanostructure that is favorable for several hybrid electronics applications.[109] The effect of the orientation and structuring of NCs in the polymer matrix can also be achieved by careful control of the material solidification conditions. Directional epitaxial solidification of poly(3-hexylthiophene) in a crystalizable solvent such as 1,3,5-trichlorobenzene leads to semicrystalline layers consisting of the periodic alternation of crystalline lamellar domains separated by amorphous interlamellar zones.[110] If this solidification is carried out in the presence of spherical or rod-like CdSe NCs of an appropriate size, the NCs are rejected to the amorphous zones. The resulting supramolecular organization is periodic and consists of alternating zones of crystalline poly(3-hexylthiophene) and zones in which NCs are embedded in the amorphous polymer matrix.

In these methods, neither ligand exchange at the NCs' surface nor conjugated oligomer (polymer) functionalization are carried out, which of course limits their applicability to selected systems for which the special processing conditions, described above, can be applied. In some procedures, an additional step is undertaken, consisting of an exchange of the initial surface ligands for more labile ones, such as pyridine. During casting, the "temporary" pyridine ligands are at least partially removed and the resulting "naked" NCs are stabilized by the poly(3-hexylthiophene) matrix.[111]

The principal interest of this procedure is to remove the initial surface ligands. The latter generally contain long alkyl chains, which act as an insulating, 1 to 2 nm thick barrier surrounding the NCs and impeding efficient charge transfer processes between the organic and the inorganic part of the hybrid. Drawbacks of this method are the reduced solubility of the NCs in most organic solvents, and the sensitivity to even small changes of the ligand exchange and processing conditions, which affects the reproducibility of the results.

5.4.1.2 Complexation of the NC Surface with Conjugated Macromolecules Bearing an Anchoring Group (See Scheme 5.2, Routes II to VI)

A versatile route to NCs–conjugated oligomers (polymers) hybrids is the exchange of the initial surface ligands with conjugated ones. This requires, of course, functionalization of the conjugated molecule (macromolecule) by introducing an appropriate anchoring group. For oligomers, the anchor functions can be introduced either as terminal or as side groups (Scheme 5.2, routes II and III). For polymers, side groups are preferable (route V), taking into account that with increasing molecular mass, the ratio of the end groups to the number of mers decreases, making the anchor functions less and less available for complexation. An instructive example of this approach is the

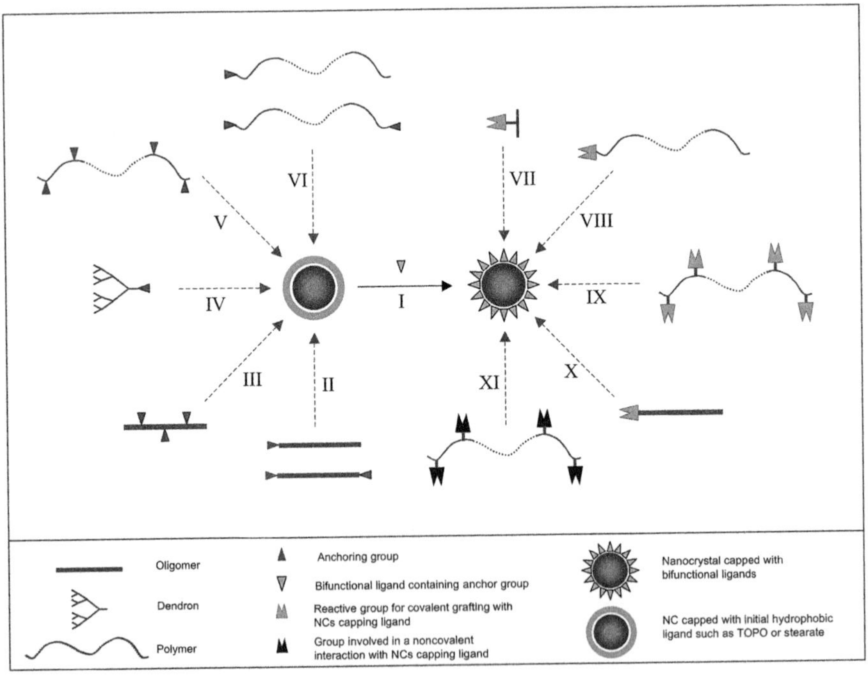

Scheme 5.2 Schematic representation of the different strategies developed to produce hybrid materials composed of semiconductor NCs and conjugated oligomers/polymers.

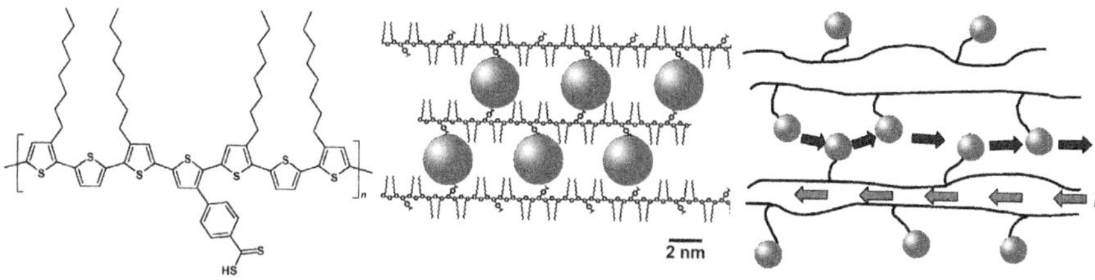

Scheme 5.3 Carbodithioate-functionalized poly(octylthiophene) (left) and its hybrid with 3 nm CdSe nanocrystals (middle). Idealized double-cable structure allowing for efficient electron (e^-) and hole (h^+) transport (right).[112] (Reprinted with permission from C. Querner et al., *Chem. Mater.* **2006**, *18*, 4817–4826. Copyright 2006 American Chemical Society.)

preparation of solution processible oligo- or poly(alkylthiophenes) containing carbodithioic anchoring functions (see Scheme 5.3).[112] The ligand exchange occurs in surprisingly mild conditions owing to the strong chelating effect of the carbodithioic complexing group.[113] In the preparation of this type of polymeric ligands, the building blocks approach is usually applied, which involves a multistep procedure.[112] In view of the potential use of these hybrids as components for electronic and optoelectronic devices, their regioregular chain microstructure without head to head couplings (*h-h*) of the alkylthiophene units is vital since the *h-h* coupling induces chain torsion and significantly lowers the charge carriers' mobility.[114,115] Enhanced resistance against photooxidation of the carbodithioic anchoring group is also technologically important, especially in photovoltaic applications.[113]

Oligo- and poly(thiophene)s with other than carbodithioic complexing groups have equally been used to prepare hybrids with CdSe NCs via ligand exchange procedures. Reported examples comprise thiol-[116] and phosphonic acid-functionalized[117] oligothiophenes as well as poly(thiophene) containing tetrahydro-4H-thiopyran-4-ylidene side groups.[118] In a somewhat similar approach, Liu et al. prepared hybrids of CdSe nanorods with amino-terminated poly(3-hexylthiophene) (route VI).[119] In this case the initial surface ligands were first replaced by pyridine, followed by the exchange of pyridine for the polymeric ligands.

Hybrids containing dendron-type oligothiophenes were also successfully fabricated (route IV).[120,121] Despite the fact that the synthesized dendrons contain a strongly complexing group (phosphonic acid function), the preparation of their hybrids with CdSe NCs requires a two-step procedure. In the first step, the initial (TOPO) ligands are exchanged for pyridine ones, which in the next step are displaced by phosphonic acid-functionalized oligothiophene dendrons.

Finally, hybrids of NCs with other families of conjugated oligomers and polymers, bearing different anchor functions, have also been prepared by the ligand exchange approach. Sudeep et al. have prepared hybrids of CdSe NCs with oligo(phenylene vinylene)s containing phosphine oxide terminal groups.[122] Javier et al. produced a hybrid of CdSe NCs and oligo-*p*-(phenylene-ethynylene)$_n$-dibenzylthiols ($n = 0$, 1, 3), in which the NCs are separated by rigid conjugated molecular spacers of 1.3 to 3.1 nm length.[123] A triblock copolymer containing a central polyfluorene-type conjugated segment linked to two poly(2-(dimethylamino)ethyl-methacrylate) units has recently been assembled with CdSe NCs.[124] Multidentate coordination assures in this case the stability of the formed hybrid. A poly(*p*-phenylene vinylene) (PPV) derivative bearing a terminal sulfide function has been synthesized by Chen et al. and introduced via ligand exchange on the surface of CdSe/ZnS NCs.[125] The hybrid material obtained was tested in a light-emitting diode configuration showing enhanced efficiency compared to the pristine polymer.

Coating of the NCs' surface with conjugated oligomers facilitates their dispersion in polymer matrices consisting of their high molecular mass analogs. This aspect can be discussed in terms of the solubility parameters, δ. To a first approximation, such dispersion can be considered as a molecular solid solution in which the polymer matrix is "a solvent" and the oligomer coated NCs "a solute." The solubility parameter of the solute is then determined by the chemical constitution of the NCs' surface and by consequence must be close to that of the solvent since both are chemically very similar. As a result, a homogeneous dispersion of the NCs in the polymer matrix can be obtained. This is demonstrated experimentally by the preparation of a hybrid prepared from poly(3,3″-dioctyl-2,2′:5′,2″-terthiophene) and carboxylic group functionalized dioctyl-2,2′:5′,2″-terthiophene coated CdSe NCs, which shows a uniform NCs distribution.[126] On the contrary, the use of TOPO-coated NCs with the same polymer results in a strong phase separation.

It turns out that in the majority of cases the exchange of the ligands with the conjugated oligomers/polymers is incomplete, which can constitute a serious problem depending on the desired application of the resulting hybrid. For this reason, NC preparation methods have been developed in which conjugated oligomers functionalized with a complexing group are used *in the reaction medium as the initial capping agents*. For example, CdS NC hybrids with a series of oligothiophene-dicarboxylic acids have been prepared.[127] Despite obvious advantages of this procedure (facile one-step, one-pot preparation) its few drawbacks should be mentioned. First, in the preparation of NCs of controlled size, shape, and low polydispersity the selection of ligands is a delicate matter of crucial importance. Thus, in the majority of cases, it is more advantageous to use nonconjugated capping ligands, known to yield good quality NCs, and then exchange them for the conjugated ones. Second, several NC fabrication procedures require the use of high temperatures and rather harsh reaction conditions. This is especially true for quasi-monodispersed samples. Although polyconjugated systems show, in general, a rather good thermal stability, some of them are not suited for the high temperature synthesis, especially after their functionalization.

5.4.1.3 Covalent Grafting of Functionalized Conjugated Macromolecules and Functionalized NCs (Routes VII to X)

5.4.1.3.1 Two-Step Procedure Comprising Preliminary Ligand Exchange (Scheme 5.2, Route I, Followed by Routes VII to X) Preparation of organic–inorganic hybrids by grafting constitutes an alternative to the aforementioned methods and allows for a very precise engineering of the hybrid components. In the most general way, the grafting approach can be described

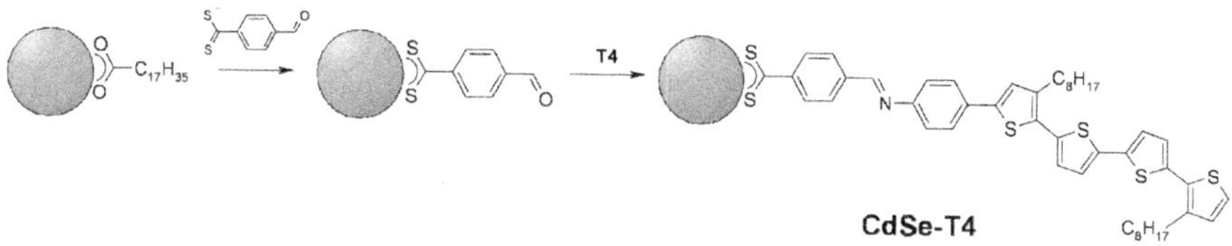

Scheme 5.4 Preparation of an organic–inorganic hybrid by grafting of aniline-terminated tetrathiophene, **T4**, onto 4-formyldithiobenzoate-capped CdSe NCs.[128]

as follows. In the first step, the initial NC surface ligands (e.g., TOPO, stearic acid) are replaced by special bifunctional linker ligands containing a complexing function and a reactive group capable of reacting with an end- or side-functionalized conjugated oligomer (polymer). In the next step, the grafting reaction is carried out, whose chemical nature depends on the type of the reactive groups in the linker ligand and in the molecule (macromolecule) being grafted. A representative example of this procedure is shown in Scheme 5.4. It starts with the exchange of the initial NCs capping ligands of stearate type for 4-formyldithiobenzoate. Due to the capability of the carbodithioate anchor group to form strong chelate-type bonds with the cadmium atoms at the NCs surface, the ligand exchange takes place under mild conditions with essentially 100% yield.[128] Next, grafting of aniline-terminated tetrathiophene proceeds via condensation between its amine group and the aldehyde group of the linker ligand, resulting in the formation of an azomethine linkage.

An identical strategy has been used for grafting aniline tetramer [a model compound mimicking the behavior of poly(aniline)] onto 4-formyldithiobenzoate-capped CdSe NCs.[113,129,130]

Zhang et al. have grafted vinyl group-terminated poly(3-hexylthiophene) onto CdSe nanorods using either *p*-bromobenzyl-di-*n*-octylphosphine oxide (DOPO-Br) or 2-(4-bromo-2,5-di-*n*-octyl-phenyl)-ethanethiol as linker ligands.[131] The proposed procedure consists of three steps. In the first one, the initial TOPO ligands are exchanged for pyridine ones. This step is then followed by the exchange of pyridine for the aforementioned brominated ligands. Finally, the vinyl group-terminated polymer is grafted by Heck coupling.

5.4.1.3.2 One-Step Procedure Using the Bifunctional Linker Ligand in the NCs Synthesis In addition to their function as linker molecules, some bifunctional ligands can also serve as surface capping ligands in the synthesis of semiconductor NCs. As a consequence, in this strategy the grafting of conjugated oligomers (polymers) does not need to be preceded by the ligand exchange step. DOPO-Br, a TOPO derivative already mentioned in the previous paragraph, readily complexes the surface of CdSe NCs *in situ* during their nucleation and growth in the reaction medium. Hybrids of poly(*p*-phenylene vinylene) derivatives and CdSe NCs have been prepared by copolymerization of 1,4-divinylbenzene and 1,4-dibromobenzene derivatives, initiated by the coupling of the former to DOPO-Br ligands on the NCs surface (Scheme 5.5).[132] However, strict control of the copolymerization

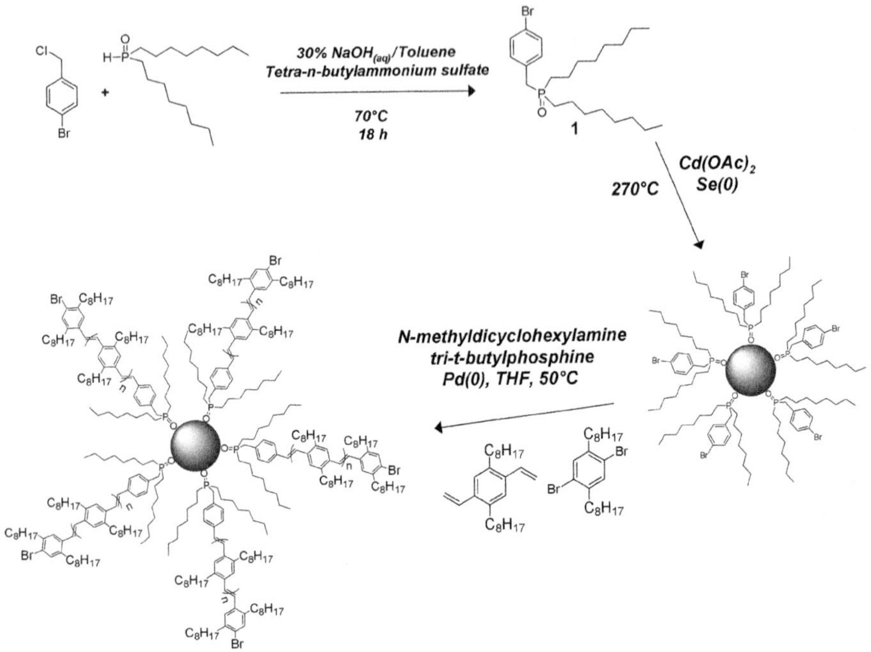

Scheme 5.5 Synthesis of DOPO-Br-capped CdSe NCs and subsequent polymerization of PPV from the NCs'surface.[132] (Reprinted with permission from H. Skaff et al., *J. Am. Chem. Soc.* **2004**, *126*, 11322–11325. Copyright 2004 American Chemical Society.)

conditions is difficult in this case and the organic part of the resulting hybrid consists of oligomers of different length. This is disadvantageous as far as electronic and optoelectronic applications are concerned because for oligomeric conjugated compounds both the position of the HOMO and LUMO levels and their electronic transitions—key parameters for their applications—strongly depend on the chain length and only monodispersed samples give strictly reproducible results.

In a similar approach, *p*-phenylene vinylene oligomers containing 3- to 6-mers have been grown on DOPO-Br capped NCs by homocoupling of 1-bromo-2,5-di-*n*-octyl-4-vinylbenzene.[133] Finally, using a one-step procedure, vinyl-terminated poly(hexylthiophene) has been grafted onto DOPO-Br coated NCs via Heck coupling.[134]

5.4.1.4 Assembly of Functionalized NCs with Functionalized Conjugated Oligomers (Polymers) by Means of Noncovalent Interactions (Scheme 5.2, Route XI)

The entropy of NCs and conjugated polymers mixing is low. Thus, in order to stabilize a homogeneous dispersion of quantum dots in the polymer matrix it is necessary to increase the interactions between the NCs' capping ligands and the polymer chains. In this respect, the application of supramolecular chemistry concepts such as, for example, molecular recognition becomes helpful. This approach has previously been used by Rotello and coworkers in the preparation of organic–inorganic hybrids consisting of diaminotriazine functionalized poly(styrene) and thymine-capped gold nanoparticles.[135,136] Here, molecular recognition between the complementary functional groups assured the uniform distribution of the inorganic component within the polymer matrix. The same strategy has been used for the fabrication of layers of self-assembled semiconductor NCs on appropriate electrodes[137] or for selective immobilization of metal NCs on functionalized surfaces.[138] The molecular recognition concept has been extended to the design and preparation of conjugated polymers/semiconductor NCs hybrids. De Girolamo et al. prepared a solution processible poly(alkylthiophene) derivative containing diaminopyrimidine-functionalized side groups, namely poly(3-hexylthiophene-*co*-3-(6-oxy-2,4-diaminopyrimidine)-hexylthiophene) (P3HT-*co*-P3(ODAP)HT).[139] This functionalized polymer is able to molecularly recognize 1-(6-mercaptohexyl)thymine (MHT) capped CdSe NCs via three hydrogen bonds, as depicted schematically in Scheme 5.6.

Direct processing of P3HT-*co*-P3(ODAP)HT/CdSe(MHT) hybrids is difficult because no common solvent can be found for the polymeric and nanocrystalline components. To overcome this problem, a special technique has been proposed. This one-pot method consists of the preparation of a chloroform solution of the molecular

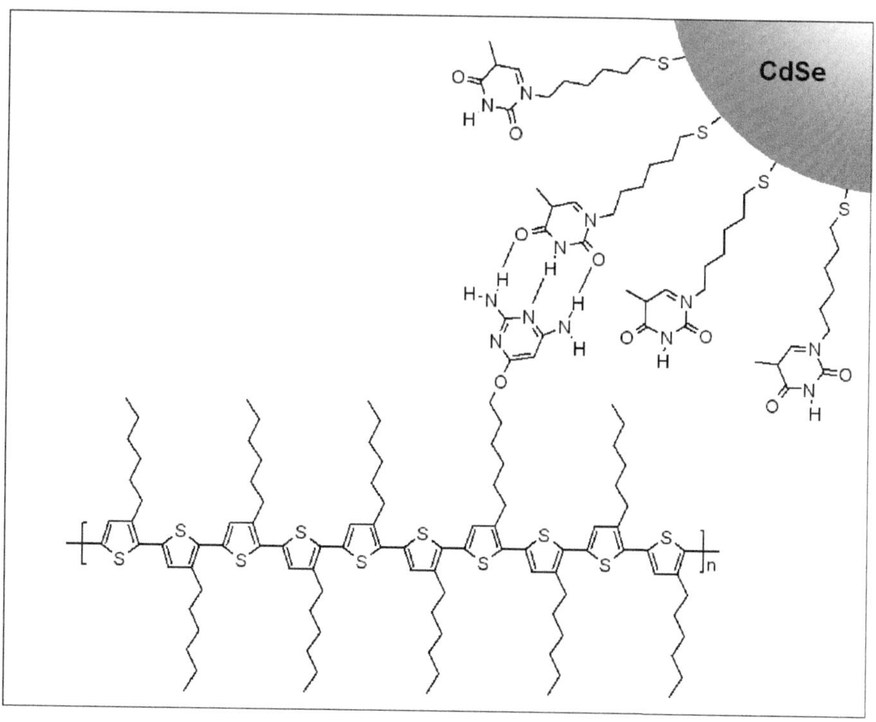

Scheme 5.6 Molecular recognition between poly(3-hexylthiophene-*co*-3-(6-oxy-2,4-diaminopyrimidine)hexylthiophene) and 1-(6-mercaptohexyl)thymine coated CdSe via hydrogen bonds.[139]

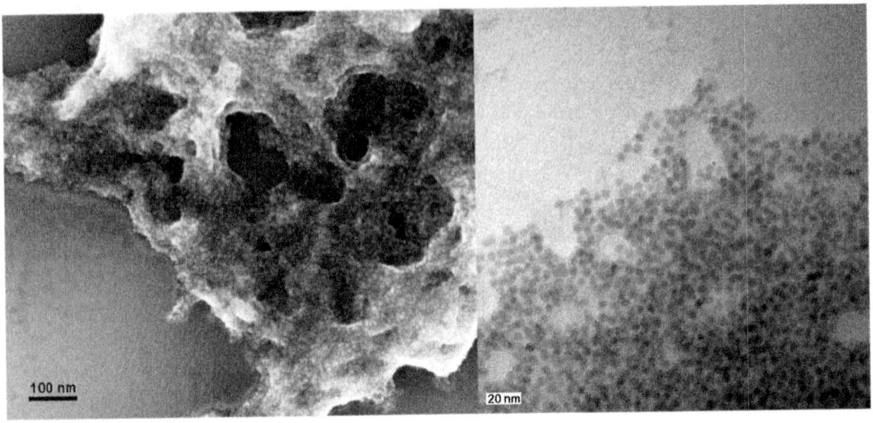

Figure 5.5 SEM (left) and TEM (right) image of the molecular hybrid consisting of poly(3-hexylthiophene-*co*-3-(6-oxy-2,4-diaminopyrimidine)hexylthiophene) and CdSe-1-(6-mercaptohexyl)thymine obtained by a one-pot processing method.[139] (Reprinted with permission from J. De Girolamo et al., *J. Phys. Chem. C* **2007**, *111*, 14681–14688. Copyright 2007 American Chemical Society.)

conjugate of P3HT-*co*-P3(ODAP)HT and MHT, to which a chloroform solution of stearate-capped CdSe NCs is added. Exchange of the stearate ligands for the conjugated ones, occurring *in situ*, followed by casting and slow solvent evaporation, results in the formation of a molecular hybrid material with uniform distribution of the NCs within the polymer matrix (Fig. 5.5).

The molecular recognition phenomenon can further be exploited as the driving force for the processing of NCs–conjugated polymer hybrids using the layer-by-layer (LbL) technique.[140] One of the appealing features of the functional copolymer P3HT-*co*-P3(ODAP)HT, which facilitates its application to this technique, is its adherence to a large variety of substrates containing oxide or hydroxide groups, with no necessity of surface modification or depositing a special anchoring layer. The adherence of the polymer originates from its capability of forming hydrogen bonds between its DAP groups and oxidized surfaces. The fabrication of hybrid films is achieved by alternating dipping the substrate into a P3HT-*co*-P3(ODAP)HT solution in chloroform and into a dispersion of CdSe(MHT) in a DMF-MeOH mixture (10 : 1) with an intermediate step of rinsing with chloroform and drying under argon flow after each dipping step.

The LbL deposition process can be controlled by UV-vis spectroscopy. Figure 5.6a shows the absorption spectra recorded during consecutive depositions of P3HT-*co*-P3(ODAP)HT and MHT-capped CdSe NCs bilayers on an ITO-coated glass substrate. The steady increase of the absorbance at both 504 nm (diagnostic of the polymer) and 622 nm (diagnostic of 5.8 nm CdSe NCs) indicates that the growth of the film is a progressive and reproducible process. From the linear increase of the polymer and NCs absorbance versus the number of bilayers (Fig. 5.6b), it can be concluded that the same amounts of polymer and NCs are deposited at each deposition step.

Detailed SEM studies of the LbL deposited films show that the surface coverage of the polymer layer by the NCs layer, deposited on it, is incomplete (approximately

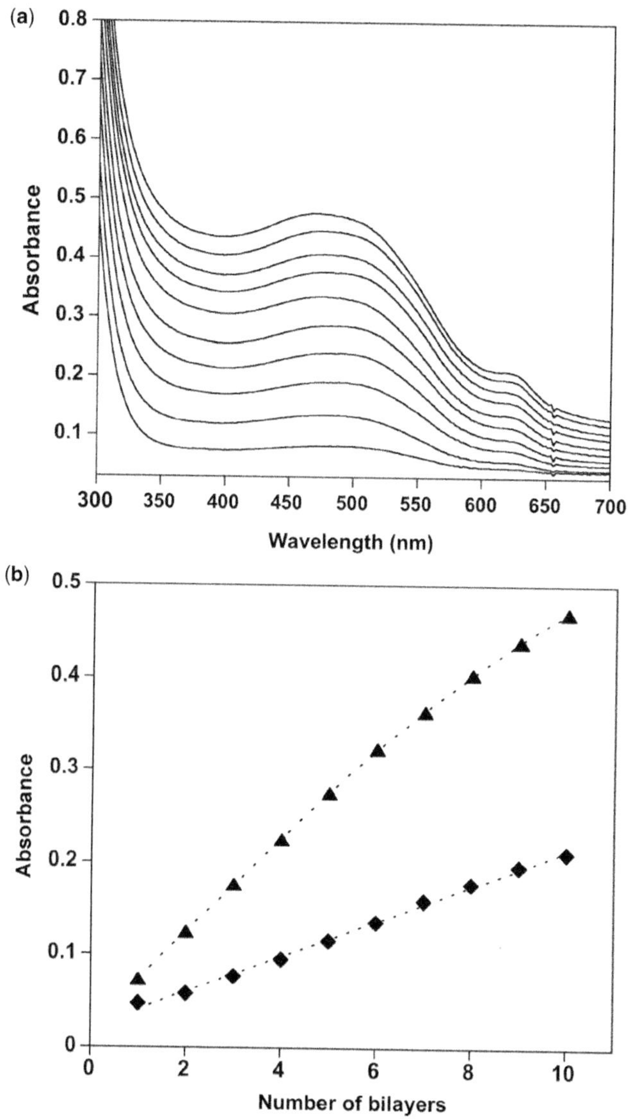

Figure 5.6 (a) UV-vis absorption spectra recorded during the successive deposition of 10 bilayers of P3HT-*co*-P3(ODAP)HT/CdSe(MHT) (NC diameter 5.8 nm), using the LbL method. In the order of increasing absorbance, each spectrum corresponds to the addition of one bilayer. (b) Absorbance measured at 504 nm and 622 nm as a function of the number of bilayers.[140] (Reprinted with permission from J. De Girolamo et al., *J. Phys. Chem. C* **2008**, *112*, 8797–8801. Copyright 2008 American Chemical Society.)

30% to 35%; see Fig. 5.7). As a result, an "interpenetrating network" morphology is found rather than a multilayer one for multibilayer films, since each consecutive polymer layer fills the voids between NCs in the NCs layer.[140] This morphology is advantageous for photovoltaic applications of the newly obtained hybrid materials (see below).

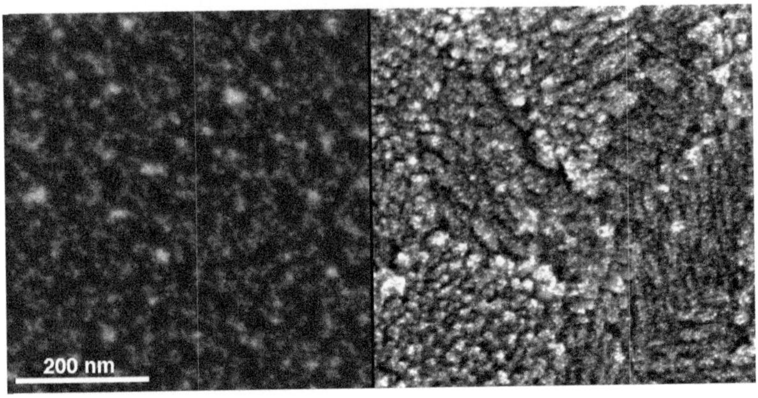

Figure 5.7 SEM images of a single bilayer of P3HT-*co*-P3(ODAP)HT/CdSe(MHT), deposited on an oxidized silicon substrate (left) or on an ITO-coated glass substrate (right).

Another type of noncovalent interaction, namely electrostatic ones, has recently been exploited by Wang et al. in the fabrication of a hybrid film containing a water-soluble poly(thiophene) derivative and CdSe NCs.[141] In this case CdSe NCs were functionalized with 11-mercaptoundecanonic acid (MUA) molecules prior to their self-assembly with side chain-functionalized poly(thiophene), that is, poly(3-(3′-thienyloxy)propyltrimethylammonium bromide) (P3TOPA).

Noncovalent bonding, $\pi-\pi$ interactions, has also been used to improve the miscibility between thiophenol-capped CdS NCs and a poly(fluorene) derivative of dendritic structure (Scheme 5.7).[142]

5.4.2 Hybrid Characterization

Taking into account the rather complex nature of conjugated oligomer (polymer)-semiconducting NCs hybrids, their detailed chemical and physicochemical characterization requires the application of several complementary techniques. In hybrids prepared via ligand exchange or by ligand exchange followed by oligomer (polymer) grafting, the estimation of the degree of the ligand exchange is of crucial importance since this process determines the chemical composition of the resulting hybrid. The changes induced in the composition of the organic part of the hybrid by the ligand exchange can be determined by elemental analysis, especially in cases where a characteristic element can be found in the new ligand, which is not present in the old one and vice versa. More frequently, the verification of the degree of the ligand exchange is carried out by ^{1}H NMR spectroscopy through the integration of the peaks corresponding to the initial ligands and the exchanged ones.[143] The NMR lines of the capping ligands in colloidal solutions of NCs are somehow broadened, as compared to those registered for the corresponding "free" ligands, because of the limited molecular mobility of the molecules complexed on the NC surface. However, in the majority of cases this phenomenon does not perturb the determination of the ligands' relative

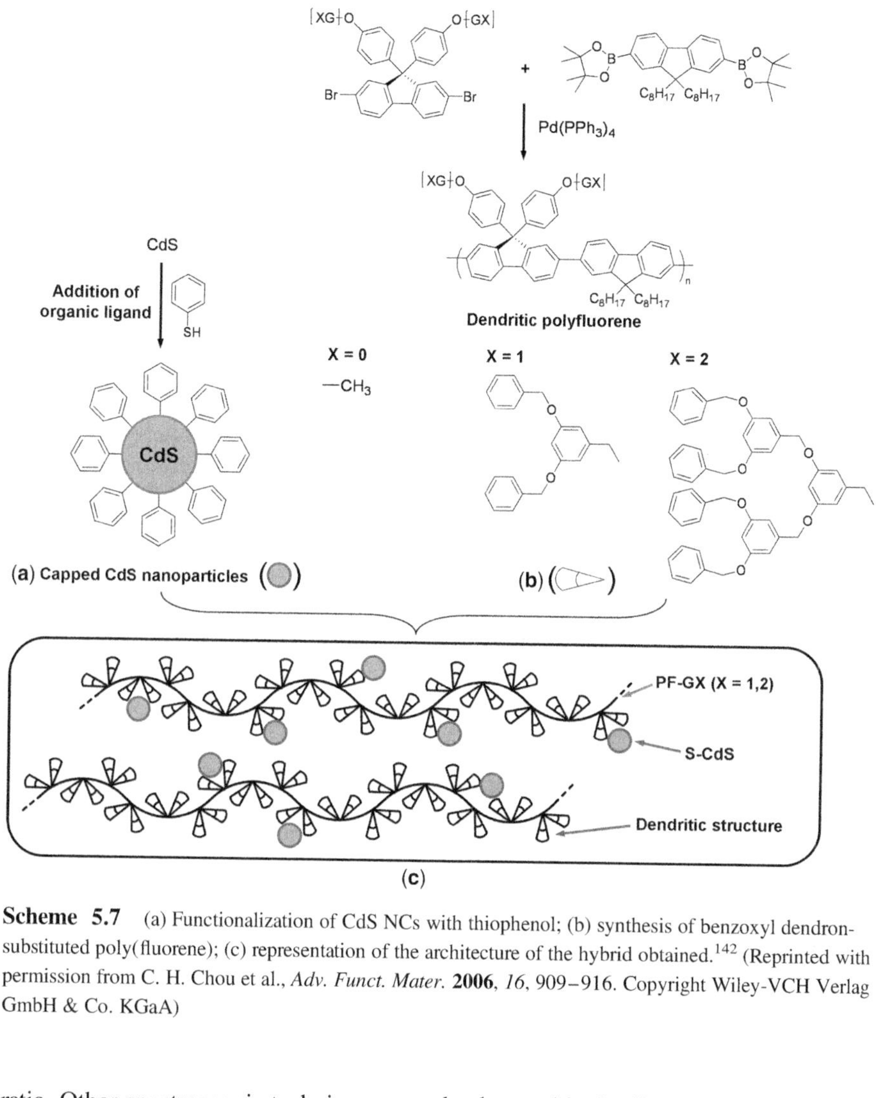

Scheme 5.7 (a) Functionalization of CdS NCs with thiophenol; (b) synthesis of benzoxyl dendron-substituted poly(fluorene); (c) representation of the architecture of the hybrid obtained.[142] (Reprinted with permission from C. H. Chou et al., *Adv. Funct. Mater.* **2006**, *16*, 909–916. Copyright Wiley-VCH Verlag GmbH & Co. KGaA)

ratio. Other spectroscopic techniques can also be used in the ligand exchange investigations (UV-vis, IR) provided that the molar absorption coefficient of the diagnostic peaks, selected for the initial and the exchanged ligands, are known.

Another important parameter characterizing the hybrids is the organic to inorganic matter ratio. This ratio can be determined from either elemental analysis or by thermogravimetric (TG) investigations. The majority of organic components of the hybrid decompose at higher temperatures yielding no residual mass. The mass measured after the organic part decomposition corresponds to the inorganic part of the hybrid (which is thermally more stable) provided that no oxidation of the inorganic core (or other type of a mass changing reaction) took place during the TG measurement.

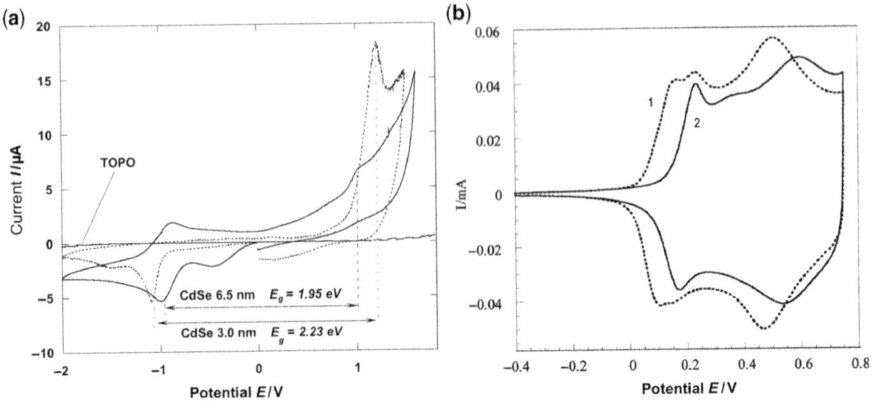

Figure 5.8 Cyclic voltammograms of (a) differently sized CdSe NCs;[129] (b) hexyl (curve 1) and dodecyl (curve 2)-substituted poly(3-alkylthiophenes).[147] (Reprinted with permission from M. Skompska and A. Szkurlat, *Electrochim. Acta* **2001,** *46*, 4007–4015. Copyright 2001, Elsevier.

Combining the NMR and TG data and knowing the size of monodispersed NCs in a given sample, it is possible to calculate the number of their surface atoms and, in the next step, to find the degree of their surface complexation or, in other words, to find the surface ligand to NCs surface atoms ratio. This is also an important parameter characterizing the hybrid.

The application of hybrids in electronics requires the determination of the HOMO and LUMO levels of their organic and inorganic components since the type of their alignment determines their possible applications. The LUMO–HOMO energy difference of the hybrid components can be extracted from the UV-vis spectroscopic data.[144,145] It is recommended to determine it first for both components separately (i.e. for free conjugated ligands and NCs capped with initial ligands), since interactions with NCs may influence the conformation of the conjugated ligands after their complexation and, by consequence, alter the LUMO–HOMO energy difference (see below).[129]

Optical studies give access to the value of the LUMO–HOMO energy difference but do not provide information concerning their position with respect to the vacuum level. Both conjugated oligomers (polymers) and semiconductor NCs are electrochemically active and in their voltammograms give a clear anodic wave upon their oxidation and a cathodic one upon their reduction (see Fig. 5.8). In the simplest reasoning the onset of the oxidation peak can be interpreted as originating from the abstraction of the first electron, that is, as corresponding to the HOMO level. By analogy, the onset of the reduction peak can be attributed to the position of the LUMO level. The calibration of the positions of these peaks with respect to ferrocene/ferrocenium (Fc/Fc^+) couple (whose position relative to the vacuum level is known and equal to -4.8 eV) allows for the determination of the position of the HOMO and LUMO levels using the following expressions:[146]

$$E_{\mathrm{HOMO}} = -(4.8 + E_{\mathrm{ox}})\,\mathrm{eV} \quad \text{and} \quad E_{\mathrm{LUMO}} = -(4.8 + E_{\mathrm{red}})\,\mathrm{eV}$$

For NCs, the electrochemically determined $E_{LUMO}-E_{HOMO}$ energy difference is, within experimental error, the same or very close to that measured optically. By both methods, the quantum confinement effect, characteristic of NCs, is evidenced.[129,148] The same applies to conjugated oligomers (polymers), whose optically and electrochemically determined $E_{LUMO}-E_{HOMO}$ energies usually agree.

In the voltammetric determination of the HOMO and LUMO levels, one must be aware of the fact that the onsets of the anodic oxidation and the cathodic reduction peaks may sometimes depend on the scan rate, which can introduce some error in the determination of the energy levels. In such cases, it is recommended to determine the position of the HOMO and LUMO levels by spectroelectrochemistry using the quasi-static approach. In this approach, the electrode is polarized to potentials close to but lower than the expected onset of the oxidation (reduction). The potential is then being raised (lowered) in small increments of approximately 10 mV. At each step the spectrum is registered (UV-vis-NIR, Raman, etc.). Since oxidation-(reduction-) induced spectral changes are very characteristic, the potential of the oxidation (reduction) onset can clearly be identified. It is illustrated in Figure 5.9, where UV-vis spectroelectrochemical investigations of poly[(4,4″-dioctyl-2,2′:5′,2″-terthiophene-3′-yl)ethyl acetate] is presented. No spectral changes are observed upon the electrode potential increase from $E = -0.2$ V to $E = +0.4$ V versus Ag/Ag^+, then upon further potential increase by 25 mV, a drastic change of the spectrum occurs which is associated with the formation of the polyradical cation form of the polymer.[149]

Organic and inorganic components of the hybrid may mutually influence the position of their HOMO and LUMO levels and, as a result, their spectroscopic and redox properties do not follow the rule of additivity. For example, grafting of aniline tetramer to NCs, via an azomethine-phenylene spacer and carbodithioate anchor function, results in a lowering of the NCs' HOMO level.[129] This is caused by the fact that the charge of the oligoaniline radical cations, which are formed at lower potentials, makes the oxidation of the NCs more difficult. On the contrary, in molecularly processed hybrids of poly(3-hexylthiophene-*co*-3-(6-oxy-2,4-diaminopyrimidine)-hexylthiophene) and mercaptohexylthymine-capped CdSe NCs, the formation of a self-assembled network makes the NCs less resistant against electrochemical oxidation. Spectroelectrochemical investigations of this hybrid show unequivocally that the oxidation of the NCs starts at lower potentials as compared to the case of stearate-capped NCs.[150]

As has already been mentioned, both colloidal NCs and conjugated oligomers (polymers) exhibit photoluminescence and electroluminescence, which reflects their semiconductor nature. In most examples of organic–inorganic molecular hybrids, prepared either by ligand exchange or by grafting of a conjugated oligomer (polymer) to a linker ligand, the luminescence is quenched efficiently via charge and/or energy transfer as a consequence of the relative energy level alignment.[132,134] Effective luminescence quenching is especially important in hybrids which are to be used as components of photovoltaic cells (see below) and can be considered as a manifestation of the exciton dissociation phenomenon in hybrid systems showing a staggered (type II) energy band alignment. The ligand exchange-induced quenching can be followed

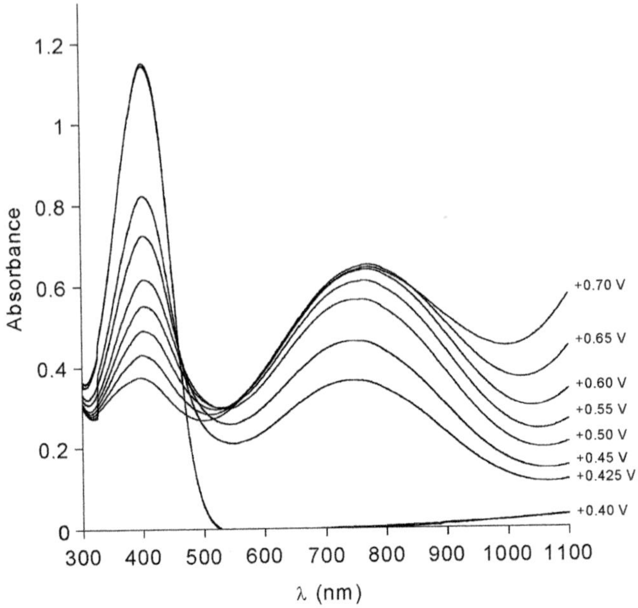

Figure 5.9 UV-vis-spectroelectrochemical determination of the oxidative doping onset directly related to the HOMO level; case of poly[(4,4″-dioctyl-2,2′:5′,2″-terthiophene-3′-yl)ethyl acetate]. Note very profound spectral changes upon increasing the electrode polarization potential from 0.400 V to 0.425 V versus Ag/Ag^{+}.[149] (Reprinted with permission from K. Buga et al., *J. Mater. Chem.* **2006**, *16*, 2150–2164. Copyright the Royal Society of Chemistry.)

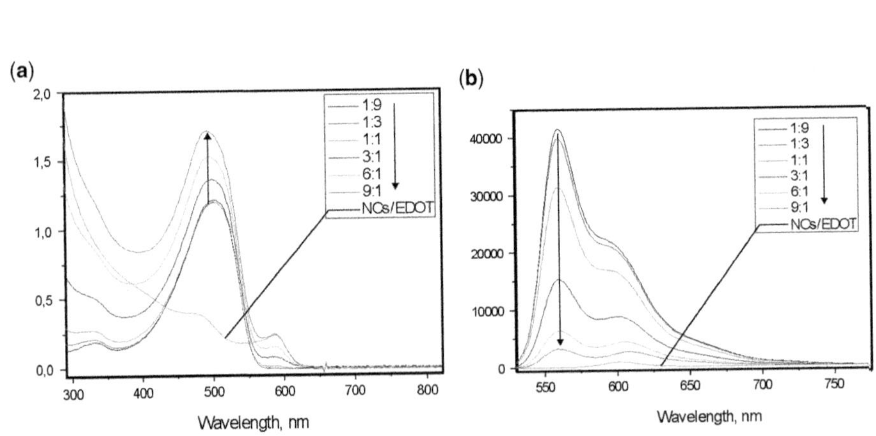

Figure 5.10 (a) UV-vis absorption; (b) photoluminescence spectra of 3,4-ethylenedioxythiophene-(EDOT) coated CdSe nanorods as a function of the quantity of poly[2-methoxy-5-(3′,7′-dimethyloctyloxy)-1,4-phenylenevinylene] (MDMO-PPV) added (solvent: $CHCl_3$).[151] (Reprinted with permission from D. Aldakov et al., *Eur. Phys. J. Appl. Phys.* **2006**, *36*, 261–265. Copyright the European Physical Journal.)

in a titration experiment where small aliquots of the oligomer (polymer) solution are added to a colloidal solution of the NCs containing the initial ligands or vice versa. Registration of the luminescence spectra after each addition allows, after their normalization, for the determination of the degree of quenching as a function of the NCs to oligomer (polymer) ratio.[117,151] The quenching can also be considered as an indirect measure of the ligand exchange effectiveness. A typical example of a titration experiment is shown in Figure 5.10.

To summarize, complete characterization of molecular hybrids of conjugated oligomers (polymers) and NCs requires the application of several complementary techniques such as elemental analysis, thermogravimetry, and NMR spectroscopy, which must be used for the determination of the hybrids' chemical composition and the degree of the complexation at the NC surface. Spectroscopic, electrochemical, and spectroelectrochemical investigations are necessary to establish their redox properties, the alignment of the HOMO and LUMO levels of the hybrid components, and the efficiency of their luminescence quenching. Finally, microscopic techniques (TEM, SEM, AFM, and STM) are vital for the elucidation of the hybrids' supramolecular organization and morphology.

5.5 HYBRID APPLICATIONS IN OPTOELECTRONICS

Two principal applications of conjugated oligomers (polymers)-NCs composites or hybrids as components of electronic devices can be envisioned: in photovoltaic cells (PCs) and in light emitting diodes (LEDs). Both applications will be discussed below.

5.5.1 Photovoltaic Cells

Photovoltaic cells are devices that transform radiation energy into electricity. If this radiation is solar light they are called solar cells. The simplest organic solar cell consists of a layer of an appropriate conjugated polymer (oligomer) sandwiched between two electrodes of different work functions. One of these electrodes must be transparent since the radiation must reach the semiconductor layer. Upon radiation, excitons are formed in this layer which then dissociate. The negative and positive charges (holes and electrons) formed in the dissociation process migrate to the electrodes, creating the difference of potentials. In this simple device, however, the photoinduced charge generation—the key parameter of the cell operation—is extremely inefficient.[152] This is caused by the fact that the conjugated polymers used for the fabrication of photovoltaic cells are good electron donors but poor electron acceptors. As a consequence, the formed excitons can dissociate to liberate the charges only in the vicinity of the electrode since the electrode is the only electron accepting area in the device. The photoinduced charge generation is facilitated if, in addition to the conjugated polymer, an electron-accepting component is dispersed in the active layer. In this case, the large interface area between the electron donating and the electron accepting components promotes the photoinduced charge generation. If both phases form interpenetrating, percolating networks, the created electrons and holes can be transported to the respective electrodes. This briefly described concept of a bulk heterojunction was first proven experimentally by Heeger and coworkers, who fabricated bulk heterojunction solar cells consisting of MEH-PPV (poly(2-methoxy-5-(2′-ethyl-hexyloxy)-1,4-phenylene-vinylene)) as an electron donating semiconductor and fullerenes (C_{60}) as electron acceptors.[153] In more recent works C_{60} was replaced by its derivative, (1-(3-methoxycarbonyl)propyl-1-phenyl[6,6]C_{61}) (PCBM). Some of the solar cells based on regioregular poly(3-hexylthiophene) and PCBM show power conversion efficiences approaching 5%.[154]

Photovoltaic cells containing NCs as electron accepting components are based on the same principle of bulk heterojunction. In addition to their appropriate electronic structure, NCs offer some advantages over other electron accepting components. First, the shape of NCs can be varied (spheres, rods, tetrapods, hyperbranched) and by changing their shape, it is possible to tune the percolation threshold since the dispersion of objects of higher aspect ratios results in its lowering. Second, the absorption spectrum of the crystals and their HOMO and LUMO levels can also be tuned by changing their size or their composition. Thus, it is possible to fabricate NCs that, in addition to the conjugated polymer, strongly contribute to the collection of photons under solar irradiation. The principles of the device functioning of photovoltaic cells with semiconductor NCs as electron acceptors are illustrated in Scheme 5.8.

For proper operation of a bulk heterojunction photovoltaic cell, a special alignment of the HOMO and LUMO levels of the bulk heterojunction components must be accomplished, compatible with the electrodes work functions, as depicted in Scheme 5.8. If an exciton is formed in the polymer phase, then the electron is transferred to the NC phase and reaches the aluminum electrode via its percolating pathway. The remaining hole is transported to the ITO electrode through the polymer phase. In the alternative case, that is, the formation of an exciton in the NCs phase, the hole is transferred to the polymer phase and then transported to the ITO electrode, whereas the electron reaches the aluminum electrode through the NCs phase.

A typical three layer conjugated polymer-semiconductor NCs bulk heterojunction-type photovoltaic cell is presented in Scheme 5.9. The two additional layers of

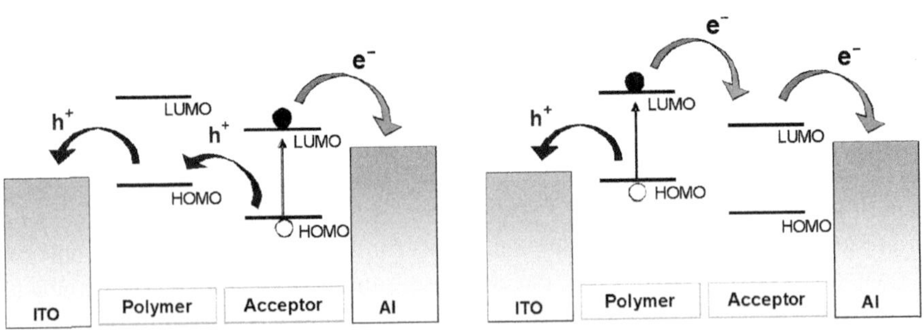

Scheme 5.8 Energy level alignment of bulk heterojunction components (conjugated polymer and semiconductor nanocrystals) facilitating the dissociation of excitons and charge separation. Left panel: Case describing excitons formed in the nanocrystal phase. Right panel: case describing excitons formed in the polymer phase.

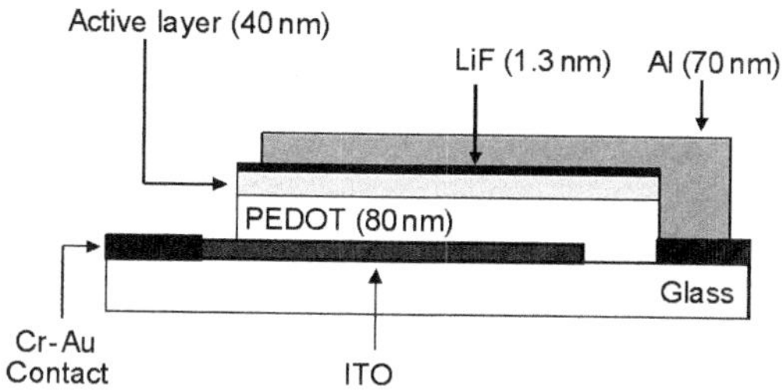

Scheme 5.9 Scheme of a hybrid photovoltaic cell with an active layer consisting of a composite of a conjugated polymer and semiconductor nanocrystals (so-called bulk heterojunction).

poly-(3,4-ethylenedioxythiophene) and polystyrene sulfonate (PEDOT : PSS and LiF) improve the contact between the active layer and the electrodes and facilitate the extraction of holes and electrons, respectively.

The first bulk heterojunction-type solar cells containing semiconductor NCs as electron accepting components were reported by Greenham et al., who used spherical CdSe or CdS NCs in nanocomposites with MEH-PPV.[111] The low efficiency of the solar cells obtained (0.2% under AM 1.5 conditions at 5 W m^{-2}; 0.1% under one sun conditions at 800 W m^{-2}; AM 1.5 = air mass 1.5 conditions = standard terrestrial solar spectral irradiance distribution on a sun-facing 37° tilted surface as defined by the American Society for Testing and Materials in document ASTM G173) is in accordance with the observed incomplete photoluminescence quenching. The morphology obtained suffered from distinct phase separation of the active layer components. In this study, it was also shown that the initial surface ligands (TOPO) constituted an insulating barrier, impeding exciton dissociation and charge transfer, and therefore these ligands were replaced by pyridine molecules.

Significant progress in the cell efficiency was achieved after replacing spherical NCs with rod-like ones[155] and by optimizing the ligand exchange with pyridine as well as other processing parameters (solvent mixture, concentrations, thermal annealing, etc.). Solar cells fabricated from 7 × 60 nm CdSe nanorods and regioregular poly(3-hexylthiophene) showed an improved power conversion efficiency of 1.7% under AM 1.5 conditions at 1 W m^{-2}.[156]

In the fabrication of hybrid, bulk heterojunction cells, special emphasis must be put on the supramolecular organization and morphology of the polymer phase. Processing of poly(3-hexylthiophene) together with rod like CdSe NCs from 1,2,4-trichlorobenzene leads to fibrilar morphology of the polymeric phase, facilitating charge carriers transport. The resulting interpenetrating network of two high aspect ratio components is very favorable for photovoltaic applications. Although for NC contents required for the functioning of the cell the fibrilar morphology is partially lost, still high cell efficiences, exceeding 2.5%, can be obtained.[157] Tetrapods as well as hyperbranched CdSe NCs were also tested as hybrid photovoltaic cells components. Cells based on tetrapods with poly(*p*-phenylene vinylene) showed the highest energy conversion efficiency reported to date, 2.8%,[158] whereas those fabricated using hyperbranched NCs and poly(3-hexylthiophene) exhibited 2.2%.[159]

Table 5.3 Energy Conversion Efficiencies Reported for Photovoltaic Devices of Various Types[a]

Material	η (AM 1.5)	Reference
Monocrystalline silicon	24.4% (12%–16%)	160
Multicrystalline silicon	20.3% (9%–12%)	161
Amorphous silicon	9.47% (4%–8%)	162
Dye-sensitized mesoporous TiO_2/redox couple in liquid electrolyte (Grätzel cell)	10.4%	163
Dye-sensitized mesoporous TiO_2/MeOTAD (solid state)	5.0%	164
MDMO-PPV/C_{70}	3.0%	165
P3HT/PCBM	4.9%	154
P3HT/CdSe nanorods	2.6%	157
MDMO-PPV/CdSe tetrapods	2.8%	158

[a]In addition to the highest reported value, the numbers in brackets indicate typical values observed for the corresponding device type (see List of Abbreviations).

It is instructive to compare the data obtained for conjugated polymers/semiconductor NCs hybrid solar cells with those reported for other types of these devices investigated in recent years (see Table 5.3). Given a large number of different parameters, which characterize the performance of photovoltaic cells, it is sometimes difficult to compare literature values. For these reasons, all data collected in Table 5.3 represent energy conversion efficiencies under AM 1.5 using an irradiation power of 100 mW cm^{-2}, according to the following relationships:

$$\text{Energy conversion efficiency:} \quad \eta = \frac{P_{max}}{P_{in}} = \frac{I_{SC} \times V_{OC} \times FF}{P_{in}}$$

$$\text{Fill factor:} \quad FF = \frac{P_{max}}{I_{SC} \times V_{OC}}$$

where P_{max} = largest power output
I_{SC} = short circuit current
V_{OC} = open circuit voltage
FF = fill factor
P_{in} = incident light power intensity

To conclude, although the energy conversion efficiency of conjugated polymer/semiconductor NCs hybrid photovoltaic cells is still significantly lower than those measured for silicon-based or Grätzel-type cells, the new nanomaterials described show several promising features. Their technologically most important advantages are as follows: (1) they are easy to process, using low-cost methods of low energy consumption; (2) they can be deposited on large surface, cheap substrates, which can partially compensate for their lower efficiency. Of course, still much research is needed to improve their efficiency and long-term stability.

5.5.2 Light-Emitting Devices

Light emitting diodes (LEDs) are electronic devices that transform electricity into light exploiting the electroluminescence phenomenon. In this respect, they can be considered as inverse to photovoltaic cells. The simplest LED consists of a single layer of a semiconductor sandwiched between two electrodes. As in the case of photovoltaic cells, one of these electrodes must be transparent to transmit the created light. These electronic devices are called diodes because they show *I-V* characteristics typical of diodes. If the quantum efficiency for electroluminescence is constant, the luminance–voltage characteristics follows the *I-V* one. This means that up to a given voltage, no current flows and no light is emitted. Above this onset voltage, both the current and the luminance quickly increase with increasing voltage. The operation of the simplest LED can be described briefly as follows. Holes and electrons are injected into the semiconductor layer from the anode and the cathode, respectively. The injected charges form either singlet or triplet excitons, of which only the former decay radiatively to generate light. In more advanced modifications of this simplest LED, additional layers are added which lower the barriers for the injection of holes and electrons (so called hole and electron transporting layers). Since the first report on polymer LEDs,[166] impressive progress in the fabrication of conjugated polymer-based LEDs has been achieved. Presently, fabricated organic LEDs combine lifetimes exceeding several thousands of hours with good external efficiencies.

On the contrary, LEDs with emissive layers consisting of semiconductor NCs exhibit much lower efficiencies to date. However, monodispersed NCs show optical properties unmatched by any class of organic chromophores. Typical linewidths of NCs' luminescence are inferior to 30 nm (FWHM),[167] whereas emission lines of conjugated polymers are much broader and frequently perturbed by a vibrational structure.[168] It can therefore intuitively be assumed that conjugated polymers/NCs composites may combine the efficiencies of polymeric LEDs with the color purity of semiconductor NCs. This is, however, only possible if an efficient energy transfer from the polymer matrix to the NCs occurs, resulting in electroluminescence essentially from NCs. Dispersion of NCs facilitates the energy transfer between the components of the composite; however, it also increases the possibility of different nonradiative relaxation pathways since binding of conjugated molecules to the NCs surface frequently quenches the luminescence of both the polymer and NCs (see above). For this reason, other device structures are more frequently used such as a bilayer NC–polymer structure or a tri- or even multilayer one, in which the NCs layer is sandwiched between two or more organic layers.[169] These systems are of course not real molecular hybrids and will therefore not be discussed here.

In a different approach, composites of semiconductor NCs with industrial insulating polymers have been used in the fabrication of so-called down-conversion LEDs.[170] The principle of their functioning can be described briefly as follows. If the light from a blue- or UV-emitting diode passes through a composite material consisting of core–shell NCs dispersed in a transparent polymer it can be absorbed and re-emitted at lower energy/higher wavelength. Thus, these devices are based on NCs' photoluminescence and not on their electroluminescence, circumventing by this means the problem of electrical charge carrier injection into the NC layer. The color of the re-emission can be tuned by the size and composition of the NCs used. Moreover, by controlling the ratio of differently sized NCs it is possible to precisely tune the re-emission color, which is of particular interest for the additive generation of white light.[171] Poly(alkyl methacrylate)s with long and branched substituents such as poly(lauryl methacrylate) are especially well suited as matrices for the NCs dispersion since they are amorphous and transparent. Their composites with NCs can be obtained by radical polymerization of the corresponding monomers in the presence of an appropriated cross-linking agent and core–shell NCs.[170]

5.6 PERSPECTIVES

Cadmium and, to a less extent, lead chalcogenide-based core and core–shell NCs are so far unmatched by any other type of semiconductor NCs as far as control of the size and shape, color purity, and fluorescence quantum yield are concerned. However, the toxicity of cadmium and lead strongly limits their technological use. Therefore, a strong need for the development of new synthesis methods for cadmium- and lead-free NCs of similar quality exists. In this context, indium phosphide, transition metal sulfides (e.g., Cu_2S, SnS) and chalcopyrite derivatives (e.g., $CuInS_2$, $CuInSe_2$) are potential candidates for the replacement of cadmium or lead chalcogenides in most applications.

The next research axis, whose exploration must be intensified in the near future, is surface ligand engineering. The organic surfactant layer strongly influences the optical and electronic properties not only of individual NCs but also of their supramolecular aggregates and of their molecular composites and hybrids with electroactive oligomers or polymers. Classical *aliphatic* surface ligands, which allow the dispersion of NCs in organic solvents, are of insulating nature, impeding efficient charge transfer. On the other hand, the electronic properties of *conjugated* ligands directly influence the electronic structure of the NCs, changing their properties frequently in an undesirable way. Taking into account the large number of potential conjugated surface ligands, molecular modeling methods capable of predicting the final properties of the NC-ligands associations must be developed to allow for a more rational surface ligand design.

New methods of molecular level processing must be developed. This implies a much wider use of supramolecular chemistry concepts. Subtle interactions like molecular recognition phenomena are very useful in this respect, because they allow for the formation of ordered supramolecular aggregates of different size and topology. Only few papers published to date explore this approach, which seems extremely versatile and promising. Appropriate selection of the complementary recognition groups should lead to a significant simplification of the processing procedures, leading in the most favorable case to one-step, one-pot preparation methods.

Applications of conjugated polymers–NCs hybrids in various (opto-)electronic devices and sensors require a deeper understanding of the relationship between their chemical and photophysical properties and their structure at different levels from the molecular to the bulk material. This requires the use of complementary methods and combined measurements, including correlation between the structural and electrical transport properties, redox and spectroelectrochemical properties, to name a few.

REFERENCES

1. C. B. Murray, C. R. Kagan, M. G. Bawendi, *Annu. Rev. Mater. Sci.* **2000**, *30*, 545–610.
2. J. Park, J. Joo, S. G. Kwon, Y. Jang, T. Hyeon, *Angew. Chem. Int. Ed.* **2007**, *46*, 4630–4660.
3. A. L. Rogach, A. Eychmüller, S. G. Hickey, S. V. Kershaw, *Small* **2007**, *3*, 536–557.
4. A. L. Rogach, D. V. Talapin, E. V. Shevchenko, A. Kornowski, M. Haase, H. Weller, *Adv. Funct. Mater.* **2002**, *12*, 653–664.
5. D. Y. Godovsky, *Adv. Polym. Sci.* **2000**, *153*, 163–205.
6. E. Holder, N. Tessler, A. L. Rogach, *J. Mater. Chem.* **2008**, *18*, 1064–1078.
7. L. E. Brus, *J. Chem. Phys.* **1984**, *80*, 4403–4409.
8. M. G. Bawendi, M. L. Steigerwald, L. E. Brus, *Annu. Rev. Phys. Chem.* **1990**, *41*, 477–496.
9. A. L. Efros, M. Rosen, M. Kuno, M. Nirmal, D. J. Norris, M. Bawendi, *Phys. Rev. B* **1996**, *54*, 4843–4856.
10. A. L. Efros, A. L. Efros, *Sov. Phys. Semicond. USSR* **1982**, *16*, 772–775.
11. A. Henglein, *Ber. Bunsen-Ges. Phys. Chem.* **1982**, *86*, 301–305.
12. R. Rossetti, L. Brus, *J. Phys. Chem.* **1982**, *86*, 4470–4472.
13. R. Rossetti, S. Nakahara, L. E. Brus, *J. Chem. Phys.* **1983**, *79*, 1086–1088.
14. C. B. Murray, D. J. Norris, M. G. Bawendi, *J. Am. Chem. Soc.* **1993**, *115*, 8706–8715.
15. D. J. Norris, A. Sacra, C. B. Murray, M. G. Bawendi, *Phys. Rev. Lett.* **1994**, *72*, 2612–2615.
16. D. V. Talapin, A. L. Rogach, A. Kornowski, M. Haase, H. Weller, *Nano Lett.* **2001**, *1*, 207–211.
17. A. I. Ekimov, F. Hache, M. C. Schanneklein, D. Ricard, C. Flytzanis, I. A. Kudryavtsev, T. V. Yazeva, A. V. Rodina, A. L. Efros, *J. Opt. Soc. Amer. B* **1993**, *10*, 100–107.
18. M. Nirmal, D. J. Norris, M. Kuno, M. G. Bawendi, A. L. Efros, M. Rosen, *Phys. Rev. Lett.* **1995**, *75*, 3728–3731.
19. S. Empedocles, M. Bawendi, *Acc. Chem. Res.* **1999**, *32*, 389–396.
20. R. G. Neuhauser, K. T. Shimizu, W. K. Woo, S. A. Empedocles, M. G. Bawendi, *Phys. Rev. Lett.* **2000**, *85*, 3301–3304.
21. M. Kuno, D. P. Fromm, H. F. Hamann, A. Gallagher, D. J. Nesbitt, *J. Chem. Phys.* **2000**, *112*, 3117–3120.
22. A. M. Munro, I. J. L. Plante, M. S. Ng, D. S. Ginger, *J. Phys. Chem.C* **2007**, *111*, 6220–6227.
23. E. Jang, S. Jun, Y. S. Chung, L. S. Pu, *J. Phys. Chem. B* **2004**, *108*, 4597–4600.
24. O. I. Micic, H. M. Cheong, H. Fu, A. Zunger, J. R. Sprague, A. Mascarenhas, A. J. Nozik, *J. Phys. Chem. B* **1997**, *101*, 4904–4912.
25. D. V. Talapin, N. Gaponik, H. Borchert, A. L. Rogach, M. Haase, H. Weller, *J. Phys. Chem. B* **2002**, *106*, 12659–12663.
26. C. Y. Yeh, Z. W. Lu, S. Froyen, A. Zunger, *Phys. Rev. B* **1992**, *46*, 10086–10097.
27. K. Jacobs, J. Wickham, A. P. Alivisatos, *J. Phys. Chem. B* **2002**, *106*, 3759–3762.
28. O. Madelung, M. Schulz, H. Weiss, *Numerical Data and Functional Relationships in Science and Technology, New Series, Group III: Crystal and Solid State Physics*, Vol. III/17b, Springer, Berlin, **1982**.
29. M. A. Hines, P. Guyot-Sionnest, *J. Phys. Chem.* **1996**, *100*, 468–471.
30. B. O. Dabbousi, J. Rodriguez Viejo, F. V. Mikulec, J. R. Heine, H. Mattoussi, R. Ober, K. F. Jensen, M. G. Bawendi, *J. Phys. Chem. B* **1997**, *101*, 9463–9475.
31. X. G. Peng, M. C. Schlamp, A. V. Kadavanich, A. P. Alivisatos, *J. Am. Chem. Soc.* **1997**, *119*, 7019–7029.
32. S. H. Wei, A. Zunger, *Appl. Phys. Lett.* **1998**, *72*, 2011–2013.
33. K. Biswas, C. N. R. Rao, *Chem. Eur. J.* **2007**, *13*, 6123–6129.
34. P. S. Shah, T. Hanrath, K. P. Johnston, B. A. Korgel, *J. Phys. Chem. B* **2004**, *108*, 9574–9587.
35. A. Fojtik, H. Weller, U. Koch, A. Henglein, *Ber. Bunsen-Ges. Phys. Chem.* **1984**, *88*, 969–977.

36. L. Spanhel, M. Haase, H. Weller, A. Henglein, *J. Am. Chem. Soc.* **1987**, *109*, 5649–5655.
37. P. Lianos, J. K. Thomas, *Chem. Phys. Lett.* **1986**, *125*, 299–302.
38. M. P. Pileni, *Langmuir* **1997**, *13*, 3266–3276.
39. A. Rogach, S. Kershaw, M. Burt, M. Harrison, A. Kornowski, A. Eychmüller, H. Weller, *Adv. Mater.* **1999**, *11*, 552–555.
40. M. T. Harrison, S. V. Kershaw, A. L. Rogach, A. Kornowski, A. Eychmüller, H. Weller, *Adv. Mater.* **2000**, *12*, 123–125.
41. M. Kuno, K. A. Higginson, S. B. Qadri, M. Yousuf, S. H. Lee, B. L. Davis, H. Mattoussi, *J. Phys. Chem. B* **2003**, *107*, 5758–5767.
42. X. Wang, Q. Peng, Y. D. Li, *Acc. Chem. Res.* **2007**, *40*, 635–643.
43. V. K. LaMer, R. H. Dinegar, *J. Am. Chem. Soc.* **1950**, *72*, 4847–4854.
44. C. D. Donega, P. Liljeroth, D. Vanmaekelbergh, *Small* **2005**, *1*, 1152–1162.
45. D. V. Talapin, A. L. Rogach, E. V. Shevchenko, A. Kornowski, M. Haase, H. Weller, *J. Am. Chem. Soc.* **2002**, *124*, 5782–5790.
46. Z. A. Peng, X. G. Peng, *J. Am. Chem. Soc.* **2001**, *123*, 183–184.
47. L. H. Qu, Z. A. Peng, X. G. Peng, *Nano Lett.* **2001**, *1*, 333–337.
48. W. W. Yu, X. G. Peng, *Angew. Chem. Int. Ed.* **2002**, *41*, 2368–2371.
49. D. G. Wu, M. E. Kordesch, P. G. Van Patten, *Chem. Mater.* **2005**, *17*, 6436–6441.
50. Y. A. Yang, H. M. Wu, K. R. Williams, Y. C. Cao, *Angew. Chem. Int. Ed.* **2005**, *44*, 6712–6715.
51. J. Jasieniak, C. Bullen, J. van Embden, P. Mulvaney, *J. Phys. Chem. B* **2005**, *109*, 20665–20668.
52. S. Sapra, A. L. Rogach, J. Feldmann, *J. Mater. Chem.* **2006**, *16*, 3391–3395.
53. N. Pradhan, D. Reifsnyder, R. G. Xie, J. Aldana, X. G. Peng, *J. Am. Chem. Soc.* **2007**, *129*, 9500–9509.
54. Y. C. Cao, J. H. Wang, *J. Am. Chem. Soc.* **2004**, *126*, 14336–14337.
55. N. Pradhan, S. Efrima, *J. Am. Chem. Soc.* **2003**, *125*, 2050–2051.
56. N. Pradhan, B. Katz, S. Efrima, *J. Phys. Chem. B* **2003**, *107*, 13843–13854.
57. J. Joo, H. B. Na, T. Yu, J. H. Yu, Y. W. Kim, F. X. Wu, J. Z. Zhang, T. Hyeon, *J. Am. Chem. Soc.* **2003**, *125*, 11100–11105.
58. D. V. Talapin, S. Haubold, A. L. Rogach, A. Kornowski, M. Haase, H. Weller, *J. Phys. Chem. B* **2001**, *105*, 2260–2263.
59. W. W. Yu, Y. A. Wang, X. G. Peng, *Chem. Mater.* **2003**, *15*, 4300–4308.
60. V. Kloper, R. Osovsky, J. Kolny-Olesiak, A. Sashchiuk, E. Lifshitz, *J. Phys. Chem. C* **2007**, *111*, 10336–10341.
61. S. L. Lin, N. Pradhan, Y. J. Wang, X. G. Peng, *Nano Lett.* **2004**, *4*, 2261–2264.
62. J. H. Yu, J. Joo, H. M. Park, S. I. Baik, Y. W. Kim, S. C. Kim, T. Hyeon, *J. Am. Chem. Soc.* **2005**, *127*, 5662–5670.
63. M. A. Hines, P. Guyot-Sionnest, *J. Phys. Chem. B* **1998**, *102*, 3655–3657.
64. P. Reiss, G. Quemard, S. Carayon, J. Bleuse, F. Chandezon, A. Pron, *Mater. Chem. Phys.* **2004**, *84*, 10–13.
65. R. G. Xie, X. H. Zhong, T. Basché, *Adv. Mater.* **2005**, *17*, 2741–2744.
66. M. Green, G. Wakefield, P. J. Dobson, *J. Mater. Chem.* **2003**, *13*, 1076–1078.
67. X. H. Zhong, Z. H. Zhang, S. H. Liu, M. Y. Han, W. Knoll, *J. Phys. Chem. B* **2004**, *108*, 15552–15559.
68. J. S. Steckel, P. Snee, S. Coe-Sullivan, J. R. Zimmer, J. E. Halpert, P. Anikeeva, L. A. Kim, V. Bulovic, M. G. Bawendi, *Angew. Chem. Int. Ed.* **2006**, *45*, 5796–5799.
69. M. Protiere, P. Reiss, *Small* **2007**, *3*, 399–403.
70. X. H. Zhong, Y. Y. Feng, W. Knoll, M. Y. Han, *J. Am. Chem. Soc.* **2003**, *125*, 13559–13563.
71. R. E. Bailey, S. M. Nie, *J. Am. Chem. Soc.* **2003**, *125*, 7100–7106.
72. O. I. Micic, C. J. Curtis, K. M. Jones, J. R. Sprague, A. J. Nozik, *J. Phys. Chem.* **1994**, *98*, 4966–4969.
73. A. A. Guzelian, J. E. B. Katari, A. V. Kadavanich, U. Banin, K. Hamad, E. Juban, A. P. Alivisatos, R. H. Wolters, C. C. Arnold, J. R. Heath, *J. Phys. Chem.* **1996**, *100*, 7212–7219.
74. D. Battaglia, X. G. Peng, *Nano Lett.* **2002**, *2*, 1027–1030.

75. S. Xu, S. Kumar, T. Nann, *J. Am. Chem. Soc.* **2006**, *128*, 1054–1055.
76. L. Li, M. Protière, P. Reiss, *Chem. Mater.* **2008**, *20*, 2621–2623.
77. A. A. Guzelian, U. Banin, A. V. Kadavanich, X. Peng, A. P. Alivisatos, *Appl. Phys. Lett.* **1996**, *69*, 1432–1434.
78. O. I. Micic, J. R. Sprague, C. J. Curtis, K. M. Jones, J. L. Machol, A. J. Nozik, H. Giessen, B. Fluegel, G. Mohs, N. Peyghambarian, *J. Phys. Chem.* **1995**, *99*, 7754–7759.
79. Y. H. Kim, Y. W. Jun, B. H. Jun, S. M. Lee, J. W. Cheon, *J. Am. Chem. Soc.* **2002**, *124*, 13656–13657.
80. M. Furis, Y. Sahoo, D. J. MacRae, F. S. Manciu, A. N. Cartwright, P. N. Prasad, *J. Phys. Chem. B* **2003**, *107*, 11622–11625.
81. K. Sardar, M. Dan, B. Schwenzer, C. N. R. Rao, *J. Mater. Chem.* **2005**, *15*, 2175–2177.
82. C. B. Murray, S. H. Sun, W. Gaschler, H. Doyle, T. A. Betley, C. R. Kagan, *IBM J. Res. Dev.* **2001**, *45*, 47–56.
83. A. J. Houtepen, R. Koole, D. L. Vanmaekelbergh, J. Meeldijk, S. G. Hickey, *J. Am. Chem. Soc.* **2006**, *128*, 6792–6793.
84. B. L. Wehrenberg, C. J. Wang, P. Guyot-Sionnest, *J. Phys. Chem. B* **2002**, *106*, 10634–10640.
85. J. M. Pietryga, R. D. Schaller, D. Werder, M. H. Stewart, V. I. Klimov, J. A. Hollingsworth, *J. Am. Chem. Soc.* **2004**, *126*, 11752–11753.
86. I. C. Baek, S. Il Seok, N. C. Pramanik, S. Jana, M. A. Lim, B. Y. Ahn, C. J. Lee, Y. J. Jeong, *J. Colloid Interface Sci.* **2007**, *310*, 163–166.
87. A. Sashchiuk, L. Amirav, M. Bashouti, M. Krueger, U. Sivan, E. Lifshitz, *Nano Lett.* **2004**, *4*, 159–165.
88. W. W. Yu, J. C. Falkner, B. S. Shih, V. L. Colvin, *Chem. Mater.* **2004**, *16*, 3318–3322.
89. M. A. Hines, G. D. Scholes, *Adv. Mater.* **2003**, *15*, 1844–1849.
90. W. G. Lu, J. Y. Fang, K. L. Stokes, J. Lin, *J. Am. Chem. Soc.* **2004**, *126*, 11798–11799.
91. J. E. Murphy, M. C. Beard, A. G. Norman, S. P. Ahrenkiel, J. C. Johnson, P. R. Yu, O. I. Micic, R. J. Ellingson, A. J. Nozik, *J. Am. Chem. Soc.* **2006**, *128*, 3241–3247.
92. P. W. Voorhees, *J. Stat. Phys.* **1985**, *38*, 231–252.
93. H. Reiss, *J. Chem. Phys.* **1951**, *19*, 482–487.
94. H. T. Liu, J. S. Owen, A. P. Alivisatos, *J. Am. Chem. Soc.* **2007**, *129*, 305–312.
95. X. B. Chen, Y. B. Lou, A. C. Samia, C. Burda, *Nano Lett.* **2003**, *3*, 799–803.
96. W. W. Yu, L. H. Qu, W. Z. Guo, X. G. Peng, *Chem. Mater.* **2003**, *15*, 2854–2860.
97. J. J. Li, Y. A. Wang, W. Z. Guo, J. C. Keay, T. D. Mishima, M. B. Johnson, X. G. Peng, *J. Am. Chem. Soc.* **2003**, *125*, 12567–12575.
98. W. C. W. Chan, S. M. Nie, *Science* **1998**, *281*, 2016–2018.
99. H. Mattoussi, J. M. Mauro, E. R. Goldman, G. P. Anderson, V. C. Sundar, F. V. Mikulec, M. G. Bawendi, *J. Am. Chem. Soc.* **2000**, *122*, 12142–12150.
100. S. Pathak, S. K. Choi, N. Arnheim, M. E. Thompson, *J. Am. Chem. Soc.* **2001**, *123*, 4103–4104.
101. J. K. Jaiswal, H. Mattoussi, J. M. Mauro, S. M. Simon, *Nature Biotechnol.* **2003**, *21*, 47–51.
102. B. Dubertret, P. Skourides, D. J. Norris, V. Noireaux, A. H. Brivanlou, A. Libchaber, *Science* **2002**, *298*, 1759–1762.
103. T. Pellegrino, L. Manna, S. Kudera, T. Liedl, D. Koktysh, A. L. Rogach, S. Keller, J. Radler, G. Natile, W. J. Parak, *Nano Lett.* **2004**, *4*, 703–707.
104. V. L. Colvin, M. C. Schlamp, A. P. Alivisatos, *Nature* **1994**, *370*, 354–357.
105. S. Gunes, H. Neugebauer, N. S. Sariciftci, *Chem. Rev.* **2007**, *107*, 1324–1338.
106. D. J. Milliron, S. M. Hughes, Y. Cui, L. Manna, J. B. Li, L. W. Wang, A. P. Alivisatos, *Nature* **2004**, *430*, 190–195.
107. L. Carbone, C. Nobile, M. De Giorg, F. D. Sala, G. Morello, P. Pompa, M. Hytch, E. Snoeck, A. Fiore, I. R. Franchini, M. Nadasan, A. F. Silvestre, L. Chiodo, S. Kudera, R. Cingolani, R. Krahne, L. Manna, *Nano Lett.* **2007**, *7*, 2942–2950.
108. M. Brinkmann, J. C. Wittmann, *Adv. Mater.* **2006**, *18*, 860–863.
109. S. Gupta, Q. L. Zhang, T. Emrick, T. P. Russell, *Nano Lett.* **2006**, *6*, 2066–2069.
110. M. Brinkmann, D. Aldakov, F. Chandezon, *Adv. Mater.* **2007**, *19*, 3819–3823.
111. N. C. Greenham, X. G. Peng, A. P. Alivisatos, *Phys. Rev. B* **1996**, *54*, 17628–17637.

112. C. Querner, A. Benedetto, R. Demadrille, P. Rannou, P. Reiss, *Chem. Mater.* **2006**, *18*, 4817–4826.
113. C. Querner, P. Reiss, J. Bleuse, A. Pron, *J. Am. Chem. Soc.* **2004**, *126*, 11574–11582.
114. H. Sirringhaus, P. J. Brown, R. H. Friend, M. M. Nielsen, K. Bechgaard, B. M. W. Langeveld-Voss, A. J. H. Spiering, R. A. J. Janssen, E. W. Meijer, *Synth. Met.* **2000**, *111*, 129–132.
115. H. Sirringhaus, P. J. Brown, R. H. Friend, M. M. Nielsen, K. Bechgaard, B. M. W. Langeveld-Voss, A. J. H. Spiering, R. A. J. Janssen, E. W. Meijer, P. Herwig, D. M. de Leeuw, *Nature* **1999**, *401*, 685–688.
116. B. C. Sih, M. Wolf, *J. Phys. Chem. C* **2007**, *111*, 17184–17192.
117. D. J. Milliron, A. P. Alivisatos, C. Pitois, C. Edder, J. M. J. Frechet, *Adv. Mater.* **2003**, *15*, 58–63.
118. R. van Beek, A. P. Zoombelt, L. W. Jenneskens, C. A. van Walree, C. D. Donega, D. Veldman, R. A. J. Janssen, *Chem. Eur. J.* **2006**, *12*, 8075–8083.
119. J. S. Liu, T. Tanaka, K. Sivula, A. P. Alivisatos, J. M. J. Frechet, *J. Am. Chem. Soc.* **2004**, *126*, 6550–6551.
120. J. Locklin, D. Patton, S. X. Deng, A. Baba, M. Millan, R. C. Advincula, *Chem. Mater.* **2004**, *16*, 5187–5193.
121. R. C. Advincula, *Dalton Trans.* **2006**, 2778–2784.
122. P. K. Sudeep, K. T. Early, K. D. McCarthy, M. Y. Odoi, M. D. Barnes, T. Emrick, *J. Am. Chem. Soc.* **2008**, *130*, 2384–2385.
123. A. Javier, C. S. Yun, J. Sorena, G. F. Strouse, *J. Phys. Chem. B* **2003**, *107*, 435–442.
124. C. Fang, X. Y. Qi, Q. L. Fan, L. H. Wang, W. Huang, *Nanotechnology* **2007**, *18*, 035704.
125. K. B. Chen, M. H. Chen, S. H. Yang, C. H. Hsieh, C. S. Hsu, C. C. Chen, H. J. Chien, *J. Pol. Sci. A Pol. Chem.* **2006**, *44*, 5378–5390.
126. R. Pokrop, PhD thesis, Warsaw University of Technology, Warsaw, Poland, **2007**.
127. T. Antoun, R. Brayner, S. Al Terary, F. Fievet, M. Chehimi, A. Yassar, *Eur. J. Inorg. Chem.* **2007**, 1275–1284.
128. D. Aldakov, C. Querner, Y. Kervella, B. Jousselme, R. Demadrille, E. Rossitto, P. Reiss, A. Pron, *Microchim. Acta* **2008**, *160*, 335–344.
129. C. Querner, P. Reiss, S. Sadki, M. Zagorska, A. Pron, *Phys. Chem. Chem. Phys.* **2005**, *7*, 3204–3209.
130. C. Querner, P. Reiss, M. Zagorska, O. Renault, R. Payerne, F. Genoud, P. Rannou, A. Pron, *J. Mater. Chem.* **2005**, *15*, 554–563.
131. Q. Zhang, T. P. Russell, T. Emrick, *Chem. Mater.* **2007**, *19*, 3712–3716.
132. H. Skaff, K. Sill, T. Emrick, *J. Am. Chem. Soc.* **2004**, *126*, 11322–11325.
133. M. Y. Odoi, N. I. Hammer, K. Sill, T. Emrick, M. D. Barnes, *J. Am. Chem. Soc.* **2006**, *128*, 3506–3507.
134. J. Xu, J. Wang, M. Mitchell, P. Mukherjee, M. Jeffries-El, J. W. Petrich, Z. Q. Lin, *J. Am. Chem. Soc.* **2007**, *129*, 12828–12833.
135. U. Drechsler, B. Erdogan, V. M. Rotello, *Chem. Eur. J.* **2004**, *10*, 5570–5579.
136. R. Shenhar, V. M. Rotello, *Acc. Chem. Res.* **2003**, *36*, 549–561.
137. R. Baron, C. H. Huang, D. M. Bassani, A. Onopriyenko, M. Zayats, I. Willner, *Angew. Chem. Int. Ed.* **2005**, *44*, 4010–4015.
138. C. R. van den Brom, I. Arfaoui, T. Cren, B. Hessen, T. T. M. Palstra, J. T. M. De Hosson, P. Rudolf, *Adv. Funct. Mater.* **2007**, *17*, 2045–2052.
139. J. De Girolamo, P. Reiss, A. Pron, *J. Phys. Chem. C* **2007**, *111*, 14681–14688.
140. J. De Girolamo, P. Reiss, A. Pron, *J. Phys. Chem. C* **2008**, *112*, 8797–8801.
141. S. B. Wang, C. Li, G. Q. Shi, *Sol. Energy Mater. Sol. Cells* **2008**, *92*, 543–549.
142. C. H. Chou, H. S. Wang, K. H. Wei, J. Y. Huang, *Adv. Funct. Mater.* **2006**, *16*, 909–916.
143. S. A. Majetich, A. C. Carter, J. Belot, R. D. McCullough, *J. Phys. Chem.* **1994**, *98*, 13705–13710.
144. W. W. Yu, L. H. Qu, W. Z. Guo, X. G. Peng, *Chem. Mater.* **2004**, *16*, 560–560.
145. M. Trznadel, A. Pron, M. Zagorska, *Macromolecules* **1998**, *31*, 5051–5058.
146. I. Polec, A. Henckens, L. Goris, M. Nicolas, M. A. Loi, P. J. Adriaensens, L. Lutsen, J. V. Manca, D. Vanderzande, N. S. Sariciftci, *J. Pol. Sci. A Pol. Chem.* **2003**, *41*, 1034–1045.
147. M. Skompska, A. Szkurlat, *Electrochim. Acta* **2001**, *46*, 4007–4015.

148. E. Kucur, J. Riegler, G. A. Urban, T. Nann, *J. Chem. Phys.* **2003**, *119*, 2333–2337.
149. K. Buga, R. Pokrop, A. Majkowska, M. Zagorska, J. Planes, F. Genoud, A. Pron, *J. Mater. Chem.* **2006**, *16*, 2150–2164.
150. J. De Girolamo, P. Reiss, M. Zagorska, R. De Bettignies, S. Bailly, J.-Y. Mevellec, S. Lefrant, J.-P. Travers, A. Pron, *Phys. Chem. Chem. Phys.* **2008**, *10*, 4027–4035.
151. D. Aldakov, F. Chandezon, R. De Bettignies, M. Firon, P. Reiss, A. Pron, *Eur. Phys. J. Appl. Phys.* **2006**, *36*, 261–265.
152. H. Antoniadis, B. R. Hsieh, M. A. Abkowitz, S. A. Jenekhe, M. Stolka, *Synth. Met.* **1994**, *62*, 265–271.
153. B. Kraabel, D. McBranch, N. S. Sariciftci, D. Moses, A. J. Heeger, *Mol. Cryst. Liq. Cryst. Sci. Technol. A* **1994**, *256*, 733–738.
154. M. Reyes-Reyes, K. Kim, D. L. Carroll, *Appl. Phys. Lett.* **2005**, *87*, 083506.
155. W. U. Huynh, X. G. Peng, A. P. Alivisatos, *Adv. Mater.* **1999**, *11*, 923–925.
156. W. U. Huynh, J. J. Dittmer, A. P. Alivisatos, *Science* **2002**, *295*, 2425–2427.
157. B. Q. Sun, N. C. Greenham, *Phys. Chem. Chem. Phys.* **2006**, *8*, 3557–3560.
158. B. Q. Sun, H. J. Snaith, A. S. Dhoot, S. Westenhoff, N. C. Greenham, *J. Appl. Phys.* **2005**, *97*, 014914.
159. I. Gur, N. A. Fromer, C. P. Chen, A. G. Kanaras, A. P. Alivisatos, *Nano Lett.* **2007**, *7*, 409–414.
160. J. H. Zhao, A. H. Wang, M. A. Green, F. Ferrazza, *Appl. Phys. Lett.* **1998**, *73*, 1991–1993.
161. O. Schultz, S. W. Glunz, G. P. Willeke, *Progr. Photovoltaics* **2004**, *12*, 553–558.
162. J. Meier, J. Spitznagel, U. Kroll, C. Bucher, S. Fay, T. Moriarty, A. Shah, *Thin Solid Films* **2004**, *451-52*, 518–524.
163. M. K. Nazeeruddin, P. Pechy, T. Renouard, S. M. Zakeeruddin, R. Humphry-Baker, P. Comte, P. Liska, L. Cevey, E. Costa, V. Shklover, L. Spiccia, G. B. Deacon, C. A. Bignozzi, M. Grätzel, *J. Am. Chem. Soc.* **2001**, *123*, 1613–1624.
164. H. J. Snaith, A. J. Moule, C. Klein, K. Meerholz, R. H. Friend, M. Grätzel, *Nano Lett.* **2007**, *7*, 3372–3376.
165. M. M. Wienk, J. M. Kroon, W. J. H. Verhees, J. Knol, J. C. Hummelen, P. A. van Hal, R. A. J. Janssen, *Angew. Chem. Int. Ed.* **2003**, *42*, 3371–3375.
166. J. H. Burroughes, D. D. C. Bradley, A. R. Brown, R. N. Marks, K. Mackay, R. H. Friend, P. L. Burns, A. B. Holmes, *Nature* **1990**, *347*, 539–541.
167. P. Reiss, J. Bleuse, A. Pron, *Nano Lett.* **2002**, *2*, 781–784.
168. A. Kraft, A. C. Grimsdale, A. B. Holmes, *Angew. Chem. Int. Ed.***1998**, *37*, 402–428.
169. Q. Sun, Y. A. Wang, L. S. Li, D. Y. Wang, T. Zhu, J. Xu, C. H. Yang, Y. F. Li, *Nature Photonics* **2007**, *1*, 717–722.
170. J. Lee, V. C. Sundar, J. R. Heine, M. G. Bawendi, K. F. Jensen, *Adv. Mater.* **2000**, *12*, 1102–1105.
171. H. S. Chen, C. K. Hsu, H. Y. Hong, *IEEE Photonics Technol. Lett.* **2006**, *18*, 193–195.

Chapter 6

Functionalized Carbon Nanotubes for Bioapplications

LINGRONG GU, FUSHEN LU, PENGJU G. LUO,
HAIFANG WANG, MOHAMMED J. MEZIANI, AND
YA-PING SUN

6.1 INTRODUCTION

Carbon nanotubes as pseudo-one-dimensional carbon allotropes of high aspect ratio, high surface area, and excellent material properties such as ultimate electrical and thermal conductivities and mechanical strength have generated much excitement in the nanoscience and nanotechnology community.[1–3] These all-carbon hollow

The Supramolecular Chemistry of Organic–Inorganic Hybrid Materials. Edited by Knut Rurack and Ramón Martínez-Máñez

graphitic nanomaterials may conceptually be viewed as rolled-up structures of one or more layers of graphene sheets for single-walled (SWNT) or multiple-walled (MWNT) carbon nanotubes, respectively, though their formation is generally spontaneous in various production methods. In recent years, the availability of carbon nanotubes has improved dramatically, both in terms of quality and quantity, which has stimulated the worldwide effort on using carbon nanotubes for technological applications.[3,4] Nevertheless, since carbon nanotubes (especially SWNTs) are still relatively expensive, their uses in technologies that offer higher value but require lower quantity are probably better justified in the near term. Obviously, one such technological area of enormous potential is the application of carbon nanotubes in biology and medicine, where most significant recent progress has been achieved.

The unique properties of carbon nanotubes offer a wide range of opportunities and potential applications. For example, the rich electronic properties of carbon nanotubes have been explored for the development of highly sensitive and specific nanoscale biosensors.[5,6] There have been promising results on the use of carbon nanotubes in various electroanalytical nanodevices,[7] and as electromechanical actuators for artificial muscles.[8] The near-infrared optical absorption of carbon nanotubes (semiconducting SWNTs) has been exploited for laser heating cancer therapy.[9] In recent years, the rapid development toward maturation of methodologies for the chemical modification and functionalization of carbon nanotubes has helped to address many biocompatibility-related issues, opening up an even wider range of bioapplication opportunities in areas such as drug delivery,[10] bioconjugation and specific recognition, including especially the use of sugar-functionalized carbon nanotubes to bind and aggregate anthrax spores.[11] The biological effects and consequences of carbon nanotubes *in vitro* and *in vivo* have also been investigated for a clear understanding of the environmental impact and toxicity issues of these newly developed nanoscale materials.[12]

In this chapter, we review the recent advances in the chemical modification of carbon nanotubes related specifically to their bioapplications, discuss various approaches and developments in the use of carbon nanotubes in biology and medicine, and provide some perspectives on further investigations and emerging opportunities.

6.2 FUNCTIONALIZATION OF CARBON NANOTUBES

In recent years, the development and implementation of various covalent and noncovalent functionalization methods for chemical modification and solubilization of carbon nanotubes have reached some degree of maturation.[13–15] The functionalization and especially the aqueous solubilization of carbon nanotubes have had significant benefits with regard to making nanotubes biocompatible for bioapplications. Among the now widely used covalent functionalization approaches are addition reactions (Scheme 6.1), derived from those traditionally used for graphite surfaces or established for fullerenes, and coupling reactions targeting oxidized defect sites on the carbon nanotube surface (typically the surface defect-derived carboxylic acid moieties to react with amines and alcohols for amide and ester linkages, respectively,

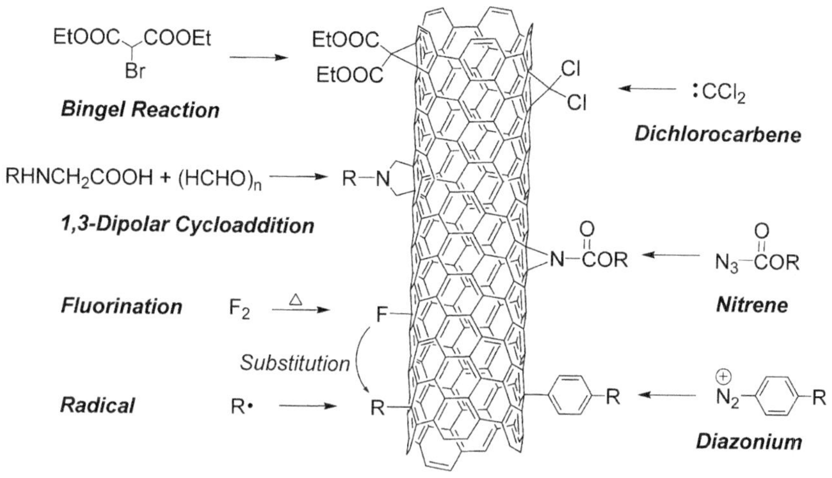

Scheme 6.1 Covalent addition reactions on the sidewall of carbon nanotubes.

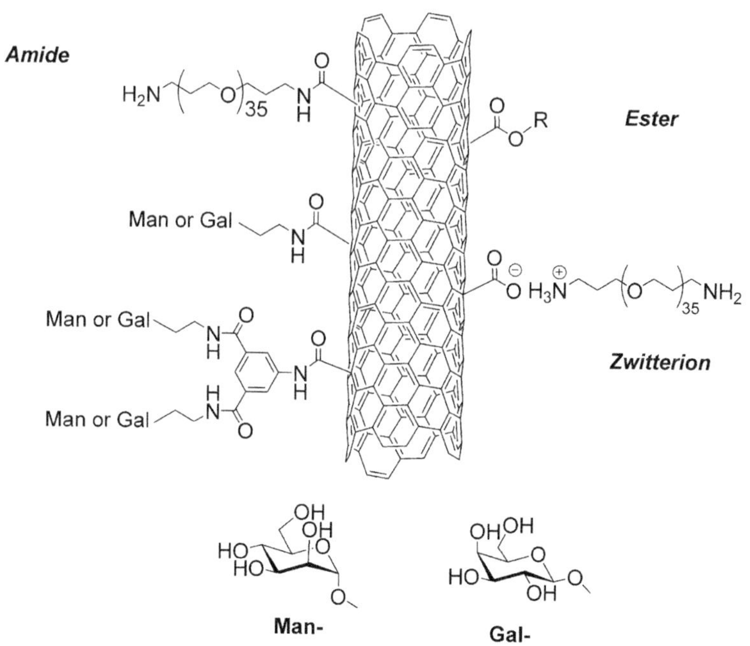

Scheme 6.2 Covalent reactions targeting carboxylic acids (derived from nanotube surface defects).

Scheme 6.2). For some functional groups such as amines, the covalent amidation functionalization is likely accompanied by significant noncovalent interactions or strong adsorption of the functional groups on the nanotube surface, thus substantially improving some properties of the functionalized nanotube samples (such as excellent water solubility for a hydrophilic functionalization agent[16]).

Noncovalent approaches can usually preserve the structures and properties of carbon nanotubes after functionalization[17] (though not necessarily the near-infrared absorption characteristics due to well-established doping effects), thus are equally important to the biocompatibilization and bioapplications of nanotubes.[15] Among commonly employed noncovalent schemes are surfactant dispersion,[18] $\pi-\pi$ stacking with aromatic compounds,[19] and polymer wrapping.[20]

Besides synthetic polymers and small molecules, biological or bioactive species are used in the functionalization of carbon nanotubes not only for water solubility but also enhanced biocompatibilities and biorecognition capabilities. Various proteins, DNAs, and carbohydrates have been covalently or noncovalently functionalized with carbon nanotubes, producing highly aqueous stable and biocompatible

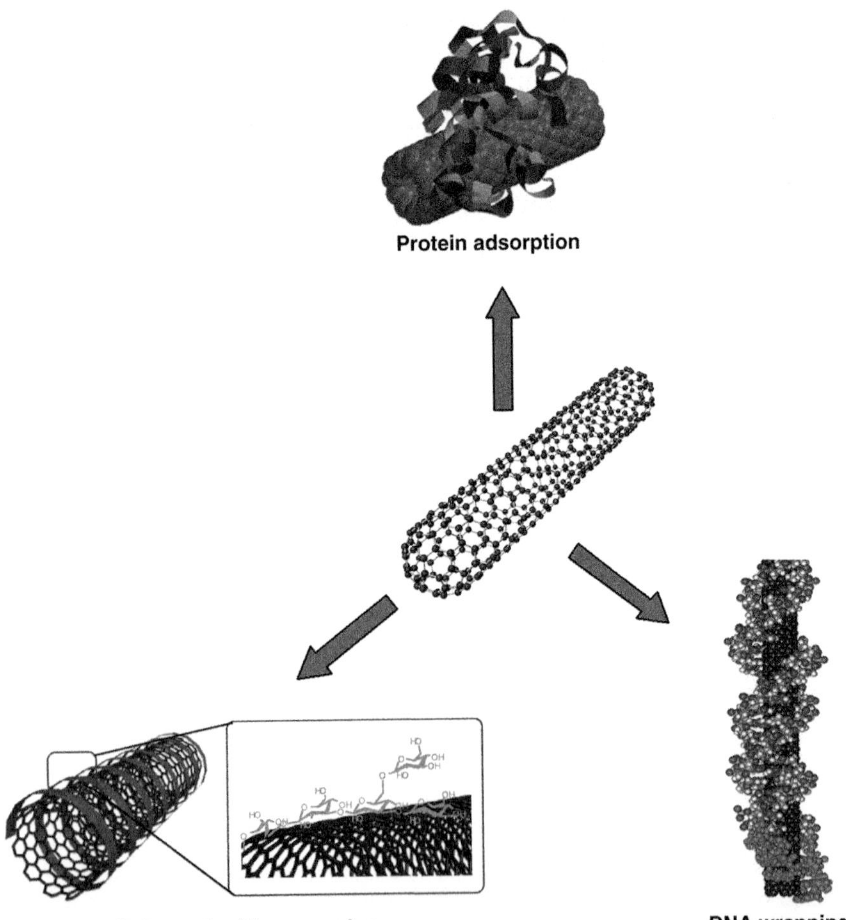

Scheme 6.3 Noncovalent biofunctionalized carbon nanotubes. (From References 21, 41, and 53. Reprinted with permission.)

biocarbon nanotube systems (Scheme 6.3). Some of the recent advances in this area are highlighted below.

6.2.1 Proteins and Peptides

Nonspecific adsorption is a commonly used method for preparing protein–carbon (and/or peptide–carbon) nanotube conjugates. For example, Nepal and Geckeler reported the dispersion and debundling of SWNTs with different proteins such as lysozyme, histone, hemoglobin, myoglobin, ovalbumin, bovine serum albumin, trypsin, and glucose oxidase.[21] The dispersion ability of the proteins depends on various factors, including their primary structure and pH. Proteins (histone and lysozyme) with higher basic residues were found to disperse carbon nanotubes better. The higher cationic (or anionic) charge density at a pH below (or above) the isoelectric point may contribute to the repulsion of protein-stabilized SWNTs, thus to less aggregation and better dispersion.[21]

Hydrophobin (HFBI) is an amphiphilic protein and can self-assemble into a monomolecular film at the air–water interface. With the aid of sonication, Kurppa et al. solubilized as-produced SWNTs with HFBI.[22] The nanotubes were found to be embedded in the HFBI film. By engineering HFBI protein with a gold nanoparticle, the same authors also prepared an interesting hybrid one-dimensional nanostructure of gold nanoparticles attached onto nanotube sidewalls in 2.6 nm spacing.[22]

Mechanistically, electrostatic interactions as well as hydrophobic interactions (especially π-stacking) are recognized to play critical roles in protein–carbon nanotube interactions.[23–26] For example, it has been suggested that the electron density of the aromatic residue on the hydrophobic face of the amphiphilic helical peptides can affect their efficiency in nanotube dispersion.[23] Mu et al. studied interactions between functionalized MWNTs (in different diameters and surface functionalities) and various proteins.[24] Their results showed that the larger-diameter MWNTs were favored in the binding with proteins, and the binding was enhanced by the negative charges associated with nanotube functionalities and also affected by the surface electrostatic and stereochemical properties of the proteins.[24]

Honek's group recently identified a SWNT-binding motif (the tryptophan residue in a peptide called UW-1) through phage-displayed peptide library screening. Their study suggested that hydrophobic as well as π-stacking interactions between the tryptophan residue in UW-1 and the SWNT sidewall may dictate the high binding affinity of UW-1 to SWNTs.[25] In a follow-up study, the tryptophan residue was altered and the effect of the alteration on the binding to SWNTs was evaluated. The experimental results combined with computational modeling indicated that the highest occupied molecular orbital of tryptophan residue interacts with the lowest unoccupied molecular orbital of SWNT.[26]

Although the noncovalent binding is a versatile way to functionalize carbon nanotubes with proteins, the interaction is rather weak in comparison to covalent binding. For example, it has been reported that the nanotube-adsorbed proteins were largely removed via dialysis against water at room temperature, while covalently attached ones were stable under the same treatment.[27]

As a classic example in covalent functionalization of carbon nanotubes with proteins, Sun and coworkers reported that carbon nanotubes could be functionalized with bovine serum albumin (BSA) or horse spleen ferritin via carbodiimide-activated amidation reaction.[28,29] The BSA-functionalized SWNTs (BSA-SWNT) were then conjugated with goat anti-*Escherichia coli* O157 to form immuno-SWNT under physiologically compatible conditions. Compared to BSA-SWNT, the immuno-SWNT were capable of recognizing pathogenic *E. coli* O157:H7 cells through specific antibody–antigen interactions, as confirmed by scanning electron microscopy (SEM) imaging results.[30] The bacterial recovery, determined by colony enumeration, was as high as 80% to 90% when a larger amount of immuno-SWNT and/or a high initial bacterial count were used.[31]

The same platform was used to prepare BSA-functionalized magnetic MWNTs.[32] After magnetic separation, those nanotubes with iron or iron oxide encapsulation were isolated and coated with anti-*E. coli* O157. The resulting immunomagnetic MWNTs were evaluated in immunomagnetic separation of *E. coli* O157:H7 cells in pure and

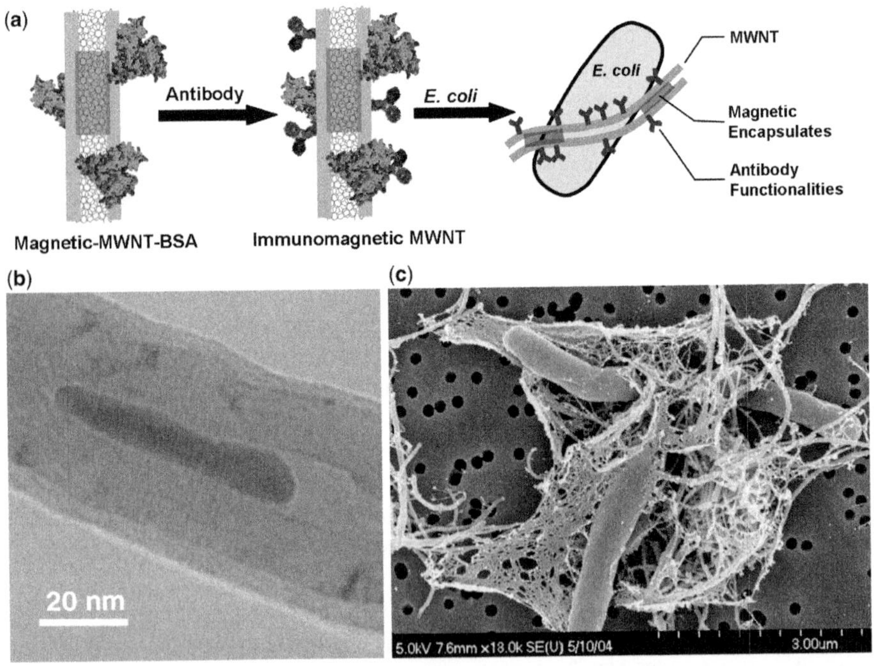

Figure 6.1 (a) A scheme on the formation of immunomagnetic MWNTs and their interaction with *E. coli* cells; (b) a high magnification TEM image of the magnetic-MWNT-BSA sample; (c) SEM images of the immunomagnetic separation sample showing *E. coli* cells bound with immunomagnetic MWNT species.[32] (Reprinted with permission from Y. Lin et al., *J. Nanosci. Nanotechnol.* **2006**, *6*, 868–871. Copyright 2006 American Scientific Publishers.)

mixed (with *Salmonella typhimurium*) cultures. The immunomagnetic MWNTs were able to efficiently capture and precipitate *E. coli* in aqueous media in a commercial magnetic separator within 5 min (Fig. 6.1). Such capture was sensitive to approximately 40 bacterial counts per 100 μL aqueous buffer, and highly selective, with no bound background bacteria even at a relatively high interfering concentration [e.g., selective toward *E. coli* O157:H7 at 800 colony forming units (CFU) 100 μL^{-1} with background flora of *S. typhimurium* DT104 cells at ~3000 CFU 100 μL^{-1}].[32]

The same amidation reaction was employed by Joshi et al. to covalently functionalize MWNTs with a specific peptide sequence, which specifically binds to the heptameric receptor-binding subunit of anthrax toxin.[33] These nanotube–peptide conjugates were found to selectively destroy anthrax toxin with the reactive oxygen species generated by the nanotubes upon near-IR radiation.[33]

In addition to direct amidation, proteins were also covalently attached to carbon nanotubes through a secondary bond which is usually cleavable under controlled conditions. You et al. recently reported a BSA-conjugated MWNTs via disulfide linkage which can be cleaved at a glutathione concentration of 20 mM, releasing BSA. In their procedure, acid-treated MWNTs were first covalently bound with pyridyldithio groups through traditional defect site amidation reaction, and the pyridyldithio-MWNT complex was then conjugated with BSA in a disulfide-exchange reaction.[34]

6.2.2 Carbohydrates

Carbohydrates are highly hydrophilic biomolecules and used extensively in the functionalization of carbon nanotubes through either defect site covalent bonding or noncovalent wrapping. In addition to their high aqueous solubility and biocompatibility, carbohydrate-functionalized carbon nanotubes represent a unique class of pseudo one-dimensional sugar arrays of unusual properties.[11,35] Many of these properties are not available to multivalent sugar ligands displayed on traditional scaffolds such as polymers, dendrimers, and proteins. For example, Sun and coworkers demonstrated that the monosaccharide (galactose or mannose)-functionalized SWNTs could effectively bind and aggregate anthrax (*Bacillus anthracis*) spores in the presence of a divalent cation like Ca^{2+} (Fig. 6.2).[11] The overall effectiveness in the aggregation of *B. anthracis* spores was assessed as up to 97.7% CFU reduction in a laboratory setting. The binding was found to be reversible by adding ethylene diamine tetraacetic acid (EDTA), which is a chelating agent for Ca^{2+}. The results suggested that the binding was likely due to Ca^{2+}-mediated carbohydrate–carbohydrate interactions between the SWNT-displayed multivalent monosaccharides and the spore surface sugars. Interestingly, experiments with polymeric nanoparticles functionalized with the same sugars (~120 nm diameter polystyrene beads) did not show any spore aggregation under similar experimental conditions, suggesting that the observed binding and spore aggregation were unique to multivalent sugar ligands displayed on SWNTs.[11]

According to an earlier report from the same research group, galactose-functionalized SWNTs (Gal-SWNT) were used to bind and agglutinate pathogenic *E. coli* O157:H7 cells.[35] The binding required no divalent cations and was specific to

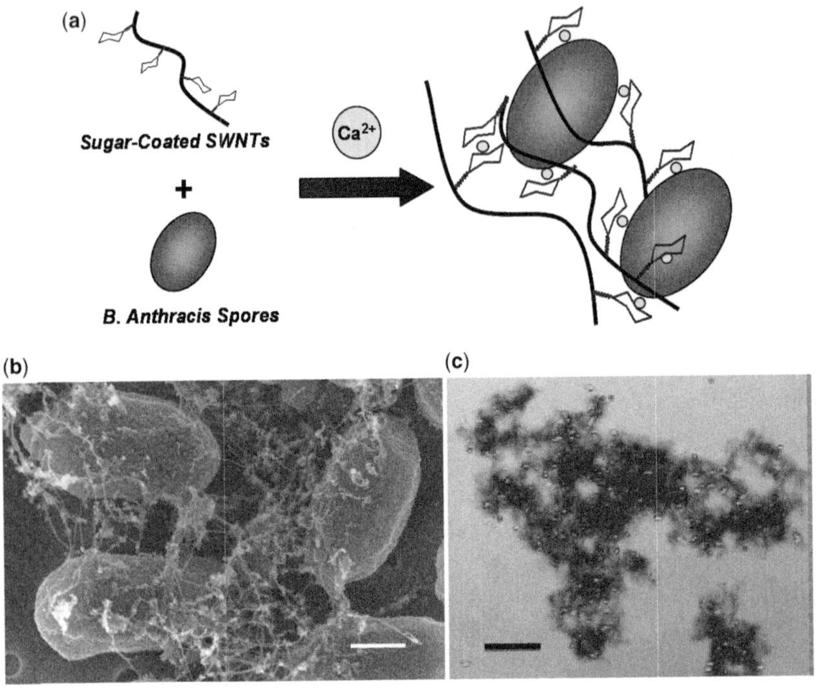

Figure 6.2 The Ca^{2+}-mediated aggregation of anthrax spores with sugar functionalized SWNTs: (a) a schematic illustration; (b) SEM image (scale bar = 500 nm); and (c) optical image (scale bar = 20 μm).[11] (Reprinted with permission from H. Wang et al., *J. Am. Chem. Soc.* **2006**, *128*, 13364–13365. Copyright 2006 American Chemical Society.)

β-D-galactose- but not α-D-mannose- or BSA-functionalized nanotubes, indicating that the multivalent interactions between the nanotube-bound galactoses and the *E. coli* cell surface galactose-binding proteins probably dictate the Gal-SWNT-mediated aggregation of *E. coli* cells.

The multivalency of carbohydrate ligands should in principle affect significantly the overall affinity of multivalent ligand–receptor interactions.[36] Thus, a series of dendritic galactopyranosides and mannopyranosides were synthesized for their functionalization of SWNTs to increase the valency of the sugar ligands displayed on each nanotube as well as to improve the aqueous solubility of the functionalized nanotubes.[37,38] According to the bioevaluation of those sugar dendron-functionalized carbon nanotubes by using established binding assays with *E. coli* cells, the β-D-galactose dendron-functionalized SWNTs (Gal_2-SWNT), compared to simple β-D-galactose-functionalized SWNTs (Gal-SWNT), had more effective CFU reduction in the agglutination of *E. coli* cells.[38]

Noncovalent wrapping with polysaccharides is another commonly used approach in the carbohydrate functionalization and solubilization of carbon nanotubes.[39–42] For example, Liu et al. used alginic acid (AA), a natural polysaccharide constituted with repeated β-D-mannuronic acid (βM) and α-L-guluronic acid (αG) segments, to solubilize raw MWNTs by sonication.[40] The transmission electron microscopy (TEM) and NMR studies of the MWNT-AA complex showed that MWNT was partially wrapped by AA possibly by M segments lying on the nanotube sidewall while part of the αG segments stand up on it. The addition of divalent or trivalent metal cations, such as Zn^{2+}, Ni^{2+}, Co^{2+}, Pb^{2+}, La^{3+}, and Eu^{3+}, resulted in precipitation of the MWNT-AA complex. The precipitation could be redispersed in water by chelating the metal cations with EDTA.[40]

Schizophyllan (SPG) is a polysaccharide with a natural triple helix (t-SPG) structure, which can be dissociated to single chains (s-SPG) in DMSO. The triple helix structure (t-SPG) can be recovered in a simple exchange of DMSO with water. Similar to that in cyclodextrin, the hollow inside of the helical column in SPG is considered hydrophobic, and thus could entrap SWNTs through hydrophobic interactions. Shinkai and coworkers prepared SPG and curdlan (an analog of SPG containing the main chain but not the side chain) wrapped SWNTs through first mixing/sonicating in DMSO and then exchanging with water.[39,41,42] The microscopy characterization showed a unique periodic stripe structure in the SWNT/s-SPG complex with two s-SPG chains right-handed helically wrapped on one nanotube. This was the first

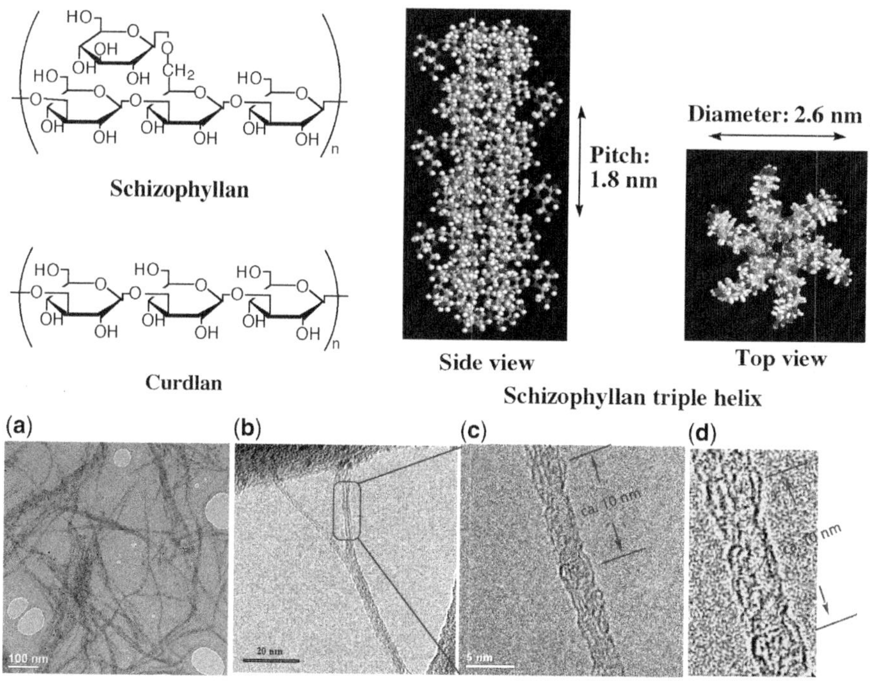

Figure 6.3 (Top) The chemical structures of schizophyllan, curdlan, and a representative model of schizophyllan triple helix. (a) TEM image of as-grown-SWNT/s-SPG composite, and (b,c) its magnified picture. (d) The original image of (c) was Fourier filtered to enhance the contrast of the composite.[41] (Reprinted with permission from M. Numata et al., *J. Am. Chem. Soc.* **2005**, *127*, 5875–5884. Copyright 2005 American Chemical Society.)

visual image that clearly demonstrated a patterned structure of water-soluble polymer-wrapped SWNTs (Fig. 6.3).[41] A similar patterned structure was observed later in the case of amylose-wrapped SWNTs.[43]

6.2.3 DNAs/RNAs

DNA- and RNA-functionalized carbon nanotubes hold interesting prospects in various fields, including biosensors[44] and gene delivery.[45] Similar to proteins and carbohydrates, DNA can be conjugated to carbon nanotubes through either noncovalent adsorption[46–48] or covalent functionalizations in terms of direct amidation[49,50] or the use of a bifunctional linker.[51] For example, Lee et al. used the bifunctional linker approach to functionalize carbon nanotubes with single-strand DNAs (ssDNA).[51] The procedure included several steps, first the sidewall functionalization of SWNTs with a 4-nitrobenzenediazonium salt, then the electrochemical reduction of $-NO_2$ to $-NH_2$ for reaction with a heterobifunctional cross-linker to introduce a maleimide group, followed by coupling with 5′-thiol-modified ssDNA (Scheme 6.4).[51] The nanotube-bound ssDNAs were able to recognize their appropriate complementary sequence with a high degree of selectivity.

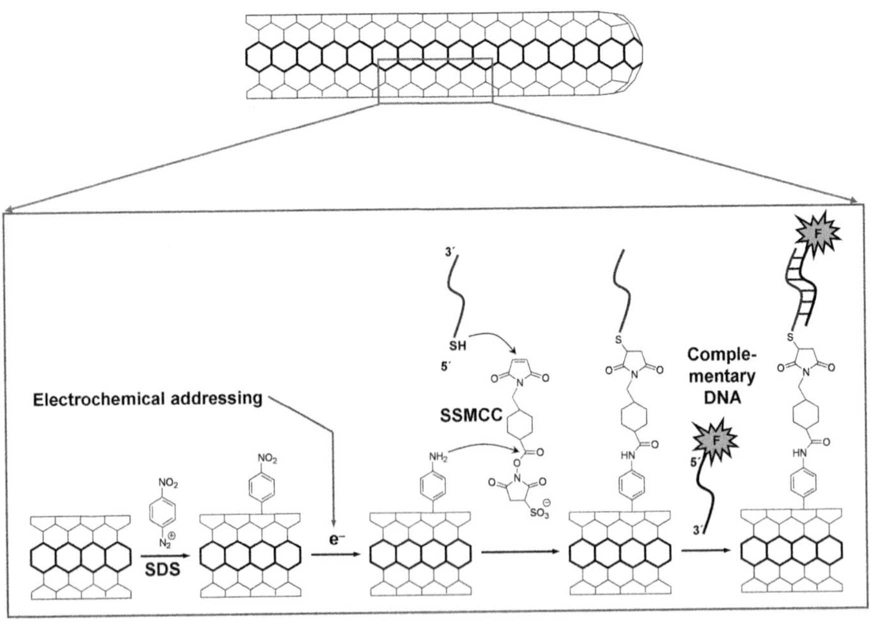

Scheme 6.4 Schematic illustration of addressable biomolecular functionalization of carbon nanotubes.[51] (Reprinted with permission from C.-S. Lee et al., *Nano Lett.* **2004**, *4*, 1713–1716. Copyright 2004 American Chemical Society.)

The specific interaction between DNA sequences on the carbon nanotube surface facilitates the assembly of nanotubes into architectures necessary for electrical circuits and molecular sensing applications. The DNA-directed multicomponent self-assembly of carbon nanotubes and gold nanoparticles has been demonstrated.[47–49] For example, Mao et al. grafted ssDNA onto the nanotube surface via noncovalent wrapping.[47] The ssDNA-SWNT complex then hybridized with complimentary DNA-coated gold nanoparticles (prepared through thiol–gold interaction). Interestingly, the DNA-directed assembly was found to be a reversible process, with formation at low temperature and dissociation at high temperature.[47] Similarly, Han et al. prepared a thiolated DNA strand $d(GT)_{29}SH$ to wrap SWNT, which was later used as a rigid template for the assembly of gold nanoparticles.[48]

The functionalization of carbon nanotubes with DNAs allows precise control of their interfacial properties. Dai and coworkers noncovalently adsorbed poly-T DNA onto SWNT sidewalls, which were well dispersed and used for ultrathin atomic layer deposition of HfO_2 for high k dielectrics.[44] With the ~3 nm HfO_2 coating, high performance SWNT field effect transistors (FETs) free of gate leakage were obtained. The same group also grafted short interfering RNA (siRNA) onto the SWNT surface.[45] In their strategy, SWNTs were first noncovalently coated with amine-terminated phospholipid-poly(ethylene glycol) (PL-PEG), and the amine attached to the nanotube was then covalently coupled with thiol-modified siRNA. The resulting siRNA-SWNT conjugates exhibited efficient silencing effects (50% to 60% knockdown) on both CD4 receptor and CXCR4 co-receptor (required for HIV entry into human T cells and infection).[45]

The DNAs attached to carbon nanotubes were shown to likely preserve their biological activity.[52] For example, luciferase DNAs containing a T7 promoter sequence required for *in vitro* transcription were adsorbed onto SWNT bucky papers. Compared to free DNAs, ~36% of the nanotube-bound DNAs remained functional in terms of RNA polymerase-catalyzed transcription/translation reaction.[52] Heller et al.[53] found that the double-stranded DNA noncovalently wrapping SWNTs preserved its property of the conformation transition from right-handed B form to the left-handed Z form on the addition of divalent cations such as Hg^{2+}, which caused a detectable red shift in band-gap emission of SWNTs. Thus, these DNA-wrapped nanotubes were used as a detecting agent for the ion in whole blood and tissue and within living mammalian cells (Fig. 6.4).[53] However, in some cases, the bounded carbon nanotubes can interfere with the normal DNA activities.[54] For instance, a study by Peng et al. also found that the attachment of SWNTs to human telomeric i-motif DNA dramatically changed its kinetics of S1 cleavage by increasing the turnover number 22-fold over that of free DNA alone.[54]

The interaction between DNA and carbon nanotubes is stable enough to allow separation of the dispersed nanotubes into well-defined subpopulations.[46,55–57] Recently, ssDNA-wrapped SWNTs were separated into multiple fractions with different length distributions, and each fraction was examined for cellular toxicity to various cell lines.[55] As demonstrated experimentally, while the longer SWNTs (>200 nm) were excluded by cells, the shorter ones were found to internalize into the cytoplasm,

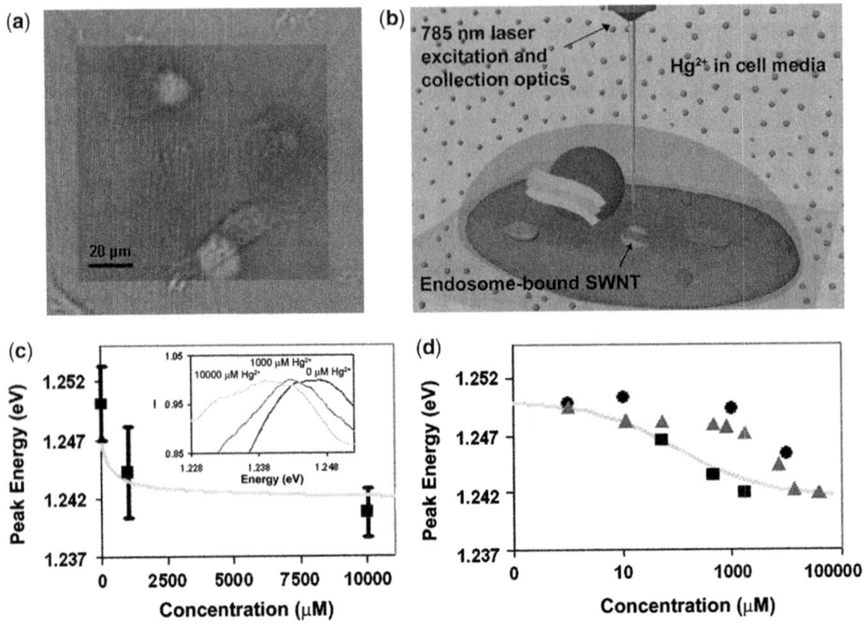

Figure 6.4 (a) Area map of the (6,5) nanotube peak fluorescence intensity of DNA-SWNTs within murine 3T3 fibroblast cells overlayed on an optical micrograph of the same region. (b) Illustration of the experimental method used for ion-binding experiments conducted in mammalian cells. A cell containing endosome-bound DNA-SWNTs undergoes 785 nm excitation through a microscope objective. (c) The (6,5) nanotube fluorescence peak energy of DNA-SWNTs in 3T3 fibroblasts plotted versus Hg^{2+} concentration in the cell medium. The fluorescence energy of a population of 8 to 10 cells was averaged for each data point. Error bars indicate 1 SD. The red line shows the model curve from original Hg^{2+} binding experiment conducted in Tris buffer. The inset shows individual spectra at each concentration. (d) The (6,5) nanotube fluorescence energy of DNA-SWNTs in the following highly absorptive and scattering media: whole rooster blood (green triangles), black dye solution (black squares), and chicken tissue (blue circles) plotted on a model curve (red) from Hg^{2+} addition to SWNTs in buffer. The ΔE of all blood and tissue data points was corrected for an initial red shift due to the cellular environment.[53] (Reprinted with permission from D. A. Heller et al., *Science* **2006**, *311*, 508–511. Copyright AAAS.) (See color insert.)

causing significantly decreased metabolic activity.[55] Therefore, shorter SWNTs may be more toxic to cells than longer ones.

The DNA–carbon nanotube interaction is a complicated and dynamic process. Many studies on this subject have been pursued through a series of techniques, including molecular dynamic simulation, microscopy, circular dichroism, and optical spectroscopy.[57,58] Although the detailed mechanism is not fully understood at present, several physical factors have been proposed to be driving DNA-carbon nanotube interactions,[46,59–61] such as entropy loss due to confinement of the DNA backbone, van der Waals and hydrophobic (π-stacking) interactions, electronic interactions between DNA and carbon nanotubes, and nanotube deformation. A recent UV optical spectroscopy study of the ssDNA-SWNT system demonstrated experimentally that π-stacking was a force behind the DNA-SWNT binding, and the bound nucleotides were probably oriented along the nanotube's long axis.[60] This was later confirmed by Meng et al. in their study based on the time-dependent density functional theory (TDDFT) and optical spectroscopy measurements of the DNA-SWNT complex.[61] Their results suggested that the DNA bases were oriented relative to the nanotube axis, and the orientation depended mainly on the base identity instead of the diameter and chirality of the nanotubes.[61]

Generally, DNA-nanotube conjugates are widely pursued for their interesting and promising prospects in a number of areas, including especially nanoscale devices, biosensors, electronic sequencing, and therapeutic delivery, taking advantage of the unique pseudo-one-dimensional cylindrical structure of carbon nanotubes and the sequence-specific pairing interaction and conformational flexibility of DNA.

6.3 BIOAPPLICATIONS OF CARBON NANOTUBES

As reviewed above, there has been significant progress in addressing some fundamental and technical issues in the functionalization and solubilization of carbon nanotubes with various biomolecules and bioactive species. It has stimulated the exploration of carbon nanotubes for applications in a variety of biological and biomedical systems and devices. Here, some interesting recent studies and achievements for the use of carbon nanotubes in biosensors, biodelivery, bioimaging, tissue engineering, and related results on toxicity and biodistribution are highlighted and discussed.

6.3.1 Biosensors

Carbon nanotubes, especially SWNTs, with their fascinating electrical properties, dimensional proximity to biomacromolecules (e.g., DNA of 1 nm in size), and high sensitivity to surrounding environments, are ideal components in biosensors not only as electrodes for signal transmission but also as detectors for sensing biomolecules and biospecies. In terms of configuration and detection mechanism, biosensors based on carbon nanotubes may be divided into two categories: electrochemical sensors and field effect transistor (FET) sensors. Since a number of recent reviews on the former have been published,[6,62,63] our focus here is mostly on FET sensors.

Generally, a carbon nanotube FET device is constructed by a substrate (gate), two microelectrodes (source and drain), and bridging material between the electrodes, which is typically an individual SWNT or a SWNT network. A SWNT FET is usually fabricated by casting a dispersion of bulk SWNTs or directly growing nanotubes on the substrate by chemical vapor deposition (CVD) either before or after the electrodes are patterned.[64] Due to the diffusive electron transport properties of semiconducting SWNTs, the current flow in SWNT FET is extremely sensitive to the substance adsorption or other related events on which the sensing is based.

A wide variety of applications for SWNT FET have been explored, especially for the detection of proteins, antibody–antigen interactions, carbohydrates, DNAs and their hybridization, and single nucleotide polymorphism.[6] The detection limit for the sensing of proteins or protein–protein interactions has generally been in the range of 100 pM to 100 nM.[65] In a recent study, however, Byon and Choi improved the sensitivity to 1 pM for both nonspecific and specific protein bindings by modifying the geometry of the SWNT FET device.[66] The substantially higher sensitivity (a factor of 10,000) was accredited mainly to the larger Schottky contact area on the electrodes, which was achieved by angled deposition of electrode metals. This work might contribute to an acceleration in the progress toward the realization of nanoscale and label-free electronic biosensor systems.

Although antibodies are often used as specific targeting agents in FET devices for highly selective biosensing,[65–67] they are typically not very stable and their large sizes (10 to 15 nm) may sacrifice the detection sensitivity in some cases. Alternatively, synthetic oligonucleotides (i.e., aptamers) have been demonstrated for their advantages (stability, small size, etc.) in the specific detection of amino acids, drugs, and proteins. For example, Maehashi et al.[68] immobilized a 5′-amino-modified aptamer onto the nanotube surface with the aid of a linker molecule, 1-pyrenebutanoic acid succinimidyl ester. The corresponding FET device exhibited a sharp decrease in the source-drain current in the presence of immunoglobulin E (IgE) target, with the detection limit down to 250 pM.[68] Compared to the IgE-mAb (monoclonal antibody against IgE)-modified carbon nanotube FET devices, the aptamer-modified ones performed better in the detection of IgE under similar experimental conditions.[68]

Star et al. reported synthetic oligonucleotide-immobilized carbon nanotube FETs as selective detectors for DNA immobilization and hybridization.[69] Their device could specifically recognize target DNA sequences, including H63D single-nucleotide polymorphism discrimination in the *HFE* gene (responsible for hereditary hemochromatosis).[69] The addition of the divalent cation Mg^{2+} dramatically increased sensitivity by three orders of magnitude to reach the detection limit of 1 pM.[69] As reported by Gui et al. synthetic oligonucleotide-immobilized SWNT networks could also be fabricated with two different metal contacts (Au and Cr) for electrical detection of DNA hybridization.[70] On DNA hybridization, the drain current (I_d) increased in a Cr-contacted FET device but decreased in an Au-contacted one, indicating that the electrode-SWNT interface played an important role in the electrical signal transmission.[70]

While it is generally acknowledged that SWNT devices are operated in terms of the Schottky barrier modulation effect,[65,71] as well as the chemical gating effect,[72,73] the sensing mechanism for nanotube-based FETs still remains a subject of debate. According to a recent report, experimentally the effect of protein adsorption on the relationship (I–V curve) between source-drain current and the liquid gate potential could be used as a tool to identify the sensing mechanism in the SWNT transistor (Fig. 6.5).[74] It was found that the sensing was indeed due to a combination of the Schottky barrier and electrostatic gating effects. The finding was consistent with another recent study of DNA immobilization on a back-gated SWNT network.[72] While earlier reports suggested that the sensitive region was limited to the nanoscale

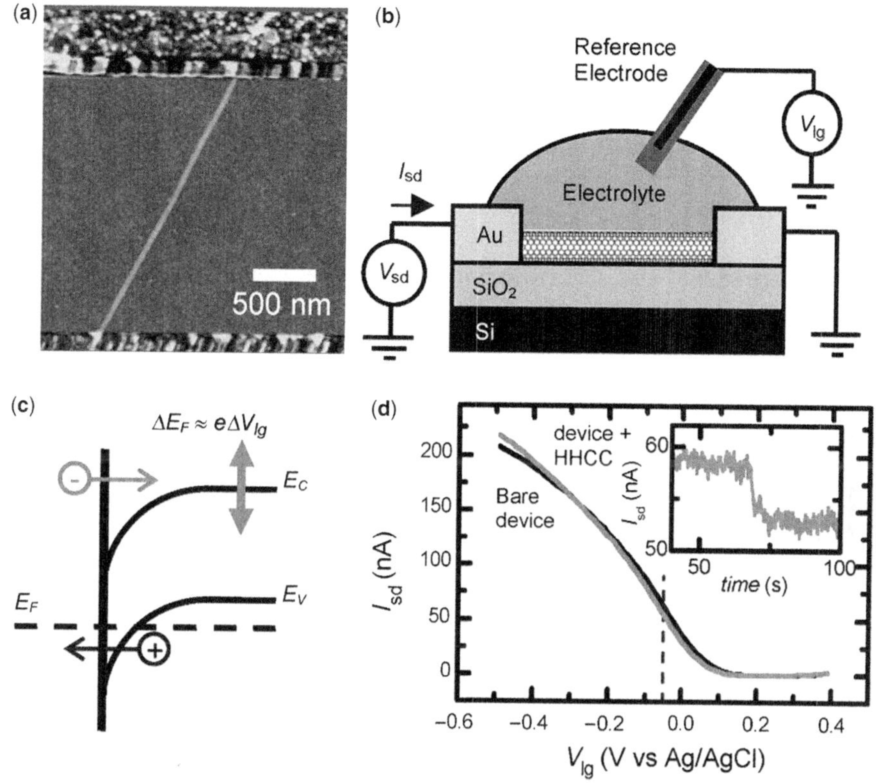

Figure 6.5 Experimental layout and results of a typical biosensing experiment. (a) Atomic force microscopy topology image of a SWNT between Cr/Au contacts on an insulating SiO_2 substrate. (b) Measurement setup, where a source-drain bias potential is applied and the device is gated through an Ag/AgCl reference electrode inserted in the electrolyte. The electrolyte is contained in a home-built flow cell (not depicted). (c) Band diagram for a hole-doped SWNT showing electron conduction through thermally activated carriers into conductance band and hole conduction through tunneling through the Schottky barrier. E_F is the Fermi energy. E_C and E_V are the energies of SWNT conductance and valence band edges, respectively. The liquid gate potential changes the doping level in the bulk of the SWNT as indicated by the double arrows. (d) Results of a typical biosensing experiment. Current versus liquid gate potential curves acquired before (black line) and after (gray line) adsorption of 1 μM HHCC in PB buffer ($V_{sd} = 10$ mV). The inset shows the real-time drop in conductance when HHCC is flushed over the device, while the gate potential is held constant at −50 mV versus Ag/AgCl, (as indicated by the dashed line).[74] (Reprinted with permission from I. Heller et al., *Nano Lett.* **2008**, *8*, 591–595. Copyright 2008 American Chemical Society.)

contacts,[65,66] these new results indicated that it could also extend to the bulk channel section of the SWNT.

As discussed earlier, semiconducting SWNTs are the ones useful in nanotube-based FET devices.[63,75] With rare exceptions,[76] however, as-produced SWNTs are mixtures of semiconducting and metallic nanotubes (in a ratio close to the statistical limit of 2 to 1).[77] Therefore, the selective removal of metallic SWNTs for purer semiconducting ones, such as by improving the postfabrication electrical breaking down technique[78] or taking advantage of the recent advances in postproduction separation of semiconducting SWNTs from the as-produced mixtures,[77,79,80] may prove highly beneficial.

6.3.2 Biodelivery

Current cancer and gene therapy require the internalization of desired drugs and genes into the target cells. However, most anticancer drugs and genes suffer from low cellular uptake. Many delivery systems, such as liposomes, polymers, dendrimers, and a variety of nanomaterials, have been developed. Carbon nanotubes are particularly interesting as potential delivery vehicles because of their unique ability to penetrate cell membranes without the need for any external transporter system. In the literature, nanotubes have been demonstrated as effective carriers for shuttling and delivering various peptides, proteins, nucleic acids, and small-molecule drugs into living cells.[10]

Methotrexate (MTX) is a widely used anticancer drug, though it has a very low cellular uptake. Pastorin et al. employed a double functionalization strategy to attach both fluorescein isothiocyanate (FITC, a fluorescent probe) and MTX onto the sidewall of MWNTs (via 1,3-cycloaddition of azomethine ylides).[81] In their procedure, carbon nanotubes were first cofunctionalized with two orthogonally protected amino groups that were selectively de-protected and subsequently derivatized with FITC and MTX. As demonstrated in the *in vitro* study with Jurkat cells, the nanotube-bound MTXs were rapidly internalized into cells and accumulated in the cell cytoplasm.[81] The same group of researchers prepared MWNT-doxorubicin (DOX, a popular anticancer drug) supramolecular complex through presumably the π-stacking interactions between the nanotube sidewall (MWNTs suspended by pluronic copolymer) and the DOX aromatic hydroxyl-anthraquinonic rings.[82] The fluorescence intensity of the MWNT-bound DOX was suppressed due to the nanotube quenching effect. The toxicological assay with MCF-7 human breast cancer cells suggested a significant enhancement in cytotoxicity of the MWNT-DOX complex as compared to the free DOX and DOX/pluronic mixture. Since pluronic copolymer-dispersed MWNTs did not exhibit any toxicity, the enhanced cytotoxicity of the MWNT-DOX complex was attributed to an effective delivery of DOX into MCF-7 cells via MWNTs.[82]

Dai and coworkers have been extremely active in studies of using carbon nanotube as delivery vehicles.[9,45,83,84] According to their 2007 report[83] on the cellular delivery of a large amount of DOX via SWNTs, two water-soluble SWNT samples were prepared, with one (PL-SWNT) from the noncovalent suspension with phospholipid-poly(ethylene glycol) (PL–PEG, ~120 PEG units) and the other (PEG-SWNT) from covalent functionalization with PEG (220 units). These samples were used for the noncovalent attachment of DOX through π-stacking. The DOX binding to and dissociation from SWNT were found to be dependent on pH and the nanotube diameter, generally lower pH and smaller diameter in favor of DOX dissociation from the nanotube.[83] The *in vitro* toxicity study with U87 human glioblastoma cancer cell showed significant cell death and cell apoptosis induced by the DOX-attached SWNTs (PL-SWNT-DOX), while PL-SWNT (no DOX loading) exhibited no toxic effect on cells. The IC_{50} (half-maximum inhibitory concentration) value for PL-SWNT-DOX was estimated to be ~8 μM, higher than that for the free DOX (~2 μM). For specifically targeting U87 cancer cells, a cyclic arginine-glycine-aspartic acid peptide (RGD, targeting integrin $\alpha_v\beta_3$ receptors upregulated in many solid tumors) was conjugated to the terminal amine groups in PL-SWNT. The bioevaluation on the effectiveness of RGD-conjugated PL-SWNT-DOX showed a lower IC_{50} value of ~3 μM for RGD-positive U87 cancer cells, but no improvement for MCF-7 cells (short of corresponding receptors, negative control). The enhanced toxicity of RGD-conjugated PL-SWNT-DOX was attributed to the targeted intracellular delivery of DOX (Fig. 6.6).[83]

Similarly, Feazell et al. tethered platinum (IV) prodrug complexes onto SWNTs through PL-PEG linkage.[84] The resulting platinum (IV) complex-SWNT conjugates [SWNT-Pt (IV)] showed dramatic enhancement of cytotoxicity to Ntera-2 cancer cells, while the free platinum (IV) complex was nearly nontoxic. The cellular platinum concentration of SWNT-Pt (IV) was six times higher than that of untethered platinum (IV) complex, suggesting the effective delivery of platinum (IV) via SWNTs.[84]

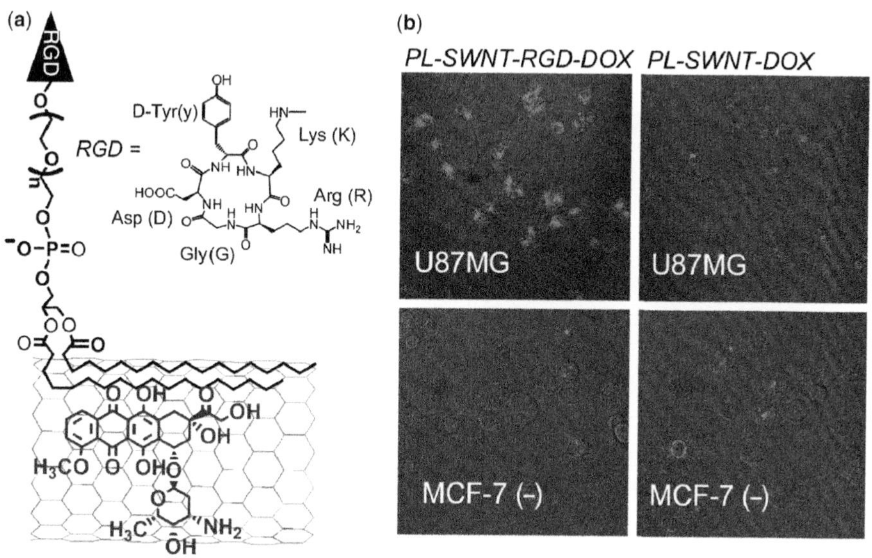

Figure 6.6 RGD peptide selectively enhances doxorubicin delivery by SWNTs and toxicity to integrin $\alpha_v\beta_3$-positive cells. (a) Schematic structure of PL-SWNT-RGD-DOX, that is, SWNT functionalized with RGD at the termini of PEG and loaded with doxorubicin on the sidewall by π-stacking. (b) Confocal fluorescence images of integrin $\alpha_v\beta_3$-U87MG cells (top) and negative MCF-7 cells (bottom) treated with either PL-SWNT-DOX (right) or PL-SWNT-RGD-DOX (left). The concentration of DOX was 2 μM in all experiments. The U87MG cells incubated with PL-SWNT-RGD-DOX showed stronger DOX fluorescence in the cells than in the other three cases.[83] (Reprinted with permission from Z. Liu et al., *ACS Nano* **2007**, *1*, 50–56. Copyright 2007 American Chemical Society.) (See color insert.)

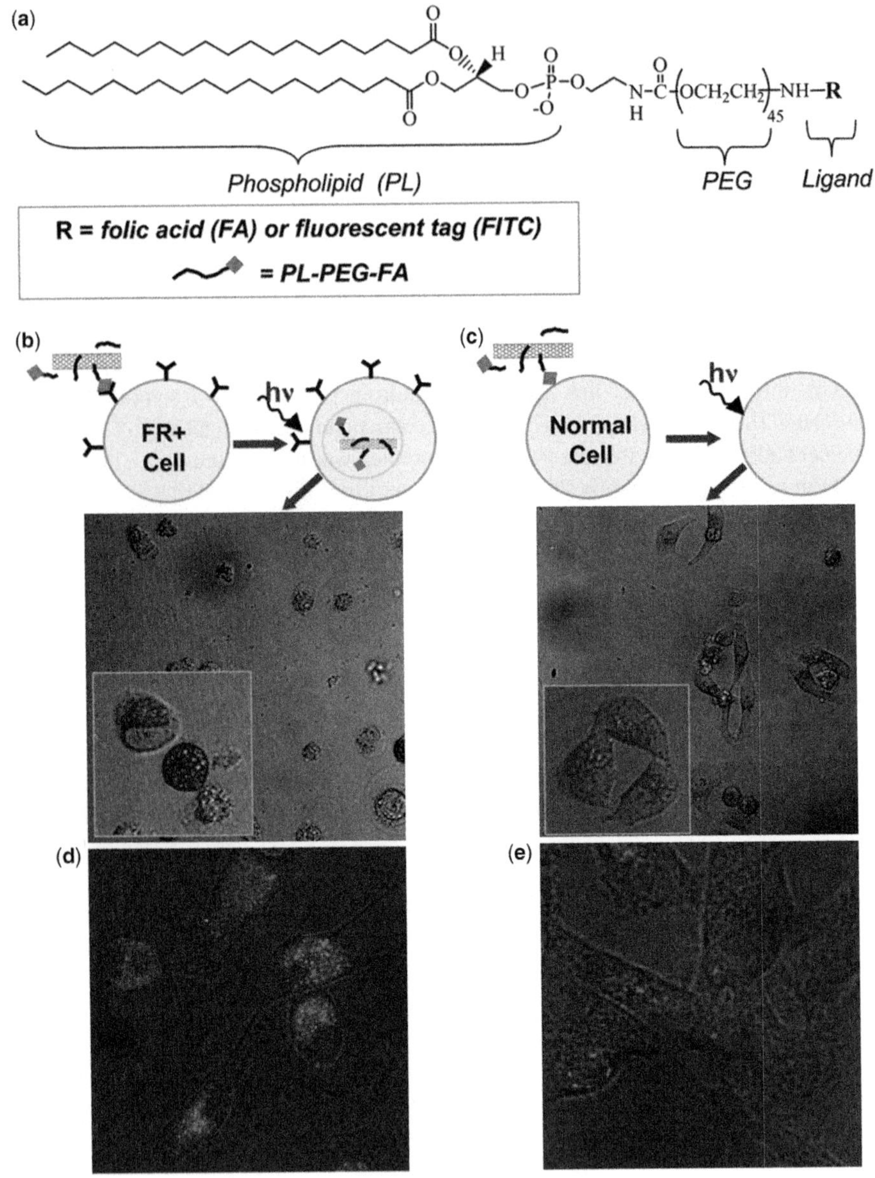

Figure 6.7 Selective targeting and killing of cancer cells. (a) Chemical structure of PL-PEG-FA and PL-PEG-FITC, for solubilizing individual SWNTs. (b) (Upper) Schematic of selective internalization of PL-PEG-FA-SWNTs into folate-overexpressing (FR^+) cells via receptor binding and then NIR 808 nm laser radiation. (Lower) Image showing death of FR^+ cells with rounded cell morphology after the process described above (808 nm laser radiation at 1.4 W cm^{-2} for 2 min). (Inset) Higher magnification image shows details of the killed cells. (c) (Upper) Schematic of no internalization of PL-PEG-FA-SWNTs into normal cells without available FRs. (Lower) Image showing normal cells with no internalized SWNTs are unharmed by the same laser radiation condition as in (b). (Inset) Higher magnification image

Beyond small-molecule drugs, other bioactive species such as nucleic acids were also successfully delivered into the cell by using carbon nanotubes.[45,85] Zhang et al. functionalized SWNTs with $-CONH-(CH_2)_6-NH_3^+Cl^-$ to mediate the conjugation of siRNAs that specifically targeted murine telomerase reverse transcriptase (mTERT) expression to form the mTERT siRNA:SWNT+ complex.[85] These specifically biofunctionalized SWNTs successfully entered tumor cell lines, silencing the expression of the targeted gene, inhibiting cell proliferation and promoting cell senescence *in vitro*, as well as suppressing tumor *in vivo*.[85] Similarly, siRNA was grafted onto carbon nanotubes for delivery into human T cells and primary cells, which exhibited superior silencing effects over conventional liposome-based nonviral agents.[45]

As mentioned earlier, the exceptional cell membrane-penetrating ability combined with the near-infrared (NIR) absorption characteristics of carbon nanotubes make them suitable materials in hyperthermia cancer therapy.[9] Kam et al. conjugated Cy3-labeled single-stranded DNA with SWNTs (DNA-SWNT), which were internalized into HeLa cells and cumulated in the cytoplasm region.[9] It was found that the DNA was dissociated from SWNTs and entered into the cell nucleus following six 10-second on-and-off pulses of 1.4 W cm^{-2} laser radiation at 808 nm, and the continued exposure under the same radiation for 2 min induced cell death. This was most likely due to the extensive local heating generated by the NIR absorption of SWNTs inside the cells.[9] To selectively target and kill cancer cells, folic acid (FA)-terminated phospholipid-poly(ethylene glycol) (PL-PEG)-functionalized SWNTs (PL-PEG-FA-SWNTs) were synthesized. The resulting conjugates were incubated with folate-overexpressing (FR^+, receptors for FA ligands) HeLa cells and also normal cells before laser radiation. Extensive death in the FR^+ cells was observed, while the normal cells maintained normal proliferations (negative control, Fig. 6.7).[9] The direct injection of SWNTs into tumor cells to induce cell death by NIR radiation represents an attractive alternative to endocytosis for applications in cancer therapeutics.[9]

For eventual *in vivo* applications of the widely pursued carbon nanotube delivery systems, a critical challenge is to be able to keep the nanotubes in the bloodstream long enough for their intended functions.[86] In a recent study by Yang et al., the oligomeric poly(ethylene glycol) functionalization (PEGylation) of SWNTs was found to be remarkably effective in achieving prolonged blood circulation (half-time on the order of 20 hours for the PEGylated nanotubes).[87] Experimentally, skeleton ^{13}C-enriched SWNTs were functionalized with the oligomeric PEGs by targeting the nanotube surface defects in well-established reactions. The plasma

Figure 6.7 (*Continued*) shows a live normal cell in stretched shape. (d) Confocal image of FR^+ cells after incubation in a solution of SWNTs with two cargoes (PL-PEG-FA and PL-PEG-FITC). The strong green FITC fluorescence inside cells confirms the SWNT uptake with FA and FITC cargoes. (e) The same as (d) for normal cells without abundant FRs on cell surfaces. There is little green fluorescence inside cells, confirming little uptake of SWNTs with FA and FITC cargoes. (Magnifications: ×20.)[9] (Reprinted with permission from N. W. S. Kam et al., *Proc. Natl. Acad. Sci. U.S.A.* **2005**, *102*, 11600–11605. Copyright 2005 National Academy of Sciences, U.S.A.) (See color insert.)

pharmacokinetic study was performed by injecting male KM mice intravenously with a solution of the PEGylated SWNTs, and by quantifying the nanotube concentrations in the blood postexposure at different time intervals in terms of the ^{13}C isotopic abundance determination (isotope ratio mass spectrometry). At one day postexposure, for example, about 30% injected dose (%ID) remained in blood circulation,[87] compared to only 0.2%ID for pristine SWNTs at the same time postexposure.[88] In a similar study by Dai and coworkers, a comparable blood circulation time was obtained for SWNTs noncovalently functionalized with branched PEGs.[89] There were also preliminary results[87] suggesting that the tumor uptake of the PEGylated SWNTs benefited from their prolonged blood circulation. For the EMF6 model (breast cancer in BABL/c mice) and the Lewis model (lung cancer in C57BL mice), the tumor uptakes of intravenously administered PEGylated SWNTs were 8%ID/g and 9%ID/g, respectively,[87] considerably higher than those for SWNTs without covalent PEGylation (in the absence of any specific targeting moieties).[90]

Despite the experimental demonstration of the efficiency of carbon nanotube-enabled or -associated delivery, the cell uptake mechanism and pathway of such delivery are still being debated. On the basis of the observed temperature-dependent cell uptake of carbon nanotubes, Dai and coworkers suggested an endocytosis mechanism.[91–93] For example, they used noncovalently DNA- and protein-functionalized SWNTs to study the uptake by HeLa (adherent) and HL60 (nonadherent) cells.[93] These functionalized nanotubes were generally short (50 to 200 nm in length), individualized or in small bundles. The *in vitro* experiments with both cell lines were consistent with the energy-dependent endocytosis mechanism for the internalization of these short nanotubes.[93] The internalization pathway could be mainly through clathron-coated pits instead of caveolae or lipid rafts. They also studied the uptake of large DNA-SWNT aggregates (200 to 2000 nm long and up to 15 nm wide); their results were similarly consistent with endocytosis being the uptake mechanism. However, because of the low solubility of these micro-scale nanotube aggregates, the uptake pathway was less certain.[93]

The group of Kostarelos, Prato, Bianco, and others disagreed with the endocytosis mechanism and proposed instead an energy-independent nonendocytotic mechanism involving insertion and diffusion of nanotubes through the lipid bilayer of the cell membrane.[94,95] In their proposed mechanism, the large biomolecules used in the solubilization of nanotubes might alter the interactions with cells and affect the intracellular transport kinetics of the nanotubes.[94,95] To elucidate such a possibility and to study the key steps in the cell uptake process, this group functionalized SWNTs and MWNTs with a wide range of functionalities, including ammonium, small fluorescent probes, anticancer drugs, and antibiotics.[95] All of these functionalized nanotubes were taken up by a wide variety of cells, and intracellularly trafficked through different cellular barriers into the perinuclear region, even under endocytosis-inhibiting conditions (Fig. 6.8).[95] These experimental results were consistent with those from molecular simulation, supporting the hypothesis of carbon nanotubes acting as nanoneedles to pierce or penetrate the plasma membrane.[96]

The apparent disagreement on the nanotube uptake mechanism could be due in part to differences in the nanotube constructs and the experimental procedures.[95]

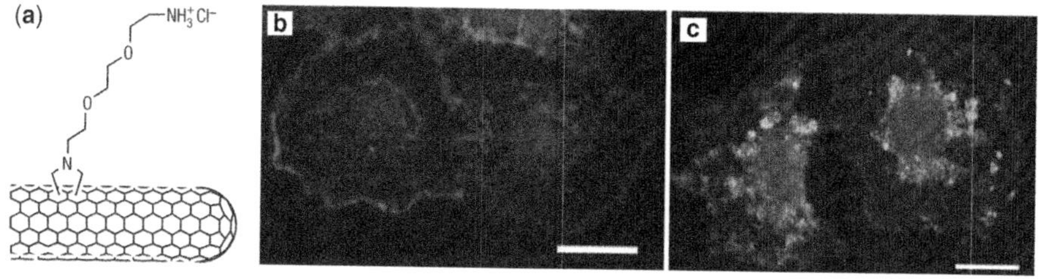

Figure 6.8 (a) A scheme of SWNT-NH_3^+ (*f*-CNT **1**) and the intracellular trafficking and perinuclear localization of *f*-CNT **1**. Triple-channel confocal image of A549 cells incubated for 2 h (b) in the absence (control) and (c) with 25 μg of *f*-CNT **1** (green). Plasma membrane stained with dye WGA-TRITC (red) and (c) nucleus counterstained with dye TO-PRO3 (blue). The scale bars correspond to 20 μm.[95] (Reprinted with permission from K. Kostarelos et al., *Nature Nanotechnol.* **2007**, *2*, 108–113. Copyright 2007 Macmillan Publishers Ltd.) (See color insert.)

More specifically, the nonendocytic internalization was found for small molecules covalently attached to carbon nanotubes, while translocation via endocytosis was found for large bio- or aqueous soluble macromolecules-conjugated SWNTs.[95] Nevertheless, debate on the mechanism is healthy and stimulating, and an improved understanding is critical to facilitate future development of carbon nanotube transporters for biodelivery applications.

Beyond the traditional delivery that relies on the cellular uptake of carbon nanotubes, Chen et al. have developed a MWNT-based cell nano-injector which delivered quantum dots (QDs) into cells through physical insertion into the cell membrane.[97] Experimentally, a biotin-tethered pyrene (via a disulfide linkage) was synthesized and attached to MWNT-atomic force microscope (AFM) tips through π-stacking. The biotin-functionalized MWNT-AFM tips were then conjugated with streptavidin-coated QDs at a ratio of 5 to 50 QDs per nanotube. The QDs were delivered into a specific cell via nano-injection and released due to the cleavage of the disulfide bond. A major advantage suggested for this technology was that the delivery-release process can be repeated many times without cell damage (Fig. 6.9).[97]

6.3.3 Bioimaging

Due to the exceptional capability of carbon nanotubes to penetrate cell membranes as well as their other unique physical and chemical properties, there is an increasing interest in their uses for cellular and subcellular imaging.[98–105] There are two major types of bioimaging with carbon nanotubes. One is direct imaging based on either the near-infrared band-gap fluorescence in semiconducting SWNTs[18] or the visible emission due to passivated surface defects in functionalized SWNTs and MWNTs.[106] The other is the indirect imaging based on associated fluorescent or radioactive agents.

In the absence of harsh chemical treatment or alteration, the well-dispersed semiconducting SWNTs exhibit band-gap fluorescence emissions in the near-IR spectral

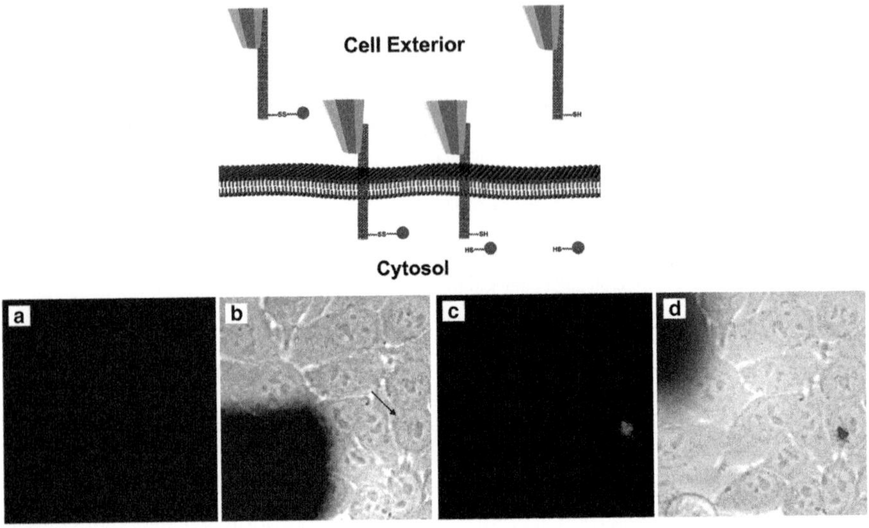

Figure 6.9 (Top) Schematic of the nanoinjection procedure and nanoinjection of QD streptavidin conjugates into a target HeLa cell. (a) Fluorescence image of the cells before nanoinjection. (b) Combined bright-field and fluorescence image of the cells before nanoinjection. The inserted arrow indicates the target cell. The dark shape in the lower left corner is the AFM cantilever. (c) Fluorescence image of the cells after the nanoinjection, showing fluorescent QD streptavidin conjugates released inside the target cell. (d) Combined bright-field and fluorescence image of the cells after the nanoinjection. The QD streptavidin conjugates are visible in red. The dark shape in the upper left corner is the retracted AFM cantilever. In all cases, fluorescence images were acquired with $\lambda_{ex} = 415$ nm and data collection with a 655 nm filter. Images are 70×70 μm.[97] (Reprinted with permission from X. Chen et al., *Proc. Natl. Acad. Sci. U.S.A.* **2007**, *104*, 8218–8222. Copyright 2007 National Academy of Sciences, U.S.A.) (See color insert.)

region (typically 900 to 1600 nm), and the fluorescence spectra are characteristic of the chiral indices (n, m) of the specific nanotubes.[18] Since most natural biomolecules are relatively transparent and nonemissive in this spectral region, the sharp emission spectra of SWNTs can be detected even in a complex biological environment. The bandgap fluorescence is quenched by surface defects in the carbon nanotubes. However, upon oxidation and chemical functionalization of carbon nanotubes at surface defect sites, both SWNTs and MWNTs exhibit strong photoluminescence in the visible and near-IR spectral regions.[106,107] Therefore, the well-functionalized carbon nanotubes are also amenable to optical bioimaging applications.

Welsher et al. used antibody-conjugated semiconducting SWNTs as near-IR fluorescent tags for selective probing of cell surface receptors and cell imaging.[105] The SWNTs were noncovalently dispersed by phospholipid-poly(ethylene glycol)-amine. The functionalized SWNTs bearing amine groups were then conjugated with thiolated antibodies (Rituxan and Herceptin) specific to different cell surface receptors. In solution, these antibody-conjugated SWNTs emitted in the 1000 to 1600 nm spectral region when excited at 785 nm, suggesting that the electronic properties of the SWNTs were preserved after the antibody conjugation.[105] The *in vitro* cell experiments and near-IR imaging showed specific binding of the nanotube-attached antibodies to the host cells. The binding specificity was evaluated in terms of the fluorescence intensity ratio between host cells and no-host cells, and the results suggested a high selectivity of 55–20 : 1 for these antibody-conjugated SWNTs.[105]

Cherukuri et al. employed pluronic copolymer-suspended pristine SWNTs ($\sim$1.0 nm average in diameter, and $\sim$1 μm average in length) for the intracellular near-IR imaging of mouse peritoneal macrophage-like cells.[98] It was found that the nanotubes were continually taken up by the cells (up to 7.3 μg/mL in SWNT concentration) without any toxic effect. The uptake was dependent on temperature, time, and SWNT concentration. Macrophage samples with internalized SWNTs exhibited characteristic nanotube fluorescence spectra (Fig. 6.10).[98] The fluorescence intensities increased smoothly with incubation time and external nanotube concentration. The macrophage-like cells maintained their normal behavior at a nanotube concentration of $\sim$4 μg/mL. In another recent study by Leeuw et al., *Drosophila melanogaster* (fruit flies) were fed with bovine serum albumin (BSA)-dispersed SWNTs.[101] The organisms and biological tissues were studied by using the intrinsic near-IR fluorescence imaging of SWNTs. It was reported that only a very small fraction ($\sim 10^{-8}$) of the ingested nanotubes became incorporated into the organs of the larvae.[101] Generally for the imaging with SWNTs in the near-IR spectral region, the signal attenuation due to water absorption could be avoided or minimized by appropriate wavelength selection, such as by using band-gap emissions of SWNTs beyond 1600 nm.

The visible photoluminescence (extending to the near-IR) in functionalized carbon nanotubes is much brighter, amenable to optical bioimaging applications. A representative example is the characteristic photoluminescence of ammonia salts-functionalized SWNTs (SWNT-NH_3^+), with the emission maximum at 485 nm (corresponding to 395 nm excitation).[99] The photoluminescence was used by Lacerda et al. to visualize the internalization and perinuclear localization of SWNT-NH_3^+ in human caucasian lung carcinoma A549 cells.[99] There was no cell plasma membrane damage even at a nanotube dose up to 500 μg/mL and 24 h post-incubation. This was the first report on the UV/vis florescence imaging of SWNTs in cells without the need for attaching a large fluorescent label.[99] By minimizing the possible alternation on the nanotube surface, the method could offer sharper images in comparison with those obtained with large fluorophore-attached SWNTs.

A representative example of indirect imaging with SWNTs labeled with fluorescent molecules was illuminated by Dai and coworkers.[102] They solubilized as-produced SWNTs via sonication with fluorescein-modified poly(ethylene glycol) (Fluor-PEG, 114 PEG units). The absorption maximum of the nanotube-attached Fluor-PEG was red shifted $\sim$3 nm from that of free fluorescein, and the fluorescence intensity was quenched about 67% in phosphate buffer solution (pH 7.4). In the bio-evaluation with BT474 breast cancer cells, substantial intracellular fluorescence was observed, suggesting cellular uptake of the Fluor-PEG-functionalized SWNTs.[102]

In addition to fluorescence, radioisotope tracing and physical methods (e.g., microscopy) are also effective techniques for the imaging of carbon nanotubes *in vitro* and *in vivo*.[100,103] For example, Porter et al. directly visualized individual SWNTs in cells through a new technique called low-loss energy-filtered transmission

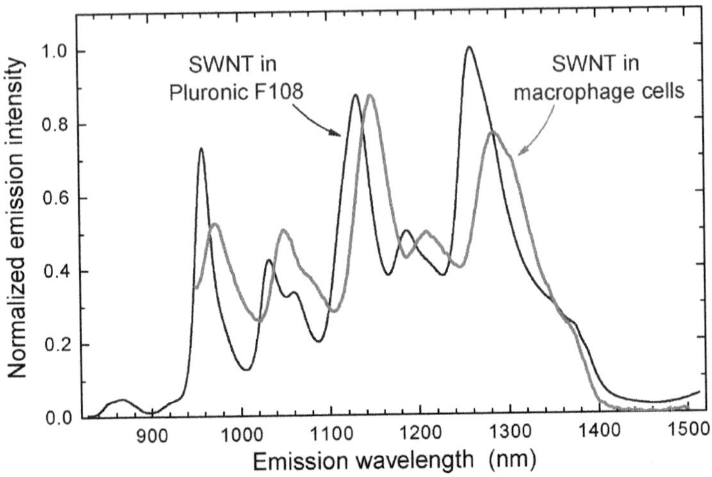

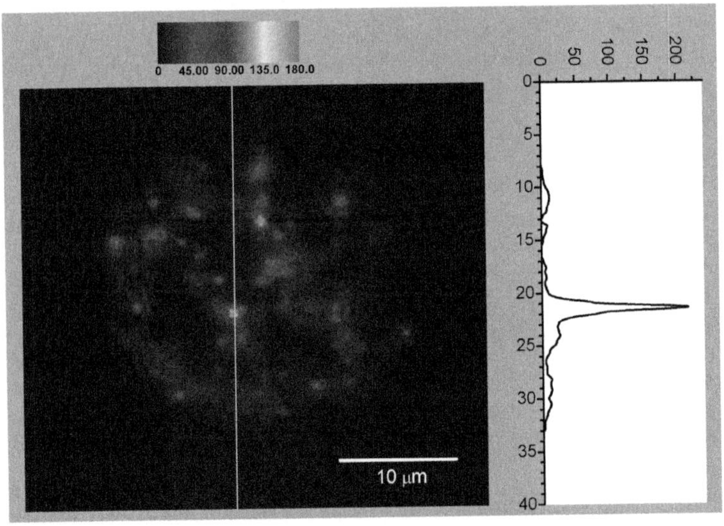

Figure 6.10 (Top) SWNT emission spectra in an aqueous pluronic F108 suspension (black trace) and in macrophage cells incubated in SWNT suspension and then washed (grey trace). Samples were excited at 660 nm. Intensities have been scaled to aid comparison. (Bottom) Fluorescence image of one macrophage-like cell incubated with SWNTs, showing emission detected from 1125 to 1600 nm with excitation at 660 nm. Intensities are coded with false color, and the image was obtained from a z-axis series by deconvolution processing. Intensity along the white vertical line is plotted on the graph to the right, showing high image contrast and localized emission sources.[98] (Reprinted with permission from P. Cherukuri et al., *J. Am. Chem. Soc.* **2004**, *126*, 15638–15639. Copyright 2004 American Chemical Society.)

electron microscopy (EFTEM), which enabled the direct determination of SWNT distribution in both stained and unstained cells.[103] The nanotubes were seen to enter the cytoplasm and localize within the cell nucleus, causing cell mortality in a dose-dependent manner.[103]

6.3.4 Tissue Engineering

As a result of the extended carbon–carbon sp^2 network structure, carbon nanotubes are of high mechanical strength and chemical stability.[108] The Young's modulus of SWNTs is close to 1 TPa, and the maximum tensile strength is about 30 GPa, which make them suitable substrates for tissue engineering.[109] For example, the as-grown MWNTs on quartz substrate from the CVD production method were used as constructs for culturing human osteoblastic cells.[110] The cells were found to be successfully attached to and grown on the MWNT constructs with higher metabolic activities than those cultured on the control surface (highly ordered pyrolytic graphite or HOPG). Compared to the well-spread cells cultured on HOPG, those grown on MWNTs exhibited a spherical morphology and a disorganized cytoskeleton.[110]

Neural stem cells are known for their sensitivity to the environment. Jan and Kotov reported the differentiation of mouse embryonic neural stem cells on SWNT-polyelectrolyte multilayer thin films [poly(ethylene imine) PEI/SWNT, six layers], which were assembled layer by layer.[111] The cells were successfully differentiated to neurons, astrocytes, and oligodendrocytes with a clear formation of neurites. Compared to the widely used poly(L-ornithine) (PLO) substrates, the six-layer PEI/SWNT thin films exhibited similar properties in terms of biocompatibility, neurites outgrowth, and the expression of neural markers.[111]

The *in vivo* biocompatibility and biodegradability of MWNT scaffold for tissue growth were also reported by Abarrategi et al. The scaffold, with a well-defined microchannel porous structure, was constructed by acid-treated MWNTs (up to 89 wt%) and chitosan (CHI).[112] The *in vitro* cell growth results showed that MWNT/CHI scaffolds were biocompatible for adhesion, spreading, proliferation, and viability of C2C12 (myoblastic mouse) cells.[112] To study the *in vivo* ectopic bone formation at a nonskeletal site, the recombinant human bone morphogenetic protein-2 (rhBMP-2, a potent osseoinductor)-adsorbed MWNT/CHI scaffolds were implanted into the mouse muscle tissue. After three weeks, bone tissue regeneration was observed with significant MWNT/CHI scaffold degradation and replacement by cells (Fig. 6.11).[112] The biocompatibility of the MWNT/CHI scaffold was confirmed by the absence of chronic inflammation during the whole implantation period.

6.3.5 Toxicity and *in vivo* Biodistribution

With the rapid advances in the development of carbon nanotube-based biomaterials and technologies, there is an urgent need for the understanding of their toxicological and pharmacological properties. In the current knowledge, the toxicity of carbon nanotubes depends on many factors, including dosage, physical form, and chemical modifications.[12,113] For example, Cui et al. studied *in vitro* toxicity of as-produced SWNTs in human HEK293 kidney cells.[114] They found that SWNTs could inhibit cell proliferation and at the same time decrease the cell's ability to adhere in a dose- and time-dependent manner. However, the observed toxicity of the pristine SWNTs

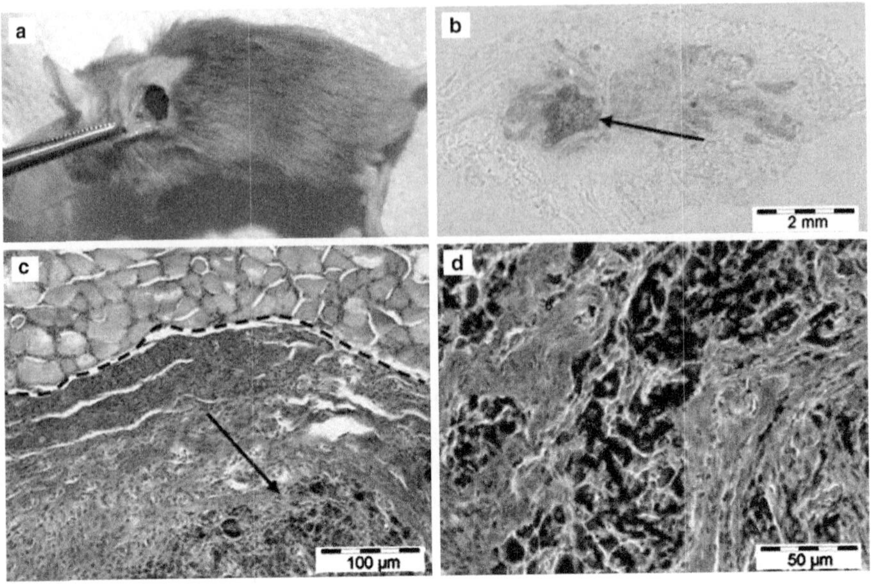

Figure 6.11 (a) Surgery implantation of rhBMP-2 adsorbed MWNT/CHI scaffolds into mouse subcutaneous muscular pocket. Optical microscope micrograph (b) shows regenerated bone tissue and a minor fraction of remaining MWNT/CHI scaffold. Optical micrograph (c) shows a detail of regenerated bone tissue (collagen expressing cells, blue-green colored) after major disassembly of the MWNT/CHI scaffold, surrounded by muscle tissue (pink colored). The well-limited interface between adjacent tissues is remarkable (see black dashed line). The remaining MWNT/CHI scaffold (black colored) is pointed out by a black arrow. Optical micrograph (d) shows detail of the remaining scaffold with plenty of fibroblasts (purple colored), prior to its disassembly and colonization by collagen expressing cells (blue-green colored).[112] (Reprinted with permission from A. Abarrategi et al., *Biomaterials* **2008**, *29*, 94–102. Copyright 2008 Elsevier.) (See color insert.)

might be due to the lack of water solubility and metal contamination, as debated in different reports.[115,116] To elucidate these issues, Isobe et al. prepared a metal-free and water-soluble carbon nanohorn aggregate (NHA) to simulate the covalently bonded class of self-aggregated carbon nanotubes.[116] The cytotoxicity study of the NHA on 3T3 and HeLa cell lines showed a very low cytotoxicity, even lower than that of the widely used quartz microparticles, though there were still doubts on the reliability of the results because of some experimental problems.[116]

In general, on chemical modification carbon nanotubes exhibit much less toxicity or nontoxicity to living cell lines that have been investigated so far.[117,118] For instance, Dumortier et al. conducted an *in vitro* cell uptake study of the functionalized SWNTs with B and T lymphocytes and macrophages.[117] Two types of functionalized SWNTs were used, one prepared via 1,3-dipolar cycloaddition reaction and the other obtained through oxidation/amidation treatment. Both types of the functionalized nanotubes were rapidly taken up by lymphocytes and macrophages without affecting the overall cell viability, and the one functionalized through the cycloaddition (thus having higher water solubility) was found not affecting the functional activity of the different types of immunoregulatory cells.[117]

Carbon nanotubes may enter the body via intravenous, dermal, subcutaneous, inhalational, intraperitoneal, or oral routes.[119] The potential *in vivo* toxicity of carbon nanotubes has been reported and discussed.[12] Largely preliminary *in vivo* toxicity studies on respiratory and skin exposure to pristine carbon nanotubes did show some harmful effects.[120,121] In a recent study reported by Poland et al.[122] pristine and BSA-dispersed MWNTs were injected intraperitoneally into mice at a dose of 50 μg per mouse. Only the long pristine MWNTs (more than 20 μm in length) showed some pathogenic effects similar to the foreign body inflammatory response caused by long asbestos (Fig. 6.12).[122] Conclusive results on the long-term exposure to these commercially supplied long MWNTs are still needed.

Schipper et al. also conducted a pilot toxicologic study of the covalently and noncovalently PEG-functionalized SWNTs in mice.[123] Their results suggested that the functionalized SWNTs persisted within liver and spleen macrophages for four months without apparent toxicity.[123]

In the *in vivo* experiments on pristine and the PEGylated SWNTs in mice, Yang et al.[87,88] did not find any of the animals exhibiting signs of acute toxicity response during the experimental period, even at a high exposure of 80 mg pristine or 24 mg PEGylated SWNTs (nanotube equivalent, both skeleton ^{13}C-enriched) per kilogram body weight, consistent with other reports.

A systematic and quantitative analysis of the metabolic process for carbon nanotubes in living body, such as their degradation, distribution, clearance, and bioaccumulation, is mandatory in order to fully understand their *in vivo* toxicity. Again according to the studies by Yang et al., the ^{13}C-enriched pristine SWNTs were cleared from the bloodstream quickly and distributed throughout most of the organs within 24 h.[88] The nanotubes accumulated primarily in the lungs, liver, and spleen and were retained at the relatively high accumulation levels over 28 days. Most noticeably and interestingly, there was around 1%ID to 3%ID accumulation of SWNTs in the brain, suggesting that the nanotubes could overcome the blood-brain barrier. While the accumulation level of SWNTs in liver was relatively constant, there was a gradual decrease in the lungs from 15%ID to 9.4%ID in the monitoring period, for which the secretion by alveolar macrophage as mucus through mucociliary transport and the translocation through lymph nodes were suggested as the possible clearance pathways.[88] Different from their modified/functionalized counterparts,[87] these pristine SWNTs could hardly be detected in urine and feces by either ^{13}C isotope ratio measurements or transmission electron microscopy (TEM) analyses, suggesting that the nanotube morphology along with surface functionalities might significantly affect the nanotube *in vivo* behaviors.[88] The PEGylated SWNTs were found to have a similar biodistribution profile to that of the pristine ones, with a rapid distribution balance among blood and tissues and also significant accumulation in liver and spleen (Fig. 6.13).[87] At day 7 postexposure, for example, ~25%ID and ~3%ID of the PEGylated SWNTs remained in liver and spleen, respectively. Nevertheless,

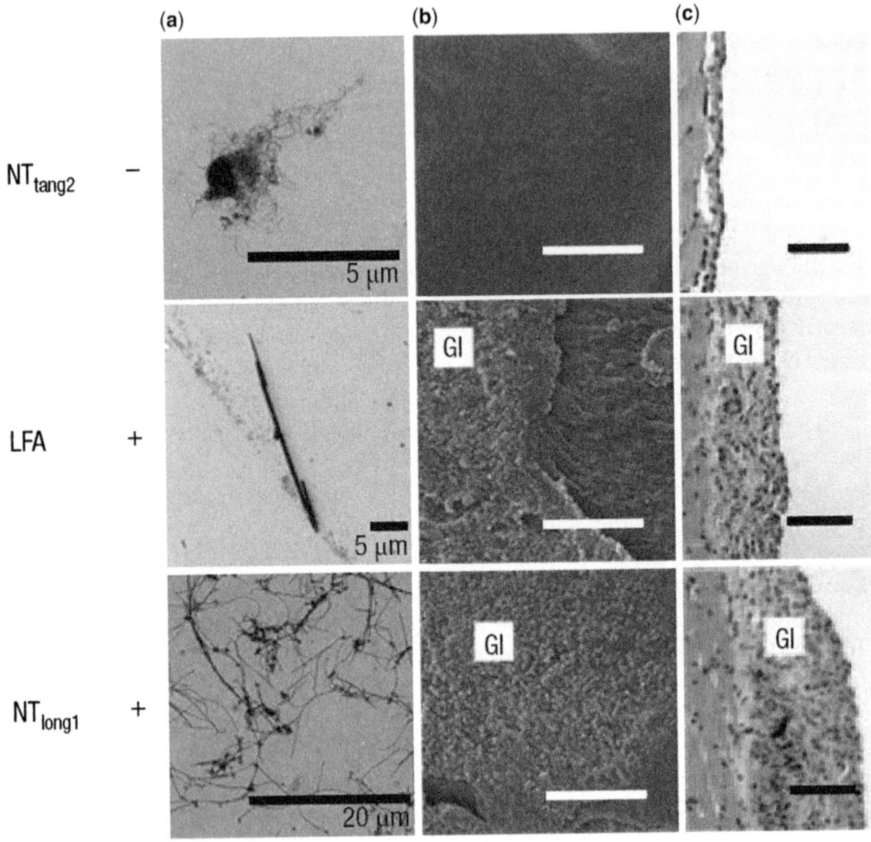

Figure 6.12 Effect of the fiber on diaphragms after 7 days. (a) TEM images of the test samples. NT_{tang2}: low-aspect-ratio tangled MWNT aggregates; LFA: long-fiber amosite; NT_{long1}: MWNTs with a substantial portion longer than 20 μm. Female C57Bl/6 mice were injected i.p. with 50 μg of sample, killed after 7 days, and the diaphragms excised and prepared for visualization. (b) SEM images and (c) hematoxylin and eosin histology sections ($n = 3$) of the diaphragms show the presence of granulomatous inflammation (GI) in mice exposed to LFA, NT_{long1}. A small granuloma response in one of the three mice treated with NT_{tang2} was observed. The muscular portion of the peritoneal diaphragm (PD) and the mesothelial layer (ML) are aligned to show granulomatous inflammation at the peritoneal aspect of the diaphragm surface. Scale bar in (b): 200 μm. Scale bar in (c): 50 μm.[122] (Reprinted with permission from C. A. Poland et al., *Nature Nanotechnol.* **2008**, *3*, 423–428. Copyright 2008 Macmillan Publishers Ltd.)

the uptake of the PEGylated SWNTs by the reticuloendothelial system (RES) was significantly reduced (28%ID)[87] in comparison with the pristine SWNTs (37%ID).[88]

Radiotracing is an efficient and broadly adopted method for *in vivo* visualization of functionalized carbon nanotubes. The radioactive labels, such as ^{125}I,[124] ^{111}In,[100,104] ^{64}Cu,[90] ^{99m}Tc,[125] and ^{14}C,[126] have made it possible to quantitatively map the location of carbon nanotubes inside animals at different doses and time points. In

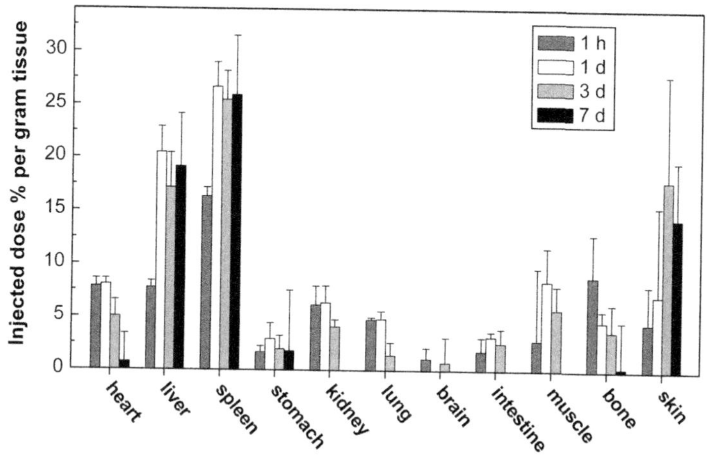

Figure 6.13 Time-dependent biodistributions of PEG-SWNTs in mice postexposure at a dose of 2.4 mg SWNTs-equivalent/kg body weight.[87] (Reprinted with permission from S.-T. Yang et al., *Small* **2008**, *4*, 940–944. Copyright Wiley-VCH Verlag GmbH & Co. KGaA.)

a study by Singh et al., SWNTs were sidewall functionalized with diethylenetriaminepentaacetic ^{111}In [(^{111}In)DTPA] and were intravenously administered into mice.[100] The conjugates were cleared rapidly from systemic blood circulation with a half-life of ~3 h through a renal excretion route, and no retaining was found in any of the RES organs (liver and spleen).[100] The same functionalization strategy was later transferred to the preparation of [^{111}In]DTPA-labeled MWNTs.[104] After the injection into the mice through the tail vein, the dynamic radioactivity tracking of the [^{111}In]DTPA-MWNT showed that the nanotubes quickly entered the systemic blood circulation and started their accumulation in kidney and bladder within 60 seconds. At 24 h, almost all MWNTs were eliminated from the body, and only a trace amount of radioactivity was detected in the kidney (Fig. 6.14).[104] The biodistributions of DTPA-functionalized MWNTs (without ^{111}In label) and serum-coated purified MWNTs were also compared by using the traditional histological examination of the major organs at 24 h after the mice were intravenously administered the MWNTs. The purified MWNTs exhibited accumulation in lung and liver, while no such accumulation was found for DTPA-functionalized MWNT, consistent with the dynamic radioactivity tracking results obtained with [^{111}In]DTPA-MWNT.[104]

Similarly, Liu et al. investigated the biodistribution and tumor targeting ability of ^{64}Cu-labeled SWNTs in mice by *in vivo* positron emission tomography (PET), *ex vivo* biodistribution analysis, and Raman spectroscopy.[90] They found that the functionalized SWNTs were surprisingly stable *in vivo*, exhibiting relatively long blood circulation time and low RES uptake. Efficient targeting of integrin positive tumor in mice was also demonstrated by tethering PL-PEG chains with RGD peptides (recognizing integrin positive tumors).[90]

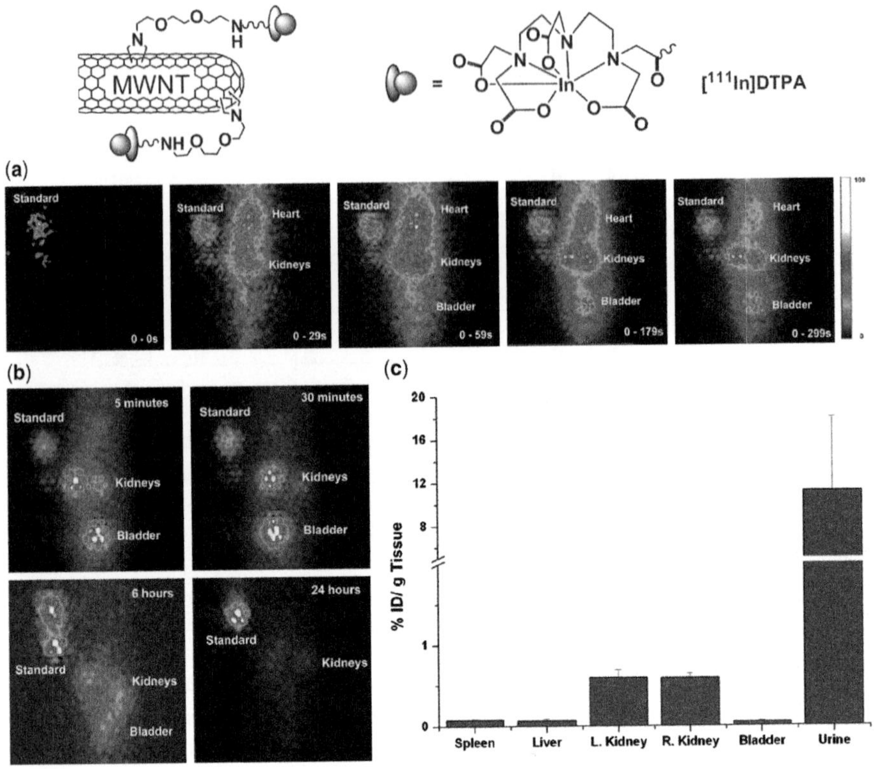

Figure 6.14 (Top) A scheme of [^{111}In]DTPA-MWNT and normal rat distribution of [^{111}In]DTPA-MWNT. (a) Dynamic anterior planar images of whole body distribution of [^{111}In]DTPA-MWNT within 5 min after intravenous administration in rats. Color scale for radioactivity levels shown in arbitrary units. (b) Static anterior planar images of whole body distribution of [^{111}In]DTPA-MWNT in rats after 5 min, 30 min, 6 h, and 24 h postinjection [difference between (a), 0 to 299 s image and (b), 5 min image is due to lag-time in camera setup). (c) Percent ID radioactivity per gram tissue at 24 h after intravenous administration of [^{111}In]DTPA-MWNT quantified by gamma counting ($n = 3$ and error bars for standard deviation).[104] (Reprinted with permission from L. Lacerda et al., *Adv. Mater.* **2008**, *20*, 225–230. Copyright Wiley-VCH Verlag GmbH & Co. KGaA.)

The chelation of radioactive elements with surface functionalities on carbon nanotubes is straightforward. However, such conjugates could suffer from decreasing or even losing activity over time due to gradual decay or dissociation of the radiolabel from the carbon nanotubes. Therefore, direct detection techniques based on the intrinsic physical properties of carbon nanotubes, such as Raman and near-IR fluorescence, might be more advantageous in the monitoring of their *in vivo* behaviors in both short and long terms. For example, the noncovalently PL-PEG (both linear and branched)-functionalized SWNTs were intravenously injected into mice for the study of their blood circulation and long-term fate in terms of the nanotubes' intrinsic Raman

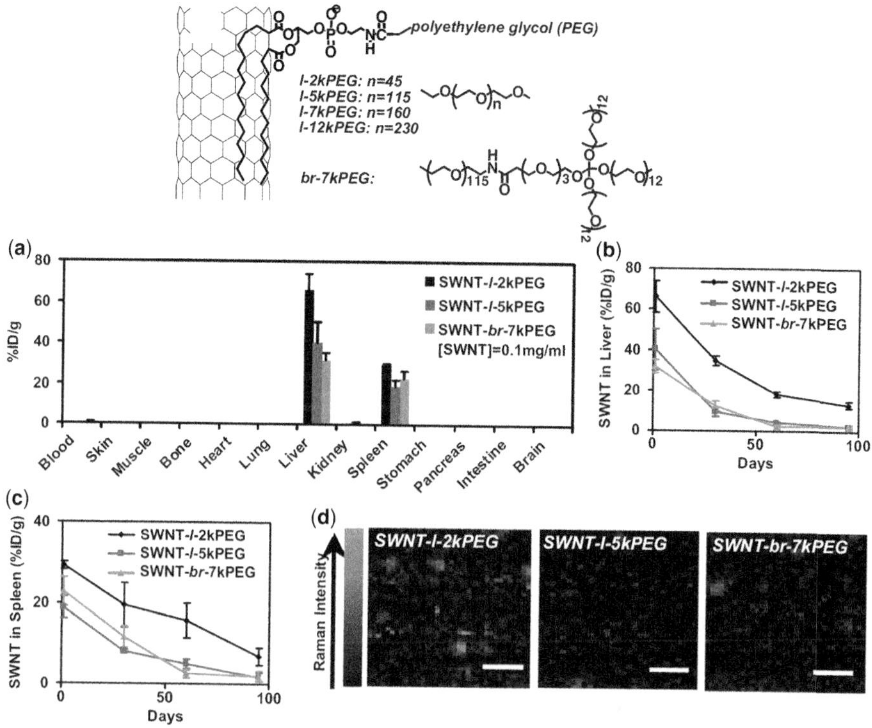

Figure 6.15 (Top) A scheme of noncovalently functionalized SWNTs by various PEGylated phospholipids, and SWNTs in mice tissues probed by *ex vivo* Raman spectroscopy after injection into mice. (a) Biodistribution of SWNT-*l*-2kPEG, SWNT-*l*-5kPEG, and SWNT-*br*-7kPEG, respectively, at 1 day postinjection measured by Raman spectroscopy. The SWNT concentrations in most organs are below detection limit. (b) and (c) Evolution of the concentrations of SWNTs retained in the liver and spleen of mice over a period of 3 months. Compared with SWNT-*l*-2kPEG, much lower concentrations of retained SWNTs in the liver and spleen were observed for SWNT-*l*-5kPEG and SWNT-*br*-7kPEG. (d) Raman mapping images of liver slices from mice treated with SWNT-*l*-2kPEG (left), SWNT-*l*-5kPEG (center), and SWNT-*br*-7kEG (right) at 3 months postinjection. More SWNT signals were observed in the SWNT-*l*-2kPEG-treated mouse sample than the other two samples under the same Raman imaging conditions (laser power, beam size, etc.). The error bars in (a) to (c) were based on three to four mice per group. Note that the injected SWNT solutions had a concentration of 0.1 mg/mL (optical density was 4.6 at 808 nm for 1 cm path).[89] (Reprinted with permission from Z. Liu et al., *Proc. Natl. Acad. Sci. U.S.A.* **2008**, *105*, 1410–1415. Copyright 2008 National Academy of Sciences, U.S.A.) (See color insert.)

signatures (Fig. 6.15).[89] Effective PEG coating (longer chain and branched structure) on SWNTs afforded longer blood circulation, lower RES uptake, and near-complete clearance from main organs via biliary and renal pathways in ~2 months, which was likely due to their improved hydrophilicity and resistance to protein nonspecific binding.[89]

Similarly, Cherukuri et al. used the near-IR fluorescence of individualized semiconducting SWNTs to study their blood elimination kinetics and biodistribution in rabbits.[127] The pluronic copolymer-coated SWNTs were intravenously administered into rabbits at a dose of ~20 μg SWNT per kilogram of body mass. The pluronic coating was found to be displaced by blood protein within seconds, as indicated by changes in the near-IR emission spectra. The nanotube concentration in the blood serum decreased exponentially with a half-life of 1 h, and no adverse effects were observed in terms of the rabbit behavior and the pathological examination. At 24 h post administration, significant concentrations of nanotubes were found only in the liver.[127]

It seems that based on the available *in vivo* biodistribution and toxicity studies there is no good evidence suggesting any significant acute toxicity or negative health effects of carbon nanotubes on the animals involved in the experiments. Obviously, further investigations are necessary and are likely already in progress in many laboratories to better address the toxicity and related issues.

6.4 CONCLUSIONS AND PERSPECTIVES

The gradually maturing research concerning the chemical modification and functionalization of carbon nanotubes has stimulated the proliferation of studies on their potential bioapplications. Here we have reviewed the recent advances in the modification and functionalization of carbon nanotubes by biomolecules (e.g., proteins, DNAs, and carbohydrates). Upon modification and/or functionalization, the toxicological impacts of carbon nanotubes are significantly reduced, offering great opportunities for the use of carbon nanotubes for diagnostic and therapeutic purposes. Many experimental investigations have demonstrated that carbon nanotubes are ideal scaffolds to carry or display multiple biomolecules or bioactive species for applications such as capturing pathogens, substance delivery, bioimaging, and high-performance biosensors.

Carbon nanotubes currently represent one of the most promising platforms for novel biomaterials, though many technological challenges have to be overcome in order to realize their full potentials. For example, the performance of carbon nanotube-based biosensors depends strongly on the structural and electrical properties of the nanotube, so that the patterned growth and controlled deposition of nanotubes at desired sites with defined structures are necessary for more effective biosensing. The establishment of more accurate mechanisms for the interactions of carbon nanotubes with cells and other biological species is fundamentally important to the development of carbon nanotube-based bio-nanotechnologies. Along the same line, in-depth and systematic studies of pharmacokinetic profiles and long-term *in vivo* toxicity of carbon nanotubes are critical to their eventual clinical applications and therefore must be vigorously and carefully pursued. Nevertheless, the obviously rapid advances in this field have led us to conclude that there is a bright future for carbon nanotubes in biology and medicine.

REFERENCES

1. L. Gu, Ph.D. Dissertation, Clemson University, Clemson, SC, **2008**.
2. (a) M. S. Dresselhaus, G. Dresselhaus, P. Avouris, Eds, *Carbon Nanotubes: Synthesis, Structure, Properties and Applications*, Berlin: Springer, **2001**; (b) Special Issue on Carbon Nanotubes, *Acc. Chem. Res.* **2002**, *35*(12).
3. P. M. Ajayan, *Chem. Rev.* **1999**, *99*, 1787–1800.
4. R. E. Smalley Memorial Issue (Eds: D. J. Nelson, M. S. Strano, W. W. Adams, R. F. Curl, R. B. Weisman), *J. Phys. Chem. C* **2007**, 111(48).
5. J. Wang, *Electroanalysis* **2005**, *17*, 7–14.
6. Y. Yun, Z. Dong, V. Shanov, W. R. Heineman, H. B. Halsall, A. Bhattacharya, L. Conforti, R. K. Narayan, W. S. Ball, M. J. Schulz, *Nano Today* **2007**, *2*, 30–37.
7. G. A. Rivas, M. D. Rubianes, M. C. Rodríguez, N. F. Ferreyra, G. L. Luque, M. L. Pedano, S. A. Miscoria, C. Parrado, *Talanta* **2007**, *74*, 291–307.
8. (a) R. H. Baughman, C. Cui, A. A. Zakhidov, Z. Iqbal, J. N. Barisci, G. M. Spinks, G. G. Wallace, A. Mazzoldi, D. De Rossi, A. G. Rinzler, O. Jaschinski, S. Roth, M. Kertesz, *Science* **1999**, *284*, 1340–1344; (b) A. M. Fennimore, T. D. Yuzvinsky, W. Q. Han, M. S. Fuhrer, J. Cumings, A. Zettl, *Nature* **2003**, *424*, 408–410.
9. N. W. S. Kam, M. O'Connell, J. A. Wisdom, H. Dai, *Proc. Natl. Acad. Sci. U.S.A.* **2005**, *102*, 11600–11605.
10. (a) N. W. S. Kam, H. Dai, *Phys. Stat. Sol. B* **2006**, *243*, 3561–3566; (b) L. Lacerda, S. Raffa, M. Prato, A. Bianco, K. Kostarelos, *Nano Today* **2007**, *2*, 38–43; (c) M. Prato, K. Kostarelos, A. Bianco, *Acc. Chem. Res.* **2008**, *41*, 60–68.
11. H. Wang, L. Gu, Y. Lin, F. Lu, M. J. Meziani, P. G. Luo, W. Wang, L. Cao, Y.-P. Sun, *J. Am. Chem. Soc.* **2006**, *128*, 13364–13365.
12. (a) C. W. Lam, J. T. James, R. McCluskey, S. Arepalli, R. L. Hunter, *Crit. Rev. Toxicol.* **2006**, *36*, 189–217; (b) J. Kolosnjaj, H. Szwarc, F. Moussa, *Adv. Exp. Med. Biol.* **2007**, *620*, 181–204.
13. Y. Lin, S. Taylor, H. Li, K. A. S. Fernando, L. Qu, W. Wang, B. Zhou, Y.-P. Sun, *J. Mater. Chem.* **2004**, *14*, 527–541.
14. (a) S. Banerjee, T. Hemraj-Benny, S. S. Wong, *Adv. Mater.* **2005**, *17*, 17–29; (b) K. Balasubramanian, M. Burghard, *Small* **2005**, *1*, 180–192.
15. D. Tasis, N. Tagmatarchis, A. Bianco, M. Prato, *Chem. Rev.* **2006**, *106*, 1105–1136.
16. W. Huang, S. Fernando, L. F. Allard, Y.-P. Sun, *Nano Lett.* **2003**, *3*, 565–568.
17. K. A. S. Fernando, Y. Lin, W. Wang, S. Kumar, B. Zhou, S. Y. Xie, L.T. Cureton, Y.-P. Sun, *J. Am. Chem. Soc.* **2004**, *126*, 10234–10235.
18. M. J. O'Connell, S. M. Bachilo, C. B. Huffman, V. C. Moore, M. S. Strano, E. H. Haroz, K. L. Rialon, P. J. Boul, W. H. Noon, C. Kittrell, J. Ma, R. H. Hauge, R. B. Weisman, R. E. Smalley, *Science* **2002**, *297*, 593–596.
19. R. J. Chen, Y. Zhang, D. Wang, H. Dai, *J. Am. Chem. Soc.* **2001**, *123*, 3838–3839.
20. (a) A. Star, J. F. Stoddart, D. Steuerman, M. Diehl, A. Boukai, E. W. Wong, X. Yang, S. W. Chung, H. Choi, J. R. Heath, *Angew. Chem. Int. Ed.* **2001**, *40*, 1721–1725; (b) V. C. Moore, M. S. Strano, E. H. Haroz, R. H. Hauge, R. E. Smalley, *Nano Lett.* **2003**, *3*, 1379–1382.
21. D. Nepal, K. E. Geckeler, *Small* **2007**, *3*, 1259–1265.
22. K. Kurppa, H. Jiang, G. R. Szilvay, A. G. Nasibulin, E. I. Kauppinen, M. B. Linder, *Angew. Chem. Int. Ed.* **2007**, *46*, 6446–6449.
23. V. Z. Poenitzsch, D. C. Winters, H. Xie, G. R. Dieckmann, A. B. Dalton, I. H. Musselman, *J. Am. Chem. Soc.* **2007**, *129*, 14724–14732.
24. Q. Mu, W. Liu, Y. Xing, H. Zhou, Z. Li, Y. Zhang, L. Ji, F. Wang, Z. Si, B. Zhang, B. Yan, *J. Phys. Chem. C* **2008**, *112*, 3300–3307.
25. Z. Su, T. Leung, J. F. Honek, *J. Phys. Chem. B* **2006**, *110*, 23623–23627.
26. Z. Su, K. Mui, E. Daub, T. Leung, J. Honek, *J. Phys. Chem. B* **2007**, *111*, 14411–14417.

27. B. R. Azamian, J. J. Davis, K. S. Coleman, C. B. Bagshaw, M. L. Green, *J. Am. Chem. Soc.* **2002**, *124*, 12664–12665.
28. (a) W. Huang, S. Taylor, K. Fu, Y. Lin, D. Zhang, T. W. Hanks, A. M. Rao, Y.-P. Sun, *Nano Lett.* **2002**, *2*, 311–314; (b) K. Fu, W. Huang, Y. Lin, D. Zhang, T. W. Hanks, A. M. Rao, Y.-P. Sun, *J. Nanosci. Nanotechnol.* **2002**, *2*, 457–461.
29. Y. Lin, L. F. Allard, Y.-P. Sun, *J. Phys. Chem. B* **2004**, *108*, 3760–3764.
30. T. Elkin, X. Jiang, S. Taylor, Y. Lin, L. Gu, H. Yang, J. Brown, S. Collins, Y.-P. Sun, *ChemBioChem* **2005**, *6*, 640–643.
31. Y. Lin, T. Elkin, S. Taylor, L. Gu, B. Chen, L. M. Veca, B. Zhou, H. Yang, J. Brown, R. Joseph, E. Jones, X. Jiang, Y.-P. Sun, *Microchim. Acta* **2006**, *152*, 249–254.
32. Y. Lin, X. Jiang, T. Elkin, K. A. S. Fernando, L. Gu, S. Taylor, H. Yang, E. Jones, W. Wang, Y.-P. Sun, *J. Nanosci. Nanotechnol.* **2006**, *6*, 868–871.
33. A. Joshi, S. Punyani, S. S. Bale, H. Yang, T. Borca-Tasciuc, R. S. Kane, *Nature Nanotechnol.* **2008**, *3*, 41–45.
34. Y.-Z. You, C.-Y. Hong, C.-Y. Pan, *J. Phys. Chem. C* **2007**, *111*, 16161–16166.
35. L. Gu, T. Elkin, X. Jiang, H. Li, Y. Lin, L. Qu, T.-R. J. Tzeng, R. Joseph, Y.-P. Sun, *Chem. Commun.* **2005**, 874–876.
36. M. Mammen, S.-K. Choi, G. M. Whitesides, *Angew. Chem. Int. Ed.* **1998**, *37*, 2754–2794.
37. L. Gu, Y. Lin, L. Qu, Y.-P. Sun, *Biomacromolecules* **2006**, *7*, 400–402.
38. L. Gu, P. G. Luo, H. Wang, M. J. Meziani, Y. Lin, L. M. Veca, L. Cao, F. Lu, X. Wang, R. A. Quinn, W. Wang, P. Zhang, S. Lacher, Y.-P. Sun, *Biomacromolecules* **2008**, *9*, 2408–2418.
39. T. Hasegawa, T. Fujisawa, M. Numata, M. Umeda, T. Matsumoto, T. Kimura, S. Okumura, K. Sakurai, S. Shinkai, *Chem. Commun.* **2004**, 2150–2151.
40. Y. Liu, P. Liang, H. Y. Zhang, D. S. Guo, *Small* **2006**, *2*, 874–878.
41. M. Numata, M. Asai, K. Kaneko, A.-H. Bae, T. Hasegawa, K. Sakurai, S. Shinkai, *J. Am. Chem. Soc.* **2005**, *127*, 5875–5884.
42. M. Numata, K. Sugikawa, K. Kaneko, S. Shinkai, *Chem. Eur. J.* **2008**, *14*, 2398–2404.
43. C. Fu, L. Meng, Q. Lu, X. Zhang, C. Gao, *Macromol. Rapid Commun.* **2007**, *28*, 2180–2184.
44. Y. Lu, S. Bangsaruntip, X. Wang, L. Zhang, Y. Nishi, H. Dai, *J. Am. Chem. Soc.* **2006**, *128*, 3518–3519.
45. Z. Liu, M. Winters, M. Holodniy, H. Dai, *Angew. Chem. Int. Ed.* **2007**, *46*, 2023–2027.
46. M. Zheng, A. Jagota, M. S. Strano, A. P. Santos, P. Barone, S. G. Chou, B. A. Diner, M. S. Dresselhaus, R. S. Mclean, G. B. Onoa, G. G. Samsonidze, E. D. Semke, M. Usrey, D. J. Walls, *Science* **2003**, *302*, 1545–1548.
47. Y. Chen, H. Liu, T. Ye, J. Kim, C. Mao, *J. Am. Chem. Soc.* **2007**, *129*, 8696–8697.
48. X. Han, Y. Li, Z. Deng, *Adv. Mater.* **2007**, *19*, 1518–1522.
49. S. Li, P. He, J. Dong, Z. Guo, L. Dai, *J. Am. Chem. Soc.* **2005**, *127*, 14–15.
50. X. Wang, F. Liu, G. T. S. Andavan, X. Jing, K. Singh, V. R. Yazdanpanah, N. Bruque, R. R. Pandey, R. Lake, M. Ozkan, K. L. Wang, C. S. Ozkan, *Small* **2006**, *2*, 1356–1365.
51. C.-S. Lee, S. E. Baker, M. S. Marcus, W. Yang, M. A. Eriksson, R. J. Hamers, *Nano Lett.* **2004**, *4*, 1713–1716.
52. K. Rege, G. Viswanathan, G. Zhu, A. Vijayaraghavan, P. M. Ajayan, J. S. Dordick, *Small*, **2006**, *2*, 718–722.
53. D. A. Heller, E. S. Jeng, T.-K. Yeung, B. M. Martinez, A. E. Moll, J. B. Gastala, M. S. Strano, *Science* **2006**, *311*, 508–511.
54. Y. Peng, X. Li, J. Ren, X. Qu, *Chem. Commun.* **2007**, 5176–5178.
55. M. L. Becker, J. A. Fagan, N. D. Gallant, B. J. Bauer, V. Bajpai, E. K. Hobbie, S. H. Lacerda, K. B. Migler, J. P. Jakupciak, *Adv. Mater.* **2007**, *19*, 939–945.
56. L. Zhang, S. Zaric, X. Tu, X. Wang, W. Zhao, H. Dai, *J. Am. Chem. Soc.* **2008**, *130*, 2686–2691.
57. X. Zhao, J. K. Johnson, *J. Am. Chem. Soc.* **2007**, *129*, 10438–10445.
58. R. R. Johnson, A. T. C. Johnson, M. L. Klein, *Nano Lett.* **2008**, *8*, 69–75.
59. S. Manohar, T. Tang, A. Jagota, *J. Phys. Chem. C* **2007**, *111*, 17835–17845.
60. M. E. Hughes, E. Brandin, J. A. Golovchenko, *Nano Lett.* **2007**, *7*, 1191–1194.

61. S. Meng, W. L. Wang, P. Maragakis, E. Kaxiras, *Nano Lett.* **2007**, *7*, 2312–2316.
62. (a) E. Katz, I. Willner, *ChemPhysChem* **2004**, *5*, 1084–1004; (b) K. Balasubramanian, M. Burghard, *Anal. Bioanal. Chem.* **2006**, *385*, 452–468; (c) B. L. Allen, P. D. Kichambare, A. Star, *Adv. Mater.* **2007**, *19*, 1439–1451; (d) S. N. Kim, J. F. Rusling, F. Papadimitrakopoulos, *Adv. Mater.* **2007**, *19*, 3214–3228.
63. P. Avouris, Z. Chen, V. Perebeinos, *Nature Nanotechnol.* **2007**, *2*, 605–615.
64. P. Stokes, S.I. Khondaker, *Nanotechnology* **2008**, *19*, 175202.
65. R. J. Chen, H. C. Choi, S. Bangsaruntip, E. Yenilmez, X. Tang, Q. Wang, Y.-L. Chang, H. Dai, *J. Am. Chem. Soc.* **2004**, *126*, 1563–1568.
66. H. R. Byon, H. C. Choi, *J. Am. Chem. Soc.* **2006**, *128*, 2188–2189.
67. C. Li, M. Curreli, H. Lin, B. Lei, F. N. Ishikawa, R. Datar, R. J. Cote, M. E. Thompson, C. Zhou, *J. Am. Chem. Soc.* **2005**, *127*, 12484–12485.
68. K. Maehashi, T. Katsura, K. Kerman, Y. Takamura, K. Matsumoto, E. Tamiya, *Anal. Chem.* **2007**, *79*, 782–787.
69. A. Star, E. Tu, J. Niemann, J.-C. P. Gabriel, C. S. Joiner, C. Valcke, *Proc. Natl. Acad. Sci. U.S.A.* **2006**, *103*, 921–926.
70. E. L. Gui, L.-J. Li, K. Zhang, Y. Xu, X. Dong, X. Ho, P. S. Lee, J. Kasim, Z. X. Shen, J. A. Rogers, S. G. Mhaisalkar, *J. Am. Chem. Soc.* **2007**, *129*, 14427–14432.
71. S. Heinze, J. Tersoff, R. Martel, V. Derycke, J. Appenzeller, P. Avouris, *Phys. Rev. Lett.* **2002**, *89*, 106801.
72. A. Star, J.-C. P. Gabriel, K. Bradley, G. Gruner, *Nano Lett.* **2003**, *3*, 459–463.
73. K. Bradley, M. Briman, A. Star, G. Gruner, *Nano Lett.* **2004**, *4*, 253–256.
74. I. Heller, A. M. Janssens, J. Mannik, E. D. Minot, S. G. Lemay, C. Dekker, *Nano Lett.* **2008**, *8*, 591–595.
75. (a) J. Kong, N. R. Franklin, C. Zhou, M. G. Chapline, S. Peng, K. Cho, H. Dai, *Science* **2000**, *287*, 622–625; (b) P. G. Collins, K. Bradley, M. Ishigami, A. Zettl, *Science* **2000**, *287*, 1801–1804.
76. (a) Y. Li, D. Mann, M. Rolandi, W. Kim, A. Ural, S. Hung, A. Javey, J. Cao, D. W. Wang, E. Yenilmez, Q. Wang, J. F. Gibbons, Y. Nishi, H. Dai, *Nano Lett.* **2004**, *4*, 317–321; (b) S. M. Bachilo, L. Balzano, J. E. Herrera, F. Pompeo, D. E. Resasco, R. B. Weisman, *J. Am. Chem. Soc.* **2003**, *125*, 11186–11187.
77. Y. Lin, K. A. S. Fernando, W. Wang, Y.-P. Sun, in *Carbon Nanotechnology* (Ed. L. Dai), Amsterdam: Elsevier, **2006**, 255–295.
78. P. G. Collins, M. S. Arnold, P. Avouris, *Science* **2001**, *292*, 706–709.
79. D. Chattopadhyay, I. Galeska, F. Papadimitrakopoulos, *J. Am. Chem. Soc.* **2003**, *125*, 3370–3375.
80. H. Li, B. Zhou, Y. Lin, L. Gu, W. Wang, K. A. S. Fernando, S. Kumar, L. F. Allard, Y.-P. Sun, *J. Am. Chem. Soc.* **2004**, *126*, 1014–1015.
81. G. Pastorin, W. Wu, S. Wieckowski, J. P. Briand, K. Kostarelos, M. Prato, A. Bianco, *Chem. Commun.* **2006**, 1182–1184.
82. H. Ali-Boucetta, K. T. Al-Jamal, D. McCarthy, M. Prato, A. Bianco, K. Kostarelos, *Chem. Commun.* **2008**, 459–461.
83. Z. Liu, X. Sun, N. Nakayama-Ratchford, H. Dai, *ACS Nano* **2007**, *1*, 50–56.
84. R. P. Feazell, N. Nakayama-Ratchford, H. Dai, S. J. Lippard, *J. Am. Chem. Soc.* **2007**, *129*, 8438–8439.
85. Z. Zhang, X. Yang, Y. Zhang, B. Zeng, S. Wang, T. Zhu, R. B. S. Roden, Y. Chen, R. Yang, *Clin. Cancer Res.* **2006**, *12*, 4933–4939.
86. V. P. Torchilin, *Adv. Drug Deliv. Rev.* **2006**, *58*, 1532–1555.
87. S.-T. Yang, K. A. S. Fernando, J.-H. Liu, J. Wang, H.-F. Sun, Y. Liu, M. Chen, Y. Huang, X. Wang, H. Wang, Y.-P. Sun, *Small* **2008**, *4*, 940–944.
88. S.-T. Yang, W. Guo, Y. Lin, X. Deng, H. Wang, H. Sun, Y. Liu, X. Wang, W. Wang, M. Chen, Y. Huang, Y.-P. Sun, *J. Phys. Chem. C* **2007**, *111*, 17761–17764.
89. Z. Liu, C. Davis, W. Cai, L. He, X. Chen, H. Dai, *Proc. Natl. Acad. Sci. U.S.A.* **2008**, *105*, 1410–1415.
90. Z. Liu, W. Cai, L. He, N. Nakayama, K. Chen, X. Sun, X. Chen, H. Dai, *Nature Nanotechnol.* **2007**, *2*, 47–52.

91. N. W. S. Kam, T. C. Jessop, P. A. Wender, H. Dai, *J. Am. Chem. Soc.* **2004**, *126*, 6850–6851.
92. N. W. S. Kam, H. Dai, *J. Am. Chem. Soc.* **2005**, *127*, 6021–6026.
93. N. W. S. Kam, Z. Liu, H. Dai, *Angew. Chem. Int. Ed.* **2006**, *45*, 577–581.
94. D. Pantarotto, J. Briand, M. Prato, A. Bianco, *Chem. Commun.* **2004**, 16–17.
95. K. Kostarelos, L. Lacerda, G. Pastorin, W. Wu, S. Wieckowski, J. Luangsivilay, S. Godefroy, D. Pantarotto, J.-P. Briand, S. Muller, M. Prato, A. Bianco, *Nature Nanotechnol.* **2007**, *2*, 108–113.
96. C. F. Lopez, S. O. Nielsen, P. B. Moore, M. L. Klein, *Proc. Natl. Acad. Sci. U.S.A.* **2004**, *101*, 4431–4434.
97. X. Chen, A. Kis, A. Zettl, C. R. Bertozzi, *Proc. Natl. Acad. Sci. U.S.A.* **2007**, *104*, 8218–8222.
98. P. Cherukuri, S. M. Bachilo, S. H. Litovsky, R. B. Weisman, *J. Am. Chem. Soc.* **2004**, *126*, 15638–15639.
99. L. Lacerda, G. Pastorin, D. Gathercole, J. Buddle, M. Prato, A. Bianco, K. Kostarelos, *Adv. Mater.* **2007**, *19*, 1480–1484.
100. R. Singh, D. Pantarotto, L. Lacerda, G. Pastorin, C. Klumpp, M. Prato, A. Bianco, K. Kostarelos, *Proc. Natl. Acad. Sci. U.S.A.* **2006**, *103*, 3357–3362.
101. T. K. Leeuw, R. M. Reith, R. A. Simonette, M. E. Harden, P. Cherukuri, D. A. Tsyboulski, K. M. Beckingham, R. B. Weisman, *Nano Lett.* **2007**, *7*, 2650–2654.
102. N. Nakayama-Ratchford, S. Bangsaruntip, X. Sun, K. Welsher, H. Dai, *J. Am. Chem. Soc.* **2007**, *129*, 2448–2449.
103. A. E. Porter, M. Gass, K. Muller, J. N. Skepper, P. A. Midgley, M. Welland, *Nature Nanotechnol.* **2007**, *2*, 713–717.
104. L. Lacerda, A. Soundararajan, R. Singh, G. Pastorin, K. T. Al-Jamal, J. Turton, P. Frederik, M. A. Herrero, S. Li, A. Bao, D. Emfietzoglou, S. Mather, W. T. Phillips, M. Prato, A. Bianco, B. Goins, K. Kostarelos, *Adv. Mater.* **2008**, *20*, 225–230.
105. K. Welsher, Z. Liu, D. Daranciang, H. Dai, *Nano Lett.* **2008**, *8*, 586–590.
106. J. E. Riggs, Z. Guo, D. L. Carroll, Y.-P. Sun, *J. Am. Chem. Soc.* **2000**, *122*, 5879–5880.
107. Y. Lin, B. Zhou, R. B. Martin, K. B. Henbest, B. A. Harruff, J. E. Riggs, Z. Guo, L. F. Allard, Y.-P. Sun, *J. Phys. Chem. B* **2005**, *109*, 14779–14782.
108. D. H. Robertson, D. W. Brenner, J. W. Mintmire, *Phys. Rev. B* **1992**, *45*, 12592–12595.
109. (a) M. F. Yu, B. S. Files, S. Arepalli, R. S. Ruoff, *Phys. Rev. Lett.* **2000**, *84*, 5552–5555; (b) B. S. Harrison, A. Atala, *Biomaterials* **2007**, *28*, 344–353.
110. S. Mwenifumbo, M. S. Shafferc, M. M. Stevens, *J. Mater. Chem.* **2007**, *17*, 1894–1902.
111. E. Jan, N. A. Kotov, *Nano Lett.* **2007**, *7*, 1123–1128.
112. A. Abarrategi, M. C. Gutiérrez, C. Moreno-Vicente, M. J. Hortigüela, V. Ramos, J. L. López-Lacomba, M. L. Ferrer, F. del Monte, *Biomaterials* **2008**, *29*, 94–102.
113. L. Lacerda, A. Bianco, M. Prato, K. Kostarelos, *Adv. Drug Deliv. Rev.* **2006**, *58*, 1460–1470.
114. D. Cui, F. Tian, C. S. Ozkan, M. Wang, H. Gao, *Toxicol. Lett.* **2005**, *155*, 73–85.
115. (a) K. Donaldson, R. Aitken, L. Tran, V. Stone, R. Duffin, G. Forrest, A. Alexander, *Toxicol. Sci.* **2006**, *92*, 5–22; (b) L. Ding, J. Stilwell, T. Zhang, O. Elboudwarej, H. Jiang, J. P. Selegue, P. A. Kooke, J. W. Gray, F. F. Chen, *Nano Lett.* **2005**, *5*, 2448–2464.
116. H. Isobe, T. Tanaka, R. Maeda, E. Noiri, N. Solin, M. Yudasaka, S. Iijima, E. Nakamura, *Angew. Chem. Int. Ed.* **2006**, *45*, 6676–6680.
117. H. Dumortier, S. Lacotte, G. Pastorin, R. Marega, W. Wu, D. Bonifazi, J.-P. Briand, M. Prato, S. Muller, A. Bianco, *Nano Lett.* **2006**, *6*, 1522–1528.
118. C. M. Sayes, F. Liang, J. L. Hudson, J. Mendez, W. Guo, J. M. Beach, V. C. Moore, C. D. Doyle, J. L. West, W. E. Billups, K. D. Ausman, V. L. Colvin, *Toxicol. Lett.* **2006**, *161*, 135–142.
119. H. C. Fischer, W. CW. Chan, *Curr. Opin. Biotech.* **2007**, *18*, 565–571.
120. J. Muller, F. Huaux, D. Lison, *Carbon* **2006**, *44*, 1048–1056.
121. N. A. Monteiro-Riviere, A. O. Inman, *Carbon* **2006**, *44*, 1070–1078.
122. C. A. Poland, R. G. Duffin, I. Kinloch, A. Maylnard, W. A. H. Wallace, A. Seaton, V. Stone, S. Brown, W. Macnee, K. Donaldson, *Nature Nanotechnol.* **2008**, *3*, 423–428.

123. M. L. Schipper, N. Nakayama-Ratchford, C. R. Davis, N. W. S. Kam, P. Chu, Z. Liu, X. Sun, H. Dai, S. S. Gambhir, *Nature Nanotechnol.* **2008**, *3*, 216–221.
124. X. Deng, S. Yang, H. Nie, H. Wang, Y. Liu, *Nanotechnology* **2008**, *19*, 075101.
125. J. Guo, X. Zhang, Q. Li, W. Li, *Nuclear Med. Biol.* **2007**, *34*, 579–583.
126. X. Deng, G. Jia, H. Wang, H. Sun, X. Wang, S. Yang, T. Wang, Y. Liu, *Carbon* **2007**, *45*, 1419–1424.
127. P. Cherukuri, C. J. Gannon, T. K. Leeuw, H. K. Schmidt, R. E. Smalley, S. A. Curley, R. B. Weisman, *Proc. Natl. Acad. Sci. U.S.A.* **2006**, *103*, 18882–18886.

Chapter 7

Metal–Organic Frameworks (MOFs) and Coordination Polymers

SHIN-ICHIRO NORO AND SUSUMU KITAGAWA

The Supramolecular Chemistry of Organic–Inorganic Hybrid Materials. Edited by Knut Rurack and Ramón Martínez-Máñez

7.1 INTRODUCTION

Over the past decade, coordination compounds with infinite structures have been studied intensively. In particular, compounds with frameworks constructed from metal ions and organic bridging ligands form a family of polymers that are known as coordination polymers or metal–organic frameworks (in this chapter, we will use the term coordination polymers). The expression coordination polymer first appeared early in the 1960s, and by the mid 1960s, this family of compounds had already been reviewed.[1] In an effort to open up this area of chemistry fully, versatile synthetic approaches involving assemblies of molecular building blocks (metal ions, bridging ligands, counter anions, and guest molecules) in solution have been pursued to achieve the formation of target structures. The key to success is the design of molecular building blocks that direct the desired architectural, chemical, and physical properties to result in solid-state materials. Coordination polymers are highly crystalline materials, and chemists need to obtain good quality single crystals to understand their structure and properties. However, in general, such polymers are insoluble in solvents, with maintenance of their original framework. Therefore, the most common crystallization method, recrystallization, is not available. In a surprisingly short period, the structural chemistry of coordination polymers has reached a mature level because of the application of useful crystallization methods, such as slow diffusion and hydrothermal and solvothermal methods. A survey of recent research work shows an extraordinary increase in the number of published articles. The next challenge in this area is to develop coordination polymers with chemical and physical functionalities by modifying the coordination framework. Unique functionalities have already begun to be found in coordination polymers. In particular, the porous properties of coordination polymers have been studied intensively.

Porous compounds have attracted the attention of chemists, physicists, and materials scientists because of scientific interest in the creation of nanometer-sized spaces and the discovery of novel phenomena in them through their characterization and processing, as well as commercial interest in their application in separation, storage, and heterogeneous catalysis. Until the mid-1990s, there were two types of porous materials: inorganic and carbon-based materials. Zeolites, which are among the best-known porous inorganic solids, are three-dimensional crystalline, hydrated alkaline or alkaline earth aluminosilicates with the general formula $(M^{I}M^{II}_{1/2})_m(Al_mSi_nO_{2(m+n)}) \cdot xH_2O$ $(n \geq m)$ (M = cations).[2,3] Their frameworks, built from corner-sharing TO_4 tetrahedra (T = Al, Si), delimit interconnected tunnels or cages into which H_2O molecules and cations are inserted. The porosity is provided by elimination of the H_2O molecules, with the framework usually remaining unaffected. The cavities, usually quantified structurally by the number of polyhedra surrounding the pore, are used for molecular sieve requirements in gas separation and catalytic operations. Activated carbon has a high porosity and high specific surface area, but has a disordered structure, the essential feature of which is a twisted network of defective hexagonal carbon layers, cross-linked by aliphatic bridging groups.[4] The width of the layer planes varies, but is typically about 5 nm. Simple functional groups and heteroelements are incorporated into the network, and are bound to the periphery of the carbon layer planes.

Recently, porous coordination polymers (PCPs) have been developed, which are beyond the scope of the former two classes of porous materials.[5–25] Werner complexes, $[M(NCS)_2(4\text{-methylpyridine})_4]$ ($M = Ni^{II}$ or Co^{II}),[26] Prussian blue compounds,[27–29] and Hofmann clathrates and their derivatives, which are built of CN linkages between square planar or tetrahedral tetracyanometallate(II) units and octahedral metal(II) units coordinated by complementary ligands,[29–31] are known as porous materials that can incorporate a variety of small molecules. Their main frameworks utilize small cyanide anions as bridging ligands. Since the early 1990s, research on the structure of porous coordination polymers with longer bridging ligands than the cyanide ligand has increased greatly, and some examples with functional pores (anion-exchange, catalytic properties, and adsorption) have begun to appear.[32–36] A survey of the research published in recent years shows an extraordinary increase in the number of articles. The following three features would provide pivotal advantages: (1) designability, (2) regularity, and (3) flexibility.

1. High designability: the key to success in obtaining highly functional materials is the design of the desired architectural, chemical, and physical properties of the resulting solid-state compounds. One can take advantage of this in the design of PCPs because the formation reactions mostly occur under mild conditions, and the choice of a certain combination of discrete building units leads to a high probability of a desired extended network.
2. Regularity: regular pore distribution in a porous solid is important for adsorption because when the size of a pore is comparable to that of a guest molecule, then the periodic potential from the pore wall can influence the form and orientation of the adsorbed guest molecule. A regular pore distribution can be readily realized for coordination polymers as well as inorganic porous materials. The pores of coordination polymers have a regular periodical structure because of their crystalline form, which affords a periodic potential on their channel surface. The structural relationships between adsorbed guest molecules and host frameworks [e.g., (a) position of the guest molecules in the pores, (b) the assembled structure of the guest molecules in the pores, and (c) the perturbation that the guest molecules receive from the pore walls] are key subjects for understanding the adsorption behavior and the physical or chemical properties of adsorbed guest molecules in the pores. In addition, molecules confined in a uniform restricted nanospace form molecular assemblies and provide unique properties that are not realized in the bulk state.
3. Flexibility: recent reports on the dynamic properties of PCPs show that they are much more flexible than generally believed. Dynamic pores can form a type of "soft" framework with bistability, whose two states oscillate back and forth between one of two counterparts. A system can exist in either of two states for different parameters of an external field. The structural rearrangement

of molecules proceeds from the "closed" phase to the "open" phase in response to a guest molecule. Some PCPs have such flexibility, and they can be developed as a special class of materials, such as highly selective gas sensors or gas separation compounds, that cannot be obtained using a rigid porous material. Dynamic structural transformations based on these flexible frameworks are among the most interesting phenomena, and are presumably a characteristic of coordination polymers, the so-called third generation PCPs.[37]

7.2 BUILDING BLOCKS

Basic building blocks to create PCPs are metal ions as connectors, bridging ligands as linkers, and guest molecules as templates. In this section, we introduce well-designed building blocks, as shown in Figure 7.1.

7.2.1 Secondary Building Units

Polynuclear clusters constructed from two or more metal ions and carboxylate units [so-called secondary building units (SBUs)] can have special coordination numbers and geometries. SBUs are conceptual units that are not employed in the synthesis as a distinct molecular building block. However, specific SBUs can be generated *in situ* under the correct reaction conditions.[6] Because the metal ions are locked into their positions by the carboxylate units, SBUs are sufficiently rigid to produce extended frameworks with a high structural stability. In SBUs with terminal guests, the reactivity of the metal site can be studied through the removal of these guests, which creates coordination sites.

7.2.2 Metalloligands

Coordinatively unsaturated metal centers (UMCs) can provide functionalities as chromic sensors and flexible frameworks, because the ligation and release of the UMCs' guests often influence the coordination geometry or ligand field. Moreover, it is well known that a variety of metal ions can act as active centers in catalyzed reactions. A combination of UMC characteristics and traditional porous properties (shape and size selectivity) can be used to create desired, highly efficient, and tailor-made functional materials. Recently, a new synthetic method has been proposed for PCPs with UMCs using metalloligands, that is, two-step self-assembly.[38] First, a metalloligand is synthesized that acts as a framework linker and a source of coordinatively unsaturated metal centers. Second, the metalloligand is added to another metal ion, which acts as a nodal unit in a framework. Consequently, two types of metal center coexist in the framework, and a larger space around the metal ion in the channel wall can be obtained, which is significant for an attack by a guest molecule. The key point of this method is a partial separation of the metal functionalities: the framework node and the coordinatively unsaturated metal centers.

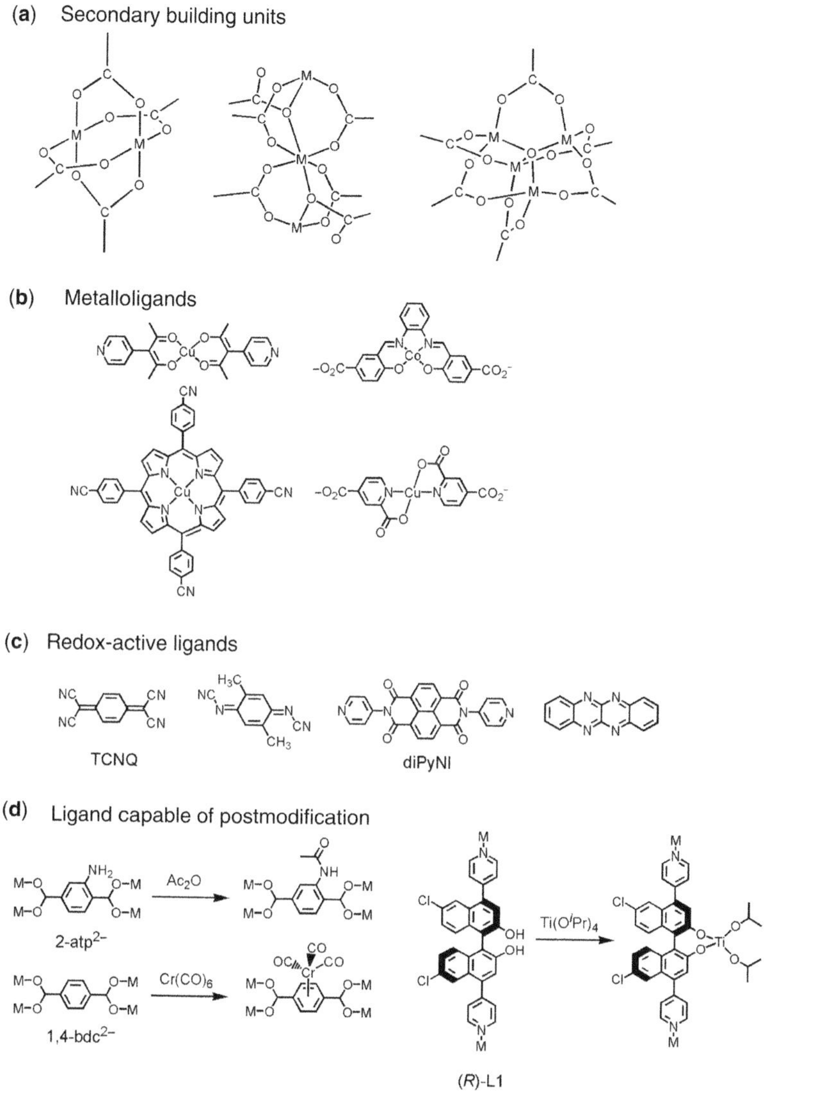

Figure 7.1 A library of building blocks.

7.2.3 Redox-Active Ligands

Charge transfer (CT) interactions play a significant role in developing novel photoactive materials, although only a few examples of this type of interaction in PCPs have been seen to date.[39–42] One available method to engender CT interactions between a host framework and a guest molecule is to use redox-active ligands, because such ligands forming the pore walls of PCPs have enough space to attract a guest molecule at a nearby site. For example, $\{[Zn(TCNQ)(4,4'\text{-bpy})]\cdot 6MeOH\}_n$

($TCNQ^{2-}$ = 7,7,8,8-tetracyano-*p*-quinodimethane dianion, 4,4′-bpy = 4,4′-bipyridine) has a three-dimensional pillared layer framework with redox-active ligands, $TCNQ^{2-}$.[41] The Zn^{II} ions are linked by $TCNQ^{2-}$ dianions to give a two-dimensional corrugated layer. The 4,4′-bpy ligands act as pillars, with the Zn^{II} ions in the adjacent layers linked to form a three-dimensional pillared layer structure. This compound possesses two-dimensional channels with cross sections of 3.4 × 5.9 $Å^2$ between the layers. As a result, a dense array of strong donor sites ($TCNQ^{2-}$) is formed on the pore surface of the framework, providing a highly electron-rich surface for the guest molecules. When $\{[Zn(TCNQ)(4,4'\text{-bpy})] \cdot 6MeOH\}_n$ is immersed in benzene, the guest MeOH can be exchanged with benzene to form $\{[Zn(TCNQ)(4,4'\text{-bpy})] \cdot 2\text{benzene}\}_n$ within a period of 10 s, producing a compound that is red in color versus the yellow color of the parent compound. The crystal structure of $\{[Zn(TCNQ)(4,4'\text{-bpy})] \cdot 2\text{benzene}\}_n$ indicates that the benzene guests interact with the most negatively charged carbon atom of the $TCNQ^{2-}$ dianions via C–H···C-type hydrogen bonds, representing the existence of an electrostatic and/or CT interaction between the $TCNQ^{2-}$ dianions and the benzene guests. For other guests (e.g., toluene, ethylbenzene, anisole, benzonitrile, and nitrobenzene), changes in color different from that of $\{[Zn(TCNQ)(4,4'\text{-bpy})] \cdot 6MeOH\}_n$, similar to that observed for benzene, were seen, except for anisole and nitrobenzene, where the occlusion of anisole and nitrobenzene forms light yellow and dark brown crystals, respectively. Such a difference in color is associated with the electron-accepting characteristics of the guest molecules.

7.2.4 Ligands Capable of Postmodification

Postmodification of the pore surface without destruction of the framework structure after formation of PCPs is the next challenging study for obtaining comfortable space for specific guests.[43–53] Recently, novel strategies have been developed to introduce a catalytic metal center by postmodification of the porous framework.[43–48] A chiral bridging ligand, L1, with orthogonal functional groups, as shown in Figure 7.1d, bridges the Cd^{II} centers of the one-dimensional $[Cd(\mu\text{-}Cl)_2]_n$ units to form an extended porous network in $\{[Cd_3Cl_6(L1)_8] \cdot 4DMF \cdot 6MeOH \cdot 3H_2O\}_n$ with very large chiral channels with an area of approximately 1.6 × 1.8 nm^2.[43] The channel walls of this porous compound have free –OH sites acting as secondary functional groups that can further interact with other metal ions. Indeed, $Ti(O^iPr)_4$ can react with the chiral dihydroxy groups in the L1 parts of this Cd^{II} network, producing an active heterogeneous catalyst for the addition of $ZnEt_2$ to aromatic aldehydes to produce chiral secondary alcohols. The performance of this heterogeneous catalyst rivals that of the homogeneous analog under similar conditions. The three-dimensional PCP $[Zn_4O(1,4\text{-bdc})_3]_n$ ($1,4\text{-bdc}^{2-}$ = 1,4-benzenedicarboxylate) can incorporate photoactive $Cr(CO)_3$ groups onto the benzene rings of $1,4\text{-bdc}^{2-}$ ligands in an η^6 fashion after construction of the framework.[44] The PCP obtained, $[Zn_4O((\eta^6\text{-}1,4\text{-bdc})Cr(CO)_3)_3]_n$, shows substitution of a single CO ligand per metal by either N_2 or H_2 under gentle photolysis conditions, forming $[Zn_4O((\eta^6\text{-}1,4\text{-bdc})Cr(CO)_2(X))_3]_n$

($X = N_2$ or H_2). These PCPs display a remarkable stability compared with their analogs $[(C_6H_5Me)Cr(CO)_2(X)]$ ($X = N_2$ or H_2) because the Cr^0 centers are pinned to the walls of the framework, which prevents reactions involving two or more Cr^0 atoms and enables the irreversible evacuation of CO.

7.2.5 Cartridges of Functional Groups

Tweaking the porous properties simply by replacing a molecular "cartridge" with a different cartridge has been developed in a new class of PCPs.[54,55] The three-dimensional PCP, $\{[Zn_3I_6(tpt)_2(triphenylene)] \cdot x(nitrobenzene) \cdot y(MeOH)\}_n$ (tpt = 2,4,6-tris(4-pyridyl)triazine), is composed of two interpenetrating networks, in which the pores are surrounded by aromatic bricks.[56] These bricks consist of alternatively layered tpt and triphenylene. The former forms an infinite three-dimensional network via coordination to ZnI_2, whereas the latter is involved in the three-dimensional network without forming any covalent or coordination bonds with other components. Therefore, the noncovalently intercalated triphenylene can be replaced with functionalized triphenylenes (Fig. 7.1e) without causing any change in the porous structure, and intercalated triphenylenes are regarded as being the cartridges of the functional groups (Fig. 7.2). For example, one of the triphenylenes directs a phenol group towards the interior of the pores. The resulting PCP selectively adsorbs alcohol guests such as propan-2-ol.

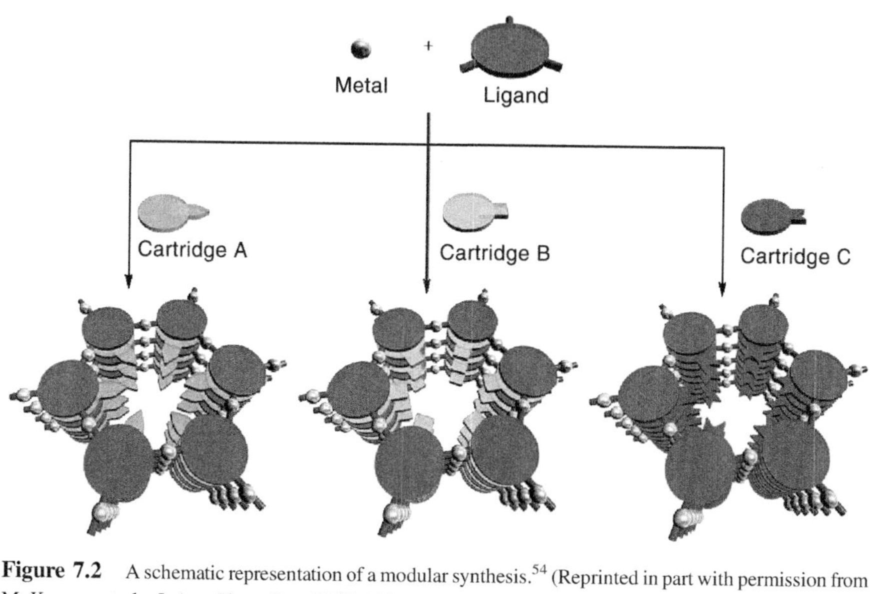

Figure 7.2 A schematic representation of a modular synthesis.[54] (Reprinted in part with permission from M. Kawano et al., *J. Am. Chem. Soc.* **2007**, *129*, 15418–15419. Copyright 2007 American Chemical Society.)

7.3 SYNTHESIS AND CHARACTERIZATION METHODS

7.3.1 Synthetic Methods

The self-assembly process for coordination polymers is a useful method for the following reasons: (1) a wide variety of frameworks can be realized just from simple building blocks of metal ions, organic bridging ligands, and counter anions; (2) easy and rational modification of organic bridging ligands is possible; (3) several interactions such as coordination bonds, hydrogen bonds, aromatic interactions, M–M bonds, and van der Waals interactions are available; and (4) the reaction can be controlled using the temperature, pH, solvent, and pressure. For coordination polymers, recrystallization cannot be used to obtain high purity compounds or large single crystals, because these compounds are insoluble in most solvents. Diffusion, hydrothermal, and solvothermal methods are common techniques used to produce pure coordination polymers. A new synthetic approach has also been developed (see below).

7.3.1.1 *In Situ Synthesis of Ligands*

The *in situ* slow hydrolysis of precursor ligands can result in less soluble phases that are not accessible directly from their corresponding hydrolyzed ligands by virtue of the presence of a large excess of metal ions under hydro(solvo)thermal conditions.[57] For example, Cd^{II} and Zn^{II} coordination polymers that show second-order nonlinear optical properties can be obtained by reaction of the metal salts with cyanopyridine, and pyridinecarboxaldehyde and its derivatives, whose cyano, carboxaldehyde, and ester substituent groups slowly hydrolyze to form the corresponding carboxylic acid.[58] The reaction of NaN_3, 3-cyanopyridine, and $ZnCl_2$ or $CdCl_2$ under hydrothermal conditions yields $[Zn(OH)(3\text{-ptz})]_n$ or $[CdN_3(3\text{-ptz})]_n$ (3-ptz$^-$ = 5-(3-pyridyl)tetrazolate).[59] For each reaction, the IR spectrum indicates the absence of the cyano group, which is consistent with a [2 + 3] cycloaddition between the cyano group and the azide anion.

7.3.1.2 *Microwave Methods*

Microwave irradiation has been developed as a novel heating technique for the hydro(solvo)thermal synthesis of organic or inorganic solid-state materials. In particular, this method has been shown to provide an efficient way to synthesize purely inorganic porous materials with short crystallization times, narrow particle size distributions, facile morphology control, phase selectivity, and efficient evaluation of process parameters. Even under ambient conditions, microwave heating of a mixture of starting materials results in an efficient reaction to provide interesting coordination polymers. In this context, the first example discussed is $[Cu_2(\text{oxalate})_2(\text{pyz})_3]_n$ (pyz = pyrazine), which has been synthesized by the author.[60] The first successful application of microwave synthesis for porous coordination polymers was achieved in the case of $\{[Cr_3FO(1,3,5\text{-btc})_2(H_2O)_3] \cdot xH_2O\}_n$ (1,3,5-btc^{3-} = 1,3,5-benzenetricarboxylate, MIL-100), which has a hierarchical pore system (microporous = 5 to 9 Å and mesoporous = 25 to 30 Å) with a very high specific surface area (~3100 $m^2 \cdot g^{-1}$).[61,62] The X-ray diffraction (XRD) pattern of a sample obtained after microwave irradiation (reaction temperature = 493 K) for a period of 4 h agrees well with the pattern of MIL-100 synthesized for a period of 4 d at 493 K using conventional hydrothermal heating. The crystal yield from microwave synthesis is 44%, which is comparable with the result of 45% from the conventional synthesis after 4 d. The physical and textural properties are also very similar to those of the crystals synthesized using conventional hydrothermal heating. Other porous coordination polymers have been synthesized using microwave heating to shorten the reaction time,[63,64] and to discover new materials not yet obtained using conventional hydrothermal synthesis.[65]

7.3.1.3 Solvent-Free Methods

Solvent-free synthesis is of interest for several reasons. For example, it can give insight into the roles of solvent molecules in templating porous structures, gives access to large-scale green production processes, and even provides more convenient laboratory-scale preparative methods. The three-dimensional coordination polymer $[Cu(isonicotinate)_2]_n$ is the first example that higher-dimensional connectivity, which is generally required to support permanent open porosity, can be produced by grinding.[66] A ball mill was used to grind together $Cu(O_2CCH_3)_2 \cdot H_2O$ and isonicotinic acid. Typical conditions involved a 20 mL steel vessel containing a steel ball bearing and approximately 0.5 g of reactants, at an oscillation rate of 25 Hz maintained for 10 min. The compound obtained has the characteristic odor of acetic acid, released as a by-product. The fully desolvated compound is obtained by heating the mixture to 473 K for a period of 3 h, and shows a similar XRD pattern to that simulated from the single-crystal data for the empty framework host that was synthesized in the solvothermal reaction between 4-cyanopyridine and $CuCl_2$ in a mixture of water and ethanol after heating to 423 K under autogenous pressure for a period of 48 h.[67,68] This solvent-free method is quick, and gives a quantitative yield without the need for solvents or external heating. Clearly, it can present higher efficiency in terms of materials, energy, and time compared with solvothermal methods.

7.3.1.4 Addition of Organic Polymers

Precise control of crystal growth of coordination polymers is currently a key challenge in their development as intelligent building blocks in optical, electronic, and catalytic applications.[69–72] There have been several advanced studies in which template molecules, such as surfactants, organic polymers, vesicles, Langmuir–Blodgett films, and self-assembled organic monolayers serve as fields for nucleation and/or growth of the crystallization of coordination polymers, while controlling the size, morphology, and orientation of the resulting crystals. However, to date, little attention has been paid to the crystal growth control of coordination polymers with organic ligands.

Functionalized organic polymers have been used widely as substances for the preparation of metal-based nanoparticles. The selection of appropriate protective polymers leads to various colloid morphologies by controlling the growth and handling of the agglomeration process, and also leads to attractive new properties based on organic–inorganic hybridizations on the nanometer scale. Recently, the control of crystal growth of the porous coordination polymer $[Cu_2(pzdc)_2(pyz)]_n$ (CPL-1, CPL = coordination pillared layer and $pzdc^{2-}$ = pyrazine-2,3-dicarboxylate) has been achieved by addition of the organic polymer poly(vinylsulfonic acid, sodium salt) (PVSA), which has the ability to complex and stabilize Cu^{II} by electrostatic interaction.[73] The addition of PVSA to the reaction mixture significantly affects the nucleation process, which determines the crystal size of this compound in the range of 1 to 100 μm. PVSA controls the crystal size and shape but also controls the preferential orientation of the plate crystals, which results in alignment of the channel direction in the bulk powder state. Nitrogen adsorption measurements show typical characteristic micropore adsorption isotherms that are independent of the crystal size. However, the adsorption speed of the samples shows a novel size-dependent feature because of changes in the diffusion length of N_2 in the one-dimensional channels.

Heterogeneous nucleation from insoluble functionalized organic polymers is also a powerful, efficient technique for discovering and selecting among the different polymorphic forms of coordination polymers. It has been proposed that the polymer serves to direct the formation of one phase over others at the time of nucleation by selectively stabilizing one mode of aggregation under conditions (e.g., solvent, temperature, and extent of supersaturation) that are otherwise identical. For example, the three-dimensional coordination polymer $\{[Zn_4O(2\text{-atp})_3] \cdot xG\}_n$ (2-atp^{2-} = 2-aminoterephthalate, IRMOF-3, isoreticular metal–organic framework No. 3) with a primitive-cubic framework structure[74] is formed in the absence of insoluble polymers. However, a new phase, PNMOF-3 (PNMOF = polymer-nucleated metal–organic framework) is additionally observed in crystallizations containing cross-linked MAA/DVB copolymers (MAA = methacrylic acid and DVB = divinylbenzene), but is not observed when 4VP/DVB copolymers (4VP = 4-vinylpyridine) or polydivinylbenzene are used.[75] The crystal structure of PNMOF-3 shows a framework with a composition of $[Zn_4(2\text{-atp})_3(NO_3)_2(H_2O)_2]_n$ with hexagonal grid layers. This result demonstrates the importance of acid functionality in the selection of this framework.

7.3.2 Characterization Methods

In addition to routinely used methods, such as elemental analysis, IR and UV-vis-NIR spectra, thermogravimetry-differential thermal analysis (TG-DTA), single-crystal X-ray diffraction, and gas adsorption, there are some important characterization methods for coordination polymers.

7.3.2.1 Synchrotron Powder XRD Measurements

The atomic-level crystal structures in coordination polymers, which are required information to fully understand the physical properties, are determined using XRD measurements. Many coordination polymers form good quality single crystals that are suitable for single-crystal XRD analysis because of their high degree of regularity. Recent advances in collimation techniques and detector technology have enabled the fast collection of highly redundant data which allow the measurement of unstable or very small single crystals in a facile way. In the case of a powder (i.e., an aggregate of microcrystals), which is unfavorable for single-crystal XRD measurements, the crystal structure can be determined using synchrotron powder XRD measurements. There are advantages to using a synchrotron light source: (1) the synchrotron radiation light source provides a much higher resolution and higher counting statistics diffraction data compared with laboratory sources; and (2) using a high brilliance light source, a diffraction experiment can be performed using a small mass of sample over very short data collection times. Such advantages enable the collection of good quality structural data that compares favorably with that obtained using single-crystal XRD analysis. The powder diffraction method has the advantage of avoiding difficulties related to the collapse of a single crystal from the large volume changes during guest adsorption and desorption.[76]

The determination of a one-dimensional ladder molecular array structure of O_2 molecules was first performed using *in situ* synchrotron powder XRD measurements on CPL-1 accommodating O_2 molecules.[77,78] The intermolecular distance of the adsorbed O_2 molecules (3.28(4) Å) is close to the nearest distance in the solid α-O_2 phase, whose close-packed structure appears below 24 K. This result indicates that O_2 molecules adsorbed in the nanochannels form van der Waals dimers, $(O_2)_2$. The X-ray structural analysis shows that O_2 molecules are in the solid state rather than the liquid state, even at 130 K and 80 kPa, which are much higher than the boiling point of bulk O_2 at atmospheric pressure, 54.4 K. This result is ascribed to the strong confinement effect of CPL-1.

7.3.2.2 Inelastic Neutron Scattering

Inelastic neutron scattering studies have been used to investigate spectroscopically the H_2–framework interaction because of the high potential of H_2 as a clean energy carrier. This technique provides a sensitive probe of the adsorptive sites for molecular H_2 because it monitors the hindered rotational transitions of bound molecules. Promising observations were made for $[Zn_4O(1,4\text{-bdc})_3]_n$ loaded with varying amounts of H_2.[79] These spectra show that at least two distinct adsorption sites exist on the framework, which are assigned to the inorganic and organic components. Neutrons are ideal scattering probes for such experiments, because X-rays are insensitive to the low electron density of hydrogen atoms.

7.3.2.3 Computational Simulation

Based on the accumulated crystallographic and adsorption data of PCPs, computational modeling studies of small-molecule adsorption have been performed (an approach that is common in carbon and inorganic materials chemistry).[80–82] PCPs have an advantage for simulations, in that their well-characterized regular structure eliminates the need to make assumptions about the host structures. Grand canonical Monte Carlo simulations of H_2, N_2, Ar, CH_4, CO_2, and *n*-butane in PCPs have been reported, and these show good agreement with measured isotherms.[83–90] These calculations were also used for proposing new PCPs with improved performance.[91–95] For example, the interaction energy between H_2 and lithium alkoxide groups in Li-modified PCPs is larger than in unmodified PCPs.[91,92,94]

7.3.2.4 Solid-State NMR Studies of Molecular Motion

NMR is an excellent method for studying the dynamics of molecules in solids. In general, all nuclear spin interactions are anisotropic, that is, they depend on the molecular orientation within the applied magnetic field of the NMR experiment. In-depth information on the dynamic behavior of the guests and/or framework in PCPs is essential for the evaluation and understanding of micropore filling and design of new functions. In particular, ^2H has been used extensively in molecular motion studies. This spin-1 nucleus gives rise to quadrupole coupling constants in the range 160 to 220 kHz in most organic solids. Therefore, its static quadrupole-broadened powder patterns are sensitive to motions with frequencies in the range 10^4 to 10^7 Hz, which makes them ideal for studying a large range of motions in solids. Molecular motions of MeOH guests adsorbed in the one-dimensional channels of CPL-1 have been studied.[96] The high degree of crystallinity of CPL-1 allows the MeOH guests to undergo only two motions, rotation and wobbling, whose aspects have not been reported in zeolites previously. The dynamic behavior of PCPs with rotational groups has also been characterized.[97,98]

7.4 DESIGN OF FRAMEWORKS

7.4.1 Frameworks Suitable for Highly Porous, Robust, Three-Dimensional PCPs

Three-dimensional PCPs tend to produce highly porous, robust frameworks because coordination bonds are distributed in a three-dimensional manner in their polymers. Three types of frameworks have been designed, systematically synthesized, and characterized to obtain three-dimensional PCPs: (1) α-Po-type frameworks, (2) zeolitic frameworks, and (3) pillared layer frameworks.

The α-Po-type framework, whose appearance is like a jungle gym, is a candidate for the construction of exceptionally rigid and highly porous structures with three-dimensional intersecting channels. A systematic design of the pore size and functionality has been achieved in three-dimensional PCPs $[Zn_4O(L)]_n$ (L = a variety of dicarboxylate ligands), as shown in Figure 7.3,[74] in which an oxide-centered Zn_4O tetrahedron is edge-bridged by six carboxylates to give the octahedron-shaped SBU that reticulates into a three-dimensional porous network. Other α-Po-type PCPs

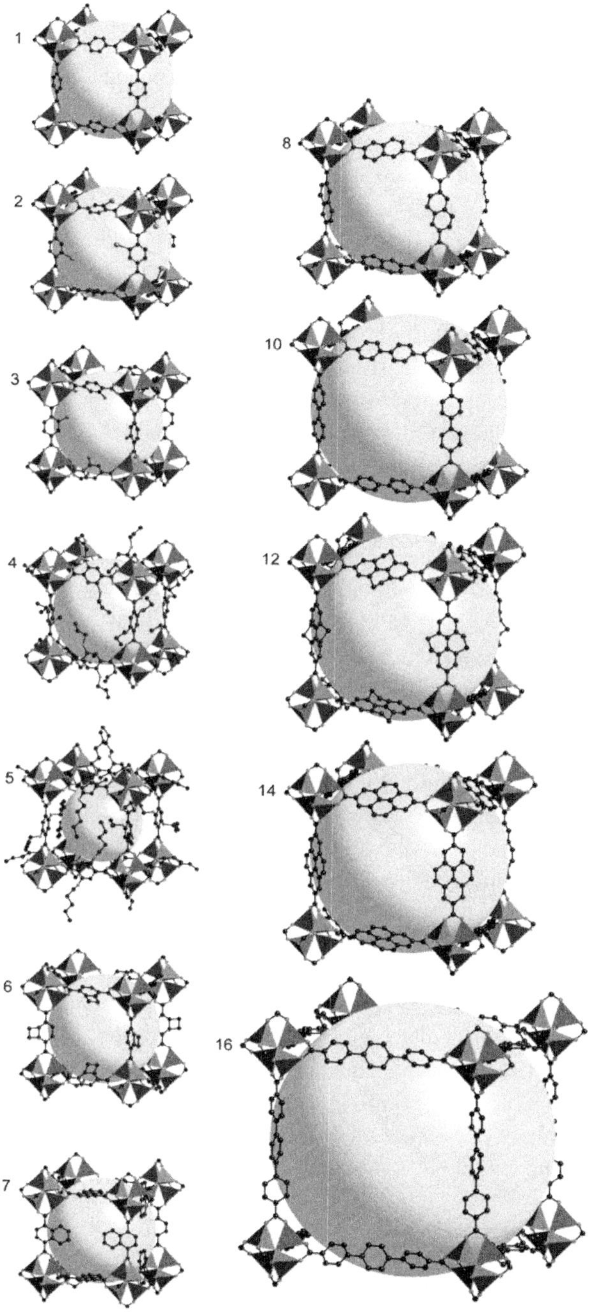

Figure 7.3 Single-crystal X-ray structures of IRMOF-*n* ($n = 1-8$, 10, 12, 14, and 16). All the hydrogen atoms have been omitted, and only one orientation of the disordered atoms is shown for clarity.[74] (Reprinted with permission from M. Eddaoudi et al., *Science* **2002**, *295*, 469–472. Copyright AAAS.)

have been systematically synthesized in $[M_2(L)_2(L')]_n$ (M = Zn^{II} and Cu^{II}, L = a variety of anionic dicarboxylate ligands, and L′ = neutral nitrogen-containing heterocyclic ligands) with two types of organic bridging ligand.[99–102] The prototype compound, $[Zn_2(1,4\text{-bdc})_2(\text{dabco})]_n$ (dabco = 1,4-diazabicyclo[2.2.2]octane), is composed of dinuclear Zn_2 units with a paddle-wheel structure, which are bridged by 1,4-bdc^{2-} dianions to form a distorted two-dimensional square-grid $[Zn_2(1,4\text{-bdc})_2]_n$.[99] The axial sites of the Zn_2 paddle wheels are occupied by dabco molecules, which act as pillars to extend the two-dimensional layers into a three-dimensional structure with a compressed primitive cubic (α-Po) net. The framework has large, three-dimensional interconnected voids. The wide, open channels (7.5 × 7.5 $Å^2$) running along the c-axis are interlinked by smaller windows (approximately 4 Å) oriented along the a- and b-axes, which are still large enough for the passage of small gas molecules. In the case of compounds containing dabco pillar ligands, noninterpenetrated frameworks are always observed, because the dabco ligand is too short to allow the insertion of another framework into the windows forming between the two-dimensional layers. On the other hand, compounds with longer 4,4′-bpy pillar ligands than dabco easily form interpenetrated frameworks.

Zeolite networks are the most significant topologies for porous materials, and are extremely scarce in porous coordination polymers. Three new zeolite-type porous coordination polymers, $\{[Zn(\text{mim})_2]\cdot 2H_2O\}_n$, $\{[Zn(\text{eim})_2]\cdot 2H_2O\}_n$, and $\{[Zn(\text{eim/mim})_2]\cdot 1.25H_2O\}_n$ (mim^- = 2-methylimidazolate, and eim^- = 2-ethylimidazolate) have been synthesized by the liquid-phase diffusion method.[103] Imidazolate (im^-) bridges make an M–im–M angle close to 145°, which is coincident with the Si—O—Si angle, which is the preferred configuration, and is commonly found in many zeolites. Therefore, under the right conditions it is possible to prepare porous zeolite-type frameworks using imidazolate-type ligands. Although all these compounds have infinite three-dimensional frameworks, their topologies are different from each other [zeolitic sodalite (SOD, 4^26^4), analcime (ANA, $4^26^28^2$), and rho (RHO, 4^36^3) topologies, respectively]. By taking advantage of simple modifications of imidazolates at a coordinatively unimportant position, the orientation of adjacent metal coordination polyhedra can be tuned. After the above report, a large number of porous metal–imidazolate frameworks, known as zeolitic imidazolate frameworks (ZIFs), have been systematically synthesized and characterized.[104–106] A study of the gas adsorption and thermal and chemical stability of two prototypical members, $[Zn(\text{mim})_2]_n$ (ZIF-8) and $[Zn(\text{phim})_2]_n$ (phim^- = benzimidazolate, ZIF-11), demonstrates their permanent porosity (Langmuir specific surface area = 1810 m^2/g), high thermal stability (up to 823 K), and remarkable chemical resistance to boiling alkaline water and organic solvents.[104]

On the macroscopic scale, pillared layer structures are frequently found in ancient buildings, such as the Parthenon in Athens. Even on the microscopic scale, the pillared layer motif is very useful for the construction of various porous frameworks, because simple modification of a pillar module can be used to control porous structures and properties. The Cu^{II} coordination polymers, $[Cu_2(\text{pzdc})_2(L)]_n$ (L = pillar ligand), known as coordination pillared layers (CPLs), have pillared layer structures and are a suitable system for the design of porous functionalities.[78,107,108] Two-dimensional layers constructed from Cu^{II} ions and anionic $pzdc^{2-}$ ligands, which do not have voids large enough for molecules to pass through, are linked by pillar ligands [pyz (CPL-1), 4,4′-bpy (CPL-2), bpethe (1,2-bis(4-pyridyl)ethene) (CPL-5), 4-pia (*N*-4-pyridylisonicotinamide) (CPL-6), and dpyg (1,2-di(4-pyridyl)glycol) (CPL-7)], providing three-dimensional porous pillared layer structures with one-dimensional channels. The channel dimensions and surface properties of CPLs can be controlled by modification of the pillar ligands (approximately 4×6 Å^2 for CPL-1, 8×6 Å^2 for CPL-2, 10×6 Å^2 for CPL-5, 10×6 Å^2 for CPL-6, and 6×4 Å^2 for CPL-7). These CPLs show characteristic and unprecedented porous functionalities.

7.4.2 Frameworks Required to Create Flexible PCPs

A guideline for rigid pores in PCPs is that stiff building units need to be linked by strong coordination bonds to form a three-dimensional framework. However, dynamic pores are subject to another constraint to give a type of flexible framework, that is, building units (or motifs) with flexible moieties need to be linked by strong bonds, or stiff building blocks (or motifs) need to be connected by weaker bonds. Another possible option is the combination of flexible building blocks (motifs) and weak linkages. The generation of a host framework that interacts with exchangeable guest species in a switchable fashion has implications for the generation of previously undeveloped advanced materials with applications in areas such as molecular sensing. For the construction of flexible PCPs, it is necessary to consider devices based on weaker interactions, such as van der Waals forces, $\pi-\pi$ stacks, and hydrogen bonds, among the macroscopic integrities. One feasible idea is that a weak interaction device is placed in between the motifs, forming a contrivance for a flexible PCP, which are grouped as shown in Figure 7.4.[109]

The one-dimensional coordination polymer, $[Rh_2(bza)_4(pyz)]_n$ ($bza^- =$ benzoate), is a flexible single-crystal adsorbent compound.[110] The component chains adopt a perfect linear geometry, with the chain skeleton bridged by the pyz ligands in the axial direction of the well-known paddle-wheel-type dinuclear metal benzoate. Each one-dimensional chain aggregates by $\pi-\pi$ stacking of the aromatic rings of the ligands. The original framework has no channel structures able to hold guest molecules, although there are empty cages with dimensions of $9 \times 4 \times 3$ Å^3, with narrow windows of approximately 1 Å at their four corners, which are formed by the benzene rings of bza^- moieties arranged parallel to the chain vector. However, on cooling samples in the presence of guest molecules, the crystal undergoes a phase transition to form a new structure, generating one-dimensional channels. The cages are transformed into channels by a slippage of the chain skeletons along the chain vector.

$\{[Ni(NO_3)_2(bpbp)_2] \cdot 4(o\text{-xylene})\}_n$ (bpbp = 4,4′-bis(4-pyridyl)biphenyl) contains two-dimensional layers that are formed by square grids with dimensions of 20×20 Å^2.[111] The layers pack on top of each other, such that they overlap in one direction and are offset in the other direction, which results in approximate channel dimensions of 10×20 Å^2 (Fig. 7.5). This compound shows guest specificity and

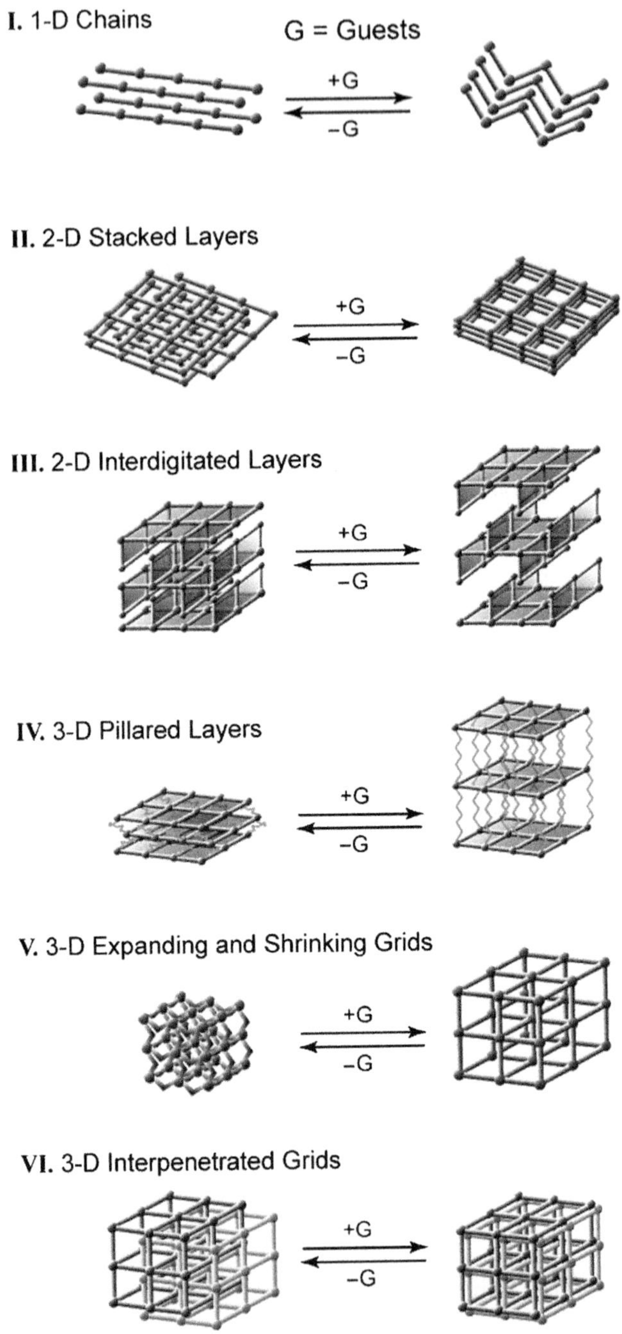

Figure 7.4 Contrivances for flexible PCPs categorized with structural dimensionalities.[109] (Reprinted with permission from S. Kitagawa and K. Uemura, *Chem. Soc. Rev.* **2005**, *34*, 109–119. Copyright the Royal Society of Chemistry.)

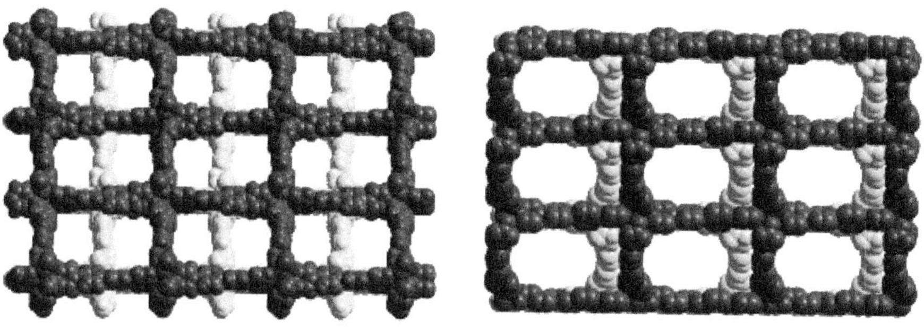

Figure 7.5 Packing of two-dimensional layers in (left) $\{[Ni(NO_3)_2(bpbp)_2] \cdot 4(o\text{-xylene})\}_n$ and (right) $\{[Ni(NO_3)_2(bpbp)_2] \cdot 1.7\text{mesitylene}\}_n$. Guest molecules are omitted for clarity.[112] (Reprinted with permission from K. Biradha et al., *Angew. Chem. Int. Ed.* **2002**, *41*, 3395–3398. Copyright Wiley-VCH Verlag GmbH Co. KGaA.)

exchanges *o*-xylene for mesitylene, but not for *m*-xylene, or 1,3- or 1,2-dimethoxybenzene.[112] Crystals of a mesitylene-inclusion network, $\{[Ni(NO_3)_2(bpbp)_2] \cdot 1.7\text{mesitylene}\}_n$, are prepared by immersion of $\{[Ni(NO_3)_2(bpbp)_2] \cdot 4(o\text{-xylene})\}_n$ in mesitylene. The crystal structure of $\{[Ni(NO_3)_2(bpbp)_2] \cdot 1.7\text{mesitylene}\}_n$ shows considerable sliding of the layers relative to the two-dimensional framework of $\{[Ni(NO_3)_2(bpbp)_2] \cdot 4(o\text{-xylene})\}_n$. The dimensions of the channels in $\{[Ni(NO_3)_2(bpbp)_2] \cdot 1.7\text{mesitylene}\}_n$ (15×20 Å^2) are much larger than those in $\{[Ni(NO_3)_2(bpbp)_2] \cdot 4(o\text{-xylene})\}_n$ as shown in Figure 7.5.

$\{[Cu(BF_4)_2(4,4'\text{-bpy})(H_2O)_2] \cdot 4,4'\text{-bpy}\}_n$ forms noninterpenetrated two-dimensional layers having grids of 7.7×11.6 Å^2.[113] The dehydration of this compound on vacuum heating affords another two-dimensional framework, $[Cu(BF_4)_2(4,4'\text{-bpy})_2]_n$.[114] Pores for the accommodation of guest molecules are not observed in this dehydrated compound because the neighboring layers obstruct the pores. However, the adsorption of N_2, Ar, CH_4, and CO_2 suddenly begins at a definite relative pressure, despite the almost zero adsorption below this threshold pressure, which is known as a gate phenomenon. The gate pressure depends on the adsorbate, which implies the possibility of separating mixed gases in this compound. Structural analysis of the compound adsorbing CO_2 gas shows that the interlayer distance changes from 0.46 nm to 0.68 nm on CO_2 adsorption. Although the details on interlayer interactions have not yet been reported, weak interlayer interactions (e.g., hydrogen bonds, π–π stacking interactions, and van der Waals interactions) may contribute to the dynamic behavior of the porous framework. A two-dimensional framework analogous to $[Cu(BF_4)_2(4,4'\text{-bpy})_2]_n$ has been synthesized.[115] The compound $[Cu(OTf)_2(4,4'\text{-bpy})_2]_n$ ($OTf^- =$ trifluoromethanesulfonate) has two-dimensional square-grid sheets with axially coordinated OTf^- anions. The stacking of these sheets creates free space (17.9%). The N_2 adsorption isotherm at 77 K shows that the adsorption branch is characterized by a definite double step, and a marked hysteresis is present. The mass of N_2 adsorbed at the first uptake is in good agreement with the void volume of $[Cu(OTf)_2(4,4'\text{-bpy})_2]_n$ calculated from the crystal structure. The second uptake may originate from an increase in the pore volume on expansion of the interlayer regions, which was verified by synchrotron XRD measurements. Compared with $[Cu(BF_4)_2(4,4'\text{-bpy})_2]_n$, which shows a single step isotherm, the difference arises from the presence of open micropores because of the pillaring role of the OTf^- anions that fix neighboring sheets at an appropriate position and prevent the clogging of the micropores.

$\{[Cu(dhbc)_2(4,4'\text{-bpy})] \cdot H_2O\}_n$ ($dhbc^- = $ 2,5-dihydroxybenzoate) is a flexible and dynamic PCP based on interdigitation.[116] The Cu^{II} ions are connected by 4,4′-bpy ligands to produce one-dimensional linear chains, and are linked by $dhbc^-$ ligands to give a two-dimensional sheet motif. This motif has interlocking ridges and hollows constructed by the $dhbc^-$ benzene planes in an upright fashion. The distance of 3.443 Å between the planes of the nearest-neighbor $dhbc^-$ ligands indicates the presence of π–π stacking interactions, and the motifs are mutually interdigitated to create one-dimensional channels along the a-axis with a cross-sectional area of 3.6×4.2 Å^2. This compound undergoes a marked crystal transformation triggered by the desorption of included H_2O and guest adsorption. A detailed structural investigation by synchrotron powder XRD measurements shows a change in cell parameters on dehydration. This change corresponds to a cell volume contraction of 27%, and is attributed to a glide motion of the two π stack ring moieties ($dhbc^-$), which results in a decrease in the channel cross-sectional area. Characteristic hysteretic adsorption isotherms that have gate-opening and gate-closing pressures for various gases (e.g., CO_2, CH_4, O_2, and N_2) are observed in the dehydrated compound. This may be attributed to expansion and contraction of the crystal structure triggered by gas adsorption and desorption, respectively.

7.5 FUNCTIONALITIES

7.5.1 Porous Fundamental Properties

7.5.1.1 Gas Adsorbency

Removal of CO_2 from the flue exhausts of power plants, currently a major source of emissions, is commonly accomplished by chilling and pressurizing the exhaust gas or by passing the fumes through a fluidized bed of aqueous amine solution, neither of which is cost effective.[117] Other methods based on chemisorption of CO_2 on oxide surfaces or adsorption within porous silicates, carbon, and membranes have been pursued as means for CO_2 uptake.[118] The room-temperature CO_2 adsorption isotherms of some PCPs ranging in specific surface area from 345 to 4508 $m^2 \cdot g^{-1}$ have been measured to examine a range of structural and porous attributes.[119,120] The voluminous space enclosed by $[Zn_4O(btb)_2]_n$ ($btb^{3-} = $ 1,3,5-benzenetribenzoate, MOF-177) with a specific surface area of 4508 $m^2 \cdot g^{-1}$, results in a CO_2 capacity of 33.5 $mmol \cdot g^{-1}$ at 3.5 MPa, which is far greater than that of any other reported porous material, and exceeds standard materials (a carbon sample of MAXSORB shows an uptake of 25 $mmol \cdot g^{-1}$ at 3.5 MPa),[121] having 150% of their capacity.[119] The three-dimensional PCP $[Cr(OH)(1,4\text{-bdc})]_n$ (MIL-53) shows unprecedented CO_2 adsorption properties.[122] Desolvated MIL-53 exhibits a CO_2 uptake of around

8.5 mmol · g^{-1} at 2.0 MPa and 304 K. The adsorption of CO_2 on hydrated MIL-53 shows very little uptake at pressures up to 1.0 MPa, while a single and distinct uptake of approximately 7.7 mmol · g^{-1} occurs in the 1.2 to 1.8 MPa pressure range (Fig. 7.6). This is not the case for other CO_2 adsorbents, such as zeolites, where H_2O often blocks specific sites. It has been shown previously that MIL-53 is stable at high humidity, and thus, a process in which a significant amount of H_2O is present may not diminish the adsorption of CO_2. Thus, a preadsorber would not be required in a pressure-swing adsorption-type process, and process design could be simplified.

Acetylene is a key starting material for the synthesis of various organic compounds, such as 1,4-butanediol, which is widely used in the preparation of polyurethane and polyester plastics. One issue associated with acetylene is its storage. Unlike other gases, acetylene cannot be stored in a steel cylinder under high pressure, because it becomes highly explosive when compressed above 0.2 MPa at room temperature, even in the absence of oxygen. The current method of storing acetylene involves dissolution of the gas in acetone or DMF placed in a steel cylinder along with an absorbent to prevent polymerization. However, acetylene stored by such a method contains volatile solvents as impurities, and the degree of its purity is not very high. Another issue in acetylene production and use is its separation from other gases, which includes CO_2, a common impurity present in many industrial processes. Therefore, high-density storage of highly pure acetylene at low pressures in a nonvolatile solid matrix with a high selectivity is a key research subject for its production, transportation, and conversion. A marked difference in the acetylene and CO_2 adsorption isotherms is observed in CPL-1.[123,124] The adsorption isotherms of acetylene show a steep rise in the very low pressure region and reach saturation (42 $cm^3 \cdot g^{-1}$), whereas those of CO_2 show a gradual adsorption. The degree of saturation of acetylene corresponds to one molecule per unit pore. The maximum ratio of adsorbed acetylene relative to that of CO_2 is 26.0 at 1.1 kPa and 270 K, indicating that CPL-1 accommodates acetylene in preference to CO_2. The isosteric heat at the fractional filling ratio of 0.2 of acetylene, 42.5 kJ · mol^{-1}, is higher than that of CO_2, 31.9 kJ · mol^{-1}. This result indicates a higher enhancement in the interaction of acetylene with CPL-1 than that of CO_2. Such a strong interaction with acetylene is attributed to the acidic hydrogen atoms at both ends of acetylene ($pK_a = 25$). Such an acidic property of acetylene makes it possible to interact with the oxygen atoms of the carboxylate parts, which was verified directly using synchrotron powder XRD measurements on CPL-1 with adsorbed acetylene. The density of adsorbed acetylene was estimated to be 0.434 g · cm^{-3}. This density is equivalent to that of an extrapolated state of acetylene of 41 MPa at room temperature, and is 200 times larger than the value of the compression limit for the safe use of acetylene at room temperature and 0.20 MPa (0.0021 g · cm^{-3}). Selective and high density acetylene adsorption is observed in other PCPs, $[M(HCOO)_2]_n$ ($M = Mg^{II}$ and Mn^{II}).[125] The amount of adsorbed acetylene at 298 K is above 60 $cm^3 \cdot g^{-1}$ for $[Mg(HCOO)_2]_n$ and 50 $cm^3 \cdot g^{-1}$ for $[Mn(HCOO)_2]_n$, whereas the uptake of other gases, including CO_2, at 298 K decreases substantially, especially at low pressure. The X-ray crystal structure of acetylene adsorbed $[M(HCOO)_2]_n$ shows that acetylene molecules are held in the zigzag channels of the frameworks through only van der

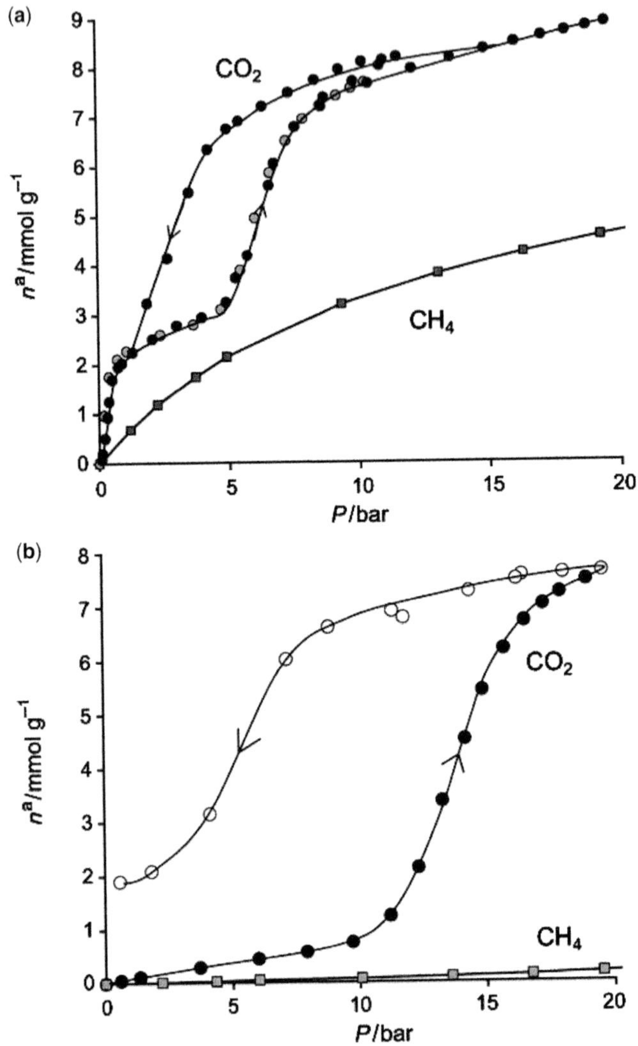

Figure 7.6 Comparison of the adsorption isotherms of carbon dioxide and methane at 304 K on (a) dehydrated and (b) hydrated forms of MIL-53. Grey- and black-filled circles represent two different experiments. n^a = amount adsorbed.[122] (Reprinted with permission from P. L. Llewellyn et al., *Angew. Chem., Int. Ed.* **2006**, *45*, 7751–7754. Copyright Wiley-VCH Verlag GmbH & Co. KGaA.)

Waals interactions, in contrast to the case with CPL-1. Therefore, such discrimination is presumably because of the slightly larger size of acetylene (5.5 Å) relative to CO_2 (5.3 Å), which provides the former with more effective van der Waals interactions with the framework walls.

Methane stands out among various alternative fuels when its profusion and availability are considered. However, the lack of an effective, economic, and safe on-board storage system is one of the major technical barriers preventing methane-driven automobiles from competing with traditional vehicles. To promote the vehicular application of methane, the U.S. Department of Energy (DOE) has set a target for methane storage of $180\ cm^3(STP) \cdot cm^{-3}$ (standard temperature and pressure equivalent volume of methane per volume of the adsorbent material) at 3.5 MPa, near to ambient temperature, with the energy density of adsorbed natural gas being comparable to that of compressed natural gas used in current practice. Although the best material to achieve high methane adsorption per mass or volume of adsorbent would have the following properties: a high specific surface area, a high free volume, a low adsorbent framework density, and a strong energetic interaction with the adsorbed methane, these are not all necessarily compatible. For example, increasing the free volume may create very wide pores, but the heat of adsorption for molecules in the centers of such wide pores may be close to zero, and so this space would be wasted. On the other hand, the heat of adsorption is higher for smaller pores, but the capacity of such pores may be smaller. Therefore, the pores of methane adsorbents must have an appropriate size and surface area. A computational study has indicated that aromatic rings in PCPs can improve the methane binding energy, enhancing both the uptake and the isosteric heat of adsorption.[85] Several PCPs have been screened for methane storage, but none have reached the DOE target.[74,120,126,127] Recently, the three-dimensional PCP, $[Cu_2(adip)]_n$ ($adip^{4-}$ = 5,5′-(9,10-anthracenediyl)di-isophthalate) with cuboctahedral nanoscopic cages, exhibited a methane adsorption capacity that is higher than the DOE target for methane storage.[128] The Langmuir specific surface area and specific pore volume are $2176\ m^2 \cdot g^{-1}$ and $0.87\ cm^3 \cdot g^{-1}$, respectively, estimated from N_2 adsorption measurements at 77 K. High pressure methane adsorption studies show that this PCP has an absolute methane adsorption capacity of $230\ cm^3$ $(STP) \cdot cm^{-3}$ (28% higher than the DOE target of $180\ cm^3\ (STP) \cdot cm^{-3}$ at ambient temperature), and an isosteric heat of adsorption of around $30\ kJ \cdot mol^{-1}$: both record high values among those reported for methane storage materials. The high isosteric heat of adsorption supports the prediction that adding more aromatic rings to the ligand and incorporating nanoscopic cages into the framework can lead to a higher affinity for methane.[85]

The development of light materials that reversibly store and release large amounts of hydrogen is essential for the use of hydrogen as a clean-burning substitute for fossil fuels in mobile systems.[129] The 2010 energy density targets for a hydrogen storage system set by the U.S. DOE are 6.0 wt% and 45 kg H_2 per m^3. The goals for 2015 are even more demanding: 9.0 wt% and 81 kg H_2 per m^3, which approach the expectations of the automotive industry. The initial finding of reversible hydrogen adsorption in $[Zn_4O(1,4\text{-bdc})_3]_n$ (MOF-5)[79,130] was followed by numerous reports of PCPs displaying significant storage capacities.[131–140] One key feature that characterizes most PCPs is the relatively low change in enthalpy of adsorption associated with H_2 uptake (typically 4 to $7\ kJ \cdot mol^{-1}$), such that cryogenic temperatures are required to observe significant H_2 uptake. Clearly, frameworks exhibiting stronger binding interactions are needed to facilitate H_2 adsorption at higher temperatures. Indeed, a binding energy of $15\ kJ \cdot mol^{-1}$ has been predicted to maximize the amount of adsorbed H_2 accessible at 298 K within the pressure range 0.15 to 2.0 MPa.[141] Two main strategies have been pursued for enhancing H_2 binding within PCPs. First, by forming frameworks with very narrow pores, wherein overlapping potentials from two or more pore walls interact with a single H_2 molecule, H_2 binding energies of 9.1 and 9.5 $kJ \cdot mol^{-1}$ for $[Zn_4O(thpdc)_3]_n$ ($thpdc^{2-}$ = tetrahydroxypyrenedicarboxylate, IRMOF-11) and $[Mg_3(ndc)_3]_n$ (ndc^{2-} = 2,6-naphthalenedicarboxylate), respectively, have been observed at low coverage.[142,143] More promisingly, it has been proposed that the presence of coordinatively unsaturated metal centers (UMCs) could increase the affinity for H_2 through strong metal–H_2 interactions.[91,92,94,129,144–150] $[NaNi_3(OH)(sip)_2]_n$ (sip^{3-} = 5-sulfoisophthalate) can be described as $Na_2Ni_6O_{34}$ clusters bridged by sip^{3-} to form a three-dimensional PCP.[149] This PCP has three crystallographically independent Ni^{II} ions, two of which give UMCs. The existence of strong Ni^{II}–H_2 (or D_2) interaction has been verified by the temperature dependence of H_2 adsorption measurements (H_2 binding energy of 10.4 $kJ \cdot mol^{-1}$) and inelastic neutron scattering spectroscopy. The three-dimensional PCP, $\{[Mn(DMF)_6]_3[Mn_{12}Cl_3(btt)_8(H_2O)_{12}]_2 \cdot 42DMF \cdot 11H_2O \cdot 20MeOH\}_n$ (btt^{3-} = 1,3,5-benzenetristetrazolate), which shows a cubic structure in which chloride-centered square-planar $[Mn_4Cl]^{7+}$ units are linked via btt^{3-} ligands to form an anionic, three-dimensional porous framework (Fig. 7.7), remains intact upon desolvation, and exhibits a total H_2 uptake of 6.9 wt% at 77 K and 9.0 MPa, which at 60 g H_2/L provides a storage density 85% of that of liquid hydrogen.[147] This material exhibits a maximum isosteric heat of adsorption of 10.1 $kJ \cdot mol^{-1}$. Neutron powder diffraction data demonstrate that this is directly related to H_2 binding at coordinatively unsaturated Mn^{II} centers within the framework. Ion exchange and metal impregnation techniques have been recently developed to achieve stronger H_2 binding. Exchange of the guest Mn^{II} ions with selected cations (e.g., Li^{I}, Cu^{I}, Fe^{II}, Co^{II}, Ni^{II}, Cu^{II}, and Zn^{II}) in the above-mentioned Mn^{II} PCP results in the formation of isostructural framework compounds.[148] Similar to the parent compound, the new porous materials are stable to desolvation and exhibit a high H_2 storage capacity ranging from 2.00 to 2.29 wt% at 77 K and 0.12 MPa. Measurements of the isosteric heat of adsorption at zero coverage show a difference of 2 $kJ \cdot mol^{-1}$ between the weakest and the strongest H_2 binding materials, which is attributed to variations in the strength of the interaction between H_2 molecules and coordinatively unsaturated metal centers within each framework. Li^{I} impregnation was performed in the three-dimensional PCP, $[Zn_2(ndc)_2(diPyNI)]_n$ (diPyNI = *N,N'*-di-(4-pyridyl)-1,4,5,8-naphthalenetetracarboxydiimide).[150] Because diPyNI is a redox-active ligand, it is reversibly reducible at −0.79 and −1.30 V versus the ferrocene couple. The redox reaction of Li metal with the as-synthesized material allows the impregnation of approximately 5 mol% of Li^+ cations, resulting in a remarkable increase in the H_2 adsorption capacity from 0.93 to 1.63 wt% at 77 K and 0.1 MPa. The zero-coverage H_2 binding energy in the Li^{I}-doped material is modest with 6.1 $kJ \cdot mol^{-1}$, which is nevertheless in good agreement with previous measurements for Li^{I}-exchanged zeolites.

7.5.1.2 Separation

Unlike traditional porous zeolites, whose pores are confined by tetrahedral oxide skeletons, and thus, are difficult to tune, the pores within PCPs can be systematically

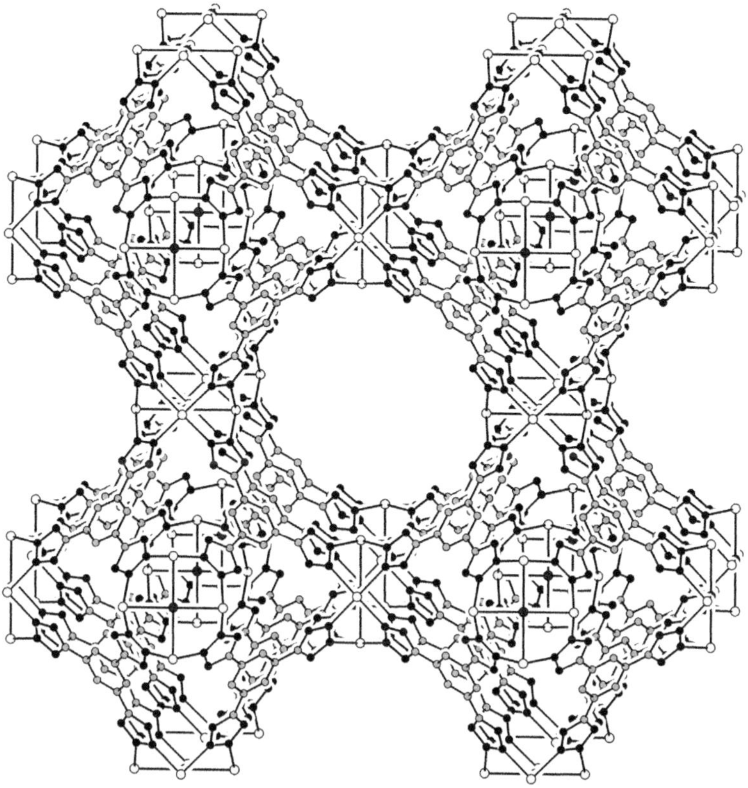

Figure 7.7 Three-dimensional porous framework of $\{[Mn(DMF)_6]_3[Mn_{12}Cl_3(btt)_8(H_2O)_{12}]_2 \cdot 42DMF \cdot 11H_2O \cdot 20MeOH\}_n$. Guest molecules are omitted for clarity.[147] (Reprinted in part with permission from M. Dincă et al., *J. Am. Chem. Soc.* **2006**, *128*, 16876–16883. Copyright 2006 American Chemical Society.)

varied by a judicious choice of building units. This superior feature is significantly important to develop novel functional materials for separation. The selective adsorption capacities of PCPs have been mainly determined by: (1) their size-exclusive effects, in which smaller molecules can go through the porous channels while larger substrates are blocked, and (2) their differential interactions with the substrates. Although the selective adsorption properties of a large number of PCPs have been estimated using single-component adsorption experiments, examples that report the separation functions of a specific guest from binary, ternary, or higher mixtures are limited.[151–157]

$[Zn(\mu_4\text{-TCNQ–TCNQ})(4,4'\text{-bpy})]_n$ shows a highly selective capacity to separate benzene from cyclohexane.[155] The Zn^{II} ions are linked by 4,4′-bpy ligands to give a one-dimensional chain. $[TCNQ–TCNQ]^{2-}$, which is derived from a σ dimerization of two $TCNQ^-$ anions, acts as a cross-linker connecting the four one-dimensional chains to form a three-dimensional porous framework (Fig. 7.8). The channels delimited by the ligands are of an undulating form, not straight, but a unique form comprising an alternating arrangement of two types of tubes of large and small

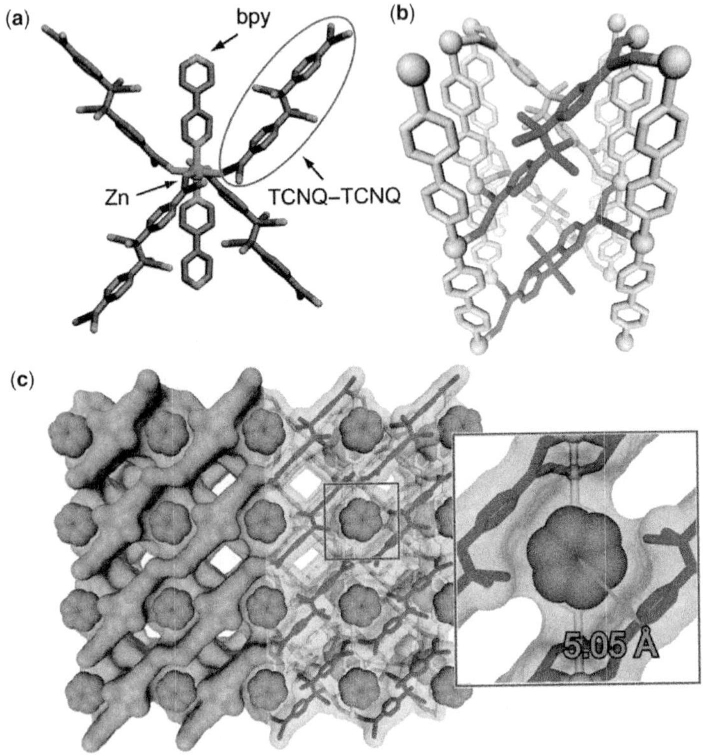

Figure 7.8 (a) Coordination environment of Zn^{II} ion of $\{[Zn(\mu_4\text{-TCNQ–TCNQ})(4,4'\text{-bpy})] \cdot 1.5\text{benzene}\}_n$. (b) TCNQ dimer connected to four one-dimensional chains of Zn^{II} and 4,4′-bpy. (c) Benzene arranged in the cage of the undulating channel. The hydrogen atoms are omitted for clarity.[155] (Reprinted in part with permission from S. Shimomura et al., *J. Am. Chem. Soc.* **2007**, *129*, 10990–10991. Copyright 2007 American Chemical Society.)

diameter. The benzene guest is accommodated firmly in the cavity by the size effect and the C—H $\cdots$ π interaction with the host framework, as shown in Figure 7.8c. In contrast, no uptake of cyclohexane into the host framework occurs, despite almost the same molecular shape and boiling point. These differ only in the position of the protons in the molecular plane, and are linked directly to the strength of the C—H $\cdots$ π interaction with the host framework.

Membrane separation technology is gaining market share from traditional separation technologies because of its low energy consumption, low capital investment, low maintenance, and flexible size. Mixed-matrix membranes are designed to combine the processability of organic polymers with the selective adsorption and diffusion properties of inorganic molecular sieves. Increasing the solubility, or limiting the diffusion of gases, in the membrane should improve the permselectivity. Highly selective, mixed-matrix membranes have been prepared using the three-dimensional PCP, $\{[Cu_2(PF_6)(NO_3)(4,4'\text{-bpy})_2] \cdot 2PF_6\}_n$,[158] confined within amorphous glassy polysulfone (PSf).[151] At loading of 5 wt% or less, the membranes are found to be homogeneous. Scanning electron microscopy images of membrane containing 5 wt% of $\{[Cu_2(PF_6)(NO_3)(4,4'\text{-bpy})_2] \cdot 2PF_6\}_n$ reveal that the PCP appear well dispersed in the PSf matrix. The ideal gas selectivity of X over CH_4 (X/CH_4, X = He, H_2, N_2, and O_2) for the mixed-matrix membranes is superior to that of the pure PSf membrane. Enhancement of the size-sieving nature of membrane by the addition of PCP increases the ideal selectivity for gases, with the dimensions defined by the pore size.

7.5.2 Reaction Vessels and Catalytic Properties

Considering the attractive features of the channels of PCPs, the utilization of PCPs should be of key importance for the creation of unique nanosized reaction vessels. In particular, the use of the PCP channels as a vessel for polymerization is an attractive idea for many reasons, and would not only allow multilevel control of polymerization (e.g., control of stereochemistry, regiochemistry, molecular weight, and helicity), but would also provide well-defined nanostructures and nanohybrids to enable the fabrication of new-generation materials.[159]

Three-dimensional PCPs, $[M_2(1,4\text{-bdc})_2(\text{dabco})]_n$ (M = Zn^{II} and Cu^{II}), exhibit the topotactic selective radical polymerization of divinylbenzenes (DVBs) inside their one-dimensional channels, as shown in Figure 7.9.[160] Polymerization of *p*-DVB in the channels of the Zn^{II} PCP affords the linear polymer, in which only one vinyl group of *p*-DVB is selectively polymerized, while hardly any undesired kink structures (e.g., branching and/or bonding through both vinyl groups) are formed. Linear polymers bearing pendant vinyl groups in their backbone are attractive in light of the potential reactivity of the olefinic double bonds, and for application as polymer precursors of functional materials. In contrast, no trace of polymeric product was observed in the Cu^{II} PCP having almost the same structure as the Zn^{II} PCP. PCPs containing Zn^{II} often show similar flexible and dynamic properties upon guest inclusion. Therefore, the flexibility of the Zn^{II} PCP allows adsorption of the relatively large *p*-DVB molecules by expansion of the host structure, such that the monomer molecules are arranged adjacent to each other, which is essential for polymerization.

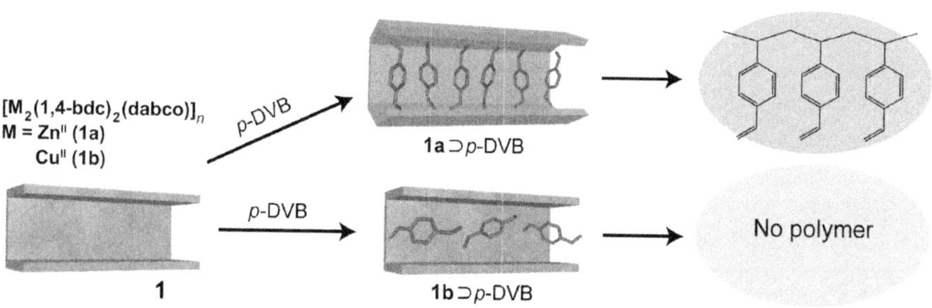

Figure 7.9 Encapsulation and polymerization of *p*-DVBs in the channels of $[M_2(1,4\text{-bdc})_2(\text{dabco})]_n$.[160] (Reprinted with permission from T. Uemura et al., *Angew. Chem. Int. Ed.* **2007**, *46*, 4987–4990. Copyright Wiley-VCH Verlag GmbH & Co. KGaA.)

PCPs with well-defined pores and surface-isolated Lewis acid sites could potentially serve as size- or shape-selective heterogeneous catalysts, in a similar manner to zeolites.[33,43,161–164] The two-dimensional PCP, $\{[Cd(4,4'\text{-bpy})_2(H_2O)_2]\cdot 2NO_3\cdot 4H_2O\}_n$, was the first example that showed catalytic properties for the cyanosilylation of aldehydes.[33] Experimental data in the case of cyanosilylation of imines, which is also performed by the same compound, led to the conclusion that hydrophobic grid cavities bind to the substrate very efficiently to promote a rapid reaction, and that the heterogeneous reaction involves the selective activation of the imino nitrogen by the weak Lewis acid Cd^{II} center.[161] In this polymer, the NO_3^- anions exist in a coordination-free state. This situation contributes to increasing the Lewis acidity of the Cd^{II} centers.

Lewis base sites in PCPs have the potential to show different catalytic properties from Lewis acid sites.[52,165,166] The three-dimensional PCP, $\{[Cd(NO_3)_2(4\text{-btapa})_2]\cdot 6H_2O\cdot 2DMF\}_n$ (4-btapa = 1,3,5-benzenetricarboxylic acid tris[*N*-(4-pyridyl)-amide]), has amide groups that act as guest interaction sites on the surfaces of channels with dimensions of 4.7 × 7.3 Å^2.[165] A Knoevenagel condensation reaction catalyzed by this PCP demonstrates its selective heterogeneous base catalytic properties, which depend on the size of the reactants. This solid catalyst maintains its crystalline framework after the reaction and is easily recycled.

Chiral PCPs can be used for heterogeneous asymmetric catalysis.[43,52,162,164,167] The chiral porous Zr^{IV} phosphonate with Ru^{II}-binap fragments (binap = 2,2′-bis(diphenylphosphanyl)-1,1′-binaphthyl) has a permanent porosity, and shows asymmetric catalytic activity in the hydrogenation of β-keto esters with enantiomeric excess values of up to 95%.[162]

7.5.3 Magnetic Properties

7.5.3.1 Porous Coordination Polymer Magnets

Porous coordination polymer magnets (PCPMs) are optimal compounds for realization of multifunctionality responding to external stimuli, such as a magnetic field and guest adsorption/desorption, because coordination polymers have metal ions as magnetic centers and have flexible characteristics inducing magnetic perturbations. In general, highly porous coordination polymers need long bridging ligands, while a strong magnetic interaction is achieved using short bridging ligands. Such a difference in structural demands makes it difficult for coordination polymers to have a high porosity and a strong magnetic interaction simultaneously.

The use of bridging ligands with a magnetic center is an effective method to overcome the above problem. Bridging ligands with a magnetic center have the characteristics of the bridging of metal centers by coordination bonds, but also enhancement of the magnetic interactions between the metal centers. $\{[Cu_3(ptmtc)_2(py)_6(H_2O)(EtOH)_2]\cdot 6H_2O\cdot 11EtOH\}_n$ (H_3ptmtc = (4,4′,4″-tricarboxydodecachlorotriphenyl)methyl radical) is a two-dimensional PCPM with one-dimensional channels oriented along the [001] and [100] directions, which incorporate guest H_2O and EtOH molecules (Fig. 7.10).[168] The ptmtc^{3-} bridging ligand

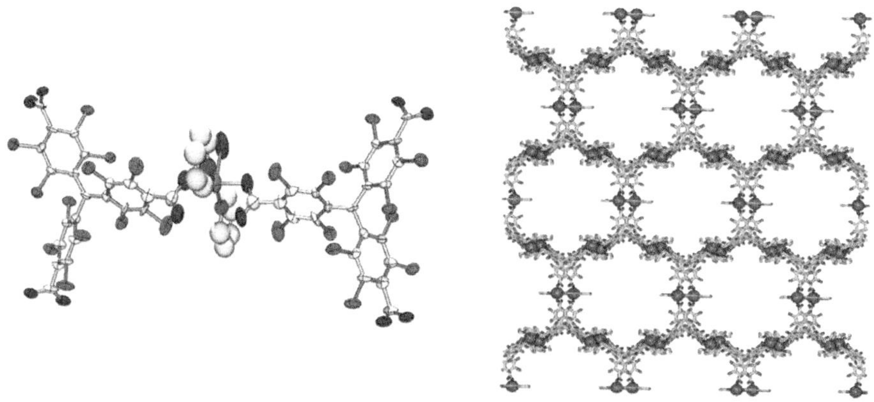

Figure 7.10 (Left) ORTEP (Oak Ridge thermal ellipsoid plot) of coordination circumstance around the Cu^{II} center; and (right) two-dimensional honeycomb framework in $\{[Cu_3(ptmtc)_2(py)_6(H_2O)(EtOH)_2] \cdot 6H_2O \cdot 11EtOH\}_n$. The Cu^{II} ions, depicted in large spheres, are located in the middle of hexagon sides. Guest molecules are omitted for clarity.[168] (Reprinted with permission from D. Maspoch et al., *Nature Mater.* **2003**, 2, 190–195. Copyright 2003 Macmillan Publishers.)

has the organic radical part (a methyl radical, $S = 1/2$) centered in the molecular structure, which enables long-range ferrimagnetic ordering below 2 K. Interestingly, after removal of all the guest molecules, the resulting desolvated compound, $[Cu_3(ptmtc)_2(py)_6]_n$, loses its crystallinity and behaves as a paramagnet in the measured temperature range. The crystallinity and magnetic properties are recovered when the desolvated compound is exposed to MeOH or EtOH vapor.

Metalloligands, which are metal complexes containing two or more Lewis base sites that are able to bridge with other metal ions, can also act as building blocks for developing PCPMs. The hexacyanometalate and octacyanometalate anions, $[M(CN)_6]^{n-}$ and $[M(CN)_8]^{n-}$, respectively, are suitable ligands, because they provide a relatively long (approximately 10 Å) inner pore, and form monodentate cyanide bridges for effective magnetic interactions and structural flexibility. The three-dimensional PCPM ferromagnet, $\{[Ni(dipn)]_2[Ni(dipn)(H_2O)][Fe(CN_6)]_2 \cdot 11H_2O\}_n$ (dipn = *N,N*-di(3-aminopropyl)amine), shows a significant magnetic conversion associated with a reversible gaseous H_2O-induced crystalline to amorphous-like phase transformation.[169] This compound has one-dimensional rectangular channels with dimensions of 3.5×8.7 Å^2 and a solvent-accessible void of 29%. A long-range ferromagnetic ordering ($T_c = 8.5$ K) is observed in the crystalline phase, $\{[Ni(dipn)]_2[Ni(dipn)(H_2O)][Fe(CN_6)]_2 \cdot 11H_2O\}_n$ with H_2O guests. Once the H_2O guests are partially removed from the channels, the resulting desolvated compound, $\{[Ni(dipn)]_2[Ni(dipn)(H_2O)][Fe(CN_6)]_2 \cdot H_2O\}_n$, shows an amorphous-like XRD pattern and exhibits an almost paramagnetic behavior, which indicates that the ferromagnetic interaction is weakened, and the magnetic domain is moderately fragmented while keeping any short-range order on dehydration. A change in magnetic properties corresponding to the capture and release of coordinated and/or hydrogen-bonded guests has been reported in some PCPMs with $[M(CN)_6]^{n-}$ or $[M(CN)_8]^{n-}$.[170–172]

Multicarboxylate ligands are used in the synthesis of PCPMs, although it is very difficult to control the coordination structures.[173–177] $[Mn_3(HCOO)_6]_n$, a three-dimensional flexible porous diamondoid framework based on Mn^{II}-centered $MnMn_4$ tetrahedral nodes, exhibits a wide spectrum of guest-inclusion behavior and long-range magnetic ordering with guest-modulated critical temperature ($T_c = 4.8$ to 9.7 K).[173,174]

7.5.3.2 Porous Spin-Crossover Coordination Polymers

Spin-crossover (SCO) centers represent a convenient physically addressable molecular-scale magnetic switch. Much current attention is directed towards the coordinative-link of SCO sites, motivated largely by the supposition that a physical connection of this type increases cooperative communication between SCO centers applicable towards developing bistable materials for binary switching applications. Recent work has shown that such SCO coordination polymers can incorporate an additional level of functionality associated with their often porous nature. Guest adsorption/desorption in such systems, and the associated perturbation of the Fe^{II} coordination environment and framework structure, provides a unique avenue for investigating SCO phenomena in the solid state.[178,179] The PCP $\{[Fe_2(NCS)_4(azpy)_4] \cdot EtOH\}_n$ (azpy = trans-4,4′-azopyridine) displays a broad half-SCO transition that depends on guest inclusion.[178] Desorption of the unbound EtOH guests from this interpenetrated compound leads to a reversible single crystal-to-single crystal transformation in which the one-dimensional channels partially collapse (Fig. 7.11). In contrast to the EtOH-loaded compound, the desolvated compound retains its high spin state down to a low temperature, a feature that may reflect a weakening of the ligand field with distortion of the coordination geometries following guest removal. Adsorption of other alcohol guests switches back on the SCO sites, with subtle

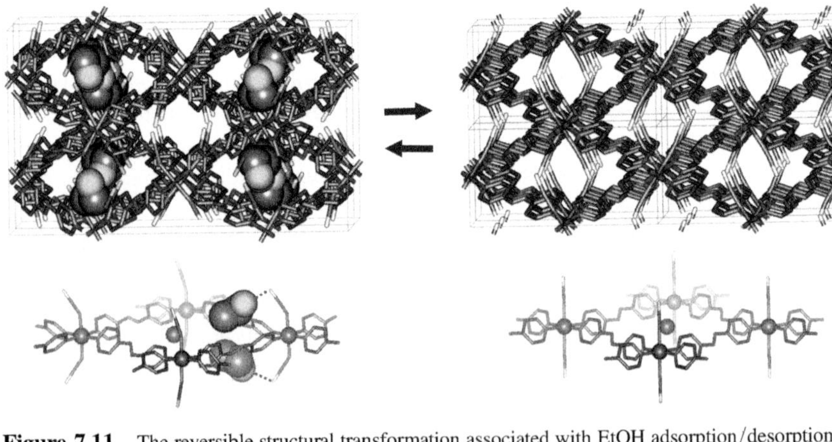

Figure 7.11 The reversible structural transformation associated with EtOH adsorption/desorption in the PCP, $\{[Fe_2(NCS)_2(azpy)_4] \cdot EtOH\}_n$.[178] (Reprinted with permission from G. J. Halder et al., *Science* **2002**, *298*, 1762–1765. Copyright AAAS.)

differences in behavior attributed to the steric influence of the guest on the local Fe^{II} coordination geometry.

7.5.4 Other Properties

The synthesis of a novel proton conductor is one of the most urgent subjects from the viewpoint of developing energy and energy conservation technologies, including photovoltaics, hydrogen storage, and fuel cells. The two-dimensional PCP $[Cu(H_2dtoa)]_n$ (H_2dtoa^{2-} = dithiooxamide anion) with pores with a diameter of approximately 6 Å exhibits a very high proton conductivity (10^{-6} S cm^{-1}) at a relative humidity of 75% and 300 K.[180] Furthermore, the proton conductivity increases by four orders of magnitude for a relative humidity up to 100%. The mechanism of proton conductivity seems to be similar to that of the proton exchange membrane Nafion, which contains many clusters of H_2O molecules in the pores of the polymer. This compound also includes many H_2O guests in its pores, which is the cause of the high proton conductivity. The proton conductivity of the derivative $[Cu(R_2dtoa)]_n$ (R = C_2H_4OH) is two orders of magnitude higher than that of the $[Cu(Et_2dtoa)]_n$ (Et = C_2H_5) derivative, which is derived from the presence of OH groups.[181]

If we can combine ferroelectrically polarizable guest molecules with PCPMs, then we can obtain new types of multiferroic materials, where ferroelectricity and ferromagnetism coexist. Such multifunctional materials have recently aroused increasing interest from the viewpoint of the development of new materials with very large magnetoelectric effects. $\{[Mn_3(HCOO)_6] \cdot EtOH\}_n$ is considered to be the first ferroelectric PCP.[182] Because the desolvated compound $[Mn_3(HCOO)_6]_n$ shows a very small and almost temperature-independent dielectric constant, irrespective of the direction of the electric field,[183] the observed ferroelectric property is considered to come mainly from the guest EtOH molecules.

7.6 PERSPECTIVES

Coordination polymers are organic–inorganic hybrid compounds formed by the self-assembly of discrete molecular building blocks. By tuning the reaction conditions, chemists can obtain desired or unprecedented coordination frameworks. Because the frameworks obtained are extremely regular and topologically interesting, a large number of coordination polymers have been synthesized and characterized crystallographically. From this review, although only a limited number of coordination polymers could be discussed, the large variety of coordination polymers with unique and unprecedented structures becomes apparent, and novel coordination polymers are being continuously synthesized. This review also shows that coordination polymers make an important contribution to developing high-performance and novel functional porous materials. In particular, coordination polymers with dynamic micropores coupled with guest occlusion are promising materials, leading to novel porous functions, such as gas separators, sensors, and actuators. The next target in this research area is to create functional porous materials to contribute to nanotechnology and environmental compliance. To achieve such a target, one needs to use building blocks with nontoxic and abundant elements for the construction of coordination polymers. We believe that coordination polymers will be recognized as useful materials in the near future.

ACKNOWLEDGMENT

This work was supported by the program Exploratory Research for Advanced Technology (ERATO), Japan Science and Technology Agency (JST).

REFERENCES

1. J. C. Bailar, Jr., *Prep. Inorg. React.* **1964**, *1*, 1–27.
2. D. W. Breck, *Zeolite Molecular Sieves: Structure, Chemistry, and Use*, New York: Wiley, **1974**.
3. W. M. Meier, D. H. Olsen, C. Baerlocher, *Atlas of Zeolite Structure Types*, London: Elsevier, **1996**.
4. R. C. Bansal, *Active Carbon*, New York: Dekker, **1988**.
5. S. Kitagwaa, R. Kitaura, S. Noro, *Angew. Chem., Int. Ed.* **2004**, *43*, 2334–2375.
6. O. M. Yaghi, M. O'Keeffe, N. W. Ockwig, H. K. Chae, M. Eddaoudi, J. Kim, *Nature* **2003**, *423*, 705–714.
7. C. Janiak, *Dalton Trans.* **2003**, 2781–2804.
8. M. J. Rosseinsky, *Microporous Mesoporous Mater.* **2004**, *73*, 15–30.
9. G. Férey, C. Mellot-Draznieks, C. Serre, F. Millange, J. Dutour, S. Surblé, I. Margiolaki, *Science* **2005**, *309*, 2040–2042.
10. G. Férey, C. Mellot-Draznieks, C. Serre, F. Millange, *Acc. Chem. Res.* **2005**, *38*, 217–225.
11. B. Moulton, M. J. Zaworotko, *Curr. Opin. Solid State Mater. Sci.* **2002**, *6*, 117–123.
12. D. G. Samsonenko, H. Kim, Y. Sun, G.-H. Kim, H.-S. Lee, K. Kim, *Chem. Asian J.* **2007**, *2*, 484–488.
13. X. Lin, J. Jia, X. Zhao, K. M. Thomas, A. J. Blake, G. S. Walker, N. R. Champness, P. Hubberstey, M. Schröder, *Angew. Chem. Int. Ed.* **2006**, *45*, 7358–7364.
14. M. Dincă, J. R. Long, *J. Am. Chem. Soc.* **2005**, *127*, 9376–9377.
15. E. Y. Lee, S. Jang, M. P. Suh, *J. Am. Chem. Soc.* **2005**, *127*, 6374–6381.
16. C. Kachi-Terajima, T. Akatsuka, M. Kohbara, S. Takamizawa, *Chem. Asian J.* **2007**, *2*, 40–50.
17. B. Kesanli, W. Lin, *Coord. Chem. Rev.* **2003**, *246*, 305–326.
18. K. Biradha, Y. Hongo, M. Fujita, *Angew. Chem. Int. Ed.* **2002**, *41*, 3395–3398.
19. B. F. Abrahams, M. Moylan, S. D. Orchard, R. Robson, *Angew. Chem. Int. Ed.* **2003**, *42*, 1848–1851.
20. M. E. Kosal, J.-H. Chou, S. R. Wilson, K. S. Suslick, *Nature Mater.* **2001**, *1*, 118–121.
21. D. Li, K. Kaneko, *J. Phys. Chem. B* **2000**, *104*, 8940–8945.
22. S. Hermes, F. Schröder, R. Chelmowski, C. Wöll, R. A. Fischer, *J. Am. Chem. Soc.* **2005**, *127*, 13744–13745.
23. X.-S. Wang, S. Ma, D. Sun, S. Parkin, H.-C. Zhou, *J. Am. Chem. Soc.* **2006**, *128*, 16474–16475.
24. B. D. Chandler, G. D. Enright, K. A. Udachin, S. Pawsey, J. A. Ripmeester, D. T. Cramb, G. K. H. Shimizu, *Nature Mater.* **2008**, *7*, 229–235.
25. C. J. Kepert, *Chem. Commun.* **2006**, 695–700.
26. R. M. Barrer, *Molecular Sieves*, Washington, DC: American Chemical Society, **1974**.
27. R. E. Wilde, S. N. Ghosh, B. J. Marshall, *Inorg. Chem.* **1970**, *9*, 2512–2516.
28. H. J. Buser, D. Schwarzenbach, W. Petter, A. Ludi, *Inorg. Chem.* **1977**, *16*, 2704–2710.
29. K. R. Dunbar, R. A. Heintz, *Prog. Inorg. Chem.* **1997**, *45*, 283–391.
30. J. L. Atwood, J. E. D. Davies, D. D. MacNicol, *Inclusion Compounds*, Vol. 1, London: Academic Press, **1984**.

31. J. L. Atwood, J. E. D. Davies, D. D. MacNicol, *Inclusion Compounds*, Vol. 5, Oxford: Oxford University Press, **1991**.
32. B. F. Hoskin, R. Robson, *J. Am. Chem. Soc.* **1994**, *116*, 1151–1152.
33. M. Fujita, Y. J. Kwon, S. Washizu, K. Ogura, *J. Am. Chem. Soc.* **1994**, *116*, 1151–1152.
34. O. M. Yaghi, G. Li, H. Li, *Nature* **1995**, *378*, 703–706.
35. D. Venkataraman, G. B. Gardner, S. Lee, J. S. Moore, *J. Am. Chem. Soc.* **1995**, *117*, 11600–11601.
36. M. Kondo, T. Yoshitomi, K. Seki, H. Matsuzaka, S. Kitagawa, *Angew. Chem. Int. Ed.* **1997**, *36*, 1725–1727.
37. S. Kitagawa, M. Kondo, *Bull. Chem. Soc. Jpn.* **1998**, *71*, 1739–1753.
38. S. Kitagawa, S. Noro, T. Nakamura, *Chem. Commun.* **2006**, 701–707.
39. H. J. Choi, M. P. Suh, *J. Am. Chem. Soc.* **2004**, *126*, 15844–15851.
40. O. Ohmori, M. Kawano, M. Fujita, *J. Am. Chem. Soc.* **2004**, *126*, 16292–16293.
41. S. Shimomura, R. Matsuda, T. Tsujino, T. Kawamura, S. Kitagawa, *J. Am. Chem. Soc.* **2006**, *128*, 16416–16417.
42. D. Tanaka, S. Horike, S. Kitagawa, M. Ohba, M. Hasegawa, Y. Ozawa, K. Toriumi, *Chem. Commun.* **2007**, 3142–3144.
43. C.-D. Wu, A. Hu, L. Zhang, W. Lin, *J. Am. Chem. Soc.* **2005**, *127*, 8940–8941.
44. S. S. Kaye, J. R. Long, *J. Am. Chem. Soc.* **2008**, *130*, 806–807.
45. S. Hermes, M.-K. Schröter, R. Schmid, L. Khodeir, M. Muhler, A. Tissler, R. W. Fischer, R. A. Fischer, *Angew. Chem. Int. Ed.* **2005**, *44*, 6237–6241.
46. S. Hermes, F. Schröder, S. Amirjalayer, R. Schmid, R. A. Fischer, *J. Mater. Chem.* **2006**, *16*, 2464–2472.
47. S. Hermes, D. Zacher, A. Baunemann, C. Wöll, R. A. Fischer, *Chem. Mater.* **2007**, *19*, 2168–2173.
48. M. J. Ingleson, J. P. Barrio, J.-B. Guilbaud, Y. Z. Khimyak, M. J. Rosseinsky, *Chem. Commun.* **2008**, 2680–2682.
49. Z. Wang, S. M. Cohen, *J. Am. Chem. Soc.* **2007**, *129*, 12368–12369.
50. J. S. Costa, P. Gamez, C. A. Black, O. Roubeau, S. J. Teat, J. Reedijk, *Eur. J. Inorg. Chem.* **2008**, 1551–1554.
51. Z. Wang, S. M. Cohen, *Angew. Chem. Int. Ed.* **2008**, *47*, 4699–4702.
52. J. S. Seo, D. Whang, H. Lee, S. I. Jun, J. Oh, Y. J. Jeon, K. Kim, *Nature* **2000**, *404*, 982–986.
53. Y.-F. Song, L. Cronin, *Angew. Chem. Int. Ed.* **2008**, *47*, 4635–4637.
54. M. Kawano, T. Kawamichi, T. Haneda, T. Kojima, M. Fujita, *J. Am. Chem. Soc.* **2007**, *129*, 15418–15419.
55. M. J. Zaworotko, *Nature* **2008**, *451*, 410–411.
56. O. Ohmori, M. Kawano, M. Fujita, *Angew. Chem. Int. Ed.* **2005**, *44*, 1962–1964.
57. X.-M. Chen, M.-L. Tong, *Acc. Chem. Res.* **2007**, *40*, 162–170.
58. O. R. Evans, W. Lin, *Acc. Chem. Res.* **2002**, *35*, 511–522.
59. R.-G. Xiong, X. Xue, H. Zhao, X.-Z. You, B. F. Abrahams, Z. Xue, *Angew. Chem. Int. Ed.* **2002**, *41*, 3800–3803.
60. S. Kitagawa, T. Okubo, S. Kawata, M. Kondo, M. Katada, H. Kobayashi, *Inorg. Chem.* **1995**, *34*, 4790–4796.
61. G. Férey, C. Serre, C. Mellot-Draznieks, F. Millange, S. Surblé, J. Dutour, I. Margiolaki, *Angew. Chem. Int. Ed.* **2004**, *43*, 6296–6301.
62. S. H. Jhung, J.-H. Lee, J.-S. Chang, *Bull. Korean Chem. Soc.* **2005**, *26*, 880–881.
63. J. Y. Choi, J. Kim, S. H. Jhung, H.-K. Kim, J.-S. Chang, H. K. Chae, *Bull. Korean Chem. Soc.* **2006**, *27*, 1523–1524.
64. S. H. Jhung, J.-H. Lee, J. W. Yoon, C. Serre, G. Férey, J.-S. Chang, *Adv. Mater.* **2007**, *19*, 121–124.
65. S. H. Jhung, J.-H. Lee, P. M. Forster, G. Férey, A. K. Cheetham, J.-S. Chang, *Chem. Eur. J.* **2006**, *12*, 7899–7905.
66. A. Pichon, A. Lazuen-Garay, S. L. James, *CrystEngComm* **2006**, *8*, 211–214.
67. Y.-H. Liu, Y.-L. Lu, H.-L. Tsai, J.-C. Wang, K. L. Lu, *J. Solid State Chem.* **2001**, *158*, 315–319.
68. J. Y. Lu, A. M. Babb, *Chem. Commun.* **2002**, 1340–1341.
69. K. Uemura, S. Kitagawa, *Chem. Lett.* **2005**, 132–137.

70. D. R. Talham, *Chem. Rev.* **2004**, *104*, 5479–5501.
71. C.-Y. Su, A. M. Goforth, M. D. Smith, P. J. Pellechia, H.-C. zur Loye, *J. Am. Chem. Soc.* **2004**, *126*, 3576–3586.
72. S. Hermes, F. Schröder, R. Chelmowski, C. Wöll, R. A. Fisher, *J. Am. Chem. Soc.* **2005**, *127*, 13744–13745.
73. T. Uemura, Y. Hoshino, S. Kitagawa, K. Yoshida, S. Isoda, *Chem. Mater.* **2006**, *18*, 992–995.
74. M. Eddaoudi, J. Kim, N. Rosi, D. Vodak, J. Wachter, M. O'Keeffe, O. M. Yaghi, *Science* **2002**, *295*, 469–472.
75. A. L. Grzesiak, F. J. Uribe, N. W. Ockwig, O. M. Yaghi, A. J. Matzger, *Angew. Chem. Int. Ed.* **2006**, *45*, 2553–2556.
76. Y. Kubota, M. Takata, T. C. Kobayashi, S. Kitagawa, *Coord. Chem. Rev.* **2007**, *251*, 2510–2521.
77. R. Kitaura, S. Kitagawa, Y. Kubota, T. C. Kobayashi, K. Kindo, Y. Mita, A. Matsuo, M. Kobayashi, H.-C. Chang, T. C. Ozawa, M. Suzuki, M. Sakata, M. Takata, *Science* **2002**, *298*, 2358–2361.
78. M. Kondo, T. Okubo, A. Asami, S. Noro, T. Yoshitomi, S. Kitagawa, T. Ishii, H. Matsuzaka, K. Seki, *Angew. Chem. Int. Ed.* **1999**, *38*, 140–143.
79. N. L. Rosi, J. Eckert, M. Eddaoudi, D. T. Vodak, J. Kim, M. O'Keeffe, O. M. Yaghi, *Science*, **2003**, *300*, 1127–1129.
80. T. Ohkubo, J. Miyawaki, K. Kaneko, R. Ryoo, N. A. Seaton, *J. Phys. Chem. B* **2002**, *106*, 6523–6528.
81. J. A. MacKinnon, J. Eckert, D. F. Coker, A. L. R. Bug, *J. Chem. Phys.* **2001**, *114*, 10137–10150.
82. L. A. Clark, S. Chempath, R. Q. Snurr, *Langmuir* **2005**, *21*, 2267–2272.
83. A. Vishnyakov, P. I. Ravikovitch, A. V. Neimark, M. Bülow, Q. M. Wang, *Nano Lett.* **2003**, *3*, 713–718.
84. A. I. Skoulidas, *J. Am. Chem. Soc.* **2004**, *126*, 1356–1357.
85. T. Düren, L. Sarkisov, O. M. Yaghi, R. Q. Snurr, *Langmuir* **2004**, *20*, 2683–2689.
86. K. S. Walton, R. Q. Snurr, *J. Am. Chem. Soc.* **2007**, *129*, 8552–8556.
87. T. Düren, R. Q. Snurr, *J. Phys. Chem. B* **2004**, *108*, 15703–15708.
88. H. Frost, T. Düren, R. Q. Snurr, *J. Phys. Chem. B* **2006**, *110*, 9565–9570.
89. N. A. Ramsahye, G. Maurin, S. Bourrelly, P. L. Llewellyn, T. Loiseau, C. Serre, G. Férey, *Chem. Commun.* **2007**, 3261–3263.
90. K. S. Walton, A. R. Millward, D. Dubbeldam, H. Frost, J. J. Low, O. M. Yaghi, R. Q. Snurr, *J. Am. Chem. Soc.* **2008**, *130*, 406–407.
91. E. Klontzas, A. Mavrandonakis, E. Tylianakis, G. E. Froudakis, *Nano Lett.* **2008**, *8*, 1572–1576.
92. A. Blomqvist, A. C. Moyses, P. Srepusharawoot, R. Ahuja, *Proc. Natl. Acad. Sci. U.S.A.* **2007**, *104*, 20173–20176.
93. T. Sagara, J. Ortony, E. Ganz, *J. Chem. Phys.* **2005**, *123*, 214707.
94. S. S. Han, W. A. Goddard III, *J. Am. Chem. Soc.* **2007**, *129*, 8422–8423.
95. S. S. Han, W.-Q. Deng, W. A. Goddard III, *Angew. Chem. Int. Ed.* **2007**, *46*, 6289–6292.
96. S. Horike, R. Matsuda, R. Kitaura, S. Kitagawa, T. Iijima, K. Endo, Y. Kubota, M. Takata, *Chem. Commun.* **2004**, 2152–2153.
97. S. L. Gould, D. Tranchemontagne, O. M. Yaghi, M. A. Garcia-Garibay, *J. Am. Chem. Soc.* **2008**, *130*, 3246–3247.
98. S. Horike, R. Matsuda, D. Tanaka, S. Matsubara, M. Mizuno, K. Endo, S. Kitagawa, *Angew. Chem. Int. Ed.* **2006**, *45*, 7226–7230.
99. D. N. Dybtsev, H. Chun, K. Kim, *Angew. Chem. Int. Ed.* **2004**, *43*, 5033–5036.
100. R. Kitaura, K. Seki, G. Akiyama, S. Kitagawa, *Angew. Chem. Int. Ed.* **2003**, *42*, 428–431.
101. B.-Q. Ma, K. L. Mulfort, J. T. Hupp, *Inorg. Chem.* **2005**, *44*, 4912–4914.
102. B. Chen, S. Ma, E. J. Hurtado, E. B. Lobkovsky, H.-C. Zhou, *Inorg. Chem.* **2007**, *46*, 8490–8492.
103. X.-C. Huang, Y.-Y. Lin, J.-P. Zhang, X.-M. Chen, *Angew. Chem. Int. Ed.* **2005**, *45*, 1557–1559.
104. K. S. Park, Z. Ni, A. P. Côté, J. Y. Choi, R. Huang, F. J. Uribe-Romo, H. K. Chae, M. O'Keeffe, O. M. Yaghi, *Proc. Natl. Acad. Sci. U.S.A.* **2006**, *103*, 10186–10191.
105. H. Hayashi, A. P. Côté, H. Furukawa, M. O'Keeffe, O. M. Yaghi, *Nature Mater.* **2007**, *6*, 501–506.
106. R. Banerjee, A. Phan, B. Wang, C. Knobler, H. Furukawa, M. O'Keeffe, O. M. Yaghi, *Science* **2008**, *319*, 939–943.

107. D. Li, K. Kaneko, *J. Phys. Chem. B* **2000**, *104*, 8940–8945.
108. R. Kitaura, K. Fujimoto, S. Noro, M. Kondo, S. Kitagawa, *Angew. Chem. Int. Ed.* **2002**, *41*, 133–135.
109. S. Kitagawa, K. Uemura, *Chem. Soc. Rev.* **2005**, *34*, 109–119.
110. S. Takamizawa, E. Nakata, H. Yokoyama, K. Mochizuki, W. Mori, *Angew. Chem. Int. Ed.* **2003**, *42*, 4331–4334.
111. K. Biradha, Y. Hongo, M. Fujita, *Angew. Chem. Int. Ed.* **2000**, *39*, 3843–3845.
112. K. Biradha, Y. Hongo, M. Fujita, *Angew. Chem. Int. Ed.* **2002**, *41*, 3395–3398.
113. A. J. Blake, S. J. Hill, P. Hubberstey, W.-S. Li, *J. Chem. Soc. Dalton Trans.* **1997**, 913–914.
114. A. Kondo, H. Noguchi, S. Ohnishi, H. Kajiro, A. Tohdoh, Y. Hattori, W.-C. Xu, H. Tanaka, H. Kanoh, K. Kaneko, *Nano Lett.* **2006**, *6*, 2581–2584.
115. A. Kondo, H. Noguchi, L. Carlucci, D. M. Proserpio, G. Ciani, H. Kajiro, T. Ohba, H. Kanoh, K. Kaneko, *J. Am. Chem. Soc.* **2007**, *129*, 12362–12363.
116. R. Kitaura, K. Seki, G. Akiyama, S. Kitagawa, *Angew. Chem. Int. Ed.* **2003**, *42*, 428–431.
117. J. Johnson, *Chem. Eng. News* **2004**, *82*, 36–42.
118. Z. Yong, V. Mata, A. E. Rodrigues, *Sep. Purif. Technol.* **2002**, *26*, 195–205.
119. A. R. Millward, O. M. Yaghi, *J. Am. Chem. Soc.* **2005**, *127*, 17998–17999.
120. S. Bourrelly, P. L. Llewellyn, C. Serre, F. Millange, T. Loiseau, G. Férey, *J. Am. Chem. Soc.* **2005**, *127*, 13519–13521.
121. S. Himeno, T. Komatsu, S. Fujita, *J. Chem. Eng. Data* **2005**, *50*, 369–376.
122. P. L. Llewellyn, S. Bourrelly, C. Serre, Y. Filinchuk, G. Férey, *Angew. Chem. Int. Ed.* **2006**, *45*, 7751–7754.
123. R. Matsuda, R. Kitaura, S. Kitagawa, Y. Kubota, R. V. Belosludov, T. C. Kobayashi, H. Sakamoto, T. Chiba, M. Takata, Y. Kawazoe, Y. Mita, *Nature* **2005**, *436*, 238–241.
124. G. Férey, *Nature* **2005**, *436*, 187–188.
125. D. G. Samsonenko, H. Kim, Y. Sun, G.-H. Kim, H.-S. Lee, K. Kim, *Chem. Asian J.* **2007**, *2*, 484–488.
126. S. Noro, S. Kitagawa, M. Kondo, K. Seki, *Angew. Chem. Int. Ed.* **2000**, *39*, 2082–2084.
127. K. Seki, *Chem. Commun.* **2001**, 1496–1497.
128. S. Ma, S. Sun, J. M. Simmons, C. D. Collier, D. Yuan, H.-C. Zhou, *J. Am. Chem. Soc.* **2008**, *130*, 1012–1016.
129. J. L. C. Rowsell, O. M. Yaghi, *Angew. Chem. Int. Ed.* **2005**, *44*, 4670–4679.
130. J. L. C. Rowsell, A. R. Millward, K. S. Park, O. M. Yaghi, *J. Am. Chem. Soc.* **2004**, *126*, 5666–5667.
131. G. Férey, M. Latroche, C. Serre, F. Millange, T. Loiseau, A. Percheron-Guegan, *Chem. Commun.* **2003**, 2976–2977.
132. D. N. Dybtsev, H. Chun, K. Kim, *Angew. Chem. Int. Ed.* **2004**, *43*, 5033–5036.
133. X. Zhao, B. Xiao, A. J. Fletcher, K. M. Thomas, D. Bradshaw, M. J. Rosseinsky, *Science* **2004**, *306*, 1012–1015.
134. L. Pan, M. B. Sander, X. Huang, J. Li, M. Smith, E. Bittner, B. Bockrath, J. K. Johnson, *J. Am. Chem. Soc.* **2004**, *126*, 1308–1309.
135. B. Kesanli, Y. Cui, M. R. Smith, E. W. Bittner, B. C. Bockrath, W. Lin, *Angew. Chem. Int. Ed.* **2005**, *44*, 72–75.
136. B. Panella, K. Hönes, U. Müller, N. Trukhan, M. Schubert, H. Pütter, M. Hirscher, *Angew. Chem. Int. Ed.* **2008**, *47*, 2138–2142.
137. Y. Liu, J. F. Eubank, A. J. Cairns, J. Eckert, V. C. Kravtsov, R. Luebke, M. Eddaoudi, *Angew. Chem. Int. Ed.* **2007**, *46*, 3278–3283.
138. X. Lin, J. Jia, X. Zhao, K. M. Thomas, A. J. Blake, G. S. Walker, N. R. Champness, P. Hubberstey, M. Schröder, *Angew. Chem. Int. Ed.* **2006**, *45*, 7358–7364.
139. H. Wu, W. Zhou, T. Yildirim, *J. Am. Chem. Soc.* **2007**, *129*, 5314–5315.
140. D. Sun, S. Ma, Y. Ke, D. J. Collins, H.-C. Zhou, *J. Am. Chem. Soc.* **2006**, *128*, 3896–3897.
141. S. K. Bhatia, A. L. Myers, *Langmuir* **2006**, *22*, 1688–1700.
142. M. Dinca, J. R. Long, *J. Am. Chem. Soc.* **2005**, *127*, 9376–9377.
143. J. L. C. Rowsell, O. M. Yaghi, *J. Am. Chem. Soc.* **2006**, *128*, 1304–1315.

144. B. Chen, N. W. Ockwig, A. R. Millward, D. S. Contreras, O. M. Yaghi, *Angew. Chem. Int. Ed.* **2005**, *44*, 4745–4749.
145. S. S. Kaye, J. R. Long, *J. Am. Chem. Soc.* **2005**, *127*, 6506–6507.
146. R. C. Lochan, M. Head-Gordon, *Phys. Chem. Chem. Phys.* **2006**, *8*, 1357–1370.
147. M. Dincă, A. Dailly, Y. Liu, C. M. Brown, D. A. Neumann, J. R. Long, *J. Am. Chem. Soc.* **2006**, *128*, 16876–16883.
148. M. Dincă, J. R. Long, *J. Am. Chem. Soc.* **2007**, *129*, 11172–11176.
149. P. M. Forster, J. Eckert, B. D. Heiken, J. B. Parise, J. W. Yoon, S. H. Jhung, J.-S. Chang, A. K. Cheetham, *J. Am. Chem. Soc.* **2006**, *128*, 16846–16850.
150. K. L. Mulfort, J. T. Hupp, *J. Am. Chem. Soc.* **2007**, *129*, 9604–9605.
151. J. Won, J. S. Seo, J. H. Kim, H. S. Kim, Y. S. Kang, S.-J. Kim, Y. Kim, J. Jegal, *Adv. Mater.* **2005**, *14*, 80–84.
152. B. Chen, C. Liang, J. Yang, D. S. Contreras, Y. L. Clancy, E. B. Lobkovsky, O. M. Yaghi, S. Dai, *Angew. Chem. Int. Ed.* **2006**, *45*, 1390–1393.
153. Y.-Y. Zhou, X.-P. Yan, K.-N. Kum, S.-W. Wang, M.-G. Liu, *J. Chromatogr. A* **2006**, *1116*, 172–178.
154. P. S. Bárcia, F. Zapata, J. A. C. Silva, A. E. Rodrigues, B. Chen, *J. Phys. Chem. B* **2007**, *111*, 6101–6103.
155. S. Shimomura, S. Horike, R. Matsuda, S. Kitagawa, *J. Am. Chem. Soc.* **2007**, *129*, 10990–10991.
156. L. Bastin, P. S. Bárcia, E. J. Hurtado, J. A. C. Silva, A. E. Rodrigues, B. Chen, *J. Phys. Chem. C* **2008**, *112*, 1575–1581.
157. Y. Zhang, I. H. Musselman, J. P. Ferraris, K. J. Balkus, Jr., *J. Membr. Sci.* **2008**, *313*, 170–181.
158. S. Noro, R. Kitaura, M. Kondo, S. Kitagawa, T. Ishii, H. Matsuzaka, M. Yamashita, *J. Am. Chem. Soc.* **2002**, *124*, 2568–2583.
159. T. Uemura, S. Horike, S. Kitagawa, *Chem. Asian J.* **2006**, *1–2*, 36–44.
160. T. Uemura, D. Hiramatsu, Y. Kubota, M. Takata, S. Kitagawa, *Angew. Chem. Int. Ed.* **2007**, *46*, 4987–4990.
161. O. Ohmori, M. Fujita, *Chem. Commun.* **2004**, 1586–1587.
162. A. Hu, H. L. Ngo, W. Lin, *Angew. Chem. Int. Ed.* **2003**, *42*, 6000–6003.
163. P. Horcajada, S. Surble, C. Serre, D. Y. Hong, Y. K. Seo, J. S. Chang, J. M. Greneche, I. Margiolaki, G. Férey, *Chem. Commun.* **2007**, 2820–2822.
164. S.-H. Cho, B. Ma, S. T. Nguyen, J. T. Hupp, T. E. Albrecht-Schmitt, *Chem. Commun.* **2006**, 2563–2565.
165. S. Hasegawa, S. Horike, R. Matsuda, S. Furukawa, K. Mochizuki, Y. Kinoshita, S. Kitagawa, *J. Am. Chem. Soc.* **2007**, *129*, 2607–2614.
166. D. M. Shin, I. S. Lee, Y. K. Chung, *Cryst. Growth Des.* **2006**, *6*, 1059–1061.
167. D. N. Dybtsev, A. L. Nuzhdin, H. Chun, K. P. Bryliakov, E. P. Talsi, V. P. Fedin, K. Kim, *Angew. Chem. Int. Ed.* **2006**, *45*, 916–920.
168. D. Maspoch, D. Ruiz-Molina, K. Wurst, N. Domingo, M. Cavallini, F. Biscarini, J. Tejada, C. Rovira, J. Veciana, *Nature Mater.* **2003**, *2*, 190–195.
169. N. Yanai, W. Kaneko, K. Yoneda, M. Ohba, S. Kitagawa, *J. Am. Chem. Soc.* **2007**, *129*, 3496–3497.
170. S. Ohkoshi, T. Tsunobuchi, H. Takahashi, T. Hozumi, M. Shiro, K. Hashimoto, *J. Am. Chem. Soc.* **2007**, *129*, 3084–3085.
171. J. Milon, M. C. Daniel, A. Kaiba, P. Guionneau, S. Brandes, J. P. Sutter, *J. Am. Chem. Soc.* **2007**, *129*, 13872–13878.
172. S. Ohkoshi, K. Arai, Y. Sato, K. Hashimoto, *Nature Mater.* **2004**, *3*, 857–861.
173. Z. M. Wang, B. Zhang, H. Fujiwara, H. Kobayashi, M. Kurmoo, *Chem. Commun.* **2004**, 416–417.
174. Z. Wang, Y. Zhang, T. Liu, M. Kurmoo, S. Gao, *Adv. Funct. Mater.* **2007**, *17*, 1523–1536.
175. B. Zhang, Z. M. Wang, M. Kurmoo, S. Gao, K. Inoue, H. Kobayashi, *Adv. Funct. Mater.* **2007**, *17*, 577–584.
176. M. Kurmoo, H. Kumagai, K. W. Chapman, C. J. Kepert, *Chem. Commun.* **2005**, 3012–3014.
177. B. Zhang, Z. Wang, M. Kurmoo, S. Gao, K. Inoue, H. Kobayashi, *Adv. Funct. Mater.* **2007**, *17*, 577–584.

178. G. J. Halder, C. J. Kepert, B. Moubaraki, K. S. Murray, J. D. Cashion, *Science* **2002**, *298*, 1762–1765.

179. S. M. Neville, G. J. Halder, K. W. Chapman, M. B. Duriska, P. D. Southon, J. D. Cashion, J.-F. Létard, B. Moubaraki, K. S. Murray, C. J. Kepert, *J. Am. Chem. Soc.* **2008**, *130*, 2869–2876.

180. H. Kitagawa, Y. Nagao, M. Fujishima, R. Ikeda, S. Kanda, *Inorg. Chem. Commun.* **2003**, *6*, 346–348.

181. Y. Nagao, T. Kubo, K. Nakasuji, R. Ikeda, T. Kojima, H. Kitagawa, *Synth. Meth.* **2005**, *154*, 89–92.

182. H. Cui, Z. Wang, K. Takahashi, Y. Okano, H. Kobayashi, A. Kobayashi, *J. Am. Chem. Soc.* **2006**, *128*, 15074–15075.

183. H. Cui, K. Takahashi, Y. Okano, H. Kobayashi, Z. Wang, A. Kobayashi, *Angew. Chem. Int. Ed.* **2005**, *44*, 6508–6512.

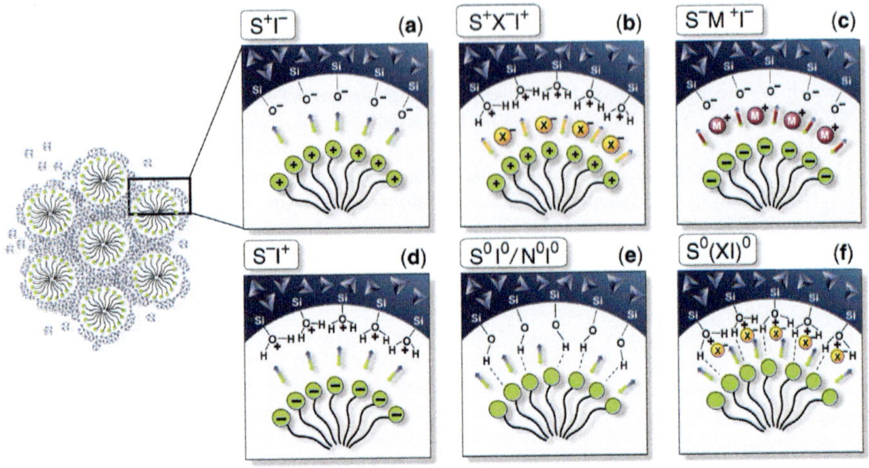

Figure 3.6 See page 51 for figure caption.

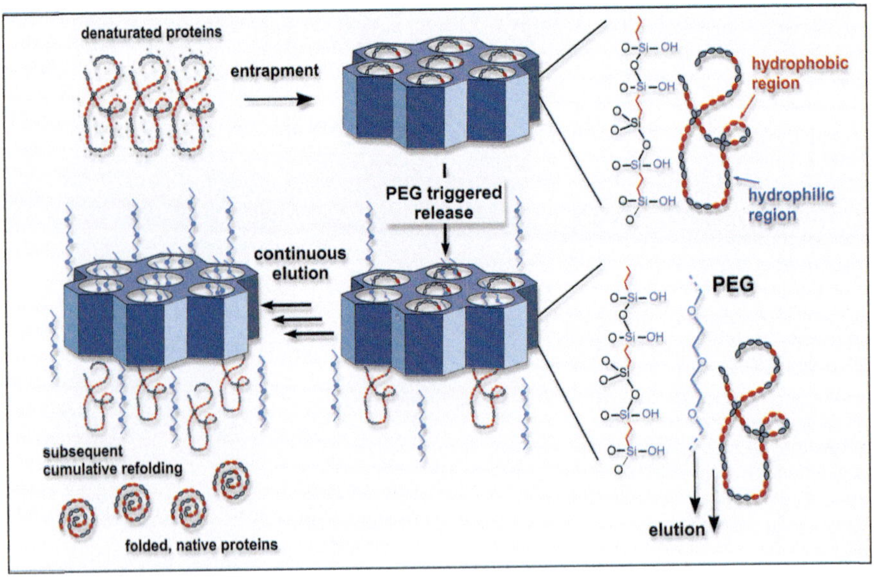

Figure 3.18 See page 73 for figure caption.

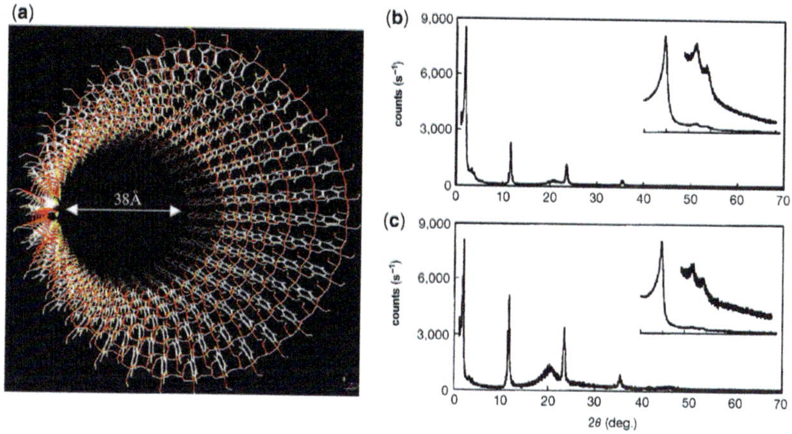

Figure 3.21 See page 76 for figure caption.

Figure 3.23 See page 81 for figure caption.

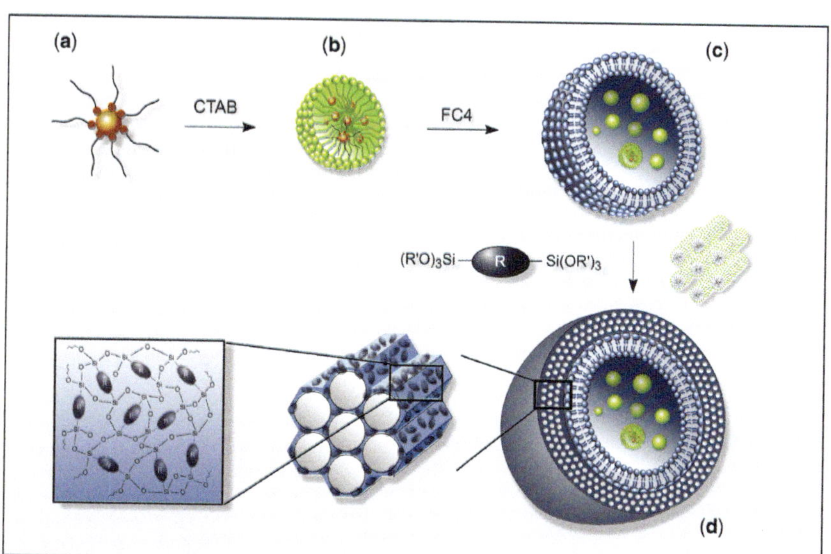

Figure 3.30 See page 101 for figure caption.

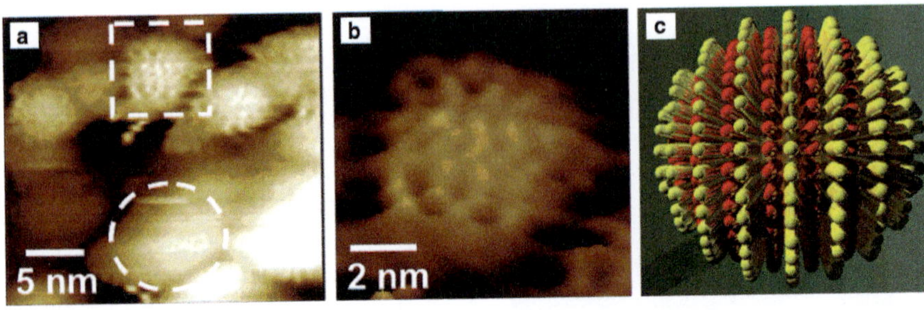

Figure 4.11 See page 140 for figure caption.

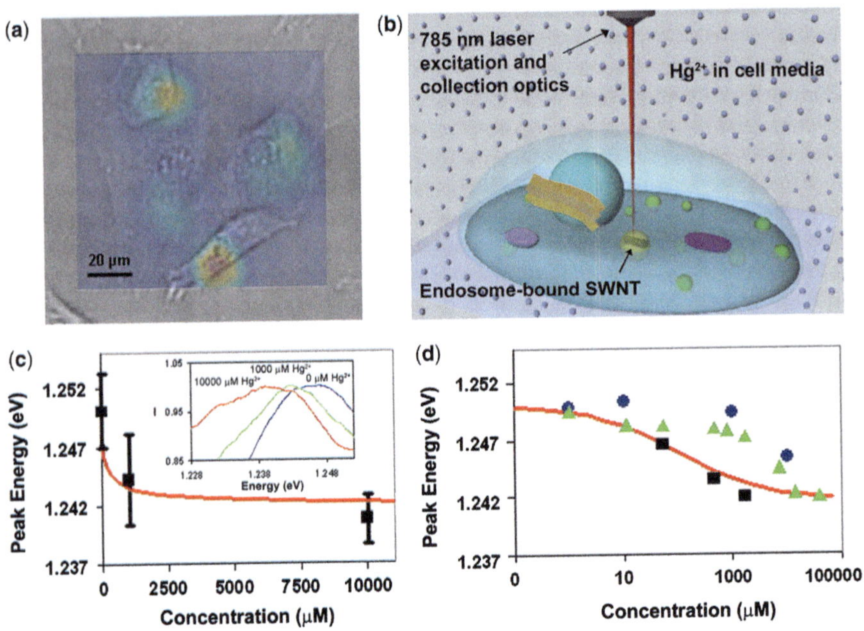

Figure 6.4 See page 208 for figure caption.

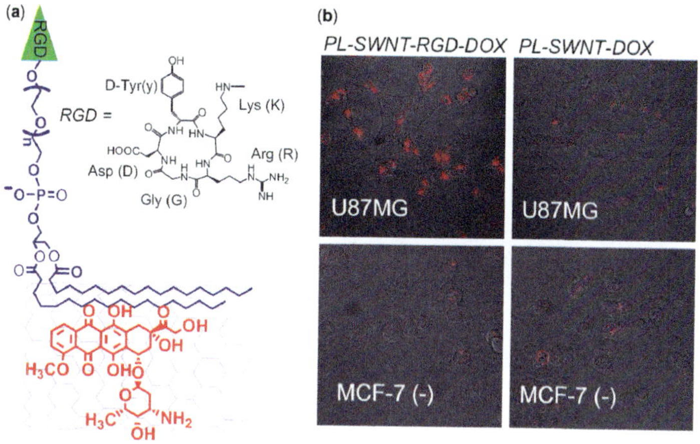

Figure 6.6 See page 213 for figure caption.

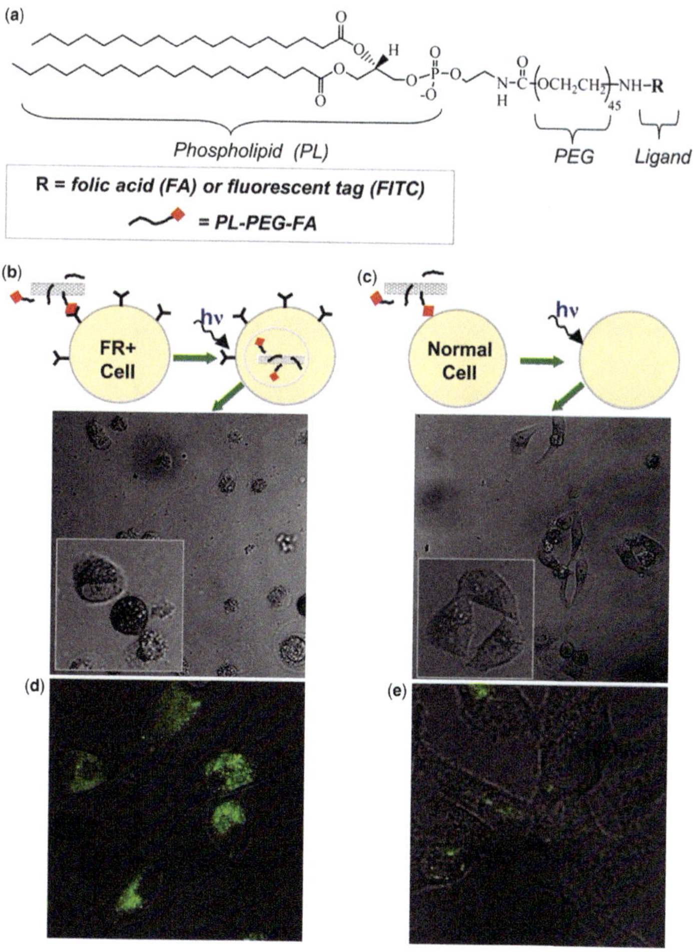

Figure 6.7 See page 214 for figure caption.

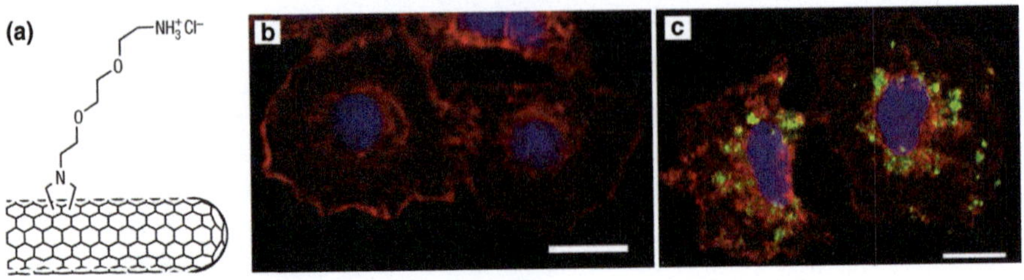

Figure 6.8 See page 217 for figure caption.

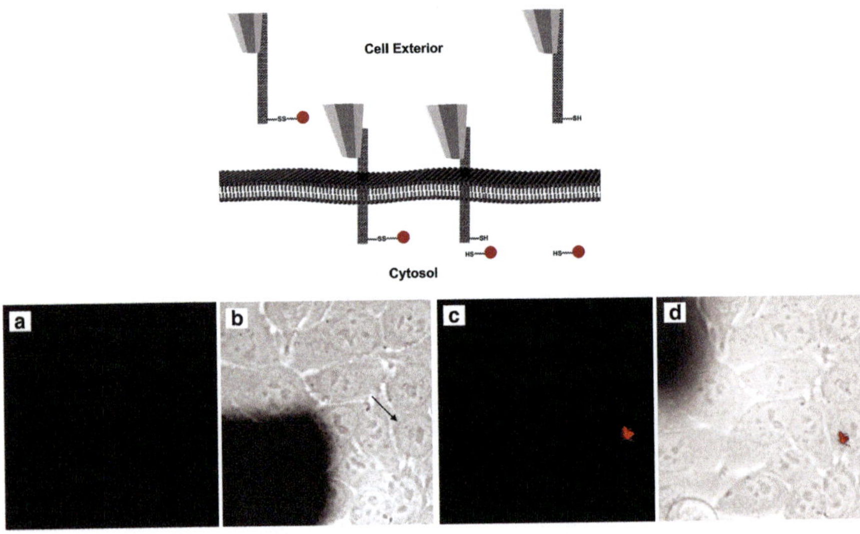

Figure 6.9 See page 218 for figure caption.

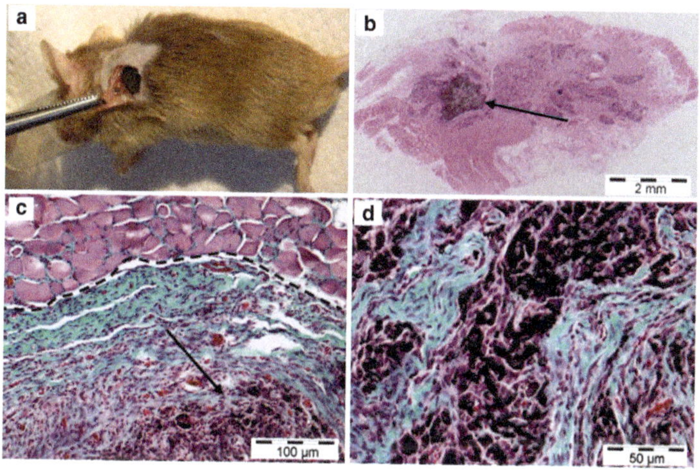

Figure 6.11 See page 223 for figure caption.

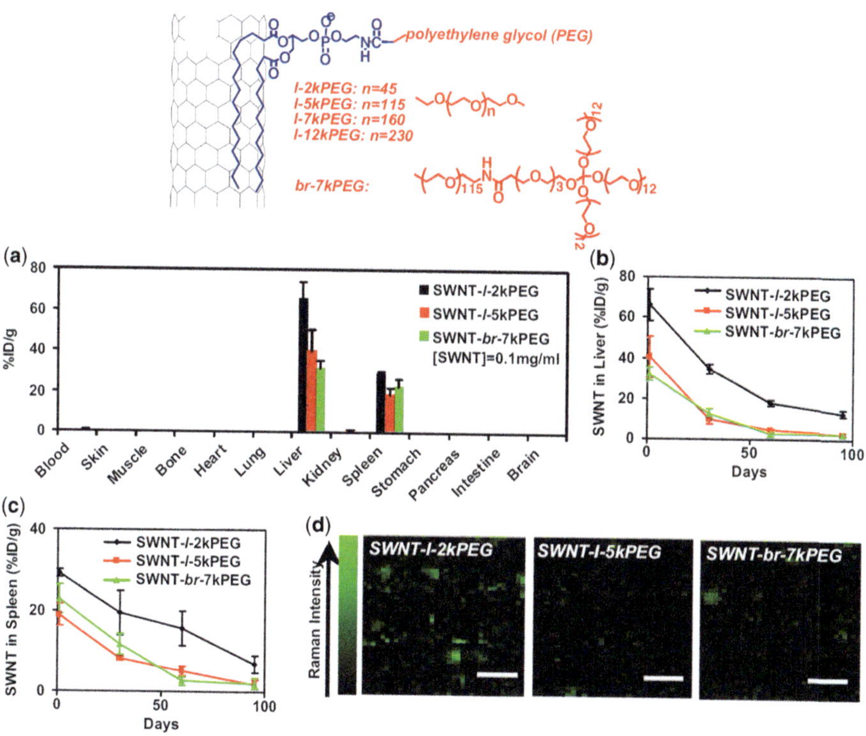

Figure 6.15 See page 228 for figure caption.

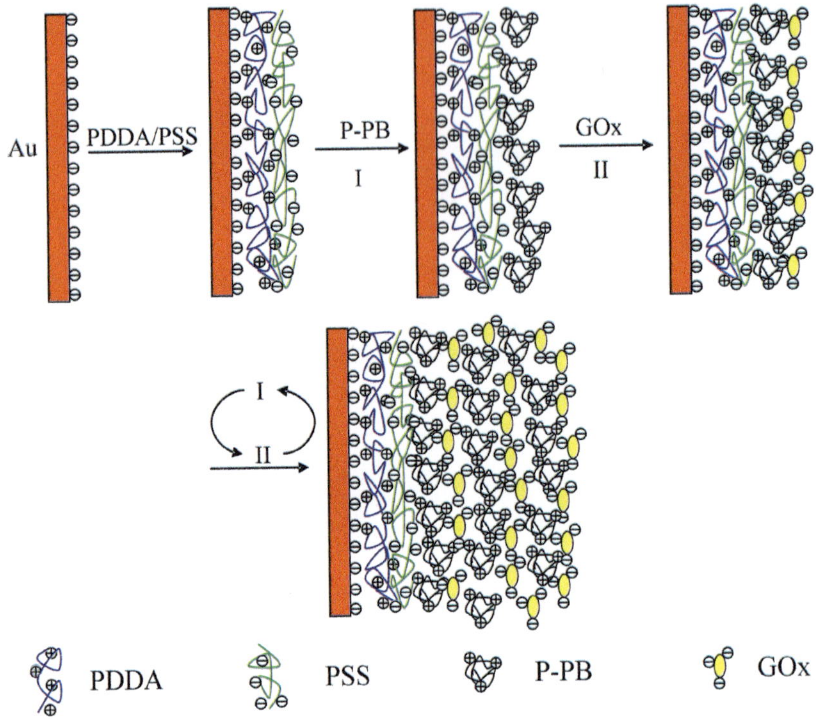

Figure 8.12 See page 287 for figure caption.

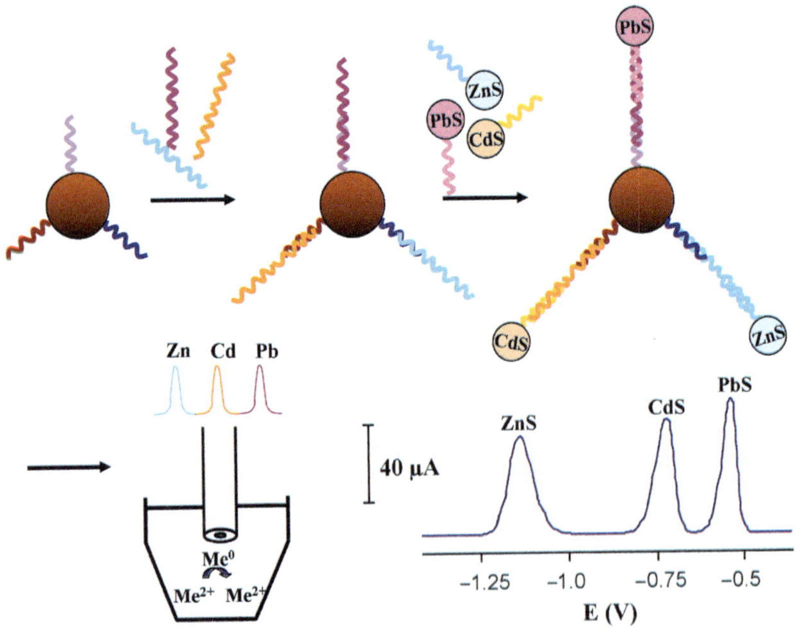

Figure 8.17 See page 293 for figure caption.

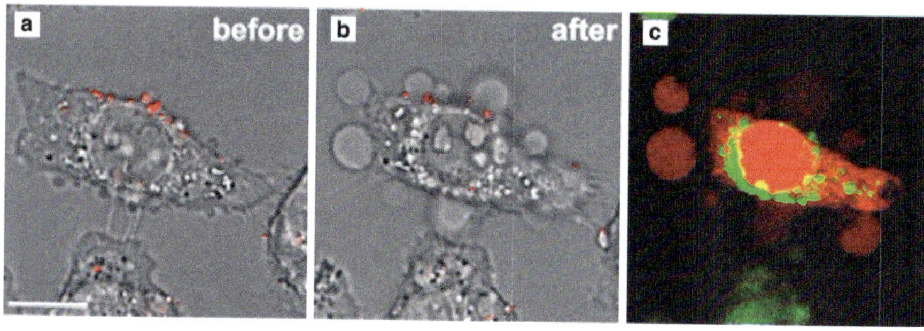

Figure 10.6 See page 328 for figure caption.

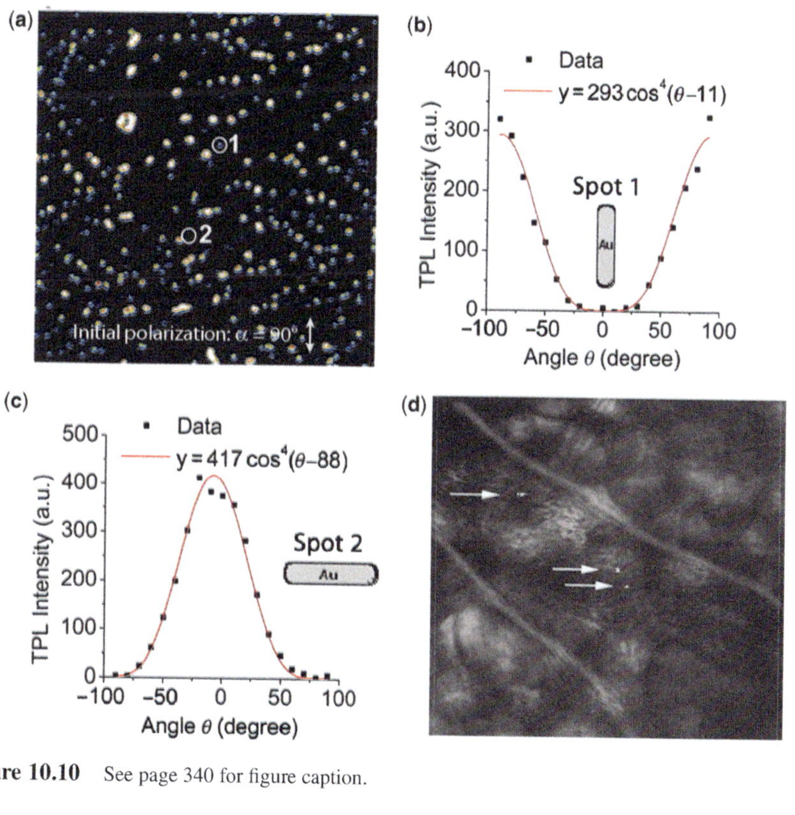

Figure 10.10 See page 340 for figure caption.

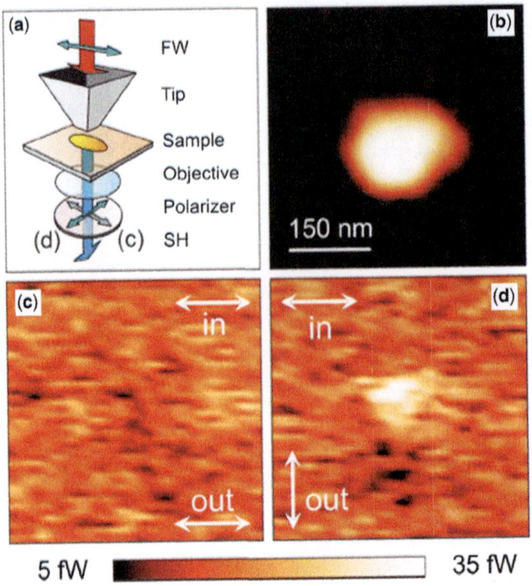

Figure 10.11 See page 342 for figure caption.

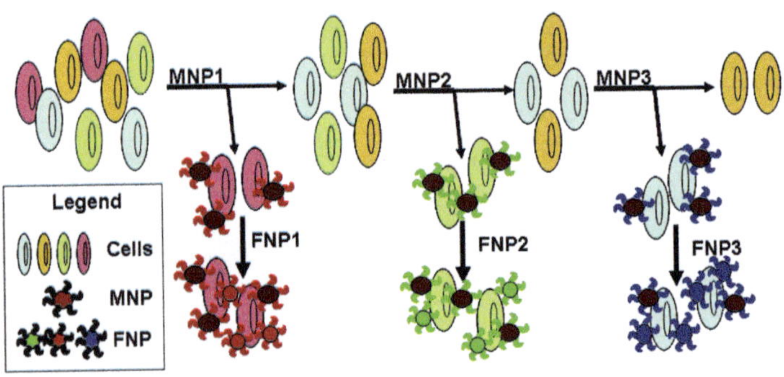

Figure 11.4 See page 355 for figure caption.

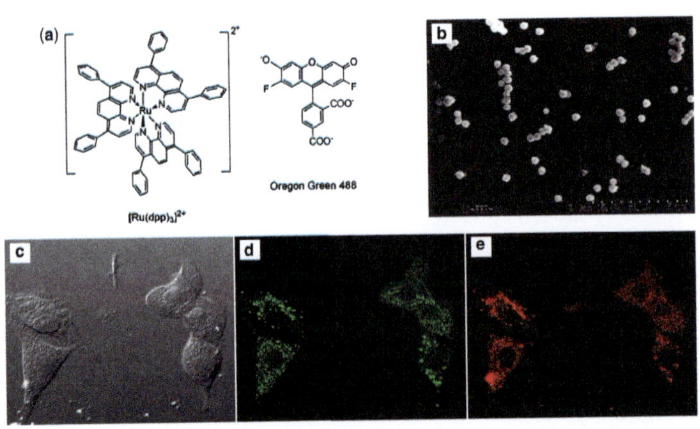

Figure 11.7 See page 358 for figure caption.

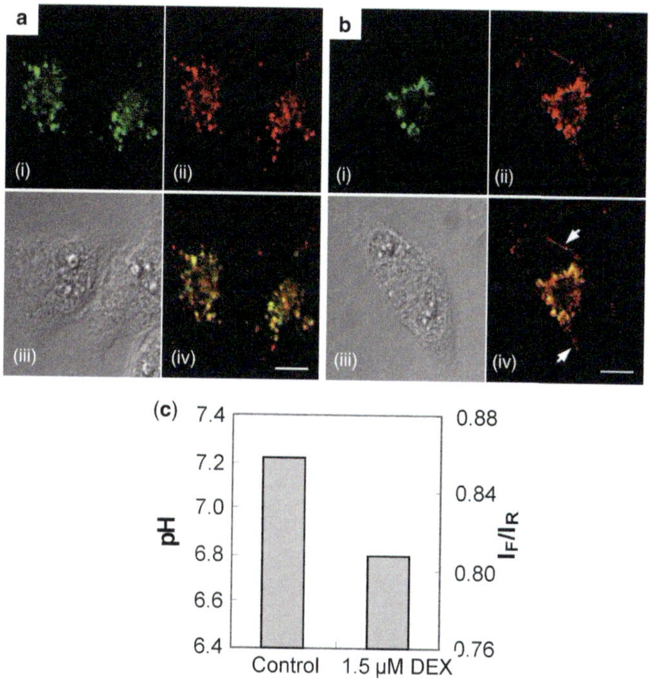

Figure 11.8 See page 360 for figure caption.

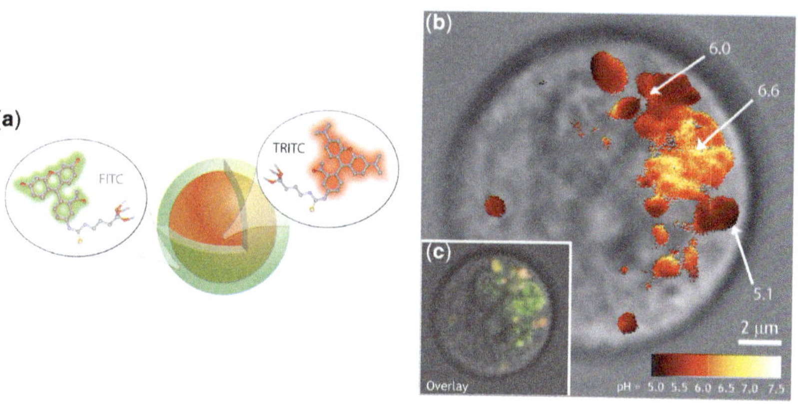

Figure 11.17 See page 372 for figure caption.

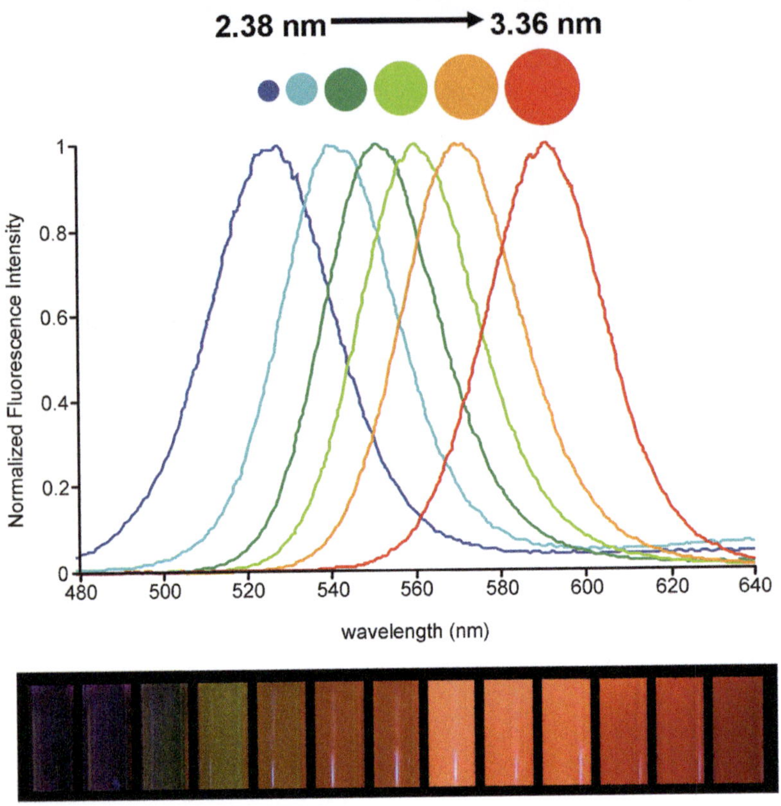

Figure 12.1 See page 380 for figure caption.

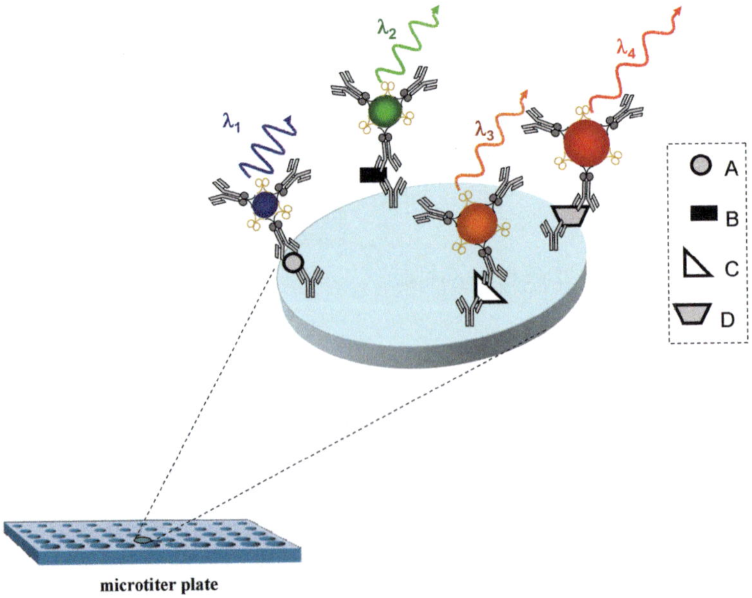

Figure 12.3 See page 386 for figure caption.

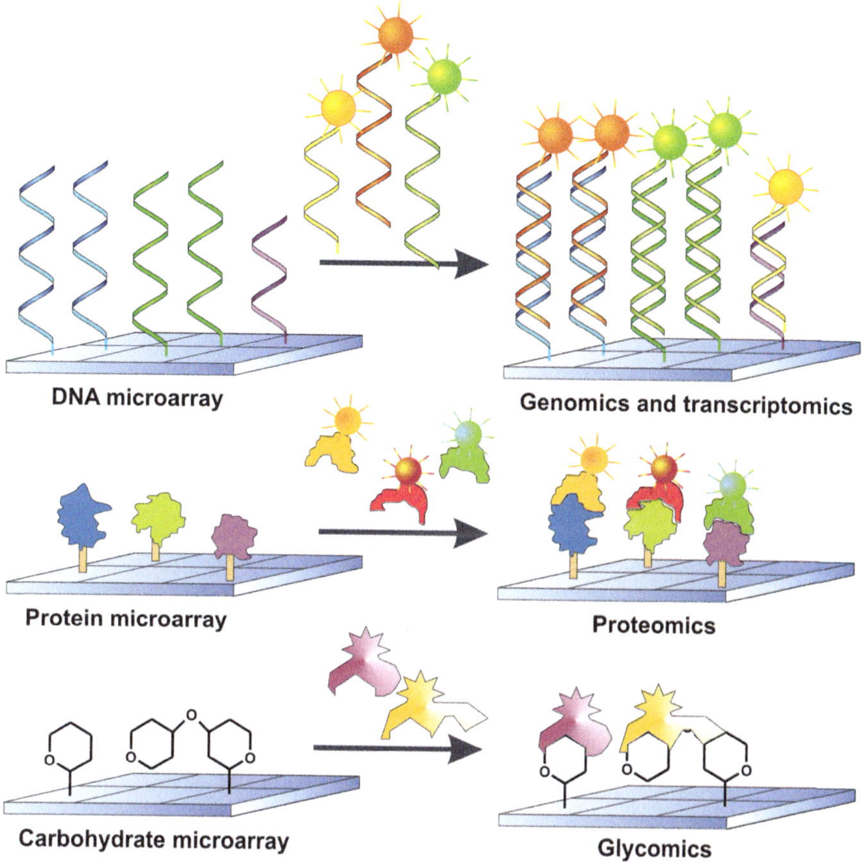

Figure 14.1 See page 435 for figure caption.

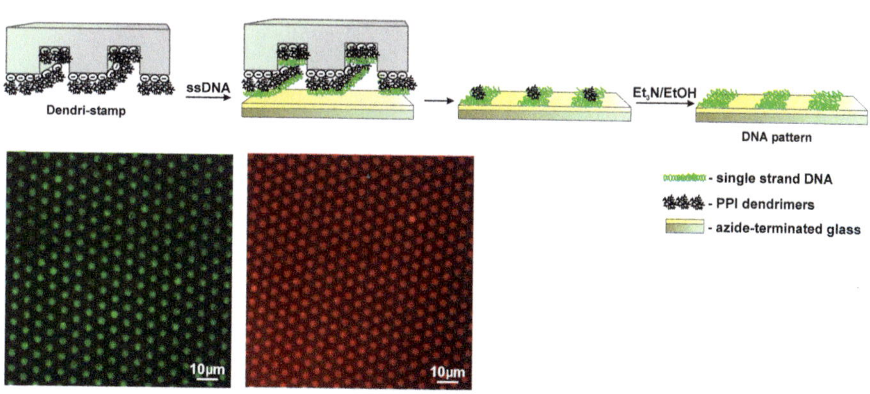

Figure 14.11 See page 451 for figure caption.

Figure 20.3 See page 587 for figure caption.

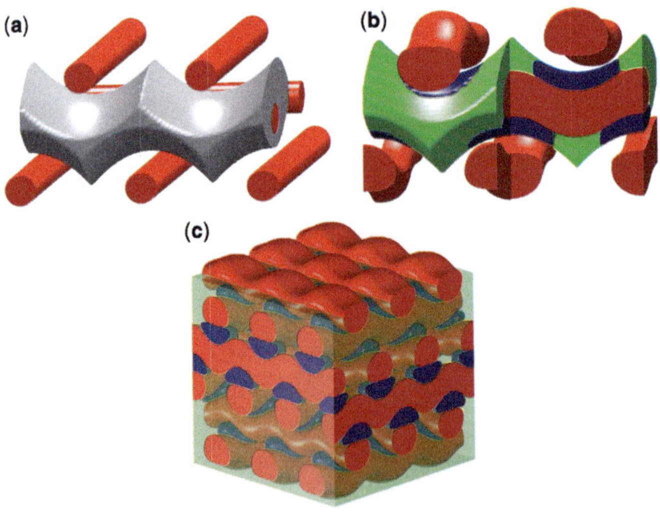

Figure 21.16 See page 618 for figure caption.

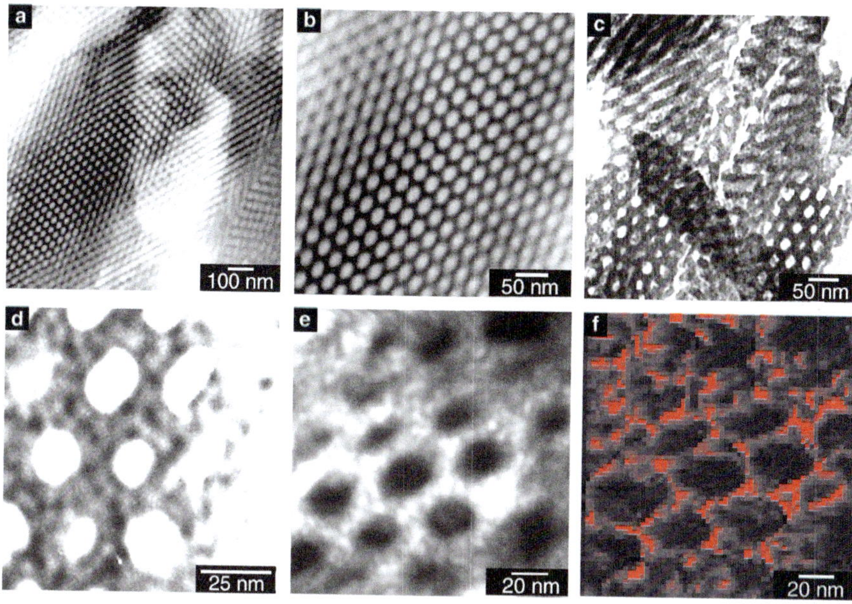

Figure 21.18 See page 622 for figure caption.

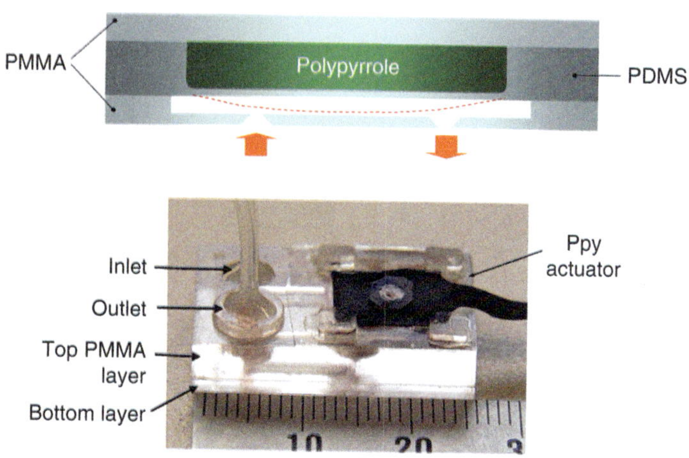

Figure 23.2 See page 659 for figure caption.

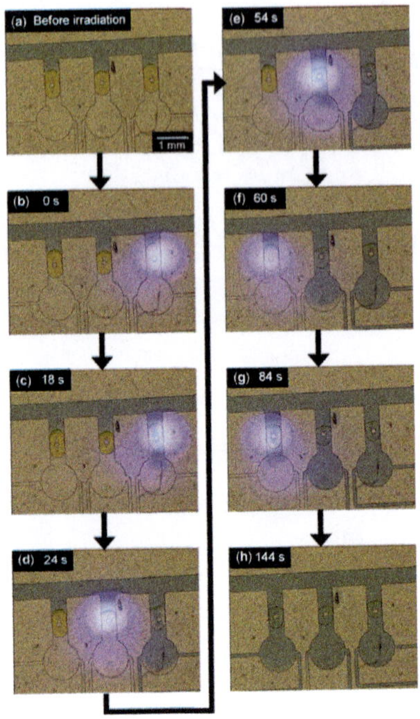

Figure 23.7 See page 664 for figure caption.

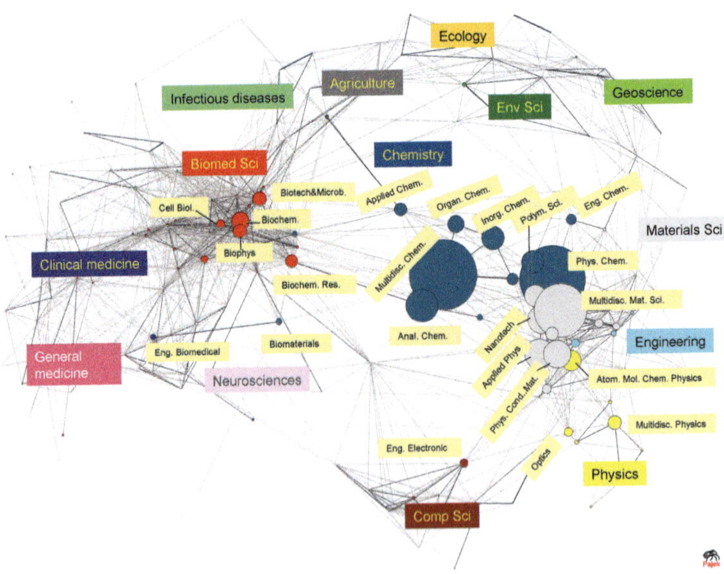

Figure 24.3 See page 682 for figure caption.

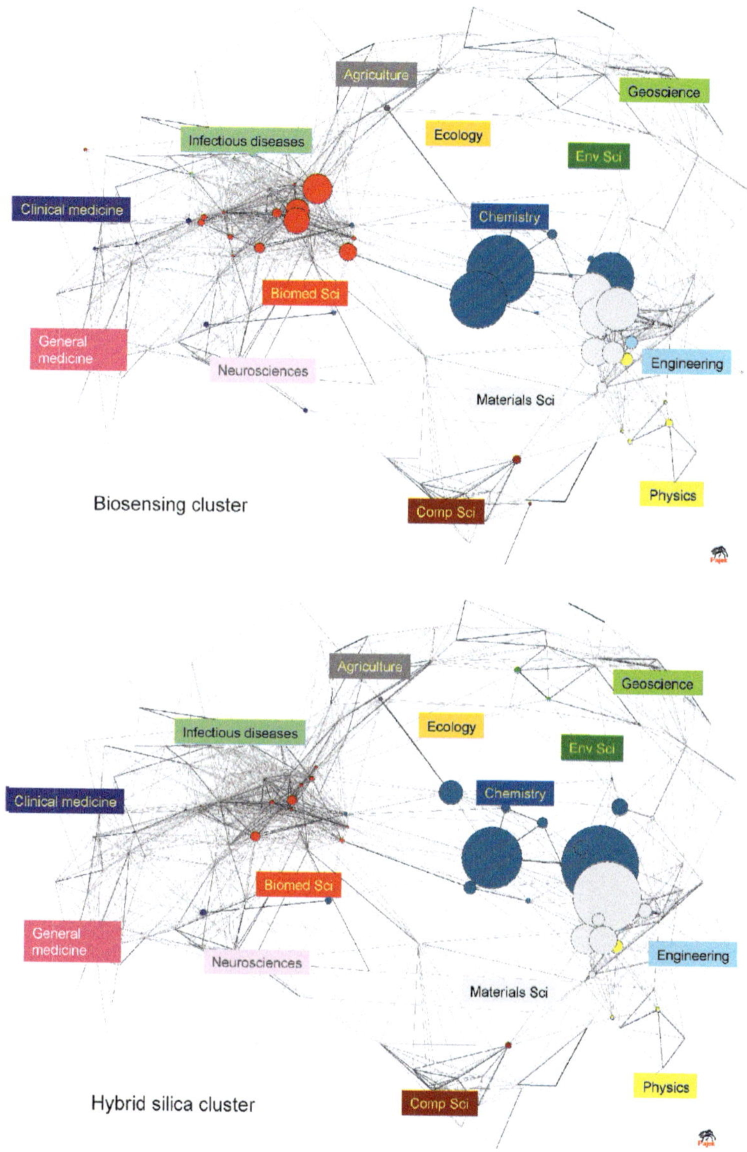

Figure 24.4 See page 684 for figure caption.

Part Two

Improvement of Signaling and Sensing by Organization on Surfaces

Chapter 8

Nanoparticle and Biomolecular–Nanoparticle Hybrid Supramolecular Complexes for Electrochemical Signaling

RONEN POLSKY, JASON C. HARPER, AND
SUSAN M. BROZIK

The Supramolecular Chemistry of Organic–Inorganic Hybrid Materials. Edited by Knut Rurack and Ramón Martínez-Máñez

8.1 INTRODUCTION

Nanoparticles have served as building blocks for supramolecular assemblies since the early 1990s and more recently have proven to be valuable in chemical and biomolecular recognition schemes as optical and electrochemical labels and for interfacing electrochemical sensing systems with electrodes. The integration of material interfaces with nanoparticles, from monolayers to thin films, that result in various supramolecular architectures has led to a new generation of hybrid materials with unique tailor-made properties impacting diverse areas concerning nanotechnology,[1] fundamental studies of nanomaterials,[2] nanoelectronics,[3] and chemical and biological sensing devices.[4] The properties associated with nanoparticle-modified surfaces depend greatly on the type of nanoparticles used, their size, shape, surface properties, and interparticle distance. Different architectures of nanometer sized particles and electrodes can be organized from highly ordered and stable arrays ideal for nanoelectronic devices to disorganized structures that can respond to internal and external stimuli suitable for the design of sensing applications.

Conductive and semiconductive nanoparticles have unique, size-dependent electronic, physical, and optical properties while biomolecules possess specific binding properties and functionalities that have been optimized over billions of years of evolution. A major technical challenge is controlling the assembly and organization of these two building blocks into two-dimensional and three-dimensional arrays in which the intrinsic properties of these two materials can be characterized and then used as functional devices (sensors, building blocks for electronic nanocircuits, nanodevices, etc.). Recent work in the conjugation, assembly, and properties of these bio-nano assemblies is discussed in the following sections as well as the electrochemical techniques used to elucidate their unique properties. Although the focus of this chapter is the organization and electrochemical characterization of these ensembles on electrode surfaces, a short discussion on the electrochemical properties of monolayer protected clusters is reviewed since capacitance measurements on three-dimensional monolayers show similar double layer charging with comparable two-dimensional self-assembled monolayers of alkanethiolates on gold.[5]

8.2 ELECTROCHEMICAL MONITORING OF MONOLAYER PROTECTED CLUSTERS

The size-dependent properties of nanoparticles differ greatly from the corresponding bulk materials. An example is the size quantization phenomenon commonly observed in II–VI and III–V inorganic semiconductor nanocrystals.[6] During the intermediate transition towards that of the bulk metal (usually between 2 and 20 nm), localization of electrons and holes in a confined volume causes an increase in its effective optical band gap as the size of the nanoparticle decreases, observed as a blue shift in its optical spectrum. Brus predicted that there should also be a dependence on the redox potential for these same classes of quantum dots.[7] Bard and coworkers showed this experimentally and have reported on the direct observation between the electrochemical band gap and the electronic spectra using electrochemical measurements on thioglycerol-capped CdS nanoparticles in *N,N*-dimethylformamide.[8] A multielectron transfer process was proposed where the electrons are consumed by fast coupled chemical reactions due to decomposition of the cluster which are limited by diffusion of the solution particles to the electrode surface and the irreversibility of the oxidation and reduction reactions. Figure 8.1 shows a typical cyclic voltammogram for the thioglycerol-capped CdS-nanoparticles at a Pt electrode. Compared to the response of the blank solution, clear oxidation and reduction peaks are observed, with a peak to peak separation of the oxidation and reduction peaks of 2.96 V (peaks A1 and C1). This is comparable to the 3.23 eV (1s-1s transition) calculated from the electronic spectra directly correlated to the electron transfer at HOMO and LUMO transitions, respectively.

Murray and coworkers have observed the charging of the double-layer capacitance of freely diffusing alkanethiolate monolayer-protected gold clusters (MPC). The nanoparticles can be treated as diffusive nanoelectrodes and behave as multivalent redox species as charge injection into the core is quantized.[5,9] A potential change of a tenth of a volt was found necessary to add a single electron to the double layer of a single cluster molecule due to the very small capacitance of the nanoparticles. For smaller voltages it was speculated that some nanoparticle-electrode collisions resulted in charging by a single electron while others did not, in a probabilistic manner.[5] Differential pulse voltammetry (DPV) on Au_{147} MPCs at 295 K also showed a similar quantization effect exhibiting 15 evenly spaced peaks characteristic of quantized

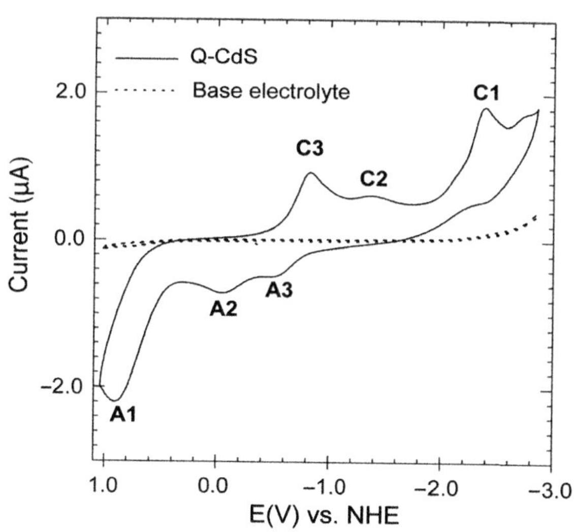

Figure 8.1 Cyclic voltammetric response in the absence and presence of thioglycol-capped CdS Q-particles (1 mg/mL of fraction IV) at a Pt electrode. 50 mM tetrahexylammonium perchlorate in DMF, sweep rate = 50 mV s^{-1}. C1, C2, and C3 refer to cathodic peaks 1, 2, and 3, respectively. A1, A2, and A3 refer to anodic peaks 1, 2, and 3 respectively.[8] (Reprinted with permission from S. K. Haram et al., *J. Am. Chem. Soc.* **2001**, *123*, 8860–8861. Copyright 2001 American Chemical Society.)

charge injection into the metal core, shown in Figure 8.2,[10] while 13 peaks were observed for Au_{140} particles in CH_2Cl_2 at 263 K.[11] Murray's group further demonstrated that by electrolytically charging solutions of MPCs with different charges (redox states), the clusters could potentially be used as nanoscale charge storage devices. The stored charges were used in a controllable and quantifiable way to drive electron transfer reactions.[12]

The self-assembling of MPCs was accomplished by ligand place-exchange with alkanedithiols followed by chemisorption, through free thiol groups, to gold electrodes allowing the electrochemical interrogation of the MPCs on an electrode surface in both organic and aqueous media.[13] The overall electrode double-layer capacitance was found to be governed mainly by the adsorbed particles as shown by collective charging of individual surface-immobilized MPCs. In aqueous solutions, the shrinking of the effective monolayer thickness (due to the hydrophobicity of the alkanethiolate layers) caused the charging to be dominated by the underlying electrode. Therefore, the filling of interparticle voids with low dielectric hexanethiol molecules allowed the observation of the discrete electron charging of the MPCs in aqueous media, although not as well defined as in organic media. Further MPC assemblies were constructed by exploiting complexation interactions between divalent metal ions and pyridine moieties revealing rectified quantized charging characteristics in aqueous solutions in the presence of hydrophobic anions, as shown schematically in Figure 8.3.[14] The hydrophobicity of the electrolyte ions produced molecular diode-like responses that evolve from conventional to quantized character through an ion-binding mechanism between the divalent transition metal ions and pyridine moieties used for the construction of the nanoparticle supramolecular structure.

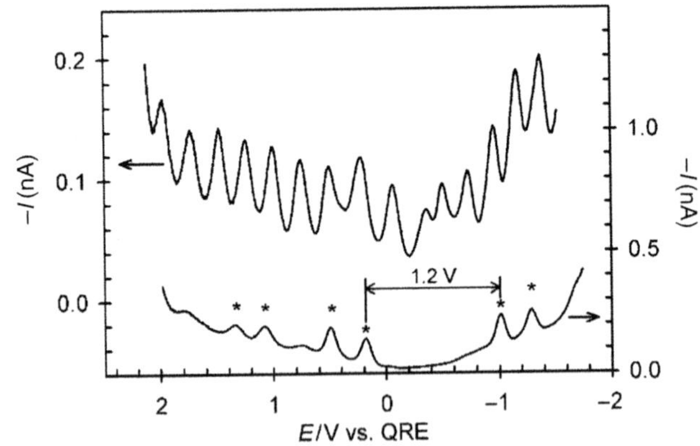

Figure 8.2 DPV responses for MPC solutions measured at a Pt microelectrode. Upper: as-prepared 177 μM C6S-Au_{147} showing 15 high-resolution quantized double layer charging (QDL) peaks. Lower: 170 μM C6S-Au_{38} showing a HOMO-LUMO gap. It can be seen that the as-prepared solution contains a residual fraction of Au_{38} that smears out the charging response in *E* regions where QDL peaks overlap. The electrode potential was scanned negative to positive.[10] (Reprinted with permission from B. M. Quinn et al., *J. Am. Chem. Soc.* **2003**, *125*, 6644–6645. Copyright 2003 American Chemical Society.)

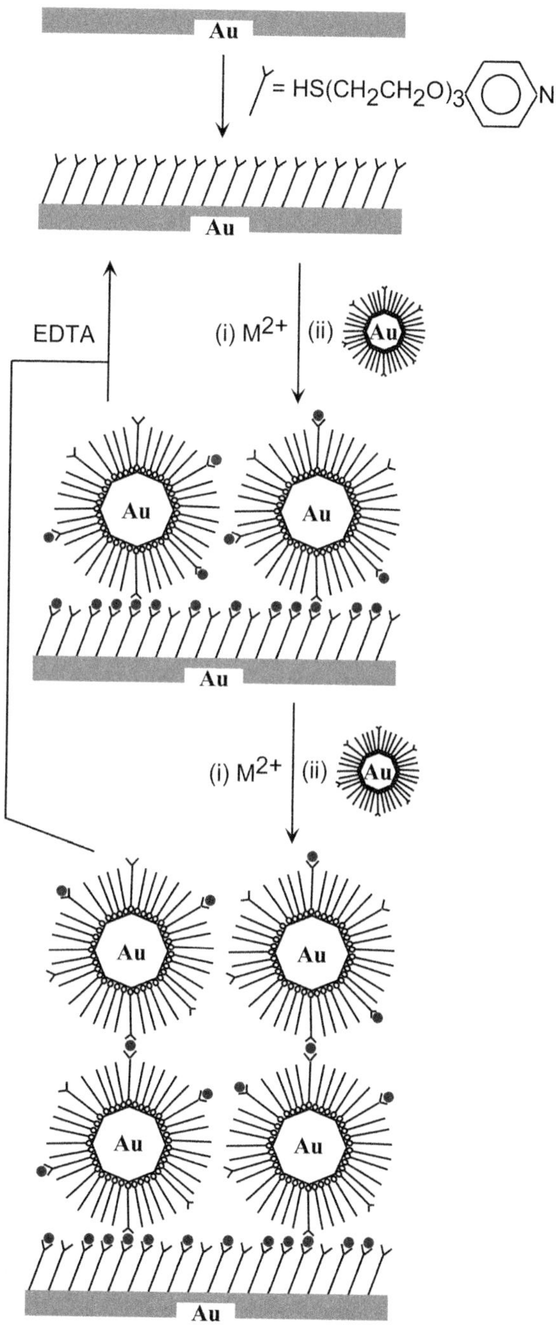

Figure 8.3 Procedure for MPC assembling at an Au electrode surface, similar to layer-by-layer deposition, by the chelating interactions between divalent (transition) metal ions and pyridine moieties.[14] (Reprinted with permission from S. Chen et al., *J. Phys. Chem. B* **2002**, *106*, 1903–1908. Copyright 2002 American Chemical Society.)

8.3 ELECTROCHEMICAL INTERROGATION OF SURFACE-BOUND NANOPARTICLES

Electrochemical analysis of nanoparticle surface ensembles can provide information on the surface area of the nanoparticles and characterization of the chemistries (functional groups) attached to the nanoparticles for further control of the unique properties of these supramolecular architectures. The potential cycling of many noble metal surfaces in acidic media leads to the formation and oxidation of both adsorbed hydrogen and oxygen and is a common method for determining the electroactive surface area of the metal.[15] This has been exploited to determine the surface area of gold nanoparticles on carbon and conducting glass substrates.[16] The charge associated with the cathodic peak for bis-bipyridinum cyclophane (**1**) cross-linked gold nanoparticles deposited onto an ITO electrode, shown in Figure 8.4, is directly related to the reduction of the oxide layer accumulated during the anodic scan and is used to determine the overall surface area of the particles. The number of nanoparticles on the electrode was then calculated based on the known particle diameter.[16d] A linear relationship between the oxidation and reduction waves was observed with increasing layers of gold nanoparticles showing that this technique is suitable to determine the number of nanoparticles in multilayer structures also, provided all nanoparticles are accessible to the background electrolyte and there is no interference from reactions at the underlying electrode. Information concerning the active surface area, the number of surface-bound nanoparticles, and density is important in determining the amount of

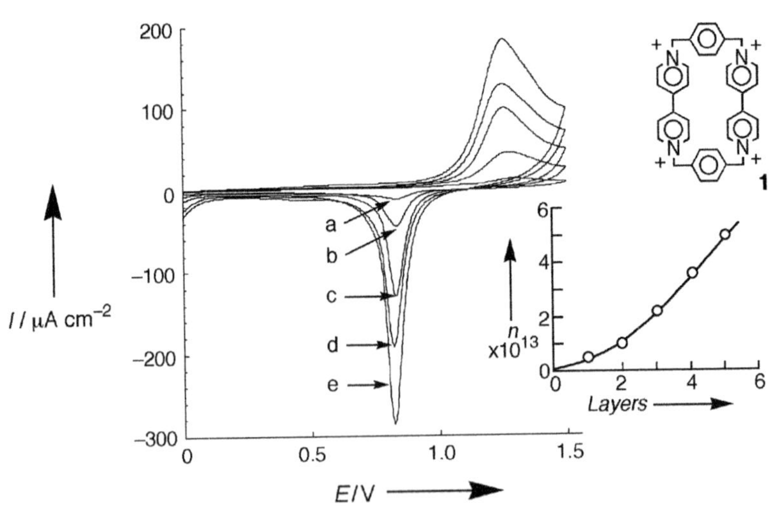

Figure 8.4 Cyclic voltammograms of one to five layers of bis-bipyridinum cyclophane (**1**) cross-linked AuNPs at an ITO electrode showing adsorption (oxidation) and desorption (reduction) of oxygen at the NP surface corresponding to surface area. 1.0 M H_2SO_4, scan rate = 50 mV s^{-1}. Inset: Calibration curve is the number of AuNPs cm^{-2} (n) versus number of layers.[4] (Reprinted with permission from A. N. Shipway et al., *ChemPhysChem* **2000**, *1*, 18–52. Copyright Wiley-VCH Verlag GmbH & Co. KGaA.)

surface modifiers to add and can play a profound role when tailoring a surface for analytical applications.

Nanoparticles can be synthesized by a variety of methods, such as electrodeposition,[17] ultrasound,[18] and gas-phase techniques,[19] while the most common method for nanoparticle synthesis is the chemical reduction of the metal precursor salt to form nanoparticle colloids. The latter method results in the capping of the nanoparticle where the passivating layer, which is responsible for controlling the nanoparticle size and stability, can also be used to add reactive groups for desired conjugation chemistries or functionalities. Deposition of mixed monolayer Au MPCs containing ferrocenyl,[20] amido-[21] or silylferrocenyl,[22] and biferrocenyl thiolate ligands have been examined electrochemically by scanning over the potential regions associated with the Fe^{II}/Fe^{III} reaction.[22a,23] Uniform redox active second oxidation processes of the biferrocenyl units of the AuNPs-biferrocenylthiolate electrode films was observed confirming that AuNPs containing multiple redox molecules could assemble according to charge accumulation. In contrast, the mono-ferrocenylthiolate protected clusters did not exhibit uniform oxidation processes, indicating the necessity for multi-electron systems for uniform deposition films[20b] allowing the construction of alternating multilayered structures of palladium and gold nanoparticles with connecting biferrocenyl groups onto an ITO electrode using this phenomenon.[24]

The supramolecular assembly of β-cyclodextrin-capped AuNPs on ITO previously coated with a monolayer of ferrocene molecules was shown to exhibit interesting redox properties, as presented in Figure 8.5.[25] This is a fine example of the functionalization of a nanoparticle with a functionalized recognition element to create a unique supramolecular scaffold with tailor-made properties. Specifically, the electrocatalytic determination of ascorbic acid was demonstrated by this surface assembly with excellent discrimination over dopamine.

An application of biomolecule–nanoparticle hybrid supramolecular assemblies is the sensitive detection of p53, a tumor suppressor protein, by AuNP streptavidin conjugates capped with multiple ferrocene molecules.[26] As shown in Figure 8.6, a mixed monolayer of double-stranded DNA and 1-hexanethiol on an Au electrode is used to capture p53, which binds to DNA containing the consensus site sequence. Subsequent conjugation of the captured p53 with biotin allowed for attachment of

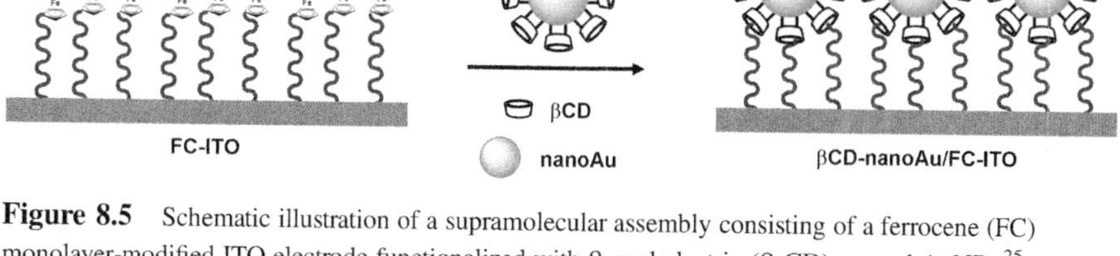

Figure 8.5 Schematic illustration of a supramolecular assembly consisting of a ferrocene (FC) monolayer-modified ITO electrode functionalized with β-cyclodextrin (β-CD)-capped AuNPs.[25] (Reprinted with permission from F. Zuo et al., *Electroanalysis* **2008**, *20*, 894–899. Copyright Wiley-VCH Verlag GmbH & Co. KGaA.)

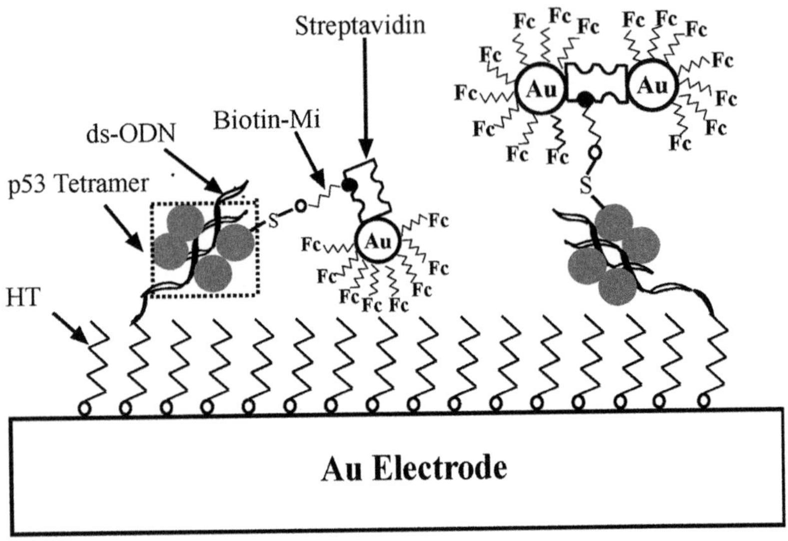

Figure 8.6 Schematic representation of a biomolecular-nanoparticle supramolecular assembly consisting of double-stranded DNA (ds-ODN) modified electrodes for the capture of p53 protein. Electrochemical signal generation and amplification occurred by subsequent labeling of the p53 tetramer with biotin (Biotin-Mi) and capture of streptavidin-modified Fc-capped AuNPs.[26] (Reprinted with permission from J. Wang et al., *Anal. Chem.* **2008**, *80*, 769–774. Copyright 2008 American Chemical Society.)

streptavidin modified ferrocene capped AuNPs. This system was capable of detecting wild-type p53 at cellular concentrations, complimenting the more common enzyme-linked immunosorbent assay (ELISA) which measures total p53 (wild-type and mutant p53), providing additional information relevant to choosing appropriate cancer therapies.

8.3.1 Scanning Probe Techniques

Scanning electrochemical microscopy (SECM) is a useful tool for studying electron transfer properties in nanoparticle films due to the micron dimensions of the SECM tip, making it ideal for localized interrogation of nanoparticle interfaces and studies of charge transfer properties. Hydrogen evolution at gold nanoparticles with a methyl viologen mediator was used to determine the electron transfer kinetics of a gold nanoparticle film on a nonconducting glass substrate using SECM.[27] The nonconducting substrate removed any possible interferences from the underlying surface, demonstrating the ability of SECM to measure rates of electron transfer reactions for immobilized nanoparticles and the possibility to investigate size effects on kinetics of reactions occurring at nanostructures. Quantitative measurements of the rate of interfacial electron transfers between Au phenylethylthiolate-capped MPCs dissolved in 1,2-dichloroethane and an aqueous redox species, $IrCl_6^{3-}$, was reported by Murray and coworkers and is presented in Figure 8.7.[28] As the SECM tip was lowered

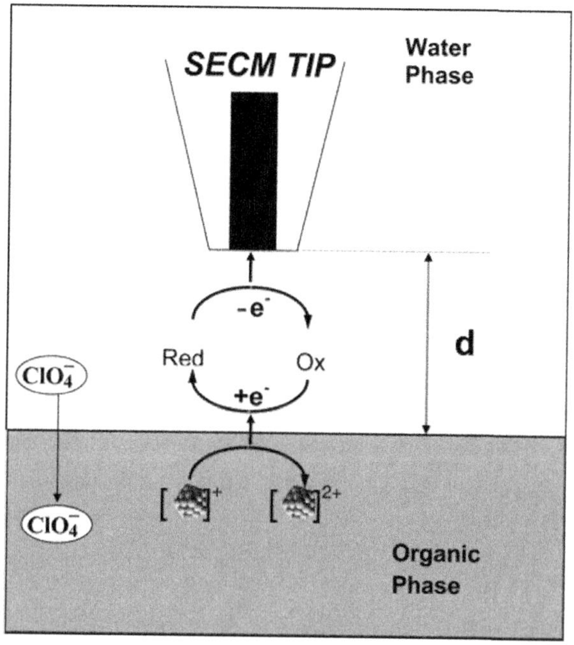

Figure 8.7 Schematic diagram of SECM approach measurement of the heterogeneous electron transfer rate between an organic-soluble MPC and an aqueous redox species, $IrCl_6^{3-}$. Electroneutrality was maintained by transfer of perchlorate ions across the interface.[28] (Reprinted with permission from D. G. Georganopoulou et al., *Nano Lett.* **2004**, *4*, 1763–1767. Copyright 2004 American Chemical Society.)

toward the dichloroethane layer containing the organic soluble MPCs, the aqueous $IrCl_6^{3-}$ oxidant began to react in an interfacial bimolecular electron transfer with the Au MPCs. Oxidation of the MPCs yielded the reduced form of the $IrCl_6^{3-}$ mediator, which was then reoxidized at the SECM tip. Determination of the apparent heterogeneous rate constant was achieved by theoretically fitting the tip current versus tip to interface distance curves.

Quinn et al. followed the compression of a monolayer of monodisperse thiol-capped Ag nanoparticles on a Langmuir trough using the SECM direct electrochemical visualization of the lateral connectivity of the particles in the film at the air–water interface with ferrocene methanol as the redox mediator.[29] The transition from negative feedback of the insulating substrate to positive feedback of the conducting substrate was monitored as the interparticle separation decreased by compressing the film. Ruiz et al. measured the lateral (in-plane) and cross-film electron transport properties in multilayer polymer–nanoparticle films shown in Figure 8.8.[30] The layer-by-layer assembly of Au MPCs containing mixed monolayers of hexanethiolate and mercaptoundecanoic acid ligands was constructed by alternating oppositely charged layers of polyelectrolytes. The conductivity of a single nanoparticle layer could be quantified as well as the film conductance as a function of the number of consecutive layers using the SECM technique.

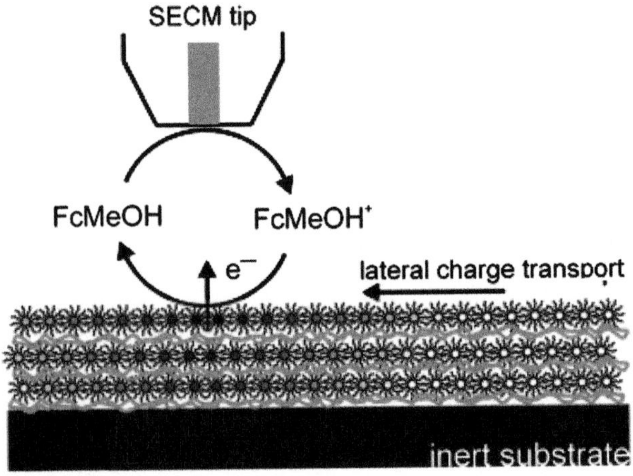

Figure 8.8 Schematic diagram showing SECM measurement of lateral (in-plane) and cross-film electron transport properties in multilayer polymer/nanoparticle films.[30] (Reprinted with permission from V. Ruiz et al., *Nano Lett.* **2003**, *3*, 1459–1462. Copyright 2003 American Chemical Society.)

Schiffrin and coworkers functionalized 6 nm gold nanoparticles with polymethylene chains containing electrochemically reversible bipyridinium units assembled onto surfaces to make a nanometer-scale electronic switch with scanning tunneling measurements.[31] Electron tunneling between the nanoparticle and the substrate was controlled by the potential applied, measured by the tip of a scanning tunneling microscope, causing electron injection to the redox center requiring less than 30 electrons to operate (with half of the redox ligands in their reduced state).

8.4 CONJUGATION OF BIOMOLECULES WITH NANOPARTICLES

The conjugation of nanoparticles with biomolecules has a profound impact on the catalytic and recognition properties of the hybrid material and the overall performance and construction of sensing devices and novel bio-nano structures. The relative size of biomolecules, ranging from nano- to micrometers, makes them ideal candidates for conjugation with nanoparticles; while specific functionality of the resulting hybrid structure is imparted from the evolutionary specificity of the biomaterial, such as DNA hybridization recognition through hydrogen bonding of complementary nucleic acid-base pairing, "lock and key" mechanisms such as antibody–antigen and substrate–enzyme recognition, and small molecule recognition resulting from (for example) various DNA aptamer– and peptide–host complexes.

The various capping agents used for the nanoparticle synthesis such as hydrophobic monolayers, positively or negatively charged hydrophilic monolayers, and polymer layers, can be exploited for biomolecular conjugation.[32] Electrostatic interactions represent the simplest form of biomolecule–nanoparticle conjugation. Many noble metal nanoparticles such as gold, silver, and platinum are routinely synthesized using a citric acid reducing agent that leaves a surface layer of carboxyl groups that are negatively charged at physiological pH and can be used to coat the nanoparticle with proteins that have positively charged amino acid side chains,[33] or the direct adsorption of heme-containing redox-enzymes.[34] The binding of basic leucine zipper proteins to lipoic acid stabilized CdSe/ZnS core–shell particles,[35] the electrostatic binding of proteins onto phospholipid coated gold nanoparticles,[36] and negatively charged DNA onto positively charged Cd^{2+}-rich CdS NPs have been reported.[37] A disadvantage of the electrostatic binding of molecules is that the reversibility of the binding can result in removal of the protein from the nanoparticle, and the possible conformational changes of proteins which result in loss of catalytic and biorecognition properties.

Chemisorption of biomolecules onto nanoparticles through ligand-exchange reactions has also been widely reported. The most common chemisorption bioconjugation method is substitution of weakly adsorbed nanoparticle stabilizers/capping agents with sulfur-containing molecules. Pioneering studies by the groups of Mirkin and Alivisatos exploited thiol-modified single-stranded DNA to prepare AuNP-DNA colloidal conjugates.[38] This procedure results in tethering of the DNA molecule onto the nanoparticle with the polyanionic strands stabilizing the structures leaving bases exposed in solution with minimal steric hindrance, ideal for hybridization to complementary strands.

The thiol-containing amino acid, cysteine, present in many peptide groups and on the surface of many proteins, can also be used to bind nanoparticles.[39] Penadés and coworkers have described several works using thiol-derivatized neoglycoconjugates to link to gold nanoparticles suitable for protein recognition and to study carbohydrate interactions.[40] Maran and coworkers recently reported the use of small α-aminoisobutyric acid (Aib)-based peptide ligands for the protection of Au MPCs with core diameters as small as 1.1 nm.[41] Unlike other MPCs employing small alkanethiol ligands, which are prone to lateral motion leading to possible solvent or electrolyte penetration through the capping layer, Aib MPCs show well-defined, rigid, and stable capping due to both intra- and interligand hydrogen bonding. The ability of the Aib peptide ligand to influence the MPC dipole moment was also exploited to significantly affect the redox properties of the MPC, shifting DPV monitored redox potentials from 33 mV to up to 800 mV depending on the number of peptide ligands on the MPC, as shown in Figure 8.9.[42] Incorporation of thiol groups by chemical means or genetic engineering into some proteins and enzymes is also possible.[43]

Nanoparticles can also be chemically coated with various oxide shells, resulting in increased stability over weakly adsorbed stabilizers allowing attachment of biomolecules. For instance, functionalization of silica-coated gold nanoparticles with 11-triethoxysilylundecanaldehyde provides a versatile method for biomolecule conjugation using bifunctional linkers.[44] Further bioconjugation of the aldehyde-modified SiO_2-coated nanoparticles with amino-terminated oligonucleotide and biotin was demonstrated.[44a] The wide variety of terminal functional groups available,

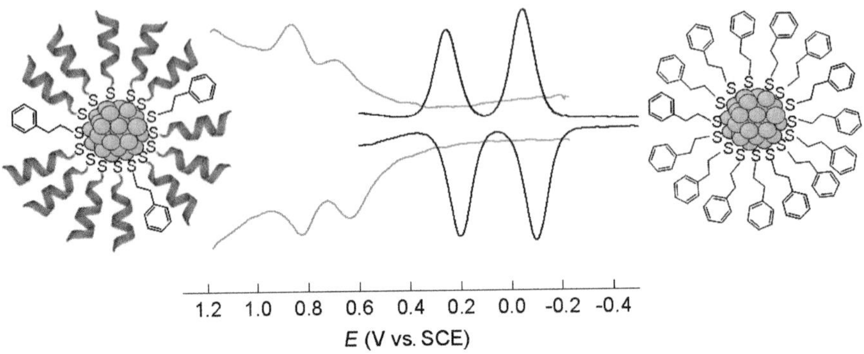

Figure 8.9 DPV of phenylethanethiolate protected Au MPCs (black), and mixed protected Au MPCs with phenylethanethiolate and Aib peptides modified by ligand exchange (gray) showing significant shift in redox potentials due to Aib peptide induced shift in particle dipole moment.[42] (Reprinted with permission from A. H. Holm et al., *Langmuir* **2006**, *22*, 10584–10589. Copyright 2006 American Chemical Society.)

that is, amine, maleimide, carboxyl, etc., make covalent cross-linking possible and increase stability of the nanoparticle–biomolecule hybrid structure.

The preparation of nanoparticles with single functional units provides an interesting approach to attach a particle to a single biomolecule, or to prevent aggregation when attempting to conjugate to a biomolecule such as a protein that has multiple binding sites. In a recent example, Brozik and coworkers used 1.4 nm Au nanoparticles with a single *N*-hydroxysuccinimide functional group to modify anti-TNF-α antibodies, resulting in a gold nanoparticle-modified antibody through conjugation to free amino groups on the antibody surface, as shown in Figure 8.10.[45] An electroless deposition of Pd using an ascorbic acid reducing agent resulted in the coating of the gold nanoparticles with a Pd shell yielding PdNP-coated antibodies that were sensitive to the oxygen reduction reaction. These antibodies proved effective as a secondary probe to create a reagent-less electrochemical immunoassay with a low detection limit comparable to typical ELISA-type assays.

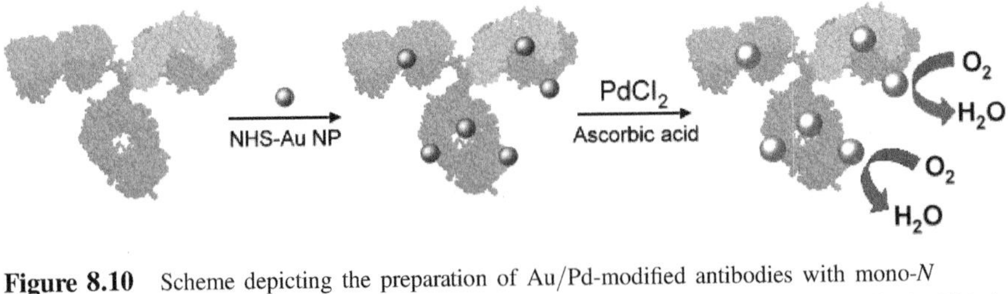

Figure 8.10 Scheme depicting the preparation of Au/Pd-modified antibodies with mono-*N* hydroxysuccinimide 1.4 nm AuNPs which covalently attached to anti-mouse TNF-α antibody. This is followed by electroless deposition of Pd onto the AuNP seeds.[45] (Reprinted with permission from R. Polsky et al., *Chem. Commun.* **2007**, 2741–2743. Copyright the Royal Society of Chemistry.)

8.5 BIOMOLECULE–NANOPARTICLE HYBRID SYSTEMS

8.5.1 Nanoparticle-Enhanced Enzyme Bioelectrocatalysis

Nanoparticles are well suited for construction of third generation enzyme biosensors that rely on the direct electron transfer between redox centers buried within the protein and electrodes. The small size of the particles provides ideal environments for contacting redox centers with little loss of bioactivity, while the conductive nature of the metal particles facilitates communication with the underlying electrode by providing a means of electron shuttling.

Gold nanoparticles have been used to immobilize micro-peroxidase 11,[46] tyrosinase,[47] and hemoglobin[48] to construct amperometric sensors, while silver nanoparticles have been used to enhance electron transfer of cytochrome c and myoglobin onto pyrolitic graphite electrodes.[49] The use of semiconductor and oxide nanoparticles has also been reported, such as horseradish peroxidase (HRP) on TiO_2 nanoparticles,[50] as well as Fe_3O_4 and MnO_4 nanoparticles to immobilize and facilitate direct electron transfer.[51]

As the previous works relied on the random and nonordered adsorption of the enzymes on the conductive nanoparticles, a large distribution of energy barriers results which can severely limit the effectiveness of direct electron transfer to the electrode. A highly sophisticated method for contacting glucose oxidase through a single 1.4 nm Au_{55} nanoparticle that was functionalized with N^6-(2-aminoethyl)flavin adenine dinucleotide (FAD cofactor, amine derivative; **1**) was reported by the Willner group, as shown in Figure 8.11.[52] The 1.4 nm Au-FAD conjugate was reconstituted with apo-glucose oxidase and assembled on a thiolated monolayer with various dithiols (**2** to **4**) as linkers onto a gold electrode. Alternatively, the gold nanoparticle-FAD layer was initially immobilized on the gold electrode with the thiolated monolayer followed by reconstitution of the apo-glucose oxidase (Fig. 8.11b). The bioelectrocatalyzed oxidation of glucose was accomplished in this fashion by the electrical contact established through the nanoparticle-reconstituted apo-glucose oxidase layer resulting in a very fast electron transfer rate of about $5000\ s^{-1}$ without interference from ascorbic acid. This electron transfer rate is about seven times faster than that between glucose oxidase and its natural cofactor, oxygen, resulting in a glucose enzyme-electrode that is insensitive to oxygen.

8.5.2 Nanoparticle–Enzyme Supramolecular Complexes

Immobilization of nanoparticles and enzymes onto electrode surfaces using various entrapment chemistries, that is, sol–gels, layered polyions, and hydrogels, has allowed for greater functionality and improved stability. Crumbliss and coworkers demonstrated the direct electron transfer between the redox center of HRP by evaporation

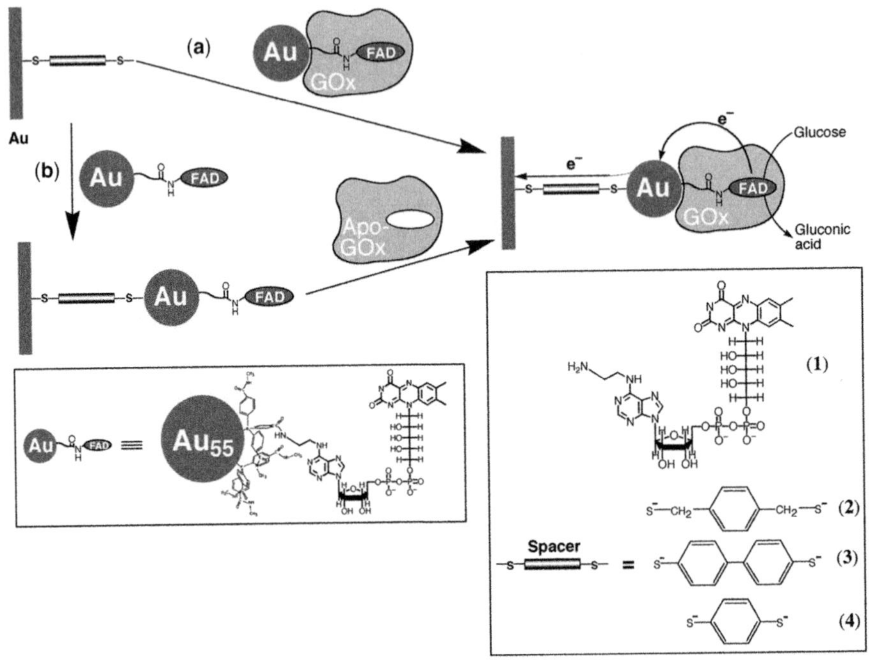

Figure 8.11 Electrical coupling of AuNP-reconstituted glucose oxidase to a Au electrode by: (a) adsorption of AuNP-reconstituted glucose oxidase to a dithiol monolayer-modified Au electrode; (b) adsorption of AuNPs functionalized with FAD on the dithiol-modified gold electrode followed by reconstitution of apo-glucose oxidase onto the functional AuNPs. Also shown are the structures for the FAD-modified AuNP and various spacer molecules used.[52] (Reprinted with permission from Y. Xiao et al., *Science* **2003**, *299*, 1877–1881. Copyright AAAS.)

of a HRP-AuNP conjugate sol–gel solution onto a glassy carbon electrode to obtain a mediator-free H_2O_2 sensor.[53] A thiol-containing AuNP sol–gel network was also used to immobilize HRP onto a gold electrode where the thiol functionality served to chemisorb the AuNPs followed by adsorption of HRP onto the AuNPs, leading to direct electron transfer and effective H_2O_2 detection.[54]

The layer-by-layer incorporation of glucose oxidase (GOx) and Prussian blue nanoparticles (PB-NP) to produce a multilayer film amperometric sensor was reported.[55] Positively charged poly(diallyldimethylammonium) chloride (PDDA) and negatively charged poly(sodium 4-styrenesulfonate) (PSS) were first deposited onto a 3-mercapto-1-propanesulfonic acid-modified gold electrode, as shown in Figure 8.12. This was followed by adsorption of PDDA modified PB-NPs (Fig. 8.12, I) and negatively charged GOx (Fig. 8.12, II). The resulting sensor's response to glucose was tailored by controlling the number of PB-NP and GOx bilayers. PB-NPs were used as catalysts for the reduction of H_2O_2 produced by the oxidation of glucose by GOx.

A biocomposite composed of a chitosan hydrogel, GOx and AuNPs immobilized onto a gold electrode via a single step electrodeposition was used to construct an

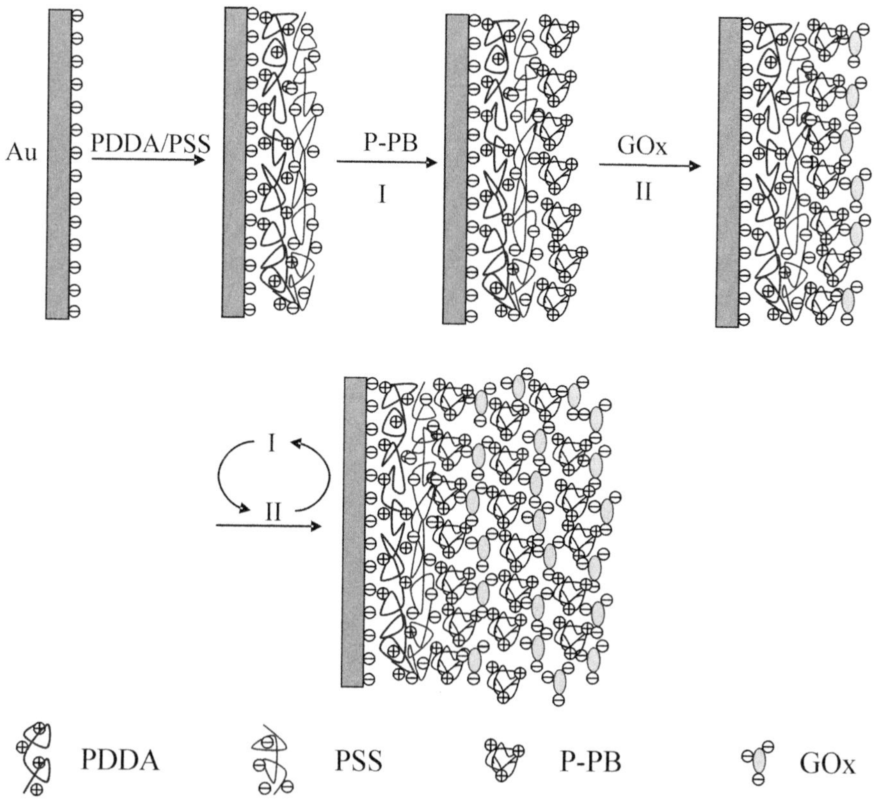

Figure 8.12 Schematic illustration of the layer-by-layer deposition on an Au electrode initially with positively charged poly(diallydimethylammonium) chloride (PDDA) and negatively charged poly(sodium 4-styrenesulfonate). Subsequent depositions entailed: (I) PDDA-modified Prussian blue nanoparticles (P-PB), followed by (II) negatively charged glucose oxidase (GOx).[55] (Reprinted with permission from W. Zhao et al., *Langmuir* **2005**, *21*, 9630–9634. Copyright 2005 American Chemical Society.) (See color insert.)

amperometric glucose sensor.[56] Electrodeposition of the chitosan hydrogel occurred via applying an oxidative potential to a solution of chitosan, GOx, and AuNPs, reducing H^+ ions to H_2, increasing the pH at the cathode until chitosan was no longer soluble ($>$ pH 6.3) and deposited on the electrode. Incorporation of AuNPs was shown to improve the stability of the immobilized GOx and the oxidation of H_2O_2 at the Au electrode.

A three-enzyme cascade employing AuNPs for the electrochemical detection of cholesterol in serum and whole blood using cholesterol oxidase (COx) and cholesterol esterase (CE) immobilized together in a carageenan hydrogel was reported by Crumbliss and coworkers.[57] Hydrolysis of cholesterol esters in whole blood was catalyzed by the entrapped CE, forming the free cholesterol which was then oxidized by COx, producing H_2O_2. The H_2O_2 then diffuses through the hydrogel to the underlying glassy carbon electrode that was modified with HRP and AuNPs. The separation of COx and CE from the underlying HRP-AuNP surface was necessary as COx adsorption onto AuNPs inactivated the enzyme. The current generated by HRP catalyzed H_2O_2 reduction was strongly correlated to total cholesterol levels in serum and whole blood.

8.5.3 DNA and Protein Biosensing

8.5.3.1 Nanoparticle-Enhanced Electron Conduction-Based Sensing

Conductimetric biosensors operate by monitoring changes in conductance, or the passage of electrons (the reciprocal of resistance), between two electrodes. The interaction or binding of biomolecules between the two electrodes may alter the measured conduction, but these subtle changes are often masked by the large background of the highly ionic electrolyte buffer. The use of conductivity combined with nanoparticles as a technique for the detection of biomolecules was introduced by Velev and Kaler[58] based on silver precipitation onto gold nanoparticles in between electrode gaps, resulting in the significant altering of resistance and the direct electric conductivity, as presented in Figure 8.13. In this initial work, the *in situ* assembly of colloidal latex particles modified with IgG protein capture antibodies onto micropatterned electrodes was used as the biosensor platform. After target protein capture and conjugation with a colloidal gold-labeled secondary antibody probe, silver precipitation onto the assembly resulted in the enhanced resistance/conductivity readout.

Mirkin's group expanded on this concept for DNA analysis by immobilizing ssDNA oligonucleotide probes between 20 μm electrode gaps. Following target DNA hybridization, another hybridization step was performed with DNA-modified colloidal gold nanoparticles, silver precipitation, and a conductimetric readout resulting in a detection limit of 500 femtomolar for a DNA target with a point mutation selectivity of ~100,000 : 1.[59] Gold nanoparticles modified with polyaniline as a secondary probe for a disposable immunochromatographic sensor on a screen printed membrane strip for the detection of human serum albumin,[60] and the use of conductive ITO nanoparticles modified with DNA oligonucleotides for the construction of a DNA sensor[61] based on similar conductimetric readouts in between electrode gaps were reported. The conductive properties of the nanoparticles in these two works obviated the need for the silver precipitation step, simplifying the protocols.

8.5.3.2 Nanoparticles as Labels for Electrochemical Biodetection

The use of gold nanoparticles as an electrochemical label for bionanalysis was introduced by Costa-Garcia and coworkers.[62] Adsorptive voltammetry was used to detect the gold signal and monitor biotin–streptavidin interactions on a pretreated carbon paste electrode with an adsorbed biotinylated albumin layer after immersion into a solution of streptavidin-coated colloidal gold. While colloidal gold had found widespread use as a histochemical label for biomolecules and an optical label based on absorption

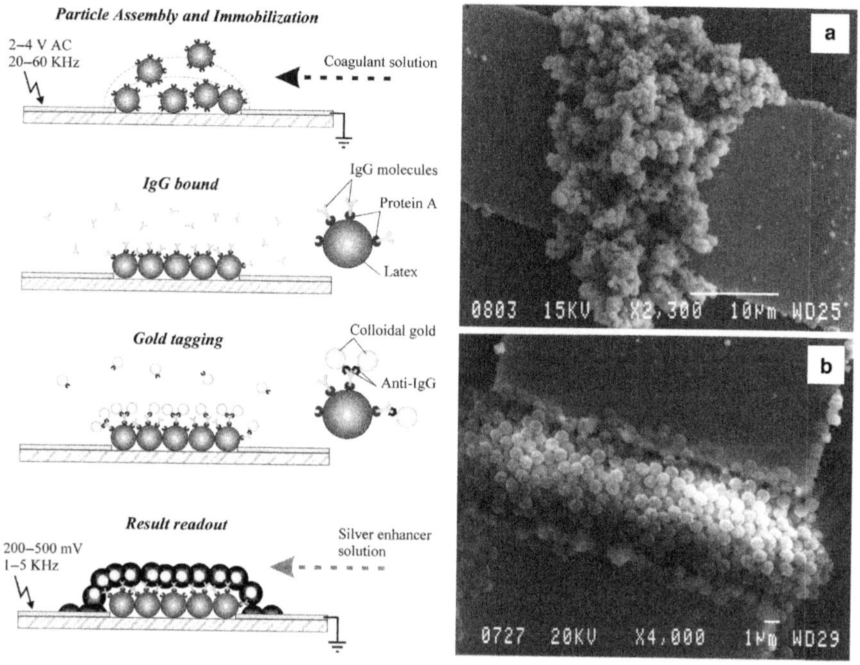

Figure 8.13 Schematic diagram illustrating assembly of the biosensor with protein A- and IgG-modified latex beads via dielectrophoresis in the micrometer-sized gaps between planar electrodes. AuNP-conjugated anti-IgG antibodies bind to the IgG-modified latex spheres facilitating Ag deposition and electron conduction, as shown in the SEM micrograph (a). A negative control sample showing minimal Ag deposition is shown in SEM micrograph (b).[58] (Adapted with permission from O. D. Velev and E. W. Kaler, *Langmuir* **1999**, *15*, 3693–3698. Copyright 1999 American Chemical Society.)

of surface plasmon bands, this work ushered in the electrochemical use of gold nanoparticles as labels for detection of biomolecules. The particular electrochemical technique involved the anodic dissolution of the gold using an applied potential followed by the monitoring of the reductive signal during a subsequent anodic scan. Limoges and coworkers expanded upon this and developed an electrochemical immunoassay for detection of immunoglobin G, as well as a DNA detection method for quantification of a human cytomegalovirus DNA sequence using an anodic stripping voltammetric technique and the dissolution of the gold nanoparticles in acidic bromine media.[63] Simultaneously, Wang and coworkers used a similar method for detection of the BRCA1 breast cancer gene via hybridization of a biotinylated oligonucleotide target to magnetic bead-conjugated oligonucleotide probes.[64] The detection of the hybridized target was accomplished by subsequent labeling with streptavidin-coated colloidal gold followed by dissolution and detection of the gold label using potentiometric stripping analysis at a thick film screen printed carbon electrode after magnetic separation. Other nanoparticles that have been used for electrochemical DNA detection in similar fashion include CdS,[65] gold-coated Fe,[66] Ag,[67] and tris(2,2′-bipyridyl) cobalt(III)-doped SiO_2 nanoparticles.[68]

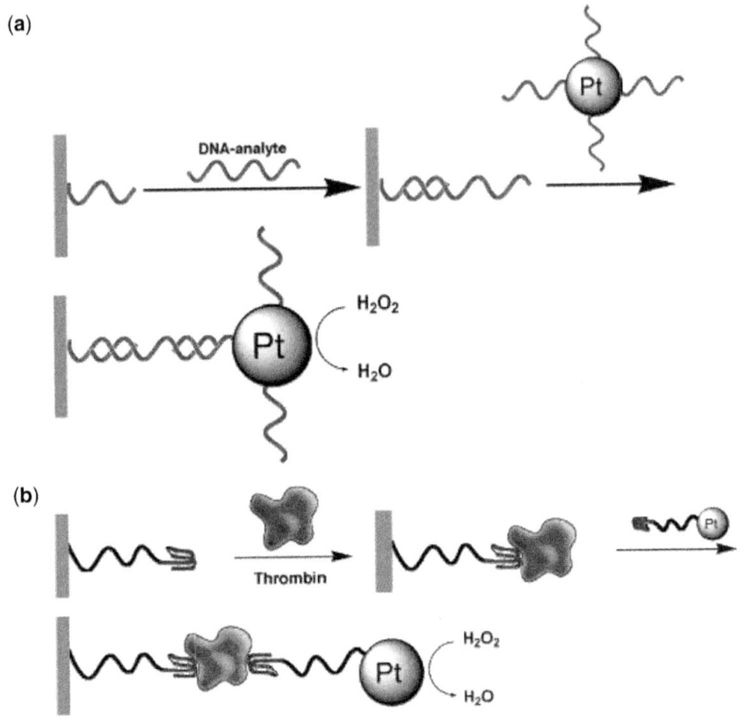

Figure 8.14 Scheme depicting the analytical procedure for biodetection based on catalytic H_2O_2 reduction at PtNPs: (a) Detection probe modified PtNPs for DNA detection; (b) thrombin aptamer DNA sequence modified PtNPs for thrombin protein detection.[69] (Reprinted with permission from R. Polsky et al., *Anal. Chem.* **2006**, *78*, 2268–2271. Copyright 2006 American Chemical Society.)

The electrocatalytic properties of nanoparticles have also been exploited for biomolecule detection. Analogous to the catalytic turnover of enzymes used as labels to produce electroactive products, metallic nanoparticles can be used as catalysts for amperometric detection without the inherent environmental instability associated with using enzyme labels. Introduced by Willner and coworkers, DNA-coated platinum nanoparticles were used to detect DNA, or the protein thrombin, as shown in Figure 8.14a, b, respectively.[69] Catalytic reduction of H_2O_2 at the PtNPs produced a significant amperometric signal providing comparable or improved sensitivity over common aptamer-based protein detection or enzymatic labeling assays. In a subsequent study, discussed in Section 8.4 of this chapter, PdNP-modified antibodies, sensitive to the oxygen reduction reaction, were used for the detection of the cytokine, TNF-α.[45] The use of Pd to catalyze oxygen reduction resulted in a pseudo-reagentless sensor as oxygen is ubiquitous in aqueous solutions.

8.5.3.3 *Signal Amplification*

One of the biggest hurdles in DNA sensing is that low copy numbers are typically present and thus preamplification using the polymerase chain reaction is required. High throughput proteomics demands large numbers of proteins to be detected at low concentrations. To overcome these obstacles several schemes have been developed to further amplify electrochemical signals. A signal amplification method was demonstrated where additional gold was deposited onto the nanoparticles from a solution of gold salt and a chemical reducing agent, leading to an enlargement of the particles and a 65 times increase of the signal resulting from DNA hybridization.[64] Similarly, silver precipitation onto the gold nanoparticles followed by nitric acid dissolution and detection of the dissolved silver using the same stripping method resulted in a detection limit in the picomolar range.[70] A triple amplification strategy yielding attomolar DNA detection was also demonstrated. As shown in Figure 8.15, streptavidin-coated polystyrene spheres bound with multiple biotinylated AuNPs were used as DNA labels.[71] Upon capture of the spheres by the probe DNA, further catalytic growth of gold onto the AuNPs, followed by Au dissolution and stripping voltammetry, provided significant signal enhancement. Tremendous sensitivity can be achieved in this fashion due to the "built-in" preconcentration step of deposited metal inherent in the electrochemical stripping technique and the catalytic enlargement of the metal nanoparticles. Particular emphasis must be placed, however, on controlling the deposition time so as to limit nonspecific deposition to minimize background signals. Additional enhancement techniques that have been reported include precipitation of Cd onto CdS nanoparticles[65] and Cu onto AuNPs for the detection of human IgG.[72]

A solid-state signal transduction was also accomplished by silver-enhanced precipitation onto DNA-coated AuNPs after hybridization to complementary DNA strand-coated magnetic beads. The silver-enhanced colloidal gold DNA-magnetic bead conjugate was collected onto an electrode by positioning a magnet behind the electrode transducer, followed by chronopotentiometric detection.[73] The precipitation of silver over the entire conjugate and the magnetic attraction contacted the whole assembly with the electrode and allowed electrochemical stripping to quantify DNA

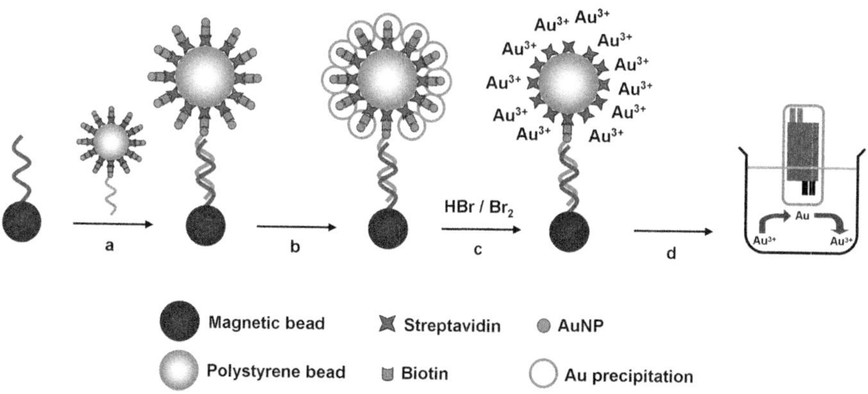

Figure 8.15 Schematic illustration of amplified DNA detection employing probe DNA modified magnetic beads. (a) Treatment of probe with target DNA modified with streptavidin-conjugated polystyrene beads loaded with biotinlyated AuNPs followed by (b) Au precipitation onto AuNP seeds, (c) dissolution of the Au, and (d) detection via electrochemical stripping.[71] (Adapted with permission from A.-N. Kawde and J. Wang, *Electroanalysis* **2004**, *16*, 101–107. Copyright Wiley-VCH Verlag GmbH & Co. KGaA.)

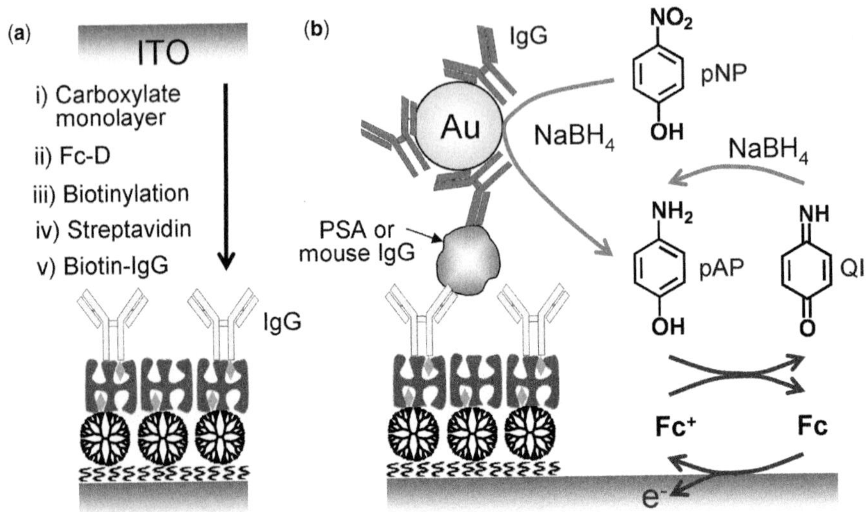

Figure 8.16 Schematic representation of (a) the preparation of the immunosensing layer and (b) the electrochemical-chemical catalytic cycle amplified detection of mouse IgG or prostate specific antigen (PSA).[75] (Reprinted with permission from J. Das et al., *J. Am. Chem. Soc.* **2006**, *128*, 16022–16023. Copyright 2006 American Chemical Society.)

hybridization without dissolution of the metal, obviating the need for the acidic dissolution media. Alegret and coworkers modified this approach by using Au_{67} nanoparticles to modify DNA in a 1 : 1 ratio, hybridizing complementary DNA, and using DNA-modified magnetic beads as a secondary probe, followed by the anchoring of the hybridized assembly onto a magnetic graphite-epoxy composite electrode for the direct electrochemical detection of the gold labels.[74]

An AuNP-mediated electrochemical signal amplification strategy that did not require precipitation of metal for enhancement was reported by Yang and coworkers.[75] In this typical sandwich-type assay, shown in Figure 8.16, a layer of IgG capture probe was immobilized to an ITO electrode with an underlying layer of ferrocenyl-tethered dendrimers (Fc-D). Following binding of the target protein, treatment with IgG-labeled 10 nm AuNPs completed the sandwich (Fig. 8.16b) and served as a catalyst for the six-electron reduction of *p*-nitrophenol (pNP) to *p*-aminophenol (pAP). Generated pAP molecules were oxidized to *p*-quinone imine (QI) by the immobilized ferrocene groups. Oxidized QI were then reduced back to pAP by $NaBH_4$, forming a fast electrochemical-chemical catalytic cycle. This system achieved an impressive detection limit of 1 fg mL^{-1} (approximately 7 aM) with a dynamic range that covered 10 orders of magnitude.

8.5.3.4 *Multianalyte Detection*

Taking advantage of the ability to detect multiple group II and group III metals at mercury film electrodes using stripping analysis, a multianalyte DNA sensor was developed using different semiconductor nanocrystals (ZnS, CdS, and PbS) for the

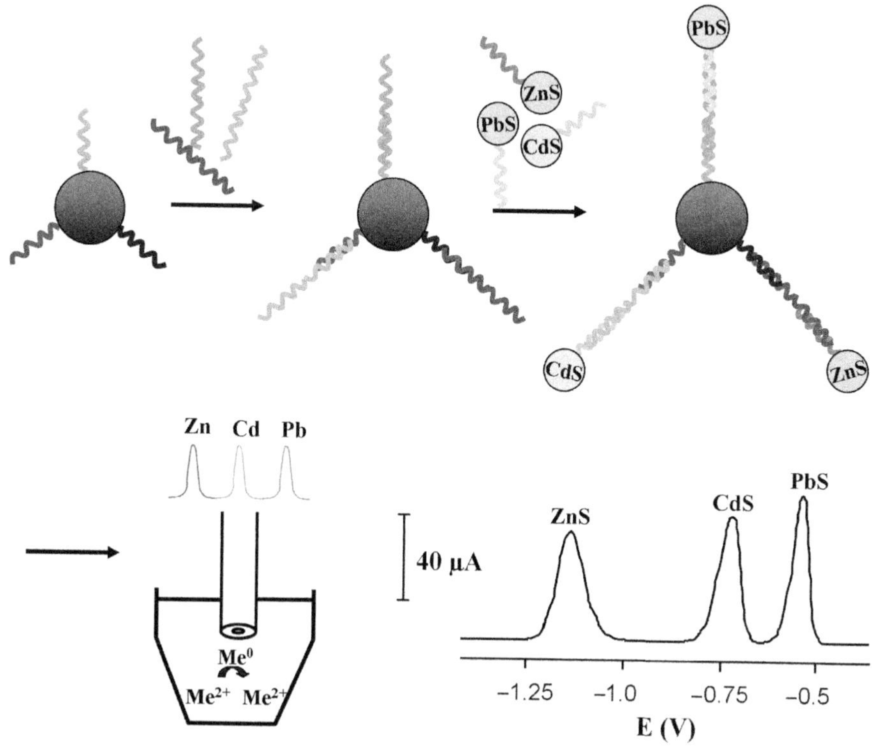

Figure 8.17 Schematic diagram for detection of three unique DNA sequences via hybridization of target DNA with magnetic bead-conjugated probe DNA followed by hybridization with three different detection probe DNA-modified semiconductor nanoparticles, ZnS, CdS, and PbS. Dissolution and electrochemical stripping yielded three well-resolved peaks corresponding to the specific metal, and hence, unique DNA sequences detected.[76] (Reprinted with permission from J. Wang et al., *J. Am. Chem. Soc.* **2003**, *125*, 3214–3215. Copyright 2003 American Chemical Society.) (See color insert.)

simultaneous detection of three different DNA sequences.[76] Dissolution of the semiconductor nanocrystals in 1 M HNO_3 followed by stripping analysis at a mercury film-coated glassy carbon electrode resulted in three well-defined stripping peaks for the Zn, Cd, and Pb tracers, as shown in Figure 8.17. This method of using multiple semiconductor nanoparticle labels was further developed to construct an electrochemical immunoassay to simultaneously detect β_2-microglobulin, IgG, bovine serum albumin, and C-reactive proteins,[77] as well as the use of DNA aptamer probes for the detection of proteins based on the competitive displacement of PbS- and CdS-modified thrombin and lysozyme, respectively.[78]

8.6 CONCLUSION

Despite a rich diversity of promising approaches reviewed in this chapter, many of the potential applications of nanoparticle and biomolecular–nanoparticle hybrid supramolecular complexes have yet to be realized. Tremendous advances have been made towards an understanding of their dynamics, including how these complex systems assemble and function. The electrochemical signaling of such systems has provided valuable insights relating how the theoretical aspects behind the size-dependent properties of nanoparticles can translate into a functional system with real world applications. Electrochemical methods can in many instances be used as an ideal tool for the manipulation and control of organized architectures and the construction of useful transduction platforms. Future prospects for electrochemical-based nanoparticle supramolecular systems are promising, with efforts being driven towards chemical and biological sensing devices that can have implications in areas such as environmental, microelectronic, defense, pharmaceutical, and health-related fields.

REFERENCES

1. (a) E. Katz, I. Willner, J. Wang, *Electroanalysis* **2004**, *16*, 19–44; (b) Y. N. Xia, X.-M. Zhao, G. M. Whitesides, *Microelectron. Eng.* **1996**, *32*, 255–268.
2. M.-C. Daniel, D. Astruc, *Chem. Rev.* **2004**, *104*, 293–346.
3. W. Lu, C. M. Lieber, *Nature Mater.* **2007**, *6*, 841–850.
4. A. N. Shipway, E. Katz, I. Willner, *ChemPhysChem* **2000**, *1*, 18–52.
5. S. J. Green, J. J. Stokes, M. J. Hostetler, J. J. Pietron, R. W. Murray, *J. Phys. Chem. B* **1997**, *101*, 2663–2668.
6. M. G. Bawendi, W. L. Wilson, L. Rothberg, P. J. Carroll, T. M. Jedju, M. L. Steigerwald, L. E. Brus, *Phys. Rev. Lett.* **1990**, *65*, 1623–1626.
7. L. E. Brus, *J. Chem. Phys.* **1983**, *79*, 5566–5571.
8. S. K. Haram, B. M. Quinn, A. J. Bard, *J. Am. Chem. Soc.* **2001**, *123*, 8860–8861.
9. (a) J. F. Hicks, F. P. Zamborini, R. W. Murray, *J. Phys. Chem. B* **2002**, *106*, 7751–7757; (b) J. F. Hicks, D. T. Miles, R. W. Murray, *J. Am. Chem. Soc.* **2002**, *124*, 13322–13328.
10. B. M. Quinn, P. Liljeroth, V. Ruiz, T. Laaksonen, K. Kontturi, *J. Am. Chem. Soc.* **2003**, *125*, 6644–6645.
11. D. T. Miles, R. W. Murray, *Anal. Chem.* **2003**, *75*, 1251–1257.
12. J. J. Pietron, J. F. Hicks, R. W. Murray, *J. Am. Chem. Soc.* **1999**, *121*, 5565–5570.
13. (a) S. Chen, *J. Phys. Chem. B* **2000**, *104*, 663–667; (b) S. Chen, *J. Am. Chem. Soc.* **2000**, *122*, 7420–7421; (c) S. Chen, R. Pei, *J. Am. Chem. Soc.* **2001**, *123*, 10607–10615; (d) F. P. Zamborini, J. F. Hicks, R. W. Murray, *J. Am. Chem. Soc.* **2000**, *122*, 4514–4515.
14. S. Chen, R. Pei, T. Zhao, D. J. Dyer, *J. Phys. Chem. B* **2002**, *106*, 1903–1908.
15. A. J. Bard, L. R. Faulkner, *Electrochemical Methods*, New York: Wiley, **2001**.
16. (a) B. E. Baker, N. J. Kline, P. J. Treado, M. J. Natan, *J. Am. Chem. Soc.* **1996**, *118*, 8721–8722; (b) M. O. Finot, G. D. Braybrook, M. T. McDermott, *J. Electroanal. Chem.* **1999**, *466*, 234–241; (c) M. Lahav, A. N. Shipway, I. Willner, *J. Chem. Soc. Perkin Trans. 2* **1999**, *9*, 1925–1931; (d) M. Lahav, A. N. Shipway, I. Willner, M. B. Nielsen, J. F. Stoddart, *J. Electroanal. Chem.* **2000**, *482*, 217–221.
17. B. Alperson, H. Demange, I. Rubinstein, G. Hodes, *J. Phys. Chem. B* **1999**, *103*, 4943–4948.
18. A. Gedanken, *Ultrason. Sonochem.* **2004**, *11*, 47–55.
19. F. E. Kruis, H. Fissan, A. Peled, *J. Aerosol Sci.* **1998**, *29*, 511–535.
20. (a) M. J. Hostetler, S. J. Green, J. J. Stokes, R. W. Murray, *J. Am. Chem. Soc.* **1996**, *118*, 4212–4213; (b) R. S. Ingram, M. J. Hostetler, R. W. Murray, *J. Am. Chem. Soc.* **1997**, *119*, 9175–9178.
21. (a) A. Labande, J. Ruiz, D. Astruc, *J. Am. Chem. Soc.* **2002**, *124*, 1782–1789; (b) A. Labande, D. Astruc, *Chem. Commun.* **2000**, 1007–1008.

22. (a) M.-C. Daniel, J. Ruiz, S. Nlate, J. Palumbo, J.-C. Blais, D. Astruc, *Chem. Commun.* **2001**, 2000–2001; (b) M.-C. Daniel, J. Ruiz, S. Nlate, J.-C. Blais, D. Astruc, *J. Am. Chem. Soc.* **2003**, *125*, 2617–2628.
23. D. Astruc, J.-C. Blais, M.-C. Daniel, S. Nlate, J. Ruiz, *C.R. Chim.* **2003**, *6*, 1117–1127.
24. M. Yamada, H. Nishihara, *Chem. Commun.* **2002**, 2578–2579.
25. F. Zuo, C. Luo, Z. Zheng, X. Ding, Y. Peng, *Electroanalysis* **2008**, *20*, 894–899.
26. J. Wang, X. Zhu, Q. Tu, Q. Guo, C. S. Zarui, J. Momand, X. Z. Sun, F. Zhou, *Anal. Chem.* **2008**, *80*, 769–774.
27. J. Zhang, R. M. Lahtinen, K. Kontturi, P. R. Unwin, D. J. Schiffrin, *Chem. Commun.* **2001**, 1818–1819.
28. D. G. Georganopoulou, M. V. Mirkin, R. W. Murray, *Nano Lett.* **2004**, *4*, 1763–1767.
29. B. M. Quinn, I. Prieto, S. K. Haram, A. J. Bard, *J. Phys. Chem. B* **2001**, *105*, 7474–7476.
30. V. Ruiz, P. Liljeroth, B. M. Quinn, K. Kontturi, *Nano Lett.* **2003**, *3*, 1459–1462.
31. D. I. Gittins, D. Bethell, D. J. Schiffrin, R. J. Nichols, *Nature* **2000**, *408*, 67–69.
32. E. Katz, I. Willner, *Angew. Chem. Int. Ed.* **2004**, *43*, 6042–6108.
33. (a) W. Shenton, S. A. Davis, S. Mann, *Adv. Mater.* **1999**, *11*, 449–452; (b) M. J. Meziani, H. W. Rollins, L. F. Allard, Y.-P. Sun, *J. Phys. Chem. B* **2002**, *106*, 11178–11182; (c) H. Larsericsdotter, S. Oscarsson, J. Buijs, *J. Colloid Interface Sci.* **2001**, *237*, 98–103.
34. (a) J. B. Broderick, M. J. Natan, T. V. O'Halloran, R. P. Van Duyne, *Biochemistry* **1993**, *32*, 13771–13776; (b) I. D. G. Macdonald, W. E. Smith, *Langmuir* **1996**, *12*, 706–713.
35. H. Mattoussi, J. M. Mauro, E. R. Goldman, G. P. Anderson, V. C. Sundar, F. V. Mikulec, M. G. Bawendi, *J. Am. Chem. Soc.* **2000**, *122*, 12142–12150.
36. D. Ibano, Y. Yokota, T. Tominaga, *Chem. Lett.* **2003**, *32*, 574–575.
37. (a) R. Mahtab, J. P. Rogers, C. J. Murphy, *J. Am. Chem. Soc.* **1995**, *117*, 9099–9100; (b) R. Mahtab, J. P. Rogers, C. P. Singleton, C. J. Murphy, *J. Am. Chem. Soc.* **1996**, *118*, 7028–7032; (c) J. R. Lakowicz, I. Gryczynski, Z. Gryczynski, K. Nowaczyk, C. J. Murphy, *Anal. Biochem.* **2000**, *280*, 128–136.
38. (a) C. A. Mirkin, R. L. Letsinger, R. C. Mucic, J. J. Storhoff, *Nature* **1996**, *382*, 607–609; (b) A. P. Alivisatos, K. P. Johnsson, X. G. Peng, T. E. Wilson, C. J. Loweth, M. P. Bruchez, Jr., P. G. Schultz, *Nature* **1996,** *382*, 609–611.
39. K. Naka, H. Itoh, Y. Tampo, Y. Chujo, *Langmuir* **2003**, *19*, 5546–5549.
40. J. M. de la Fuente, A. G. Barrientos, T. C. Rojas, J. Rojo, J. Canada, A. Fenández, S. Penadés, *Angew. Chem. Int. Ed.* **2001**, *40*, 2257–2261.
41. L. Fabris, S. Antonello, L. Armelao, R. L. Donkers, F. Polo, C. Tonioto, F. Maran, *J. Am. Chem. Soc.* **2006**, *128*, 326–336.
42. A. H. Holm, M. Ceccato, R. L. Donkers, L. Fabris, G. Pace, F. Maran, *Langmuir* **2006**, *22*, 10584–10589.
43. (a) S. S. Ghosh, P. M. Kao, A. W. McCue, H. L. Chappelle, *Bioconjugate Chem.* **1990**, *1*, 71–76; (b) M. A. Firestone, M. L. Shank, S. G. Sligar, P. W. Bohn, *J. Am. Chem. Soc.* **1996**, *118*, 9033–9041.
44. (a) S. H. Liu, M. Y. Han., *Adv. Funct. Mater.* **2005**, *15*, 961–967; (b) E. Dujardin, S. Mann, *Adv. Mater.* **2002**, *14*, 775–788.
45. R. Polsky, J. C. Harper, D. R. Wheeler, S. M. Dirk, J. A. Rawlings, S. M. Brozik, *Chem. Commun.* **2007**, 2741–2743.
46. F. Patolsky, T. Gabriel, I. Willner, *J. Electroanal. Chem.* **1999**, *479*, 69–73.
47. Z. M. Liu, H. Wang, Y. Yang, H. F. Yang, S. Q. Hu, G. L. Shen, R. Q. Yu, *Anal. Lett.* **2004**, *37*, 1079–1091.
48. H.-Y. Gu, A.-M. Yu, H.-Y. Chen, *J. Electroanal. Chem.* **2001**, *516*, 119–126.
49. T. Liu, J. Zhong, X. Gan, C. Fan, G. Li, N. Matsuda, *ChemPhysChem* **2003**, *4*, 1364–1366.
50. Y. Zhang, P. L. He, N. F. Hu, *Electrochim. Acta* **2004**, *49*, 1981–1988.
51. (a) D. F. Cao, P. L. He, N. F. Hu, *Analyst* **2003**, *128*, 1268–1274; (b) Y. Lvov, B. Munge, O. Giraldo, I. Ichinose, S. L. Suib, J. F. Rusling, *Langmuir* **2000**, *23*, 8850–8857.
52. Y. Xiao, F. Patolsky, E. Katz, J. F. Hainfeld, I. Willner, *Science* **2003**, *299*, 1877–1881.

53. J. G. Zhao, R. W. Henkens, J. Stonehuerner, J. P. O'Daly, A. L. Crumbliss, *J. Electroanal. Chem.* **1992**, *327*, 109–119.
54. J. B. Jia, B. Q. Wang, A. G. Wu, G. J. Cheng, Z. Li, S. J. Dong, *Anal. Chem.* **2002**, *74*, 2217–2223.
55. W. Zhao, J.-J. Xu, C.-G. Shi, H.-Y. Chen, *Langmuir* **2005**, *21*, 9630–9634.
56. X.-L. Luo, J.-J. Xu, Y. Du, H.-Y. Chen, *Anal. Biochem.* **2004**, *334*, 284–289.
57. A. L. Crumbliss, J. G. Stonehuerner, R. W. Henkens, J. Zhao, J. P. O'Daly, *Biosens. Bioelectron.* **1993**, *8*, 331–337.
58. O. D. Velev, E. W. Kaler, *Langmuir* **1999**, *15*, 3693–3698.
59. S.-J. Park, T. A. Taton, C. A. Mirkin, *Science* **2002**, *295*, 1503–1506.
60. J.-H. Kim, J.-H. Cho, G. S. Cha, C.-W. Lee, H.-B. Kim, S.-H. Paek, *Biosens. Bioelectron.* **2000**, *14*, 907–915.
61. Y. Fan, X. Chen, J. Kong, C.-H. Tung, Z. Gao, *Angew. Chem. Int. Ed.* **2007**, *46*, 2051–2054.
62. M. B. González-García, C. Fernández-Sánchez, A. Costa-García, *Biosens. Bioelectron.* **2000**, *15*, 315–321.
63. (a) M. Dequaire, C. Degrand, B. Limoges, *Anal. Chem.* **2000**, *72*, 5521–5528; (b) L. Authier, C. Grossiord, P. Brossier, B. Limoges, *Anal. Chem.* **2001**, *73*, 4450–4456.
64. J. Wang, D. Xu, A.-N. Kawde, R. Polsky, *Anal. Chem.* **2001**, *73*, 5576–5581.
65. J. Wang, G. Liu, R. Polsky, A. Merkoci, *Electrochem. Commun.* **2002**, *4*, 722–726.
66. J. Wang, G. D. Liu, A. Merkoci, *Anal. Chim. Acta* **2003**, *482*, 149–155.
67. H. Cai, Y. Xu, N. Zhu, P. He, Y. Fang, *Analyst* **2002**, *127*, 803–808.
68. N. Zhu, H. Cai, P. He, Y. Fang, *Anal. Chim. Acta* **2003**, *481*, 181–189.
69. R. Polsky, R. Gill, L. Kaganovsky, I. Willner, *Anal. Chem.* **2006**, *78*, 2268–2271.
70. J. Wang, R. Polsky, D. K. Xu, *Langmuir* **2001**, *17*, 5739–5741.
71. A.-N. Kawde, J. Wang, *Electroanalysis* **2004**, *16*, 101–107.
72. X. Mao, J. Jiang, Y. Luo, G. Shen, R. Yu, *Talanta* **2007**, *73*, 420–424.
73. J. Wang, D. K. Xu, R. Polsky, *J. Am. Chem. Soc.* **2002**, *124*, 4208–4209.
74. M. Pumera, M. T. Castaneda, M. I. Pividori, R. Eritja, A. Merkoci, S. Alegret, *Langmuir* **2005**, *21*, 9625–9629.
75. J. Das, M. A. Aziz, H. Yang, *J. Am. Chem. Soc.* **2006**, *128*, 16022–16023.
76. J. Wang, G. D. Liu, A. Merkoci, *J. Am. Chem. Soc.* **2003**, *125*, 3214–3215.
77. G. D. Liu, J. Wang, J. Kim, M. R. Jan, and G. E. Collins, *Anal. Chem.* **2004**, *76*, 7126–7130.
78. J. A. Hansen, J. Wang, A.-N. Kawde, Y. Xiang, K. V. Gothelf, G. Collins, *J. Am. Chem. Soc.* **2006**, *128*, 2228–2229.

Chapter 9

Modified Nanoparticles as Nanoelectrocatalysts and Amplifying Sensors

SHAOJUN GUO, ERKANG WANG, AND XIURONG YANG

9.1 INTRODUCTION

Advanced functional inorganic nanoparticles (NPs) have been studied intensively in the last couple of decades due to their unusual chemical and physical properties compared with their bulk materials, which enable them to be promising in applications as diverse as electronics,[1] optics,[2] optoelectronics,[3] nanosensors,[4] information storage,[5] fuel cells,[6] biomedicine,[7] biological labeling,[8] gene delivery,[9] electrocatalysis,[10] and surface enhanced Raman scattering (SERS).[11,12] For instance, metal NPs with

The Supramolecular Chemistry of Organic–Inorganic Hybrid Materials. Edited by Knut Rurack and Ramón Martínez-Máñez

size-, morphology-, and composition-dependent electronic and optical properties have been increasingly synthesized[13,14] and employed as excellent building blocks for the construction and assembly of electrochemical devices. Quantum dots (QDs), developed recently as a new class of fluorescent probes for replacing traditional organic dyes, have been widely employed for applications in biomedicine, bioassays, and bioimaging.[15,16] Superparamagnetic NPs, which are of great interest for researchers from a wide range of disciplines, including biotechnology/biomedicine and magnetic resonance imaging, have been facilely prepared by a number of suitable methods.[17,18]

On the other hand, a rising interdisciplinary field (nanoelectrochemistry) which combines the simplicity, rapidness, high selectivity, and high sensitivity of electrochemistry with the unique electronic, optical, and magnetic properties of NPs has been the focus of intense research. Why is nanoelectrochemistry so popular for the electroanalytical chemist and electrochemist? This can be explained as follows.

1. Inorganic NPs exhibit higher surface to volume ratios than their bulk counterparts. Thus, electrochemical interfaces modified with inorganic NPs will provide a higher electrochemically active area and therefore lead to higher detection sensitivity.
2. Some novel NPs, particularly metal NPs, can facilely act as enhancing agents for effectively accelerating the electron transfer between electrode and probe molecules, which will lead to a more rapid current response.
3. Inorganic NPs can be adopted as supramolecular assembly units with advanced functional properties for constructing a variety of composite architectures on the surface of electrodes and for the further tailoring of electrochemical sensing interfaces.
4. Inorganic NPs can be conjugated with important biomolecules such as redox enzymes and further act as nano-connectors that activate redox enzymes or electrical labels for biorecognition events.
5. Electrochemical interfaces modified with inorganic NPs behave as nanoelectrode ensembles. In principle, the electroanalytical detection limit at a nanoelectrode ensemble can be much lower than that at an analogous macrosized electrode because the ratio between the faradaic and capacitive currents is higher.[19–23]

Inspired by these important features, nanoelectrochemistry has been flourishing recently and many research papers have reported the preparation of inorganic NPs for the modification of interfaces for enhanced electrochemical signaling. In this chapter, we will address recent research accomplishments that have led to powerful inorganic NP-based electrochemical or electrochemiluminescence (ECL) devices and examine future prospects and challenges. The approaches described include inorganic NPs as enhancing elements for electrochemical sensors, for example, for small molecule and biomolecule sensing, and inorganic NPs as advanced nanoelectrocatalysts for fuel cells as well as modified inorganic NPs for ECL amplified detection.

However, due to the explosion of publications in this field, we do not claim that this chapter includes all of the published work about modified NPs as nanoelectrocatalysts and amplifying sensors. We apologize to the authors of much excellent research work that, due to the large activity in this important field, we have unintentionally left out.

9.2 INORGANIC NPs AS ENHANCING ELEMENTS IN SENSING

Inorganic NPs have received considerable interest in the past decades due to their interesting optical, electronic, and magnetic properties and because of their suitability for many applications in fields such as sensors, optics, electronics, SERS, biomedicine, information storage, etc. For the electroanalytical chemist, it is primarily interesting how electrochemical sensing platforms can be constructed for ever improved sensitivity and selectivity by using inorganic NPs as the enhancing elements. Fortunately, many inorganic materials such as metals, semiconductors, carbon nanomaterials, and magnetic NPs provide good opportunities for completing this aim. Various methodologies have been used for the tailoring of inorganic NPs on electrode surfaces for electrocatalytic applications, which include the anchoring by electrostatic interaction, covalent linkage and electrochemical deposition, etc. Thus, electrodes modified with inorganic NPs that behave as nanoelectrode ensembles have been used widely as enhancing electrocatalytic interfaces for the development of electrochemical sensors and also for applications in fuel cells.

9.2.1 Inorganic NPs as Enhancing Elements for Small Molecule Sensing

Inorganic NPs play a key role in catalytic and electrocatalytic reactions. Generally, the catalytic efficiency of inorganic NPs depends on three main factors: (1) the surface area to volume ratio of the NPs, (2) the nature of electronic interactions between the NPs and the reactant molecules, and (3) the effective and uniform immobilization of NPs on the surface of an electrode. In the last years, many groups have designed different electrochemical sensing platforms for electrocatalyzing different target molecules by satisfying the important factors mentioned before.

It was found that electrodes modified with inorganic NPs exhibit higher electrocatalytic activities than other traditionally modified electrodes for the detection of several small molecules such as hydrazine,[24–29] H_2O_2,[30–32] nitric oxide,[33–35] nitrite,[36] trinitrotoluene (TNT),[37] or various poisonous substances.[38–43] For example, several electrodes modified with inorganic NPs have been used successfully for the detection of hydrazine.[24–29] Unwin et al.[24] reported that a polyaniline (PANI)/PtNP composite thin film electrode, which was prepared by the spontaneous redox reaction of K_2PtCl_4 with PANI, showed relatively high electrocatalytic properties for hydrazine in both acidic and neutral aqueous media. However, the overpotential for hydrazine oxidation is still high. Jena and Raj[25] demonstrated the ultrasensitive electrochemical detection of hydrazine using AuNPs self-assembled on a sol–gel-derived three-dimensional silicate network, followed by seed-mediated growth of gold. This nanostructured platform was highly sensitive toward the electrochemical oxidation of hydrazine. A very large decrease in the overpotential (~800 mV) and significant enhancement in the peak currents with respect to the bulk Au electrode were observed without using any redox mediator. The nanostructured platform showed excellent sensitivity with an experimental detection limit of 200 pM. Zinc oxide nanonails have recently been used to prepare a hydrazine electrochemical sensor.[26] Figure 9.1a shows

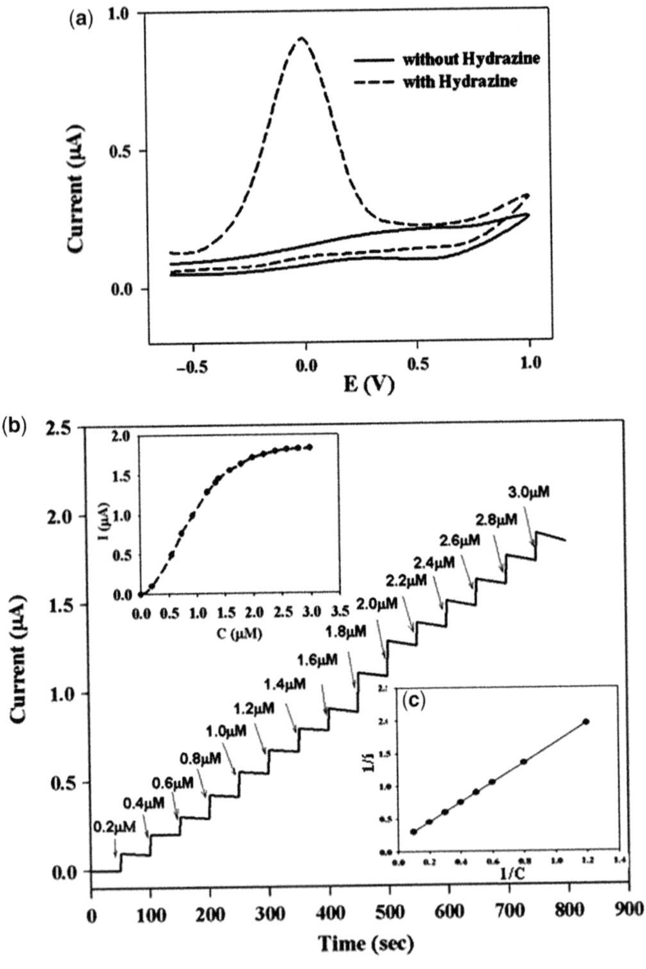

Figure 9.1 (a) Cyclic voltammetric sweep curve for the Nafion/ZnO/Au electrode without hydrazine (solid line) and with 1 mM N_2H_4 (dashed line) in 0.01 M PBS (pH = 7.4). The scan rate was 100 mV/s. (b) Amperometric response of the Nafion/ZnO/Au electrode with successive addition of hydrazine into 0.01 M PBS buffer solution (pH = 7.4). The applied potential was in the range of −0.5 to +0.4 V. (c) Plot of 1/i vs. 1/C exhibiting a linear relationship for the steady state current and hydrazine concentration.[26] (Reprinted with permission from A. Umar et al., *Chem. Commun.* **2008**, 166–168. Copyright the Royal Society of Chemistry.)

cyclic voltammograms (CVs) for the ZnO nanonail-modified gold (Nafion/ZnO/Au) electrode without hydrazine (solid line) and with 1 mM N_2H_4 (dashed line) in 0.01 M phosphate buffer solution (PBS, pH = 7.4) at a scan rate of 100 mV s^{-1}. It can be seen that in contrast to the Nafion/ZnO/Au electrode without hydrazine (solid line), a well-defined oxidation peak with 1 mM N_2H_4 in 0.01 M PBS (dashed line) is observed. Figure 9.1b shows a typical amperometric response of the Nafion/ZnO/Au electrode on the successive addition of hydrazine (from 0.2 to 3 mM) into continuously stirred 0.01 M PBS solution at an applied potential in the range of –0.5 to 0.4 V. The amperometric sensor exhibited a rapid and sensitive response to the change of hydrazine concentration and an obvious increase in the oxidation current upon successive addition of hydrazine was obtained. An inset in Figure 9.1 (b) shows the corresponding calibration curve of the fabricated amperometric hydrazine sensor. A linear range from 0.1 to 1.2 μM could be obtained.

The highly sensitive detection of other small molecules such as nitrite, NO, and H_2O_2 could also be carried out on electrochemical interface modified with inorganic NPs. As is known, the determination of nitrite is very important in various fields, such as environmental and food chemistry and the life sciences. However, a fast and reliable sensing method is still a challenge for its detection at low concentrations.[44] To solve this problem, Lin et al.[36] reported a novel approach to the electrochemical oxidation of nitrite by using a glassy carbon electrode (GCE) with electrochemically codeposited PtNPs and Fe^{III} on its surface. A detection limit of 4.7×10^{-7} M was reached. NO is an important short-lived intermediate in many biological systems and the direct detection of low-level NO remains a challenge today. Interestingly, electrochemical sensing interfaces modified with inorganic NPs provide feasible access to completing this aim. For instance, polyelectrolyte/AuNP hybrid films,[33] Fe^{III}-PtNPs[34] and TiO_2-Au nanocomposite films[35] have been employed to prepare novel electrochemical sensors for NO with high sensitivity. In addition, the determination of H_2O_2 using silver-DNA hybrid NPs,[30] platinum NP-embedded carbon films,[31] and magnetic Fe_3O_4-NPs[32] was demonstrated by several groups.

Poisonous substances such as arsenic, antimony, and mercury are widely distributed in nature and occur in the form of inorganic or organic compounds. The presence of these poisonous substances in food and water above a certain level presents a serious threat to public health. Therefore, it is of utmost importance to develop analytical methods for the detection of such poisonous substances in water and food.[38] Recently, several excellent papers have been published that combine nanotechnology and electrochemical techniques such as stripping voltammetry and in which the detection of poisonous substances with high sensitivity and selectivity could be achieved. For instance, Luong's group[38] developed an electrochemical flow-analysis system for the determination of arsenite. They found that the detection sensitivity could be significantly improved by the electrochemical deposition of gold NPs on a GCE. The electrochemical system was linear up to 15 ppb with a detection limit of 0.25 ppb. Later, they prepared a boron-doped diamond (BDD) microelectrode modified with platinum NPs for the oxidative determination of As^{III} at levels below 1 ppb.[39] More interestingly, this PtNP-modified electrode could well overcome severe interferences usually caused by chloride and copper. In addition, AuNP-modified carbon screen-printed and GC electrodes were used successfully to determine antimony[42] and mercury,[43] respectively.

9.2.2 Inorganic NPs as Enhancing Elements for Biomolecule Sensing

The development of sensing devices for the fast and reliable monitoring of glucose and other carbohydrates for the treatment and control of diabetes has been a focal subject in analytical chemistry for the past few decades. Interestingly, this aim could also be achieved using electrochemical interfaces modified with inorganic NPs.[45] Although these enzyme-based sensors showed high selectivity and good sensitivity for the detection of glucose, they suffered from the lack of stability due to the intrinsically labile nature of the enzymes, the requirement of a high overpotential for the oxidation of enzymatically generated H_2O_2 and their proneness to interference due to other redox-active molecules such as ascorbic acid.[45] Therefore, the enzyme-free amperometric sensing of glucose by using inorganic NPs as electron accelerating agents provides an alternative. Recently, it was shown that electrodes modified with Pt nanotube arrays and three-dimensional macroporous Pt films, obtained by electrodeposition of Pt on a three-dimensional ordered array of silica NPs and subsequent removal of the silica template with HF, exhibit good sensitivity towards glucose.[46,47] However, these electrodes required electrochemical cleaning before each measurement to renew the electrode surface. Gold NPs can overcome the above poisoning problem caused by the reaction intermediates of sugars.[45,48,49] For instance, Jena and Raj[45] reported a nonenzymatic electrochemical method for the detection of glucose by using AuNPs self-assembled on a three-dimensional silicate network obtained by the sol–gel process. The 3-mercaptopropyltrimethoxysilane (MPTMS) three-dimensional network was first self-assembled on a polycrystalline Au electrode, then the AuNPs were chemisorbed onto the thiol groups present inside and on the surface of the silicate network (MPTMS-AuNP) and finally they were enlarged by gold deposition on the surface of the self-assembled AuNPs through the reduction of $HAuCl_4$ using hydroxylamine as reductant (MPTMS-eAuNP). Figure 9.2 compares the CVs obtained for the oxidation of glucose at the MPTMS-modified polycrystalline Au electrode, the MPTMS-AuNP electrode, and the MPTMS-eAuNP electrode. The MPTMS-modified polycrystalline Au electrode does not show any characteristic response (Fig. 9.2a) for the oxidation of glucose. However, the MPTMS-AuNP electrode shows a well-defined voltammetric peak at 0.24 V (Fig. 9.2b), whereas the MPTMS-eAuNP electrode shows the oxidation peak already at much less positive potential ($\sim$0.16 V; Fig. 9.2c). These results show that the AuNPs on the silicate network efficiently catalyze the oxidation process. This novel nonenzymatic glucose sensor showed excellent sensitivity with a detection limit of 50 nM. Furthermore, Mirsky et al.[48] prepared a nonenzymatic electrochemical sensor for the detection of sugars by layer-by-layer (LbL) deposition of gold NPs on gold electrodes.

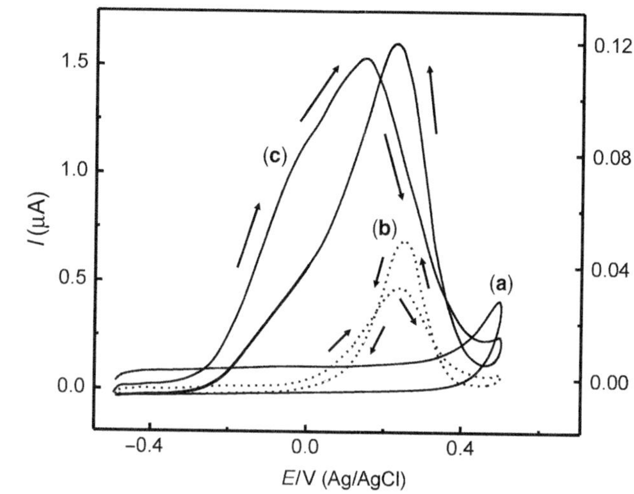

Figure 9.2 Cyclic voltammograms for the oxidation of glucose (1 mM) at: (a) MPTMS-modified polycrystalline Au electrode; (b) MPTMS-AuNP electrode; (c) MPTMS-eAuNP electrode in 0.1 M PBS. Scan rate: $10\ mV\ s^{-1}$.[45] (Reprinted with permission from B. K. Jena and C. R. Raj, *Chem. Eur. J.* **2006**, *12*, 2702–2708. Copyright Wiley-VCH-Verlag GmbH & Co. KGaA.)

In fact, inorganic NPs could also be employed as enhancing materials for the electrochemical investigation of cells[50] and for electrocatalyzing various important small biomolecules such as norepinephrine,[51] dopamine,[52–56] catechol,[57] epinephrine,[58] and ascorbic acid.[56,59,60] For instance, a novel electrochemical sensor has been constructed by use of the GCE coated with gold NPs/choline.[53] The as-modified electrode exhibited strong electrocatalytic activity toward the oxidation of dopamine, ascorbic acid, and uric acid, with obvious reduction of overpotentials. At the same time, Dong's group[59] reported a simple and effective method to detect ascorbic acid using surface plasmon resonance (SPR) spectroscopy and electrochemistry by catalyzed deposition of silver ions on an SPR gold film. Figure 9.3a depicts the fabrication scheme of the biosensor for detecting ascorbic acid on an SPR gold film. At first, a monolayer of cysteamine was assembled on the surface of the SPR gold film by adding 0.01 M cysteamine solution to the Teflon cell for 12 h. After the amino adhesion agent was deposited, AuNPs with a diameter of ~2.5 nm were assembled on the surface of the gold film. Subsequently, Ag atoms were deposited on the surface of the AuNP-modified SPR gold film by adding a solution including 6 mM $AgNO_3$ and different concentrations of ascorbic acid to the Teflon cell for 200 s. It was found that the depth change of the SPR reflectance intensity and the minimum reflectivity angle curves mainly depend on the amount of Ag atoms deposited on the gold film, which in turn was controlled by the concentration of ascorbic acid (Fig. 9.3b). By monitoring the deposition of Ag atoms on the gold film, ascorbic acid was detected in the concentration range of 2×10^{-5} M to 1×10^{-3} M.

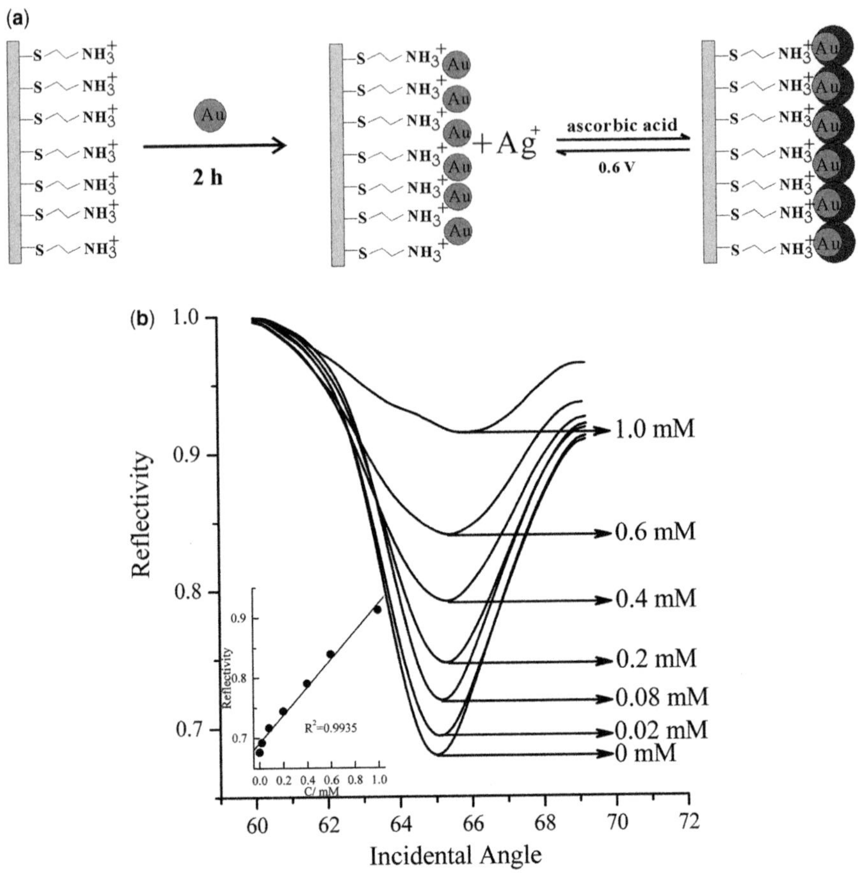

Figure 9.3 (a) Assembly of AuNPs on SPR gold film and the biocatalytic Ag atom growth in the presence of ascorbic acid. (b) SPR reflectance intensity and minimum reflectivity angle curves resulting from the biocatalytic deposition of Ag atom in the presence of different concentrations of ascorbic acid as indicated. Insert: A calibration curve to convert the change of reflectivity at the minimum of SPR spectrum to different ascorbic acid concentrations.[59] (Reprinted with permission from J. Wang et al., *Electrochem. Commun.* **2007**, *9*, 343–347. Copyright 2007 Elsevier.)

9.2.3 Inorganic NPs as Advanced Nanoelectrocatalysts for Fuel Cells

Research on the direct conversion of chemical energy to electricity via fuel cells has received considerable attention in the past decades. Fuel cells are indeed attractive alternatives to combustion engines for electrical power generation in transportation applications and also as promising future power sources, especially for mobile and portable applications. Thus, the search for excellent electrocatalysts for the electrocatalytic oxygen reduction and methanol oxidation reactions, which are the two important cathodic and anodic reactions in fuel cells, is intensively pursued by scientists around the world today. Inorganic NPs, particularly metal NPs, are still the best choice as advanced nanoelectrocatalysts for fuel cells. In the last years, research interests have been directed towards synthesizing and designing advanced metal nanomaterials with novel morphologies and towards intensively investigating their applications in fuel cells. In this section, we will discuss recent advances in designing different inorganic NPs as advanced nanoelectrocatalysts for fuel cells.

Generally, gold NPs are good nanoelectrocatalysts for fuel cells because of their large surface to volume ratios, good stability, excellent conductivity, and rather high inertness against poisoning caused by reaction intermediates. In the last years, a large number of articles has been published that study the oxygen reduction reaction (ORR) and methanol oxidation reaction (MOR) involving AuNPs as electrocatalysts.[61–73] El-Deab and Ohsaka[61] reported that gold NPs electrodeposited on a gold electrode can act as good electrocatalysts for the cathodic reduction of oxygen. Two well-defined reduction peaks were observed for such AuNP-modified electrodes, indicating a 2-step 4-electron reduction pathway for oxygen in a strongly acidic medium. Our group[65] prepared 2D and three-dimensional AuNPs/[tetrakis(*N*-methylpyridyl) porphyrinato]cobalt (CoTMPyP) nanostructured materials by a "bottom-up" self-assembly technique, as shown in Figure 9.4. This novel AuNPs/CoTMPyP self-assembled nanostructured material exhibited tunable electrocatalytic activity for oxygen reduction. The electrochemical reduction of oxygen has also been studied

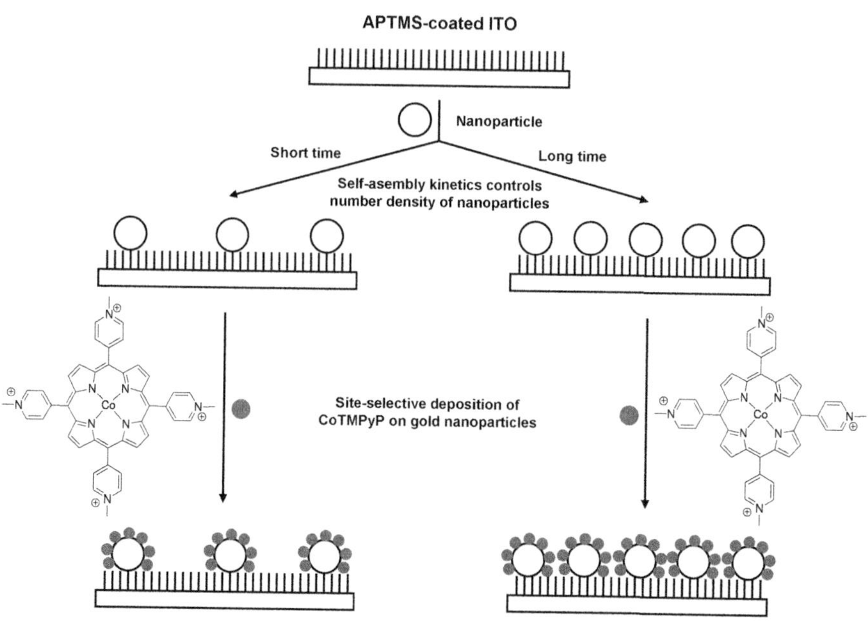

Figure 9.4 Schematic illustration of site-selective deposition of CoTMPyP on gold NP surfaces and tuning the quantities of the as-anchored electrocatalysts by number density of gold NPs.[65] (Reprinted with permission from W. Cheng et al., *J. Phys. Chem. B* **2004**, *108*, 19146–19154. Copyright 2004 American Chemical Society.)

by Tammeveski's group[69] on gold NP/multiwalled carbon nanotube (AuNP/MWNT)-modified GCE in 0.5 M H_2SO_4 solution using the rotating disk electrode (RDE) technique. Electrochemical studies indicated that the AuNP/MWNT catalyst showed a remarkable electrocatalytic activity towards oxygen reduction in acidic media. To further increase the effectively active area of gold NPs, Jena and Raj[70] described a simple environmentally benign route for the synthesis of flower-like gold NPs and investigated their electrocatalytic activity in MOR and ORR. It was found that the electrode modified with flower-like gold NPs showed excellent electrocatalytic activity in these reactions. Some bimetallic materials containing gold have also been employed in electrocatalytic fields. For instance, Zhou and Lee[71] demonstrated that core–shell structured Au@PdNPs could significantly improve the catalytic activity and stability of Pd in the electrooxidation of formic acid at room temperature.

Although gold NPs exhibited high electrocatalytic activities for ORR or MOR, it is very hard to achieve four-electron reduction of oxygen or higher current density of methanol oxidation on gold NP-modified electrodes. For numerous nanocatalysts, Pt and Pt-based nanomaterials are still indispensable and the most effective catalysts. A large number of papers have reported the design of Pt catalysts.[74–77] For example, Ye and Crooks[75] demonstrated that dendrimer-encapsulated PtNPs of uniform size could be immobilized on GCE, which was electrocatalytically active for the ORR. Dong's group[77] alternately deposited negatively charged citrate-stabilized platinum NPs and positively charged [tetrakis(*N*-methylpyridyl)porphyrinato] cobalt on a 4-aminobenzoic acid-modified glassy carbon electrode through electrostatic LbL assembly, directly forming a three-dimensional nanostructured material. The multilayer films containing PtNPs exhibited high electrocatalytic activity for the reduction of dioxygen and possessed high stability. For instance, RDE voltammetry and rotating ring-disk electrode voltammetry demonstrated that the PtNP-containing multilayer films could catalyze an almost four-electron reduction of O_2 to H_2O in an air-saturated 0.5 M H_2SO_4 solution.

Although PtNPs have been proven to be very efficient electrocatalysts, one important issue puzzling scientists is the limited supply and high cost of Pt. The search for other highly active electrocatalysts thus continues to be a key scientific and technological challenge in the area of electrochemical energy (fuel cells). Recent work by Narayanan and El-Sayed correlating the catalytic activity of PtNPs with the number of surface atoms indicated that a large number of edge and corner atoms hold the key to improving their catalytic performance.[78] Generally, high-index planes have a greater density of unsaturated atomic steps, ledges, and kinks which can serve as active sites for effectively breaking chemical bonds, thus leading to electrocatalysts with higher efficiency.[79–81] For instance, Sun et al.[80] reported a simple high-temperature organic phase synthesis of monodisperse Pt nanocubes (Fig. 9.5) and studied their catalysis for oxygen reduction under fuel cell reaction conditions. They found that the activity of the Pt nanocubes was more than twice as high as that of a commercial Pt catalyst, indicating that the nanocubes indeed offered an enhancement in oxygen reduction. In addition, typical research in this direction reported by Tian and coworkers[81] demonstrated that PtNPs of unusual tetrahexahedral (THH) shape could be prepared at high yield by an electrochemical treatment of Pt nanospheres supported

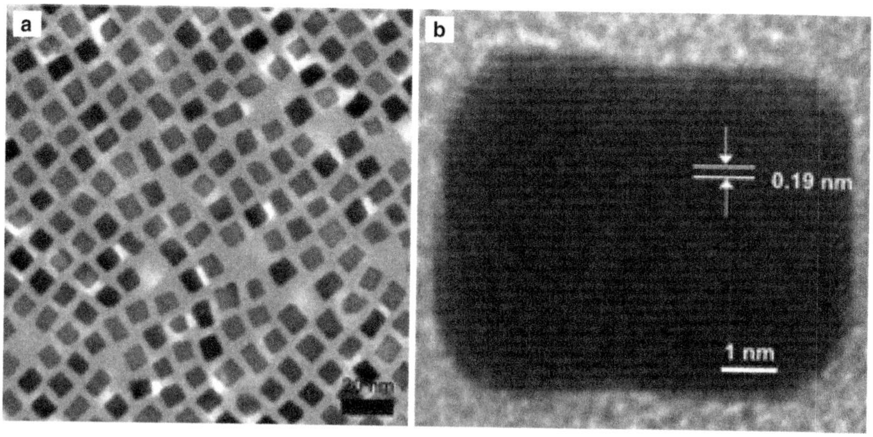

Figure 9.5 (a) TEM image of the 8 nm Pt nanocubes; (b) HRTEM image of a single Pt nanocube.[80] (Reprinted with permission from C. Wang et al., *J. Am. Chem. Soc.* **2007**, *129*, 6974–6975. Copyright 2007 American Chemical Society.)

on glassy carbon by a square-wave potential. In their synthesis, the size of THH nanocrystals could be tailored from 20 to 240 nm by varying the number of cycles of square-wave treatment. The single-crystal THH NPs are enclosed by 24 high-index facets such as {730}, {210}, and/or {520} surfaces that have a large density of atomic steps and dangling bonds. These high-energy surfaces were thermally (up to 800°C) and chemically stable and exhibited much higher (up to 400%) catalytic activity for the electrooxidation of small organic fuels such as formic acid and ethanol than equivalent Pt surface areas.

Although the maximization of high-index surfaces and abundant corner and edge sites should be the criterion of choice for selection of an excellent nanocatalyst,[79] changing the solid Pt nanocatalyst to hollow nanostructures will greatly reduce the amount of Pt and improve the efficient utility of Pt. Several groups have reported the synthesis of hollow Pt-based nanospheres[82–89] for electrocatalytic applications. For instance, hollow PtNPs have been prepared in three steps through a core–shell–hydrolysis route.[82] First, Ag^+ was reduced to yield AgNP cores, followed by reduction of Pt^{2+}, forming a Pt shell around the AgNPs. The remaining AgNP templates are subsequently removed by hydrolysis with bis(*p*-sulfonatophenyl)phenylphosphine. Alternatively, when for instance Co is used as core template, first CoNPs are reductively synthesized. Change of the reaction conditions for reductive Pt shell formation then leads to the simultaneous oxidation of the Co core and Co^{2+} dissolution. Depending on the protocol chosen, hollow Pt[83] or bimetallic Pt-CoNPs[84] can be obtained. Hollow Pt-CoNPs of varying composition can also be synthesized directly by refluxing the Pt and Co precursors.[85] According to a similar route as described in Reference 83, that is, by sacrificial CoNP core templating, bimetallic Au-PtNPs can be prepared.[86] However, all of these hollow Pt-based nanospheres possess smooth surfaces, which is less advantageous for their use as high-performance electrocatalysts. To further increase the surface to volume ratios of Pt, reduce the usage of expensive Pt and enhance the activity of Pt, it is necessary to change the Pt shell to a more complex morphology (e.g., an urchinlike nanostructure).[87,88] Recently, our group[87] demonstrated that a high-efficiency and low-cost spongelike Au/Pt core–shell electrocatalyst with hollow cavity could be facilely obtained via a simple two-step wet-chemical process. It was found that this spongelike hybrid nanomaterial exhibited much higher catalytic activity for the ORR and MOR than the common Pt electrode.

Another important approach to reduce the high cost of Pt is underpotential deposition (UPD) of a Pt monolayer on the surface of inorganic NPs.[90–96] For instance, Adzic's group reported a new class of ORR electrocatalysts based on supported Pt monolayers. Pt was deposited in a monolayer amount on the surfaces of carbon-supported core–shell NPs.[93] Dong et al.[94] modified GCEs with films of multiply deposited Pt overlayers on PtNPs by alternately depositing Ag monolayers through underpotential deposition (UPD) and replacing the Ag monolayer by Pt in a redox process involving K_2PtCl_4. The Pt films thus prepared behaved as novel nanostructured electrocatalysts for dioxygen reduction and hydrogen evolution reaction with enhanced electrocatalytic activities when compared to that of the bulk polycrystalline Pt electrode.

Another important question that has to be faced in fuel cell development is related to the Pt poisoning caused by reaction intermediates in the direct methanol oxidation. This is principally due to self-poisoning of the surface caused by reaction intermediates such as CO and CO-like species, which are strongly adsorbed on the Pt catalyst during the electrooxidation of methanol and drastically reduce the electrocatalytic activity (catalyst poisoning) and performance. To minimize poisoning, many researchers have used Pt-based alloys, especially Pt/Ru and Pt/Au, as bifunctional catalysts in which the platinum activates the fuel to dehydrogenation and the other metals are thought to provide the necessary oxygen for complete oxidation of the fuel to CO_2.[97–101] Yang et al.[99] developed a new synthetic method for making Pt/Ru alloy nanodendrites with defined three-dimensional shapes. Furthermore, Xu et al.[100] presented a novel approach to prepare uniform bimetallic Au/Pt alloy NPs, using an *N,N*-dimethylformamide (DMF) coordinated Au-Pt complex as the precursor. The corresponding novel Pt-based alloy NPs exhibited higher catalytic property and better catalyst tolerance than the traditional catalysts. Another new approach in the search for Pt-based electrocatalysts was to replace Ru with less expensive materials such as Bi,[101] Fe,[102–104] Pb,[105] Cu,[106,107] Sn,[108] and Cu/Co.[109] In addition, other alloy NPs that contain no Pt have also been designed to further reduce the high cost of Pt-based materials and at the same time improve the electrocatalytic activity.[110,111] For instance, Bard et al.[110] proposed guidelines for the design of improved bimetallic (and related) electrocatalysts for the ORR in acidic media.[110] Adzic's group[111] reported on a new Pd-Fe alloy nanoelectrocatalyst, which has a very high activity for the ORR. Figure 9.6 displays the ORR polarization curves for Pd_3Fe/C (line c), Pt/C (line b), and Pd/C (line a) in an O_2-saturated 0.1 M $HClO_4$ solution obtained using an RDE. The curve of Pd_3Fe/C exhibits a 70 mV shift to more positive potentials compared with Pd/C. Furthermore, the activity of Pd_3Fe/C is higher than that of

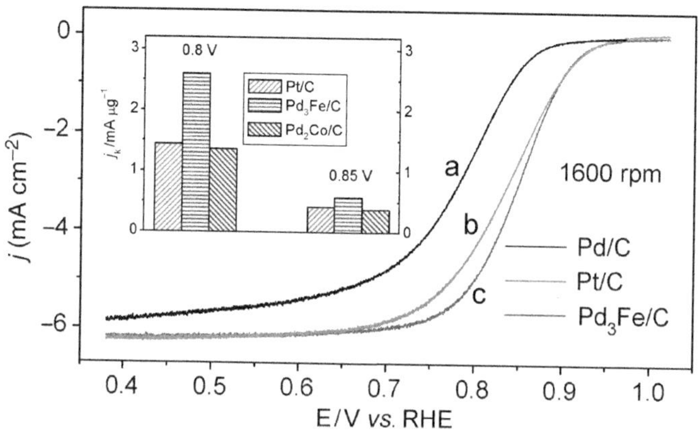

Figure 9.6 Polarization curves for the ORR on Pd/C (ETEK, line a), Pt/C (ETEK, line b), and Pd_3Fe/C NPs (line c) in 0.1 M $HClO_4$. Sweep rate: 10 mV s^{-1}; room temperature. The Pd or Pt loading is 10 μg cm^{-2}. Inset: Comparison of mass activity for Pt/C (ETEK), Pd_3Fe/C, and Pd_2Co/C at 0.8 and 0.85 V.[111] (Reprinted with permission from M. Shao et al., *J. Am. Chem. Soc.* **2006**, *128*, 3526–3527. Copyright 2006 American Chemical Society.)

the state-of-the-art commercial Pt/C electrocatalyst. The inset in Figure 9.6 shows their mass activity, which is a better indicator of an electrocatalyst's quality. The mass activity of Pd_3Fe is 1.4 and 1.9 times higher than that of a commercial Pt/C electrocatalyst at 0.85 and 0.8 V, respectively. This indicates that Pd-Fe alloy nanoelectrocatalysts possess higher electrocatalytic activities than commercial electrocatalysts, thus offering the potential for considerable cost reduction in fuel cell manufacturing.

Actually, on the basis of economic concerns, inorganic NPs, particularly Au, Pt, and alloy NPs mentioned above, could also be attached to the surface of supporting materials such as carbon nanotubes to further increase the effective electrochemically active area of the electrode. In the recent past, many methodologies have been exploited to enhance the utilization of Pt by dispersing NPs onto different supporting materials such as carbon,[112–117] carbon nanotubes,[118–127] silica,[128] graphite nanofibers,[129,130] TiO_2 nanotubes,[131] and silica/MWNT coaxial nanocables.[132] For instance, Yang et al.[122] demonstrated that three-dimensional flowerlike PtNPs could be electrodeposited onto MWNTs by using a three-step protocol. More interestingly, the resulting hybrid nanomaterials displayed very high electrocatalytic activity. Hsin et al.[129] synthesized a PtRu/herringbone graphitic carbon nanofiber (GCNF) nanocomposite in which the GCNF support has a herringbone-like atomic structure. When these hybrid nanomaterials were used as the anode catalyst for fuel cells, the fuel cell performance could be enhanced by ~50% relative to that of an unsupported PtRu colloid anode catalyst. In addition, Guo et al.[132] reported that high-density gold/platinum hybrid NPs could be attached to the surface of 1D silica/MWNT coaxial nanocables with high surface to volume ratios (Fig. 9.7). It was found that this hybrid nanomaterial exhibited a high electrocatalytic activity for enhancing oxygen reduction (low overpotential associated with the oxygen reduction reaction and almost four-electron electroreduction of dioxygen to water). Actually, other

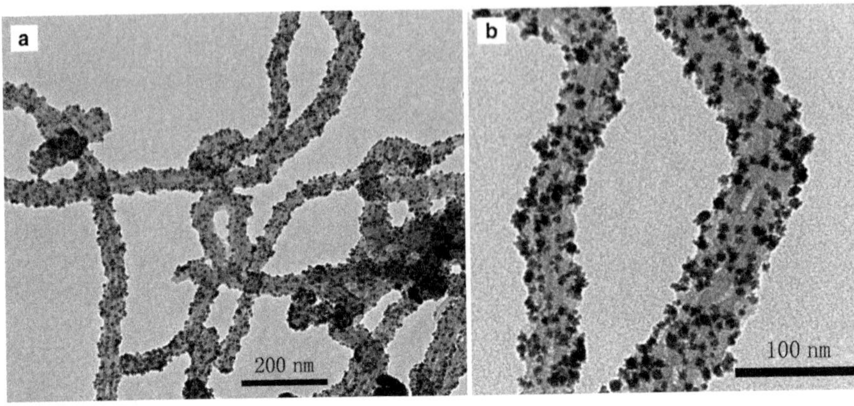

Figure 9.7 Typical TEM images (a) and (b) of the as-prepared gold/platinum hybrid NPs supported coaxial nanocables at different magnifications.[132] (Reprinted with permission from S. Guo et al., *J. Phys. Chem. C*. **2008**, *112*, 2389–2393. Copyright 2008 American Chemical Society.)

publications have also shown successfully that supporting materials, particularly conducting nanomaterials, could obviously increase the effective electrochemically active area and thus improve electrocatalytic activities for the ORR and MOR. Space limitations preclude an exhaustive review here.

9.3 MODIFIED INORGANIC NPs FOR ELECTROCHEMILUMINESCENCE (ECL) DETECTION

Electrogenerated chemiluminescence (ECL) has attracted considerable attention during the past decades due to its versatility, good temporal and spatial control, very low background signal, and simplified optical setup. It has become an important and valuable detection method in analytical chemistry. Recently, several novel ECL systems based on inorganic NPs have been developed due to their high sensitivity and rapid response caused by excellent catalytic properties of NPs. In this section, we will provide some advances on ECL amplifying systems with modified NPs.

Gold NPs with unique electrochemical, catalytic, and chemical properties have been applied to luminol and lucigenin ECL systems, which were reported by Cui et al.[133–137] For example, they demonstrated that on a gold electrode with self-assembled 16 nm AuNPs, two anodic and one cathodic ECL peaks of luminol in neutral aqueous solution were greatly enhanced and a new cathodic ECL peak was initiated compared with a bare gold electrode; in alkaline solution, two anodic ECL peaks much stronger than those with a bare gold electrode were obtained.[133] Later, they[136] studied ECL of luminol on an Au nanorod-modified gold electrode. It was found that one new strong ECL peak was obtained at −0.28 V (vs. SCE) on the Au nanorod-modified gold electrode in both neutral and alkaline solutions and enhanced with an increase in pH. In contrast, four peaks are usually observed on gold NP-modified electrodes.

Among all ECL systems, $Ru(bpy)_3^{2+}$ is one of the most extensively studied compounds due to its stability, reversibility, sensitivity, and capability of undergoing ECL at room temperature in aqueous solution. To develop regenerative ECL-based sensors and detection devices, to reduce the cost of expensive $Ru(bpy)_3^{2+}$ and simplify the detection system, considerable attention has been paid to the immobilization of $Ru(bpy)_3^{2+}$ on the electrode surface. To date, many methods for the effective immobilization of $Ru(bpy)_3^{2+}$ on solid electrode surfaces have been developed.[138–142] Prominent examples include the immobilization of $Ru(bpy)_3^{2+}$ in polymer[138] or Nafion films,[139] TiO_2-Nafion composite films,[140] carbon nanotube-based nanocomposite films[141] and sol–gels.[142] Recently, inorganic NPs could also provide a new platform for the effective immobilization of $Ru(bpy)_3^{2+}$ on electrodes based on different strategies.[143–146] Our group[143] has devised a simple self-assembly method for the effective immobilization of $Ru(bpy)_3^{2+}$ on an electrode surface using AuNPs as amplifying materials. The whole preparation process is shown in Figure 9.8. First, the

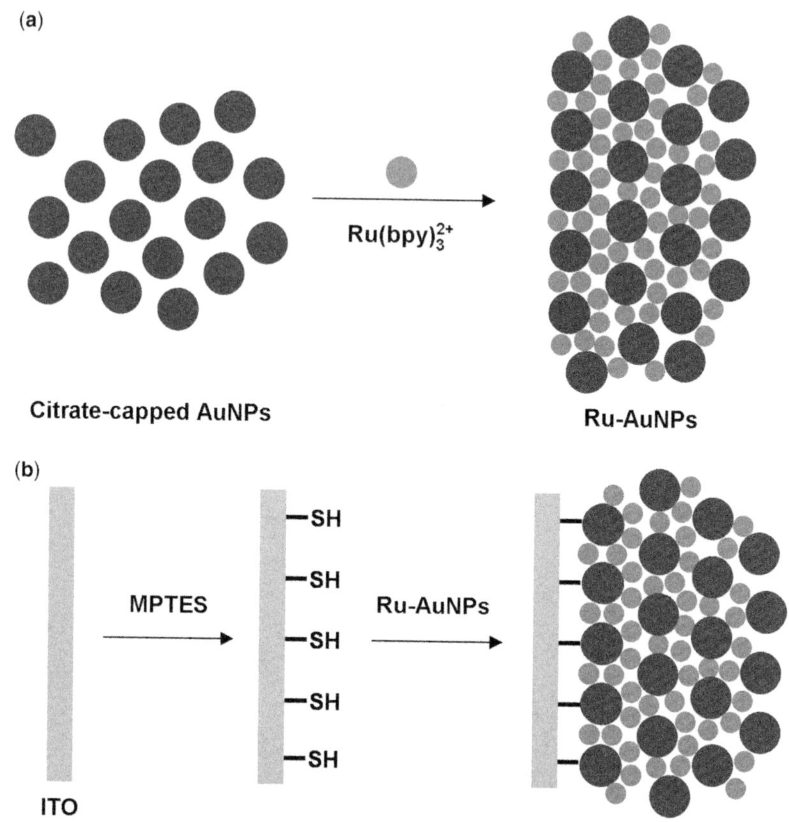

Figure 9.8 Scheme illustrating (a) the formation of Ru-AuNPs in aqueous medium due to electrostatic interactions between $Ru(bpy)_3^{2+}$ and citrate-capped AuNPs and (b) the immobilization of Ru-AuNPs on a sulfhydryl-derivatized ITO electrode surface.[143] (Reprinted with permission from X. Sun et al., *Anal. Chem.* **2005**, *77*, 8166–8169. Copyright 2005 American Chemical Society.)

electrostatic interactions between citrate-capped AuNPs and $Ru(bpy)_3Cl_2$ in aqueous medium were used to fabricate the $Ru(bpy)_3^{2+}$-AuNPs aggregates (Ru-AuNPs). Then Au-S interactions between as-formed Ru-AuNPs and sulfhydryl groups were used to effectively immobilize the Ru-AuNPs on a sulfhydryl-derivatized ITO electrode surface. The as-prepared ITO electrode showed excellent stability, and the ECL active species ($Ru(bpy)_3^{2+}$) contained therein exhibited excellent ECL behaviors.

Later, Dong's group[146] developed an attractive alcohol dehydrogenase biosensor based on the above-mentioned self-assembly strategy. This biosensor displayed a wide linear working range, good stability, and high sensitivity, with a detection limit of 3.33×10^{-6} M. In addition, the immobilization of $Ru(bpy)_3^{2+}$ on PtNPs by self-assembly has also been well demonstrated by our group.[144,145] Silica NPs could also immobilize $Ru(bpy)_3^{2+}$ molecules because of the electrostatic attraction between the negatively charged SiO_2 NPs and positively charged $Ru(bpy)_3^{2+}$. Such hybrids have also been used to construct an electrochemical platform for ECL amplifying sensing.[147–153] For instance, Dong's group[147] reported that positively charged $Ru(bpy)_3^{2+}$ and negatively charged SiO_2 NPs could be assembled on ITO electrodes by an LbL method. The resulting multilayer films containing $Ru(bpy)_3^{2+}$ were used for the ECL determination of tripropylamine (TPA). Figure 9.9 shows cyclic voltammograms (CVs) of the $\{SiO_2/Ru(bpy)_3^{2+}\}_{10}$ multilayer films acquired at 100 mV s^{-1} in phosphate buffer (pH 8.2) with and without 1.3 μM TPA. In the presence of TPA, the oxidation current increased significantly while the cathodic current decreased. This ECL sensor showed high sensitivity and good stability. Later, they[149] developed a novel ECL sensor based on $Ru(bpy)_3^{2+}$-doped silica NPs conjugated with a

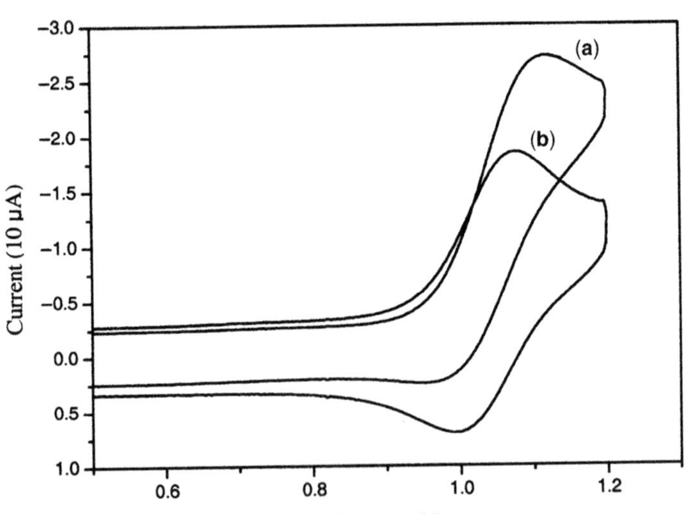

Figure 9.9 CVs of $\{SiO_2/Ru(bpy)_3^{2+}\}_{10}$ films with (a) and without (b) 1.3 μmol L^{-1} TPA (pH 8.2). Scan rate, 100 mV s^{-1}.[147] (Reprinted with permission from Z. Guo et al., *Anal. Chem.* **2004**, *76*, 184–191. Copyright 2004 American Chemical Society.)

biopolymer chitosan membrane. This sensor showed a detection limit of 2.8 nM for TPA, which is three orders of magnitude lower than that observed for a Nafion-based ECL sensor.

More recently, Dong and coworkers[151] synthesized a new kind of nanoarchitecture containing a magnetic Fe_3O_4 core and a luminescent, $Ru(bpy)_3^{2+}$-encapsulating silica shell and further explored the feasibility of applying the as-prepared nanostructure to fabricating an ECL sensor. Consequently, the as-prepared ECL sensor showed a wide linear working range, high sensitivity, and good stability. Followed by these novel immobilization strategies, Fang's group has successfully developed novel biosensors for the detection of thrombin[152] and DNA using $Ru(bpy)_3^{2+}$-doped silica NPs.[153]

After Bard's research group reported ECL from octanol-, octane-, and octanethiol-capped Si[154] and trioctylphosphine oxide (TOPO)-capped CdSe[155] or CdTe[156] nanocrystals in organic solvents, more people in recent years have devoted their attention to studying ECL systems based on QDs in aqueous solution. A few reports have been published on the ECL of QDs[157] and on the development of novel electrochemical sensors based on QD ECL systems.[158–166] For instance, Zhang et al.[157] first observed a band gap ECL of ZnS NPs in alkaline aqueous solution at a Pt electrode during the potential applied between −2.0 V (vs. Ag/AgCl, saturated KCl) and +0.86 V. Novel types of QD-based ECL sensors were also developed to detect the concentration of H_2O_2 in aqueous solution.[158–161] For example, Zou and Ju[158] reported the first QD-based ECL sensor in aqueous systems, which was based on the electron-transfer reaction between reduced nanocrystal species formed in CdSe NP thin films and hydrogen peroxide. The prepared ECL sensor showed high sensitivity, good stability, and reproducibility, indicating that QDs could be a new

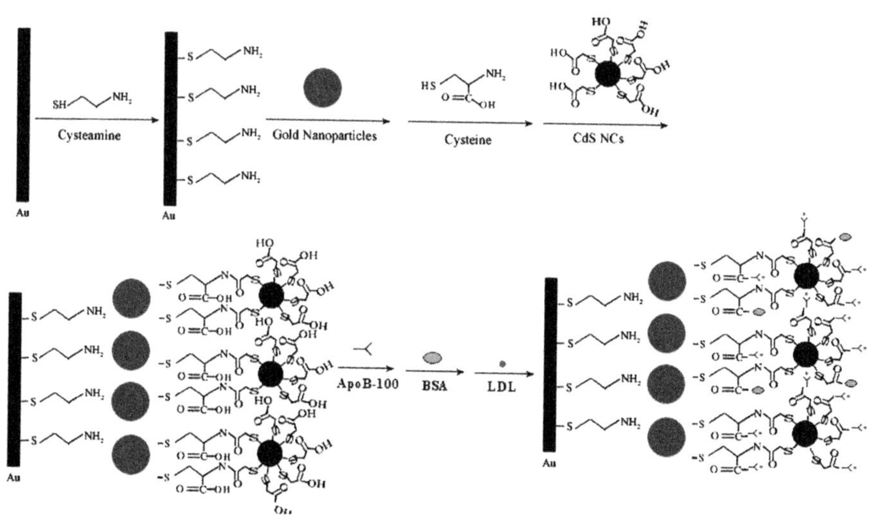

Figure 9.10 Schematic diagram for the biosensor fabrication and LDL binding.[163] (Reprinted with permission from G. Jie et al. *Anal. Chem.* **2007**, *79*, 5574–5581. Copyright 2007 American Chemical Society.)

and promising material for ECL applications by electron-transfer reactions between electrochemically formed QDs and coreactants such as H_2O_2. According to this principle, several novel biosensors for the detection of antigens,[162] low density lipoprotein (LDL),[163] catechol derivatives,[164] thiol compounds,[165] and amines[166] have been successfully realized by several groups. For instance, a versatile immunosensor using CdTe QDs as electrochemical and fluorescent labels has been developed by Zhu et al.[162] for sensitive protein detection. They also reported a novel label-free ECL biosensor based on the ECL of CdS NPs for the detection of LDL using self-assembly and gold NPs for the construction of an amplifying device.[163] The biosensor was prepared as follows (Fig. 9.10): The gold NPs were first assembled onto a cysteamine monolayer on the gold electrode surface. This gold NP-modified electrode was next treated with cysteine and then reacted with CdS NPs to afford a CdS NP-modified electrode. Then, apoB-100 (ligand of LDL receptor) was covalently conjugated to the CdS NP-modified electrode. After the specific binding of apoB-100 with LDL, the biosensor underwent less ECL reaction due to the increased steric hindrance. It was found that the ECL peak intensity of the biosensor decreased linearly with LDL concentration in the range of 0.025 to 16 ng mL^{-1} with a detection limit of 0.006 ng mL^{-1}.

9.4 SUMMARY, CONCLUSIONS, AND OUTLOOK

In this chapter, recent significant advances in the use of modified NPs as nanoelectrocatalysts and amplifying sensors have been addressed. The introduction of inorganic NPs into electrochemistry implants their unique properties into electrochemical sensing interfaces, resulting in many novel strategies tested and some highly sensitive systems fabricated. This rapidly expanding interdisciplinary area has attracted great research efforts from chemists, physicists, biologists, and materials scientists and brought great opportunities to electroanalytical chemists in particular. However, this area is still on the horizon from the viewpoint of applied research, and the widespread practical application of inorganic NP-based electroanalytical devices is not yet possible. Several important factors need to be overcome for the deeper development of nanoelectrochemistry.

1. More novel and sensitive electrochemical sensing interfaces still need to be constructed.
2. Novel and simple methods for the immobilization of NPs need to be developed to further reduce the electron transfer resistance and improve the stability, sensitivity, selectivity, and life-span of the electrodes.
3. Other attractive nanostructured materials in the form of disks, plates, sponges, stars, flakes, urchins, prisms, wires, and rods have been the focus of intense research and have been primarily synthesized by many strategies. The electrochemical signaling with these modified NPs with novel morphologies needs to be further studied.
4. Generally, inorganic NPs assembled on electrode surfaces, which behave as nanoelectrode ensembles, are not diffusionally isolated and the overall current

is smaller than the sum of currents passing through the nanoelectrode when they operate independently, due to the overlap of the diffusion layer of the individual nanoelectrodes.[167] This will result in significant reduction of detection sensitivity. So, it is still very much necessary to develop well-defined nanoelectrode arrays for a further enhancement of the sensitivity and a reduction of the background signal. However, major efforts are necessary to meet these challenges in the nearer future.

ACKNOWLEDGMENT

This work was supported by the National Science Foundation of China (Nos. 20675076 and 20820102037) and the Chinese Academy of Sciences (No. $KJCX_2$.YW.H11) and 973 Project (2007CB714500 and 2009CB930100).

REFERENCES

1. M. A. El-Sayed, *Acc. Chem. Res.* **2001**, *34*, 257–264.
2. C. J. Murphy, N. R. Jana, *Adv. Mater.* **2002**, *14*, 80–82.
3. P. V. Kamat, *J. Phys. Chem. B.* **2002**, *106*, 7729–7744.
4. J. Zhang, J. Liu, Q. Peng, X. Wang, Y. Li, *Chem. Mater.* **2006**, *18*, 867–871.
5. A. Lu, E. L. Salabas, F. Schüth, *Angew. Chem. Int. Ed.* **2007**, *46*, 1222–1244.
6. R. Srivastava, P. Mani, N. Hahn, P. Strasser, *Angew. Chem. Int. Ed.* **2007**, *46*, 8988–8991.
7. J. Chen, D. Wang, J. Xi, L. Au, A. Siekkinen, A. Warsen, Z.-Y. Li, H. Zhang, Y. Xia, X. Li, *Nano Lett.* **2007**, *7*, 1318–1322.
8. J. K. Jailswail, H. Mattoussi, J. M. Mauro, S. M. Simon, *Nature Biotechnol.* **2003**, *21*, 47–51.
9. Y. Huh, E. Lee, J. Lee, Y. Jun, P. Kim, C. Yun, J. Kim, J. Suh, J. Cheon, *Adv. Mater.* **2007**, *19*, 3109–3112.
10. S. Guo, Y. Fang, S. Dong, E. Wang, *J. Phys. Chem. C* **2007**, *111*, 17104–17109.
11. W. E. Doering, M. E. Piotti, M. J. Natan, R. G. Freeman, *Adv. Mater.* **2007**, *19*, 3100–3108.
12. S. Guo, L. Wang, E. Wang, *Chem. Commun.* **2007**, 3163–3165.
13. D. Seo, J. C. Park, H. Song, *J. Am. Chem. Soc.* **2006**, *128*, 14863–14870.
14. E. Katz, I. Willner, *Angew. Chem. Int. Ed.* **2004**, *43*, 6042–6108.
15. Z. Zhelev, H. Ohba, R. Bakalova, *J. Am. Chem. Soc.* **2006**, *128*, 6324–6325.
16. J. Yuan, W. Guo, E. Wang, *Anal. Chem.* **2008**, *80*, 1141–1145.
17. J. Ge, Y. Hu, Y. Yin, *Angew. Chem. Int. Ed.* **2007**, *46*, 7428–7431.
18. J. Park, J. Joo, S. G. Kwon, Y. Jang, T. Hyeon, *Angew. Chem. Int. Ed.* **2007**, *46*, 4630–4660.
19. R. M. Penner, C. R. Martin, *Anal. Chem.* **1987**, *59*, 2625–2630.
20. A. Escorcia, A. A. Dhirani, *J. Electroanal. Chem.* **2007**, *601*, 260–268.
21. W. Cheng, S. Dong, E. Wang, *Anal. Chem.* **2002**, *74*, 3599–3604.
22. W. Cheng, S. Dong, E. Wang, *Langmuir* **2002**, *18*, 9947–9952.
23. F. Wang, J. Wang, H. Chen, S. Dong, *J. Electroanal. Chem.* **2007**, *600*, 265–274.
24. A. P. O'Mullane, S. E. Dale, J. V. Macpherson, P. R. Unwin, *Chem. Commun.* **2004**, 1606–1607.
25. B. K. Jena, C. R. Raj, *J. Phys. Chem. C* **2007**, *111*, 6228–6232.
26. A. Umar, M. M. Rahman, S. H. Kim, Y. Hahn, *Chem. Commun.* **2008**, 166–168.
27. C. Yang, A. S. Kumara, M. Kuo, S. Chien, J. Zen, *Anal. Chim. Acta* **2005**, *554*, 66–73.
28. J. Li, X. Lin, *Sens. Actuators B* **2007**, *126*, 527–535.
29. C. Batchelor-McAuley, C. E. Banks, A. O. Simm, T. G. J. Jones, R. G. Compton, *Analyst* **2006**, *131*, 106–110.
30. S. Wu, H. Zhao, H. Ju, C. Shi, J. Zhao, *Electrochem. Commun.* **2006**, *8*, 1197–1203.

31. T. You, O. Niwa, Masato Tomita, S. Hirono, *Anal. Chem.* **2003**, *75*, 2080–2085.
32. H. Wei, E. Wang, *Anal. Chem.* **2008**, *80*, 2250–2254.
33. A. Yu, Z. Liang, J. Cho, F. Caruso, *Nano Lett.* **2003**, *3*, 1203–1207.
34. S. Wang, X. Lin, *Electrochim. Acta* **2005**, *50*, 2887–2891.
35. E. V. Milsom, J. Novak, M. Oyama, F. Marken, *Electrochem. Commun.* **2007**, *9*, 436–442.
36. S. Wang, Y. Yin, X. Lin, *Electrochem. Commun.* **2004**, *6*, 259–262.
37. S. Hrapovic, E. Majid, Y. Liu, K. Male, J. H. T. Luong, *Anal. Chem.* **2006**, *78*, 5504–5512.
38. E. Majid, S. Hrapovic, Y. Liu, K. B. Male, J. H. T. Luong, *Anal. Chem.* **2006**, *78*, 762–769.
39. S. Hrapovic, Y. Liu, J. H. T. Luong, *Anal. Chem.* **2007**, *79*, 500–507.
40. X. Dai, R. G. Compton, *Electroanalysis* **2005**, 1325–1330.
41. Y. Song, G. Muthuraman, Y. Chen, C. Lin, J. Zen, *Electroanalysis* **2006**, *18*, 1763–1770.
42. O. D. Renedo, M. J. A. Martínez, *Anal. Chim. Acta* **2007**, *589*, 255–260.
43. O. Abollino, A. Giacomino, M. Malandrino, G. Piscionieri, E. Mentasti, *Electroanalysis* **2008**, *20*, 75–83.
44. M. J. Moorcroft, J. Davis, R. G. Compton, *Talanta* **2001**, *54*, 785–803.
45. B. K. Jena, C. R. Raj, *Chem. Eur. J.* **2006**, *12*, 2702–2708.
46. J. Yuan, K. Wang, X. H. Xia, *Adv. Funct. Mater.* **2005**, *15*, 803–809.
47. Y. Y. Song, D. Zhang, W. Gao, X. H. Xia, *Chem. Eur. J.* **2005**, *11*, 2177–2182.
48. F. Kurniawan, V. Tsakova, V. M. Mirsky, *Electroanalysis* **2006**, *18*, 1937–1942.
49. M. Tominaga, T. Shimazoe, M. Nagashima, I. Taniguchi, *Electrochem. Commun.* **2005**, *7*, 189–193.
50. L. Ding, C. Hao, Y. Xue, H. Ju, *Biomacromolecules* **2007**, *8*, 1341–1346.
51. L. P. Lu, S. Q. Wang, X. Q. Lin, *Anal. Chim. Acta* **2004**, *519*, 161–166.
52. C. R. Raj, T. Okajima, T. Ohsaka, *J. Electroanal. Chem.* **2003**, *543*, 127–133.
53. P. Wang, Y. Li, X. Huang, L. Wang, *Talanta* **2007**, *73*, 431–437.
54. E. B. Bustos, M. G. G. Jiménez, B. R. Díaz-Sánchez, E. Juaristi, T. W. Chapman, L. A. Godínez, *Talanta* **2007**, *72*, 1586–1592.
55. J. Mathiyaras, S. Senthilkumar, K. L. N. Phani, V. Yegnaraman, *Mater. Lett.* **2008**, *62*, 571–573.
56. S. Thiagarajan, S. Chen, *Talanta* **2007**, *74*, 212–222.
57. L. Su, L. Q. Mao, *Talanta* **2006**, *70*, 68–74.
58. L. Wang, J. Bai, P. Huang, H. Wang, L. Zhang, Y. Zhao, *Electrochem. Commun.* **2006**, *8*, 1035–1040.
59. J. Wang, F. Wang, X. Zou, Z. Xu, S. Dong, *Electrochem. Commun.* **2007**, *9*, 343–347.
60. F. Zuo, C. Luo, Z. Zheng, X. Ding, Y. Peng, *Electroanalysis* **2008**, *20*, 894–899.
61. M. S. El-Deab, T. Ohsaka, *Electrochem. Commun.* **2002**, *4*, 288–292.
62. M. S. El-Deab, T. Ohsaka, *J. Electroanal.Chem.* **2003**, *553*, 107–115.
63. J. M. Macak, F. Schmidt-Stein, P. Schmuki, *Electrochem. Commun.* **2007**, *9*, 1783–1787.
64. C. R. Raj, A. I. Abdelrahman, T. Ohsaka, *Electrochem. Commun.* **2005**, *7*, 888–893.
65. W. Cheng, S. Dong, E. Wang, *J. Phys. Chem. B* **2004**, *108*, 19146–19154.
66. F. N. Crespilho, F. C. Nart, O. N. Oliveira, Jr., C. M. A. Brett, *Electrochim. Acta* **2007**, *52*, 4649–4653.
67. M. S. El-Deab, T. Sotomura, T. Ohsaka, *Electrochim. Acta* **2006**, *52*, 1792–1798.
68. Y. Zhang, S. Asahina, S. Yoshihara, T. Shirakashi, *Electrochim. Acta* **2003**, *48*, 741–747.
69. N. Alexeyeva, T. Laaksonen, K. Kontturi, F. Mirkhalaf, D. J. Schiffrin, K. Tammeveski, *Electrochem. Commun.* **2006**, *8*, 1475–1480.
70. B. K. Jena, C. R. Raj, *Langmuir* **2007**, *23*, 4064–4070.
71. W. Zhou, J. Y. Lee, *Electrochem. Commun.* **2007**, *9*, 1725–1729.
72. K. Matsuoka, K. Miyazaki, Y. Iriyama, K. Kikuchi, T. Abe, Z. Ogumi, *J. Phys. Chem. C* **2007**, *111*, 3171–3174.
73. J. Hernández, J. Solla-Gullón, E. Herrero, A. Aldaz, J. M. Feliu, *Electrochim. Acta* **2006**, *52*, 1662–1669.
74. L. Wang, S. Guo, L. Huang, S. Dong, *Electrochem. Commun.* **2007**, *9*, 827–832.
75. H. Ye, R. M. Crooks, *J. Am. Chem. Soc.* **2005**, *127*, 4930–4934.
76. Z. G. Estephan, L. Alawieh, L. I. Halaoui, *J. Phys. Chem. C* **2007**, *111*, 8060–8068.
77. M. Huang, Y. Shao, X. Sun, H. Chen, B. Liu, S. Dong, *Langmuir* **2005**, *21*, 323–329.
78. R. Narayanan, M. A. El-Sayed, *J. Phys. Chem. B* **2005**, *109*, 12663–12676.

79. Y. Xiong, B. J. Wiley, Y. Xia, *Angew. Chem. Int. Ed.* **2007**, *46*, 7157–7159.
80. C. Wang, H. Daimon, Y. Lee, J. Kim, S. Sun, *J. Am. Chem. Soc.* **2007**, *129*, 6974–6975.
81. N. Tian, Z. Zhou, S. Sun, Y. Ding, Z. L. Wang, *Science* **2007**, *316*, 732–735.
82. J. Yang, J. Y. Lee, H. Too, S. Valiyaveettil, *J. Phys. Chem. B* **2006**, *110*, 125–129.
83. H. P. Liang, H. M. Zhang, J. S. Hu, Y. G. Guo, L. J. Wan, C. L. Bai, *Angew. Chem. Int. Ed.* **2004**, *43*, 1540–1543.
84. J. Zhai, M. Huang, Y. Zhai, S. Dong, *J. Mater. Chem.* **2008**, *18*, 923–928.
85. G. Chen, D. Xia, Z. Nie, Z. Wang, L. Wang, L. Zhang, J. Zhang, *Chem. Mater.* **2007**, *19*, 1840–1844.
86. H. Liang, Y. Guo, H. Zhang, J. Hu, L. Wan, C. Bai, *Chem. Commun.* **2004**, 149–151.
87. S. Guo, S. Dong, E. Wang, *J. Phys. Chem. C* **2007**, *111*, 17104–17109.
88. S. Guo, S. Dong, E. Wang, *Chem. Eur. J.* **2008**, *14*, 4689–4695.
89. L. Lu, G. Sun, S. Xi, H. Wang, H. Zhang, *Langmuir* **2003**, *19*, 3074–3077.
90. S. Park, P. X. Yang, P. Corredor, M. J. Weaver, *J. Am. Chem. Soc.* **2002**, *124*, 2428–2429.
91. J. Lee, S. Hwang, H. Lee, J. Kwak, *J. Phys. Chem. B* **2004**, *108*, 5372–5379.
92. Y. D. Jin, Y. Shen, S. J. Dong, *J. Phys. Chem. B* **2004**, *108*, 8142–8147.
93. J. Zhang, F. H. B. Lima, M. H. Shao, K. Sasaki, J. X. Wang, J. Hanson, R. R. Adzic, *J. Phys. Chem. B* **2005**, *109*, 22701–22704.
94. M. Huang, Y. Jin, H. Jiang, X. Sun, H. Chen, B. Liu, E. Wang, S. Dong, *J. Phys. Chem. B* **2005**, *109*, 15264–15271.
95. J. L. Zhang, M. B. Vukmirovic, Y. Xu, M. Mavrikakis, R. R. Adzic, *Angew. Chem., Int. Ed.* **2005**, *44*, 2132–2135.
96. J. Zhai, M. Huang, S. Dong, *Electroanalysis* **2007**, *19*, 506–509.
97. P. Costamanga, S. Srinivisan, *J. Power Sources* **2001**, *102*, 242–245.
98. T. J. Schmidt, H. A. Gasteiger, R. J. Behm, *J. Electrochem. Soc.* **1999**, *146*, 1296–1304.
99. X. Teng, S. Maksimuk, S. Frommer, H. Yang, *Chem. Mater.* **2007**, *19*, 36–41.
100. J. Xu, T. Zhao, Z. Liang, L. Zhu, *Chem. Mater.* **2008**, *20*, 1688–1690.
101. C. Roychowdhury, F. Matsumoto, V. B. Zeldovich, S. C. Warren, P. F. Mutolo, M. Ballesteros, U. Wiesner, H. D. Abruna, F. J. DiSalvo, *Chem. Mater.* **2006**, *18*, 3365–3372.
102. M. Chen, J. P. Liu, S. Sun, *J. Am. Chem. Soc.* **2004**, *126*, 8394–8395.
103. M. Chen, J. Kim, J. P. Liu, H. Fan, S. Sun, *J. Am. Chem. Soc.* **2006**, *128*, 7132–7133.
104. M. Chen, T. Pica, Y. B. Jiang, P. Li, K. Yano, J. P. Liu, A. K. Datye, H. Fan, *J. Am. Chem. Soc.* **2007**, *129*, 6348–6349.
105. S. Maksimuk, S. Yang, Z. Peng, H. Yang, *J. Am. Chem. Soc.* **2007**, *129*, 8684–8685.
106. S. Koh, P. Strasser, *J. Am. Chem. Soc.* **2007**, *129*, 12624–12625.
107. P. Mani, R. Srivastava, P. Strasser, *J. Phys. Chem. C* **2008**, *112*, 2770–2778.
108. Z. Liu, D. Reed, G. Kwon, M. Shamsuzzoha, D. E. Nikles, *J. Phys. Chem. C* **2007**, *111*, 14223–14229.
109. R. Srivastava, P. Mani, N. Hahn, P. Strasser, *Angew. Chem., Int. Ed.* **2007**, *46*, 8988–8991.
110. J. L. Fernandez, D. A. Walsh, A. J. Bard, *J. Am. Chem. Soc.* **2005**, *127*, 357–365.
111. M. Shao, K. Sasaki, R. R. Adzic, *J. Am. Chem. Soc.* **2006**, *128*, 3526–3527.
112. J. Luo, P. N. Njoki, Y. Lin, L. Wang, C. J. Zhong, *Electrochem. Commun.* **2006**, *8*, 581–587.
113. Q. Huang, H. Yang, Y. Tang, T. Lu, D. L. Akins, *Electrochem. Commun.* **2006**, *8*, 1220–1224.
114. J. T. Moore, D. Chu, R. Jiang, G. A. Deluga, C. M. Lukehart, *Chem. Mater.* **2003**, *15*, 1119–1124.
115. J. Tian, G. Sun, L. Jiang, S. Yan, Q. Mao, Q. Xin, *Electrochem. Commun.* **2007**, *9*, 563–568.
116. K. Matsuoka, K. Miyazaki, Y. Iriyama, K. Kikuchi, T. Abe, Z. Ogumi, *J. Phys. Chem. C* **2007**, *111*, 3171–3174.
117. Y. Song, Y. Li, X. Xia, *Electrochem. Commun.* **2007**, *9*, 201–205.
118. E. Frackowiak, G. Lota, T. Cacciaguerra, F. Beguin, *Electrochem. Commun.* **2006**, *8*, 129–132.
119. Y. Liu, J. Chen, W. Zhang, Z. Ma, G. F. Swiegers, C. O. Too, G. G. Wallace, *Chem. Mater.* **2008**, *20*, 2603–2605.
120. J. Xu, K. Hua, G. Sun, C. Wang, X. Lv, Y. Wang, *Electrochem. Commun.* **2006**, *8*, 982–986.
121. L. Cao, F. Scheiba, C. Roth, F. Schweiger, C. Cremers, U. Stimming, H. Fuess, L. Chen, W. Zhu, X. Qiu, *Angew. Chem. Int. Ed.* **2006**, *45*, 5315–5319.
122. Y. Zhao, L. Fan, H. Zhong, Y. Li, S. Yang, *Adv. Funct. Mater.* **2007**, *17*, 1537–1541.

123. A. Kongkanand, K. Vinodgopal, S. Kuwabata, P. V. Kamat, *J. Phys. Chem. B* **2006**, *110*, 16185–16188.
124. Y. Mu, H. Liang, J. Hu, L. Jiang, L. Wan, *J. Phys. Chem. B* **2005**, *109*, 22212–22216.
125. M. C. Gutiérrez, M. J. Hortigüela, J. M. Amarilla, R. Jiménez, M. L. Ferrer, F. del Monte, *J. Phys. Chem. C* **2007**, *111*, 5557–5560.
126. H. J. Wang, H. Yu, F. Peng, P. Lv, *Electrochem. Commun.* **2006**, *8*, 499–504.
127. V. Selvaraj, M. Alagar, *Electrochem. Commun.* **2007**, *9*, 1145–1153.
128. S. Guo, J. Zhai, Y. Fang, S. Dong, E. Wang, *Chem. Asian J.* **2008**, *3*, 1156–1162.
129. Y. L. Hsin, K. C. Hwang, C. Yeh, *J. Am. Chem. Soc.* **2007**, *129*, 9999–10010.
130. E. S. Steigerwalt, G. A. Deluga, D. E. Cliffel, C. M. Lukehart, *J. Phys. Chem. B* **2001**, *105*, 8097–8101.
131. L. Yang, W. Yang, Q. Cai, *J. Phys. Chem. C* **2007**, *111*, 16613–16617.
132. S. Guo, S. Dong, E. Wang, *J. Phys. Chem. C* **2008**, *112*, 2389–2393.
133. H. Cui, Y. Xu, Z. Zhang, *Anal. Chem.* **2004**, *76*, 4002–4010.
134. H. Cui, Y. Dong, *J. Electroanal. Chem.* **2006**, *595*, 37–46.
135. Y. Dong, H. Cui, Y. Xu, *Langmuir* **2007**, *23*, 523–529.
136. Y. Dong, H. Cui, C. Wang, *J. Phys. Chem. B* **2006**, *110*, 18408–18414.
137. H. Cui, W. Wang, C. Duan, Y. Dong, J. Guo, *Chem. Eur. J.* **2007**, *13*, 6975–6984.
138. C. H. Lyons, E. D. Abbas, J.-K. Lee, M. F. Rubner, *J. Am. Chem. Soc.* **1998**, *120*, 12100–12107.
139. W.-Y. Lee, T. A. Nieman, *Anal. Chem.* **1995**, *67*, 1789–1796.
140. H. N. Choi, S.-H. Cho, W.-Y. Lee, *Anal. Chem.* **2003**, *75*, 4250–4256.
141. S. Guo, E. Wang, *Electrochem. Commun.* **2007**, *9*, 1252–1257.
142. M. Sykora, T. J. Meyer, *Chem. Mater.* **1999**, *11*, 1186–1189.
143. X. Sun, Y. Du, S. Dong, E. Wang, *Anal. Chem.* **2005**, *77*, 8166–8169.
144. X. Sun, Y. Du, L. Zhang, S. Dong, E. Wang, *Anal. Chem.* **2006**, *78*, 6674–6677.
145. Y. Du, B. Qi, X. Yang, E. Wang, *J. Phys. Chem. B* **2006**, *110*, 21662–21666.
146. L. Zhang, Z. Xu, X. Sun, S. Dong, *Biosens. Bioelectron.* **2007**, *22*, 1097–1100.
147. Z. Guo, Y. Shen, M. Wang, F. Zhao, S. Dong, *Anal. Chem.* **2004**, *76*, 184–191.
148. L. Zhang, S. Dong, *Electrochem. Commun.* **2006**, *8*, 1687–1691.
149. L. Zhang, S. Dong, *Anal. Chem.* **2006**, *78*, 5119–5123.
150. H. Wei, J. Liu, L. Zhou, J. Li, X. Jiang, J. Kang, X. Yang, S. Dong, E. Wang, *Chem. Eur. J.* **2008**, *14*, 3687–3693.
151. L. Zhang, B. Liu, S. Dong, *J. Phys. Chem. B* **2007**, *111*, 10448–10452.
152. X. Wang, J. Zhou, W. Yun, S. Xiao, Z. Chang, P. He, Y. Fang, *Anal. Chim. Acta* **2007**, *598*, 242–248.
153. Z. Chang, J. Zhou, K. Zhao, N. Zhu, P. He, Y. Fang, *Electrochim. Acta* **2006**, *52*, 575–580.
154. Z. Ding, B. M. Quinn, S. K. Haram, L. E. Pell, B. A. Korgel, A. J. Bard, *Science* **2002**, *296*, 1293–1297.
155. N. Myung, Z. Ding, A. J. Bard, *Nano Lett.* **2002**, *2*, 1315–1319.
156. Y. Bae, N. Myung, A. J. Bard, *Nano Lett.* **2004**, *4*, 1153–1161.
157. L. Shen, X. Cui, H. Qi, C. Zhang, *J. Phys. Chem. C* **2007**, *111*, 8172–8175.
158. G. Zou, H. Ju, *Anal. Chem.* **2004**, *76*, 6871–6876.
159. G. Jie, B. Liu, J. Miao, J. Zhu, *Talanta* **2007**, *71*, 1476–1480.
160. Z. Dai, J. Zhang, J. Bao, X. Huang, X. Mo, *J. Mater. Chem.* **2007**, *17*, 1087–1093.
161. S. Ding, J. Xu, H. Chen, *Chem. Commun.* **2006**, 3631–3633.
162. R. Cui, H. Pan, J. Zhu, H. Chen, *Anal. Chem.* **2007**, *79*, 8494–8501.
163. G. Jie, B. Liu, H. Pan, J. Zhu, H. Chen, *Anal. Chem.* **2007**, *79*, 5574–5581.
164. X. Liu, H. Jiang, J. Lei, H. Ju, *Anal. Chem.* **2007**, *79*, 8055–8060.
165. H. Jiang, H. Ju, *Anal. Chem.* **2007**, *79*, 6690–6696.
166. L. Zhang, X. Zou, E. Ying, S. Dong, *J. Phys. Chem. C* **2008**, *112*, 4451–4454.
167. K. Stulik, C. Amatore, K. Holub, V. Marecek, W. Kutner, *Pure Appl. Chem.* **2000**, *72*, 1483–1492.

Chapter 10

Signal Generation with Gold Nanoparticles: Photophysical Properties for Sensor and Imaging Applications

QINGSHAN WEI AND ALEXANDER WEI

The Supramolecular Chemistry of Organic–Inorganic Hybrid Materials. Edited by Knut Rurack and Ramón Martínez-Máñez

10.1 INTRODUCTION

Nanostructured materials are well suited to support the diverse activities currently being explored in contemporary supramolecular science. Nanomaterials and supramolecular systems operate on similar length scales, and in recent years, chemists have been successful in controlling the number and types of molecules presented on the surfaces of nanoparticles, ranging from single-molecule functionalization to multilamellar structures. Like molecules, nanoparticles can be prepared in a wide variety of shapes and sizes, and in many cases exhibit tunable physical properties that can be exploited for applications in analytical chemistry and information technology. Nanoparticles are also building blocks in their own right, and can be directed by various modes of self-assembly to create unique architectures with novel materials properties.

Here, we focus on Au nanoparticles as complementary functional materials for supramolecular systems, and discuss recent demonstrations of hybrid nanosystems in chemical and biomolecular sensing, biomedical imaging, photothermal actuation, or a combination of these. Nearly all of the properties to be discussed are derived from *surface plasmons*, an electrodynamic phenomenon involving the collective excitation of conduction electrons at specific wavelengths of light. Plasmon-resonant nanomaterials can amplify optical signals by many orders of magnitude, and the resonance conditions are highly sensitive to both structural and surface effects. This tunability creates a wealth of opportunities for coupling the optical properties of Au nanoparticles to supramolecular function.

In this chapter, we provide a brief discussion on surface plasmons, intended for the nonspecialist. This is followed by a closer examination of some photophysical properties intrinsic to Au nanostructures, as well as examples of their application to supramolecular or bioanalytical systems. Readers interested in a deeper understanding of the underlying physics of surface plasmons are directed to the monographs or reviews in References 1 to 13. With respect to supramolecular chemistry, it must be emphasized that a successful fusion with metal nanoparticles requires a good working knowledge of surface chemistry and the various chemisorptive strategies for introducing ligands onto metal surfaces. A comprehensive treatise on this subject can be found in Chapter 4. Lastly, due to space limitations and the enormous number of publications involving the application of Au nanoparticles in chemical systems, we regret that only a selected number of examples can be presented, to illustrate some of the many possible applications supported by these fascinating materials. Indeed, we have intentionally left out some of the more obvious applications, such as the changes in optical absorbance caused by the aggregation of gold nanoparticles, in order to focus on some more recent (and possibly underexploited) photophysical effects.

10.2 SURFACE PLASMONS

Coinage metals such as Au and Ag are well known for their ability to support surface plasmons. For bulk metal with smooth surfaces, the conduction electrons can be excited as a plasmon "wave" by incident photons at the surface (metal–dielectric interface), and can propagate for several microns before the energy is reemitted (or scattered) as light. This effect, which is commonly referred to as surface plasmon resonance (SPR), also involves a strong attenuation in reflectance at a specific angle of incidence. The resonance condition is highly sensitive to the local environment and is useful for detecting changes in molecular surface adsorption, but its quantitative analysis requires precise alignment of the collection optics (Fig. 10.1).

Metal nanoparticles with dimensions below the wavelength of light are also highly responsive to light; however, the conduction electrons are trapped within the particle and thus forced to oscillate as induced dipoles in a localized fashion. Here, the resonance condition is defined by the intrinsic material properties (dielectric function) of the metal as well as the surrounding medium, and also by nanoparticle size and shape. In the case of spherical Au nanoparticles in the size range of 5 to 40 nm, the localized surface plasmon resonance (LSPR) is strongest at green wavelengths

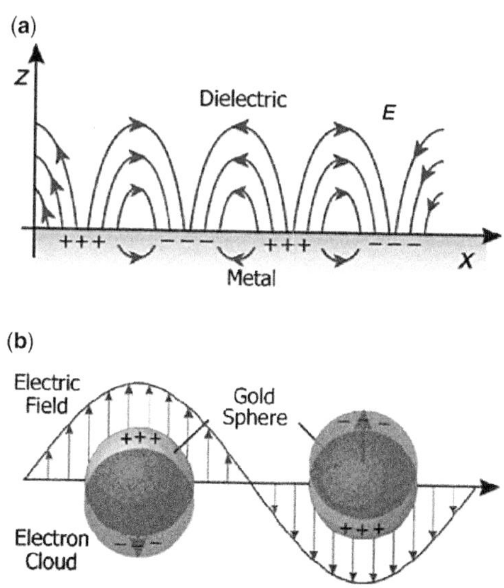

Figure 10.1 (a) Surface plasmon wave propagating (in the x direction) along a metal–dielectric interface; (b) localized surface plasmon resonance (oscillating induced dipole) produced by Au nanospheres.[7] (Reprinted with permission from Y. Xia and N. J. Halas, *MRS Bull.* **2005**, *30*, 338–343.)

($\lambda_{SPR} = 520$ to 540 nm), with colloidal dispersions producing a characteristic wine-red color. If the Au nanoparticles are smaller than 5 nm, the LSPR peak is broadened due to surface scattering.[14] If the nanospheres are much greater than 40 nm (the approximate mean free path of an electron in Au), the LSPR shifts to longer wavelengths due to retardation effects, and secondary peaks may appear due to multipolar plasmon modes.

The intense optical properties of colloidal gold have been appreciated for a very long time, with famous examples dating back to the days of the Roman empire, but the physical origin of "red gold" was first established by Faraday in the nineteenth century.[15] The electrodynamic nature of surface plasmons was first described by Mie for spherical Au particles at the beginning of the twentieth century,[16] then extended by Gans toward ellipsoidal particles.[17] In the simple case of small metal nanoparticles ($r < 20$ nm) the LSPR can be attributed solely to the dipolar oscillation of conduction electrons, whereas higher-order plasmon modes increase in significance at the expense of dipolar modes for large or anisotropic particles.[18]

The extinction cross-section C_{ext} for spherical particles can be approximated by using Mie theory:

$$C_{ext} = \frac{24\pi^2 r^3 \varepsilon_m^{3/2}}{\lambda} \left| \frac{\varepsilon''}{(\varepsilon' + 2\varepsilon_m)^2 + (\varepsilon'')^2} \right| \quad (10.1)$$

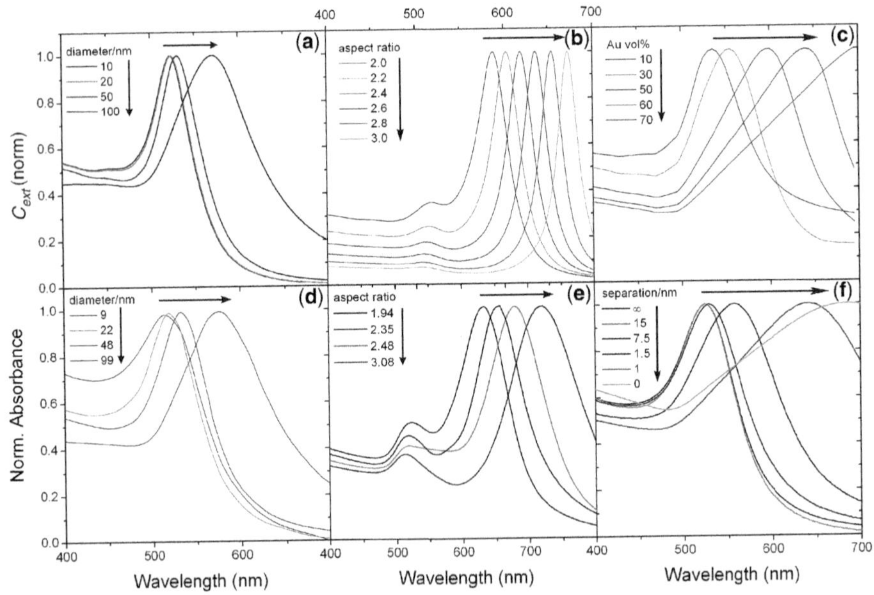

Figure 10.2 Simulated and normalized experimental LSPR spectra from different Au nanostructures. (a, d) Au nanospheres with varying diameters; (b, e) Au nanorods with varying aspect ratio; (c, f) nanospheres loaded onto thin glass films with varying surface coverages.[8] Horizontal and vertical arrows correspond to the progressions in λ_{LSPR} with increasing size or aspect ratio. (Reprinted with permission from L. M. Liz-Marzán, *Langmuir* **2006**, 22, 32–41. Copyright 2006 American Chemical Society.)

where ε_m is the dielectric constant of the local medium, and ε' and ε'' are real and imaginary components of the complex dielectric function $\varepsilon(\omega)$. A plasmon resonance occurs when the condition $\varepsilon' = -2\varepsilon_m$ is satisfied; in the case of spherical Au nanoparticles, ε' becomes negative at visible wavelengths; if the particles are anisotropic, ε' becomes negative at red to near-infrared (NIR) wavelengths. The LSPR is also sometimes described in terms of extinction efficiency Q_{ext}, which is defined as the ratio of C_{ext} to geometrical cross-section area (πr^2 in the case of spheres).

The extinction cross-section C_{ext} can be further divided into absorption and scattering components, C_{abs} and C_{sca}. In principle, these terms can be solved analytically for any Au nanostructure by Mie theory; in practice, exact solutions are only available for a few ideal shapes, such as solid spheres,[16] spheroids,[19] concentric shells, and infinite cylinders (wires).[20] However, the LSPR of anisotropic nanoparticles (e.g., nanorods, nanoprisms, and nanocages) can be reasonably approximated using numerical methods such as discrete dipole approximation (DDA)[21] and the finite difference time domain (FDTD) method.[22] Figure 10.2 reveals a good agreement between DDA simulations and the actual extinction spectra for Au nanospheres, nanorods, and nanoparticle films on thin glass substrates.[8] The DDA method can also be used to compare the relative contributions of absorption, scattering, and

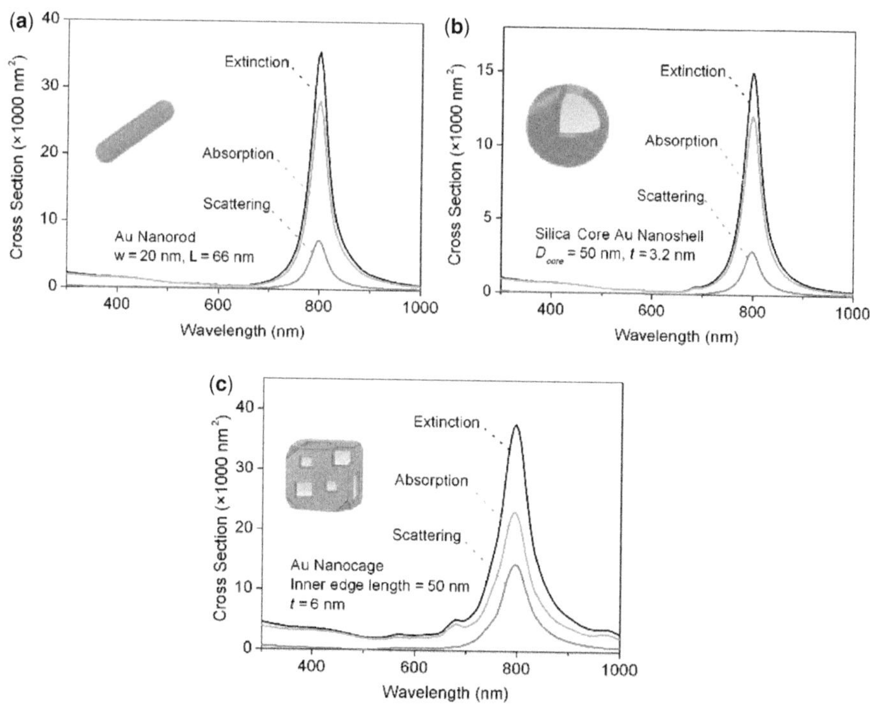

Figure 10.3 Comparison of optical spectra (C_{abs}, C_{sca}, and C_{ext}) calculated using DDA for (a) Au nanorods, (b) Au nanoshells, and (c) Au nanocages.[23] (Reprinted with permission from M. Hu et al., *Chem. Soc. Rev.* **2006**, *35*, 1084–1094. Copyright the Royal Society of Chemistry.)

extinction of nanostructures having similar LSPRs but different shapes, such as NIR-resonant Au nanorods, nanocages, and nanoshells (Fig. 10.3).[23] For example, Au nanorods have relatively high C_{abs} whereas nanocages have greater C_{sca}, but the C_{ext} of both nanorods and nanocages at their respective resonances are more than twice that of hollow Au nanoshells having the same size. Therefore, the combination of Mie theory and numerical calculations provides a strong theoretical foundation for designing plasmonic nanoparticles with tunable optical properties.

10.2.1 Factors Affecting LSPR

As mentioned above, the localized plasmon resonance wavelength (λ_{SPR}) and extinction cross-section are strongly influenced by physical factors such as particle size and shape, and also by chemical factors such as local dielectric effects on the particle surface.[3,8,9,12,24] Nanoparticle shape is an important parameter which affects both λ_{SPR} and the range of wavelengths over which it can vary. Although Au nanospheres have been used widely, the adjustable range of their LSPR band is in fact quite limited.[25] Anisotropic nanostructures such as nanorods, nanoshells, and nanoprisms have access to a much larger range of wavelengths. These often have more than one resonance peak, introducing some possibilities for multiplexing; for example, Au nanorods have two plasmon bands associated with the short and long axes, with the longitudinal plasmon resonance (λ_{LPR}) correlating with nanorod aspect ratio in a quasi-linear fashion (Fig. 10.2b, e).[26–28]

The LSPR response can also be sensitive to the spacing between particles, providing a particularly fertile opportunity to support chemical or biological sensing applications. In the case of single nanoparticles, the electromagnetic fields generated by the localized plasmon oscillations support near-field effects, but these decrease rapidly from the particle surface.[10] However, when two nanoparticles are in close proximity (separated by less than a particle diameter), their plasmon resonances can be coupled and the near-field effects between the particles can be greatly enhanced.[29] Experimental and computational studies have also shown that the aggregation or assembly of Au nanoparticles induces a red shift in wavelength, relative to that of isolated particles (Fig. 10.2c, f). A famous example involving interparticle plasmons has been provided by Mirkin and coworkers, in which the DNA-mediated aggregation of Au nanoparticles produces an easily detectable change in color from red to blue.[30] Variations on this theme have produced systems that can recognize complementary DNA strands, using as little as 50 femtomoles.[31] More recently, a "plasmonic ruler" based on the distance-dependent coupling of Au nanoparticles was described as an alternative to Förster resonance energy transfer (FRET), and has been shown to measure interparticle separations of up to 70 nm.[32]

Interparticle plasmon coupling is important not only for modulating the optical response of chemically induced aggregation, but also for generating intense local electromagnetic fields for surface-enhanced Raman scattering (SERS) and other plasmon-enhanced spectroscopies. The coupling effects strengthen with increasing particle diameter, and also with decreasing interparticle spacing. These tunable

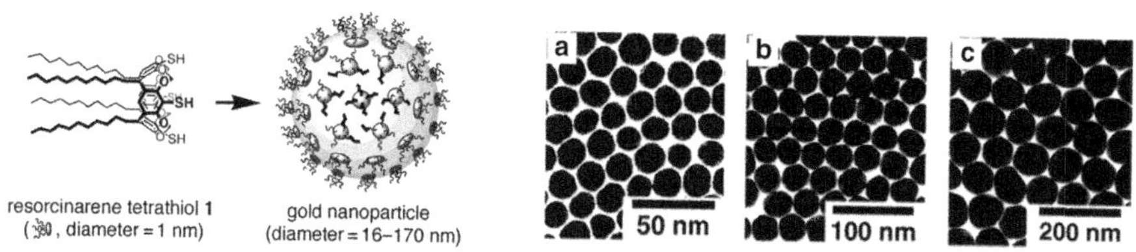

Figure 10.4 Resorcinarene-encapsulated colloidal Au nanoparticles (left) and their self-assembly into two-dimensional arrays (right). Particle sizes: (a) 16 ± 3 nm; (b) 34 ± 2 nm; (c) 87 ± 7 nm.[33] (Reprinted with permission from B. Kim et al., *J. Am. Chem. Soc.* **2001**, *123*, 7955–7956. Copyright 2001 American Chemical Society.)

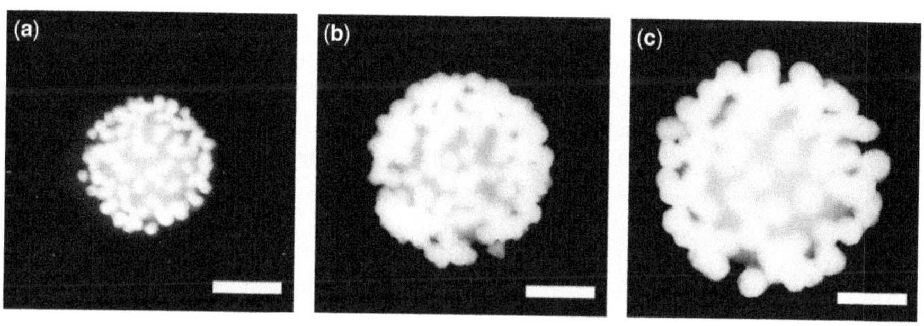

Figure 10.5 Spherical ensembles of citrate-stabilized Au nanoparticles onto polymer-coated SiO_2 cores. Au/SiO_2 particle sizes: (a) 40/330 nm; (b) 60/460 nm; (c) 80/550 nm. Scale = 200 nm.[34] (Reprinted with permission from B. Sadtler and A. Wei, *Chem. Commun.* **2002**, 1604–1605. Copyright the Royal Society of Chemistry.)

parameters have stimulated efforts to design plasmonic substrates based on nanoparticle self-assembly; two examples shown here include the preparation of planar two-dimensional nanoparticle arrays by self-organization of resorcinarene-encapsulated Au nanoparticles at the air–water interface (Fig. 10.4),[33] and spherical arrays of Au nanoparticles on silica microspheres by electrostatic self-assembly (Fig. 10.5).[34] Applications of localized field effects for supramolecular sensing will be discussed in Sections 10.4 and 10.5.

10.2.2 Plasmonic Nanomaterials as Refractive Index Sensors

The sensitivity of the LSPR wavelength of Au nanoparticles to local environmental effects renders them useful for a variety of chemical and biological sensing applications.[35–39] SPR-like sensors based on nanoparticles have been used to report molecular recognition events, by monitoring refractive index (RI) changes induced by surface adsorption.[40–45] For example, the surface conjugation of biomolecules such as proteins,[46] DNA or carbohydrates,[47] and the formation of silica coatings[8,48] or polymeric shells[49] all produce a change in the local RI due to the modulation of surface dielectric and can be monitored by tracking shifts in λ_{SPR}. Many of these nanosensors can be applied toward the detection of analytes under aqueous physiological conditions; for example, antibody-functionalized Au nanoshells have been used to detect proteins in saline, serum, and whole blood.[50]

Anisotropic Au nanostructures have demonstrated even greater potential as ultrasensitive RI probes. Van Duyne and coworkers have conducted a systematic investigation of surface effects on the LSPR of triangular metal nanoprisms and other anisotropic nanostructures.[51–58] The LSPR shifts to longer wavelength with the adsorption of alkanethiols, and increases further with chain length.[51] The RI sensitivity of individual Ag nanoparticles is also quite shape dependent. Rod-shaped and triangular Ag nanoparticles exhibit LSPR shifts of 235 nm and 197 nm, respectively, per refractive index unit (RIU^{-1}), greater than that observed for spherical Ag nanoparticles (161 nm RIU^{-1}).[51] The greater RI sensitivity of anisotropic plasmon-resonant nanostructures is theoretically validated by the plasmon resonance condition $\varepsilon' = -\chi\varepsilon_m$, where χ is a shape factor that increases with particle aspect ratio (for spherical nanoparticles, $\chi = 2$). DDA calculations on Au nanorods and nanocages confirm their high sensitivity to RI effects, in accord with experimental studies. Nanorods with an aspect ratio of 3.5 were predicted to be three times more sensitive to RI effects than spherical particles (aspect ratio 1),[59] and hollow Au nanocages (50 nm diameters, with wall thickness of 4.5 nm) were predicted to be nearly seven times more sensitive than solid nanoparticles of the same diameter.[60]

10.3 ABSORPTION PROPERTIES OF GOLD NANOPARTICLES

The optical properties of Au nanoparticles less than 40 nm are dominated by their absorption cross-section (C_{abs}). The absorbance of a colloidal solution containing N particles per unit volume is readily defined by the Beer–Lambert Law:

$$A = NC_{abs}L/\ln 10 \tag{10.2}$$

where L is the optical path length (in cm). The C_{abs} of a 5 nm Au particle is approximately 3 nm^2 at λ_{SPR}, which is over two orders of magnitude larger than that of a typical organic dye at room temperature.[61] The absorption cross-section increases linearly with particle volume, so a 25 nm Au particle has a C_{abs} of roughly 375 nm^2. However, the plasmon-resonant absorption cross-section is even larger in anisotropic Au nanoparticles, and scales nonlinearly with size; for example, the C_{abs} for Au nanorods with aspect ratio 2 (25 × 50 nm) has been estimated to be 5×10^3 nm^2, much more than double that of a 25 nm Au sphere.[62] Anisotropic Au nanoparticles are also much more strongly absorbing than semiconductor quantum-dot nanoparticles, when compared on a per volume basis.[6]

10.3.1 Plasmon-Enhanced Heating: Nanoparticles as Photothermal Agents

The very large absorption cross-sections of Au nanoparticles give rise to photothermal effects, that is, an increase in local temperature upon plasmon resonance excitation. The concept of Au nanoparticles as photothermal transducers has inspired numerous efforts to develop biomedical applications such as localized hyperthermal therapy,[63–65] photothermal contrast agents,[66] and optothermally controlled drug release,[67–69] and also novel platforms for information storage.[70]

The photophysical origin of plasmonic heating can be attributed to the nonradiative relaxation of the optically excited conduction electrons, whose energies then diffuse into the metal lattice (electron–phonon coupling) and are dissipated as heat into the surrounding medium (phonon–phonon interactions).[71–73] Under ideal conditions, the nanoparticle temperature can be estimated by the following equation:[73]

$$\Delta T = \frac{E_{abs}}{mc_p} \tag{10.3}$$

where E_{abs} is the absorbed photon energy, m represents the mass of the Au nanoparticle, and c_p is the heat capacity of Au. Based on this simple expression, a 5 nm Au sphere should experience a 15 K increase in surface temperature when saturated at plasmon resonance by 20 mW laser.[74] The rate of energy dissipation from hot Au nanoparticles to their surroundings has been investigated experimentally by a number of groups using ultrafast laser spectroscopy.[71] The photothermal relaxation times have been found to be quadratically proportional to the particle diameter (10 ps for 4 nm particles, compared to near 400 ps for 50 nm particles), which means that smaller particles cool down faster because of their larger surface to volume ratios.[71]

With respect to heat released to the immediate surroundings, the temperature distribution from the surface of Au nanoparticles has been calculated using a simple heat transfer model.[75,76] The local temperature increase $\Delta T(r)$ around a single nanoparticle is described as:

$$\Delta T(r) = \frac{V_{NP}Q}{4\pi k_0 r} \tag{10.4}$$

where V_{NP} is the nanoparticle volume, Q is photothermal heat, k_0 is the thermal conductivity of the surrounding medium, and r is the distance from the nanoparticle surface. The model suggests a thermal gradient with an approximately $1/r$ dependence, with a maximum surface temperature $\Delta T_{\max}$ proportional to R_{NP}^2, the nanoparticle radius squared. For instance, in the case of a 5 nm Au sphere in water, the surface temperature rise is 15 K but drops to 3 K just 10 nm from the surface.[74]

Plasmon-resonant Au nanoparticles are attractive as contrast agents for photothermal imaging.[74,77–79] Unlike fluorescent probes, Au nanoparticles are stable against photobleaching, transient signal quenching (blinking), and photosaturation effects. The conversion of light energy to heat gives rise to a localized temperature gradient as described above, resulting in a refractive index change in the surrounding media. Boyer et al. demonstrated that Au nanoparticles as small as 2.5 nm could be detected via photothermal imaging, even in a high background scattering environment.[74] In an example of a biological application, 10 nm Au particles coated with anti-IgG were used as photothermal imaging probes for single-molecule detection in cells.[80]

The high photothermal efficiency of Au nanoparticles and their well-established biocompatibility also make them excellent candidates for targeted photothermal therapy. A number of groups have provided *in vitro* demonstrations of cell-selective targeting, using Au nanoparticles conjugated to biomolecular recognition elements. For example, Au nanospheres labeled with antibodies specific for CD8 or epidermal growth factor receptors were able to selectively destroy cancer cells expressing those surface proteins, when irradiated with a 532 nm laser.[81,82]

For *in vivo* applications, visible wavelengths are less effective due to their limited penetration depth into biological tissue, whereas a "biological window" exists for NIR wavelengths between 750 nm and 1.3 microns.[83] Therefore, anisotropic nanoparticles with large NIR absorption cross-sections are considered the photothermal agents of choice for *in vivo* photothermal applications. A seminal report on the use of NIR-active Au nanostructures for *in vivo* photothermal imaging and therapy was reported in 2003 by Hirsch et al.[84] Core–shell silica@Au nanoparticles (Au nanoshells) were coated with poly(ethylene glycol) (PEG), an antifouling polymer, and investigated both as a contrast agent for photothermal magnetic resonance imaging (MRI) and as a device for localized photothermal ablation. Nanoshells deposited in the vicinity of tumors could induce irreversible tissue damage after a brief exposure to an NIR laser, which increased the bulk temperature by nearly 40 K.[84] Nanoshells were also conjugated with anti-HER2 for the targeted photothermal ablation of breast cancer cells.[85] Au nanocages, a similar NIR-absorbing agent with a hollow interior, have also demonstrated a capacity for optical imaging and therapy.[86–90]

Au nanorods are particularly noteworthy as photothermal agents, due to their large and wavelength-selective absorption cross-sections and synthetic accessibility.

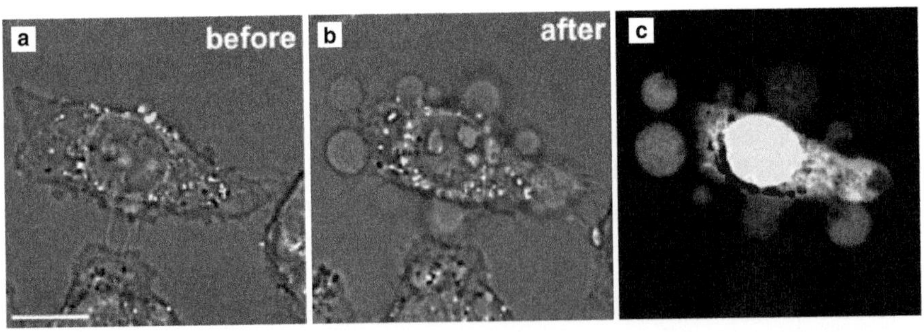

Figure 10.6 (a, b) Folate-conjugated Au nanorods (red) targeted to the membranes of KB cells, before and after exposure to a brief NIR laser pulse. (c) Evidence of "optoporation" was revealed by using ethidium bromide (red) and a fluorescent dye to indicate the presence of intracellular Ca^{2+} (green).[94] (Reprinted with permission from L. Tong et al., *Adv. Mater.* **2007**, *19*, 3136–3141. Copyright Wiley-VCH Verlag GmbH & Co. KGaA.) (See color insert.)

Like the nanoshells and nanocages, the photothermal effects of Au nanorods have been applied in a targeted fashion toward tumor cells,[91–94] parasitic protozoans,[95] macrophage,[96] and pathogenic bacteria.[97] In one recent study, it has been shown that Au nanorods can be targeted to tumor cell membranes and photothermally activated as "optoporation" agents, and can induce cell necrosis by enabling an intracellular influx of Ca^{2+} ions (Fig. 10.6).[94] With respect to nonbiological applications, Au nanorods have been blended with polymers and investigated as materials for photothermal recording.[98] These examples represent just a slice from the portfolio of potential applications for Au nanorods and other NIR-active nanoparticles. Further developments will depend on the chemist's ability to develop appropriate surface functionalization schemes to render such nanoparticles compatible with the application at hand; this topic will be revisited in a later section.

10.3.2 Photoacoustic Effects

The photoacoustic properties of plasmon-resonant Au nanoparticles are in fact derived from photothermal effects. The heat released produces variations in local pressure, resulting in audible acoustic waves that can be detected by ultrasonic transducers. Not suprisingly, Au nanoparticles have been employed as contrast agents for photoacoustic tomography (PAT), an emerging noninvasive imaging technique based on NIR-induced photoacoustic effects.[99] Images are obtained by the reconstruction of the optical energy absorption distributed across an array of acoustic transducers.

The advantage of PAT over pure optical or ultrasonic imaging techniques relies on the combined merits of optical irradiation and acoustic detection. On one hand, laser pulse irradiation provides a site-specific illumination compared with ultrasonic signal generation, with the benefit of high spatial resolution. On the other hand, the low diffusion of propagating acoustic waves is advantageous over reflected optical signals, which suffer from scattering by biological tissue.[100] The depth resolution of PAT can be several centimeters in biological tissue,[101] whereas pure optical imaging is limited to penetration depths of 1 mm or less.[102]

Plasmon-resonant Au nanoparticles have only recently been employed as exogenous contrast agents for PAT. Normally, the PAT relies on differences in optical absorption by different tissue components for contrast. A NIR light source provides the benefit of greater penetration depth, because absorption (and also tissue photodamage) is minimal in normal tissue, but requires the appropriate contrast agent. Both Au nanoshells ($>$100 nm) and nanocages of more compact size ($<$50 nm) have been detected in rat brains by PAT after intravenous injection, with observable vascular contrast within minutes.[103,104] Au nanorods are also appealing as contrast agents for PAT; in one recent example, Au nanorods could be detected with concentration differences as low as 1.25 pM.[105,106] This sensitivity is 75 times better than the detection of superparamagnetic iron oxide nanoparticles by MRI.[105] Other examples involving PAT include nanorod-based quantitative flow analysis for biological tissues,[107] and the distribution kinetics of drug delivery systems labeled with Au nanorods.[108]

10.4 SCATTERING PROPERTIES OF GOLD NANOPARTICLES

Up to this point, we have been considering the absorption properties of Au nanoparticles and the significance of C_{abs} in colorimetric or photothermal applications. However, colloidal Au particles are also famous for resonant light scattering, which depends on C_{sca}.[109,110] The earlier description of C_{ext} is dominated by C_{abs} and intended only for smaller nanospheres ($r < 20$ nm). C_{ext} is more correctly described as the sum of $C_{abs} + C_{sca}$:

$$C_{abs} = -\frac{8\pi^2}{\lambda} r^3 \mathrm{Im}\left(\frac{m-1}{m+2}\right) \tag{10.5}$$

$$C_{sca} = \frac{128\pi^5}{3\lambda^4} r^6 \left|\frac{m-1}{m+2}\right|^2 \tag{10.6}$$

where m is $\varepsilon(\omega)/\varepsilon_m$, the ratio of the frequency-dependent dielectric function of the metal particle and the dielectric constant of the surrounding medium, respectively.[3,111] Whereas C_{abs} scales approximately linearly with particle volume, C_{sca} scales quadratically with particle volume and becomes significant for spherical particles above 40 nm, to the extent that single-particle scattering is easily detected by dark-field microscopy.[112–116] Numerous single-particle scattering studies have investigated the influence of particle size and shape on the optical response, particular spectral frequency,[117–119] peak intensity,[111] and resonance linewidth.[120–125]

When excited with a broadband (white light) source, Au nanoparticles will scatter light at the resonance frequency corresponding to λ_{SPR}, often giving rise to a brilliant iridescence.[39,126,127] In most cases, the nanoparticle is much smaller than λ_{SPR}, such that elastic (Rayleigh) scattering is in effect. The relative efficiency of Rayleigh scattering versus absorption (Q_{sca} versus Q_{abs}) is illustrated in Figure 10.7 as a function of size for 20, 40, and 80 nm Au particles. The scattering efficiency is only significant for the largest particles, meaning that changes based on size are modest in comparison to those based on shape anisotropy, as observed earlier with NIR-resonant nanoparticles (Fig. 10.3).[128] Nevertheless, the scattering efficiencies of Au nanoparticles at λ_{SPR} are two orders of magnitude larger than nonmetallic particles,[109,110] and their molar scattering coefficients are four to five orders of magnitude greater than that of most fluorophores.[128]

10.4.1 Optical Contrast for Two-Dimensional Imaging: Dark-Field Microscopy

The study of plasmon-resonant nanoparticles as light scattering agents has a long history, starting with Zsigmondy's original demonstration in 1914.[129] Recent advances in dark-field optics and microscopy have led to a resurgence of interest in resonant light scattering, and its applications in biological detection and imaging.[110,130,131] Two forms of dark-field illumination are traditionally used for elastic light scattering: (1)

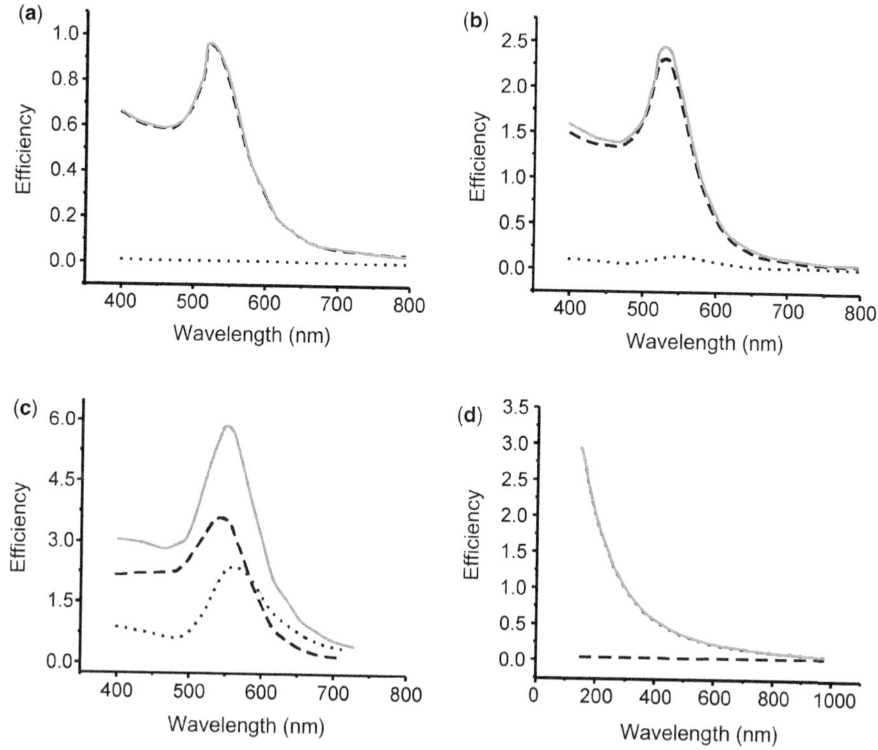

Figure 10.7 Calculated spectra of extinction, absorption, and scattering efficiencies for different sized Au nanoparticles: (a) $d = 20$ nm, (b) $d = 40$ nm, (c) $d = 80$ nm. (d) Nonresonant (polystyrene) particles, $d = 300$ nm.[128] Q_{ext}, Q_{abs}, and Q_{sca} are represented as solid, dashed, and dotted lines, respectively. (Reprinted with permission from P. K. Jain et al., *J. Phys. Chem. B* **2006**, *110*, 7238–7248. Copyright 2006 American Chemical Society.)

a wide-angle reflection mode in which the incident and collected light are both set overhead at oblique angles, and (2) an evanescent-wave mode in which nanoparticles are illuminated underneath by an electromagnetic field produced under total-internal reflection (TIR) conditions, with scattered light collected overhead. The evanescent field declines exponentially in intensity over a few tens of nanometers from the interface, coupling only with particles on or very close to the substrate.[132] For this reason, illumination by evanescent-wave coupling tends to be much more localized than by wide-angle incidence, with higher signal-to-noise ratios and greater imaging contrast.[132,133]

The strong plasmon-resonant scattering of Au nanoparticles makes them attractive as bright, unbleachable markers for chemical and bioanalytical sensing applications, with the caveat that a size of at least 40 nm is necessary to produce an easily detectable optical signal.[74] Despite this, plasmon-resonant scattering has been used with great success for the detection of single-stranded oligonucleic acids,[134–136] virus coat proteins,[137] and cancer cell biomarkers,[138] and has also been used to support RI-based chemical sensing (see Section 10.2.2)[139] and intracellular imaging studies in real time using dark-field microscopy.[140,141]

10.4.2 Optical Contrast for Three-Dimensional Imaging: Optical Coherence Tomography

Optical coherence tomography (OCT) is a cross-sectional tomographic imaging technique that has received increasing amounts of attention in recent years for subsurface tissue imaging.[142] OCT is typically performed using Michelson interferometry,[143] and produces morphology-dependent contrast using reflected or backscattered light from structures within the imaged sample. In OCT imaging, the axial resolution (Δz_{FWHM}) is inversely dependent on the bandwidth of the light source and can be as low as 1 μm, which is one to two orders of magnitude higher resolution than ultrasound imaging, magnetic resonance imaging (MRI), or X-ray computed tomography (CT).[144] The OCT imaging depth is determined by the relative transmittivity of incident light as well as the intrinsic scattering properties of the sample; when NIR light is used, a penetration depth of 2 to 3 mm into biological tissues can be achieved.[145]

NIR-resonant Au nanoparticles have shown exciting potential as optical contrast agents, and can greatly enhance the low contrast profiles associated with many applications of interest involving OCT.[146] Plasmon-resonant probes have so far been employed in at least two types of OCT imaging modalities. The most common type is based on backscattering contrast, but modalities that exploit high absorption contrast have also benefited. For conventional backscattering OCT, Au nanoshells (>100 nm in diameter) have been developed as contrast agents with high backscattering coefficients by adjusting the ratio of the core diameter and shell thickness.[147,148] For example, Au nanoshells with 291 nm SiO_2 cores and 25 nm thick Au coatings can produce a detectable (2 dB) OCT signal at a concentration of 10^9 nanoshells mL^{-1}.[149] Highly scattering Au nanoshells have recently been demonstrated for *in vivo* OCT imaging, with noticeable enhancement of OCT signal in the mice dorsal skin following a tail vein injection.[148] Similarly, PEG-modified "stealth" Au nanoshells can be passively concentrated in tumor tissue due to the leakiness of the tumor vasculature, commonly referred to as enhanced permeability and retention (EPR).[150]

Au nanorods and cubelike Au nanocages are even more appealing as contrast agents because of their smaller overall size ($\leq$50 nm) and narrow plasmon linewidths. Like spherical Au nanoparticles, the C_{abs} of nanorods and nanocages is proportional to particle volume, whereas their scattering cross-sections vary quadratically.[61] The structural anisotropy of Au nanorods also has great potential for polarization-sensitive OCT modalities, as the LPR mode can be physically aligned with the polarized incident beam. Au nanorods and nanocubes have been employed in conventional backscattering OCT, but require a very high concentration to produce detectable contrast.[151,152] Instead, the size-dependent absorption and scattering properties of these nanoparticles are better suited to support OCT modalities based on differential absorption or backscattering albedo (the ratio of backscattering to total extinction), which have the advantage of producing contrast in tissues with intrinsically high scattering cross-sections. An OCT modality based on low backscattering albedo has been demonstrated with Au nanorods having strong absorption at NIR wavelengths; the detection limit of Au nanorods was estimated to be as low as 30 ppm in a highly backscattering tissue phantom.[153] Au nanocages (<100 nm) have been used to support a spectroscopic mode of OCT, in which contrast was generated by the significant attenuation of backscattering signal at nanocage concentrations above 500 pM.[154] Au nanorods have also been found to be useful for spectroscopic OCT, and were investigated as contrast agents in excised human breast tumors.[155]

10.5 SURFACE-ENHANCED RAMAN SCATTERING (SERS)

Among the various forms of plasmon-enhanced spectroscopies, SERS is easily the most famous. Born 30 years ago from the description of the electromagnetic field enhancement effect by Jeanmaire and van Duyne,[156] SERS has become the method of choice for demonstrating the existence of local (near-field) effects in plasmonic nanostructures, and a popular analytical tool for surface spectroscopy and chemical sensing.[157–161] SERS signals are based on the Raman vibrational modes of chemical species adsorbed onto the surfaces of plasmonic nanomaterials. These Raman intensities are intrinsically weak, but can be amplified many orders of magnitude by local electromagnetic fields generated from coupled plasmon modes. As its name implies, SERS requires a large scattering cross-section, and is thus intimately dependent on the plasmon resonance of the supporting metal nanostructure. Unlike elastic light scattering, however, SERS signals are Stokes-shifted and require a spectrometer for signal deconvolution. Chemical (i.e., surface charge-transfer) effects are also known to contribute toward the SERS signal,[162] but are not discussed in detail here.

The enduring interest in SERS as a plasmon-enhanced sensing modality is largely based on two factors. First, SERS emissions are spectroscopic in nature, and can be used to characterize or detect adsorbates based on their Raman vibrational patterns. SERS is thus often considered as a label-free method of detecting chemical analytes, and is appealing as an alternative to methods based on fluorescence, which requires labeling and produces a single wavelength of emission. Second, several reports over the last decade have demonstrated the possibility of acquiring SERS signals from single molecules or single nanoparticles.[163–169] This usually involves coupling electromagnetic amplification effects with resonance Raman signals, and is thus especially pertinent to dye-labeled nanoparticles, often referred to as SERS tags. The structural features that generate hot spots or hot particles for single-molecule SERS are still rather poorly defined; nevertheless, advancements in this area continue to be made, most recently in the form of PEG-coated SERS tags for *in vivo* imaging,[169] which the authors claim to be brighter than quantum-dot probes under comparable excitation conditions.

SERS is considered an inelastic scattering effect because the signals of interest are Stokes-shifted from the incident excitation. This means that the field-enhancement effect can involve more than one frequency. The overall SERS signal $S_R(\nu_S)$ can be expressed by the equation below:

$$S_R(\nu_S) = N'\sigma_{ads}^{R}|A(\nu_L)|^2|A(\nu_S)|^2 I(\nu_L) \quad (10.7)$$

where ν_L and ν_S are the incident and Stokes-shifted frequencies respectively and $A(\nu_L)$ and $A(\nu_S)$ represent the frequency-dependent field enhancement factors, N' is the number of SERS-active molecules in the excitation volume, $I(\nu_L)$ is the intensity of incident light at ν_L, and σ_{ads}^{R} is the increased Raman cross-section due to chemical enhancement effects.[158,159] Overall, the local field factors amplify the Raman scattering of probe molecules by roughly to the fourth power, resulting in enormous signal amplification. In comparison, Rayleigh scattering is only enhanced by field effects to the second power.

Conventional SERS substrates typically have disordered nanoscale features, such as that found in electrochemically roughened Au or Ag surfaces.[170] SERS substrates can also be rationally designed by lithography, by self-assembly, or a combination of the two. Both supramolecular and nanoscale self-assembly methods can be used to fabricate two-dimensional Au nanoparticle arrays with tunable optical properties.[33,171,172] For example, macrocyclic surfactants based on resorcinarenes (a subclass

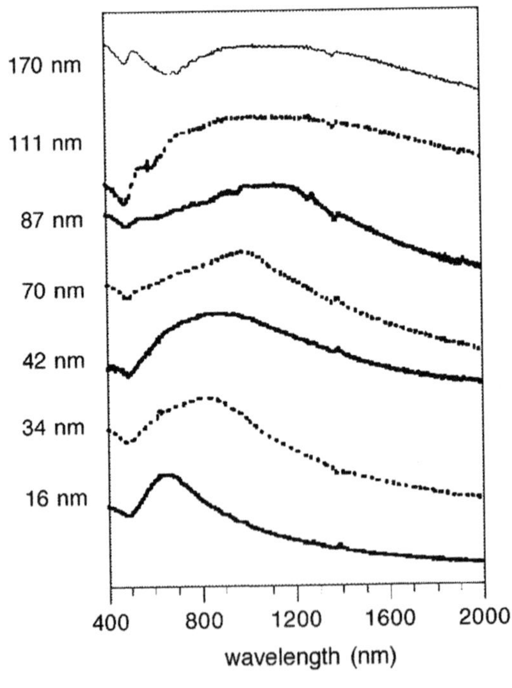

Figure 10.8 Optical extinction of two-dimensional gold nanoparticle arrays with different unit particle sizes.[33] (Reprinted with permission from B. Kim et al., *J. Am. Chem. Soc.* **2001**, *123*, 7955–7956. Copyright 2001 American Chemical Society.)

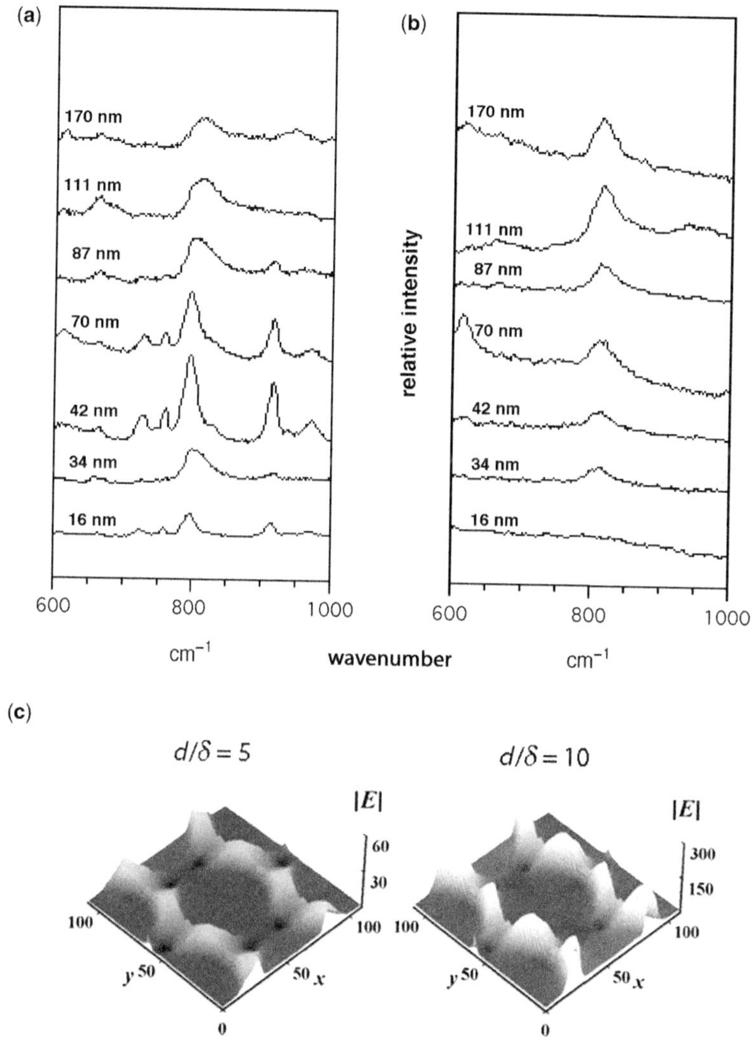

Figure 10.9 SERS signals from hexagonal close-packed Au nanoparticle arrays, excited at (a) 647 nm and (b) 1064 nm. Raman signals are produced by encapsulating surfactant (resorcinarene tetrathiol) between nanoparticles.[178] (c) Localized electromagnetic field factors within two-dimensional hexagonal arrays of metal nanoparticles at two different d/δ ratios.[180] [(a) and (c) adapted and reprinted with permission from A. Wei et al., *ChemPhysChem* **2001**, *2*, 743–745. Copyright Wiley-VCH Verlag GmbH & Co. KGaA. Part (c) reprinted with permission from D. A. Genov et al., *Nano Lett.* **2004**, *4*, 153–158. Copyright 2004 American Chemical Society.]

of calixarenes) have been found to have excellent properties for dispersing colloidal metal nanoparticles,[173] and can mediate the self-assembly of Au nanoparticles in the 20 to 200 nm range into two-dimensional arrays at biphasic interfaces (Fig. 10.4).[33,174–177] The self-assembled metal nanoparticle films exhibit collective optical properties that vary as a function of particle size, with plasmon-resonant extinctions in the visible and NIR region (Fig. 10.8). In addition, the Au nanoparticle arrays vary in their optical reflectivities, with 70 nm particle films producing the maximum specular reflectance.[178] The apparently tunable optical responses are due to variations in particle size and also to interparticle coupling effects.

The self-assembled Au nanoparticle arrays exhibit strong SERS properties, with surface-averaged signal enhancement factors ranging from 10^4 to 10^7 depending on the excitation wavelength. The enhancements are correlated with the periodicity and collective plasmonic response of the nanoparticle arrays: For example, 42 nm arrays produced the largest empirical enhancement factor upon excitation at 647 nm, whereas 111 nm arrays produced the strongest signal enhancements at 1064 nm (Fig. 10.9). The wavelength-dependent enhancement of SERS signal is strongest when the wavelengths of incident light excitation and Stokes-shifted Raman emissions overlap with the plasmon resonance band of the nanoparticle array. The two-dimensional arrays also support highly localized electromagnetic field factors in between neighboring particles. These interparticle gaps are just a few nanometers wide, and are considered to be responsible for producing most of the SERS signal intensities.[179] Theoretical and semiempirical studies on two-dimensional arrays of metal nanoparticles indicate that the surface-averaged SERS enhancement is highly sensitive to the particle diameter to spacing ratio (d/δ).[180] These findings suggest a straightforward approach for engineering nanostructured arrays with maximum amplification of SERS signal.

Localized plasmonic enhancements can also support novel optical detection modalities based on SERS-active nanoparticles, including *in vivo* applications.[169,181–184] The Raman reporters (usually dye molecules) are immobilized on the surface of plasmon-resonant nanoparticles with hot spots, which can be in the form of single nanoparticles or nanoaggregates.[181–183] Only those Raman-active dyes that are very close to these highly active regions will produce SERS signals, as the effect is extremely distance dependent. The enhancement factors are even more strongly attenuated if the molecules of interest are not directly adsorbed onto the surface. In this scenario, the SERS signal $S_R(\nu_S)$ can be expressed as:

$$|A(\nu_L)|^2|A(\nu_S)|^2 \approx \left|\frac{\varepsilon(\nu_L) - \varepsilon_0}{\varepsilon(\nu_L) + 2\varepsilon_0}\right|^4 \left(\frac{r}{r+d}\right)^{12} \tag{10.8}$$

where enhancement factors $A(\nu_L)$ and $A(\nu_S)$ describe $S_R(\nu_S)$ [see Equation (10.7)], $\varepsilon(\nu_L)$ is the permittivity of the metal at ν_L, ε_0 is that of vacuum, r is the effective particle diameter, and d is the distance between the Raman-active molecule and metal surface.[158] According to Equation (10.8), the SERS intensity produced by a probe molecule decreases to the 12th power with distance d. This implies that the SERS signal decreases rapidly for molecules positioned outside of hot spots, and is dominated by a few molecules that are optimally positioned on the surface within the plasmon-enhanced electromagnetic field. A recent study involving a two-dimensional array of submicron Ag hemispheres established that such hot spots can contribute to nearly a quarter of the overall SERS signal.[185]

NIR-absorbing metal nanostructures are appealing for biomedical imaging applications for reasons discussed previously, and this includes biological applications of SERS. For example, NIR-active core–shell superparticles have been prepared by the electrostatic assembly of densely packed Au nanoparticles on submicron silica spheres.[34] Such superparticle probes can be implanted into mammalian cells by cationic transfection,[186] and have produced SERS signals from absorbed DNA.[187] Biocompatible SERS nanoparticle tags can also be used as contrast agents for *in vivo* detection, as previously discussed.[169]

10.6 SURFACE-ENHANCED FLUORESCENCE

Plasmon-amplified electromagnetic fields on the surfaces of metal nanostructures can also support surface-enhanced fluorescence (SEF), although the conditions are much more stringent than for SERS. SEF is also highly sensitive to the distance between the metal nanoparticle and fluorescent molecule: If separated by less than 5 nm, fluorescence quenching is significant due to resonant energy or electron transfer. However, emissions can be amplified by as much as 100-fold if the molecules are separated from the metal surface at distances on the order of 10 nm.[188] A detailed investigation of electronic interactions between photoactive molecules and Au nanoparticles can be found in a review by Thomas and Kamat.[189] More information about SEF is available through some topical reviews.[188,190]

For fluorophores, the quantum yield Q is given by:

$$Q = \frac{k_r}{k_r + k_{nr}} \tag{10.9}$$

where k_r and k_{nr} are radiative and nonradiative decay rates, respectively. The fluorescence lifetime τ is expressed in the following equation:

$$\tau = \tau_r + \tau_{nr} = \frac{1}{k_r} + \frac{1}{k_{nr}} \tag{10.10}$$

An increase in fluorescence quantum yield is possible due to the suppression of the nonradiative decay of fluorophores upon binding to metallic nanoparticles. However, radiative decay is dramatically increased near the surface of metal nanoparticles, whereas changes in k_{nr} are less sensitive to distance.

Recent developments of SEF include high spatial selectivity of the fluorescence enhancement via lithographic nanopatterning[191] or near-field tip enhancement.[192] In addition to organic dyes, quantum dots (QDs) attached to Au nanostructures provide another robust SEF system where the interparticle separation can be controlled by the length of polymer linkers[193] or the thickness of silica shells.[194] Anisotropic Au nanoparticles and Au nanoshells can be employed to obtain plasmon-enhanced emissions in the NIR regime.[195] On the other hand, Au nanoparticle-mediated fluorescence quenching can also be applied toward biomolecular detection schemes, as demonstrated with avidin–biotin,[196] breast cancer biomarkers,[197] glucose,[198] and matrix metalloproteinases.[199]

10.7 PHOTOLUMINESCENCE

The discovery of photoluminescence from noble metals dates back nearly 40 years ago, when Au films were first observed to produce weak but detectable emissions upon photoexcitation.[200] Photoluminescence from metals has been described as a three-step process: (1) excitation of electrons from the *d*- to the *sp*-band to generate electron-hole pairs, (2) scattering of electrons and holes on the picosecond timescale with partial energy transfer to the phonon lattice, and (3) electron-hole recombination with subsequent photoemission. However, the quantum efficiency of linear (single-photon) photoluminescence from smooth Au films is low, on the order of 10^{-10}. More recently, metal nanoclusters were found to have discrete electronic states, and can exhibit a strong photoluminescence in the visible to infrared range.[201,202] The quantum efficiency has been measured to be in the range of 10^{-4} to 10^{-5}, five to six orders of magnitude larger than that produced from bulk gold.[203,204] Nowadays, metal nanoclusters are often referred to as artificial atoms, and show size-dependent fluorescence transitions analogous to that observed in molecular dyes.[205] Fluorescent Au nanoclusters have been prepared by various methods, such as electrochemical etching,[206] dendrimer-templated synthesis,[207] and wet-chemical reduction in the presence of alkanethiol ligands.[208] Such nanoclusters have also been shown to be sensitive to Hg^{2+} ions, and therefore useful as chemosensors.[208]

Linear photoluminescence is much less efficient in larger spherical particles (>5 nm), but is strong again for anisotropic Au nanostructures such as nanorods. In some cases, the photoluminescence efficiency appears to be strongly correlated with the plasmon resonance modes supported by the metal nanostructures. The quantum efficiency of single-photon luminescence from Au nanorods has been reported to be more than six orders of magnitude (10^{-3} to 10^{-4}) higher than that of bulk gold metal due to resonant coupling with localized plasmons.[209,210] The peak position of single-photon emission from Au nanorods appears to increase linearly with nanorod aspect ratio, while the quantum efficiency has a quadratic relationship.[209] However, the photoluminescence decreases and broadens in the case of longer rods (aspect ratio >3.5), possibly due to reduced coupling with the plasmon band or interband transitions.[211] Furthermore, longer Au nanorods produce a second emission peak centered at 740 nm, close to the band edge of Au, whose intensity increases with nanorod aspect ratio and lower excitation energies.[212] Regardless of their origins, the fluorescence properties of Au nanorods are clearly useful and have been demonstrated for the detection of DNA and proteins.[213]

10.8 NONLINEAR OPTICAL PROPERTIES

Gold nanoparticles have large second- and third-order nonlinear susceptibilities and are therefore a promising class of nonlinear optical materials.[214] We will briefly discuss several nonlinear optical processes from metal nanoparticles, such as multiphoton luminescence, hyper-Rayleigh scattering, and multiharmonic generation.

10.8.1 Multiphoton Luminescence

Two-photon luminescence (TPL) from Au nanoparticles has been proposed to originate from the radiative recombination of holes in the *d* band with electrons in the conduction band after two-photon absorption, a similar mechanism as single-photon luminescence.[200] However, the relatively weak TPL signal can be enormously amplified under plasmon-resonant conditions, to the extent that single particles are visible. In some cases, the TPL emission can be amplified by plasmon band coupling; for example, 100 nm Au particles can be excited at NIR wavelengths to produce TPL spectra whose emission peaks coincide with plasmon resonance.[215] TPL signals can also be enhanced with plasmon-resonant two-photon absorption: near-field imaging revealed strong TPL emissions in the hot spot produced between two Au nanoparticles, whose coupled plasmon modes overlap with the NIR excitation wavelength.[216] NIR-resonant nanoparticles such as nanorods also have large two-photon absorption cross-sections, and are particularly appealing as TPL-based contrast agents (see below).[217] It is worth mentioning that besides TPL, whose emissions are in the visible region, two-photon NIR emission has also been observed from both roughened Au surfaces[215] and Au nanoclusters.[218] This NIR emission is believed to originate from intraband transitions with the conduction band[215] and may also have promise for biological imaging applications, given sufficient sensitivity for NIR photodetection.

The intensity of TPL has a quadratic dependence with respect to excitation power, as expected for two-photon absorption. The TPL spectra produced by Au nanostructures are in the visible range (500 to 700 nm) and are generally quite broad, due to the multiple optical transitions between the *d* band and the conduction band.[215,219] However, the autofluorescence background under multiphoton imaging conditions is very low, and the TPL signals do not photobleach and can be continuously generated over the lifetime of the Au nanoparticles. This optical stability, in conjunction with gold's well-known biocompatibility, make Au nanoparticles very attractive as biological imaging agents based on TPL signaling. Spherical Au nanoparticles have been targeted to cells, and can be detected by TPL microscopy using pulsed NIR laser excitation.[220] However, anisotropic Au nanostructures are capable of producing much stronger TPL signals due to resonant NIR absorption, and their use is strongly favored in current biophotonic imaging applications. These include Au nanorods (see below), nanoplates,[221] nanoshells,[222] and closely spaced Au nanoparticle dimers.[223,224]

Au nanorods are especially appealing as TPL probes due to their high quality factors at NIR wavelengths and their sensitivity to incident polarization, which can provide additional orientational information.[217,225] The TPL intensities of Au nanorods have been shown to have a $\cos^4$ relation with respect to the polarization of incident light, and are at a maximum when aligned with their longitudinal axes (Fig. 10.10).[217,225] The two-photon absorption cross-section of individual Au nanorods is much larger than that of organic dye molecules ($>$2000 GM, compared to 1–300 GM for organic fluorophores).[226] The brightness of Au nanorods is even sufficient for *in vivo* TPL imaging, and Au nanorods have been detected at subpicomolar

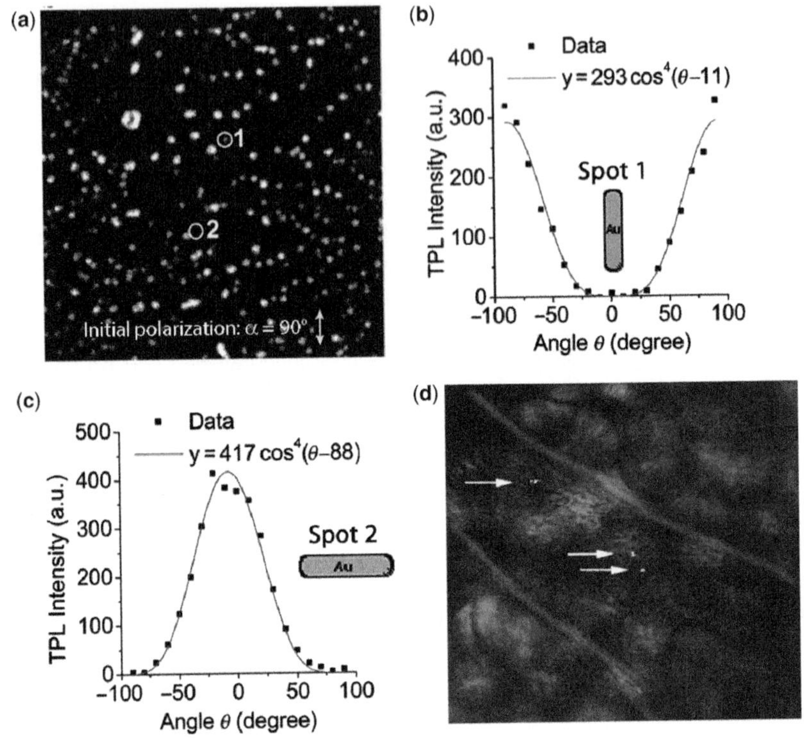

Figure 10.10 (a) TPL image of Au nanorods dispersed on a glass cover slip, with random orientations. (b, c) Polarization-dependent TPL signals from single Au nanorods [spots 1 and 2 in (a)], illustrating the $\cos^4$ relationship between the nanorod axis and incident polarization. (d) Au nanorods (marked with white arrows) detected *in vivo*, passing through a mouse ear blood vessel (outlined for clarity).[217] (Reprinted with permission from H. Wang et al., *Proc. Natl. Acad. Sci. U.S.A.* **2005**, *102*, 15752–15756. Copyright 2005 National Academy of Sciences, U.S.A.) (See color insert.)

concentrations while flowing through the blood vessels of a mouse earlobe.[217] It should be noted that multiphoton imaging is typically performed using confocal microscopy techniques, which provides axial as well as lateral focusing for three-dimensional spatial resolution. This has enabled the TPL imaging of Au nanorods in a three-dimensional tissue phantom with penetration depths on the order of 75 μm.[227]

While Au nanorods and other NIR-active nanoparticles have untapped potential for clinical *in vivo* imaging applications, the focus at present is on preclinical *in vitro* studies to better ascertain the biological effects of these nanomaterials. Surface chemistry becomes a critical issue, as it determines the biocompatibility, dispersion stability, and site-directed targeting of nanoparticles. For example, Au nanorods coated with cetyltrimethylammonium bromide (CTAB), a cationic surfactant used during nanorod synthesis, are internalized by KB cells via a nonspecific uptake pathway within hours of its addition to the culture medium, and transported toward the perinuclear region.[228] Nonspecific cell uptake can be dramatically reduced by exhaustive removal of CTAB and substitution by methyl-terminated PEG chains, which can be introduced by chemisorption using the standard alkanethiol approach[229,230] or via *in situ* dithiocarbamate (DTC) formation,[228,231] which conveniently converts alkylamines into strongly chemisorptive ligands for metal surfaces. Chemisorptive ligands are also useful for functionalizing nanoparticle surfaces with biorecognition elements such as homing oligopeptides or folic acid conjugates, and targeting them toward cell-surface biomarkers or cognate receptors.[93] The delivery of ligand-functionalized nanorods can be monitored by TPL microscopy, both for site-directed targeting and also during their simultaneous application as photothermal agents.[94] Such capabilities present many exciting opportunities for using NIR-active Au nanostructures as multifunctional "theragnostic" agents, in which potential therapies are guided by real-time imaging modalities.

Three-photon luminescence (3PL) and multiphoton luminescence (MPL) from Au nanostructures has also been reported. 3PL was first reported by Shen and coworkers,[232] then later studied systemically as a function of particle size from 2.5 to 125 nm using pulsed NIR laser excitation.[233] 3PL-active Au nanoparticles can also be prepared in a glassy matrix, to support photonic applications such as optical memory and data storage.[234] Plasmon-resonant nanostructures can support both two- and three-photon excitation processes, which can be preferentially activated simply by adjusting the excitation intensity.[235] MPL has recently been generated by Au nanowire arrays, and shown to be 10^4 times brighter than 3PL.[236] However, as the nonlinearity of the photoemission process increases, it becomes increasingly sensitive to polarization effects: in the case above, the ratio of intensity to excitation power could be as large as 18.3, corresponding to a $\cos^{50}$ relationship with incident polarization.

10.8.2 Nonlinear Scattering and Second Harmonic Generation

Hyper-Rayleigh scattering (HRS) and second harmonic generation (SHG) are both nonlinear optical responses characterized by a doubling of the incident frequency. Both require the absence of centrosymmetric inversion, but only SHG is produced by spatially ordered molecules, for example, a lipid monolayer or bilayer assembly on a glass surface. HRS is produced as an isotropic and incoherent scattering response, whereas SHG is coherent with predominantly forward emission. While SHG tends to be surface-selective, HRS can be generated by the transient loss of centrosymmetry in a bulk liquid phase (an otherwise isotropic medium) by random fluctuations in molecular or particle orientations. The generality of this condition has enabled HRS to become a common technique for studying the nonlinear properties of Au nanoparticles in liquid suspensions.[237,238] HRS signals are typically obtained by placing the detector 90° with respect to the incident beam at NIR wavelengths; for example, the fundamental wavelength produced by an Nd:YAG laser (1064 nm) can generate a strong, plasmon-resonant HRS response at 532 nm from Au nanospheres. The HRS signal is proportional to the number density of Au nanoparticles as well as the intensity of incident light.[237] For anisotropic Au nanorods, HRS signal can also be modulated by polarization effects.[239]

SHG signals have been observed from isolated Au nanoparticles at a biphasic interface,[240,241] as well as nanoparticle dimers[242] and two-dimensional arrays,[243] protein-mediated Au nanoparticle aggregates,[244] and nanoparticle multilayers.[245] SHG signals collected from dispersed Au nanoparticles under wavelength scanning conditions (800 to 1200 nm) will produce a maximum intensity at 520 nm, consistent with their plasmon absorption. The strong SHG signal produced from centrosymmetric Au nanoparticles may be counterintuitive at first, but it is the nanoparticle surface that provides the source of broken symmetry,[246] which can then be amplified by dipolar plasmon resonances in smaller nanoparticles (<50 nm)[247] as well as by multipolar contributions in larger particles (>50 nm).[248,249] SHG signals from single Au nanoparticles have been imaged by scanning near-field optical microscopy (SNOM), providing direct evidence of plasmon-resonant SHG (Fig. 10.11).[250,251] SHG from Au nanorods has also been investigated and found to be greatest when the excitation wavelength overlapped with the longitudinal plasmon resonance band for maximum absorption.[252]

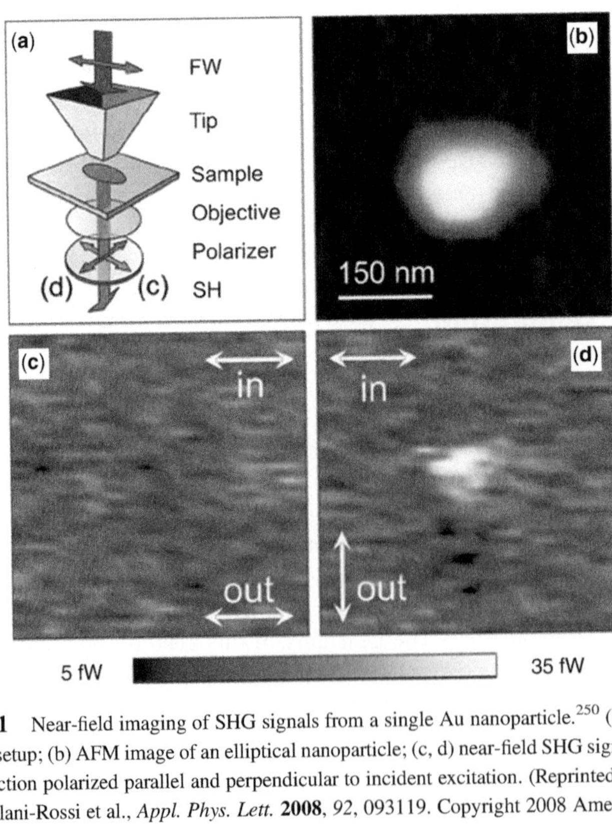

Figure 10.11 Near-field imaging of SHG signals from a single Au nanoparticle.[250] (a) Schematic of experimental setup; (b) AFM image of an elliptical nanoparticle; (c, d) near-field SHG signal mapping with emission detection polarized parallel and perpendicular to incident excitation. (Reprinted with permission from M. Zavelani-Rossi et al., *Appl. Phys. Lett.* **2008**, *92*, 093119. Copyright 2008 American Institute of Physics.) (See color insert.)

Third-harmonic generation (THG) has also emerged as an attractive imaging modality for biological and materials research. THG signals are usually only observed at interfaces and are considered to have a limited capacity for molecular imaging applications. However, in the case of metal nanoparticles, the THG signals are amplified by plasmon-resonant excitation, enabling their employment as exogenous contrast agents based on this modality. THG signals from Au nanoparticles have been demonstrated, but only for larger colloidal nanoparticles (>40 nm)[253] or for particles coated with Ag, which greatly increases their ability to enhance THG signals.[220] The THG activity of Ag nanoparticles is strongest when the fundamental wavelength is in the NIR range (1230 nm). Like NIR-resonant Au nanostructures, these particles have demonstrated their utility as plasmon-resonant nonlinear optics (NLO) contrast agents for biophotonic imaging.[254]

10.9 CONCLUDING REMARKS

The examples provided in this chapter are intended to alert the reader to the diversity of opportunities available for employing gold nanoparticles in chemistry and biology. These represent just a small fraction of the rapidly growing body of research that exploits the optical properties of gold nanoparticles and other plasmon-resonant nanostructures for chemical and biological applications. By emphasizing several recent developments in the field, we hope to encourage the creative expansion of supramolecular chemistry and materials science into this fertile research area.

REFERENCES

1. U. Kreibig, M. Vollmer, *Optical Properties of Metal Clusters*, New York: Springer, **1995**.
2. S. Link, M. A. El-Sayed, in *Semiconductor and Metal Nanocrystals: Synthesis and Electronic and Optical Properties* (Ed: V. I. Klimov), New York: Routledge, **2003**, 421–452.
3. K. L. Kelly, E. Coronado, L. L. Zhao, G. C. Schatz, *J. Phys. Chem. B* **2003**, *107*, 668–677.
4. I. O. Sosa, C. Noguez, R. G. Barrera, *J. Phys. Chem. B* **2003**, *107*, 6269–6275.
5. A. Wei, in *Nanoparticles: Scaffolds and Building Blocks* (Ed: V. M. Rotello), New York: Kluwer Academic, **2004**, 173–200.
6. N. Halas, *MRS Bull.* **2005**, *30*, 362–367.
7. Y. Xia, N. J. Halas, *MRS Bull.* **2005**, *30*, 338–343.
8. L. M. Liz-Marzán, *Langmuir* **2006**, *22*, 32–41.
9. A. Moores, F. Goettmann, *New J. Chem.* **2006**, *30*, 1121–1132.
10. P. Jain, X. Huang, I. El-Sayed, M. El-Sayed, *Plasmonics* **2007**, *2*, 107–118.
11. W. A. Murray, W. L. Barnes, *Adv. Mater.* **2007**, *19*, 3771–3782.
12. C. Noguez, *J. Phys. Chem. C* **2007**, *111*, 3806–3819.
13. C. L. Nehl, J. H. Hafner, *J. Mater. Chem.* **2008**, *18*, 2415–2419.
14. M. C. Daniel, D. Astruc, *Chem. Rev.* **2004**, *104*, 293–346.
15. M. Faraday, *Philos. Trans. R. Soc. London* **1857**, *147*, 145–181.
16. G. Mie, *Ann. Phys.* **1908**, *25*, 377–445.
17. R. Gans, *Ann. Phys.* **1912**, *37*, 881–900.
18. J. E. Millstone, S. Park, K. L. Shuford, L. Qin, G. C. Schatz, C. A. Mirkin, *J. Am. Chem. Soc.* **2005**, *127*, 5312–5313.
19. S. Asano, G. Yamamoto, *Appl. Opt.* **1980**, *14*, 29–49.

20. A. C. Lind, J. M. Greenberg, *J. Appl. Phys.* **1966**, *37*, 3195–3203.
21. B. T. Draine, P. J. Flatau, *J. Opt. Soc. Am. A* **1994**, *11*, 1491–1499.
22. C. Oubre, P. Nordlander, *J. Phys. Chem. B* **2004**, *108*, 17740–17747.
23. M. Hu, J. Chen, Z.-Y. Li, L. Au, G. V. Hartland, X. Li, M. Marquez, Y. Xia, *Chem. Soc. Rev.* **2006**, *35*, 1084–1094.
24. S. Link, M. A. El-Sayed, *J. Phys. Chem. B* **1999**, *103*, 8410–8426.
25. P. N. Njoki, I. I. S. Lim, D. Mott, H. Y. Park, B. Khan, S. Mishra, R. Sujakumar, J. Luo, C. J. Zhong, *J. Phys. Chem. C* **2007**, *111*, 14664–14669.
26. B. Nikoobakht, M. A. El-Sayed, *Chem. Mater.* **2003**, *15*, 1957–1962.
27. T. K. Sau, C. J. Murphy, *Langmuir* **2004**, *20*, 6414–6420.
28. D. A. Zweifel, A. Wei, *Chem. Mater.* **2005**, *17*, 4256–4261.
29. S. K. Ghosh, T. Pal, *Chem. Rev.* **2007**, *107*, 4797–4862.
30. R. Elghanian, J. J. Storhoff, R. C. Mucic, R. L. Letsinger, C. A. Mirkin, *Science* **1997**, *277*, 1078–1081.
31. T. A. Taton, C. A. Mirkin, R. L. Letsinger, *Science* **2000**, *289*, 1757–1760.
32. C. Sönnichsen, B. M. Reinhard, J. Liphardt, A. P. Alivisatos, *Nature Biotechnol.* **2005**, *23*, 741–745.
33. B. Kim, S. L. Tripp, A. Wei, *J. Am. Chem. Soc.* **2001**, *123*, 7955–7956.
34. B. Sadtler, A. Wei, *Chem. Commun.* **2002**, 1604–1605.
35. J. Zhao, X. Zhang, C. R. Yonzon, A. J. Haes, R. P. Van Duyne, *Nanomedicine* **2006**, *1*, 219–228.
36. K. A. Willets, R. P. Van Duyne, *Annu. Rev. Phys. Chem.* **2007**, *58*, 267–297.
37. M. E. Stewart, C. R. Anderton, L. B. Thompson, J. Maria, S. K. Gray, J. A. Rogers, R. G. Nuzzo, *Chem. Rev.* **2008**, *108*, 494–521.
38. C. J. Murphy, A. M. Gole, S. E. Hunyadi, J. W. Stone, P. N. Sisco, A. Alkilany, B. E. Kinard, P. Hankins, *Chem. Commun.* **2008**, 544–557.
39. N. L. Rosi, C. A. Mirkin, *Chem. Rev.* **2005**, *105*, 1547–1562.
40. Y. Niidome, H. Takahashi, S. Urakawa, K. Nishioka, S. Yamada, *Chem. Lett.* **2004**, *33*, 454–455.
41. C.-D. Chen, S.-F. Cheng, L.-K. Chau, C. R. C. Wang, *Biosens. Bioelectron.* **2007**, *22*, 926–932.
42. S. M. Marinakos, S. Chen, A. Chilkoti, *Anal. Chem.* **2007**, *79*, 5278–5283.
43. C. Wang, Y. Chen, T. Wang, Z. Ma, Z. Su, *Chem. Mater.* **2007**, *19*, 5809–5811.
44. C. Yu, J. Irudayaraj, *Anal. Chem.* **2007**, *79*, 572–579.
45. K. M. Mayer, S. Lee, H. Liao, B. C. Rostro, A. Fuentes, P. T. Scully, C. L. Nehl, J. H. Hafner, *ACS Nano* **2008**, *2*, 687–692.
46. C.-C. You, M. De, V. M. Rotello, *Curr. Opin. Chem. Biol.* **2005**, *9*, 639–646.
47. W. B. Tan, Y. Zhang, *J. Biomed. Mater. Res. Part A* **2005**, *75A*, 56–62.
48. L. M. Liz-Marzán, P. Mulvaney, *J. Phys. Chem. B* **2003**, *107*, 7312–7326.
49. J. Shan, H. Tenhu, *Chem. Commun.* **2007**, 4580–4598.
50. L. R. Hirsch, J. B. Jackson, A. Lee, N. J. Halas, J. L. West, *Anal. Chem.* **2003**, *75*, 2377–2381.
51. A. D. McFarland, R. P. Van Duyne, *Nano Lett.* **2003**, *3*, 1057–1062.
52. C. R. Yonzon, E. Jeoung, S. Zou, G. C. Schatz, M. Mrksich, R. P. Van Duyne, *J. Am. Chem. Soc.* **2004**, *126*, 12669–12676.
53. A. J. Haes, W. P. Hall, L. Chang, W. L. Klein, R. P. Van Duyne, *Nano Lett.* **2004**, *4*, 1029–1034.
54. A. J. Haes, L. Chang, W. L. Klein, R. P. Van Duyne, *J. Am. Chem. Soc.* **2005**, *127*, 2264–2271.
55. A. J. Haes, S. Zou, J. Zhao, G. C. Schatz, R. P. Van Duyne, *J. Am. Chem. Soc.* **2006**, *128*, 10905–10914.
56. W. P. Hall, J. N. Anker, Y. Lin, J. Modica, M. Mrksich, R. P. Van Duyne, *J. Am. Chem. Soc.* **2008**, *130*, 5836–5837.
57. J. Sung, E. M. Hicks, R. P. Van Duyne, K. G. Spears, *J. Phys. Chem. C* **2007**, *111*, 10368–10376.
58. J. Sung, M. Sukharev, E. M. Hicks, R. P. Van Duyne, T. Seideman, K. G. Spears, *J. Phys. Chem. C* **2008**, *112*, 3252–3260.
59. K. S. Lee, M. A. El-Sayed, *J. Phys. Chem. B* **2006**, *110*, 19220–19225.
60. Y. Sun, Y. Xia, *Anal. Chem.* **2002**, *74*, 5297–5305.
61. C. F. Bohren, D. R. Huffman, *Absorption and Scattering of Light by Small Particles*, New York: Wiley, **1983**.

62. A. Gulati, H. Liao, J. H. Hafner, *J. Phys. Chem. B* **2006**, *110*, 22323–22327.
63. X. Huang, P. Jain, I. El-Sayed, M. El-Sayed, *Lasers Med. Sci.* **2007**, *23*, 217–228.
64. X. Huang, P. K. Jain, I. H. El-Sayed, M. A. El-Sayed, *Nanomedicine* **2007**, *2*, 681–693.
65. P. K. Jain, I. H. El-Sayed, M. A. El-Sayed, *Nano Today* **2007**, *2*, 18–29.
66. L. Cognet, S. Berciaud, D. Lasne, B. Lounis, *Anal. Chem.* **2008**, *80*, 2288–2294.
67. A. G. Skirtach, C. Dejugnat, D. Braun, A. S. Susha, A. L. Rogach, W. J. Parak, H. Möhwald, G. B. Sukhorukov, *Nano Lett.* **2005**, *5*, 1371–1377.
68. A. G. Skirtach, A. M. Javier, O. Kreft, K. Köhler, A. P. Alberola, H. Möhwald, W. J. Parak, G. B. Sukhorukov, *Angew. Chem. Int. Ed.* **2006**, *45*, 4612–4617.
69. O. Kreft, A. G. Skirtach, G. B. Sukhorukov, H. Möhwald, *Adv. Mater.* **2007**, *19*, 3142–3145.
70. J. W. M. Chon, C. Bullen, P. Zijlstra, M. Gu, *Adv. Funct. Mater.* **2007**, *17*, 875–880.
71. M. Hu, G. V. Hartland, *J. Phys. Chem. B* **2002**, *106*, 7029–7033.
72. S. Link, M. A. El-Sayed, *Int. Rev. Phys. Chem.* **2000**, *19*, 409–453.
73. C. Burda, X. Chen, R. Narayanan, M. A. El-Sayed, *Chem. Rev.* **2005**, *105*, 1025–1102.
74. D. Boyer, P. Tamarat, A. Maali, B. Lounis, M. Orrit, *Science* **2002**, *297*, 1160–1163.
75. A. O. Govorov, H. H. Richardson, *Nano Today* **2007**, *2*, 30–38.
76. H. H. Richardson, Z. N. Hickman, A. O. Govorov, A. C. Thomas, W. Zhang, M. E. Kordesch, *Nano Lett.* **2006**, *6*, 783–788.
77. S. Berciaud, L. Cognet, G. A. Blab, B. Lounis, *Phys. Rev. Lett.* **2004**, *93*, 257402.
78. S. Berciaud, L. Cognet, P. Tamarat, B. Lounis, *Nano Lett.* **2005**, *5*, 515–518.
79. S. Berciaud, D. Lasne, G. A. Blab, L. Cognet, B. Lounis, *Phys. Rev. B* **2006**, *73*, 045424.
80. L. Cognet, C. Tardin, D. Boyer, D. Choquet, P. Tamarat, B. Lounis, *Proc. Natl. Acad. Sci. U.S.A.* **2003**, *100*, 11350–11355.
81. C. M. Pitsillides, E. K. Joe, X. Wei, R. R. Anderson, C. P. Lin, *Biophys. J.* **2003**, *84*, 4023–4032.
82. I. H. El-Sayed, X. Huang, M. A. El-Sayed, *Cancer Lett.* **2006**, *239*, 129–135.
83. R. Weissleder, *Nature Biotechnol.* **2001**, *19*, 316–317.
84. L. R. Hirsch, R. J. Stafford, J. A. Bankson, S. R. Sershen, B. Rivera, R. E. Price, J. D. Hazle, N. J. Halas, J. L. West, *Proc. Natl. Acad. Sci. U.S.A.* **2003**, *100*, 13549–13554.
85. C. Loo, A. Lowery, N. Halas, J. West, R. Drezek, *Nano Lett.* **2005**, *5*, 709–711.
86. J. Chen, B. Wiley, Z. Y. Li, D. Campbell, F. Saeki, H. Cang, L. Au, J. Lee, X. Li, Y. Xia, *Adv. Mater.* **2005**, *17*, 2255–2261.
87. H. M. Chen, H. C. Peng, R. S. Liu, K. Asakura, C. L. Lee, J. F. Lee, S. F. Hu, *J. Phys. Chem. B* **2005**, *109*, 19553–19555.
88. S. E. Skrabalak, J. Chen, L. Au, X. Lu, X. Li, Y. Xia, *Adv. Mater.* **2007**, *19*, 3177–3184.
89. S. E. Skrabalak, L. Au, X. Lu, X. Li, Y. Xia, *Nanomedicine* **2007**, *2*, 657–668.
90. J. Chen, D. Wang, J. Xi, L. Au, A. Siekkinen, A. Warsen, Z. Y. Li, H. Zhang, Y. Xia, X. Li, *Nano Lett.* **2007**, *7*, 1318–1322.
91. X. Huang, I. H. El-Sayed, W. Qian, M. A. El-Sayed, *J. Am. Chem. Soc.* **2006**, *128*, 2115–2120.
92. H. Takahashi, T. Niidome, A. Nariai, Y. Niidome, S. Yamada, *Chem. Lett.* **2006**, *35*, 500–501.
93. T. B. Huff, L. Tong, Y. Zhao, M. N. Hansen, J.-X. Cheng, A. Wei, *Nanomedicine* **2007**, *2*, 125–132.
94. L. Tong, Y. Zhao, T. B. Huff, M. N. Hansen, A. Wei, J.-X. Cheng, *Adv. Mater.* **2007**, *19*, 3136–3141.
95. D. Pissuwan, S. M. Valenzuela, C. M. Miller, M. B. Cortie, *Nano Lett.* **2007**, *7*, 3808–3812.
96. D. Pissuwan, S. Valenzuela, M. Killingsworth, X. Xu, M. Cortie, *J. Nanopart. Res.* **2007**, *9*, 1109–1124.
97. R. S. Norman, J. W. Stone, A. Gole, C. J. Murphy, T. L. Sabo-Attwood, *Nano Lett.* **2008**, *8*, 302–306.
98. P. Zijlstra, J. W. M. Chon, M. Gu, *Opt. Express* **2007**, *15*, 12151–12160.
99. T. Sun, G. J. Diebold, *Nature* **1992**, *355*, 806–808.
100. X. Wang, Y. Pang, G. Ku, X. Xie, G. Stoica, L. V. Wang, *Nature Biotechnol.* **2003**, *21*, 803–806.
101. G. Ku, L. V. Wang, *Opt. Lett.* **2005**, *30*, 507–509.
102. H. F. Zhang, K. Maslov, G. Stoica, L. V. Wang, *Nature Biotechnol.* **2006**, *24*, 848–851.
103. Y. Wang, X. Xie, X. Wang, G. Ku, K. L. Gill, D. P. O'Neal, G. Stoica, L. V. Wang, *Nano Lett.* **2004**, *4*, 1689–1692.

104. X. Yang, S. E. Skrabalak, Z. Y. Li, Y. Xia, L. V. Wang, *Nano Lett.* **2007**, *7*, 3798–3802.
105. M. Eghtedari, A. Oraevsky, J. A. Copland, N. A. Kotov, A. Conjusteau, M. Motamedi, *Nano Lett.* **2007**, *7*, 1914–1918.
106. K. Kim, S.-W. Huang, S. Ashkenazi, M. O'Donnell, A. Agarwal, N. A. Kotov, M. F. Denny, M. J. Kaplan, *Appl. Phys. Lett.* **2007**, *90*, 223901.
107. C.-K. Liao, S.-W. Huang, C.-W. Wei, P.-C. Li, *J. Biomed. Opt.* **2007**, *12*, 064006.
108. D. L. Chamberland, A. Agarwal, N. Kotov, J. B. Fowlkes, P. L. Carson, X. Wang, *Nanotechnology* **2008**, *19*, 095101.
109. J. Yguerabide, E. E. Yguerabide, *Anal. Biochem.* **1998**, *262*, 137–156.
110. J. Yguerabide, E. E. Yguerabide, *Anal. Biochem.* **1998**, *262*, 157–176.
111. M. A. van Dijk, A. L. Tchebotareva, M. Orrit, M. Lippitz, S. Berciaud, D. Lasne, L. Cognet, B. Lounis, *Phys. Chem. Chem. Phys.* **2006**, *8*, 3486–3495.
112. S. Schultz, D. R. Smith, J. J. Mock, D. A. Schultz, *Proc. Natl. Acad. Sci. U.S.A.* **2000**, *97*, 996–1001.
113. C. Sönnichsen, S. Geier, N. E. Hecker, G. von Plessen, J. Feldmann, H. Ditlbacher, B. Lamprecht, J. R. Krenn, F. R. Aussenegg, V. Z. H. Chan, J. P. Spatz, M. Moller, *Appl. Phys. Lett.* **2000**, *77*, 2949–2951.
114. C. Sönnichsen, T. Franzl, T. Wilk, G. von Plessen, J. Feldmann, *New J. Phys.* **2002**, *4*, 93.
115. J. Muller, C. Sonnichsen, H. von Poschinger, G. von Plessen, T. A. Klar, J. Feldmann, *Appl. Phys. Lett.* **2002**, *81*, 171–173.
116. C. Sönnichsen, T. Franzl, T. Wilk, G. von Plessen, J. Feldmann, O. Wilson, P. Mulvaney, *Phys. Rev. Lett.* **2002**, *88*, 077402.
117. J. J. Mock, M. Barbic, D. R. Smith, D. A. Schultz, S. Schultz, *J. Chem. Phys.* **2002**, *116*, 6755–6759.
118. H. Kuwata, H. Tamaru, K. Esumi, K. Miyano, *Appl. Phys. Lett.* **2003**, *83*, 4625–4627.
119. C. L. Nehl, H. Liao, J. H. Hafner, *Nano Lett.* **2006**, *6*, 683–688.
120. M. Z. Liu, P. Guyot-Sionnest, *J. Phys. Chem. B* **2004**, *108*, 5882–5888.
121. X. Wang, Z. Zhang, G. V. Hartland, *J. Phys. Chem. B* **2005**, *109*, 20324–20330.
122. C. L. Nehl, N. K. Grady, G. P. Goodrich, F. Tam, N. J. Halas, J. H. Hafner, *Nano Lett.* **2004**, *4*, 2355–2359.
123. C. Novo, D. Gomez, J. Perez-Juste, Z. Zhang, H. Petrova, M. Reismann, P. Mulvaney, G. V. Hartland, *Phys. Chem. Chem. Phys.* **2006**, *8*, 3540–3546.
124. M. Hu, H. Petrova, A. R. Sekkinen, J. Chen, J. M. McLellan, Z. Y. Li, M. Marquez, X. Li, Y. Xia, G. V. Hartland, *J. Phys. Chem. B* **2006**, *110*, 19923–19928.
125. M. Hu, C. Novo, A. Funston, H. Wang, H. Staleva, S. Zou, P. Mulvaney, Y. Xia, G. V. Hartland, *J. Mater. Chem.* **2008**, *18*, 1949–1960.
126. R. Jin, Y. Cao, C. A. Mirkin, K. L. Kelly, G. C. Schatz, J. G. Zheng, *Science* **2001**, *294*, 1901–1903.
127. C. J. Orendorff, T. K. Sau, C. J. Murphy, *Small* **2006**, *2*, 636–639.
128. P. K. Jain, K. S. Lee, I. H. El-Sayed, M. A. El-Sayed, *J. Phys. Chem. B* **2006**, *110*, 7238–7248.
129. R. A. Zsigmondy, *Colloids and the Ultramicrocope: A Manual of Colloid Chemistry and Ultramicroscopy*, New York: Wiley, **1914**.
130. J. Yguerabide, E. E. Yguerabide, *J. Cell. Biochem. Suppl.* **2001**, *84*, 71–81.
131. D. A. Schultz, *Curr. Opin. Biotechnol.* **2003**, *14*, 13–22.
132. J. A. Steyer, W. Almers, *Nature Rev. Mol. Cell Biol.* **2001**, *2*, 268–275.
133. D. Toomre, D. J. Manstein, *Trends Cell Biol.* **2001**, *11*, 298–303.
134. T. A. Taton, G. Lu, C. A. Mirkin, *J. Am. Chem. Soc.* **2001**, *123*, 5164–5165.
135. J. J. Storhoff, A. D. Lucas, V. Garimella, Y. P. Bao, U. R. Muller, *Nature Biotechnol.* **2004**, *22*, 883–887.
136. G. L. Liu, Y. Yin, S. Kunchakarra, B. Mukherjee, D. Gerion, S. D. Jett, D. G. Bear, J. W. Gray, A. P. Alivisatos, L. P. Lee, F. F. Chen, *Nature Nanotechnol.* **2006**, *1*, 47–52.
137. B. Dragnea, C. Chen, E. S. Kwak, B. Stein, C. C. Kao, *J. Am. Chem. Soc.* **2003**, *125*, 6374–6375.
138. I. H. El-Sayed, X. Huang, M. A. El-Sayed, *Nano Lett.* **2005**, *5*, 829–834.
139. G. Raschke, S. Kowarik, T. Franzl, C. Sonnichsen, T. A. Klar, J. Feldmann, A. Nichtl, K. Kurzinger, *Nano Lett.* **2003**, *3*, 935–938.

140. S. Kumar, N. Harrison, R. Richards-Kortum, K. Sokolov, *Nano Lett.* **2007**, *7*, 1338–1343.
141. J. Aaron, E. de la Rosa, K. Travis, N. Harrison, J. Burt, M. José-Yacamán, K. Sokolov, *Opt. Express* **2008**, *16*, 2153–2167.
142. J. G. Fujimoto, M. E. Brezinski, G. J. Tearney, S. A. Boppart, B. Bouma, M. R. Hee, J. F. Southern, E. A. Swanson, *Nature Med.* **1995**, *1*, 970–972.
143. D. Huang, E. A. Swanson, C. P. Lin, J. S. Schuman, W. G. Stinson, W. Chang, M. R. Hee, T. Flotte, K. Gregory, C. A. Puliafito, *Science* **1991**, *254*, 1178–1181.
144. J. G. Fujimoto, *Nature Biotechnol.* **2003**, *21*, 1361–1367.
145. A. F. Low, G. J. Tearney, B. E. Bouma, I.-K. Jang, *Nature Clin. Pract. Cardiovasc. Med.* **2006**, *3*, 154–162.
146. S. A. Boppart, A. L. Oldenburg, C. Xu, D. L. Marks, *J. Biomed. Opt.* **2005**, *10*, 041208.
147. C. Loo, A. Lin, L. Hirsch, M. H. Lee, J. Barton, N. Halas, J. West, R. Drezek, *Technol. Cancer Res. Treat.* **2004**, *3*, 33–40.
148. J. K. Barton, N. J. Halas, J. L. West, R. A. Drezek, *Proc. SPIE Int. Soc. Opt. Eng.* **2004**, *5316*, 99–106.
149. A. Agrawal, S. Huang, A. W. H. Lin, M.-H. Lee, J. K. Barton, R. A. Drezek, T. J. Pfefer, *J. Biomed. Opt.* **2006**, *11*, 041121.
150. A. M. Gobin, M. H. Lee, N. J. Halas, W. D. James, R. A. Drezek, J. L. West, *Nano Lett.* **2007**, *7*, 1929–1934.
151. T. S. Troutman, J. K. Barton, M. Romanowski, *Opt. Lett.* **2007**, *32*, 1438–1440.
152. J. Chen, F. Saeki, B. J. Wiley, H. Cang, M. J. Cobb, Z. Y. Li, L. Au, H. Zhang, M. B. Kimmey, X. D. Li, Y. Xia, *Nano Lett.* **2005**, *5*, 473–477.
153. A. L. Oldenburg, M. N. Hansen, D. A. Zweifel, A. Wei, S. A. Boppart, *Opt. Express* **2006**, *14*, 6724–6738.
154. H. Cang, T. Sun, Z.-Y. Li, J. Chen, B. J. Wiley, Y. Xia, X. Li, *Opt. Lett.* **2005**, *30*, 3048–3050.
155. A. L. Oldenburg, M. N. Hansen, A. Wei, S. A. Boppart, *Proc. SPIE Int. Soc. Opt. Eng.* **2008**, *6867*, 68670E.
156. D. L. Jeanmaire, R. P. Van Duyne, *J. Electroanal. Chem.* **1977**, *84*, 1–20.
157. K. Kneipp, H. Kneipp, I. Itzkan, R. R. Dasari, M. S. Feld, *Chem. Rev.* **1999**, *99*, 2957–2976.
158. G. A. Baker, D. S. Moore, *Anal. Bioanal. Chem.* **2005**, *382*, 1751–1770.
159. K. Kneipp, H. Kneipp, J. Kneipp, *Acc. Chem. Res.* **2006**, *39*, 443–450.
160. Z.-Q. Tian, B. Ren, J.-F. Li, Z.-L. Yang, *Chem. Commun.* **2007**, 3514–3534.
161. W. E. Doering, M. E. Piotti, M. J. Natan, R. G. Freeman, *Adv. Mater.* **2007**, *19*, 3100–3108.
162. A. Campion, P. Kambhampati, *Chem. Soc. Rev.* **1998**, *27*, 241–250.
163. S. Nie, S. R. Emory, *Science* **1997**, *275*, 1102–1106.
164. K. Kneipp, Y. Wang, H. Kneipp, L. T. Perelman, I. Itzkan, R. R. Dasari, M. S. Feld, *Phys. Rev. Lett.* **1997**, *78*, 1667–1670.
165. H. Xu, E. J. Bjerneld, M. Käll, L. Börjesson, *Phys. Rev. Lett.* **1999**, *83*, 4357–4360.
166. W. E. Doering, S. Nie, *J. Phys. Chem. B* **2002**, *106*, 311–317.
167. W. Gu, H. Choi, K. Kim, *J. Phys. Chem. A* **2007**, *111*, 8121–8125.
168. X. M. Qian, S. M. Nie, *Chem. Soc. Rev.* **2008**, *37*, 912–920.
169. X. Qian, X.-H. Peng, D. O. Ansari, Q. Yin-Goen, G. Z. Chen, D. M. Shin, L. Yang, A. N. Young, M. D. Wang, S. Nie, *Nature Biotechnol.* **2008**, *26*, 83–90.
170. P. Gao, D. Gosztola, L.-W. H. Leung, M. J. Weaver, *J. Electroanal. Chem.* **1987**, *233*, 211–222.
171. B. Kim, M. A. Carignano, S. L. Tripp, A. Wei, *Langmuir* **2004**, *20*, 9360–9365.
172. B. Kim, R. Balasubramanian, W. Pérez-Segarra, A. Wei, B. Decker, J. Mattay, *Supramol. Chem.* **2005**, *17*, 173–180.
173. A. Wei, *Chem. Commun.* **2006**, 1581–1591.
174. K. B. Stavens, S. V. Pusztay, S. Zou, R. P. Andres, A. Wei, *Langmuir* **1999**, *15*, 8337–8339.
175. R. Balasubramanian, B. Kim, S. L. Tripp, X. Wang, M. Lieberman, A. Wei, *Langmuir* **2002**, *18*, 3676–3681.
176. S. V. Pusztay, A. Wei, K. B. Stavens, R. P. Andres, *Supramol. Chem.* **2002**, *14*, 291–294.
177. R. Balasubramanian, Y.-G. Kwon, A. Wei, *J. Mater. Chem.* **2007**, *17*, 105–112.

178. A. Wei, B. Kim, B. Sadtler, S. L. Tripp, *ChemPhysChem* **2001**, *2*, 743–745.
179. K. Imura, H. Okamoto, M. K. Hossain, M. Kitajima, *Nano Lett.* **2006**, *6*, 2173–2176.
180. D. A. Genov, A. K. Sarychev, V. M. Shalaev, A. Wei, *Nano Lett.* **2004**, *4*, 153–158.
181. J. Ni, R. J. Lipert, G. B. Dawson, M. D. Porter, *Anal. Chem.* **1999**, *71*, 4903–4908.
182. J. D. Driskell, K. M. Kwarta, R. J. Lipert, M. D. Porter, J. D. Neill, J. F. Ridpath, *Anal. Chem.* **2005**, *77*, 6147–6154.
183. J. Kneipp, H. Kneipp, M. McLaughlin, D. Brown, K. Kneipp, *Nano Lett.* **2006**, *6*, 2225–2231.
184. M. Y. Sha, H. Xu, S. G. Penn, R. Cromer, *Nanomedicine* **2007**, *2*, 725–734.
185. Y. Fang, N.-H. Seong, D. D. Dlott, *Science* **2008**, *321*, 388–392.
186. Y. Zhao, B. Sadtler, M. Lin, G. H. Hockerman, A. Wei, *Chem. Commun.* **2004**, 784–785.
187. A. Wei, *e-J. Surf. Sci. Nanotechnol.* **2006**, *4*, 9–18.
188. J. Lakowicz, C. Geddes, I. Gryczynski, J. Malicka, Z. Gryczynski, K. Aslan, J. Lukomska, E. Matveeva, J. Zhang, R. Badugu, J. Huang, *J. Fluoresc.* **2004**, *14*, 425–441.
189. K. G. Thomas, P. V. Kamat, *Acc. Chem. Res.* **2003**, *36*, 888–898.
190. E. Fort, S. Grésillon, *J. Phys. D Appl. Phys.* **2008**, *41*, 013001.
191. P. P. Pompa, L. Martiradonna, A. Della Torre, F. Della Sala, L. Manna, M. De Vittorio, F. Calabi, R. Cingolani, R. Rinaldi, *Nature Nanotechnol.* **2006**, *1*, 126–130.
192. P. Bharadwaj, P. Anger, L. Novotny, *Nanotechnology* **2007**, *18*, 044017.
193. J. Lee, A. O. Govorov, N. A. Kotov, *Angew. Chem. Int. Ed.* **2005**, *44*, 7439–7442.
194. N. Liu, B. S. Prall, V. I. Klimov, *J. Am. Chem. Soc.* **2006**, *128*, 15362–15363.
195. F. Tam, G. P. Goodrich, B. R. Johnson, N. J. Halas, *Nano Lett.* **2007**, *7*, 496–501.
196. E. Oh, M. Y. Hong, D. Lee, S. H. Nam, H. C. Yoon, H. S. Kim, *J. Am. Chem. Soc.* **2005**, *127*, 3270–3271.
197. C. C. Huang, C. K. Chiang, Z. H. Lin, K. H. Lee, H. T. Chang, *Anal. Chem.* **2008**, *80*, 1497–1504.
198. B. Tang, L. Cao, K. Xu, L. Zhuo, J. Ge, Q. Li, L. Yu, *Chem. Eur. J.* **2008**, *14*, 3637–3644.
199. S. Lee, E.-J. Cha, K. Park, S.-Y. Lee, J.-K. Hong, I.-C. Sun, S. Y. Kim, K. Choi, I. C. Kwon, K. Kim, C.-H. Ahn, *Angew. Chem. Int. Ed.* **2008**, *47*, 2804–2807.
200. A. Mooradian, *Phys. Rev. Lett.* **1969**, *22*, 185–187.
201. T. P. Bigioni, R. L. Whetten, O. Dag, *J. Phys. Chem. B* **2000**, *104*, 6983–6986.
202. S. Link, A. Beeby, S. FitzGerald, M. A. El-Sayed, T. G. Schaaff, R. L. Whetten, *J. Phys. Chem. B* **2002**, *106*, 3410–3415.
203. J. P. Wilcoxon, J. E. Martin, F. Parsapour, B. Wiedenman, D. F. Kelley, *J. Chem. Phys.* **1998**, *108*, 9137–9143.
204. H. Y. Lin, Y. F. Chen, *Appl. Phys. Lett.* **2006**, *88*, 101914.
205. T. H. Lee, J. I. Gonzalez, J. Zheng, R. M. Dickson, *Acc. Chem. Res.* **2005**, *38*, 534–541.
206. H. Duan, S. Nie, *J. Am. Chem. Soc.* **2007**, *129*, 2412–2413.
207. Y. Bao, C. Zhong, D. M. Vu, J. P. Temirov, R. B. Dyer, J. S. Martinez, *J. Phys. Chem. C* **2007**, *111*, 12194–12198.
208. C.-C. Huang, Z. Yang, K.-H. Lee, H.-T. Chang, *Angew. Chem. Int. Ed.* **2007**, *46*, 6824–6828.
209. M. B. Mohamed, V. Volkov, S. Link, M. A. El-Sayed, *Chem. Phys. Lett.* **2000**, *317*, 517–523.
210. O. P. Varnavski, M. B. Mohamed, M. A. El-Sayed, T. Goodson, *J. Phys. Chem. B* **2003**, *107*, 3101–3104.
211. S. Eustis, M. El-Sayed, *J. Phys. Chem. B* **2005**, *109*, 16350–16356.
212. C.-Z. Li, K. B. Male, S. Hrapovic, J. H. T. Luong, *Chem. Commun.* **2005**, 3924–3926.
213. B. Pan, D. Cui, P. Xu, Q. Li, T. Huang, R. He, F. Gao, *Colloid Surf. A* **2007**, *295*, 217–222.
214. Y. R. Shen, *The Principles of Nonlinear Optics*, New York: Wiley, **1984**.
215. M. R. Beversluis, A. Bouhelier, L. Novotny, *Phys. Rev. B* **2003**, *68*, 115433.
216. A. Bouhelier, M. R. Beversluis, L. Novotny, *Appl. Phys. Lett.* **2003**, *83*, 5041–5043.
217. H. Wang, T. B. Huff, D. A. Zweifel, W. He, P. S. Low, A. Wei, J.-X. Cheng, *Proc. Natl. Acad. Sci. U.S.A.* **2005**, *102*, 15752–15756.
218. G. Ramakrishna, O. Varnavski, J. Kim, D. Lee, T. Goodson, *J. Am. Chem. Soc.* **2008**, *130*, 5032–5033.
219. G. T. Boyd, Z. H. Yu, Y. R. Shen, *Phys. Rev. B* **1986**, *33*, 7923–7936.
220. D. Yelin, D. Oron, S. Thiberge, E. Moses, Y. Silberberg, *Opt. Express* **2003**, *11*, 1385–1391.
221. K. Imura, T. Nagahara, H. Okamoto, *Appl. Phys. Lett.* **2006**, *88*, 023104.

222. J. Park, A. Estrada, K. Sharp, K. Sang, J. A. Schwartz, D. K. Smith, C. Coleman, J. D. Payne, B. A. Korgel, A. K. Dunn, J. W. Tunnell, *Opt. Express* **2008**, *16*, 1590–1599.
223. P. J. Schuck, D. P. Fromm, A. Sundaramurthy, G. S. Kino, W. E. Moerner, *Phys. Rev. Lett.* **2005**, *94*, 017402.
224. K. Ueno, S. Juodkazis, V. Mizeikis, K. Sasaki, H. Misawa, *Adv. Mater.* **2008**, *20*, 26–30.
225. A. Bouhelier, R. Bachelot, G. Lerondel, S. Kostcheev, P. Royer, G. P. Wiederrecht, *Phys. Rev. Lett.* **2005**, *95*, 267405.
226. W. R. Zipfel, R. M. Williams, W. W. Webb, *Nature Biotechnol.* **2003**, *21*, 1369–1377.
227. N. J. Durr, T. Larson, D. K. Smith, B. A. Korgel, K. Sokolov, A. Ben-Yakar, *Nano Lett.* **2007**, *7*, 941–945.
228. T. B. Huff, M. N. Hansen, Y. Zhao, J. X. Cheng, A. Wei, *Langmuir* **2007**, *23*, 1596–1599.
229. H. Liao, J. H. Hafner, *Chem. Mater.* **2005**, *17*, 4636–4641.
230. T. Niidome, M. Yamagata, Y. Okamoto, Y. Akiyama, H. Takahashi, T. Kawano, Y. Katayama, Y. Niidome, *J. Control. Release* **2006**, *114*, 343–347.
231. Y. Zhao, W. Perez-Segarra, Q. Shi, A. Wei, *J. Am. Chem. Soc.* **2005**, *127*, 7328–7329.
232. Y.-C. Shen, Z. Tang, M. Gui, J. Cheng, X. Wang, Z. Lu, *Chem. Lett.* **2000**, *29*, 1140–1141.
233. R. A. Farrer, F. L. Butterfield, V. W. Chen, J. T. Fourkas, *Nano Lett.* **2005**, *5*, 1139–1142.
234. M. Eichelbaum, B. E. Schmidt, H. Ibrahim, K. Rademann, *Nanotechnology* **2007**, *18*, 355702.
235. S. Xiao, H. Gong, X. Su, J. Han, Y. Han, M. Chen, Q. Wang, *J. Phys. Chem. C* **2007**, *111*, 10185–10189.
236. Q. Q. Wang, J. B. Han, D. L. Guo, S. Xiao, Y. B. Han, H. M. Gong, X. W. Zou, *Nano Lett.* **2007**, *7*, 723–728.
237. P. Galletto, P. F. Brevet, H. H. Girault, R. Antoineb, M. Broyer, *Chem. Commun.* **1999**, 581–582.
238. I. Russier-Antoine, C. Jonin, J. Nappa, E. Benichou, P.-F. Brevet, *J. Chem. Phys.* **2004**, *120*, 10748–10752.
239. J. Nappa, G. Revillod, J.-P. Abid, I. Russier-Antoine, C. Jonin, E. Benichou, H. H. Girault, P. F. Brevet, *Faraday Discuss.* **2004**, *125*, 145–156.
240. R. Antoine, P. F. Brevet, H. H. Girault, D. Bethell, D. J. Schiffrin, *Chem. Commun.* **1997**, 1901–1902.
241. R. Antoine, M. Pellarin, B. Palpant, M. Broyer, B. Prevel, P. Galletto, P. F. Brevet, H. H. Girault, *J. Appl. Phys.* **1998**, *84*, 4532–4536.
242. M. L. Sandrock, C. D. Pibel, F. M. Geiger, C. A. Foss, *J. Phys. Chem. B* **1999**, *103*, 2668–2673.
243. T. Hayakawa, Y. Usui, S. Bharathi, M. Nogami, *Adv. Mater.* **2004**, *16*, 1408–1412.
244. I. Russier-Antoine, J. Huang, E. Benichou, G. Bachelier, C. Jonin, P.-F. Brevet, *Chem. Phys. Lett.* **2008**, *450*, 345–349.
245. M. Ishifuji, M. Mitsuishi, T. Miyashita, *Chem. Commun.* **2008**, 1058–1060.
246. J. I. Dadap, J. Shan, T. F. Heinz, *J. Opt. Soc. Am. B* **2004**, *21*, 1328–1347.
247. J. Nappa, G. Revillod, I. Russier-Antoine, E. Benichou, C. Jonin, P. F. Brevet, *Phys. Rev. B* **2005**, *71*, 165407.
248. J. Nappa, I. Russier-Antoine, E. Benichou, C. Jonin, P. F. Brevet, *J. Chem. Phys.* **2006**, *125*, 184712–184716.
249. I. Russier-Antoine, E. Benichou, G. Bachelier, C. Jonin, P.-F. Brevet, *J. Phys. Chem. C* **2007**, *111*, 9044–9048.
250. M. Zavelani-Rossi, M. Celebrano, P. Biagioni, D. Polli, M. Finazzi, L. Duò, G. Cerullo, M. Labardi, M. Allegrini, J. Grand, P. M. Adam, *Appl. Phys. Lett.* **2008**, *92*, 093119.
251. M. Celebrano, M. Zavelani-Rossi, D. Polli, G. Cerullo, P. Biagioni, M. Finazzi, L. Duò, M. Labardi, M. Allegrini, J. Grand, P. M. Adam, *J. Microsc.* **2008**, *229*, 233–239.
252. C. Hubert, L. Billot, P. M. Adam, R. Bachelot, P. Royer, J. Grand, D. Gindre, K. D. Dorkenoo, A. Fort, *Appl. Phys. Lett.* **2007**, *90*, 181105.
253. M. Lippitz, M. A. van Dijk, M. Orrit, *Nano Lett.* **2005**, *5*, 799–802.
254. S.-P. Tai, Y. Wu, D. B. Shieh, L. J. Chen, K. J. Lin, C. H. Yu, S. W. Chu, C. H. Chang, X. Y. Shi, Y. C. Wen, K. H. Lin, T. M. Liu, C. K. Sun, *Adv. Mater.* **2007**, *19*, 4520–4523.

Chapter 11

Optical Signaling with Silica Nanoparticles

FABRIZIO MANCIN, PAOLO TECILLA, AND
UMBERTO TONELLATO

11.1 INTRODUCTION

Many years before nanotechnology established itself as a scientific discipline, nanoparticles were already being produced and used.[1] Among them, silicon dioxide (silica) colloids (Fig. 11.1) played a prominent role. The preparation of silica nanoparticles was first reported in the nineteenth century and commercial production of aqueous silica sol started in 1933.[2] Since then, colloidal silica particles have been used in a wide range of applications, including investment casting, silicon wafer polishing, composite materials or beverage clarification, to cite a few.

In contrast to many other nanomaterials, silica nanoparticles do not acquire any peculiar property from their submicrometric size, except for the corresponding increase of the surface area. As a matter of fact, they can simply be regarded as extremely small and highly porous glass spheres. What makes silica nanoparticles very interesting from the supramolecular point of view is the presence of a well-defined structure with compartments (bulk, surface, pores, shells, etc.) that can be rather

The Supramolecular Chemistry of Organic–Inorganic Hybrid Materials. Edited by Knut Rurack and Ramón Martínez-Máñez

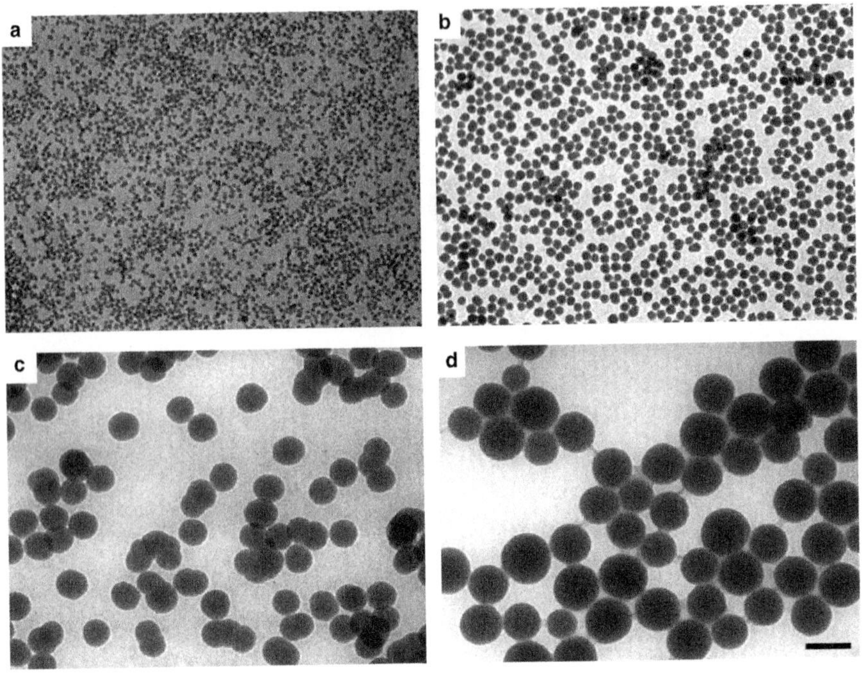

Figure 11.1 TEM images of silica nanoparticles of different sizes: (a) 20 nm, (b) 50 nm, (c) 200 nm, (d) 290 nm, prepared with the Stöber method [scale bar at the corner of (d) corresponds to 300 nm].

easily engineered with the desired properties and functionalized or doped with organic species.[3,4] In this way, it is possible to design and develop new recognition motifs, to modify the properties of the appended molecules, to spatially organize different species with different roles. In addition, silica nanoparticles are cheap and easy to prepare, transparent to light, relatively chemically inert, biocompatible, and water dispersible.

Silica nanoparticles are commonly prepared by polymerization of appropriate precursors such as silicates, silicon alkoxides, or chlorides (Fig. 11.2).[2] Besides the industrial methods, which rely mainly on condensation of sodium silicate in water induced by sodium removal through ion exchange, three different synthetic methods are currently used in research labs to prepare silica nanoparticles loaded with organic molecules. In the first method, proposed by Kolbe in 1956[5] and developed by Stöber and coworkers in the late 1960s,[6] the particles are formed via hydrolysis and

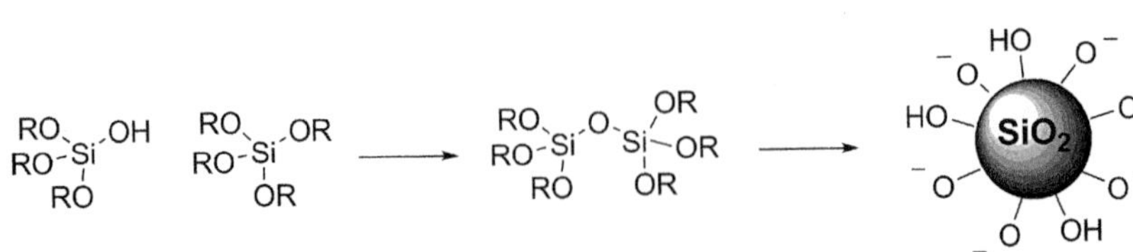

Figure 11.2 General synthesis of silica nanoparticles (R = H, alkyl group).

condensation of a silicon alkoxide (usually tetraethoxysilane, TEOS) in ethanol solutions containing water and ammonia. Final dimensions are mainly dependent on the amount of water and ammonia added and diameters ranging from 20 up to hundreds of nanometers can be obtained. Nanoparticles prepared with this method are less dense, and hence more porous, than amorphous silica.[7,8] They are relatively monodisperse and their shape changes from irregular, in the case of smaller particles, to almost perfectly spherical, in the case of the largest ones (Fig. 11.1). Extensive investigations by van Blaaderen proved that addition of organoalkoxysilane derivatives to the reaction mixture allows incorporation of organic species in the silica network.[9,10] Moreover, surface functionalization can be achieved by reacting preformed silica nanoparticles with organoalkoxysilane derivatives.[11] Finally, multishell structures can be easily produced with a one-pot procedure by subsequent additions of silane precursors to the reaction mixture.[9]

The second synthetic method was developed in the early 1990s by Arriagada and Osseo-Asare and involves the ammonia-catalyzed polymerization of TEOS in reverse microemulsion.[12,13] The water pools of the microemulsion act as "nanoreactors" where the particle growth occurs and the final size is controlled by the water/organic solvent ratio. With this methodology, highly monodisperse and perfectly spherical particles are obtained with sizes ranging from 20 to 100 nm. Hydrophilic molecules, or even nanoparticles, added to the reaction mixture are entrapped in the silica matrix allowing the preparation of doped silica particles.[14] Surface functionalization and bioconjugation are achieved either by reaction with organoalkoxysilane derivatives or by direct modification of the surface silanol groups.[15]

Finally, polymerization of lipophilic organosilane derivatives, such as vinyltriethoxysilane (VTES), in micellar aqueous solutions has been recently proposed by Prasad and coworkers.[16,17] Again, highly monodisperse particles with 10 to 80 nm diameters are obtained and lipophilic molecules added to the reaction mixture are entrapped in the resulting particles.

As soon as synthetic protocols to obtain silica nanoparticles doped with organic molecules became available, embedding of fluorescent dyes into such nanosystems was explored and several applications in the field of sensing were envisaged. In a first instance, the sensitivity of several fluorescence-based assays such as ELISA (enzyme linked immunosorbent assay), FIA (fluoroimmunoassay), and microarray-based assays was improved by the use of dye-doped particles. Moreover, nanoparticles conjugated with biologically active molecules, such as targeting agents or antibodies, were employed both in fluorescence assays and imaging.[15,18]

Further progress in the development of sensing systems based on silica nanoparticles was achieved by doping with fluorescent probes. In these systems, the nanoparticles act as transporters delivering the probe across membranes into the cell, thus improving both performance and protection from matrix interferences.[19–22] As we will see later, the possibility to confine several probes and dye molecules within the same particle also allows the design of more complex sensing schemes.

Finally, silica nanoparticles have been used recently as scaffolds to self-organize the sensing units or even their separate components and allow complex functions originating from the spatial organization of different molecular entities in the particle compartments.[3,4,23,24]

11.2 BIOSENSING AND BIOMARKING WITH SILICA NANOPARTICLES

Inclusion of fluorescent dyes within silica nanoparticles can significantly affect their properties. In fact, the microenvironment they experience is different from that in solution. The solvent is at least partially excluded by the particle interior and this may reduce the importance of collisional quenching and increase the fluorescence quantum yield.[25] Oxygen exclusion from the hydrophilic particle cores results in increased dye photostability.[15] Finally, a single nanoparticle is doped with a great number of dye molecules and is of course much easier to detect than a single molecule.[26] The result of these different factors is that the brightness of dye-doped silica nanoparticles can be compared in some cases with that of semiconductor nanocrystals or quantum dots (Fig. 11.3).[27] Moreover, their preparation, surface modification, and bioconjugation are much easier than in the case of quantum dots and even polymer nanoparticles.[15]

The sensitivity of fluorescence-based assays can hence be greatly improved by the use of dye-doped silica nanoparticles and this approach has been pioneered and subsequently deeply investigated by Tan and coworkers.[15] Their luminophore of choice was the water-soluble, positively charged tris(2,2′-bipyridyl)dichlororuthenium(II) $[Ru(bpy)_3]^{2+}$ hydrochloride that can be easily incorporated into silica nanoparticles prepared using the reverse microemulsion method. The charge complementarity between the dye and the silica matrix prevents leaching from the particles.[15]

The silanol groups on the particle surface can then be functionalized following several different methods. Synthesized nanoparticles can be reacted with commercially available alkoxysilane derivatives bearing thiol or amino groups.[28,29] Alternatively, they can be added to the microemulsion reaction mixture to obtain surface

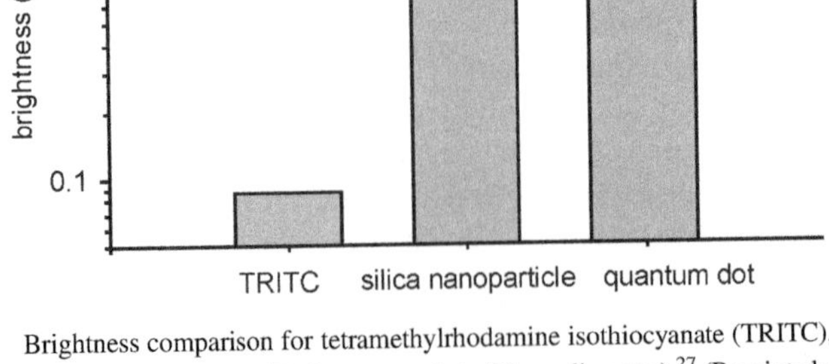

Figure 11.3 Brightness comparison for tetramethylrhodamine isothiocyanate (TRITC), TRITC-doped silica nanoparticles (20 nm), and CdSe/ZnS quantum dots (28 nm diameter).[27] (Reprinted with permission from H. Ow et al., *Nano Lett.*, **2005**, *5*, 113–117. Copyright 2005 American Chemical Society.)

functionalized nanoparticles with a one-pot procedure.[30] The amino functions can then be converted into carboxylic or aldehyde groups by reaction with succinic anhydride or glutaraldehyde.[28,31] Finally, conjugation with biomolecules can be achieved by standard coupling reactions with the amino groups present on such species.[29] Recently, the direct preparation of carboxy functionalized nanoparticles using carboxylethylsilanetriol has been reported.[30]

A different approach to surface functionalization involves direct activation of the silanol groups by reaction with cyanogen bromide. The resulting isocyanate groups can be immediately exploited for covalent immobilization of biomolecules.[32]

Antibody surface-modified nanoparticles can be employed for biomarking or bioimaging. Several applications for the detection of leukemia and other cancer cells have been reported.[32–34] In very recent examples, small magnetic nanoparticles and/or quantum dots have been incorporated into larger silica nanoparticles for multimodal imaging (dual band emission, fluorescence, and magnetic resonance).[35–38] In a similar approach, again proposed by Tan and coworkers, magnetic nanoparticle-doped and dye-doped silica nanoparticles, both conjugated with antibodies, have been used for subsequent separation, through application of a magnetic field, and detection, by fluorescence staining, of cancer cells (Fig. 11.4).[39,40]

Bioconjugated dye-doped nanoparticles can be also very useful as luminescent probes in immunoassays, with the sandwich fluoroimmunoassay (FIA) method being one of the most popular ones. According to this protocol (Fig. 11.5), a solution containing the antigen is introduced into the analysis well, whose surface is coated with the corresponding antibody. Antibody-conjugated nanoparticles are then added and form sandwich complexes with the surface-bound antigen. After final washing, the residual fluorescence emission is measured in the washing solution and converted into analyte concentration. The high photostability and brightness of dye-doped

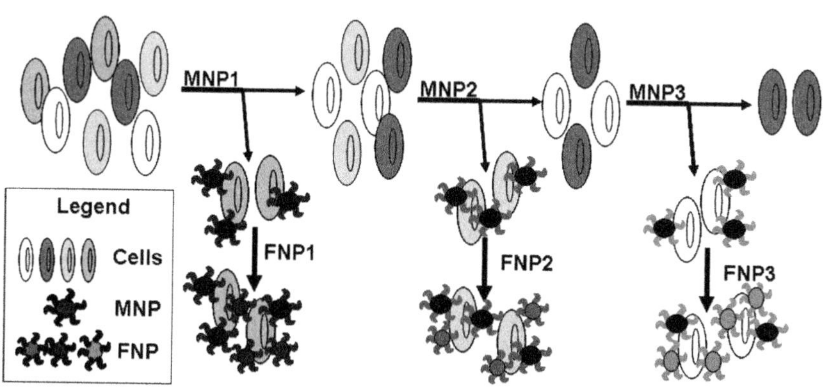

Figure 11.4 Schematic representation of the multiple extraction procedure with the antibody-conjugated silica nanoparticles doped with magnetic nanoparticles (MNP) being added and extracted stepwise and the corresponding antibody-conjugated silica nanoparticles doped with fluorescent dyes (FNP) being added post-magnetic extraction of cell samples.[40] (Reprinted with permission from J. E. Smith et al., *Anal. Chem.*, **2007**, *79*, 3075–3082. Copyright 2007 American Chemical Society.) (See color insert.)

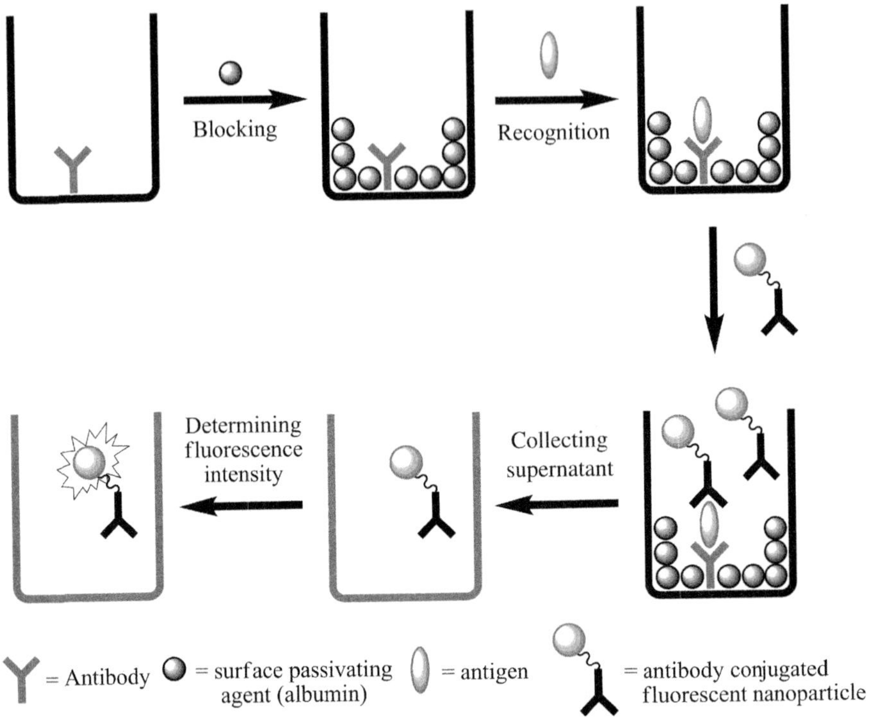

Figure 11.5 Scheme of sandwich fluoroimmunoassay for the determination of human IgG (immunoglobulin G) with goat anti-human IgG antibody-conjugated silica nanoparticles ($Ab\text{-}NH_2\text{-}AlC_4Pc\text{-}SiO_2$ NPs).[41]

nanoparticles, compared with those of conventional dyes, greatly improves the assay sensitivity up to the detection of analytes at trace levels.[29,31,41–44]

The use of silica nanoparticles doped with lanthanide chelates of antenna ligands can bring about further advantages. In fact, the long lifetimes of these inorganic fluorophores allow for better separation between signal and noise when detected in a time-resolved mode. In addition, the large apparent Stokes shift and sharp emission spectra further eliminate background interferences. Fluorescence emission of lanthanide chelates is, however, generally weaker than the emission of fluorescent organic dyes, preventing their direct use in highly sensitive bioassays. The use of doped nanoparticles, each containing a large number of individual chelates, provides a first-rate solution to this problem.[45–48]

When single-strand DNA oligonucleotides are grafted to the surface of dye-doped nanoparticles, DNA detection through hybridization assays can be achieved. Several sensing schemes have been explored, ranging from electrogenerated chemiluminescence after hybridization on gold electrodes[49] to Förster resonance energy transfer (FRET)-enhanced detection of hybridization on nanoparticles[50] and sandwich assays on microarrays (Fig. 11.6).[26,51,52]

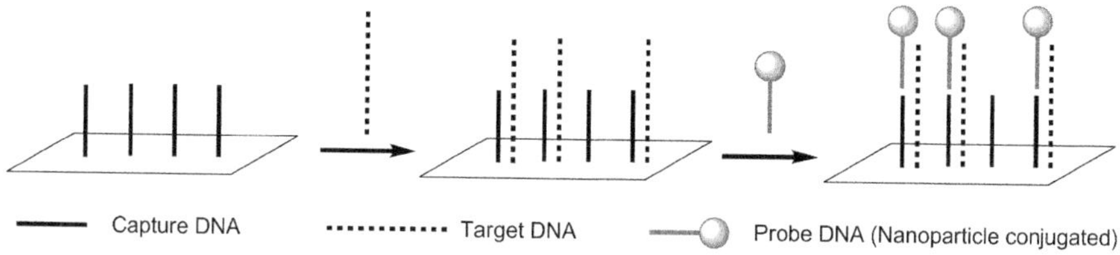

Figure 11.6 Scheme of sandwich assay for the detection of gene products using DNA-conjugated dye-doped silica nanoparticles. The combined sequences of DNA1 and DNA3 are complementary to that of the target DNA2. Capture DNA1 is immobilized on a glass substrate, target DNA2 and DNA3 conjugated dye-doped nanoparticles are added for hybridization. Detection of the DNA2 is done by monitoring fluorescence emission on the glass surface after washing.[51] (Reprinted with permission from X. Zhao et al., *J. Am. Chem. Soc.*, **2003**, *125*, 11474–11475. Copyright 2003 American Chemical Society.)

11.3 SILICA PEBBLEs

In medical and biochemical research, as the size of the sample is reduced down to the dimensions of living cells or their subcompartments, the real-time measurements of chemical and physical parameters with high spatial resolution and negligible perturbation of the sample becomes extremely important and challenging. Conventional probes are based either on chemosensors,[53,54] molecules that can selectively recognize and signal the presence of a specific analyte, or chemically modified pulled optical fibers.[55] Chemosensors, with their molecular dimensions, avoid physical perturbation of the cell but are exposed to chemical interferences, such as protein or membrane binding, which may result in alteration of the response, sequestration of the probe, and/or toxicity. On the other hand, in chemically modified pulled optical fibers, the dye is protected from undesired chemical interferences but, due to their size, the optical fibers are strongly invasive. In the late 1990s, Kopelman and coworkers[56] proposed a new approach to intracellular sensing, which combines the physical noninvasiveness of free chemosensors with the chemical noninvasiveness of pulled fiber optical sensors. These new sensors are dubbed PEBBLEs, an acronym of *probes encapsulated by biologically located embedding* or, more recently, *photonic explorers for bioanalysis with biologically localized embedding*.[20,21]

PEBBLEs are water-soluble nanoparticles based on biologically inert matrices of cross-linked polymers, typically poly(acrylamide), poly(decylmethacrylate), silica, or organically modified silicates (ORMOSILs), which encapsulate a fluorescent chemosensor and, often, a reference dye. These matrices have been used to make sensors for pH, metal ions, as well as for some nonionic species. The small size of the PEBBLE sensors (from 20 to 600 nm) enables their noninvasive insertion into a living cell, minimizing physical interference. The semipermeable and transparent nature of the matrix allows the analyte to interact with the indicator dye that reports the interaction via a change in the emitted fluorescence. Moreover, when compared to "naked" chemosensors, nanoparticles can protect the indicator from chemical interferences and minimize its toxicity. Another important feature of PEBBLEs, particularly valuable in intracellular sensing applications, is that the polymer matrix creates a separate sensing phase in which multiple dyes and ionophores can be combined, resulting in complex sensing schemes. These include reference dyes for ratiometric sensing or ionophore/chromoionophore combinations that allow for the use of nonfluorescent highly selective ionophores, although such ion-correlation PEBBLEs use preferentially poly(decylmethacrylate) and not silica matrices.

Silica-based PEBBLEs are typically prepared following a modification of the Stöber protocol by condensation of TEOS directly in the presence of the sensing components, which are physically trapped in the nanoparticle matrix. Sol–gel PEBBLEs are then coated with poly(ethylene glycol) (PEG) in order to increase the biocompatibility.[57] A variation of these nanoparticles are the ORMOSIL PEBBLEs which are prepared by growing an ORMOSIL layer of methyltrimethoxysilane on a preformed nanoparticle obtained by condensation of phenyltrimethoxysilane. The sensing components are added before the formation of the second layer and are trapped in this external shell. Both types of PEBBLEs are particularly well suited for the realization of ratiometric sensors typically obtained by entrapping an analyte sensitive dye and a reference dye that is not sensitive toward the analyte. For example, Kopelman and coworkers[58] have reported the realization of a ratiometric sensor for intracellular oxygen obtained by including a ruthenium complex $[Ru(dpp)_3]^{2+}$ (dpp = 4,4′-diphenylphenanthroline) and the dye Oregon Green 488 in silica PEBBLEs with diameters ranging from 100 to 400 nm (Fig. 11.7). In the presence of oxygen, the fluorescence

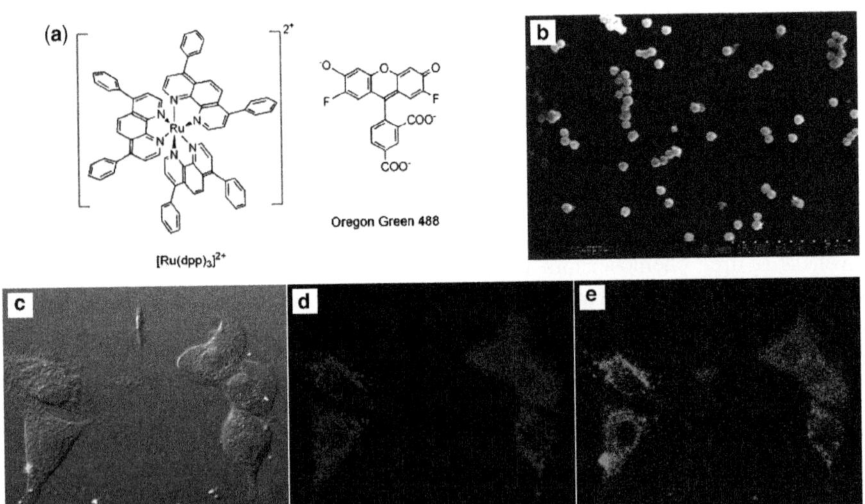

Figure 11.7 (a) Chemical structures of the dyes used for the preparation of the PEBBLE oxygen sensor reported by Kopelman and coworkers.[58] (b) Scanning electron micrographs showing the size distribution of sol–gel-PEG particles produced by the modified Stöber method. Scale bar 5 μm. (c) to (e) Confocal images of rat C6 glioma cells loaded with sol–gel PEBBLEs by gene gun injection showing the colocalization of the two dyes. (c) Nomarski illumination. (d) Oregon Green fluorescence of PEBBLEs inside cell. (e) $[Ru(dpp)_3]^{2+}$ fluorescence of PEBBLEs inside cells.[58] (Reprinted with permission from H. Xu et al., *Anal. Chem.*, **2001**, *73*, 4124–4133. Copyright 2001 American Chemical Society.) (See color insert.)

emission of $[Ru(dpp)_3]^{2+}$ is strongly quenched while the emission of Oregon Green is not affected, thus allowing ratiometric quantification of the level of oxygen. These PEBBLEs have been inserted into living cells using gene gun delivery techniques and have been used for the monitoring of variations of the oxygen level in the cytosol. Similar results have been reported by the same research group[59] using 120 nm diameter ORMOSIL PEBBLEs trapping inside a platinum porphyrin (platinum(II) octaethylporphine ketone) as oxygen-sensitive dye and octaethylporphine as reference dye. Compared to the sol–gel PEBBLE, this ORMOSIL PEBBLE sensor has higher sensitivity, better linearity, and longer excitation and emission wavelengths, resulting in reduced background noise for cellular measurement. Moreover, these PEBBLE sensors show excellent reversibility and stability against leaching as well as in long-term storage.

The versatility of PEBBLEs was demonstrated targeting several different analytes. pH is probably one of the most obvious targets, perhaps also because of the large number of pH-sensitive dyes that are available. Wang and coworkers[60] described a pH sensor made by encapsulating the pH-sensitive rhodamine-β-isothiocyanate dye in 60 nm silica nanoparticles. The PEBBLEs were prepared under reverse microemulsion conditions by polymerization of TEOS in the presence of the dye. The nanoparticles suspended in water by ultrasonication detect fluorimetrically the pH with a dynamic range in the interval 5.0 to 10.0, good reproducibility and stability, and a relatively short response time (4 min). In a later report, the same group reported dual-fluorophore-doped silica nanoparticles suited for ratiometric pH sensing.[61] In this case, the pH-sensitive dye fluorescein and the pH-insensitive dye phenosafranine were trapped into 30 nm silica nanoparticles obtained again by TEOS polymerization under reverse microemulsion condition. The silica shell prevents the leaching of the dyes when the nanoparticles are dispersed in water and the fluorescence intensity ratio of the two dyes varies linearly as a function of pH in the range from 4.0 to 8.0, allowing ratiometric pH detection. The results in realistic water samples taken from different sources are in good agreement with those obtained with the conventional glass electrode, demonstrating the real applicability of these PEBBLE pH sensors.

PEBBLEs are designed for biological applications and, in particular, for intracellular imaging. A nice example of noninvasive monitoring of intracellular pH changes induced by drug stimulation has been reported by Wang and coworkers.[62] The pH sensor is a dual-fluorophore-doped silica nanoparticle which contains a pH-sensitive indicator (fluorescein isothiocyanate, FITC) and a reference dye ($[Ru(bpy)_3]^{2+}$ as the hexahydrate). The 42 nm diameter nanoparticles were obtained first by reacting the fluorescein isothiocyanate dye with 3-aminopropyl triethoxysilane (APTES) and then by polymerizing the resulting FITC-APTES conjugate with TEOS in the presence of $[Ru(bpy)_3]^{2+}$ under reverse microemulsion conditions so that the FITC dye is covalently cross-linked with the silica matrix while the $[Ru(bpy)_3]^{2+}$ is simply trapped inside by electrostatic interactions, as discussed in the previous section. The sensor is able to ratiometrically report pH with a dynamic range of pH 4 to 7, well centered for biological studies, and shows fast response times (<1 s), reversibility, limited leakage of the luminophores, and enhanced stability to photobleaching, especially in the case of FITC. The sensor nanoparticles are readily taken up by cells through endocytosis, and intracellular pH imaging was demonstrated with two examples. In the first one, the lysosomal pH change in murine macrophages stimulated by chloroquine, a drug known for this effect, was studied. In the second one, the intracellular pH change occurring during apoptosis of HeLa cells was investigated (Fig. 11.8). The apoptosis was induced by treatment of the HeLa cells containing the sensor nanoparticles with dexamethasone, a glucocorticoid drug commonly used for such studies. Upon treatment with the drug, HeLa cells underwent apoptosis associated with intracellular acidification with a drop in pH from 7.2 to 6.5 which was monitored through fluorescence imaging and flow cytometry. Due to relevance of these phenomena for several pathologies, the potentialities of such nanosized sensor appear really interesting.

Other examples of the detection of biologically relevant targets under physiological conditions are represented by calcein-doped silica nanoparticles for the determination of calcium in blood serum[63] and PEBBLEs that report redox processes.[64–66]

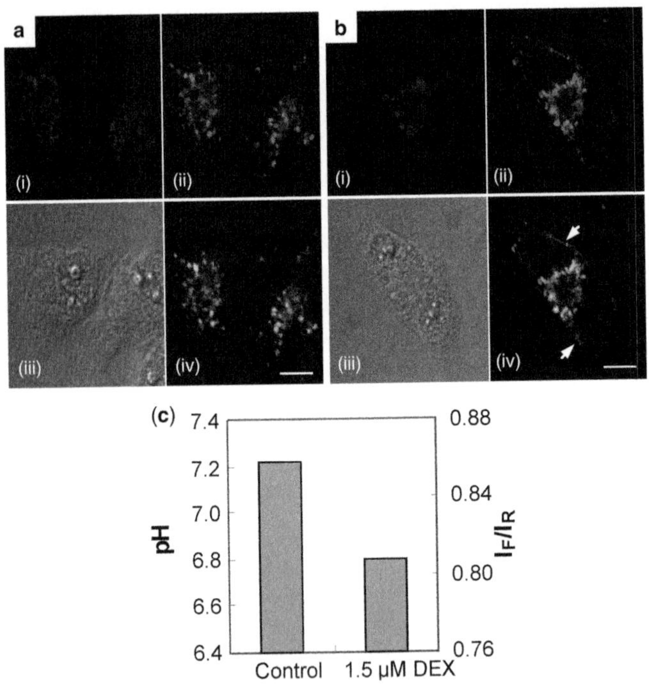

Figure 11.8 Change in fluorescence in HeLa cells loaded with Wang's PEBBLEs after treatment with 1.5 μmol L^{-1} dexamethasone. (a) and (b) represent HeLa cells treated with and without dexamethasone, respectively. (c) Shows the change in the I_F/I_R ratio (F = FITC, R = $[Ru(bpy)_3]^{2+}$) and the pH changes in (a) and (b). (i) and (ii) represent fluorescence images of FITC and $[Ru(bpy)_3]^{2+}$ in PEBBLEs internalized by HeLa cells, respectively. Images labeled (iii) represent bright-field images of HeLa cells, and those labeled (iv) are (i) and (ii) merged. White arrows indicate the fluorescence of PE-AnnexinV combined with phosphatidylserine on the cell membrane in (ii) and (iv). Scale bar, 10 μm.[62] (Reprinted with permission from J. F. Peng et al., *Anal. Bioanal. Chem.* **2007**, *388*, 645–654. Copyright Springer.) (See color insert.)

This last application appears quite appealing because biological processes involving radical species have a putative role in pathological phenomena such as aging, cancer, and other degenerative processes. For example, lipid peroxidation is a process of great interest associated with the damage of biological membranes. The process is a radical chain reaction initiated by the abstraction of hydrogen from unsaturated fatty acid by reactive oxygen species (ROS) such as hydroxyl radicals. The resulting lipid radicals react with atmospheric oxygen to form lipid peroxyl radicals.[67] The process is associated with a weak emission of chemiluminescence which is thought to arise from triplet carbonyls and singlet oxygen derived from hydroperoxides decomposition. The signal is very weak and needs to be amplified using fluorescent dyes such as coumarin derivatives in order to allow imaging of the process.[68] Greenway and coworkers reported a silica PEBBLE sensor able to enhance the weak chemiluminescence associated with lipid peroxidation based on coumarin 343.[64] To prevent dye leaching, the coumarin 343 was covalently linked, via amide bond formation, to aminopropyl surface-coated silica nanoparticles (15 nm diameter). The resulting dye-functionalized nanoparticles were subsequently incorporated in 100 nm diameter sol–gel silica nanoparticle obtained by polymerization of TEOS in the presence of the above nanoparticles under reverse microemulsion conditions (Fig. 11.9). The sensor was tested using a lipid peroxidation model in which cumene hydroperoxide was used to initiate lipid peroxydation of L-α-phosphatidylcholine

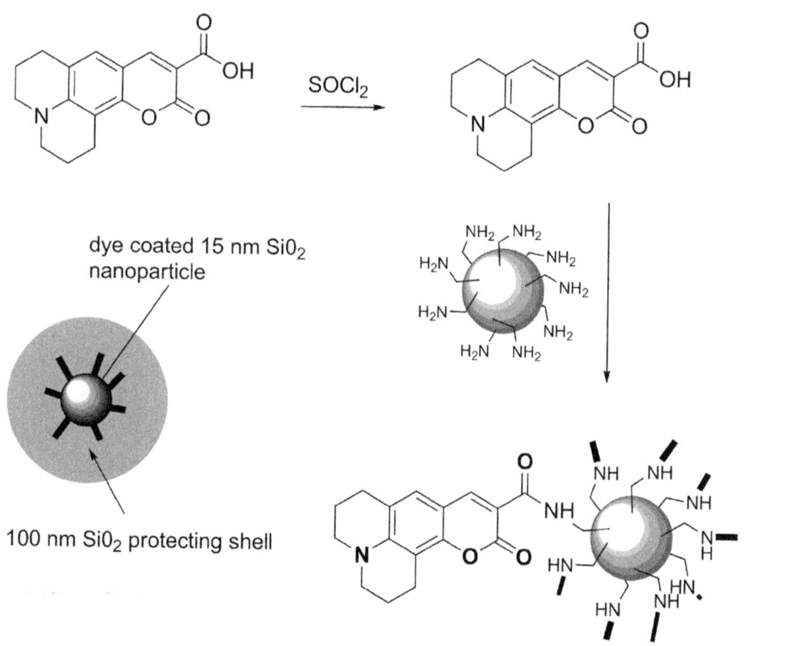

Figure 11.9 Schematic representation of the SiO_2 nanoparticles prepared by Greenway and coworkers[64] for the detection of lipid peroxidation; 15 nm SiO_2 nanoparticles covalently coated with the coumarin dye are embedded in a 100 nm silica protecting shell.

in water. The results indicate that the PEBBLEs are able to enhance the low-level chemiluminescence emitted by the model system by approximately two times. The effect is lower than that observed with the free dye (enhancement factor of approximately 10) probably due to the shielding effect of the silica shell but, in any case, these PEBBLEs represent the first example where the emitted signal arises from chemiluminescence.

A similar approach was used by the same research group for the design of PEBBLEs sensitive to ROS.[65] In this case, the dye is 5(6)-carboxyfluorescein which is first reacted with APTES and then used to decorate the surface of 15 nm diameter silica nanoparticles. As before, the functionalized nanoparticles thus obtained are then buried inside larger (150 nm diameter) silica PEBBLEs. The sensor reports the presence of ROS (hydrogen peroxide, hydroxyl anions, nitric oxide, peroxynitrile, and superoxide anions) by a bleaching of the dye fluorescence with detection limits in the nanomolar concentration range. With respect to the free dye and to the functionalized 15 nm silica nanoparticles, the incorporation in the sol–gel matrix reduces photobleaching, dye leaching, and interference from proteins such as bovine serum albumin (BSA). The PEBBLEs were successfully introduced into bovine oviduct cells using a lipid transfection reagent but no data were reported on the detection of ROS inside the cells.

A different application but again based on redox chemistry has been reported by Zou and Chen,[66] who prepared PEBBLEs containing tetra-carboxyl-substituted iron phthalocyanine (TCFePc), a compound that exhibits peroxidase activity. The system was prepared by copolymerization of a TCFePc-APTES conjugate with TEOS under reverse microemulsion conditions, leading to 60 nm diameter silica nanoparticles with covalently linked peroxidase mimic. In water and in the presence of hydrogen peroxide, the TCFePc containing nanoparticles catalyze the oxidation of thiamine (vitamin B_1) yielding thiochrome, a strongly fluorescent product. Thiamine can, therefore, be detected with a linearity range of 0.005 to 1.0 μM and a detection limit of 2.0 nM, and the validity of the analytical method was demonstrated with the direct analysis of two kinds of pharmaceutical tablets. With respect to the free TCFePC, the catalytic PEBBLEs show higher activity and this was explained with a stronger affinity of the substrate for the silica nanoparticles and a lower propensity of the immobilized catalyst to form aggregates in water which are known to decrease the efficiency of the catalyst. In any case, this system represents a new application of PEBBLEs in a sensing process where the nanoparticles have the role to carry and stabilize the catalyst involved in the chemical transformation of the substrate.

The confinement of a relatively large number of dye molecules in the small volume of a nanoparticle may trigger collective phenomena otherwise not observable in bulk solution. This has been demonstrated by Prasad and coworkers in the case of an ORMOSIL pH sensor.[69] The PEBBLEs contain a naphthalenylvinylpyridine derivative (NVP) as pH-sensitive fluorescent dye which has been functionalized with a triethoxysilane anchor by reaction with an excess of (3-isocyanatopropyl)triethoxysilane (ICTES). The sol–gel polymerization in aqueous micellar solution of the NVP-ICTES derivative with VTES gives spherically shaped 33 nm silica nanoparticles in which the dye is covalently linked to the silica matrix and uniformly distributed in the nanoparticle volume. The NVP dye responds ratiometrically to protons, with a fluorescence red shift from a high-energy blue neutral form to a low-energy yellow protonated form. Moreover, the red-shifted absorption by protonation partially overlaps with the shorter-wavelength side of the neutral blue emission. Therefore, energy transfer is possible between the neutral and protonated form if the two states exist in proximity. In the NVP-loaded PEBBLEs, the protonation of the dye occurs mainly on the surface of the nanoparticles and therefore FRET from the neutral dyes of the core to the protonated dyes on the surface is observed. This FRET process, made possible by the organization of the dye in the nanospace, amplifies the emitted signal, increasing the sensitivity of the sensor with respect to the free dye in bulk solution for which FRET is not observed. Moreover, this energy transfer process increases the apparent pK_a of the dye to 6.4 for the nanoparticle-encorporated dye, well centered for biological application. A similar behavior was also reported under two-photon excitation conditions.

A second example of a collective process made possible by the organization of nanoparticles was reported by Montalti and coworkers.[70] They prepared silica nanoparticles (30 nm diameter) in which a dansylated polyamine, obtained by reaction of dansyl chloride with 3-[2-(2-aminoethylamino)ethylamino]propyltrimethoxysilane, is covalently included in the silica matrix by copolymerization with TEOS under basic conditions. Due to the presence of the polyamine receptor, the nanoparticles bind metal ions with relatively high affinity and the addition of Cu^{2+}, Co^{2+}, and Ni^{2+} results in the quenching of the fluorescence emission of the dansyl unit. Interestingly, the quenching observed per single analyte ion is much stronger than that theoretically expected for conventional 1:1 complex formation, allowing the detection of nanomolar concentrations of the metal ions. This effect has been interpreted with the occurrence of multiple simultaneous quenching processes favored by the dense packing of the fluorescent dyes in the nanoparticle. As a consequence, a single quenching ion is able, when complexed by the amine in the silica matrix, to interact with many or even all of the fluorophores located nearby, causing their nonradiative deactivation. Due to the results obtained the investigators estimated that a single copper ion is able to quench up to 13 dansyl units, leading to strong signal amplification (Fig. 11.10). However, even in the presence of a large excess of Cu^{2+} (10 equivalents with respect to the dansyl dye) the quenching of the luminescence is not complete. This residual fluorescence has been attributed to the presence of very compact, nonpermeable regions in the nanoparticle structure where the fluorophores are completely shielded from the quenching metal ion.

A further peculiar and valuable feature of PEBBLEs is the possibility of external modification of the nanoparticle, which may allow, for example, signal enhancement, as elegantly demonstrated by the group of Kopelman with MagMOONs.[71] In biological samples, the background fluorescence often constitutes a real problem for the sensitive detection of the signal from fluorescent dye-containing sensors and background subtraction can be an important source of experimental error. To overcome this problem, Kopelman and coworkers have introduced magnetically modulated optical nanoprobes (MagMOONs), which are nanoparticles designed to emit a fluorescence signal and to blink in a rotating magnetic field, allowing rapid *in situ* background subtraction.

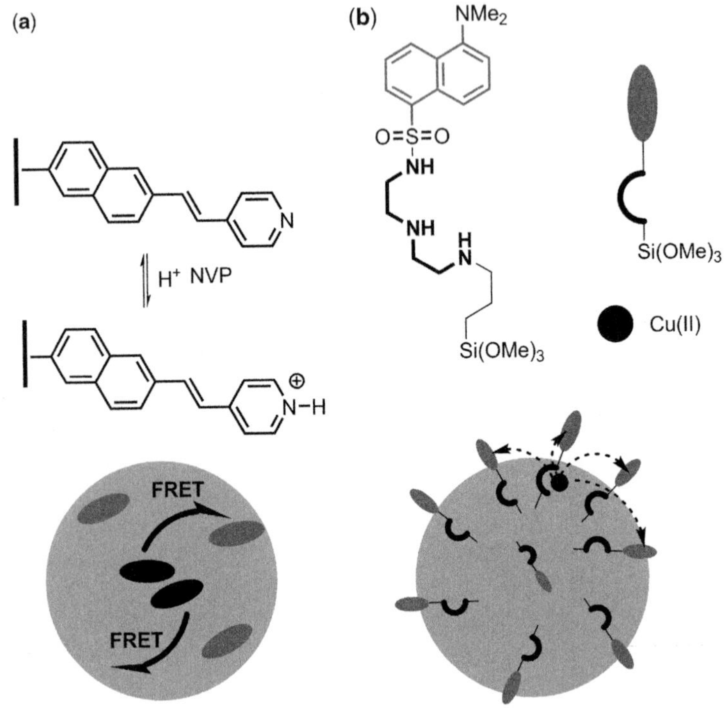

Figure 11.10 Examples of collective effects in silica nanoparticles. (a) FRET processes from the inner neutral to the protonated NVP dye on the surface amplify the emitted signal; (b) the complexation of a single Cu^{2+} ion (black circle) quenches the emission of several (up to 13) dansylamide dyes located in the surrounding leading to improved sensitivity.[69,70]

MagMOONs are generated by vapor depositing a layer of aluminum over one hemisphere of a silica nanosphere in which a magnetic material and fluorescent dyes are entrapped. When a rotating magnetic field is applied to MagMOONs in solution, the microspheres rotate, seemingly passing through the phases of the moon (Fig. 11.11a). Due to the presence of the aluminum coating, which prevents excitation light from entering and fluorescence from leaving the nanoparticles, MagMOONs appear to blink when they are rotating (Fig. 11.11b). Demodulating the blinking MagMOON signal dramatically increases the signal-to-noise ratio. Combining PEBBLEs with MagMOONs can, therefore, increase the sensitivity to intracellular analytes, allowing the use of fewer probes and/or weaker fluorescent dyes, and permitting extension of the range of applications to samples with high fluorescent background such as, for example, stained cells.

Following this approach, Kopelman and coworkers have prepared a pH sensor[72] using as magnetic material barium ferrite nanocrystals (BaM) and as dye the dextran-linked carboxy SNARF-1, a commercially available fluorescein derivative with a convenient pK_a of 7.5 and dual excitation and emission, which allows for ratiometric pH sensing. The two components were incorporated in silica nanospheres of 300 nm

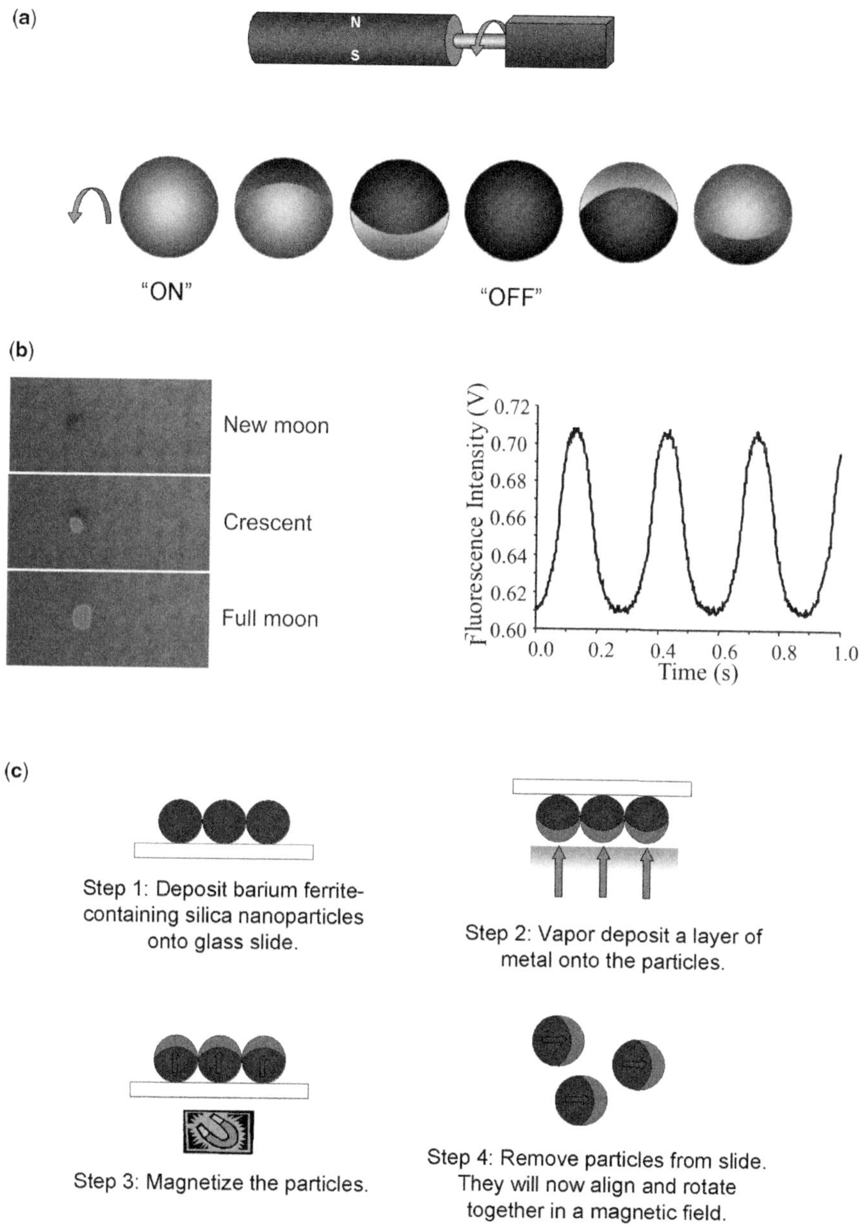

Figure 11.11 Drawing of the MagMOONs developed by Kopelman. (a) In a rotating magnetic field MagMOONs appear to blink as they rotate through the phases of the moon; (b) fluorescence emission during the rotation; (c) schematic illustration of the preparation of aluminum capped MagMOONs.[71,72] (Reprinted with permission from J. N. Anker et al., *J. Magn. Magn. Mater.* **2005**, *293*, 655–662; and T. G. Roberts et al., *J. Magn. Magn. Mater.* **2005**, *293*, 715–724. Copyright 2005 Elsevier.)

prepared by the Stöber method. The size of the nanoparticles is of particular importance because MagMOONs must be heavily loaded with magnetic material and fluorescent dyes to operate as described above. The actual 300 nm size is a compromise between the estimated optimal size (1 μM) and the nucleation effect exerted by the dextran polymer, which limits the growth of the nanoparticle in the Stöber synthesis. Silica nanospheres containing BaM and the dye were coated on one hemisphere with aluminum metal with the procedure described in Figure 11.11c. In brief, the nanoparticles were deposited on a microscope glass slide and aluminum was deposited onto the surface of the slide to a thickness of approximately 20 nm by vapor deposition. Before being removed from the plate, the MagMOONs were magnetized by placing the slide in a strong magnetic field so that the uncoated hemisphere was oriented toward the magnetic north pole. Finally, the MagMOONs were removed from the slide using a soft brush and dispersed in water by sonication. The MagMOONs thus obtained, after pH calibration, were tested in buffer at pH 5.5 in the presence of an external rotating magnetic field. The result clearly indicates the ability of MagMOONs to detect a very weak signal in the presence of a background 250 times higher, thanks to the modulation of the fluorescence emitted. These results are clearly encouraging and MagMOONs promise to enhance the signal-to-noise ratio of a variety of applications based on fluorescence spectroscopy, including biomedical immunoassays, intracellular chemical sensors, cellular labels or tags, and protein folding studies.

11.4 SILICA NANOPARTICLES BASED SELF-ORGANIZED SENSING AGENTS

The study of some of the systems described in the previous paragraph indicated that the spatial confinement of several photoactive units within the small nanoparticle volume can be exploited in order to obtain cooperative or collective effects, such as ratiometric sensing or signal amplification, which significantly improve the performance of the sensing agent. However, the degree of cooperation between the components embedded into a sensing nanoparticle can be further increased by taking advantage of the dense packing of active units achievable by surface functionalization and the possibility to compartmentalize them in different particle domains.

In 2001, McBranch, Whitten and coworkers showed that adsorption of the cationic cyanine **1** (Fig. 11.12) on the surface of silica nanoparticles leads to the formation of extended J-aggregates.[73] Such "self-assembed polymers" detect the presence of the negatively charged anthraquinone disulfonate (**2**, Fig. 11.12) through fluorescence quenching, and the measured quenching constant is four orders of magnitude larger than in the case of the free dye in solution. Such an increment is due to both enhanced binding of the substrate to the polycationic cyanine assembly on the particle and to the amplified quenching of an exciton delocalized over many dye units.

A similar, but covalent, approach was employed by Raymo and Cejas, who functionalized the surface of commercially available silica nanoparticles (20 nm diameter) with 2,7-diazapyrenium dications and used them for the detection of dopamine in

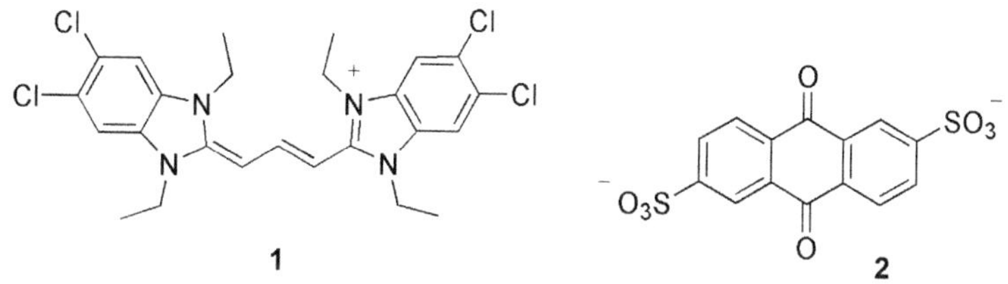

Figure 11.12 Cyanine dye **1** and antraquinone disulfonate **2** used by McBranch and Whitten.[73]

aqueous solutions.[74,75] The nanoparticles were functionalized following two different protocols (Fig. 11.13): (1) a two-step procedure involving first the passivation of the nanoparticles with a trialkoxysilane derivative bearing a halo-alkyl moiety and then the coupling with *N*-benzyl-2,7-diazapyrenium monocation; (2) a one-step procedure involving the direct passivation of the silica surface with a 2,7-diazapyrenium dication derivative bearing a trialkoxysilane group. The one-step procedure yielded a much higher surface coverage but the dense packing of the fluorescent units results in a substantial decrease of the fluorescence quantum yield. Addition of dopamine or catechol

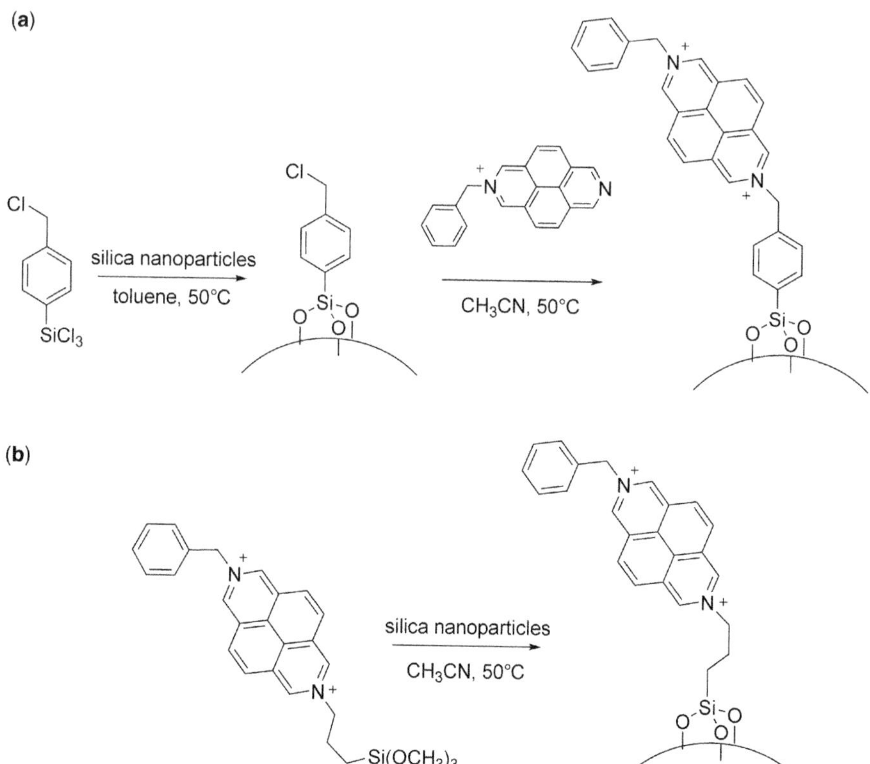

Figure 11.13 Two-step (a) and one-step (b) protocols to prepare diazapyrenium coated nanoparticles.[75]

induces emission quenching as a consequence of the formation of donor-acceptor complexes between the dye and the substrate. Particles prepared by the one-step procedure are significantly more sensitive than those prepared using the two-step protocol. Most likely, the spatial proximity of the 2,7-diazapyrenium units on the particle surface allows their cooperation in substrate binding, favoring the formation of acceptor-donor-acceptor sandwich complexes.

The main difference between these two last examples and the PEBBLE approach is in the cooperation between the active units in the binding of the substrate, which ensures an increased affinity and sensitivity. Still, individual sensing units are linked to the particles and the working scheme of the nanosensor is not different from that of the individual units. In 1999, we explored a new approach to the realization of self-organized fluorescence chemosensors. In this method, fluorescent dyes and receptors are synthesized separately and then assembled on the surface of a proper template.[76,77] In the assembly, the two sensor components are close enough to actively interact and convert the analyte recognition by the receptor units into a variation of the emission properties of the dyes. In the first example, we used surfactant aggregates as templates and we showed that formation of co-micelles containing an inert surfactant, a lipophilic fluorescent dye, and a Cu^{2+}-selective ligand, results in the self-organization of a fluorescence chemosensor capable of detecting Cu^{2+} ions with high sensitivity and selectivity.[76,78] The main advantages of such a system are (1) selectivity, which is mainly due to the ligand choice; (2) simplicity: the sole mixing of the components (most of them are commercially available) in water is required to prepare the sensor; (3) the possibility to tune the detection range just by the modification of the components ratio; and (4) modularity, which allows the modification or the optimization of the system by simply substituting one of the components.

However, surfactant aggregates are delicate objects due to their dynamic nature: they form only above the critical micellar concentration and they are very sensitive to environmental conditions, such as temperature and ionic strength. As seen before, the structure of surface-functionalized nanoparticles closely resembles that of a frozen co-micelle. On the basis of these premises, commercially available particles (20 nm diameter) were functionalized with the triethoxysilane derivatives of the Cu^{2+} selective ligand picolinamide (**3**, Fig. 11.14) and the fluorophore dansylamide (**4**, Fig. 11.14).[79,80] The grafting of the sensor components to the particle surface ensures the spatial proximity required to signal Cu^{2+} by quenching of the dye fluorescence emission. In 9 : 1 DMSO/water solution, the coated silica nanoparticles selectively detect copper ions down to micromolar concentrations and the operative range of the sensor can be tuned by the simple modification of the components ratio. Moreover, it was possible to obtain clear evidence of the cooperation among the ligand subunits in the formation of binding sites with an increased affinity for the substrate. The preparation of a small library of coated silica nanoparticles with the same ligand and different fluorophores proved the versatility of this approach. The emission spectra of these particles span over a large wavelength interval from 300 to 600 nm (correspondingly, excitation wavelengths are in the range 285 to 466 nm), allowing the choice of the most suitable sensor for the desired application. Using a stronger ligand for the Cu^{2+} ion, we were able to exploit collective processes, similar to

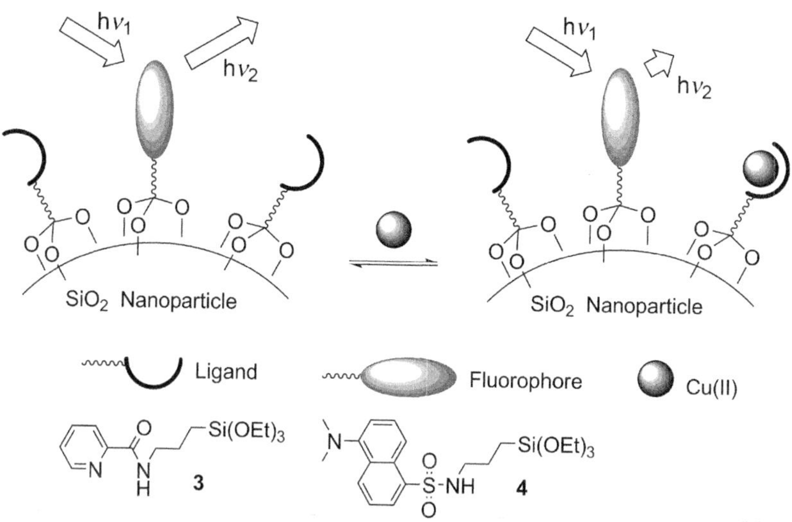

Figure 11.14 Working scheme of Cu^{2+} sensor self-organized on silica nanoparticles.[79] (Reprinted with permission from E. Brasola et al., *Chem. Commun.* **2003**, 3026–3027. Copyright the Royal Society of Chemistry.)

those reported by Montalti[70] discussed in the previous paragraph, in which one single metal ion is capable of quenching the emission of about 10 surrounding dyes, thus lowering the detection limit of the sensor to the nanomolar range.

An important evolution of this approach has been described by Martínez-Máñez, Sancenón and coworkers, who realized a nanoparticle-based colorimetric sensor for organic carboxylates (Fig. 11.15).[81] The sensor working scheme relies on the polarity controlled merocyanine-spiropyran equilibrium (Fig. 11.15): the prevailing species in polar environments is the red merocyanine form, whereas in nonpolar media, the pale yellow spiropyran form dominates. A thiourea derivative and a spirobenzopyran dye bearing the usual trialkoxysilane moieties were anchored to the surface of 20 nm diameter commercial silica nanoparticles. Addition of organic carboxylates to a nanoparticle suspension in water at pH 7 leads to a clear color change from red to pale yellow. The interaction of the carboxylates with the thiourea groups induces a polarity change of the environment surrounding the dye units causing the cyclization of the dye to the spiropyran form (Fig. 11.15). No effect is observed after the addition of other inorganic anions. It must be underlined that this is one of the very few examples of self-organized sensors where a detection mechanism different from quenching of the emission by the substrate is at play.

One of the main features of silica nanoparticles is that the particle interior, and not only the surface, can be functionalized. In the previous examples, the silica nanoparticles act just as mere scaffolds where the two sensor components, the signaling dye and the receptor, are assembled. A careful analysis of the properties of silica gels, and particularly of their metal sorption ability, led us to imagine that the sensor scheme could be further simplified and the degree of self-organization increased.

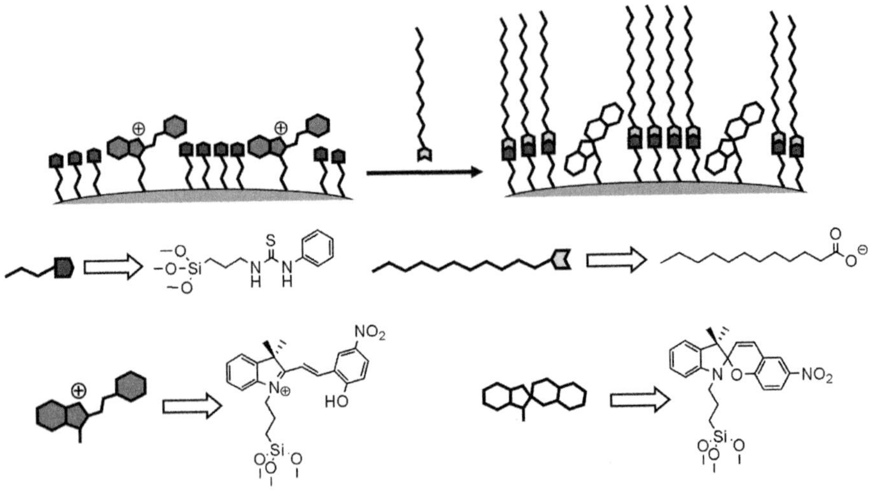

Figure 11.15 Working scheme of silica nanoparticle-based sensor for organic carboxylates. The nanoparticle surface is functionalized with anion binding sites (thiourea, dark gray) and signaling units (merocyanine, gray). Addition of lipophilic carboxylates (light gray) results in the coordination to the binding sites and causes a change in polarity of the environment around the signaling group, inducing cyclization to the spirocyclic form and a color modulation from pink-red to pale yellow.[81] (Reprinted with permission from P. Calero et al., *Chem. Commun.* **2008**, 1668–1670. Copyright the Royal Society of Chemistry.)

Accordingly, we prepared and studied[82] silica particles containing the dansylamide dye, by copolymerization of TEOS and the dansyl triethoxysilane derivative **4** (Fig. 11.14) following the Stöber–van Blaaderen method. Addition of Cu^{2+} ions to solutions of such dye-doped silica nanoparticles resulted in a strong fluorescence quenching, allowing the determination of metal ion concentration in the micromolar range. The sensor is remarkably selective and addition of Co^{2+} or Ni^{2+} to a particle solution resulted in very small changes in the emission intensity. In this case, silica itself provides the receptor units of the sensor: the network of acidic silanol groups on the particle's surface forms the binding sites for metal ions. Once bound to the silica network, the Cu^{2+} ions are close enough to the fluorescent units entrapped in the particle to quench their emission, thus producing a measurable signal. Hence, the formation of the particles leads, at the same time, to the formation of metal ion binding sites as well as to the linking of a fluorescent reporter in their proximity, thus giving rise to a simple conversion of fluorescent dyes into sensors. Interestingly, the particle size is an important parameter that influences the efficiency of the sensors. Smaller particles show an almost complete quenching of the emission, while the larger ones undergo a smaller fluorescence variation upon Cu^{2+} addition. This was explained by assuming that the particles are accessible to the solvent and the analyte only up to a certain depth.[70] When the particle is small, the solvent permeable layer is deep enough to make the whole body of the particle accessible to the Cu^{2+} ions so that all the fluorophores are effectively quenched. As the diameter of the colloid increases, an increasing fraction of the dyes, located in the particle core, remains unaffected by the binding of the metal ions, occurring only in the outer shell.

This approach can be extended to the detection of other transition metals if the nature of the functional groups present on the particle surface is changed. Fifty nanometer dansylamide-doped silica nanoparticles coated with a thin layer of thiols signal the presence of Pb^{2+} ions in millimolar concentration in water, by quenching the fluorescence emission.[83] Moreover, the sensing system can be converted from an on-off probe to a more valuable ratiometric probe (Fig. 11.16). As previously discussed, this is quite easily achieved with nanoparticle-based chemosensors by the addition of a second, substrate-insensitive dye that provides the reference signal. However, in the case of quenching substrates, such strategy could have no effect since the quencher would not distinguish between sensitive and reference dyes and ultimately quench both. Differentiation between the responses of the nanoparticle-embedded dyes can be obtained by shell structuring of the nanoparticle itself. We prepared 50 nm silica nanoparticles organized with a core−shell structure in the following way: a 40 nm diameter core doped with a methoxynaphthalene dye, a 7 nm thick insulating silica shell, a second 3 nm thick shell doped with a dansylamide derivative, and finally a thiol functionalized surface layer (Fig. 11.16). When Pb(II) is bound to the surface of the nanoparticles in the binding sites spontaneously formed in the thiol layer, it quenches the emission of the neighboring dansylamide dyes in the outer shell, while the methoxynaphthalene dyes, buried in the nanoparticle core and further protected by the insulating shell, remain unaffected (Fig. 11.16). This produces a ratiometric response to the presence of Pb^{2+} ions.

In a previous example, core−shell silica nanoparticles where employed by Wiesner and coworkers for ratiometric pH sensing *in vitro* (Fig. 11.17).[84]

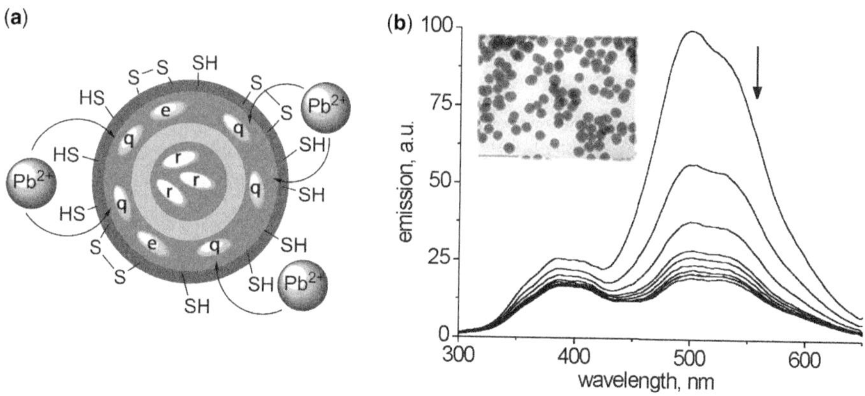

Figure 11.16 Multishell silica nanoparticle for ratiometric fluorescence sensing of Pb^{2+} ions. (a) Schematic working scheme of the sensor: binding of the Pb^{2+} ions to the thiol groups present on the surface causes the quenching of the emission of the dansylamide dyes (labeled **e** when unperturbed and **q** when quenched) in the outer shell but not of the methoxynaphthalene dyes in the core (labeled **r**). (b) TEM micrograph of the particles (inset) and ratiometric behavior of the fluorescence emission at different Pb^{2+} concentrations.[83] (Reprinted with permission from M. Arduini et al., *Langmuir*, **2007**, *23*, 8632–8636. Copyright 2007 American Chemical Society.)

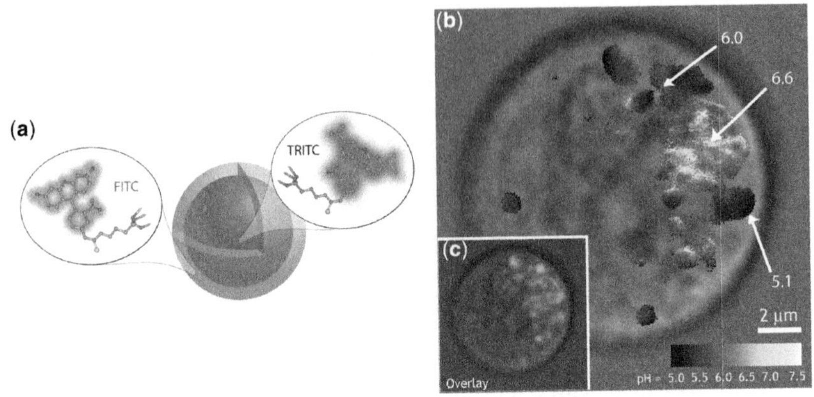

Figure 11.17 (a) Schematic diagram showing the core–shell architecture of the pH sensing nanoparticles: the reference dye tetramethylrhodamine isothiocyanate (TRITC) is sequestered in the core and the sensor-dye fluorescein isothiocyanate (FITC) is confined in the outer shell. (b, c) Confocal fluorescence microscopy images (overlaid on bright field) of pH sensors in RBL mast cells showing: overlaid images of reference dye and sensor dye channels (c), and false-color ratiometric imaging of pH in various intracellular compartments (d).[84] (Reprinted with permission from A. Burns et al., *Small*, **2006**, *2*, 723–726. Copyright Wiley-VCH Verlag GmbH & Co. KGaA.) (See color insert.)

Tetramethylrhodamine, which is pH insensitive, was covalently linked to 50 nm silica particles that were coated with a 10 nm thick shell doped with the pH responsive fluorescein. In this case, shell structuring is not needed to obtain the ratiometric response but it produces an improvement of the sensor performance. In fact, the presence of sensitive units in the solvent-inaccessible core of the nanoparticle, which would result in a less pronounced ratiometric effect, is avoided.

A completely different and elegant approach for the realization of silica nanoparticle-based chemosensors was explored by Lin and coworkers in 2004 (Fig. 11.18).[85] Here, the nanoparticles were designed to act as a nanochromatographic system with fluorescence detection. Mesoporous silica nanospheres with average pore diameter of 2.5 nm were prepared following a procedure set up by the authors, which involves the polymerization of 3-mercaptopropyl-trimethoxysilane in the presence of the template surfactant cetyltrimethylammonium bromide (CTAB). Before the removal of the surfactant, which acts also as protecting agent for the thiol groups located in the mesopores, the nanoparticle surface was functionalized with 5,6-epoxyhexyltriethoxysilane (EHTES). Treatment with boiling HCl/MeOH solution leads to conversion of the epoxyhexyl groups to 5,6-dihydroxyhexyl groups and to surfactant removal as well. At the end, the nanosphere surface is coated with an 11 nm thick poly(lactic acid) (PLA) shell and the thiol groups present in the pores react with *o*-phthalaldehyde to obtain pore surface-anchored *o*-phthalic hemithioacetal (OPTA) groups. Addition of amines to a suspension of these nanospheres originates in a fluorescence emission as a consequence of their reaction with the *o*-phthalic hemithioacetal groups located in the particles' pores to form fluorescent isoindole derivatives. Moreover, different substrates penetrate through the PLA shell at different rates depending on their

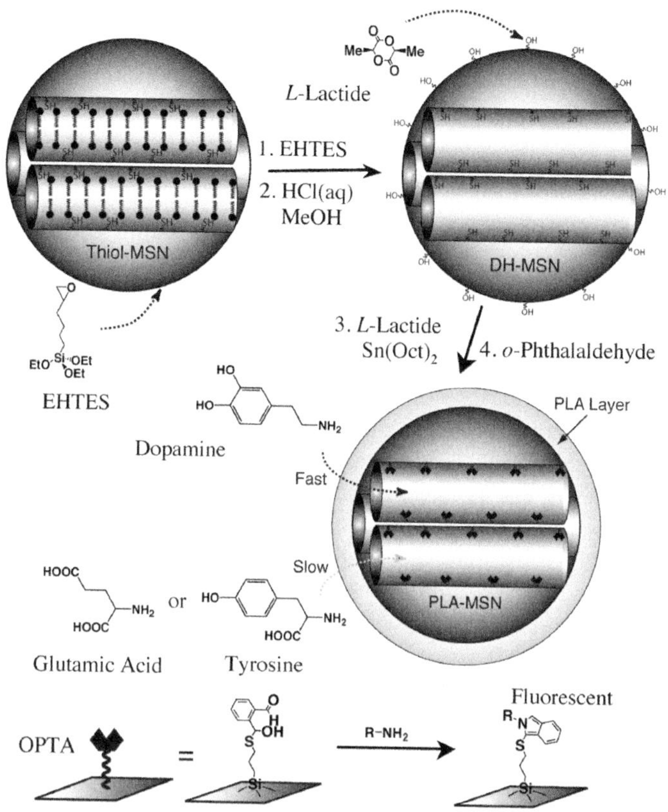

Figure 11.18 Schematic representation of the synthesis and working scheme of Lin's fluorescence sensor system for the detection of amine-containing neurotransmitters. EHTES: 5,6-epoxyhexyltriethoxy-silane), •—— CTAB surfactant.[85] (Reprinted with permission from D. R. Radu et al., *J. Am. Chem. Soc.*, **2004**, *126*, 1640–1641. Copyright 2004 American Chemical Society.)

polarity. As a consequence, addition of the biologically relevant, amine containing neurotransmitters dopamine and glutamic acid results in different rates of fluorescence increase which allows selective detection of dopamine.

11.5 CONCLUSIONS AND OUTLOOK

Optical sensing has always attracted a great interest for the advantages connected to light absorption and emission measurements. The use of simple and inexpensive instrumentation, high sensitivity, spatial and temporal resolution, and low interference with the analyzed sample are among the most valuable features of optical sensing. Nanosystems, since their appearance, offered new opportunities and perspective to this field. The peculiar emission features of quantum dots as well as the strong and aggregation-sensitive plasmonic absorption of metal nanoparticles stimulated the design and realization of new sensing schemes. Dye-doped silica nanoparticles are another major player in this field. The reasons for such interest are not in their peculiar properties, which are not much different from those of the embedded dyes, but in their ability to organize organic and inorganic dyes to obtain complex behaviors. The immediate advantage over other fluorescent, or optically active, nanosystems is the simplicity of preparation, which involves one-pot reactions under mild conditions without particular precautions. Embedded dyes undergo a substantial improvement of their properties, such as photostability and emission quantum yield, while concentration of a high number of units in a single object leads to high brightness and hence facilitates detection.

On this basis, improvements in fluorescence-based bioassays, probes, and stains appear logical for these extremely important applications. Moreover, the number and type of different species that can be embedded into a particle are limited mainly by the imagination and synthetic ability of the chemist. Multiple fluorophore nanoparticles allow dual band emission, ratiometric sensing, and the exploitation of cooperative and collective effects. On the other hand, organization of binding units on the particle's surface leads to spontaneous organization of high affinity binding sites for the substrate. In this view, self-assembled and self-organized systems appear capable of opening new perspectives to a wider application of such systems.

REFERENCES

1. D. L. Feldheim, C. A. Foss, Jr., in *Metal Nanoparticles: Synthesis, Characterization and Applications* (Eds. D. L. Feldheim, C. A. Foss, Jr.), New York: Dekker, **2002**, 1–14.
2. A. Yoshida, in *Colloidal Silica: Fundamentals and Applications* (Eds. H. E. Bergna, W. O. Roberts), Boca Raton, FL: CRC Press, **2006**, 48–56.
3. A. Burns, H. Ow, U. Wiesner, *Chem. Soc. Rev.* **2006**, *35*, 1028–1042.
4. I. I. Slowing, B. G. Trewyn, S. Giri, V. S. Y. Lin, *Adv. Funct. Mater.* **2007**, *17*, 1225–1236.
5. G. Kolbe, Das komplexchemischeVerhalten der Kieselsaure, Ph.D. Dissertation, Jena, Germany, **1956**.
6. W. Stöber, A. Fink, E. Bohn, *J. Colloid. Interface Sci.* **1968**, *26*, 62–69.
7. A. van Blaaderen, A. P. M. Kentgens, *J. Non-Cryst. Solids* **1992**, *149*, 161–178.
8. A. van Blaaderen, J. Vangeest, A. Vrij, *J. Colloid Interface Sci.* **1992**, *154*, 481–501.
9. A. van Blaaderen, A. Vrij, *Langmuir* **1992**, *8*, 2921–2931.
10. N. A. M. Verhaegh, A. van Blaaderen, *Langmuir* **1994**, *10*, 1427–1438.
11. A. van Blaaderen, A. Vrij, *J. Colloid Interface Sci.* **1993**, *156*, 1–18.
12. F. J. Arriagada, K. Osseo-Asare, *J. Colloid Interface Sci.* **1999**, *211*, 210–220.
13. K. Osseo-Asare, F. J. Arriagada, *J. Colloid Interface Sci.* **1999**, *218*, 68–76.
14. X. J. Zhao, R. P. Bagwe, W. H. Tan, *Adv. Mater.* **2004**, *16*, 173–176.
15. S. Santra, J. S. Xu, K. M. Wang, W. H. Tan, *J. Nanosci. Nanotechnol.* **2004**, *4*, 590–599.
16. I. Roy, T. Y. Ohulchanskyy, H. E. Pudavar, E. J. Bergey, A. R. Oseroff, J. Morgan, T. J. Dougherty, P. N. Prasad, *J. Am. Chem. Soc.* **2003**, *125*, 7860–7865.
17. I. Roy, T. Y. Ohulchanskyy, D. J. Bharali, H. E. Pudavar, R. A. Mistretta, N. Kaur, P. N. Prasad, *Proc. Natl. Acad. Sci. U.S.A.* **2005**, *102*, 279–284.
18. L. He, C. S. Toh, *Anal. Chim. Acta* **2006**, *556*, 1–15.
19. J. Z. Lu, Z. Rosenzweig, *Fresenius J. Anal. Chem.* **2000**, *366*, 569–575.
20. S. M. Buck, Y. E. L. Koo, E. Park, H. Xu, M. A. Philbert, M. A. Brasuel, R. Kopelman, *Curr. Opin. Chem. Biol.* **2004**, *8*, 540–546.
21. S. M. Buck, H. Xu, M. Brasuel, M. A. Philbert, R. Kopelman, *Talanta* **2004**, *63*, 41–59.

22. L. Prodi, *New J. Chem.* **2005**, *29*, 20–31.
23. F. Mancin, E. Rampazzo, P. Tecilla, U. Tonellato, *Chem. Eur. J.* **2006**, *12*, 1844–1854.
24. L. Basabe-Desmonts, D. N. Reinhoudt, M. Crego-Calama, *Chem. Soc. Rev.* **2007**, *36*, 993–1017.
25. M. Montalti, L. Prodi, N. Zaccheroni, G. Battistini, S. Marcuz, F. Mancin, E. Rampazzo, U. Tonellato, *Langmuir* **2006**, *22*, 5877–5881.
26. X. C. Zhou, J. Z. Zhou, *Anal. Chem.* **2004**, *76*, 5302–5312.
27. H. Ow, D. R. Larson, M. Srivastava, B. A. Baird, W. W. Webb, U. Wiesner, *Nano Lett.* **2005**, *5*, 113–117.
28. M. Qhobosheane, S. Santra, P. Zhang, W. H. Tan, *Analyst* **2001**, *126*, 1274–1278.
29. W. Lian, S. A. Litherland, H. Badrane, W. H. Tan, D. H. Wu, H. V. Baker, P. A. Gulig, D. V. Lim, S. G. Jin, *Anal. Biochem.* **2004**, *334*, 135–144.
30. L. Wang, C. Y. Yang, W. H. Tan, *Nano Lett.* **2005**, *5*, 37–43.
31. W. Yang, C. G. Zhang, H. Y. Qu, H. H. Yang, J. G. Xu, *Anal. Chim. Acta* **2004**, *503*, 163–169.
32. S. Santra, P. Zhang, K. M. Wang, R. Tapec, W. H. Tan, *Anal. Chem.* **2001**, *73*, 4988–4993.
33. X. X. He, J. Y. Chen, K. M. Wang, D. L. Qin, W. H. Tan, *Talanta* **2007**, *72*, 1519–1526.
34. J. F. Peng, K. M. Wang, W. H. Tan, X. X. He, C. M. He, P. Wu, F. Liu, *Talanta* **2007**, *71*, 833–840.
35. S. T. Selvan, P. K. Patra, C. Y. Ang, J. Y. Ying, *Angew. Chem. Int. Ed.* **2007**, *46*, 2448–2452.
36. N. R. Jana, C. Earhart, J. Y. Ying, *Chem. Mater.* **2007**, *19*, 5074–5082.
37. W. J. Rieter, J. S. Kim, K. M. L. Taylor, H. Y. An, W. L. Lin, T. Tarrant, W. B. Lin, *Angew. Chem. Int. Ed.* **2007**, *46*, 3680–3682.
38. C. W. Lu, Y. Hung, J. K. Hsiao, M. Yao, T. H. Chung, Y. S. Lin, S. H. Wu, S. C. Hsu, H. M. Liu, C. Y. Mou, C. S. Yang, D. M. Huang, Y. C. Chen, *Nano Lett.* **2007**, *7*, 149–154.
39. J. K. Herr, J. E. Smith, C. D. Medley, D. H. Shangguan, W. H. Tan, *Anal. Chem.* **2006**, *78*, 2918–2924.
40. J. E. Smith, C. D. Medley, Z. W. Tang, D. Shangguan, C. Lofton, W. H. Tan, *Anal. Chem.* **2007**, *79*, 3075–3082.
41. X. L. Chen, J. L. Zou, T. T. Zhao, Z. B. Li, *J. Fluoresc.* **2007**, *17*, 235–241.
42. X. Hun, Z. J. Zhang, *Talanta* **2007**, *73*, 366–371.
43. X. Hun, Z. J. Zhang, *Food Chem.* **2007**, *105*, 1623–1629.
44. X. Hun, Z. J. Zhang, *Biosens. Bioelectron.* **2007**, *22*, 2743–2748.
45. M. Q. Tan, G. L. Wang, X. D. Hai, Z. Q. Ye, J. L. Yuan, *J. Mater. Chem.* **2004**, *14*, 2896–2901.
46. M. Q. Tan, Z. Q. Ye, G. L. Wang, J. L. Yuan, *Chem. Mater.* **2004**, *16*, 2494–2498.
47. Y. Xu, Q. G. Li, *Clin. Chem.* **2007**, *53*, 1503–1510.
48. H. Zhang, Y. Xu, W. Yang, Q. G. Li, *Chem. Mater.* **2007**, *19*, 5875–5881.
49. Z. Chang, J. M. Zhou, K. Zhao, N. N. Zhu, P. G. He, Y. Z. Fang, *Electrochim. Acta* **2006**, *52*, 575–580.
50. Y. S. Wang, B. Liu, *Chem. Commun.* **2007**, 3553–3555.
51. X. Zhao, R. Tapec-Dytioco, W. H. Tan, *J. Am. Chem. Soc.* **2003**, *125*, 11474–11475.
52. Y. Chen, Z. H. Lu, *Anal. Chim. Acta* **2007**, *587*, 180–186.
53. D. W. Domaille, E. L. Que, C. J. Chang, *Nature Chem. Biol.* **2008**, *4*, 168–175.
54. J. F. Callan, A. P. de Silva, D. C. Magri, *Tetrahedron* **2005**, *61*, 8551–8588.
55. O. S. Wolfbeis, *Anal. Chem.* **2000**, *72*, 81R–89R.
56. H. A. Clark, S. L. R. Barker, M. Brasuel, M. T. Miller, E. Monson, S. Parus, Z.-Y. Shi, A. Song, B. Thorsrud, R. Kopelman, A. Ade, W. Meixner, B. Athey, M. Hoyer, D. Hill, R. Lightle, M. A. Philbert, *Sens. Actuators B* **1998**, *51*, 2–16.
57. H. Xu, F. Yan, E. F. Monson, R. Kopelman, *J. Biomed. Mater. Res. A* **2003**, *66*, 870–879.
58. H. Xu, J. W. Aylott, R. Kopelman, T. J. Miller, M. A. Philbert, *Anal. Chem.* **2001**, *73*, 4124–4133.
59. Y.-E. L. Koo, Y. Cao, R. Kopelman, S. M. Koo, M. Brasuel, M. A. Philbert, *Anal. Chem.* **2004**, *76*, 2498–2505.
60. F. Gao, L. Wang, L. J. Tang, C. Q. Zhu, *Microchim. Acta* **2005**, *152*, 131–135.
61. F. Gao, L. J. Tang, L. Dai, L. Wang, *Spectrochim. Acta A* **2007**, *67*, 517–521.
62. J. F. Peng, X. X. He, K. M. Wang, W. H. Tan, Y. Wang, Y. Liu, *Anal. Bioanal. Chem.* **2007**, *388*, 645–654.
63. X. Hun, Z. J. Zhang, *Microchim. Acta* **2007**, *159*, 255–261.

64. N. Baker, G. M. Greenway, R. A. Wheatley, C. Wiles, *Analyst* **2007**, *132*, 104–106.
65. V. J. Hammond, J. W. Aylott, G. M. Greenway, P. Watts, A. Webster, C. Wiles, *Analyst* **2008**, *133*, 71–75.
66. J. L. Zou, X. L. Chen, *Microchem. J.* **2007**, *86*, 42–47.
67. X. Shen, J. Tian, Z. Zhu, X. Li, *Biophys. Chem.* **1991**, *40*, 161–167.
68. Y. A. Vladimirov, A. Arroyo, J. M. Taylor, Y. Y. Tyurina, T. Matsura, V. A. Tyurin, V. E. Kagan, *Arch. Biochem. Biophys.* **2000**, *384*, 154–162.
69. S. Kim, H. E. Pudavar, P. N. Prasad, *Chem. Commun.* **2006**, 2071–2073.
70. M. Montalti, L. Prodi, N. Zaccheroni, *J. Mater. Chem.* **2005**, *15*, 2810–2814.
71. J. N. Anker, C. J. Behrend, H. Huang, R. Kopelman, *J. Magn. Magn. Mater.* **2005**, *293*, 655–662.
72. T. G. Roberts, J. N. Anker, R. Kopelman, *J. Magn. Magn. Mater.* **2005**, *293*, 715–724.
73. R. M. Jones, L. D. Lu, R. Helgeson, T. S. Bergstedt, D. W. McBranch, D. G. Whitten, *Proc. Natl. Acad. Sci. U.S.A.* **2001**, *98*, 14769–14772.
74. F. M. Raymo, M. A. Cejas, *Org. Lett.* **2002**, *4*, 3183–3185.
75. M. A. Cejas, F. M. Raymo, *Langmuir* **2005**, *21*, 5795–5802.
76. P. Grandini, F. Mancin, P. Tecilla, P. Scrimin, U. Tonellato, *Angew. Chem. Int. Ed.* **1999**, *38*, 3061–3064.
77. M. Arduini, E. Rampazzo, F. Mancin, P. Tecilla, U. Tonellato, *Inorg. Chim. Acta* **2007**, *360*, 721–727.
78. M. Berton, F. Mancin, G. Stocchero, P. Tecilla, U. Tonellato, *Langmuir* **2001**, *17*, 7521–7528.
79. E. Brasola, F. Mancin, E. Rampazzo, P. Tecilla, U. Tonellato, *Chem. Commun.* **2003**, 3026–3027.
80. E. Rampazzo, E. Brasola, S. Marcuz, F. Mancin, P. Tecilla, U. Tonellato, *J. Mater. Chem.* **2005**, *15*, 2687–2696.
81. P. Calero, E. Aznar, J. M. Lloris, M. D. Marcos, R. Martinez-Manez, J. V. Ros-Lis, J. Soto, F. Sancenon, *Chem. Commun.* **2008**, 1668–1670.
82. M. Arduini, S. Marcuz, M. Montolli, E. Rampazzo, F. Mancin, S. Gross, L. Armelao, P. Tecilla, U. Tonellato, *Langmuir* **2005**, *21*, 9314–9321.
83. M. Arduini, F. Mancin, P. Tecilla, U. Tonellato, *Langmuir* **2007**, *23*, 8632–8636.
84. A. Burns, P. Sengupta, T. Zedayko, B. Baird, U. Wiesner, *Small* **2006**, *2*, 723–726.
85. D. R. Radu, C. Y. Lai, J. W. Wiench, M. Pruski, V. S. Y. Lin, *J. Am. Chem. Soc.* **2004**, *126*, 1640–1641.

Chapter 12

Organically Modified Quantum Dots in Chemical and Biochemical Analysis

MARÍA TERESA FERNÁNDEZ ARGÜELLES,
JOSÉ M. COSTA-FERNÁNDEZ, ROSARIO PEREIRO, AND
ALFREDO SANZ-MEDEL

12.1 INTRODUCTION

Research in semiconductor quantum dots (QDs) started two decades ago with the realization that the optical and electronic properties of such semiconductor

The Supramolecular Chemistry of Organic–Inorganic Hybrid Materials. Edited by Knut Rurack and Ramón Martínez-Máñez

nanocrystals were strongly dependent on their particle size, due to quantum confinement of the charge carriers in small spaces.[1] During the following two decades, extensive research was carried out dealing with fundamental studies on the synthesis and photophysical properties of QDs.[2–5] So far, tremendous research effort has been made also to search for potential applications of these semiconductor nanocrystals mainly for the development of novel optoelectronic devices, quantum-dot lasers, and high density memories.[6,7] Other promising applications in optoelectronics are photovoltaic cells and light emitting devices which are discussed in more detail in Chapter 5.

In two breakthrough papers published in 1998, one from Alivisatos's group at the University of California Berkeley[8] and the other from Chan and Nie (then at Indiana University–Bloomington),[9] the authors simultaneously demonstrated that highly luminescent quantum dots can be made water soluble and biocompatible by surface modification and bioconjugation. After such preliminary developments, their use as biological probes started to attract considerable attention. Some other key advances enabling emerging practical applications of quantum dots in biochemistry and medicine include the synthesis of high-quality colloidal QDs in large quantities[10] and the control of the surface chemistry of QDs via conjugation with appropriate functional molecules.[11] Proper QD surface modification can result in increased luminescent quantum yields and improved stability of the semiconductor nanocrystals, preventing them from aggregation and allowing them to interact with target analytes, all of these features being of crucial interest for chemical analysis and sensing applications.[12] In fact, only a few years after biocompatible QDs were first described,[8,9] they have already been employed in many biochemical applications and assays, including DNA sorting, fluoroimmunoassays, as Förster resonance energy transfer (FRET) tools for microenvironmental sensing, in biooptical coding, cellular and whole animal imaging, and, more recently, in chemical analysis, targeting classic organic and inorganic analytes.[13,14]

This chapter focuses on analytical applications of surface-modified QDs and general approaches to develop novel QD methodologies for chemical and biochemical analysis, an area of growing interest in the last few years. A brief discussion of the attractive luminescent properties of QDs, as compared with the more conventional organic dyes, is mandatory to understand such developments in (bio)chemical analysis.

Fluorescence spectroscopy has become an invaluable analytical tool for bioanalytical research.[15] A high number of fluorescence-based analytical methodologies have been employed successfully for commercial applications in bioscience. The amazing development of such methods over the past few decades has been, and continues to be, mainly a result of the tremendous progress in the basic understanding of the photophysical and recognition properties of organic luminescent molecules. Indeed, sensitive compounds with fluorescence responses tailored to specific analytes can now be designed and synthesized relying on established strategies. Decades of basic research on the photophysical properties of organic dyes in conjunction with the advances in modern chemical synthesis methodologies have afforded a wide number of sensitive fluorophores.[16] However, organic chromophores typically suffer from important drawbacks, including poor photobleaching resistances, short fluorescence lifetimes or inappropriate excitation and emission wavelengths (they have relatively broad emission spectra which often results in signal overlap from different dyes) and are sometimes very sensitive toward slight changes in their environment. Such drawbacks often limit the high potential and applicability of fluorescence in chemical analysis.[17] Recent investigations on the special photophysical properties of organically modified quantum dots indicate that QDs can overcome some of the typical problems associated with conventional organic dyes, making them preferable in many analytical applications.[13,18,19]

As other semiconductors, QDs are characterized by a certain band gap between their valence and conduction electron bands.[20] When a photon having an excitation energy exceeding the semiconductor band gap is absorbed by a QD, electron–hole pairs are generated (electrons are excited from the valence to the conduction band) and the recombination of electron–hole pairs (the relaxation of the excited state) results in the emission of the measured fluorescence light.

These semiconductor nanocrystals exhibit atom-like size-dependent energy states due to the confinement of the charge carriers (electrons and holes) in three dimensions.[21,22] Quantum confinement here means that the particle size is well below the involved wavelength. A result of such quantum confinement is that the energy states of electron and hole within the semiconductor nanocrystals are discrete and the electron and hole energy levels (and therefore the band gap) are a function of the QD diameter:[23–25] the smaller the diameter of the quantum dot is, the larger becomes the energy gap and vice versa. As wavelength (frequency) depends on the band gap energy, longer emission wavelengths result for bigger QDs.[5,26] For example, small CdSe QDs (about 2 nm in diameter) fluoresce in the green, whereas bigger ones (about 5 nm in diameter) fluoresce in the red. Figure 12.1 shows the fluorescence spectra of CdSe QDs with different nanoparticle diameters. As can be seen, differently sized CdSe QDs can be tuned in the 500 to 700 nm spectral range. Moreover, considering that each material has tunability limits (which depend on the physical limitations of the dot's size), access to other wavelength ranges is possible by choosing different chemical materials; for example, Zn-based QDs emit below 400 nm and Pb-based QDs have an emission in the near-infrared spectral region.

In order to reduce nonradiative relaxation on the nanoparticle surface, which is equivalent to recombination of electron–hole pairs without emitting fluorescence light, a second shell of a semiconductor material with higher band gap can be grown around the semiconductor core.[27] Chan and Nie estimated that single ZnS-capped CdSe QDs are about 20 times brighter than single rhodamine 6G molecules.[9] Furthermore, as a result of their discrete atomic-like electronic structure, the full width at half maximum (FWHM) of the emission band of the quantum dots (typically about 15 to 40 nm)[28] is smaller than that of organic dyes. Thus, detection of quantum dots suffers less from cross-talk that might result from the emission of a different fluorophore bleeding into the detection channel of the fluorophore of interest (e.g., from the fluorescent analyte itself or from a labeled analyte).

Several other QD properties have significant practical implications for their use as fluorescent indicators in chemical analysis. Their comparatively low photodegradation

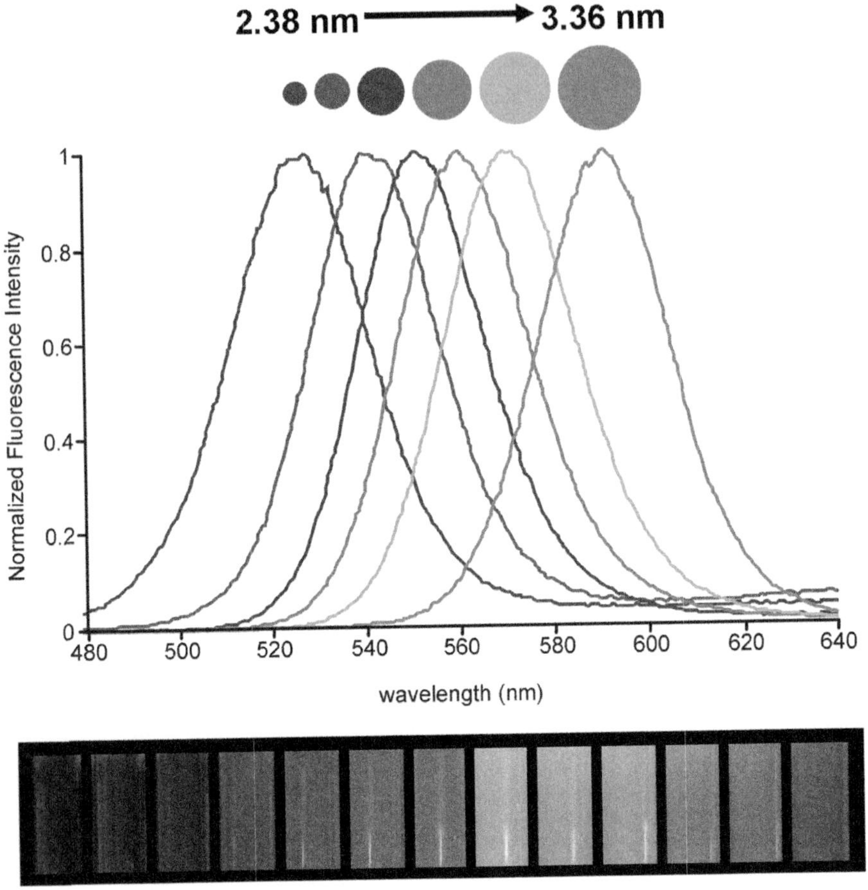

Figure 12.1 Size-tunable fluorescence spectra of CdSe QDs with different diameter sizes. At the bottom, a picture of the different glass vials containing the colloidal suspensions of the differently sized QDs under irradiation with UV light is shown. (See color insert.)

rates make continuous or long-term monitoring of slow biological processes possible. In fact, QDs were observed to be 100 times more stable against photobleaching than conventional organic fluorophores.[9] Moreover, several studies have demonstrated that the photoluminescence properties of CdSe QDs (including quantum yields, band maxima, and FWHM) did not show any detectable change upon aging in air for several months.[29]

In addition, the luminescence lifetime of quantum dots is approximately one order of magnitude longer than that of their organic counterparts.[30] Thus, the autofluorescence of biological samples can effectively be filtered, for example, with time-gated measurements, in the case of these inorganic luminophores. Moreover, QDs are also particularly interesting for multiplexed analysis because of two of their inherent properties: (1) they have narrow and tunable emission spectra and hence there is no risk of signal overlap from different QDs, if different QDs are made selective for different analytes; and (2) different QDs can be suitably excited by a single wavelength.

However, QDs are not free from shortcomings. Some disadvantages of QDs include the size, which is bigger than conventional organic dyes, and might hamper some applications, blinking processes (where the luminescence switches on and off), which can make very sensitive measurements not very reliable, especially when single molecule spectroscopy needs to be carried out, and the typically non-exponential decays, which limit their use in lifetime multiplexing. Moreover, the toxicity of QDs is still an important issue that so far has restricted their large-scale production for the market.

In any case, the ability to tune their luminescence characteristics by particle size control, combined with relatively high quantum yields, narrow fluorescence emission, very broad absorption spectra and photostability provide new solutions to many of the problems associated with traditional luminescence sensors and hold a strong promise for a completely new set of analytical applications with QDs.[18]

12.2 STRATEGIES FOR CHEMICAL ANALYSIS

The appealing optical properties of QDs make these semiconductor nanocrystals very attractive for a wide variety of applications. In the field of (bio)chemical analysis, the ability to tune the QDs' optical properties and to tailor the chemical and biological characteristics of their surfaces are enabling a growing number of creative analytical strategies.

After the pioneer work of Alivisatos's[8] and Nie's[9] groups, many authors have highlighted the interesting advantages of quantum dot bioconjugates over conventional organic dyes as labels in bioanalysis and diagnostics, as tags for protein and DNA immunoassays, or as biocompatible labels for *in vivo* imaging studies.[13,23,30–32] However, in contrast to the fast development and widespread use of QDs in bioimaging and medical applications, the use of such semiconductor nanocrystals as luminescent indicators in chemical analysis for more classical organic and inorganic analytes is showing a somewhat slow start and until recently, only very few reports were published in this field.

QDs may act as an "active" sensory element (for chemical analysis applications) when the intrinsic fluorescence properties of the QDs are changed upon reaction with the analyte (e.g., by surface modification). In probes with QDs as "passive" labels, however, selective receptor molecules (such as antibodies) are conjugated to the surface of QDs. In this latter approach, QD-labeled antibodies are used to recognize the analyte and, in this way, "visualize" the eventually bound antigen.

QDs have been also used extensively as efficient donors in the development of Förster resonance energy transfer (FRET) systems. The key developments and most recent applications in chemical analysis using such QDs-based strategies are reviewed below.

12.2.1 Analytical Methodologies Based on a Direct Interaction of the Analyte with the QD ("Active" Sensors)

The luminescence of QDs is very sensitive to their surface state. Therefore, it is reasonable to expect that eventual chemical or physical interactions between a given chemical species in the surrounding environment a nanoparticle (NP) with the surface of the NP would result in changes of the efficiency of the core electron–hole recombination,[33] affecting the QD's luminescence. This concept has been demonstrated in some chemical analysis applications. The desired selectivity can be somewhat controlled by chemically tailoring the surface of the semiconductor nanocrystals. Depending on the functionalization strategy, the presence of the analyte interacting with the surface of the QD can quench or enhance the quantum dot luminescence. Following this approach, QDs have been reported for "active" fluorescence sensing of small molecules and ions.[14]

In a representative work, Chen and Rosenzweig[34] were able to alter the selectivity of CdS nanoparticles to respond either to Zn^{2+} or Cu^{2+} simply by changing the capping layer. They showed that, while polyphosphate-capped CdS QDs responded to almost all mono- and divalent metal cations (thus showing no ion selectivity), luminescence emission from thioglycerol-capped CdS QDs was quenched only by Cu^{2+} and Fe^{3+}, but was not affected by other ions at similar concentrations. On the other hand, the luminescence emission of L-cysteine-capped CdS quantum dots was enhanced in the presence of zinc ions, but was not affected by cations like Cu^{2+}, Ca^{2+}, and Mg^{2+}. Using this set of QD probes, the authors described the selective detection of zinc and copper in physiological buffered samples, with detection limits of 0.8 μM and 0.1 μM for Zn^{2+} and Cu^{2+}, respectively. This was claimed to be the first use of semiconductor nanocrystals as ion probes in aqueous samples.

Other systems, based on the quenching of the luminescence of the QDs by the analyte, were proposed for the determination of small cationic species. For example, Isarov and Chrysochoos[35] observed that the addition of copper(II) perchlorate in 2-propanol to CdS nanoparticles leads to copper ions binding onto the QD surface, accompanied by rapid reduction of Cu^{2+} to Cu^{+}, this resulting in a quenching of the luminescence that could be used to detect Cu ions in solution. It was proposed that copper ions bound onto the surface of the QDs facilitate nonradiative electron–hole annihilation, thus resulting in a quenching of the luminescence from the nanoparticles. Measurement of the luminescence deactivation of peptide-coated CdS quantum dots has been proposed for the optical sensing of Cu^{2+} and Ag^{+}.[36] Moreover, water-soluble CdSe QDs with their surface modified with 2-mercaptoethane sulfonic acid were used for the sensitive and selective determination of Cu^{2+} in aqueous solutions based on fluorescence quenching measurements.[37]

Following rather similar approaches, CdSe-ZnS QDs modified with bovine serum albumin (BSA),[38] CdSe/CdS QDs modified by mercaptoethanol,[39] and CdTe QDs treated with mercaptopropionic acid[40] were proposed for the determination of Cu^{2+}, CdSe QDs modified with mercaptoacetic acid and BSA were assayed for the analysis of silver,[41] CdSe/ZnS QDs capped with sulfur calixarene for the determination of mercury ions in acetonitrile,[42] and thioglycolic acid-capped CdTe quantum dots for Hg^{2+} determination after further coating of the semiconductor nanocrystals with denatured BSA to achieve the desired selectivity.[43]

The same principle was applied for the detection of inorganic anions. The use of surface-modified CdSe quantum dots functionalized with tert-butyl-*N*-(2-mercaptoethyl)-carbamate groups has been proposed for the selective and sensitive determination of free cyanide in methanol. In a further work,[44] the authors reported the synthesis of water-soluble luminescent CdSe quantum dots (surface-modified with 2-mercaptoethane sulfonate), for the sensitive (detection limit of 1.1×10^{-6} M) and selective (over many other anionic species) determination of free cyanide in aqueous solution (Fig. 12.2). The addition of surfactant agents to the aqueous solution stabilized the QDs. Other authors evaluated polyphosphate-stabilized CdS QDs for optical sensing of iodide,[45] observing a strong luminescence intensity decay (decay times of 10 μs), brought about by the analyte. Such quenching effects were attributed to inner filter effects, nonradiative recombination pathways, and electron transfer processes.

Analytical strategies based on the activation effect caused by the analyte on the QD luminescence emission also have been proposed. In a pioneering work, the addition of Zn and Mn ions to colloidal solutions of CdS or ZnS QDs resulted in an important enhancement of the luminescence quantum yield of the nanoparticles. This effect was attributed to the passivation of surface trap sites that are either being "filled" or energetically moved closer to the band edges.[33,46] This behavior provided the basis for the optical sensing of such metal cations with QDs. Chen and Zhu[47] proposed a method for the determination of trace levels of silver ions based

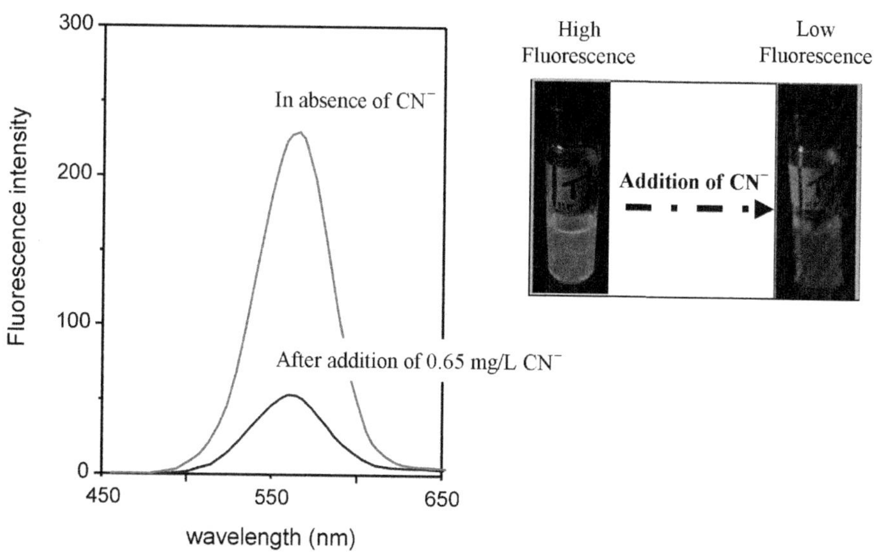

Figure 12.2 Effect of the addition of 0.65 mg L^{-1} cyanide on the luminescence emission spectra of surface-modified CdSe QDs.

on luminescence enhancement of water-soluble CdS quantum dots modified with L-cysteine. The authors attributed the fluorescence enhancement effect to the complex formation between silver ions and the RS-groups adsorbed on the surface of the modified QDs, resulting in the creation of radiative centers at the CdS/Ag-SR complex.

Surface modification can also promote different photoluminescence responses to chemical species present in the environment. Li et al.[48] synthesized water-soluble luminescent thiol-capped CdTe QDs and nanorods and investigated the effect of divalent metal ions on their photoluminescence behavior. They found that the trends of photoluminescence changes were almost similar for both, CdTe QDs and nanorods. Moreover, zinc ions enhanced the luminescence emission of the QDs while other metal ions (e.g., calcium, magnesium, manganese, nickel, and cadmium) caused luminescence quenching.

More complex chemical species have been determined using different cappings on the QDs' surfaces. Shi et al. have recently demonstrated that QDs coated with oleic acid were efficiently quenched by a diversity of nitroaromatic explosives, such as 2,4,6-trinitrotoluene (TNT) or nitrobenzene.[49] Different quenching behaviors were observed for the different molecules. Nevertheless, modified Stern–Volmer equations could be used, in most cases, to provide linear calibration curves. Time domain measurements showed that static quenching was the dominant process as no change in luminescence lifetime was observed in the presence of the analyte.

The sensing of reactive gas molecules using polymer-embedded CdSe QDs has also been demonstrated.[50] The authors found that the photoluminescence of the semiconductor nanocrystals responded to the environment in a reversible, rapid, and species-specific fashion. After QD synthesis, a photostimulation was found to be necessary to obtain a stable emission profile and to provide reliable responses to the presence of the gases. The photostimulated response was thought to be related to photon–phonon coupling of the optical absorption and emission processes occurring in the semiconductor nanocrystal. Based on the same principle, differently sized trioctylphosphine oxide (TOPO) capped CdSe QDs were incorporated into a polymeric film and the photoactivated CdSe QDs unexpectedly exhibited distinct photoluminescence response patterns upon exposure to polar and nonpolar vapors in air.[51] On the basis of principal component analyses of the spectral response of a multi-QD doped film, the authors proposed that a selective gas sensor could be obtained, introducing the possibility of multiparameter gas sensing. The principle was demonstrated with a polymethylmethacrylate (PMMA) film doped with CdSe QDs with distinct average sizes (2.8 nm and 5.6 nm) and different concentrations of methanol and toluene.

Other approaches already realized include the development of a simple and rapid method for spironolactone determination based on the quenching of the fluorescence of CdSe quantum dots by the analyte[52] or the determination of anthracene with a detection limit of 1.6×10^{-8} M based on the quenching effect of this polycyclic aromatic hydrocarbon on water-soluble CdSe/ZnS quantum dots, prepared via a simple sonochemical procedure using β-cyclodextrin (β-CD) as surface coating agent.[53]

All these analytical methods based on chemical or physical interactions between target chemical species and the surface of the quantum dots appear to be simple, easy to develop, and some of them have shown high sensitivity. Unfortunately, they appear to be restricted to just a few reactive small molecules or ions that are able to interact directly with the QD's surface. Moreover, the reactions are usually not specific, and they also depend strongly on the surface properties of the QDs (including their coating). Therefore, a limited use for practical analytical applications can be envisaged for such strategies, that is, the use of QDs as "active" elements in sensors.

12.2.2 Molecular Recognition Based on QD Immunoassays

QDs have rapidly emerged as interesting fluorescent labels able to overcome chemical and spectral limitations of traditional organic fluorophore molecules. Although analytical applications using QDs developed only rather slowly in the beginning because of the difficulties encountered to efficiently process these inorganic semiconductors into aqueous or biocompatible probes,[54] immunoassay-based applications have rapidly increased in recent years. It is well known that in fluorescence immunoassays a target molecule (analyte) can be detected by its specific binding to a fluorophore-labeled antibody. After an incubation period, the unbound antibody can be washed away and the quantity of bound antibody can be determined.[18] Using this general approach, QD-labeled antibodies have been employed as fluorescent labels for the detection of different analytes.

The majority of the assays are based on the use of CdSe/ZnS QDs capped with carboxylic functional groups, such as mercaptoacetic acid (TA), mercaptosuccinic acid, or dihydrolipoic acid (DHLA), which provide negative charge to the QD surface. The carboxylic group can be used, for example, to conjugate rabbit anti-human immunoglobulin G via ethyl-3-(dimethyl aminopropyl) carbodiimide for the detection of human immunoglobulin G (IgG), achieving a detection limit of 2 mg mL^{-1} on a microspotted glass chip.[55] Moreover, the negative charge on the nanoparticle surface has allowed creating QD bioconjugates with mixed-protein surfaces, where each protein imparts a different functionality.[56] In this way, the positively charged engineered adaptor protein G (PG) designed to bind antibodies (by electrostatic forces) through their Fc domain (PG-zb) can be conjugated to QDs and used in fluorescence-based immunoassays for the detection of proteins, small molecules, or multianalyte systems. Engineered maltose binding protein dimer appended with a positive leucine zipper domain (MBP-zb) binds tightly to the DHLA coating through electrostatic interactions. This bioconjugation provides surface passivation, which gives rise to an increase in the fluorescence quantum yield, while maintaining its maltose-binding capability. Moreover, this protein enables rapid and easy removal of excess unconjugated antibodies after coupling to QDs by purification with an amylose resin column.

Detection of staphylococcal enterotoxin B (SEB), a causative agent of food poisoning, was achieved by QDs conjugated with polyclonal sheep anti-SEB antibody.[57] Moreover, this approach also harbors the possibility of a multiplexed immunoassay (see Fig. 12.3), which was first reported by Goldman et al.[58] in 2004: four toxins of interest in food- or water-borne illnesses (cholera toxin, ricin,

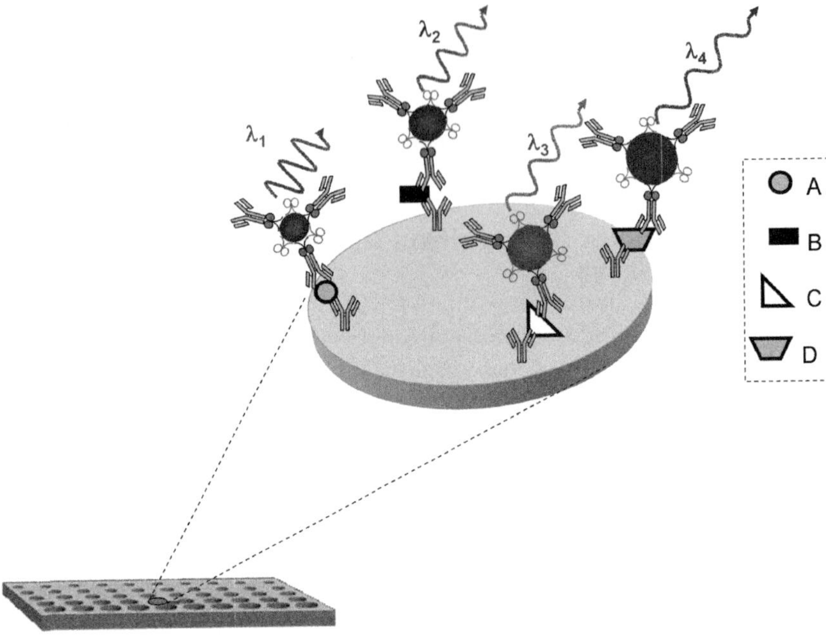

Figure 12.3 Scheme of multianalyte immunoassay. Four analytes (A, B, C, and D) can be detected simultaneously in a sandwich assay format in a single well of a microtiter plate. Antibodies against all four analytes are adsorbed on the wells of the plate, exposed to a mix of all four analytes, and analytes are detected by anti-toxin antibodies conjugated to various QDs. (See color insert.)

shiga-like toxin 1, and SEB) were detected simultaneously with a sandwich immunoassay in single wells of a microtiter plate by using QDs with emission wavelengths of 510, 555, 590, and 610 nm. In this work, CdSe/ZnS QDs capped with DHLA were conjugated to MBP-zb and to the antitoxin antibodies using a PG-zb as a bridge between the QDs and the antibodies. With the designed conjugation and purification protocols, QDs with an average of three antibodies per QD were obtained. The simultaneous detection of the four analytes was demonstrated at concentrations of 1000 and 30 ng mL^{-1} of each toxin in the mixture; the four spectra could be resolved with a simple deconvolution scheme. However, cross-reactivity problems were observed.

More functionalities could be achieved by the bioconjugation of avidin to CdSe/ZnS QDs capped with DHLA and MBP-zb. Such an approach has been exploited in the development of a fluorescence immunoassay for the detection of toxins.[59] Avidin, a highly positively charged tetramer, is here employed as a bridge between the QDs and the specific receptors of the analytes (biotinylated antibodies) to form QD-antibody conjugates for the simultaneous detection of two protein toxins (SEB and cholera toxin). The avidin–biotin system has been widely used for simple bioconjugations in biomedical research. Avidin interacts stoichiometrically with high affinity with biotin, binding one biotin per subunit. In such binding assays, it is important to avoid cross-linking between different antibody-QD conjugates because this would result in an undesirable precipitation of the aggregated conjugates. Therefore, it is necessary to immobilize the mixed-surface QD intermediate on a column before adding biotinylated antibody. These conjugated antibody-QDs were first used in direct binding experiments where antibody-QDs were able to bind goat IgG, immobilized in microtiter plate wells. Moreover, the authors demonstrated the application of such avidin-conjugated DHLA-QDs in sandwich fluoroimmunoassays for the detection of the two toxins. Although the detection limits achieved with this methodology were not superior to conventional immunoassays, these experiments show that QDs can be used as fluorescent labels. Moreover, biotin-modified antibodies, proteins, and DNAs are commercially available or can be easily prepared. Hence, these avidin-QD conjugates are promising versatile fluorophores, which might allow the attachment of QDs to any biotinylated protein or DNA.

Based again on the well-characterized and highly specific biotin–avidin interactions, novel biotinylated QD conjugates were designed by chemically attaching biotin groups onto the functional domain of MBP-zb.[60] These biotinylated QD-MBP-zb ensembles enabled the formation of QD immunocomplexes via avidin and neutravidin bridges. This methodology was applied to the detection of SEB in a sandwich assay format. Nitrotyrosine-modified neutravidin (nitro-neutravidin) was employed as bridging protein between QD-MBP-zb-biotin conjugates and biotinylated rabbit anti-SEB antibody in order to decrease the nonspecific adsorptions observed. High signal-to-background ratios were obtained, giving rise to a detection limit of 10 ng mL^{-1} for SEB.

The approaches described in the previous paragraphs were based on the conjugation of proteins to QD-DHLA by electrostatic assembly. As shown above, this methodology requires a positively charged linker protein, such as avidin, or protein engineering. Such engineering schemes can be complex and not always result in a biomolecule with the desired attributes. In 2005, Mattoussi et al. developed a method for conjugating QDs to proteins based on the coordination of histidine (His) residues (in a polyHis tag) to Zn ions on the shell of CdSe/ZnS nanoparticles.[61] It is known that His coordinates to immobilized metal ions such as Cu^{2+}, Ni^{2+}, Zn^{2+}, and Co^{2+} (this is a key factor because many fusion proteins have been engineered with an artificial polyHis tag incorporated on the N- or C-terminus of the protein to facilitate single-step purification using the technique of immobilized metal ion affinity chromatography, IMAC). The detection of TNT in a competitive immunoassay has been achieved in such a way.[61]

Moreover, not only QDs capped with carboxylic functional groups have been employed for immunoassay applications. It is known that the use of amphiphilic polymers can provide nanoparticles with very good colloidal stability, rendering them water soluble. In this procedure, hydrophobic alkyl chains interact with the alkyl chains of QD ligands, and the hydrophilic groups (Fig. 12.4), disposed to the external part, impart water solubility to the nanoparticles.[62,63] This procedure often results in QDs with final sizes three to four times larger (~20 nm) than the original particles,[64] a disadvantage for live cell and *in vivo* applications. However, polymer-coated QDs are brighter and more stable than surface-exchanged QDs due to the retention of the

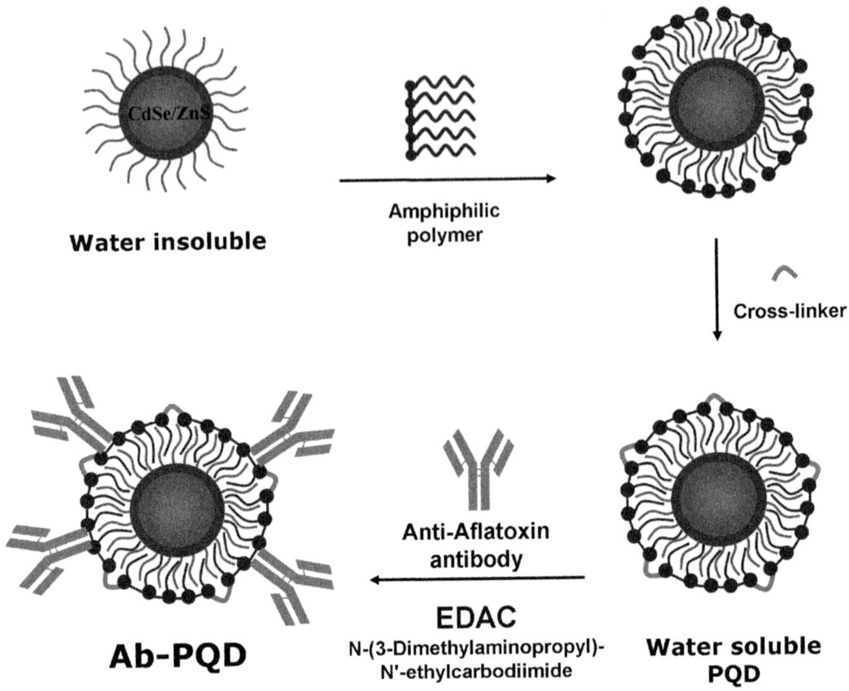

Figure 12.4 Polymer coating and QD-antibody bioconjugation scheme.

native TOPO ligands, and they can also be conjugated to biological probes such as nucleic acids, proteins, or small molecules. In fact, antibodies can be very easily conjugated to polymer-coated QDs, then purified via size exclusion chromatography and finally characterized with fluorescence and mass spectrometric measurements. The usefulness of such conjugated QDs for the detection of aflatoxin B1 has been demonstrated most recently.[65]

Some methodologies employ new matrices to carry out immunoassays. For instance, SEB and cholera toxin were immobilized within hydrogels and detected by direct binding to biotinylated antibodies with streptavidin-coated QDs.[66] Hydrogel-based three-dimensional microarrays are an increasingly attractive format for biological assays because they offer benefits over traditional two-dimensional (2D) immobilization protocols for proteins. When immobilized on 2D substrates, some proteins are prone to denaturation at the interface between the solution and solid support. Hydrogels on the other hand create a medium that is stable, biocompatible, and simulates a solution phase system, in contrast to the limited binding kinetics of the solid–liquid phase interaction evident on a 2D surface. Moreover, sugar polyacrylate hydrogel possesses accessible carbohydrate moieties that have the characteristic advantage of very large pores allowing for diffusion of the nanoparticles into the gel. Streptavidin-coated QDs were able to diffuse into the gel and bind to a protein immobilized within the hydrogel, demonstrating that QDs can be advantageously used in this new matrix. Another interesting approach, assayed for the detection of human prostate specific antigen, is based on the alternation of amphiphilic copolymers consisting of hydrocarbons and maleic anhydride groups. In this way, QD-polymer complexes grow epitaxially into nanobeads of narrow size dispersity.[67] As a result of this nanostructure formation mechanism, more than 250 QDs can be loaded into a nanobead of 100 nm of diameter and so higher sensitivity is expected with this kind of nanostructures.

In order to facilitate the separation of QD-labeled antibodies from those bound to the corresponding antigen (and, so, retained on a microwell plate) the use of magnetic microspheres has been proposed. The major problem using magnetic beads for fluorescence immunoassays is the autofluorescence of the bead, which can interfere with the final analyte detection signal.[68,69] This spectral overlap can be overcome by using QDs with a fluorescence emission shifted away from the bead autofluorescence. These microspheres have been already applied for the detection of mixtures of explosives and biological antigens such as tumor necrosis factor.[70]

A novel trend is the lab-on-a-chip immunoassay, based on the use of a configuration that some authors also call nano-on-micro or NOM,[71] which consists of microspheres or microbeads that incorporate a high number of QDs in their structure and which could be used for several purposes such as increased sensitivity, multiplexed analysis,[72] or reduction of photobleaching. QDs were enveloped within a silica shell to obtain particles with an average diameter of around 170 nm, and an enhancement in anti-photobleaching was observed due to the inert barriers generated for the QDs.[73] Then, the surface of the QD-containing particle was functionalized with carboxylic groups and subsequently with human IgG and the final design was applied to the recognition of goat-anti-human IgG in solution.

Also, through the convergence of nano- and microtechnologies (i.e., QDs and microfluidics), Chan et al. have created a diagnostic system capable of multiplexed, high-throughput analysis of infectious agents in human serum samples.[74] As a proof-of-concept, the authors have demonstrated the ability to detect serum biomarkers of the most globally prevalent blood-borne infectious diseases (e.g., hepatitis B and C) with low sample volume ($<100\ \mu L$), rapidity (<1 h), and 50 times higher sensitivity than that of the currently available methods approved by the FDA (U.S. Food and Drug Administration). This promising device will most probably be developed further into a portable handheld point-of-care diagnostic system.

Currently, QD-labeled antibodies have been employed for the determination of different analytes in matrices of diverse nature, such as the determination of sulfamethazine in milk with very good sensitivity ($0.6\ ng\ mL^{-1}$ detection limit in milk),[75] for simultaneous biomonitoring of exposure to agrochemicals such as 3-phenoxybenzoic acid and atrazine-mercapturate, employed as insecticide and herbicide, respectively,[76] for the simultaneous detection of viruses like H9 avian influenza virus and MHV68 virus,[77] or for the detection of the important cancer marker total prostate-specific antigen on a disposable carbon substrate surface, providing a very sensitive tool for its detection in undiluted human serum samples with a detection limit of $0.25\ ng\ mL^{-1}$.[78] The feasibility of multiplexed labeling was also demonstrated for the detection of tumor biomarkers in cells and tissues via conjugation of five different QDs to primary antibodies against significant tumor markers in breast cancer cells.[79]

To sum up, QDs have finally taken the step from pure demonstration experiments to real applications and, although QDs will probably never replace organic fluorophores as such, they could become the dominant fluorescence labels in certain types of immunoassays in the near future.[31]

12.2.3 Biosensor Nanoassemblies Based on FRET

FRET occurs when the electronic excitation energy of a donor chromophore is transferred to an acceptor molecule nearby via a through-space dipole–dipole interaction between the donor–acceptor pair.[80] The strong dependence of the FRET efficiency on donor–acceptor distance has been widely exploited in studying the structure and dynamics of proteins and nucleic acids, in the detection and visualization of intermolecular association, and in the development of intermolecular binding assays.[81,82]

A critical feature for FRET biosensors development involves a careful selection of energy donor and acceptor chromophores to function in concert with desired recognition elements. Although organic dyes are available to achieve this, FRET-based studies involving pairs of organic dye molecules (as the donor–acceptor complexes) are often limited by cross-talk caused by spectral overlap of the donor and acceptor emission. The broad emission spectrum of the donor, with its long red tail, can often overlap significantly with the emission spectrum of the acceptor. In this sense, the broad absorption spectra of semiconductor QDs allow for a greater flexibility in the selection of the donor excitation wavelength without excitation of the acceptor. Moreover, the narrow emission spectra of QDs minimize the spectral overlap of the donor QD and the acceptor emission spectrum and further reduce cross-talk. In addition, the emission spectra of semiconductor QDs are readily tunable through the dot's size. This will enable tuning the emission spectrum of the donor into resonance with the absorption spectrum of the acceptor, to enhance the energy transfer efficiency and to optimize the spectral resolution of the donor and acceptor emission. Thus, QDs offer an enormous potential to develop new FRET-based nanoscale assemblies for continuous monitoring of target (bio)chemical species in diverse environments.*

In 2001, Van Orden et al. reported the first biologically relevant FRET investigation using QDs as energy donors.[83] They conjugated biotinylated bovine serum albumin (bBSA) to water-soluble CdSe–ZnS-TA QDs via a thiol linkage with 2-iminothiolane. The acceptor was streptavidin covalently labeled with tetramethylrhodamine (TMR). Once the avidin–biotin interaction took place, a decrease in the QD luminescence and an increase in the TMR luminescence were observed, which was attributed to FRET between QD and TMR. Although no time-resolved fluorescence experiments were carried out and no estimates for the separation distance

*It should be considered that, in general, quantum dots are not appropriate FRET acceptors. The broad semiconductor-type absorption bands of QDs typically overlap with the absorption spectra of conventional donors and, therefore, the potential acceptor QD would be always co-excited, see, for example, Reference 129.

were provided for this QD–dye pair, experimental data suggested that low FRET efficiencies would be expected with this QD–protein–dye configuration. Also in 2001, Studer et al.[84] studied BSA–QD conjugates made by using glutaric dialdehyde as a cross-linking agent in conjunction with L-cystein capping of CdTe QDs. They demonstrated the efficient excitation energy transfer between the protein units and core levels of QDs by associating a luminescence enhancement effect with the concentration of adsorbed protein units.

The ability to conjugate each QD donor with a relatively large number of quasi-symmetrically arranged specific receptors offers an exciting opportunity for construction and optimization of new types of optical nanosensors. These include assemblies with mixed surfaces for multiplexed sensing, improved optical sensors for intracellular monitoring, with important applications in a variety of areas, including healthcare, environmental monitoring, and biodefense. However, there are some limitations when longer wavelength emitting QDs are employed as FRET donors, due to larger core sizes and lower quantum yields. As FRET is generally most efficient when the distance between donors and acceptors is in the 20 to 60 Å range, strategies to anchor the protein receptor as close as possible to energy-donating QD cores to maximize FRET have been studied. Mauro et al. developed the first well-characterized QD-protein/receptor nanosensor assemblies. In these assemblies, MBP coordinates to DHLA-CdSe/ZnS QDs.[85] The designed QD–MBP sugar-sensing nanoassemblies have their saccharide binding sites occupied by the displaceable signal transducer consisting of energy acceptor dye QSY9 conjugated to β-CD (Fig. 12.5). The excellent overlap between QD emission and QSY9 absorption allows quenching of the QD emission by ~50%. As MBP has a similar binding affinity for β-CD and maltose, the addition of maltose readily competes with and displaces the β-CD-QSY9 quencher, resulting in a concentration-dependent increase in QD photoluminescence. In other words, a quantitative maltose sensing system is gained by such assembly.

A potential limitation of using QD donors for FRET-based optical sensors involves the bigger physical dimensions of semiconductor nanocrystals compared to organic dyes. To overcome this problem, a two-step FRET QD-MBP sugar-sensing assembly containing a fluorescent dye acting as a relay station was also successfully evaluated.[85]

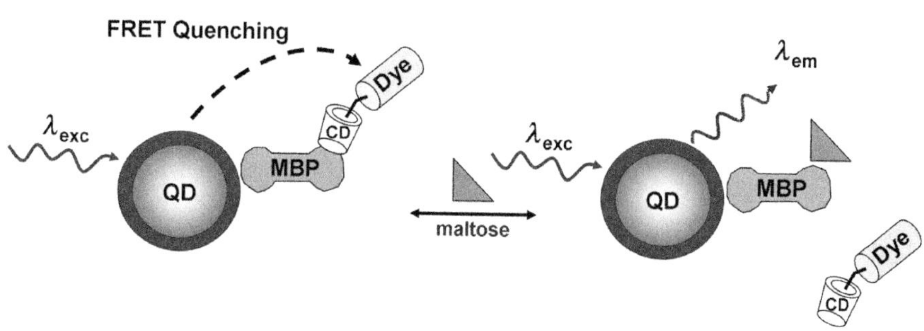

Figure 12.5 Sugar-sensing nanoassembly based on a FRET process. (Adapted from Reference 85.)

Other useful receptors for building nanoassemblies that have already been reported and use QD FRET donors include DNA,[86,87] RNA,[88] peptides,[89] recombinant proteins,[90] antibody fragments,[91] and small molecules, such as drugs. Whole IgG antibodies may be difficult to use directly due to their large size in the light of donor–acceptor distance constraints. Thus, the use of an adequate antibody fragment instead of a full antibody provides a more compact QD conjugate and is better suited for FRET. In this line, a nanoscale sensing assembly based on FRET was developed by Mattoussi et al. for the specific detection of the explosive TNT in aqueous environments.[91] In this approach, anti-TNT specific antibody fragments appended with an oligohistidine sequence were specifically immobilized on the surface of CdSe/ZnS QDs. A dye-labeled TNT analog prebound in the binding/recognition site quenches the QD photoluminescence via proximity-induced FRET. Addition of soluble TNT displaces the dye-labeled analog, eliminating FRET and resulting in a concentration-dependent recovery of QD photoluminescence. Such assembly proved to be an effective sensor for TNT and was tested in artificial seawater and contaminated soil, with 20 ng mL^{-1} being the lowest reported value to observe a measurable signal.

Nanometer-sized gold particles (AuNPs), acting as acceptors, have been used as well for the development of FRET systems, with QDs as donors.[92] When QDs are bound to AuNPs by peptide sequences specific for collagenase a suppression of luminescence, due to energy transfer, occurs. These QD–peptide–AuNPs imaging probes are activated upon peptide cleavage by collagenase, resulting in the release of AuNPs and the subsequent restoration of the QD fluorescence emission. These probes can be customized by changing the sequence of the peptide linker for applications ranging from imaging proteolytic activity of cells to assessing the metastatic potential of cancerous lesions.

One of the advantages of FRET-based sensors is that a single interaction with the acceptor can produce a big change in FRET signal. In this sense, Zhou et al. directly coupled an Alexa 594 fluorophore-labeled double-stranded DNA acceptor to a CdSe/ZnS/3-mercaptopropionic acid QD donor through a C6-thiol linker.[93] This strategy reduces the donor–acceptor distance, leading to a greatly improved FRET efficiency. The FRET between the QD and dye-labeled DNA was studied in both bulk solution and at the single-molecule level. The QD emission (558 nm) overlaps the absorption spectrum of the dye. Thus, conjugation of the dye-labeled DNA to the QD quenches the QD fluorescence, while it enhances the acceptor fluorescence at 618 nm through FRET. Unlike the QD–protein conjugates, for which a number of acceptors are needed for each QD to achieve moderate FRET, Zhou et al. showed that highly efficient FRET (88%) can be obtained at low acceptor to donor ratios (i.e., 2) at the single-molecule level with little aggregation.

In 2005, imaging of single QDs to study the internal motion of a biological molecule was employed for the first time.[94] For this purpose, FRET between a single streptavidin-coated QD (585 nm) acting as a donor, and a single organic Cy5 fluorophore as acceptor was measured. Moreover, this process was employed to measure the conformational changes of individual biomolecules using the DNA four-way junction as a model system. Two complementary DNA single strands, one containing Cy5 at the 5′ end and the other with biotin at the 3′ end, were employed. QD/DNA composites were immobilized on a quartz surface coated with biotinylated

BSA. Thus, the streptavidin-coated QD acts as the donor and as a bridge to the surface. Although single molecule conformational changes could be measured by the procedure described above, some relevant aspects (e.g., the final bioconjugate size), should be improved in order to achieve better yields of the FRET process.

There are some advantages connected to the use of hydrophobic QDs, including higher quantum yield than the water soluble ones and better stability. Moreover, many biological processes occur within lipid membranes. Their natural hydrophobicity makes QDs obvious candidates for sensing transmembrane or intermembrane processes. Kloepfer et al. employed hydrophobic QDs as FRET donors, within a model membrane system using biomimetic lipid vesicles.[95] The FRET process occurs from hydrophobic QDs, embedded into lipid membranes, to both a hydrophilic dye that is able to float free in solution around the membrane (Cy3.5) and a hydrophobic dye that inserts into the lipid layer in a similar manner to the QDs. Higher FRET efficiencies were achieved for the hydrophobic dye. This work opens the way towards using QDs within the membrane leaflets of biological cells, where they may act as FRET donors to naturally occurring molecules, to genetically engineered assemblies or to organic dyes added to the membrane or to the inside or outside of the cell.

In the last years, the number of publications related to QD-FRET-based systems has increased continuously. In 2008, a FRET-based nanosensor was developed for the rapid detection of human cardiac troponin I, which is a key factor for the early detection of myocardial infarction.[96] In this system, a donor(QD)-labeled protein A is bound to an acceptor-labeled capture antibody and the presence of troponin I antigen generates a conformational change within the structure of the antibody. This results in a change of the distance between the donor and acceptor and, therefore, a shift in energy transfer is observed. A limit of detection of 55 nM of troponin in human plasma and a very short time of analysis (1 minute) were reported using this biosensor.

In a similar approach, a system based on QDs acting as donor and an acceptor consisting of rhodamine isothiocyanate (RITC) conjugated to bovine serum albumin was developed for the detection of trypsin (Try) and trypsin inhibitor (TI) in urine samples with high sensitivity and low matrix interferences.[97] When Try was used to digest BSA, the FRET efficiency decreased, allowing the detection of Try at concentrations as low as 110 pM. On the other hand, the digestion activity of Try was inhibited by the presence of TI, and a limit of detection of 250 pM was achieved for TI. Multiplexed detection has been demonstrated as well with QD-based FRET biosensors used for the determination of protease and trypsin.[98]

Several FRET strategies for analyte-mediated distance changes have been demonstrated (Fig. 12.6):

1. The analyte acts as a competitive ligand for receptor binding by displacing the original fluorescent ligand.[85,91]
2. The analyte specifically cleaves the linkage between the NC donor and the acceptor.[92]
3. The analyte mediates the binding of the acceptor to the donor.[99]
4. The analyte changes the conformation of the linkage between the donor and the acceptor.[100]

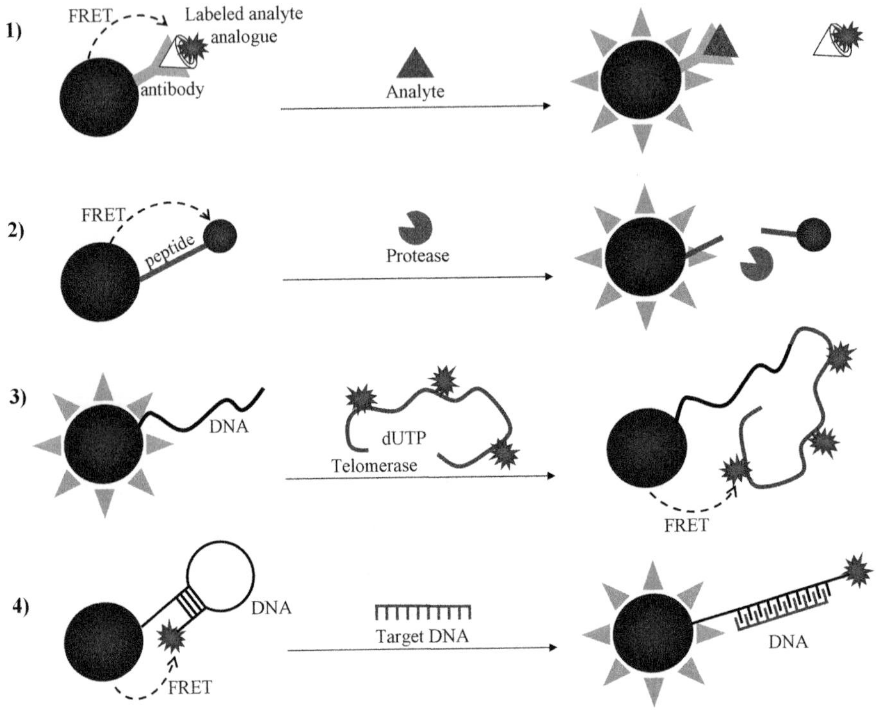

Figure 12.6 FRET strategies for analyte-mediated distance changes. (1) Analyte displaces fluorescent ligand. (2) Analyte cleaves donor-acceptor linkage. (3) Analyte mediates donor-acceptor binding. (4) Analyte changes the conformation of acceptor-donor linkage.

These sensing architectures have in common that the FRET contrast is based on (reversibly or irreversibly) displacing the acceptor from the NC donor.

Another configuration in which the donor constantly transfers energy to the acceptor and its fluorescence emission is changed through a modulation of its quantum efficiency upon ligand binding has also been reported. In this example, the analyte affects both the optical properties of the acceptor and the FRET efficiency measured. For such static systems to provide sufficient signal and stability the acceptor should be linked as close to the donor as possible and the linkage should be stable. In this line, a FRET geometry in which the acceptor is directly incorporated into the polymer shell used to provide colloidal stability to the QD donor has been developed. A very stable linkage of the acceptor to the QD, excellent colloidal stability of the whole assembly, a reduced donor–acceptor distance, which allows for a FRET efficiency of 84%, and facile postmodification steps are the characteristics of such polymer-coated particles.[101] These features make QDs promising candidates for the investigation of cellular traffic at the single particle level.

Triplet exciton energy transfer from a thin film of phosphorescent molecules (Ir complexes) to a monolayer of CdSe/ZnS core-shell QDs acting as acceptor was also demonstrated,[102] with increases in the corresponding photoluminescence intensity and in the photoluminescence lifetime for QDs. The triplet exciton energy transfer is manifested in time-resolved photoluminescence measurements as the QD photoluminescence time constant increases from 40 ns to 400 ns, and a concomitant 55% increase of the time-integrated QD photoluminescence intensity is measured. One important feature of this study is that it confirms definitively that CdSe/ZnS core–shell QDs can efficiently accept energy from a metallo-organic donor, demonstrating effective energy transfer from a phosphorescent dye to QD luminophores. The latter could benefit the development of QD optoelectronic technologies, such as QD-light emitting diodes.

The literature reveals that FRET methodologies can be employed not only for detection, but also to help drug delivery and therapy. One example of this important application is the PSMA biosensor: the surface of a QD was functionalized with an aptamer (QD-Apt) which recognizes the extracellular domain of the prostate specific membrane antigen (PSMA), allowing for the uptake (and imaging) of prostate cancer cells that express the PSMA protein.[103] Additionally, the intercalation of doxorubicin (DOX), a widely used antineoplastic anthracycline drug with fluorescence properties, in the double-stranded stem of the aptamer results in a targeted QD-Apt(DOX) conjugate with reversible self-quenching properties. This multifunctional nanoparticle system can deliver DOX to the targeted prostate cancer cells and sense the delivery of DOX by activating the fluorescence of QD, which concurrently facilitates imaging of the cancer cells. Thus, cancer imaging, therapy, and sensing in vitro are enabled by such a QD-based system.

Of course, there are still some remaining problems, including aggregation, bioconjugate sizes, and pH stability, that need to be improved,[104] but highly relevant applications of QD-based FRET methodologies have been shown in this chapter. It can be envisaged that a refinement of FRET-based QD sensing assemblies will take place in the near future. Therefore, once the conjugate's architecture is more controlled, their practical applications in analytical chemistry, bioscience, and nanobiotechnology will be more reliable.

12.3 FUTURE DIRECTIONS

As shown in the previous sections, QDs are being used successfully nowadays in increasing analytical applications. In fact, QDs have already fulfilled some of their original great promise as "ground-breaking labels" in chemical analysis. The present tremendous interest in QDs guarantees to extend previous applications to real life problems and inspire new creative ones. Thus, new developments and advances in the use of QDs for (bio)chemical analysis are expected in the near future. In our opinion, future developments and applications could be related to the combination of QDs with different detection principles, the coupling and bioconjugation of these nanoparticles to novel recognition elements or their immobilization in appropriate solid supports for new sensing materials development.

12.3.1 Novel Detection Modes

Research on the luminescence properties of QDs has slowly expanded to using detection principles alternative to fluorescence. In fact, quantum dots have been already used in combination with surface plasmon resonance (SPR) measurements. Preliminary studies reported a clear exciton peak due to the quantum confinement effect in uncoated CdSe/ZnS quantum dots, and surface plasmon resonance and a red-shifted exciton peak in Ag-coated CdSe/ZnS composite quantum dot structures.[105] Redox transformations occurring on chemically modified surfaces of QDs may significantly alter the refractive index of the interface and thus induce changes in the plasmon angle of the SPR spectra.[106] This approach was the basis of an SPR sensor for acetylcholine esterase inhibitors based on the photoelectrochemical charging effect of Au nanoparticles in an AuNP/CdS-QD array (coating an Au/glass surface) which was followed by means of SPR changes upon continuous irradiation of the sample.[106] Similarly, the use of SPR spectroscopy combined with surface-modified QDs coupled to other enzymes could provide an alternative tool for new SPR (bio)sensor probes. On the other hand, the combination of surface plasmon enhanced fluorescence spectroscopy (SEF) with a fluorescent analyte tagged by CdSe/ZnS QDs has already been successfully demonstrated for detection of single-stranded DNA sequences by SPR and SEF.[107] The spectral resolution of the multicolor images obtained with a spectrograph demonstrates the potential of the combination of QD-DNA conjugates with SEF for future applications in DNA chip analytics.

More recently, investigation of the luminescence properties of QDs is slowly expanding into phosphorescence, a detection principle that may provide advantages for the design of luminescence analytical methodologies.[108,109] For example, doping sol–gel porous matrices with Tb_2S_3 QDs produces photoluminescent materials with an emission at 600 nm with typical characteristics of room temperature phosphorescence, although the origin of the luminescence and the emission mechanism is not understood yet.[110] A phosphorescence emission from Ir complex functionalized CdSe/ZnS quantum dots has been reported more recently.[111] For Ir–CdSe/ZnS in aerated methanol, the quantum yield of the 590 nm CdSe/ZnS phosphorescent emission was determined to be 0.4, which is sufficiently high for potential applications of such systems to detect specific target agents in imaging and perhaps as new room-temperature phosphorescence (RTP) analytical sensors.

Although very preliminary, the above studies seem to open the door to novel transduction schemes and applications of QDs for (bio)chemical analysis.

12.3.2 Bioconjugation to Highly Specific Receptors

As detailed previously, so far QDs have been bioconjugated to different antibodies in an assortment of fluoroimmunoassays aiming at selective detection of proteins and small molecules. Such studies have documented that QDs may be used as "generalized" reporters in immunoassays. More recently some researchers have started to investigate alternative analyte receptors to be conjugated to such reporter QDs.

In particular, molecularly imprinted polymers (MIPs) and aptamers have been invoked to develop novel analytical methodologies.

MIPs offer a cheap and versatile platform for molecule-specific recognition.[112,113] Literature available on the subject frequently underlines the biomimetic properties exhibited by these polymers, with substrate-selective mechanisms mimicking those found in natural entities such as antibodies and enzymes. In this line, incorporation of QDs into MIPs (Fig. 12.7) has been investigated for caffeine detection.[114] The determination of the target species was accomplished by measuring the quenching of the emission from the QDs included in the polymeric structure due to energy transfer processes between the QDs and the template molecules. It was reported that the presence of analogous molecular structures like theophylline and theobromine had no effect on the signals and a control polymer (i.e., without template) showed no change in the QDs photoluminescence. These important analytical results demonstrated that it is possible to couple semiconductor nanocrystals with the selective recognition capability of MIPs, opening new possibilities for the development of QD-based analytical assays.

On the other hand, very recently the possibility to resort to aptamers as a new type of recognition elements that can be used in combination with QDs for analytical applications has been demonstrated.[115] Aptamers, that is, single-stranded DNA or RNA molecules with a ligand-binding ability, exhibit many important advantages as recognition elements in biosensing when compared to traditional antibodies. They are small in size, chemically stable, and cost effective. More importantly, aptamers offer remarkable flexibility and convenience in the design of their structures, which has led to novel biosensors that have exhibited high sensitivity and selectivity. Since their discovery in the early 1990s, many aptamers have been evolved *in vitro* to bind any class of molecule, ranging from small organic molecules, peptides, and proteins to intact viral particles and even cells.

In a recent preliminary paper, QDs have been used to encode aptamer-linked nanostructures sensitive to adenosine and cocaine, respectively.[115] The nanostructures also contain gold nanoparticles that serve as quenchers. Addition of analytes

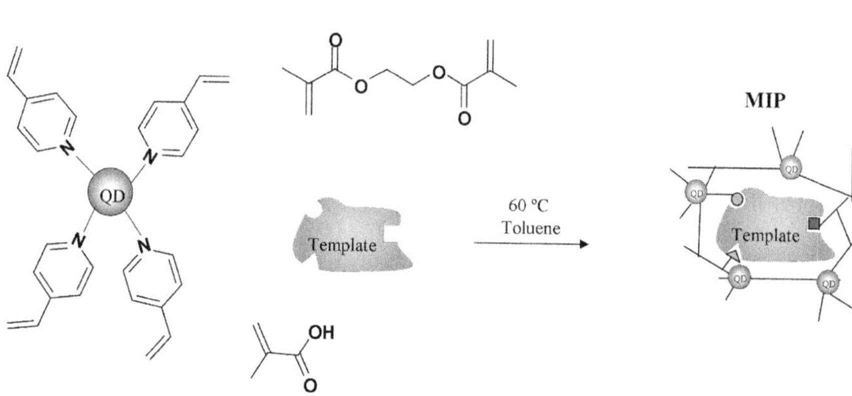

Figure 12.7 Scheme for the synthesis of CdSe/ZnS incorporated MIP. (Adapted from Reference 114.)

disassembles the nanostructures and results in increased emission from QDs allowing simultaneous detection of both analytes. Considering that aptamers could be ideally selected against essentially any desired target molecule, such multiplex detection systems could be useful for simultaneous determination of multiple analytes.

12.3.3 Immobilization of QDs in Solid Supports: Preparation of Sensing Materials

All the reported approaches have demonstrated the feasibility of using organically modified QDs as excellent analytical probes for the determination of a wide variety of biochemical species. Nevertheless, in most cases colloidal QDs dispersed in aqueous media have been utilized. An attractive further step to develop useful optical sensor approaches[116] consists of the immobilization of such conjugated QDs in appropriate solid supports to fabricate "active" solid phases.[109] In an ideal solid support the conjugated nanoparticles should be retained while maintaining their sensing and optical properties, and avoid leaching into the solution. This is a challenging task today and very often some compromises must be accepted.

In this context, sol–gel materials are very versatile and have been demonstrated to be suited for the development of luminescent optical sensors by trapping indicator molecules inside the inorganic structure during the polymerization process.[117] Several approaches have been already proposed for the synthesis of sol–gel materials doped with QDs.[118–120] Most of the reported synthetic routes prepare the QDs first and subsequently modify their surface adequately in a solution phase. This is followed by the sol–gel processing in order to obtain an inorganic material doped with the luminescent QDs. However, achieving immobilization while maintaining the QDs unique optical properties is not an easy task. In early attempts, the sensitivity of QDs to the surrounding environment often led to broad size distributions and poor stability. In order to overcome such limitations, several strategies have been investigated, including the use of alkyl amines as a bifunctional aid in QD-glass synthesis (amines act as gelation catalysts and stabilizers)[118] or the use of supercritical CO_2 drying, allowing for highly porous aerogel structures which are thus amenable to further functionalization and liquid phase interactions.[121] An efficient way to obtain concentrated and monodispersed quantum dots embedded in a solid host is to prepare, in a previous step, the organically modified QDs followed by their addition to the sol before the aging. This allows proper passivation and functionalization of the dots to be performed in advance and has proved to result in highly photostable doped glasses.[122]

Most of the reported applications of doping a sol–gel host with QDs are related to the use of photoluminescent QDs for optoelectronic applications, for example, solar concentrators or active media for tunable lasers.[123] To date, work on implementation of a chemical or biochemical sensor using sol–gel QD doped materials has been very limited. Nevertheless, taking into consideration the variety of chemical and biological sensing applications using sol–gel immobilized organic dyes, such implementation can be expected for QDs too. In fact, recently, simultaneous determination of oxygen and temperature based on the use of luminescent QDs in combination with a ruthenium dye, both immobilized in a sol–gel host, has been demonstrated.[124]

Due to the versatility of QDs and sol–gel processes, it can be envisaged that such concepts hold a great potential to develop new "active" phases for luminescence-based (bio)sensors.

Conversely, an important number of organic polymer encapsulated QD applications for chemical sensing has been reported so far. Some of the more representative approaches include PMMA CdSe films for temperature sensing,[125] QDs encapsulated in acrylic nanospheres for ion sensing,[126] or a variety of polymer beads for imaging and labeling applications,[127] among others. Alternative immobilization approaches include the deposition of QDs onto hydrophilic substrates using the Langmuir–Blodgett technique[128] or the use of layer-by-layer electrostatic self-assembly. The latter technique is particularly attractive as it allows a very fine control of film thickness and deposition on surfaces of complex shape. In addition, it is compatible with QD functionalization and can also be employed in combination with sensory dyes.

12.4 CONCLUDING REMARKS

The use of organically modified QDs has seen considerable progress during the last few years, as a new class of highly valuable luminophores, and novel analytical methodologies and a wide range of (bio)chemical applications have already been demonstrated. Conjugation of selective recognition elements (such as antibodies) to organically modified QDs is already a well-established strategy to develop novel immunoassays and labeling applications. QDs have been used also as FRET donors in the development of fluorescent probes and immunoassays for varied applications. Using different conjugation schemes of the nanoparticles (donors) with the acceptor molecules led to improved efficiency of FRET measurements, as well as to novel applications of this technique.

Many research efforts have been devoted to improve the synthesis of QDs (searching for nanoparticles with higher quality and stability) and to their chemical surface modifications. As a result, QD synthesis in organic solvents is well established nowadays and size, shape, and composition control is today possible. Of course, QD-based analytical methodologies will benefit from such continuous basic advances. In this context, although nowadays the use of QDs as "active" probes is limited to a few applications, a great increase in the number of luminescent methodologies based on QDs for chemical analysis is to be expected in the near future. In particular, the use of QD-acceptor pairs in FRET-based assays is being exploited in the development of multiplex configurations, specifically targeted to simultaneous multianalyte determinations. Other promising approaches being investigated include the use of QDs in combination with alternative detection principles (e.g., phosphorescence) or, more generally, novel bioconjugations with highly selective recognition elements (such as MIPs or aptamers).

However, there are still limitations of QDs that need to be overcome, such as the dependence of the luminescence on surface modification or chemical environment. In addition, the not fully understood blinking process of QDs is another limitation when single molecule spectroscopy needs to be approached.

Last, but not least, full application of QDs in chemical sensors would require the immobilization of the nanoparticles into appropriate solid supports in order to develop reliable "active" phases (able to provide, for instance, convenient fiber optic-based sensing applications). Although only a few reports have been published so far regarding the trapping of the QDs in solid matrices, some important steps have already started towards the realization of the potential of these technologies. There is still plenty of room for further development in all those directions.

To conclude, we may say that the unique photophysical properties of organically modified semiconductor quantum dots make these nanoparticles very promising in the field of luminescent analytical methodologies. In fact, it is already apparent that these inorganic luminophores can complement (if not replace at times) conventional organic fluorophores in a diversity of practical sensing applications. The future of QDs for chemical and biochemical analysis looks bright as they are already playing a key role in the development of a new generation of luminescence-based analytical methodologies, particularly for (bio)chemical sensors.

ACKNOWLEDGMENTS

Financial support from project CTQ2006-02309 (Ministerio de Educación y Ciencia, Spain) is gratefully acknowledged.

REFERENCES

1. A. I. Ekimov, A. A. Onushchenko, *Sov. Phys. Semicond.* **1982**, *16*, 775–778.
2. L. E. Brus, *J. Chem. Phys.* **1983**, *79*, 5566–5571.
3. L. E. Brus, *J. Chem. Phys.* **1984**, *80*, 4403–4409.
4. L. Spanhel, M. Haase, H. Weller, A. Henglein, *J. Am. Chem. Soc.* **1987**, *109*, 5649–5655.
5. C. B. Murray, D. J. Norris, M. G. Bawendi, *J. Am. Chem. Soc.* **1993**, *115*, 8706–8715.
6. S. Chaudhary, M. Ozkan, W. C. W. Chan, *Appl. Phys. Lett.* **2004**, *84*, 2925–2927.
7. V. Colvin, M. C. Schlamp, A. P. Alivisatos, *Nature* **1994**, *370*, 354–357.
8. M. Bruchez, M. Moronne, P. Gin, S. Weiss, A. P. Alivisatos, *Science* **1998**, *281*, 2013–2016.
9. W. C. W. Chan, S. M. Nie, *Science* **1998**, *281*, 2016–2018.
10. Z. A. Peng, X. G. Peng, *J. Am. Chem. Soc.* **2001**, *123*, 1389–1395.
11. X. Peng, M. C. Schlamp, A. V. Kadavanich, A. P. Alivisatos, *J. Am. Chem. Soc.* **1997**, *119*, 7019–7029.
12. C. J. Murphy, *Anal. Chem.* **2002**, *74*, 520A–526A.
13. A. M. Smith, S. Nie, *Analyst* **2004**, *129*, 672–677.
14. J. M. Costa-Fernández, R. Pereiro, A. Sanz-Medel, *Trends Anal. Chem.* **2006**, *25*, 207–218.
15. G. Orellana, *Anal. Bioanal. Chem.* **2004**, *379*, 344–346.
16. R. P. Haugland, Ed., *The Handbook: A Guide to Fluorescent Probes and Labeling Technologies*, Eugene, OR: Molecular Probes, **2005**.
17. J. R. Lakowicz, Ed., *Principles of Fluorescence Spectroscopy*, New York: Kluwer Academic, **1999**.
18. R. E. Bailey, A. M. Smith, S. Nie, *Physica E* **2004**, *25*, 1–12.
19. U. Resch-Genger, M. Grabolle, S. Cavaliere-Jaricot, R. Nitschke, T. Nann, *Nature Methods* **2008**, *5*, 763–775.
20. M. G. Lagally, *J. Chem. Educ.* **1998**, *75*, 277–279.
21. M. G. Bawendi, M. L. Steigerwald, L. E. Brus, *Annu. Rev. Phys. Chem.* **1990**, *41*, 477–496.
22. A. P. Alivisatos, *Science* **1996**, *271*, 933–937.

23. W. J. Parak, T. Pellegrino, C. Plank, *Nanotechnology* **2005**, *16*, R9–R25.
24. L. E. Brus, *J. Phys. Chem.* **1986**, *90*, 2555–2560.
25. B. M. Hutchins, T. T. Morgan, M. G. Ucak-Astarlioglu, M. E. Williams, *J. Chem. Educ.* **2007**, *84*, 1301–1303.
26. J. E. Bowen-Katari, V. L. Colvin, A. P. Alivisatos, *J. Phys. Chem.* **1994**, *98*, 4109–4117.
27. B. O. Dabbousi, J. Rodriguez-Viejo, F. V. Mikulec, J. R. Heine, H. Mattoussi, R. Ober, K. F. Jensen, M. G. Bawendi, *J. Phys. Chem. B* **1997**, *101*, 9463–9475.
28. M. Tomasulo, I. Yildiz, S. L. Kaanumalle, F. M. Raymo, *Langmuir* **2006**, *24*, 10284–10290.
29. L. Qu, X. Peng, *J. Am. Chem. Soc.* **2002**, *124*, 2049–2055.
30. F. M. Raymo, I. Yildiz, *Phys. Chem. Chem. Phys.* **2007**, *9*, 2036–2043.
31. C.-A. J. Lin, T. Liedl, R. A. Sperling, M. T. Fernández-Argüelles, J. M. Costa-Fernández, R. Pereiro, A. Sanz-Medel, W. H. Chang, W. J. Parak, *J. Mater. Chem.* **2007**, *17*, 1343–1346.
32. P. Alivisatos, *Nature Biotechnol.* **2004**, *22*, 47–52.
33. D. E. Moore, K. Patel, *Langmuir* **2001**, *17*, 2541–2544.
34. Y. Chen, Z. Rosenzweig, *Anal. Chem.* **2002**, *74*, 5132–5138.
35. A. V. Isarov, J. Chrysochoos, *Langmuir* **1997**, *13*, 3142–3149.
36. K. M. Gatas-Asfura, R. M. Leblanc, *Chem. Commun.* **2003**, 2684–2685.
37. M. T. Fernández-Argüelles, W. J. Jin, J. M. Costa-Fernández, R. Pereiro, A. Sanz-Medel, *Anal. Chim. Acta* **2005**, *549*, 20–25.
38. H. Y. Xie, J. G. Liang, Z. L. Zhang, Y. Liu, Z. K. He, D. W. Pan, *Spectrochim. Acta A* **2004**, *60*, 2527–2530.
39. Y. Lai, Y. Yu, P. Zhong, J. Wu, Z. Long, C. Liang, *Anal. Lett.* **2006**, *39*, 1201–1209.
40. C. Bo, Z. Ping, *Anal. Bioanal. Chem.* **2005**, *381*, 986–992.
41. J. G. Liang, X. P. Ai, Z. K. He, D. W. Pang, *Analyst* **2004**, *129*, 619–622.
42. H. Li, Y. Zhang, X. Wang, D. Xiong, Y. Bai, *Mater. Lett.* **2007**, *61*, 1474–1477.
43. Y.-S. Xia, C.-Q. Zhu, *Talanta* **2008**, *75*, 215–221.
44. W. J. Jin, M. T. Fernández-Argüelles, J. M. Costa-Fernández, R. Pereiro, A. Sanz-Medel, *Chem. Commun.* **2005**, 883–885.
45. J. R. Lakowicz, I. Gryczynski, Z. Gryczynski, C. J. Murphy, *J. Phys. Chem. B* **1999**, *103*, 7613–7620.
46. K. Sooklal, B. S. Cullum, S. M. Angel, C. J. Murphy, *J. Phys. Chem.* **1996**, *100*, 4551–4555.
47. J.-L. Chen, C.-Q. Zhu, *Anal. Chim. Acta* **2005**, *546*, 147–153.
48. J. Li, D. Bao, X. Hong, D. Li, J. Li, Y. Bai, T. Li, *Colloid Surf. A* **2005**, *257–258*, 267–271.
49. G. H. Shi, Z. B. Shang, Y. Wang, W. J. Jin, T. C. Zhang, *Spectrochim. Acta A* **2008**, *70*, 247–252.
50. A. Y. Nazzal, L. Qu, X. Peng, M. Xiao, *Nano Lett.* **2003**, *3*, 819–822.
51. R. A. Potyrailo, A. M. Leach, *Appl. Phys. Lett.* **2006**, *88*, 134110.
52. J. Liang, S. Huang, D. Zeng, Z. He, X. Ji, X. Ai, H. Yang, *Talanta* **2006**, *69*, 126–130.
53. C. P. Han, H. B. Li, *Chin. Chem. Lett.* **2008**, *19*, 215–218.
54. F. Pinaud, X. Michalet, L. A. Bentolila, J. M. Tsay, S. Doose1, J. J. Li, G. Iyer, S. Weiss, *Biomaterials* **2006**, *27*, 1679–1687.
55. B. Sun, W. Xie, G. Yi, D. Chen, Y. Zhou, J. Cheng, *J. Immunol. Methods* **2001**, *249*, 85–89.
56. E. R. Goldman, I. L. Medintz, H. Mattoussi, *Anal. Bioanal. Chem.* **2006**, *384*, 560–563.
57. E. R. Goldman, G. P. Anderson, P. T. Tran, H. Mattoussi, P. T. Charles, J. M. Mauro, *Anal. Chem.* **2002**, *74*, 841–847.
58. E. R. Goldman, A. R. Clapp, G. P. Anderson, H. T. Uyeda, J. M. Mauro, I. L. Mendizt, H. Mattoussi, *Anal. Chem.* **2004**, *76*, 684–688.
59. E. R. Goldman, E. D. Balighian, H. Mattoussi, M. K. Kuno, J. M. Mauro, P. T. Tran, G. P. Anderson, *J. Am. Chem. Soc.* **2002**, *124*, 6378–6382.
60. B. M. Lingerfelt, H. Mattoussi, E. R. Goldman, J. M. Mauro, G. P. Anderson, *Anal. Chem.* **2003**, *75*, 4043–4049.
61. E. R. Goldman, I. L. Medintz, A. Hayhurst, G. P. Anderson, J. M. Mauro, B. L. Iverson, G. Georgiou, H. Mattoussi, *Anal. Chim. Acta* **2005**, *534*, 63–67.
62. X. Y. Wu, H. J. Liu, J. Q. Liu, K. N. Haley, J. A. Treadway, J. P. Larson, N. F. Ge, F. Peale, M. P. Bruchez, *Nature Biotechnol.* **2003**, *21*, 41–46.

63. B. Dubertret, P. Skourides, D. J. Norris, V. Noireaux, A. H. Brivanlou, A. Libchaber, *Science* **2002**, *298*, 1759–1762.
64. T. Pellegrino, L. Manna, S. Kudera, T. Liedl, D. Koktysh, A. L. Rogach, S. Keller, J. Rädler, G. Natile, W. J. Parak, *Nano Lett.* **2004**, *4*, 703–707.
65. M. T. Fernández-Argüelles, J. M. Costa-Fernández, R. Pereiro, A. Sanz-Medel, *Analyst* **2008**, *133*, 444–447.
66. E. R. Goldman, T. J. O'Shaughnessy, C. M. Soto, C. H. Patterson, Jr., C. R. Taitt, M. S. Spector, P. T. Charles, *Anal. Bioanal. Chem.* **2004**, *380*, 880–886.
67. J. Yang, S. R. Dave, X. Gao, *J. Am. Chem. Soc.* **2008**, *130*, 5286–5292.
68. R. Wilson, D. G. Spiller, I. A. Prior, R. Bhatt, A. Hutchinson, *J. Mater. Chem.* **2007**, *17*, 4400–4406.
69. M. I. Santori, C. Gonzalez, L. Serrano, M. Isalan, *Nature Protoc.* **2006**, *1*, 526–531.
70. A. Agrawal, T. Sathe, S. Nie, *J. Agric. Food Chem.* **2007**, *55*, 3778–3782.
71. L. J. Lucas, J. N. Chesler, J. Y. Yoon, *Biosens. Bioelectron.* **2007**, *23*, 675–681.
72. Q. Ma, X. Wang, Y. Li, X. Su, Q. Jin, *Luminescence* **2007**, *22*, 438–445.
73. X.-X. Zhu, Y.-C. Cao, X. Jin, J. Yang, X.-F. Hua, H.-Q. Wang, B. Liu, Z. Wang, J.-H. Wang, L. Yang, Y.-D. Zhao, *Nanotechnology* **2008**, *19*, 025708.
74. J. M. Klostranec, Q. Xiang, G. A. Farcas, J. A. Lee, A. Rhee, E. I. Lafferty, S. D. Perrault, K. C. Kain, W. C. W. Chan, *Nano Lett.* **2007**, *7*, 2812–2818.
75. J. Shen, F. Xu, H. Jiang, Z. Wang, J. Tong, P. Guo, S. Ding, *Anal. Bioanal. Chem.* **2007**, *389*, 2243–2250.
76. M. Nichkova, D. Dosev, A. E. Davies, S. J. Gee, I. M. Kennedy, B. D. Hammock, *Anal. Lett.* **2007**, *40*, 1423–1433.
77. Z. Deng, Y. Zhang, J. Yue, F. Tang, Q. Wei, *J. Phys. Chem. B* **2007**, *111*, 12024–12031.
78. K. Kerman, T. Endo, M. Tsukamoto, M. Chikae, Y. Takamura, E. Tamiya, *Talanta* **2007**, *71*, 1494–1499.
79. M. V. Yezhelyev, A. Al-Hajj, C. Morris, A. I. Marcus, A. T. Liu, M. Lewis, C. Cohen, P. Zrazhevskiy, J. W. Simons, A. Rogatko, S. Nie, X. Gao, R. M. O'Regan, *Adv. Mater.* **2007**, *19*, 3146–3151.
80. C. R. Kagan, C. B. Murray, M. Nirmal, M. G. Bawendi, *Phys. Rev. Lett.* **1996**, *76*, 1517–1520.
81. C. R. Kagan, C. B. Murray, M. G. Bawendi, *Phys. Rev. B* **1996**, *54*, 8633–8643.
82. K. M. Sanchez, D. E. Schlamadinger, J. E. Gable, J. E. Kim, *J. Chem. Educ.* **2008**, *85*, 1253–1256.
83. D. M. Willard, L. L. Carillo, J. Jung, A. Van Orden, *Nano Lett.* **2001**, *1*, 469–474.
84. N. N. Mamedova, N. A. Kotov, A. L. Rogach, J. Studer, *Nano Lett.* **2001**, *1*, 281–286.
85. I. L. Medintz, A. R. Clapp, H. Mattoussi, E. R. Goldman, B. Fisher, J. M. Mauro, *Nature Mater.* **2003**, *2*, 630–638.
86. T. C. Lim, V. J. Bailey, Y.-P. Ho, T.-H. Wang, *Nanotechnology* **2008**, *19*, 075701.
87. J. H. Kim, S. Chaudhary, M. Ozkan, *Nanotechnology* **2007**, *18*, 195105.
88. C.-H. Zhang, L. W. Johnson, *Anal. Chem.* **2007**, *79*, 7775–7781.
89. A. R. Clapp, T. Pons, I. L. Medintz, J. B. Delehanty, J. S. Melinger, T. Tiefenbrunn, P. E. Dawson, B. R. Fisher, B. O'Rourke, H. Mattoussi, *Adv. Mater.* **2007**, *19*, 1921–1926.
90. H. Mattoussi, J. M. Mauro, E. R. Goldman, G. P. Anderson, V. C. Sundar, F V. Mikulec, M. G. Bawendi, *J. Am. Chem. Soc.* **2000**, *122*, 12142–12150.
91. E. R. Goldman, I. L. Medintz, J. L. Whitley, A. Hayhurst, A. R. Clapp, H. T. Uyeda, J. R. Deschamps, M. E. Lassman, H. Mattoussi, *J. Am. Chem. Soc.* **2005**, *127*, 6744–6751.
92. E. Chang, J. S. Miller, J. Sun, W. W. Yu, V. L. Colvin, R. Drezek, J. L. West, *Biochem. Biophys. Res. Comm.* **2005**, *334*, 1317–1321.
93. D. Zhou, J. D. Piper, C. Abell, D. Klenerman, D. Kang, L. Ying, *Chem. Commun.* **2005**, *38*, 4807–4809.
94. S. Hohng, T. Ha, *ChemPhysChem* **2005**, *6*, 956–960.
95. J. A. Kloepfer, N. Cohen, J. L. Nadeau, *J. Phys. Chem. B* **2004**, *108*, 17042–17049.
96. R. C. Stringer, D. Hoehn, S. A. Grant, *IEEE Sens. J.* **2008**, *8*, 295–300.
97. C.-W. Liu, H.-T. Chang, *Open Anal. Chem. J.* **2007**, *1*, 1–6.

98. M. Suzuki, Y. Husimi, H. Komatsu, K. Suzuki, K. T. Douglas, *J. Am. Chem. Soc.* **2008**, *130*, 5720–5725.

99. F. Patolsky, R. Gill, Y. Weizmann, T. Mokari, U. Banin, I. Willner, *J. Am. Chem. Soc.* **2003**, *125*, 13918–13919.

100. I. L. Medintz, A. R. Clapp, J. S. Melinger, J. R. Deschamps, H. Mattoussi, *Adv. Mater.* **2005**, *17*, 2450–2455.

101. M. T. Fernández-Argüelles, A. Yakovlev, R. A. Sperling, C. Luccardini, S. Gaillard, A. Sanz-Medel, J. M. Mallet, J. C. Brochon, A. Feltz, M. Oheim, W. Parak, *Nano Lett.* **2007**, *7*, 2613–2617.

102. P. O. Anikeeva, C. F. Madigan, S. A. Coe-Sullivan, J. S. Steckel, M. G. Bawendi, V. Bulović, *Chem. Phys. Lett.* **2006**, *424*, 120–125.

103. V. Bagalkot, L. Zhang, E. Levy-Nissenbaum, S. Jon, P. W. Kantoff, R. Langer, O. C. Farokhzad, *Nano Lett.* **2007**, *7*, 3065–3070.

104. A. R. Clapp, I. L. Medintz, H. Mattoussi, *ChemPhysChem* **2006**, *7*, 47–57.

105. J. H. Song, T. Atay, S. F. Shi, H. Urabe, A. V. Nurmikko, *Nano Lett.* **2005**, *5*, 1557–1561.

106. M. Zayats, A. B. Kharitonov, S. P. Pogorelova, O. Lioubashevski, E. Katz, I. Willner, *J. Am. Chem. Soc.* **2003**, *125*, 16006–16014.

107. R. Robelek, L. Niu, E. L. Schmid, W. Knoll, *Anal. Chem.* **2004**, *76*, 6190–6165.

108. A. Sanz-Medel, *Anal. Chim. Acta* **1993**, *283*, 367–378.

109. J. Kuijt, F. Ariese, U. A. Th. Brinkman, C. Gooijer, *Anal. Chim. Acta* **2003**, *488*, 135–171.

110. P. Yang, M. K. Lu, C. F. Song, G. J. Zhou, D. Xu, D. R. Yan, *Inorg. Chem. Commun.* **2002**, *5*, 187–191.

111. J.-M. Hsieh, M.-L. Ho, P.-W. Wu, P.-T. Chou, T.-T. Tsaib, Y. Chi, *Chem. Commun.* **2006**, 615–617.

112. B. Sellergren, Ed., *Molecularly Imprinted Polymers: Man-Made Mimics of Antibodies and Their Application in Analytical Chemistry*, Amsterdam: Elsevier, **2001**.

113. J. O. Mahony, K. Nolan, M. R. Smyth, B. Mizaikoff, *Anal. Chim. Acta* **2005**, *534*, 31–39.

114. C. I. Lin, A. K. Joseph, C. K. Chang, Y. D. Lee, *J. Chromatogr. A* **2004**, *1027*, 259–262.

115. J. Liu, J. H. Lee, Y. Lu, *Anal. Chem.* **2007**, *79*, 4120–4125.

116. J. Ruzicka, E. M. Hansen, *Anal. Chim. Acta* **1985**, *173*, 3–21.

117. J. M. Costa Fernández, A. Sanz Medel, *Anal. Chim. Acta* **2000**, *407*, 61–69.

118. S. T. Selvan, C. Bullen, M. Ashokkumar, P. Mulvaney, *Adv. Mater.* **2001**, *13*, 985–989.

119. C. Bullen, P. Mulvaney, C. Sada, M. Ferrari, A. Chiasera, A. Martucci, *J. Mater. Chem.* **2004**, *14*, 1112–1116.

120. H. Jiang, X. Y. J. Che, M. Wang, F. Kong, *Ceram. Int.* **2004**, *30*, 1685–1689.

121. I. U. Arachchige, S. L. Brock, *Acc. Chem. Res.* **2007**, *40*, 801–809.

122. R. Reisfeld, T. Saraidarov, *Opt. Mater.* **2006**, *28*, 64–70.

123. R. Reisfeld, *Opt. Mater.* **2001**, *16*, 1–7.

124. P. A. S. Jorge, C. Maule, A. J. Silva, R. Benrashid, J. L. Santos, F. Farahi, *Anal Chim. Acta* **2008**, *606*, 223–229.

125. V. C. Sundar, C. M. Rudzinski, A. W. Wun, M. G. Bawendi, D. G. Nocera, *Appl. Phys. Lett.* **2003**, *83*, 3555–3557.

126. M. J. Ruedas-Rama, X. Wang, E. A. H. Hall, *Chem. Commun.* **2007**, 1544–1546.

127. H.-Q. Wang, J.-H. Wang, Y.-Q. Li, X.-Q. Li, T.-C. Liu, Z.-L. Huang, Y.-D. Zhao, *J. Colloid Interface Sci.* **2007**, *316*, 622–627.

128. J. Xu, X. Ji, K. M. Gattas-Asfura, C. Wang, R. M. Leblanc, *Colloid Surf. A* **2006**, *284–285*, 35–42.

129. A. R. Clapp, I. L. Medintz, B. R. Fisher, G. P. Anderson, H. Mattoussi, *J. Am. Chem. Soc.* **2005**, *127*, 1242–1250.

Part Three

Control of Supramolecular Nanofabrication, Motion, and Morphology

Chapter 13

Chemically Directed Self-Assembly of Nanoparticle Structures on Surfaces

XING YI LING, DAVID N. REINHOUDT, AND
JURRIAAN HUSKENS

13.1 INTRODUCTION

The self-assembly of ordered nanostructures consisting of nanoparticles with sizes between 1 and 1000 nm has attracted a lot of attention owing to their unique optical, electronic, magnetic, catalytic, and other physical properties.[1] The ability to attach nanoparticles onto planar surfaces in a well-defined, controllable, and reliable

The Supramolecular Chemistry of Organic–Inorganic Hybrid Materials. Edited by Knut Rurack and Ramón Martínez-Máñez

manner is an important prerequisite for the fabrication of micro- or nanostructured devices suitable for the application in the field of nano(bio)technology.[2]

In general, there are two approaches to assemble nanostructured materials, namely, physical assembly and chemical assembly. Physical assembly techniques are based on the assembly of nonfunctionalized nanoparticles on surfaces by physical forces, which include convective or capillary assembly,[3,4] spin coating,[5] and sedimentation.[6] The physical assembly of nanoparticles generally results in relatively simple, closely packed two- or three-dimensional particle arrays. In addition, the physically assembled nanoparticle structures lack long term stability because they were deposited at relatively low surface pressures.[7]

Chemical assembly utilizes coupling chemistries to direct and control the deposition of nanoparticles with surface functionalities onto a functionalized substrate. The control over the surface functionalities of nanoparticles allows tailoring of the nanostructures in a predictable manner and thus the formation of functional, more complex nanostructured architectures on surfaces to meet the needs for specific applications such as molecular electronics and biosensing.[8] Various chemical interaction types, such as covalent bonding,[9] electrostatic forces,[10] and host–guest interactions,[11,12] have been employed to chemically govern the self-assembly of nanoparticles onto surfaces. Cross-linking of the neighboring particles with chemical forces by selective binding further enhances the stability of nanoparticle assemblies.[13] These methods are anticipated to directly control the spatial distribution of nanoparticles across a large area in more complex patterns when combined with nanopatterning schemes.

In this review, we describe the recent developments of chemically directed self-assembly of nanoparticle structures on surfaces. The first part focuses on the chemical interactions used to direct the assembly of nanoparticles on surfaces. The second part highlights a few major top-down patterning techniques employed in combination with chemical nanoparticle assembly in manufacturing two- or three-dimensional nanoparticle structures. The combination of top-down and bottom-up techniques is essential in the fabrication of nanoparticle structures of various kinds to accommodate the need for device applications.

13.2 CHEMICAL NANOPARTICLE ASSEMBLY ON NONPATTERNED SURFACES

Recent advances in nanotechnology have led to well-defined nanoparticles and self-assembled monolayers (SAMs) with desired surface functionalities. The chemically directed assembly of functionalized nanoparticles onto SAMs utilizes the specific binding affinity between organic head groups of the SAMs with the functional groups on the surface of the nanoparticles. In this section, the use of covalent bonding[14] and noncovalent interactions, including electrostatic interactions, metal coordination,[15] hydrogen bonding,[13] host–guest interactions,[12] and biomolecular interactions,[16] for the construction of functional nanoparticle architectures is discussed.

13.2.1 Covalent Bonding

The coupling chemistries that have been widely used in organic chemistry for producing chemical bonds have been applied to form irreversible and stable nanoparticle arrays on surfaces. Eychmüller et al. used the carbodiimide coupling chemistry to covalently assemble carboxylate-functionalized CdTe nanocrystals (NCs) onto amino-terminated glass surfaces, which resulted in densely covered nanoparticle films.[14] The same principle was also applied to coat SiO_2 microparticles by CdTe NCs.

The assembly of monolayers of alkylbromide-functionalized Co nanoparticles onto amino-terminated silicon surfaces through direct nucleophilic substitution was reported by Kim et al.[17] The nanoparticle density on the surface can be controlled by changing the immersion time of the silicon surface in the nanoparticle solution. Directed assembly of nanoparticles was observed on a chemically patterned surface.

Yang et al. utilized the diazo-nucleophile covalent bonding to assemble poly(methacrylic acid) (PMAA)-capped Fe_3O_4 nanoparticles onto 2-nitro-*N*-methyl-4-diazonium-formaldehyde resin (NDR) in a layer-by-layer (LbL) manner to form a multilayered photosensitive precursor film.[18] The assembly was initiated by the electrostatic attraction between the negatively charged carboxylate groups of the PMAA surface and the cationic diazoresin of NDR. The multilayer films formed via electrostatic interactions are less stable in polar solvents or aqueous electrolyte solutions, which limits their application range. To circumvent the limited stability of the electrostatic LbL multilayer films, a photo-cross-linking reaction between the diazoresin and carboxylate groups was performed by UV irradiation to convert the ionic bonds to covalent bonds. The stability of the LbL films was evaluated by solvent etching in a mixture of polar solvents, which indicated stable films with no obvious desorption of nanoparticles.

Sun et al. extended the diazo-carboxylate covalent bonding to the assembly between diazo-resins and gold nanoparticles.[19] The ionic LbL assembly was first achieved by using diazo resins (DAR) and citrate-capped gold nanoparticles. Under UV irradiation, diazonium groups decomposed to phenyl cations that reacted with the nucleophilic carboxylate groups on the gold nanoparticles via an S_N1 reaction.

p-Aminothiophenol-capped cadmium sulfide (CdS) nanoparticles were electrochemically cross-linked onto *p*-aminothiophenol-functionalized Au electrode surfaces.[20] The covalent LbL assembly of nanoparticles to the electrode was monitored by electrochemistry and atomic force microscopy (AFM) imaging, which revealed a random and densely packed CdS nanoparticle array, with surface coverage of 65% of the theoretical coverage of a dense monolayer of nanoparticles. The dianiline-bridged CdS nanoparticles assembled on the Au electrode revealed highly efficient photoelectrochemical properties in the presence of triethanolamine as a sacrificial electron donor.

Paraschiv et al. demonstrated a functionalized surface with locally isolated functional groups (Fig. 13.1).[9] Au nanoparticles ($d \sim 3$ nm) stabilized with propanethiol and monofunctionalized with (mercaptopropyl)methyldimethoxysilane (MPMD) were first prepared by a place-exchange reaction. The Au nanoparticles were covalently assembled onto the surface via a single methyldimethoxysilane unit by reaction with Si-OH surface groups. AFM images revealed the attachment of Au nanoparticles

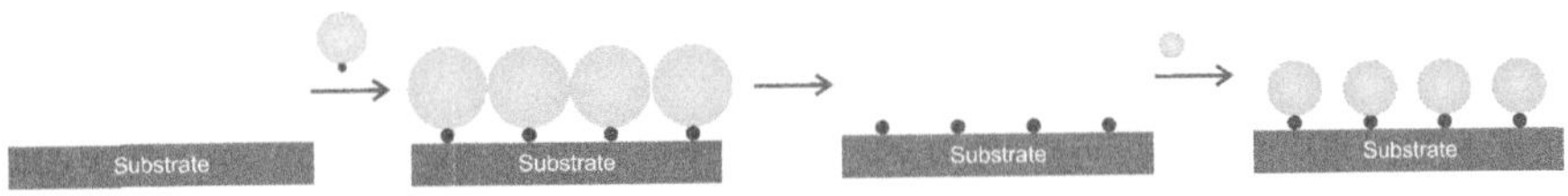

Figure 13.1 The preparation of surfaces with spaced single functional groups and (re)attachment of entities to these functional groups via specific interactions.

to the surface in a dense but disordered fashion. The Au nanoparticles were then removed to set free the single functional units, spaced by a minimum distance governed by the nanoparticle size. This allowed reattachment of the other entities, but the subsequent attachment of nanoparticles via thiol place-exchange was found to be much slower than the initial covalent assembly reaction.

13.2.2 Noncovalent Interactions

Despite the high stability of the covalently bound nanoparticle films, the order is often lacking. This is attributed to the rapid and strong chemical reaction that occurs between the functional groups of the nanoparticles and SAMs, leading to irreversible anchoring of the nanoparticles onto the surface. Hence, fine-tuning of the coupling chemistry is needed to obtain a balance between the ordering and the stability of the nanoparticle array.

The use of noncovalent interactions has been exploited for the synthesis of, for example, receptor-functionalized nanoparticles and/or SAMs with molecular recognition abilities at the interface. The advantage of noncovalent interactions over covalent bonding is that the former offer the possibility for error correction.[21]

13.2.2.1 Electrostatic Interactions

Electrostatic assembly, which involves attractive forces between two oppositely charged entities (polymers, nanoparticles, and substrates), has been proposed in the pioneering work of Iler for the assembly of two- and three-dimensional structures.[22] The LbL assembly of charged polyelectrolytes was later reported by Decher et al. for the fabrication of multilayer films of polyelectrolytes.[10,23] Their technique is based on the consecutive adsorption of polyanions and polycations from dilute aqueous solutions onto a charged substrate (Fig. 13.2).

Owing to its simplicity, the LbL assembly of multilayer films by electrostatic interactions has been widely extended to the assembly of chemically functionalized particles onto surfaces. The LbL assembly of multilayer nanoparticle films of metallic,[24,25] semiconductor,[26,27] inorganic,[28] and polymeric nanoparticles[28–30] has been demonstrated by using polyelectrolytes as a sandwich layer between the nanoparticles.[31,32] Sandwich structures of negatively charged CdS and TiO_2 nanoparticles and positively charged polyelectrolytes have been fabricated (Fig. 13.3).[26] The fluorescence emission intensity was found to increase linearly with increasing number of polyelectrolyte-CdS bilayers. Composite films of polyelectrolyte-TiO_2–polyelectrolyte-CdS behave like an n-type semiconductor under irradiation.[26]

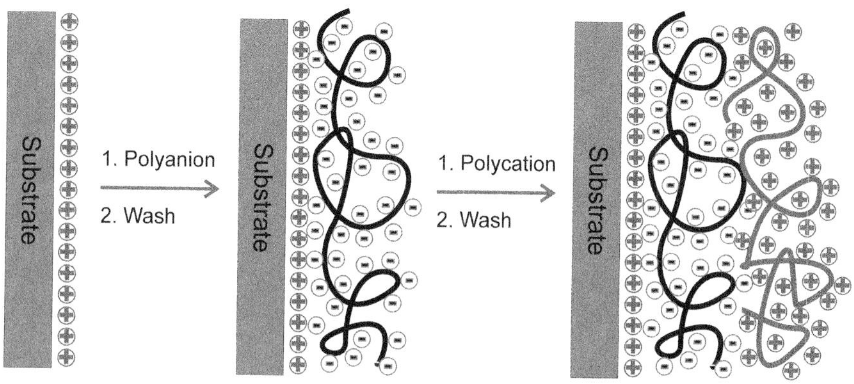

Figure 13.2 Layer-by-layer (LbL) assembly of oppositely charged polyelectrolyte films via electrostatic interactions.[10]

Similar methodologies were adopted by others. For instance, Akashi et al.[33] and Kunitake et al.[34] studied, independently, the adsorption of multilayers of nanoparticles and polyelectrolytes onto surfaces and its kinetics. Their studies revealed that the LbL films of nanoparticles and polyelectrolytes were molecularly flat after each bilayer assembly. However, both reports highlight the requirement of achieving an adsorption and desorption balance of the polyelectrolytes at the solid–liquid interface. The growth and properties of the multilayer films are sensitive to chemical factors such as polyelectrolyte and nanoparticle concentrations, ionic strength, pH, and hydrogen bonding, more pronounced than observed in the LbL assembly of polyelectrolytes. In particular, an increase in ionic strength enhanced not only the extent of adsorption but also the nanoparticles polyelectrolyte mass ratio per adsorption cycle, as a result of excess and free cationic sites on the polyelectrolytes. Murray et al. noticed the formation of multiple layers of poly(styrene sulfonate) and arylamine-functionalized nanoparticles incorporated during a single adsorption cycle in an LbL process, implying looping/entanglement of charged polymer chains with charged nanoparticles.[35] Quantized double layer charging in an LbL assembled film was observed, indicating

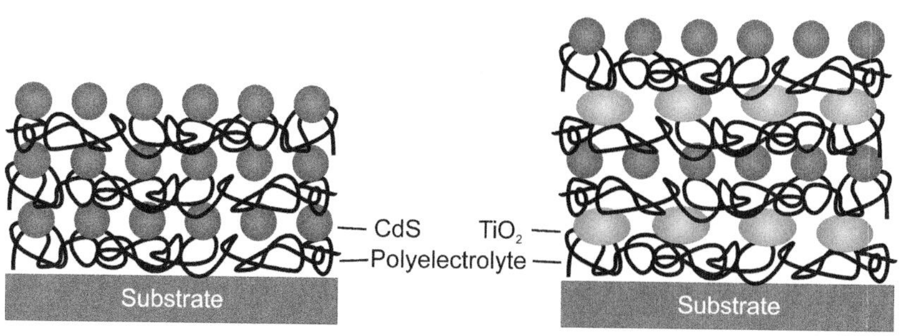

Figure 13.3 The LbL assembly of negatively charged CdS and TiO_2 nanoparticles, and positively charged polyelectrolytes to form nanoparticle-composite films .[26]

a substantial microscopic mobility of both the polyelectrolytes and the nanoparticles within the film. These observations indicated the structural differences between LbL films of pure polyelectrolytes and of polyelectrolytes and nanoparticles. The differences in size, morphology, and effective charge density between nanoparticles and polyelectrolytes significantly affect their adsorption behavior.

The direct LbL assembly of oppositely charged nanoparticles, which did not involve polyelectrolytes, was also examined. Sastry et al. demonstrated the formation of alternating layers of gold and silver nanoparticles via sequential electrostatic assembly.[36] In the absence of polyelectrolytes, the effective charging of gold and silver nanoparticles was accomplished by the adsorption of 4-aminothiophenol and 4-carboxythiophenol molecules on the nanoparticle surfaces, respectively. The multilayer films were stable up to 100°C.

Willner and coworkers demonstrated three-dimensional networks of Au, Ag, and mixed composites of Au and Ag nanoparticles assembled on a conductive (indium-doped tin oxide) glass support by stepwise LbL assembly with *N,N'*-bis(2-aminoethyl)-4,4'-bipyridinium as a redox-active cross-linker.[8,37] The electrostatic attraction between the amino-bifunctional cross-linker and the citrate-protected metal particles led to the assembly of a multilayered composite nanoparticle network. The surface coverage of the metal nanoparticles and bipyridinium units associated with the Au nanoparticle assembly increased almost linearly upon the formation of the three-dimensional (3D) network. A coulometric analysis indicated an electroactive 3D nanoparticle array, implying that electron transport through the nanoparticles is feasible. A similar multilayered nanoparticle network was later used in a study on a sensor application by using bis-bipyridinium cyclophane as a cross-linker for Au nanoparticles and as a molecular receptor for π-donor substrates.[8]

13.2.2.2 Metal–Ligand Coordination

Metal coordination chemistry enables ligand-bearing components to be assembled into supramolecular structures using appropriate metals or metal ions. The strong coordinative bonding between sulfur groups and transition metal surfaces, for example, gold[38] and silver,[39] has been exploited in the assembly of thiol-functionalized nanoparticles onto surfaces. Rizza et al. studied the thiol-based self-assembly of TiO_2 NCs on a substrate by two related methods.[40] In the first method, bare TiO_2 NCs were self-assembled on a thiol-functionalized gold substrate, forming a NC SAM. In the second method, thiol-functionalized TiO_2 NCs were assembled on annealed gold surfaces.

Kiely et al.[41] and Brust et al.[42] adapted an LbL approach to form a 3D multilayered nanoparticle assembly. A substrate with free thiols was subjected to a gold nanoparticle solution. After washing, the substrate was rinsed with a dithiol solution, providing new surface thiol groups for attachment of additional layers of particles, forming multilayered Au nanoparticle films (Fig. 13.4). Their study indicated a linear relationship between the number of adsorption layers and the apparent thickness of the nanoparticle assemblies.[42] Optical investigation of the Au nanoparticle films revealed that the individual nanoparticles did not fuse into larger units because of the protection by the dithiol ligand shells. Study of the optical and electronic properties of LbL films of 6 nm Au nanoparticles and dithiols revealed nonmetallic properties.[42]

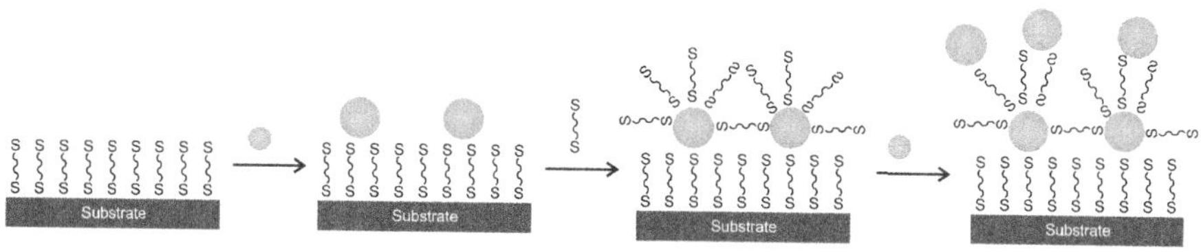

Figure 13.4 The construction of multilayered nanomaterials from LbL assembly of Au nanoparticles and dithiol ligands.[41]

The temperature dependence of the conductivity behavior of the nanoparticle films predicted that electronic conduction occurred via an electron hopping mechanism.

Further characterization of 3D Au and Ag nanoparticle multilayered films, with a short dithiol as a cross-linker, was studied by Natan et al.[43] They confirmed the porous, discontinuous morphology of the LbL films. Changes in electrical and optical film properties were reported when bifunctional dithiols of different lengths were used. Multilayer films assembled from more than six LbL cycles exhibited high conductivity and resembled bulk Au. In contrast, films of similar particle coverage generated using a longer cross-linker (1,6-hexanedithiol) exhibited a higher transmission in the near-infrared region and showed a reduced conductivity. The presence of conductive and insulating regions consistent with the metal–insulator transition was observed. On conducting substrates, Au or Ag monolayers were electrochemically addressable and behaved like a collection of closely spaced microelectrodes.[40] Dithiols have been further reported as cross-linkers for the multilayer fabrication of CdS nanoparticles[27] and alternating layers of Pt and CdS nanoparticles[44] on gold substrates.

Bharathi et al. reported the construction of an electrochemical interface with a tunable kinetic barrier by a mercaptopropyltrimethoxysilane-modified silica network. The thiol served as a matrix for the encapsulation of gold nanoparticles.[45] The silica network without Au nanoparticles exhibited a kinetic barrier for the electron transfer between the electrode surface and the electroactive species (ferrocyanide) in solution. However, with increasing immersion time of the silica network in an Au nanoparticle solution, the voltammetric response of ferrocyanide was gradually restored. This illustrated that the gold nanoparticles provided a conductive pathway to facilitate normal electron transfer. Wang and Wang utilized a similar gold nanoparticle-modified electrode to inhibit the adsorption of cytochrome c onto a bare electrode and to act as a bridge for electron transfer between protein and electrode.[46]

Rubinstein et al. constructed monolayers and multilayers (via LbL assembly) of bishydroxamate-functionalized Au nanoparticles onto bishydroxamate disulfide SAMs using Zr^{4+} as binding ions (Fig. 13.5).[15] The thickness of the densely packed Au nanoparticle layers increased regularly with the number of nanoparticle layers assembled, indicating an LbL growth of a monolayer of nanoparticles at a time, as a result of the highly specific metal coordination binding. Controlled spacing of nanoparticle layers from the surface was accomplished by binding of the Au nanoparticles to an organic multilayer spacer. The spacer comprised a Zr^{4+}-coordinated monolayer of a bishydroxamate disulfide ligand on gold and multilayers of branched hexahydroxamate ligands. The electrical behavior of the coordinated nanoparticle layers separated from the Au substrate by the organic spacer showed an ohmic

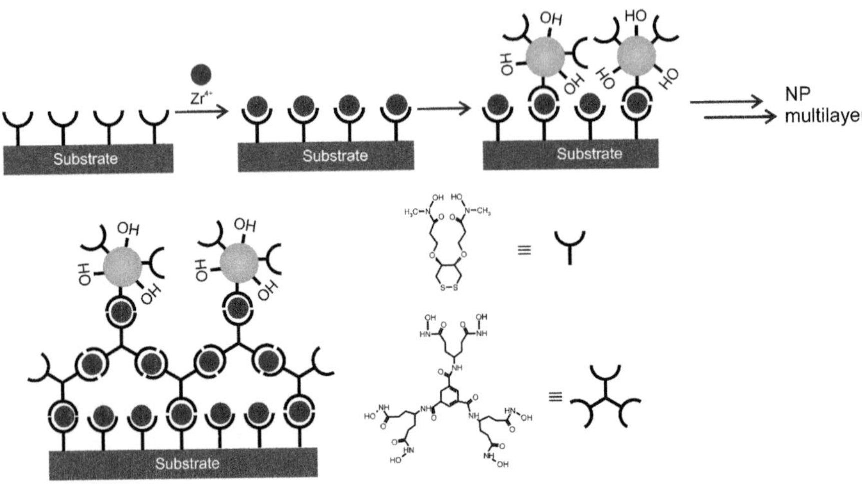

Figure 13.5 LbL assembly of bishydroxamate-functionalized Au nanoparticles onto the bishydroxamate disulfide SAMs by Zr^{4+} as binding ions. Controlled spacing of nanoparticle layer was achieved using multilayer branched hexahydroxamate ligands.[15]

resistance that increased with the number of nanoparticle layers, with a larger resistance observed when an organic multilayer spacer was used.

Murray et al. developed Cu^{2+}- and Zn^{2+}-carboxylate linker chemistry to prepare monolayer and multilayer films of Au alkanethiolate-monolayer-protected clusters (MPCs) onto a mercaptoundecanoic acid monolayer.[47] [(2-mercaptopropanoyl)-amino]acetic acid-(tiopronin)-functionalized Au MPCs were subsequently attached via Cu^{2+}-carboxylate chemistry. Quantized double-layer charging was observed in these films.[47] Attachment of additional layers of tiopronin-MPCs was demonstrated by repeating the sequential adsorption steps of Cu^{2+} ions and tiopronin-MPCs, resulting in a high-yield surface attachment. The Cu^{2+}-carboxylate chemistry was also used to induce the reversible aggregation of tiopronin-MPCs. The aggregation was controlled by changing the pH of the Cu^{2+} solution, with increased aggregation at lower pH. The reversible formation of transiently soluble alkanethiolate- and tiopronin-MPCs was demonstrated by treatment with sodium acetate solution or concentrated acetic acid. In conclusion, the use of highly specific coordination chemistry offers a convenient tool for the construction of multilayer and multicomponent nanostructures on surfaces.

13.2.2.3 Hydrogen Bonding

Binder et al.[13] utilized the intrinsic multivalent hydrogen bonding between "Hamilton-type" receptors[48] and barbiturates through six hydrogen bonds to assemble nanoparticles onto surfaces. On a monolayer of Hamilton-type receptors with an intrinsic binding strength of $\sim 10^5\ M^{-1}$, barbituric acid-functionalized Au nanoparticles were selectively adsorbed. The density of the receptors on the surface was varied by adjusting the receptor concentration during preparation. In addition, the Hamilton receptors were incorporated into one block of microphase-separated block copolymer thin films.[13] Highly selective binding of barbituric acid-functionalized Au nanoparticles onto the specific block copolymer phases on the surfaces was observed.

Hao and Lian reported hydrogen bonding-based routes for LbL assembly of polymer/Au nanoparticle multilayer thin films.[49] Au nanoparticles modified with carboxylate or pyridine groups were adsorbed onto poly(4-vinylpyridine) (P4VP) and carboxylate-functionalized Au nanoparticle surfaces, respectively. Alternating deposition of poly(acrylic acid) (PAA) and Au nanoparticles with pyridine groups resulted in a multilayer buildup, which showed a linear increase of the film thickness with the number of adsorbed Au nanoparticle layers. Fourier transform infrared spectroscopy verified the hydrogen bonding between the pyridine and carboxylate groups, which is the driving force for the formation of the polymer/Au multilayer thin films.

Similar polymer/Au nanoparticle multilayer thin films were made by Wu et al. in a study of pH-sensitive dissociation behavior of poly(3-thiophene acetic acid) (PTAA) and PAA in a LbL film (of 8 bilayers).[50] Unlike the pure polymer LbL film, the Au nanoparticles-containing LbL films were difficult to be released from the substrate by varying the pH. It was suggested that the gold particles act as a cross-linker in between the multilayers, thus further enhancing the stability of the LbL films.

Barbiturate-triaminodiazine hydrogen bonding was used to assemble CdS nanoparticles and gold/CdS nanoparticle hybrids onto gold surfaces.[51] A CdS nanoparticle-modified electrode generated a light-induced photocurrent in the system. A 2.6-fold enhancement in photocurrent was observed for the gold/CdS composite-modified electrode. The enhancement was attributed to the charge separation of the electron-hole pair that is generated upon the photochemical excitation of the CdS nanoparticles.

13.2.2.4 Host–Guest Interactions

Host–guest chemistry involves the complexation of two or more molecules that are held together in a unique structure via specific interactions, for example, hydrophobic interactions, Van der Waals forces, or hydrogen bonding. Cyclodextrin is an interesting host molecule because it is a natural receptor that forms stable and specific inclusion complexes with a variety of organic guest molecules in aqueous media.[52,53] The bonds between hosts and guests are continuously broken and formed. Hence, the use of molecules with multiple binding sites is used to enhance the binding affinities (multivalency).[54]

Different cyclodextrin monolayers have been synthesized and their surface properties have been characterized. Kaifer et al. prepared per-6-thiol-cyclodextrins, and described the interfacial monovalent ferrocene complexation at the monolayer.[55] Mittler-Neher et al. studied the kinetics of the adsorption of mono- and multithiolate-functionalized CD SAMs.[56,57] Beulen et al.,[58] Huskens et al.,[59] Auletta et al.,[60] and Nijhuis et al.[61] reported on the concept of molecular printboards, that is, β-cyclodextrin (β-CD) SAMs on gold or silicon oxide substrates, onto which complementary multivalent guest-functionalized dendrimer molecules were adsorbed, resulting in the formation of kinetically stable supramolecular assemblies. With the aid of adamantyl- or ferrocenyl-functionalized poly(propylene imine) dendrimers as a noncovalent supramolecular glue, CD-functionalized nanoparticles were assembled onto CD SAMs (Fig. 13.6).[12,62,63] Ferrocenyl-functionalized silica nanoparticles were also directly adsorbed onto CD SAMs via host–guest complexation.[11] All of these host- or guest-functionalized nanoparticle layers bind strongly at the interface owing to the formation of multivalent interactions.[64]

The supramolecular interaction is also appealing for its highly tunable binding strength between a nanoparticle and an interface. Different guest moieties (e.g., ferrocene, adamantane) of different binding strength can be selected to bind to the CD SAMs via different binding affinity.[52] It has been demonstrated that the total number of interactions involved in the nanoparticle assembly can be easily varied by using dendrimers of different generations and/or by the introduction of native CD molecules to the nanoparticle solution to induce competition so that CD hosts on the particle surface have to compete with the native CD in binding to the surface with pre-adsorbed guest-functionalized dendrimers.[63] Tuning the binding strength during the nanoparticle assembly enables ordering of nanoparticles, such that a nearly hexagonally close packed nanoparticle array with high stability was obtained.

Reversible attachment of nanostructures at molecular printboards was exemplified by the adsorption and desorption of CD-functionalized nanoparticles onto and from stimuli-responsive pre-adsorbed ferrocenyl-dendrimers at a CD SAM (Fig. 13.7).[65] Electrochemical oxidation of the ferrocenyl endgroups was employed to induce desorption of the nanostructure from the CD SAM. An *in situ* adsorption and desorption of ferrocenyl dendrimers and CD-functionalized Au nanoparticles ($d \sim 3$ nm) onto and from the molecular printboard was observed by a combination of surface plasmon resonance spectroscopy (SPR) and electrochemistry. Similar behavior was observed when larger CD-functionalized silica nanoparticles ($d \sim 60$ nm) were desorbed from the surface with the aid of ultrasonication.

The stepwise construction of self-assembled organic/inorganic multilayers based on multivalent supramolecular interactions between guest-functionalized dendrimers and nanoparticles and host-modified gold nanoparticles has been developed, yielding supramolecular LbL assembly (Fig. 13.8).[12] Multilayer thin films composed of

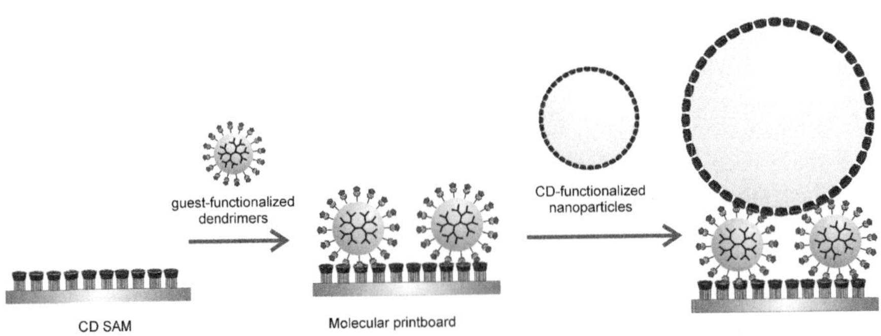

Figure 13.6 The adsorption of multivalent guest-functionalized dendrimers onto a CD SAM and the subsequent assembly of complementary CD-functionalized nanoparticles.

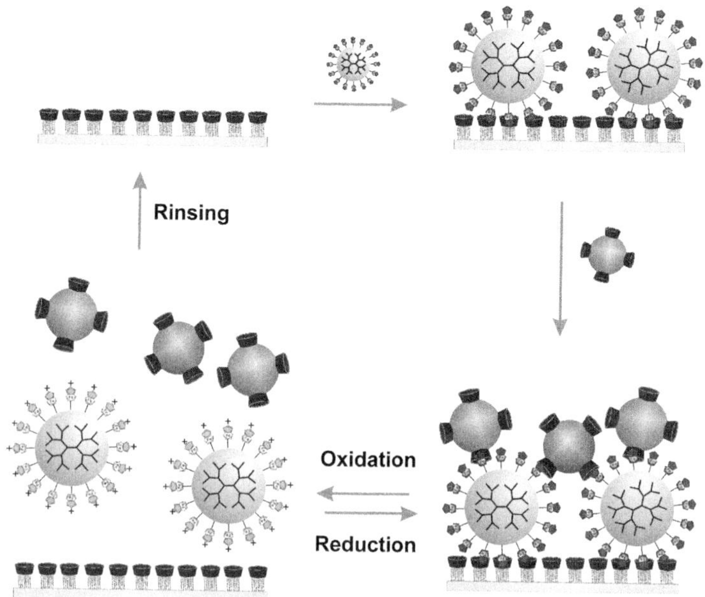

Figure 13.7 The adsorption and desorption of CD-functionalized nanoparticles onto and from a CD SAM with ferrocenyl dendrimers as a reversible supramolecular glue.

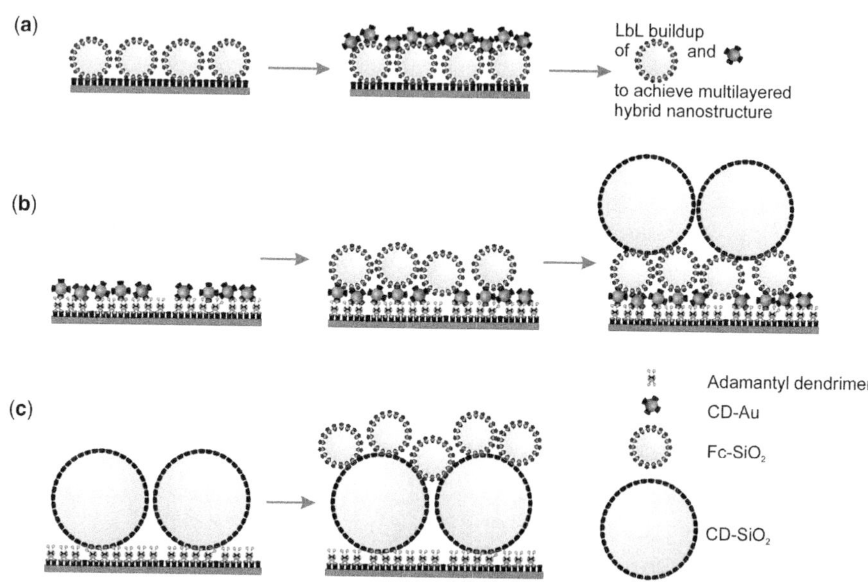

Figure 13.8 The assembly of multicomponent nanostructures by using complementary guest- and host-functionalized nanoparticles (a). The nanoparticle assembly from small to large (b) and large to small nanoparticles (c) by supramolecular interactions.

CD-functionalized gold nanoparticles and adamantyl-terminated dendrimers have been prepared on CD SAMs, whereby the thickness was controlled at the nanometer level. Densely packed and multilayered nanoparticle structures were obtained by alternating assembly steps of complementary guest- and host-functionalized nanoparticles (Fig. 13.8a).[66] The effects of order of the nanoparticle assembly steps from large to small and from small to large nanoparticles (Fig. 13.8b, c) revealed that the specific supramolecular assembly of nanoparticles was self-limited, that is, one nanoparticle layer per assembly step, allowing the control over the thickness of the supramolecular hybrid nanostructure by choosing the size of the nanoparticles, irrespective of the core material of the nanoparticles. The roughness of the top nanoparticle layer of the structure was directly influenced by the size of the underlying nanoparticles.

13.2.2.5 Biomolecular Interactions

In recent years, there has been a lot of research in the utilization of biomolecule-functionalized nanoparticles for the formation of hybrid nanostructures. While much effort was spent on solution systems, there are some examples that exploit the binding specificity of the biomolecules, for example, proteins, peptides, and DNA to assemble nanoparticles on surfaces.

The strong and specific biotin–streptavidin binding was used to assemble biomolecule-functionalized nanoparticles in multilayered structures.[67] Application of an electrical field allowed the assembly of multilayer structures by using extremely low concentrations of nanoparticles with minimal nonspecific binding. A microelectrode array was used to facilitate the rapid parallel electrophoretic transport and binding of biotin- and streptavidin-functionalized fluorescent nanoparticles to specific sites. By controlling the current, voltage, and activation time at each nanoparticle adsorption step, the directed assembly of more than 50 layers of nanoparticles was accomplished within an hour.

Taton et al. used gold nanoparticles functionalized with thiol-modified oligonucleotides to detect the presence of its complementary sequence hybridized on a transparent substrate (Fig. 13.9).[16] In comparison to conventional fluorophore probes, this technique is three times more sensitive in discriminating an oligonucleotide sequence with a single basepair mismatch. In addition, signal amplification by reduction of silver ions on the nanoparticles drastically increased the sensitivity of

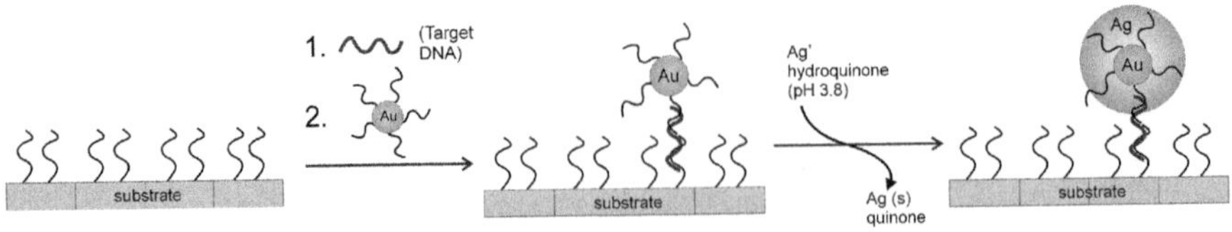

Figure 13.9 The assembly of oligonucleotide-functionalized Au nanoparticles on an Au surface via DNA hybridization and signal enhancement by reduction of silver ions on the nanoparticles for DNA array detection.[16]

this detection system, exceeding that of the fluorophore system by two orders of magnitude.

Alternatively, a DNA array based on electrical detection was reported to detect a target oligonucleotide at a concentration as low as 500 fM. Oligonucleotide-functionalized gold nanoparticles were locally and specifically assembled between an electrode gap, functionalized with complementary oligonucleotides. The silver deposition on these nanoparticles resulted in conductivity changes, which allow detection of target oligonucleotides.[68]

13.3 CHEMICAL NANOPARTICLE ASSEMBLY ON PATTERNED SURFACES

In device fabrication, the location of functional materials is as important as their properties. The integration of solid particles into devices usually requires placing them in specific positions. Hence, the combination of top-down patterning techniques and bottom-up self-assembly is crucial in obtaining (submicron) patterned functional nanostructures on surfaces.

The introduction of SAMs on localized areas of a substrate allows straightforward further functionalization and directed assembly of nanoparticles. By using chemistry, specific binding can be introduced, allowing the control of nanoparticle assembly onto lithographic patterns. Wet-chemical self-assembly of nanoparticles is particularly attractive for the fabrication of nanoparticle-based nanostructures because of its compatibility with various kinds of substrates with complex shapes. In this section, conventional and nonconventional patterning techniques for the chemical assembly of nanoparticles will be highlighted.

13.3.1 Patterning by Photolithography

Conventional patterning techniques, for example, photolithography and electron beam lithography, are frequently used for fabrication of patterned substrates owing to their capability to produce nanometer features with remarkable perfection. However, the slow process of electron beam lithography has limited most of its application fabricating high-end devices.[69,70]

Photolithography is one of the most widely implemented fabrication techniques. It involves an exposure of a resist on an inert surface to an irradiation source (e.g., UV, X-ray) through a mask with a prefabricated pattern to induce chemical changes to the exposed areas of the substrate, yielding a replica of the pattern of the mask.[71] Development of the exposed resist, for example, by chemical etching, results in topographically or chemically patterned substrates. In general, chemical patterns with different wettability, owing to the relative simplicity, have been most widely employed for the assembly of multicomponent and 3D nanoparticle crystals.

Heath et al. exploited photolithography to generate chemical patterns for the assembly of different nanocrystals (Fig. 13.10).[72] Organic monolayers of an adsorbate functionalized with the photolabile protection group nitroveratryloxycarbonyl

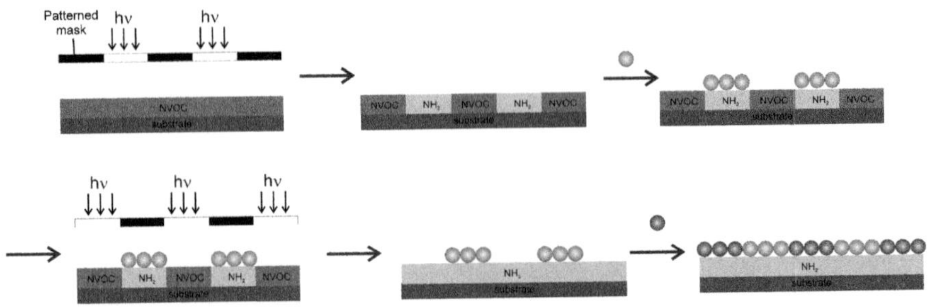

Figure 13.10 Reaction scheme for the stepwise preparation of multicomponent particle arrays by photolithography.[72] Chemically patterned amino-functionalized surfaces were formed by selectively exposing the NVOC substrate by UV exposure. Multicomponent amino-functionalized nanocrystals were subsequently adsorbed via ligand exchange.

(NVOC) were deprotected locally by UV exposure through a mask, resulting in spatially and chemically distinct areas on the substrate. On patterns of an adsorbate functionalized with the photolabile NVOC and amino-functionalized regions, multiple types of metal and semiconductor nanocrystals were produced. The adsorption of amine-functionalized CdSe/CdS core–shell nanocrystals, Au, and Pt nanoparticles onto the pattern was achieved by the ligand exchange between the amine groups of nanocrystals with the substrate-bound amino groups. A wide range of fluorescence intensities (from 100/1 to 8/1 signal-to-noise ratio) of the CdSe/CdS core–shell nanocrystals on the surface was observed, indicating that the binding selectivity of the nanocrystals is dependent on the type of particle, the particle concentration, the chemical composition of the NC solution, and the immersion time of the pattern in the particle solution. The assembly of nanocrystals is characterized by strong interparticle and particle–substrate dispersion attractions that scale geometrically with the size of the particle. Such interactions can compete effectively with ligating particle–substrate interactions, and thus decrease the binding selectivity.

Similar NVOC deprotection chemistry was used by Jonas et al. to study the selective assembly of polybutylmethacrylate particles onto chemically patterned silane layers via electrostatic interactions.[73] Substrates functionalized with triethoxysilane with photoprotected NVOC amino groups were patterned by photolithography. Site-specific nanoparticle adsorption was observed on the photo-deprotected layers after local conversion to amino groups. A three-step mechanism was suggested for the site-selective particle adsorption. The positioning and adhesion of the nanoparticles in liquid suspension are controlled by electrostatic attraction and polar interactions (e.g., hydrogen bonding) between the substrate and the particle surfaces. They depend on the solution pH that governs the pK_a of the carboxylic acid on the nanoparticles. Capillary forces between particles and the surface laterally rearrange the particles during the drying process. Third, an irreversible reorganization of the particle–substrate interface occurs after complete evaporation.

Rotello et al. patterned silica substrates with thymine and positively charged poly(vinyl-*N*-methylpyridinium) (PVMP) polymers by photolithography.[74] By using the triple hydrogen bonding diamidopyridine-thymine motif and pyridinium-carboxylate electrostatic interactions, selective self-assembly of diaminopyridine-functionalized polystyrene and carboxylate-derivatized CdSe/ZnS core–shell nanoparticles onto the complementary domains of the patterned substrate was observed. Owing to the specificity and selectivity of the interactions involved, the recognition-directed orthogonal self-assembly of multicomponent nanoparticle arrays can be performed in one step.

The combination of patterning and chemical assembly has also been extended to the formation of patterned 3D nanoparticle crystals. Parikh et al. fabricated patterned substrates with different chemical functionalities and wettability by photolithography (Fig. 13.11).[75] A concentrated nanoparticle solution was injected into the gap between the patterned substrate and a hydrophilic glass substrate. After solvent evaporation, the structure was peeled off, leaving cleaved face-centered cubic nanoparticle crystals on each of the two surfaces, corresponding to the substrate hydrophilicity. The crystal thickness was controlled by the spacing between the substrates, and by the amount and concentration of the nanoparticle solution. The measured photonic stop gaps for the nanoparticle crystals were slightly lower than the theoretically predicted values, probably due to nanoparticle shrinking as a result of dehydration.

In addition to the manipulation of wettability of the patterned substrate, Masuda et al. utilized the dynamic interactions between particles, substrate, and solution, and the shrinkage of nanoparticle droplets to form spherical particle assemblies (Fig. 13.12).[76,77] A droplet of a SiO_2 nanoparticle solution in methanol was placed on photolithographically patterned hydrophilic and hydrophobic SAMs. The substrate was then immersed in decalin. The nanoparticles selectively contacted the hydrophilic regions and bound to the droplet interfaces by surface tension of the emulsified droplets. The interfacing between two solutions and shrinkage of the particle-containing droplets resulted in rearrangement of the nanoparticles to form close-packed spherical particle assemblies.

The selective adsorption of catalytic nanoparticles onto patterned substrates has been demonstrated by Akamatsu et al. to allow direct metallization on insulating substrates.[78] TiO_2 nanocrystals were selectively adsorbed onto a hydrophobic region of lithographically patterned glass substrate via electrostatic interactions. TiO_2 nanocrystals, known for their strong oxidizing ability, were used as photocatalyst to oxidize methanol in solution to produce formic acid. This led to the reduction of copper ions to produce metallic copper films. The thickness of the deposited copper films

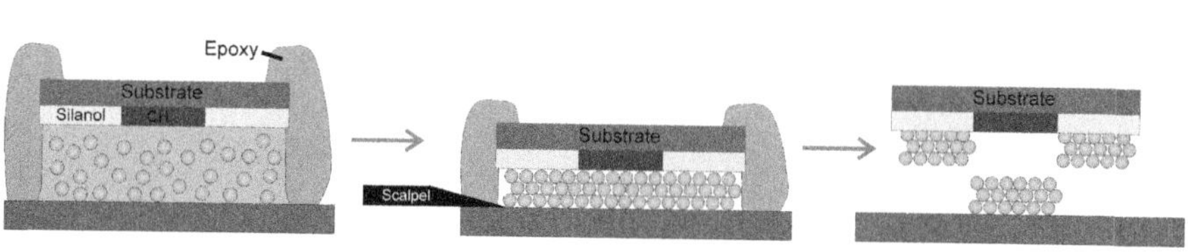

Figure 13.11 The patterning of nanoparticle crystals by combination of photolithography and surface wettability.[75] A concentrated nanoparticle solution slowly evaporated in the gaps between the patterned substrate and a glass substrate. When the structure was peeled off, patterned nanoparticle crystals on both surfaces were formed.

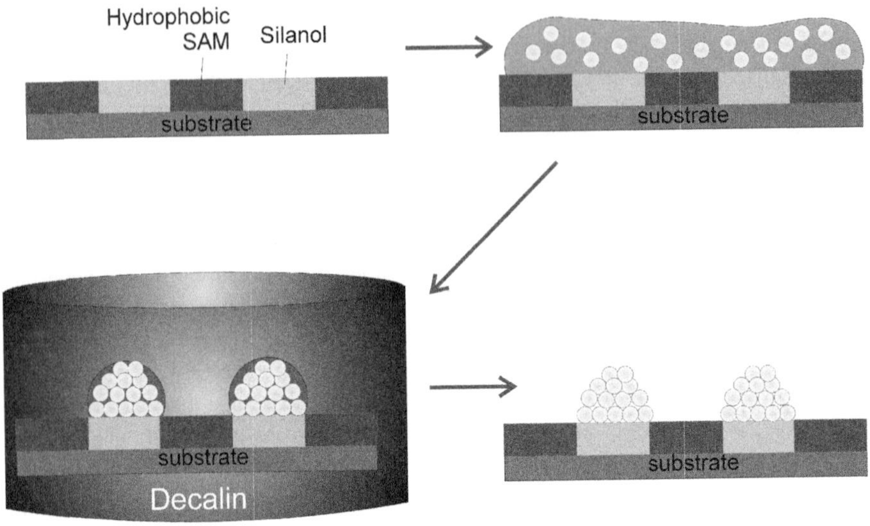

Figure 13.12 The preparation of spherically shaped nanoparticle crystals. A chemically patterned SAM with attached droplets of a SiO_2 nanoparticle solution was immersed in decalin. The interfacing between two solutions and shrinkage of the particle droplets resulted in rearrangement of the nanoparticles to form close-packed spherical particle assemblies.

was controlled by varying the irradiation time and power, and by the initial concentration of methanol as a hole scavenger. The deposited copper thin films exhibited electrical conductivity slightly lower than bulk copper, probably due to the formation of relatively large grains of copper that were loosely connected to each other.

13.3.2 Patterning by Soft Lithography

Unconventional patterning techniques, such as soft lithography (e.g., microcontact printing), nanoimprint lithography, and scanning probe lithography, are increasingly used for the cost-effective fabrication of nanostructures. They are particularly attractive because they can be performed even without cleanroom facilities.

Microcontact printing (μCP) utilizes an elastomeric stamp that can be molded from a patterned substrate prefabricated by photolithography or e-beam lithography.[30,79] The elastomeric stamp, most commonly made of poly(dimethylsiloxane) (PDMS), is employed for transferring and fabricating features at the intended target surface in a nondestructive manner, making it particularly suitable for patterning of SAMs, nanoparticles, biomolecules, and nanostructures.

There are two μCP methodologies for the assembly of nanoparticles into patterned nanoparticle arrays on surfaces: (1) the direct use of nanoparticles as the "ink" in μCP and thus the transfer of the nanoparticle ink to the substrate by the stamp, and (2) the preparation of patterned monolayers on a substrate to direct the adsorption of nanoparticles from solution. Whitesides and coworkers pioneered the μCP of SAMs and nanostructures. In their earlier work, micropatterns of palladium nanoparticles were prepared by μCP, which served as a catalyst for electroless deposition of copper (Fig. 13.13).[80] A PDMS stamp was soaked in a Pd nanoparticle solution and stamped onto amino-functionalized substrates. The electroless deposition of copper only happened at the patterned Pd nanoparticle regions. Printing on curved substrates and the fabrication of free-standing copper and multilevel metal structures with variable thickness in different regions were also demonstrated.

Andres and coworkers demonstrated the μCP of densely packed alkanethiolate-functionalized Au nanoparticle arrays in monolayer and multilayer structures.[81,82] Dense and hexagonally packed monolayers of nanoparticles were first assembled on a water surface. By using the Langmuir–Schäfer technique, the Au nanoparticle monolayer was transferred to a PDMS stamp, and printed onto a substrate. Multilayers were prepared by repeating the printing process in an LbL scheme, in which subsequent particle layers may be made up of the same or different types of particles. Similarly, the assembly of irregular, densely packed monolayers of polystyrene nanoparticles on μCP substrates via carbodiimide coupling was reported.[83] The conformal contact of the carbodiimide-functionalized polystyrene particles resulted in the covalent attachment of the nanoparticles at a carboxylate-functionalized surface.

Owing to the flexibility of PDMS stamps, nanocomposites were also feasible for printing. Wang et al. reported the μCP of a polymeric/inorganic nanocomposite of hydrolyzed poly(styrene-*alt*-maleic anhydride) (HSMA) and TiO_2 on substrates from a dispersion of TiO_2 nanoparticles in an HSMA solution.[84] The printed composite layers had a dish shape, with more nanoparticles accumulated in the periphery of the dishes. The size of the rim and the thickness of the printed dish depended on the TiO_2 particle concentration. Calcination of the composite dishes removed the polymer and resulted in nanostructured TiO_2 layers. In addition, Wu et al. reported the μCP of CdS/dendrimer nanocomposites on hydroxy-terminated silicon surfaces.[85] Dendrimers were used as hosts for CdS nanoparticles, and facilitated the adsorption of the nanoparticles to the surface via electrostatic forces, hydrogen bonds, and/or Van der Waals interactions.

Wolf et al. introduced the self-assembly, transfer, and integration (SATI) of nanoparticles with high placement accuracy.[86,87] Silica and polymer nanoparticles were positioned on a PDMS stamp through convective assembly (Fig. 13.14). By controlling the printing temperature or by using a thin polymer layer as an adhesion layer, nanoparticles of different shapes and sizes were printed onto the target substrate.

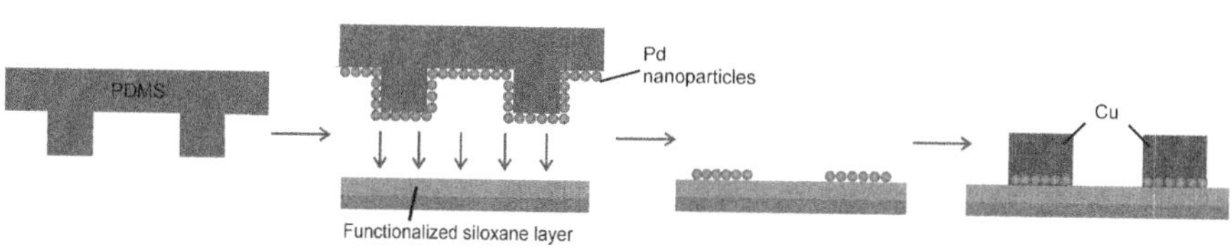

Figure 13.13 The μCP of Pd nanoparticles on an amino-functionalized substrate and the subsequent electroless deposition of copper.[80]

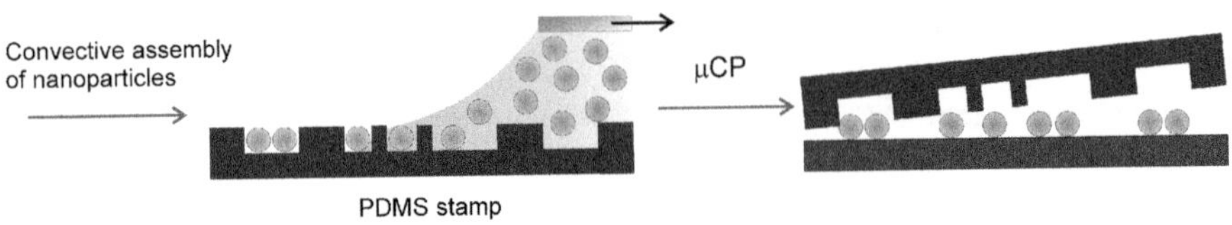

Figure 13.14 The use of convective assembly to control the arrangement of nanoparticles on a patterned PDMS stamp and the subsequent printing of the nanoparticles onto a substrate with single-particle resolution.[87]

Using this technique, the authors have demonstrated the printing of a 60 nm Au nanoparticle array with single-particle resolution.

Huskens et al. exploited host–guest interactions between dendritic guest molecules and CD-modified nanoparticles for the formation of organic/metal nanoparticle multilayers on a PDMS stamp (Fig. 13.15).[88] The multilayer stacks were transferred to a complementary host surface, while no materials remained on the protruding areas of the PDMS stamp. These multilayers showed a well-defined thickness control of 2 nm per bilayer.

In the indirect printing approach, organic molecules were first printed onto substrates. The μCP of organic adsorbates serves to direct the assembly of nanoparticles for the formation of ordered 2D arrays of particles. Whitesides et al. prepared patterned surfaces with grids of hydrophobic (CH_3-terminated) and hydrophilic (COOH-terminated) SAMs of alkanethiols on a gold substrate.[89] The nanoparticle solution wetted exclusively the hydrophilic regions of surface by controlling the substrate withdrawing speed from the nanoparticle solution. The dimensions of these particle patterns can be controlled by the concentration and composition of the solution, and shape and area of the hydrophilic regions.

Hammond et al. reported the self-organization of SiO_2 and polystyrene nanoparticles on a μCP patterned polyelectrolyte substrate (Fig. 13.16).[28] The multicomponent nanoparticle assembly was driven by spatial electrostatic and hydrophobic

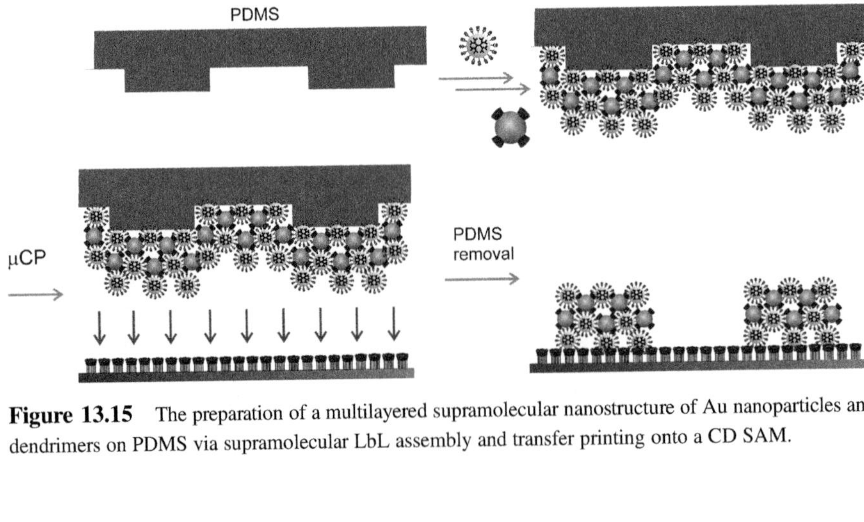

Figure 13.15 The preparation of a multilayered supramolecular nanostructure of Au nanoparticles and dendrimers on PDMS via supramolecular LbL assembly and transfer printing onto a CD SAM.

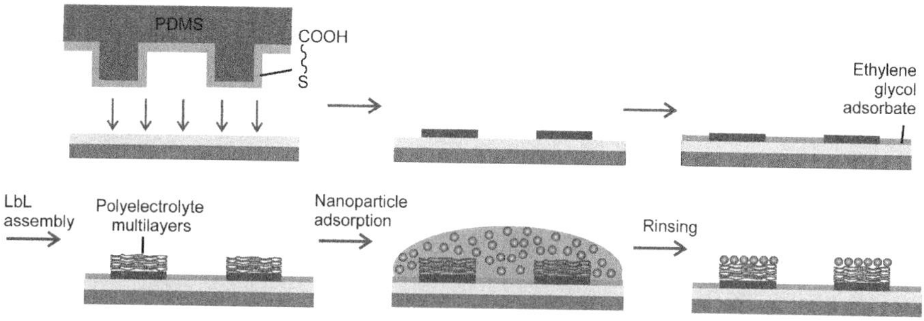

Figure 13.16 Chemical patterning by μCP, LbL assembly of polyelectrolytes, and nanoparticle assembly by electrostatic interactions.[28]

interactions between the nanoparticles and the polyelectrolyte substrate. The surface charge density was modulated by pH, ionic strength, and effective surface charge of the polyelectrolyte.[90] However, only a moderate packing density was achieved due to the repulsive forces between the particles. In addition, a balance between the interactions involved during the nanoparticle assembly is needed to obtain an optimum adsorption strength and deposition selectivity.

Zhou et al. studied the selective LbL assembly of NCs on microcontact printed carboxylate-functionalized SAMs.[91] The LbL assembly was achieved by alternating adsorption of NCs functionalized with hydrophobic trioctylphosphine oxide and 2-mercaptoethanesulfonic acid, and positively charged linear poly(ethyleneimine) (LPEI). By adsorption of 11-mercaptoundecylhexa(ethylene glycol) in the noncontacted areas, nonspecific interactions were minimized. A uniform and linear growth of the polymer-NC layers was observed.

Combinatorially selected peptides and peptide–organic conjugates were used as linkers to direct the attachment of NCs on a microcontact printed carboxylate SAM.[92] The use of genetically engineered peptides for inorganics (GEPIs) allowed control over structural and organizational conformations during the nanoparticle assembly. In addition, the spatial configurations of the NCs at the surface were varied by tailoring GEPIs with π-conjugated functional molecules. An increase in the average NC attachment density was observed when peptide–organic conjugates were employed in the formation of hybrid nanostructures.

13.3.3 Patterning by Nanoimprint Lithography

Nanoimprint lithography (NIL) is an embossing method for fabricating patterns by mechanical deformation of an imprint resist. Unlike μCP, it offers three-dimensional patterning with high resolution features down to 6 nm,[93] making it a low cost, high throughput, and high resolution technique. Two general methods are used to pattern the imprint resist, hot embossing of a thermoplastic polymer and UV imprint lithography of a photocurable monomer.

Huskens et al. utilized NIL as a tool to pattern SAMs on silicon substrates.[94] As shown in Figure 13.17, a prefabricated silicon wafer with a pattern was pressed against

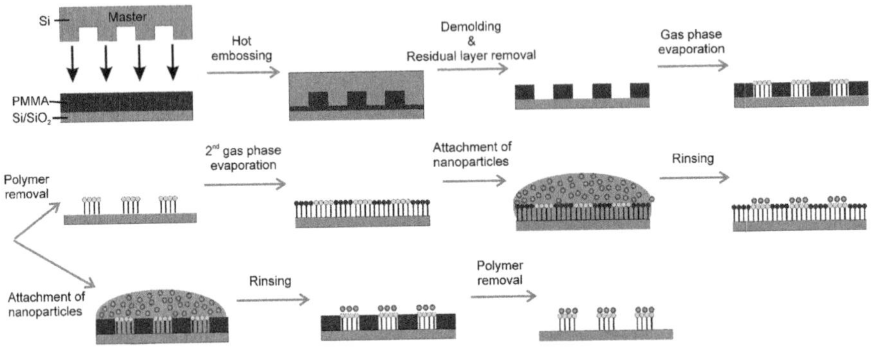

Figure 13.17 Schematic description of the NIL process to form chemical or topographically patterned SAMs. The nanoparticles were subsequently attached specifically onto the SAM by electrostatic interactions.

a thin layer of poly(methyl methacylate) PMMA on a silicon substrate above the glass transition temperature. The system was then cooled and the thin polymer residual layer was removed. SAMs of aminoalkylsilanes were formed on the uncovered regions. Carboxylate-functionalized polystyrene nanoparticles were electrostatically attached to the amino-functionalized surface. Alternatively, chemical patterns were created by removing the patterned polymer layers, and adsorbing a second silane on the remaining areas. Selective attachment of carboxylate-functionalized silica nanoparticles on complementary amino regions was observed in this case.

NIL patterns were also used for the assembly of nanoparticles via supramolecular host–guest interactions.[95] The NIL-patterned substrate was functionalized with CD SAMs via a three-step synthesis process. The fabrication of 3D nanostructures was achieved by the alternating assembly of multivalent guest-functionalized dendrimers and CD-functionalized Au nanoparticles.[88] This methodology can be applied to various nanoparticles, regardless of their size and core material. For instance, CD-functionalized silica and polystyrene nanoparticles were adsorbed onto NIL-patterned CD SAMs with preadsorbed guest-functionalized dendrimers.[60,92] Recently, Huskens et al. demonstrated the supramolecular LbL assembly of 3D multicomponent nanostructures of nanoparticles by alternating assembly steps of complementary ferrocenyl-functionalized silica nanoparticles and different kinds of host-functionalized nanoparticles (see Fig. 13.8).[66]

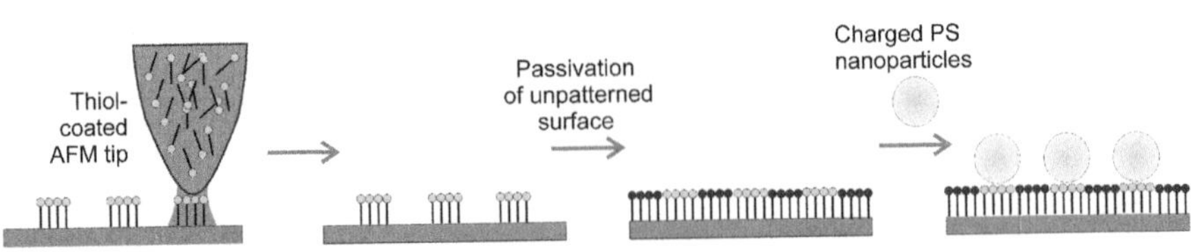

Figure 13.18 The formation of chemically patterned substrates by DPN, passivation of the unpatterned surface and subsequent attachment of nanoparticles via electrostatic interactions.

13.3.4 Patterning by Scanning Probe Lithography

The application of scanning probe lithography (SPL) has been widespread owing to its ability to modify substrates with very high resolution and ultimate pattern flexibility.[96] Dip-pen nanolithography (DPN),[97] high contact force AFM,[98] and constructive nanolithography[99] are some of the most commonly employed techniques, all of which aim to control the position and directed assembly of molecules and nanoparticles.

DPN is a scanning probe nanopatterning technique developed by Rosi and Mirkin.[97] An AFM tip is used to deliver molecules to a surface through a water meniscus, which naturally forms in ambient atmosphere. DPN can also be used to generate many customized templates from the same or different chemical inks. Making use of this idea, Demers and Mirkin employed the DPN-based strategy for generating charged chemical templates to study the assembly of single particles into 2D lattices (Fig. 13.18).[100] They used 16-mercaptohexadecanoic acid (MHDA) to make templates, and positively charged protonated amino-modified polystyrene particles ($d \sim 930$ nm) were then electrostatically assembled onto the MHDA surface.[100] The nonpatterned regions were passivated by a hydrophobic alkanethiol, which prevented undesired particle diffusion. They also extended the DPN-driven nanoparticle assembly to magnetic nanoparticles (e.g., Fe_2O_3, $MnFe_2O_4$).[101] Detailed experimental studies showed that the size of the nanoparticle dots or lines correlated with the size of the dots or lines of the MHDA patterns generated by DPN. The template dot diameter could be controlled by tuning the tip–substrate contact time, whereas the line width could be controlled by adjusting the scan speed.

Sagiv et al. introduced constructive nanolithography, a surface patterning process utilizing conductive AFM tips as nanoelectrochemical pens.[99] The top surface of an organosilane monolayer on silicon was electrochemically transformed and inscribed by electrical pulses delivered via a conductive AFM tip. Further surface chemical derivatization and guided self-assembly resulted in hierarchical LbL assembly. Utilizing this strategy, they reported the assembly of Au clusters on patterned silicon substrates. As shown in Figure 13.19, alkylsilane monolayer regions were electrochemically transformed to carboxylate-functionalized monolayers.[102,103] Subsequent exposure

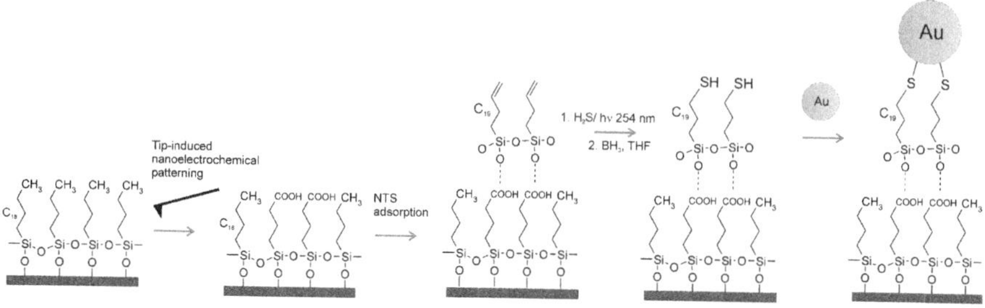

Figure 13.19 The template-directed self-assembly of Au clusters on silicon substrate patterned by constructive nanolithography.

to nonadecenyltrichlorosilane (NTS) resulted in patterned monolayers with terminal vinyl groups (—CH=CH_2). Photochemical radical addition of H_2S to these vinyl moieties and further reduction (with $BH_3 \cdot THF$) of a fraction of the disulfide groups produced a thiol-functionalized layer. Phosphine ligands on Au clusters were partially lost due to the exchange process during gold cluster attachment to the thiol-coated surface.

Liu et al. utilized AFM-based nanooxidation to fabricate Au nanoparticle arrays on silicon substrates.[98] An octadecyltrichlorosilane (OTS) monolayer on silicon was first subjected to localized chemical oxidation by using a conductive AFM tip to form silicon oxide. The oxide region was then further modified to an amino-terminated silane monolayer via selective chemical adsorption. The patterned substrate was exposed to negatively charged Au nanoparticles which resulted in the formation of nanoparticle arrays via electrostatic interactions in the amino-terminated silane regions. By optimizing the humidity, applied voltage, and the pulse duration of the system, oxidized areas as small as 15 nm were fabricated. A very regular nanoparticle array, with only one nanoparticle per oxide dot was obtained.[98] However, the assembly efficiency of nanoparticles onto the guiding templates was not perfect (70%), and showed a decrease in efficiency when the oxide dots were smaller. The authors attributed this to smaller effective areas to adsorb nanoparticles compared to the actual geometrical areas. The difference in effective areas and geometrical areas were attributed to imperfectness of the AFM nanooxidation process.

13.4 CONCLUSIONS

Parallel advances in the development of the self-assembly of nanoparticles by chemical means and of different patterning strategies have enabled the creation of two- and three-dimensional nanoparticle crystals. In particular, the use of noncovalent chemistry for the directed assembly of functionalized nanoparticles onto desired patterned substrates has resulted in highly specific assembly of nanoparticles with controlled affinity and reversibility. However, it is clear that the reported nanoparticle structures are still relatively simple in structure, and that the intriguing properties of the nanoparticle structures have not been extensively studied. Hence, the development of the integration of self-assembly of nanoparticles into nanofabrication schemes still requires more efforts to fully extend its potential to the fabrication of devices that target specific applications. It is anticipated, using chemically-directed self-assembly as a nanofabrication tool, that more ordered and complex arrays of chemically, optically, and geometrically tunable nanostructured building blocks can be realized. These will one day be of benefit for the development of novel technologies such as optical data processing, molecular electronics, and quantum computing.

REFERENCES

1. M. C. Daniel, D. Astruc, *Chem. Rev.* **2004**, *104*, 293–346.
2. A. Arsenault, S. Fournier-Bidoz, B. Hatton, H. Miguez, N. Tetreault, E. Vekris, S. Wong, S. M. Yang, V. Kitaev, G. A. Ozin, *J. Mater. Chem.* **2004**, *14*, 781–794.

3. N. D. Denkov, O. D. Velev, P. A. Kralchevsky, I. B. Ivanov, H. Yoshimura, K. Nagayama, *Nature* **1993**, *361*, 26.
4. L. Malaquin, T. Kraus, H. Schmid, E. Delamarche, H. Wolf, *Langmuir* **2007**, *23*, 11513–11521.
5. G. A. Ozin, S. M. Yang, *Adv. Funct. Mater.* **2001**, *11*, 95–104.
6. J. Wijnhoven, W. L. Vos, *Science* **1998**, *281*, 802–804.
7. S. Chen, *Langmuir* **2001**, *17*, 2878–2884.
8. M. Lahav, A. N. Shipway, I. Willner, M. B. Nielsen, J. F. Stoddart, *J. Electroanal. Chem.* **2000**, *482*, 217–221.
9. V. Paraschiv, S. Zapotoczny, M. R. de Jong, G. J. Vancso, J. Huskens, D. N. Reinhoudt, *Adv. Mater.* **2002**, *14*, 722–726.
10. G. Decher, *Science* **1997**, *277*, 1232–1237.
11. X. Y. Ling, D. N. Reinhoudt, J. Huskens, *Langmuir* **2006**, *22*, 8777–8783.
12. O. Crespo-Biel, B. Dordi, D. N. Reinhoudt, J. Huskens, *J. Am. Chem. Soc.* **2005**, *127*, 7594–7600.
13. R. Zirbs, F. Kienberger, P. Hinterdorfer, W. H. Binder, *Langmuir* **2005**, *21*, 8414–8421.
14. A. Shavel, N. Gaponik, A. Eychmuller, *ChemPhysChem* **2005**, *6*, 449–451.
15. M. Wanunu, R. Popovitz-Biro, H. Cohen, A. Vaskevich, I. Rubinstein, *J. Am. Chem. Soc.* **2005**, *127*, 9207–9215.
16. T. A. Taton, C. A. Mirkin, R. L. Letsinger, *Science* **2000**, *289*, 1757–1760.
17. J. I. Park, W. R. Lee, S. S. Bae, Y. J. Kim, K. H. Yoo, J. Cheon, S. Kim, *J. Phys. Chem. B* **2005**, *109*, 13119–13123.
18. H. Zhang, R. Wang, G. Zhang, B. Yang, *Thin Solid Films* **2003**, *429*, 167–173.
19. Y. Bai, S. Zhao, K. Zhang, C. Sun, *Colloid Surf. A* **2006**, *281*, 105–112.
20. E. Granot, F. Patolsky, I. Willner, *J. Phys. Chem. B* **2004**, *108*, 5875–5881.
21. D. N. Reinhoudt, M. Crego-Calama, *Science* **2002**, *295*, 2403–2407.
22. R. K. Iler, *J. Colloid Interface Sci.* **1966**, *21*, 569–594.
23. G. Decher, J. D. Hong, J. Schmitt, *Thin Solid Films* **1992**, *210*, 831–835.
24. J. W. Ostrander, A. A. Mamedov, N. A. Kotov, *J. Am. Chem. Soc.* **2001**, *123*, 1101–1110.
25. N. Malikova, I. Pastoriza-Santos, M. Schierhorn, N. A. Kotov, L. M. Liz-Marzan, *Langmuir* **2002**, *18*, 3694–3697.
26. N. A. Kotov, I. Dekany, J. H. Fendler, *J. Phys. Chem.* **1995**, *99*, 13065–13069.
27. T. Nakanishi, B. Ohtani, K. Uosaki, *J. Phys. Chem. B* **1998**, *102*, 1571–1577.
28. I. Lee, H. P. Zheng, M. F. Rubner, P. T. Hammond, *Adv. Mater.* **2002**, *14*, 572–577.
29. P. Maury, M. Escalante, D. N. Reinhoudt, J. Huskens, *Adv. Mater.* **2005**, *17*, 2718–2723.
30. M. A. Meitl, Z. T. Zhu, V. Kumar, K. J. Lee, X. Feng, Y. Y. Huang, I. Adesida, R. G. Nuzzo, J. A. Rogers, *Nature Mater.* **2006**, *5*, 33–38.
31. D. L. Feldheim, K. C. Grabar, M. J. Natan, T. E. Mallouk, *J. Am. Chem. Soc.* **1996**, *118*, 7640–7641.
32. J. Schmitt, G. Decher, W. J. Dressick, S. L. Brandow, R. E. Geer, R. Shashidhar, J. M. Calvert, *Adv. Mater.* **1997**, *9*, 61–65.
33. T. Serizawa, H. Takeshita, M. Akashi, *Langmuir* **1998**, *14*, 4088–4094.
34. Y. Lvov, K. Ariga, M. Onda, I. Ichinose, T. Kunitake, *Langmuir* **1997**, *13*, 6195–6203.
35. J. F. Hicks, Y. Seok-Shon, R. W. Murray, *Langmuir* **2002**, *18*, 2288–2294.
36. A. Kumar, A. B. Mandale, M. Sastry, *Langmuir* **2000**, *16*, 6921–6926.
37. R. Blonder, L. Sheeney, I. Willner, *Chem. Commun.* **1998**, 1393–1394.
38. R. G. Nuzzo, D. L. Allara, *J. Am. Chem. Soc.* **1983**, *105*, 4481–4483.
39. H. Sellers, A. Ulman, Y. Shnidman, J. E. Eilers, *J. Am. Chem. Soc.* **1993**, *115*, 9389–9401.
40. R. Rizza, D. Fitzmaurice, S. Hearne, G. Hughes, G. Spoto, E. Ciliberto, H. Kerp, R. Schropp, *Chem. Mater.* **1997**, *9*, 2969–2982.
41. C. J. Kiely, J. Fink, J. G. Zheng, M. Brust, D. Bethell, D. J. Schiffrin, *Adv. Mater.* **2000**, *12*, 640–643.
42. M. Brust, D. Bethell, C. J. Kiely, D. J. Schiffrin, *Langmuir* **1998**, *14*, 5425–5429.
43. M. D. Musick, C. D. Keating, M. H. Keefe, M. J. Natan, *Chem. Mater.* **1997**, *9*, 1499–1501.
44. K. V. Sarathy, P. J. Thomas, G. U. Kulkarni, C. N. R. Rao, *J. Phys. Chem. B* **1999**, *103*, 399–401.
45. S. Bharathi, M. Nogami, S. Ikeda, *Langmuir* **2001**, *17*, 1–4.
46. L. Wang, E. Wang, *Electrochem. Commun.* **2004**, *6*, 49–54.

47. A. C. Templeton, F. P. Zamborini, W. P. Wuelfing, R. W. Murray, *Langmuir* **2000**, *16*, 6682–6688.
48. S. K. Chang, A. D. Hamilton, *J. Am. Chem. Soc.* **1988**, *110*, 1318–1319.
49. E. C. Hao, T. Q. Lian, *Chem. Mater.* **2000**, *12*, 3392–3396.
50. Y. Jiang, Y. Shen, P. Y. Wu, *J. Colloid Interface Sci.* **2008**, *319*, 398–405.
51. R. Baron, C. H. Huang, D. M. Bassani, A. Onopriyenko, M. Zayats, I. Willner, *Angew. Chem. Int. Ed.* **2005**, *44*, 4010–4015.
52. M. V. Rekharsky, Y. Inoue, *Chem. Rev.* **1998**, *98*, 1875–1917.
53. K. A. Connors, *Chem. Rev.* **1997**, *97*, 1325–1357.
54. A. Mulder, J. Huskens, D. N. Reinhoudt, *Org. Biomol. Chem.* **2004**, *2*, 3409–3424.
55. M. T. Rojas, R. Koniger, J. F. Stoddart, A. E. Kaifer, *J. Am. Chem. Soc.* **1995**, *117*, 336–343.
56. M. Weisser, G. Nelles, P. Wohlfart, G. Wenz, S. Mittler-Neher, *J. Phys. Chem.* **1996**, *100*, 17893–17900.
57. G. Nelles, M. Weisser, R. Back, P. Wohlfart, G. Wenz, S. Mittler-Neher, *J. Am. Chem. Soc.* **1996**, *118*, 5039–5046.
58. M. W. J. Beulen, J. Bügler, B. Lammerink, F. A. J. Geurts, E. Biemond, K. G. C. van Leerdam, F. C. J. M. van Veggel, J. F. J. Engbersen, D. N. Reinhoudt, *Langmuir* **1998**, *14*, 6424–6429.
59. J. Huskens, M. A. Deij, D. N. Reinhoudt, *Angew. Chem. Int. Ed.* **2002**, *41*, 4467–4471.
60. T. Auletta, B. Dordi, A. Mulder, A. Sartori, S. Onclin, C. M. Bruinink, M. Peter, C. A. Nijhuis, H. Beijleveld, H. Schonherr, G. J. Vancso, A. Casnati, R. Ungaro, B. J. Ravoo, J. Huskens, D. N. Reinhoudt, *Angew. Chem. Int. Ed.* **2004**, *43*, 369–373.
61. C. A. Nijhuis, J. Huskens, D. N. Reinhoudt, *J. Am. Chem. Soc.* **2004**, *126*, 12266–12267.
62. V. Mahalingam, S. Onclin, M. Peter, B. J. Ravoo, J. Huskens, D. N. Reinhoudt, *Langmuir* **2004**, *20*, 11756–11762.
63. X. Y. Ling, L. Malaquin, D. N. Reinhoudt, H. Wolf, J. Huskens, *Langmuir* **2007**, *23*, 9990–9999.
64. M. J. W. Ludden, D. N. Reinhoudt, J. Huskens, *Chem. Soc. Rev.* **2006**, *35*, 1122–1134.
65. X. Y. Ling, D. N. Reinhoudt, J. Huskens, *Chem. Mater.* **2008**, *20*, 3574–3578.
66. X. Y. Ling, I. Y. Phang, D. N. Reinhoudt, G. J. Vancso, J. Huskens, *Int. J. Mol. Sci.* **2008**, *9*, 486–497.
67. D. A. Dehlinger, B. D. Sullivan, S. Esener, M. J. Heller, *Small* **2007**, *3*, 1237–1244.
68. S. J. Park, T. A. Taton, C. A. Mirkin, *Science* **2002**, *295*, 1503–1506.
69. J. H. Moon, J. Ford, S. Yang, *Polym. Adv. Technol.* **2006**, *17*, 83–93.
70. T. Ito, S. Okazaki, *Nature* **2000**, *406*, 1027–1031.
71. Y. Xia, J. A. Rogers, K. E. Paul, G. M. Whitesides, *Chem. Rev.* **1999**, *99*, 1823–1848.
72. T. Vossmeyer, S. Jia, E. DeIonno, M. R. Diehl, S. H. Kim, X. Peng, A. P. Alivisatos, J. R. Heath, *J. Appl. Phys.* **1998**, *84*, 3664–3670.
73. C. A. Fustin, G. Glasser, H. W. Spiess, U. Jonas, *Langmuir* **2004**, *20*, 9114–9123.
74. H. Xu, R. Hong, T. Lu, O. Uzun, V. M. Rotello, *J. Am. Chem. Soc.* **2006**, *128*, 3162–3163.
75. A. M. Brozell, M. A. Muha, A. N. Parikh, *Langmuir* **2005**, *21*, 11588–11591.
76. Y. Masuda, T. Itoh, K. Koumoto, *Adv. Mater.* **2005**, *17*, 841–845.
77. Y. Masuda, T. Itoh, K. Koumoto, *Langmuir* **2005**, *21*, 4478–4481.
78. K. Akamatsu, A. Kimura, H. Matsubara, S. Ikeda, H. Nawafune, *Langmuir* **2005**, *21*, 8099–8102.
79. B. D. Gates, Q. B. Xu, M. Stewart, D. Ryan, C. G. Willson, G. M. Whitesides, *Chem. Rev.* **2005**, *105*, 1171–1196.
80. P. C. Hidber, W. Helbig, E. Kim, G. M. Whitesides, *Langmuir* **1996**, *12*, 1375–1380.
81. V. Santhanam, R. P. Andres, *Nano Lett.* **2004**, *4*, 41–44.
82. V. Santhanam, J. Liu, R. Agarwal, R. P. Andres, *Langmuir* **2003**, *19*, 7881–7887.
83. M. Himmelhaus, H. Takei, *Phys. Chem. Chem. Phys.* **2002**, *4*, 496–506.
84. M. Wang, H.-G. Braun, E. Meyer, *Chem. Mater.* **2002**, *14*, 4812–4818.
85. X. C. Wu, A. M. Bittner, K. Kern, *Adv. Mater.* **2004**, *16*, 413–417.
86. T. Kraus, L. Malaquin, E. Delamarche, H. Schmid, N. D. Spencer, H. Wolf, *Adv. Mater.* **2005**, *17*, 2438–2442.
87. T. Kraus, L. Malaquin, H. Schmid, W. Riess, N. D. Spencer, H. Wolf, *Nature Nanotechnol.* **2007**, *2*, 570–576.

88. O. Crespo-Biel, B. Dordi, P. Maury, M. Péter, D. N. Reinhoudt, J. Huskens, *Chem. Mater.* **2006**, *18*, 2545–2551.
89. D. Qin, Y. N. Xia, B. Xu, H. Yang, C. Zhu, G. M. Whitesides, *Adv. Mater.* **1999**, *11*, 1433–1437.
90. H. P. Zheng, I. Lee, M. F. Rubner, P. T. Hammond, *Adv. Mater.* **2002**, *14*, 569–572.
91. D. J. Zhou, A. Bruckbauer, C. Abell, D. Klenerman, D. J. Kang, *Adv. Mater.* **2005**, *17*, 1243–1248.
92. M. T. Zin, A. M. Munro, M. Gungormus, N. Y. Wong, H. Ma, C. Tamerler, D. S. Ginger, M. Sarikaya, A. K. Y. Jen, *J. Mater. Chem.* **2007**, *17*, 866–872.
93. L. J. Guo, *Adv. Mater.* **2007**, *19*, 495–513.
94. P. Maury, M. Péter, V. Mahalingam, D. N. Reinhoudt, J. Huskens, *Adv. Funct. Mater.* **2005**, *15*, 451–457.
95. P. Maury, M. Péter, O. Crespo-Biel, X. Y. Ling, D. N. Reinhoudt, J. Huskens, *Nanotechnology* **2007**, *18*, 044007.
96. D. Wouters, U. S. Schubert, *Angew. Chem. Int. Ed.* **2004**, *43*, 2480–2495.
97. N. L. Rosi, C. A. Mirkin, *Chem. Rev.* **2005**, *105*, 1547–1562.
98. J. C. Garno, Y. Yang, N. A. Amro, S. Cruchon-Dupeyrat, S. Chen, G. Y. Liu, *Nano Lett.* **2003**, *3*, 389–395.
99. R. Maoz, S. R. Cohen, J. Sagiv, *Adv. Mater.* **1999**, *11*, 55–61.
100. L. M. Demers, C. A. Mirkin, *Angew. Chem. Int. Ed.* **2001**, *40*, 3069–3071.
101. X. G. Liu, L. Fu, S. H. Hong, V. P. Dravid, C. A. Mirkin, *Adv. Mater.* **2002**, *14*, 231–234.
102. S. T. Liu, R. Maoz, J. Sagiv, *Nano Lett.* **2004**, *4*, 845–851.
103. S. T. Liu, R. Maoz, G. Schmid, J. Sagiv, *Nano Lett.* **2002**, *2*, 1055–1060.

Chapter 14

Immobilization and Patterning of Biomolecules on Surfaces

DOROTA I. ROŻKIEWICZ, BART JAN RAVOO, AND
DAVID N. REINHOUDT

The Supramolecular Chemistry of Organic–Inorganic Hybrid Materials. Edited by Knut Rurack and Ramón Martínez-Máñez

14.1 INTRODUCTION

"Biochips" and "bioplatforms" that aim at the simultaneous analysis of thousands of oligonucleotides,[1] proteins,[2] carbohydrates,[3] drug-like molecules,[4] cells,[5] or even tissues[6] in a microarray have attracted a lot of attention during the last ten years. Currently, there is a high demand for methods to produce substrates for biomolecule immobilization that provide a high probe loading density, maintain biological activity, and are easy to prepare and use. The development of a general method for biomolecule immobilization will reduce the differences in chip production and increase the reproducible comparison between chips fabricated by many research groups and companies.

Surface chemistry is one of the keys to successful development of reliable biochips. Surfaces must be designed and prepared to optimize the immobilization of probe biomolecules and also should resist nonspecific binding of target species. In terms of immobilization method, the chemical and biochemical differences between different classes of biomolecules have to be taken into account. For any given probe molecule, there is likely to be an optimal surface immobilization technique that will allow for attachment at the highest possible concentration and with preservation of required activity. The biomolecules of interest can be broadly grouped into proteins (antibodies, enzymes, and receptors), nucleic acids [DNA, RNA, and peptide nucleic acids (PNA)], carbohydrates and glycoconjugates, lipids, and various small molecules (e.g., metabolites, peptides, and hormones). For instance, proteins are chemically and structurally much more complex and heterogeneous than nucleic acids, and they can easily lose their structure and biochemical activity due to denaturation, dehydration, or oxidation. The detection of proteins by antibody–antigen interactions is characterized by a broad range of specificity and affinity while the hybridization of two DNA strands is limited by the degree of complementarity of two strands and a few other physical parameters like melting temperature (T_m) or ionic strength.[7] Since DNA is uniformly negatively charged, the spontaneous adsorption to the substrate is much easier to exclude in comparison with protein adsorption due to electrostatic, coordinative, van der Waals, and hydrophobic interactions, as well as conformational changes.[8] The immobilization of the biomolecules should be conducted in such a manner that the background signal is minimal. Achieving a low degree of unspecific binding is a very important factor for the quality of the microarray, signal intensity, and quantification of binding events. Another important factor that should be taken into consideration when designing a biochip is the lateral interaction that can affect the density of surface-bound biomolecules. These interactions can result from either electrostatic repulsion between molecules with like charges, or dipole–dipole interactions, which can be repulsive or attractive, depending on the position and ordering of molecules on the surface. Taking into account the above mentioned issues, the surface chemistry must be carefully tailored depending on the application of each biochip.

The performance of the immobilized biomolecules is strongly dependent on a number of parameters related to the immobilization process. These include: (1) the chemical and physical properties of the surface, since they can influence nonspecific binding of target and nontarget molecules; (2) the distance between the immobilized

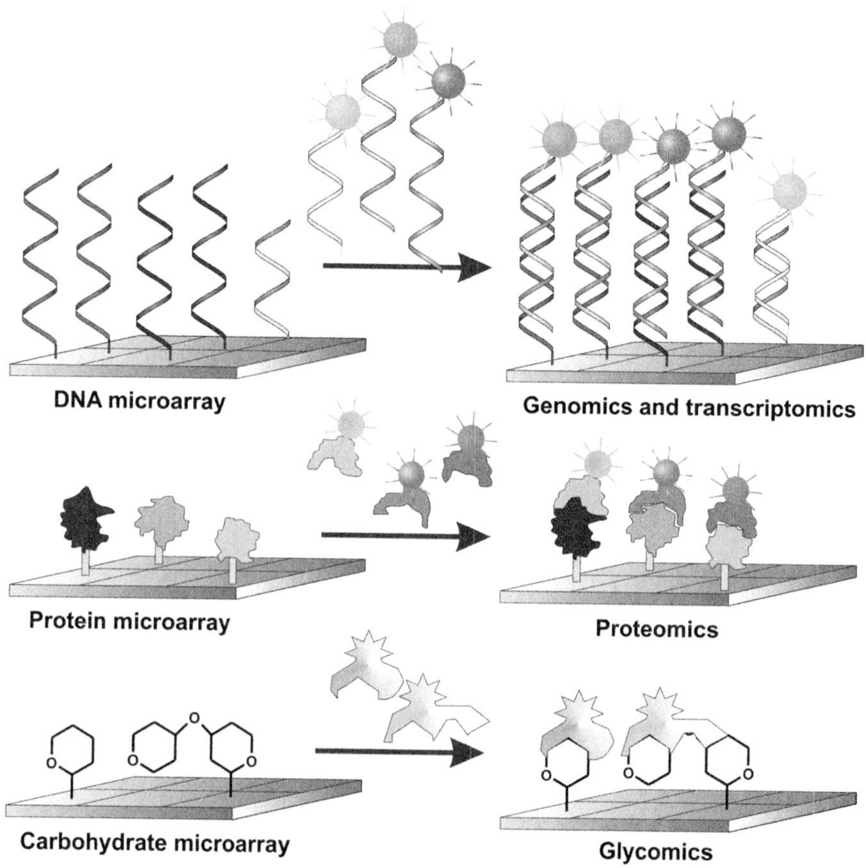

Figure 14.1 Immobilization of biomolecules in microarray format for the investigation of biological processes. Molecular recognition is the key event in all microarrays. (Adapted from Reference 11.) (See color insert.)

biomolecules and the solid surface and the orientation of the immobilized molecules; (3) the density of the biomolecules on the surface, which can limit the sensitivity; and (4) the chemical composition and structure of the biomolecule. Overall, it is complex to predict or model all effects of the above interactions and the behavior of biomolecule adsorption on a particular surface (Fig. 14.1).

The surface and type of selected immobilization method will affect the bioactivity, the concentration, and the target-binding ability of bound probe molecules. Gold and glass substrates are good candidates for the immobilization of biomolecules. These substrates have a number of favorable characteristics: (1) they are chemically homogeneous and stable, (2) surface properties like wettability can be fine-tuned, (3) these substrates can be modified with a wide range of chemical functionalities, (4) the surface modification processes are generally highly reproducible, and (5) in case of glass, the substrates have low intrinsic fluorescence. Glass substrates are very commonly used for microarray application due to their chemical resistance against solvents, low fluorescent noise, optical transparency, and flatness. In addition they are easy to handle and inexpensive. The immobilization of biomolecules onto gold and/or glass supports involves three general steps: (1) the chemical modification of the biomolecule in such a manner that it can interact with complementary functionalities present on the substrate to form a stable bond, (2) the modification of the support surface with sufficient functional groups to allow specific binding and prevent nonspecific adsorption of the biomolecules, and (3) the use of a transport-delivery system (e.g., a stamp, a tip, or a microfluidic channel) that brings small quantities of the biomaterial to specific positions on the surface.

Specific groups involved in covalent immobilization of biomolecules to the surface are very often introduced into the structure of this molecule, especially in oligonucleotides and carbohydrates. Carboxylic and amino functionalities are the modifications most frequently used, although aldehyde, thiol, and oxyamino are useful alternatives. In some cases coupling molecules (cross-linkers) are necessary to increase the affinity (or reactivity) of the interacting groups.

Direct attachment of biomolecules to the surface can introduce a steric constraint to reactivity of the molecule which is not encountered when considering molecules free in a solution. This effect can be minimized if, for example, a spacer is introduced between the biomolecule and the linking group. The spacer can be of nearly any desired length and possess a variety of chemical characteristics, that is, it can be rigid or flexible, hydrophilic or hydrophobic, charged or neutral.[1,9]

Depending on the type of biomolecule, different approaches in attachment to the substrate are preferred. Proteins usually are bound to the surface via amino groups of lysines or at the termini of the peptide chain. Antibodies can be bound through thiol groups, with the most reactive of them being positioned either in the hinge region or between light and heavy chains. Alternatively antibodies can be bound via carbohydrate residues of the Fc region and coupled with protein A or protein G.[7,10] Oligonucleotides can be bound to the surface through the reactive linker introduced into the 5′-terminus of the strand. Chemically modified or unmodified carbohydrates can be attached covalently to functionalized supports or can be nonspecifically and noncovalently immobilized to the underivatized surfaces.[11] Lipids and lipid bilayers can be physisorbed onto hydrophilic surfaces such as mica or amino-modified substrates.

The covalent immobilization of biomolecules seems to be the best method for the attachment since physical adsorption of molecules, albeit the simplest strategy, can be rather hard to control. In case of adsorption, the close proximity between the molecules and the surface could have an influence on the reactivity of that molecule. In addition the surface can be susceptible to exchanging adsorbed molecules with others, for example, adsorbed proteins can be exchanged with the proteins from the solution.[12] Covalent attachment of biomolecules to a surface generally has more advantages. It can be designed in more controlled fashion when the immobilization takes place by exposing the active sites on the surface and binding occurs through specific sites that are distributed elsewhere in the biomolecule. Covalent immobilization could also be random when the binding occurs through the active functional groups that are arbitrarily or widely distributed over the structure of the molecule. Nevertheless, the advantages of covalent immobilization depend strongly on surface chemistry and analytical techniques used for detection.

Another issue in the immobilization of biomolecules on solid supports is spatial control and diversity or loading capacity of different molecules on one substrate. Many techniques have been developed to pattern DNA, proteins, carbohydrates, and other biomolecules on planar substrates. Soft lithography opened the access to simple and cost-effective methods to create patterns of biomolecules by microcontact printing (μCP), microfluidic networks, and microwells. Dip-pen nanolithography (DPN) enables production of submicrometer features and when conducted using the multiple-pen array it can result in thousands of features produced simultaneously. Ink-jet technology and contact printing systems have been developed especially for microarray purposes. These massively parallel methods can fabricate thousands of spots of different oligonucleotides or proteins on one chip. In this chapter, methods and aspects of immobilization and patterning of biomolecules are described. The main focus is the immobilization of biomolecules and soft lithography as a universal technique for fabrication of functional biomolecular patterns on solid surfaces.

14.2 SUBSTRATES FOR THE IMMOBILIZATION OF BIOMOLECULES

Glass slides are one of the most popular substrates for the immobilization of biomolecules, since they possess many advantages such as availability, low cost, flatness, rigidity, and optical transparency and the possibility of the introduction of diverse reactive chemical groups.[13–16] The main method for the functionalization of glass slides for covalent attachment of biomolecules is the reaction between the surface silanols (Si–OH) (obtained after activation of the surface with, for example, "piranha" solution or oxygen plasma) and alkyltrichlorosilanes, trialkoxy(alkyl)silanes, or chlorodimethyl alkylsilanes with a terminal functional group.[17] Monolayers of organosilanes with a terminal functional group are robust and versatile substrates for the immobilization of biomolecules.

The density of active surface groups is an important factor that can influence the immobilization of biomolecules. One example for enhancing surface density with reactive groups on the glass surface is modification with dendrimers or dendritic structures. These nanoscopic polymers are characterized by regular dendritic branching, radial symmetry, ease of control of surface functionality, and uniform size.[18] Covalent attachment of dendrimers to a support builds a three-dimensional (3D) structure that can consequently yield a higher density of reactive groups and as a result of that a higher density of immobilized molecules. Dendrimer-modified substrates were previously used for the attachment of DNA molecules.[19,20] The method was based on modification of glass with aminosilanes and subsequent activation with a homobifunctional linker, such as disuccinimidylglutarate (DSG) or 1,4-phenylenediisothiocyanate (PDITC). The activated amine groups were further reacted with a starburst dendrimer that contains amino groups in the outer sphere [poly(amidoamine) dendrimers,

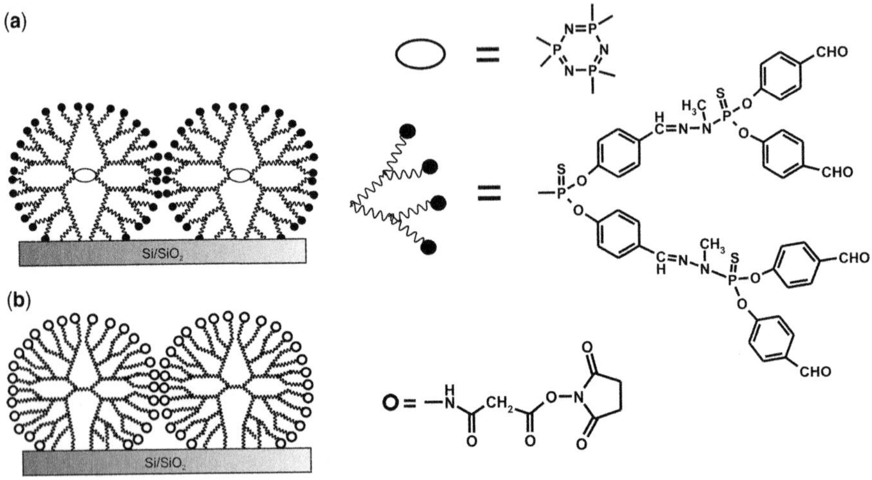

Figure 14.2 Schematic illustration of dendrimer-modified glass support. (a) Aldehyde-terminated dendrimers attached onto amino-modified glass slides (dendri-slides). (b) PAMAM dendrimers activated with glutaric anhydride and *N*-hydroxysuccinimide.

PAMAM]. The final step of this functionalization relied on activation and cross-linking of attached dendrimers with a homobifunctional spacer (DSG or PDITC). Alternatively, after attachment of dendrimers to the surface, glutaric anhydride activated with *N*-hydroxysuccinimide can be used. This surface modification yields a thin, chemically reactive polymer film, which is covalently attached to the glass support and can be directly used for the covalent attachment of amino-modified components, such as DNA or peptides (Fig. 14.2b).

Dendron-modified glass slides were used to obtain controlled spacing between immobilized DNA molecules.[21] Aldehyde-functionalized dendrimers were attached covalently to the amino-silanized glass substrates generating a reactive layer that can bind efficiently amino-functionalized DNA. These "dendrislides" can create high surface coverage with oligonucleotides that can be further hybridized with high yield (Fig. 14.2a).[22] Grafting of DNA on such a surface resulted in uniform, homologous, and also high density layers. The modification with dendritic molecules can be achieved by either *in situ* synthesis or by direct surface attachment of presynthesized branched structures.

Another interesting modification of glass surfaces was introduced by Beier and Hoheisel.[23] They synthesized a flexible, dendritic linker system that enables covalent immobilization of oligonucleotides and PNAs with high loading capacity in a controlled manner. This method facilitates the modulation of surface properties such as hydrophobicity and charge. The synthesis of the linker system involves two consecutive reactions: an acylation of surface-bound amine groups with acid chloride (4-nitrophenyl-chloroformate or acryloylchloride) and subsequent reaction with an amine. A bis-amine results in a linker system, while a polyamine produces a dendritic structure (Fig. 14.3). Because polyamines possess primary and secondary amine

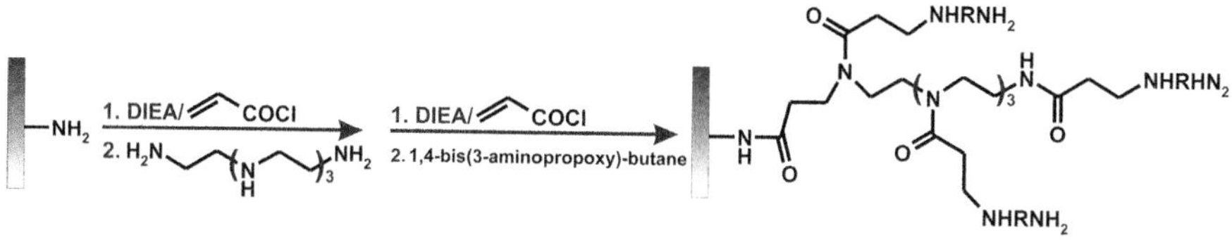

Figure 14.3 Dendritic linker structure synthesized on an amino-functionalized substrate. DIEA = diisopropylethylamine.

groups, not one unique product is generated in such a reaction, but a mixture of compounds.

The density of immobilized proteins was successfully improved when a dendrimer-activated surface was used.[24] Poly(propyleneimine) dendrimers were bound to the Si and glass surface that was modified with 1,1′-carbonyldiimidazole under anhydrous conditions. Due to the high concentration of charged amine groups of dendrimers, this method provides a fine platform for high density protein immobilization.

Another method to increase the number of reactive groups on the glass surface is modification of the substrates with colloidal silica films. These films are usually spin-coated onto a glass substrate and stabilized by thermal curing.[25] The "three-dimensional," porous, colloidal silica coatings of surfaces exhibit better performance than flat glass substrates in terms of the efficiency of chemical synthesis. Glazer et al. showed immobilization of DNA probe arrays onto the substrates modified with colloidal silica films.[26] Subsequent hybridization with the target sequence revealed 20 times higher signal intensity than the one obtained on modified, flat glass. Cunningham and Campbell[27] have established that colloidal silica films can be attached to plastic supports and used for immobilization of antibodies with higher protein loading per centimeter square than conventional two-dimensional surfaces.

Self-assembled monolayers (SAMs) on gold and other metal surfaces have been extensively studied in biosensors applications.[28–30] The thiol–gold chemistry is well known and is much easier controlled than organosilane chemistry (Fig. 14.4).[31–34] The

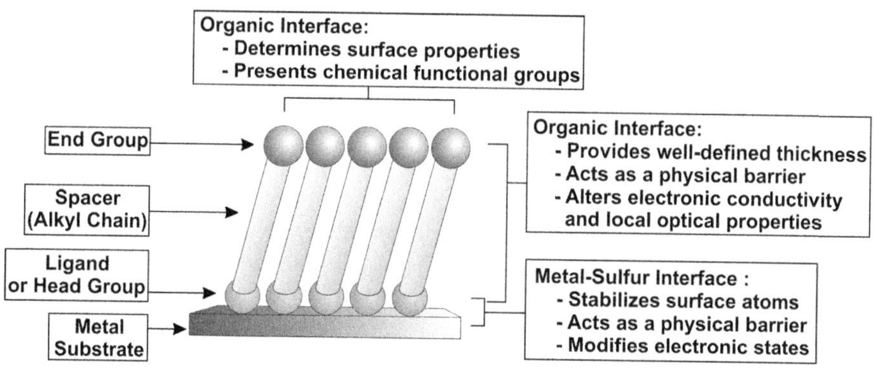

Figure 14.4 Schematic diagram of a self-assembled monolayer of *n*-alkylthiols on a gold substrate. (Adapted from Reference 31.)

functionality of the thiol SAM can be tailored by choosing a reactive end group of the thiol (carboxylic, hydroxyl, or amine are the most frequently used examples). Biomolecules can be covalently attached to such monolayers in two manners: directly to the gold surface (when possessing linkers with thiol termination or when possessing thiol residues in its structure) or by the reaction with an end group of a thiol SAM. Monolayers can also be introduced as a mixture of usually two components, thiols with different functionalities as end groups (mixed monolayers). This influences the spacing between attachment points for biomolecules and prevents steric hindrance between them, protecting the loss of their full reactivity. Mixed SAMs for the covalent attachment of biomolecules possess utility in minimizing nonspecific binding or affinity of biomolecules to the surface when the thiols with reactive groups are mixed with thiols that can prevent nonspecific absorption (mainly thiols modified with polyethylene glycol, polyvinylpyrrolidone, polyvinyl alcohol chains, and others).[35] Even though gold showed the best properties for the functionalization with SAMs, other metal surfaces also have been used for the attachment of biomolecules. Oligonucleotides and proteins have been immobilized onto SAMs on metal substrates such as titanium, aluminum, and silver.[36–39]

A number of papers have been published that demonstrate that biomolecules can also be immobilized in molecular monolayers on silicon,[40] silicon nitride,[41] and amorphous carbon[42] surfaces. It is evident that these new methods will further increase the scope of substrates for biological microarrays.

14.3 IMMOBILIZATION OF BIOMOLECULES WITH COVALENT AND NONCOVALENT LINKERS

The reactivity of the solid substrates is determined by the surface functional groups. Immobilization of the biomolecules to the substrates can be performed via several routes. Immobilization can be done by a direct attachment of the molecule to the functionalized or nonfunctionalized surface, or by the employment of a cross-linker (homobifunctional, heterobifunctional, or multifunctional) between the functionalized surface and the biomolecule. Generally, the selection of the substrate and the chemistry is crucial for the successful immobilization of biomolecules and application of that substrate. Here, the reaction between the biomolecule and the reactive surface groups is described for both types of substrates: silicon oxide/glass and gold.

14.3.1 Surface Activation for Covalent Attachment

Surface activation compounds are bound to the substrate to change the functionality of the substrate. Gold and silicon oxide or glass surfaces can be modified with reactive monolayers of thiols and alkylsilanes, respectively. These monolayers should bear a functionality that can react with biomolecules via a covalent bond. The monolayers may be modified by the introduction of another reactive group or not modified, keeping their native reactive groups. The preparation of clean gold or glass surfaces in order to remove any organic surface contaminations that could possibly interfere with subsequent monolayer formation involves cleaning with detergents, strong oxidants ("piranha" solution, H_2SO_4/H_2O_2, NH_3/H_2O_2), UV-plasma cleaning, or sonication. Typical chemistry involved in binding biomolecules relies on the reaction between epoxide-, aldehyde-, or isothiocyanate-terminated reactive monolayers and amino-modified or amino-containing biomolecules. Examples of such surface reactions are listed in Table 14.1.

Reaction of surface hydroxyl groups with 2,2,2-trifluoroethanesulfonyl chloride (tresyl chloride) yields sulfonate esters which in a subsequent step can be reacted with amino- or mercapto-modified biomolecules for covalent immobilization

Table 14.1 Characteristic Surface Reactions to Couple Biomolecules with Solid Supports

Surface chemistry	Modification of biomolecule	References
$Si\text{-}SiO_2$/glass-silane-epoxide	Amino-DNA	43, 44
$Si\text{-}SiO_2$/glass-silane-PDI-PAMAM-PDI	Amino-DNA	20
$Si\text{-}SiO_2$/glass-silane-aldehyde	Amino-DNA	45
$Si\text{-}SiO_2$/glass-silane-sulfhydryl	Disulfide-DNA	46
$Si\text{-}SiO_2$/glass-silane-sulfhydryl	Maleimide-DNA, Maleimide-carbohydrates	9, 47, 48
$Si\text{-}SiO_2$/glass-silane-maleimide	Thiol-DNA	49
$Si\text{-}SiO_2$/glass-silane-bromoacetamide	Phosphorothiolate-DNA	50
$Si\text{-}SiO_2$/glass-silane-sulfhydryl	Acrylamide-DNA	51
$Si\text{-}SiO_2$/glass-silane-phosphane derivative	Azide-carbohydrates	52
$Si\text{-}SiO_2$/glass-silane-bromoacetamide	(Internal)phosphorothiolate-DNA	16
$Si\text{-}SiO_2$/glass	Silanized-DNA	53
Au-thiol/amine-maleimide	Thiol-DNA	54
Au-thiol/amine-disulfide	Thiol-DNA	55
Au-thiol/azide	Acetylene-DNA	56
Au-thiol/benzoquinone	Cyclopentadiene-carbohydrates	57
Au-thiol/NHS ester	Amino-proteins	58
Au	Thiol-DNA or cysteine-containing peptides or proteins	51

(Fig. 14.5a).[59] Aldehyde-modified surfaces can be reacted with amino groups of biomolecules to form imine linkages which can be subsequently reduced to secondary amines by the reaction with sodium borohydride (Fig. 14.5b).[60] Disulfide-modified biomolecules can be immobilized onto mercapto-silanized glass supports by a thiol/disulfide exchange reaction (Fig. 14.5c).[46] Isothiocyanate functionalized substrates can be used for the immobilization of amino-functionalized molecules (Fig. 14.5d) similarly to epoxy-terminated SAMs (Fig. 14.5e).[61] Azide-functionalized surfaces can be used to immobilize alkyne-bearing molecules by triazole formation usually using copper(I) catalyst (Fig. 14.5g). This chemistry is highly reliable, fast, and resistant to side reactions.[56] Active sulfhydryl-functionalized surfaces can

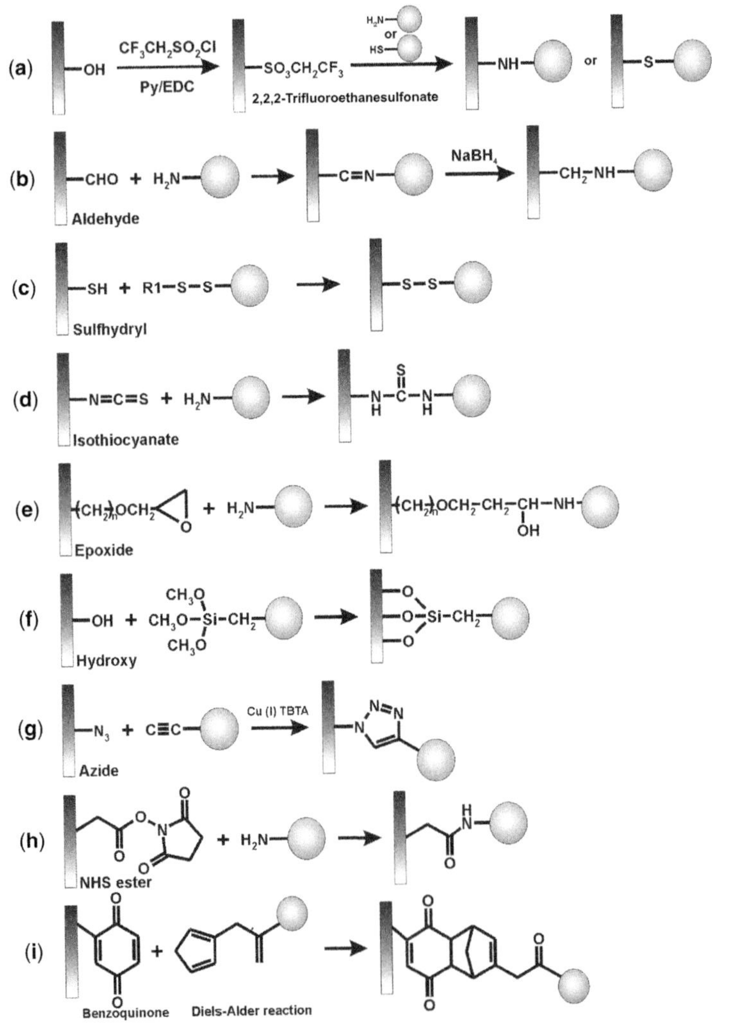

Figure 14.5 Surface activation for direct immobilization of biomolecules.

be easily reacted with maleimide-modified biomolecules.[55] Biomolecules containing free amino groups can be coupled to a monolayer bearing *N*-hydroxysuccinimidyl (NHS) ester groups on the surface through an amide bond. The NHS-ester coupling reaction is easy to carry out and proceeds in good yields (Fig. 14.5h).[58]

Silicon substrates can be modified with 3-mercaptopropyltrimethoxysilane (MPTMS) and reacted with biomolecules functionalized with acrylamide groups. The acrylamide moieties react with surface thiol groups by Michael addition forming covalent links between biomolecule and the surface. In terms of using this chemistry to pattern the surface, remaining thiol surface groups can be passivated by the reaction with acrylic acid monomers at pH 10.[51] Unmodified microscope slides may be used for direct covalent conjugation with an activated, silanized oligonucleotide (Fig. 14.5f).[59] The advantage of this system is that the immobilization process is very simple and functionalization of the surface is not necessary.

A clean gold surface can react directly with thiol-containing molecules. Direct immobilization of proteins on gold surfaces can be achieved simply by affinity of cysteine residues in the protein to the gold surface resulting in a strong adsorption.[62,63] After immobilization, a passivation should be carried out in order to block the remaining gold areas. The benefit of that method is that no surface modification is needed in order to attach the molecule.[51] The introduction of a diene-functionality into a protein is a one-step reaction by acylation of the amino group of lysine from the protein and a bifunctional cross-linker which possess a succinimidyl ester on one end and a hexadienyl functionality on the other end. This dienyl conjugate can then be immobilized onto maleimide-modified substrates via Diels–Alder cycloaddition. The Diels–Alder reaction was shown to be compatible in the bioconjugation process and immobilization of oligonucleotides and other biomolecules.[64,65] The reaction is highly selective and can be carried out in water with a higher rate than in organic solvents. Another example of the application of Diels–Alder reactions for the immobilization of biomolecules was presented by Houseman and Mrksich.[57] Cyclopentadiene-modified carbohydrates were reacted with a benzoquinone-functionalized monolayer on gold. This reaction was found to be highly efficient and selective for the immobilization of carbohydrates on the surface (Fig. 14.5i).

14.3.2 Introduction of Homobifunctional Cross-Linkers on the Surface

It is essential that after immobilization of biomolecules on the surface, the biomolecule will retain its activity. In order to prevent the loss of bioactivity, the molecule should be separated from the surface to avoid steric clashes with incoming molecules. One way to do so is to introduce a linker molecule to create a certain distance between the surface and biomolecule. Homobifunctional surface-activating molecules possess reactive groups on both terminal ends to conjugate the attached biomolecule and the (modified) surface through two identical functional groups. Usually the homobifunctional cross-linker is first reacted with the surface and then the biomolecule is attached. The requirement for this linker is that it should not connect two neighboring groups on the surface blocking the reactivity for the further attachment of other molecules. The linker should be rigid and chemically stable. The most commonly used homobifunctional cross-linkers are glutaraldehyde, 1,4-butanediol diglycidyl ether, 1,4-phenylene diisothiocyanate, dimethylsuberimidate, and terephthaldialdehyde (Fig. 14.6).[60] Amino-modified surfaces can be applied for attachment of molecules that contain free amino groups when the substrate is reacted with a homobifunctional cross-linker like bis(sulfosuccinimidyl)suberate (BS^3) in acetate buffer,[66] terephthaldialdehyde, disuccinimidyl carbonate, or others (Fig. 14.6). A sulfhydryl-terminated monolayer on gold or on silicon oxide surface can be reacted with 2,2′-dipyridyl disulfide to form disulfide bonds on the surface.[55] These disulfide bonds can then be used

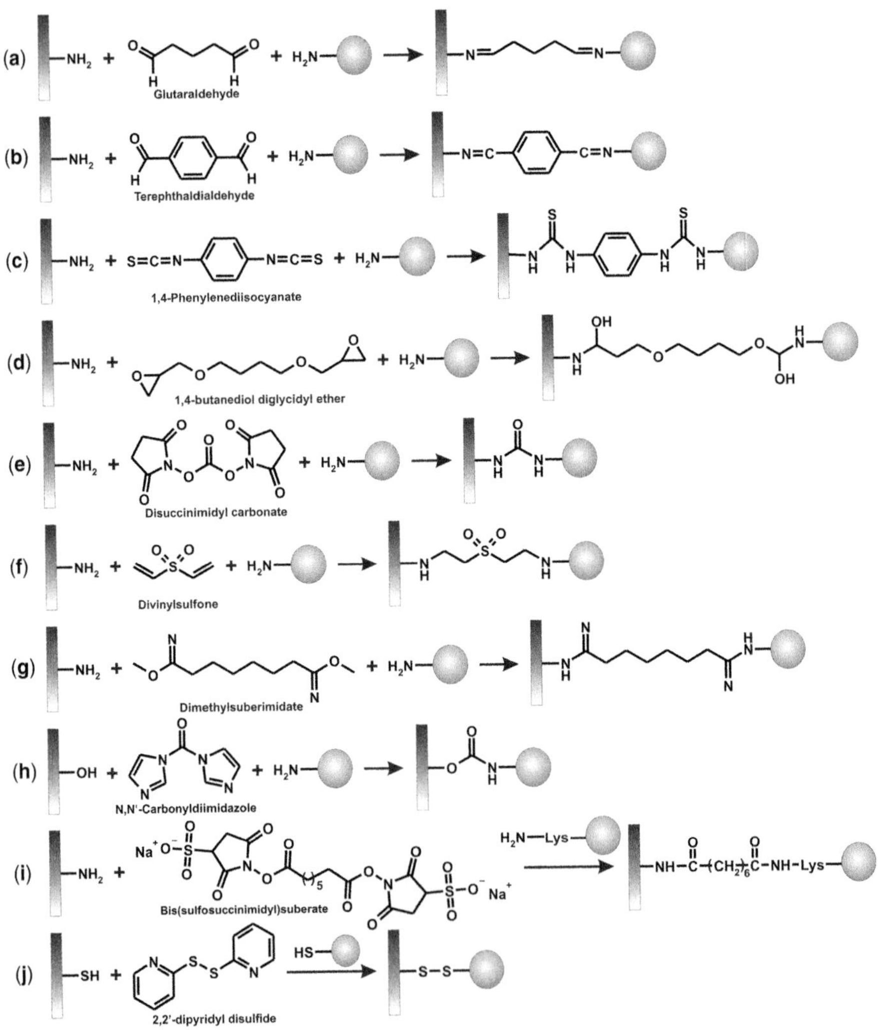

Figure 14.6 Homobifunctional cross-linkers to connect identical functional groups to surfaces.

in a thiol-disulfide exchange reaction with free sulfhydryls in order to attach biomolecules, such as thiol-modified DNA or cysteine-containing polypeptides, onto the surface (Fig. 14.6j).

14.3.3 Introduction of Heterobifunctional Cross-Linkers on the Surface

Heterobifunctional cross-linkers are employed to couple two different functional groups between the monolayer and the molecule for the subsequent immobilization. The wide range of these cross-linkers and their specific reactions are presented in Figure 14.7. One example in which the use of heterobifunctional cross-linkers was applied for immobilization of biomolecules was the work of Corn and coworkers where the sulfosuccinimidyl 4-(*N*-maleimidomethyl)cyclohexane-1-carboxylate (SSMCC) linker was applied to link thiol-modified DNA molecules with an amino-functionalized surface.[54]

Mercapto-functionalized surfaces can be modified with a heterobifunctional linker in order to bind amine groups from biomolecules like, for instance, antibodies.[67] Amino-group-containing biomolecules can be bound to that surface via maleimido-R-*N*-succinimidylester, which contains a sulfhydryl reactive maleimide group at one end

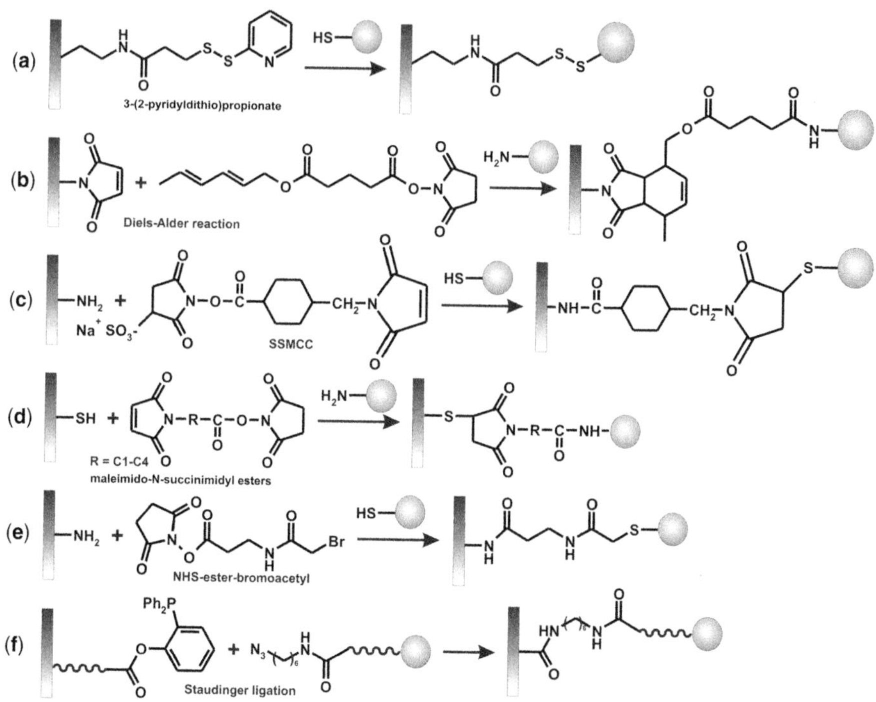

Figure 14.7 Heterobifunctional cross-linkers for the immobilization of biomolecules.

and an NHS ester at the other end (Fig. 14.7d). Similarly, the maleimide-NHS cross-linker can bind mercapto-functionalized biomolecules in reverse orientation on the amino-functionalized surface (Fig. 14.7c). Another example of coupling two molecules where one contains an amino functionality and the other a sulfhydryl can involve succinimdyl 3-(bromoacetamido)propionate. The NHS ester group will bind the linker to amino-functionalized surfaces, exposing the bromoacetyl group for the subsequent reaction with a sulfhydryl containing molecule. A useful review about heterobifunctional cross-linkers is available.[68] An additional advantage of using an NHS-maleimide cross-linker (e.g., aminocaproic acid linker) is that once it is bound to NH_2-surface groups exposing a maleimid functionality it can react with diene-modified biomolecules (for instance proteins).[69] Waldmann et al. have explored the Staudinger ligation to immobilize azide-containing carbohydrates onto phosphane-derivatized glass slides and fabricate small-molecule microarrays (Fig. 14.7f).[52] The glass substrate was functionalized with PAMAM dendrimers to increase the number of reactive sites on the surface.

14.3.4 Supramolecular Immobilization of Biomolecules

An interesting alternative to the covalent immobilization of biomolecules is the use of specific noncovalent interactions to attach biomolecules to surfaces. Tampé et al. described multivalent Ni^{2+} nitrilotriacetate (NTA) thiols that bind histidine-tagged proteins on SAMs on gold substrates.[70] In contrast to the mono-NTA/His_6-tag interaction, which has drawbacks because of its low affinity and fast dissociation, drastically improved stability of protein binding by the multivalent chelator surfaces was observed. The immobilized His_6-tagged proteins are uniformly oriented and retain their function. At the same time, proteins can be removed from the chip surface under mild conditions by adding competing ligands. Exploiting another noncovalent binding motif, Huskens et al. showed that streptavidin (SAv) is attached to β-cyclodextrin (β-CD) SAMs on gold via orthogonal host–guest and SAv–biotin interactions.[71] The orthogonal linkers combine a biotin functionality to bind SAv and adamantyl functionalities to bind β-CD SAMs through host–guest interaction. The two supramolecular approaches were combined in the multivalent binding of His_6-tagged proteins to β-CD SAMs on gold and on glass by using the Ni^{2+} complex of an adamantyl NTA linker.[72] His_6-tagged maltose binding protein, His_6-tagged DsRed and the α-proteasome could be attached to the β-CD SAMs in a specific manner.

14.4 SOFT LITHOGRAPHY WITH BIOMOLECULES

Soft lithography is an emerging method for micro- and nanofabrication of two- or three-dimensional structures on surfaces.[73–78] This method developed by Whitesides and coworkers uses elastomeric stamps, molds, and conformable photomasks for creating patterns as small as tens of nanometers on substrates. Depending on the

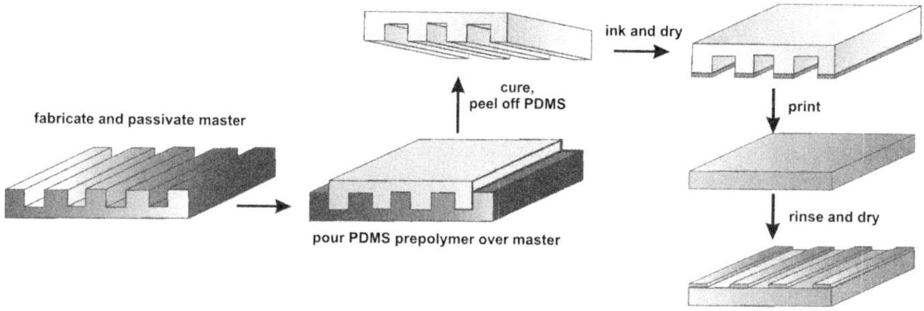

Figure 14.8 Schematic representation of the fabrication of PDMS stamps and the process of microcontact printing.

way that molds are used there are different techniques that belong to soft lithography: replica molding (REM), micromolding in capillaries (MIMIC), microtransfer molding (μTM), solvent-assistant microcontact molding (SAMIM), and microcontact printing (μCP). The main advantage of these techniques lies in the ease and simplicity of their use, which makes them attractive for a wide range of applications. μCP is one of the simplest methods for fabrication of patterns on a surface. An elastomeric stamp with a chemical ink is contacted with the target substrate and forms a SAM pattern, as in Figure 14.8.

Soft lithography, in particular μCP, relies on the fabrication of a master that is used to produce replicas in an elastomeric polymer. The master can be fabricated by any technique that is capable of producing well-defined relief structures on a surface. The most commonly used technique for silicon master fabrication is photolithography. Flexible stamps are generated by casting a liquid elastomer, poly(dimethylsiloxane) (PDMS) on the top of the mold and subsequent curing for at least one hour at 60°C. Since the PDMS stamp is flexible, it is important that its pattern is well defined by means of lateral and vertical control of the chemical pattern. The ratio of width to depth of the pattern should be carefully designed so deformations like pairing, buckling, or collapsing will be avoided.[78] Another drawback of the PDMS stamps is contamination of low molecular mass unpolymerized alkylsiloxanes that leach from the stamp during the printing process. This problem can be avoided by extraction of the stamp with apolar solvents.[79] Swelling during inking of the stamp can cause pattern enlargement during printing.[80,81] A pattern that is increased in size can also be made by applying too much pressure on the stamp during printing. Furthermore, diffusion of the ink due to the low molecular mass of the ink or too high concentration of the ink, or excess of the ink can affect the quality of the pattern. The diffusion of ink that is bound to the surface noncovalently can also occur.[82] Even though μCP has its limitations in resolution, edge definition, and deformation of the stamp, as well as diffusion of inks, this technique is still very attractive for printing biomolecules since it is biocompatible and does not influence or damage the molecule. Biomolecules generally have a higher molecular mass which limits diffusion of molecules during the printing process. The surface of the PDMS stamp can be tailored by depositing other layers, for example, functional silanes, to transfer the ink to the surface with higher yield and better compatibility. The surface chemistry plays an important role since it can attract, bind, or adhere the ink molecules.[74,83] Although μCP was initially used for printing alkylthiols on gold substrates this method was extended to alkylsiloxanes on glass and silicon oxide surfaces,[84] which resulted in additional applications in nanobiotechnology, such as patterned surfaces for local cell immobilizations, fabrication of microarrays or for biosensor and microfluidic purposes. The scope of ink molecules has widened from alkylthiols and alkylsiloxanes to nanoparticles, inorganic inks, organic molecules such as peptides,[85] proteins,[86] or DNA.[87,88]

14.4.1 Microcontact Printing of Biomolecules

The transfer of biomolecules from the stamp to the substrate by μCP generally depends on how the surfaces of the stamp and the substrate are chemically, physically, or biochemically modified. The simplest μCP approach for patterning of biomolecules relies on the direct transfer of the ink molecules, which are adsorbed on the stamp, to a target surface by conformal contact. Nevertheless, depending on the properties of the biomolecule, the surface of the stamp needs to be modified in terms of wettability, charge distribution, introduction of reactive groups, etc. Generally biomolecules such as proteins, lipids, or oligonucleotides are suitable for μCP due to their large molecular weight, which helps in formation of well-defined, high-contrast patterns since the diffusion is limited.

In terms of printing of biomolecules, there are several important factors that have to be considered. The affinity of the biomolecule to the stamp and to the substrate must be tailored so that it is higher for the substrate than for the stamp. The binding of the biomolecule should not cause denaturation (if applied, e.g., in case of proteins) and therefore should not affect the secondary or tertiary structure that can cause the unfolding of the molecule. The biomolecule should be attached to the substrate in a way that will expose all the active sites to the target molecules.

Surface modification of PDMS stamps plays an important role in transfer and printing of biological material. Although the mechanism of protein transfer by μCP is not totally understood, Tan et al. have demonstrated that both stamp and substrate wettability is crucial for biomolecule transport.[89] They showed that a minimum wettability of the substrate is required for successful μCP of proteins, and this minimum wettability can be decreased if the wettability of the stamp is decreased. Their findings also revealed that the mechanism of μCP of proteins is different from protein adsorption because (1) surfaces that are resistant to protein adsorption in aqueous environment are susceptible to μCP under ambient conditions and (2) the amount of immobilized protein varies gradually with the wettability of the substrate for the adsorption process, yet in μCP only negligible protein is immobilized below a certain threshold substrate wettability. It was shown in the literature that many proteins (e.g., immunoglobulin G, IgG) readily adsorb to uncharged PDMS surfaces through van der Waals interactions under physiological conditions even though electrostatic interactions can play a more important role at lower ionic strengths on charged surfaces.[90]

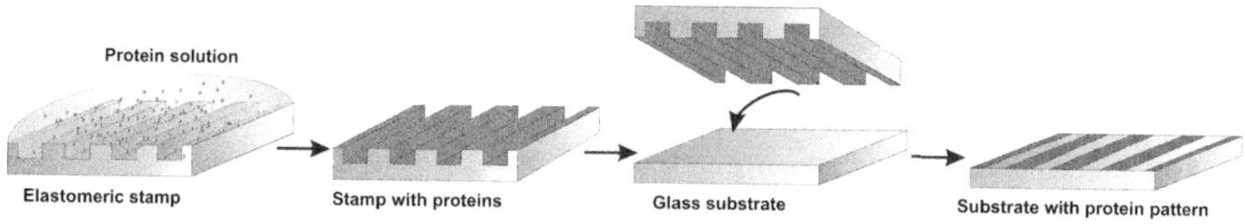

Figure 14.9 Microcontact printing of proteins.[93] A protein solution is incubated on the top of an elastomeric stamp. After drying, the stamp is brought into conformal contact with the glass substrate and transfer of proteins occurs only in the places of contact between the stamp and the substrate.

Applied strategies of pattern formation of proteins or other biomolecules on surfaces relies on: inking of the nonmodified stamp with biomolecule solution, incubation, drying the stamp and bringing the stamp into conformal contact with the (modified) substrate.[90,91] Bernard et al. demonstrated that direct transfer of proteins from nonmodified PDMS stamps to a target glass surface results in patterns with high surface coverage (Fig. 14.9).[87]

Bovine serum albumin (BSA) protein adsorbs readily from solution to hydrophobic surfaces including PDMS and can be easily printed on different types of modified surfaces. For instance, Ross et al. printed BSA from nonmodified PDMS stamp to substrate modified with planar supported lipid bilayer.[92] The deposition of proteins by μCP was demonstrated on different types of substrates, including oxides, metals, and polymers.[87,93] The concept of direct μCP of protein onto a glass substrate via nonmodified PDMS stamps was further extended to the fabrication of single protein molecules such as antibodies (e.g., IgG) and green fluorescent proteins (GFPs) on glass surface.[93] Gold binding polypeptide (GBP) is an interesting example of a protein that is applicable in direct μCP on gold surfaces. This protein does not contain cysteine residues, which are generally known to form a covalent bond with gold. The binding of this protein is independent of thiol linkages and offers therefore a new way of interaction between the biomolecule and the surface. GBP-GFP-His_6 fusion protein was printed directly onto a gold surface in a mixture with BSA, Tween 20, sodium phosphate, and NaCl. The protein was immobilized by chemisorption onto the substrate and later applied for high throughput assays of protein–protein as well as DNA–DNA interactions in microfluidic devices.[94] Kwak et al. patterned cytochrome C onto gold surfaces using a nonmodified PDMS stamp.[95] Since cytochrome C has the ability to transfer an electron, this protein is often used in studies on biomolecular photodiode systems. Cytochrome C was used as an ink to wet the surface of PDMS stamps and the protein arrays were transferred directly from the stamp to a carboxyl-terminated SAM (16-mercaptohexadecanoic acid, MHDA) on gold. After successive washing with detergent and water, the pattern of protein stayed intact on the surface. Active enzymes were also patterned successfully using SAMs on gold surfaces.[96] Metalloprotein (azurin, Az) was printed on a glass substrate modified with mercaptosilane that allows a site-specific binding of the protein. The pattern was assessed by immunofluorescence experiments with anti-Az serum.[97]

The hydrophobic nature of the PDMS stamp can be adjusted by oxidation in O_2-plasma, which produces a thin, glassy silicate layer on the stamp. This layer is brittle and loses its hydrophilic character unless the stamp is kept under water.[66] Another method for changing the surface properties of the stamp is modification with siloxanes, for example, APTES or poly(ethylene glycol) siloxane. Polylysine was microcontact printed on clean, nonmodified glass surfaces via an oxidized PDMS stamp using electrostatic interactions between the positively charged polypeptide and the negatively charged glass surface.[98]

Other biomolecules that are of interest in μCP are lipids and lipid bilayers. Supported lipid bilayers are very fragile assemblies that are formed by lipids that are organized into two opposing leaflets on hydrophilic surfaces, such as glass or mica substrates. These structures can be also patterned on solid substrates but the μCP technique differs slightly from the ones that were applied for proteins or DNA. First, the bilayer has to be formed on the oxidized PDMS stamp from the buffer solution by lipid vesicle fusion. Second, printing has to be carried out in water, otherwise the bilayer will lose its structure.[99] This method allows efficient and reliable transfer of membrane patches to glass surfaces.

The noncovalent adsorption of proteins by μCP is experimentally simple, but suffers from the disadvantage that the attachment can be reversible by rinsing the pattern with certain buffers and detergents or replacement by other proteins in solution. Moreover, the orientation of the deposited protein is not controlled. Delamarche et al. proposed the use of stamps modified with poly(ethylene oxide) silanes.[100] The modification was conducted by oxidation of the PDMS stamp and reaction with APTES to yield an amino-functionalized surface. The next step was the reaction with homobifunctional cross-linker BS^3 to bind surface amino groups with poly(ethylene glycol) (PEG) chains (Fig. 14.10).

When poly(ethylene oxide) (PEO) silane was grafted onto oxidized PDMS stamps it acts as a protein repellent layer. This property was utilized to design a flat stamp with regions that can attract proteins (nonmodified PDMS) and regions modified with PEO that have protein-repellent properties. The local modification of native PDMS was conducted by oxidation in O_2-plasma with the application of a metal mask (areas that were covered by the mask were not oxidized and not modified). Proteins (immunoglobulin G, IgG) were transferred successfully to the glass substrates

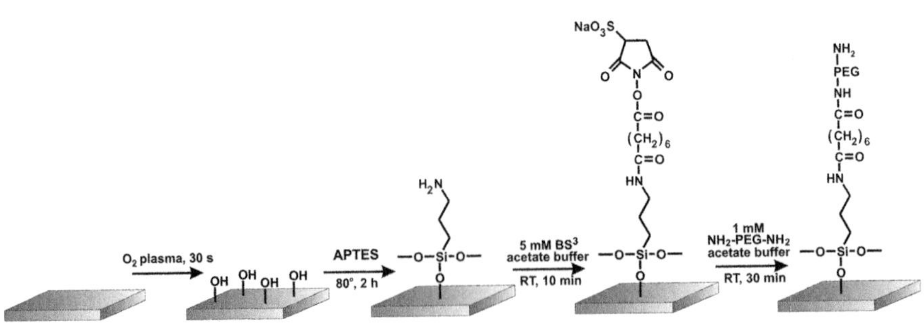

Figure 14.10 Modification of a PDMS stamp for microcontact printing of polar inks. An oxidized PDMS stamp is reacted with APTES and subsequently with BS^3. Finally, the stamp is reacted with poly(ethylene glycol).

and immobilized in a well-defined pattern with high accuracy and contrast. When proteins are applied to such a stamp they are directed to the hydrophobic parts of the stamp. A different approach has also been presented. When the PEO-modified stamp (according to the procedure mentioned above) is contacted with another flat, dry, nonmodified PDMS stamp (ink pad) that was incubated with IgG buffer solution, a homogeneous layer of proteins is transferred to the PEO regions of the other stamp. This stamp can be subsequently contacted with a glass substrate and used to pattern IgG proteins.

An additional example where the modification of the stamp surface was an important factor is μCP of DNA molecules.[87] To attract DNA molecules to the stamp, the surface of that stamp was modified with APTES resulting in positively charged amino group functionalization. In that experiment, electrostatic interactions play an important role in transfer and delivery of DNA (considering DNA as a negatively charged polyelectrolyte). A much more efficient way to transfer micropatterns of DNA and RNA to a surface was obtained by modification of the PDMS stamp with positively charged dendrimers such as poly(propyleneimine) (PPI; "dendri stamp") in a layer-by-layer arrangement.[101,102] The electrostatic interactions between dendrimers and oligonucleotides ensure successful transfer of DNA or RNA to the target surface. Imine chemistry or "click" chemistry can be applied to bind covalently modified DNA and RNA molecules to a chemically functionalized substrate (Fig. 14.11).[100,102] A different approach to pattern DNA molecules on the surface was proposed by Xu et al. DNA-surfactant molecules were prepared by attaching a hydrophobic alkyl chain to the 3′ or 5′ end.[88] The hydrophobic tail allows for the appropriate adsorption to the hydrophobic PDMS stamp. This method allows for efficient transfer and delivery of DNA to the surface.

Another way to overcome problems with wettability and compatibility with aqueous solutions of PDMS stamp is simply to use other materials for fabrication

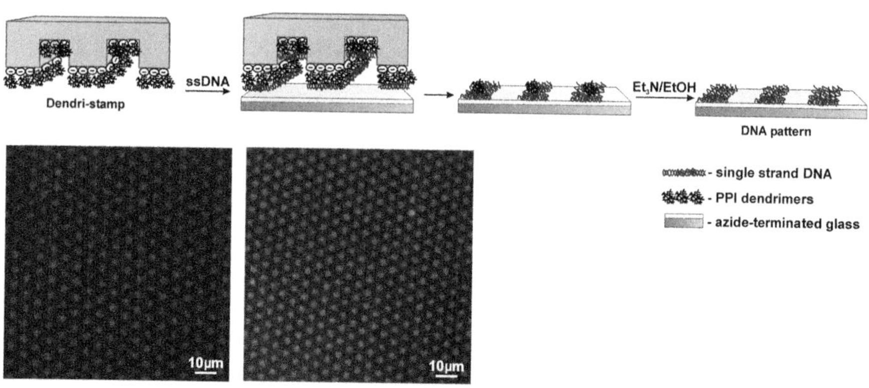

Figure 14.11 Transfer printing of DNA and RNA using dendri-stamps.[101,102] A PDMS stamp is oxidized and coated with a PPI dendrimer. DNA or RNA binds to the dendrimer-coated stamp in a layer-by-layer arrangement. If the DNA (or RNA) is functionalized with an alkyne, it can be printed on an azide-terminated self-assembled monolayer. If the DNA (or RNA) is functionalized with an amine, it can be printed on an aldehyde-terminated self-assembled monolayer. (See color insert.)

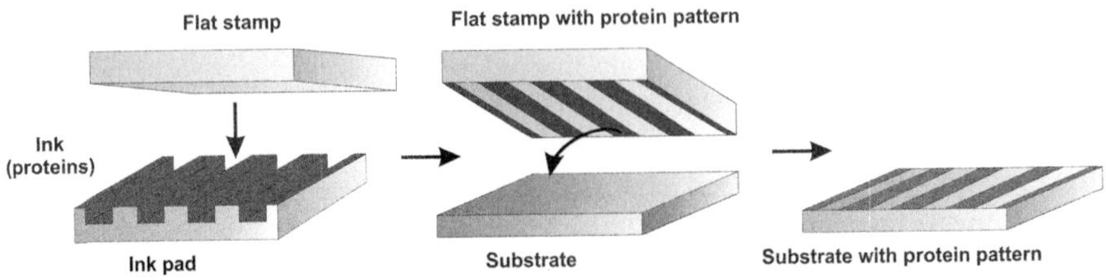

Figure 14.12 Pattern transfer with a flat PDMS stamp. A flat stamp is inked with a patterned ink pad. The ink is then printed on the substrate. (Adapted from Reference 105.)

of the stamp. There are many approaches and many examples have been presented in the literature but two of them were very successful. Agarose stamps[103] are suitable as a mold for transferring water-soluble biomolecules due their high permeability to water. Multiple stamping is possible without the need for intermediate re-inking of the stamp. Spencer and coworkers introduced polyolefin plastomers (POPs) as a stamp material for μCP of proteins (fibrinogen and poly-L-lysine-*g*-poly(ethylene glycol).[104] By using the POP elastomer, a higher resolution of the printed pattern can be achieved and also possible contamination from the stamp (which sometimes can be observed in case of PDMS stamp) does not occur.

Flat PDMS stamps can be substituted for PDMS stamps possessing protruding features.[105] The method was presented by Geissler et al. and it showed improved contrast and resolution of the pattern. The technique relies on introducing a pattern to the flat stamp by contacting with another PDMS stamp (with features; Fig. 14.12) or by using microfluidic networks or microwells. Subsequently, the patterned flat stamp can be contacted with glass or another substrate to transfer the pattern. This method can be applied specifically for high molecular weight inks.

14.4.2 Reactive Microcontact Printing of Biomolecules

μCP was not only used to position (bio)molecules on the surface but also to synthesize them on planar supports.[106–108] Huck et al. introduced peptide synthesis by μCP (Fig. 14.13).[85] An oxidized, hydrophilic PDMS stamp was inked with a solution of {2-[2-(Fmoc-amino)ethoxy]ethoxy}acetic acid (Fmoc = fluorenylmethoxycarbonyl) to activate an amino-functionalized surface. In the second step, the elastomeric stamp was inked with *tert*-butoxycarbonyl (Boc) protected amino acid and brought into conformal contact with the activated surface to form a covalent amide bond without the use of a catalyst. Similarly, a 20-mer peptide nucleic acid (PNA) was synthesized by printing commercially available nucleotides in 20 steps by using the same activation cycle. The PNA strand was finally hybridized with complementary and noncomplementary DNA.[85]

Covalent immobilization of cytophilic proteins by μCP can be also used to pattern cells on substrates. Cytophilic proteins can be printed in micropatterns on top of reactive SAMs using, for example, imine chemistry.[109] Aldehyde-functionalized

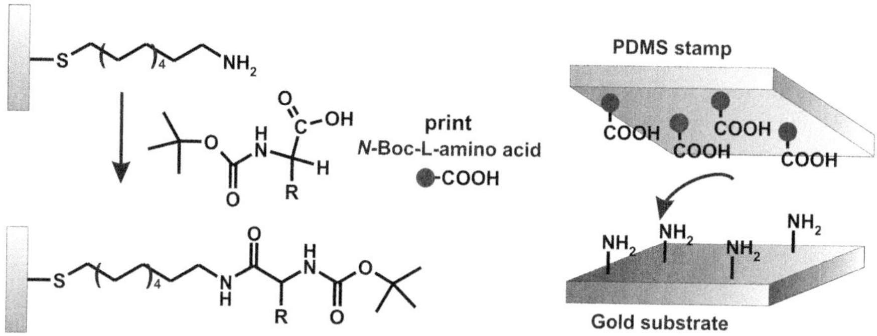

Figure 14.13 Peptide synthesis by μCP.[85] An oxidized PDMS stamp is inked with an *N*-Boc-L-amino acid and pressed into a contact against an amino-functionalized gold substrate to yield a covalent peptide bond.

SAMs can be used as a substrate for the μCP of collagen-derived proteins using an oxidized PDMS stamp. After immobilization of the proteins into adhesive "islands", the remaining areas need to be blocked with amino-poly(ethylene glycol) in order to prevent cell adhesion in those regions. It was found that cells such as HeLa cells adhere and spread selectively on the protein islands, while avoiding the PEG zones. These findings illustrate the importance of μCP as a method for positioning proteins and cells at surfaces.

Another example where the PDMS stamp was used as a tool to locally transfer and covalently bind molecules was reported by Wilhelm and Wittstock.[96] The enzyme glucose oxidase (GOx) was mixed together with a coupling agent (EDAC, *N*-(3-dimethylaminopropyl)-*N*-ethylcarbodiimide hydrochloride) in PBS (phosphate buffered saline) buffer and printed directly onto an amino-terminated glass substrate. During contact time, the activated enzyme was allowed to react covalently with the amino-terminated monolayer on the glass. The activity of the immobilized enzyme was confirmed by scanning electrochemical microscopy (SECM) measurements.

By using a pattern of two different types of biomolecules on the surface, it is possible to tailor the surface regions that are cell adhesive and cell resistant. Feng et al. showed that by covalent μCP of chitosan onto an aldehyde-enriched glass surface followed by incubation in BSA solution, the surface will have properties for cell localization and cell growth guidance.[110]

Matthews and coworkers[15] used covalent μCP for the attachment of glycosaminoglycan (GAG) polysaccharides and heparan sulfate proteoglycan. The GAG solution was mixed with a reducing agent ($NaBH_3CN$) and used as an ink. The ink was incubated on a nonmodified PDMS stamp and printed onto an amino-functionalized glass substrate. Similarly, a solution of heparan sulfate proteoglycans was mixed with the homobifunctional cross-linker (BS^3) and incubated on a hydrophilic oxidized PDMS stamp. Finally, the stamp was contacted with an APTES-modified glass slide to immobilize the heparan sulfate covalently.

"Click" chemistry can also be efficiently combined with μCP. 1,3-Dipolar cycloadditions where alkynes and azides react to give triazoles can serve as a good example for the reaction conducted under stamp confinement.[111] Usually this reaction needs a Cu^I catalyst to accelerate the rate of reaction. Synthesis under nanoscale confinement between an elastomeric stamp and a reactive substrate leads to the desired product within a short period of time, without a catalyst, and under mild conditions. A high concentration of reagents in the contact area of the elastomeric stamp and the monolayer surface is sufficient to obtain full conversion reaction within minutes of contact time, even without a Cu catalyst. In particular for the immobilization of biomolecules it is advantageous to exclude the toxic Cu catalysts (Fig. 14.11).[102,112]

14.4.3 Affinity Contact Printing of Biomolecules

Affinity contact printing (αCP)[113] relies on inking the surface of a PDMS stamp with antigens as capture molecules, which allows subsequent binding of selected antibodies from a solution containing mixtures of proteins (Fig. 14.14). Affinity stamps were prepared by modification of the PDMS stamp with aminosilanes following the reaction with a homobifunctional cross-linker (BS^3) to produce the activated, hydrophilic surface. This activated stamp was used to couple antigens to small areas using microwells, microfluidic networks, and μCP. Repeating this procedure with a different type of antigen, the stamp could be functionalized with a pattern of various antigens which is valuable for microarray applications (A). When several types of antigens are immobilized on an activated stamp they can be further exposed to a solution of different antibodies (B) to extract and immobilize a "matching partner" (C). The captured antibodies can then be printed onto a glass substrate (D) forming microarrays of antibodies (E).

A different example of affinity contact printing was shown by Jang et al.[114] This multistep approach relies on the modification of a PDMS stamp with aminosilanes and

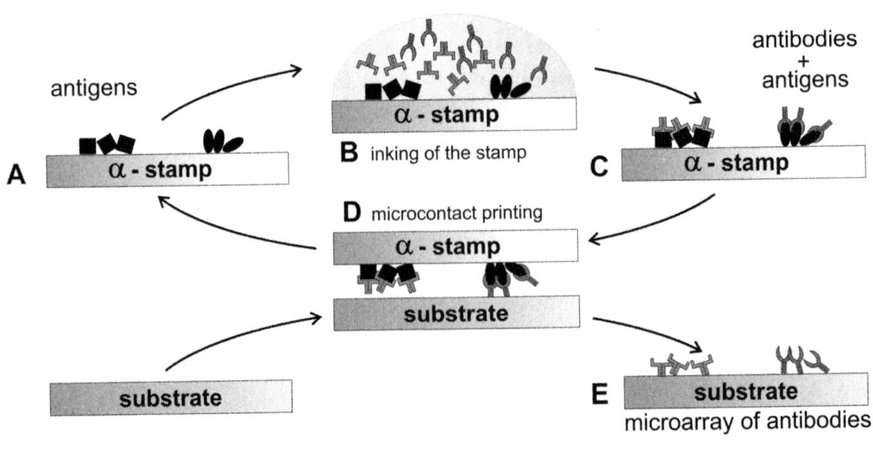

Figure 14.14 Affinity contact printing (αCP)[113] relies on inking the surface of a PDMS stamp with antigens as capture molecules, which allows subsequent binding of selected antibodies from a solution containing mixtures of proteins.

succinic anhydride to introduce carboxylic acid groups on the surface, followed by covalent immobilization of a monoclonal antibody to epidermal growth factor receptor (EGFR). The EGFR-modified stamp was subsequently incubated with a solution of membrane proteins from cell membrane extracts and crude cell lysates. Finally, the stamp was contacted with a gold substrate that was modified with an amino-terminated monolayer covered with a micrometer-thick nematic liquid crystal (LC) film of 4-cyano-4′-pentylbiphenyl. The orientation of the LC film is different on the amino-terminated surface and the regions of the surface presenting EGFR, providing a facile label-free method to optically detect the presence of the EGFR on the surface. Similarly, the group of Abbott used affinity contact printing to immobilize proteins that subsequently can be imaged with LCs.[115] This method relies on covalent modification of a PDMS stamp with biotinylated BSA. The BSA-functionalized stamp was inked with anti-biotin IgG and brought into conformal contact with amino-modified gold surface. After printing, the protein pattern was imaged by spreading an LC film on the surface.

An interesting concept of pattern transfer of nucleic acids using μCP was introduced by Kim and Crooks[116] and Stellacci and coworkers.[117] Their strategies rely on the fabrication of DNA arrays onto flat solid supports by immobilization via the 5′ end functionalized with, for example, amine linkers. The DNA array is further hybridized with complementary strands that possess a capturing group at the 5′ terminus. These groups can be reacted with functionalized surfaces simply by bringing the surface into conformal contact with the DNA array. After reaction, the two surfaces can be separated and the pattern of single-stranded DNA is transferred to the surface that was in the contact with the patterned hybridized array. The strands are mechanically separated and new arrays of patterned DNA can be used for the next transfer of microarrays.

14.4.4 Patterning of Biomolecules via Microchannels and Microwells

In order to overcome the typical limitations of μCP, such as the low surface density of immobilized biomolecules (in particular on polymeric surfaces), due to the small amount of reactant absorbed on the stamp, Hyun and Chilkoti developed a new patterning technique using elastomeric microwells.[118] This method relies on the fabrication of hydrophilic microwells in a PDMS stamp and involves the oxidation of the PDMS stamp, the modification of the stamp areas between the wells with hydrophobic monolayers (hexadecanethiol), the inking of the stamp with protein solution that goes directly into hydrophilic wells avoiding hydrophobic areas, the drying of the stamp, and the printing of the pattern onto the target substrate (Fig. 14.15). A cell-adhesive peptide was covalently printed on NHS-ester-functionalized poly(ethylene terephthalate) substrates.

Similarly, substrates can be patterned with biomolecules by means of a microfluidic network (μFN).[119] Bernard et al.[87] presented patterns of antigen made by flow of solution in a microfluidic network. After a blocking step with BSA, the solution with

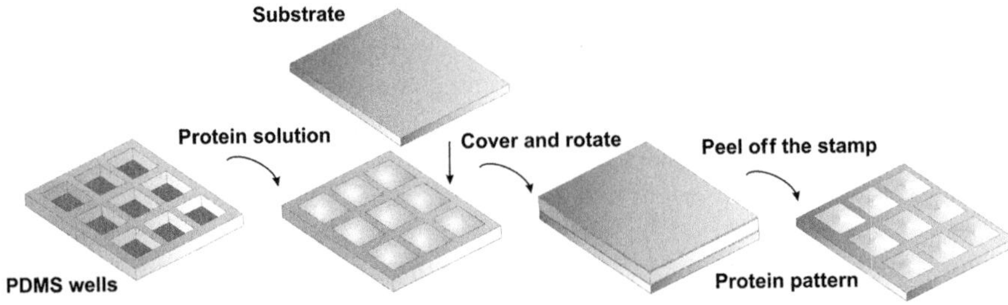

Figure 14.15 Micropatterning of biological molecules on a polymer surface using elastomeric microwells.

antibodies was delivered in a second μFN across the pattern of antigens. Binding between antigen and antibody resulted in a pattern of cross-reacted zones. This method has many advantages, such as a high density of parallel patterns that are precise and well defined due to the fact that microfluidic networks have high resolution and high-contrast capabilities.

μCP of proteins was also combined with microfluidic networks (μFNs) by printing fluorescently labeled antibodies inside microstructures (independent microchannels) coated with Au and a thiolated poly(ethylene glycol) (PEG-SH).[120] The modified, hydrophilic channels repel proteins from solution but promote their transfer from a stamp during printing. Such patterned channels might find application in miniaturization of fluorescence surface immunoassays for *in vitro* diagnostics, drug discovery, or environmental monitoring. Similarly, Geissler et al. used IgG proteins to locally deposit them from the solution onto planar PDMS by μFNs.[105] Subsequently, the PDMS stamp was brought into conformal contact with a clean (nonmodified) glass substrate. The proteins were chemisorbed onto this support in a well-defined manner to give a pattern with high contrast and resolution. The same protein (IgG) was efficiently microcontact printed when the substrate was more hydrophilic than the PDMS stamp whose surfaces were modified with $-CF_3$ or $-NH_2$ functionalities.[89]

Whitesides and coworkers combined microfluidic networks with a PDMS platform to create patterned gradients of biomolecules on a surface.[121] This method involves a two-step process: (1) formation of a gradient of avidin within well-defined patterns by use of microfluidic channels and (2) specific interaction between the avidin gradient pattern and biotin. Such patterns with a density gradient of immobilized biomolecules may find application in studies on cell development and function.

14.5 SPOTTING OF BIOMOLECULES ON SURFACES

In the microarray industry, the most commonly used techniques to pattern biologically active molecules are robotic contact printing or ink-jet printing. In general, these methods can fabricate biochips with thousands of different compounds, which can be analyzed simultaneously. The most commonly spotted biomolecules include oligonucleotides,[1] proteins,[2] and carbohydrates.[3] Microarray-based platforms have been commonly used in sequencing, single nucleotide polymorphism (SNP) detection, characterization of protein–DNA interactions, screening antigen–antibody and carbohydrate–protein interactions, detection of biomarkers, in disease proteomics, and others. Various robotic systems deposit biomolecules into an accurately designed grid made of spots with an average size of 100 μm.

Contact printing relies on dispensing pico- or nanoliters of a biomolecule solution from a set of metallic pins which physically contact the substrate onto which the biomolecules should be immobilized (Fig. 14.16). A defined probe volume is loaded into the pin cavity via capillary forces. Small drops of biomolecule solution are deposited by surface tapping (Fig. 14.17). Pins for contact printing are available in a wide assortment of tip and channel sizes. The tip size determines the spot size and the smaller the tip the smaller is the spot. The smallest tip size that is currently commercially available is 50 μm in diameter and can produce spots about 65 μm in diameter with the sample channel size from 0.25 μL to 1.25 μL. Usually the spot size is approximately 30% larger than the diameter of the tip capillary with the standard sample channel of 0.25 μL. The advantage of this system is that the drop is placed in an exactly defined position on the substrate; there are no shifts of the drop, which is very often observed in noncontact methods. The disadvantage of this system is that pins may damage the coating of the slide while contacting the substrate.

Robotic contact printing can be combined with μCP to spot different biomolecules onto the stamp and transfer the biomolecule array several times with very good spot morphology and homogeneous probe density. This concept was utilized to transfer DNA and RNA arrays.[101] From only one spotted stamp four arrays of good quality can be obtained. The advantage of this system is that diverse probes can be spotted onto the stamp in contrast to simple μCP where only one type of molecule can be used as an ink per single printing step.

Ink-jet printing is a noncontact method; the sample solution is dispensed without direct contact between the print head and the substrate. In modern ink-jet technology, a

Figure 14.16 Set of pins built within a print head for contact printing in microarray applications. (Image courtesy of The Netherlands Cancer Institute, W. Brugman.)

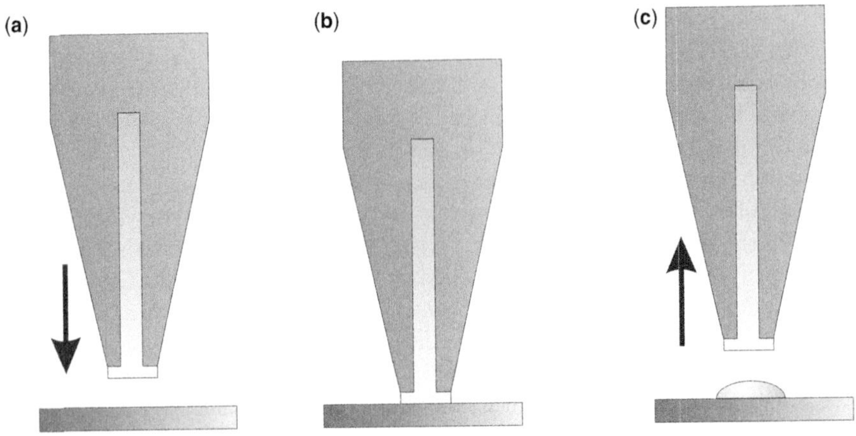

Figure 14.17 Printing mechanism based on surface tension and adhesion. The sample is first loaded into a sample channel by capillary action. The horizontal terminus of the tip ensures that a thin layer of a sample solution is accumulated at the end of the tip. The loaded tip contacts the printing surface depositing a droplet between the substrate and the pin. After the contact time (0.05 s), the pin is retracted and the droplet is held by strong adhesive forces to the substrate.

piezoelectric print head is used to print the arrays. The print head contains a set of nozzles in a linear arrangement controlled by a computer which directs them where to deposit ink, how much to use, and which biomolecules to spot. The nozzles can be employed in subsets for printing microarrays of a variety of sample types. There are many different types of nozzles based on piezoelectric, bubble-generated (thermal printers), or microsolenoid (magnetic) actuators. The volume of the dispensed drops is usually around 100 pL, which produces spots of sizes 90 to 120 μm in diameter, depending on the buffer and the surface chemistry of the substrate. Ink-jet technology is especially suitable for protein microarrays since proteins are very fragile and this method is less invasive than contact printing. The disadvantage of this method is the accuracy of drop positioning and delivery. Ink-jet methods can be used to print DNA, proteins, cells, and other biomolecules.

14.6 DIP-PEN NANOLITHOGRAPHY OF BIOMOLECULES

One of the challenges in microarray technology is the spot size and spot density on the substrate. Miniaturization in the form of nanoarrays that can allow fitting many more genes or proteins on one chip. Several methods, including dip-pen nanolithography (DPN), nanografting, nanoshaving, or focus ion beam lithography have been developed to fabricate nanoscale patterns of biomolecules on the surface. DPN is a promising tool that has perspectives to generate massive, parallel biomolecule arrays in the nanoscale regime. DPN is a scanning-based lithography method for fabrication of nanometer-size patterns where the AFM tip is used to transport and deliver ink containing chemical reagents to the target surface (Fig. 14.18).[122]

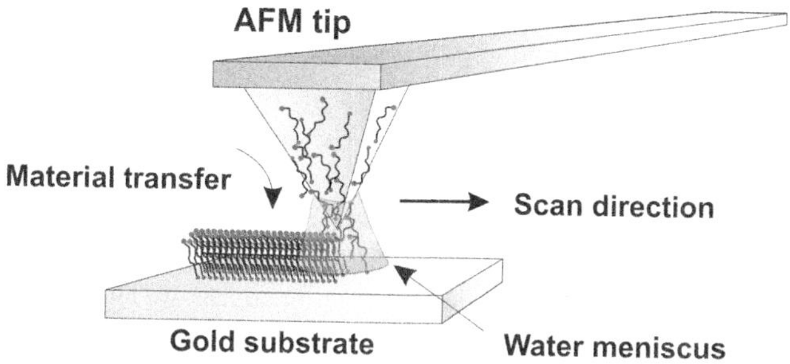

Figure 14.18 Dip-pen nanolithography.[122] The thiol ink is transferred from the tip to the gold substrate through the water meniscus between the tip and the substrate according to the direction of scan.

DPN was used initially to pattern gold surfaces with octadecanethiol (ODT) or 16-mercaptohexadecanoic acid (MHDA) as nanoscale etch resists to produce metal and semiconductor nanostructures. Later, the technique was expanded to a variety of inks (i.e., oligonucleotides,[51] proteins,[123,124] silanes,[125,126] and colloidal particles[127]) and substrates (polymers, glass, and silicon oxide). The minimal size of the features that can be generated by this method is in the sub-100 nm range. There are many factors like humidity, set-point, tip modification, the chemistry involved in binding between the ink and the substrates, the chemical purity of ink, tip and substrate, the tip shape and size, the properties of the ink and the substrate that play an important role in a successful nanopatterning by DPN. Humidity is a critical factor that should be taken into account when DPN is used to pattern biomolecules.[122] For instance, protein patterning was conducted with a relative humidity of 80% to 90% at room temperature. Very often, these experiments are conducted in a glove box or environmental chamber to control the humidity. Another important factor controlling the transport of ink molecules from the tip to the surface is the set-point since large set-points can lead to larger features and contact area. The transport of the ink to the substrate is a complex process. This transport occurs through the meniscus of water which is condensed between the tip and the surface. The water meniscus influences the transport of the ink and this is the reason why humidity control and water solubility of ink is crucial in successful delivery of molecules.[122]

DPN was applied for the fabrication of DNA patterns on gold and silicon oxide surfaces.[51,128] For the successful transfer and delivery of oligonucleotides, the AFM silicon nitride tip was modified with aminosilanes, which promotes adhesion of the negatively charged DNA ink to the tip surface. When gold substrates were patterned, thiol-modified DNA was used as an ink. After drawing an array of DNA spots the remaining bare gold area was passivated with hydrophobic ODT. The same tip inked with 5-acrylamide-modified DNA can be used to pattern silicon oxide substrates activated with MPTMS. The acrylamide moieties can react with thiol groups from the substrate by Michael addition to link covalently DNA molecules. After patterning the substrate, the remaining thiol groups can be blocked with acrylic acid monomer. The biological activity of immobilized DNA can be investigated by subsequent hybridization with fluorescently labeled complementary and noncomplementary oligonucleotides. Another experiment relied on the employment of DPN combined with wet etching techniques to generate arrays of gold nanostructures that were functionalized with oligonucleotides.[129] The fabrication of the gold nanostructure arrays was conducted by drawing MHDA patterns by DPN on a gold surface which served as an etch-resistant layer, etching the remaining gold areas, passivating exposed silicon oxide areas with octadecyltrimethoxysilane and removing the alkanethiol layer from the remaining gold regions by photooxidiation of the substrate with a UV lamp. The final step involved immobilization of disulfide-modified oligonucleotides.

There is another approach to generate protein patterns by DPN.[63] Since proteins can adsorb spontaneously to the gold surface, the gold AFM cantilever was first protected by the deposition of 11-mercaptoundecylpenta(ethylene glycol)disulfide (PEG-SH) in order to preserve the reflective properties of Au. The subsequent step involved coating of the AFM tip with gold and reaction with thioctic acid to obtain a hydrophilic, carboxylic acid-functionalized surface that facilitates protein adsorption. The tip was consequently covered with proteins (lysozyme and rabbit IgG). The experiment was carried out at 80% to 90% relative humidity at room temperature. DPN was carried out directly on gold surfaces since cysteine residues in the proteins possess strong affinity to gold. After patterning, the gold regions surrounding the pattern were reacted with PEG-SH to form a protein-resistant layer. A similar experiment was also carried out on modified silicon oxide surfaces.[63] To transfer successfully anti-rabbit IgG and other proteins, the AFM tip was modified with 2-[methoxypoly(ethyleneoxy)propyl]-trimethoxysilane (PEG-Si) to make the tip surface hydrophilic and biocompatible (Fig. 14.19). This modification also causes inhibition of protein adsorption, allowing

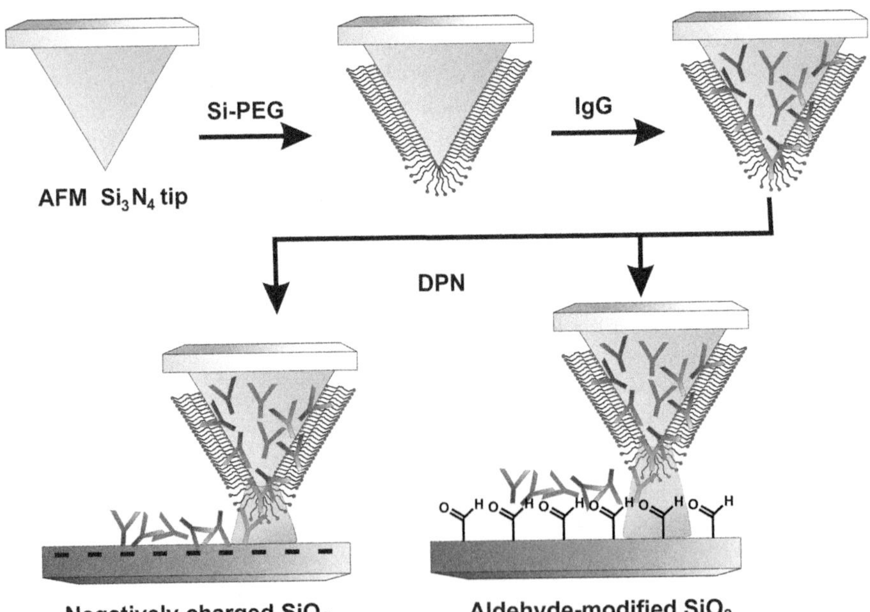

Figure 14.19 Schematic representation of DPN of protein IgG. The pattern formation can be conducted by two methods, using negatively charged SiO_2 or aldehyde-modified SiO_2.

the proteins to be transferred from the tip to the target surface. To immobilize the proteins two surfaces were chosen, negatively charged silicon obtained by treatment with base and aldehyde-terminated silicon. In the first case, the proteins can be attached by electrostatic interactions between the positively charged residues of the protein and the negatively charged surface. In the second case, the proteins can be attached covalently to the aldehyde-modified surface by reaction between amino groups from the lysine residues or amines at the N-termini in the proteins and aldehyde surface groups, resulting in the formation of Schiff bases. To expand the array to multiple components, two different fluorophore-labeled proteins, anti-rabbit IgG and anti-human IgG, were deposited on aldehyde-modified silicon oxide surface by DPN in a serial fashion.

Considering the affinity between metal ions (i.e., Zn^{2+}, Cu^{2+}, Ni^{2+}, and Co^{2+}) and proteins another method to pattern proteins on gold surfaces by DPN was developed.[130] This method utilizes Zn^{2+} ions ($Zn(NO_3)_2 \cdot 6H_2O$) that can be coordinated with MHDA, which can be patterned on gold by an AFM tip after passivation of the remaining gold areas with PEG-SH. The Zn-modified features can be subsequently reacted with the desired antibody. Biorecognition properties can be further studied by interaction with other proteins.

More complex biomolecules were used to generate patterns on gold surfaces by DPN. Tobacco mosaic virus (TMV) patterns were fabricated by the following method. First, patterns of MHDA on gold were fabricated using DPN.[131] The remaining gold areas were reacted with PEG-SH to minimize nonspecific binding of the virus particles to the unpatterned regions. To bridge the virus particles and patterned carboxylic groups from the surface, Zn^{2+} ions were added to the substrate and TMV particles were then immobilized. The protein capsid of the virus has metal-binding groups for Zn^{2+} and other metals and this finding was utilized to immobilize these particles to the patterned surface. The TMV attachment was confirmed by the highly specific reaction with TMV antiserum which, by binding to the virus particles, increases the height of each virus.

Classic DPN, which involves just one tip for writing, is not a high-speed and efficient method in patterning over large, centimeter-wide areas. The parallelization of the writing process is an essential issue for enhancing the acceleration of the ink deposition process. The major challenges in performing parallel-probe lithography include the simultaneous engagement of arrays of tips to the surface and also the fabrication of large arrays of sharp tips, the control over tip feedback, and the introduction of a wide variety of inks to the tip array as well as the characterization of the drawn features.[131–133] Fifty-five thousand pens, writing simultaneously, have been used to draw patterns in the nanometer regime on substrates.[134] This setup can generate 88 million features (dots) with a diameter of 100 ± 20 nm, spaced by 400 nm, and was used to write patterns of ODT and MHDA. The array of 55,000 tips was also used to simultaneously generate phospholipid patterns over a surface area of 1 cm^2.[135] Parallel DPN patterning was used in an indirect approach to generate templates on the surface for protein immobilization.[136] Smaller pen arrays were also used to design various patterns. A 26-pen array was applied to draw dot features of reactive 11-mercaptoundecanoyl-*N*-hydroxysuccinimide ester ($NHSC_{11}SH$) on a gold surface.

After drawing an array of $NHSC_{11}SH$ dots on a gold surface, the substrate was passivated with 11-mercaptoundecyltri(ethylene glycol) and reacted with protein A/G (an engineered protein containing domains of protein A and domains of protein G) that was selectively immobilized only on the dots that were written by DPN through covalent coupling. To extend this method to antibody-based arrays, the substrate with patterns of A/G proteins was reacted with human IgG antibody. The biological activity of that array was demonstrated by the reaction with antigens (β-galactosidase and ubiquitin).

Conventional silicon nitride tips used in DPN can absorb chemical inks poorly. This can be detected since during the writing process in DPN the spreading of ink is commonly observed. To enhance the ink absorption to the tip surface, two important factors should be considered: (1) tip coating or fabrication of a tip from a material with better absorption than Si_3N_4, and (2) ink-diffusion from the tip to the surface. Wang et al. have developed a new tip made entirely of PDMS to write efficiently patterns using ODT as ink.[137] This method is called scanning probe contact printing (SP-CP) and combines the advantages of μCP and DPN to generate monolayer features chemisorbed on a gold surface at a submicron scale. To overcome the limitation related to AFM feedback control and performance due to the nonreflective PDMS material, an improved method of SP-CP was developed.[121] A clean, Si_3N_4 tip was immersed in liquid PDMS in the inkwell to coat the tip with the polymer using an AFM to control the movement of the tip. The PDMS coating acts as an ink reservoir that easily absorbs different types of ink. This tip is biocompatible with proteins, oligonucleotides, and other biomolecules. Different patterns can be generated using the stamp tip, including hollow dots and lines, and the minimum width of the features can reach sub-100 nm. The stamp tip can serve not only as a tool to create nanopatterns but also to image the generated patterns.

14.7 CONCLUDING REMARKS

With the increasing interest in and application of DNA and protein microarrays, biosensors, cell surface interactions, and biomedical implants, various systems and strategies for the immobilization and patterning of biomolecules have been developed and some have become well established. A wide diversity of chemical methods for biomolecule immobilization on inorganic substrates has been implemented by different research groups. Often, the choice of the method is a compromise between effectiveness, cost, and technology. In consideration of stability and durability of the attached biomolecules, certainly, the covalent attachment has to be preferred.

μCP and its variations provide a useful toolbox to generate patterns of biomolecules. The applications of biomolecular patterns can be found in many areas of (bio)-nanotechnology, molecular biology, cell engineering, microarrays, and others. The most efficient printing would be achieved when the ink could form covalent or multivalent noncovalent bonds with the support and when the stamp would carry an array of different molecules so that each print would result in the replication of an array of biomolecules. An attractive solution seems to be the combination of μCP with microfluidic systems where many inks can be introduced in parallel to the stamp although the density of channels and the conformal contact with the substrate is limited and can be an obstacle in achieving the delivery of a high diversity of molecules. The surface synthesis under PDMS stamp confinement opens up the opportunity of conducting chemical reactions much faster on the surface than in the solution.

To decrease the size of the pattern on the surface, DPN can offer ample possibilities. The newest technology introduced 55,000 tip pens working in parallel, which can create 88 million features on an area of 1 cm^2. In this field, the biggest challenge still remains to introduce a corresponding 55,000 different inks to the array of tips.

REFERENCES

1. C. Heise, F. F. Bier, *Top Curr. Chem.* **2006**, *261*, 1–25.
2. B. Schweitzer, S. F. Kingsmore, *Curr. Opin. Biotechnol.* **2002**, *13*, 14–19.
3. K. R. Love, P. H. Seeberger, *Angew. Chem. Int. Ed.* **2002**, *41*, 3583–3586.
4. K. S. Lam, M. Renil, *Curr. Opin. Chem. Biol.* **2002**, *6*, 353–358.
5. R. Z. Wu, S. N. Bailey, D. Sabatini, *Trends Cell Biol.* **2002**, *12*, 485–488.
6. M. S. Fejzo, D. J. Slamon, *Am. J. Pathol.* **2001**, *159*, 1645–1650.
7. W. Kusnezow, J. D. Hoheisel, *J. Mol. Recognit.* **2003**, *16*, 165–176.
8. V. V. Hlady, J. Buijs, *Curr. Opin. Biotechnol.* **1996**, *7*, 72–77.
9. A. del Campo, I. J. Bruce, *Topics Curr. Chem.* **2005**, *260*, 77–111.
10. R. Horejsi, G. Kollenz, F. Dachs, H. M. Tillian, K. Schauenstein, E. Schauenstein, W. Steinschifter, *J. Biochem. Biophys. Methods* **1997**, *34*, 227–236.
11. I. Shin, S. Park, M.-R. Lee, *Chem. Eur. J.* **2005**, *11*, 2894–2901.
12. V. Ball, P. Huetz, A. Elaissari, J. P. Cazenave, J. C. Voegel, P. Schaaf, *Proc. Natl. Acad. Sci. U.S.A.* **1994**, *91*, 7330–7334.
13. P. J. Hergenrother, K. M. Depew, S. L. Schreiber, *J. Am. Chem. Soc.* **2000**, *122*, 7849–7850.
14. M. C. Pirrung, C.-Y. Huang, *Bioconjugate Chem.* **1996**, *7*, 317–321.
15. A. Peramo, A. Albritton, G. Matthews, *Langmuir* **2006**, *22*, 3228–3234.
16. X. Zhao, S. Nampalli, A. J. Serino, S. Kumar, *Nucleic Acids Res.* **2001**, *29*, 955–959.
17. S. Onclin, B. J. Ravoo, D. N. Reinhoudt, *Angew. Chem. Int. Ed.* **2005**, *44*, 6282–6304.
18. W. C. Heiser, Ed., *Gene Delivery to Mammalian Cells. Vol. 1. Nonviral Gene Transfer Techniques*, Totowa, NJ: Humana, **2004**.
19. R. Benters, C. M. Niemeyer, D. Wöhrle, *ChemBioChem* **2001**, *2*, 686–694.
20. R. Benters, C. M. Niemeyer, D. Drutschmann, D. Blohm, D. Wöhrle, *Nucleic Acids Res.* **2002**, *30*, e10.
21. B. J. Hong, S. J. Oh, T. O. Youn, S. H. Kwon, J. W. Park, *Langmuir* **2005**, *21*, 4257–4261.
22. V. Le Berre, E. Trévisiol, A. Dagkessamanskaia, S. Sokol, A.-M. Caminade, J.-P. Majoral, B. Meunier, J. François, *Nucleic Acids Res.* **2003**, *31*, e88.
23. M. Beier, J. D. Hoheisel, *Nucleic Acids Res.* **1999**, *27*, 1970–1977.
24. S. Pathak, A. K. Singh, J. R. McElhanon, P. M. Dentinger, *Langmuir* **2004**, *20*, 6075–6079.
25. S. Wong, V. Kitaev, G. A. Ozin, *J. Am. Chem. Soc.* **2003**, *125*, 15589–15598.
26. M. Glazer, J. Fidanza, G. McGall, C. Frank, *Chem. Mater.* **2001**, *13*, 4773–4782.
27. E. Cunningham, C. J. Campbell, *Langmuir* **2003**, *19*, 4509–4511.
28. F. Xu, G. Zhen, F. Yu, E. Kuennemann, M. Textor, W. Knoll, *J. Am. Chem. Soc.* **2005**, *127*, 13084–13085.
29. J. J. Gooding, V. G. Praig, E. A. H. Hall, *Anal. Chem.* **1998**, *70*, 2396–2402.
30. C. Boozer, J. Ladd, S. Chen, Q. Yu, J. Homola, S. Jiang, *Anal. Chem.* **2004**, *76*, 6967–6972.
31. J. C. Love, L. A. Estroff, J. K. Kriebel, R. G. Nuzzo, G. M. Whitesides, *Chem. Rev.* **2005**, *105*, 1103–1170.

32. J. J. Gooding, F. Mearns, H. Yang, J. Liu, *Electroanalysis* **2003**, *15*, 81–96.
33. H. Schönherr, F. J. B. Kremer, S. Kumar, J. A. Rego, H. Wolf, H. Ringsdorf, M. Jaschke, H.-J. Butt, E. Bamberg, *J. Am. Chem. Soc.* **1996**, *118*, 13051–13057.
34. D. K. Schwartz, *Annu. Rev. Phys. Chem.* **2001**, *52*, 107–137.
35. T. P. Sullivan, W. T. S. Huck, *Eur. J. Org. Chem.* **2003**, 17–29.
36. K. M. Pertays, G. E. Thompson, M. R. Alexander, *Surf. Interface Anal.* **2004**, *36*, 1361–1366.
37. A. Nanci, J. D. Wuest, L. Peru, P. Brunet, V. Sharma, S. Zalzal, M. D. McKee, *J. Biomed. Mater. Res.* **1998**, *40*, 324–335.
38. A. Klinger, D. Steinberg, D. Kohavi, M. N. Sela, *J. Biomed. Mater. Res.* **1997**, *36*, 387–392.
39. J. P. Lee, Y. J. Jang, M. M. Sung, *Adv. Funct. Mater.* **2003**, *13*, 873–876.
40. L. C. P. M. de Smet, G. A. Stork, G. H. F. Hurenkamp, Q. Y. Sun, H. Topal, P. J. E. Vronen, A. B. Sieval, A. Wright, G. M. Visser, H. Zuilhof, E. J. R. Sudholter, *J. Am. Chem. Soc.* **2003**, *125*, 13916–13917.
41. A. Arafat, M. Giesbers, M. Rosso, E. J. R. Sudholter, K. Schroen, R. G. White, L. Yang, M. R. Linford, H. Zuilhof, *Langmuir* **2007**, *23*, 6233–6244.
42. B. Sun, P. E. Colavita, H. Kim, M. Lockett, M. S. Marcus, L. M. Smith, R. J. Hamers, *Langmuir* **2006**, *22*, 9598–9605.
43. W. G. Beattie, L. Meng, S. L. Turner, R. S. Varma, D. D. Dao, K. L. Beattie, *Mol. Biotechnol.* **1995**, *4*, 213–225.
44. J. B. Lamture, K. L. Beattie, B. E. Burke, M. D. Eggers, D. J. Ehrlich, R. Fowler, M. A. Hollis, B. B. Kosicki, R. K. Reich, S. R. Smith, R. S. Varma, M. E. Hogan, *Nucleic Acids Res.* **1994**, *22*, 2121–2125.
45. N. Zammatteo, L. Jeanmart, S. Hamels, S. Courtois, P. Loutte, J. Hevesi, J. Remacle, *Anal. Biochem.* **2000**, *280*, 143–150.
46. H.-Y. Rogers, P. Jiang-Baucom, Z.-J. Huang, V. Bogdanov, S. Anderson, M. T. Boyce-Jacino, *Anal. Biochem.* **1999**, *266*, 23–30.
47. A. Steel, M. Torres, J. Hartwell, Y.-Y. Yu, N. Ting, G. Hoke, H. Yang, in *Microarray Biochip Technology* (Ed. M. Schena), Natick, MA: Eton, **2000**, 87–117.
48. S. Park, M.-R. Lee, S.-J. Pyo, I. Shin, *J. Am. Chem. Soc.* **2004**, *126*, 4812–4819.
49. T. Okamoto, T. Suzuki, N. Yamamoto, *Nature Biotechnol.* **2000**, *18*, 438–441.
50. M. C. Pirrung, J. D. Davis, A. L. Odenbaugh, *Langmuir* **2000**, *16*, 2185–2191.
51. L. M. Demers, D. S. Ginger, S.-J. Park, Z. Li, S.-W. Chung, C. A. Mirkin, *Science* **2002**, *296*, 1836–1838.
52. M. Köhn, R. Wacker, C. Peters, H. Schröder, L. Soulere, R. Breinbauer, C. M. Niemeyer, H. Waldmann, *Angew. Chem. Int. Ed.* **2003**, *42*, 5830–5834.
53. A. Kumar, O. Larsson, D. Parodi, Z. Liang, *Nucleic Acids Res.* **2000**, *28*, e71.
54. J. M. Brockman, A. G. Frutos, R. M. Corn, *J. Am. Chem. Soc.* **1999**, *121*, 8044–8051.
55. E. A. Smith, M. J. Wanat, Y. Cheng, S. V. P. Barreirs, A. G. Frutos, R. M. Corn, *Langmuir* **2001**, *17*, 2502–2507.
56. N. K. Devaraj, G. P. Miller, W. Ebina, B. Kakaradov, J. P. Collman, E. T. Kool, C. E. D. Chidsey, *J. Am. Chem. Soc.* **2005**, *127*, 8600–8601.
57. B. T. Houseman, M. Mrksich, *Chem. Biol.* **2002**, *9*, 443–454.
58. E. Delamarche, A. Bernard, H. Schmid, A. Bietsch, B. Michel, H. Biebuyck, *J. Am. Chem. Soc.* **1998**, *120*, 500–508.
59. A. Kumar, Z. Liang, *Nucleic Acids Res.* **2001**, *29*, e2.
60. D. I. Rożkiewicz, B. J. Ravoo, D. N. Reinhoudt, *Langmuir* **2005**, *21*, 6337–6343.
61. K. Lindroos, U. Liljedahl, M. Raitio, A. C. Syvänen, *Nucleic Acids Res.* **2001**, *29*, e69.
62. D. L. Wilson, R. Martin, S. Hong, M. Cronin-Golomb, C. A. Mirkin, D. L. Kaplan, *Proc. Natl. Acad. Sci. U.S.A.* **2001**, *98*, 13660–13664.
63. J.-H. Lim, D. S. Ginger, K.-B. Lee, J. Heo, J.-M. Nam, C. A. Mirkin, *Angew. Chem. Int. Ed.* **2003**, *42*, 2309–2312.
64. G. M. Husar, D. J. Anziano, M. Leuck, D. P. Sebesta, *Nucleosides Nucleotides Nucleic Acids* **2001**, *20*, 559–566.

65. H. A. Latham-Timmons, A. Wolter, J. S. Roach, R. Giare, M. Leuck, *Nucleosides Nucleotides Nucleic Acids* **2003**, *22*, 1495–1497.
66. C. Donzel, M. Geissler, A. Bernard, H. Wolf, B. Michel, J. Hilborn, E. Delamarche, *Adv. Mater.* **2001**, *13*, 1164–1167.
67. W. Kusnezow, A. Jacob, A. Walijew, F. Diehl, J. D. Hoheisel, *Proteomics* **2003**, *3*, 254–264.
68. G. T. Hermanson, *Bioconjugate Techniques*, San Diego, CA: Academic Press, **2008**.
69. A. D. de Araujo, J. M. Palomo, J. Cramer, M. Köhn, H. Schröder, R. Wacker, C. M. Niemeyer, K. Alexadrov, H. Waldmann, *Angew. Chem. Int. Ed.* **2006**, *45*, 296–301.
70. A. Tinazli, J. L. Tang, R. Valiokas, S. Picuric, S. Lata, J. Piehler, B. Liedberg, R. Tampé, *Chem. Eur. J.* **2005**, *11*, 5249–5259.
71. M. J. W. Ludden, M. Peter, D. N. Reinhoudt, J. Huskens, *Small* **2006**, *2*, 1192–1202.
72. M. J. W. Ludden, X. Y. Ling, T. Gang, W. P. Bula, H. L. G. E. Gardeniers, D. N. Reinhoudt, J. Huskens, *Chem. Eur. J.* **2008**, *14*, 136–142.
73. X.-M. Zhao, Y. Xia, G. M. Whitesides, *J. Mater. Chem.* **1997**, *7*, 1069–1074.
74. Y. Xia, G. M. Whitesides, *Angew. Chem. Int. Ed.* **1998**, *37*, 550–575.
75. Y. Xia, G. M. Whitesides, *Annu. Rev. Mater. Sci.* **1998**, *28*, 153–184.
76. R. S. Kane, S. Takayama, E. Ostuni, D. E. Ingber, G. M. Whitesides, *Biomaterials* **1999**, *20*, 2363–2376.
77. J. L. Wilbur, A. Kumar, E. Kim, G. M. Whitesides, *Adv. Mater.* **1994**, *6*, 600–604.
78. J. A. Rogers, K. E. Paul, G. M. Whitesides, *J. Vac. Sci. Technol. B* **1998**, *16*, 88–97.
79. D. J. Graham, D. D. Price, B. D. Ratner, *Langmuir* **2002**, *18*, 1518–1527.
80. Y. Xia, G. M. Whitesides, *Langmuir* **1997**, *13*, 2059–2067.
81. J. A. Helmuth, H. Schmid, R. Stutz, A. Stemmer, H. Wolf, *J. Am. Chem. Soc.* **2006**, *128*, 9296–9297.
82. R. B. Bass, A. W. Lichtenberger, *Appl. Surf. Sci.* **2004**, *226*, 335–340.
83. A. P. Quist, E. Pavlovic, S. Oscarsson, *Anal. Bioanal. Chem.* **2005**, *381*, 591–600.
84. N. L. Jeon, K. Finnie, K. Branshaw, R. G. Nuzzo, *Langmuir* **1997**, *13*, 3382–3391.
85. T. P. Sullivan, M. L. van Poll, P. Y. W. Dankers, W. T. S. Huck, *Angew. Chem. Int. Ed.* **2004**, *43*, 4190–4193.
86. K. E. Schmalenberg, H. M. Buettner, K. E. Uhrich, *Biomaterials* **2004**, *25*, 1851–1857.
87. S. A. Lange, V. Benes, D. P. Kern, J. K. H. Hober, A. Bernard, *Anal. Chem.* **2004**, *76*, 1641–1647.
88. C. Xu, P. Taylor, M. Ersoz, P. D. J. Fletcher, V. Paunov, *J. Mater. Chem.* **2003**, *13*, 3044–3048.
89. J. L. Tan, J. Tien, C. S. Chen, *Langmuir* **2002**, *18*, 519–523.
90. J. R. LaGraff, Q. Chu-LaGraff, *Langmuir* **2006**, *22*, 4685–4893.
91. D. J. Graber, T. J. Zieziulewicz, D. A. Lawrence, W. Shain, J. N. Turner, *Langmuir* **2003**, *19*, 5431–5434.
92. E. E. Ross, J. R. Joubert, R. J. Wysocki, K. Nebesny, T. Spratt, D. F. O'Brien, S. S. Saavedra, *Biomacromolecules* **2006**, *7*, 1393–1398.
93. J. P. Renault, A. Bernard, A. Bietsch, B. Michel, H. R. Bosshard, E. Delamarche, M. Kreiter, B. Hecht, U. P. Wild, *J. Phys. Chem. B* **2003**, *107*, 703–711.
94. T. J. Park, S. Y. Lee, S. J. Lee, J. P. Park, K. S. Yang, K.-B. Lee, S. Ko, J. B. Park, T. Kim, S. K. Kim, Y. B. Shin, B. H. Chung, S.-J. Ku, D. H. Kim, I. S. Choi, *Anal. Chem.* **2006**, *78*, 7197–7205.
95. S. K. Kwak, G. S. Lee, D. J. Ahn, J. W. Choi, *Mater. Sci. Eng. C* **2004**, *24*, 151–155.
96. T. Wilhelm, G. Wittstock, *Langmuir* **2002**, *18*, 9485–9493.
97. A. Biasco, D. Pisignano, B. Krebs, R. Cingolani, R. Rinaldi, *Synth. Methods* **2005**, *153*, 21–24.
98. C. D. James, R. C. Davis, L. Kam, H. G. Craighead, M. Isaacson, J. N. Turner, W. Shain, *Langmuir* **1998**, *14*, 741–744.
99. J. S. Hovis, S. G. Boxer, *Langmuir* **2001**, *17*, 3400–3405.
100. E. Delamarche, C. Donzel, F. S. Kamounah, H. Wolf, M. Geissler, R. Stutz, P. Schmid-Winkel, B. Michel, H. J. Mathieu, K. Schaumburg, *Langmuir* **2003**, *19*, 8749–8758.
101. D. I. Rozkiewicz, W. Brugman, R. M. Kerkhoven, B. J. Ravoo, D. N. Reinhoudt, *J. Am. Chem. Soc.* **2007**, *129*, 11593–11599.
102. D. I. Rożkiewicz, J. Gierlich, G. A. Burley, K. Gutsmiedl, T. Carell, B. J. Ravoo, D. N. Reinhoudt, *ChemBioChem* **2007**, *8*, 1997–2002.

103. C. J. Campbell, S. K. Smoukov, K. J. M. Bishop, B. A. Grzybowski, *Langmuir* **2005**, *21*, 2637–2640.
104. G. Csusc, T. Künzler, K. Feldman, F. Robin, N. D. Spencer, *Langmuir* **2003**, *19*, 6104–6109.
105. M. Geissler, A. Bernard, A. Bietsch, H. Schmid, B. Michel, E. Delamarche, *J. Am. Chem. Soc.* **2000**, *122*, 6303–6304.
106. J. Lahiri, E. Ostuni, G. M. Whitesides, *Langmuir* **1999**, *15*, 2055–2060.
107. K. B. Lee, D. J. Kim, Z. W. Lee, S. I. Woo, I. S. Choi, *Langmuir* **2004**, *20*, 2531–2535.
108. J. H. Hyun, H. W. Ma, P. Banerjee, J. Cole, K. Gonsalves, A. Chilkoti, *Langmuir* **2002**, *18*, 2975–2979.
109. D. I. Rożkiewicz, Y. Kraan, V. Subramaniam, M. W. T. Werten, A. F. de Wolf, B. J. Ravoo, D. N. Reinhoudt, *Chem. Eur. J.* **2007**, *12*, 6290–6297.
110. J. Feng, C. Gao, B. Wang, J. Shen, *Thin Solid Films* **2004**, *460*, 286–290.
111. D. I. Rożkiewicz, D. Jańczewski, W. Verboom, B. J. Ravoo, D. N. Reinhoudt, *Angew. Chem. Int. Ed.* **2006**, *45*, 5292–5296.
112. O. Michel, B. J. Ravoo, *Langmuir* **2008**, *24*, 12116–12118.
113. J. P. Renault, A. Bernard, D. Juncker, B. Michel, H. R. Bosshard, E. Delamarche, *Angew. Chem. Int. Ed.* **2002**, *41*, 2320–2324.
114. C.-H. Jang, M. L. Tingey, N. L. Korpi, G. J. Wiepz, J. H. Schiller, P. J. Bertics, N. L. Abbott, *J. Am. Chem. Soc.* **2005**, *127*, 8912–8913.
115. M. L. Tingey, S. Wilyana, E. J. Snodgrass, N. L. Abbott, *Langmuir* **2004**, *20*, 6818–6826.
116. J. Kim, R. M. Crooks, *Anal. Chem.* **2007**, *79*, 7267–7274.
117. A. A. Yu, T. A. Savas, G. S. Taylor, A. G. Elie, H. I. Smith, F. Stellacci, *Nano Lett.* **2005**, *5*, 1061–1064.
118. J. Hyun, A. Chilkoti, *J. Am. Chem. Soc.* **2001**, *123*, 6943–6944.
119. A. Papra, A. Bernard, D. Juncker, N. B. Larsen, B. Michel, E. Delamarche, *Langmuir* **2001**, *17*, 4090–4095.
120. J. Foley, H. Schmid, R. Stutz, E. Delamarche, *Langmuir* **2005**, *21*, 11296–11303.
121. X.-Y. Jiang, Q. Xu, K. W. Dertinger, A. D. Stroock, T. Fu, G. M. Whitesides, *Anal. Chem.* **2005**, *77*, 2338–2347.
122. D. S. Ginger, H. Zhang, C. A. Mirkin, *Angew. Chem. Int. Ed.* **2004**, *43*, 30–45.
123. A. Noy, A. E. Miller, J. E. Klare, B. L. Weeks, B. W. Woods, J. J. De Yoreo, *Nano Lett.* **2002**, *2*, 109–112.
124. K.-B. Lee, S.-J. Park, C. A. Mirkin, J. C. Smith, M. Mrksich, *Science* **2002**, *295*, 1702–1705.
125. H. Jung, R. Kulkarni, C. P. Collier, *J. Am. Chem. Soc.* **2003**, *125*, 12096–12097.
126. D. J. Pena, M. P. Raphael, J. M. Byers, *Langmuir* **2003**, *19*, 9028–9032.
127. M. Ben Ali, T. Ondarcuhu, M. Brust, C. Joachim, *Langmuir* **2002**, *18*, 872–872.
128. S.-W. Chung, D. S. Ginger, M. W. Morales, Z. Zhang, V. Chandrasekhar, M. A. Ratner, C. A. Mirkin, *Small* **2005**, *1*, 64–69.
129. H. Zhang, Z. Li, C. A. Mirkin, *Adv. Mater.* **2002**, *20*, 1472–1474.
130. R. A. Vega, D. Maspoch, C. K.-F. Shen, J. J. Kakkassery, B. J. Chen, R. A. Lamb, C. A. Mirkin, *ChemBioChem* **2006**, *7*, 1653–1657.
131. R. A. Vega, D. Maspoch, K. Salaita, C. A. Mirkin, *Angew. Chem. Int. Ed.* **2005**, *44*, 6013–6015.
132. S. W. Lee, R. G. Sanedrin, B.-K. Oh, C. A. Mirkin, *Adv. Mater.* **2005**, *17*, 2749–2753.
133. S. Hong, C. A. Mirkin, *Science* **2000**, *288*, 1808–1811.
134. K. Salaita, Y. Wang, J. Fragala, R. A. Vega, C. Liu, C. A. Mirkin, *Angew. Chem. Int. Ed.* **2006**, *45*, 7220–7223.
135. S. Lenhert, P. Sun, Y. Wang, H. Fuchs, C. A. Mirkin, *Small* **2007**, *3*, 71–75.
136. S. W. Lee, B.-K. Oh, R. G. Sanedrin, K. Salaita, T. Fujigaya, C. A. Mirkin, *Adv. Mater.* **2006**, *18*, 1133–1136.
137. X. Wang, K. S. Ryu, D. A. Bullen, J. Zou, H. Zhang, C. A. Mirkin, C. Liu, *Langmuir* **2003**, *19*, 8951–8955.

Chapter 15

Switchable Host–Guest Chemistry on Surfaces

JILIE KONG, CHUNMING JIANG, AND LI MU

15.1 INTRODUCTION

There has been widespread interest in the past decade in the development of smart or intelligent hybrid functional materials that incorporate switchable host and guest components. These materials respond to slight changes in the external medium, including temperature, electric potential, or photoirradiation. Such external stimuli usually trigger changes in the structure, conformation, or redox states of the materials, which modifies their surface, and thus results in changes of macroscopic properties of the surface through the stimulation of switchable host–guest chemistry. To date, various types of materials have been employed to establish this kind of switchable host–guest surface chemistry and some reviews have been published.[1–3] Here, we report representative examples from two categories based on the method by which the switchable functions are immobilized on the surface of the materials. One strategy is self-assembly, which is normally based on the dissociation/reassociation, conformational transition, or reversible attachment/detachment of host and guest in response to external stimulation. The second method is based on the modification of solid

The Supramolecular Chemistry of Organic–Inorganic Hybrid Materials. Edited by Knut Rurack and Ramón Martínez-Máñez

substrates by the grafting of certain photo-, electro-, and thermo-switchable polymer hybrids.

15.2 SWITCHABLE HOST–GUEST CHEMISTRY ON SURFACES

15.2.1 Self-Assembly

15.2.1.1 Electrochemically Driven Self-Assembly

Electrochemical stimulation is a convenient method widely used for controlling surface properties. Kong et al. have introduced a strategy of controlling protein assembly on switchable low-density self-assembled monolayers (LD-SAMs), which may establish a new way to design functional hybrid composite films that undergo reversible conformational transitions stimulated by electrical potential.[4–7] The LD-SAM surface was generated by attachment of an inclusion complex between 16-mercaptohexadecanoic acid (MHDA) encapsulated by cyclodextrin (CD) on an Au surface, followed by the dissociation of CD. The MHDA molecules of the resulting LD-SAM undergo conformational transitions between bent and straight states, which represent the two states in which either the ionic carboxyl groups or the alkyl chains of MHDA face the outside as a function of the applied potentials, respectively. As a consequence, the LD-SAM can exhibit hydrophobic or hydrophilic properties which can reversibly be interconverted at will by simply changing the applied potential, to enable the controlled assembly of avidin as a model of a charged protein on this smart surface

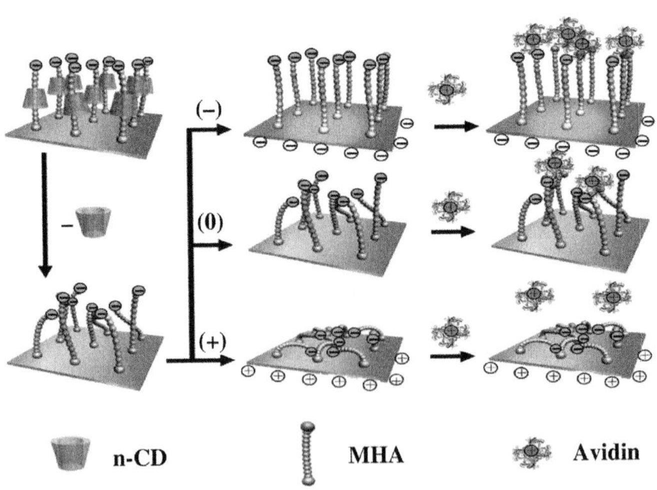

Figure 15.1 The idealized illustration for the preparation of LD-MHDA-SAMs, the transition for anchored MHDA at applied potentials, and the subsequent protein assembly at negative (−) and positive (+) potential, and the open circuit state (0).[4] (Reprinted with permission from Y. Liu et al., *Chem. Commun.* **2004**, 1194–1195. Copyright the Royal Society of Chemistry.)

(Fig. 15.1).[4] Experimental results illustrated that the fluorescence emission for avidin-LD-MHDA-SAM after applying a negative potential for 30 min was about 4.9 times higher than that for conditioning the same components of the assembly at positive potential. Furthermore, these findings have been extended successfully to the application of the LD-SAM in the fabrication of smart microfluidic chips for protein separation (Fig. 15.2).[5] The inner surface of the chip's channel was first coated with a thin layer of Au through sputtering and was subsequently modified with loosely packed SAMs of thiols with terminal carboxylic or amino groups. Accordingly, the microchips prepared in this way can reversibly adsorb and release different charged proteins under electrical control. It was demonstrated that two model proteins, avidin and streptavidin, were readily adsorbed by the smart chips under negative and positive potential, respectively. In addition, more than 90% of the adsorbed proteins could be released upon an electrical command. Such microfluidic systems would be useful for controlled on-chip capture of target proteins directly from a complex mixture and would provide a new method for fabricating miniaturized bioanalytical devices for protein separation (Fig. 15.3).[5]

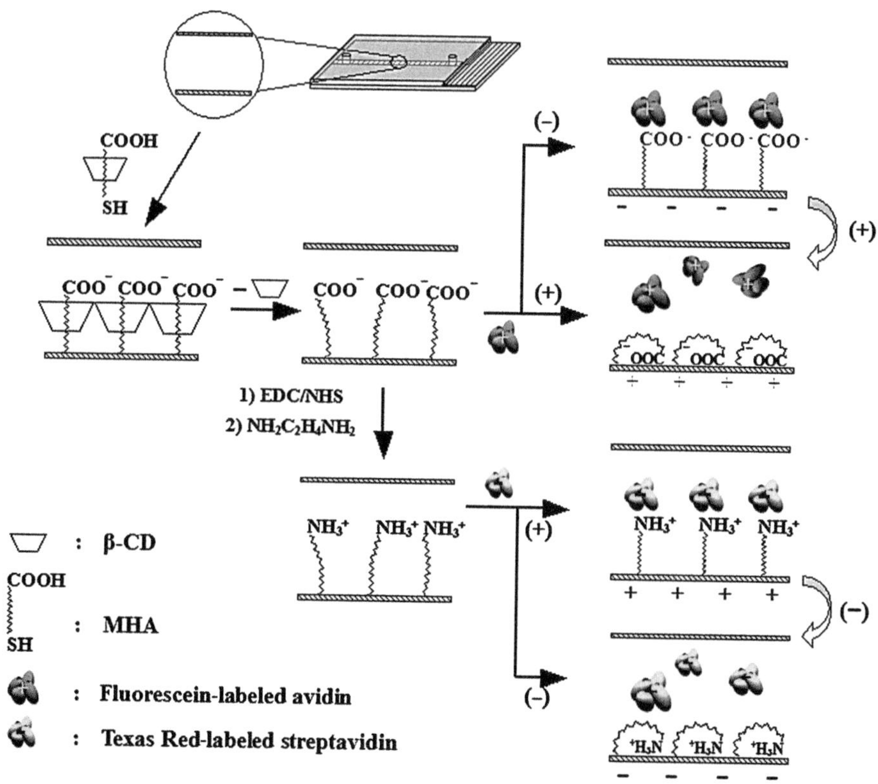

Figure 15.2 Preparative route to smart microfluidic chips and illustration of the electrically controlled adsorption/release of avidin and streptavidin.[5] (Reprinted with permission from L. Mu et al., *Chem. Eur. J.* **2007**, *13*, 5113–5120. Copyright Wiley-VCH Verlag GmbH & Co. KGaA.)

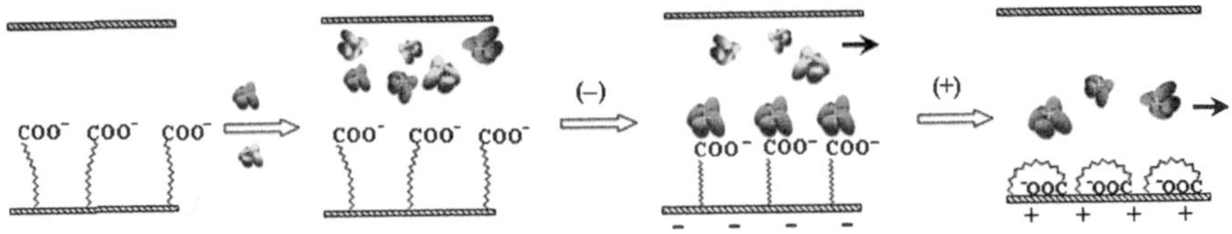

Figure 15.3 Principle of operation of the smart microfluidic chips for protein separation.[5] (Reprinted with permission from L. Mu et al., *Chem. Eur. J.* **2007**, *13*, 5113–5120. Copyright Wiley-VCH Verlag GmbH & Co. KGaA.)

On the other hand, controlling different potentials can strongly affect the stability of CD host–guest complexes. One of the examples of a redox-controlled host–guest complex uses inclusion complexes between ferrocene-appended nanotubes and β-CDs (Fig. 15.4).[8] The template nanotubes, self-assembled from small peptide molecules, incorporate free amide groups on the tubular surface as binding sites. Functionalization of the peptide nanotubes was achieved by assembly of functional molecules to the binding sites of the nanotubes. The ferrocene nanotubes in particular were fabricated by binding carboxylic acid-derivatized ferrocenes onto the template peptide nanotubes via hydrogen bonding. When these ferrocene-functionalized nanotubes were incubated with β-CD SAMs coated on patterned Au substrates, the ferrocene nanotubes recognized and attached onto the β-CD SAMs via host–guest molecular recognition. The binding affinity of the host–guest pair can now be controlled with external stimuli such as an electric field because the electrochemically oxidized form of ferrocene is not effectively bound by CD.[9] In this way, the attachment or detachment of ferrocene nanotubes to or from the β-CD SAMs could be easily controlled electrochemically by tuning the redox state of the ferrocene units of the nanotubes. This control of attachment and detachment of guest nanotubes to and from host surfaces may be applied to build nanometer-sized switching components that can be turned on and off by tuning the redox state of the guest with external electric fields.

A monomeric compound composed of both metallocenes (cobaltocenium and ferrocene) linked by a small positively charged tether resulted in a three-state

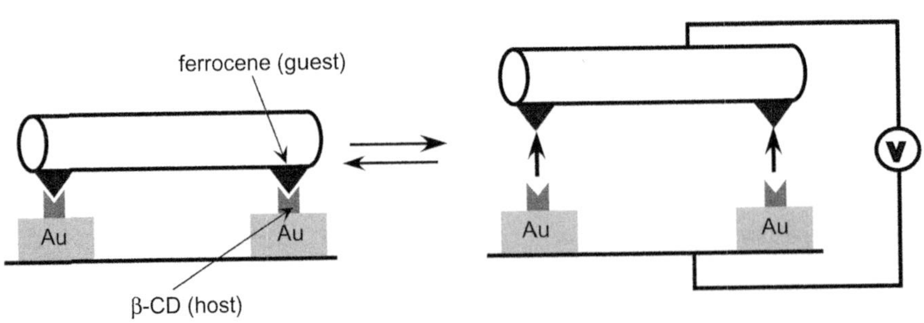

Figure 15.4 Attachment and detachment of ferrocene nanotubes on the β-CD SAMs/Au substrates controlled by electric fields.[8] (Reprinted with permission from Y. F. Chen et al., *Langmuir*, **2004**, *20*, 8409–8413. Copyright 2004 American Chemical Society.)

host–guest system the binding of which to β-CD can also be controlled electrochemically.[10] Whereas the fully oxidized form is not able to bind to β-CD, the fully reduced form binds to two β-CDs. Accordingly, the molecule in the mixed ferrocene-cobaltocenium redox state binds to one β-CD unit only.

The electrochemically controlled delivery of redox-active ferrocene dendrimers (Fc-D) to a so-called molecular printboard (Fig. 15.5) was also reported.[11] The molecular printboard is comprised of heptathioether-functionalized β-CDs which form a densely packed SAM on Au. β-CD monolayers on SiO_2 were also prepared.[12] The ferrocene dendrimer–β-CD assemblies form stable monolayers on the β-CD SAMs due to the formation of multivalent interactions and the stoichiometry of the immobilized dendrimers could be quantified. Electrochemical conversion of the ferrocene moieties resulted in an effective desorption of the dendrimers from the molecular printboard. The immobilization of Fc-Ds at molecular printboards on SiO_2 could also be controlled by scanning electrochemical microscopy (SECM).[13] Biferrocene (BFc) consists of two directly linked ferrocene moieties which are strongly electrochemically communicating, and has three different oxidations states.[14] BFc also forms strong inclusion complexes with β-CD when incorporated into dendrimers, for example, in the form of BFc-decorated poly(propylene imine) dendrimers.[15] The dendrimers could be immobilized at the β-CD host surface and oxidation of the BFc moiety to the mixed valence state caused effective desorption of the dendrimers from the molecular printboard.[15] In a related work, a custom-designed electrochemical apparatus with variable temperature control was used to investigate the redox-switching behavior of a linear motor-molecule confined on an Au surface in the fashion of a disulfide-tethered bistable rotaxane.[16] For a more detailed account of this and related systems, the reader is referred to Chapter 17.

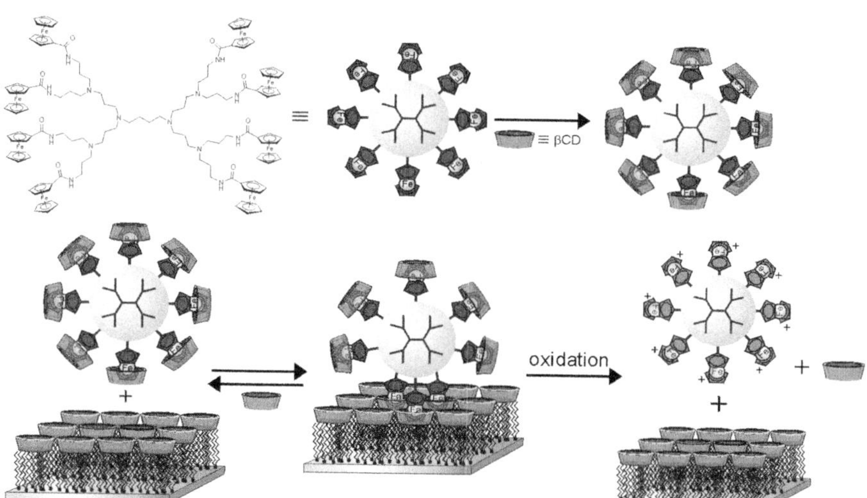

Figure 15.5 Formation of a dendrimer–β-CD assembly (top) and the electrochemically controlled adsorption at the β-CD surface.[11] (Reprinted with permission from C. A. Nijhuis et al., *Langmuir* **2005**, *21*, 7866–7876. Copyright 2005 American Chemical Society.)

15.2.1.2 Light-Driven Self-Assembly

Azobenzene can be considered as a molecular-sized switch that can be reversibly interconverted from an on state (the *trans* form) to an off state (the *cis* form), both thermodynamically rather stable under normal conditions, by external stimuli (e.g., photochemical and/or thermal). In other words, in systems using azobenzene an input from the macroscopic world generates a response at the molecular (microscopic) level.[17–19] Indeed, light-driven photochromism of azobenzene resulting from the photoisomerization of azobenzene molecules has been known for years. On this basis, the control of surface properties has been studied extensively. These machine-like switching motions have been applied to molecular shuttles, motors, and information storage in various matrices. For example, when azobenzenes are attached to nanotubes as switchable recognition components, nanotube-based photoactive switches can be fabricated on α-CDs/Au substrates. CDs and their derivatives have been known to form inclusion complexes with a number of complementary azo compounds via host–guest recognition, and photoisomerization of the azo compound controls the inclusion or exclusion of the guest through the stability of the host–guest complexes. As a proof-of-principle, Matsui et al. demonstrated that azobenzene-coated nanotubes could recognize and attach onto well-defined regions of thiolated α-CD SAMs on patterned Au substrates via host–guest molecular recognition.[20] The reversible attachment and detachment of azobenzene-appended nanotubes on the α-CD SAMs is illustrated in Figure 15.6. The binding affinity between the azobenzene-functionalized nanotubes and the α-CD SAMs was controlled by UV irradiation. In other words, upon isomerization of the azobenzene units to the *cis* form after UV irradiation, the modified nanotubes were detached from the α-CD hosts and removed from the support. On the other hand, the conformation of the azobenzenes on the nanotubes returned to the *trans* form after keeping the material in the dark for 48 h at 4°C, reactivating the switchable nanotubes for the next cycle. This smart approach may be applied to build photoactive nanometer-scale mechanical switches in electronics.

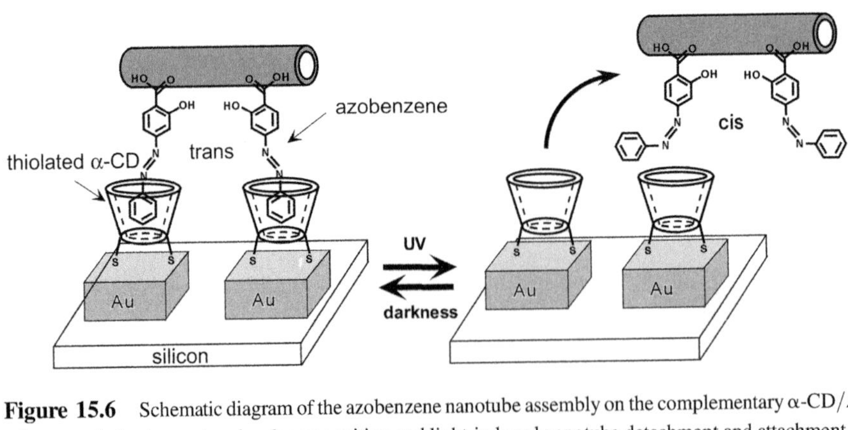

Figure 15.6 Schematic diagram of the azobenzene nanotube assembly on the complementary α-CD/Au substrates via host–guest molecular recognition and light-induced nanotube detachment and attachment on the α-CD surfaces.[20] (Reprinted with permission from I. A. Banerjee et al., *J. Am. Chem. Soc.* **2003**, *125*, 9542–9543. Copyright 2003 American Chemical Society.)

Diarylethenes are promising synthetic photoswitchable molecules because of their outstanding fatigue-resistance in reversible light-induced transformation between two isomers with different absorption spectra.[21,22] The π-conjugation extends over the entire molecular in the closed form, whereas it is restricted to each half of the molecule in the open form. As a consequence, the closed form is expected to exhibit better electrical conductance than the open form. The two forms are referred to as the on and off states of the switch, respectively. The UV-vis spectra of this molecular switch in toluene show that the wavelengths that can be used for on to off switching are $420 < \lambda < 650$ nm, and $300 < \lambda < 350$ nm for the reverse operation.

Photochemical host–guest systems on the basis of photoswitchable macrocycles have also been investigated most recently in a number of publications. Schäfer et al.[23] and Anselmetti et al.[24] modified a bifunctional resorcin[4]arene with two anthracene units and proved its switching function in photochemical induction and heating cycles. Moreover, they modified the lower rim of the resorcin[4]arene with four 10-(decylthio)decyl linkers for an oriented surface immobilization on gold via molecular self-assembly (Fig. 15.7). The affinity modulation of this optical switch was

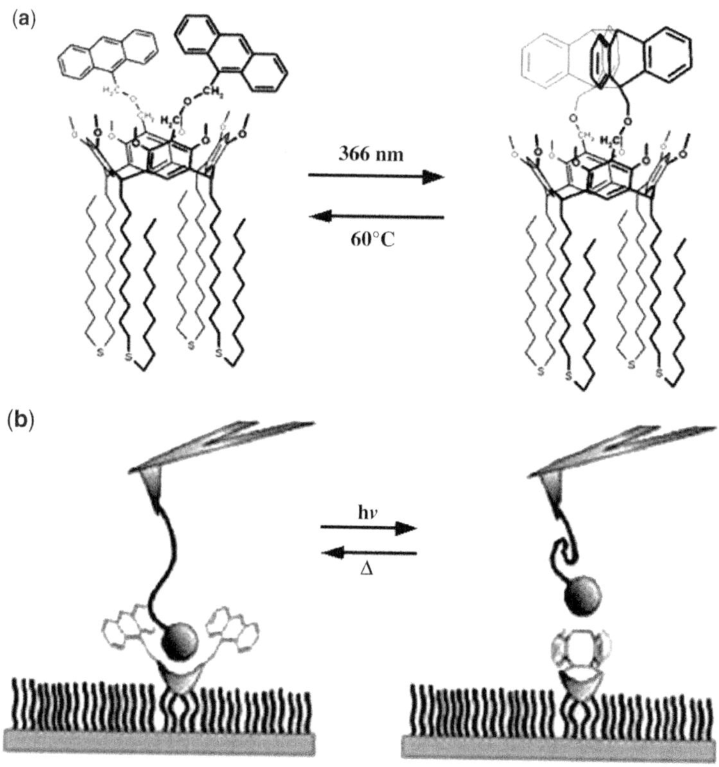

Figure 15.7 Resorcin[4]arene photoswitch for self-assembly (a) and pictograms visualizing the diluted self-assembled monolayer of this photoswitch for single-molecule affinity studies by AFM (b).[23] (Reprinted with permission from C. Schäfer et al., *J. Am. Chem. Soc.* **2007**, *129*, 1488–1489. Copyright 2007 American Chemical Society.)

investigated by single-molecule force spectroscopy[25–27] similar to recent experiments where an affinity ranking of resorcin[4] arene cavitand complexes with different guests was reported.[28] Briefly, a resorcin[4] arene self-assembled monolayer was immobilized on a gold surface through a 10-(decylthio)decyl linker. The guest molecule, an ammonium ion, was immobilized via an organic residue and a poly(ethylene glycol) linker to an atomic force microscopy (AFM) tip. The switching from closed to open isomers was performed by heating the sample to 60°C for 2 h. The cavity was subsequently closed by irradiation with a UV lamp at 368 ± 7 nm for 5 min.

15.2.1.3 Dissociation and Reassociation of Assembled Molecules

The dissociation and reassociation of the components of the molecules that assemble in the SAM triggered by external stimulation bring remarkable changes for the properties of the resulting SAM, as reviewed by our previous work.[7] Many SAMs consist of supermolecules. One of the best investigated classes of supramolecular machines is that based on complexes known as pseudorotaxanes. The reason for this is that the dethreading and rethreading movements of a thread through the center of a ring are reminiscent of the action of a linear motor. The reversible association of various pseudorotaxanes is controlled by light and can subsequently be operated with machine-like motion.[29–31] Besides light, pH can also induce the reversible formation of pseudorotaxanes. Figure 15.8[32] depicts a SAM assembled with pseudorotaxanes containing cucurbit[6]uril (CB) threaded onto a molecular string with a thioctic acid group[33] as an anchoring point toward a gold surface. The pseudorotaxane molecules in the SAM dissociate under alkaline conditions, while they reform under acidic conditions. Consequently, the pseudorotaxane SAM obtained may behave as an ion gate

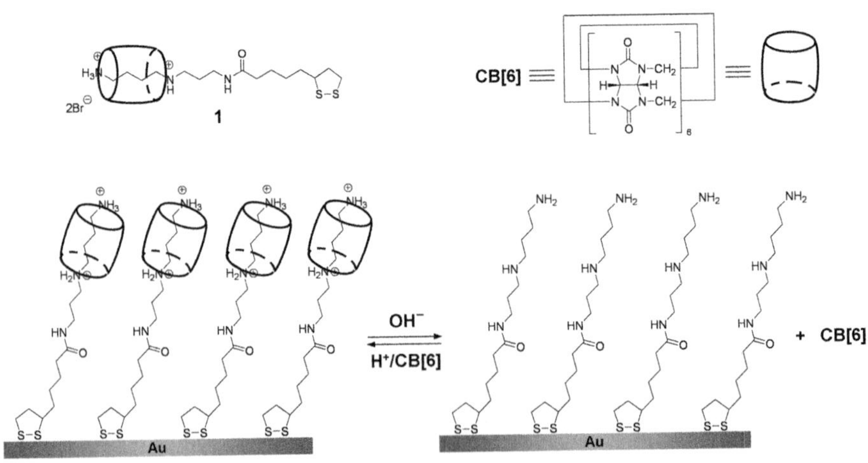

Figure 15.8 Schematic diagram of a SAM of a pseudorotaxane on Au and the dethreading and rethreading of CB molecules.[32] (Reprinted with permission from K. Kim et al., *Angew. Chem. Int. Ed.* **2003**, *42*, 2293–2296. Copyright Wiley-VCH Verlag GmbH & Co. KGaA.)

when the SAM is formed on the surface of an electrode. Depending on the presence or absence of the threaded CB, it may block or allow the access of electroactive species to the electrode surface (see the section on ion channel sensors in Chapter 19). In other words, the conductance of the pseudorotaxane SAM may be switched under the control of pH. These switching processes are reversible and have been repeated up to three cycles with about 10% to 20% decrease in the amount of CB molecules rethreading on the molecule string in each cycle. Like pseudorotaxanes, certain oligomer molecules can also perform similar switching processes, taking advantage of their cleaving and reformation, and hence giving switchable wettability.[34]

15.2.2 Grafted Modification

When a molecular recognition host and a stimuli-responsive polymer are integrated in the same material, a novel function can be achieved in the combined compound that is different from the component materials.

One typical example is poly(*N*-isopropyl acrylamide) (PNIPAAm) which exhibits a dramatic change in volume over a small variation in temperature on either side of its lower critical solution temperature (LCST). Nozaki et al. have investigated the formation of a complex conjugate that consisted of a molecular recognition compound, CD, which was modified with PNIPAAm.[35] They reported that the binding constant of the complex showed a drastic decrease in passing from below to above the LCST in an aqueous solution. Another interesting practical application was demonstrated by Yanagioka and coworkers.[36] By introducing a single vinyl group into a CD moiety, they polymerized the complex using PNIPAAm to maintain a high water diffusivity rate and a high mobility of the polymer. This copolymer was fixed onto the pore surface of a porous polyethylene substrate using plasma-graft polymerization. The objective was to develop a thermosensitive polymer membrane that exhibits high selectivity under a moderate environmental change and one in which the CD moieties could recognize a specific guest molecule, with the PNIPAAm controlling the molecular recognition ability. The concept of the solid separation membrane is illustrated schematically in Figure 15.9. When the part of the guest molecule that is outside the cavity is introduced into the cavity, the binding

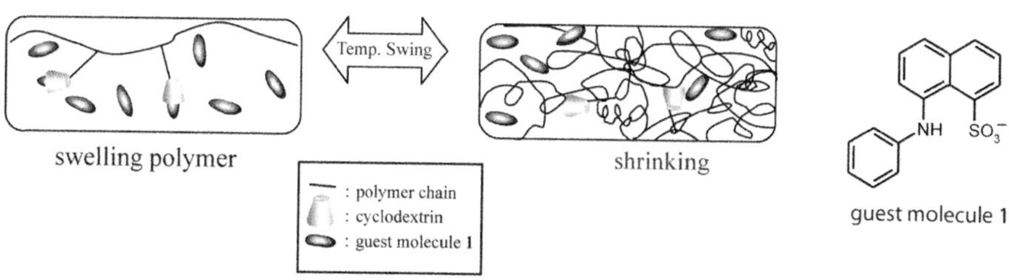

Figure 15.9 Proposed separation mechanism to control the binding constant between the CD moieties and guest molecules. Right: Chemical structure of guest molecules **1**.[36] (Reprinted with permission from M. Yanagioka et al., *Ind. Eng. Chem. Res.* **2003**, *42*, 380–385. Copyright 2003 American Chemical Society.)

constant for the CD cavity changes because of the volume change of the copolymer at temperatures around the LCST. At temperatures below the LCST, the copolymer swells and the steric hindrance exerted by the polymer near the cavity is small; a guest molecule can fit into the cavity. At temperatures above the LCST, the copolymer shrinks and the copolymer chains agglomerate around the cavity, increasing steric hindrance, which leads to a smaller binding constant. Furthermore, the copolymer chains can be fixed at the surface of a porous substrate, that is, at the outlets of the pores, and a polymer can be grafted on the inside of the pores. This provides two advantages from both fundamental and applied aspects. One advantage is that a copolymer fixed to the surface of a substrate can be applied either as a separation material or as a molecular sensing material. Another advantage is that this copolymer system shows a much faster response to an environmental stimulus.

In addition, temperature-responsive properties of PNIPAAm have also been used in fabricating molecular ion gating membranes.[37,38] In these studies, the concept of an ion gating molecular recognition membrane using synthetic host substances and a thermosensitive polymer was proposed. Figure 15.10[38] illustrates the concept. The membrane was prepared by plasma graft copolymerization, which filled the pores of a porous polyethylene film with a copolymer of *N*-isopropyl

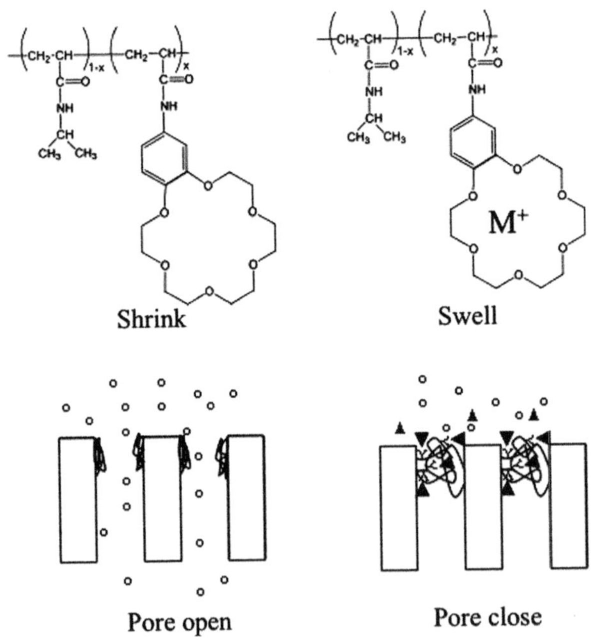

Figure 15.10 Schematic representation of a molecular recognition ion gating membrane. The copolymer of thermosensitive polymer *N*-isopropylacrylamide, and crown ether polymer, BCAm, is grafted on the pore surface. The membrane senses a specific ion M^{n+}, by the crown receptor, and closes its pores rapidly.[38] (Reprinted with permission from T. Ito et al., *J. Am. Chem. Soc.* **2002**, *124*, 7840–7846. Copyright 2002 American Chemical Society.)

acrylamide and benzo[18]crown-6-acrylamide (BCAm). *N*-Isopropyl acrylamide, a thermosensitive polymer, exhibits a dramatic gating effect due to gel swelling, and the crown receptor of the BCAm allows ionic molecular sensing, which causes a shift in the LCST when the receptor captures a specific ion.[39,40] The polymer chain incorporating crown receptor units behaves like an ionic polymer chain upon host (crown)–guest (ion) complex formation, resulting in a change of the LCST to a higher temperature because of the osmotic pressure. As a consequence, the gel can isothermally phase change as the result of a specific ion signal at a temperature between the two LCSTs. Ca^{2+} and Ba^{2+} ions were chosen as the test substances because ions are important for chemical signals in biomembranes. This membrane is of potential use for artificial internal organs, drug delivery systems, or the treatment of wastewater.

15.3 CONCLUSIONS

Here, we have reviewed reversibly switchable surfaces based on host–guest chemistry that are generated by using different types kinds of versatile switching methods, such as electric field, photoirradiation, or thermal treatment, with two different ways of immobilization of the addressable functional units, including self-assembly and modification by grafting. The intelligent host–guest switching on surfaces driven by external stimuli holds great promise for designing bioanalytical methods, microfluidic devices, biochips, controllable drug release systems, and smart material devices.

ACKNOWLEDGMENTS

This work was supported by the National Nature Science Foundation of China (NSFC) (Grant No: 20525519, 90606014, 20890020), and Shanghai Leading Academic Discipline Project B109 & Shanghai Nano-08XD14010.

REFERENCES

1. S. Saha, K. C.-F. Leung, T. D. Nguyen, J. F. Stoddart, J. I. Zink, *Adv. Funct. Mater.* **2007**, *17*, 685–693.
2. C. A. Nijhuis, B. J. Ravoo, J. Huskens, D. N. Reinhoudt, *Coord. Chem. Rev.* **2007**, *251*, 1761–1780.
3. H. Tian, Q.-C. Wang, *Chem. Soc. Rev.* **2006**, *35*, 361–374.
4. Y. Liu, L. Mu, B. H. Liu, S. Zhang, P. Y. Yang, J. L. Kong, *Chem. Commun.* **2004**, 1194–1195.
5. L. Mu, Y. Liu, S. Y. Cai, J. L. Kong, *Chem. Eur. J.* **2007**, *13*, 5113–5120.
6. L. Mu, Y. Liu, S. Zhang, B. H. Liu, J. L. Kong, *New J. Chem.* **2005**, *29*, 847–852.
7. Y. Liu, L. Mu, B. H. Liu, J. L. Kong, *Chem. Eur. J.* **2005**, *11*, 2622–2631.
8. Y. F. Chen, I. A. Banerjee, L. Yu, R. Djalali, H. Matsui, *Langmuir*, **2004**, *20*, 8409–8413.
9. A. E. Kaifer, *Acc. Chem. Res.* **1999**, *32*, 62–71.
10. B. González, I. Cuadrado, B. Alonso, C. M. Casado, M. Morán, A. E. Kaifer, *Organometallics* **2002**, *21*, 3544–3551.
11. C. A. Nijhuis, F. Yu, W. Knoll, J. Huskens, D. N. Reinhoudt, *Langmuir* **2005**, *21*, 7866–7876.
12. S. Onclin, A. Mulder, J. Huskens, B. J. Ravoo, D. N. Reinhoudt, *Langmuir* **2004**, *20*, 5460–5466.
13. C. A. Nijhuis, J. K. Sinha, G. Wittstock, J. Huskens, B. J. Ravoo, D. N. Reinhoudt, *Langmuir* **2006**, *22*, 9770–9775.

14. H. Nishihara, M. Murata, *J. Inorg. Organomet. Polym. Mater.* **2005**, *15*, 147–156.
15. C. A. Nijhuis, K. A. Dolatowska, B. J. Ravoo, J. Huskens, D. N. Reinhoudt, *Chem. Eur. J.* **2007**, *13*, 69–80.
16. H.-R. Tseng, D. Wu, N. X. Fang, X. Zhang, J. F. Stoddart, *ChemPhysChem* **2004**, *5*, 111–116.
17. M. D. Ward, *Chem. Ind.* **1997**, 640–645.
18. C. J. Barrett, J. Mamiya, K. G. Yager, T. Ikeda, *Soft Matter* **2007**, *3*, 1249–1261.
19. S. Yagai, A. Kitamura, *Chem. Soc. Rev.* **2008**, *37*, 1520–1529.
20. I. A. Banerjee, L. T. Yu, H. Matsui, *J. Am. Chem. Soc.* **2003**, *125*, 9542–9543.
21. N. Katsonis, T. Kudernac, M. Walko, S. Jan van der Molen, B. J. van Wees, B. L. Feringa, *Adv. Mater.* **2006**, *18*, 1397–1400.
22. T. Kudernac, S. J. van der Molen, B. J. van Wees, B. L. Feringa, *Chem. Commun.* **2006**, 3597–3599.
23. C. Schäfer, R. Eckel, R. Ros, J. Mattay, D. Anselmetti, *J. Am. Chem. Soc.* **2007**, *129*, 1488–1489.
24. D. Anselmetti, F. W. Bartels, A. Becker, B. Decker, R. Eckel, M. McIntosh, J. Mattay, P. Plattner, R. Ros, C. Schäfer, N. Sewald, *Langmuir* **2008**, *24*, 1365–1370.
25. G. U. Lee, L. A. Chrisey, R. J. Colton, *Science* **1994**, *266*, 771–773.
26. U. Dammer, O. Popescu, P. Wagner, D. Anselmetti, H. J. Guntherodt, G. N. Miservic, *Science* **1995**, *267*, 1173–1175.
27. M. Rief, M. Gautel, F. Oesterheit, J. M. Fernandez, H. E. Gaub, *Science* **1997**, *276*, 1109–1112.
28. R. Eckel, R. Ros, B. Decker, J. Mattay, D. Anselmetti, *Angew. Chem. Int. Ed.* **2005**, *44*, 484–488.
29. S. Chia, J. Cao, J. F. Stoddart, J. I. Zink, *Angew. Chem. Int. Ed.* **2001**, *40*, 2447–2451.
30. V. Balzani, A. Credi, F. M. Raymo, J. F. Stoddart, *Angew. Chem. Int. Ed.* **2000**, *39*, 3348–3391.
31. R. Ballardini, V. Balzani, A. Credi, M. T. Gandolfi, M. Venturi, *Acc. Chem. Res.* **2001**, *34*, 445–455.
32. K. Kim, W. S. Jeon, J. K. Kang, J. W. Lee, S. Y. Jon, T. Kim, K. Kim, *Angew. Chem. Int. Ed.* **2003**, *42*, 2293–2296.
33. J. Madoz, B. A. Kuznetzov, F. J. Medrano, J. L. Garcia, V. M. Fernandez, *J. Am. Chem. Soc.* **1997**, *119*, 1043–1051.
34. S. Abbott, J. Ralston, G. Reynolds, R. Hayes, *Langmuir* **1999**, *15*, 8923–8928.
35. T. Nozaki, Y. Maeda, K. Ito, H. Kitano, *Macromolecules* **1995**, *28*, 522–524.
36. M. Yanagioka, H. Kurita, T. Yamaguchi, S. Nakao, *Ind. Eng. Chem. Res.* **2003**, *42*, 380–385.
37. T. Yamaguchi, T. Ito, T. Sato, T. Shinbo, S. Nakao, *J. Am. Chem. Soc.* **1999**, *121*, 4078–4079.
38. T. Ito, T. Hioki, T. Yamaguchi, T. Shinbo, S. Nakao, S. Kimura, *J. Am. Chem. Soc.* **2002**, *124*, 7840–7846.
39. M. Irie, Y. Misumi, T. Tanaka, *Polymer* **1993**, *34*, 4531–4535.
40. M. Irie, *Adv. Polym. Sci.* **1993**, *110*, 49–65.

Chapter 16

Nanogated Mesoporous Silica Materials

IGOR I. SLOWING, BRIAN G. TREWYN, AND
VICTOR S.-Y. LIN

16.1 INTRODUCTION

Inorganic materials such as silica can constitute excellent scaffolds for a variety of chemical and biochemical processes to occur, not only because of their mechanical stability or physicochemical properties, but also because of their exquisite structural

The Supramolecular Chemistry of Organic–Inorganic Hybrid Materials. Edited by Knut Rurack and Ramón Martínez-Máñez

features. Those features can be rationally designed and can eventually shape the interactions among other species in their surroundings. Mesoporous silica materials exhibit an internal space—the mesopores—that can be separated from the exterior by means of selective gating structures, thus allowing specific processes to take place within nanoscale volumes. Many applications can benefit from such a selective confinement, including catalysis, sensing, separation, and drug delivery.

16.2 MESOPOROUS SILICA MATERIALS: SYNTHESIS AND PROPERTIES

16.2.1 Principle of the Synthesis

The synthesis of mesoporous silica materials is a good example of supramolecular chemistry involving organic and inorganic species. Mesoporous silica materials were discovered independently by scientists led by Kresge at the Mobil Oil Corporation and by a research group led by Kuroda at Waseda University in the early 1990s.[1–4] Even though the synthetic approaches of both groups were different, top-down for Kuroda and bottom-up for the Mobil researchers, the key principle they applied was the same: the organization of the silica polymer directed by alkyltrimethylammonium micelles. Upon postsynthesis removal of the alkyltrimethylammonium surfactants by calcination, the researchers were able to show a network of mesopores in the spaces previously occupied by the surfactant micelles. As opposed to individual molecule templation as in zeolites, the templation in mesoporous silicas is performed by supramolecular entities, that is, micelles. A remarkable result in the early syntheses of mesoporous silica was the observation that the material displayed X-ray diffraction patterns that did not arise from crystallinity of the silica but from a highly ordered arrangement of the pores. This observation was confirmed by transmission electron microscopy (Fig. 16.1).

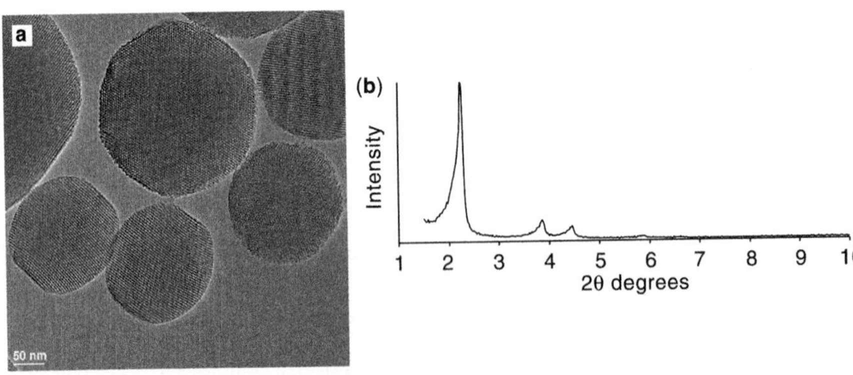

Figure 16.1 (a) Transmission electron micrograph and (b) small angle X-ray diffraction pattern of a mesoporous silica material with a two-dimensional hexagonal array of pores.

The current methods for the synthesis of mesoporous silica materials can be divided into three main types: bottom-up, based on the polymerization of orthosilicic acid derivatives; top-down, based on the reorganization of silica; and evaporation-induced self-assembly,[5,6] all of them promoted by surfactant templation.

Upon investigating the mechanism of the synthesis of mesoporous silica, Kresge and coworkers questioned whether the surfactant formed the ordered mesophase independently or if the ordered state was a result of the interaction between the surfactant and the silica precursor (Fig. 16.2).[7] Some observations suggested that the second mechanism most accurately described what actually occurred. For instance, mesoporous silica could be prepared at concentrations far below the secondary critical micelle concentration (CMC) of the surfactant.[8] Furthermore, different pore structures could be achieved upon variation of the amount of silica precursor at a constant surfactant concentration, indicating that any preexisting organization of the surfactant had no effect on the final architecture of the material.[7] An electron paramagnetic resonance study by Ottaviani and coworkers followed the kinetics of the mesophase formation in a surfactant-silicate solution, and demonstrated that immediately after the addition of the silicate the reaction mixture consisted of a disordered aggregate which then gradually progressed to a hexagonal mesophase.[9] A model to explain the contribution of the silica precursors to the formation of the ordered mesophase was proposed by

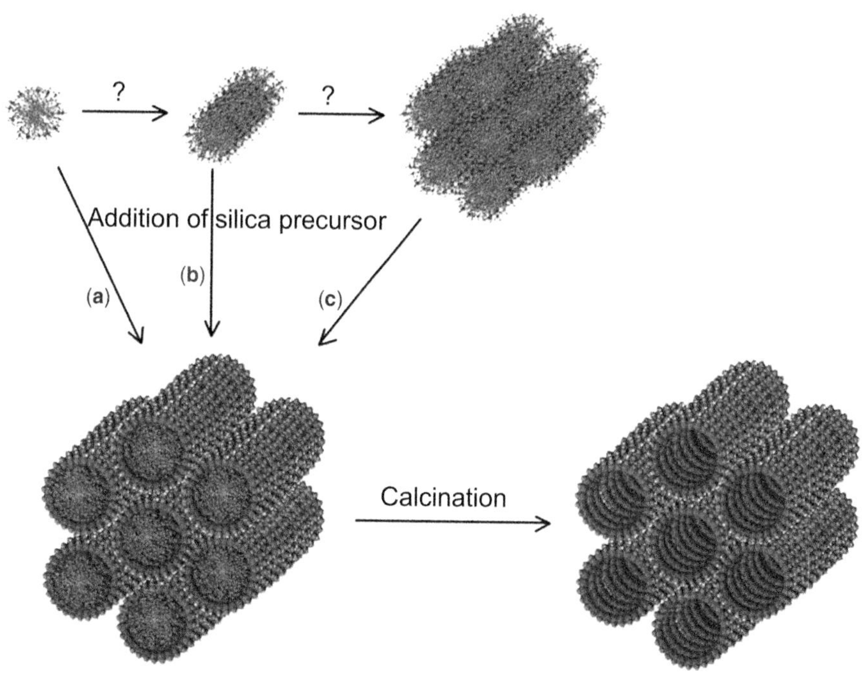

Figure 16.2 Possible mechanisms of the formation of mesoporous silica materials, by generation of hexagonal silica-surfactant mesophases from the interaction of silica precursor with: (a) spherical micelles, (b) rod-like micelles, or (c) a preformed liquid crystalline phase of the surfactant.

Stucky et al.[10] and Monnier and coworkers.[11] A more recent study by Gov and coworkers showed that, on addition of the silica source, the interactions between surfactant molecules are changed to reduce their curvature leading to structural changes in the mesophase.[12] However, when the syntheses are performed at high surfactant concentrations, they are considered to proceed through a preexisting liquid crystal template (route c in Fig. 16.2).[13]

Mesoporous silicas are characterized by surface areas 10 to 20 times larger than that of zeolites, typically around 1000 $cm^2\ g^{-1}$. The pores can be adjusted in a range between 2 and 30 nm, and the pore volumes range between 0.5 and 2.5 $cm^3\ g^{-1}$. The mesopore sizes and volumes are ideal for chemical reactions to take place in their interior, therefore the ability to block the pore entrances with stimuli responsive gates enables a control in chemical reactivity suitable for a variety of applications, including selective sensing, catalysis, and delivery, among others.

16.2.2 Surfactant Types and Pore Architectures

After the initial success in employing cationic alkyltrimethylammonium salts as templates in basic media for the synthesis of mesoporous silica, other surfactants were investigated for similar properties. A major contribution in this respect was the introduction by Stucky and coworkers of a series of block-copolymer surfactants at acidic pH that expanded the attainable pore sizes from the original range of 2 to 10 nm to values reaching 30 nm.[14,15] Other surfactants that have been employed include polyethylene oxide,[16] amines at neutral pH,[17] room temperature ionic liquids,[18] anionic species as co-templating agents,[19,20] fluorinated hydrocarbons,[21,22] carbohydrates,[23] chiral amino acid-based surfactants,[24] and long chain derivatized siloxanes.[25] The use of different surfactants enables the synthesis to be performed under a variety of conditions, including a wide range of pH, temperatures, and solvents. This flexibility has allowed for the incorporation of other species (e.g., metal oxides) into the framework that may be sensitive to a particular set of conditions. The properties of the surfactant employed in the synthesis, especially the CMC and packing parameter, strongly influence the pore architecture of the material.[26]

16.2.3 Manipulation of Pore Size and Particle Morphology

Achieving control of the pore size is relevant not only because it provides a size and shape restriction to the molecules that can penetrate and react in the interior of the material,[27,28] but also because the presence of multiple active sites localized within the pores increases the reactivity of functionalized mesoporous materials as compared to nonporous substrates.[29] The pore size of mesoporous silica can be increased from the base value of approximately 2 nm up to 30 nm by a variety of strategies, which include controlling the length of the hydrocarbon chain of small surfactant templates,[1,30] varying the molecular weights of the hydrophobic blocks in the case of

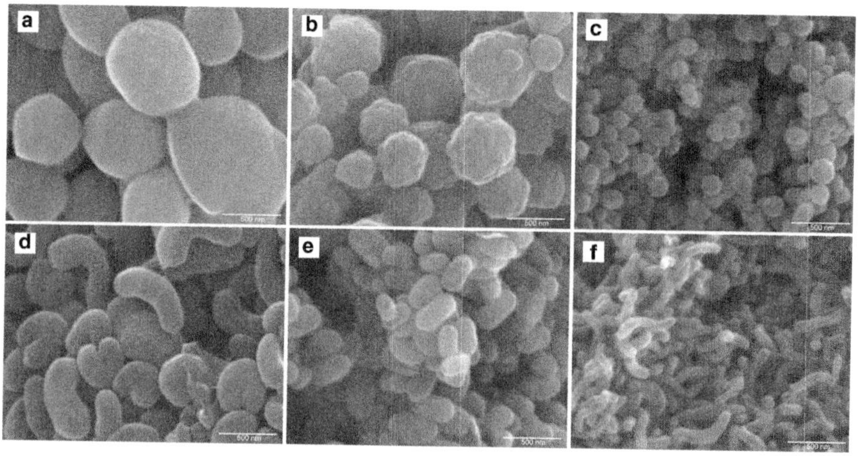

Figure 16.3 Some morphologies and sizes of mesoporous silica particles: (a) to (c) spherical with different sizes and textures; (d) to (f) rod-like with different sizes.

block copolymer surfactants,[31] hydrothermal treatment,[32] or the use of swelling agents like mesitylene or amines.[1,33,34]

Particle size and morphology are relevant properties for separation,[26] catalysis,[35] and drug delivery purposes.[36] Chan et al.[37] have suggested that for cetyltrimethylammonium bromide (CTAB) templated synthesis, mesoporous silica tends to form spherical particles, in order to minimize the surface free energy. This spherical shape can be altered, however, depending on the relative rates of silica polymerization versus mesophase formation.[37] Control of the relative rates can be attained by modifying the properties of the solution. One alternative is the use of cosolvents that slow down the formation of the mesophase by reducing the condensation rates of the silica precursors.[37,38] Another option is the simultaneous use of various surfactants that differ in the magnitude of their affinity for silica, creating discontinuities or different domains within the mesophase.[39–41] The co-condensation of organic groups can also lead to specific morphologies which include spheres, rods, fibers, helical rods, or aggregates of those shapes (Fig. 16.3).[42–44] The particle sizes can be controlled by the abovementioned conditions and by the concentrations of the reagents in the system, and can be adjusted with a good degree of monodispersity from 20 nm up to tens of microns.[41–43,45]

16.3 INCORPORATION OF ORGANIC MOIETIES WITHIN MESOPOROUS SILICA

Similar to higher organisms combining muscles with skeletons to attain mobility, the addition of flexible organic functionalities to the rigid mesoporous silica structures enables the formation of gated systems. Organic species can be covalently attached to silica with relative ease. The main sources for the functionalization of silica are

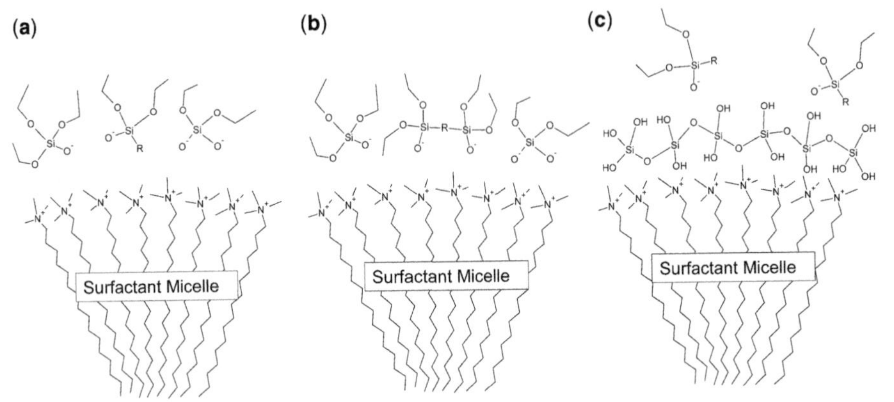

Figure 16.4 Incorporation of organic functionalities into mesoporous silica by (a) co-condensation of organotrialkoxysilanes, (b) co-condensation of bis(trialkoxysilyl)organics, and (c) postsynthesis grafting of organotrialkoxysilanes.

organotrichlorosilanes or organotrialkoxysilanes, a large variety of which are commercially available. Incorporating organic functionalities into mesoporous silica leads to an immense set of highly versatile hybrid materials. Organic groups can be introduced directly during the synthesis of the materials by co-condensation of either organotrialkoxysilanes to yield groups that emerge from the surface of silica (Fig. 16.4a) or bis(trialkoxysilyl)organics to yield groups that are embedded within the walls of the mesoporous material (Fig. 16.4b); the latter are known as periodic mesoporous organosilicas (PMO).[46–48] An alternative consists in grafting the organotrialkoxysilanes postsynthesis of the mesoporous material (Fig. 16.4c). Each method has advantages and drawbacks that have to be weighted according to the application for which the material is designed.

16.3.1 Incorporation During Synthesis by Co-Condensation

In this method an organotrialkoxysilane is introduced in the reaction mixture along with the silica precursor (commonly tetraethoxysilane, TEOS).[49] This mode of functionalization provides the highest homogeneity of the organic group within the mesoporous structure. The organic group ends up incorporated at concentrations around 2 mmol g^{-1} without loss of order and stability of the final mesoporous material.[43] Such a loading corresponds to surface densities of approximately 1 group nm^{-2}. The maximum load depends, however, on the nature of the functional group that is introduced. By using a series of anionic functionalized organosilanes, Lin and coworkers[50] were able to demonstrate that the efficiency of loading a co-condensing organosilane is proportional to the degree of electrostatic matching of the functional group with the surfactant. They demonstrated that the least hydrated functional groups were the most efficient in replacing bromide from the Stern layer of the CTAB micelles, thus providing the highest loads. After forming a disulfide bond between mercaptopropyltrimethoxysilane and 2-mercaptoethanesulfonic acid by means of dipyridyldisulfide chemistry, the authors were able to co-condense the product to yield up to three times more accessible groups than without the sulfonic acid functionality. Mercaptopropionic acid coupled to the mercaptopropyltrimethoxysilane gave only two times the amount of accessible groups than the silane itself. Based on these results and on the possibilities that disulfide chemistry offers, the authors proposed mercaptopropyltrimethoxysilane as a scaffold for tuning the degree of accessible functionalities in a silica matrix and for achieving high loading of a variety of functional groups.[50]

16.3.2 Hybrid Mesoporous Frameworks: Periodic Mesoporous Organosilicas

The co-condensation of bis(trialkoxysilyl)organics with TEOS or any other silica source leads to materials in which the organic group becomes a constitutive part of the walls of the structure. These materials were termed periodic mesoporous organosilicas and were discovered simultaneously by three independent groups led by Ozin,[46] Inagaki,[47] and Stein.[48] This type of functional materials has the advantage of a high homogeneity in the distribution of the organic groups and avoiding pore blocking by the organics. A wide variety of functional groups has been incorporated with this approach.

16.3.3 Postsynthesis Incorporation

The postsynthesis functionalization of mesoporous silicas, commonly known as grafting, is a versatile method of derivatizing the surface with organic groups.[51] The main advantage of this method is the selective functionalization of the external and internal (pore) surfaces with different groups.[52,53] Another advantage is that the structure is more stable and usually has a higher degree of order than the co-condensed material.[54] The drawbacks, however, are that the degree of functionalization is lower than that of the two previously mentioned methods; and that the groups cannot be introduced homogeneously. The main reason for these disadvantages is that the functional groups react initially at the external surface and the pore entrances, thus blocking the penetration of the reactants to the innermost sites of the pores; this problem is especially relevant when the pores are 3 nm or smaller.[55] An interesting application of this technique is the preparation by Burleigh and coworkers[56] of a mesoporous silica material first functionalized with *N*-[3-(trimethoxysilyl)-propyl]ethylenediamine, followed by adsorption of copper, and then followed by a second grafting of the *N*-[3-(trimethoxysilyl)-propyl]ethylenediamine, which gave ligands that were grafted within the material in an adequate disposition to recognize copper ions. This variant was designated as a stepwise imprint functionalization.[56]

16.4 GATED MESOPOROUS SILICA

As mentioned before, the large interior volumes of mesoporous materials and their pore sizes are ideal for promoting molecular interactions, and confinement and selectivity can be achieved upon introduction of gates. This feature has led to the quest for smart designs of nanoscopic gates and their immobilization within the entrances to the mesopores of the materials. As will be seen, many research groups have responded successfully to that challenge.

16.4.1 Nanoparticle Gated Mesoporous Silica

Solid gates can form composites that minimize the leaching of molecules previously loaded into the pores of mesoporous silica up to zero measurable release, an ideal characteristic for drug delivery applications. Based on this concept, Lin and coworkers[57,58] have designed a series of nanoparticle gated systems (Fig. 16.5). Taking advantage of a synthesis that allows the formation of mesoporous silica particles with regular morphologies and sizes between 150 and 300 nm, they have demonstrated that these materials can be employed as intracellular delivery agents.[57,58]

Lin and coworkers demonstrated not only the possibility to effectively cap the channels of mesoporous silica, but also the application of the composite to a biological system. Mesoporous silica nanoparticles (MSN) were functionalized by co-condensation with mercaptopropyl groups, and the mercapto functionality was then reacted with 2-(pyridyldisulfanyl)ethylamine to yield 3-(propyldisulfanyl)aminopropyl-functionalized MSN. The MSN was allowed to take up vancomycin and ATP by diffusion from a high concentration solution, and was then reacted with mercaptoacetic acid-functionalized cadmium sulfide nanocrystals to give a cadmium sulfide pore-capped material. The product was washed until vancomycin and ATP were undetectable by HPLC. The final loads of the materials were 25 and 47 $\mu mol\ g^{-1}$ MSN, respectively. The material exhibited less than 1% leaching after prolonged periods of suspension in physiological buffer. Efficient loading and capping of the material was further confirmed by decreased surface areas and decreased X-ray

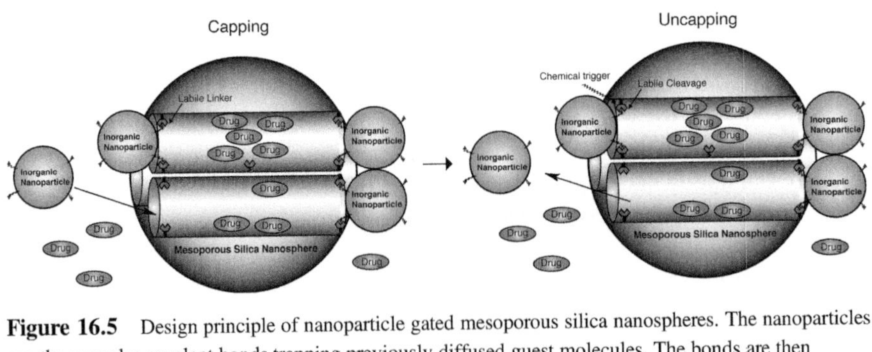

Figure 16.5 Design principle of nanoparticle gated mesoporous silica nanospheres. The nanoparticles cap the pores by covalent bonds trapping previously diffused guest molecules. The bonds are then broken under specific stimuli allowing the guest molecules to diffuse away.

diffraction signal intensity of the loaded material. The uncapping of the system was achieved by addition of disulfide reducing agents such as dithiothreitol (DTT) or mercaptoethanol, which promoted a rapid release of the entrapped substances: after 24 hours 85% of the loaded species were released into the solution. A practical application of the gated system was demonstrated when ATP loaded MSN capped with the CdS gate was introduced into a culture of astrocytes. CdS gated MSN were immobilized in a flow cell in the presence of astrocytes preloaded with a calcium sensitive dye. Upon perfusion of mercaptoethanol, the fluorescence due to CdS was gradually lost from the site of immobilization of MSN; at the same time the dye showed an increase in the intracellular concentration of calcium, as a result of ATP stimulation.[59]

A further refinement of the system was achieved when the CdS gates were replaced by the more biocompatible superparamagnetic iron oxide Fe_3O_4 nanoparticles, and was effectively endocytosed by mammalian cells. As a proof of principle 3-(propyldisulfanyl)propionic acid-functionalized MSN was loaded with fluorescein and then reacted with aminopropyl surface derivatized iron oxide nanoparticles. The capped material showed no leaching of the fluorophore after 5 days in suspension; however, on addition of disulfide reducing agents DTT or dihydrolipoic acid (DHLA) the loaded fluorophore was efficiently released within 48 hours. After adding a suspension of fluorescein loaded Fe_3O_4 gated nanoparticles to a HeLa cell culture and incubating overnight, cells that endocytosed the material were isolated magnetically. The magnetically isolated cells were labeled with nuclear stain, and observed under confocal fluorescence microscope. The microscopy images showed green fluorescent cell bodies, suggesting the material was uncapped after internalization by the cells, thus demonstrating the ability of the material to act as an intracellular controlled release device. The authors believe antioxidants naturally present in the cell such as DHLA were responsible for breaking the linking disulfide bonds and triggering the release of the fluorescein cargo.[60]

After demonstrating efficient delivery of gated mesoporous silica into animal cells, Lin and coworkers employed a different nanoparticle gated material to perform delivery of functional biochemical species into plant cells.[61] Gold nanoparticle capped MSN were prepared using the same disulfide linking strategy as above. The endocytosis by plant cells is more difficult than by animal cells, and protoplasts showed no uptake of MSN. A previous study, however, had shown that endocytosis of MSN by animal cells is affected by the external functionalization of the particles.[62] Based on that information, the authors modified the external surface of MSN by incorporating triethyleneglycol (TEG) functionality on the external surface, allowing the material to be taken up by the protoplasts. Furthermore, the TEG-functionalized MSN could be coated with a green fluorescent protein (GFP) promoting plasmid and protect it from digestion by restriction enzymes. Upon endocytosis of the composite the protoplasts were shown to express the encoded protein with efficiency comparable to that of standard polyethyleneglycol transfection vehicles. β-estradiol loaded gold-capped MSN were applied to transgenic tobacco plantlets containing an inducible promoter GFP gene. The gold nanoparticle caps increase the density of MSN greatly, allowing for gene gun bombardment to be employed as the mechanism of delivery into plant cells. After bombardment of the plantlets with the β-estradiol

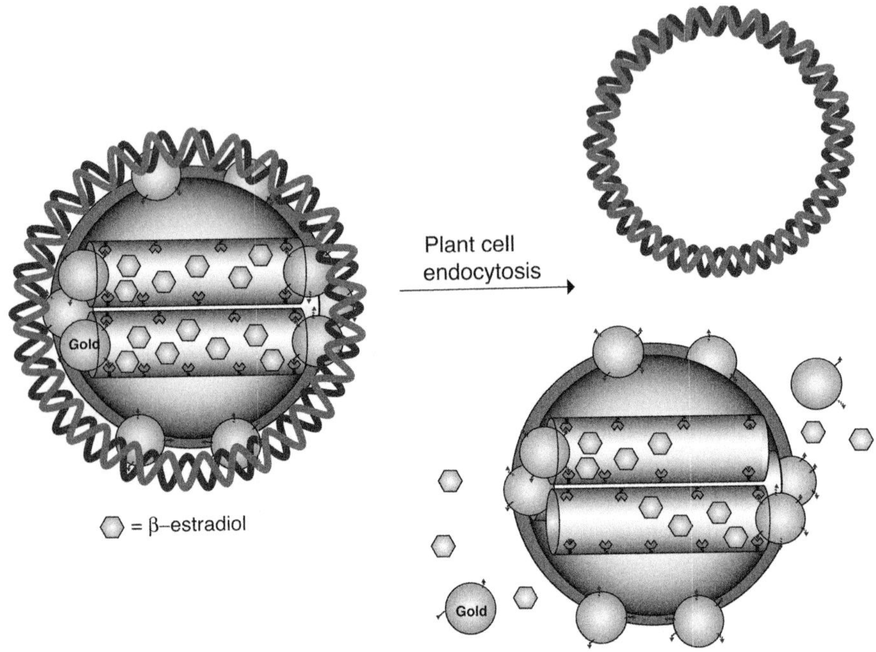

Figure 16.6 DNA-coated gold-capped mesoporous silica nanoparticle loaded with β-estradiol, employed for the simultaneous delivery of a gene and its promoter into plant cells.

loaded gold gated MSN and perfusion of DTT, it was possible to observe the expression of the fluorescent protein. Furthermore, the bombardment of nontransgenic plantlets with GFP encoding plasmid covered gold nanoparticle gated MSN containing β-estradiol also showed expression of GFP upon germination in DTT containing medium (Fig. 16.6). This proved that the gated MSN is an ideal vehicle to attain both spatial and temporal control in the delivery of various substances required to perform a biologically critical function.[61]

16.4.2 Organic Molecule Gated Mesoporous Silica

Even if small organic molecules are not able to completely block the pore entrances as solid nanoparticle caps do, they can still prevent the transport of molecules that are either larger than the openings that remain in the gate, or whose chemical properties tend to keep them away from the gating molecules. A large number of systems designed by several groups exemplify the efficiency of this approach.

Considering that the goal of a gating system is to provide some selectivity in the access to the pores of the material, the pore size itself can be considered as a gate. It has been demonstrated that even if proteins of different sizes are strongly adsorbed onto the surfaces of mesoporous silica, only the smaller ones are able to penetrate inside of the pores,[63] this is especially relevant when the proteins are to be immobilized by chemisorption rather than by physisorption within the pores.[27] This selectivity has been applied to the isolation and analysis of peptides from human plasma in the investigation of biomarkers.[64] The first study in which mesoporous silica was proposed as a candidate for drug delivery consisted in investigating the effect of pore size on the diffusion of drugs preloaded on the pores of the material.[65]

As mentioned above, gates do not necessarily need to completely block pore entrances as caps do, but they can limit the diffusion of undesired species while stimulating the diffusion of molecules of interest into the pores by chemical means. In this way, Lin and coworkers showed the first example of selective detection of closely related species within mesoporous silica.[66] The authors immobilized the amine-sensitive *o*-phthalic hemithioacetal by reaction of its aldehyde with mercaptopropyl groups in the mesopores of the material. To attain selectivity the authors grafted the material with different secondary groups: propyl, phenyl, and pentafluorophenyl. The materials were tested for reaction with the neurotransmitter dopamine and glucosamine to yield a fluorescent isoindole. Both substances were able to react with the hemithioacetal at the same rate within solution and have similar molecular sizes so pore size discrimination could not be used to distinguish them. The authors found that the sensor was not efficient for dopamine detection when no secondary functional group was present on the surface of silica, or when the secondary group was the hydrophobic propyl. However, good selectivity was obtained when the material was functionalized with the aromatic groups, especially with the pentafluorophenyl group that was able to form the best donor/acceptor $\pi-\pi$ stacking interactions with the dopamine. Glucosamine, to the contrary, was able to give a good response only when the material had no secondary functional groups. In this way it was demonstrated that selectivity could be attained in the mesoporous material by surface functionalization.[66] The sensor was further refined by polymerizing a layer of poly(lactic acid) on the exterior surface of the MSN, which acted as an external gate to selectively prevent the diffusion of species to the interior of the material (Fig. 16.7). When comparing the ability of three different neurotransmitters to diffuse into the pores and react with the hemithioacetal, they discovered that dopamine was able to penetrate faster than glutamic acid and tyrosine into the material at pH 7.4. The observed selectivity was attributed to the fact that while dopamine (isoelectric point 9.7) is positively charged at physiological pH, the other two species are negatively charged (isoelectric points: 5.7 for tyrosine and 3.2 for glutamic acid), so the acid polymeric layer tended to repel the latter two species while letting the dopamine diffuse through.[53]

The use of a polymer species as a way to control diffusion to the inside of mesoporous silica was also employed by López and coworkers.[67] In this work the researchers polymerized *N*-isopropyl acrylamide on mesoporous silica by atom transfer radical polymerization, and took advantage of the changes the polymer experiences upon thermal treatment. The authors discovered that the hybrid material could take up more fluorescein than nonfunctionalized material at temperatures above 45°C. At that temperature the polymer is in a collapsed hydrophobic state and partially covers the negatively charged surface of silica that otherwise repels the negatively charged fluorescein dye. At temperatures below 30°C the polymer exists in a hydrated state in which the chains are expanded. Interestingly, the fluorescein loaded hybrid particles were

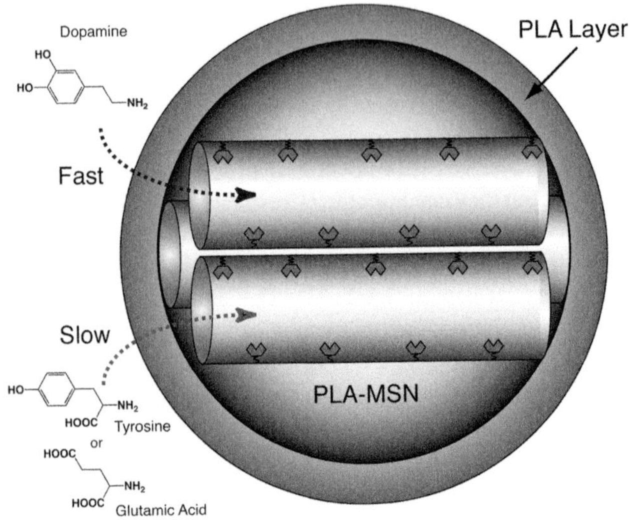

Figure 16.7 Schematic representation illustrating the selective diffusion of dopamine over glutamic acid and tyrosine into the mesopores of PLA-coated MSN-based fluorescence sensor system.

able to release the dye faster when resuspended at 50°C than when resuspended at 25°C. This suggested that the polymer molecules extended at low temperatures were able to partially block the diffusion of the dye outside of the pores.[67]

The use of photochemistry as a way to control the diffusion in and out of the pores of mesoporous materials was introduced by Mal and coworkers,[68] when they grafted propoxycoumarin at the pore entrances of the material. After removal of the surfactant template, the propoxycoumarin grafted material was loaded in organic media with the steroid cholestane followed by irradiation at wavelengths above 310 nm which induced the dimerization of the coumarin, thus closing the gate (Fig. 16.8). The material was able to retain 28 weight% of the steroid; to the contrary, material that was irradiated before being set in contact with cholestane was unable to take up the steroid. After washing the cholestane loaded material and resuspending it in solvent, the researchers irradiated it with a wavelength around 250 nm cleaving the coumarin dimer and allowing the steroid to release. Around 80% of the loaded cholestane was recovered by washing the 250 nm irradiated material with hexane. In this way the authors demonstrated full control of the opening and closing of the gate by photochemical stimulation.[68]

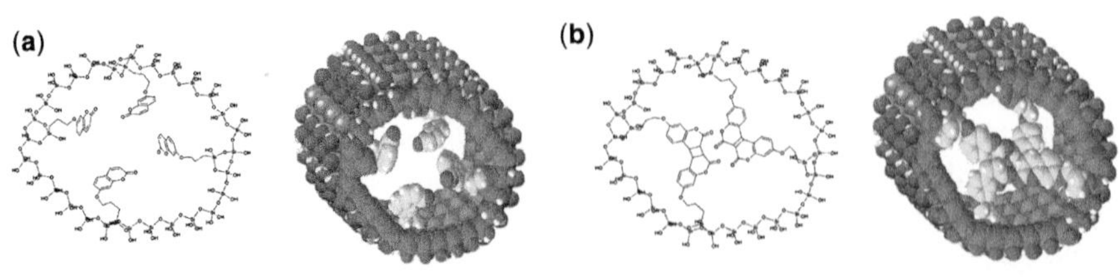

Figure 16.8 Photochemical gate based on coumarin dimerization, (a) in its open and (b) in its closed state.

Another photochemically based gating system is the one by Brinker and coworkers in which the cis-trans isomerization of azobenzene was applied.[69] (Triethoxysilylpropylureido)azobenzene was co-condensed with TEOS by evaporation-induced self-assembly to yield a functionalized mesoporous silica thin film. The propylureidoazobenzene groups inside of the mesopores could then be isomerized from their stable elongated trans structure to the more compact cis geometry upon UV irradiation. The trans geometry could be recovered by thermal treatment. The authors suggested that the results of the transitions from trans to cis could lead to an increase in pore size of approximately 0.7 nm and an increase of the polarity from 0 to 3 D. This could be applied to regulation of ion transport; however, they did not confirm such expectations.[69]

A third photochemically controlled gate for mesoporous silica materials was reported recently by Martínez-Máñez and coworkers. In this work the researchers synthesized a spirobenzopyran derivative with a silica anchoring triethyoxysilyl group, and they reacted it with a suspension of mesoporous silica material containing tris(2,2′-bipyridyl)ruthenium chloride dye. Since the spirobenzopyran can be converted into a merocyanine positively charged at neutral pH, a negatively charged G1.5 poly(amidoamine) (PAMAM) dendrimer was added to the suspension in order to electrostatically bind the merocyanine photoproduct, thereby capping the mesopores of the material. The researchers were able to demonstrate that PAMAM was able to cap the pores of the material and prevent the escape of the dye only when the pH of the suspension was high enough for the merocyanine form to be protonated and when the solution had been previously exposed to UV irradiation. In this way, the authors proposed the gated material could be employed as a NAND-type pH- and irradiation-controlled logic gate.[70]

The use of electrostatic interactions was also employed by Martínez-Máñez and coworkers to develop amine gated mesoporous silica as chemosensors for anions. Mesoporous silica was grafted with aminopropyltrimethoxysilane, and the product was employed for the reductive amination of 9-anthraldehyde to yield a fluorescent aminoanthracene-functionalized mesoporous silica. The fluorescence of a suspension of the aminoanthracene-functionalized material was significantly quenched by the presence of adenosine triphosphate (ATP) anions in the solution, but to a lesser extent by adenosine diphosphate (ADP) or adenosine monophosphate (AMP). Furthermore, small anions like chloride, bromide, or even phosphate were unable to induce quenching of the fluorophore-functionalized material. Interestingly, homogeneous solutions of the fluorophore gave a response to the anions only at much higher concentrations. The selectivity of the aminoanthracene receptor was explained in terms of hydrogen bonding of the ammonium and the anions and of π stacking interactions. Immobilizing the aminoanthracene moiety on nonporous silica gave a less sensitive material than the one formed by the grafting onto mesoporous silica. The enhanced response of the mesoporous silica immobilized aminoanthracene receptor was explained by the fitting of the ATP to the geometrical distribution of the fluorophore molecules which was in turn determined by the pore size and structure of the material.[71,72]

The effect of pH on the interaction between triamine groups was used by Martínez-Máñez and coworkers to create a pH and anion controlled gated mesoporous silica material. The material was prepared by co-condensation of mercaptopropyltriethoxysilane with TEOS followed by grafting of *N*-(3-trimethoxysilylpropyl-2-aminoethyl)ethylenediamine on the external surface of the material. The open/close state of the triamine gates was monitored by the bleaching of squaraine dye produced by the mercaptopropyl groups present inside of the pores of the material. No bleaching of the dye was observed at acidic pH in the presence of anions, with the largest and most efficiently coordinating anions providing the highest protection against bleaching. Upon an increase in pH the bleaching of the dye was observed. These observations were explained by the interactions occurring between the amines as a function of pH. At high pH values, the amines are deprotonated and are able to interact with each other through hydrogen bonding (Fig. 16.9a). At low pH, however, the protonated ammonium groups have a repulsion effect and orient themselves to close the gate (Fig. 16.9b). The closure was notably improved by electrostatic interactions with anionic species, which acted as seals for the gate (Fig. 16.9c), demonstrating a synergistic effect in the hybrid material.[73] Further insight on the anion-pH responsive triamine system was attained by molecular dynamics simulations and by experimental study of a larger number of anions. The controlled release of tris(2,2′-bipyridylruthenium)chloride was also used to experimentally evaluate the system. It was found that the anions that tend to form hydrogen bonding networks like phosphate were particularly efficient for closing the gates.[74] This gated material was recently employed for the controlled release of vitamin B2.[75]

An approach that could be considered complementary to the triamine grafted material is that of Xiao and coworkers,[76] who grafted the ethyl ester of 11-(triethoxysilyl)undecanoic acid on the surface of mesoporous silica, followed by hydrolysis to produce carboxylic acid-functionalized material. After suspending the functionalized material in a solution of the drug vancomycin, the researchers added the cationic polyelectrolyte poly(dimethyldiallylammonium) chloride (PDDA), which electrostatically associated to the carboxylate groups of the material thus closing the entrance to the pores. In this way, the authors were able to load up to 36 wt% of the drug into the material. Upon suspension in aqueous media, the hybrid material released the loaded drug in a pH-dependent manner. While too little release was observed at pH 6.5, a large amount of the drug was able to escape at pH 4.5, and an even larger amount (90 wt% after 30 minutes) was detected at pH 2.0.[76]

Shi and Zhu also used a polyelectrolyte to achieve a mesoporous silica-based, pH-responsive drug release system. Their approach, however, was different from the abovementioned triamine- and carboxylate-based methods. In this study, mesoporous silica spheres were coated with a layer of poly(allylamine hydrochloride) followed by a second layer of poly(styrene sulfonate) and repeated eight times. The coated hybrid material was then suspended in an aqueous solution of the antibiotic gentamicin at pH 2. After stirring the suspension for 4 hours, the pH of the suspension was raised to 8, and the material was isolated and washed with pH 8 solution. The drug could be loaded up to 34 wt%. The solid was then dispersed in a solution at pH 8 and the leaching of gentamicin was monitored for 12 hours reaching a maximum value of 5%. After that time, the pH of the solution was decreased to 2, and a fast release of the drug could be observed reaching a value of 70% over 12 hours. The detailed mechanism of both

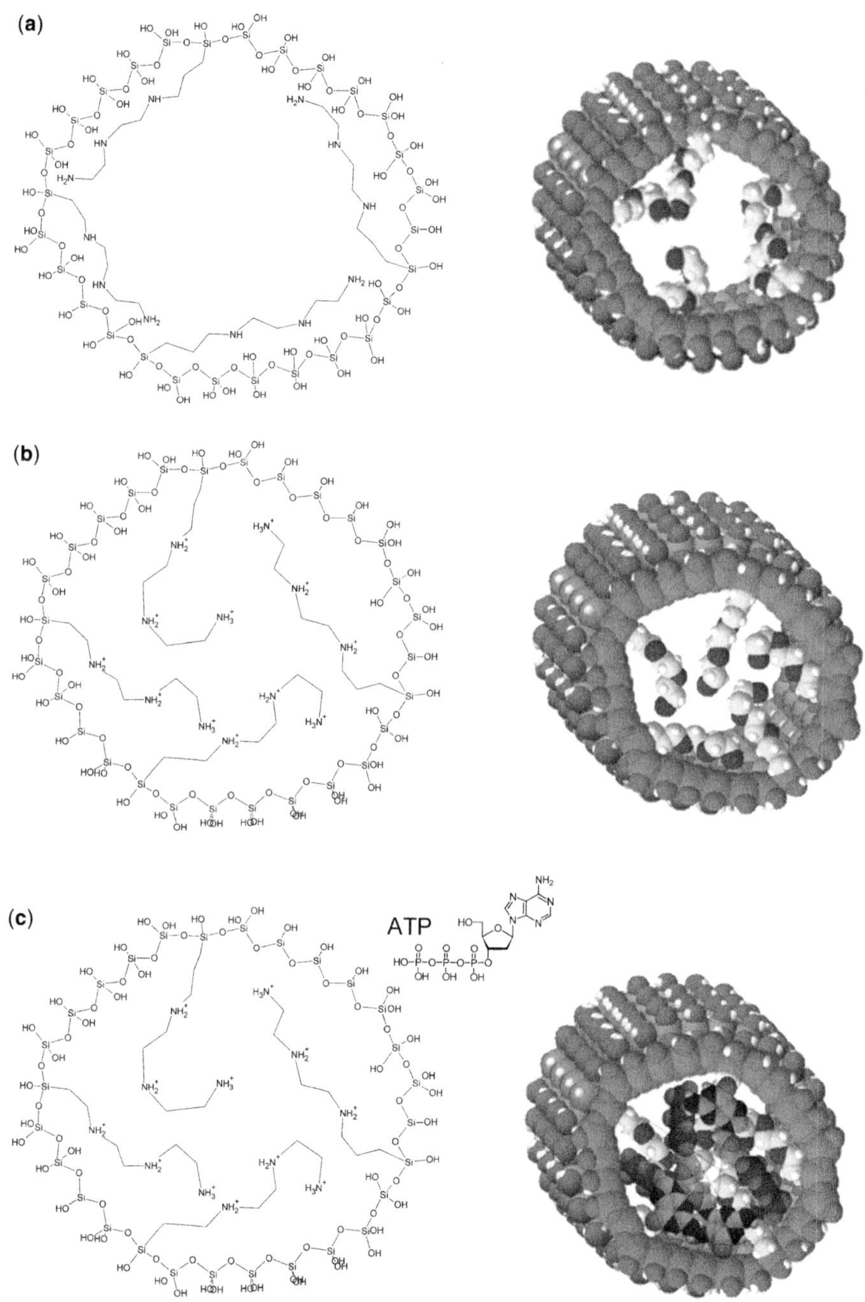

Figure 16.9 Triamine-based gating strategy: (a) at high pH the amine groups interact with each other and the pore wall through hydrogen bonding and the gate is open; (b) at low pH the protonated ammonium ions repel each other, closing the gate; (c) in the presence of bulky polyvalent counterions the pores are effectively sealed.

uptake and release was not explained by the researchers, who described the process in terms of an adjustable permeability of the multilayers as a function of pH.[77]

Covalently linked PAMAM was utilized as a gate for mesoporous silica materials by Lin and coworkers. This material was used to prove that mesoporous silica nanoparticles (MSN) can be employed for the delivery of two different species simultaneously. MSN grafted with isocyanatopropyl triethoxysilane were suspended in an ethanolic solution of the fluorescent dye Texas Red, and then reacted with a solution of G-2 PAMAM dendrimer to generate a urea bond and cap the pores. The PAMAM capped material was complexed to DNA, as evidenced by agarose gel electrophoresis, and protected the DNA against enzymatic cleavage by restriction endonucleases (Fig. 16.10). The researchers formed a complex of the material with plasmid DNA encoding for enhanced green fluorescent protein (EGFP) and added a suspension of it to cultures of human cervical cancer (HeLa), Chinese hamster ovary (CHO) and astrocyte cells. After doing so the authors demonstrated for the first time the endocytosis of MSN by animal cells by fluorescence confocal microscopy of the Texas Red encapsulated within the pores, and by transmission electron microscopy. Furthermore, the authors demonstrated the functionality of the hybrid material, by observing the cellular expression of the EGFP vector delivered under fluorescent microscope and through flow cytometry.[57]

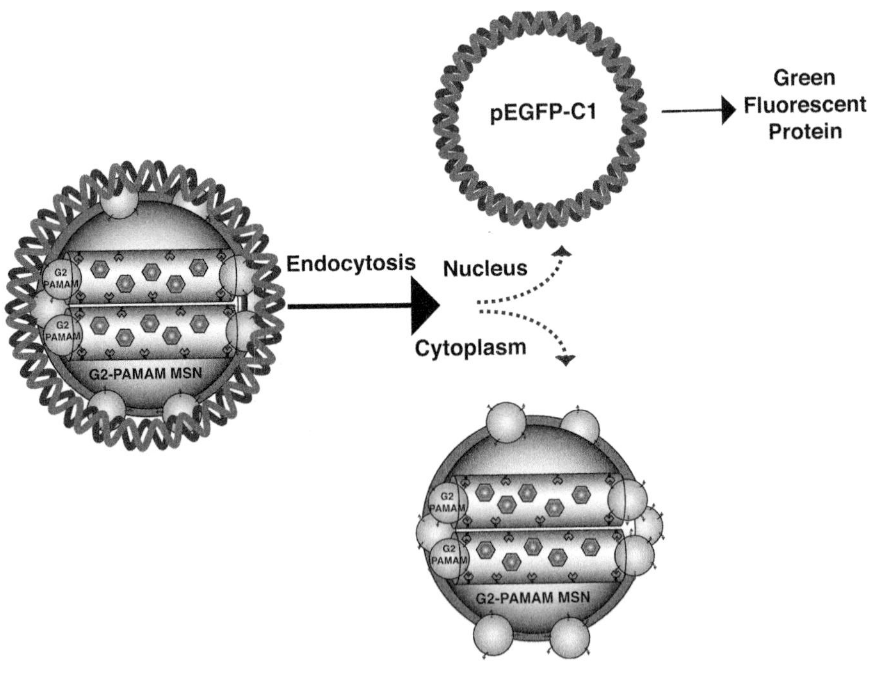

Figure 16.10 Schematic representation illustrating the cell uptake of the nonviral gene transfection system based on DNA-coated PAMAM-capped MSN and the dissociation of the plasmid DNA from the MSN.

By using disulfide bonds to cap MSN with PAMAM, Yeung and coworkers[78] were able to compare in real time and at the microscopic level the kinetics of uncapping the "soft" organic PAMAM gates versus the "hard" nanoparticle CdS gates previously reported by Lin et al.[59] For that purpose, 2-(propyldisulfanyl)ethylamine-functionalized MSN material was prepared as described above, and suspended in a buffered aqueous solution of ATP. The different caps—mercaptopropionic acid-functionalized CdS nanoparticles and two different generations of PAMAM (G 2.5 and G 4.5)—were added to the suspensions containing ATP and functionalized MSN and formed amide bonds between the caps and the MSN through carbodiimide chemistry. After isolating the loaded and gated MSN materials they were resuspended in buffered solution and the release of ATP triggered by disulfide reducing agent was monitored by its activation of luciferase chemiluminescence. The release was observed under a microscope equipped with an iCCD chip with 0.80 Hz data collection. No significant release of ATP was found in the absence of disulfide reducing trigger. Interestingly, the researchers found that upon application of the trigger, the nanoparticle gated MSN released faster than both of the dendrimer gated systems (2, 5, and 20 minutes for CdS, G 4.5, and G 2.5 PAMAM, respectively). The results were interpreted in terms of the different flexibilities that the caps had, which resulted in different numbers of cap-MSN bonds. More flexible caps, such as the smaller PAMAM dendrimer, were able to form a larger number of bonds with the MSN and fit more tightly within the pores of the material than the rigid nanoparticle caps.[78]

The use of disulfide bond as a gate was also recently employed by Feng and coworkers. In their approach, the authors grew poly-(*N*-acryloxysuccinimide) through reversible addition-fragmentation chain transfer (RAFT) polymerization. After loading the mesopores with a fluorophore, the polymer was cross-linked by reaction of the succinimidyl moieties with cystamine. The release of a fluorescent guest was observed upon breaking the disulfide with DTT.[79]

A gating approach that resembles the coumarin dimerization was attempted by Fujiwara and coworkers by using disulfide bonds as gates. The goal of this research was to prepare a selective catalyst in which the promoted reaction took place inside the mesopores. Calcined mesoporous silica was treated with aluminum acetoacetate to yield an aluminum-doped mesoporous silica material, known to catalyze the dimerization of α-methylstyrene (AMS). The researchers confirmed the activity of their material by obtaining a 99% yield of the mentioned reaction with the mesoporous catalyst. A disilane compound was prepared by reaction of 6-hydroxy-2-naphthyl disulfide with 3-(triethoxysilyl)propyl isocyanate, and then grafted onto the surface of the aluminum-doped material. No product was obtained when the disulfide-functionalized material was employed in the catalytic dimerization of AMS. However, upon cleaving the disulfide gate with DTT, the material became catalytic for the AMS dimerization, giving a high yield (97%). After regeneration of the disulfide bridge by addition of iodine, the material once again became an ineffective catalyst for the reaction. After performing a series of control experiments the researchers were able to conclude that the incorporated disulfide moiety was an efficient gate to prevent the diffusion of reactants into the pore, and that the catalytic activity of the system can be reversibly controlled by redox chemistry.[80]

A series of gates based on rotaxane motifs in which the threads were anchored to the openings of mesoporous silica was developed by Zink and coworkers.[81–84] The first system the researchers developed was a pseudorotaxane consisting of a derivatized 1,5-dioxynaphthalene (DON) thread bound to the mesoporous silica surface and a cyclobis(paraquat-*p*-phenylene) ($CBPQT^{4+}$) ring that was able to noncovalently bind DON. The DON-functionalized mesoporous material was loaded with the dye tris(2,2′-phenylpyridyl)iridium(III) and then the $CBPQT^{4+}$ was added to the system threading itself into the DON and consequently closing the entrance to the pores. The pores were reopened by addition of a reducing agent ($NaCNBH_3$) after which the release of the entrapped dye was observed by luminescence.[81] The system was further developed by including a second recognition site based on tetrathiafulvalene (TTF) and turning the functionality into a rotaxane by incorporating a bulky group at the outer end of the axis (Fig. 16.11a). When the TTF unit was in the oxidized state (TTF^{2+}) the $CBPQT^{4+}$ had no affinity for it so it remained interacting with the DON unit, which was close to the pore entrance, keeping the gate closed (Fig. 16.11b). Upon reduction of the TTF^{2+} moiety with ascorbic acid, the $CBPQT^{4+}$ unit shuttled to the TTF position, opening the gate (Fig. 16.11c). The authors were able to demonstrate that the gate was successful by spectroscopically determining the position of the $CBPQT^{4+}$ and by observing the loading and release of dyes into and out of the pores of the hybrid material.[82] The authors optimized their system later, by studying the effects of chain length, relative positions of the recognition sites within the chain, methods of surfactant removal from the material, and methods of functionalizing the material. They established, for example, that it was necessary to shorten the thread in order to prevent leaking of guests with sizes smaller than 1×1 nm.[83]

The system was modified by Zink and coworkers in order to trigger the uncapping of the gate by photochemical activation. In order to attain this goal, the axis of the pseudorotaxane was replaced by 1,5-bis[2-(2-(2-hydroxyethoxy)ethoxy)ethoxy]-naphthalene (BHEEEN) which was attached to mesoporous silica by reaction with an isocyanatopropyl triethoxysilane grafted group. The functionalized material was then loaded with the dye tris(2,2′-phenylpyridyl)iridium(III) ($Ir(ppy)_3$) and then immersed in a solution of $CBPQT^{4+}$ chloride for capping. As in the previously mentioned system, the $CBPQT^{4+}$ affinity for the central DON group of BHEEEN led to the formation of the pseudorotaxane and closed the pore of the mesoporous material. The uncapping could be performed by addition of the reducing agent $NaCNBH_3$. However, the reduction of $CBPQT^{4+}$ could also be performed photolytically with the aid of a photosensitizer. The researchers grafted 9-anthracenecarboxylic acid or tris(2,2′-phenylpyridyl)ruthenium(II) to the surface of the mesoporous material. These groups were able to promote the reduction and unthreading of $CBPQT^{4+}$ upon irradiation with a 364 nm and a 488 nm laser, respectively. By turning on and off the laser at various time points, the researchers were able to observe changes in the release profile of the dye, which demonstrated the ability that the system had to provide a temporal control in the dye release. The system, however, required the addition of triethanolamine reducing agent to regenerate the sensitizers.[84]

The pseudorotaxane approach was also applied by Zink and coworkers to create gates controlled by pH.[85] In this work, the researchers grafted

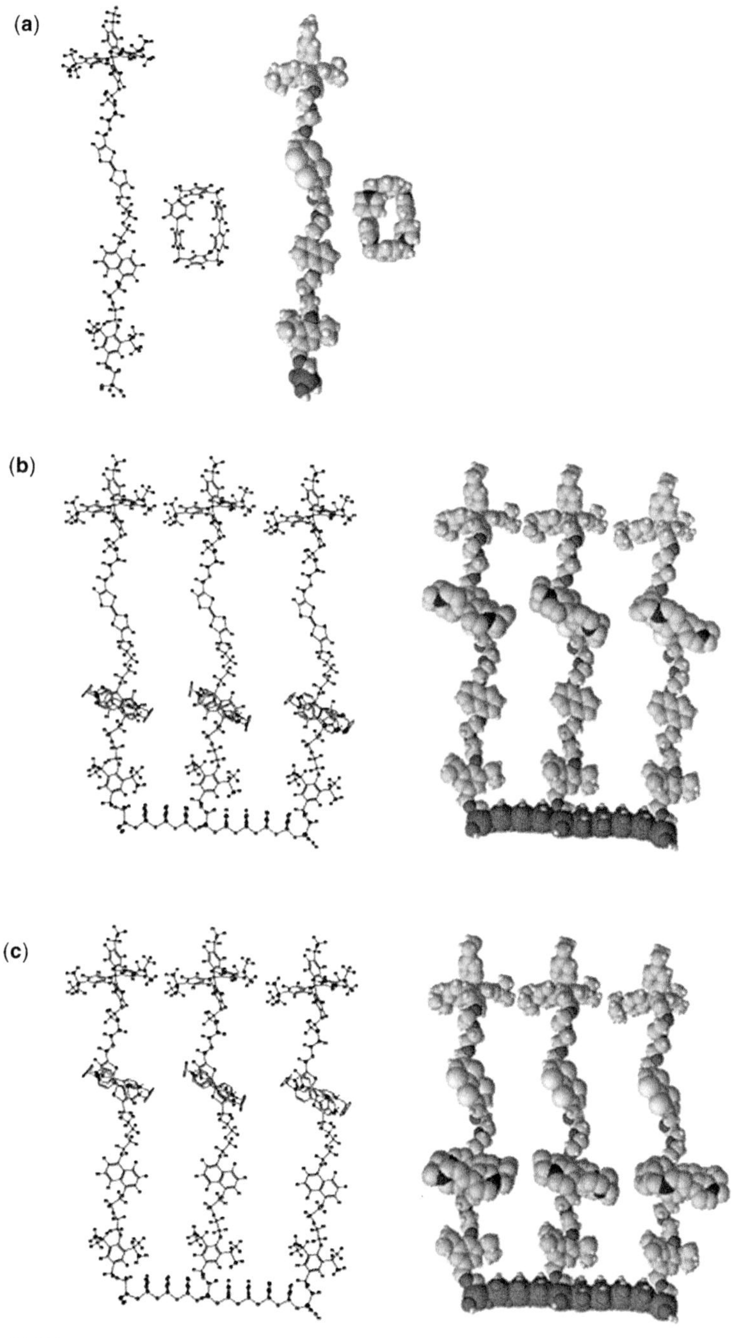

Figure 16.11 (a) Redox-controlled rotaxane gate consisting of a DON–TTF derivatized axis and a movable $CBPQT^{4+}$ unit. (b) In its reduced form $CBPQT^{4+}$ binds the TTF, opening the gate. (c) In its oxidized form $CBPQT^{4+}$ binds the DON, closing the gate.

isocyantopropyltrimethoxysilane on the pore openings of mesoporous silica particles and used this reactive group to synthesize naphthalene-containing dialkylammonium threads. The dye, coumarin 460, was then loaded into the material followed by addition of a dibenzo[24]crown-8 (DB24C8) unit which interacted by hydrogen bonding with the positive ammonium groups of the thread to generate a pseudorotaxane supramolecular complex. The complex formation enabled closing of the entrance to the pore, trapping the previously loaded coumarin 460 with approximately 20% leaching as determined by luminescence spectrometry. The researchers demonstrated that the addition of a base could trigger the dethreading of the DB24C8 and induce the release of the dye. It was determined that the efficiency of the basic triggers was affected by steric constraints, in that the smallest bases were the most efficient to open the gate.[85] Since it is desirable to have systems that work in aqueous media, and the systems developed by Zink and coworkers were functional only in organic media, they modified their pH responsive system to be able to use it in water. In their new system aminoethyl grafted MSN was reacted with propargyl bromide, and the resulting alkyne-functionalized material was suspended in a solution of rhodamine B for absorption of the dye into the mesopores. The researchers then used cucurbit[6]uril (CB[6]) as a catalyst for performing an interfacial Huisgen reaction between the alkyne-functionalized MSN and 2-azidoethylamine. The synthetic design allowed the CB[6] moiety to be concurrently threaded into the resulting chain forming the pseudorotaxane and capping the MSN pores. Hydrogen bonding interactions between the threaded CB[6] and two amine groups above and below it prevented the macrocycle from sliding out of the pseudorotaxane. Upon deprotonation of the amines by addition of sodium hydroxide up to pH 10, the hydrogen bonding interactions were lost and the CB[6] was able to diffuse away from the MSN, thus enabling the release of the loaded dye. The researchers recognized that subjecting silica material to such a high pH value could be detrimental for the material; however, they reported that no significant changes in the X-ray diffraction pattern and the particle morphology were observed after the release of the dye.[86]

A pH-responsive pseudorotaxane operational in aqueous solution was also reported by Kim and coworkers.[87] In their system the thread consisted of a low molecular weight amino terminal polyethyleneimine (PEI, $M_n = 1100$), and it was grafted by a 1,1′-carbonyldiimidazol catalyzed reaction with a carboxylic acid-functionalized mesoporous silica particle. The hybrid material was then suspended in a solution of the dye calcein in physiological buffer. The pseudorotaxane was then formed by addition of cyclodextrin (CD), which at basic pH is able to form hydrogen bonds with the deprotonated nitrogen atoms of the polyethyleneimine thus threading into the chains and closing the pores. Different CD were tested for their ability to be threaded by the PEI axis; αCD and γCD proved to yield the desired pseudorotaxane. The addition of CD was performed at pH 11, and the authors evaluated the dethreading of the CD and the consequent uncapping of the system upon acidification of the solution to pH 5.5. A fast release of the loaded dye was observed upon acidification of the suspension. The system, composed of biocompatible elements, proved to constitute an efficient, pH-regulated gate; however, as mentioned above, the stability of the mesoporous silica at such high pH values remains a practical concern.[87]

An αCD containing rotaxane was recently reported by Stoddart and coworkers. In this system the threading axis is a polyethyleneglycol formed by reaction of aminopropyl-functionalized mesoporous silica with triethyleneglycol monoazide monotosylate. After loading the material in a solution of rhodamine B, the CD was added to the solution and threading was promoted by setting the solution at 5°C. A bulky, adamantane ester containing stopper was linked to the azide terminated axis by Huisgen cycloaddition. After capping the release was verified by adding porcine liver esterase which by breaking the ester bond to the adamantyl terminal group enabled the CD to escape from the thread and uncap the pores (Fig. 16.12). The release of the loaded dye was monitored by luminescence spectroscopy.[88]

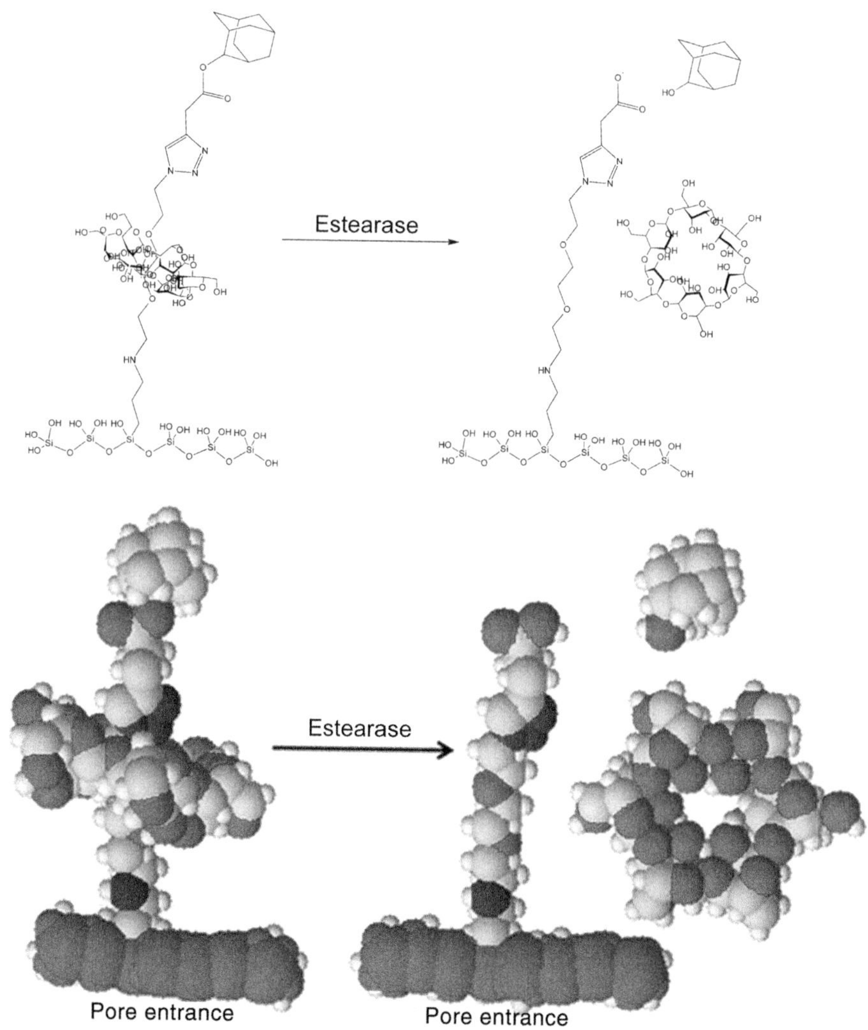

Figure 16.12 Estearase cleavage mediated opening of an αCD-based rotaxane gate.

16.5 OUTLOOK

As can be appreciated in this brief review, despite the short age, the field of gated mesoporous silicas is gradually approaching maturity. The chemical inventive nature of the researchers in the field is evidenced by the existence of a relatively large and fast growing number of ingenious capping systems and uncapping mechanisms. It must be recognized, however, that despite the claims of relevance and applicability of the developed systems, most of the published work to date has been limited to merely demonstrating the principles of operation. At this stage of development, it is necessary to make a call for practical applications of all these highly creative systems. It has to be recognized also, that there are still too few examples of higher order gating related to selective permeability for the penetration of chemical species into the mesopores, even if achieving such a goal for a variety of substances could be of high relevance. Therefore, it is possible to say that even if the field is becoming increasingly solid, there are still plenty of challenges that have not been addressed and are waiting for the contribution of new generations of daring and insightful scientists.

REFERENCES

1. J. S. Beck, J. C. Vartuli, W. J. Roth, M. E. Leonowicz, C. T. Kresge, K. D. Schmitt, C. T. W. Chu, D. H. Olson, E. W. Sheppard, *J. Am. Chem. Soc.* **1992**, *114*, 10834–10843.
2. C. T. Kresge, M. E. Leonowicz, W. J. Roth, J. C. Vartuli, J. S. Beck, *Nature* **1992**, *359*, 710–712.
3. T. Yanagisawa, T. Shimizu, K. Kuroda, C. Kato, *Bull. Chem. Soc. Jpn.* **1990**, *63*, 988–992.
4. S. Inagaki, Y. Fukushima, K. Kuroda, *Chem. Commun.* **1993**, 680–682.
5. Y. Lu, R. Ganguli, C. A. Drewien, M. T. Anderson, C. J. Brinker, W. Gong, Y. Guo, H. Soyez, B. Dunn, M. H. Huang, J. I. Zink, *Nature* **1997**, *389*, 364–368.
6. C. J. Brinker, Y. Lu, A. Sellinger, H. Fan, *Adv. Mater.* **1999**, *11*, 579–585.
7. J. C. Vartuli, K. D. Schmitt, C. T. Kresge, W. J. Roth, M. E. Leonowicz, S. B. McCullen, S. D. Hellring, J. S. Beck, J. L. Schlenker, *Chem. Mater.* **1994**, *6*, 2317–2326.
8. N. Li, S. Liu, H. Luo, *Anal. Lett.* **2002**, *35*, 1229–1238.
9. A. Galarneau, F. Di Renzo, F. Fajula, L. Mollo, B. Fubini, M. F. Ottaviani, *J. Colloid Interface Sci.* **1998**, *201*, 105–117.
10. G. D. Stucky, A. Monnier, F. Schüth, Q. Huo, D. Margolese, D. Kumar, M. Krishnamurty, P. Petroff, A. Firouzi, M. Janicke, B. F. Chmelka, *Mol. Cryst. Liq. Cryst. Sci. Technol. Sect. A Mol. Cryst. Liq. Cryst.* **1994**, *240*, 187–200.
11. A. Monnier, F. Schüth, Q. Huo, D. Kumar, D. Margolese, R. S. Maxwell, G. D. Stucky, M. Krishnamurty, P. Petroff, A. Firouzi, M. Janicke, B. F. Chmelka, *Science* **1993**, *261*, 1299–1303.
12. N. Gov, I. Borukhov, D. Goldfarb, *Langmuir* **2006**, *22*, 605–614.
13. G. S. Attard, J. C. Glyde, C. G. Goltner, *Nature* **1995**, *378*, 366–368.
14. D. Zhao, Q. Huo, J. Feng, B. F. Chmelka, G. D. Stucky, *J. Am. Chem. Soc.* **1998**, *120*, 6024–6036.
15. D. Zhao, J. Feng, Q. Huo, N. Melosh, G. H. Frederickson, B. F. Chmelka, G. D. Stucky, *Science* **1998**, *279*, 548–552.
16. S. A. Bagshaw, E. Prouzet, T. J. Pinnavaia, *Science* **1995**, *269*, 1242–1244.
17. P. T. Tanev, T. J. Pinnavaia, *Science* **1995**, *267*, 865–867.
18. B. G. Trewyn, C. M. Whitman, V. S. Y. Lin, *Nano Lett.* **2004**, *4*, 2139–2143.
19. S. Che, A. E. Garcia-Bennett, T. Yokoi, K. Sakamoto, H. Kunieda, O. Terasaki, T. Tatsumi, *Nature Mater.* **2003**, *2*, 801–805.
20. D. Chen, Z. Li, C. Yu, Y. Shi, Z. Zhang, B. Tu, D. Zhao, *Chem. Mater.* **2005**, *17*, 3228–3234.

21. Y. Han, J. Y. Ying, *Angew. Chem. Int. Ed.* **2005**, *44*, 288–292.
22. Y. Han, D. Li, L. Zhao, J. Song, X. Yang, N. Li, Y. Di, C. Li, S. Wu, X. Xu, X. Meng, K. Lin, F.-S. Xiao, *Angew. Chem. Int. Ed.* **2003**, *42*, 3633–3637.
23. Y. Wei, K.-Y. Qiu, in *Nanoporous Materials: Science and Engineering* (Eds. G. Q. Lu, X. S. Zhao), Singapore: World Scientific, **2004**, 873–892.
24. S. Che, Z. Liu, T. Ohsuna, K. Sakamoto, O. Terasaki, T. Tatsumi, *Nature* **2004**, *429*, 281–284.
25. A. Shimojima, Z. Liu, T. Ohsuna, O. Terasaki, K. Kuroda, *J. Am. Chem. Soc.* **2005**, *127*, 14108–14116.
26. Q. Huo, D. I. Margolese, G. D. Stucky, *Chem. Mater.* **1996**, *8*, 1147–1160.
27. H. H. P. Yiu, C. H. Botting, N. P. Botting, P. A. Wright, *Phys. Chem. Chem. Phys.* **2001**, *3*, 2983–2985.
28. C. Zapilko, Y. Liang, W. Nerdal, R. Anwander, *Chem. Eur. J.* **2007**, *13*, 3169–3176.
29. M. H. Lee, S. J. Lee, J. H. Jung, H. Lim, J. S. Kim, *Tetrahedron* **2007**, *63*, 12087–12092.
30. M. Widenmeyer, R. Anwander, *Chem. Mater.* **2002**, *14*, 1827–1831.
31. C. Yu, J. Fan, B. Tian, G. D. Stucky, D. Zhao, *J. Phys. Chem. B* **2003**, *107*, 13368–13375.
32. A. Sayari, P. Liu, M. Kruk, M. Jaroniec, *Chem. Mater.* **1997**, *9*, 2499–2506.
33. A. Sayari, *Angew. Chem. Int. Ed.* **2000**, *39*, 2920–2922.
34. A. Sayari, Y. Yang, M. Kruk, M. Jaroniec, *J. Phys. Chem. B* **1999**, *103*, 3651–3658.
35. T. M. Suzuki, T. Nakamura, E. Sudo, Y. Akimoto, K. Yano, *Microporous Mesoporous Mater.* **2008**, *111*, 350–358.
36. B. G. Trewyn, J. A. Nieweg, Y. Zhao, V. S. Y. Lin, *Chem. Eng. J.* **2008**, *137*, 23–29.
37. H. B. S. Chan, P. M. Budd, T. D. V. Naylor, *J. Mater. Chem.* **2001**, *11*, 951–957.
38. H. Zhang, J. Sun, D. Ma, G. Weinberg, D. S. Su, X. Bao, *J. Phys. Chem. B* **2006**, *110*, 25908–25915.
39. L. Qi, J. Ma, H. Cheng, Z. Zhao, *Chem. Mater.* **1998**, *10*, 1623–1626.
40. M. Gruen, I. Lauer, K. K. Unger, *Adv. Mater.* **1997**, *9*, 254–257.
41. K. Suzuki, K. Ikari, H. Imai, *J. Am. Chem. Soc.* **2004**, *126*, 462–463.
42. S. Huh, J. W. Wiench, B. G. Trewyn, S. Song, M. Pruski, V. S. Y. Lin, *Chem. Commun.* **2003**, 2364–2365.
43. S. Huh, J. W. Wiench, J.-C. Yoo, M. Pruski, V. S. Y. Lin, *Chem. Mater.* **2003**, *15*, 4247–4256.
44. S. Sadasivan, D. Khushalani, S. Mann, *J. Mater. Chem.* **2003**, *13*, 1023–1029.
45. D. Zhao, J. Sun, Q. Li, G. D. Stucky, *Chem. Mater.* **2000**, *12*, 275–279.
46. T. Asefa, M. J. MacLachlan, N. Coombs, G. A. Ozin, *Nature* **1999**, *402*, 867–871.
47. S. Inagaki, S. Guan, Y. Fukushima, T. Ohsuna, O. Terasaki, *J. Am. Chem. Soc.* **1999**, *121*, 9611–9614.
48. B. J. Melde, B. T. Holland, C. F. Blanford, A. Stein, *Chem. Mater.* **1999**, *11*, 3302–3308.
49. S. L. Burkett, S. D. Sims, S. Mann, *Chem. Commun.* **1996**, 1367–1368.
50. D. R. Radu, C.-Y. Lai, J. Huang, X. Shu, V. S. Y. Lin, *Chem. Commun.* **2005**, 1264–1266.
51. X. Feng, G. E. Fryxell, L. Q. Wang, A. Y. Kim, J. Liu, K. M. Kemner, *Science* **1997**, *276*, 923–926.
52. F. De Juan, E. Ruiz-Hitzky, *Adv. Mater.* **2000**, *12*, 430–432.
53. D. R. Radu, C.-Y. Lai, J. W. Wiench, M. Pruski, V. S. Y. Lin, *J. Am. Chem. Soc.* **2004**, *126*, 1640–1641.
54. M. H. Lim, A. Stein, *Chem. Mater.* **1999**, *11*, 3285–3295.
55. J. Liu, Y. Shin, Z. Nie, J. H. Chang, L.-Q. Wang, G. E. Fryxell, W. D. Samuels, G. J. Exarhos, *J. Phys. Chem. A* **2000**, *104*, 8328–8339.
56. M. C. Burleigh, S. Dai, E. W. Hagaman, C. E. Barnes, Z. L. Xue, *ACS Symp. Ser.* **2001**, *778*, 146–158.
57. D. R. Radu, C.-Y. Lai, K. Jeftinija, E. W. Rowe, S. Jeftinija, V. S. Y. Lin, *J. Am. Chem. Soc.* **2004**, *126*, 13216–13217.
58. I. I. Slowing, B. G. Trewyn, V. S. Y. Lin, *J. Am. Chem. Soc.* **2007**, *129*, 8845–8849.
59. C.-Y. Lai, B. G. Trewyn, D. M. Jeftinija, K. Jeftinija, S. Xu, S. Jeftinija, V. S. Y. Lin, *J. Am. Chem. Soc.* **2003**, *125*, 4451–4459.
60. S. Giri, B. G. Trewyn, M. P. Stellmaker, V. S. Y. Lin, *Angew. Chem. Int. Ed.* **2005**, *44*, 5038–5044.
61. F. Torney, B. G. Trewyn, V. S. Y. Lin, K. Wang, *Nature Nanotechnol.* **2007**, *2*, 295–300.
62. I. Slowing, B. G. Trewyn, V. S. Y. Lin, *J. Am. Chem. Soc.* **2006**, *128*, 14792–14793.
63. A. Katiyar, N. G. Pinto, *Small* **2006**, *2*, 644–648.

64. R. Tian, H. Zhang, M. Ye, X. Jiang, L. Hu, X. Li, X. Bao, H. Zou, *Angew. Chem. Int. Ed.* **2007**, *46*, 962–965.
65. M. Vallet-Regi, A. Rámila, R. P. del Real, J. Pérez-Pariente, *Chem. Mater.* **2001**, *13*, 308–311.
66. V. S. Y. Lin, C.-Y. Lai, J. Huang, S.-A. Song, S. Xu, *J. Am. Chem. Soc.* **2001**, *123*, 11510–11511.
67. Q. Fu, G. V. R. Rao, L. K. Ista, Y. Wu, B. P. Andrzejewski, L. A. Sklar, T. L. Ward, G. P. López, *Adv. Mater.* **2003**, *15*, 1262–1266.
68. N. K. Mal, M. Fujiwara, Y. Tanaka, *Nature* **2003**, *421*, 350–353.
69. N. Liu, Z. Chen, D. R. Dunphy, Y.-B. Jiang, R. A. Assink, C. J. Brinker, *Angew. Chem. Int. Ed.* **2003**, *42*, 1731–1734.
70. E. Aznar, R. Casasús, B. García-Acosta, M. D. Marcos, R. Martínez-Máñez, F. Sancenón, J. Soto, P. Amorós, *Adv. Mater.* **2007**, *19*, 2228–2231.
71. A. B. Descalzo, D. Jiménez, M. D. Marcos, R. Martínez-Máñez, J. Soto, J. El Haskouri, C. Guillem, D. Beltrán, P. Amorós, M. V. Borrachero, *Adv. Mater.* **2002**, *14*, 966–969.
72. A. B. Descalzo, M. D. Marcos, R. Martínez-Máñez, J. Soto, D. Beltrán, P. Amorós, *J. Mater. Chem.* **2005**, *15*, 2721–2731.
73. R. Casasús, M. D. Marcos, R. Martínez-Máñez, J. V. Ros-Lis, J. Soto, L. A. Villaescusa, P. Amorós, D. Beltrán, C. Guillem, J. Latorre, *J. Am. Chem. Soc.* **2004**, *126*, 8612–8613.
74. R. Casasús, E. Climent, M. D. Marcos, R. Martínez-Máñez, F. Sancenón, J. Soto, P. Amorós, J. Cano, E. Ruiz, *J. Am. Chem. Soc.* **2008**, *130*, 1903–1917.
75. A. Bernardos, E. Aznar, C. Coll, R. Martínez-Máñez, J. M. Barat, M. D. Marcos, F. Sancenón, A. Benito, J. Soto, *J. Control. Release* **2008**, *131*, 181–189.
76. Q. Yang, S. Wang, P. Fan, L. Wang, Y. Di, K. Lin, F.-S. Xiao, *Chem. Mater.* **2005**, *17*, 5999–6003.
77. Y. Zhu, J. Shi, *Microporous Mesoporous Mater.* **2007**, *103*, 243–249.
78. J. A. Gruenhagen, C.-Y. Lai, D. R. Radu, V. S. Y. Lin, E. S. Yeung, *Appl. Spectrosc.* **2005**, *59*, 424–431.
79. R. Liu, X. Zhao, T. Wu, P. Feng, *J. Am. Chem. Soc.* **2008**, *130*, 14418–14419.
80. M. Fujiwara, S. Terashima, Y. Endo, K. Shiokawa, H. Ohue, *Chem. Commun.* **2006**, 4635–4637.
81. R. Hernandez, H.-R. Tseng, J. W. Wong, J. F. Stoddart, J. I. Zink, *J. Am. Chem. Soc.* **2004**, *126*, 3370–3371.
82. T. D. Nguyen, H.-R. Tseng, P. C. Celestre, A. H. Flood, Y. Liu, J. F. Stoddart, J. I. Zink, *Proc. Natl. Acad. Sci. U.S.A.* **2005**, *102*, 10029–10034.
83. T. D. Nguyen, Y. Liu, S. Saha, K. C. F. Leung, J. F. Stoddart, J. I. Zink, *J. Am. Chem. Soc.* **2007**, *129*, 626–634.
84. T. D. Nguyen, K. C. F. Leung, M. Liong, Y. Liu, F. Stoddart, J. I. Zink, *Adv. Funct. Mater.* **2007**, *17*, 2101–2110.
85. T. D. Nguyen, K. C. F. Leung, M. Liong, C. D. Pentecost, J. F. Stoddart, J. I. Zink, *Org. Lett.* **2006**, *8*, 3363–3366.
86. S. Angelos, Y.-W. Yang, K. Patel, J. F. Stoddart, J. I. Zink, *Angew. Chem. Int. Ed.* **2008**, *47*, 2222–2226.
87. C. Park, K. Oh, S. C. Lee, C. Kim, *Angew. Chem. Int. Ed.* **2007**, *46*, 1455–1457.
88. K. Patel, S. Angelos, W. R. Dichtel, A. Coskun, Y.-W. Yang, J. I. Zink, J. F. Stoddart, *J. Am. Chem. Soc.* **2008**, *130*, 2382–2383.

Chapter 17

Building Molecular Machines on Surfaces

ALBERTO CREDI, SERENA SILVI, AND
MARGHERITA VENTURI

17.1 INTRODUCTION

The design, synthesis, and operation of multicomponent molecular systems capable of performing specific, directional mechanical movements under the action of a defined energy input—namely, molecular machines—constitute a fascinating challenge in the field of nanoscience. Such a daring goal finds its scientific origin in the existence of

The Supramolecular Chemistry of Organic–Inorganic Hybrid Materials. Edited by Knut Rurack and Ramón Martínez-Máñez

natural molecular machines.[1,2] Owing to the progress made in several branches of chemistry, physics, biology, and materials science, various types of artificial or semi-synthetic molecular machines have been developed. Among the systems reported there are molecular tweezers,[3] adaptable receptors,[4] propellers,[5] rotors,[6] turnstiles,[7] gyroscopes,[8,9] gears,[10] brakes,[11] a molecular pedal,[12] ratchets,[13] switches,[14] shuttles,[15] elevators,[16] muscles,[17–19] valves,[20] processive artificial enzymes,[21] walkers,[22,23] and catalytic self-propelled micro- and nanorods.[24] Several excellent reviews[25–36] and a monograph[37] dealing with artificial molecular machines and motors are available.

Most of the studies on molecular machines have been performed in solution. Under such conditions, the systems investigated contain a huge number of molecules that behave independently from one another because they cannot be addressed individually and hence controlled. In spite of this limitation, these systems illustrate a number of very interesting concepts concerning molecular-level movements controlled by external inputs. Incoherence, however, remains a significant impediment to designing and realizing systems capable of performing useful functions. It seems therefore reasonable that for several real applications the molecular machines have to be interfaced with the macroscopic world by ordering them in some way at interfaces or on surfaces.[38] The problem of obtaining ordered arrays of molecular devices can be addressed by a variety of techniques.

Realizing the summation of motion at the molecular level[39] to achieve macroscopic motion,[40] for instance mimicking the operation of miosyn/actin driven movement of muscles, is a real challenge. The approaches taken include changes in surface properties (e.g., surface tension), direct translation of molecular to macroscopic motion, and signal collection by suitably designed architectures.

In the last few years, the development of scanning probe techniques[41–43] has enabled direct observation and manipulation of single molecules on surfaces. Techniques of this kind have opened novel directions to the study of molecular machines and have also contributed to better understand the differences between movement at the macroscopic and at the molecular level.

Studies on molecular machines on surfaces and at interfaces have been recently carried out by several research groups. We will briefly describe here only a small number of paradigmatic examples. Interested readers can consult recent review articles[27,29,30,33–36,44–49] and the original papers.

17.2 ROTARY MOTORS

In a macroscopic rotary machine, the turning part is called the rotor and the stationary part is called the stator. In the molecular world, any part of a rotary motor that is rigidly attached to the surface can be considered as the stator, while that which is not is called the rotor (or rotator).

Surface-mounted molecular rotary motors are extremely interesting from a basic viewpoint.[33] They could also find applications in a variety of molecular-sized devices and machines, for example, in the fields of nanoelectronics, nanophotonics, and nanofluidics.[50] Two different types of surface-mounted molecular rotors can be

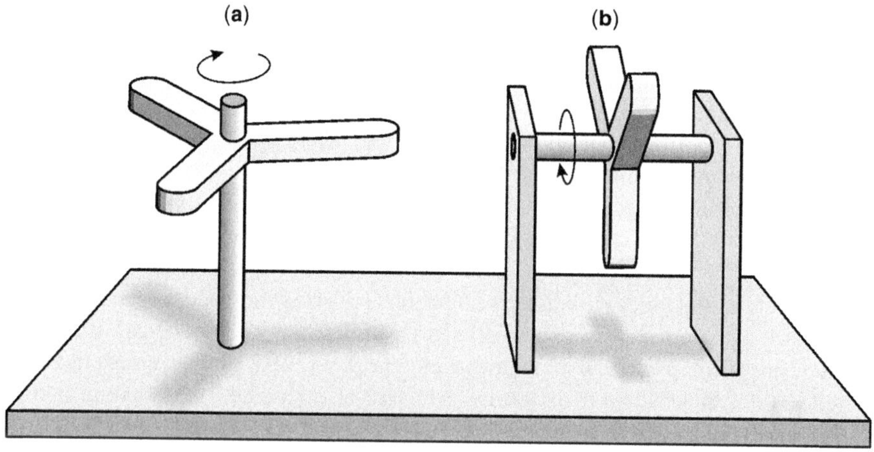

Figure 17.1 Azimuthal (a) and altitudinal (b) surface-mounted molecular rotors.[33] (Reprinted with permission from V. Balzani et al., *ChemPhysChem* **2008**, *9*, 202–220. Copyright Wiley-VCH Verlag GmbH & Co. KGaA.)

distinguished. In an azimuthal rotor (Fig. 17.1, left), the axis of rotation is perpendicular to the surface, whereas in an altitudinal rotor (Fig. 17.1, right), it is parallel to the surface.

Light-driven unidirectional molecular rotary motors based on photoisomerization around a C=C bond[51] have been successfully immobilized on nanoparticle gold surfaces by two thiol-functionalized "legs" to yield an azimuthal rotary motor (compound **1** in Fig. 17.2).[52] Repetitive and unidirectional 360-degree rotary motion of the rotor moiety with respect to the surface-immobilized stator part was observed to occur on

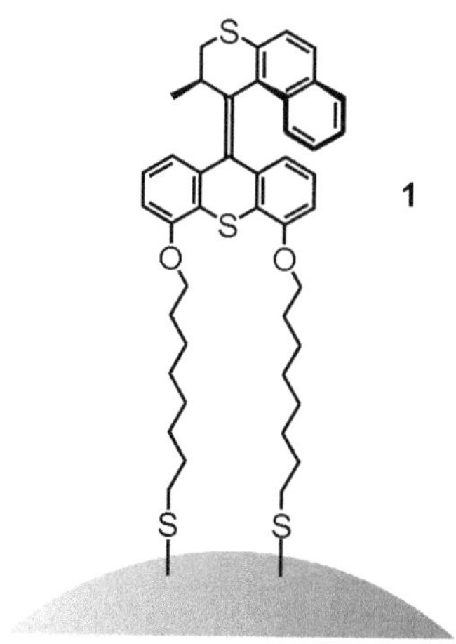

1

Figure 17.2 A light-driven molecular azimuthal rotary motor immobilized on a nanoparticle gold surface.[52] (Reprinted with permission from V. Balzani et al., *ChemPhysChem* **2008**, *9*, 202–220. Copyright Wiley-VCH Verlag GmbH & Co. KGaA.)

irradiation. The use of two attachment points per molecule prevents uncontrolled thermal rotation of the entire system with respect to the surface. This study, however, demonstrated also the difficulties that can be encountered in the immobilization process. The length and nature of the connecting units is, indeed, a critical issue in the attachment of photoactive components on metal surfaces as it can affect excited state properties in a dramatic way.[53] More recently, photoinduced controlled rotation was obtained with a modified second-generation motor attached to a quartz surface which does not interfere with photochemical processes and circular dichroism analysis of the system.[54]

Molecular dynamics simulations on artificial surface-mounted molecular rotors have been performed and extensively reviewed.[33,55] The theoretical rotational motion, driven by a circularly polarized electric field, of a dipolar chiral rhenium complex rotor attached to a molecular grid was also studied.[56] In vacuum and at

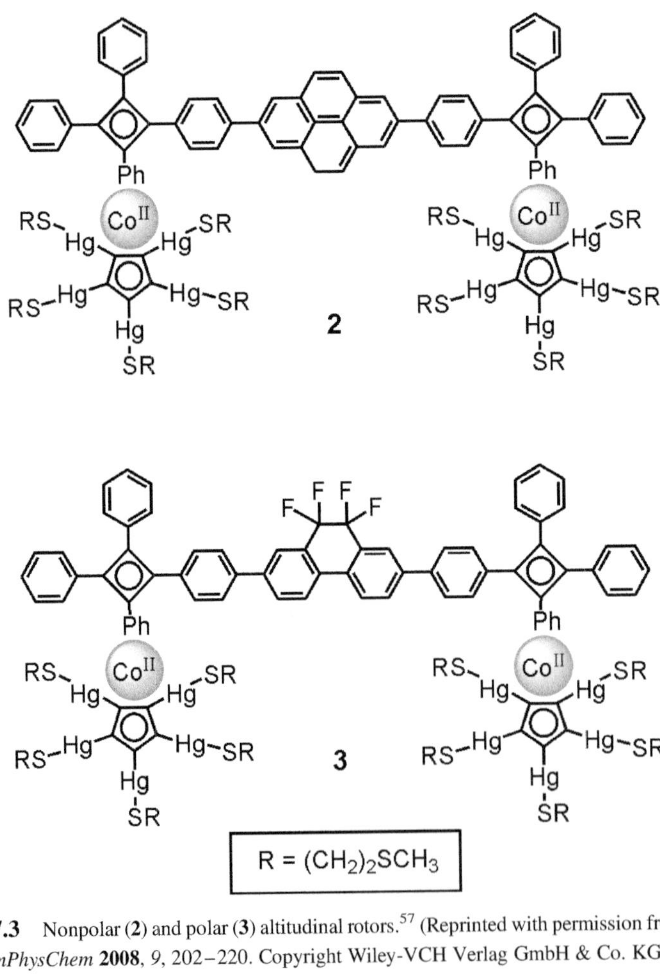

Figure 17.3 Nonpolar (**2**) and polar (**3**) altitudinal rotors.[57] (Reprinted with permission from V. Balzani et al., *ChemPhysChem* **2008**, 9, 202–220. Copyright Wiley-VCH Verlag GmbH & Co. KGaA.)

low temperature, the simulations suggest that at low field frequencies the major restriction to unidirectional rotation is the thermal energy. At higher frequencies, friction is the dominant hindrance to directional motion.

Nonpolar and dipolar altitudinal rotors (compounds **2** and **3** in Fig. 17.3) have been synthesized. ^{19}F NMR spectroscopy showed that the barrier to rotation in **3** was extremely low in solution. Both systems have then been immobilized on Au(111) surfaces and studied with a variety of techniques.[57] The results obtained indicated that for a fraction of molecules the static electric field from the scanning tunneling microscopy (STM) tip could induce an orientation change in the dipolar rotor but not in the nonpolar analog (for a recent example of an azimuthal molecular rotor controlled by the STM tip, see Reference 58). Compound **3** can exist as three pairs of helical enantiomers because of the propeller-like conformation of the tetraarylcyclobutadienes. For at least one out of the three diastereomers, an asymmetric potential energy surface can be predicted by molecular dynamics simulations on application of an alternating electric field.[55]

A family of molecular rotors (e.g., compound **4** in Fig. 17.4 a) has been designed to perform rotation under electrochemical stimulation.[59,60] The molecules have a piano-stool structure with a stator meant to be grafted on an oxide surface and a rotor bearing redox-active groups, so that addressing the molecule with nanoelectrodes would trigger rotation (Fig. 17.4 b). To avoid intramolecular electron transfer between two electroactive units, which would compete with rotation, insulating

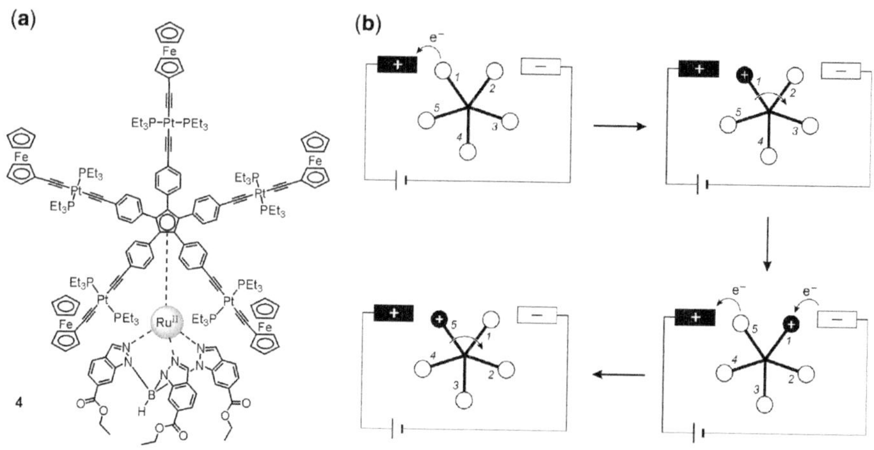

Figure 17.4 (a) Molecular rotor **4** designed to perform rotation under electrochemical stimulation in a nanoelectrode junction. (b) Schematic representation of the proposed operation mechanism.[60] A potential difference applied across the nanojunction results in oxidation of the ferrocene group nearest the anode, which is then electrostatically repelled away towards the cathode. The oxidized unit close to the cathode is reduced while a new ferrocene unit is oxidized at the anode. Overall, the passage of each electron across the nanojunction results in the clockwise rotation of the upper part of the molecule by one-fifth of a turn. (Reprinted with permission from V. Balzani et al., *ChemPhysChem* **2008**, *9*, 202–220. Copyright Wiley-VCH Verlag GmbH & Co. KGaA.)

spacers based on platinum acetylide units were inserted into the structure, but several difficulties remain to be overcome.

17.3 SURFACE ROVING MOLECULES DRIVEN BY STM TIP

A first problem for molecular-level movement generated by an STM tip is related to the fact that the molecule to be moved lies on—and therefore interacts with—a surface. Such an interaction can prevent the movement and, in some cases, it can also happen that the molecule, subjected to the contrasting forces due to the interactions with surface and tip, breaks down into pieces.

The system must be carefully designed to minimize the interaction of the moving part with the surface. Molecules called "molecular landers"[61] such as compound **5** (Fig. 17.5)[62] have been proposed in which interaction with the surface is through the orthogonally oriented di-*tert*-butyl "legs." Such an arrangement should reduce the interaction with the surface of the central phenyl rotor whose angle could be manipulated by an STM tip to control electrical conduction between the polyaromatic boards.[62] This system would constitute a single-molecule nanomechanical electrical switch. Although reversible conformation switching of biphenyl molecules adsorbed on a Si(100) surface has been obtained through resonance electronic excitation,[63] studies on a triphenyl derivative (1,4″-paratriphenyldimethylacetone) on a silicon surface have evidenced problems connected with the manipulation of intramolecular conformations with an STM tip.[64]

An even more difficult task in the field of single-molecule manipulation is that of causing a combined translational and rotational movement. For this purpose,

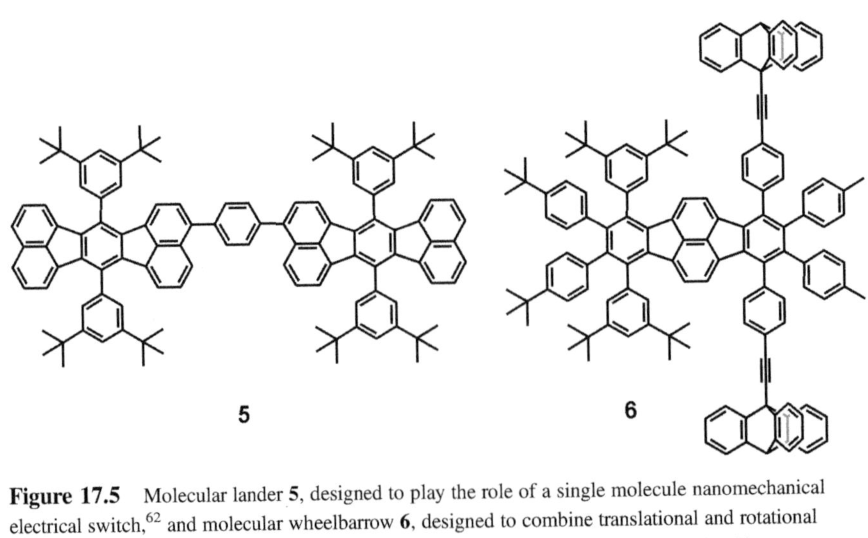

5

6

Figure 17.5 Molecular lander **5**, designed to play the role of a single molecule nanomechanical electrical switch,[62] and molecular wheelbarrow **6**, designed to combine translational and rotational movements on a surface.[65] (Adapted with permission from V. Balzani et al., *ChemPhysChem* **2008**, *9*, 202–220. Copyright Wiley-VCH Verlag GmbH & Co. KGaA.)

molecular "wheelbarrows" should be constructed. Compound **6** (Fig. 17.5), which corresponds to a dissymmetrized lander, was indeed prepared to play a wheelbarrow function.[59,65] Its skeleton is made of polycyclic aromatic hydrocarbons connected with two legs (3,5-di-*tert*-butyl phenyl groups) and two wheels (ethynyl triptycene groups). After deposition on a Cu(100) surface, STM images were obtained and lateral manipulation was attempted. Motions could not be observed, however, because the interaction of the wheels with the metal surface is very strong and the applied force leads to fragmentation of the molecule.[66]

In a recent experiment, translational and rotational motion of a single molecule was driven and monitored by the STM tip at 7 K in a sort of rack-and-pinion device.[67] The pinion consists of a 1.8 nm diameter molecule acting as a six-toothed wheel interlocked at the edge of a self-assembled molecular island functioning as a rack. The rotation of the pinion molecule tooth by tooth along the rack could be monitored owing to the presence of a chemical tag attached to one of its cogs.

Fullerenes, on account of their spherical and cylindrical shapes, appear to be more suitable than trypticene wheels for rolling translations on surfaces. Indeed, it has been demonstrated that fullerenes can undergo tip-induced surface translations through a rolling mechanism.[68] Accordingly, a nanocar has recently been constructed (compound **7** in Fig. 17.6).[69] This molecule consists of an oligo(phenylethylene ethynylene) chassis connected with four C_{60} wheels by means of freely rotating alkynyl axles. On Au(111) surface a four lobe image is observed which is stationary at room temperature. Sequential STM images show that on increasing temperature a translational motion occurs perpendicular to the axles, with pivoting around one wheel accounting for changes in direction. The behavior of three-wheel analogs confirms a thermally driven rolling motion of the nanocar. Movement of the nanocar was

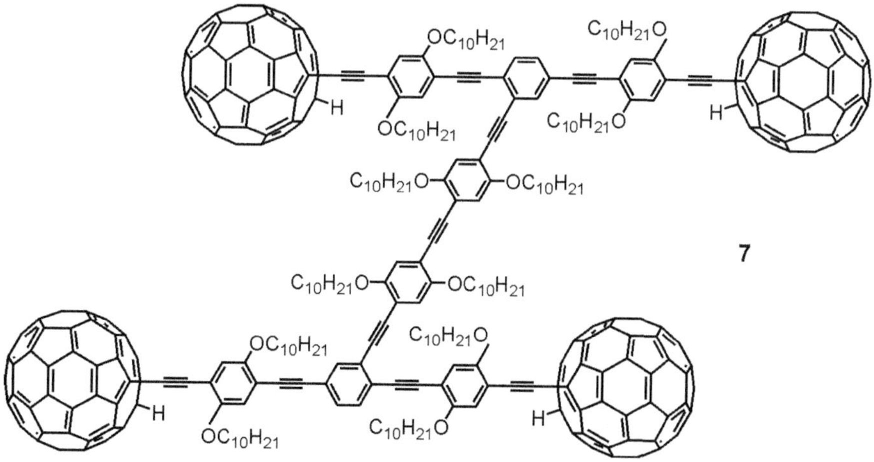

7

Figure 17.6 Nanocar **7** based on an oligo(phenylethylene ethynylene) chassis connected with four C_{60} wheels.[69] (Adapted with permission from V. Balzani et al., *ChemPhysChem* **2008**, *9*, 202–220. Copyright Wiley-VCH Verlag GmbH & Co. KGaA.)

also obtained by manipulation with the STM tip. Again, the most frequent movement was perpendicular to the axles, with the molecule apparently pulled by the tip.

Prototypes of light-powered nanocars have recently been constructed.[70,71] One of them consists[70] of a light-driven unidirectional motor[51] like that shown in Figure 17.2, an oligo(phenylene ethynylene) chassis, and an axle system with four carborane wheels. Substitution of C_{60} by carborane was necessary to prevent quenching of the motor excited state by fullerene. These systems do work in solution, but light-driven movement across a surface poses several problems and has not yet been demonstrated.

17.4 MOLECULAR VALVES

A macroscopic valve is a machine constructed by combining a controllable component that regulates the flow of gases or liquids with a reservoir, which can also serve as a supporting platform for the movable element. Construction of such a device at the molecular scale requires the integration of a stable and inert nanocontainer with an appropriate moving part that can act as a gatekeeper to regulate the molecular transport into and out of the container. The construction of functioning nanovalves can be useful for controlling drug delivery, signal transduction, nanofluidic systems, and sensors.

A light-actuated nanovalve derived from a channel protein was constructed to control photochemical transport of solutes across a lipid bilayer[72,73] (for a related system, see Reference 73). The valve consists of a channel protein modified with photochromic spiropyran[74] components. On irradiation with UV light, the neutral spiropyran units convert into a zwitterionic form. The change in hydrophobicity results in the channel opening. Visible light reverses the process and closes the channel again. The hybrid protein valve is compatible with a liposome encapsulation system and allows for external photocontrol of transport through the channel, demonstrated by the controlled release of an encapsulated fluorescent compound.

Allosteric control of a ligand-gated ion channel[75] and light-controlled expulsion of molecules from mesostructured silica nanoparticles,[76] both based on the photoinduced isomerization of azobenzene, have also been reported.

An alternative approach to the construction of nanovalves is based on the controlled stimulation of mechanical movements in rotaxanes.[20] The redox switchable bistable rotaxane $\mathbf{8}^{4+}$ (Fig. 17.7) has been attached to mesoporous silica nanoparticles. As shown in the illustration, the silica nanopores can be closed and opened by moving the mechanically interlocked ring component of rotaxane $\mathbf{8}^{4+}$ closer to and away from the pores' orifices, respectively.[77] When the ring sits on the preferred tetrathiafulvalene (TTF) station, access to the interior of the nanoparticles is unrestricted and diffusion of solutes from the surrounding solution can occur. Chemically induced oxidation of the TTF unit results in a shuttling of the ring closer to the solid surface, thereby blocking the access to the pores and trapping any solute molecule inside. Reduction of the TTF unit reverses the mechanical motion, thus releasing the guest molecules and returning the system to its initial state. More recently, it has been shown that the properties of these rotaxane-based valves can be fine-tuned by

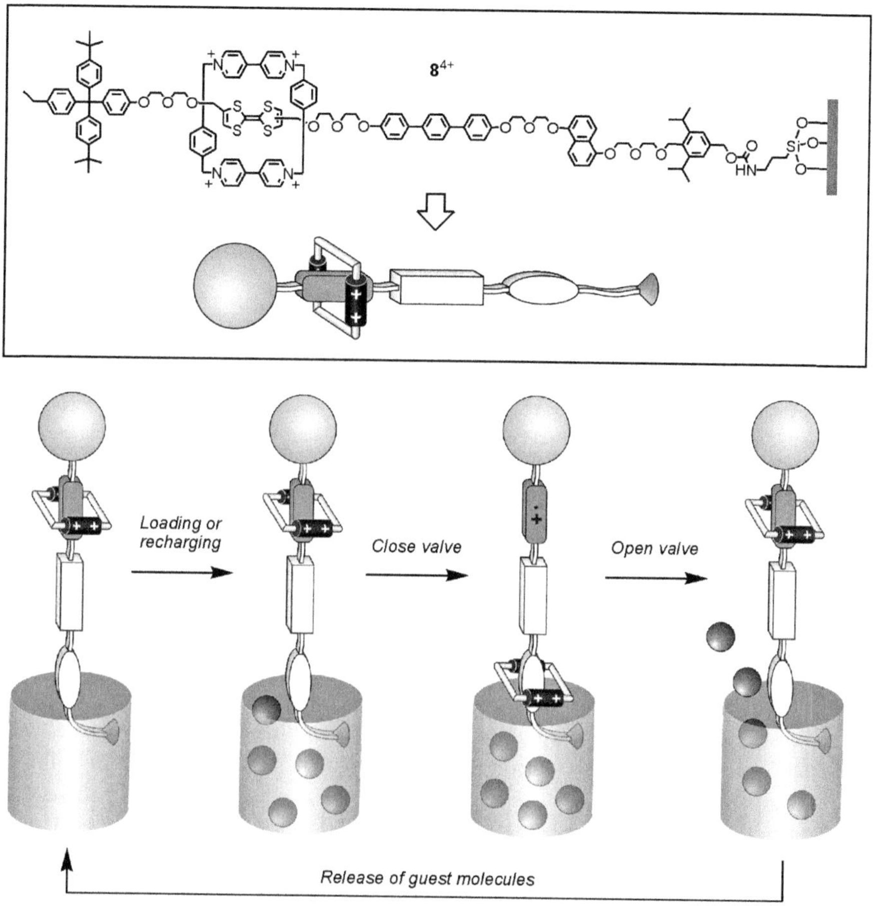

Figure 17.7 A molecular valve based on redox switchable bistable rotaxane **8**$^{4+}$.[77] (Adapted with permission from V. Balzani et al., *ChemPhysChem* **2008**, *9*, 202–220. Copyright Wiley-VCH Verlag GmbH & Co. KGaA.)

changing the length of the linkers between the surface and the rotaxane molecules and the location of the movable components on the nanoparticles.[78] pH-driven[79] and enzyme-responsive[80] nanovalves have also been constructed by threading/dethreading a surface-bonded pseudorotaxane.

17.5 MOLECULAR MUSCLES

An exciting development in the field of molecular machines has been the construction of a rudimentary molecular-scale muscle[17–19,31] based on the topology of a rotaxane dimer, which can undergo contraction and stretching movements in solution.

Cumulative nanoscale movements within surface-bound molecular muscles based on another type of rotaxanes have been harnessed to perform large-scale

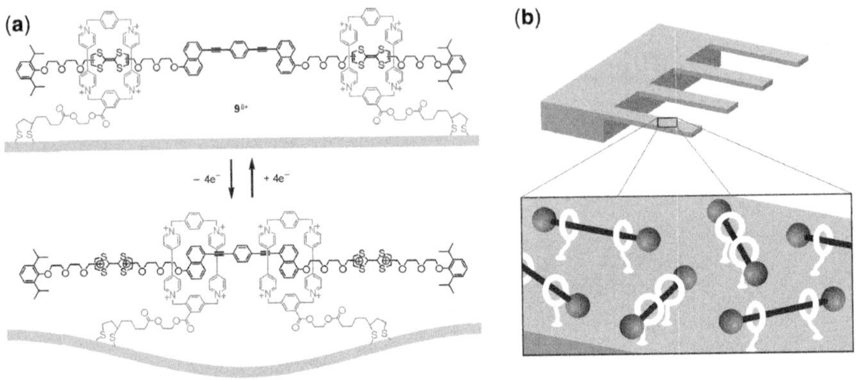

Figure 17.8 A molecular muscle based on a self-assembled monolayer of the palindromic bistable rotaxane $\mathbf{9}^{8+}$ (a), anchored on the gold surface of microcantilever beams (b).[81] (Adapted with permission from V. Balzani et al., *ChemPhysChem* **2008**, *9*, 202–220. Copyright Wiley-VCH Verlag GmbH & Co. KGaA.)

mechanical work. The investigated system[81] consists of the palindromic bistable rotaxane $\mathbf{9}^{8+}$ with two rings and four stations, namely two TTF and two oxynaphthalene (ON) moieties (Fig. 17.8). Tethers attached to each ring anchor the whole rotaxane system as a self-assembled monolayer to the gold surface of microcantilever beams. Each microcantilever is covered with approximately 6 billion randomly oriented rotaxane molecules. The set up was then inserted into a fluid cell. Oxidation of the ring-preferred TTF station with ferric perchlorate results in shuttling of the two rings onto the ON stations, significantly shortening the inter-ring separation (from 4.2 to 1.4 nm). This change in the inter-ring distance induces a tensile stress on the gold surface of the microcantilever which causes an upward bending of the beam by about 35 nm, whereas a monolayer of the dumbbell alone is inactive. Reduction with ascorbic acid returns the cantilever to its original position. The process can be repeated over several cycles, with gradually decreasing amplitude. Stress-induced bending of microcantilevers has also been observed for molecular recognition processes,[82] conformational changes of DNA molecules,[83] and redox[84] and photochemical[85] reactions.

17.6 THREADED MOLECULAR STRUCTURES

The self-assembly of threaded supramolecular complexes on solid surfaces can be obtained in systems in which either the acyclic or the macrocyclic component of a pseudorotaxane is immobilized on an appropriate surface.[86]

A general approach is provided by the emerging field of heterosupramolecular chemistry, where molecular or supramolecular species are linked to nanoparticles.[87] Confinement of molecular-level devices and machines in restricted environments such as those offered by porous materials[88] (e.g., zeolites[89–91]) have also been investigated.

Interesting systems have been reported in which a β-cyclodextrin monolayer has been immobilized on gold electrodes to act as an active interface for the electrochemical and microgravimetric transduction of optical signal recorded by a bipyridinium-azobenzene guest.[92] If the surface is an electrode, it may be possible to induce a coherent dethreading-threading movement under electrochemical control.[86,93] Sensitization of light-induced processes could be possible on a semiconductor surface.

The light-driven dethreading of a pseudorotaxane whose thread component is immobilized on the surface of a silica sol–gel film was investigated.[94] A molecular triad for photoinduced charge separation was attached to a gold electrode and used to drive the dethreading of a pseudorotaxane in a photoelectrochemical setup.[95]

The threading of the dibenzylammonium cation into dibenzo[24]crown[8] has been exploited to assemble pseudorotaxanes at the surface of a nanocrystal[96] and to induce and control formation of nanoparticle aggregates.[97] However, the possibility of pH-triggered aggregation, related to the acid-base controlled dethreading of hydrogen-bonded pseudorotaxanes, was not explored.

Other stimulating experiments arise from the possibility of investigating supramolecular host–guest interactions by force spectroscopy on single molecules.[98] For example, the tip of an atomic force microscope (AFM) could be chemically modified with a thread-like guest, and moved towards a surface bearing suitable macrocyclic hosts, therefore acting as a nanoscopic fishing rod. As the surface is approached, recognition between the host and the guest would lead, in an ideal case, to formation of a single pseudorotaxane structure. At this point, retraction of the tip is expected to stretch the supramolecular system and eventually produce dethreading of the pseudorotaxane. The disruption of individual host–guest complexes of several different types has been investigated with this technique.[99–101] For instance, in the case of the decomplexation between ferrocene moieties immobilized on an AFM tip and β-cyclodextrin (β-CD) receptors in highly ordered self-assembled monolayers on Au(111), the rupture force of individual ferrocene–cyclodextrin complexes was estimated as 56 ± 10 pN. From the correlation among the forces measured at the single-molecule level and the bulk thermodynamic and kinetic parameters, interesting mechanistic information on the binding-unbinding process can be inferred. Among the many factors that must be taken into account in the design of such experiments, the molecular connections of the guest to the tip and of the host to the surface are of great importance to obtain ordered monolayers and avoid direct interactions between the tip and the surface. Moreover, to increase the chances of having a very small number of molecular contacts, or even a single one, between the tip and the substrate, the chemical functionalities on the two approaching surfaces should be appropriately diluted by careful engineering of the monolayers.

17.7 INTERLOCKED COMPOUNDS

17.7.1 Interlocked Compounds On Surfaces

Interlocked molecular species such as rotaxanes and catenanes can be appended to a surface after suitable functionalization.[20,77,78,81,102] Alternatively, the molecular components can become mechanically bound as a consequence of surface immobilization.[103–106] One example is the surface-bound rotaxane *trans*-**10** (Fig. 17.9), consisting of a ferrocene-functionalized β-CD macrocycle threaded on a molecule containing a photoisomerizable azobenzene unit and a long alkyl chain.[104] A monolayer of *trans*-**10** was self-assembled on a gold electrode. Therefore, the ring component is prevented from dethreading at either end of the axle by a bulky anthracene stopper group and the presence of the gold surface. The azobenzene unit in the *trans* configuration is complexed by β-CD; photoisomerization to the *cis* form renders complexation sterically impossible, so that the β-CD ring is displaced to the alkyl component. Back-photoisomerization restores the original *trans* configuration. The position of the β-CD-tethered ferrocene unit was determined by chronoamperometry. A fast current decay ($k = 65\ s^{-1}$) was observed for the *trans* isomer, implying that the ring component is close to the electrode surface (Fig. 17.9). Photoisomerization of the

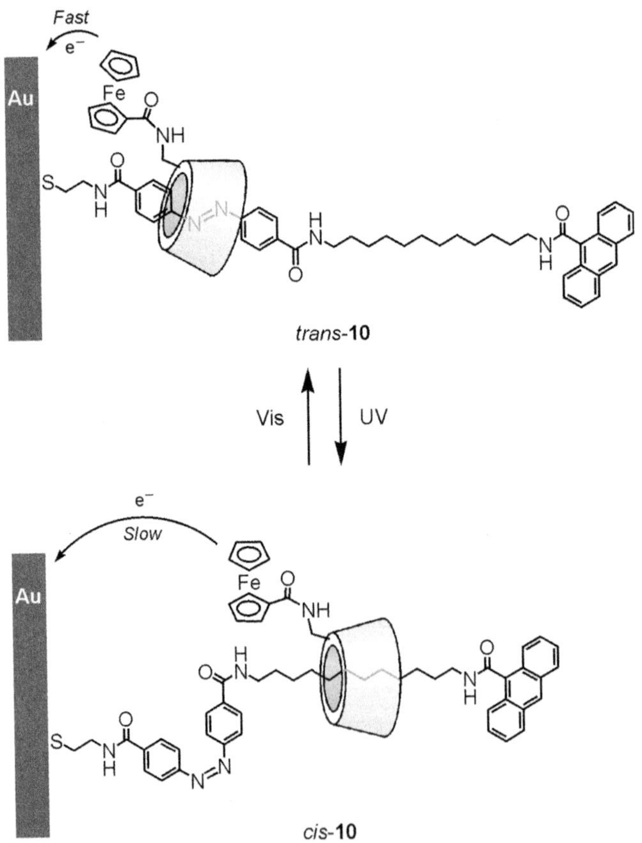

Figure 17.9 Schematic representation of the surface-bound photoswitchable rotaxane **10**.[104] Such a device is capable of transducing an optical signal into an electronic signal by means of the photocontrolled ring shuttling in the rotaxane molecules. (Adapted with permission from V. Balzani et al., *ChemPhysChem* **2008**, 9, 202–220. Copyright Wiley-VCH Verlag GmbH & Co. KGaA.)

monolayer to the *cis* state resulted in a chronoamperometric transient characterized by a substantially lower electron-transfer rate ($k = 15\ s^{-1}$). This result indicates that in *cis*-**10** the β-CD ring is more distant from the electrode surface. Owing to the reversibility of azobenzene photoisomerization, a cyclic pattern for the rate constant of the heterogeneous electron-transfer process was observed. In this optoelectronic system, optical information is transduced by a mechanical shuttling to an electronic signal.[104]

Changes in molecular properties of bistable species mounted on surfaces can strongly affect the macroscopic properties of the surface itself. It has been shown[107] that light excitation of the two-station rotaxane **11** (Fig. 17.10a) physisorbed on a chemically modified gold plate (Fig. 17.10b) can cause the displacement of a liquid droplet across the surface. The system comprises a ring that can move along a thread that contains a photoactive fumaramide (FUM) station and a tetrafluorosuccinimide (TFS) station. The former station has a high binding affinity for the ring, which can be switched off by photoisomerization to the maleamide (MAL) isomeric form. Therefore, in the dark the ring surrounds the fumaramide station, but on light excitation it moves to surround the tetrafluorosuccinimide station (Fig. 17.10a). The shuttle movement is thus used to expose or conceal the fluorinated moiety of the dumbbell,

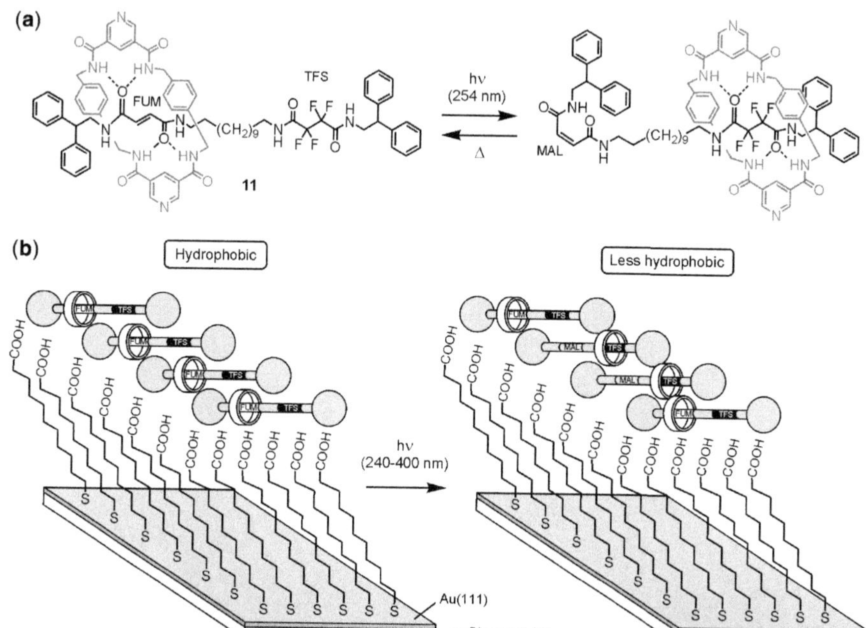

Figure 17.10 (a) Chemical formula of rotaxane **11** and its operation as a light-controlled molecular shuttle. (b) Schematic representation of the photoinduced change of the hydrophobic/hydrophilic character of a gold surface covered with a self-assembled monolayer of rotaxane **11** and 11-mercaptoundecanoic acid.[107] (Adapted with permission from V. Balzani et al., *ChemPhysChem* **2008**, *9*, 202–220. Copyright Wiley-VCH Verlag GmbH & Co. KGaA.)

thereby changing the surface tension (Fig. 17.10b). When an area besides that occupied by the droplet of liquid is switched, the liquid is attracted to the irradiated zone and therefore moves across the surface. The collective action of a monolayer of these molecular shuttles has sufficient power to move a microliter droplet of diiodomethane on a millimeter scale up to a 12° incline. Liquid transportation using photoresponsive surfaces may prove useful for delivering analytes in lab-on-a-chip environments, or for performing chemical reactions on a tiny scale without a reaction vessel by bringing individual drops containing different reactants together.

Catenanes in which one of the macrocyclic rings contains a disulfide bridge that allows adsorption of the molecule on the gold surface of, for example, an electrode, have been prepared.[105] The adsorbed species can be viewed as catenanes in which the gold atoms of the surface constitute an integral part of one of the two rings. Such systems yield clear electrochemical responses, but the electrochemically driven ring switching motions observed in solution[108] are no longer observed in the monolayer. These experiments indicate that intramolecular motions in the surface-confined catenanes are severely slowed down, if not totally frozen, in comparison with the species in solution.

The catenane bis-phenanthroline ligand **12** (Fig. 17.11) has been deposited on an Ag(111) surface by vacuum sublimation.[109] STM measurements show that the deposited catenane molecules self-organize as ordered dimer chain structures at the surface, presumably via intermolecular π–π stacking. Upon an *in situ* addition of Cu atoms to the surface-adsorbed catenanes, the dimer chain structures are disrupted and isolated species are observed instead. X-ray photoelectron spectroscopy (XPS) experiments suggest that the phenanthroline units of **12** coordinate with copper, indicating that the catenane ligand is transformed into a Cu-complexed catenane.[109] Since it is known[110] that the two interlocked macrocyclic rings of **12** circumrotate upon complexation to Cu^+ ions in solution (Fig. 17.11), the results reveal that, when this catenane is adsorbed on the silver surface, its two macrocyclic rings can glide within one another so as to generate the corresponding Cu-complexed species. This is indeed the

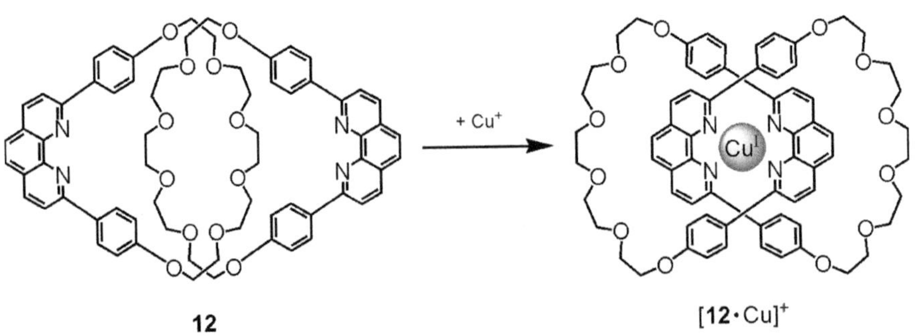

Figure 17.11 Chemical formula of the bis-phenathroline catenane ligand **12** that was adsorbed on a Ag(111) surface.[109] In solution, complexation of **12** by Cu^+ ions is accompanied by conformational changes involving the circumrotation of the macrocyclic components through each other's cavity.[110]

first time that the chemically driven motion of an artificial molecular machine has been imaged at the single-molecule level.

Ordering suitably tailored molecules in two dimensions at interfaces and then transferring them onto surfaces is a widely used technique. A series of investigations[111–113] have shown that cationic rotaxanes and catenanes such as $\mathbf{13}^{4+}$ (Fig. 17.12) can be self-organized at the air–water interface when the PF_6^-

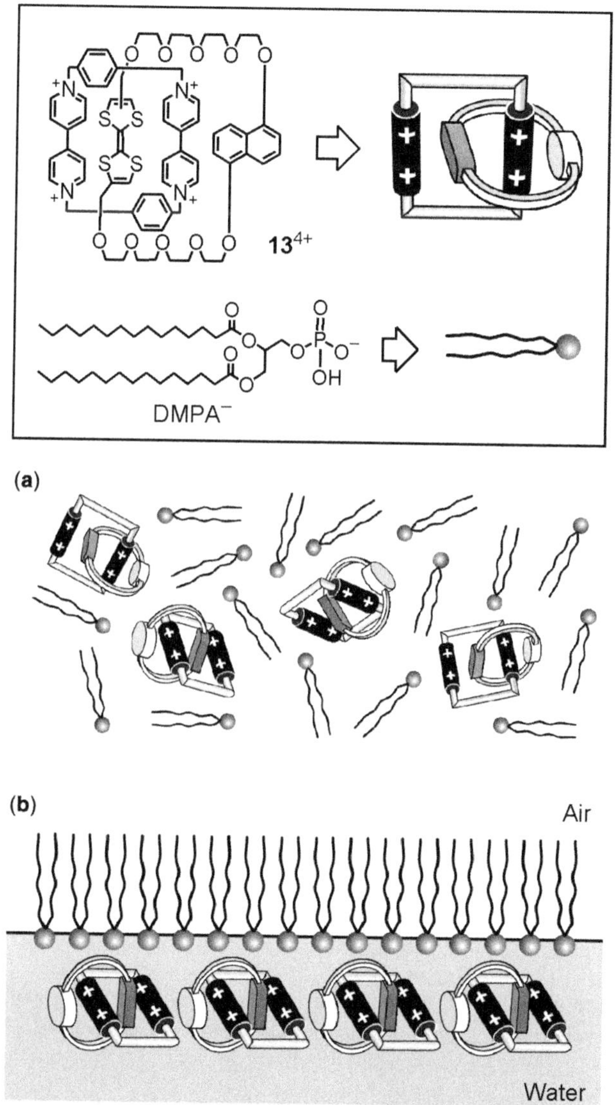

Figure 17.12 (a) Disordered state in solution and (b) ordered monolayer of the tetracationic catenane $\mathbf{13}^{4+}$.[113] (Adapted with permission from V. Balzani et al., *ChemPhysChem* **2008**, *9*, 202–220. Copyright Wiley-VCH Verlag GmbH & Co. KGaA.)

counteranions are replaced by amphiphilic anions such as the dimyristoylphosphatidyl anion ($DMPA^-$). With these counterions, aligned in a parallel manner at a surface, the phosphonate portion points toward the aqueous substrate. To balance this high negative charge density, a layer of cationic rotaxanes and catenanes associates near the surface as a highly ordered, reproducible monolayer, possibly stabilized by $\pi-\pi$ stacking interactions between adjacent molecules. For instance catenane $\mathbf{13}^{4+}$ formed stable monolayers (Fig. 17.12b) that could be transferred to hydrophobized quartz by use of the Langmuir–Blodgett technique.[113] The area per catenane tetracation is around 1.2 nm^2, with the four $DMPA^-$ counterions believed to be clustered in a layer above the tetracations. When catenane $\mathbf{13}^{4+}$ was oxidized to its hexacationic state, $\mathbf{13}^{6+}$, prior to Langmuir film deposition, and then deposited as its hexakis $DMPA^-$ salt, a different behavior was observed. Two stable films were obtained, with areas per catenane of 2.7 and 1.5 nm^2, respectively. These areas were interpreted in terms of the anions dictating the area occupied when the surface pressure is below 20 mN m^{-1}, whereas at higher surface pressures the area occupied is determined by the $\pi-\pi$ stacking interactions. Both states of the catenane were transferred (below 20 mN m^{-1}) onto freshly cleaved mica and the (111) face of gold. When the mica-bound monolayers were examined by tapping-mode atomic force microscopy (AFM), it was observed that, while the monolayer containing the catenane $\mathbf{13}^{4+}$ is extremely flat, that of catenane $\mathbf{13}^{6+}$ is very bumpy, with roughly 25 nm wide and 2 nm high hills. When the corresponding gold-supported monolayers were examined by scanning tunneling spectroscopy (STS), important differences were noted. The monolayer containing $\mathbf{13}^{4+}$ only conducted current above +0.3 V, whereas that containing $\mathbf{13}^{6+}$ exhibited a linear $I-V$ curve reminiscent of that of a metal.[113]

17.7.2 Interlocked Compounds in Solid-State Devices

The results reported above encouraged attempts to incorporate rotaxanes and catenanes in electrically addressable solid-state devices. In a first study,[114] the V-shaped rotaxanes $\mathbf{14}^{4+}$ and $\mathbf{15}^{4+}$ (Fig. 17.13) were used in the form of molecular monolayers sandwiched between lithographically fabricated metal wires. The switches were read by monitoring current flow at reducing voltages. In the closed state, current flow was dominated by resonant tunneling through the electronic states of the molecules. The switches were irreversibly opened by applying an oxidizing voltage across the device. The V-shaped dumbbell component of the rotaxanes exhibits similar properties. The results obtained showed that these systems behave as singly configurable logic gates.

More interesting observations were made on incorporation of the bistable catenane $\mathbf{13}^{4+}$ (Fig. 17.12) into a solid-state device.[115] A monolayer of the catenane was transferred onto a photolithographically patterned polycrystalline silicon electrode. The patterning was such that the Langmuir film was deposited along several parallel lines of poly-Si on the electrode. A second set of orthogonally oriented wires was then deposited on top of the first set such that a crossbar architecture is obtained. This second set of electrodes consisted of a 5 nm thick layer of Ti, followed

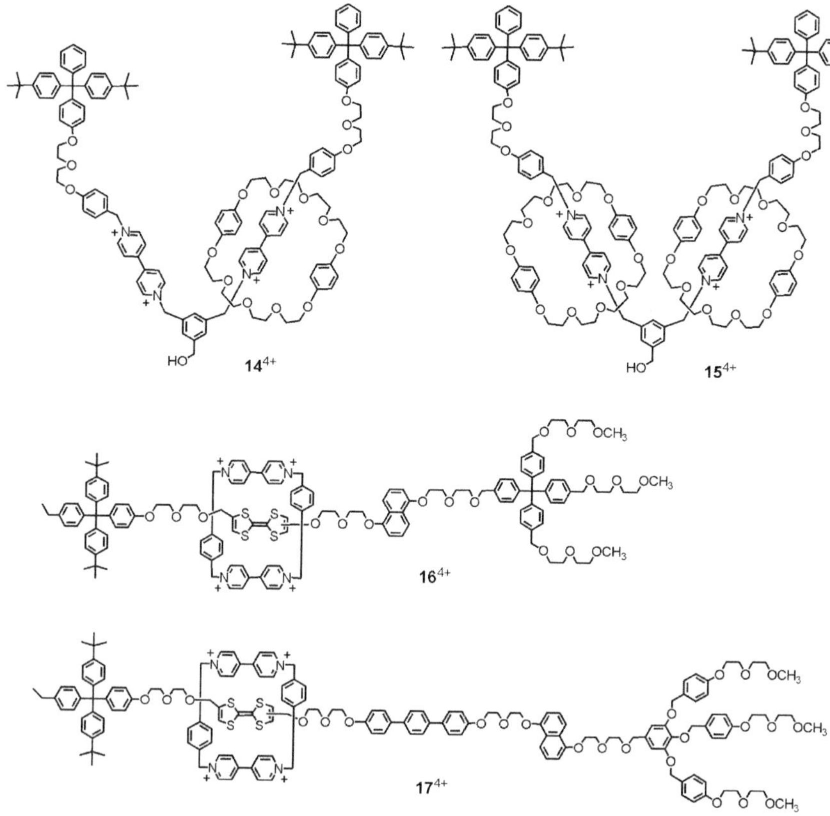

Figure 17.13 Structure formulas of V-shaped rotaxanes $\mathbf{14}^{4+}$ and $\mathbf{15}^{4+}$, and bistable rotaxanes $\mathbf{16}^{4+}$ and $\mathbf{17}^{4+}$, used to construct switchable electronic junctions for memory and logic function purposes.[114,118,119] (Adapted with permission from V. Balzani et al., *ChemPhysChem* **2008**, *9*, 202–220. Copyright Wiley-VCH Verlag GmbH & Co. KGaA.)

by a 100 nm thick layer of Al. The bottom poly-Si electrode of the molecular sandwich was 7 μm wide and the top Ti/Al electrode 10 μm wide. By this approach, an array of junctions, each one addressable individually, was constructed (Fig. 17.14). The mechanism for conduction is by electron tunneling through the single-molecule thick layer between the junction electrodes. Thus, any change in the electronic characteristics of the interelectrode medium is expected to affect the tunneling efficiency and change the resistance of the junction. It should be noticed that such devices are conductors, not capacitors. Experiments were carried out by applying a series of voltage pulses (between +2.0 and −2.0 V) and reading, after each pulse, current through the device at 200 mV, a potential that does not affect switching. The current(read)-voltage(write) curve displays a highly hysteretic profile, making the catenane junction device interesting for potential use in random access memory (RAM) storage.

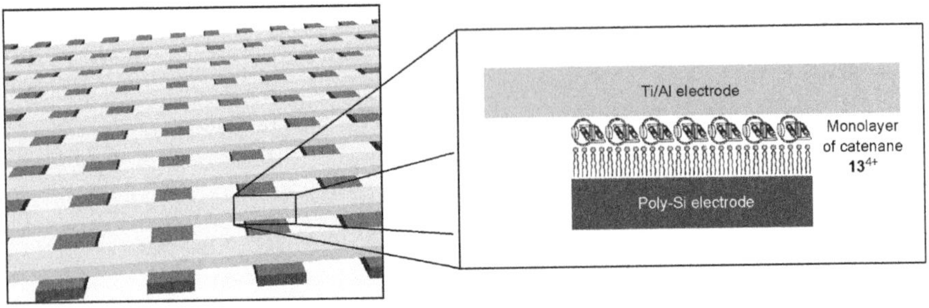

Figure 17.14 Schematic representation of a solid-state device based on junctions consisting of a monolayer of catenane $[\mathbf{13}^{4+}][DMPA^-]_4$ sandwiched between polysilicon and Ti/Al electrodes.[115] The structures of $\mathbf{13}^{4+}$ and $DMPA^-$ are shown in Fig. 17.12. (Adapted with permission from V. Balzani et al., *ChemPhysChem* **2008**, *9*, 202–220. Copyright Wiley-VCH Verlag GmbH & Co. KGaA.)

The current-voltage curve was interpreted on the basis of the mechanism illustrated in Figure 17.15a, which is derived from the behavior of the same catenane $\mathbf{13}^{4+}$ in solution.[116,117] Conformation *I* is the "switch open" state and conformation *IV* the "switch closed" state of the device. When $\mathbf{13}^{4+}$ is oxidized (+2 V), the TTF unit is ionized in state *II* and experiences a Coulombic repulsion inside the tetracationic cyclophane component, resulting in circumrotation of the crown ether and formation of conformation *III* (note that in solution at +2 V TTF undergoes two-electron oxidation and the dioxynaphthalene unit is also oxidized).[116] When the voltage is reduced to near-zero bias, a metastable conformation *IV* is obtained

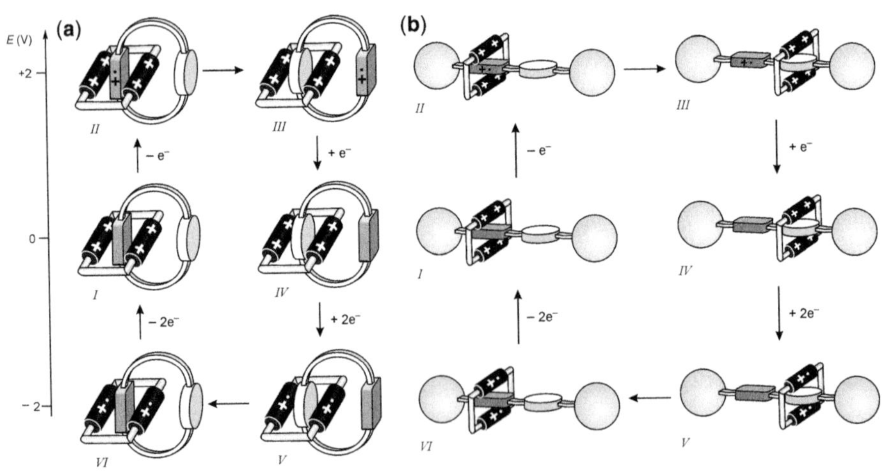

Figure 17.15 Proposed molecular-level mechanism for the operation of solid-state electronic devices based on (a) catenanes like $\mathbf{13}^{4+}$ [115–117] and (b) rotaxanes like $\mathbf{16}^{4+}$ and $\mathbf{17}^{4+}$.[118,119] (Adapted with permission from V. Balzani et al., *ChemPhysChem* **2008**, *9*, 202–220. Copyright Wiley-VCH Verlag GmbH & Co. KGaA.)

which, however, does not return to conformation *I*. The initial conformation can in fact be restored only via states *V* and *VI* in which the bipyridinium units of the cyclophane component are reduced (in solution, at the potential value used, −2 V, the bipyridinium units undergo two-electron reduction).[116] Most likely, the reduction of the bipyridinium units weakens the charge-transfer interaction with the dioxynaphthalene unit, thereby decreasing the barrier for its escape from the cyclophane cavity. The same behavior was observed for solid-state devices containing TTF-based bistable rotaxanes such as $\mathbf{16}^{4+}$ and $\mathbf{17}^{4+}$ (Fig. 17.13),[118,119] and interpreted on the basis of an analogous mechanism (Fig. 17.15b).

Because these results and their interpretation gave rise to some controversy,[120,121] intensive efforts were undertaken to characterize the mechanical rearrangement associated with redox switching in bistable catenanes like $\mathbf{13}^{4+}$ and related bistable rotaxanes,[122] as well as their structural organization in monolayers[123] and the molecule–electrode interface.[124] The metastable state corresponding to conformation *IV* (Fig. 17.15) was in fact observed for a number of different bistable rotaxanes and catenanes in a variety of environments (solution, self-assembled monolayer, and solid-state polymer matrix).[125] The conductance switching in these systems was also extensively investigated by theoretical calculations.[126] Furthermore, the color change associated with the redox-driven switching of TTF-based bistable catenanes and rotaxanes in a polymer matrix can be exploited to make solid-state electrochromic devices.[127]

Using the paradigms described above, a molecular electronic memory with an amazingly high density of 10^{11} bits cm^{-2} was constructed by sandwiching a monolayer of the bistable rotaxane $\mathbf{17}^{4+}$ (Fig. 17.13) between arrays of nanoelectrodes in a crossbar arrangement.[128] The realization of this device relies on a novel method for producing ultra-dense, highly aligned arrays and crossbars of metal or semiconductor nanowires with high aspect ratios.[129] It was estimated that each junction acting as a memory element consists of approximately 100 rotaxane molecules. For practical reasons, only 128 (16 × 8 contacts) of the 160,000 memory cells (400 × 400 nanowires) contained in the circuit were tested.[128] The measurements showed that 25% of the tested cells displayed good and reproducible switching, whereas 35% failed because of bad contacts or shorts, and the remaining 40% showed poor switching. This work is a compelling demonstration that the combination of top-down and bottom-up nanofabrication methods can lead to outstanding technological achievements. However, several aspects, such as stability, reliability, and ease of fabrication, need to be optimized before these systems can find real industrial applications.[130]

17.8 HYBRID BIO-NANOMACHINES

Looking to biology for design inspiration and harnessing its enzymatic toolbox is essential to create new nanoscale materials.[1,2] Recent scientific advances in both molecular biology and supramolecular chemistry have opened up the possibility of building functional hybrid devices based on natural motors. Combination of natural and artificial components is indeed leading to a great variety of nano- or micro-devices.

One long-term objective of this research is to utilize the finest attributes associated with the worlds of both biological and synthetic materials to create nanomechanical systems powered by biological motors. Important fields of application include miniaturized (nanofluidic) analytical systems,[131] molecular sorting,[132] controlled adaptation of materials on a molecular to mesoscopic scale,[133] and engineering lipid and polymer membrane systems with cellular processes.[134]

17.8.1 Linear Motors

Enzymes such as myosin and kinesin and their relatives are linear motors that move along polymer substrates converting the energy of ATP (adenosine triphosphate) hydrolysis into mechanical work; myosin moves along actin filaments in muscle and other cells, and kinesin along microtubules.[135] Motion is derived from a mechanochemical cycle during which the motor protein binds to successive sites along the substrate in such a way as to move forward on the average.

Linear motor proteins generate force and transport cargo unidirectionally. The movement of single kinesin molecules on microtubule tracks under variable ATP concentrations and loads has been studied.[136] Reconstituting motor proteins in their active state *ex vivo* was initially pursued to study the mechanism by which they generate force. More recently, biomolecular motors have been used extensively to power nanodevices in hybrid systems.[44,45] Molecular shuttles based on kinesin and microtubules have been constructed by using two different approaches: either the microtubules are fixed to a surface and kinesin is moving, similar to cars driving on a highway and transporting kinesin-coated objects,[137] or the kinesin is bound to the surface and the microtubules are propelled by the kinesin, analogous to a linear motor.[138] An interesting application of the second technique is a statistical approach to surface imaging by use of fluorescent microtubules moving as probe robots across a surface coated with kinesin.[138] Kinesin-motor proteins can find application in artificial microfluidic systems,[132,139] for example, driving into either one of two arms of a Y junction the direct transport of microtubules under control of an external electric field. The energy of molecular motors can also be used to promote self-organization processes of nanoparticles.[133]

17.8.2 Rotary Motors

The most important and best known natural rotary motor is ATP synthase,[135] the ubiquitous enzyme that manufactures ATP. The mechanochemistry of ATP synthase has been studied in great detail.[140–143]

This protein consists of two rotary molecular motors, F_0 and F_1, attached to a common shaft (known as the γ subunit), each attempting to rotate in the opposite direction. The F_1 motor uses the free energy of ATP hydrolysis to rotate in one direction whereas the F_0 motor uses the energy stored in a transmembrane electrochemical gradient to turn in the opposite direction. Which motor "wins" (i.e., develops more torque) depends on cellular conditions. When F_0 takes over, which is the normal situation (e.g., bacterial photosynthesis), it drives the F_1 motor in reverse whereupon it synthesizes ATP from its constituents, adenosine diphosphate, ADP, and inorganic phosphate. When F_1 dominates, it hydrolyzes ATP and drives the F_0 motor in reverse, turning it into an ion pump that moves ions across the membrane against the electrochemical gradient.

Hybrid nanomechanical devices powered by ATP synthase have been reported.[144] That shown in Figure 17.16 consists of three elements:[145] (1) nanofabricated substrates of Ni posts, each 50 to 120 nm in diameter and 200 nm high; (2) F_1-ATPase molecules, specifically modified for selective interfacing with the nanofabricated structures; (3) nanofabricated Ni rods (150 nm in diameter and 750 to 1500 nm long). The F_1-ATPase molecules were attached to the Ni posts by means of histidine tags introduced into the β-subunits.[146] Streptavidin was bound to the biotin residue on the tip of the γ shaft, and the Ni nanorods, coated with biotinylated histidine-rich peptides, were then attached to the substrate-mounted F_1-ATPase motors by means of a biotin-streptavidin linkage. Rotation of the nanopropellers, which was observed in a flow cell by means of a CCD camera, was initiated by addition of ATP, and inactivated

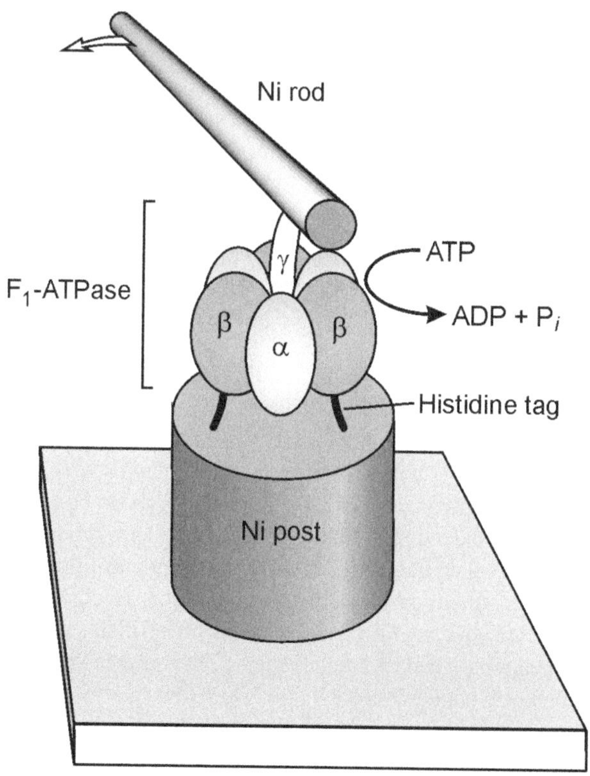

Figure 17.16 Schematic representation of a nanomechanical device powered by the F_1-ATPase molecular motor.[145] (Adapted with permission from V. Balzani et al., *ChemPhysChem* **2008**, *9*, 202–220. Copyright Wiley-VCH Verlag GmbH & Co. KGaA.)

by addition of NaN_3, an F_1-ATP synthase inhibitor.[145] Although only five of 400 total observed propellers were found to rotate, probably because of incorrect assembly of the components, these experiments demonstrate the possibility of integrating biomolecular motors with nanoengineered systems to produce nano- or micro-mechanical machines. It should be noted, however, that this device does not contain any artificial active component.

In most cases, as in the example described above, hybrid bio-nanomachines transform chemical energy into mechanical work. Particularly interesting is the reverse conversion of mechanical energy into chemical energy, obtained with an ATP-based system.[147] A magnetic bead was attached to the γ shaft of isolated F_1-ATP synthase molecules on a glass surface, and the bead was induced to rotate using electrical magnets. Rotation in the appropriate direction resulted in the appearance of ATP in the medium, as detected by the luciferase-luciferin reaction. This outstanding result shows that a vectorial force (torque) working at one particular point on a protein machine can drive a chemical reaction, occurring in physically remote catalytic sites, far from equilibrium.

Biomimetic systems based on vesicles in which either natural[148] or artificial[149] light-driven proton pumping molecular devices are used to power ATP synthase have been constructed. These devices couple a transmembrane electrochemical potential generated by a light-driven process with the chemical potential associated with the ADP-ATP conversion, thereby mimicking the entire process of bacterial photosynthesis. They constitute synthetic biological motors that could not only enable *in vitro* investigations of cellular metabolism, but also be used to power anything that requires a proton gradient or ATP to work, for example, to pump calcium ions across a lipid bilayer membrane, or even future nanomachines.

17.9 CONCLUSION

Investigations on molecular machines in solution are of fundamental importance to understand their operation mechanisms and for some use (e.g., drug delivery). It seems reasonable, however, that before such systems can find applications in many fields of technology, they have to be interfaced with the macroscopic world by ordering them in some way so that they can behave coherently and can be addressed in space. Viable possibilities include deposition on surfaces, incorporation into polymers, organization at interfaces, or immobilization into membranes or porous materials.

The recent achievements in this direction, some of which have been reviewed here, seem to indicate that useful devices based on artificial molecular machines will see the light in a not too distant future. Apart from foreseeable applications related to the development of nanotechnology, investigations on molecular machines in heterogeneous environments are important to increase the basic understanding of the processes that determine the behavior of nanoscale objects, as well as to develop reliable theoretical models. This research has also the important merit of stimulating the ingenuity of chemists, thereby conveying new incitements to chemistry as a scientific discipline.

ACKNOWLEDGMENTS

Financial support from the Ministero dell'Università e della Ricerca (PRIN 2006034123_003) and Università di Bologna is gratefully acknowledged.

REFERENCES

1. R. A. L. Jones, *Soft Machines, Nanotechnology and Life*, Oxford: Oxford University Press, **2004**.
2. D. S. Goodsell, *Bionanotechnology: Lessons from Nature*, Hoboken, NJ: Wiley, **2004**.
3. S. Shinkai, T. Nakaji, T. Ogawa, K. Shigematsu, O. Manabe, *J. Am. Chem. Soc.* **1981**, *103*, 111–115.
4. J. Frey, C. Tock, J.-P. Collin, V. Heitz, J.-P. Sauvage, *J. Am. Chem. Soc.* **2008**, *130*, 4592–4593.
5. H. Iwamura, K. Mislow, *Acc. Chem. Res.* **1988**, *21*, 175–182.
6. S. Shinkai, M. Ikeda, A. Sugasaki, M. Takeuchi, *Acc. Chem. Res.* **2001**, *34*, 494–503.
7. T. C. Bedard, J. S. Moore, *J. Am. Chem. Soc.* **1995**, *117*, 10662–10671.
8. M. A. Garcia-Garibay, *Proc. Natl. Acad. Sci. U.S.A.* **2005**, *102*, 10771–10776.
9. T. Shima, F. Hampel, J. A. Gladysz, *Angew. Chem. Int. Ed.* **2004**, *43*, 5537–5540.
10. N. Koga, Y. Kawada, H. Iwamura, *J. Am. Chem. Soc.* **1983**, *105*, 5498–5499.
11. T. R. Kelly, M. C. Bowyer, K. V. Bhaskar, D. Bebbington, A. Garcia, F. Lang, M.-H. Kim, M. P. Jette, *J. Am. Chem. Soc.* **1994**, *116*, 3657–3658.
12. T. Muraoka, K. Kinbara, T. Aida, *Nature* **2006**, *440*, 512–515.
13. T. R. Kelly, *Acc. Chem. Res.* **2001**, *34*, 514–522.
14. B. L. Feringa, Ed., *Molecular Switches*, Weinheim: Wiley-VCH, **2001**.
15. V. Balzani, M. Clemente-León, A. Credi, B. Ferrer, M. Venturi, A. H. Flood, J. F. Stoddart, *Proc. Natl. Acad. Sci. U.S.A.* **2006**, *103*, 1178–1183.
16. J. D. Badjic, V. Balzani, A. Credi, S. Silvi, J. F. Stoddart, *Science* **2004**, *303*, 1845–1849.
17. M. C. Jiménez-Molero, C. Dietrich-Buchecker, J.-P. Sauvage, *Angew. Chem. Int. Ed.* **2000**, *39*, 3284–3287.
18. J. Wu, K. C.-F. Leung, D. Benitez, J.-Y. Han, S. J. Cantrill, L. Fang, J. F. Stoddart, *Angew. Chem. Int. Ed.* **2008**, *47*, 7470–7474.
19. R. E. Dawson, S. F. Lincoln, C. J. Easton, *Chem. Commun.* **2008**, 3980–3982.
20. S. Saha, K. C.-F. Leug, T. D. Nguyen, J. F. Stoddart, J. I. Zink, *Adv. Funct. Mater.* **2007**, *17*, 685–693.
21. P. Thordarson, E. J. A. Bijsterveld, A. E. Rowan, R. J. M. Nolte, *Nature* **2003**, *424*, 915–918.
22. W. B. Sherman, N. C. Seeman, *Nano Lett.* **2004**, *4*, 1203–1207.
23. F. C. Simmel, W. U. Dittmer, *Small* **2005**, *1*, 284–299, and references therein.
24. W. F. Paxton, A. Sen, T. E. Mallouk, *Chem. Eur. J.* **2005**, *11*, 6462–6470, and references therein.
25. V. Balzani, A. Credi, F. M. Raymo, J. F. Stoddart, *Angew. Chem. Int. Ed.* **2000**, *39*, 3348–3391.
26. V. Balzani, A. Credi, F. M. Raymo, J. F. Stoddart, *Acc. Chem. Res.* **2001**, *34*(6); special issue on Molecular Machines (Ed.: J. F. Stoddart).
27. V. Balzani, A. Credi, F. M. Raymo, J. F. Stoddart, *Struct. Bond.* **2001**, *99*; special volume on Molecular Machines and Motors (Ed.: J.-P. Sauvage).
28. A. Credi, in *Handbook of Photochemistry and Photobiology*, Vol. 3 (Ed. H. S. Nalwa), Stevenson Ranch, CA: American Scientific, **2003**, 319–353.
29. A. H. Flood, R. J. A. Ramirez, W. Q. Deng, R. P. Muller, W. A. Goddard, J. F. Stoddart, *Austr. J. Chem.* **2004**, *57*, 301–322.
30. A. H. Flood, R. J. A. Ramirez, W. Q. Deng, R. P. Muller, W. A. Goddard, J. F. Stoddart, *Top. Curr. Chem.* **2005**, *262*; special volume on Molecular Machines (Ed.: T. R. Kelly).
31. J.-P. Sauvage, *Chem. Commun.* **2005**, 1507–1510.
32. K. Kinbara, T. Aida, *Chem. Rev.* **2005**, *105*, 1377–1400.
33. G. S. Kottas, L. I. Clarke, D. Horinek, J. Michl, *Chem. Rev.* **2005**, *105*, 1281–1376.
34. W. R. Browne, B. L. Feringa, *Nature Nanotechnol.* **2006**, *1*, 25–35.

35. E. R. Kay, D. A. Leigh, F. Zerbetto, *Angew. Chem. Int. Ed.* **2007**, *46*, 72–191.
36. E. R. Kay, D. A. Leigh, F. Zerbetto, *Adv. Funct. Mater.* **2007**, *17*(5); special issue on Molecular Machines and Switches (Eds. A. Credi, H. Tian).
37. V. Balzani, A. Credi, M. Venturi, *Molecular Devices and Machines: Concepts and Perspectives for the Nanoworld*, 2nd edition, Weinheim: Wiley-VCH, **2008**.
38. D. L. Allara, *Nature* **2005**, *437*, 638–639.
39. V. Ferri, M. Elbing, G. Pace, M. D. Dickey, M. Zharnikov, P. Samorì, M. Mayor, M. A. Rampi, *Angew. Chem. Int. Ed.* **2008**, *47*, 3407–3410.
40. S. Masiero, S. Lena, S. Pieraccini, G. P. Spada, *Angew. Chem. Int. Ed.* **2008**, *47*, 3184–3187.
41. P. Samorì, Ed., *Scanning Probe Microscopies Beyond Imaging: Manipulation of Molecules and Nanostructures*, Weinheim: Wiley-VCH, **2006**.
42. L. Grill, *J. Phys. Cond. Matter* **2008**, *20*, 053001.
43. L. Grill, *Science* **2007**, *316*, 1143–1158; special section on Single Molecules.
44. H. Hess, G. D. Bachand, V. Vogel, *Chem. Eur. J.* **2004**, *10*, 2110–2116.
45. M. G. L. van den Heuvel, C. Dekker, *Science* **2007**, *317*, 333–336.
46. N. Koch, *ChemPhysChem* **2007**, *8*, 1438–1455.
47. T. Y. Huang, *MRS Bull.* **2008**, *33*, 226–231.
48. V. Balzani, A. Credi, M. Venturi, *ChemPhysChem* **2008**, *9*, 202–220.
49. S. Silvi, M. Venturi, A. Credi, *J. Mater. Chem.* **2009**, *19*, 2279–2294.
50. B. Wang, P. Král, *Phys. Rev. Lett.* **2007**, *98*, 266102.
51. N. Koumura, R. W. J. Zijlstra, R. A. van Delden, N. Harada, B. L. Feringa, *Nature* **1999**, *401*, 152–155.
52. R. A. van Delden, M. K. J. ter Wiel, M. M. Pollard, J. Vicario, N. Koumura, B. L. Feringa, *Nature* **2005**, *437*, 1337–1340.
53. K. G. Thomas, B. I. Ipe, P. K. Sudeep, *Pure Appl. Chem.* **2002**, *74*, 1731–1738.
54. M. M. Pollard, M. Lubonska, P. Rudolf, B. L. Feringa, *Angew. Chem. Int. Ed.* **2007**, *46*, 1278–1280.
55. J. Vacek, J. Michl, *Adv. Funct. Mater.* **2007**, *17*, 730–739.
56. J. Vacek, J. Michl, *Proc. Natl. Acad. Sci. U.S.A.* **2001**, *98*, 5481–5486.
57. X. L. Zheng, M. E. Mulcahy, D. Horinek, F. Galeotti, T. F. Magnera, J. Michl, *J. Am. Chem. Soc.* **2004**, *126*, 4540–4542.
58. N. Wintjes, D. Bonifazi, F. Cheng, A. Kiebele, M. Stohr, T. Jung, H. Spillmann, F. Diederich, *Angew. Chem. Int. Ed.* **2007**, *46*, 4089–4092.
59. G. Rapenne, *Org. Biomol. Chem.* **2005**, *3*, 1165–1169.
60. G. Vives, A. Gonzales, J. Jaud, J.-P. Launay, G. Rapenne, *Chem. Eur. J.* **2007**, *13*, 5622–5631.
61. F. Moresco, A. Gourdon, *Proc. Natl. Acad. Sci. U.S.A.* **2005**, *102*, 8809–8814.
62. C. Viala, A. Secchi, A. Gourdon, *Eur. J. Org. Chem.* **2002**, 4185–4189.
63. M. Lastapis, M. Martin, D. Riedel, L. Hellner, G. Comtet, G. Dujardin, *Science* **2005**, *308*, 1000–1003.
64. L. Soukiassian, A. J. Mayne, G. Comtet, L. Hellner, G. Dujardin, A. Gourdon, *J. Chem. Phys.* **2005**, *122*, 134704.
65. G. Jimenez-Bueno, G. Rapenne, *Tetrahedron Lett.* **2003**, *44*, 6261–6263.
66. L. Grill, K.-H. Rieder, F. Moresco, G. Jimenez-Bueno, C. Wang, G. Rapenne, C. Joachim, *Surf. Sci.* **2005**, *584*, L153–L158.
67. F. Chiaravallotti, L. Gross, K.-H. Rieder, S. M. Stojkovic, A. Gourdon, C. Joachim, F. Moresco, *Nature Mater.* **2007**, *6*, 30–33.
68. D. L. Keeling, M. J. Humphry, R. H. J. Fawcett, P. H. Beton, C. Hobbs, L. Kantorovich, *Phys. Rev. Lett.* **2005**, *94*, 146104.
69. Y. Shirai, A. J. Osgood, Y. M. Zhao, Y. X. Yao, L. Saudan, H. B. Yang, Y. H. Chiu, L. B. Alemany, T. Sasaki, J. F. Morin, J. M. Guerrero, K. F. Kelly, J. M. Tour, *J. Am. Chem. Soc.* **2006**, *128*, 4854–4864.
70. J.-F. Morin, Y. Shirai, J. M. Tour, *Org. Lett.* **2006**, *8*, 1713–1716.
71. T. Sasaki, J. M. Tour, *Org. Lett.* **2008**, *10*, 897–900.

72. A. Kocer, M. Walko, W. Meijberg, B. L. Feringa, *Science* **2005**, *309*, 755–758.
73. I. Vlassiouk, C. D. Park, S. A. Vail, D. Gust, S. Smirnov, *Nano Lett.* **2006**, *6*, 1013–1017.
74. G. Berkovic, V. Krongauz, V. Weiss, *Chem. Rev.* **2000**, *100*, 1741–1754.
75. M. Volgraf, P. Gorostiza, R. Numano, R. H. Kramer, E. Y. Isacoff, D. Trauner, *Nature Chem. Biol.* **2006**, *2*, 47–52.
76. S. Angelos, E. Choi, F. Vögtle, L. De Cola, J. I. Zink, *J. Phys. Chem. C* **2007**, *111*, 6589–6592.
77. T. D. Nguyen, H.-R. Tseng, P. C. Celestre, A. H. Flood, Y. Liu, J. F. Stoddart, J. I. Zink, *Proc. Natl. Acad. Sci. U.S.A.* **2005**, *102*, 10029–10034.
78. T. D. Nguyen, Y. Liu, S. Saha, K. C.-F. Leug, J. F. Stoddart, J. I. Zink, *J. Am. Chem. Soc.* **2007**, *129*, 626–634.
79. T. D. Nguyen, K. C.-F. Leug, M. Liong, C. D. Pentecost, J. F. Stoddart, J. I. Zink, *Org. Lett.* **2006**, *8*, 3363–3366.
80. K. Patel, S. Angelos, W. R. Dichtel, A. Coskun, Y. W. Yang, J. I. Zink, J. F. Stoddart, *J. Am. Chem. Soc.* **2008**, *130*, 2382–2383.
81. Y. Liu, A. H. Flood, P. A. Bonvallet, S. A. Vignon, B. H. Northrop, H.-R. Tseng, J. A. Jeppesen, T. Y. Huang, B. Brough, M. Ballaer, S. Magonov, S. D. Solares, W. A. Goddard III, C.-M. Ho, J. F. Stoddart, *J. Am. Chem. Soc.* **2005**, *127*, 9745–9759.
82. M. Watari, J. Galbraith, H.-P. Lang, M. Sousa, M. Hegner, C. Gerber, M. A. Horton, R. A. McKendry, *J. Am. Chem. Soc.* **2007**, *129*, 601–609.
83. W. Shu, D. Liu, M. Watari, C. K. Riener, T. Strunz, M. E. Welland, S. Balasubramanian, R. A. McKendry, *J. Am. Chem. Soc.* **2005**, *127*, 17054–17060.
84. W. Shi, M. I. Giannotti, X. Zhang, M. A. Hempenius, H. Schonherr, G. J. Vancso, *Angew. Chem. Int. Ed.* **2007**, *46*, 8400–8404.
85. T. Hugel, N. B. Holland, A. Cattani, L. Moroder, M. Seitz, H. E. Gaub, *Science* **2002**, *296*, 1103–1106.
86. B. C. Bunker, D. L. Huber, J. G. Kushmerick, T. Dunbar, M. Kelly, C. Matzke, J. Cao, J. O. Jeppesen, J. Perkins, A. H. Flood, J. F. Stoddart, *Langmuir* **2007**, *23*, 31–34.
87. T. Niazov, S. Shlyahovsky, I. Willner, *J. Am. Chem. Soc.* **2007**, *129*, 6374–6375.
88. P. J. Langley, J. Hulliger, *Chem. Soc. Rev.* **1999**, *28*, 279–291.
89. J. R. Kincaid, *Chem. Eur. J.* **2000**, *6*, 4055–4061, and references therein.
90. M. Álvaro, B. Ferrer, H. García, E. J. Palomares, V. Balzani, A. Credi, M. Venturi, J. F. Stoddart, S. Wenger, *J. Phys. Chem. B* **2003**, *107*, 14319–14325.
91. Z. Popović, M. Otter, G. Calzaferri, L. De Cola, *Angew. Chem. Int. Ed.* **2007**, *46*, 6188–6191.
92. I. Willner, B. Willner, *J. Mater. Chem.* **1998**, *8*, 2543–2556.
93. G. Cooke, F. M. A. Duclairoir, V. M. Rotello, J. F. Stoddart, *Tetrahedron Lett.* **2000**, *41*, 8163–8166.
94. S. Chia, J. Cao, J. F. Stoddart, J. I. Zink, *Angew. Chem. Int. Ed.* **2001**, *40*, 2447–2451.
95. S. Saha, E. Johansson, A. H. Flood, H.-R. Tseng, J. I. Zink, J. F. Stoddart, *Chem. Eur. J.* **2005**, *11*, 6846–6858.
96. D. Fitzmaurice, S. N. Rao, J. A. Preece, J. F. Stoddart, S. Wenger, N. Zaccheroni, *Angew. Chem. Int. Ed.* **1999**, *38*, 1147–1150.
97. D. Ryan, S. N. Rao, H. Rensmo, D. Fitzmaurice, J. A. Preece, S. Wenger, J. F. Stoddart, N. Zaccheroni, *J. Am. Chem. Soc.* **2000**, *122*, 6252–6257.
98. B. Samorì, G. Zuccheri, P. Baschieri, *ChemPhysChem* **2005**, *6*, 29–34.
99. T. Auletta, M. R. de Jong, A. Mulder, F. C. J. M. van Veggel, J. Huskens, D. N. Reinhoudt, S. Zou, S. Zapotoczny, H. Schonherr, G. J. Vancso, L. Kuipers, *J. Am. Chem. Soc.* **2004**, *126*, 1577–1584.
100. S. Kado, K. Yamada, T. Murakami, K. Kimura, *J. Am. Chem. Soc.* **2005**, *127*, 3026–3030.
101. C. Schafer, R. Eckel, R. Ros, J. Mattay, D. Anselmetti, *J. Am. Chem. Soc.* **2007**, *129*, 1488–1489.
102. G. Fioravanti, N. Haraszkiewicz, E. R. Kay, S. M. Mendoza, C. Bruno, M. Marcaccio, P. G. Wiering, F. Paolucci, P. Rudolf, A. M. Brouwer, D. A. Leigh, *J. Am. Chem. Soc.* **2008**, *130*, 2593–2601.
103. T. Lu, L. Zhang, G. W. Gokel, A. E. Kaifer, *J. Am. Chem. Soc.* **1993**, *115*, 2542–2543.
104. I. Willner, V. Pardo-Yssar, E. Katz, K. T. Ranjit, *J. Electroanal. Chem.* **2001**, *497*, 172–177.

105. L. Raehm, J.-M. Kern, J.-P. Sauvage, C. Hamann, S. Palacin, J.-P. Bourgoin, *Chem. Eur. J.* **2002**, *8*, 2153–2162.
106. E. Katz, O. Lioubashevsky, I. Willner, *J. Am. Chem. Soc.* **2004**, *126*, 15520–15532.
107. J. Berna, D. A. Leigh, M. Lubomska, S. M. Mendoza, E. M. Pérez, P. Rudolf, G. Teobaldi, F. Zerbetto, *Nature Mater.* **2005**, *4*, 704–710.
108. A. Livoreil, J.-P. Sauvage, N. Armaroli, V. Balzani, L. Flamigni, B. Ventura, *J. Am. Chem. Soc.* **1997**, *119*, 12114–12124.
109. D. Payer, S. Rauschenbach, N. Malinowski, M. Konuma, C. Virokanadara, U. Starke, C. Dietrich-Buchecker, J.-P. Collin, J.-P. Sauvage, N. Lin, K. Kern, *J. Am. Chem. Soc.* **2007**, *129*, 15662–15667.
110. C. O. Dietrich-Buchecker, J.-P. Sauvage, J.-M. Kern, *J. Am. Chem. Soc.* **1984**, *106*, 3043–3045.
111. R. C. Ahuja, P.-L. Caruso, D. Möbius, G. Wildburg, H. Ringsdorf, D. Philp, J. A. Preece, J. F. Stoddart, *Langmuir* **1993**, *9*, 1534–1544.
112. C. L. Brown, U. Jonas, J. A. Preece, H. Ringsdorf, M. Seitz, J. F. Stoddart, *Langmuir* **2000**, *16*, 1924–1930.
113. M. Asakawa, M. Higuchi, G. Mattersteig, T. Nakamura, A. R. Pease, F. M. Raymo, T. Shimidzu, J. F. Stoddart, *Adv. Mater.* **2000**, *12*, 1099–1102.
114. C. P. Collier, E. W. Wong, M. Belohradsky, F. M. Raymo, J. F. Stoddart, P. J. Kuekes, R. S. Williams, J. R. Heath, *Science* **1999**, *285*, 391–394.
115. C. P. Collier, G. Mattersteig, E. W. Wong, Y. Luo, K. Beverly, J. Sampaio, F. M. Raymo, J. F. Stoddart, J. R. Heath, *Science* **2000**, *289*, 1172–1175.
116. M. Asakawa, P. R. Ashton, V. Balzani, A. Credi, C. Hamers, G. Mattersteig, M. Montalti, A. N. Shipway, N. Spencer, J. F. Stoddart, M. S. Tolley, M. Venturi, A. J. P. White, D. J. Williams, *Angew. Chem. Int. Ed.* **1998**, *37*, 333–337.
117. M. Ceccarelli, F. Mercuri, D. Passerone, M. Parrinello, *J. Phys. Chem. B* **2005**, *109*, 17094–17099.
118. D. R. Stewart, D. A. A. Ohlberg, P. A. Beck, Y. Chen, R. S. Williams, J. O. Jeppesen, K. A. Nielsen, J. F. Stoddart, *Nano Lett.* **2004**, *4*, 133–136.
119. Y. Luo, C. P. Collier, J. O. Jeppesen, K. A. Nielsen, E. Delonno, G. Ho, J. Perkins, H.-R. Tseng, T. Yamamoto, J. F. Stoddart, J. R. Heath, *ChemPhysChem* **2002**, *3*, 519–525.
120. R. F. Service, *Science* **2003**, *302*, 556–559.
121. J. R. Heath, J. F. Stoddart, R. S. Williams, *Science* **2004**, *303*, 1136–1137.
122. S. Nygaard, K. C.-F. Leung, I. Aprahamian, T. Ikeda, S. Saha, B. W. Laursen, S. Y. Kim, S. W. Hansen, P. C. Stein, A. H. Flood, J. F. Stoddart, J. O. Jeppesen, *J. Am. Chem. Soc.* **2007**, *129*, 960–970.
123. S. S. Jang, Y. H. Jang, Y.-H. Kim, W. A. Goddard III, J. W. Choi, J. R. Heath, B. W. Laursen, A. H. Flood, J. F. Stoddart, K. Nørgaard, T. Bjørnholm, *J. Am. Chem. Soc.* **2005**, *127*, 14804–14816.
124. M. R. Diehl, D. W. Steuerman, H.-R. Tseng, S. A. Vignon, A. Star, P. C. Celestre, J. F. Stoddart, J. R. Heath, *ChemPhysChem* **2003**, *4*, 1335–1339.
125. J. W. Choi, A. H. Flood, D. W. Steuerman, S. Nygaard, A. B. Braunschweig, N. N. P. Moonen, B. W. Laursen, Y. Luo, E. DeIonno, A. J. Peters, J. O. Jeppesen, K. Xu, J. F. Stoddart, J. R. Heath, *Chem. Eur. J.* **2006**, *12*, 261–279.
126. Y. H. Kim, W. A. Goddard III, *J. Phys. Chem. C* **2007**, *111*, 4831–4837.
127. D. W. Steuerman, H.-R. Tseng, A. J. Peters, A. H. Flood, J. O. Jeppesen, K. A. Nielsen, J. F. Stoddart, J. R. Heath, *Angew. Chem. Int. Ed.* **2004**, *43*, 6486–6491.
128. J. E. Green, J. W. Choi, A. Boukai, Y. Bunimovich, E. Johnston-Halperin, E. DeIonno, Y. Luo, B. A. Sheriff, K. Xu, Y. S. Shin, H.-R. Tseng, J. F. Stoddart, J. R. Heath, *Nature* **2007**, *445*, 414–417.
129. N. A. Melosh, A. Boukai, F. Diana, B. Gerardot, A. Badolato, P. M. Petroff, J. R. Heath, *Science* **2003**, *300*, 112–115.
130. P. Ball, *Nature* **2007**, *445*, 362–363.
131. G. D. Bachand, S. B. Rivera, A. Carroll-Portillo, H. Hess, M. Bachand, *Small* **2006**, *2*, 381–385.
132. M. G. L. van den Heuvel, M. P. de Graaff, C. Dekker, *Science* **2006**, *312*, 910–914.
133. H. Hess, *Soft Matter* **2006**, *2*, 669–677.

134. D. W. Wendell, J. Patti, C. D. Montemagno, *Small* **2006**, *2*, 1324–1329.
135. M. Schliwa, Ed., *Molecular Motors*, Weinheim: Wiley-VCH, **2003**.
136. M. Reconditi, M. Linari, L. Lucil, A. Stewart, Y.-B. Sun, P. Boesecke, T. Narayanan, R. F. Fischetti, T. Irving, G. Piazzesi, M. Irving, V. Lombardi, *Nature* **2004**, *428*, 578–581.
137. G. Muthukrishnan, B. M. Hutchins, M. E. Williams, W. O. Hanckock, *Small* **2006**, *2*, 626–630.
138. S. Ramachandran, K.-H. Ernst, G. D. Bachand, V. Vogel, H. Hess, *Small* **2006**, *2*, 330–334.
139. C.-T. Lin, M.-T. Kao, K. Kurabayashi, E. Meyhöfer, *Small* **2006**, *2*, 281–287.
140. P. D. Boyer, *Angew. Chem. Int. Ed.* **1998**, *37*, 2296–2307.
141. D. Stock, A. G. W. Leslie, J. E. Walker, *Science* **1999**, *286*, 1700–1705.
142. H. Noji, R. Yasuda, M. Yoshida, K. J. Kinosita, Jr., *Nature* **1997**, *386*, 299–302.
143. C. Bustamante, D. Keller, G. Oster, *Acc. Chem. Res.* **2001**, *34*, 412–420, and references therein.
144. D. W. Wendell, J. Patti, C. D. Montemagno, *Small* **2006**, *2*, 1324–1329.
145. R. K. Soong, G. D. Bachand, H. P. Neves, A. G. Olkhovets, H. G. Craighead, C. D. Montemagno, *Science* **2000**, *290*, 1555–1558.
146. G. D. Bachand, R. K. Soong, H. P. Neves, A. Olkhovets, H. G. Craighead, C. D. Montemagno, *Nano Lett.* **2001**, *1*, 42–44.
147. H. Itoh, A. Takahashi, K. Adachi, H. Noji, R. Yasuda, M. Yoshida, K. Kinosita, Jr., *Nature* **2004**, *427*, 465–468.
148. T.-J. M. Luo, R. Soong, E. Lan, B. Dunn, C. Montemagno, *Nature Mater.* **2005**, *4*, 220–224.
149. G. Steinberg-Yfrach, P. A. Liddell, S.-C. Hung, A. L. Moore, D. Gust, T. A. Moore, *Nature* **1997**, *385*, 239–241.

Chapter 18

Control of Morphology in Mesoporous and Mesostructured Hybrid Materials

DARREN R. DUNPHY, BERND SMARSLY, AND
C. JEFFREY BRINKER

18.1 INTRODUCTION

Arguably one of the greatest advancements in materials science and chemistry over the last two decades has been the development of porous inorganic materials with narrow pore size distributions and high surface area produced through the co-assembly of molecular inorganic precursors (i.e., sol–gel chemistry and processing)[1] with self-assembled lyotropic surfactant mesophases.[2,3] Both pore size (from about 2 to 30 nm) and pore network morphology are readily engineered through choice of surfactant as well as processing conditions; common pore geometries are given in Figure 18.1, along with examples of common types of surfactant templates. Although originally demonstrated for porous materials of pure silica, this templating strategy has been extended to many other metal oxides (e.g., TiO_2 or Al_2O_3).[4] In addition to mesoscale

The Supramolecular Chemistry of Organic–Inorganic Hybrid Materials. Edited by Knut Rurack and Ramón Martínez-Máñez

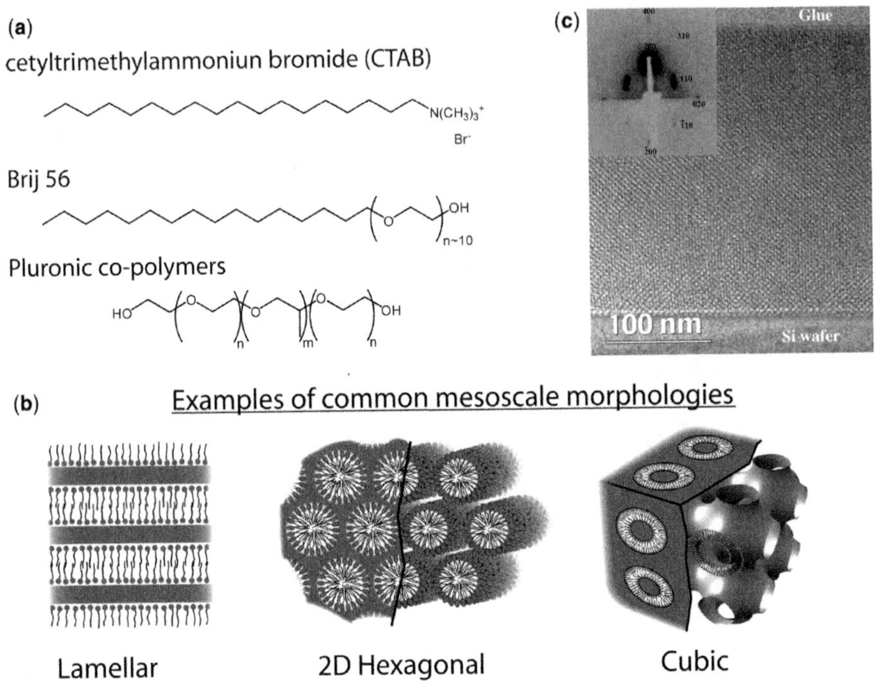

Figure 18.1 (a) Examples of surfactant types used to template mesoporous and mesostructured materials. (b) Common network morphologies for inorganic materials templated by self-assembled surfactant mesophases (the cubic structure shown, one of many types observed experimentally, is based on a surfactant bilayer draped over a bicontinuous minimal surface). (c) Transmission electron microscope cross-section of a Brij 56-templated film co-condensed with an azobenzene pendant silsesquioxane along with a 2D X-ray scattering pattern for this film (inset).[57] (Reprinted with permission from N. Liu et al., *Nano Lett.* **2004,** *4*, 551–554. Copyright 2004 American Chemical Society.)

(pore network) morphology, both atomic and macroscopic structures (i.e., crystallinity and particle shape and size, respectively) are readily tailored for specific applications.

The range of functionalities and properties achievable for inorganic materials is often limited when contrasted with organic counterparts, however. For example, the diversity of forces between organic molecules and inorganic surfaces is narrow in comparison with the types of interactions seen between organic molecules.[5] Furthermore, organic materials often exhibit superior hydrolytic stability as well as unique optical, electrochemical, surface, and mechanical properties. Nevertheless, the processing complexity for organic materials (i.e., polymers) is generally higher than that for inorganic oxides synthesized through low-temperature sol–gel chemistry. Also, inorganic networks can be more physically durable than organic materials.

Combining the processability and sturdiness of inorganic materials with the functionality of organic chemistry is made possible through the use of hybrid organic–inorganic molecular precursors containing robust Si—C bonds. These precursors are generally classified into two categories (Fig. 18.2): those with one reactive silicon

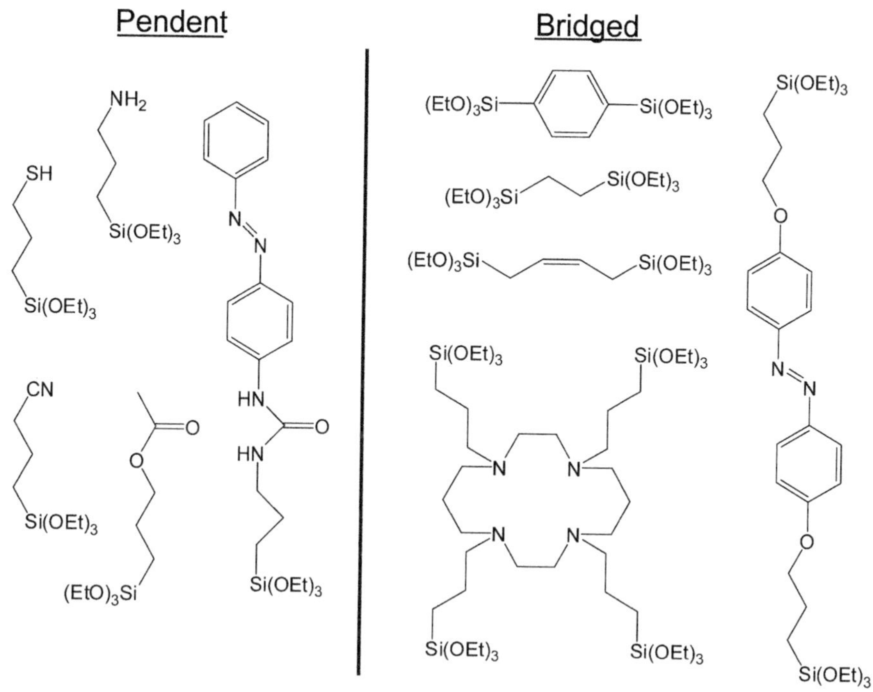

Figure 18.2 A sample of pendant and bridged silsesquioxane precursors suitable for creating hybrid nanostructured porous materials.

center (so-called pendant silsesquioxanes) and those with two or more silicon groups (bridged silsesquioxanes). Replacement of some or all of the purely inorganic molecular precursors in the synthesis of surfactant mesophase-templated mesoporous materials produces hybrid materials with covalently linked inorganic and organic components. A key property of these materials is that the location of the organic component within the overall mesostructure is directed by partitioning during the self-assembly process between the inorganic and surfactant phases (forming the wall framework and pore volumes, after surfactant extraction, respectively), with the partitioning behavior of the precursor depending on the precise molecular structure of that precursor (Fig. 18.3).[6] The ability to locate organic functionality within the overall mesostructure represents yet another level of structural control available in mesoporous materials, enabling the optimization of many functional properties such as adsorption selectivity[7,8] or energy transfer efficiency.[9,10]

The partitioning of organosilane precursors is exemplified by the case of pendant silsesquioxanes with amphiphile-like structure co-condensed into the mesostructure during self-assembly.[11] As shown in Figure 18.3a, the organic functionality extends into the surfactant phase, leaving the reactive silicon alkoxide group exposed at the surface of the micelle to form an extended network with other silica precursors both at the micelle–silica interface as well as in the inorganic phase. Extraction of the surfactant template leaves the organic functionality covalently attached to the silica

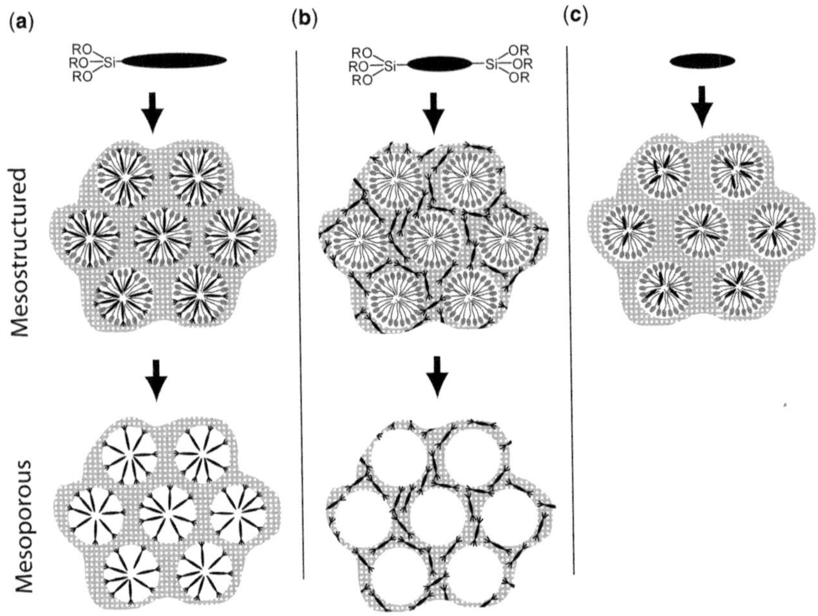

Figure 18.3 Control of hybrid precursor or monomer/polymer location within self-assembled porous materials controlled by partitioning of precursors between silica and surfactant phases. (a) Partitioning of amphiphilic pendant silsesquioxanes at the interface between the silica and surfactant phases; (b) integration of bridged silsesquioxanes within the silica framework; and (c) localization of hydrophobic monomers and polymers with the core of the surfactant mesophase.

framework, yielding pore surfaces homogeneously covered with organic functionality after surfactant extraction. Careful selection of the processing conditions as well as overall composition is often required to achieve high (>30%, approximately) surface coverage,[12] however, as the inclusion of organic functionality in the organic phase can impact the quality of both the mesoscale order and macroscale homogeneity. Although postextraction derivatization can also be used to decorate the surface of mesoporous silica with organic functionality, this route adds processing complexity, and may not yield uniform surface coverage throughout the pore network.[13] Derivatization will remain an important complimentary tool for producing hybrid materials, however, as some pendant functionality may not partition into the surfactant phase, or may disrupt the self-assembly process. Also, nonuniform coating can be exploited to construct hybrid materials with unique properties; for example, functional nanovalves can be placed around pore openings on particle or film surfaces to control the release of molecules loaded within the mesopore network.[14]

For nonampliphilic precursors (as well as other non-silane-modified organic molecules), partitioning into either the surfactant or silica phases is largely controlled by the overall hydrophobicity of the molecule. Most notably, bridged silsesquioxanes become homogeneously dispersed within the inorganic framework, producing, after surfactant removal, what have been termed *periodic mesoporous organosilicas*

(PMOs).[15–17] Both silanols produced from the condensation of the organosilica precursor as well as the organic functionality of the bridging group are accessible at the pore surface.[18] Through the addition of both pendant and bridged silsesquioxanes to the precursor solution, PMOs can also be synthesized with organic functionality in both the framework and at the pore surface.[19] Recently, PMO materials were reported that utilize functionalized C_{60} fullerenes as the framework material,[20] creating a new class of porous inorganic–organic–inorganic hybrids.

Organic molecules without reactive silicon groups also partition between the wall and surfactant phases; addition of hydrophobic species such as reactive monomers,[21] whole polymers,[22] or other components, including swelling agents or photoacid generators,[23] yields mesostructured *nanocomposite* materials with the organic phase homogeneously dispersed within the overall mesostructure, surrounded by the inorganic wall material. Although these materials are not technically porous, we include the discussion of these mesostructured composites as the final material morphology is achieved through the same strategy as is used for producing hybrid mesoporous networks with integral Si—C bonding.

A fourth class of related hybrid materials is briefly mentioned here for sake of completeness; instead of using liquid crystalline surfactant phases to template the overall mesostructure, intermolecular interactions such as hydrogen bonding or π stacking between organosilica precursors directs the formation of long-range order.[24–26] Materials formed through this route are crystalline, with pore dimensions in the micropore size regime.[26] Noncrystalline lamellar and two-dimensional hexagonal structures have also been formed through the self-organization of alkyl precursors incorporating preformed siloxane oligomers.[27] The structural control of materials synthesized using these precursors is often not limited to the nanoscale; as an example, the self-assembly of hollow tubes with micron-scale diameters has been realized using chiral bridged silsesquioxanes.[28]

The greatest number of literature reports regarding the synthesis and characterization of hybrid porous materials have been for hydrothermally produced powders.[29,30] The range of precursor structures integrated into these materials is extensive; although Figure 18.2 illustrates a few examples of both pendant and bridged silsesquioxanes that have appeared in the literature, the reader is referred to several excellent reviews for a more systematic listing of material compositions that have been synthesized.[3,29,30] In general, the organic component can be divided into passive and active functionality. The most prominent examples of the former are simple alkyl chains, while the latter include organic groups capable of function such as metal-ion binding rings,[7,8] energy transfer,[9,10] and photoisomerization (see later). PMO materials have also been reported with catalytically active metal ion dopants,[31] producing an added level of structural complexity. Although most metal-carbon bonds exhibit little or no stability under the conditions used to produce PMO materials, alkenediphosphonates with integral P—$(CH_2)_n$—P bonds have been reported.[32]

Of special significance in the development of PMO materials was the synthesis of porous materials with crystalline order of the bridged silsesquioxane inside the framework walls, initially achieved through the use of π–π interactions between 1,4-phenylene-bridged silsesquioxane precursors.[33] Molecular scale periodicity in the framework walls has also been obtained using larger bridging groups (e.g., biphenylene)[34] as well as nonlinear phenylene precursors.[35] Although nanoscale morphology has largely been limited to two-dimensional hexagonal pore networks, a lamellar structure with a crystalline framework has also been reported.[36] Chemical modification of the phenyl groups within these materials with sulfuric acid[37] or amino groups[38] has been demonstrated without loss of framework periodicity.

As many potential applications of mesoporous materials (sensors, separation membranes, etc.) require porous layers, considerable effort has been directed toward the development of hybrid mesoporous films.[3,17] To this end, evaporation-induced self-assembly (EISA)[39] has proven to be a convenient route for the rapid synthesis of oriented hybrid mesoporous and mesostructured films of up to approximately 1 μm in thickness. In EISA, a homogeneous solution of a soluble oxide precursor and surfactant, prepared in a mixed alcohol (or other polar solvent)–water solvent system with an initial surfactant concentration normally less than the critical micelle concentration, undergoes preferential evaporation of organic solvent during film deposition, thus concentrating the depositing film in water, surfactant, and inorganic species. The progressively increasing surfactant concentration drives the organization of the surfactant into lyotropic liquid crystalline mesophases. Subsequent inorganic condensation freezes in the final nanoscale morphology; the surfactant template can then be removed by solvent extraction or calcination; UV exposure has also been used in this regard for purely inorganic films,[40] but lack of selectivity toward organic functionality reduces the usefulness of this approach for hybrid films.

Although hybrid films have been synthesized using many of the same precursors and synthesis strategies used for powdered materials, development of film composition or morphology has lagged behind that of powders. For example, crystalline order of the PMO organosilica frameworks has been difficult to achieve for films synthesized using an EISA process, presumably due to kinetic trapping of less ordered framework structures during the rapid self-assembly process. Recently, however, phenylene-bridged PMO films with crystalline order within the wall phase were obtained by careful control of the degree of precursor condensation.[41]

An aspect of PMO research that is of critical importance in understanding the relationship between synthesis conditions, material structure, and functional properties is the fundamental characterization of the hybrid microstructure and chemical state. In general, many recent advances in the overall characterization of surfactant-templated materials (for example, tools for analysis of X-ray scattering data for oriented films)[42,43] are applicable to organosilica materials. For hybrid materials, however, special emphasis has been given to the chemistry of the surface of the hybrid material as well as the organization of templated silsesquioxane frameworks. Of special note is the use of vibrational spectroscopy to investigate framework structure. As an example, IR spectroscopy, combined with *ab initio* molecular modeling, has examined the surface structure of a phenyl-bridged PMO (with both crystalline[18] and amorphous[44] wall frameworks) along with the interaction of the PMO surface with small probe molecules. In another study, IR spectroscopy was used to examine the chemical environment of hydroxyl groups within the framework of a P123-templated phenyl-bridged PMO.[45] Raman scattering has also proven to be useful for investigations of hybrid materials.[46] Significantly, vibrational spectroscopy was found to be an unreliable indicator of Si—C bond integrity in PMOs; only NMR spectroscopy can definitively monitor the state of this bond.[47] New NMR techniques have also been developed to investigate the surface structure of mesoporous silicas modified with surface functionality through the co-condensation route.[48]

Although PMOs and related materials have many potential applications, including adsorbants,[49,50] sensors,[51] and catalysts,[52,53] we focus here on two specific classes of functional materials, nanovalves and polymer–silica nanocomposites formed via the co-assembly of hybrid precursors (or polymer–monomer species) with surfactant mesophases.

18.2 CONTROLLED RELEASE AND VALVING

For controlled release and molecular transport, the inorganic framework provides ordered monosized porosity with controlled connectivity as well as a rigid platform to support environmentally responsive organic functionality that modulates mass transport out of or through the pore network. Applications of these hybrid nanovalves may include anticorrosion agents triggered by changes in pH or redox potential, localized therapeutic delivery of drugs such as anticancer compounds,[54–56] and control of flow in microfluidic devices. Two strategies have been employed to produce these functional materials; co-condensation of the organic component during synthesis of the material via self-assembly,[57] and postsynthesis grafting of the organic component onto the surface of the preformed inorganic framework.[58] Co-condensation produces a hybrid with homogeneous dispersal throughout the material, while grafting conditions can be tailored to selectively attach functional molecules around external pore openings or over the entire pore network surface.

An example of a hybrid porous material with nanovalve functionality using the grafting route is illustrated in Figure 18.4.[58] The exterior of preformed mesoporous silica powder was derivatized using (3-isocyanatopropyl)triethoxysilane (ICTES); the ICTES serves as a linking point for a dialkylammonium pseudorotaxane through the formation of an amide linkage. After loading of the MCM-41 with a fluorescent dye, dibenzo[24]crown-8 was used to cap the pseudorotaxane, with the cyclic ether bonding to the attached stem of the nanovalve through $[N-H—O]^+$ hydrogen bonding. Addition of base to the modified MCM-41 suspension deprotonates the ammonium center, releasing the ether ring and subsequently releasing the contents of the silica nanopore network. Alternatively, competitive binding of the ether ring to free fluorodialkylammonium or metal ions can be used to open the nanovalve construct. Redox-actuated valves,[59] as well as fully reversible valves formed through the use of a bistable redox-active rotaxane,[14] have also been fabricated using a similar strategy as that described in Figure 18.4.

An alternative route to hybrid nanostructures with controlled release or mass transfer properties entails the modification of the pore network surface with photoisomerizable molecules such as azobenzene (Fig. 18.5a). Azobenzene exists in two conformations, *trans* (the most stable isomer) and *cis*; irradiation with light

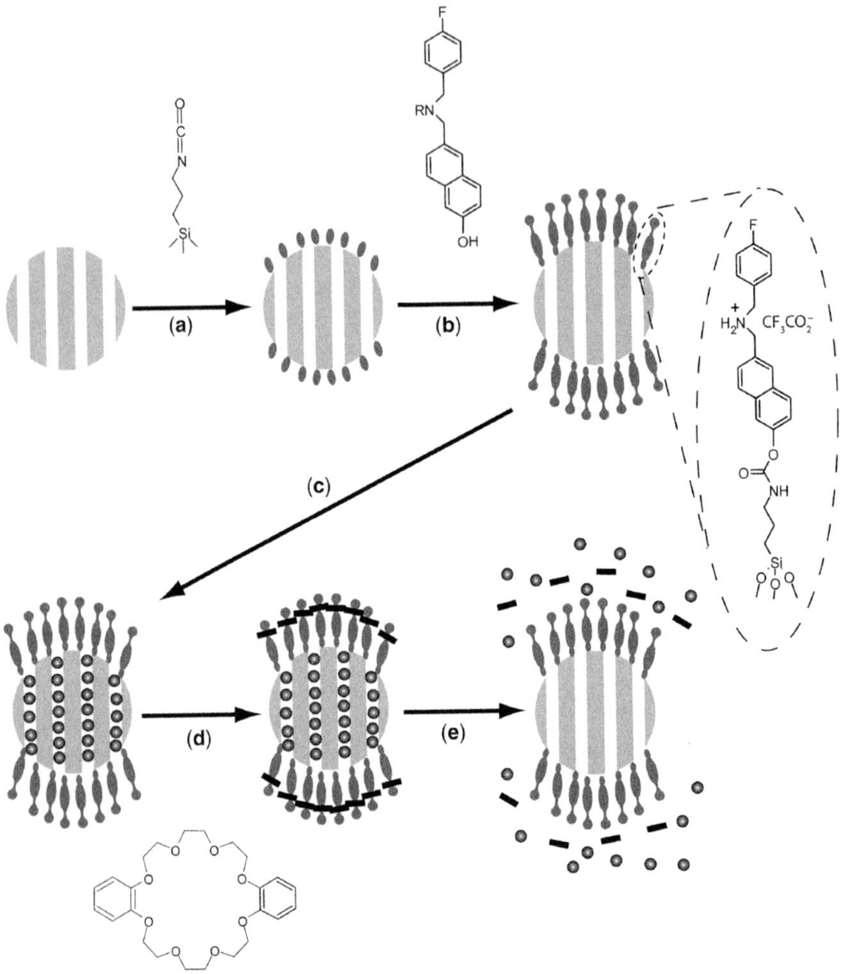

Figure 18.4 Fabrication and operation of a hybrid organic–inorganic nanovalve based on grafting of valve functionality to preformed nanoporous silica. (a) Grafting of ICTES to the surface of the mesoporous silica powders; (b) linkage of a dialkylammonium pseudorotaxane to the surface of the silica through reaction with ICTES; (c) loading of the pore volume with molecules; (d) closing of the nanovalve by complexation of the dialkylammonium pseudorotaxane with a crown ether; and (e) release of the pore volume contents by disruption of the crown ether-ammonium complex.

of approximately 350 nm induces a *trans* to *cis* isomerization while light at 450 nm (or heat) recycles the molecule back to the *trans* form. During the *trans* to *cis* isomerization, the length of the azobenzene unit decreases by approximately 3.4 Å, effectively increasing the nanopore diameter; the dipole moment of the molecule also increases from 0.5 to 3.1 debye. Using an azobenzene-modified silsesquioxane as a cosurfactant with Brij 56, Liu et al. demonstrated the formation of ordered hybrid films with 3D-accessible porosity (Fig. 18.1c) and a pore surface

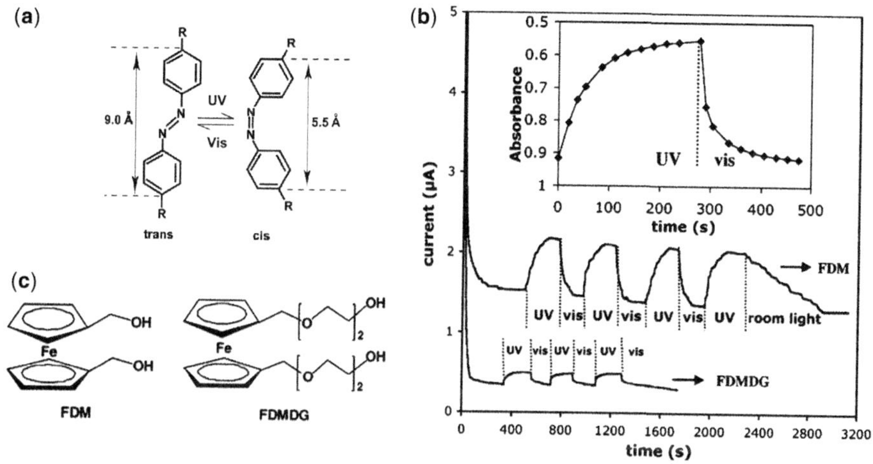

Figure 18.5 Control of mass transfer through an azobenzene-modified film. (a) Structure and photoisomerization of azobenzene; (b) structures of ferrocene derivatives used to monitor anodic current at the film–electrode interface; and (c) current as a function of light exposure for FDM and FDMDG redox probes.[57] (Reprinted with permission from N. Liu et al., *Nano Lett.* **2004,** *4*, 551–554. Copyright 2004 American Chemical Society.)

coverage of 1.2 azobenzene molecules nm^{-2}.[57] Following removal of the primary surfactant to establish the necessary critical isomerization volume, UV/Vis spectroscopy was used to follow the reversibility of the conformation change within the nanopores (Fig. 18.5b, inset). Figure 18.5b shows that changes in molecular conformation are directly correlated with modulation of mass transfer through the film. Plotted are the anodic currents obtained by the oxidation of either ferrocene dimethanol (FDM) or ferrocene dimethanol diethylene glycol (FDMDG) (Fig. 18.5c) at an electrode modified with a thin mesoporous film. Because the measurement was performed with vigorous stirring of the ferrocene solution (thus ensuring a steady-state concentration of the redox probe at the film–solution interface), all observed changes in electrode current are due to changes in mass transfer (bulk and surface diffusion) rates through the azobenzene-modified pore network. As the film is illuminated with UV light, *trans* to *cis* isomerization is accompanied by an increase in current, consistent with an increase in pore diameter (although surface polarity effects may also be a contribution to this change in flux). The current increase is completely reversed upon relaxation of the azobenzene back to the *trans* isomer, either by visible light or heat. Both the baseline anodic current as well as the total change in current upon film irradiation is significantly less for the FDMDG probe, an effect of its larger size.

Azobenzene has also been added to the internal surface of mesoporous films and powders through postsynthesis grafting.[58,60–62] The effect of oligomer size on azobenzene isomerization kinetics was studied for branched oligomer-modified azobenzene derivatives in solution and covalently grafted to nanoporous silica;[61] the size of the dendrimer had little effect on the thermal *cis* to *trans* isomerization rate in solution but slowed this transition considerably for grafted molecules as the dendrimer size was increased. Also, illumination of azobenzene-modified nanoporous silica at a wavelength where both the *cis* and *trans* isomers absorb was found to induce rapid switching between *trans* and *cis* isomers, creating a molecular impeller that can be used to expel molecules adsorbed within the mesopore network.[62] This effect was used to selectively deliver antitumor agents to cancer cells in a demonstration of a promising therapeutic technology.[58]

In addition to azobenzene, coumarin-modified mesoporous networks have been used to create light-induced controlled release.[63,64] The action of coumarin differs from that of azobenzene in that illumination drives a reversible bimolecular coupling reaction, creating a cyclobutane dimer that physically blocks pore egress.

18.3 NANOSTRUCTURED POLYMER–INORGANIC COMPOSITES

The ordered co-arrangement of functional organic polymers and an inorganic framework with nanometer-scale integration of the two phases through surfactant-mediated self-assembly (e.g., by EISA) has emerged as an attractive strategy to combine the unique properties of polymers with the rigidity of metal oxides. Of particular interest are macromolecules that respond predictably to environmental conditions such as pH and temperature (using polymeric hydrogels to endow the nanostructured oxide with sensor properties), and conducting or semiconducting polymers (enabling the synthesis of nanostructured electronic or optical devices).

The major challenge in the synthesis of nanostructured polymer–silica hybrids arises out of thermodynamics: the entropic driving force for nanoscaled mixing between polymers and silica precursors is smaller than that between small organic molecules and silica precursors, promoting undesired macroscopic phase separation between the polymer and silica networks that frustrates the formation of a well-ordered nanostructured material. However, three suitable strategies have been developed to overcome this fundamental limitation and permit the incorporation of polymeric organics into highly regular silica nanostructures using a standard EISA process:[21,65–67] (1) the addition of monomers to the EISA precursor solution followed by post-film synthesis polymerization (illustrated in Fig. 18.6a); (2) the design of surfactant templates possessing a polymerizable unit, again followed by polymerization after the completion of the self-assembly process (Fig. 18.6b); and (3) direct addition of polymers to the EISA precursor solution (Fig. 18.6c).[22]

An example of the first approach is the integration of hydrogels into nanostructured silica films by addition of a suitable monomer (e.g., methyl methacrylate, *N*-isopropyl acrylamide, etc.) and an initiator for radical polymerization to a solution containing a structure-directing surfactant and a prehydrolyzed silica precursor. During self-assembly, the monomers partition within the hydrophobic core of the surfactant mesophase; postsynthesis polymerization (for instance, by UV treatment) followed by solvent washing to remove the surfactant template yields a polymer–silica nanohybrid.

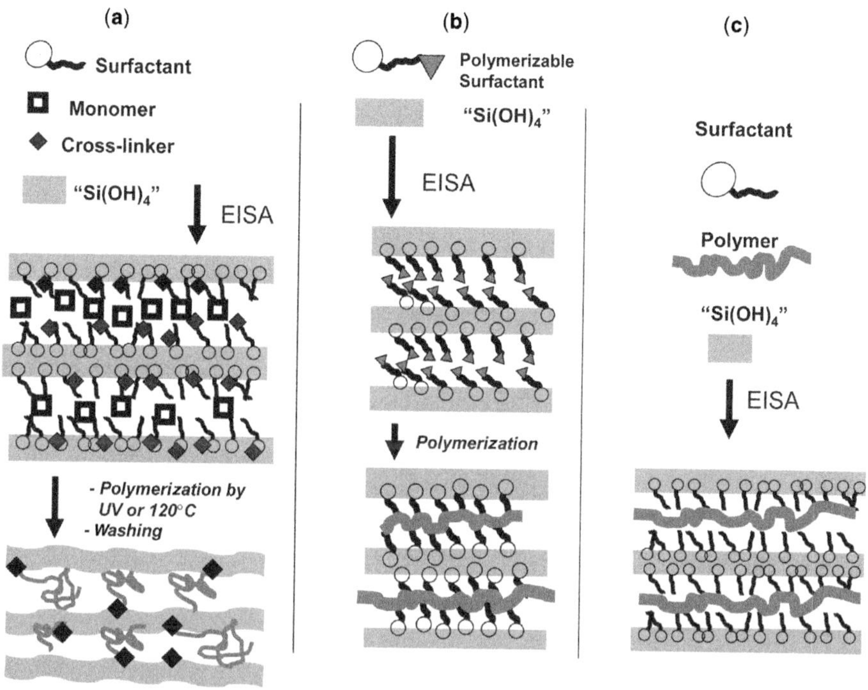

Figure 18.6 Fabrication of polymer–silica mesostructured materials through a self-assembly process by (a) addition of reactive monomers to the precursor solution, followed by postdeposition polymerization; (b) addition of a polymerizable surfactant; and (c) direct addition of a soluble polymer to the synthesis solution.

With this procedure, temperature-sensitive polymers such as poly(*N*-isopropyl acryl amide) (PNIPAAm) have been incorporated into thin silica films.[21,66] PNIPAAm has a lower critical solution temperature (LCST), undergoing a reversible swelling transition at approximately 32°C in water. PNIPAAm has been incorporated into clays, producing nanoscaled hydrogel–inorganic intercalation compounds with temperature-dependent swelling of the clay lamellae. However, these materials suffer from weak interactions between the polymer and the inorganic components. In contrast, an EISA process has been developed that includes a covalent linkage between polymer and silica, achieved by adding a siloxane featuring a polymerizable double bond [e.g., trimethoxy(7-octen-1-yl)silane] which becomes located between the silica and surfactant phase during the self-assembly of the hybrid material.[21,64] After heat-activated polymerization of the PNIPAAm network and extraction of the surfactant template, the polymer phase was found to maintain its hydrogel properties within the thin silica-PNIPAAm films, showing reversible swelling and deswelling of the hybrid film upon heat treatment and cooling above and below approximately 32°C, that is, no change in LCST.[66] Temperature-dependent *in situ* SAXS experiments showed that these films, composed of alternating layers of silica and PNIPAAm, undergo reversible changes in the PNIPAAm layer thickness by a factor of up to two, corresponding to an equivalent absolute increase in the total film thickness.[66] Hydrogel-silica hybrid films of this type could potentially be used to modulate flow in permeable membranes or in microfluidic devices through control of temperature.

In the second approach to self-assembly of polymer–silica nanocomposites, polymerizable surfactants are employed both to direct self-assembly into ordered mesophases and to serve as organic monomers which can be subsequently polymerized. For instance, Brinker et al. developed nonionic surfactants incorporating diacytelene groups which could be polymerized upon irradiation by UV light.[68,69] Similarly, poly(thiophene)[70] and poly(pyrrole)[71] have been successfully integrated into silica nanostructures in this manner. Importantly, this *in situ* polymerization produces isolated molecular wires, rather than clustered bundles of conductive polymer.[70]

Finally, direct addition of functional polymer to the EISA precursor solution has been found to be a viable route to well-ordered silica–polymer hybrids; polymers containing phenylenevinylene or fluorenyl units were incorporated into organic light-emitting diode structures.[22] Although conjugated polymers can also be added to preformed mesoporous silica networks,[72] direct addition has the advantage of a one-pot synthesis approach as well as the ability to incorporate polymers into inaccessible mesophase networks. However, the polymer must be soluble in the EISA precursor solution, a substantial limitation when working with macromolecules.

18.4 FUTURE PROSPECTS

The range of organosilane precursor structures potentially available for the synthesis of hybrid mesoporous (and mesostructured) materials is extremely large; development of new material compositions will continue to be a fertile area of research.

Many organosilane molecules are difficult to synthesize and purify, however. Also, the solubility of these precursors in the polar solvents commonly used in sol-gel chemistry can be low, and care must be taken to design the molecular structure to be compatible with typical self-assembly conditions (i.e., acidic pH). For these reasons, new synthetic routes to hybrid materials will be required to expand the range of accessible material formulations. As an example, allylorganosilanes can be used in place of difficult-to-purify alkoxy- or chloro-species to synthesize phenyl-bridged PMOs under acidic and basic conditions, yielding a material identical to that synthesized using standard alkoxysilane precursors. Both 2D hexagonal[73] and 3D cubic (*Pm*3*n*)[74] materials have been produced in this manner.

Another productive direction for future research is the addition of environmentally or chemically responsive functionality to the hybrid framework of PMO materials. Already, PMOs have been synthesized that incorporate a light-responsive azobenzene-bridged silsesquioxane diluted with tetraethylorthosilicate; reversible photoisomerization was observed that was accompanied by changes in adsorption properties toward polydisperse Au nanoparticles.[75] Molecule- (or ion-) specific binding sites could also be incorporated into the pore walls using appropriate hybrid precursors. Although examples of PMO compositions that include a metal chelating site within the inorganic framework have been limited, hybrid mesoporous materials with specific metal adsorption properties have been synthesized with covalently incorporated cyclam rings,[7] as well as amine binding sites.[8] Similarly, the inclusion of metal-organic framework (MOF) architectures in the framework phase could be used to add a second scale or type of porosity to hybrid materials for applications such as hydrogen storage or sorbents for environmental remediation.[76] Engineered binding sites could also be introduced into PMO frameworks through the use of silsesquioxane precursors that interact, in a manner analogous to that of amino acid fragments within peptides or proteins, via noncovalent interactions such as hydrogen bonding.[77]

For nanostructured hybrid thin films, development of new methods to combine bottom-up fabrication (nanoscale self-assembly) with top-down processing (film patterning) will be required to incorporate these materials into functional devices. For example, Doshi et al. added an amphiphilic photoacid generator (PAG) to an EISA precursor solution; during self-assembly, the photoacid served as a cosurfactant and was uniformly dispersed throughout the film.[23] Masked irradiation with UV light activated the PAG, inducing patterned changes in film nanostructure as well as control of silica etching susceptibility. Hybrid materials could be synthesized that incorporate this type of chemistry (or other photoactive system) directly in the silsesquioxane framework.

Recently, living cells were reported to actively direct the assembly of lipid–silica nanostructures;[78] extension of this cell-directed assembly (CDA) to hybrid architectures will add another level of organizational control to these materials.

Finally, structural characterization of hybrid materials must be refined; in many cases, it is unclear where the organic component is located within the overall mesostructure of the material. Coupled with experimental data, simulations of the self-assembly process in hybrid materials[79] should yield greater insight into the local structure of hybrid precursors.

REFERENCES

1. C. J. Brinker, G. W. Scherer, *Sol-Gel Science: The Physics and Chemistry of Sol-Gel Processing*, New York: Academic Press, **1990**.
2. Y. Wan, D. Zhao, *Chem. Rev.* **2007**, *107*, 2821–2860.
3. C. Sanchez, C. Boissière, D. Grosso, C. Laberty, L. Nicole, *Chem. Mater.* **2008**, *20*, 682–737.
4. S. W. Boettcher, J. Fan, C. K. Tsung, Q. Shi, G. D. Stucky, *Acc. Chem. Res.* **2007**, *40*, 784–792.
5. J. Israelachvili, *Intermolecular and Surface Forces*, New York: Academic Press, **1991**.
6. P. M. Minoofar, B. S. Dunn, J. I. Zink, *J. Am. Chem. Soc.* **2005**, *127*, 2656–2665.
7. J. Alauzun, A. Mehdi, C. Reyé, R. P. J. Corriu, *J. Mater. Chem.* **2007**, *17*, 349–356.
8. M. C. Burleigh, S. Dai, E. W. Hagaman, J. S. Lin, *Chem. Mater.* **2001**, *13*, 2537–2546.
9. M. Álvaro, B. Ferrer, V. Fornes, H. García, *Chem. Commun.* **2001**, 2546–2547.
10. M. Álvaro, M. Benitez, J. F. Cabeza, H. García, A. Leyva, *J. Phys. Chem. C* **2007**, *111*, 7532–7538.
11. N. Liu, Z. Chen, D. R. Dunphy, Y. Jiang, R. A. Assink, C. J. Brinker, *Angew. Chem. Int. Ed.* **2003**, *42*, 1731–1734.
12. S. Huh, J. W. Wienchg, J. Yoo, M. Pruski, V. S. Y. Lin, *Chem. Mater.* **2003**, *15*, 4247–4256.
13. M. H. Lim, A. Stein, *Chem. Mater.* **1999**, *11*, 3285–3295.
14. T. D. Nguyen, H. Tseng, P. C. Celestre, A. H. Flood, Y. Liu, J. F. Stoddart, J. I. Zink, *Proc. Nat. Acad. Sci. U.S.A.* **2005**, *109*, 10029–10034.
15. T. Asefa, M. J. MacLachlan, N. Coombs, G. A. Ozin, *Nature* **1999**, *402*, 867–871.

16. S. Inagaki, S. Guan, Y. Fukushima, T. Ohsuna, O. Terasaki, *J. Am. Chem. Soc.* **1999**, *121*, 9611–9614.
17. Y. Lu, H. Fan, N. Doke, D. A. Loy, R. A. Assink, D. A. LaVan, C. J. Brinker, *J. Am. Chem. Soc.* **2000**, *122*, 5258–5261.
18. B. Onida, L. Borello, L. Busco, P. Ugliengo, Y. Goto, S. Inagaki, E. Garrone, *J. Phys. Chem. B* **2005**, *109*, 11961–11966.
19. L. Zhang, J. Liu, J. Yang, Q. Yang, C. Li, *Microporous Mesoporous Mater.* **2008**, *109*, 172–183.
20. W. Whitnall, L. Cademartiri, G. A. Ozin, *J. Am. Chem. Soc.* **2007**, *129*, 15644–15649.
21. G. Garnweitner, B. Smarsly, R. Assink, W. Ruland, E. Bond, C. J. Brinker, *J. Am. Chem. Soc.* **2003**, *125*, 5626–5627.
22. E. Dovgolevsky, S. Kirmayer, E. Lakin, Y. Yang, C. J. Brinker, G. L. Frey, *J. Mater. Chem.* **2008**, *18*, 423–436.
23. D. A. Doshi, N. K. Huesing, M. C. Lu, H. Y. Fan, Y. F. Lu, K. Simmons-Potter, B. G. Potter, A. J. Hurd, C. J. Brinker, *Science* **2000**, *290*, 107–111.
24. N. G. Liu, K. Yu, B. Smarsly, D. R. Dunphy, Y.-B. Jiang, C. J. Brinker, *J. Am. Chem. Soc.* **2002**, *124*, 14540–14541.
25. J. J. E. Moreau, B. P. Pichon, M. Wong Chi Man, C. Bied, H. Pritzkow, J.-L. Bantignies, P. Dieudonné, J.-L. Sauvajol, *Angew. Chem. Int. Ed.* **2004**, *43*, 203–206.
26. H. Peng, Y. Lu, *Adv. Mater.* **2008**, *20*, 797–800.
27. A. Shimojima, K. Kuroda, *Angew. Chem. Int. Ed.* **2003**, *42*, 4057–4060.
28. J. J. E. Moreau, L. Vellutini, M. Wong Chi Man, C. Bied, *Chem. Eur. J.* **2003**, *9*, 1594–1599.
29. B. Hatton, K. Landskron, W. Whitnall, D. Perovic, G. A. Ozin, *Acc. Chem. Res.* **2005**, *38*, 305–312.
30. S. Fujita, S. Inagaki, *Chem. Mater.* **2008**, *20*, 891–908.
31. M. P. Kapoor, S. Inagaki, A. Bhaumik, K. Kuraoka, T. Yazawa, *J. Mater. Chem.* **2002**, *12*, 3078–3083.
32. T. Kimura, K. Kato, *J. Mater. Chem.* **2007**, *17*, 559–566.
33. S. Inagaki, S. Guan, T. Ohsuna, O. Terasaki, *Nature* **2002**, *416*, 304–307.
34. M. P. Kapoor, Q. Yang, S. Inagaki, *J. Am. Chem. Soc.* **2002**, *124*, 15176–15177.
35. M. P. Kapoor, Q. Yang, S. Inagaki, *Chem. Mater.* **2004**, *16*, 1209–1213.
36. K. Okamoto, M. P. Kapoor, S. Inagaki, *Chem. Commun.* **2005**, 1423–1425.
37. Q. Yang, M. P. Kapoor, S. Inagaki, *J. Am. Chem. Soc.* **2002**, *124*, 9694–9695.
38. M. Ohashi, M. P. Kapoor, S. Inagaki, *Chem. Commun.* **2008**, 841–843.
39. Y. F. Lu, R. Ganguli, C. A. Drewien, M. T. Anderson, C. J. Brinker, W. L. Gong, Y. X. Guo, H. Soyez, B. Dunn, M. H. Huang, J. I. Zink, *Nature* **1997**, *389*, 364–368.
40. T. Clark, J. D. Ruiz, H. Y. Fan, C. J. Brinker, B. I. Swanson, A. N. Parikh, *Chem. Mater.* **2000**, *12*, 3879–3884.
41. S.-Y. Wu, H.-S. Hsueh, M. H. Huang, *Chem. Mater.* **2007**, *19*, 5986–5990.
42. M. P. Tate, V. N. Urade, J. D. Kowalski, T. C. Wei, B. D. Hamilton, B. W. Eggiman, H. W. Hillhouse, *J. Phys. Chem. B* **2006**, *110*, 9882–9892.
43. M. P. Tate, H. W. Hillhouse, *J. Phys. Chem. C* **2007**, *111*, 7645–7654.
44. B. Onida, B. Camarota, P. Ugliengo, Y. Goto, S. Inagaki, E. Garrone, *J. Phys. Chem. B* **2005**, *109*, 21732–21736.
45. B. Camarota, B. Onida, Y. Goto, S. Inagaki, E. Garrone, *Langmuir* **2007**, *23*, 13164–13168.
46. Ö. Dag, G. A. Ozin, *Adv. Mater.* **2001**, *13*, 1182–1185.
47. F. Hoffman, M. Güngerich, P. J. Klar, M. Fröba, *J. Phys. Chem. C* **2007**, *111*, 5648–5660.
48. J. Trebosc, J. W. Wiench, S. Huh, V. S.-Y. Lin, M. Pruski, *J. Am. Chem. Soc.* **2005**, *127*, 7587–7593.
49. X. Feng, G. E. Fryxell, L. Q. Wang, A. Y. Kim, J. Liu, K. M. Kemmer, *Science* **1997**, *276*, 923–926.
50. C. Li, J. Liu, X. Shi, J. Yang, Q. Yang, *J. Phys. Chem. C* **2007**, *111*, 10948–10954.
51. D. Chandra, T. Yokoi, T. Tatsumi, A. Bhaumik, *Chem. Mater.* **2007**, *19*, 5347–5354.
52. S. Huh, H.-T. Chen, J. W. Wiench, M. Pruski, V. S. Y. Lin, *Angew. Chem. Int. Ed.* **2005**, *44*, 1826–1830.
53. J. Liu, Q. Yang, M. P. Kapoor, N. Setoyama, S. Inagaki, J. Yang, L. Zhang, *J. Phys. Chem. B* **2005**, *109*, 12250–12256.
54. B. G. Trewyn, S. Giri, I. I. Slowing, V. S. Y. Lin, *Chem. Commun.* **2007**, 3236–3245.

55. I. I. Slowing, B. G. Trewyn, V. S. Y. Lin, *J. Am. Chem. Soc.* **2007**, *129*, 8845–8849.
56. J. Lu, E. Choi, F. Tamanoi, J. I. Zink, *Small* **2008**, *4*, 421–426.
57. N. Liu, D. R. Dunphy, P. Atanassov, S. D. Bunge, Z. Chen, G. P. López, T. J. Boyle, C. J. Brinker, *Nano Lett.* **2004**, *4*, 551–554.
58. K. C. Leung, T. D. Nguyen, J. F. Stoddart, J. I. Zink, *Chem. Mater.* **2006**, *18*, 5919–5928.
59. R. Hernandez, H. Tseng, J. W. Wong, J. F. Stoddart, J. I. Zink, *J. Am. Chem. Soc.* **2004**, *126*, 3370–3371.
60. K. Meada, T. Nishiyama, T. Yamazaki, T. Suzuki, T. Skei, *Chem. Lett.* **2006**, *35*, 736–737.
61. P. Sierocki, H. Mass, P. Dragit, G. Richardt, F. Vögtle, L. De Cola, F. A. M. Brouwer, J. I. Zink, *J. Phys. Chem. B* **2006**, *110*, 24390–24398.
62. S. Angelos, E. Choi, F. Vögtle, L. De Cola, J. I. Zink, *J. Phys. Chem. C* **2007**, *111*, 6589–6592.
63. N. K. Mal, M. Fujiwara, Y. Tanaka, T. Taguchi, M. Matsukata, *Chem. Mater.* **2003**, *15*, 3385–3394.
64. Y. Zhu, M. Fujiwara, *Angew. Chem. Int. Ed.* **2007**, *46*, 2241–2244.
65. A. Sellinger, P. M. Weiss, A. Nguyen, Y. Lu, R. A. Assink, W. Gong, C. J. Brinker, *Nature* **1998**, *394*, 256–260.
66. G. Garnweitner, B. Smarsly, R. Assink, D. R. Dunphy, C. Scullin, C. J. Brinker, *Langmuir* **2004**, *20*, 9811–9820.
67. B. Smarsly, G. Garnweitner, R. Assink, C. J. Brinker, *Prog. Org. Coat.* **2003**, *47*, 393–400.
68. Y. Lu, Y. Yang, A. Sellinger, M. Lu, J. Huang, H. Fan, R. Haddad, G. López, A. R. Burns, D. Y. Sasaki, J. Shelnutt, C. J. Brinker, *Nature* **2001**, *410*, 913–917.
69. Y. Yang, Y. Lu, M. Lu, J. Huang, R. Haddad, G. Xomeritakis, N. Liu, A. P. Malanoski, D. Sturmayr, H. Fan, D. Y. Sasaki, R. A. Assink, J. A. Shelnutt, F. van Swol, G. P. López, A. R. Burns, C. J. Brinker, *J. Am. Chem. Soc.* **2003**, *125*, 1269–1277.
70. Y. Li, S. Bhosale, T. Wang, Y. Zhang, H. Zhu, J. H. Fuhrhop, *Angew. Chem. Int. Ed.* **2003**, *42*, 3818–3821.
71. D. Franke, C. Egger, B. Smarsly, C. F. J. Faul, G. J. Tiddy, *Langmuir* **2005**, *21*, 2704–2712.
72. T. Q. Nguyen, J. Wu, S. H. Tolbert, B. J. Schwartz, *Adv. Mater.* **2001**, *13*, 609–611.
73. M. P. Kapoor, S. Inagaki, S. Ikeda, K. Kakiuchi, M. Suda, T. Shimada, *J. Am. Chem. Soc.* **2005**, *127*, 8174–8178.
74. M. P. Kapoor, M. Yanagi, Y. Kasama, T. Yokoyama, S. Inagaki, T. Shimada, H. Nanbu, L. R. Juneja, *J. Mater. Chem.* **2006**, *16*, 3305–3311.
75. M. Álvaro, M. Benitez, D. Das, H. García, E. Peris, *Chem. Mater.* **2005**, *17*, 4958–4964.
76. S. L. James, *Chem. Soc. Rev.* **2003**, *32*, 276–288.
77. G. Arrachart, C. Carcel, J. J. E. Moreau, G. Hartmeyer, B. Alonso, D. Massiot, G. Creff, J.-L. Bantignies, P. Dieudonne, M. Wong Chu Man, G. Althoff, F. Babonneau, C. Bonhomme, *J. Mater. Chem.* **2008**, *18*, 392–399.
78. H. K. Baca, C. Ashley, E. Carnes, D. Lopez, J. Flemming, D. Dunphy, S. Sing, Z. Chen, N. Liu, H. Fan, G. P. López, S. M. Brozik, M. Werner-Washburne, C. J. Brinker, *Science* **2006**, *313*, 337–341.
79. A. Patti, A. D. Mackie, F. R. Siperstein, *Langmuir* **2007**, *23*, 6771–6780.

Part Four

Biomimetic Chemistry

Chapter 19

Biomimetically Inspired Signaling

KNUT RURACK, RAMÓN MARTÍNEZ-MÁÑEZ,
FÉLIX SANCENÓN, AND ANA B. DESCALZO

19.1 INTRODUCTION

Living organisms created by nature over the course of evolution host complex assemblies of organic, inorganic, and mixed or hybrid organic–inorganic structures. These structures are the basis for a multitude of active functions, ranging from vital cellular processes to, for instance, the autonomous movement of creatures. In microscopic dimensions, such structures constitute supra- or macromolecular as well as biopolymeric assemblies. These entities in turn are built from very simple molecules such as amino acids, sugars, nucleobases, and certain inorganic materials such as silica, calcium carbonate, or iron oxide. Given the striking diversity, efficiency, and success of this biomaterials-based chemistry, it is not surprising that chemists have attempted to

The Supramolecular Chemistry of Organic–Inorganic Hybrid Materials. Edited by Knut Rurack and Ramón Martínez-Máñez

use concepts and strategies from nature to develop "bottom-up" approaches for complex nanoscopic ensembles that are able to mimic biological structures and functions.[1] Otto H. Schmitt coined the term *biomimetics* in 1969,[2] for the science that studies and imitates nature's methods, mechanisms, and processes.[3,4] Apparently, nature's diversity in this respect is reflected by a large variety of different chemical fields that contribute to biomimetic research. For example, materials chemists have used molecular and supramolecular template-directed processes to prepare materials that simulate solid biological structures with increasingly complex hierarchical organization.[5] Work in other directions has dealt with the design and development of biosynthetic inorganic chemistry compounds,[6] biomaterials for tissue engineering and related clinical applications,[7] the synthesis and characterization of biologically inspired solids,[8] or biomimetic strategies for molecular recognition in catalysis.[9] A further level of sophistication in the sense of supramolecular chemistry, however, can be reached if such usually complex and nanoscopically structured solids are functionalized with organic or bio-organic molecules that can bind to, signal, or process chemical species, resulting in a new generation of organic–inorganic hybrid materials.[10,11] In other words, such systems are particularly attractive when they incorporate addressable active units that can be triggered by simple external stimuli to exert a directed function at will.

Naturally, the supramolecular chemistry of biologically inspired or biomimetic hybrid materials itself is a rather diverse field as reflected by several other chapters of this book.[12] In this chapter, we will thus focus our attention on certain examples of hybrid materials with three-dimensional[13] or quasi-3D structures that have been developed in the context of chemical recognition and signaling. The term signaling here refers to a physico-chemical process that is generated in response to the recognition of, for instance, a particular target species and that can be detected by measuring the change of a certain parameter such as a current or a light intensity at a certain wavelength. This facile concept of assembling tailored 3D inorganic scaffolds and rather simple (functional) organic groups in a modular fashion according to bio-inspired strategies allows the creation of hybrid sensing ensembles for analytes for which selectivity is hard to achieve with conventional methods or for the indication of analytes by signal amplification strategies. In fact, the examples discussed below are far from the classical "binding site–signaling unit" approach that is followed for most of the molecular chemosensors described in the literature. In this section, we pay special attention to cases where the combination of traditional supramolecular chemistry tools such as the topology or *shape* (or the complementarity of shape) and the *arrangement* of (often very different and multiple) functional groups with preorganized solid structures results in new sensing paradigms for enhanced performance. This chapter will not attempt comprehensiveness, but will focus on selected examples that show remarkable supramolecular synergic effects in relation to signaling events.

19.2 BINDING POCKETS

One of the ideas of utilizing hybrid frameworks for biomimetic signaling is inspired by the way nature deals with such issues; in particular, how proteins bind substrates.

Proteins are basic building blocks in living organisms able to bind a single substrate or an entire class of substrates tightly by using rather simple and conventional weak forces (e.g., hydrogen bonds, π stacking, etc.) in an aqueous environment in which these binding forces usually suffer from strong competition by water molecules, hydrating host, and guest. The key step proteins use to achieve this tight binding in water is to extract the substrate into a hydrophobic pocket where the active site–substrate complex is shielded. Since such active sites of proteins are usually embedded in a complex yet flexible (super)structure, upon entering of the substrate, the latter "induces the fit": the binding site is reoriented, the pocket is closed and remaining water is squeezed out.

The concept of mimicking hydrophobic pockets has been utilized in coordination chemistry for the enhancement of selectivity of supramolecular receptors and for the development of artificial metalloenzyme mimics.[14] Moreover, the research field of molecular imprinting approximates this concept from an organic chemistry point of view, mainly through molecularly imprinted polymers.[15,16] In addition, imprinted functionalized silica has become an attractive alternative to entirely organic systems in recent years; see Chapter 20 of this book for a detailed account. On the other hand, the principle of operation of enzymatic chemistry has inspired the design of hybrid systems in which the supramolecular recognition chemistry is combined with a protective action in preorganized 3D nanoscopic porous solids in analogy to the formation of the shielded enzymatic host–guest complex (Fig. 19.1). For these purposes, two different organic groups are commonly employed, recognition centers and additional groups that fine-tune the pore's polarity. Based on the facile handling of signal read-out with optical measurement equipment, such ensembles mostly rely on a suitable spectrophoto- or fluorometric event to indicate the presence of a guest. These bio-inspired signaling systems show advanced features because in addition to conventional recognition at the binding site there is a further supramolecular control,

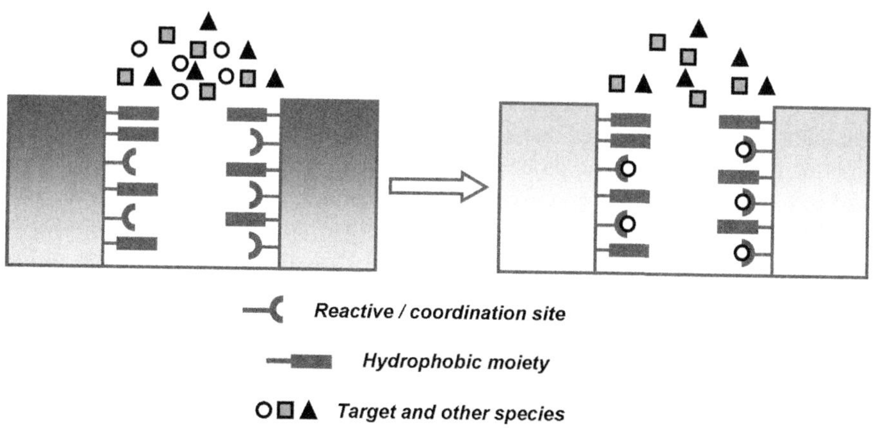

Figure 19.1 Hydrophobic binding pockets in mesoporous materials. This biomimetic approach takes advantage from both the extraction of the guest from water by finely tuned polarity and size, and the selective reaction of the guest with the binding site.

governed by the polarity, pore size, and structural features of the nanoscopic porous hosts. The results are often remarkably enhanced sensing responses and selectivities with respect to comparable indicator or probe molecules.

The first example shown here is based on the use of an MCM-41-type silica derivative as the porous 3D host that was organically modified to selectively recognize the anions of fatty acids, fatty carboxylates, in aqueous media. In a first step, the solid mesoporous scaffolding was functionalized with a urea-phenoxazinone derivative. This chromophore has two functions. Its urea moiety acts as anion receptor via hydrogen bond-mediated coordination of the carboxylate head group of the fatty acid anions. Second, the dye reports the binding event through shifts in the phenoxazine's absorption spectrum and a change in fluorescence intensity. These signal changes are a consequence of the modulated electron distribution in the ground and excited states of the phenoxazinone–fatty carboxylate complex. In a second synthetic procedure, the inner surface of the MCM-41 support was further functionalized with trimethylsilane. This chemistry transforms the hydrophilic inner walls of the silica skeleton into hydrophobic ones, containing also the indicator molecules for carboxylate recognition; hydrophilic voids are converted into hydrophobic ones (Fig. 19.2).[17] This solid **M1** displays a selective response to fatty (long-chain) carboxylates, manifested in a bathochromic shift of the absorption band and an enhancement of the fluorescence. The system is highly selective. Short-chain carboxylates, inorganic cations and anions as well as biological species such as triglycerides, cholesterol, bile acids, or organic phosphates gave only negligible responses. The solid thus only recognizes analytes that are lipophilic enough to enter the hydrophobic pores and that carry a head group that is capable of binding to the urea receptor and triggering the optical signals (Fig. 19.2). For instance, glycerophospholipids are lipophilic enough to enter the pores but are unable to induce the signaling reaction most likely due to poor coordination with the urea binding sites. As stated above, one of the most appealing characteristics of these biomimetic hybrid materials is that they usually display enhanced responses in comparison to their parent molecules, the neat molecular indicators. This fact was clearly observed here as well. The molecular urea-phenoxazinone probe **P1** (Fig. 19.2) did not respond to any of the above-mentioned guests in water, but bound only unspecifically to carboxylates and $H_2PO_4^-$ in polar organic solvents. The authors also demonstrated that a hybrid material **M2** that was monofunctionalized with the urea-phenoxazinone probe but was not hydrophobized with trimethylsilane moieties did not display any sensing features in water, presumably due to the fact that hydrophobic molecules do not readily diffuse into the pores. Furthermore, indication of small inorganic species such as $H_2PO_4^-$ is hampered because in the solvent-filled pores, water hydrates the anion so strongly that it cannot form hydrogen bonds with the urea moiety. To understand the operational mechanism, it is important to know that, whereas **M2** is readily suspended in water, hydrophobic **M1** is not suspended, that is, floats on the water surface. Only in the presence of the target analytes, suspension occurs. Its performance is thus based on an extraction of the fatty carboxylates into the pores. Moreover, although counter-cations and water molecules are co-extracted, the water content at the hydrophobic layer of the inner wall is presumably reduced so that hydrogen bonding between carboxylate and urea can occur (Fig. 19.2).

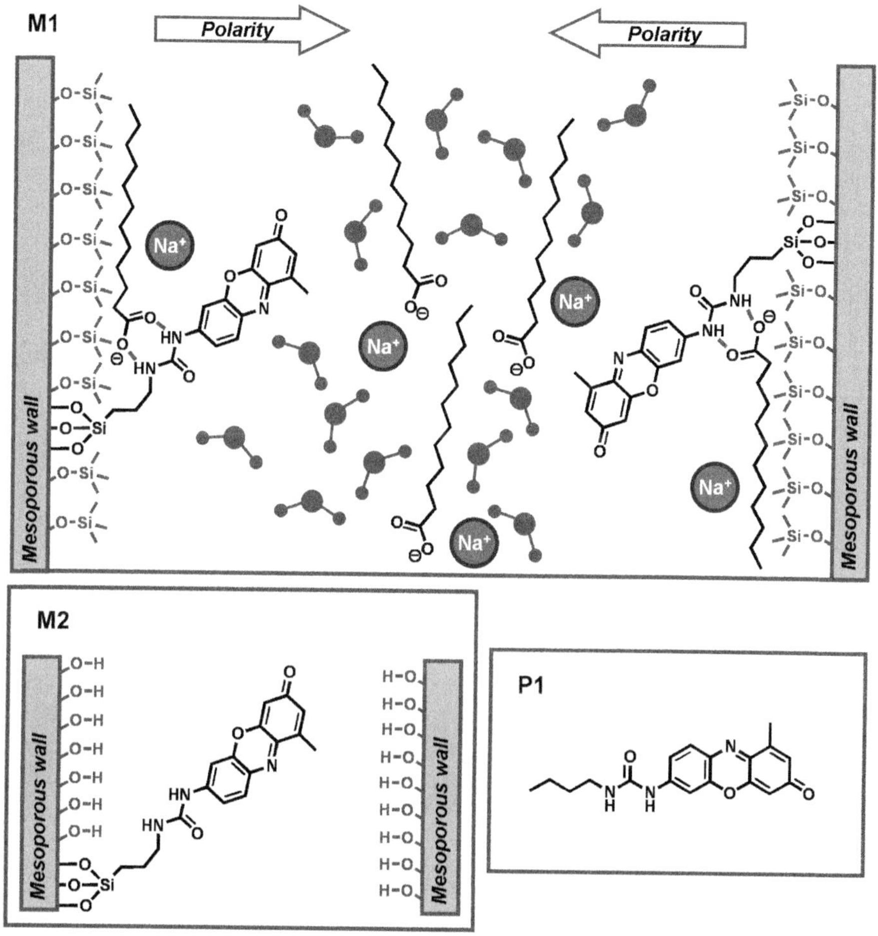

Figure 19.2 Schematic model of the biomimetic material **M1** for the sensing of long-chain carboxylates. Hydrophobic forces bind the tail to the wall, enabling hydrogen bonding interactions of the analyte's head group with the urea group of the phenoxazinone dye. The result is a color and fluorescence modulation. Lower left: inactive nonhydrophobized material **M2**. Lower right: inactive molecular probe **P1**.

The previous example took inspiration from earlier work of Lin et al.[18] In their seminal work, Lin and coworkers functionalized the inner pores of MCM-41 with *o*-phthalic hemithioacetal moieties that are able to react with amines to produce a highly fluorescent isoindole derivative. In order to enhance selectivity in the sense we have discussed above, the solids were also hydrophobized with different groups such as propyl, phenyl, and pentafluorophenyl in a second step. Interestingly, some of these solids displayed a remarkably selective and differentiable response to dopamine versus the less lipophilic glucosamine. The authors also demonstrated that this selectivity was not observed when amorphous (nonporous) silica functionalized with the same organic groups was used, stressing the importance of the 3D mesoporous structure versus locally flat 2D systems for the sensing protocol. In further studies, Lin's group modulated the approach to gain selectivity for dopamine versus tyrosine and glutamic acid by coating the mesoporous particles with poly(lactic acid).[19] The coating acts as a molecular gatekeeper that is able to regulate the penetration of certain amines into the nanoscopic pores by coulombic forces. The amines able to penetrate the poly(lactic acid) are then signaled by fluorescence changes similar to those described before. A more detailed discussion of this gated chemistry of hybrid silica nanomaterials can be found in Chapter 16 of this book.

Bis-functionalization of the pores of mesoporous silicas with signaling units and hydrophobic groups has also been employed successfully for the detection of other analytes. For instance, the discrimination of certain guests within a class of amines has been achieved with siliceous solids that contain an anilinopyrylium derivative as reactive indicator. In order to control selectivity, the solid was additionally trimethylsilylated, rendering the pores highly lipophilic (Fig. 19.3).[20] Anilinopyrylium dyes are known to react with primary amines to the corresponding pyridinium salts with a concomitant color change from magenta to bright yellow. This reaction is not specific and the chromogenic pyrylium dye reacts with a whole family of primary amines in solution. However, when placed in the hydrophobized nanoscopic pores, a very selective response is obtained. A family of linear primary aliphatic amines was tested from which only relatively short but sufficiently hydrophobic medium-chain amines (such as *n*-octylamine) induced a chromogenic reaction in water. Hydrophilic (e.g., *n*-propylamine) or long-chain amines (e.g., *n*-dodecylamine) remained silent. Apparently, short amines are too hydrophilic to enter into the mesopores and induce the chromogenic reaction. Moreover, long-chain amines appear to close the pores to such an extent that diffusion of following analytes is hampered by steric crowding. The first amines that reach the porous material **M3** react primarily with the pyrylium groups close to the pore openings so

M3

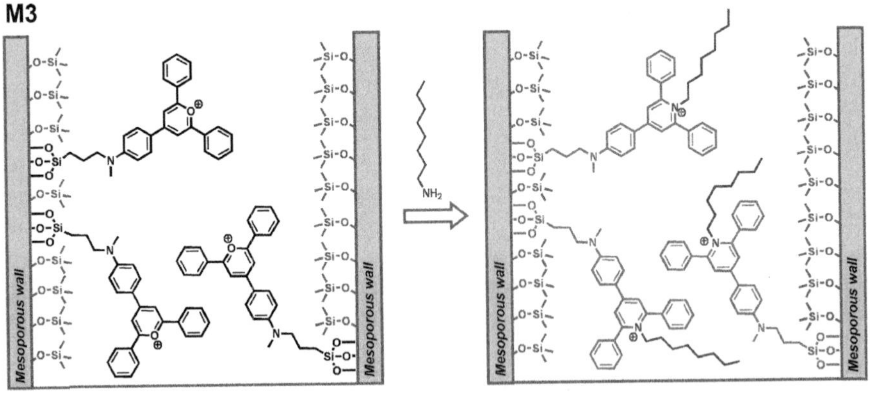

Figure 19.3 Schematic model of biomimetic material **M3** for amine signaling. Changes in color are due to the reaction between the pyrylium ring and the primary amine. Short amines are too hydrophilic and do not enter into the hydrophobic pores, whereas long-chain amines tend to clog in the pore openings.

that their long tails literally clog the pore openings. This interpretation was supported by the fact that simple 2D flat surfaces (such as silica gel or fumed silica lacking the mesopores) functionalized with the chromogenic pyrylium dye and additionally passivated with trimethylsilane reacted with both medium and long-chain primary amines in the same fashion.

This concept of reactive amine sensing appears to be general and has also been realized with MCM-41-type solids containing a dicyanomethylene-2-chloro-3-amino-indene[21] or a styrylpyrylium dye[22] as signaling units and trimethylsilyl-passivated walls. The latter hybrid material absorbs and emits in the vis/NIR spectral region and primary amines are indicated by a change in color from blue to red upon reaction. As an enhanced feature, the hydrophobized mesoporous solid containing the styrylpyrylium sensing groups was able to react selectively with biogenic amines such as histamine from fish extracts and remained silent in the presence of amino acids or long-chain fatty amines. Amine discrimination can also be tuned by changing the size of the pore openings. For instance, anilinopyrylium derivatives have also been anchored to the inner walls of the pores of zeolite Beta. Whereas the typical MCM-41-type mesopore has a diameter of approximately 2 to 3 nm, access to the pore voids of the 3D channel system of zeolite Beta is limited by a window of 6.8 Å in diameter. This hybrid material also showed size discrimination with only amines having one longitudinal section (at least in two directions) smaller than the openings of zeolite Beta being able to enter into the pores and induce color modulations. A discriminatory effect of polarity was also observed, that is, due to the hydrophobic environment of the inner walls of the zeolite's pores, color changes were only found for lipophilic amines.[23]

These results demonstrate that selective molecular recognition can be achieved not only by synthesizing complex hosts based on recognition in the traditional supramolecular sense, but also by using nanoscopic preorganized solids containing a certain pore aperture and a suitable binding site in an appropriate (usually hydrophobic) environment. These structures bear a certain resemblance to binding pockets of more complex biological systems. However, they do not constitute enzyme mimics in a classical, but in a more figurative sense: these hybrids are based on rigid inorganic supports so that a rearrangement according to an "induced fit" is not possible to shield the host–guest interaction from water molecules. Instead, principles from extraction have to be invoked. Within a biomimetic perspective, the confinement of a signaling group in a modified, periodically ordered solid structure harbors the following advantages: discrimination can be accomplished through (1) size exclusion phenomena, (2) hydrophilic/lipophilic partitioning, and (3) the selective reactivity of the guest with the binding site. These sensing benefits are especially evident when the responses of the bio-inspired hybrids are compared with those of the analogous molecular indicators or those of flat non-(meso)porous supports (see above).

An interesting aspect related to the design of these ensembles is the fine-tuning of the polarity of the inner pores. In this respect, Inumaru et al. reported the adsorption behavior of alkylphenols and alkylanilines by MCM-41-type solids doped with different degrees of Al^{3+} and equipped with alkyl chains of different length grafted onto the surface.[24,25] The authors found two well defined trends: (1) lipophilic guests were preferentially adsorbed when the chain length of the anchored functional group increases and (2) alkylamines that contain a hydrophilic head group and therefore are able to give stronger hydrogen bonding and/or weak acid–base interactions with the inorganic walls are better adsorbed than alkylphenols.

Besides controlling the hydrophilicity and size of mesoporous structures, other reports suggest that enhanced signaling features can also be achieved by anchoring a large number of coordination and/or signaling units in the confined space of pores. Such a preorganization of receptors on the surface of a mesopore reduces the conformational flexibility (entropic contributions) and increases the effective concentration of binding sites at the surface, creating a self-assembled monolayer (SAM)-like receptor environment. The latter can result in improved recognition characteristics, presumably also enhanced by the concave geometry of the mesopores. Translated into the framework of this chapter, the lack of flexibility of the artificial systems compared with the active centers of enzymes, in which a limited number of binding sites is achieving optimum complexation of the guest because of a rearrangement of the entire neighborhood, the binding "pocket," is accounted for by offering an excess of receptor units and further reducing the entropic freedom of the resulting complexes. Such a chemical amplification by coordination has also been observed for functionalized nanoparticles and in coordinating dendrimers for which the so-called "positive dendritic effect" refers to the ability of dendrimers to show enhanced recognition/coordination of guests as the generation of the dendrimer increases.[26] The strategy is illustrated by the following example of a mesoporous silica support **M4** that is functionalized with a large number of a single type of chemosensor molecule, a simple alkylaminoanthracene, for the fluorogenic sensing of ATP.[27,28] The anchored unit contains a secondary amino group that is protonated at acidic pH and that acts as the coordination site for the phosphate residues of ATP. Additionally, the amino group is attached to an anthracene chromophore that constitutes the signaling subunit and also provides additional π-stacking interactions with the adenine moiety of the target (Fig. 19.4). The addition of ATP to acidic aqueous suspensions of the solid resulted in a remarkable fluorescence quenching. The detection limit for ATP was found to be two orders of magnitude lower and the association constant for ATP–**M4** interaction two orders of magnitude larger, respectively, than those of the corresponding probe **P2** in solution (Fig. 19.4). The enhanced response of **M4** to ATP reveals a cooperative effect that is related to an effective concentration enhancement favored by the proximity of binding/indication sites on the MCM-41 solid. Again, as discussed for several of the previous examples, the mesoporous hybrid also displays remarkably enhanced sensing features when compared to a chemically similar solid that does not contain ordered mesopores but only a locally flat surface such as fumed silica (**M5**, Fig. 19.4). Apparently, only the dense coating of alkylaminoanthracene groups on the concave inner surface of the mesoporous support material leads to a confinement effect that facilitates the binding process. Such an improved performance is not possible when using 2D materials with rather flat surfaces, despite the fact that their surface can locally also be very inhomogeneous with flat, convex and concave microdomains. A recent study suggests that besides ordered porous structures and tailored pore diameter and functionalization, the size of the particulate material as such

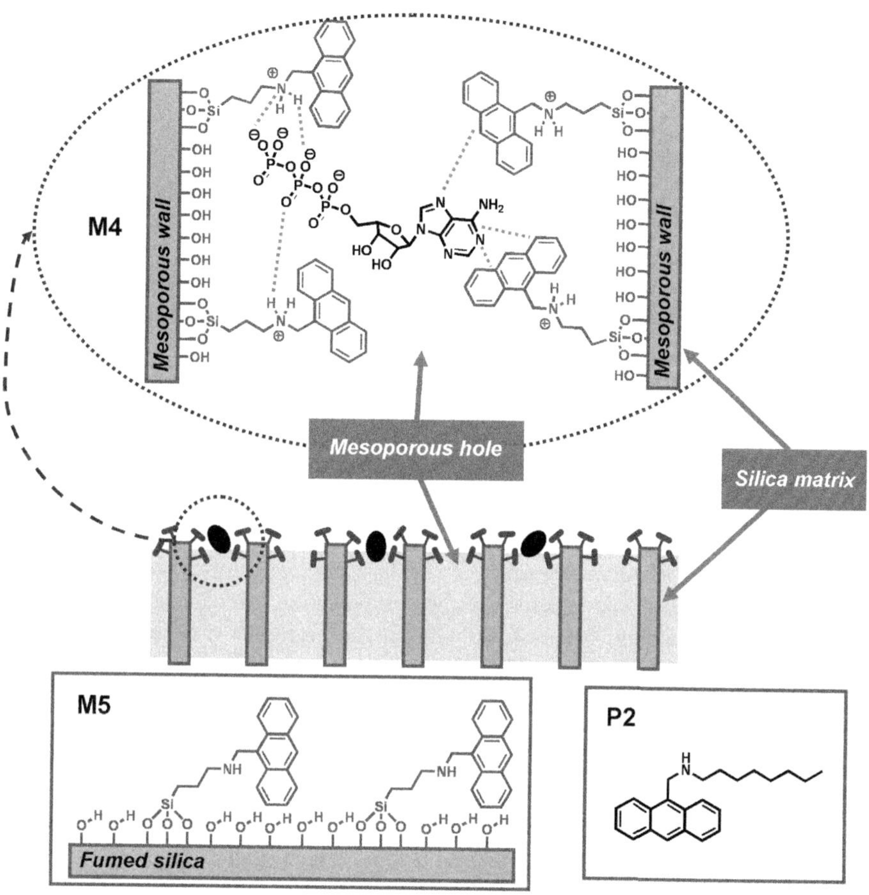

Figure 19.4 Schematic view of the surface of the ATP sensing solid **M4** showing the possible coordination between the aminoanthracene groups and the ATP anion. Lower left: weakly active nonporous material **M5**. Lower right: inactive molecular probe **P2**.

also plays an important role for optimum performance, with nanometric particles outcompeting micrometric ones in that case.[29] Moreover, these results do not only seem to be valid for larger (organic) species but also for small inorganic metal ions as analytes.

The crucial influence of the density of the signaling units on the surface of a hybrid material for a detection method that can principally suffer from auto-induced modulations such as fluorometry is also evident from the results in References 27 and 28. An increase in the number of alkylaminoanthracene units on the MCM-41 surface from a mean distance between two anthracenes of 33 or 23 nm to 10 nm led to the appearance of a significant amount of excimer fluorescence, generated from an anthracene in the excited state and a neighboring one in the ground state in the absence of an analyte. Similar observations have been made for hybrid optical materials that contain fluorophores but do not exhibit any other explicit function; see, for example, Ganschow et al.[30]

19.3 METALLOREGULATOR MIMICS

The regulation of metal ion uptake, storage, and distribution is an essential task for organisms. Many transition metal ions are essential for vital bodily or cell functions, for example iron, zinc, copper, or manganese ions. However, at elevated concentrations these metal ions can exert toxic effects. Regulatory cycles are thus required, and cells usually possess a series of proteins for this task, the most important ones being perhaps the metallothioneins.[31] These proteins are rich in cysteine and can incorporate a considerably large amount of metals, especially when forming higher-order sulfur-based protein-metal clusters. With respect to essential elements, metallothioneins are the key players in the trace metals' homeostasis. On the other hand, these proteins have the important function to sequester nonessential and basically toxic trace metals such as mercury or cadmium. Whereas metallothioneins are primarily responsible for metal storage, another class of ubiquitous metalloproteins, the ferric uptake regulator (Fur) and related proteins, are especially involved in regulation.[32] Regulation here means that these proteins first of all have to sense the concentration of a metal (ion) before they start to activate either release of the species from a reservoir or sequestration of an excess of a metal. A third strategy comes into play when taking a look at the detoxification mechanisms of biological systems. Here, organisms often utilize single amino acids or small linear peptides, generally rich in histidine, cysteine, glutathione, and imidazole groups, to create nanoscale bio-inorganic clusters of reduced metals (Au^0, Ag^0, etc.), thus withdrawing these harmful species by biomineralization.[33,34] In analogy to the model of the induced fit as sketched out above, these metalloregulatory examples have in common that rather low molecular weight proteins occur in a predominantly random and disordered structure in the metal-free state. However, upon complexation of metal ions, these proteins fold into a well-defined spatial structure, efficiently fulfilling their task.

Within a biomimetic framework, metalloregulatory proteins are important for two different applications. First, artificial analogs of the biomacromolecular metal reservoirs could function as excellent sorption materials in remediation processes. Second, nanoscopic ensembles that possess a large number of binding sites can constitute powerful sensing materials for metal ions when coupled with a suitable signaling mechanism. Besides attempts to synthesize complex ferrichrome or siderophore analogs that allow specific binding of, for instance, Fe^{3+} at higher stoichiometries,[35] especially the use of hybrid mesoporous silicas, partly containing additional chromogenic or fluorogenic groups, for metal ion sorption or sensing has received considerable attention. Again, as discussed above for the artificial analogs of binding pockets, most of the attempts reported so far try to compensate for the unique target-induced conformational changes of the flexible biological system by endowing suitable rigid nanoscopic porous inorganic networks with (several and/or large amounts of) tailored organic functions to accomplish the desired task. Since this chapter is focused on biomimetic signaling, we will not discuss organic–inorganic hybrid materials for sorption purposes,[36–42] but will devote our attention to sensing materials.

Most of the examples reported to date involve the use of MCM-41, MCM-48, SBA-15, or silica nanotube matrices for which certain selective indicator dyes are covalently anchored to the pore walls. Following this approach, sensing materials for the detection of Hg^{2+},[43–46] Cu^{2+},[47–49] Fe^{3+},[50] and Zn^{2+}[47,51] have been described. The synthesis of indicator dye derivatives with reactive groups that allow covalent attachment to silica walls, however, is sometimes rather complex. Balaji et al.[52,53] and El-Safty et al.[54] thus used a different strategy for the preparation of a family of optical sensing materials for metal ion detection. They anchored the signaling units electrostatically to the porous support. In a first example, mesoporous SBA-15 possessing pores of 5 to 30 nm in diameter was used as support and was functionalized with an organic spacer carrying a trimethylammonium end group. In a second step, the 4-(2-pyridylazo)-resorcinol (PAR) probe molecules were electrostatically anchored to the pores (**M6**, Fig. 19.5). Coordination of Cd^{2+} to the PAR units resulted in a color change from yellowish orange to violet. The system is not selective but the authors demonstrated that the cross sensitivity against Co^{2+}, Ni^{2+}, Cu^{2+}, and Fe^{3+} can be minimized by employing masking reagents.

In a further work, new nanostructured cage materials containing several colorimetric reactants for the selective chromogenic detection of selected metal ions were prepared (Fig. 19.6). The authors used the nanosized cavities contained in cubic *Fm3m* nanocage monoliths and incorporated the dyes again through electrostatic and hydrogen bonding interactions to the silica surface.[53] For instance, *N*-trimethoxysilylpropyl-*N*,*N*,*N*-trimethylammonium groups were first attached to the cavity walls and cationic dyes such as pyrogallol red (PR, **M7**) or tetraphenylporphine tetrasulfonic acid (TPPS, **M8**) were electrostatically fixed subsequently. In a related procedure, cationic dyes such as α,β,γ,δ-tetrakis(1-methylpyridinium-4-yl)porphine (TMPyP, **M9**) were attached in an electrostatic manner by first grafting SO_3^--terminated spacers onto the surface of the nanocages. Attachment of the neutral dye dithizone could be accomplished through hydrogen bonding interactions with the silanol groups at the surface (**M10**, Fig. 19.6). It was found that these systems could be used for the selective and

M6

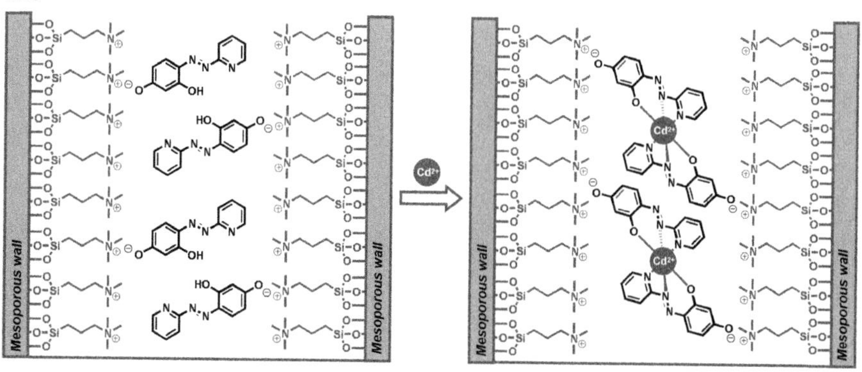

Figure 19.5 SBA-15 derivative **M6** containing a dye anchored through electrostatic interactions in the pores for the colorimetric recognition of Cd^{2+}.

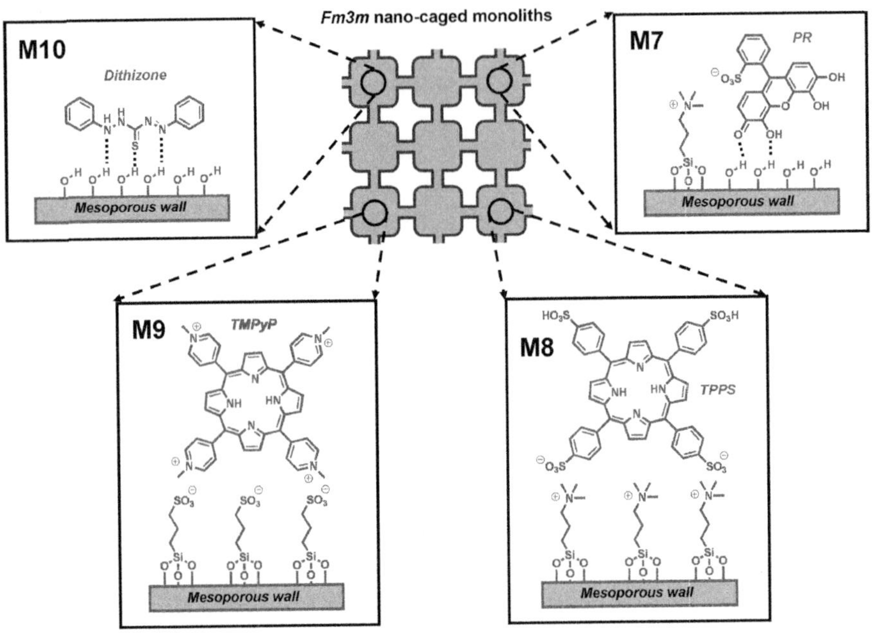

Figure 19.6 Nanocaged monoliths **M7** to **M10** containing selected dyes for the colorimetric signaling of various metal ions.

sensitive detection of Sb^{3+} (with **M7**), Hg^{2+} (**M8**), Cd^{2+} (**M9**), and Pb^{2+} (**M10**) at subnanomolar levels (Fig. 19.7). In a later work, the same authors reported the effect of the pore's geometry and shape and the particles' morphology on the colorimetric detection of Sb^{3+}.[55] They prepared similar nanocage monoliths and used PR as the dye trapped in the nano-sized cavities.

Different sensor materials for Cd^{2+} detection in water were prepared by the same authors using TMPyP, diphenlcarbazide (DPC), 4-*n*-dodecyl-6-(2-thiazolylazo) resorcinol (DTAR) and 4-*n*-dodecyl-6-(2-pyridylazo)phenol (DPAR) as reporter modules and a mesoporous solid as support. Whereas TMPyP was immobilized in a similar fashion as described above, DTAR and DPAR were anchored simply by van der Waals and hydrogen bonding interactions with the abundant hydroxyl groups of the support's surface. Additionally, the long alkyl chains on these dyes offer adequate hydrophobicity to prevent leaching of the probes in aqueous solution. DPC was finally immobilized with the aid of the cationic surfactant dilauryldimethylammonium bromide (DDAB) that is assumed to form a bilayer type of coating on the negatively charged surface of the inorganic support and thus facilitates efficient adsorption of the dye DPC onto the bilayer-coated pores.[55] Similar mesocaged nanosensors have been prepared for Bi^{3+} signaling, employing a similar inorganic matrix and diphenylthiocarbazone (DZ) as suitable dye.[56] Recent advances in the modular building-block design of such hybrid sensor materials led to the development of a family of nanoscale hybrid sensor materials that can be used for the indication of multiple metal ions.[57]

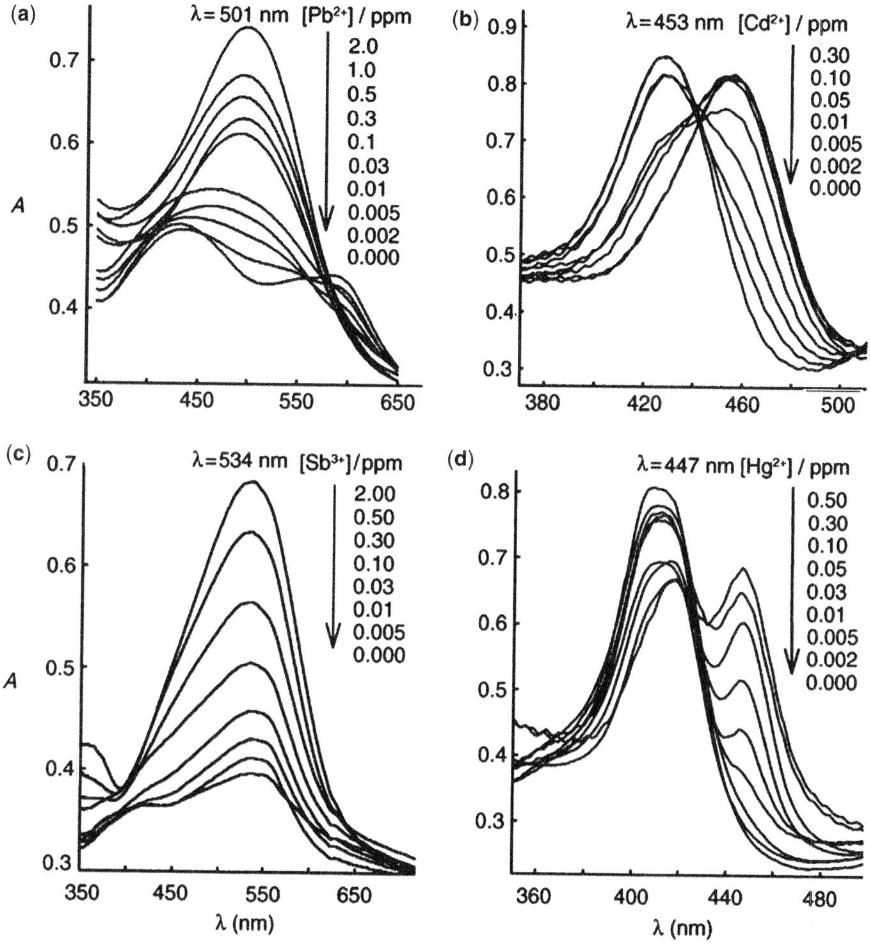

Figure 19.7 Concentration-dependent changes of UV-visible absorption spectra of sensors **M10** (a), **M9** (b), **M7** (c), and **M8** (d) after addition of Pb^{2+}, Cd^{2+}, Sb^{3+}, and Hg^{2+}, respectively.[53] (Reprinted with permission from T. Balaji et al., *Angew. Chem. Int. Ed.*, **2006**, *45*, 7202–7208. Copyright Wiley-VCH Verlag GmbH & Co. KGaA.)

The last two examples in this section refer to current attempts to merge the functions of metal ion storage and sensing. The aim is to have a material in hand that can still report the concentration of a harmful metal ion yet is also able to sequester the toxin even when present at higher concentrations. In a biological sense, such a system mimics a typical regulator protein, noticing for instance the amount of ingested Cd^{2+} or Hg^{2+}, while also constituting the active component of a detoxification or biomineralization mechanism. From a practical point of view, the implementation of adequate sorption functions is the key task for such biomimetic hybrids. In their attempt to develop a potent sorption and sensing material for Hg^{2+}, Ros-Lis et al.[29] synthesized a mesoporous silica support with a high degree of covalently attached thiol groups by

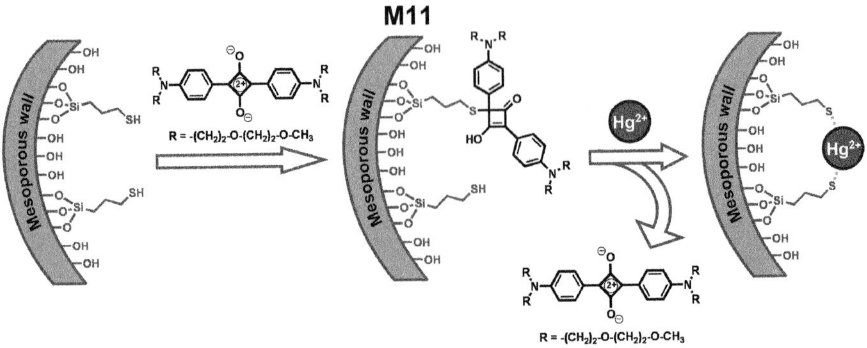

Figure 19.8 Route for the preparation of the dual sorption and sensing material **M11** (first step) and analytical as well as sequestration reaction with Hg^{2+} (second step).

reacting mercaptopropyltriethoxysilane together with triethanolamine, tetraethyl orthosilicate, and cetyltrimethylammonum bromide in a first step to yield an MCM-41-type support with a thiol-covered surface (Fig. 19.8). The resulting hybrid material was then reacted with a squaraine (SQ) derivative in basic solution. Bleaching of the solution that contained the dye indicated that, in accordance with a previous report on a chemodosimetrical SQ-thiol sensing system for Hg^{2+},[58] nucleophilic attack of the deprotonated thiol led to an addition reaction, producing a 2,4-bis(4-dialkylaminophenyl)-3-hydroxy-4-alkylsulfanylcyclobut-2-enone (APC) derivative that is covalently anchored to the silica matrix (**M11**). Signaling is then accomplished by reaction of the thiophilic target ion with the sulfur atoms, liberating and restoring the chromophore, which is released into the solution and can be quantified by absorption or fluorescence measurements. Important for the design of a potent sensing and sorption material in this case was the ordered mesoporous structure of the support with a suitable pore diameter, the optimum incorporation of propanethiol moieties into the pores, and the nanometric dimensions of the individual particles. Porous disordered or even nonporous silica materials showed significantly inferior performance, as did mesoporous silica particles of micrometric size. Optimum adjustment of the thiol group coverage of the material's pore surface is necessary to achieve best adsorption of Hg^{2+} per S atom and a low limit of detection, which is only obtained at certain thiol-APC ratios.

In a second example of a dual sensing and sorption hybrid material published recently by Jung's group, the architecture of the system was simpler.[59] These researchers only used a single organic function, a reactive derivative of the classic fluorescent ligand phenanthroline, and attached it to the surface of porous silica supports. Jung et al. targeted Cu^{2+} and showed that these hybrids were promising in discriminating Cu^{2+} against a series of other heavy and transition metal ions.

19.4 ION CHANNEL SENSORS

The processing of substrates in the binding pockets of enzymes or the storage of metal ions in reservoir proteins are perhaps the most obvious biological processes that chemists try to adopt or mimic for sensing purposes. However, since the middle of the 1980s, another very interesting and multifaceted biological processing system inspired chemists to develop artificial, that is, biomimetic analogs: ion channels. Ion channels are ubiquitously found in cell or organelle membranes and are important pathways for inter- and intracellular signaling and transport.[60–63] Their architecture basically consists of channel proteins, which are mostly symmetric assemblies organized around a central aqueous pore, immersed in a biological membrane. Many attempts have been made in the past 20 years to exploit their basic principle of operation for technological applications, ranging from the use of (modified) biological channels[64] or protein pores[65] via natural ionophores such as, for example, valinomycin in conjunction with biomimetic membranes[66,67] or under selected artificial conditions[68] to the combination of artificial ionophores with lipid membranes[69–71] and the development of entirely artificial systems such as ionophore-polymer hybrids[72] or ion-imprinted polymers.[73]

Traditionally, ion channel sensors (ICSs) are electrochemical sensors that mimic biological ion channels by employing receptor-modified electrodes, with the organic molecules usually being arranged as SAMs on the electrode surface.[74,75] In general, binding of analytes to the receptors at the electrode surface controls the reduction or oxidation rate of redox-active probes, commonly ions or small molecules. This control is either based on physical exclusion or on electrostatic attraction or repulsion between receptor and probe, which is modulated in the presence of the analyte. One of the advantages of ion channel sensors is the fact that signal amplification is an inherent feature: in the ideal case, a single analyte can close a channel through which many probes can diffuse in the open state. The net effect is an analyte-controlled switching effect.[76,77] In a representative example, gold electrodes were modified with SAMs of thioctic acid (**M12**) for the indirect detection of protamine down to 0.5 $\mu g\ mL^{-1}$ or 2.0 $\mu g\ mL^{-1}$ when $[Ru(NH_3)_6]^{3+}$ or $[Fe(CN)_6]^{3-}$ were used as redox probes.[78] Protamine (a polycation) binds to the thioctic acid SAMs and controls the reduction rate at the electrode surface by electrostatic repulsion between the guest-covered SAM and the probe (Fig. 19.9). Umezawa and coworkers found that blood electrolytes do not interfere at physiological concentrations and that the electrodes can be easily recycled by washing with acidic KCl.[78] Moreover, one of the advantages of such ICSs is that modulation of the scan rate usually allows optimization of either the LOD (limit of detection) or the dynamic range of detection. Another important issue for such electrochemical sensors is that the length of the thioctic acid derivative used for the SAM plays a role because it determines the topology of the SAM: receptors with long chains tend to form tight monolayers that can completely block the electrode surface. On the other hand, molecules with shorter chains are better suited, provided that the resulting SAMs are sufficiently stable. Depending on the probe chosen, the effect of protamine is as follows. In the ICS free state, $[Fe(CN)_6]^{3-}$ is repelled by the SAM. Upon binding of the polycation, the negative excess charge on the monolayer is reduced, leading eventually to an excess positive charge at the electrode surface at higher protamine concentrations. Once the negative charges are compensated for, $[Fe(CN)_6]^{3-}$ can easily access the electrode surface and is reversibly reduced (Fig. 19.9). However, these researchers preferred the cationic $[Ru(NH_3)_6]^{3+}$ probe because the anionic probe showed some interferences due to interaction with the polycationic analyte, resulting in a steady increase in the reduction current at

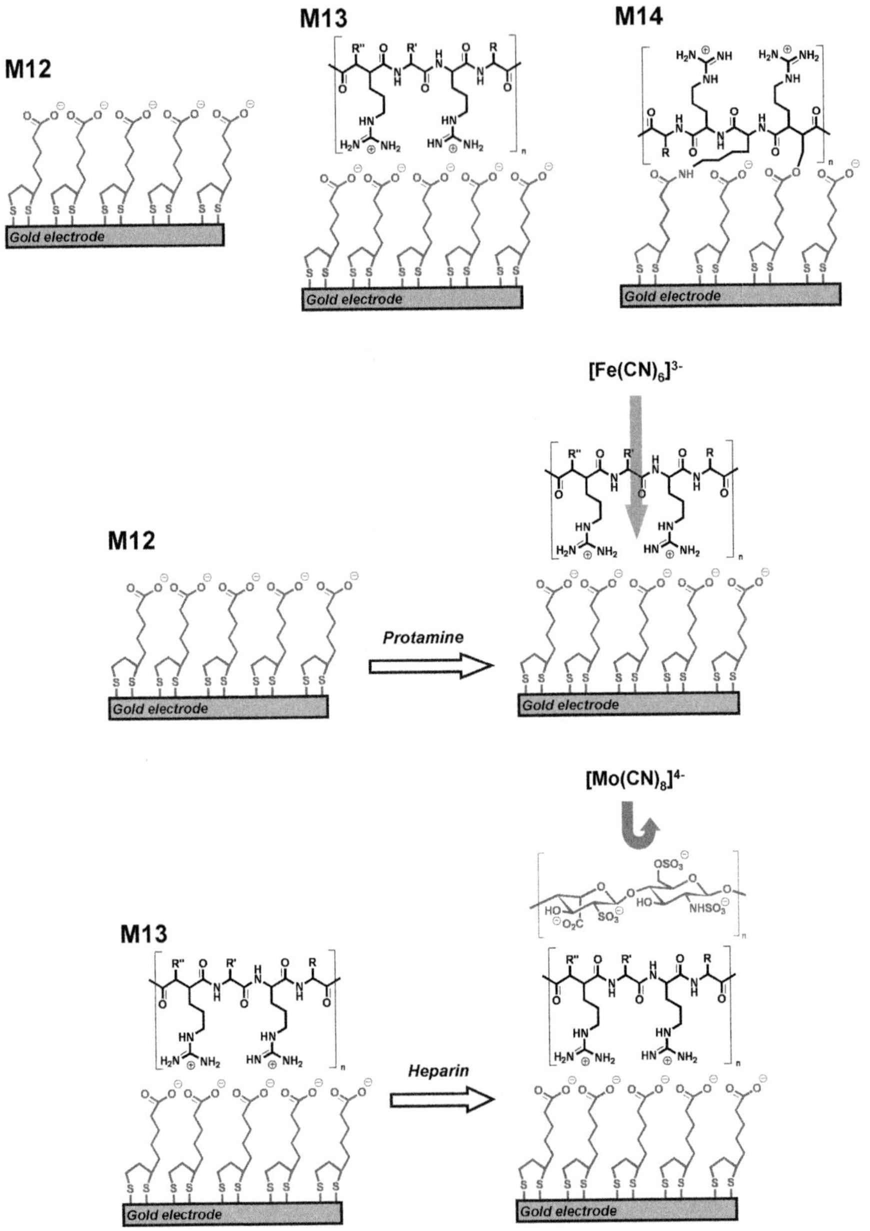

Figure 19.9 Top: Architectures of ion channel sensor materials **M12**, **M13**, and **M14**. Middle: Scheme for protamine detection with **M12** and $[Fe(CN)_6]^{3-}$ probe. Bottom: Scheme for heparin detection with the protamine–**M12** conjugate **M13** and $[Mo(CN)_8]^{4-}$ probe.

constant analyte concentration. Accordingly, when exchanging a negatively charged for a positively charged probe, the detection scheme is reversed, that is, $[Ru(NH_3)_6]^{3+}$ has free access to the electrode in the absence of the analyte and protamine addition reduces the current. A general advantage of the sensor is that the cyclovoltammogram is reversible even at high protamine concentrations so that no unspecific blocking of the electrode surface occurs.

Since protamine is routinely used to neutralize heparin in blood at the end of surgery, the detection of heparin is also possible with the previous ICS when protamine is attached to the SAM-covered electrodes. In a continuation of their protamine work, Umezawa et al. tested two different approaches and attached protamine electrostatically to the thioctic acid SAM (**M13**) as well as covalently through amide bonds (**M14**), and used the anionic probes $[Fe(CN)_6]^{3-}$ and $[Mo(CN)_8]^{4-}$ for detection (Fig. 19.9).[79] The authors found that in the covalent case, the surface coverage of protamine is higher than in the electrostatic case; however, the former electrodes were not reusable. In addition, they also used a quaternary ammonium salt as receptor instead of the thioctic acid/protamine pair and employed $[Mo(CN)_8]^{4-}$ and $[Ru(NH_3)_6]^{3+}$ as probes. It was found that the system worked well and could be readily reused. However, **M14** was found to be not as sensitive as **M13** at higher heparin concentrations.

In analogy to the examples described, many sensors have been reported for a wide range of analytes from simple metal ions[80–84] via simple anions[85] to complex guests.[86–88] Organization of the molecular recognition agents or receptors on the surface includes mostly the formation of SAMs via covalent attachment of receptor molecules.

Besides these classic ICSs, another label-free electrochemical method has become popular in recent years, sensing with biomolecule or organically modified nanotube membranes.[89] Here, specific receptors are not immobilized on the electrode surface, but in the channels of a porous membrane. An advantage of this method is that nanoporous membrane sensors can be used for highly selective electroanalytical detection that can dispense with specific redox probes. For instance, Gyurcsányi et al. prepared polymer membranes from polycarbonate, track-etched nanochannels into them and coated the pores with gold.[90] Further functionalization of the Au layer by SAM formation with a thiol-appended linker and a reactive biotin derivative in a two-step procedure and blocking of the outer surface by poly(dimethylsiloxane) (PDMS) stamping and soft lithography with a biocompatible, thiol-terminated polymer (Fig. 19.10) yielded a hybrid material **M15** with a ratio of inner surface to total surface area of 85%. As the measurement mode, these researchers employed the potentiometric monitoring of Ca^{2+} as the indicator ion, whose zero-current flux through the chemically modified nanochannels is altered by biorecognition events. Compared with the neat Au-coated nanotubules, biotin-functionalization reduces the flux of Ca^{2+} by approximately 16%. Binding of avidin then further decreases the flow dramatically.

Whereas the previous example relies on a series of channels etched into the membrane, typical single-channel stochastic sensors can be created in a closely related way.[91] For such sensors, usually one protein (e.g., α-hemolysin) is introduced into a lipid bilayer. However, the problem here is that lipid bilayers are rather fragile objects. As an alternative, Martin and coworkers embedded a single gold nanotube

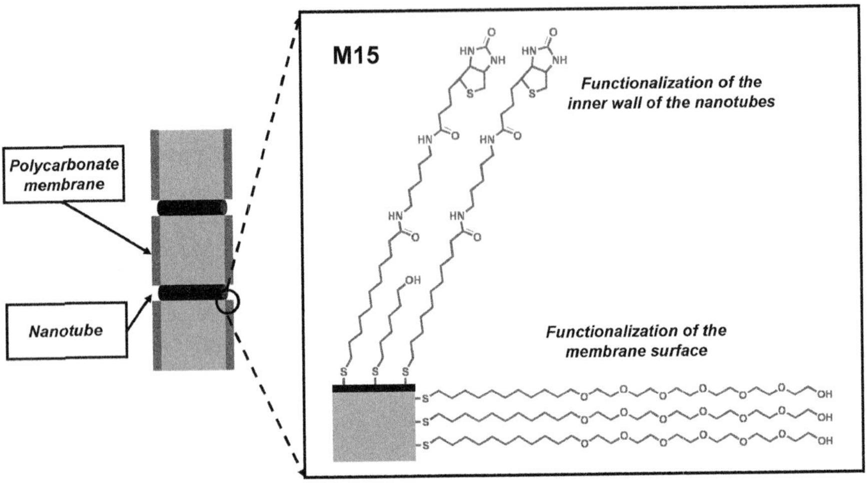

Figure 19.10 Architecture of the hybrid nanotube membrane **M15**.

into a polymer layer and attached biochemical recognition sites at the pore openings.[92] As above, the analyte (e.g., a protein) then blocks a pore. The authors etched asymmetric, conical pores with the small opening being about 30 nm in diameter into a poly(ethylene terephthalate) membrane. Decoration of the channels with Au by electroless gold plating fulfils two purposes,[93] activation of the material for functionalization and reduction (tailoring) of the pore size. In this example, Au-covered pores with a 5 to 9 nm small opening were obtained. Biochemical functionalization finally yielded pore openings of 4 to 5 nm (**M16**, Fig. 19.11). Martin et al.[94] tested different pairs of analytes/recognition elements with this concept: biotin/streptavidin, protein

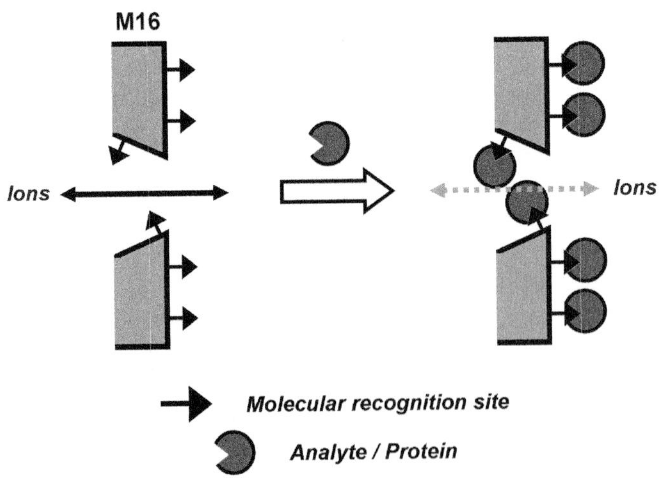

Figure 19.11 Architecture of the hybrid material **M16** containing a single nanopore and mode of operation by inhibition of ion flux.

G/immunoglobulin G (IgG), ricin/anti-ricin antibody. All the tests revealed that a protein that does not bind to the receptors does not change the flux and a protein that binds reduces flux. Since the blockage time is inversely proportional to analyte concentration, it can be used for quantification. Following this strategy, Martin's group could show that electrochemical single-molecule detection is possible.[94]

Although playing a much less important role in the field of ion channel sensors, porous silica supports such as MCM-41 have also been employed in some cases, basically due to their well-known advantages such as a well-defined porous network and facile external control of size and surface energy of the pores. In the first example discussed here, López and coworkers used atom transfer radical polymerization (ATRP) to coat the pores of MCM-41 uniformly with a stimuli-responsive polymer, poly(*N*-isopropyl acrylamide) or PNIPAAm.[95] The modulation of the transport of aqueous solutes was monitored by the uptake and release of fluorescent dyes into and from the particles, employing flow cytometry and confocal laser scanning microscopy. At low temperatures, PNIPAAm is hydrated and extended and the pores are closed. At high temperatures, PNIPAAm is hydrophobic and collapsed; the pores are open. In their studies the authors observed that the negatively charged dye fluorescein was taken up by ungrafted MCM-41 only very slowly, presumably due to electrostatic repulsions between the silanol groups of the framework and the negatively charged dye. The PNIPAAm-grafted hybrid then shows much better uptake at higher temperatures (>45°C) while at low temperatures, the hybrid shows no transport. Ion channel sensors with analyte-controlled trafficking of fluorescent probes have also been reported for entirely organic supramolecules; see, for example, Sakai et al.[96]

The second example used biotinylated MCM-41 that was incorporated into lipid bilayer membranes.[97] Ion transport through the membrane was then controlled by avidin as the analyte. Although the authors experienced some of the disadvantages of lipid bilayers, that is, unwanted transport through membrane voids (leakage current), because random physical embedding of MCM-41 particles into the bilayer produces a disordered structure, they could block a large amount of the pores (approximately 80 %) by avidin. In contrast, bovine serum albumin (BSA) as a negative control did not modulate the ion flux. The potential of mesoporous hybrid silica materials is apparent when considering that this rather imperfect membrane allowed the authors to reach an LOD of 2.9 pM.

19.5 GATED SIGNALING

As mentioned above, channels and pumps of biological systems are characterized by proteins or parts of proteins that form the pore. In addition, these proteins are commonly not flush with the membrane surface, but stick out on both sides of the membrane. These parts of the channel proteins, however, are not loosely dangling ends, but constitute (the basis for) a control station that regulates the transport through the pores, that is, they often operate as a gate.[98,99] Depending on the type of channel, either the flow of a single specific ion is regulated or, as for many ion pumps, the transport of different kinds of ions in opposite directions across cell membranes is controlled by these gates. The principle of operation of a gated ion channel or pump is basically controlled by two features: (1) coordination of a molecular stimulus at the receptors of the gate, leading to an opening or closing of the pore; and (2) the nature of the inner chemical texture of the channels that additionally determines important parameters for size-, hydrophilicity-, or charge-selective transport or cotransport. Well-known chemical gates in cells are, for instance, sited on the dendrites of neurons, where the interaction of specific neurotransmitters with their designated external receptors in the control region of the gated channel regulates ion flux.[100] Inspired by these examples from nature, chemists have embarked on the development of abiotic molecular gated devices in which (usually organic) groups, that are responsible for the gating function and that can be triggered by an external stimulus, are situated at the pore openings of inorganic scaffoldings to control the state of the gate and thus the uptake or release of guests. Accordingly, gated nanochemistry is a very fertile and promising area of research that is bringing traditional aspects of coordination and supramolecular chemistry up to the frontiers of nanoscience.

The majority of bio-inspired gated hybrid nanomaterials has been developed for cargo delivery or in the context of molecular machines and is discussed in detail in Chapters 16 and 17 of this book. Few works have demonstrated that nanoscopic gate-like scaffoldings can also be employed successfully in sensing protocols, and some of these examples are described in this section. Besides the biologically inspired architecture of these devices—gating functions being situated at pore openings—also their sensing modus operandi is often adopted from biological systems. Signaling in cells is generally connected to effector concentrations, that is, the presence or a threshold concentration of a certain substance induces the release or uptake of another species.[101,102] Translated to the world of the analytical chemist this means that a porous support material is equipped with specific binding sites at the pore openings and is loaded with indicator molecules such as dyes in the open state. Once applied

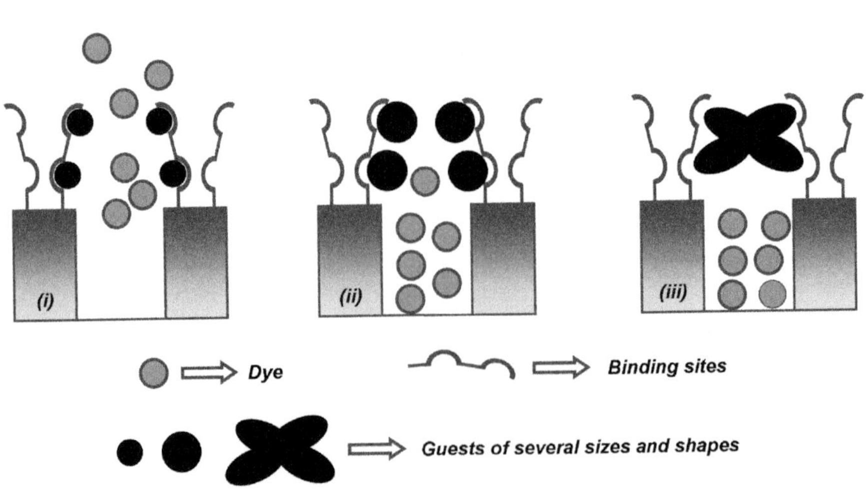

Figure 19.12 Control of the pore aperture by interaction of molecular gates with guests of different size and shape.

in the analytical sample, two scenarios are possible. In the absence of the designated analyte, the pores remain open and the indicator can diffuse into the solution (Fig. 19.12). In the presence of the target, however, this molecule or ion can bind to the receptors at the pore outlets and close the gate; diffusion of the indicator into the solution is inhibited and only the hybrid material remains colored. The examples realized so far utilize mesoporous MCM-41 supports containing suitable coordination sites for the "key" guests at the pore outlets and an easily measurable dye in the pore voids, relying on a chromofluorogenic indication through the inhibition of dye delivery. These systems are completely different in concept to the classical "binding site–signaling unit" paradigm because of two important characteristics. First, as mentioned above, a few guests are sufficient to close the gates and lock a large number of dyes in the mesoporous reservoir, thus showing clear features of signal amplification. Second, this approach detaches the recognition protocol from the signaling event, making signaling independent of the stoichiometry of the guest–receptor complex.

A first proof-of-concept was reported for an ensemble consisting of an MCM-41 mesoporous support, $[Ru(bpy)_3]^{2+}$ as the mobile indicator and polyamine receptors as gates, the single amino groups acting as the receptors (**M17**, Fig. 19.13). Like in a previously discussed example, polyamine/polyammonium receptors are well known to

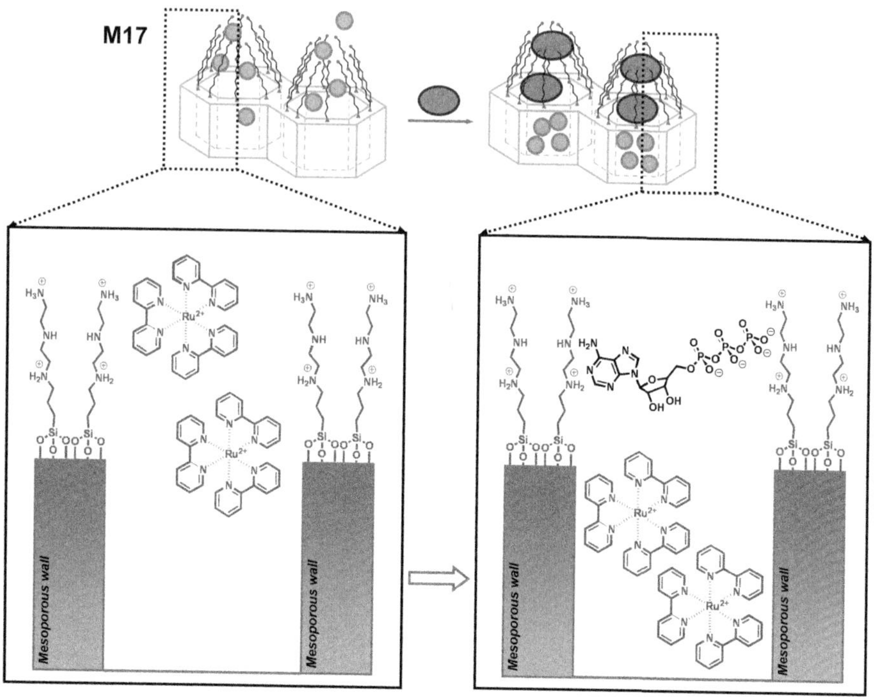

Figure 19.13 ATP signaling protocol via inhibition of dye delivery using nanoscopic gate-like scaffoldings. The mesopores of the system are loaded with a $[Ru(bpy)_3]^{2+}$ dye and polyamines are covalently anchored to the external surface of **M17** (left: the gate is open). Addition of ATP results in the inhibition of dye delivery due to strong ATP-polyamine interactions (right: the gate is closed).

bind anions through a combination of hydrogen bonding and electrostatic interactions. Among several anions tested, ATP and ADP are selectively able to close the gate, whereas other anions such as Cl^-, SO_4^{2-}, or guanosine monophosphate (GMP) are too small or form complexes that are too weak to effectively close the pore openings and stop leaching of the $[Ru(bpy)_3]^{2+}$ indicator (Figs. 19.13 and 19.14).[103]

A related approach has been reported for the selective colorimetric detection of fatty acid anions, using imidazolium-functionalized mesoporous MCM-41 with pores of 2.5 nm in diameter and again $[Ru(bpy)_3]^{2+}$ as the indicator.[104] In the absence of carboxylates, the gate is open and dye delivery is observed. Addition of linear carboxylates of different chain length ($CH_3(CH_2)_nCOO^-$, $n = 2, 4, 6, 8, 10$) resulted in different blocking behaviors. However, it was observed that pore blockage did not follow a gradual trend, but an on-off behavior was observed: complete delivery of the dye in the presence of small carboxylates from acetate to octanoate and complete pore blockage for longer chain carboxylates. This peculiar behavior is most likely due to a combination of two effects, the size of the carboxylate and the hydrophobic nature of the blocking layer formed over the pore outlets upon binding of the carboxylate moieties to the imidazolium groups. Fatty carboxylates with a long hydrocarbon chain presumably interact with themselves through van der Waals forces, leading to a hydrophobic cork on the pore openings. Besides analyte-induced pore blockage, a pore-opening protocol can also be envisioned. The latter for instance could take profit of a displacement reaction in which the indicator-filled pore is closed by a large, multitopic gatekeeper molecule, for example, a functionalized dendrimer

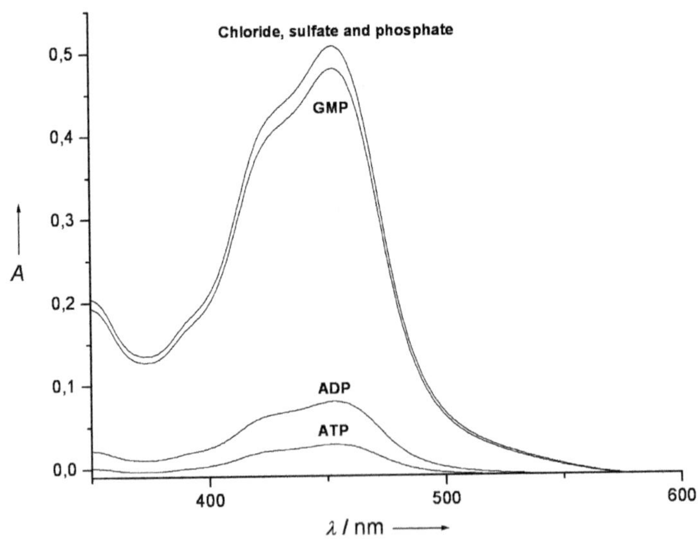

Figure 19.14 Absorbance of solutions containing solid **M17** (10 mg) and different anions (10^{-4} mol dm^{-3}) in water at pH 7.8 (Tris buffer 10^{-3} mol dm^{-3}). The band centered at 454 nm is due to the delivery of the $[Ru(bpy)_3]^{2+}$ complex from the pores to the aqueous solution.[103] (Reprinted with permission from R. Casasús et al., *Angew. Chem. Int. Ed.*, **2006**, *45*, 6661–6664. Copyright Wiley-VCH Verlag GmbH & Co. KGaA.)

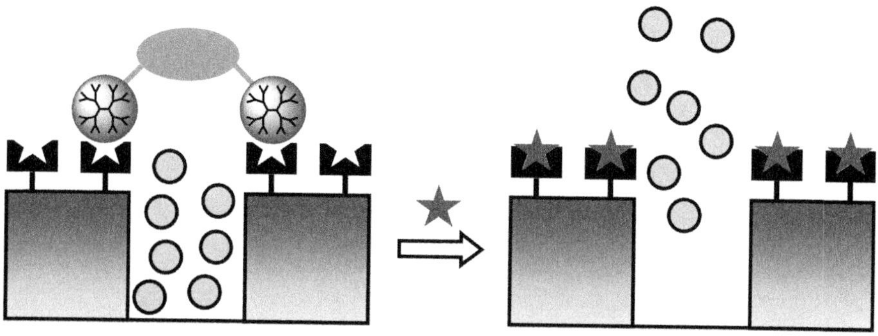

Figure 19.15 Sketch of gated closed-to-opened signaling utilizing the displacement of functionalized dendrimer "corks" by a stronger binding analyte and subsequent release of an indicator.

(Fig. 19.15). The system then would have to be designed in such a way that the binding constants between the functional groups on the dendrimer and the gating receptors are distinctly smaller than the binding constants of the target analyte and the gating receptors. In addition, the analyte should also have a significantly smaller size than the dendrimer. The presence of the target species would then lead to a favorable competition for the gating receptors, detaching the large dendrimer cap and opening the pores. However, such a system has not yet been realized. Due to the possible use of different porous 3D supports, diverse guest-selective gate-like systems and a wide range of indicator dyes, this strategy harbors an enormous potential for the development of novel signaling systems.

19.6 DISPLACEMENT ASSAYS

Having touched the key step in classic displacement assays already with the last hypothetical example in the previous section, distinct examples of the use of artificial binding pockets bearing receptors in selective colorimetric displacement assays are described in this section. In analogy to Figure 19.15, the protocol involves several steps: (1) the pores are functionalized with adequate binding sites; (2) the latter can be loaded with a dye that coordinates to these anchored sites; (3) in the presence of a target species that forms a stronger complex with the inner binding sites, displacement of the dyes and diffusion into the bulk solution takes place; (4) enabling the indirect colorimetric detection of the guest (Fig. 19.16).[105] As in traditional displacement assays,[106–108] the selectivity of the response relies on an adequate choice of the binding sites and the dye. Such a protocol is successful when the stability constant between receptor and dye (K_{dye}) is lower than that between receptor and target (K_{target}) and larger than that between receptor and potentially interfering species ($K_{interference}$); that is $K_{target} > K_{dye} > K_{interference}$.

A first example reports a mesoporous MCM-41-type solid containing guanidinium groups within the mesopores that act as binding sites and methylthymol blue as the dye (**M18**, Fig. 19.17).[105] Guanidinium groups are known to undergo hydrogen

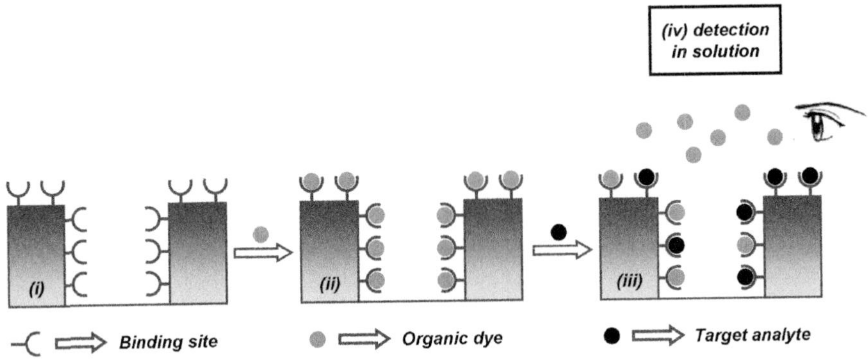

Figure 19.16 Optical detection protocol using displacement assays in bio-inspired mesoporous systems.

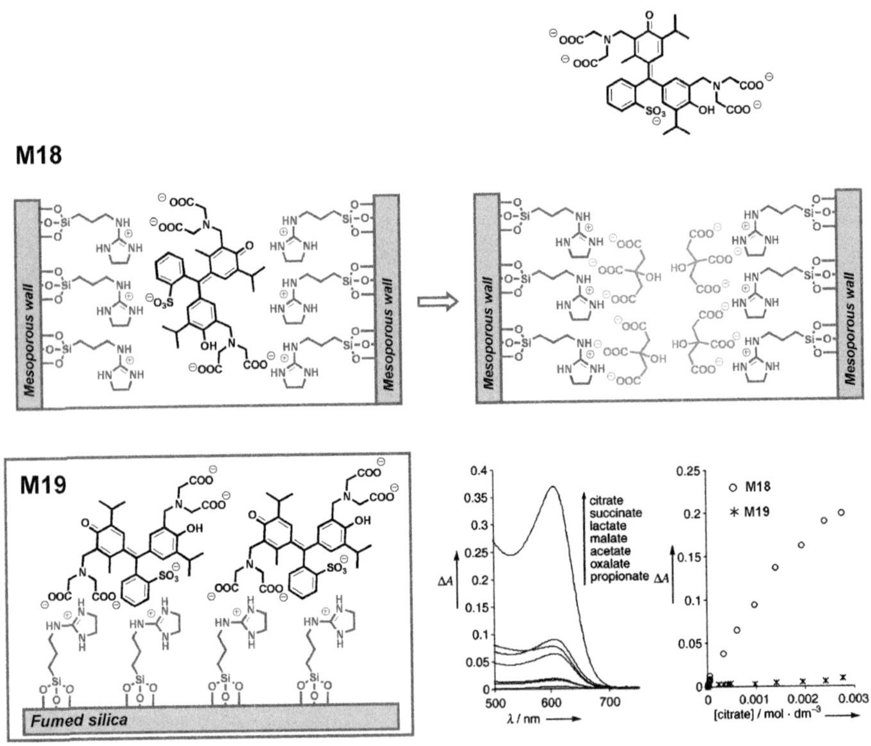

Figure 19.17 Top: protocol for citrate signaling in water by indicator displacement with the mesoporous solid **M18** containing nanoscopic binding pockets and methylthymol blue as indicator. Bottom: (left side) architecture of the weakly responding 2D analog **M19**, (right side) absorption spectra of **M18** in the presence of different carboxylates and citrate calibration curves for **M18** and **M19**.[105] (Reprinted with permission from M. Comes et al., *Angew. Chem. Int. Ed.*, **2005**, *44*, 2918–2922. Copyright Wiley-VCH Verlag GmbH & Co. KGaA.)

bonding interactions with carboxyl groups and the prepared material was tested in water in the presence of a number of carboxylates. A remarkable citrate-selective response was found, indicating that the binding pockets are able to recognize this anion through favorable coordination with respect to other carboxylates (Fig. 19.17). In analogy to other cases discussed before, functionalization of a 2D material that lacks the homogeneous porosity of mesoporous silicas with the same receptors and the same dye (attachment of the latter occurs only by electrostatic interactions, **M19**) leads to a very poor response only. In the same paper, similar protocols were carried out for the chromogenic sensing of borate. In this case, the surface of the MCM-41 solid was reacted with a mannose derivative that was further treated with a boronate-functionalized dye to bind to the sugar derivative through formation of a boronic ester bond. In the sensing experiments, the addition of borate in water at concentrations of parts per million led to a partial displacement of the boronate-functionalized dye, resulting in a simple and selective colorimetric assay for borate in aqueous solution. The solid did not show any cross-reactivity toward other compulsory anions and cations present in water.

A hybrid material for the colorimetric sensing of phosphate in water using displacement assays has also been reported.[109] The sensing system consists again of a mesoporous MCM-41-type support, the surface of which was simply functionalized with amino binding sites and the latter were subsequently filled with carboxyfluorescein. Addition of phosphate to a suspension of the hybrid resulted in a preferential interaction of phosphate with the tethered amines and displacement of the dye from the pores to the solution. Addition of sulfate also led to some displacement, but to a remarkably lesser degree. Nitrate and halide anions on the other hand remained inactive. This sensing protocol is undemanding and opens the possibility of developing new chromofluorogenic sensor materials by introduction of simple coordination sites and the noncovalent anchoring of dyes into the mesoporous materials.

19.7 CHEMICAL CHAMELEONS

An important ability of various amphibians, reptiles, crustaceans, and fish is to rapidly change their body color as a response to environmental stimuli such as light intensity or background color.[110] Such a collective change of the body color takes place in pigment cells, so-called chromatophores, and is usually orchestrated by hormones, cyclic AMP (cAMP), or other key mediators involved in the signaling pathways of the responsible parts of the nervous or the endocrine system.[111] Depending on the type of animal and chromatophore, regulatory cycles and color changing mechanisms can differ considerably. For example, whereas melanophores possess black and brown granules of melanin, the reddish colored erythrophores contain carotenoid vesicles and pteridin granules. The mechanisms of color change include the translocation of granules or microplatelets in the pigment cells, their dispersion or aggregation.[112] Besides light or background color as the trigger for chromatic responses, other stimuli have also frequently been reported. For instance, the pigment translocation in certain

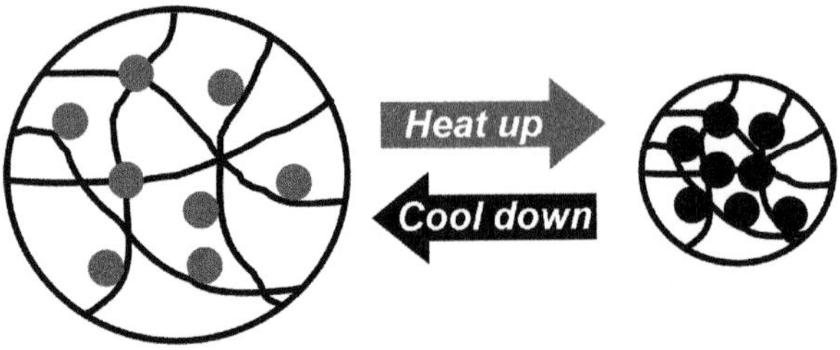

Figure 19.18 Thermosensitive hybrid microgel containing gold nanoparticles. The color changes originate from nanoparticle aggregation induced by the swelling/deswelling of the thermosensitive microgel.

shrimp species was found to depend on Ca^{2+} concentration.[113] Thermoregulation or temperature dependence of color changes has been reported for lizards and frogs.[114,115] The latter behavior, that is temperature as the stimulus for a color change, leads us to the next examples of biomimetic signaling chemistry.

Figure 19.18 illustrates the first approach reported by Suzuki and Kawaguchi.[116] These researchers immobilized gold nanoparticles (AuNPs) or Au/AgNPs in a polymer microgel and managed to swell or to shrink the particles in a controlled fashion by tuning of the temperature. In analogy to aggregation of granules in chromatophores, the shrinking of the polymer network of the hybrid material for instance leads to an aggregation of the nanoparticles, entailing a color change due to a shift of the plasmon absorption band. As for the natural systems, the color changes in this artificial ensemble are fully reversible. An elaboration of this concept permits the development of a reversible nanothermometer, enabling the precise measurement of temperature down to the nanoscale regime. Kotov et al.[117,118] realized the task by connecting two different NPs, an AuNP and a CdTe quantum dot (QD), through a poly(ethylene glycol) (PEG) polymer as a "molecular spring" (Fig. 19.19). The system is designed in such a way that the resonance conditions between AuNP plasmon and CdTe exciton lead to luminescence enhancement (instead of the commonly observed quenching). The temperature-based tuning of the distance of the two NPs then modulates the luminescence intensity through the degree of coupling.

Recalling the fact that hormones are often involved in the triggering of color changes that take place in pigment cells, recent work by Andersson et al.[119] nicely complement the chemistry described in this chapter, returning with the last example to the signaling of actual target compounds. The Swedish research group developed a bio-inorganic hybrid system for hormone detection in which they actually employed real chromatophores (frog melanophores) in combination with fluorescent polymer microbeads.[119] The sensing protocol starts with the culturing of melanophore cells in the presence of the polymer beads, leading to so-called "bead-cell solutions" that contain a large number of approximately stoichiometric 1 : 1 agglomerates of a melanophore and a 40 μm microbead. The researchers chose frog melanophores that were known to exhibit hormone-induced aggregation of their melanosomes (the melanin-containing granules) with melatonin acting as the trigger hormone. After preparing

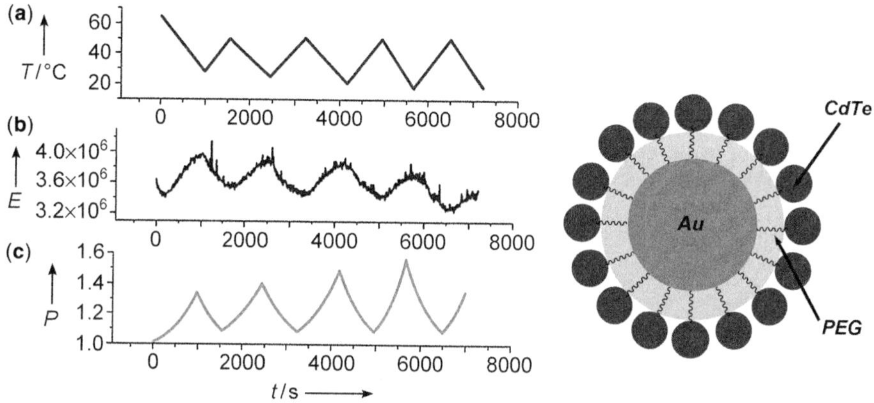

Figure 19.19 Left side: Variation of the photoluminescence intensity E (b) of the PEG-functionalized Au and CdTe nanoparticles depending on the temperature (a); (c) shows the calculated photon-field enhancement factor P of the CdTe nanoparticles as a function of time. Right side: Schematic representation of a dynamic nanothermometer based on a nanoparticle superstructure. This superstructure consists of two types of nanoparticles (gold and CdTe) connected by polymeric spacers.[118] (Reprinted with permission from J. Lee et al., *Angew. Chem. Int. Ed.*, **2005**, *44*, 7439–7442. Copyright Wiley-VCH Verlag GmbH & Co. KGaA.)

the biosensory suspension, melatonin detection was measured by an increase in fluorescence signal upon melatonin-induced aggregation of the melanophores, the optical signal arising from the interplay of suspension absorption, scattering, and fluorescence. Andersson et al. used fluorescence microplate readers as well as the very simple computer screen photoassisted technique for signal acquisition. With this technique, it was possible to discriminate between the target analyte melatonin and another hormone involved in other chromatic responses, α-melanocyte stimulating hormone (MSH). While this proof-of-principle is very convincing, future research has to elucidate the possibilities of how to transfer the protocol to the sensing of medically more relevant hormones.

19.8 CONCLUSIONS

In this chapter, we have discussed representative examples published in the field of functional biomimetic chemistry of hybrid materials for signaling applications. The contribution has shown how novel biomimetic functions can arise from synergisms between specifically designed hybrid materials and the implementation of selected supramolecular concepts. The first sections of the chapter focused on the idea of using hybrid frameworks for biomimetic signaling inspired by the way nature has created binding pockets and the concept of induced fit as well as metalloregulatory cycles. Selected examples show how the combination of certain inorganic scaffoldings with supramolecular concepts results in enhanced molecular recognition and signaling. The very wide range of solids organized at the nanometric level, the many possibilities of

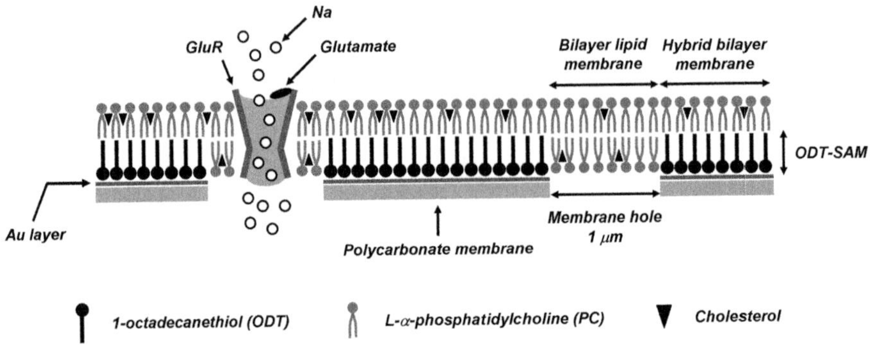

Figure 19.20 Sketch of a mixed hybrid bilayer lipid membrane incorporating a glutamate receptor.[121] (Adapted with permission from G. Favero et al., *J. Am. Chem. Soc.* **2005**, *127*, 8103–8111. Copyright 2005 American Chemical Society.)

modifying and functionalizing their structures with a multitude of possible recognition sites and supramolecular recognition protocols anticipate promising new results in the near future with particular impact on analytical applications. In a related sense, exploitation of the gating and trafficking strategies of ion channels has led to a variety of exciting applications in electrochemical sensing. We assume that especially further research on nanoporous membranes such as single-nanopore architectures will create highly sensitive analytical devices, possessing all the advantages of stochastic sensors.[120] In addition, recent progress in the stabilization of lipid bilayers[121,122] by using, for example, mixed hybrid bilayers with embedded biological receptors on gold supports is expected to lead to advances in biosensing (Fig. 19.20).[121] Since the gated porous objects retain their biological function in such a membrane with distinctly improved stability, exploitation of the large variety of biological receptors becomes possible.

The examples in Sections 19.5 and 19.6 have impressively demonstrated the unique features that can be introduced to chemical sensing with the use of gate-like ensembles. Since these systems intrinsically have (at least) three important parameters that can almost limitlessly be adjusted—pore size and morphology, inner pore texture, and gating functions—we expect a strongly growing impact of such systems for future miniaturized analytical devices. Finally, the "chemical chameleon" hybrids suggest that there is still plenty of room for the design of novel hetero-supramolecular ensembles for specific applications. If one further considers that novel hybrid material combinations are constantly prepared, such as for instance mesostructured silica–block-copolymer[123] or hydroxyapatite–carbon nanotube–polylysine hybrids,[124] the impact of the field on chemical signaling can be imagined. The common feature of all this exciting chemistry presented here is the synergistic use of concepts from nanotechnology, bio- and supramolecular chemistry that create properties and signaling functions that transfer known functions to a higher level of sophistication, often showing signaling outcomes that are hardly observable in nanoscopic solids or the analogous molecular-based systems alone.

ACKNOWLEDGMENTS

Financial support from the Ministerio de Ciencia y Tecnología and CIBER BBN (projects CTQ2006-15456-C04-01, MAT2009-14564-C04-01 and CB07/01/2012) and the European Commission (Marie Curie Intra-European Fellowship of the Human Resources & Mobility Program and EFRE/ProFIT) is gratefully acknowledged.

REFERENCES

1. P. Fratzl, *J. R. Soc. Interface* **2007**, *4*, 637–642.
2. O. H. Schmitt, *Proc. 3rd Intl. Biophysics Congress*, Boston, MA, Aug 29–Sep 03, **1969**, 297.
3. Y. Bar-Cohen, *Bioinsp. Biomim.* **2006**, *1*, P1–P12.
4. J. F. V. Vincent, O. A. Bogatyreva, N. R. Bogatyrev, A. Bowyer, A.-K. Pahl, *J. R. Soc. Interface* **2006**, *3*, 471–482.
5. C. Sanchez, H. Arribart, M. M. G. Guille, *Nature Mater.* **2005**, *4*, 277–288.
6. Y. Lu, *Angew. Chem. Int. Ed.* **2006**, *45*, 5588–5601.
7. R. Chen, J. Hunt, *J. Mater. Chem.* **2007**, *17*, 3974–3979.
8. S. Mann, *Chem. Commun.* **2004**, 1–4.
9. S. Das, G. W. Brudvig, R. H. Crabtree, *Chem. Commun.* **2008**, 413–424.
10. A. B. Descalzo, R. Martínez-Máñez, F. Sancenón, K. Hoffmann, K. Rurack, *Angew. Chem. Int. Ed.* **2006**, *45*, 5924–5948.
11. K. Rurack, R. Martínez-Máñez, in *Nanomaterials: Inorganic and Bioinorganic Perspectives* (Eds. C. M. Lukehart, R. A. Scott), Chichester: John Wiley & Sons, **2008**, 789–812.
12. K. Rurack, R. Martínez-Máñez, in *Nanomaterials: Inorganic and Bioinorganic Perspectives* (Eds. C. M. Lukehart, R. A. Scott), Chichester: John Wiley & Sons, **2008**, 31–50.
13. K. Ariga, A. Vinu, J. P. Hill, T. Mori, *Coord. Chem. Rev.* **2007**, *251*, 2562–2591.
14. E. Rizzarelli, G. Vecchio, *Coord. Chem. Rev.* **1999**, *188*, 343–364.
15. K. Haupt, K. Mosbach, *Chem. Rev.* **2000**, *100*, 2495–2504.
16. G. Wulff, *Chem. Rev.* **2002**, *102*, 1–27.
17. A. B. Descalzo, K. Rurack, H. Weisshoff, R. Martínez-Máñez, M. D. Marcos, P. Amorós, K. Hoffmann, J. Soto, *J. Am. Chem. Soc.* **2005**, *127*, 184–200.
18. V. S.-Y. Lin, C.-Y. Lai, J. Huang, S.-A. Song, S. Xu, *J. Am. Chem. Soc.* **2001**, *123*, 11510–11511.
19. D. R. Radu, C.-Y. Lai, J. W. Wiench, M. Pruski, V. S.-Y. Lin, *J. Am. Chem. Soc.* **2004**, *126*, 1640–1641.
20. M. Comes, M. D. Marcos, R. Martínez-Máñez, F. Sancenón, J. Soto, L. A. Villaescusa, P. Amorós, D. Beltrán, *Adv. Mater.* **2004**, *16*, 1783–1786.
21. S. Basurto, T. Torroba, M. Comes, R. Martínez-Máñez, F. Sancenón, L. Villaescusa, P. Amorós, *Org. Lett.* **2005**, *7*, 5469–5472.
22. B. García-Acosta, M. Comes, J. L. Bricks, M. A. Kudinova, V. V. Kurdyukov, A. I. Tolmachev, A. B. Descalzo, M. D. Marcos, R. Martínez-Máñez, A. Moreno, F. Sancenón, J. Soto, L. A. Villaescusa, K. Rurack, J. M. Barat, I. Escriche, P. Amorós, *Chem. Commun.* **2006**, 2239–2241.
23. M. Comes, M. D. Marcos, R. Martínez-Máñez, M. C. Millán, J. V. Ros-Lis, F. Sancenón, J. Soto, L. A. Villaescusa, *Chem. Eur. J.* **2006**, *12*, 2162–2170.
24. K. Inumaru, Y. Inoue, S. Kakii, T. Nakano, S. Yamanaka, *Chem. Lett.* **2003**, *32*, 1110–1111.
25. K. Inumaru, Y. Inoue, S. Kakii, T. Nakano, S. Yamanaka, *Phys. Chem. Chem. Phys.* **2004**, *6*, 3133–3139.
26. (a) P. D. Beer, D. P. Comode, J. J. Davis, *Chem. Commun.* **2004**, 414–415; (b) T. R. Tshikhudo, D. Demuru, Z. Wang, M. Brust, A. Secchi, A. Arduini, A. Pochini, *Angew. Chem. Int. Ed.* **2005**, *44*, 2913–2916; (c) D. Astruc, M.-C. Daniel, J. Ruiz, *Chem. Commun.* **2004**, 2637–2649.

27. A. B. Descalzo, D. Jiménez, M. D. Marcos, R. Martínez-Máñez, J. Soto, J. El Haskouri, C. Guillem, D. Beltrán, P. Amorós, M. V. Borrachero, *Adv. Mater.* **2002**, *14*, 966–969.
28. A. B. Descalzo, M. D. Marcos, R. Martínez-Máñez, J. Soto, D. Beltrán, P. Amorós, *J. Mater. Chem.* **2005**, *15*, 2721–2731.
29. J. V. Ros-Lis, R. Casasús, M. Comes, C. Coll, M. D. Marcos, R. Martínez-Máñez, F. Sancenón, J. Soto, P. Amorós, J. El Haskouri, N. Garró, K. Rurack, *Chem. Eur. J.* **2008**, *14*, 8267–8278.
30. M. Ganschow, M. Wark, D. Wöhrle, G. Schulz-Ekloff, *Angew. Chem. Int. Ed.* **2000**, *39*, 160–163.
31. M. Vasak, *J. Trace Elem. Med. Biol.* **2005**, *19*, 13–17.
32. J.-W. Lee, J. D. Helmann, *Biometals* **2007**, *20*, 485–499.
33. J. Ziegler, R. T. Chang, D. W. Wright, *J. Am. Chem. Soc.* **1999**, *121*, 2395–2400.
34. J. M. Slocik, D. W. Wright, *Biomacromolecules* **2003**, *4*, 1135–1141.
35. Y. Hara, M. Akiyama, *J. Am. Chem. Soc.* **2001**, *123*, 7247–7256.
36. M. Etienne, S. Sayen, B. Lebeau, A. Walcarius, *Stud. Surf. Sci. Catal.* **2002**, *141*, 615–622.
37. V. Antochshuk, O. Olkhovyk, M. Jaroniec, I.-S. Park, R. Ryoo, *Langmuir* **2003**, *19*, 3031–3034.
38. Y. Kim, B. Lee, J. Yi, *Sep. Sci. Technol.* **2004**, *39*, 1427–1442.
39. S. G. Wang, J. L. Li, *Chin. Chem. Lett.* **2006**, *17*, 221–224.
40. K. F. Lam, K. L. Yeung, G. McKay, *J. Phys. Chem. B* **2006**, *110*, 2187–2194.
41. D. Pérez-Quintanilla, I. del Hierro, M. Fajardo, I. Sierra, *J. Mater. Chem.* **2006**, *16*, 1757–1764.
42. R. C. Schroden, M. Al-Daous, S. Sokolov, B. J. Melde, J. C. Lytle, A. Stein, M. C. Carbajo, J. T. Fernández, E. Rodríguez, *J. Mater. Chem.* **2002**, *12*, 3261–3267.
43. M. H. Lee, S. J. Lee, J. H. Jung, H. Lim, J.S. Kim, *Tetrahedron* **2007**, *63*, 12087–12092.
44. R. Métivier, I. Leray, B. Lebeau, B. Valeur, *J. Mater. Chem.* **2005**, *15*, 2965–2973.
45. S. J. Lee, J.-E. Lee, J. Seo, I. Y. Jeong, S. S. Lee, J. H. Jung, *Adv. Funct. Mater.* **2007**, *17*, 3441–3446.
46. E. Palomares, R. Vilar, J. R. Durrant, *Chem. Commun.* **2004**, 362–363.
47. H. Zhang, P. Zhang, K. Ye, Y. Sun, S. Jiang, Y. Wang, W. Pang, *J. Lumin.* **2006**, *117*, 68–74.
48. L.-L. Li, H. Sun, C.-J. Fang, J. Xu, J.-Y. Jin, C.-H. Yan, *J. Mater. Chem.* **2007**, *17*, 4492–4498.
49. L. Gao, J. Q. Wang, L. Huang, X. X. Fan, J. H. Zhu, Y. Wang, Z. G. Zou, *Inorg. Chem.* **2007**, *46*, 10287–10293.
50. J.-Q. Wang, L. Huang, M. Xue, Y. Wang, L. Gao, J. H. Zhu, Z. Zou, *J. Phys. Chem. C* **2008**, *112*, 5014–5022.
51. L. Gao, Y. Wang, J. Wang, L. Huang, L. Shi, X. Fan, Z. Zou, T. Yu, M. Zhu, Z. Li, *Inorg. Chem.* **2006**, *45*, 6844–6850.
52. T. Balaji, M. Sasidharan, H. Matsunaga, *Anal. Bioanal. Chem.* **2006**, *384*, 488–494.
53. T. Balaji, S. A. El-Safty, H. Matsunaga, T. Hanaoka, F. Mizukami, *Angew. Chem. Int. Ed.* **2006**, *45*, 7202–7208.
54. S. A. El-Safty, A. A. Ismail, H. Matsunaga, F. Mizukami, *Chem. Eur. J.* **2007**, *13*, 9245–9255.
55. S. A. El-Safty, D. Prabhakaran, A. A. Ismail, H. Matsunaga, F. Mizukami, *Adv. Funct. Mater.* **2007**, *17*, 3731–3745.
56. S. A. El-Safty, A. A. Ismail, H. Matsunaga, H. Nanjo, F. Mizukami, *J. Phys. Chem. C* **2008**, *112*, 4825–4835.
57. S. A. El-Safty, A. A. Ismail, H. Matsunaga, T. Hanaoka, F. Mizukami, *Adv. Funct. Mater.* **2008**, *18*, 1485–1500.
58. J. V. Ros-Lis, M. D. Marcos, R. Martínez-Máñez, K. Rurack, J. Soto, *Angew. Chem. Int. Ed.* **2005**, *44*, 4405–4407.
59. S. J. Lee, D. R. Bae, W. S. Han, S. S. Lee, J. H. Jung, *Eur. J. Inorg. Chem.* **2008**, 1559–1564.
60. M. Montal, *Annu. Rev. Biophys. Biomol. Struct.* **1995**, *24*, 31–57.
61. T. J. Jentsch, V. Stein, F. Weinreich, A. A. Zdebik, *Physiol. Rev.* **2002**, *82*, 503–568.
62. S. Morbach, R. Krämer, *ChemBioChem* **2002**, *3*, 385–397.
63. J. A. Wemmie, M. P. Price, M. J. Welsh, *Trends Neurosci.* **2006**, *29*, 578–586.
64. H. Bayley, C. R. Martin, *Chem. Rev.* **2000**, *100*, 2575–2594.
65. Y. Jung, H. Bayley, L. Movileanu, *J. Am. Chem. Soc.* **2006**, *128*, 15332–15340.
66. A. T. A. Jenkins, N. Boden, R. J. Bushby, S. D. Evans, P. F. Knowles, R. E. Miles, S. D. Ogier, H. Schönherr, G. J. Vancso, *J. Am. Chem. Soc.* **1999**, *121*, 5274–5280.

67. L. Rose, A. T. A. Jenkins, *Bioelectrochemistry* **2007**, *70*, 387–393.
68. H. Tsukube, K. Takagi, T. Higashiyama, T. Iwachido, N. Hayama, *Inorg. Chem.* **1994**, *33*, 2984–2987.
69. J. H. van Zanten, H. G. Monbouquette, *Biotechnol. Prog.* **1992**, *8*, 546–552.
70. O. Murillo, I. Suzuki, E. Abel, C. L. Murray, E. S. Meadows, T. Jin, G. W. Gokel, *J. Am. Chem. Soc.* **1997**, *119*, 5540–5549.
71. N. Sakai, K. C. Brennan, L. A. Weiss, S. Matile, *J. Am. Chem. Soc.* **1997**, *119*, 8726–8727.
72. M. F. M. Roks, R. J. M. Nolte, *Macromolecules* **1992**, *25*, 5398–5407.
73. P. Metilda, K. Prasad, R. Kala, J. M. Gladis, T. P. Rao, G. R. K. Naidu, *Anal. Chim. Acta* **2007**, *582*, 147–153.
74. L. M. Goldenberg, M. R. Bryce, M. C. Petty, *J. Mater. Chem.* **1999**, *9*, 1957–1974.
75. Y. Umezawa, H. Aoki, *Anal. Chem.* **2004**, *76*, 320A–326A.
76. P. Bühlmann, H. Aoki, K. P. Xiao, S. Amemiya, K. Tohda, Y. Umezawa, *Electroanalysis* **1998**, *10*, 1149–1158.
77. M. Sugawara, A. Hirano, P. Bühlmann, Y. Umezawa, *Bull. Chem. Soc. Jpn.* **2002**, *75*, 187–201.
78. V. P. Y. Gadzekpo, K. P. Xiao, H. Aoki, P. Bühlmann, Y. Umezawa, *Anal. Chem.* **1999**, *71*, 5109–5115.
79. V. P. Y. Gadzekpo, P. Bühlmann, K. P. Xiao, H. Aoki, Y. Umezawa, *Anal. Chim. Acta* **2000**, *411*, 163–173.
80. M. Takaya, P. Bühlmann, Y. Umezawa, *Mikrochim. Acta* **1999**, *132*, 55–60.
81. K. Bandyopadhyay, H. Liu, S.-G. Liu, L. Echegoyen, *Chem. Commun.* **2000**, 141–142.
82. S. Flink, H. Schönherr, G. J. Vancso, F. A. J. Geurts, K. G. C. van Leerdam, F. C. J. M. van Veggel, D. N. Reinhoudt, *J. Chem. Soc. Perkin Trans. 2* **2000**, 2141–2146.
83. T. Ito, *J. Electroanal. Chem.* **2001**, *495*, 87–97.
84. H. Aoki, Y. Umezawa, A. Vertova, S. Rondinini, *Anal. Sci.* **2006**, *22*, 1581–1584.
85. H. Aoki, K. Hasegawa, K. Tohda, Y. Umezawa, *Biosens. Bioelectron.* **2003**, *18*, 261–267.
86. H. Aoki, Y. Umezawa, *Analyst* **2003**, *128*, 681–685.
87. H. Aoki, Y. Umezawa, *Electroanalysis* **2002**, *14*, 1405–1410.
88. Y. Katayama, Y. Ohuchi, H. Higashi, Y. Kudo, M. Maeda, *Anal. Chem.* **2000**, *72*, 4671–4674.
89. P. Kohli, M. Wirtz, C. R. Martin, *Electroanalysis* **2004**, *16*, 9–18.
90. R. E. Gyurcsányi, T. Vigassy, E. Pretsch, *Chem. Commun.* **2003**, 2560–2561.
91. L. T. Sexton, L. P. Horne, C. R. Martin, *Mol. BioSyst.* **2007**, *3*, 667–685.
92. Z. Siwy, L. Trofin, P. Kohli, L. A. Baker, C. Trautmann, C. R. Martin, *J. Am. Chem. Soc.* **2005**, *127*, 5000–5001.
93. Z. Siwy, E. Heins, C. C. Harrell, P. Kohli, C. R. Martin, *J. Am. Chem. Soc.* **2004**, *126*, 10850–10851.
94. E. A. Heins, Z. S. Siwy, L. A. Baker, C. R. Martin, *Nano Lett.* **2005**, *5*, 1824–1829.
95. Q. Fu, G. V. R. Rao, L. K. Ista, Y. Wu, B. P. Andrzejewski, L. A. Sklar, T. L. Ward, G. P. López, *Adv. Mater.* **2003**, *15*, 1262–1266.
96. N. Sakai, J. Mareda, S. Matile, *Acc. Chem. Res.* **2005**, *38*, 79–87.
97. K. Nozawa, C. Osono, M. Sugawara, *Sens. Actuators B* **2007**, *126*, 632–640.
98. J. T. Finn, M. E. Grunwald, K. W. Yau, *Annu. Rev. Physiol.* **1996**, *58*, 395–426.
99. I. B. Levitan, *Nature Neurosci.* **2006**, *9*, 305–310.
100. L.-L. Yuan, X. Chen, *Progr. Neurobiol.* **2006**, *78*, 374–389.
101. T. C. Rich, J. W. Karpen, *Ann. Biomed. Eng.* **2002**, *30*, 1088–1099.
102. M. Oh-hora, A. Rao, *Curr. Opin. Immunol.* **2008**, *20*, 250–258.
103. R. Casasús, E. Aznar, M. D. Marcos, R. Martínez-Máñez, F. Sancenón, J Soto, P. Amorós, *Angew. Chem. Int. Ed.* **2006**, *45*, 6661–6664.
104. C. Coll, R. Casasús, E. Aznar, M. D. Marcos, R. Martínez-Máñez, F. Sancenón, J. Soto, P. Amorós, *Chem. Commun.* **2007**, 1957–1959.
105. M. Comes, G. Rodríguez-López, M. D. Marcos, R. Martínez-Máñez, F. Sancenón, J. Soto, L. A. Villaescusa, P. Amorós, D. Beltrán, *Angew. Chem. Int. Ed.* **2005**, *44*, 2918–2922.
106. R. Martínez-Máñez, F. Sancenón, *Chem. Rev.* **2003**, *103*, 4419–4476.

107. B. T. Nguyen, E. V. Anslyn, *Coord. Chem. Rev.* **2006**, *250*, 3118–3127.
108. L. Fabbrizzi, N. Marcotte, F. Stomeo, A. Taglietti, *Angew. Chem. Int. Ed.* **2002**, *41*, 3811–3814.
109. M. Comes, M. D. Marcos, R. Martínez-Máñez, F. Sancenón, J. Soto, L. A. Villaescusa, P. Amorós, *Chem. Commun.* **2008**, 3639–3641.
110. L. E. M. Nery, A. M. L. Castrucci, *Comp. Biochem. Physiol.* **1997**, *118A*, 1135–1144.
111. M. A. Visconti, G. C. Ramanzini, C. R. Camargo, A. M. L. Castrucci, *J. Exp. Zool.* **1999**, *284*, 485–491.
112. N. Oshima, *Pigment Cell Res.* **2001**, *14*, 312–319.
113. J. C. McNamara, M. R. Ribeiro, *Biol. Bull.* **2000**, *198*, 357–366.
114. W. C. Sherbrooke, A. M. D. Castrucci, M. E. Hadley, *Physiol. Zool.* **1994**, *67*, 659–672.
115. R. B. King, S. Hauff, J. B. Phillips, *Copeia* **1994**, 422–432.
116. D. Suzuki, H. Kawaguchi, *Langmuir* **2006**, *22*, 3818–3822.
117. J. Lee, N. A. Kotov, *Nano Today* **2007**, *2*, 48–51.
118. J. Lee, A. O. Govorov, N. A. Kotov, *Angew. Chem. Int. Ed.* **2005**, *44*, 7439–7442.
119. T. P. M. Andersson, D. Filippini, A. Suska, T. L. Johansson, S. P. S. Svensson, I. Lundström, *Biosens. Bioelectron.* **2005**, *21*, 111–120.
120. J. Schmidt, *J. Mater. Chem.* **2005**, *15*, 831–840.
121. G. Favero, L. Campanella, S. Cavallo, A. D'Annibale, M. Perrella, E. Mattei, T. Ferri, *J. Am. Chem. Soc.* **2005**, *127*, 8103–8111.
122. T. Wilkop, D. Xu, Q. Cheng, *Langmuir* **2008**, *24*, 5615–5621.
123. S. C. Warren, F. J. DiSalvo, U. Wiesner, *Nature Mater.* **2007**, *6*, 156–161.
124. Y. Ding, J. Liu, X. Jin, H. Lu, G. Shen, R. Yu, *Analyst* **2008**, *133*, 184–190.

Chapter 20

Imprinted Functionalized Silica

MARYANNE M. COLLINSON

20.1 INTRODUCTION

Molecular recognition, defined as the favored binding of a molecule (i.e., a substrate) to a specific site in a receptor over other structurally and chemically related molecules, is at the forefront of science.[1–8] Long before man walked on this earth, nature had succeeded in the creation of a series of biologically based recognition elements with unmatched specificity: antibodies, enzymes, and receptors. Perhaps the simplest well-known example of this concept is the lock and key hypothesis that has been used to describe protein–substrate interactions in biological systems.[5–7]

Interest in the development of synthetic schemes and approaches to create molecular recognition elements has blossomed during the past half century for a number of reasons. In contrast to biologically based receptors, artificial mimics have the potential advantages of being less costly, more stable, and better able to withstand harsher conditions.[7–12] Furthermore, synthetic methodologies can be used to create receptors for molecules for which an artificial receptor does not exist. These "designer" materials have enormous potential in catalysis, clinical and pharmaceutical applications, chemical sensors, separation science, and electronics.[7–12]

The Supramolecular Chemistry of Organic–Inorganic Hybrid Materials. Edited by Knut Rurack and Ramón Martínez-Máñez

One promising means to create synthetic recognition systems involves an approach termed molecular imprinting.[5–23] In molecular imprinting, a polymeric framework is assembled around a suitable imprint (template) molecule, which upon removal yields diffusional pathways and/or microcavities with a specific size, shape, and/or chemical functionality in the cross-linked material. Such molecularly designed cavities show an affinity for the imprinted molecule over other structurally and chemically related compounds. A simple representation of this process is shown in Figure 20.1. To impart selectivity, many different types of intermolecular interactions can be exploited, including covalent bonds, hydrogen bonds, Van der Waals interactions, ionic interactions, metal ion coordination, and hydrophobic forces.[5–23] Not only are the number and strength of interactions important, but the porosity of the polymeric matrix as the imprinted molecule must be able to be removed and reintroduced in a timely fashion.

One powerful means to prepare such designer materials, popularized by the groups of Wulff[9] and Mosbach[6,15] relies on organic polymerization tools such as free radical polymerization. One example first described by the Mosbach group involves combining a functionalized monomer (i.e., methacrylic acid, 4-vinylbenzoic acid, 4-vinylpyridine, and styrene) with a cross-linker [i.e., ethylene glycol dimethacrylate (EGDMA)] in the presence of a porogen (solvent), a free radical initiator, and the template molecule.[8] After the rigid polymer forms, the material can be ground, sieved, and washed extensively in an appropriate solvent to remove the template.[8] The resulting materials then contain nanosized pores that are complementary in shape, size, and functionality to the template molecule. When properly designed

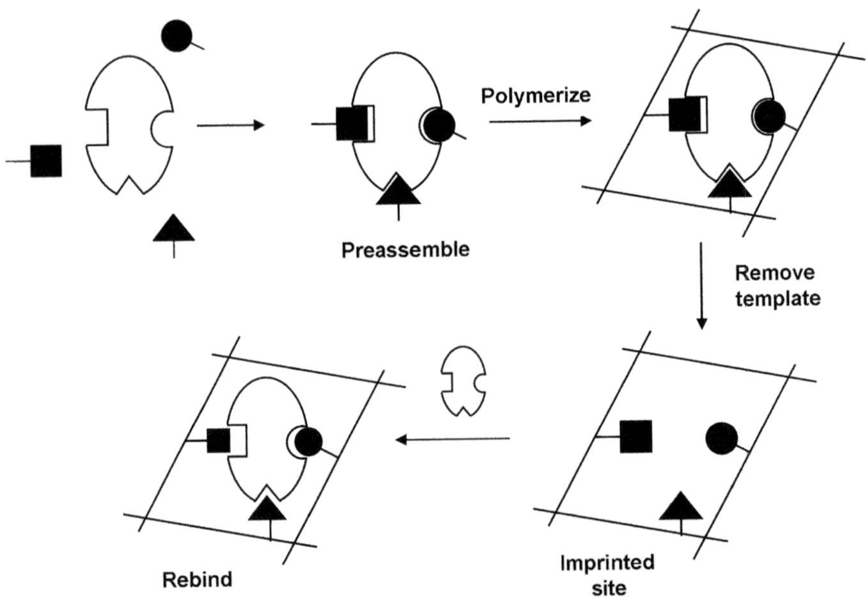

Figure 20.1 Simplified diagram of molecular imprinting.

and made, these materials show a distinct affinity for the template molecule compared to other structurally and chemically related analogs. A number of excellent review papers have appeared on this subject.[5–23]

A second, equally powerful means to prepare such materials relies on traditional inorganic polymerization tools, most notably sol–gel polymerization.[24,25] A number of excellent reviews have appeared on this subject as well.[5,12,17] In sol–gel processing, the functional monomer [i.e., an organoalkoxysilane such as 3-aminopropyltrimethoxysilane (APTMS)] is combined with the cross-linking agent [i.e., a tetrafunctional alkoxysilane such as tetramethoxysilane (TMOS) or tetraethoxysilane (TEOS)], a catalyst (such as hydrochloric acid or ammonia), and the template molecule. The resultant sol can be left to gel to form a monolith, which can then be dried, sieved, and extensively washed to remove the template. Alternatively, the sol can be spin coated, dip coated, or electrodeposited on a surface to yield a thin film, which can be subsequently washed with a solvent to remove the template and yield the imprinted cavities.

During the past decade, sol–gel chemistry as a means to prepare imprinted materials has seen a renewed interest relative to the time it was first used in the 1900s. Compared to organic polymerization tools, sol–gel processing has some unique advantages.[5,26,27]

1. Greater flexibility in terms of material preparation: thin films, powders, monoliths, and fibers can easily be made. The thin film configuration provides a means of reducing mass transfer problems common in bulk materials. Because of their thin path length for diffusion ($\ll 1$ μm thickness), response times tend to be shorter. However, concerns do arise about the thin film configuration because of their limited capacity.
2. Greater flexibility in terms of the processing conditions. Because sol–gel polymerization involves mild reaction conditions, biological molecules, water soluble molecules, and thermally sensitive molecules can be utilized as templates. Such molecules are often difficult to use as templates when traditional free radical polymerization in nonaqueous solvents is employed.
3. A large number of organoalkoxysilanes can be purchased or synthesized to help impart selectivity to the matrix. These functional monomers can be selected to complement specific functional groups on the template molecule to provide the materials with chemical specificity in addition to shape selectivity.
4. Often improved optical transparency, rigidity, chemical inertness, and porosity.

These two methods (sol–gel and free-radical polymerization) to form the polymeric framework have been compared. Marx and Liron specifically looked at thin films on glass using the template propranolol.[28] They showed that the acrylic thin film had higher uptake but also higher nonspecific adsorption, whereas the sol–gel derived thin film had a lower overall uptake, but faster uptake kinetics and a lower degree of nonspecific binding.[28] Cummins and coworkers also compared the two methods of polymerization, with a focus on bulk materials and rebinding in organic solvents using 2-aminopyridine as the template.[29] The results obtained depended on the solvent system used.[29] In acetonitrile, they showed that the sol–gel derived imprinted polymer "had a greater ability to maintain recognition" compared to the acrylic polymer.[29]

Regardless of the method by which they are made, molecularly imprinted polymers have found many applications in analytical and biochemical science.[5–23] Three of the most common applications of molecularly imprinted polymers have been chemical analysis, solid phase extraction, and chemical separations. The selective extraction, separation, detection, and/or quantification of imprint molecules are vitally important in pharmaceutical analysis, clinical chemistry, and any application that involves separations and/or chemical/biochemical sensing. Chiral stationary phases, materials for immunoassays, and sensor configurations based on mass sensing, fluorescence and other spectroscopic methods, and electrochemistry have been reported.[21] Also, a Swedish based company (www.miptechnologies.com) sells via Sigma-Aldrich (Supelco) solid-phase extraction cartridges based on molecular-imprinting technology for sample clean up.

In this chapter, the focus will be on the use of sol–gel chemistry to create imprints in functionalized silica. In the first part of this chapter, some background information and a brief history of imprinting using sol–gel chemistry will be presented. This will be followed by specific sections on (1) small molecule imprinting (covalent versus self-assembly), (2) chiral imprinting, (3) imprinting in or on novel supports, and (4) large molecule imprinting. It is important to note that this chapter is not meant to be a comprehensive compilation of the work in the field, but instead will highlight recent papers and techniques. The reader can find a number of review papers or research articles on imprinting depending on his or her interests: bioimprinting,[8,11,12,17,20] imprinted sensors,[6,13,14,18,30] imprinted stationary phases,[13,19] electrochemical sensors,[16] chiral imprinting,[22,31] and imprinting in titania.[32–34] In addition, Table 20.1 lists a few selected applications of imprinted functionalized silica and their corresponding references.

Table 20.1 Selected Examples of Applications Involving Imprinted Functionalized Silica

Reference	Application	Transducer	Template
58	Catalysts		Pyrenebutyric acid
59	Catalysts, chiral		Proline
75,79	Sorbents		Metal ions
83	Solid-phase extraction		Hg(II)
92	Extraction/detection	Electrochemical	Trinitrotoluene (TNT)
60	Sensing	Fluorescence	Fluorene
61	Sensing	Fluorescence	9-Anthrol
62	Sensing	ATR spectrometry	TNT
64,65	Sensing	Electrochemical	Dopamine
68	Sensing, chiral	Electrochemical	Catechol and ferrocene derivatives
91	Sensing	Piezoelectric	Yeasts
93	Sensing	Piezoelectric	L-histidine

20.2 BACKGROUND

Sol-gel processing provides a relatively facile means to prepare porous silica materials through the hydrolysis and condensation of alkoxysilanes.[24,25] In a typical procedure, the alkoxysilane is mixed with water in a mutual solvent. During sol–gel formation, the viscosity of the solution increases and eventually the sol becomes interconnected to form a rigid, porous structure—the gel. The time for gelation depends strongly on the sol–gel processing conditions and can be seconds to days to months (or longer). A simplified reaction scheme is depicted below for the hydrolysis and condensation of tetramethoxysilane (TMOS), a common precursor used to prepare silica materials.[24,25]

$$Si(OCH_3)_4 + H_2O \longrightarrow {\equiv}Si\text{-}OH + CH_3OH \qquad (hydrolysis) \qquad (20.1)$$

$$\equiv Si\text{-}OH + HO\text{-}Si{\equiv} \longrightarrow {\equiv}Si\text{-}O\text{-}Si{\equiv} + H_2O \qquad (condensation) \qquad (20.2)$$

Prior to gelation, the sol can be doped with a suitable molecule, which then becomes entrapped in a hydrated stable matrix upon gelation.[35] Since the sol–gel process is relatively mild compared to other polymerization procedures, temperature-sensitive and chemically fragile molecules can be incorporated in glass, including proteins and enzymes.[36–39] As a host matrix, sol–gel derived materials have many advantages relative to organic polymers, including ease of preparation and modification, photochemical and electrochemical stability, optical transparency, and good chemical stability.[36–39]

The chemical and physical properties (surface area, pore size, and pore size distribution) of the resultant materials strongly depend on how they are made. Important variables include the type and amount of the alkoxysilanes, the water to silane ratio, and the type and concentration of the catalyst (typically, hydrochloric acid, ammonia, or fluoride).[24,25] In general, low pH and/or low water to Si ratios give rise to materials that are denser, with smaller average pore sizes. Under high pH and/or high H_2O to Si ratios, more particulate sols are formed that give rise to gels that have larger interstitial porosity and larger average pore sizes.[24,25,38] The ability to tune the porosity of these high surface area materials is an important advantage when being used to prepare imprinted materials. High porosity is important when it comes down to the removal of the template from the functionalized cavities as well as when the template (or structurally related molecules) is reintroduced into the material.

Organically modified silicates (commonly termed ORMOSILs) are prepared by co-hydrolyzing and co-condensing organoalkoxysilanes [typically, R_x-$Si(OR')_{4-x}$] with the tetrafunctional silanes (i.e., TMOS), for example.[27,40–45] The trialkoxysilanes [R-$Si(OR')_3$] are often used as network modifiers to add specific properties to the material such as hydrophobicity, flexibility, and selectivity.[27,40–45] Examples of commonly used R groups are $-CH_3$, $-C_2H_5$, $-C_6H_5$, $-CH_2CH_2CH_2NH_2$, $-CH_2CH_2CH_2CN$, and $-CH_2CH_2CH_2SH$. A simplified reaction scheme is shown in Scheme 20.1 for the reaction of CH_3-$Si(OCH_3)_3$ with TMOS.

The ability to change the functional group, R, via the judicious choice of the organoalkoxysilane (many of which are commercially available) is particularly

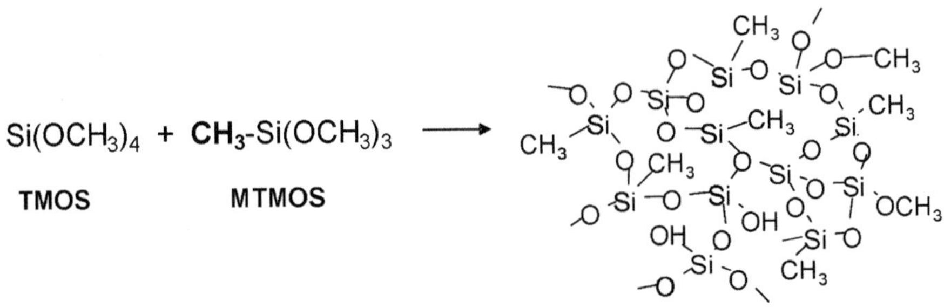

Scheme 20.1 Simplified diagram of ORMOSIL formation.

important in the preparation of materials that exhibit chemical selectivity toward a particular template. Organoalkoxysilanes with acidic, basic, and/or hydrophobic groups can be utilized depending on what type of interaction is desired, charged and/or hydrophobic. Some common examples of organoalkoxysilanes that have been used to create functionalized, imprinted silica are shown in Figure 20.2.

It is important to note that conceptually it seems easy enough to mix the different precursors together, but chemically it's complicated. The individual silanes have different rates of hydrolysis and condensation that can lead to materials that are phase

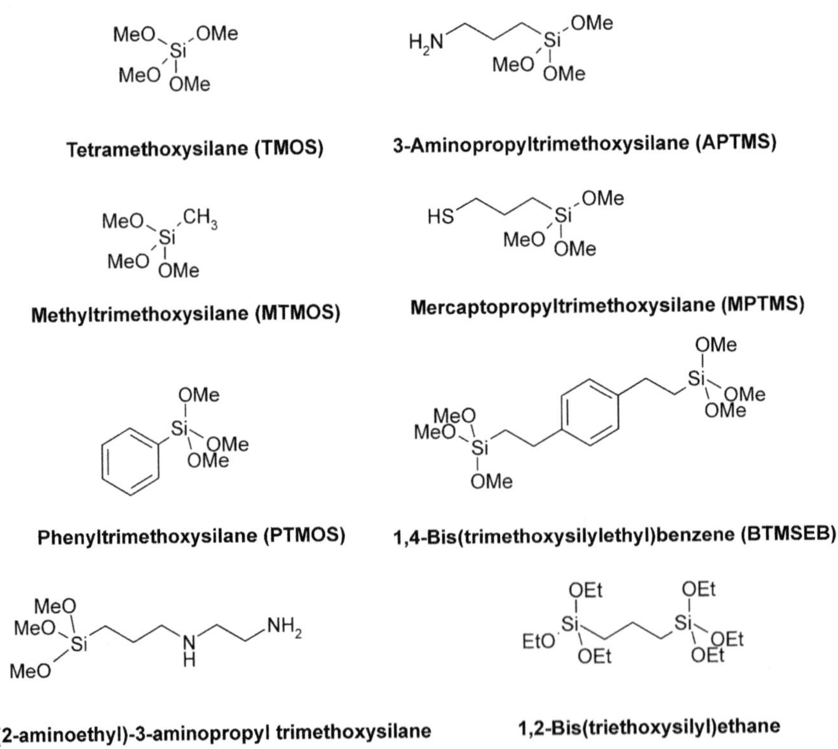

Figure 20.2 Selected examples of organoalkoxysilanes used to create imprinted silica.

separated (microscopically and/or macroscopically).[43,45] The rates of hydrolysis/condensation of the individual organoalkoxysilanes are influenced by steric and inductive factors and also the acidic/basic properties of the functional group since these groups can catalyze hydrolysis and condensation.[24,43,45] In general, under acidic conditions, organoalkoxysilanes [R-Si(OR′)$_3$] that have "simple" functional groups hydrolyze faster than the tetrafunctional silanes (i.e., TMOS). Under basic conditions, the reverse is true.[24]

Three of the most common configurations used in imprinting applications are monoliths, thin films, and powders.[24,38] Monolithic gels are easy to prepare by pouring the doped sol into an appropriate container and allowing the sol to gel and then dry. The shape of the monolith corresponds to the shape of the container used to house the sol–gel. The monoliths can be dried to form a xerogel (dried gel) and then crushed and sieved to form a powder-like material. Monoliths have the advantage of porosity, but the disadvantage of a long path length for diffusion (either out of the gel when the template needs to be removed or into the gel when the template is

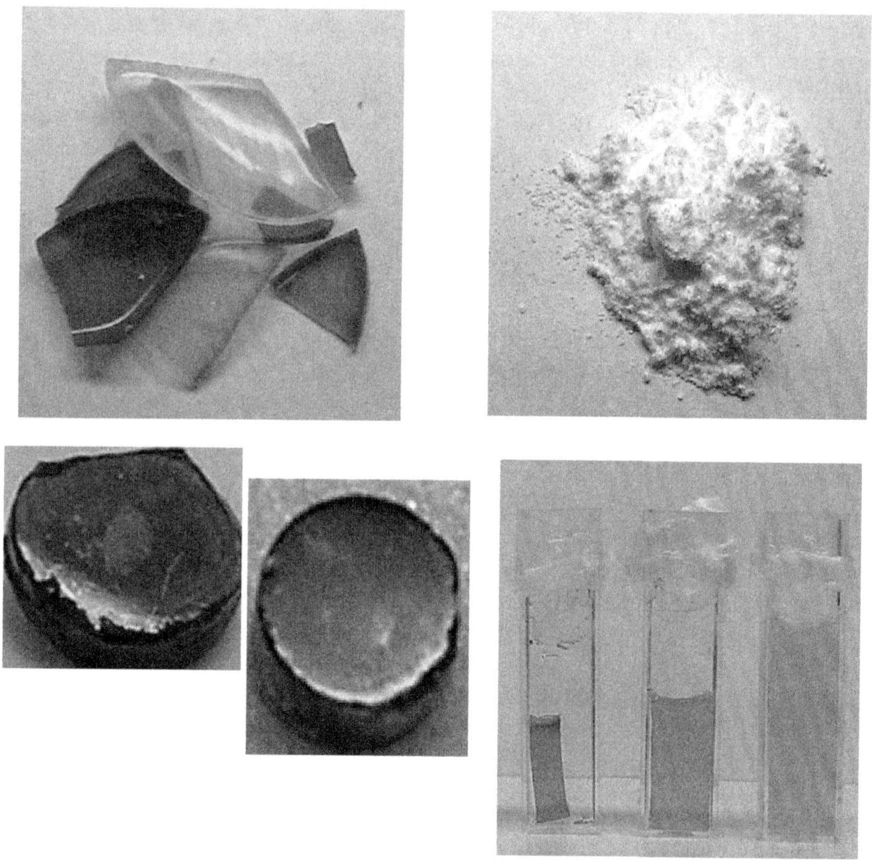

Figure 20.3 Digital photographs of sol–gel-derived materials: monoliths, powders, and films. (See color insert.)

reintroduced). The thin film configuration lessens this disadvantage due to their relatively short path length for diffusion, which is often hundreds of nanometers. Thin films can be made via the traditional methods[24,46,47] of dip coating or spin coating or more recent methods using aerosol[48] or electrodeposition techniques.[49,50] Because gelation and evaporation tend to occur simultaneously or near simultaneously in the thin film preparation methods (especially in spin-coating),[24,46,47] these materials tend to have lower porosity relative to a monolith prepared from an identical sol that may take days or weeks to age and dry. Figure 20.3 shows digital photographs of these three different configurations. The monolithic pieces contain entrapped dyes, which give rise to their distinct colors. The thin films (approximately 100 to 200 nm thick) were grown on glassy carbon via electrogeneration of hydroxide ions. Their color, which results from an optical interference effect, depends on refractive index and film thickness.[51,52] The mesoporous powders shown were prepared in a manner similar to that described by Beck and coworkers.[53,54]

An early, important example of silica imprinting was that of Dickey's work in the late 1940s and early 1950s.[55,56] In this work, the silica material was prepared from sodium silicate in the presence of a dye molecule. Upon removal of the dye via methanol extraction, the material showed an increased adsorption capacity for that particular dye. Example of dye molecules utilized included methyl orange, ethyl orange, propyl orange, and butyl orange.[55,56] Dickey argued that the mechanism for selectivity involved a "foot print" mechanism. Hodgson and coworkers later argued that it may involve an association (or recrystallization) type mechanism whereby the encapsulated molecules (unremoved template) acted as "nucleation centers" to enhance the readsorption of original dye molecule.[57] Regardless, these and other early studies introduced scientists to the notion that it may be possible to design materials that contain molecular recognition sites for a given species.

20.3 SMALL MOLECULE IMPRINTING

Small molecule imprinting in sol–gel matrices has received considerable interest in recent years, undoubtedly due to the flexibility offered by the sol–gel process.[5] Two different approaches have been utilized: covalent assembly and noncovalent self-assembly.[9] In the covalent assembly approach, the polymerizable functional group (i.e., the silicon alkoxide group) is covalently attached to the imprint molecule. The functionalized imprint molecule is then mixed with appropriate monomers (i.e., TMOS) to form the imprinted materials. After polymerization, the covalent bonds are cleaved to release the template and leave the molecular recognition pocket. Figure 20.4 shows a diagram of this process.

One of the first examples of the covalent imprinting of bulk silica was described by Katz and Davis.[58] In this study, an imprint molecule containing an aromatic core and protected 3-aminopropyltrimethoxysilane groups was synthesized. When hydrolyzed and condensed with TEOS followed by removal of the aromatic core, a "shape-selective base catalyst" was prepared.[58] Katz and Defreese also used a similar approach to create proline-amide imprints in bulk microporous silica.[59] Additional

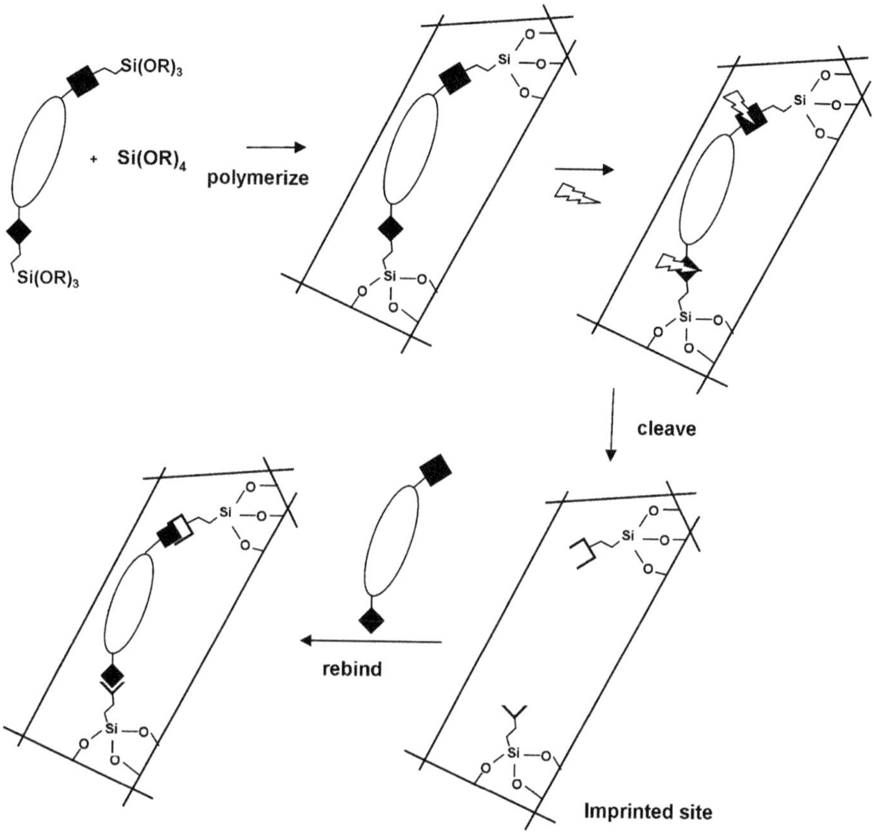

Figure 20.4 Simplified diagram of covalent imprinting in silica.

examples of the covalent approach have been reported, some of which are presented in more detail below.[60–63] The covalent approach has the advantage of producing more uniform and homogeneous imprinted sites,[8,15,58] but the disadvantage of complexity. Not only does it often require the synthesis of the template, but more elaborate and specific methods to break the covalent bond to release the template without destroying the molecular recognition pocket or the polymeric framework.[8]

The noncovalent self-assembly method relies on less specific types of interactions, including electrostatic, hydrogen bonding, metal ion coordination, Van der Waals forces, and/or hydrophobic attractions.[8] In this approach, the template and the polymerizable monomer (i.e., an organoalkoxysilane) are judiciously chosen so that they will have complementary interactions. Removal of the template is easier than in the covalent assembly approach and often involves simply washing the materials in a suitable solvent.[8]

A simple drawing of the sol–gel noncovalent imprinting scheme described by us many years ago is shown in Figure 20.5.[64,65] In this work, we were interested in making a thin film that was selective for dopamine over other common

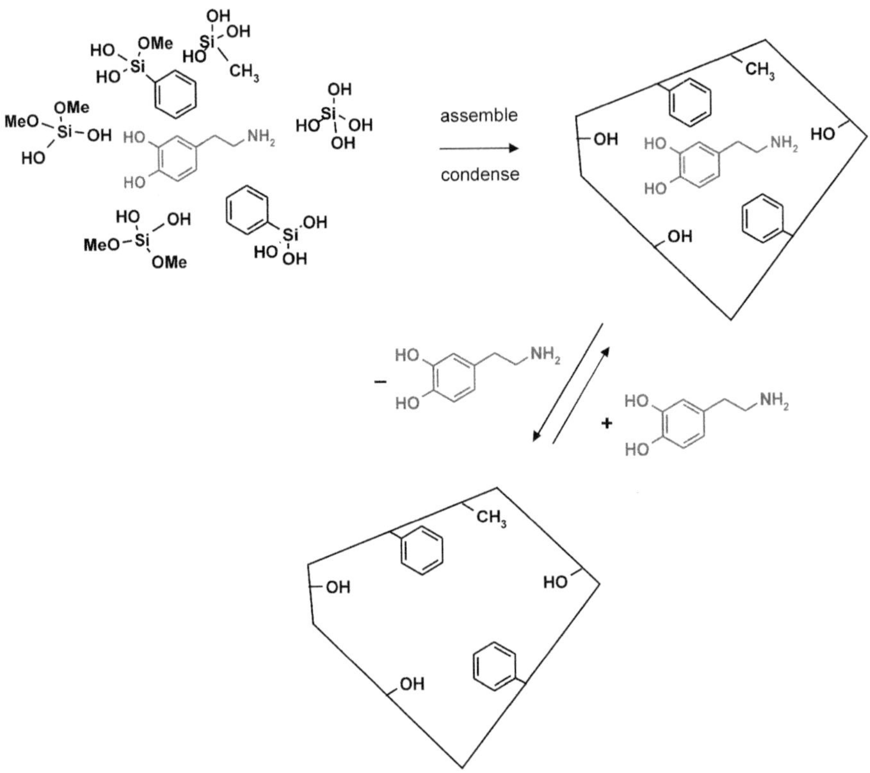

Figure 20.5 Simplified drawing depicting the noncovalent imprinting of dopamine in hybrid silica thin film.

neurotransmitters using molecular imprinting technology.[64,65] In this example, the sol was prepared by mixing the functional monomers [phenyltrimethoxysilane (PTMOS) and methyltrimethoxysilane (MTMOS)] with TMOS, which served as the primary cross-linking agent. These monomers were chosen because they would interact with the template (dopamine) using hydrophobic forces and π–π interactions. Dopamine was added to the sol and then the doped sol was spin coated on an electrode surface to form a thin film. Dopamine was removed from the matrix by soaking in phosphate buffer for an extended period of time.[64,65] The affinity of these films for dopamine was characterized using cyclic voltammetry, a common electroanalytical technique.

While technically simpler than the covalent approach, the self-assembly approach creates more heterogeneous sites and also requires templates with specific functional groups.[8] Since sol–gel chemistry is aqueous based, H-bonding interactions are significantly weaker compared to the conventional organic polymerization methods. Often, hydrophobic effects and ionic interactions are utilized. A number of other examples of the noncovalent approach to imprinting in sol–gel-derived materials are provided in recent reviews.[5,17] In the sections below, the focus will be on some of the newer aspects of small molecule imprinting in silica that involve the use of chiral templates to create materials that demonstrate enantioselectivity and the use of nanostructured materials as supports to improve diffusional transport.

20.3.1 Chiral Imprinting

Chiral porous materials have many applications in separation science and catalysis.[22] Marx and Avnir have described chirality in sol–gel materials, with a particular focus on their work.[31] In this section, an overview of some of this work is given, with emphasis on the use of chiral templates to create materials that demonstrate enantioselectivity.

One of the first mentions of an imprinted sol–gel-derived silica thin film that was able to discriminate against enantiomers was that of Marx and Liron, who created a chiral cavity for propranolol.[66] In a note added to the proof, the authors reported on data that showed a cavity templated with the pure enantiomer, *S*-propranolol, that discriminated between it and the R-enantiomer.[66] In a later publication, more details were provided on this method and two additional systems: (*R*)- and (*S*)-2,2,2-trifluoro-1-(9-anthryl) ethanol and D- and L-3,4-dihydroxyphenylalanine.[67]

In another report published near the same time, this group demonstrated the electrochemical detection of electroactive enantiomers, D- and L-3,4-dihydroxyphenylalanine (DOPA) and (*R*)- and (*S*)-*N*,*N*′-dimethylferrocenylethylamine (FcN) in imprinted sol–gel-derived thin films.[68] To improve response times, the imprinted films were made to be very thin, ~70 nm, which is about 3 to 10 times thinner than typical sol–gel-derived films. The functional monomer used to imprint DOPA was PTMOS (Fig. 20.2) because it was expected to exhibit $\pi-\pi$ and hydrophobicity interactions. For the ferrocene derivative, both PTMOS (Fig. 20.2) and carboxyethylsilanetriol sodium salt (CTES) were used for hydrophobicity and $\pi-\pi$ interactions and electrostatic interactions between the amine and carboxy groups, respectively.[68]

An intriguing report later followed these that reported on the ability of a templated cavity to selectively recognize the chirality of a molecule that is different than the template molecule.[69] In this study, Marx and coworkers created the host by hydrolyzing and co-condensing PTMOS and TMOS in the presence of a surfactant that contains two chiral centers, (1*R*,2*S*)-(-)-*N*-dodecyl-*N*-methylephedrinium bromide (DMB). The enantioselectivity of the DMB-imprinted films toward propranolol and 2,2,2-trifluoro-1-(9-anthryl) ethanol were examined. The imprinted films showed a preferential readsorption factor of *R*-propranolol over *S*-propranolol and (*R*)-2,2,2-trifluoro(anthryl)ethanol over the S enantiomer.[69] In more recent work, they described a similar approach whereby the chiral surfactant induces what this group termed "chiral porosity" in bulk silica materials when the surfactant was removed.[70] Additionally, they also described how the chiral surfactant that was doped into the silica material acted as a chiral center. The pairs of enantiomers used included (*R*)- and (*S*)-propranolol, (*R*)- and (*S*)-binaphthyl-2,2-diyl hydrogen phosphate, and (*R*)- and (*S*)-naproxen. Enantioselective preference was reversed upon going from the matrix that still contained the chiral surfactant compared to the material where the surfactant was removed. Furthermore, they showed selectivity in aqueous and organic solvents.[70] The ability to create a selective cavity for a molecule that is *not* the template is an intriguing concept. Clearly, the size of the cavity is not critical to selectivity but rather the spatial orientation of specific groups.

20.3.2 Surface Imprinting on Novel Supports

An overriding concern with imprinted materials, whether they are made with conventional organic polymerization routes or via the sol–gel process, is that of the accessibility of the imprinted sites.[17] Not only must one be able to remove the template deep within the center of the polymer, but also be able to get the template back to these same sites. The binding capacity of the imprinted polymer and the binding kinetics are also affected by accessibility. Historically, a number of different methods have been used to tackle this problem. The polymeric matrix can be ground into small particles to expose more sites and a greater surface area or a porogen (i.e., an inert solvent) can be incorporated into the polymer during formation and later washed away leaving pores. Alternatively surface imprints[23] can be created on a suitable support such as mesoporous silica particles or crushed and sieved xerogels.[71–73] Conventional silane coupling chemistry can be used to derivatize the particle with an appropriate functional group such as that shown by Mosbach and coworkers many years ago.[73] Alternatively, free radical organic polymerization can be used to form the polymeric network on the surface of this functionalized particle.[74]

Several different approaches utilize nanostructured materials as the support, including ordered mesoporous materials (i.e., MCM-41),[54] silica nanotubes and wires formed from nanopore alumina membranes, and silica nanoparticles. The advantage the mesoporous support provides is an open porous framework that facilitates removal and readsorption of the template. Dai and coworkers were among the first to describe how the inner surfaces of ordered mesoporous powders could be imprinted and used in metal ion recognition.[75] In one example, calcined ordered mesoporous silica with a high surface area and an average pore size of 25 Å was used as the support. A metal ion complex formed between a bifunctional metal ion binding ligand that contains a silicon alkoxide group (*N*-(2-aminoethyl)-3-aminopropyltrimethoxysilane; Fig. 20.2) and Cu^{2+} was coated on the inner surfaces of the support. The silicon alkoxide groups on the bifunctional ligand hydrolyzed and then condensed to the mesoporous support. The metal ion was removed by soaking the material in acid. The distribution coefficient for the imprinted material was significantly larger relative to control samples.[75] In later work, this group also showed how their approach can be extended to other ordered supports, most notably L-zeolite.[76] The internal surfaces of the zeolite were functionalized by first loading Cu^{2+} into the pores followed by the bifunctional silane ligand, (*N*-(2-aminoethyl)-3-aminopropyltrimethoxysilane; Fig. 20.2), and then condensation between ligand and surface silanol groups.[76]

Yang and coworkers also used a surface imprinting approach to prepare nanotube membranes that exhibit selectivity for the molecule estrone.[77] In this study, the silica nanotubes with pore diameters of 100 nm were synthesized within the cylindrical pores of nanopore alumina membranes. A covalent assembly approach was used to prepare the imprint.[77] Zhang and coworkers have also recently shown how silica nanotubes can be used as an imprinting scaffold.[78] In this example, trinitrotoluene (TNT) was used as the template and imprinted sites were formed on the walls of a silica nanotube. An alumina membrane was first derivatized with APTMS (Fig. 20.2) and then immersed in a sol containing TEOS, APTMS, cetyltrimethylammonium bromide (CTAB), and TNT. The membrane was dried in an oven and then dissolved in acid to create the individual silica nanotubes with the imprinted pockets on their walls. The authors showed that the uptake capacity of imprinted TNT was larger by a factor of 3.6 compared to bulk particles. In addition, uptake rates were significantly shorter using the imprinted nanotubes.[78]

A recent example of how silica nanoparticles prepared by the Stöber method can be used as supports was the work of Gao et al., who derivatized a silica nanoparticle first with APTMS and then with acryloyl chloride to form reactive vinyl groups.[74] A polymer shell with sites imprinted with the template, TNT, was then formed around the silica nanoparticle using conventional acrylic organic polymerization procedures. The capacity and binding kinetics were shown to be significantly better than traditional imprinted particles.[74]

A second method that has been described as a means to deal with the limited accessibility issue involves a copolymerization route. In this approach, the porous matrix is copolymerized with the template. Burleigh and coworkers, for example, described the preparation of imprinted polysilsesquioxanes for the recognition of metal ions.[79] Polysilsesquioxanes are hybrid porous materials synthesized from bridged alkoxide precursors such as shown in Figure 20.2.[80] In this example, the bridged silsesquioxane precursors were copolymerized with the metal ion complex in the presence of surfactant. Once the surfactant and metal ion were removed, a porous network was formed that showed a high affinity for the metal ion.[79]

In another example, a material selective for fluorene was prepared using the bridged silane precursor, bis(trimethoxysilylethyl) benzene (Fig. 20.2).[60] In this example, a covalent assembly approach was used whereby the template was covalently attached to a silicon alkoxide group. It was hypothesized that the bridging aromatic group on the bridged precursor would exhibit π–π stacking with the fluorene in addition to help create a more porous material.[60] A somewhat related approach was used by the same group to prepare materials selective toward gas-phase TNT.[62] Three different TNT templates containing one or two silicon alkoxide groups were synthesized and the imprinted polymer prepared by mixing one of these templates with the bis-functionalized precursor as the cross-linker.[62]

Dai and coworkers also described an approach that they termed hierarchically double imprinting.[81,82] In this approach, two templates are used: (1) one that produces pores on the mesoporous level (25 Å) and (2) one that produces pores on the microporous level (1 to 2 Å). The procedure Dai and coworkers used to prepare hierarchically imprinted sorbents for metal ions involved mixing CTAB with the metal ion complex (the silicon alkoxide modified bifunctional ligand coordinated with the metal ion) and a silica source.[81,82] Soxhlet extraction removed the surfactant template while extensive acid washing removed the metal ion. The resulting materials have high surface areas (200 to 600 $m^2 g^{-1}$) and showed a significant improvement in selectivity compared to control samples. In a more recent example, He and coworkers also used this same double imprinting approach to create materials selective for mercury ions.[83]

20.4 LARGE MOLECULE IMPRINTING

As described in the previous section, small molecule molecular imprinting has been well studied. In contrast, biomolecule (i.e., proteins, enzymes, and viruses) imprinting has been less developed due to the challenges that face researchers in this field, including:[11,12,17,20]

1. Harsh synthesis conditions, particularly for biomolecules.
2. Biomolecule complexity, particularly as it pertains to size, conformation flexibility, surface functionalities, and stability.
3. Limited solubility and stability in solvents commonly used in imprinting procedures.
4. Difficulties associated with the removal of the biomolecule and its re-incorporation into the molecularly designed pores due to its shear size and/or limited porosity of the polymeric framework.

Sol–gel processing, being relatively mild and more aqueous based compared to conventional organic polymerization, provides a means to imprint large biomolecules, such as proteins, enzymes, bacteria, and viruses. It still, however, is not without its challenges. Some of the early investigations focused on bulk imprinting whereby a monolithic xerogel is made and then washed appropriately to remove the imprint, dried, and crushed. Mosbach and coworkers, for example, have described approaches for molecular imprinting and enzyme entrapment based on the condensation of organosilanes on the surface of porous silica particles.[73] In addition, Venton and Gudipati prepared xerogels by combining TEOS with APTMS (Fig. 20.2) in the presence of the proteins urease and bovine serum albumin (BSA).[84] They were able to show that urease preferentially bound to the xerogel prepared when urease was used as the imprint and BSA preferentially bound to the xerogel prepared using BSA as the imprint.[84]

The approach that has been used nearly exclusively now is some form of surface imprinting.[23] The one significant advantage that surface imprinting provides is that the binding sites are more accessible, thus making it easier to remove the protein from the imprinted pocket and get it back in. Considering the shear sizes of biomolecules and their relatively slow diffusion, restricted access to and from imprinted pockets can be a significant limitation.[11,17] Surface imprinting provides one solution to this problem. Surface imprinting strategies have involved the use of thin films coated on porous particles, thin films coated on planar substrates, and stamping.

In an interesting illustration of the first strategy, Sakaguchi and coworkers covalently attached hemoglobin to an aminopropyl silica particle and then polymerized organoalkoxysilanes on the surface of the hemoglobin-modified silica particle.[85] The template was removed via treatment with oxalic acid.[85] In more recent work, Zhang and coworkers utilized a similar approach. In their case, the sphere was made from the functionalized biopolymer, chitosan.[86] The model template protein, bovine serum albumin, was covalently attached to the chitosan microsphere and then coated with a composite sol prepared from TEOS and an aminosilane.[86]

Along similar lines, the protein ricin and its A and B chains were imprinted on a surface of a silica particle using a composite sol prepared from TEOS and bis-(2-hydroxy-ethyl)–aminopropyltriethoxysilane.[87] The protein was removed by washing in buffer. The advantages this strategy holds are that the support has a high surface area, thus providing potentially more imprinted sites, it is stable, and does not require grinding.

An example of the thin film method was illustrated by Yao and coworkers.[88] In this study, human serum albumin was added to a composite sol prepared by mixing TEOS with MTMOS (Fig. 20.2) and PTMOS (Fig. 20.2). A gold coated quartz crystal previously self-assembled with thioglycolic acid was dipped into this composite sol several times to form a film (approximately 150 to 500 nm thick). The protein was removed by repeatedly washing the film in hot double-distilled water. Binding was detected using the piezoelectric quartz crystal impedance and the electrochemical impedance technique.[88] Another approach was demonstrated by Bright and coworkers, which they termed PIXIES, for protein imprinted xerogels with integrated emission sites.[18,89] The target protein was ovalbumin and the approach involved several steps: (1) determination of the appropriate formulation of sol via pin printing; (2) formation of the imprinted pocket by mixing ovalbumin and the composite sol, spin coating it on microscope slides, and removing the protein by washing the xerogel films with urea or phosphoric acid; and (3) incorporation of the fluorescent probe by tagging the imprinted pockets within the xerogel film with a luminophore.[89]

In the third case, described by Dickert and coworkers, the imprint of a biomolecule (in this case, microorganisms) was formed on the surface of suitable substrate using a stamp.[90] The microorganism was first coated on a surface and then pressed into a polymer coating on a transducer to leave the imprint of the microorganism behind. This process was demonstrated for various microorganisms stamped into polymer surfaces and also coatings prepared by the sol–gel process.[91] Mass sensitive techniques were used to detect binding.

20.5 CONCLUDING REMARKS

Imprinted functionalized silica has enormous potential to create the next generation of molecular recognition materials. Their ease of preparation coupled with processing flexibility allows them to be used in a variety of different applications, including those that involve fragile biomolecules. Integrating nanostructured supports into molecular imprinting design provides a powerful means to overcome limitations associated with mass transport into and out of these materials. Likewise, merging sol–gel-based imprinting strategies with the more traditional free radical-based organic polymerization provides a means to obtain the best of both polymerization worlds. Chiral imprinting in functionalized silica provides an important approach to create supports for the often difficult separation of enantiomers. The broad and powerful field of molecular imprinting will likely continue to develop in these directions and play an important role in sensor development, separation science, nanopore filtration, and pharmaceutical applications.

REFERENCES

1. P. D. Beer, P. A. Gale, *Angew. Chem. Int. Ed.* **2001**, *40*, 486–516.
2. L. Fabbrizzi, A. Poggi, *Chem. Soc. Rev.* **1995**, *24*, 197–202.
3. J.-M. Lehn, *Angew. Chem. Int. Ed.* **1988**, *27*, 89–112.
4. T. E. Mallouk, J. A. Gavin, *Acc. Chem. Res.* **1998**, *31*, 209–217.
5. M. E. Díaz-García, R. B. Laino, *Microchim. Acta* **2005**, *149*, 19–36.
6. D. Kriz, O. Ramstrom, K. Mosbach, *Anal. Chem.* **1997**, *69*, 345A–349A.
7. B. Sellergren, *Trends Anal. Chem.* **1997**, *16*, 310–320.
8. L. Ye, K. Mosbach, *Chem. Mater.* **2008**, *20*, 859–868.
9. G. Wulff, *Angew. Chem. Int. Ed.* **1995**, *34*, 1812–1832.
10. S. T. Wei, M. Jakusch, B. Mizaikoff, *Anal. Chim. Acta* **2006**, *578*, 50–58.
11. D. E. Hansen, *Biomaterials* **2007**, *28*, 4178–4191.
12. A. Bossi, F. Bonini, A. P. F. Turner, S. A. Piletsky, *Biosens. Bioelectron.* **2007**, *22*, 1131–1137.
13. K. Ensing, T. De Boer, *Trends Anal. Chem.* **1999**, *18*, 138–145.
14. K. Haupt, K. Mosbach, *Chem. Rev.* **2000**, *100*, 2495–2504.
15. A. G. Mayes, K. Mosbach, *Trends Anal. Chem.* **1997**, *16*, 321–332.
16. A. Merkoci, S. Alegret, *Trends Anal. Chem.* **2002**, *21*, 717–725.
17. N. W. Turner, C. W. Jeans, K. R. Brain, C. J. Allender, V. Hlady, D. W. Britt, *Biotechnol. Progr.* **2006**, *22*, 1474–1489.
18. E. L. Holthoff, F. V. Bright, *Acc. Chem. Res.* **2007**, *40*, 756–767.
19. V. T. Remcho, Z. J. Tan, *Anal. Chem.* **1999**, *71*, 248A–255A.
20. D. S. Janiak, P. Kofinas, *Anal. Bioanal. Chem.* **2007**, *389*, 399–404.
21. C. Alexander, H. S. Andersson, L. I. Andersson, R. J. Ansell, N. Kirsch, I. A. Nicholls, J. O'Mahony, M. J. Whitcombe, *J. Mol. Recognit.* **2006**, *19*, 106–180.
22. N. M. Maier, W. Lindner, *Anal. Bioanal. Chem.* **2007**, *389*, 377–397.
23. C. J. Tan, Y. W. Tong, *Anal. Bioanal. Chem.* **2007**, *389*, 369–376.
24. J. Brinker, G. Scherer, *Sol-Gel Science*, New York: Academic Press, **1990**.
25. L. L. Hench, J. K. West, *Chem. Rev.* **1990**, *90*, 33–72.
26. M. M. Collinson, *Crit. Rev. Anal. Chem.* **1999**, *29*, 289–311.
27. A. Walcarius, D. Mandler, J. Cox, M. M. Collinson, O. Lev, *J. Mater. Chem.* **2005**, *15*, 3663–3689.
28. S. Marx, Z. Liron, *Chem. Mater.* **2001**, *13*, 3624–3630.
29. W. Cummins, P. Duggan, P. McLoughlin, *Anal. Chim. Acta* **2005**, *542*, 52–60.
30. E. L. Holthoff, F. V. Bright, *Anal. Chim. Acta* **2007**, *594*, 147–161.
31. S. Marx, D. Avnir, *Acc. Chem. Res.* **2007**, *40*, 768–776.
32. S. W. Lee, D. H. Yang, T. Kunitake, *Sens. Actuators B* **2005**, *104*, 35–42.
33. T. Kunitake, S. W. Lee, *Anal. Chim. Acta* **2004**, *504*, 1–6.
34. S. W. Lee, I. Ichinose, T. Kunitake, *Langmuir* **1998**, *14*, 2857–2863.
35. D. Avnir, D. Levy, R. Reisfeld, *J. Phys. Chem.* **1984**, *88*, 5956–5959.
36. D. Avnir, *Acc. Chem. Res.* **1995**, *28*, 328–334.
37. B. C. Dave, B. Dunn, J. S. Valentine, J. I. Zink, *Anal. Chem.* **1994**, *66*, 1120A–1127A.
38. O. Lev, M. Tsionsky, L. Rabinovich, V. Glezer, S. Sampath, I. Pankratov, J. Gun, *Anal. Chem.* **1995**, *67*, 22A–30A.
39. D. Avnir, S. Braun, O. Lev, M. Ottolenghi, *Chem. Mater.* **1994**, *6*, 1605–1614.
40. M. M. Collinson, *Trends Anal. Chem.* **2002**, *21*, 30–38.
41. J. Wen, G. L. Wilkes, *Chem. Mater.* **1996**, *8*, 1667–1681.
42. S. Sánchez, F. Ribot, *New J. Chem.* **1994**, *18*, 1007–1047.
43. U. Schubert, N. Husing, A. Lorenz, *Chem. Mater.* **1995**, *7*, 2010–2027.
44. M. Pagliaro, R. Ciriminna, M. W. C. Man, S. Campestrini, *J. Phys. Chem. B* **2006**, *110*, 1976–1988.
45. H. Schmidt, *J. Sol-Gel Sci. Technol.* **1994**, *1*, 217–231.
46. C. J. Brinker, G. C. Frye, A. J. Hurd, C. S. Ashley, *Thin Solid Films* **1991**, *201*, 97–108.
47. C. J. Brinker, A. J. Hurd, P. R. Schunk, G. C. Frye, C. S. Ashley, *J. Non-Cryst. Solids* **1992**, *147–148*, 424–436.

48. J. D. Jordan, R. A. Dunbar, F. V. Bright, *Anal. Chim. Acta* **1996**, *332*, 83–91.
49. M. M. Collinson, *Acc. Chem. Res.* **2007**, *40*, 777–783.
50. A. Walcarius, E. Sibottier, M. Etienne, J. Ghanbaja, *Nature Mater.* **2007**, *6*, 602–608.
51. J. C. Manifacier, J. Gasiot, J. P. Fillard, *J. Phys. E. Sci. Instrum.* **1976**, *9*, 1002–1004.
52. C. L. Schauer, M. S. Chen, M. Chatterley, K. Eisemann, E. R. Welsh, R. R. Price, P. E. Schoen, F. S. Ligler, *Thin Solid Films* **2003**, *434*, 250–257.
53. C. T. Kresge, M. E. Leonowicz, W. J. Roth, J. C. Vartuli, J. S. Beck, *Nature* **1992**, *359*, 710–712.
54. J. S. Beck, J. C. Vartuli, W. J. Roth, M. E. Leonowicz, C. T. Kresge, K. D. Schmitt, C. T.-W. Chu, D. H. Olson, E. W. Sheppard, S. B. McCullen, J. B. Higgins, J. L. Schlenker, *J. Am. Chem. Soc.* **1992**, *114*, 10834–10843.
55. F. H. Dickey, *Proc. Natl. Acad. Sci. U.S.A.* **1949**, *35*, 227–229.
56. F. H. Dickey, *J. Phys. Chem.* **1955**, *59*, 695–707.
57. J. L. Morrison, M. Worsley, D. R. Shaw, G. W. Hodgson, *Can. J. Chem.* **1959**, *37*, 1986–1995.
58. A. Katz, M. E. Davis, *Nature* **2000**, *403*, 286–289.
59. J. L. Defreese, A. Katz, *Chem. Mater.* **2005**, *17*, 6503–6506.
60. C. A. Carlson, J. A. Lloyd, S. L. Dean, N. R. Walker, P. L. Edmiston, *Anal. Chem.* **2006**, *78*, 3537–3542.
61. E. L. Shughart, K. Ahsan, M. R. Detty, F. V. Bright, *Anal. Chem.* **2006**, *78*, 3165–3170.
62. N. R. Walker, M. J. Linman, M. M. Timmers, S. L. Dean, C. M. Burkett, J. A. Lloyd, J. D. Keelor, B. M. Baughman, P. L. Edmiston, *Anal. Chim. Acta* **2007**, *593*, 82–91.
63. H.-H. Yang, S.-Q. Zhang, W. Yang, X.-L. Chen, Z.-X. Zhuang, J.-G. Xu, X.-R. Wang, *J. Am. Chem. Soc.* **2004**, *126*, 4054–4055.
64. R. Makote, M. M. Collinson, *Chem. Mater.* **1998**, *10*, 2440–2445.
65. R. Makote, M. M. Collinson, *Chem. Commun.* **1998**, 425–426.
66. S. Marx, Z. Liron, *Chem. Mater.* **2001**, *13*, 3624–3630.
67. S. Fireman-Shoresh, D. Avnir, S. Marx, *Chem. Mater.* **2003**, *15*, 3607–3613.
68. S. Fireman-Shoresh, I. Turyan, D. Mandler, D. Avnir, S. Marx, *Langmuir* **2005**, *21*, 7842–7847.
69. S. Fireman-Shoresh, I. Popov, D. Avnir, S. Marx, *J. Am. Chem. Soc.* **2005**, *127*, 2650–2655.
70. S. Fireman-Shoresh, S. Marx, D. Avnir, *Adv. Mater.* **2007**, *19*, 2145–2150.
71. J. D. Bass, A. Katz, *Chem. Mater.* **2006**, *18*, 1611–1620.
72. D. Y. Sasaki, T. M. Alam, *Chem. Mater.* **2000**, *12*, 1400–1407.
73. M. Glad, O. Norrlow, B. Sellergren, N. Siegbahn, K. Mosbach, *J. Chromatogr.* **1985**, *347*, 11–23.
74. D. M. Gao, Z. P. Zhang, M. H. Wu, C. G. Xie, G. J. Guan, D. P. Wang, *J. Am. Chem. Soc.* **2007**, *129*, 7859–7866.
75. S. Dai, M. C. Burleigh, Y. Shin, C. C. Morrow, C. E. Barnes, Z. Xue, *Angew. Chem. Int. Ed.* **1999**, *38*, 1235–1239.
76. Z. H. Zhang, S. Dai, R. D. Hunt, Y. Wei, S. Qiu, *Adv. Mater.* **2001**, *13*, 493–496.
77. H.-H. Yang, S.-Q. Zhang, W. Yang, X.-L. Chen, Z.-X. Zhuang, J.-G. Xu, X.-R. Wang, *J. Am. Chem. Soc.* **2004**, *126*, 4054–4055.
78. C. G. Xie, B. H. Liu, Z. Y. Wang, D. M. Gao, G. J. Guan, Z. P. Zhang, *Anal. Chem.* **2008**, *80*, 437–443.
79. M. C. Burleigh, S. Dai, E. W. Hagaman, J. S. Lin, *Chem. Mater.* **2001**, *13*, 2537–2546.
80. K. J. Shea, D. A. Loy, O. Webster, *J. Am. Chem. Soc.* **1992**, *114*, 6700–6710.
81. S. Dai, M. C. Burleigh, Y. H. Ju, H. J. Gao, J. S. Lin, S. J. Pennycook, C. E. Barnes, Z. L. Xue, *J. Am. Chem. Soc.* **2000**, *122*, 992–993.
82. S. Dai, *Chem. Eur. J.* **2001**, *7*, 763–768.
83. G. H. Wu, Z. Q. Wang, J. Wang, C. Y. He, *Anal. Chim. Acta* **2007**, *582*, 304–310.
84. D. L. Venton, E. Gudipati, *Biochim. Biophys. Acta* **1995**, *1250*, 126–136.
85. T. Shiomi, M. Matsui, F. Mizukami, K. Sakaguchi, *Biomaterials* **2005**, *26*, 5564–5571.
86. F. Li, J. Li, S. S. Zhang, *Talanta* **2008**, *74*, 1247–1255.
87. M. F. Lulka, S. S. Iqbal, J. P. Chambers, E. R. Valdes, R. G. Thompson, M. T. Goode, J. J. Valdes, *Mater. Sci. Eng. C* **2000**, *11*, 101–105.
88. Z. H. Zhang, Y. M. Long, L. H. Nie, S. Z. Yao, *Biosens. Bioelectron.* **2006**, *21*, 1244–1251.

89. Z. Y. Tao, E. C. Tehan, R. M. Bukowski, Y. Tang, E. L. Shughart, W. G. Holthoff, A. N. Cartwright, A. H. Titus, F. V. Bright, *Anal. Chim. Acta* **2006**, *564*, 59–65.
90. O. Hayden, P. A. Lieberzeit, D. Blaas, F. L. Dickert, *Adv. Funct. Mater.* **2006**, *16*, 1269–1278.
91. F. L. Dickert, O. Hayden, *Anal. Chem.* **2002**, *74*, 1302–1306.
92. S. A. Trammell, M. Zeinali, B. J. Melde, P. T. Charles, F. L. Velez, M. A. Dinderman, A. Kusterbeck, M. A. Markowitz, *Anal. Chem.* **2008**, *80*, 4627–4633.
93. Z. Zhang, H. Liao, H. Li, L. Nie, S. Yao, *Anal. Biochem.* **2005**, *336*, 108–116.

Chapter 21

Bioinspired Block Copolymer-Based Hybrid Materials

MARLEEN KAMPERMAN AND ULRICH WIESNER

21.1 INTRODUCTION

Organisms use three conceptually different strategies to build their skeletal parts. The easiest approach to the filling of a given shape with solid material applies when there is no internal order or structure, that is, when the material is amorphous. A beautiful example is found within the *Euplectella* species of marine sponge, which comprises at least seven hierarchical levels within its silica skeleton, the lowest level consisting of silica nanoparticles.[1] A second approach involves building skeletal materials

The Supramolecular Chemistry of Organic–Inorganic Hybrid Materials. Edited by Knut Rurack and Ramón Martínez-Máñez

as polycrystalline assemblies formed inside self-assembled matrices composed of biological macromolecules such as proteins and polysaccharides. Control of the microenvironment of crystal formation in these systems, which include bone, teeth, and mollusk shells, is crucial. A third solution involves building a material as a single crystal, deposited from an amorphous phase that very slowly transforms in a controlled manner. The regulation of the transformation to the crystalline state is crucial to the shape, structure, and microtexture of the final material. Examples of single crystal skeletal elements are the calcitic sea urchin larval spicules and adult spines. The striking calcitic microlenses found in the dorsal arm plates of light-sensitive brittle stars also belong to this group.[2] Each of these approaches provides inspiration for the design and construction of technologically important materials based on new concepts. Thus, they open opportunities beyond those arising by simply mimicking specific materials.

As biology evolved to remarkable and complex designs, synthetic bioinspired materials are evolving towards new levels of complexity achieving larger combinations of properties within one material. This development can easily lead to long and complex, and therefore costly, synthetic procedures with multiple steps. In order to make substantial progress in this field toward translating research results into industrial applications it is necessary, however, to keep the number of required synthesis steps to a minimum. To this end, in particular "one-pot" approaches, in which the entire synthesis can be carried out in a single dish from mixing the precursors to the final product, are highly desirable. In these approaches, structure-directing agents may direct multiple components through a self-assembly process into structures with nanoscale precision to achieve desired combinations. If required, final processing, for example, through annealing, may transform the composite into final composition, structure, form, and shape.

In this chapter, recent progress in developing such synthetic one-pot approaches toward organic–inorganic hybrid materials, inspired by nature, is described. Several review articles on (bioinspired) organic–inorganic hybrid materials have appeared in the past few years.[3,4] Therefore, rather than attempting a general overview of this rapidly growing and exciting research area, general features, challenges, and future trends will be highlighted through examples of work of the Wiesner research group at Cornell University. The chapter is structured into two parts related to two of the biomineralization approaches described above. In the first part, work inspired by nature's approach to structure amorphous materials is highlighted. The second part of this chapter focuses on recent progress towards nanostructured polycrystalline assemblies from block copolymer mesophases.

21.2 NANOSTRUCTURED AMORPHOUS HYBRID MATERIALS

The first part on nanostructured amorphous hybrid materials is divided in two sections. First, new findings on coassembly mechanisms of mesostructured block copolymer–silica-type hybrids are described. More specifically, the influence of nanoparticle characteristics on structure control in a particular block copolymer–silica-type inorganic hybrid material is discussed. It will become clear that with the emerging understanding refined control is established over nanoparticle placement in segregated mesostructures that may go well beyond the silica model system. In the second section it is shown how this control, in combination with the exploration of different polymer architectures like AB dendritic and ABC triblock copolymers as structure-directing agents, offers tremendous opportunities for the generation of nanostructured materials with complex property profiles.

21.2.1 Influence of Nanoparticle Characteristics on the Assembly of Mesostructured Silica

Silica nanoparticles have been implicated as the fundamental structural unit in a number of silica-producing organisms as studies probe these structures at the molecular level.[5,6] Despite a growing recognition of silica nanoparticles' widespread occurrence in nature and the variety of silica mesostructures that have been synthesized,[7–9] the influence of nanoparticle characteristics in the assembly of mesostructured silica remains an important issue. Recently, the Wiesner group investigated the influence of nanoparticle size effects on mesostructure assembly, by using high loadings of silica-type nanoparticles in combination with block copolymers.[10,11] Although recent experiments have highlighted the importance of nanoparticle size on the miscibility of nanoparticles with homopolymers,[12,13] these experiments have only focused on the regime of low nanoparticle density—materials with less than 10 wt% of nanoparticles. In contrast, the regime of high nanoparticle density was examined by the Wiesner group, because it is most relevant for the assembly of mesostructured materials.

Using an amphiphilic AB block copolymer that self-assembles with silica-type nanoparticles, we showed that $d/R_{0,B}$ (where d is the nanoparticle diameter and $R_{0,B}$ is the root-mean-square end-to-end distance of the B block) is a critical design parameter for silica mesostructures. When the particles are small in comparison to the root-mean-square end-to-end distance of the polymer chain, they can swell the polymer domain without significantly perturbing the polymer chain conformations. In this case, the translational entropy of the particles dominates the behavior, leading to a dispersion of the nanoparticles within the B domain, as depicted on the left of Figure 21.1. In contrast, when the particle size is of the order of or larger than the root-mean-square end-to-end distance of the chains, incorporation into the B domain significantly perturbs the chain conformations since B chains have to stretch to get around the spheres, resulting in a loss in chain conformational entropy. Larger particles then segregate into a particle rich core which prevents the chain entropy penalty (at the expense of translational entropy of the particles), as shown on the right of Figure 21.1.

With this deepened understanding we showed that $d/R_{0,B}$ can be used as a design parameter to switch among different morphologies (Fig. 21.2). Therefore, considering size-dependent segregation, in addition to changes in volume fraction, is essential to morphology control.

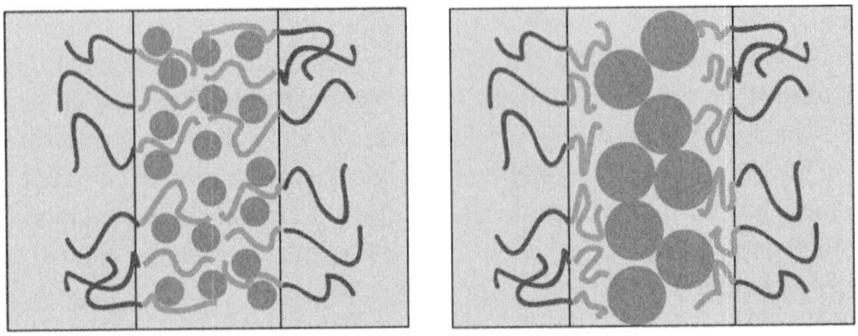

Figure 21.1 Block copolymer–nanoparticle co-assemblies as a function of nanoparticle size. The left shows nanoparticles smaller than the root-mean-square end-to-end distance of the block copolymer leading to particle distribution throughout the domain. The right scenario shows much larger nanoparticles segregating out from the block copolymer.[11] (Reprinted with permission from A. Jain and U. Wiesner, *Macromolecules* **2004**, *37*, 5665–5670. Copyright 2004 American Chemical Society.)

Silica-type nanoparticles and poly(isoprene-*block*-ethylene oxide) (PI-*b*-PEO) block copolymers (for polymer characteristics see Table 21.1, A and B) were selected as a model system for this study as it is well characterized[14] and both nanoparticle and block copolymer sizes can be independently tuned. Dipole–dipole interactions and hydrogen bonding drive mixing of the nanoparticles (NP) and PEO (B) and the Flory–Huggins interaction parameter, χ_{BP}, is negative. This allows high loadings of

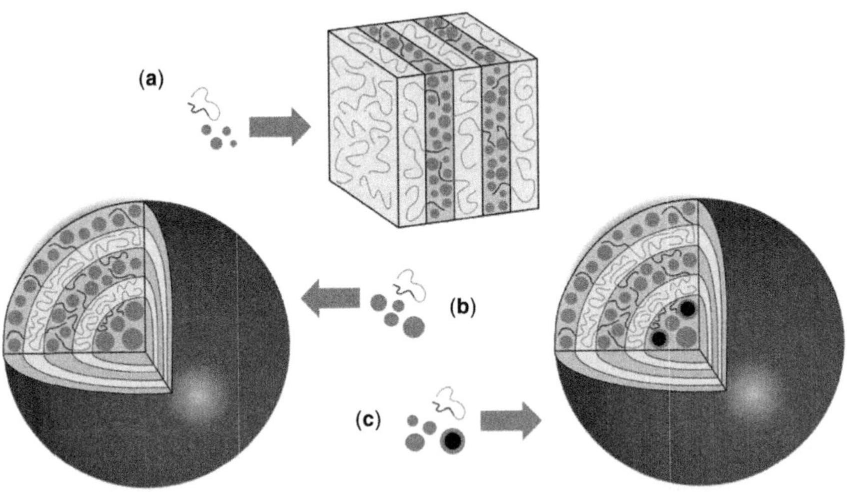

Figure 21.2 Assembly of mesostructured hybrids. (a) Nanoparticles smaller than the darker gray block's R_0 are miscible and assemble into a lamellar structure. (b) Nanoparticles larger than the darker gray block's R_0 segregate, forming a nanoparticle-rich core around which lamellae assemble into an onion-type structure. (c) This can be used to generate compositionally heterogeneous nanostructures from tailored nanoparticle size distributions.[10] (Reprinted with permission from S. C. Warren et al., *Nature Mater.* **2007**, *6*, 156–161. Copyright 2007 Macmillan Publishers Ltd.)

Table 21.1 Characterization of Polymers

	Polymer	M_w (g mol^{-1})	M_w/ M_n	wt%	wt%	wt%
A	PI-*b*-PEO	13,400	1.03	PI 85.4	PEO 14.6	–
B	PI-*b*-PEO	27,800	1.03	PI 83.6	PEO 16.4	–
C	Extended amphiphilic dendron	6000	1.05	Docosyl periphery 47[a]	Dendritic core + PEO 53[a]	–
D	PEP-*b*-PEO-*b*-PHMA	48,520	1.13	PEP 15[a]	PEO 11[a]	PHMA 74[a]
E	PEP-*b*-PEO-*b*-PHMA	25,100	1.08	PEP 22[a]	PEO 20[a]	PHMA 58[a]
F	PI-*b*-PEO	22,400	≤1.05	PI 85	PEO 15	–
G	PI-*b*-PEO	38,600	≤1.05	PI 68	PEO 32	–
H	PI-*b*-PEO	22,700	≤1.05	PI 86.6	PEO 13.4	–
I	PI-*b*-PEO	16,200	≤1.05	PI 87.5	PEO 12.5	–
J	PI-*b*-PEO	22,900	≤1.05	PI 83.9	PEO 16.1	–
K	PI-*b*-PEO	27,220	1.02	PI 83.3	PEO 16.7	–
L	PI-*b*-PEO	33,500	1.03	PI 77	PEO 23	–
M	PI-*b*-PDMAEMA	31,100	1.05	PI 67	PDMAEMA 33	–
N	PI-*b*-PDMAEMA	27,800	1.04	PI 85	PDMAEMA 15	–

[a]Volume fractions.

nanoparticles in the B block and makes the entire block copolymer phase space experimentally accessible.[3,15]

Using silica-type nanoparticles with narrow size distributions (ranging in size from 0.5 to 6.0 nm) a range of $d/R_{0,\mathrm{PEO}}$ from 0.1 to 1.4 was covered (Fig. 21.3). Nanoparticles were synthesized by adding water at varying rates to a 90:10 mol%

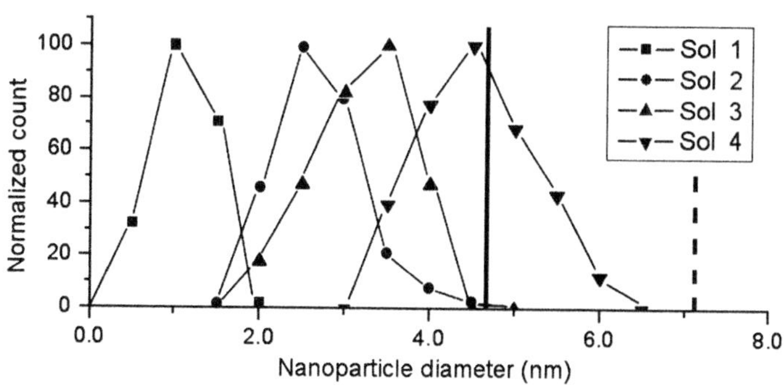

Figure 21.3 Size distributions for different sols. For their syntheses, water was added over a period of 7 min (sol 1), 2 min (sol 2), 30 s (sol 3), and 5 s (sol 4) to a mixture of silicon and aluminum alkoxides. The vertical lines denote the R_0s of the two copolymers used.[10] (Reprinted with permission from S. C. Warren et al., *Nature Mater.* **2007**, *6*, 156–161. Copyright 2007 Macmillan Publishers Ltd.)

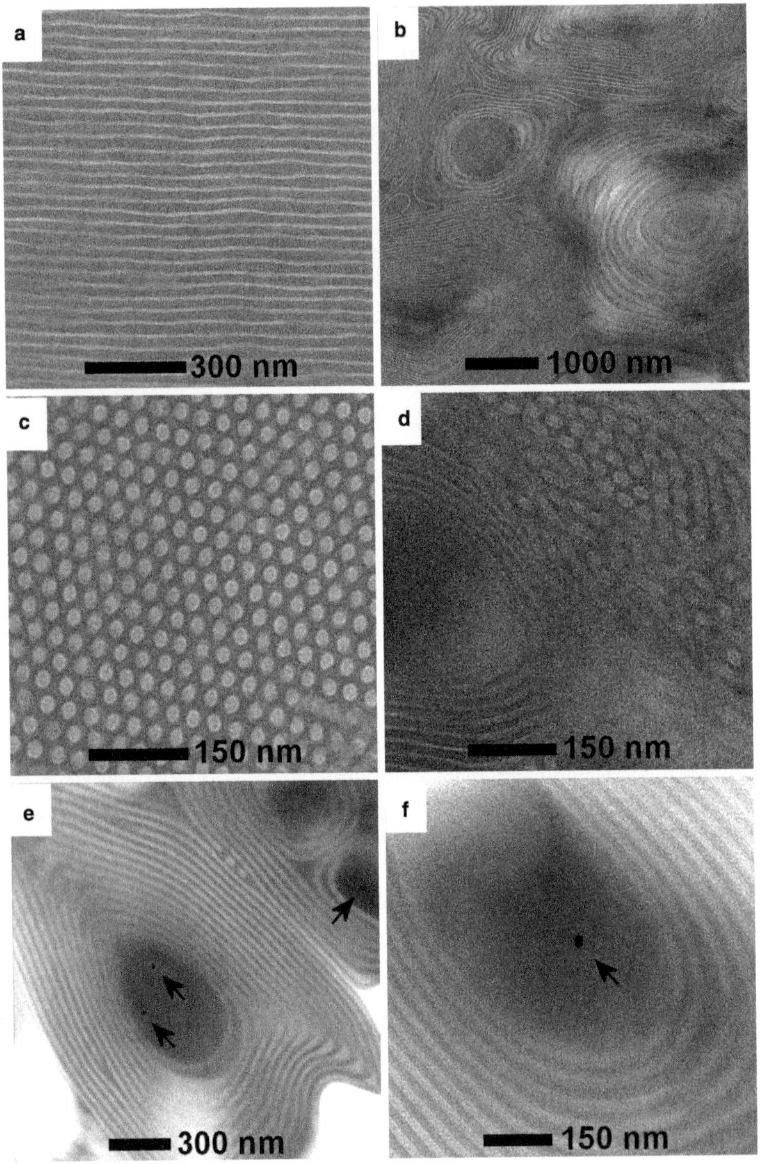

Figure 21.4 Influence of nanoparticle size on silica-type mesostructure. (a, b) TEM micrographs showing mesostructures with a nanoparticle volume fraction $\varphi_P = 0.49$ and PEO volume fraction $\varphi_{PEO} = 0.06$. (c, d) TEM micrographs showing assemblies with $\varphi_P = 0.64$ and $\varphi_{PEO} = 0.05$. (a, c) When sol 3 was combined with PI-*b*-PEO, $d/R_{0,PEO}$ ranged from 0.3 to 1.0. (b, d) When sol 4 was combined with PI-*b*-PEO, $d/R_{0,PEO}$ ranged from 0.6 to 1.4. (e, f) Gold–silica core–shell nanoparticles were directed to the core of the onion-type structures (indicated by arrows) because their diameters (14.5 ± 1.5 nm) exceeded $R_{0,PEO}$ by a factor of two.[10] (Reprinted with permission from S. C. Warren et al., *Nature Mater.* **2007**, *6*, 156–161. Copyright 2007 Macmillan Publishers Ltd.)

mixture of a silicon alkoxide, 3-glycidyloxypropyltrimethoxysilane (GLYMO), and an aluminum alkoxide, aluminum-tri-*s*-butoxide. Figure 21.3 shows the size distributions for the four sols studied as derived from quantitative analysis of atomic force microscopy (AFM) height data. Both particle size and size distribution can be controlled by altering the period of time during which water is added, with longer periods leading to smaller particles, as the results clearly demonstrate.

Polymer–inorganic composites were prepared using a solution of the structure-directing PI-*b*-PEO diblock copolymer mixed with the nanoparticles. For systems where $d/R_{0,\mathrm{PEO}}$ ranged from 0.1 to 1.0 results suggested that the sol nanoparticles do not significantly alter morphology formation other than increasing the effective PEO volume fraction and the expected lamellar or inverse hexagonal silica-type mesostructures are obtained [see transmission electron microscopy (TEM) images in (Fig. 21.4a, c)]. In contrast, when $d/R_{0,\mathrm{PEO}}$ ranged from 0.6 to 1.4 TEM images revealed onion-type morphologies (Fig. 21.4b, d). These results suggest that the largest nanoparticles segregate out, forming a silica-rich core around which a lamellar or lamellar/hexagonal structure grows, as corroborated by the darker (more electron dense) silica core in the TEM images in Figure 21.4b, d.

We expect size-dependent segregation to occur in self-assembled macromolecular systems in which there is competition between molecular interactions that favor mixing of one block with the nanoparticles (i.e., χ_{BP} is negative) and significant conformational entropy losses of the macromolecular chains wrapping around the particles. In contrast, for nonionic surfactant systems composed of short chains, such as the Brij surfactants that have only a few monomers of ethylene oxide, conformational entropy contributions to the free energy will be small. Accordingly, for systems with favorable interactions, nanoparticles that are larger than $R_{0,\mathrm{PEO}}$ may still mix with these surfactants. In this context, it is interesting to note that silaffins that have a role in the biomineralization of silica have molecular weights between 3 and 17 kg mol^{-1},[6] similar molecular weights as the copolymers used for this model study.

Onion-type silica mesostructures may be technologically useful. Similar materials, termed mesoporous multilamellar vesicles, have been the synthetic target of numerous investigations, as the structures could serve as efficient catalyst supports, adsorbents or controlled-release materials for drug delivery.[9,16] For most of these applications, it is desirable that the vesicle core and shell have different compositions, which is usually difficult to achieve. The present approach opens a simple pathway to the design of such compositionally heterogeneous structures by tailoring the nanoparticle size distributions to segregate particles with distinct interiors but similar surfaces into precisely controlled locations in the bulk structure. Large nanoparticles ($d/R_{0,\mathrm{PEO}} > 1$) of one composition can be segregated into the core, whereas smaller nanoparticles ($d/R_{0,\mathrm{PEO}} < 1$) of a different composition can be directed into the shell. Based on this self-assembly concept, the Wiesner group successfully directed gold nanoparticles into onion-type cores. A solution of 14.5 ± 1.4 nm core–shell nanoparticles with a gold core and a 1.0 ± 0.2 nm silica shell was added to hydrolyzed silicon and aluminum alkoxides and the resulting sol was mixed in the PI-*b*-PEO solution. Figure 21.4e, f shows representative TEM micrographs of the resulting self-assembled hybrids. The images reveal that the core–shell gold nanoparticles are indeed localized in the onion-type cores (use the arrows in Fig. 21.4e, f as guides). However, when 2.9 ± 1.1 nm palladium nanoparticles with hydrophilic ligands were added, the nanoparticles were dispersed evenly between the silica-rich core and shells. These results reveal the power of working with appropriately designed nanoparticle size distributions rather than average particle size alone for controlled nanoparticle placement in segregated mesostructures. This concept may apply to a wide range of nanoparticle-derived mesostructures.

21.2.2 Novel Blocked Macromolecular Amphiphiles as Structure-Directing Agents for Silica-Type Materials

Synthetic macromolecules with AB or ABA block architectures have been used to direct silica-type materials into ordered structures with lamellar, hexagonal, or cubic symmetry.[8,14,17] However, a wide gulf remains between these synthetic composites and the complex structure of biological silica. Bridging this divide will require the use of structure-directing organic molecules with more complex interactions and phase behavior. In fact the interface between polymer science and inorganic chemistry is where a lot of materials innovation can be expected in the future. For example, polymer chemistry has developed to a point where even nonpolymer synthesis experts can use the tool-box of synthetic techniques available to prepare polymers of complex chemical and structural architecture. As examples, the Wiesner group recently developed novel families of blocked extended amphiphilic dendrons[18,19] and ABC triblock copolymers,[20] see Figure 21.5, thereby significantly expanding the available architectures of structure-directing agents.

21.2.2.1 Mesostructured Block Dendron–Silica Hybrids

Amphiphilic dendrimers/dendrons present a distinct class of materials due to their unique structure and properties. The dense distribution of functional end groups for higher generations at the exterior of the dendrimer/dendron affords tremendous functionality, whereas the special molecular architecture is promising for obtaining novel well-defined structures at the nanoscale.[21] Since the pioneering work of Tomalia et al.[22] dendrimers have been shown to possess fascinating phase behavior both in solution and bulk previously unknown in soft matter.[18,23] Similar to the dendrimer/dendron core, convergent[24] as well as divergent[25] (starburst) approaches have been developed for the synthesis of linear-dendritic diblock copolymers, which promise to further enrich the phase behavior of this class of materials. Despite their tremendous potential to structure direct, however, dendrimers have been used primarily as "porogens" in inorganic matrices, where they form spherical aggregates employed to generate well-defined pores in the surrounding continuous matrix.[26] The Wiesner group recently showed that AB-type linear-dendron block copolymers, analogous to simple AB diblock copolymers, could be used to structure direct silica-type materials into structures with different morphologies.[27] The specific molecular architecture of an extended amphiphilic dendron results in unique structural features in

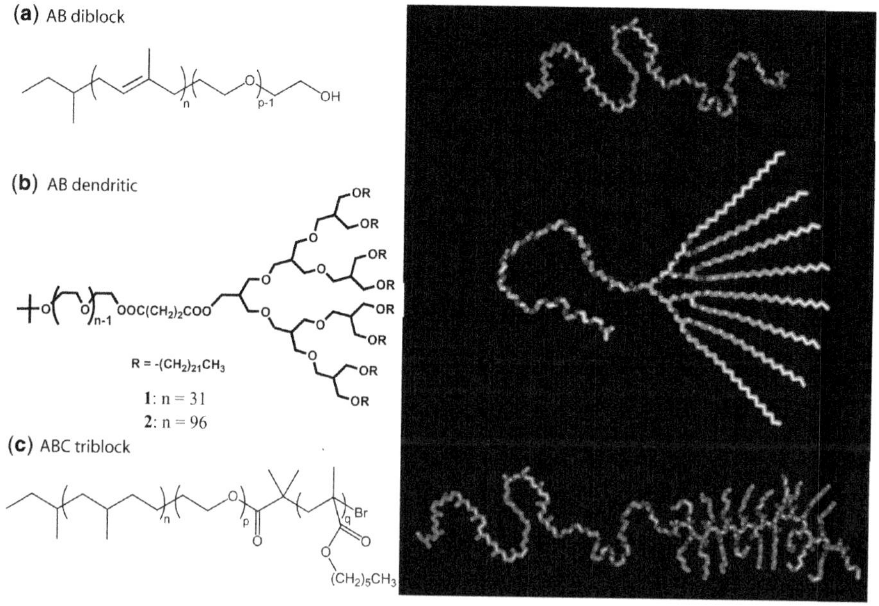

Figure 21.5 Molecular architectures and models of (a) AB diblock copolymer, (b) extended amphiphilic dendron, and (c) ABC triblock copolymer.[28] (Reprinted with permission from B. K. Cho et al., *Chem. Mater.* **2007**, *19*, 3611–3614. Copyright 2007 American Chemical Society.)

these nanostructured dendron–inorganic hybrids.[28] More specifically, an inorganic nanoparticle-induced packing transition in dendron-based nanocomposites was investigated.

The parent extended amphiphilic dendron used as a structure-directing agent consists of a hydrophilic fraction, a linear PEO attached to a PEO-like dendritic core, and a hydrophobic fraction, composed of docosyl branches.[29] The interface of a phase segregated bulk phase of this compound runs through the middle of the dendrimer part of the molecule, and is expected to have an inherent curvature, which is unique compared to conventional linear amphiphiles (see Fig. 21.5b and Table 21.1, C). On the basis of the primary *d*-spacing, obtained from small angle X-ray scattering (SAXS) of the bulk material at 55°C, and the length of a docosyl chain of 3.0 nm from a CPK model,[19] the hydrophobic docosyl peripheries are expected to self-assemble into a bilayer packing, see schematics in Figure 21.6a, b. By loading a prehydrolyzed sol to this extended dendron, using procedures similar to the one described in the previous section, two nanostructured hybrid materials were prepared. SAXS analysis suggested lamellar morphologies for both hybrids. This interpretation was supported by AFM studies of the morphology of individual nanoparticles obtained from dissolution of the bulk materials in organic solvents (see Fig. 21.7).

In contrast to the bilayer lamellar structure in the parent extended amphiphilic dendron, structural data on these hybrids suggested lamellar morphologies with an interdigitated monolayer packing of hydrophobic peripheries (see Fig. 21.6c). The

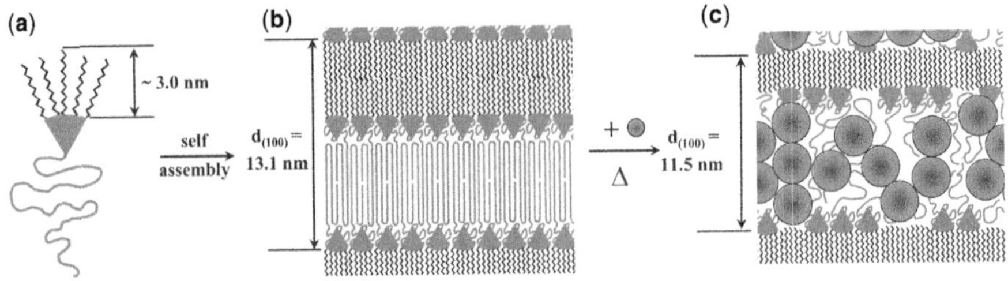

Figure 21.6 Schematic illustration of the packing transition induced by addition of the sol nanoparticles.[11] Grey and black regions represent hydrophilic and hydrophobic domains, respectively. Spheres indicate sol nanoparticles. (a) Structural schematic of the parent extended dendron; (b) bilayer packing of docosyl peripheries in the crystalline states of the parent extended dendron; (c) molecular organization of sol particles and dimensions in hybrid material.[28] (Reprinted with permission from B. K. Cho et al., *Chem. Mater.* **2007**, *19*, 3611–3614. Copyright 2007 American Chemical Society.)

transformation from a bilayer to an interdigitated monolayer arrangement of the peripheries upon addition of inorganic sol nanoparticles can be explained by relieving an interfacial steric barrier. In the crystalline state of the parent extended amphiphilic dendron, the molecular packing is dominated by the minimization of interfacial area because crystallization leaves the coil conformational freedom small, inducing the bilayer packing structure (Fig. 21.6b). As shown in the previous section, the sol–gel

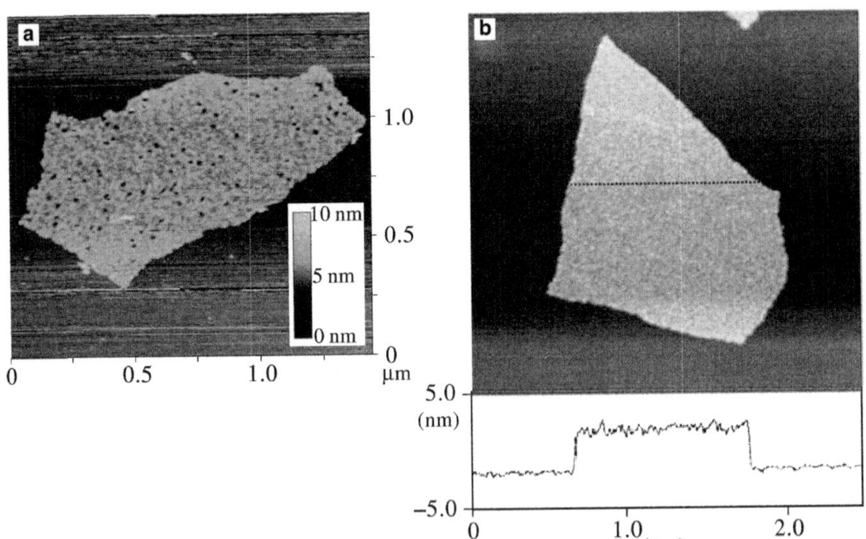

Figure 21.7 (a) and (b) show AFM images of individual nanoplates obtained after dissolution of the bulk hybrids. The height profile in (b) was recorded along the dashed line in the image. Before AFM imaging the materials were heat treated to remove any organic material.[28] (Reprinted with permission from B. K. Cho et al., *Chem. Mater.* **2007**, *19*, 3611–3614. Copyright 2007 American Chemical Society.)

process proceeds through assembly of small silica-type nanoparticles (sol) within the PEO domains. These sol particles may bring about a significant steric hindrance between PEO chains in a bilayer packing model. On the contrary, the transformation into an interdigitated monolayer arrangement provides a way to reduce the interfacial steric barrier, because the cross-sectional area occupied by a PEO chain is larger than that in a bilayer packing [compare diagram (b) with diagram (c) in Fig. 21.6].

21.2.2.2 ABC Copolymer–Silica-Type Hybrid Materials

The introduction of a third block in ABC triblock copolymers leads to a diverse range of complex morphologies.[30–32] Using ABC block copolymers as structure-directing agents promises numerous benefits, including easier access to continuous network structures,[33] novel helical[34] and noncentrosymmetric[35] structures, and the possibility of independently structuring two or more types of inorganic material. To date, however, using the rich assembly behavior of ABC block copolymers to mesostructure silica and other inorganic materials remains a challenge. In this section, new amorphous aluminosilicate mesostructures formed with the amphiphilic ABC triblock copolymer, poly(ethylene-*alt*-propylene-*block*-ethylene oxide *block*-*n*-hexyl methacrylate) (PEP-*b*-PEO-*b*-PHMA)[20] recently developed in the Wiesner group are described.[36,37] While many examples of the preparation of block copolymers built from PEO macroinitiators appear in the literature[38] this polymer constitutes a rare example in which a PEO block is incorporated in a nonterminal position.[39] For this polymer, the Flory–Huggins interaction parameters favor optional contact between PEP and PHMA domains over the obligatory PEP–PEO and PEO–PHMA interfaces. ABC copolymer–aluminosilicate composites were prepared following the same procedure described in the previous sections (and described in detail in References 11 and 14). The aluminosilicate precursors partition into the hydrophilic PEO middle block of the copolymer.[11,14] While a range of compositions were studied, only the results for two ABC copolymer–aluminosilicate hybrids will be discussed here. Table 21.2 summarizes the block volume fractions of the hybrid compounds. For polymer compositions see Table 21.1, D and E for hybrid A and B, respectively. The small differences in hybrid composition and the large resulting differences in hybrid morphologies provide a glance at the wide range of complex hybrid morphologies that can be expected in the future from the use of ABC copolymers and higher order blocked macromolecular architectures as structure-directing agents.

Table 21.2 Composition of ABC Copolymer–Aluminosilicate Composites[a]

Compound	f_{PEP}	$f_{PEO+aluminosilicate}$	f_{PHMA}
Hybrid A	0.11	0.34	0.55
Hybrid B	0.19	0.32	0.49

[a] Domain volume fractions were calculated assuming room temperature densities of $f_{PEP} = 0.855\ g\ cm^{-3}$,[40] $f_{PEO+aluminosilicate} = 1.4\ g\ cm^{-3}$,[11] $f_{PHMA} = 1.007\ g\ cm^{-3}$.[41]

21.2.2.2.1 Hexagonally Patterned Lamellae Hybrid A represents a triblock copolymer–aluminosilicate composite with the composition shown in Table 21.2. In this hybrid, the volume fraction of the PEP block was significantly smaller (11%) than that of the PEO–aluminosilicate mixed domain (34%) and PHMA block (55%). This composition ($f_A \ll f_B, f_C$) corresponds to the small end-block regime[42] of interest for studying the transition from two-domain diblock to three-domain triblock morphologies. In a diblock PEO-*b*-PHMA copolymer–aluminosilicate composite material,[43] the PHMA and PEO–aluminosilicate domains can form a simple lamellar morphology for $f_{PHMA} \approx 0.50$, as illustrated in Figure 21.8a. However, in the triblock copolymer, the enthalpy for mixing the PEP block with the PEO–aluminosilicate domain is truly prohibitive. These unfavorable interactions can be reduced by the formation of micellar PEP domains as in the balls-in-lamellae[32] (Fig. 21.8b), cylinders-in-lamellae,[44] dimple (Fig. 21.8c), and pillared-lamellae[42] (Fig. 21.8d) structures. The optimal position and shape of PEP domains depends on a trade-off between chain stretching and interfacial area. In the balls-in-lamellae[32] structure shown in Figure 21.8b, the burial of the PEP domain within the PEO–aluminosilicate layer incurs a large enthalpic penalty because the surface tension of the PEP/PEO–aluminosilicate interface is larger than that of a PEP/PHMA interface. As shown in Figure 21.8c, moving the PEP micelle to one side of the PEO–aluminosilicate sheet reduces the area of the PEP/PEO–aluminosilicate interface at the cost of forming a PEP/PHMA interface. Alternatively, as proposed by Bailey et al.[42] the PEP domain can form a pillar spanning the PEO–aluminosilicate domain as shown in Figure 21.8d. Although both structures reduce the PEP/PEO–aluminosilicate interface, the dimple structure is likely to be favored for smaller PEP micelles while the pillared structure may suit larger PEP micelles.

SAXS and electron microscopy were applied to elucidate which of these morphologies was formed.[37] The parent PEP-*b*-PEO-*b*-PHMA copolymer (Table 21.1, D) was examined with SAXS first. The large volume fraction of the PHMA domain ($f_{PHMA} = 0.74 > f_{PEP} = 0.15$, $f_{PEO} = 0.11$) should favor morphologies in which the PEP and PEO blocks form micellar or cylindrical domains[32] surrounded by a matrix of PHMA. Since the mixing enthalpy of the A and C blocks is relatively small ($\chi_{PEP/PHMA}N < \chi_{PEP/PEO}N$, $\chi_{PEO/PHMA}N$), morphologies with optional PEP/PHMA domain interfaces should be favored. Earlier experimental studies of ABC copolymers in this regime have reported a number of morphologies including the spheres-on-spheres,[45] core-shell cylinders,[46] rings-at-cylinders,[46] and helices-around-cylinders[34] structures.

Samples were aligned using reciprocating shear[47] and were prepared using a small, home-built shear cell. Specimens were pressed into the shear cell and sheared (~150% shear at ~0.5 Hertz) for 5 minutes under rough vacuum at 75°C. SAXS data taken with the X-ray beam directed perpendicular to the shear axis (horizontal) show alignment perpendicular to the shear direction, as depicted in Figure 21.9a. SAXS data taken with the X-ray beam parallel to the shear axis (Fig. 21.9c) exhibited sixfold symmetry, consistent with a hexagonal lattice. The plots of radial averaged scattering intensity (Fig. 21.9b, d) showed a broad second peak at $3^{\frac{1}{2}} \times s_0$ and shoulder at $7^{\frac{1}{2}} \times s_0$, consistent with the allowed reflections for a hexagonal unit cell. Thus,

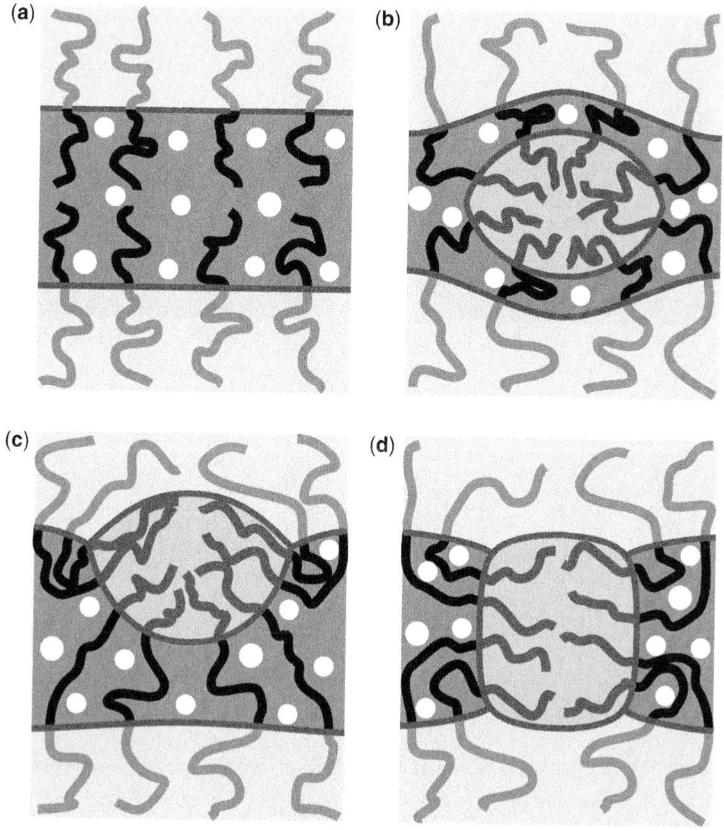

Figure 21.8 Structural models for lamellar PEP-*b*-PEO-*b*-PHMA block copolymer–aluminosilicate composite morphologies with a small PEP block. In the absence of the PEP block, the PEO (black) and PHMA (light grey) chains stretch into their respective domains while the aluminosilicate particles (white) partition into the hydrophilic PEO domain (a). Possible domain structures discussed in the text are illustrated as follows: In the balls-in-lamellae structure the small PEP block (dark grey) forms round micellar domains (b). Dimple structure with PEP micelles at the PHMA/PEO-aluminosilicate interface (c). In the pillared-lamellae structure the PEP domains form pillars spanning across the PEO–aluminosilicate domain (d).[37] (Reprinted with permission from G. E. S. Toombes et al., *Chem. Mater.* **2008**, *20*, 3278–3287. Copyright 2008 American Chemical Society.)

morphologies in which the PEP and PEO domains form a cylindrical core are consistent with SAXS from shear-aligned samples.

The PEO–aluminosilicate domains in the hybrid material have a larger volume fraction and incompatibility with the PEP and PHMA blocks than the PEO domains in the parent ABC copolymer. Thus, the hybrid materials and the parent copolymer can have different morphologies. Figure 21.10a shows a two-dimensional SAXS pattern from hybrid A in which the sample was oriented so that the normal to the film surface was directed along the y-axis (vertical). For this orientation, scattering along

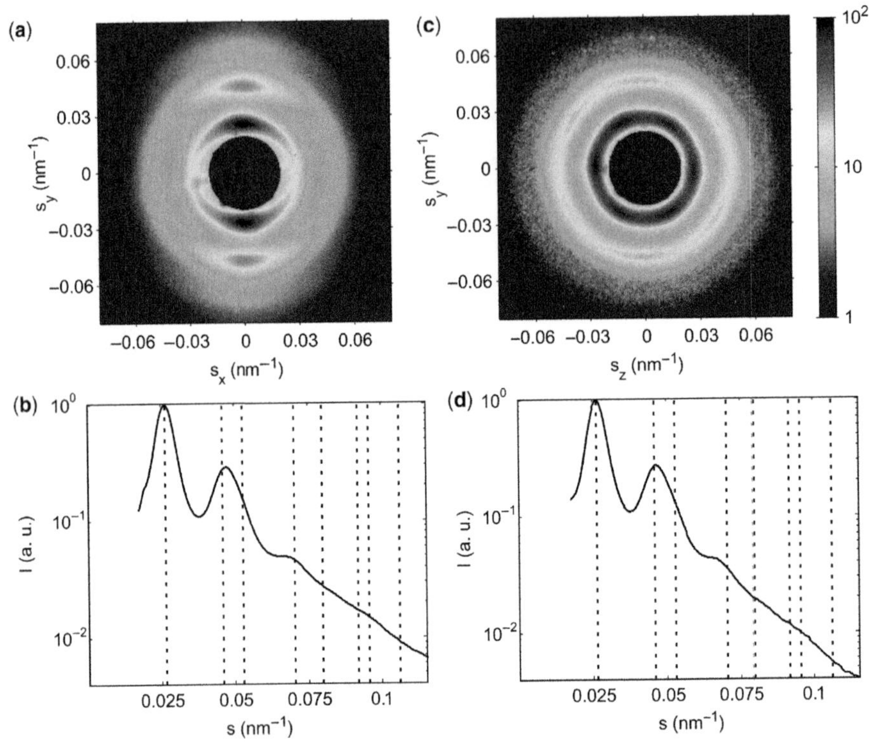

Figure 21.9 2D SAXS (logarithmic scale) from a shear-aligned specimen of the parent ABC block copolymer (a, c). The shear direction is horizontal in (a) and along the X-ray beam direction in (c), while the surface normal of the sample is vertical in both images. Radial averages (b, d) in both cases show a main peak at $s = (2.63 \pm 0.05) \times 10^{-2}\ nm^{-1}$ (repeat spacing of 38.0 ± 0.7 nm) and the dotted vertical lines indicate the allowed reflections for a hexagonal lattice.[37] (Reprinted with permission from G. E. S. Toombes et al., *Chem. Mater.* **2008**, *20*, 3278–3287. Copyright 2008 American Chemical Society.)

the y-axis (vertical) reflects order along the direction of the film normal while structure within the plane of the film leads to scattering along the x-axis (horizontal). The scattering pattern did not change when the sample was rotated about the film normal (y-axis), indicating simple fiber-type alignment of the structure with respect to the film surface.[48]

The three orders of Bragg spots along the vertical axis correspond to the lamellar stacking of aluminosilicate sheets oriented parallel to the film surface strongly aligned during the solvent-casting process.[49] The intense, in-plane scattering along the vertical row lines at $|s_x| = 0.046 \pm 0.004\ nm^{-1}$ indicates periodic structure within the sheets. From calculated structure factors the observed intensity for the outer row lines is largely consistent with the intensity expected for a pillared lamellae structure.

The response of the hybrid materials to changes in temperature and solvent content provided further support for this interpretation of the SAXS data. A lamellar block copolymer–aluminosilicate material should preferentially deform along its lamellar

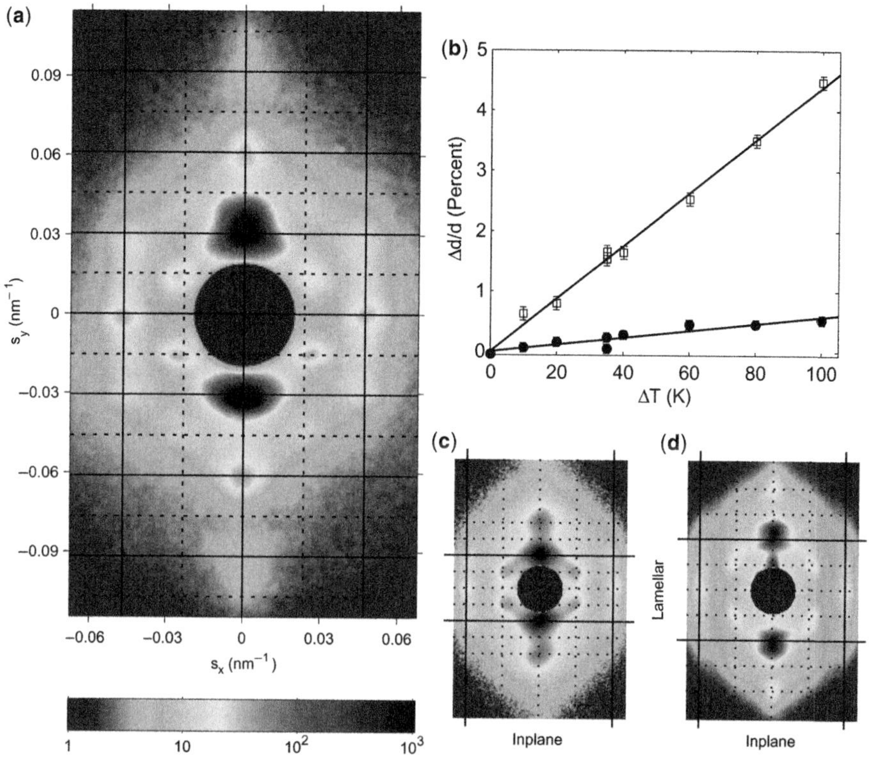

Figure 21.10 2D SAXS (logarithmic scale) with the sample's surface normal directed along the vertical axis (a). The solid layer (horizontal) and row (vertical) lines mark repeat spacings of $d_l = 33.0 \pm 3.3$ nm ($\Delta s_y = 0.030 \pm 0.003\ \text{nm}^{-1}$) and $d_r = 21.7 \pm 1.9$ nm ($\Delta s_x = 0.046 \pm 0.004\ \text{nm}^{-1}$), respectively. Hybrid material anisotropy (b to d). Percent change in lamellar (open squares) and in-plane (closed circles) repeat spacings of compound hybrid A as a function of temperature (263 K to 363 K) (b). Along the lamellar direction the linear thermal expansion coefficient ($4.33 \pm 0.11 \times 10^{-4}\ \text{K}^{-1}$) is 7.5 ± 1 times the expansion coefficient in the in-plane direction ($5.8 \pm 0.8 \times 10^{-5}\ \text{K}^{-1}$). 2D SAXS (logarithmic scale) from compound hybrid A swollen by cyclohexane ($d_l = 50 \pm 1$ nm; $d_r = 21.7 \pm 1.9$ nm) (c) and following re-evaporation of solvent ($d_l = 32.8 \pm 0.5$ nm; $d_r = 21.7 \pm 1.9$ nm) (d).[37] (Reprinted with permission from G. E. S. Toombes et al., *Chem. Mater.* **2008**, *20*, 3278–3287. Copyright 2008 American Chemical Society.)

axis because the covalent bonding network within each PEO–aluminosilicate layer constrains in-plane deformations. Because the middle PEO block of each chain is embedded within the covalent aluminosilicate network, shape changes in the sample should be directly reflected in the unit cell dimensions. Thus, following thermal expansion or swelling due to solvent uptake, the layer lines (lamellar ordering) should shift while the row lines (in-plane ordering) should remain fixed. Figure 21.10b shows the change in lamellar and in-plane cell dimensions as a function of temperature. The linear thermal expansion coefficient ($4.33 \pm 0.11 \times 10^{-4}\ \text{K}^{-1}$) along the lamellar direction was 7.5 ± 1 times the expansion coefficient in the in-plane direction ($5.8 \pm 0.8 \times 10^{-5}\,K^{-1}$), confirming the anisotropic thermal expansion of the structure along the lamellar axis. Similarly, when hybrid A was exposed to a nonpolar solvent (cyclohexane), the structure within the sheets remained essentially unchanged ($d_r = 21.7 \pm 1.9$ nm) while the adjoining PHMA domains swelled, as illustrated by the 50% ± 3% increase in lamellar spacing shown in Figure 21.10c. Remarkably, following evaporation of the solvent, the structure relaxed to the initial state, as shown in Figure 21.10d (compare to Fig. 21.10a). The extreme anisotropy of the hybrid materials provides strong support for a hexagonally patterned lamellar structure.

To further resolve the morphology electron microscopy was performed. A scanning electron microscopy (SEM) image of a cross-section perpendicular to the film surface (surface normal vertical) shows aluminosilicate layers running parallel to the surface of the film (Fig. 21.11a). Bright-field TEM imaging of the in-plane structure showed a well-ordered 2-D hexagonal mesh (Fig. 21.11b), which agrees well with the strong in-plane ordering seen in the SAXS pattern. The structure of the PEO–aluminosilicate domain within individual sheets was examined using the atomic number sensitivity of annular dark-field scanning transmission electron microscopy (STEM) imaging. In Figure 21.11c–f (aluminosilicate bright), the 2-D hexagonal pattern is readily apparent and the enrichment of silicon within the mesh

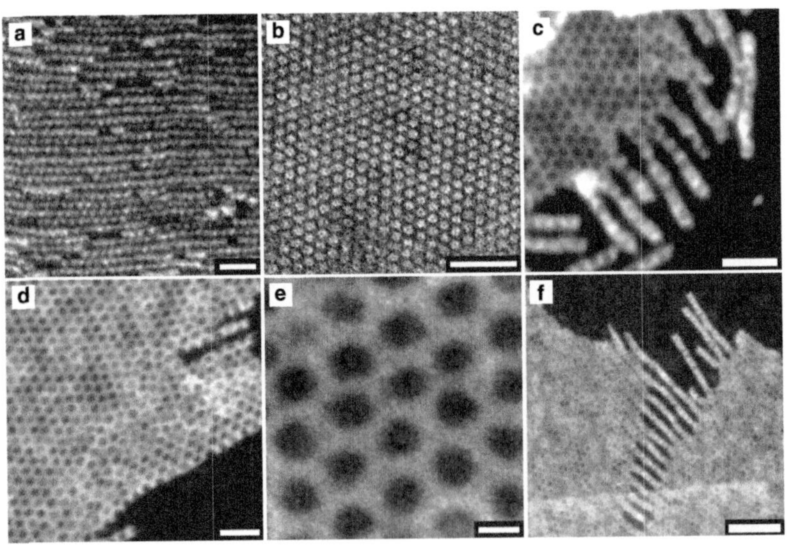

Figure 21.11 Electron micrographs of compound hybrid A. SEM image (aluminosilicate bright) of the bulk material for a cross-section perpendicular to the surface (surface normal vertical) (a, 100 nm scale bar). Bright-field TEM (aluminosilicate dark) of a thin (~50 nm) section cut parallel to the surface (b, 100 nm scale bar). Dark-field STEM images (aluminosilicate bright) of individual sheets isolated by dispersion in solvent (c, d, 100 nm scale bar, f, 200 nm scale bar) and a higher magnification image of the 2D mesh structure (e, 20 nm scale bar).[37] (Reprinted with permission from G. E. S. Toombes et al., *Chem. Mater.* **2008**, *20*, 3278–3287. Copyright 2008 American Chemical Society.)

framework was confirmed using parallel electron energy loss spectroscopy (PEELS). The dark regions in the hexagonal mesh could correspond to dimples in the PEO–aluminosilicate domain (Fig. 21.8c), or holes where the PEO–aluminosilicate phase was completely excluded (Fig. 21.8d). In Figure 21.11d, some of the spots are considerably darker than others. This variation is readily accounted for if the darker spots correspond to holes through the PEO–aluminosilicate and the brighter spots correspond to dimples in the PEO–aluminosilicate layer. This hypothesis was validated by field-emission SEM experiments on isolated sheets using both dark-field (STEM) and secondary electron (SE) signals. Level set modeling confirmed that this hexagonal patterning of the PEO–aluminosilicate sheets is consistent with micellar PEP domains forming either pillars through the sheet, as in the proposed pillared-lamellae structure,[42] as shown in Figure 21.12, or dimples at the surface of the sheet.

21.2.2.2.2 Four-Layer Woodpile Structure from Zigzag Concertinas Thin sections (~60 nm) of hybrid B (see Table 21.2) were examined via dark-field TEM. Although this hybrid has only small variations in composition as compared to hybrid A, its morphology turned out to be quite different.[36] Figure 21.13a shows the periodic character of the aluminosilicate (bright) structure. The projection gives the appearance of layers of strands running in two almost perpendicular directions, but the detailed structure of the strands and their arrangement within the bulk material remained unclear from such TEM micrographs. Three-dimensional imaging of hybrid B using scanning transmission electron tomography[50] finally revealed a complex morphology not previously achieved using AB or ABA block copolymers. The tomographic reconstruction of an isolated strand obtained through dissolution of bulk material in an organic solvent is shown in Figure 21.13b. The

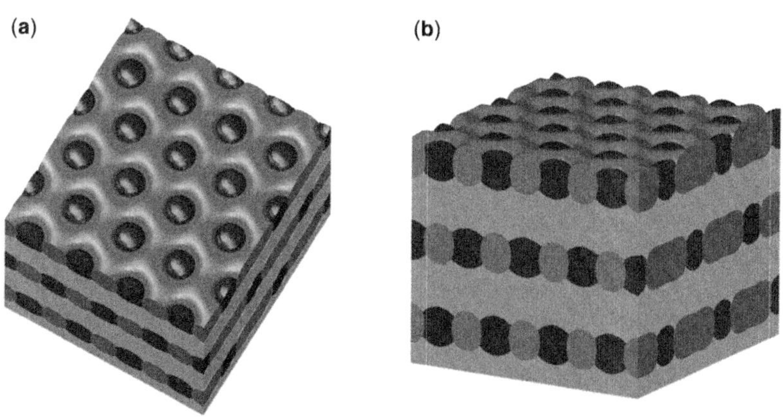

Figure 21.12 Structural models for PEP-*b*-PEO-*b*-PHMA block copolymer–aluminosilicate lamellar morphologies with a small PEP block. Top (a) and side (b) views of the pillared-lamellae structure.[37] (Reprinted with permission from G. E. S. Toombes et al., *Chem. Mater.* **2008**, *20*, 3278–3287. Copyright 2008 American Chemical Society.)

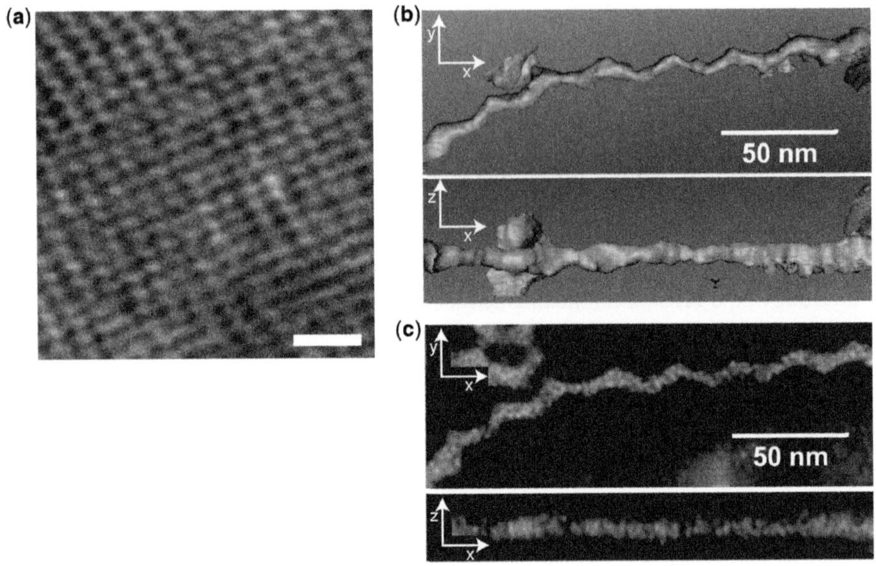

Figure 21.13 Dark-field TEM (a) of a thin section of bulk material (aluminosilicate bright) (scale bar 100 nm). Two iso-surface renders (b) showing the surface of the aluminosilicate strand from top and edge obtained from the tomographic reconstruction of an individual hybrid strand. The strand shows a clear concertina structure. The thickness of strands is approximately 10 nm, while the wiggles along each strand have a period of $d_w \sim 24$ nm and a peak-to-peak amplitude of $\sim$5 nm. Corresponding slices (c) through the center of the reconstruction show individual sol particles (diameter $\sim$2 nm) within the body of the concertina.[36] (Reprinted with permission from G. E. S. Toombes et al., *Macromolecules* **2008**, *41*, 852–859. Copyright 2008 American Chemical Society.)

PEO–aluminosilicate domains formed layers of zigzag-shaped strands with wiggles like the bellows of a concertina. The resolution of the reconstruction is sufficient to resolve the internal structure of the concertina, Figure 21.13c. Sections through the reconstruction shown in Figure 21.13c reveal a distinctly bimodal distribution of intensity, with bright and dark regions approximately 1 to 3 nm in size. The bright regions correspond to aluminosilicate-rich sol particles (high atomic number, i.e., high-Z) while the polymer-rich regions (low-Z) are darker. Please note that this experimental result is entirely consistent with the model presented in Figure 21.1, corroborating the mechanistic suggestions made in the first section.

A tomographic reconstruction was performed upon a thin section of the bulk material to determine the arrangement of individual concertinas as shown in Figure 21.14. Remarkably, the unit cell consists of a four layer woodpile[51,52] (Fig. 21.15) in which the direction of concertinas in successive layers alternates. The first and third layers of concertinas are directed along the [110] diagonal while the second and fourth layers are run along the $[1\underline{1}0]$ diagonal. Furthermore, layers are staggered, with the third and fourth layers offset by $(\underline{a} + \underline{c})/2$. The alternating direction of concertinas is most evident in slices taken through consecutive layers

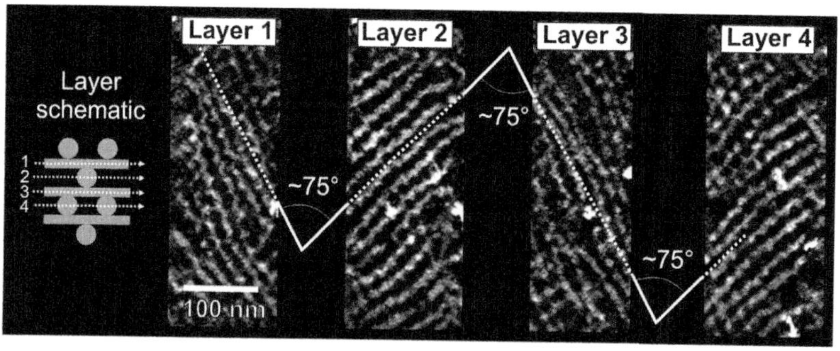

Figure 21.14 STEM tomographic reconstruction of the bulk specimen (aluminosilicate bright). Sections taken through the strand layers at depths of 0, 11, 22, and 33 nm, respectively, show successive layers of strands running in alternate directions. The spacing between strands within each layer is $d_{ip} \approx 23.3$ nm while the angle between the direction of strands in successive layers is $\phi \approx 75°$.[36] (Reprinted with permission from G. E. S. Toombes et al., *Macromolecules* **2008**, *41*, 852–859. Copyright 2008 American Chemical Society.)

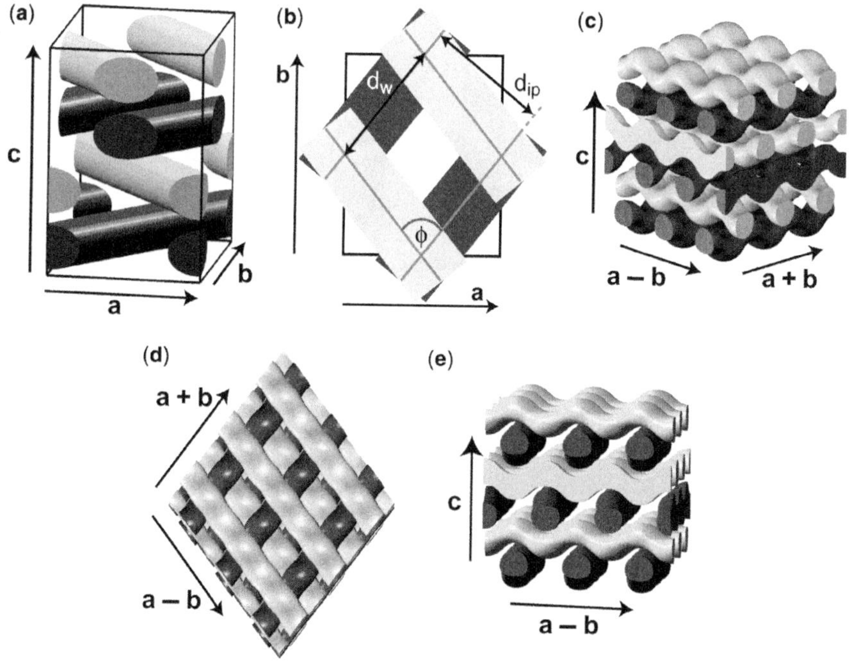

Figure 21.15 Four-layer woodpile structure. In the face-centered orthorhombic unit cell (side view a, top view b), strands in odd/even layers (dark gray/light gray) are directed along the $(\mathbf{a}+\mathbf{b})/(\mathbf{a}-\mathbf{b})$ diagonals while the third and fourth layers are offset by $(\mathbf{a}+\mathbf{c})/2$. The angle between the two strand directions ($\phi \approx 75°$), spacing between strands within each plane (d_{ip}), and period of wiggles along each strand (d_w) are indicated in the overhead view (b). Models of the undulating structure viewed from the front (c), overhead (d), and in the direction of the strands (e).[36] (Reprinted with permission from G. E. S. Toombes et al., *Macromolecules* **2008**, *41*, 852–859. Copyright 2008 American Chemical Society.)

in the sequence. In Figure 21.14, the strands run in alternate directions (**a** + **b**, **a** − **b**, **a** + **b**, ...) in successive layers. Within each layer the average distance between strands is $d_{ip} \approx 23.3$ nm while the distance between layers is ≈ 11 nm. Strands in successive layers are found to cross at $\phi \approx 75°$.

While the overall energy landscape of the ABCD woodpile lattice structure is certainly quite complicated there are some features of this morphology that can be rationalized using relatively simple geometrical arguments. The zigzag, concertina shape of the aluminosilicate strands and the alternating direction of strands within the woodpile lattice are two striking features of this material. While both features can be explained by geometrical models[36] here we will only discuss the concertina shape. Rods formed in AB–ABA block copolymers[31] and copolymer–silica materials[43] have a uniform cross-section and pack into a parallel, hexagonal array. Thus, the more complicated self-assembly behavior of ABC triblock copolymers is probably responsible for the structure of this material. Previous studies of ABC

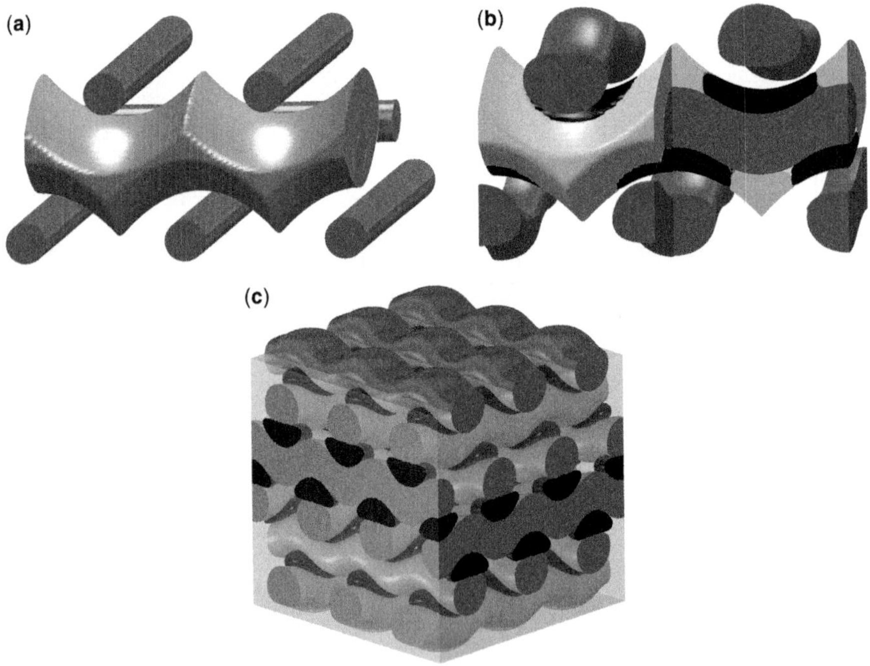

Figure 21.16 (a) Generalized Voronoi cell for the four-layer woodpile structure. The region of space closest to the central rod is shown in gray. The Voronoi cell wiggles under the rods in the layer above and over the rods in the layer below leading to a zigzag, concertina shape. The distance from the rod to the cell surface is smallest where strands cross and largest between crossings. (b, c) Model distribution of PHMA (green) and PEP domains (blue) surrounding the central PEO/aluminosilicate (red) core of the concertina. The smaller PEP domains are positioned where concertinas cross as shown in the cut-away view of a single strand (b) and for the full structure (c).[36] (Reprinted with permission from G. E. S. Toombes et al., *Macromolecules* **2008**, *41*, 852–859. Copyright 2008 American Chemical Society.) (See color insert.)

copolymer systems have reported both strands with nonuniform cross-sections[34,46] and strands packed in nonhexagonal (but parallel) arrays.[53] However, although examples of the woodpile structure are known at the molecular scale,[52,54] it has not been observed in ABC block copolymers before.

It is important to consider the PEP and PHMA domains surrounding the PEO–aluminosilicate core of each concertina. Because chain stretching is energetically costly,[55] PEP and PHMA chains tend to stretch to the nearest PEO–aluminosilicate domain. Thus, the shape of the polymer sheath is approximately the region of space closest to the core of the concertina[55,56] (a generalized Voronoi cell). Figure 21.16a shows the generalized Voronoi cell for straight rods arranged in a four-layer woodpile lattice. The region of space closest to the central rod wiggles under the rods in the layer above and over the rods in the layer below, leading to an overall concertina shape. The shape of the Voronoi cell and thus a simple space-filling requirement accounts for both the wiggling of the PEO–aluminosilicate core and for the relative offset of the layers above and below any given layer (Fig. 21.15e). Figure 21.16b, c show a plausible arrangement of the PEP (blue) and PHMA (green) domains consistent with the block volume fractions and Voronoi cell shape. The smaller PEP block forms micellar domains bridging the short gaps where concertinas cross while the PHMA fills the rest of the space.

These results demonstrate that, in analogy to the multidomain proteins in biosilicification, triblock copolymers can direct the assembly of silica into complex structures. Furthermore, combining complex ABC copolymer architectures with the physical, electrical, and optical properties of inorganic materials holds considerable promise for functional materials.

21.3 NANOSTRUCTURED POLYCRYSTALLINE HYBRID MATERIALS

So far only nanostructuring of amorphous materials with the help of block copolymers as structure-directing agents has been discussed. As described in the introduction this constitutes the simplest level of structuring of inorganic materials. When crystalline materials are considered the degree of difficulty increases substantially. This is due to the fact that crystallization energies are typically much larger than the energies associated with self-assembly, which usually is the result of weak interaction forces. In order to preserve structure control from self-assembly processes specific "tricks" usually have to be applied.

In the second part of this chapter recent progress towards nanostructured polycrystalline assemblies from block copolymer mesophases will be described. First, developments in the approaches presented so far toward nanostructured materials with crystalline nanoparticles embedded in an amorphous silica-type material will be discussed. The second section will describe a new assembly approach for the development of nanostructured materials with highly crystalline walls. The last part highlights achievements of the self-assembly of block copolymers with ligand-stabilized metal nanoparticles, leading to ordered mesoporous metal structures that are polycrystalline in nature.

21.3.1 Nanostructured Aluminosilicate Materials with Crystalline γ-Fe_2O_3 Particles Embedded in Amorphous Silica-Type Materials

Expanding the functionality of silica-based mesostructured and mesoporous materials by the incorporation of functional organic compounds,[57] substitution or addition of other inorganic materials,[58] or templating into carbon-based materials[59] has sparked much interest among researchers. In this section, we will focus on recent work of the Wiesner group on nanostructured materials with crystalline iron oxide nanoparticles embedded in amorphous silica-type materials.[60,61] Although research had been performed on surfactant-based templating and sonochemical approaches to layered iron oxide–oxyhydroxide mesophases,[62] polymer-derived magnetic bulk ceramics,[63] and bulk iron oxide silicates prepared through sol–gel techniques,[64,65] research on mesostructured iron containing silicates was scarce and limited to surfactant-based systems where the iron compound was added in a postsynthesis loading step.[66,67] Backfilling the pores is a common technique to functionalize mesoporous materials,[68] but it requires more synthesis and characterization steps and, more importantly, risks clogging the pore structure. In contrast, we presented a simple block copolymer-based, one-pot self-assembly approach to multifunctional γ-iron oxide–aluminosilicates that are mesoporous and exhibit superparamagnetic behaviour.[60] Nanoscopic iron oxide particles are incorporated in the walls of the aluminosilicate matrix; therefore, blocking of the pores as observed in earlier studies on backfilled materials was overcome even for high iron loadings.[67] This simple and versatile block copolymer-directed approach enabling large pores may lead to new techniques for the separation of magnetically labelled biological macromolecules that combine size exclusion as well as magnetic interactions. Also, the robust matrix with thick walls (>10 nm) makes the material stable to temperatures as high as 800°C allowing for catalytic applications at elevated temperatures. The approach can also be extended to other transition metal oxide systems. Iron oxide was used in the study for its potential magnetic properties, but also serves as an example of what should be possible from a wide range of commercially available metal alkoxides. The actual composition of the resulting materials can be tailored according to the application, which becomes particularly important in catalyst technology.[69]

Polymer–inorganic composites with 25 mol% Fe on a cation basis were prepared using a sol of GLYMO and $Al(O^sBu)_3$ mixed with a solution of the structure-directing PI-*b*-PEO diblock copolymer (see Table 21.1, F and G) and iron(III) ethoxide powder. For the production of a crystalline magnetic iron-oxide phase, the as-made composites were calcined at elevated temperatures in air. SAXS and TEM were performed to determine the structure and order in the materials. Figure 21.17a shows SAXS data for the pure diblock copolymer (1), the as-made polymer-inorganic composite (2), and material calcined to 750°C (3).

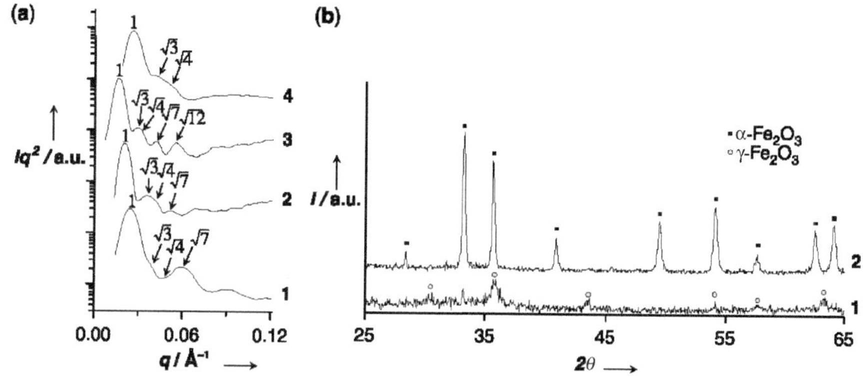

Figure 21.17 (a) SAXS on the pure block copolymer (1), the as-made 25 mol% Fe containing polymer–inorganic composite (2), and material calcined to 750°C (3). Positions of the higher order peaks suggest a hexagonal structure in all cases. (b) XRD of the bulk ceramics (prepared without block copolymer) calcined to 550°C. (1) Pure $Fe(OH)_3$ and (2) 50 mol% Fe containing aluminosilicate prepared using the same sol–gel synthesis procedure as for the composites. Filled and open symbols indicate expected peaks from α- and γ-Fe_2O_3, respectively.[60] (Reprinted with permission from C. Garcia et al., *Angew. Chem. Int. Ed.* **2003**, *42*, 1526–1530. Copyright Wiley-VCH Verlag GmbH & Co. KGaA.)

All traces, including the calcined material, exhibit higher order peaks indicative of cylinders in a hexagonal array. The data suggests that the mesoscopic order of the as-made material is preserved upon calcination. TEM corroborates these assignments (Fig. 21.18). The higher resolution image in Figure 21.18d shows homogeneously distributed dark spots within the matrix. A histogram of the size distribution of these precipitates revealed an average particle diameter of approximately 5 nm. Considering that TEM is a local technique, this number was in excellent agreement with a value of 5.6 nm calculated for the γ-Fe_2O_3 particle size from SQUID data (data not shown), which represented measurements integrated over macroscopic sample dimensions. This remarkable result indicated that the particle size distribution reflected from the TEM micrograph in Figure 21.18d was representative of the entire sample, suggesting the excellent structural control of the present approach. To directly show that these precipitates contain iron, elemental mapping was performed on a scanning transmission electron microscope (STEM) using electron energy loss spectroscopy (EELS). The dark field image of the pyrolyzed sample is shown in Figure 21.18e and the corresponding elemental iron map in Figure 21.18f. Iron is clearly abundant and clustered together consistent with iron oxide precipitates.

X-ray diffraction (XRD) on model materials suggested that in a bulk aluminosilicate matrix γ-Fe_2O_3 is the only crystalline phase detected after heat treatment at 550°C (Fig. 21.17b). Comparable results, that is, the formation of only γ-Fe_2O_3 in iron containing silicate materials prepared by similar means, were also obtained by other groups.[64,65,70] Although pure γ-Fe_2O_3 usually decomposes to nonmagnetic α-Fe_2O_3 at approximately 350°C, the remaining amorphous aluminosilicate matrix apparently hinders the transformation, thus stabilizing the γ-phase even when

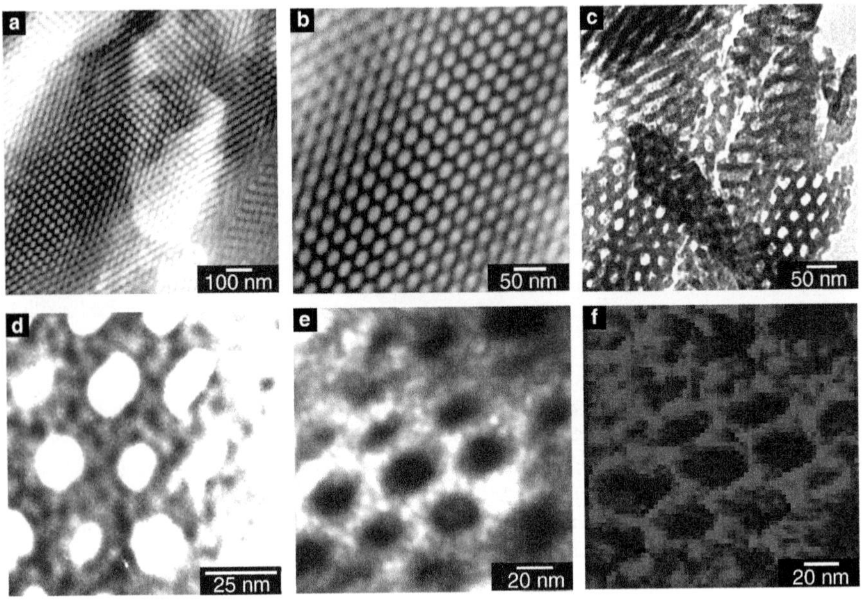

Figure 21.18 Bright-field TEM images of (top row) the as-made sample at two different magnifications revealing the inverse hexagonal morphology (a, b); after calcination to 750°C at two different magnifications (c, d). The image in (c) demonstrates that the hexagonal order is preserved after calcination. At higher magnification (d), the iron oxide precipitates show up as dark spots within the matrix of the inverse hexagonal morphology. Annular dark-field STEM images of the sample calcined to 750°C (e) together with the corresponding energy filtered iron distribution map (f).[60] (Reprinted with permission from C. Garcia et al., *Angew. Chem. Int. Ed.* **2003**, *42*, 1526–1530. Copyright Wiley-VCH Verlag GmbH & Co. KGaA.) (See color insert.)

heated to 750°C.[65] This is fortunate since one can now exploit the magnetic behavior of γ-Fe_2O_3.

These data demonstrated that, first, a ceramic with ordered mesoscale pores is obtained on calcination to temperatures as high as 750°C and, second, nucleation and growth of γ-Fe_2O_3 precipitates occurs inside the amorphous aluminosilicate matrix walls. This latter result is consistent with the phase separation of Fe_2O_3 and SiO_2 observed in their bulk phase diagram. The result is a mesoporous material with open and accessible pores that is superparamagnetic, that is, for which magnetic behavior can be switched on and off by applying an external magnetic field (for magnetic characterization the interested reader is referred to Reference 60).

The synthesis is unique in allowing for precise control over the structure and composition of the final materials. Next to mesoporous materials we used the same approach to make magnetic nanoparticles with different shapes.[61] By simply changing the ratio of inorganic material to polymer (see Table 21.1: polymers H, I, and J) different mesophases could be obtained, as shown in Figure 21.19. The SAXS traces of the as-made composites in Figure 21.19 are consistent with sphere (a), hexagonal cylinder (b), and lamellar hybrid morphologies (c). Solutions of dispersed nanoparticles with different shape and size were generated by simple dissolution and ultrasonication of

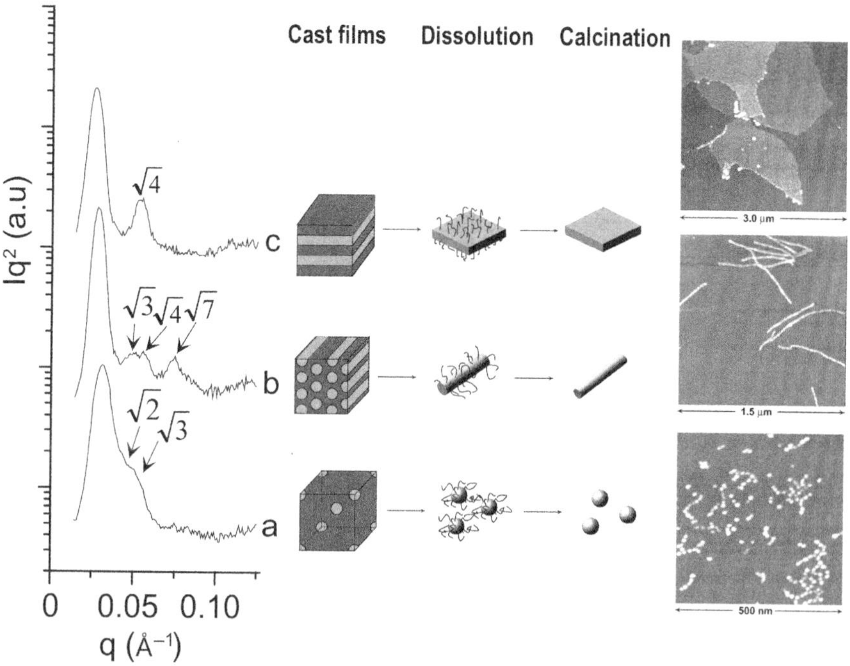

Figure 21.19 SAXS traces with mesophase assignment of the polymer–inorganic composite films and SFM images of individual nanoparticles: (a) sphere, (b) hexagonal cylinder, and (c) lamellar morphologies. The schematic in the middle shows nanoparticle shapes obtained from each phase through dissolution of the films and subsequent calcinations.[61] (Reprinted with permission from C. B. W. Garcia et al., *J. Am. Chem. Soc.* **2003**, *125*, 13310–13311. Copyright 2003 American Chemical Society.)

the bulk composite films in organic solvents and characterized by scanning force microscopy (SFM) after organic components were calcined off. The right side of Figure 21.19 shows images of individual nanoparticles originating from the sphere, cylinder, and lamellar morphologies after dissolution, deposition on a substrate, and calcination.

These results pave the way to functional nanoparticles where the size, shape, and iron oxide concentration can be controlled leading to tunable magnetic and optical properties of nanostructures on surfaces and interesting research in catalysis, molecular labeling, and detection, as well as controlled drug delivery using external magnetic fields.

21.3.2 Thermally Stable and Highly Crystalline Mesoporous Transition Metal Oxides

Mesoporous materials with a transition metal oxide framework have immense potential for applications in catalysis, photocatalysis, sensors, and electrode materials because of their characteristic catalytic, optical, and electronic properties. However, for some applications, this potential can only be maximized in the highly crystalline state. For example, when mesoporous titanium dioxide is applied in photocatalysis, the amorphous regions are known to be trap sites for the recombination of photoexcited electrons and holes,[71] limiting the device efficiency. High thermal and mechanical stability are also associated with the crystalline state. However, the direct synthesis of highly crystalline mesoporous transition metal oxides that are thermally stable and well-ordered still constitutes a major challenge.[72,73] While various soft and hard templating approaches have been developed in the past, they usually suffer from multiple, tedious steps and often result in poor structure control.[58,68,72,74] To this end, the Wiesner group recently reported the synthesis of well-organized, highly crystalline mesoporous transition metal oxides using an amphiphilic diblock copolymer as the structure-directing agent that contains a hydrophilic block and an sp^2 hybridized carbon containing hydrophobic block.[75] The approach combines advantages of both soft structure-directing assemblies and hard templating chemistries. We therefore refer to it as combined assembly by soft and hard (CASH) chemistries or simply CASH. The hybrid materials were produced by combining the amphiphilic diblock copolymer, poly(isoprene-*block*-ethylene oxide) (PI-*b*-PEO) (see Table 21.1, K and L for polymer composition), with sols of groups IV and V transition metal oxides, which selectively swell the hydrophilic PEO block of the block copolymer. Titanium oxide, TiO_2, and niobium oxide, Nb_2O_5, were chosen because of their potential use in photovoltaic cells[76] and fuel cells,[77] respectively. PEO is easily decomposed on heating while the more thermally stable PI, containing two sp^2 carbons per monomer unit, is converted to a sturdy, amorphous carbon material when heat treated under an inert environment. This *in situ* carbon is sufficient to act as a rigid support to the mesostructured oxide walls, preventing collapse when heat treated to temperatures required for obtaining highly crystalline materials. Employing an sp^2 carbon containing hydrophobic block in the structure-directing agent as the carbon source overcomes the disadvantages of other methods where additional carbon precursors are added to the synthesized mesoporous materials for the subsequent carbonization process, often leading to bimodal pore size distributions.[78] In contrast, using the CASH method the materials retain the original mesoporous structure, even after heat treatment to temperatures as high as 1000°C. In this way, highly crystalline mesoporous materials can be obtained.

Polymer–inorganic composites were made by mixing a sol prepared by a nonhydrolytic sol–gel route with a solution of the structure-directing PI-*b*-PEO diblock copolymer. To obtain highly crystalline materials, the as-made films were subsequently heat treated under argon to 700°C. During this process, PEO is burnt off and the metal oxide crystals nucleate, grow, and sinter into wall material. At the same time, part of the PI is converted to an amorphous carbon that lines the walls of the resulting cylindrical pores. This carbon can subsequently be removed by heating the material in air to 450°C, leaving behind a well-organized, highly crystalline mesoporous transition metal oxide.

To support this picture, characterization of the as-made, heat treated under argon (CCM-MO_x-C) and calcined in air (CCM-MO_x-*CASH*) materials was obtained from a variety of techniques. Results of SAXS and TEM experiments on TiO_2 and Nb_2O_5 samples are depicted in Figure 21.20. For the as-synthesized TiO_2, the SAXS pattern

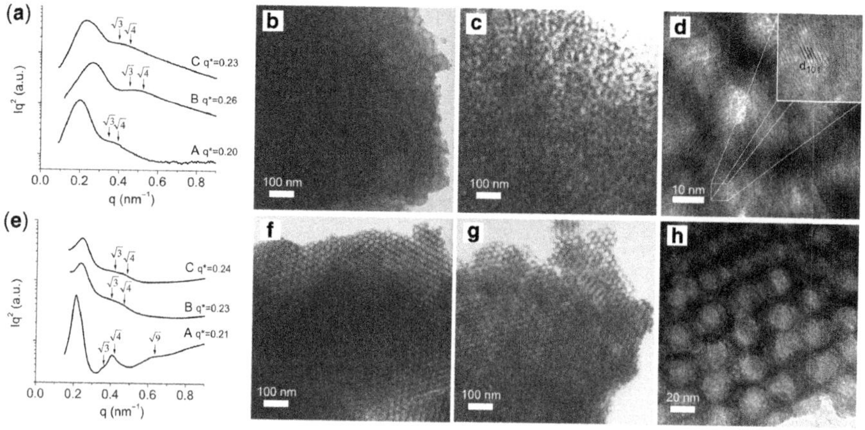

Figure 21.20 Materials characterization by X-ray scattering and electron microscopy. (a) SAXS traces of mesostructured titanium oxide: as-made composite (**A**), CCM-TiO_2-C (**B**), CCM-TiO_2-CASH (**C**); (b) TEM image of as-made TiO_2; (c) TEM image of CCM-TiO_2-CASH; (d) High resolution TEM image of CCM-TiO_2-CASH; Inset: image showing d_{101} spacing, $d_{101} = 3.48$ Å, of the anatase structure consistent with literature values ($d_{101} = 3.52$ Å from JCDPS); (e) SAXS traces of mesostructured niobium oxide: as-made composite (**A**), CCM-Nb_2O_5-C (**B**), CCM-Nb_2O_5-CASH (**C**); (f) TEM image of as-made Nb_2O_5; (g) TEM image of CCM-Nb_2O_5-C; (h) TEM image of CCM-Nb_2O_5-CASH.[75] (Reprinted with permission from J. Lee et al., *Nature Mater.* **2008**, *7*, 222–228. Copyright 2008 Macmillan Publishers Ltd.)

and the TEM results are typical for wormhole-like or short-range ordered hexagonal structures. For Nb_2O_5, the results are consistent with hexagonally packed domains. The SAXS diffractograms of both TiO_2 and Nb_2O_5 after heat treatment under argon and air together with TEM images clearly show the preservation of these structures.

The mesoporous nature of the crystallized materials is confirmed by nitrogen physisorption. CCM-TiO_2–C exhibits a type-IV nitrogen sorption isotherm with specific surface area of 75 $m^2 g^{-1}$. The sharp adsorption at $\sim 0.9P/P0$ indicates that uniform large pores are dominant. The pore size estimate based on the Barrett–Joyner–Halenda (BJH) method is 23.6 nm, in agreement with pore size estimates from TEM. Similarly, CCM-Nb2O5–C exhibits a type-VI nitrogen sorption isotherm with specific surface area of 69 $m^2 g^{-1}$. The pore size estimate based on the BJH method is 24 nm, also in agreement with analysis of microscopy (TEM) images. Raman spectroscopy was used to establish the presence of the *in situ* formed carbon and its removal in the final material. Carbon removal was also corroborated by the fact that after subsequent calcination in air at 450°C, the materials became white, see Figure 21.21.

The formation of crystalline oxides is evident from XRD analysis of the resulting materials. As-made TiO_2 heat treated under Ar (CCM-TiO_2-C) shows several well-resolved, sharp peaks on a flat baseline (Fig. 21.22a, **A**), suggesting that the amorphous as-synthesized TiO_2 was successfully converted to anatase. The crystallite size as calculated from the Scherrer equation is 10 nm. Similarly, the XRD pattern of CCM-Nb_2O_5-C (Fig. 21.22b, **A**) matched that of the hexagonal (primitive)

Figure 21.21 Photograph of CCM-Nb_2O_5-C on the left and CCM-Nb_2O_5-CASH on the right.[75] (Reprinted with permission from J. Lee et al., *Nature Mater.* **2008**, *7*, 222–228. Copyright 2008 Macmillan Publishers Ltd.)

Nb_2O_5 TT phase (PDF# 28-0317) and also showed sharp peaks on a flat baseline. The domain size of the corresponding crystallites was calculated to be 35 nm along one hkl plane and on average 12 nm in other directions. Since the wall thickness of the mesoporous material is estimated to be ~10 nm [from a combination of pore diameter from Brunauer–Emmett–Teller (BET) data and d-spacing from SAXS analysis], this suggests that the Nb_2O_5 crystallites grew parallel to the pores. Analysis after heat treatment in air showed that crystallinity and ordering of the parent materials were preserved (Figs. 21.20 and 21.22). Crystallite sizes estimated from the Scherrer equation showed no changes over the parent materials.

Representative high resolution TEM imaging (Fig. 21.20d) and selected area electron diffraction patterns (SAED, Fig. 21.22c) of various local sections of CCM-TiO_2-*CASH* confirm that the walls are indeed highly crystalline and that anatase is the crystalline form as suggested by the XRD pattern. For CCM-Nb_2O_5-*CASH*, convergent beam electron diffraction patterns (CBED, Fig. 21.22d) of various local sections also confirm that the walls are indeed highly crystalline. Since, Nb_2O_5 has a lower melting point than TiO_2 (1520°C compared to 1870°C), heat treatment to high temperatures resulted in larger crystallite sizes (35 nm oblong-shaped) than for TiO_2 (10 nm). This is evident from electron diffraction analysis as one can obtain diffraction spots for the crystalline CCM-Nb_2O_5-*CASH* with CBED (much smaller beam size as compared to SAED) rather than diffraction rings (as seen for CCM-TiO_2-*CASH* with SAED; compare Fig. 21.22c, d, respectively).

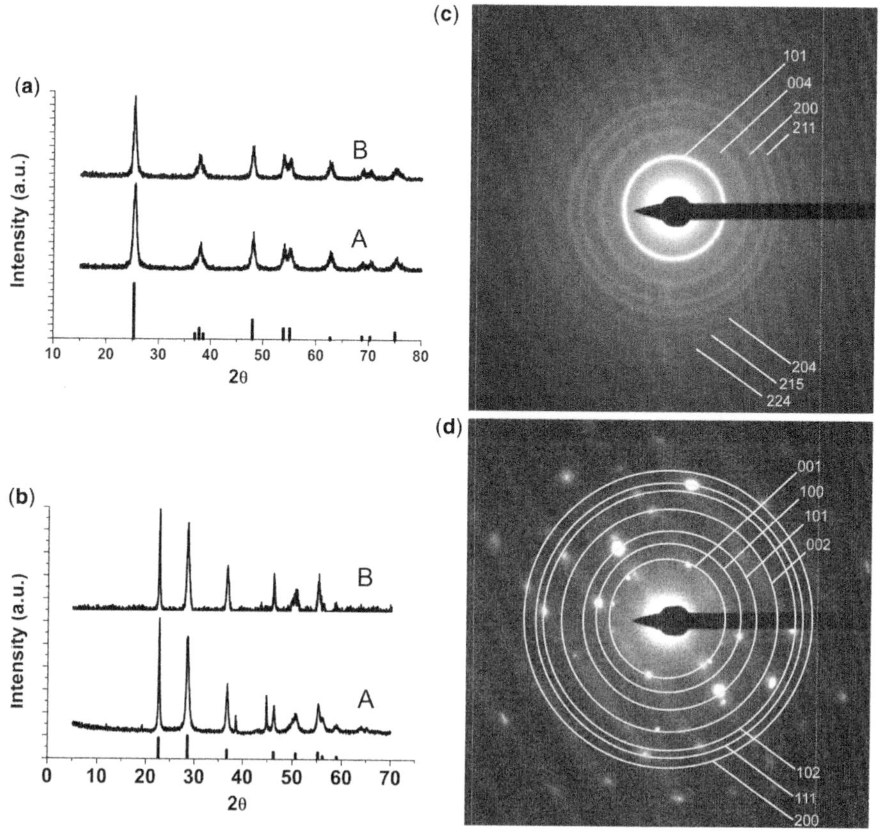

Figure 21.22 Materials characterization by X-ray scattering and electron diffraction. Powder XRD pattern of (a) crystalline TiO_2: CCM-TiO_2-C (**A**), CCM-TiO_2-CASH (**B**); bar markers signify the peaks for anatase TiO_2 structure (PDF # 21-1272); (b) crystalline Nb_2O_5: CCM-Nb_2O_5-C (**A**), CCM-Nb_2O_5-CASH (**B**); bar markers signify the peaks for crystalline Nb_2O_5 structure (PDF # 28-0371). (c) Selected area electron diffraction (SAED) pattern of CCM-TiO_2-CASH showing a sequence of diffraction rings consistent with what is expected for anatase TiO_2 (PDF # 21-1272) as the crystal structure of the mesoporous material; (d) Convergent beam electron diffraction (CBED) pattern of CCM-Nb_2O_5-CASH showing a sequence of diffraction spots consistent with what is expected for crystalline Nb_2O_5 (the hkl planes for PDF # 28-0371 are indicated by circles).[75] (Reprinted with permission from J. Lee et al., *Nature Mater.* **2008**, *7*, 222–228. Copyright 2008 Macmillan Publishers Ltd.)

In order to elucidate the importance of the role of *in situ* formed carbon in the formation of well-organized, highly crystalline mesoporous transition metal oxides, as-synthesized TiO_2 was *directly* calcined under *air* to 700°C while keeping all other conditions the same as for the CASH method. As expected, the BET surface area of the resulting material was only 0.2 $m^2\,g^{-1}$ and no porous structure could be detected by TEM imaging. This implies that the mesostructure completely collapsed. The crystallite size of this sample, heat treated to 700°C in air is 31.5 nm (calculated from XRD pattern), which is much larger than that of our sample heat treated to 700°C under argon (10 nm). All the organics from the structure-directing block copolymer PI-*b*-PEO are removed under air below 500°C and further crystallization without any carbon support causes collapse of the pores. In contrast, when the as-made TiO_2 was heated to temperatures as high as 1000°C under argon (i.e., prepared by the CASH method), the black color and mesoporosity were preserved. The pore size and surface area of CCM-TiO_2-C in this instance were 17.4 nm and 29 $m^2\,g^{-1}$, respectively. For these annealing conditions, the walls of the mesopores are composed of a mixture of anatase and rutile crystals as anatase is usually converted to rutile above 700°C.[79] Reduction in pore size and surface area, relative to the material heated to only 700°C, can be rationalized considering the changes in crystal structure occurring at such extreme temperatures, which are expected to have an effect on structural characteristics of the resulting material. Most importantly, however, even for this very high temperature the porosity of the material was retained, further demonstrating the efficiency of the CASH method.

To show the versatility of our method in terms of achieving highly crystalline, mesoporous transition metal oxides and the importance of heat treating to high temperatures, as-made Nb_2O_5 was heat treated to 400°C in air, like most reported work using soft-templating methods suggest. The resulting XRD pattern reveals that the material is still amorphous (see Fig. 21.23, **A**). Nb_2O_5 is crystallized at a higher temperature than 400°C but most previous methods only allow heat treatment to this temperature before the mesostructure collapses. Therefore the CASH method facilitates the synthesis of highly crystalline (see Fig. 21.23, **B**), mesoporous Nb_2O_5 by virtue of heat treating to a temperature as high as 700°C under an inert atmosphere. This simple yet facile method can therefore be utilized to obtain highly crystalline, mesoporous

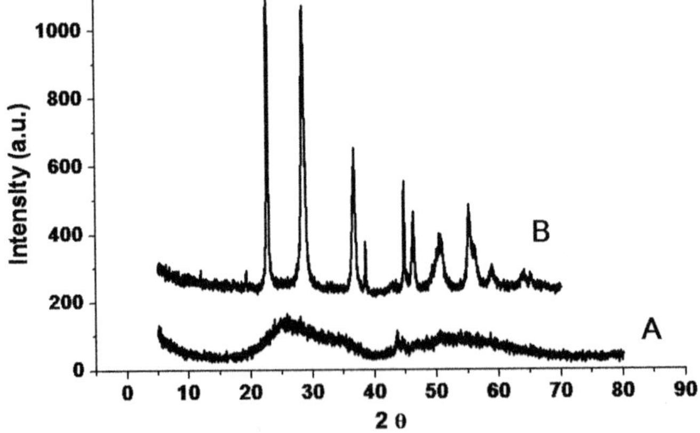

Figure 21.23 Powder XRD pattern of (**A**) Nb_2O_5 heat treated to 400 °C in air and (**B**) CCM-Nb_2O_5-CASH heat treated to 700°C.[75] (Reprinted with permission from J. Lee et al., *Nature Mater.* **2008**, 7, 222–228. Copyright 2008 Macmillan Publishers Ltd.)

transition metal oxides through high temperature heat treatments that were otherwise not possible without collapse of the mesostructure.

21.3.3 Ordered Mesoporous Metals

Controlling the structure of metals at the mesoscale is crucial for the development of improved fuel cell electrodes and may also assist in the miniaturization of optical and electronic materials for data transmission, storage, and computation.[4] Despite considerable progress in the field of metal–polymer hybrids,[80–82] their conversion into mesoporous materials with ordered and large pores ($\geq$10 nm) has not been accomplished, in part because of the low volume fraction of metals in most hybrids and the widespread use of gold, which has a high diffusion coefficient and therefore retains its mesostructure only at low temperatures.[83–85] This section describes a route to mesostructured platinum nanoparticle–block copolymer hybrids with exceptionally high nanoparticle (NP) loadings and tunable phase-separated morphologies with feature sizes >10 nm, recently developed in the Wiesner group.[86] Metal-rich NPs with a thin organic shell made from ionic liquid-type ligands, combined with NP loadings as high as 79 vol% in the hydrophilic domains of the hybrid, ensure that mesostructure order is retained upon conversion to mesoporous metal–carbon composites. The carbon can be removed from the nanocomposites to produce ordered platinum (Pt) mesostructures with ordered and large uniform pores.

The hybrid materials were produced by combining amphiphilic diblock copolymers, poly(isoprene-*block*-dimethylaminoethyl methacrylate) (PI-*b*-PDMAEMA) (see Table 21.1, M and N for polymer composition), with *N*,*N*-di-2-propoxyethyl-*N*-3-mercaptopropyl-*N*-methylammonium chloride-stabilized PtNPs. The following criteria were identified as being crucial for successful structure formation: First, the NPs should exhibit high solubility in organic solvents. When NPs would have low solubility, some fraction of the particles macroscopically precipitates during solvent evaporation and fails to mix with the block copolymer. We recently reported a ligand for metal NPs based on a thiol-containing ionic liquid[87] that imparted liquid-like behavior to the NPs, even in the absence of a solvent. This result suggested that the use of certain ionic liquids as NP ligands provides a route to high solubility. Second, the ligand-stabilized NPs should have at least a modest metal volume fraction. Although metal NPs with high solubility have been reported,[87,88] the metal volume fraction was between 0.6% and 3%.[87] Higher metal volume fractions are needed to prevent structural collapse upon removal of the organic components. Conventional short ligands can increase the metal volume fraction but usually result in insufficient solubility. Third, the NPs should be highly dispersible in just one block of the block copolymer.[4,89] Mixing should be driven by favorable enthalpic interactions between that block and the NPs, which can be achieved through ionic interactions, hydrogen bonding, and dipole–dipole interactions, among many possible routes. Fourth, as described in an earlier section, the diameter of the NPs should be below a critical limit relative to the size of the block with which they mix, approximately the root mean square end-to-end distance of the relevant block of the copolymer.[10]

Although the precise cutoff may vary depending on enthalpic considerations and NP concentration, this heuristic provides a guideline for NP size.

Casting a film of a macroscopically homogeneous solution of aged NPs and block copolymer resulted in a homogeneous, shiny black solid, which was weak and brittle (Fig. 21.24a). Analysis of samples by TEM (Fig. 21.24b) revealed that a mesostructure had formed. The order could be improved, however, by annealing at 130°C for 2 days under vacuum, as confirmed by TEM (Fig. 21.24c). The representative TEM image in Figure 21.24c revealed an inverse hexagonal mesostructure with grain sizes on the

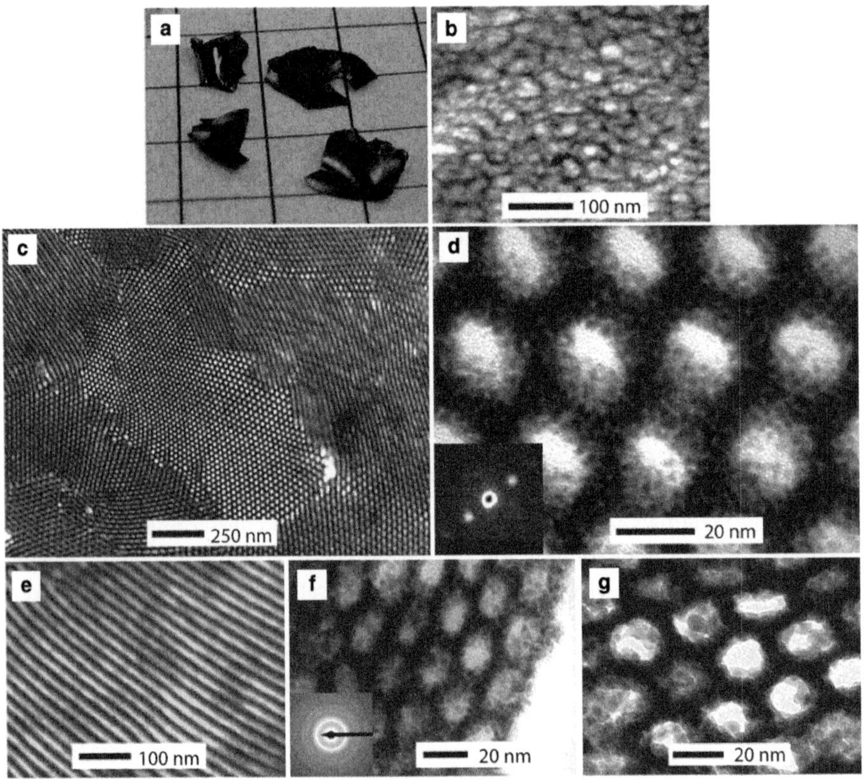

Figure 21.24 Photograph (a) and bright-field TEM images (b) to (g) of materials produced after each stage of the synthesis. (a) Pieces of unannealed inverse hexagonal hybrid film. The grid paper has 5 mm markings. (b) Unannealed inverse hexagonal hybrid. (c) An annealed inverse hexagonal hybrid. (d) Examination of the hybrid from (c) at higher magnification resolved individual PtNPs, seen as dark spots in the bright-field image. (Inset) A typical convergent-beam electron diffraction pattern (seen with an ultra-high-vacuum scanning transmission electron microscope) from a single PtNP, demonstrating its crystallinity. (e) An annealed lamellar hybrid. (f) Pyrolysis of an annealed inverse hexagonal hybrid yields a mesoporous Pt-C composite. (Inset) Selected area electron diffraction, showing Pt expected face-centered cubic scattering profile. (g) Removal of carbon with an Ar-O plasma yielded mesoporous inverse hexagonal Pt.[86] (Reprinted with permission from S. C. Warren et al., *Science* **2008**, *320*, 1748–1752. Copyright AAAS.)

order of a few micrometers. Examination of the mesostructure at higher magnification (Fig. 21.24d) revealed that individual PtNPs composed the walls of the mesostructure, with three to five NPs spanning the thickness of the wall. Besides inverse hexagonal mesostructures, samples with lamellar morphology were produced. Annealing at 130°C led to a well-developed mesostructure as confirmed by TEM (Fig. 21.24e). These results suggest that as in the cases discussed earlier of oxide structures, metal NP–block copolymer hybrid morphologies can be tailored by simply adjusting the NP volume fraction.[14]

A rapid pyrolysis process[75,90] was used to convert inverse hexagonal hybrids to ordered mesoporous Pt-carbon composites. Samples were heated at 10°C min^{-1} under N_2 or Ar to at least 410°C, followed by immediate cooling. Under these conditions, the sp^2-hybridized carbons of the PI block decompose into an amorphous carbon-rich material with slight graphitic character.[75] Characterization of the resulting materials by TEM (Fig. 21.24f) indicated that the inverse hexagonal structure was preserved. Analysis of a particular sample by PXRD and application of the Scherrer equation indicated that the Pt nanocrystals' domain size was 4.1 ± 0.4 nm, representing a substantial increase from the parent aged NPs. Nitrogen physisorption (data not shown) on this sample indicated that the mesopores were open and that 26% of the sample's volume (micropores and mesopores) was open space, as expected for an inverse hexagonal nanocomposite that has pores lined with carbon. The BET surface area was 18 $m^2\,g^{-1}$, and the pore diameter 17 nm.

For many applications, such as fuel cells, it is desirable for the metal surface to be completely exposed. The carbon was removed from microtomed thin films (~50 nm thick) of the nanocomposites using an Ar-oxygen plasma. The mesoporous Pt was structurally similar to the Pt-carbon nanocomposites, as determined by TEM (Fig. 21.24g). Close inspection of the pores in TEM images revealed that the grainy texture indicative of the carbon had disappeared. Alternatively, for thicker films (10 to 100 mm thick), a sulfuric acid:nitric acid 3:1 (v/v) etch at 70°C[91] successfully removed most carbon. Electrochemical data from acid-etched samples indeed confirmed that the metal surface was exposed, showing current densities nearly identical to that of bulk Pt (Fig. 21.25a). Energy-dispersive spectroscopy (EDS) on the metal-carbon nanocomposites (Fig. 21.25b) showed a composition of 74 wt% Pt, 18 wt% C, 7 wt% O, and 1 wt% S. In contrast, after the plasma treatment, EDS revealed that >98 wt% of the sample was Pt, with only trace contributions from carbon and oxygen (Fig. 21.25c).

We measured the electrical conductivity of Pt-C nanocomposites using two-point measurements. In a representative example the NP-polymer hybrid had a conductivity of 2.5 mS cm^{-1}, which increased to 400 S cm^{-1} upon pyrolysis. Despite the presence of carbon, to the best of our knowledge this value represented the highest electrical conductivity yet measured for ordered mesoporous materials derived from block copolymers. This discovery creates a potential pathway to a new class of ordered mesoporous metals made from nanoparticles of different elements and/or distinct compositions. Such nano-heterogeneous mesoporous metals may have a range of exceptional electrical, optical, and catalytic properties.

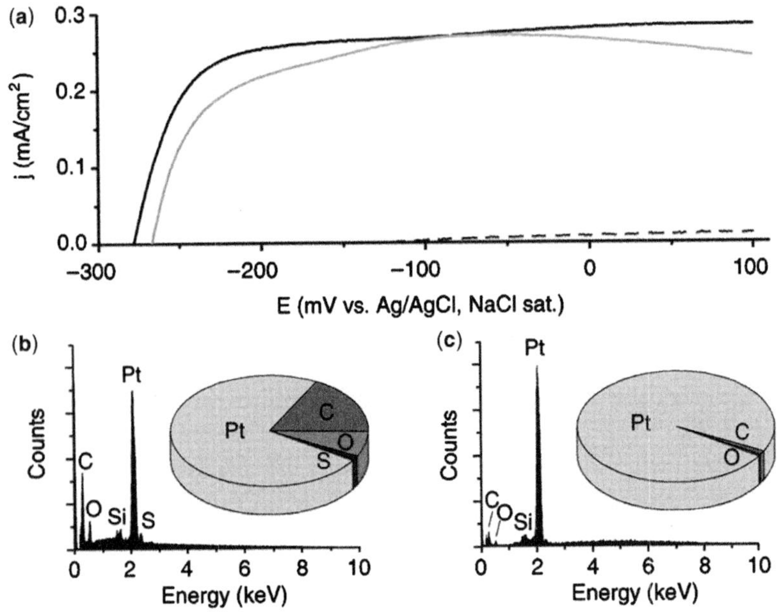

Figure 21.25 (a) Polarization curves of the H_2 oxidation reaction in H_2-saturated 0.1 M H_2SO_4 solution (at 2000 rpm and 10 mV s^{-1}). Dashed line, mesoporous Pt-C nanocomposite; grey curve, mesoporous Pt; black curve, planar Pt electrode. E, potential; sat., saturated. (b) EDS of pyrolized sample. The pie chart displays elemental weight fractions. The sample was on a Si substrate and the primary energy was 10 keV. Pt = 74 wt%, C = 18 wt%, O = 7 wt%, and S = 1 wt%. (c) EDS of acid-treated sample. Pt = 98 wt%, C = 1 wt%, and O = 0.5 wt%.[86] (Reprinted with permission from S. C. Warren et al., *Science* **2008**, *320*, 1748–1752. Copyright AAAS.)

21.4 CONCLUSIONS

Copying strategies used in nature to structure organic (polymer)–inorganic hybrids is a rich and fascinating area. In this chapter we have only discussed entirely synthetic material approaches that did not include any bio(macro)molecules. In particular, we have focused on the field of nanoparticle–block copolymer hybrid systems, which has tremendous scientific as well as technological promise. Since the start of the use of block copolymers to structure direct amorphous silica-type materials, we have gained a lot of fundamental understanding about these systems and their formation mechanisms. In the meantime, the concept has been successfully generalized towards other macromolecular architectures of the structure-directing agents, for example, dendritic blocks and ABC triblock copolymers elevating the complexity of the resulting structures significantly. Moreover, on the inorganic side, beyond amorphous materials powerful approaches towards building mesoporous polycrystalline structures have been developed.

Through the choice of the appropriate synthetic block copolymer systems as well as nanoparticles, unprecedented morphological control on the nano- and mesoscales is obtained in the block copolymer-directed synthesis of hybrid materials. For systems with favorable enthalpic interactions between one block of the copolymer and the inorganic, that is, with a negative interaction parameter, χ, experiments suggest a critical size of the nanoparticles where the self-assembly behavior of polymers and particles changes fundamentally. This critical size is the result of competing entropic effects (translation versus conformation). Competing interactions is a feature also found in complex biological systems. Below the critical size, the nanoparticles can be regarded as small molecular weight additives or solvents that are selectively added to one of the blocks, thereby swelling it, resulting in rational design criteria for controlled hybrid mesostructure formation.

The potential of this approach for new materials, with multiple nanoscale functional features for applications in nanotechnology and beyond, lies in the versatility of the polymer chemistry as well as that of the nanoparticle chemistry that can be exploited in the materials synthesis. Experiments described in this chapter and work of many others in the field so far probably only scratch the surface of things to come. In the long run, such powerful bottom-up strategies will develop into alternatives to, or converge with, various top-down approaches, thereby revolutionizing the precision with which materials will be fabricated over length scales from the molecular all the way to the macroscopic length scales. This may lead to the design of entirely new classes of materials with properties that have no analog in the natural world.

REFERENCES

1. J. Aizenberg, J. C. Weaver, M. S. Thanawala, V. C. Sundar, D. E. Morse, P. Fratzl, *Science* **2005**, *309*, 275–278.
2. J. Aizenberg, A. Tkachenko, S. Weiner, L. Addadi, G. Hendler, *Nature* **2001**, *412*, 819–822.
3. M. Kamperman, U. Wiesner, in *Block Copolymers in Nanoscience* (Eds. M. Lazzari, G. Liu, S. Lecommandoux), Weinheim: Wiley-VCH, **2006**.
4. (a) C. Sanchez, C. Boissiere, D. Grosso, C. Laberty, L. Nicole, *Chem. Mater.* **2008**, *20*, 682–737; (b) Y. Wan, D. Y. Zhao, *Chem. Rev.* **2007**, *107*, 2821–2860; (c) S. W. Boettcher, J. Fan, C. K. Tsung, Q. H. Shi, G. D. Stucky, *Acc. Chem. Res.* **2007**, *40*, 784–792; (d) A. C. Balazs, T. Emrick, T. P. Russell, *Science* **2006**, *314*, 1107–1110; (e) A. Haryono, W. H. Binder, *Small* **2006**, *2*, 600–611; (f) M. R. Bockstaller, R. A. Mickiewicz, E. L. Thomas, *Adv. Mater.* **2005**, *17*, 1331–1349; (g) M. Pagliaro, R. Ciriminna, M. W. C. Man, S. Campestrini, *J. Phys. Chem. B* **2006**, *110*, 1976–1988; (h) D. A. Olson, L. Chen, M. A. Hillmyer, *Chem. Mater.* **2008**, *20*, 869–890; (i) J. Fan, S. W. Boettcher, C. K. Tsung, Q. Shi, M. Schierhorn, G. D. Stucky, *Chem. Mater.* **2008**, *20*, 909–921.
5. (a) K. Shimizu, J. Cha, G. D. Stucky, D. E. Morse, *Proc. Natl. Acad. Sci. U.S.A.* **1998**, *95*, 6234–6238; (b) H. Cölfen, S. Mann, *Angew. Chem. Int. Ed.* **2003**, *42*, 2350–2365; (c) N. Kröger, R. Deutzmann, M. Sumper, *Science* **1999**, *286*, 1129–1132; (d) E. G. Vrieling, T. P. M. Beelen, R. A. van Santen, W. W. C. Gieskes, *Angew. Chem. Int. Ed.* **2002**, *41*, 1543–1546; (e) M. Sumper, *Science* **2002**, *295*, 2430–2433.
6. N. Kroger, S. Lorenz, E. Brunner, M. Sumper, *Science* **2002**, *298*, 584–586.
7. (a) T. Yanagisawa, T. Shimizu, K. Kuroda, C. Kato, *Bull. Chem. Soc. Jpn.* **1990**, *63*, 1535–1537; (b) C. T. Kresge, M. E. Leonowicz, W. J. Roth, J. C. Vartuli, J. S. Beck, *Nature* **1992**, *359*, 710–712; (c) Q. Huo, D. I. Margolese, U. Ciesla, P. Feng, T. E. Gier, P. Sieger, R. Leon, P. M. Petroff, F. Schueth, G. D. Stucky, *Nature* **1994**, *368*, 317–321; (d) I. A. Aksay, M. Trau, S. Manne, I. Honma, N. Yao, L. Zhou, P. Fenter, P. M. Eisenberger, S. M. Gruner, *Science* **1996**, *273*, 892–898; (e) H. Yang, N. Coombs, G. A. Ozin, *Nature* **1997**, *386*, 692–695; (f) G. J. de A. A. Soler-Illia, C. Sanchez, B. Lebeau, J. Patarin, *Chem. Rev.* **2002**, *102*, 4093–4138.

8. (a) S. A. Bagshaw, E. Prouzet, T. J. Pinnavaia, *Science* **1995**, *269*, 1242–1244; (b) D. Y. Zhao, J. L. Feng, Q. S. Huo, N. Melosh, G. H. Fredrickson, B. F. Chmelka, G. D. Stucky, *Science* **1998**, *279*, 548–552.
9. Y. Lu, H. Fan, A. Stump, T. L. Ward, T. Rieker, C. J. Brinker, *Nature* **1999**, *398*, 223–226.
10. S. C. Warren, F. J. DiSalvo, U. Wiesner, *Nature Mater.* **2007**, *6*, 156–161.
11. A. Jain, U. Wiesner, *Macromolecules* **2004**, *37*, 5665–5670.
12. S. Gupta, Q. L. Zhang, T. Emrick, A. C. Balazs, T. P. Russell, *Nature Mater.* **2006**, *5*, 229–233.
13. M. E. Mackay, A. Tuteja, P. M. Duxbury, C. J. Hawker, B. van Horn, Z. B. Guan, G. H. Chen, R. S. Krishnan, *Science* **2006**, *311*, 1740–1743.
14. M. Templin, A. Franck, A. DuChesne, H. Leist, Y. M. Zhang, R. Ulrich, V. Schadler, U. Wiesner, *Science* **1997**, *278*, 1795–1798.
15. P. F. W. Simon, R. Ulrich, H. W. Spiess, U. Wiesner, *Chem. Mater.* **2001**, *13*, 3464–3486.
16. J. Huh, V. V. Ginzburg, A. C. Balazs, *Macromolecules* **2000**, *33*, 8085–8096.
17. M. Antonietti, C. Göltner, *Angew. Chem. Int. Ed.* **1997**, *36*, 910–928.
18. B. K. Cho, A. Jain, S. M. Gruner, U. Wiesner, *Science* **2004**, *305*, 1598–1601.
19. B. K. Cho, A. Jain, J. Nieberle, S. Mahajan, U. Wiesner, S. M. Gruner, S. Turk, H. J. Rader, *Macromolecules* **2004**, *37*, 4227–4234.
20. S. Mahajan, B. K. Cho, A. Allgaier, L. J. Fetters, G. W. Coates, U. Wiesner, *Macromol. Rapid Commun.* **2004**, *25*, 1889–1894.
21. C. J. Hawker, J. M. J. Frechet, *J. Am. Chem. Soc.* **1990**, *112*, 7638–7647.
22. D. A. Tomalia, H. Baker, J. Dewald, M. Hall, G. Kallos, S. Martin, J. Roeck, J. Ryder, P. Smith, *Polymer J.* **1985**, *17*, 117–132.
23. (a) S. M. Grayson, J. M. J. Frechet, *Chem. Rev.* **2001**, *101*, 3819–3867; (b) A. W. Bosman, H. M. Janssen, E. W. Meijer, *Chem. Rev.* **1999**, *99*, 1665–1688; (c) X. B. Zeng, G. Ungar, Y. S. Liu, V. Percec, S. E. Dulcey, J. K. Hobbs, *Nature* **2004**, *428*, 157–160; (d) J. C. M. Vanhest, D. A. P. Delnoye, M. W. P. L. Baars, M. H. P. Vangenderen, E. W. Meijer, *Science* **1995**, *268*, 1592–1595.
24. I. Gitsov, K. L. Wooley, C. J. Hawker, P. T. Ivanova, J. M. J. Frechet, *Macromolecules* **1993**, *26*, 5621–5627.
25. J. Iyer, K. Fleming, P. T. Hammond, *Macromolecules* **1998**, *31*, 8757–8765.
26. (a) J. L. Hedrick, C. J. Hawker, R. D. Miller, R. Twieg, S. A. Srinivasan, M. Trollsas, *Macromolecules* **1997**, *30*, 7607–7610; (b) J. W. Kriesel, T. D. Tilley, *Chem. Mater.* **1999**, *11*, 1190–1193; (c) E. Ruckenstein, W. S. Yin, *J. Polym. Sci. Part A Polym. Chem.* **2000**, *38*, 1443–1449; (d) R. Velarde-Ortiz, G. Larsen, *Chem. Mater.* **2002**, *14*, 858–866.
27. B. K. Cho, A. Jain, S. Mahajan, H. Ow, S. M. Gruner, U. Wiesner, *J. Am. Chem. Soc.* **2004**, *126*, 4070–4071.
28. B. K. Cho, A. Jain, S. M. Gruner, U. Wiesner, *Chem. Mater.* **2007**, *19*, 3611–3614.
29. M. Jayaraman, J. M. J. Frechet, *J. Am. Chem. Soc.* **1998**, *120*, 12996–12997.
30. (a) Y. Matsushita, H. Choshi, T. Fujimoto, M. Nagasawa, *Macromolecules* **1980**, *13*, 1053–1058; (b) C. Auschra, R. Stadler, *Macromolecules* **1993**, *26*, 2171–2174.
31. F. S. Bates, G. H. Fredrickson, *Phys. Today* **1999**, *52*, 32–38.
32. Z. Wei, Z. G. Wang, *Macromolecules* **1995**, *28*, 7215–7223.
33. T. H. Epps, E. W. Cochran, T. S. Bailey, R. S. Waletzko, C. M. Hardy, F. S. Bates, *Macromolecules* **2004**, *37*, 8325–8341.
34. U. Krappe, R. Stadler, I. Voigtmartin, *Macromolecules* **1995**, *28*, 4558–4561.
35. T. Goldacker, V. Abetz, R. Stadler, I. Erukhimovich, L. Leibler, *Nature* **1999**, *398*, 137–139.
36. G. E. S. Toombes, S. Mahajan, M. Weyland, A. Jain, P. Du, M. Kamperman, S. M. Gruner, D. A. Muller, U. Wiesner, *Macromolecules* **2008**, *41*, 852–859.
37. G. E. S. Toombes, S. Mahajan, M. Thomas, P. Du, M. W. Tate, S. M. Gruner, U. Wiesner, *Chem. Mater.* **2008**, *20*, 3278–3287.
38. R. B. Grubbs, *Macromol. Chem. Phys.* **2005**, *206*, 625–627.
39. (a) J. Z. Du, Y. M. Chen, *J. Polym. Sci. Part A Polym. Chem.* **2004**, *42*, 2263–2271; (b) Y. L. Cai, S. P. Armes, *Macromolecules* **2004**, *37*, 7116–7122; (c) B. Angot, D. Taton, Y. Gnanou, *Macromolecules*

2000, *33*, 5418–5426; (d) K. Matyjaszewski, J. H. Xia, *Chem. Rev.* **2001**, *101*, 2921–2990; (e) M. Kamigaito, T. Ando, M. Sawamoto, *Chem. Rev.* **2001**, *101*, 3689–3745.
40. L. J. Fetters, D. J. Lohse, D. Richter, T. A. Witten, A. Zirkel, *Macromolecules* **1994**, *27*, 4639–4647.
41. S. S. Rogers, L. Mandelkern, *J. Phys. Chem.* **1957**, *61*, 985–990.
42. T. S. Bailey, H. D. Pham, F. S. Bates, *Macromolecules* **2001**, *34*, 6994–7008.
43. S. Renker, S. Mahajan, D. T. Babski, I. Schnell, A. Jain, J. Gutmann, Y. M. Zhang, S. M. Gruner, H. W. Spiess, U. Wiesner, *Macromol. Chem. Phys.* **2004**, *205*, 1021–1030.
44. S. Ludwigs, A. Boker, V. Abetz, A. H. E. Muller, G. Krausch, *Polymer* **2003**, *44*, 6815–6823.
45. U. Breiner, U. Krappe, T. Jakob, V. Abetz, R. Stadler, *Polymer Bull.* **1998**, *40*, 219–226.
46. U. Breiner, U. Krappe, V. Abetz, R. Stadler, *Macromol. Chem. Phys.* **1997**, *198*, 1051–1083.
47. (a) A. Keller, E. Pedemont, F. M. Willmout, *Nature* **1970**, *225*, 538–539; (b) R. M. Kannan, J. A. Kornfield, *Macromolecules* **1994**, *27*, 1177–1186; (c) H. Leist, D. Maring, T. Thurn-Albrecht, U. Wiesner, *J. Chem. Phys.* **1999**, *110*, 8225–8228; (d) S. Stangler, V. Abetz, *Rheol. Acta* **2003**, *42*, 569–577.
48. V. L. Finkenstadt, R. P. Millane, *Acta Crystallogr. Sect. A* **1998**, *54*, 240–248.
49. (a) G. Coulon, T. P. Russell, V. R. Deline, P. F. Green, *Macromolecules* **1989**, *22*, 2581–2589; (b) K. Fukunaga, H. Elbs, R. Magerle, G. Krausch, *Macromolecules* **2000**, *33*, 947–953.
50. P. A. Midgley, M. Weyland, *Ultramicroscopy* **2003**, *96*, 413–431.
51. H. S. Sozuer, J. P. Dowling, *J. Mod. Opt.* **1994**, *41*, 231–239.
52. M. Okeeffe, S. Andersson, *Acta Crystallogr. Sect. A* **1977**, *33*, 914–923.
53. (a) Y. Mogi, H. Kotsuji, Y. Kaneko, K. Mori, Y. Matsushita, I. Noda, *Macromolecules* **1992**, *25*, 5408–5411; (b) S. Brinkmann, R. Stadler, E. L. Thomas, *Macromolecules* **1998**, *31*, 6566–6572.
54. (a) S. V. Meille, S. Bruckner, W. Porzio, *Macromolecules* **1990**, *23*, 4114–4121; (b) N. L. Rosi, J. Kim, M. Eddaoudi, B. L. Chen, M. O'Keeffe, O. M. Yaghi, *J. Am. Chem. Soc.* **2005**, *127*, 1504–1518.
55. E. L. Thomas, D. J. Kinning, D. B. Alward, C. S. Henkee, *Macromolecules* **1987**, *20*, 2934–2939.
56. G. M. Grason, *Phys. Rep.* **2006**, *433*, 1–64.
57. (a) Y. F. Lu, Y. Yang, A. Sellinger, M. C. Lu, J. M. Huang, H. Y. Fan, R. Haddad, G. Lopez, A. R. Burns, D. Y. Sasaki, J. Shelnutt, C. J. Brinker, *Nature* **2001**, *410*, 913–917; (b) T. Asefa, M. J. MacLachlan, N. Coombs, G. A. Ozin, *Nature* **1999**, *402*, 867–871; (c) S. Inagaki, S. Guan, T. Ohsuna, O. Terasaki, *Nature* 2002, *416*, 304–307.
58. P. Yang, D. Zhao, D. I. Margolese, B. F. Chmelka, G. D. Stucky, *Nature* **1998**, *396*, 152–155.
59. S. H. Joo, S. J. Choi, I. Oh, J. Kwak, Z. Liu, O. Terasaki, R. Ryoo, *Nature* **2001**, *412*, 169–172.
60. C. B. W. Garcia, Y. M. Zhang, F. DiSalvo, U. Wiesner, *Angew. Chem. Int. Ed.* **2003**, *42*, 1526–1530.
61. C. B. W. Garcia, Y. M. Zhang, S. Mahajan, F. DiSalvo, U. Wiesner, *J. Am. Chem. Soc.* **2003**, *125*, 13310–13311.
62. (a) G. Wirnsberger, K. Gatterer, H. P. Fritzer, W. Grogger, B. Pillep, P. Behrens, M. F. Hansen, C. B. Koch, *Chem. Mater.* **2001**, *13*, 1453–1466; (b) Y. Q. Wang, L. X. Yin, A. Gedanken, *Ultrason. Sonochem.* **2002**, *9*, 285–290.
63. M. J. MacLachlan, M. Ginzburg, N. Coombs, T. W. Coyle, N. P. Raju, J. E. Greedan, G. A. Ozin, I. Manners, *Science* **2000**, *287*, 1460–1463.
64. (a) F. delMonte, M. P. Morales, D. Levy, A. Fernandez, M. Ocaña, A. Roig, E. Molins, K. Ogrady, C. J. Serna, *Langmuir* **1997**, *13*, 3627–3634; (b) G. Ennas, A. Musinu, G. Piccaluga, D. Zedda, D. Gatteschi, C. Sangregorio, J. L. Stanger, G. Concas, G. Spano, *Chem. Mater.* **1998**, *10*, 495–502; (c) C. Cannas, M. F. Casula, G. Concas, A. Corrias, D. Gatteschi, A. Falqui, A. Musinu, C. Sangregorio, G. Spano, *J. Mater. Chem.* **2001**, *11*, 3180–3187.
65. M. F. Casula, A. Corrias, G. Paschina, *J. Non-Cryst. Solids* **2001**, *293*, 25–31.
66. M. Froba, R. Kohn, G. Bouffaud, O. Richard, G. van Tendeloo, *Chem. Mater.* **1999**, *11*, 2858–2865.
67. L. Zhang, G. C. Papaefthymiou, J. Y. Ying, *J. Phys. Chem. B* **2001**, *105*, 7414–7423.
68. X. He, D. Antonelli, *Angew. Chem. Int. Ed.* **2002**, *41*, 214–229.
69. C. Roychowdhury, F. Matsumoto, V. B. Zeldovich, S. C. Warren, P. F. Mutolo, M. Ballesteros, U. Wiesner, H. D. Abruna, F. J. DiSalvo, *Chem. Mater.* **2006**, *18*, 3365–3372.
70. D. Niznansky, N. Viart, J. L. Rehspringer, *J. Sol-Gel Sci. Technol.* **1997**, *8*, 615–618.
71. B. Ohtani, Y. Ogawa, S. Nishimoto, *J. Phys. Chem. B* **1997**, *101*, 3746–3752.

72. D. M. Antonelli, J. Y. Ying, *Angew. Chem. Int. Ed.* **1996**, *35*, 426–430.
73. U. Ciesla, S. Schacht, G. D. Stucky, K. K. Unger, F. Schueth, *Angew. Chem. Int. Ed.* **1996**, *35*, 541–543.
74. B. Lee, D. Lu, J. N. Kondo, K. Domen, *J. Am. Chem. Soc.* **2002**, *124*, 11256–11257; (b) C.-W. Wu, T. Ohsuna, M. Kuwabara, K. Kuroda, *J. Am. Chem. Soc.* **2006**, *128*, 4544–4545; (c) F. Jiao, P. G. Bruce, *Angew. Chem. Int. Ed.* **2004**, *43*, 5958–5961; (d) D. M. Antonelli, J. Y. Ying, *Angew. Chem. Int. Ed.* **1995**, *34*, 2014–2017; (e) F. Jiao, A. Harrison, J.-C. Jumas, A. V. Chadwick, W. Kockelmann, P. G. Bruce, *J. Am. Chem. Soc.* **2006**, *128*, 5468–5474; (f) F. Jiao, J.-C. Jumas, M. Womes, A. V. Chadwick, A. Harrison, P. G. Bruce, *J. Am. Chem. Soc.* **2006**, *128*, 12905–12909; (g) X. Lai, X. Li, W. Geng, J. Tu, J. Li, S. Qiu, *Angew. Chem. Int. Ed.* **2007**, *46*, 738–741; (h) J. Roggenbuck, M. Tiemann, *J. Am. Chem. Soc.* **2005**, *127*, 1096–1097; (i) W.-C. Li, A.-H. Lu, C. Weidenthaler, F. Schueth, *Chem. Mater.* **2004**, *16*, 5676–5681; (j) J. Lee, J. Kim, T. Hyeon, *Adv. Mater.* **2006**, *18*, 2073–2094; (k) R. Ryoo, S. H. Joo, S. Jun, *J. Phys. Chem. B* **1999**, *103*, 7743–7746; (l) J. Lee, S. Yoon, T. Hyeon, S. M. Oh, K. B. Kim, *Chem. Comm.* **1999**, 2177–2178.
75. J. Lee, M. C. Orilall, S. C. Warren, M. Kamperman, F. J. Disalvo, U. Wiesner, *Nature Mater.* **2008**, *7*, 222–228.
76. M. Zukalova, A. Zukal, L. Kavan, M. K. Nazeeruddin, P. Liska, M. Graetzel, *Nano Lett.* **2005**, *5*, 1789–1792.
77. K. Drew, G. Girishkumar, K. Vinodgopal, Prashant V. Kamat, *J. Phys. Chem. B* **2005**, *109*, 11851–11857.
78. T. Katou, B. Lee, D. Lu, J. N. Kondo, M. Hara, K. Domen, *Angew. Chem. Int. Ed.* **2003**, *42*, 2382–2385.
79. P. K. Chinthala, O. G. Neeruganti, C. W. Ting, M.-S. Wong, C. K. Shyue, *J. Phys. Chem. B* **2006**, *110*, 5223–5229.
80. Y. N. C. Chan, R. R. Schrock, R. E. Cohen, *Chem. Mater.* **1992**, *4*, 24–27.
81. D. E. Fogg, L. H. Radzilowski, R. Blanski, R. R. Schrock, E. L. Thomas, *Macromolecules* **1997**, *30*, 417–426.
82. B. J. Kim, G. H. Fredrickson, C. J. Hawker, E. J. Kramer, *Langmuir* **2007**, *23*, 7804–7809.
83. P. Buffat, J.-P. Borel, *Phys. Rev. A* **1976**, *13*, 2287–2298.
84. R. Li, K. Sieradzki, *Phys. Rev. Lett.* **1992**, *68*, 1168–1171.
85. J. Erlebacher, M. J. Aziz, A. Karma, N. Dimitrov, K. Sieradzki, *Nature* **2001**, *410*, 450–453.
86. S. C. Warren, L. C. Messina, L. S. Slaughter, M. Kamperman, Q. Zhou, S. M. Gruner, F. J. Disalvo, U. Wiesner, *Science* **2008**, *320*, 1748–1752.
87. S. C. Warren, M. J. Banholzer, L. S. Slaughter, E. P. Giannelis, F. J. Disalvo, U. Wiesner, *J. Am. Chem. Soc.* **2006**, *128*, 12074–12075.
88. S. Sivaramakrishnan, P.-J. Chia, Y.-C. Yeo, L.-L. Chua, P. K. H. Ho, *Nature Mater.* **2007**, *6*, 149–155.
89. R. B. Thompson, V. V. Ginzburg, M. W. Matsen, A. C. Balazs, *Science* **2001**, *292*, 2469–2472.
90. C. Liang, K. Hong, G. A. Guiochon, J. W. Mays, S. Dai, *Angew. Chem. Int. Ed.* **2004**, *43*, 5785–5789.
91. A. C. Ferrari, J. Robertson, *Phys. Rev. B* **2000**, *61*, 14095–14107.

Part Five

Interfacial Chemistry, Multifunctionality, and Interdisciplinarity

Chapter 22

Emerging Concepts in Interfacial Chemistry of Hybrid Materials: Nanocontainer-Based Self-Healing Coatings

DMITRY G. SHCHUKIN, DARIA V. ANDREEVA, KATJA SKORB, AND HELMUTH MÖHWALD

The Supramolecular Chemistry of Organic–Inorganic Hybrid Materials. Edited by Knut Rurack and Ramón Martínez-Máñez

22.1 INTRODUCTION

The development of a new generation of hybrid coatings, which act at the same time as passive matrix material and can actively respond to changes in the local environment of functional inclusions inside the matrix, is attracting a great deal of interest in the field of materials sciences nowadays. Among other types of coatings, hybrid self-healing anticorrosion coatings are extremely important for the aerospace, maritime, and automotive industries. Active corrosion protection aims to restore material properties (functionality) if the passive coating matrix is penetrated and corrosive species come into contact with the substrate. In addition, the partial recovery of the main functionality of a material can also be considered as self-healing ability. The main function of anticorrosion coatings is protection of an underlying metallic substrate against environmentally induced corrosion attacks. Thus, it is not obligatory to recuperate all properties of the film; only the protection of the substrate has to be guaranteed. Consequently, the coatings have to release the active and repairing material within a short time after changes in the coating's integrity.

Recent developments in surface science and technology offer several general approaches to fabricate hybrid self-healing coatings. In the following section, self-healing coatings with pH-triggered inhibitor release are discussed. In Section 22.3, we report examples where light is used for annealing/repairing. These systems are self-annealing/repairing if the light acts only at defect sites because of their peculiar properties, not if, for example, the light is merely focused. The container-based coating employed for photoinduced autorepairing of polymers and polymer coatings consist of micrometer-scale containers incorporated into the polymer body.[1] These containers are filled with monomers that can be used for repairing of the coating and with appropriate catalysts or UV-sensitive agents to initiate the polymerization of the monomer released at the damaged spot of the polymer coating. In general, mechanical stress exerted on the coating leads to mechanical deformation of the microcontainers embedded therein and starts the release of the monomer from the latter, leading to a sealing of the defects.[2] Furthermore, repair of macroscopic adhesion defects formed throughout the service life of composites can be carried out by hollow fibers that are filled with a polymer resin; the fibers then fracture under excessive service life of the composites.[3] In a related sense, Lee et al. developed a protective ceramic composite coating based on sodium-clay and silica multilayers and epoxy-amine microcapsules containing $MgSO_4$.[4] Self-repairing is achieved here by the sealing of defects with magnesium hydroxide formed after release of magnesium ions from the capsules. Moreover, electroactive conductive polymers were found to be effective coating systems for corrosion control due to their ability to trap oxygen reduction that follows the corrosion process at a remote site with respect to the metal surface.[5]

Another approach for hybrid self-healing coatings is based on the use of chemical inhibitors that can be released from the coating system. However, the direct introduction of inhibitor components to the protective coatings very often leads to deactivation of a corrosion inhibitor and degradation of the polymer matrix.[6] To avoid this drawback, research activities have been shifted to the development of containers that can

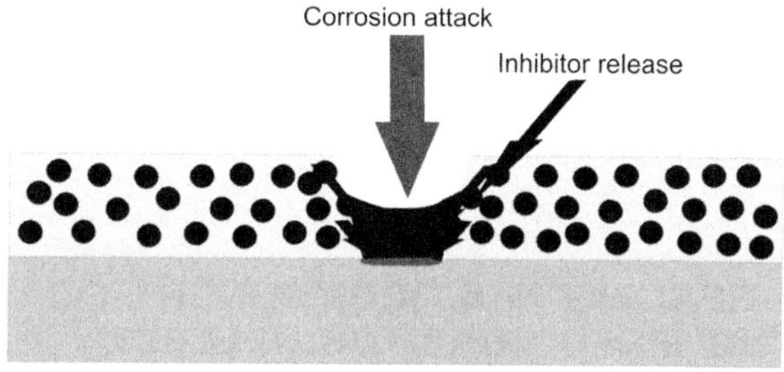

Figure 22.1 Schematic illustration of the operation of nanocontainer-based coatings.

isolate the inhibitor from the coating matrix and, simultaneously, provide active protection in the corroded area (see schematic illustration of the concept in Fig. 22.1). A rather simple approach of inhibitor entrapment is based on complexation of organic molecules by cyclodextrin.[7] Another entrapment approach is based on the use of oxide nanoparticles (e.g., zirconia particles) which can play the role of nanocarriers for corrosion inhibitors adsorbed on their surface. Generally, most of the oxide nanoparticles by themselves are known reinforcements for the coating formulations, as their addition leads to enhanced barrier properties.[8] Immobilization of an inhibitor, for example, Ce^{3+} ions, on the surface of the zirconia nanoparticles can be achieved during the synthesis of the nano-sol by controlled hydrolysis of the precursors in a Ce^{3+}-containing aqueous solution.[9] Furthermore, high loading capacities can be achieved when porous fillers with hollow cellular structure are loaded with organic and/or inorganic inhibitors.[10] The hollow cellular-structured material may be represented by diatomaceous earth, zeolite, or carbon. Inhibiting inorganic ions can also be incorporated as exchangeable ions associated with cation- and anion-exchange solids. For instance, Ca^{2+} and Ce^{3+} cation-exchanged bentonite anticorrosion pigments can be prepared by exhaustive ion exchange of naturally occurring Wyoming Bentonite.[11] For anion-exchange solids, release of inhibitor anions can be triggered by aggressive chloride ions.[12]

The most important aspects in the design of new active coatings are the development of nanocontainers with good compatibility with the composite coating matrix, the possibility to encapsulate and retain active material in the containers and to equip them with a shell, the permeability of which is sensitive to an external stimulus. These tasks can be achieved by employing the layer-by-layer (LbL) assembly approach[13] for the formation of the nanocontainer shell. The procedure of construction of the nanocontainer shell by LbL deposition of charged species (e.g., polyelectrolytes, nanoparticles, and biomaterials) consists of the templating of LbL-assembled films on the surface of containers. The shell of the resulting polyelectrolyte containers is semipermeable and sensitive to a variety of changes in physical and chemical conditions (like a shift in pH or mechanical impact), taking place in the surrounding media. Nanocontainers with regulated storage/release of inhibitors can be constructed with nanometer-scale precision. In this chapter, we demonstrate several examples of a novel generation of hybrid self-healing anticorrosion coatings based on inhibitor-loaded nanocontainers amenable to the activation of the release of the inhibitor either by a pH shift or by external electromagnetic irradiation. Aluminum alloy 2024 (AA2024) used in aerospace applications was chosen as a model metal substrate for corrosion protection.

22.2 SELF-HEALING MULTICOMPONENT COATINGS WITH pH-TRIGGERED INHIBITOR RELEASE

Polyelectrolytes are macromolecules carrying a relatively large number of functional groups that either are charged or, under suitable conditions, can become charged. The molecules may be either polycations or polyanions, or both, depending on the functional groups present.[14–16] Polyelectrolyte films are able to change their chemical composition with pH since the degree of dissociation of the polyelectrolytes is influenced by the pH value. Active species deposited as a component of the polyelectrolyte film are able to be released on demand. External stimuli such as pH, temperature, light, etc., change the degree of dissociation of polyelectrolytes and, therefore, control the concentration of bonded and nonbonded species in the film. Furthermore, coatings with pronounced buffering activity could stabilize the pH between 5 and 7.5 and, therefore, retard undesirable surface degradation which usually follows pH changes (corrosion degradation).[17] As a result, polyelectrolyte coatings exhibit very promising self-healing or self-repairing properties due to their ability to maintain integrity.

22.2.1 Mesoporous Particles with Polyelectrolyte Shell as Nanocontainers for Inhibitors

Mesoporous silica containers can be used as inhibitor hosts with controlled release properties triggered at the beginning of the corrosion process in response to local pH changes. For instance, mesoporous silica nanoparticles covered with polyelectrolyte layers can be loaded with an inhibitor (2-(benzothiazol-2-ylsulfanyl)-succinic acid) prior to introduction into a hybrid zirconia-silica sol–gel film. This hierarchical design avoids spontaneous release of the inhibitor by the formation of a polyelectrolyte shell over the container's outermost surface.

In this approach, mesoporous particles with an average pore size of about 3 to 4 nm were prepared through sol–gel synthesis by adding tetraethoxysilane to a solution containing ethanol, water, and cetyltrimethylammonium bromide as surfactant. 2-(Benzothiazol-2-ylsulfanyl)-succinic acid was loaded into the pores of the containers by incubating them in 10 mg mL^{-1} solution of corrosion inhibitor in vacuum. The polyelectrolyte shell was then formed by successive deposition of two positively charged poly(ethylene imine) (PEI) and negatively charged polystyrene sulfonate (PSS) layers. The inhibitor content in the nanocontainers was equal to 85 mg g^{-1} of the initial SiO_2 particles. The hybrid $SiO_x:ZrO_x$ sol–gel film with the inhibitor-loaded silica containers was prepared by adding mesoporous silica into the mixture of sols followed by deposition onto the metal aluminum alloy by dip-coating. The latter process yielded a uniform distribution of silica containers with an average distance between the containers of approximately 15 to 20 nm.

The release of the 2-(benzothiazol-2-ylsulfanyl)-succinic acid from the polyelectrolyte-shelled containers is triggered at pH 10.1, similar to the local pH value during the localized corrosion process of aluminum alloy. This local pH change influences the polyelectrolyte shell, in particular, its opening. Consequently, the inhibitor is released and the healing of the corrosion area is achieved. This effect is self-regulated and the inhibitor release takes place just in the region with high pH, triggering the self-healing process. When the pH returns to the initial (near neutral) value, the shell is closed and the release of the inhibitor stops.

The scanning vibrating electrode technique (SVET) permits the mapping of the current density in an electrolyte close to the substrate surface. Figure 22.2 reveals that the corrosion activity is negligible in the beginning of the measurements (Fig. 22.2a, d). Then it appears in both cases after 42 h (Fig. 22.2b, e). Simultaneously, the anodic activity of the nondoped sol–gel film is more than twice as high as that of the sol–gel film with embedded containers. Moreover, the SVET measurements 60 h after the start (Fig. 22.2c, f) show the following tendency. The anodic activity increases dramatically in the case of the plain sol–gel coating (Fig. 22.2c). In contrast, corrosion is virtually absent for the surface with the container-doped coating (Fig. 22.2f). Thus, the combination of sol–gel network stability and mesoporous inhibitor-loaded SiO_2 containers with a polyelectrolyte shell allows one to provide a high self-healing corrosion protection over long time duration.

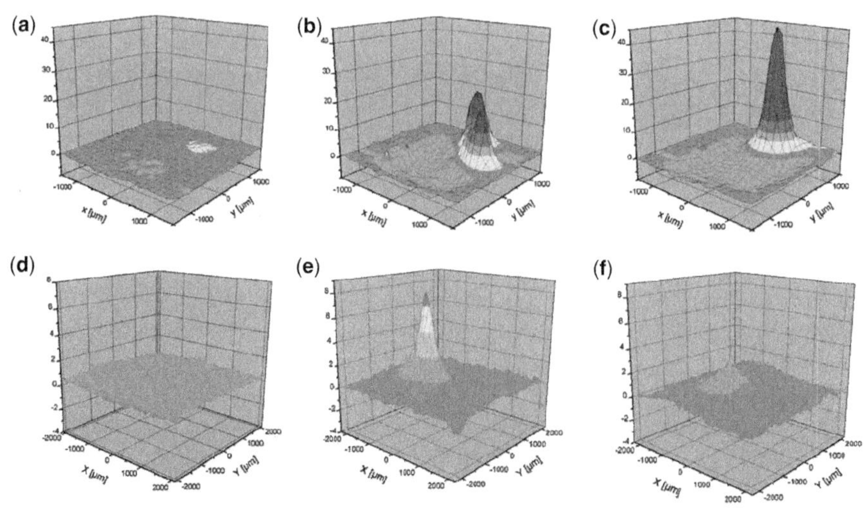

Figure 22.2 Scanning vibrating electrode (SVET) measurements of the ionic currents above the surface of the inhibitor-free SiO_x:ZrO_x hybrid film (a) to (c) and of the SiO_x:ZrO_x hybrid film with inhibitor-loaded SiO_2 nanocontainers (d) to (f). (a, d) at the beginning; (b, e) after 42 hours of corrosion; (c, f) after 60 hours. Scale units: $\mu A\ cm^{-2}$, spatial resolution 150 μm. Solution: 0.1 M NaCl.

22.2.2 Solid Particles with Polyelectrolyte Shell as Nanocontainers for Inhibitors

Another example of hybrid self-healing coatings is the use of solid, nonporous silica nanoparticles as templates for the formation of the pH-sensitive polyelectrolyte-inhibitor complex on their surface. To produce the inhibitor-loaded polyelectrolyte shell around the 70 nm silica nanoparticles, the LbL deposition procedure was followed for both the large polyelectrolyte molecules and the small molecules of the inhibitor benzotriazole. The polyelectrolyte shell is formed as described above in Section 22.2.1 Since benzotriazole is only slightly soluble in water at neutral pH, the adsorption of the third, inhibitor layer was accomplished from an acidic medium at pH 3, together with PSS. Moreover, PSS/benzotriazole adsorption was repeated to increase the inhibitor loading in the LbL film structure surrounding the silica nanoparticles. The benzotriazole content in the resulting SiO_2-based nanocontainers is approximately 95 mg g^{-1} of SiO_2 nanoparticles. In the final step, the suspension of benzotriazole-loaded nanocontainers was mixed with ZrO_2 and organosiloxane sols and deposited onto the aluminum alloy by dip-coating. As a result, the uniformly distributed nanoparticles are impregnated into the sol–gel film. AFM investigations did not show any signs of container agglomeration, confirming the high stability of the container suspension used to dope the hybrid sol–gel film. The thickness of the film measured by SEM is approximately 1.6 to 2 μm.

Electrochemical impedance measurements were performed to estimate the corrosion protection performance of the hybrid sol–gel films. Figure 22.3 demonstrates typical Bode plots of aluminum alloy coated with different hybrid films: a pure sol–gel film, a sol–gel film with inhibitor-loaded nanocontainers, and a sol–gel film modified by direct addition of benzotriazole in two different concentrations. NaCl 0.005 M was used as the corrosive medium. The first high frequency maximum observed at 10^4 to 10^5 Hz is characteristic for the capacitance of the film. Another time-dependent process at medium frequencies between 0.1 and 10 Hz can be clearly ascribed to the capacitance of the intermediate oxide film formed by both the native Al_2O_3 layer and chemical Al-O-Si bonds. The first well-defined signs of corrosion appear in the impedance spectra at low frequencies at about 0.01 Hz for the sol–gel films with benzotriazole directly introduced into the hybrid matrix. In contrast, no signs of corrosion were found in the impedance spectra for the nondoped hybrid film and for films loaded with nanocontainers, indicating an effective barrier against corrosive species. The active corrosion processes on alloy surfaces coated with sol–gel films with directly introduced benzotriazole indicate the absence of any inhibition effect. The resistances of the nondoped sol–gel coating and that of the film containing the nanocontainers show high values of about 10^4 ohm · cm^2. On the other hand, the coatings with benzotriazole directly introduced into the sol–gel matrix exhibit significantly lower resistances (one and two orders of magnitude in the case of 0.13% and 0.63% of benzotriazole, respectively). This decrease of sol–gel film resistance with increase of free benzotriazole concentration evidently indicates a strongly adverse effect on the weathering stability of the sol–gel matrix when the inhibitor is added directly into the sol–gel coating (without encapsulation in nanocontainers); an

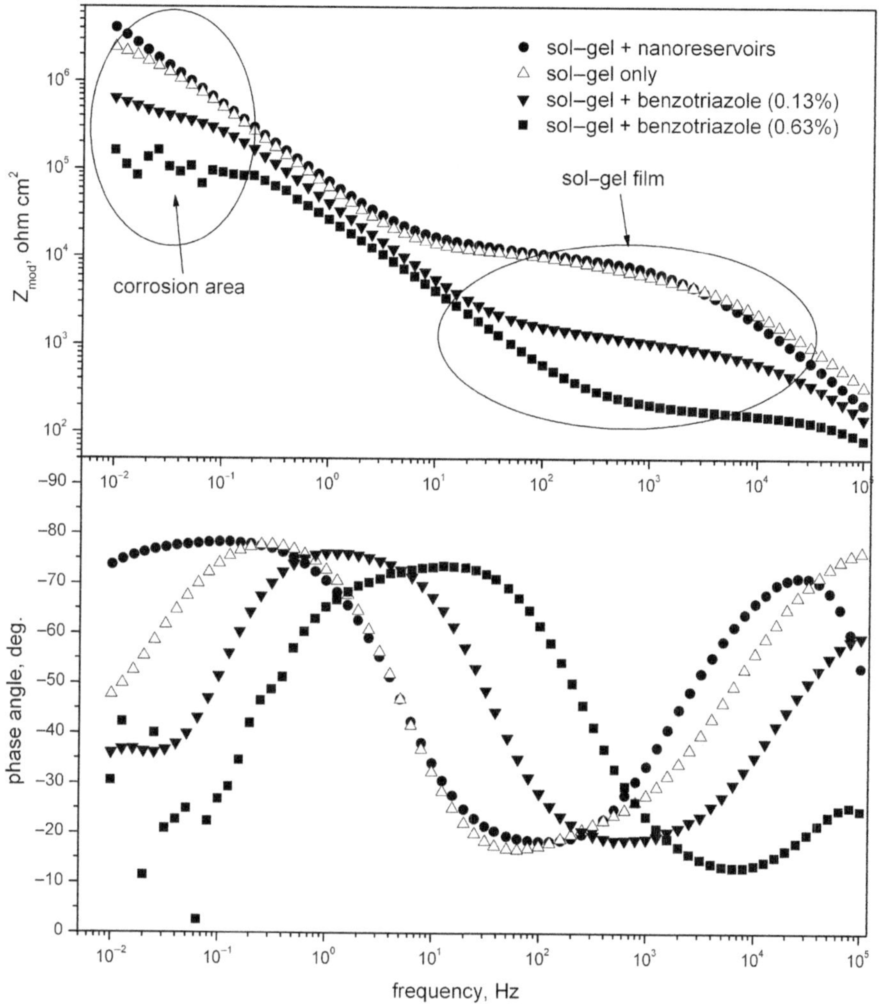

Figure 22.3 Bode plots of aluminum alloy coated with different sol–gel films after 48 h of immersion in 0.005 M NaCl solution.[6] (Reprinted with permission from D. G. Shchukin et al., *Adv. Mater.* **2006**, *18*, 1672–1678. Copyright Wiley-VCH Verlag GmbH & Co. KGaA.)

intact and normally anticorrosive sol–gel coating is even turned into a corrosion-promoting layer when doped directly with the designated inhibitor. After storage for 1 d in the case of the nondoped sol–gel film and even 14 d for the nanocontained doped film, no signs of corrosion could be detected.

An additional experiment was performed for the two films that stayed intact during the experiments described in the previous paragraph in order to provide evidence of the self-healing effect of nanoreservoirs added to the coating. The nondoped and the nanocontainer-doped sol–gel films were removed from the NaCl bath after the above indicated storage period, defects were made artificially into the films in order to provide direct access for the corrosive medium to the alloy surface and the samples were reimmersed in 0.005 M NaCl. The impedance spectra immediately after the beginning of the immersion of the defected nondoped sample show a decrease of impedance and scattering of the data at low frequencies, originating from active processes in the damaged zone. In contrast, for the composite film with nanocontainers a fast decrease of oxide resistance is only recorded for a short period of time after the artificial induction of defects. After this initial drop of impedance, a very important recovery of oxide film resistance occurs during a further 60 h of immersion for the nanocontainer-doped sol–gel coating. The increase of resistance clearly indicates self-healing of the defects in this type of coating. Additionally, self-healing of the hybrid coating with inhibitor-loaded containers was revealed by the SVET technique. Again, defects about 200 μm in diameter were artificially formed in the sol–gel coating of the substrate. A high cathodic current density appeared immediately at the origin of the defect when the nondoped coating was immersed in 0.005 M NaCl, showing well-defined corrosion activity. The defects also remained active with time. On the other hand, the sample coated with the hybrid film doped with the nanocontainers behaved completely differently. During the first 10 h, no remarkable currents were detected at all in the defect zone. Only after about 24 h, the cathodic current appeared. However, after 26 h, the effective suppression of corrosion took place, decreasing the local current density.

22.2.3 Halloysite Nanotubes as Nanocontainers for Inhibitors

Another type of container that could be successfully loaded with active compounds and incorporated into the sol–gel coatings are nanotubes. Halloysite material is aluminosilicate clay, up to 3.0 μm long and with up to 0.3 μm outer diameter.[18] The inner and outer surfaces of the tubule walls behave as a polyvalent anion, whereas their edges are amphoteric, with negative charges at high pH and positive charges at low pH. Due to the active surfaces and the special shape, halloysite can be easily loaded with cationic compounds. For instance, halloysites were used for trapping of a range of biocides and subsequent incorporation into marine antifouling paints.[19] To load the nanotubes with an inhibitor, dispersed halloysite powder was mixed with a 10 mg mL^{-1} solution of benzotriazole in ethanol at pH 7 under slightly negative pressure. The maximum content of benzotriazole in halloysite tubes was estimated as 5 wt%, which is close to the theoretically estimated free volume of halloysite. Then the suspension of benzotriazole-loaded halloysite was mixed with a sol–gel solution and deposited onto an aluminum alloy by dip-coating. The loading of inhibitor into halloysite thus provides maximal distribution of the corrosion inhibitor in the sol–gel coating. The SEM image (Fig. 22.4) demonstrates that halloysites completely cover the surface of the aluminum alloy.

The anticorrosion activity of the halloysite–sol–gel hybrid coating was tested with SVET. A small anodic activity was observed in the first 10 min (scanning

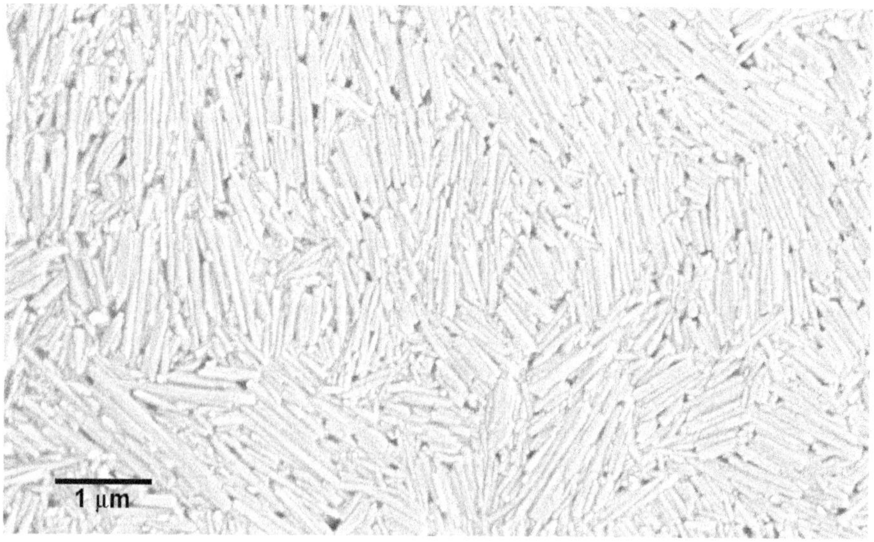

Figure 22.4 Scanning electron microscopy image of halloysite nanotubes deposited onto an aluminum alloy surface.

time) after defect formation. After 3 h, the surface emitted negligible ion flux, indicating the absence of corrosion processes. Thus, a halloysite-containing hybrid coating, when deposited on the surface of aluminum alloy, can heal defected corrosive zones of the sample and terminate corrosion propagation.

The pH sensitivity of halloysite can be enhanced by using retardant polymers, a cationic coating for the formation of a pH-sensitive polyelectrolyte shell on the nanotubes after their saturation with corrosion inhibitor. To equip the halloysite nanotubes with controlled release properties, the surface of the nanotubes was first loaded with benzotriazole and subsequently modified by LbL deposition of two polyelectrolyte bilayers, thus blocking the tubes' openings with polyelectrolytes.

The release of the entrapped inhibitor into water at neutral and alkaline pH was investigated. Approximately 10% of the loaded material is released from halloysite nanotubes without a polyelectrolyte shell after six days of storage in neutral water. On the contrary, the polyelectrolyte shell terminates the spontaneous release of the inhibitor by blocking the permeability of halloysite nanotubes. The polyelectrolyte shell can be opened by increasing the pH value, inducing the formation of defects between the polyelectrolyte multilayers. The release of benzotriazole from halloysite nanotubes with a polyelectrolyte shell is triggered at pH 10, similar to the pH value during the localized corrosion process on aluminum. Thus, the modification of the halloysite nanotubes with a polyelectrolyte shell is an important issue to achieve controlled release of the encapsulated inhibitor as well as to prevent its undesirable leakage from the coating.

22.3 SELF-HEALING MULTICOMPONENT COATINGS WITH LIGHT-INDUCED INHIBITOR RELEASE

A very useful alternative to pH-sensitive films are photosensitive coatings prepared by incorporation of photo-destructible pigments into the sol–gel matrix.[20] An external stimulus, light, can then be used for regular and fast activation of the repairing mechanisms. The surface could be periodically exposed to laser light, which stimulates release of the healing species entrapped in the photosensitive containers. If a scratch or another defect occurs and the surface becomes prone to corrosion, then illumination with laser light would assure that these defects would be cured. In this regard, such coatings behave as "smart" materials because they initiate the triggered release of the inhibitor only if corrosion occurs; otherwise, the protective nanocontainers just remain inactivated in the coating and do not initiate any action due to the blocked access of water molecules. The latter acts as intermediate species in the activation of the shell opening by absorption of remote light.

22.3.1 Activation of the Inhibitor Release by UV Light

The structure and composition of the nanocontainers are defined by the chemical origin of the inhibitors to be introduced and by the environment in which the nanocontainers should provide controlled loading/release properties. One of the interesting approaches to the fabrication of nanocontainers is based on the use of porous titania as a host for the incorporation of low molecular weight species, coated with a polyelectrolyte shell to block uncontrolled leakage from the mesopores. It is well known that under UV irradiation titanium dioxide generates photoholes with very high oxidation ability and can thus oxidize adsorbed organic materials, leading to the materials' partial degradation and the possible release of incorporated corrosion inhibitor.[20] In this case study, again, benzotriazole was used as corrosion inhibitor and was loaded into titania to investigate the effect of UV-induced shell oxidation and inhibitor release on the restoration of the anticorrosion coating.

Mesoporous titania was used in the form of nanosized particles. The titanium dioxide was synthesized by oxidation of titanium carbide in 5 M nitric acid. The resultant particles were approximately 100 nm in size, possessed 5 nm pores, and had an anatase crystal modification. The 365 nm line of a 120 W high pressure Hg lamp was used for UV activation of the titania containers. The inhibitor benzotriazole was loaded into pores of titania from a 10 mg mL^{-1} solution of benzotriazole in vacuum. The maximum benzotriazole content that could be loaded into the titania interior was 73 mg g^{-1} of the porous material, which coincides with the complete filling of the inner porous space of the TiO_2. Employing the LbL deposition technique (subsequent polyelectrolyte layers were alternately deposited from 2 mg mL^{-1} corresponding polyelectrolyte solutions starting with PEI as the first layer), the thickness of the PEI/PSS polyelectrolyte shell generated around the particles was approximately 10 nm. TiO_2 containers protected by this shell did not show any release of benzotriazole if the surrounding environment remained unchanged, that is, the inhibitor was kept inside the containers until the moment of UV-irradiation.

Model films with TiO_2 nanocontainers that were loaded with a luminescent dye (rhodamine 6G) and embedded into an epoxy-functionalized SiO_x:ZrO_x sol–gel film were used for investigating the possibility of container opening under UV illumination by confocal fluorescence microscopy. The atomic force microscopy (AFM) image of the resultant TiO_2-loaded film revealed an almost uniform distribution of meso-TiO_2-based containers over the surface of the coating. The opening of the container upon exposure to a UV lamp is not due to thermal shock since no release of the luminescent dye is observed for similarly polyelectrolyte-coated containers with a core made of mesoporous silica. The light-driven release of the dyes entrapped in the polyelectrolyte containers can thus only be attributed to the partial photodegradation of the polyelectrolyte shell deposited on the titania core as a consequence of the local pH lowering due to a photocatalytic process involving the adsorbed water. This is why this system, although it needs an additional stimulus (light), is also self-repairing.

The UV-activated opening of the TiO_2 containers and release of the inhibitor results in the suppression of the corrosion process as manifested by SVET measurements. It is important to note that photoholes are only generated on the outer TiO_2 surface and can then react directly with the polyelectrolyte shell, oxidizing it and thus making it porous to release the inhibitor. The inhibitor itself is entrapped deep enough inside the mesoporous particles that external UV light cannot penetrate and induce photodegradation of this active compound. Upon several hours of storage in a solution of sodium chloride, anodic activity was observed in the SVET image, which could be cured after five minutes of irradiation. After local UV irradiation, the total suppression of on-going corrosion was achieved. In summary, for coatings for which very fast and regulated release of the inhibitor at a damaged corrosive area is required, the use of UV-irradiation as an external trigger has distinct advantages.

22.3.2 Activation of the Inhibitor Release by IR Light

The sensitivity of functional nanocontainers to external stimuli also depends on the nature of the components of the nanocontainer shell. For instance, the use of other light-sensitive components provides access to photoactivation with light sources operating in other wavelength regions. For example, silver nanoparticles (AgNPs) can be used as light absorption centers because of their intense surface plasmon resonance bands.[21] The surface plasmon resonance of silver nanoparticles is located at ~420 nm. However, upon aggregation of AgNPs sufficient absorption can be achieved throughout the visible and into the near-infrared (NIR) wavelength range. These features open new possibilities for remote regulation of corrosion healing, for example, by using various lasers for controlled remote activation. For instance, whereas visible laser light can be used directly in the absorption maximum of the nanoparticles, also the widely used NIR lasers can serve as irradiation sources, rendering the method very flexible. Remote laser-induced release is a facile method of controlling the properties of substrate coatings and improves performance. Upon laser light illumination, the absorption centers locally disrupt the shells of the nanocontainers. The loaded inhibitor is then released, covering and healing the corrosion area.

To make the containers sensitive to IR laser light, preformed silver nanoparticles were directly incorporated into the polyelectrolyte shell. For this purpose, AgNPs were added into solutions of polyelectrolytes for LbL deposition and hence fixed between the polyelectrolyte layers. The AFM images of the resultant nanocontainer-impregnated film showed a uniform distribution of the containers over the coating, with the concentration of the silica containers equalling $\sim 10^7$ containers per meter squared.

Opening of the polyelectrolyte–silver SiO_2 containers loaded with benzotriazole was performed in two modes. First, a home-made laser setup was used in which a collimated pulsed laser (time duration 700 ps, and power 3 μJ, 820 nm) was focused onto the sample through a microscope objective (100 × magnification). Images were recorded by a charge-coupled device camera connected to a computer in transmission mode using a 150 W light source. In the second mode, an unfocused CW frequency-doubled Nd:YAG laser (operating at 532 nm) with an average incident power up to 150 mW illuminated the sample directly for ~30 seconds.

The effects observed during these studies can be attributed to the photothermal change of the conformation of the polyelectrolyte shell as a consequence of particle heating induced by laser light absorption, resulting in a switching of the shell to the open state. The rupture of the containers demonstrates that laser–nanoparticle interaction through thermal processes[22] is responsible for container activation. Other processes, for instance transport of protons or electron redistribution around the nanoparticles, do not activate the containers, because the polyelectrolyte multilayers were shown to be permeable for protons[23] and the local redistribution of electrons cannot cause the rupture of entire containers. When illumination is conducted in the NIR, the concentration of nanoparticles and their aggregation state in particular[21] play an important role. Since only NP aggregates show NIR absorption, closely located nanoparticles have to interact to open this wavelength range for use in self-healing applications. Therefore, higher concentrations of aggregated nanoparticles are needed for NIR-responsive nanocontainers. The local temperature rise on nanoparticles can be rather high, more than 100 degrees, but it can be controlled and adjusted (e.g., to just several degrees) simply by tuning the incident light intensity, allowing induction of truly controlled release of the encapsulated material. Most importantly, the presence of absorption centers in the shell of a nanocontainer assures that the temperature rise occurs only locally in the vicinity of the nanoparticle, because the temperature decreases rapidly on the nanometer scale with increasing distance from the particle surface,[24] assuring that the surrounding environment is unaffected.

Figure 22.5 shows the current map above the surface of the sol–gel film loaded with NIR-sensitive containers recorded after 24 h of immersion before and after irradiation with an IR-laser. The corrosion disappears immediately after irradiation of the local defect area. Thus, the healing of the corrosion defect is induced by remote IR opening of the inhibitor-loaded containers near the defects. Moreover, these films are characterized by high anticorrosion ability during aging.

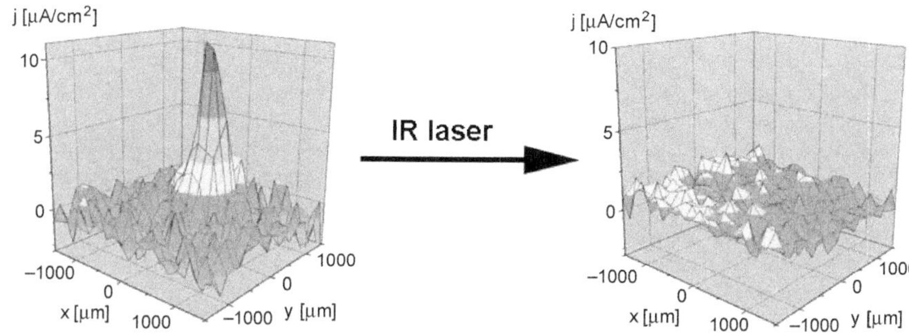

Figure 22.5 Scanning vibrating electrode (SVET) measurements of the ionic currents above the surface of a SiO_x:ZrO_x sol–gel film with inhibitor-loaded containers sensitive to IR irradiation upon 24 hours of immersion in 0.1 M NaCl before (left) and after (right) local IR-laser irradiation of the corrosion area.

22.4 CONCLUSIONS

Nanocontainers possessing the ability to release encapsulated active materials in a controlled way can be employed to develop a new family of feedback active, in particular self-healing, anticorrosion coating. Several approaches to fabricate such self-healing coatings by impregnation of SiO_x:ZrO_x sol–gel films, with inhibitor-loaded mesoporous SiO_2 or TiO_2 containers, halloysites or dense SiO_2 nanoparticles on aluminum alloy substrates were demonstrated. All containers were modified with layer-by-layer assembled polyelectrolyte or polyelectrolyte/nanoparticle shells in order to provide them permeability properties controlled by changes of pH in a local area or by external electromagnetic irradiation. Nanomaterials are most useful in this respect because of their high specific surface area that is available for loading, little mechanical disturbance, and advantageous optical properties, qualifying them especially for applications that utilize remote light sources. The release of the anticorrosion materials occurs only when triggered by the corrosion processes themselves, which prevents leakage of the active component out of the intact coating. Achievements for better control of the encapsulation/release parameters have been brought about by the progress in the fabrication of the functional nanocontainers. Moreover, the concept presented in this chapter can also be transferred to active coatings with several active functionalities (e.g., antibacterial, anticorrosion, and antistatic) when several types of nanocontainers loaded with the corresponding active agents are incorporated simultaneously into a coating matrix. This will surely be a matter of future intense research, which, as a result, may lead to highly sophisticated hybrid structures.

ACKNOWLEDGMENTS

The authors would like to thank Dr. A. Skirtach and D. Fix for carrying out experimental work presented in this chapter. The research was supported by the NanoFutur program of the German Ministry for Education and Research (BMBF) and the Volkswagen Foundation.

REFERENCES

1. S. R. White, N. R. Sottos, P. H. Geubelle, J. S. Moore, M. R. Kessler, S. R. Sriram, E. N. Brown, S. Viswanathan, *Nature* **2001**, *409*, 794–797.
2. W. Feng, S. H. Patel, M-Y. Young, J. L. Zunino, M. Xanthos, *Adv. Polym. Technol.* **2007**, *26*, 1–13.
3. W. Li, L. M. Calle, *Smart Coating for Corrosion Sensing and Protection, Proceedings of the US Army Corrosion Summit 2006*, Clearwater Beach, FL, Feb. 14–16, **2006**.
4. S. Lee, M. Müller, R. Heeb, S. Zürcher, S. Tosatti, M. Heinrich, F. Amstad, S. Pechmann, N. D. Spencer, *Tribol. Lett.* **2006**, *24*, 217–223.
5. B. Wessling, *Adv. Mater.* **1994**, *6*, 226–228.
6. D. G. Shchukin, M. Zheludkevich, K. Yasakau, S. Lamaka, M. Ferreira, H. Möhwald, *Adv. Mater.* **2006**, *18*, 1672–1678.
7. A. N. Khramova, N. N. Voevodin, V. N. Balbyshev, M. S. Donley, *Thin Solid Films* **2004**, *447–448*, 549–557.
8. Y. Chen, L. Jin, Y. Xie, *J. Sol-Gel. Sci. Technol.* **1998**, *13*, 735–738.
9. M. L. Zheludkevich, R. Serra, M. F. Montemor, K. A. Yasakau, I. M. M. Salvado, M. G. S. Ferreira, *Electrochim. Acta* **2005**, *51*, 208–217.
10. C. Schmidt, *Anti-Corrosive Coating Including a Filler with a Hollow Cellular Structure*, U.S. Patent 6383271, May 7, **2002**.
11. R. G. Buchheit, S. B. Mamidipally, P. Schmutz, H. Guan, *Corrosion* **2002**, *58*, 3–14.
12. R. B. Leggat, W. Zhang, R. G. Buchheit, S. R. Taylor, *Corrosion* **2002**, *58*, 322–328.
13. G. Schneider, G. Decher, *Nano Lett.* **2004**, *4*, 1833–1839.
14. S. S. Shiratori, M. F. Rubner, *Macromolecules* **2000**, *33*, 4213–4219.
15. S. T. Dubas, J. B. Schlenoff, *Langmuir* **2001**, *17*, 7725–7727.
16. D. M. DeLongchamp, P. T. Hammond, *Chem. Mater.* **2003**, *15*, 1165–1173.
17. D. V. Andreeva, D. Fix, D. G. Shchukin, H. Möhwald, *J. Mater. Chem.* **2008**, *18*, 1738–1740.
18. S. R. Levis, P. B. Deasy, *Int. J. Pharm.* **2002**, *243*, 125–134.
19. R. R. Price, B. P. Gaber, Y. Lvov, *J. Microencapsul.* **2001**, *18*, 713–722.
20. X. Luo, C. Zha, B. Luther-Davies, *J. Sol-Gel Sci. Technol.* **2004**, *32*, 297–301.
21. A. G. Skirtach, A. Muñoz Javier, O. Kreft, K. Köhler, A. Piera Alberola, H. Möhwald, W. J. Parak, G. B. Sukhorukov, *Angew. Chem. Int. Ed.* **2006**, *45*, 4612–4617.
22. R. E. Holmin, R. F. Ismagilov, R. Haag, V. Mujica, M. A. Ratner, M. A. Rampi, G. M. Whitesides, *Angew. Chem. Int. Ed.* **2001**, *40*, 2316–2320.
23. R. von Klitzing, H. Möhwald, *Langmuir* **1995**, *11*, 3554–3559.
24. A. G. Skirtach, C. Déjugnat, D. Braun, A. S. Susha, A. L. Rogach, G. B. Sukhorukov, *J. Phys. Chem. C* **2007**, *111*, 555–564.

Chapter 23

Molecular Schizophrenics: Switchable Materials with Multiple Functions

ROBERT BYRNE AND DERMOT DIAMOND

23.1 INTRODUCTION

Digital communication networks are at the heart of modern society. The digitization of communications, development of the Internet, and the availability of relatively inexpensive but powerful mobile computing technologies have established a global communications network capable of linking billions of people, places, and objects.

The Supramolecular Chemistry of Organic–Inorganic Hybrid Materials. Edited by Knut Rurack and Ramón Martínez-Máñez

The move from traditional analog landline to digital mobile phones has been an important part of this revolution. Inexpensive mobile phones and other wireless technologies, coupled with palmtop PCs and personal digital assistants (PDAs), provide individuals with communications capabilities unimaginable a decade ago. The exchange of files containing text, graphics, and media to multiple remote locations and websites provides a platform for instantaneous notification and dissemination of information globally. This technology is now pervasive, and those in research and business have multiple interactions with this digital world every day. However, this technology might simply be the foundation for the next wave of development that will provide a seamless interface between the real and digital worlds.

Sensornets are large-scale distributed sensing networks made up of many small sensing devices equipped with memory, processors, and short-range wireless communications capabilities.[1] These devices, known as Motes can gather and share sensor data from multiple locations through in-built wireless communications capabilities. The vision of incorporating chemical and biological sensing dimensions into these platforms is very appealing, and the potential applications in areas critical to society are truly revolutionary.[2] For example,

- Healthcare: personalized access for individuals, relatives, care givers, and other specialists to real-time or historical information generated by wearable sensors, implantable devices, or home-based diagnostics units will facilitate the movement towards home- or community-based healthcare rather than the current, unsustainable, hospital-centric model in the developed world. In addition, access to low cost communications and diagnostics will also provide a means to rapidly improve the delivery of healthcare in less well-developed regions.
- Environment: Sensors monitoring air and water quality will be able to provide early warning of pollution events arising at industrial plants, landfill sites, reservoirs, and water distribution systems at remote locations. The "environmental nervous system" concept likens the rapid access and response capabilities of widely distributed sensor networks to the human nervous system; that is, it is able to detect and categorize events as they happen, and organize an appropriate response.
- Emergency/disaster and threat detection: Given the increased concern over terrorist incidents involving chemical, biological, or radiological threats, this is a major driving force for the development of sensornets, so that such events can be quickly identified and appropriate action taken to minimize the impact. However, at the moment, chemical and biological measurements are overwhelmingly post-event, and related to gathering remedial and forensic information.[3]

The crucial missing part in this scenario is the gateway through which these worlds will communicate; how can the digital world sense and respond to changes in the real world? Unfortunately, it would appear from the lack of field deployable devices in commercial production that attempts to integrate molecular sensor science into portable devices have failed to bear the fruits promised; this problem is what we call *the chemo-/biosensing paradox*.[4]

23.1.1 Chemical Sensing Paradox

Chemo/biosensors must have an *active* surface incorporating sites that are predesigned to bind with specific target species in order to generate the chemically or biologically inspired signal; at the same time, these surfaces must be *passive* to changes that cause signal drift and loss of sensitivity.

The interactions involved in these binding events can be very subtle, and even slight changes in the surface or bulk characteristics through processes like leaching, fouling, or decomposition can have a significant effect on the output signal, and the overall performance of the device. This is in contrast to physical transducers like thermistors that can be completely enclosed in a tough protective coating without inhibiting their ability to function. Hence chemo-/biosensors suffer from baseline drift and variations (usually reduction) in sensitivity, as well as cross-response to interferents that may be present in the sample.

Consequently, chemo-/biosensors and analytical instruments must be regularly calibrated, meaning that the sensing surface is periodically removed from the sample and exposed to standards, the response characteristics checked, and any baseline drift or change in sensitivity compensated. Therefore autonomous analytical devices typically incorporate liquid handling for sampling, reagents, and waste, which requires pumps, valves, and liquid storage. This drives up the complexity, price, and power requirements, which makes the realization of small, autonomous, reliable, chemical sensing/biosensing devices impractical at present.

Therefore, the concept of the chemo-/biosensor as a device with an active membrane attached to a pen-like probe is outdated, and needs to be completely rethought. The key to progress will require breakthroughs arising from new concepts in fundamental materials science, such as the development of *adaptive materials* that have externally or locally controllable characteristics. In particular, the issue of how to predict and control surface characteristics at the interface between the device and the real world needs fresh thinking, as this is where the molecular interactions that generate the observed sensor signal happen. In a sense, these materials could be regarded as having the capability to switch between several completely different "personalities"—schizophrenics at the molecular level!

This means that analytical and material scientists, working together, can realize one of the most exciting challenges facing science today, the bridging of the digital and molecular worlds through the generation of new types of chemo/biosensing platforms. The fundamental requirement of all chemo/biosensors is the need to generate a selective response to a particular analyte, for example, by means of a selective binding event as happens in host–guest complexation, enzyme–substrate reactions, antibody–antigen interactions, or other forms of molecular recognition. Much research focuses on developing a fuller understanding of the basis for molecular recognition, which may ultimately lead to more selective devices that will be required in futuristic applications. Coupled with this attention to selectivity is the need to provide a transduction mechanism, so that the binding event can be observed via an electronic or optical signal. Researchers typically look to electrochemistry (potentiometry, voltammetry, and amperometry) or spectroscopy (visible absorbance and fluorescence) for this signal, which is generated by appropriate redox-active sites or chromophores or fluorophores, either as part of the molecular sensor itself or as part of a sensing cocktail. Success depends on the molecular binding event triggering transduction of the signaling moiety without adversely affecting the overall selectivity of the binding process. In parallel with sensor research, tremendous advances have been made in the development of compact, portable analytical instruments. For example, lab-on-a-chip devices enable complex bench processes (sampling, reagent addition, temperature control, and analysis of reaction products) to be incorporated into a compact, low-power format to provide reliable analytical information within controlled internal environments.

In this chapter, we shall discuss how sensors and sensing systems are likely to develop in the coming years, with a particular focus on the critical importance of new concepts in fundamental materials science to the realization of these futuristic chemo-/biosensing systems. Due to space limitations, we will focus on a small number of fundamental challenges, such as the ability to control the characteristics and behavior of solids (polymers) and fluids, and processes occurring at solid–liquid interfaces. In particular, we will highlight the key role that stimuli-responsive materials can play in producing new "adaptive" materials capable of exhibiting dramatic changes in properties by external stimuli, such as temperature,[5,6] an electric or magnetic field,[7–10] photon irradiation,[11,12] or specific chemicals (pH, ionic strength).[13,14] For more details, the reader should consult sources like the excellent recent special issue of *Langmuir* on Stimuli-Responsive Materials.[15]

23.2 FLUID CONTROL USING STIMULI-RESPONSIVE MATERIALS

Conventional solid-state actuators and valves such as peristaltic pumps and solenoid valves require external power and complex fabrication schemes which limit their use in many microfluidic applications. Of particular interest at the moment is the development of low-cost, efficient, polymer-based actuators and valves for sample handling in microfluidics systems.

23.2.1 Electrorheological Fluids

Electrorheological (ER) fluids are materials whose rheological properties (viscosity, yield stress, shear modulus, etc.) can be readily controlled using an external electric field. For example, in some cases, they can switch from a liquid-like material to a solid-like material within a millisecond with the aid of an electric field, by means of the so-called ER effect.[16,17] The unique feature of the ER effect is that ER fluids can reversibly and continuously change from a liquid state to a solid state. ER fluid research is focused mainly on the automotive and robotics industry as electrical and mechanical interfaces for applications such as clutches, brakes, damping devices, fuel injection, and hydraulic valves. However, more recently, there is growing realization of the potential impact of ER fluids in analytical science, and in particular, the area of microfluidics.[18,19]

Three types of ER effects have been observed so far:

- The positive ER effect—occurs when the rheological properties of a fluid increase with applied electric field;
- The negative ER effect—occurs when the rheological properties decrease with applied electric field;
- The photo-electrorheological (PER) effect—both the positive and negative ER effect can be enhanced by UV illumination in some ER systems.

Much effort has been focused on developing high-performance positive ER materials. A good positive ER fluid should have the following features: (1) a high yield stress preferably equal to or larger than 5 kPa under an electric field of 2 kV mm^{-1}; (2) a low current density passing through the ER fluid, preferably less than 20 $\mu A\ cm^{-2}$; (3) a strong ER effect within a wide temperature range from $-30°C$ to 120°C; (4) a short response time, usually less than 10^{-3} s; (5) a high stability and no particle sedimentation or material degradation problems.[17,20]

Amorphous silicate ceramics are very important ER materials and constitute a large number of ER fluids.[21,22] Aluminosilicates are popularly used to make hydrous and anhydrous ER fluids with a strong ER effect, the yield stress of such an ER material can easily reach 10 kPa at 2.5 kV mm^{-1} and a particle loading of 45 wt%. Recently, Wen et al.[23] developed electrorheological suspensions of urea-coated barium titanyl oxalate nanoparticles that show electrically reversible liquid–solid transitions. The solid state can reach yield strengths in the region of 130 kPa, breaking the theoretical upper boundary of conventional ER static yield stress.[23] For more details on the structure and mechanisms of ER fluids, readers should consult the recent excellent review by Wen et al.[24]

Organic and polymeric semiconductive materials also show a strong ER effect. They are generally electronic conductive materials with a π-conjugated bond structure. It is believed that they have better dispersing ability compared to inorganic materials. However, the ER effect of organic and polymeric ER fluids is relatively weak compared to that of aluminosilicate materials.

23.2.2 Conducting Polymer Actuators

Conjugated polymer (CPs) actuators are based on conducting polymers that undergo a volume change on oxidation or reduction of the material. Multiple mechanisms are involved in the physical swelling of CPs. The macroscopic volume change of the polymer is due mainly to the combined effect of the insertion of ions and their solvent/hydration shell, which occupies free space within the polymer matrix. For small and mobile counteranions, polymer oxidation predominantly results in the penetration of negatively charged anions from the surrounding electrolyte into the positively charged polymer, causing the polymer to swell. On reduction, charge on the polymer backbone is neutralized, electrostatic forces between the polymer–solvent and polymer–counterions become negligible, and polymer–polymer interactions become strong. As a result, counteranions can diffuse freely across the polymer–electrolyte interface to electrolyte phase, resulting in polymer shrinkage. At the opposite extreme, if the counteranion is bulky and immobile, on polymer reduction, these counterions would be entangled within the polymer network and remain inside the neutralized polymer. Consequently, charge compensation is dominated by mobile cations from the electrolyte to balance the entrapped negatively charged counterions, resulting in polymer expansion. On oxidation, the polymer backbone again becomes positively charged, promoting the expulsion of these mobile cations from the polymer and resulting in polymer shrinkage. Moreover, there is also an inflow of solvent, governed by osmosis, which acts to balance the altered ionic concentration inside the polymeric matrix. Third, conformational changes and coulombic repulsion acts on the existing polymer backbones to provide an additional contribution to the structural changes. The degree of volume expansion and its speed is thus dependent on a number of factors. These include ion types and sizes, ionic concentration, the solvent, the thickness of the CP film, and the external voltage applied. Polyanilines (PANI) and polypyrroles (PPY), Figure 23.1, and their derivatives, are popular CP actuators, as they can be prepared easily using simple procedures. These organic polymers are soft and non-abrasive, and have attracted a great deal of attention over the past few years, resulting in a large number of publications.[25,26]

Actuators based on low-voltage electrochemical systems, such as CPs, are convenient and safe, and power inputs are potentially low (55 mW).[27] One deficiency of conducting polymers compared to skeletal muscle is their low actuation strain, less than 15% in the case of CPs. It has become obvious that force generation is limited by the breaking strength of the actuator material. It has been reported that the maximum stress that can be generated by an actuator is typically 50% of the breaking stress, so that for PANI fibers, stresses on the order of 190 MPa should be achievable. However, in practice the breaking strain of CP fibers when immersed in electrolyte and operated electromechanically are significantly lower than their dry-state strengths.

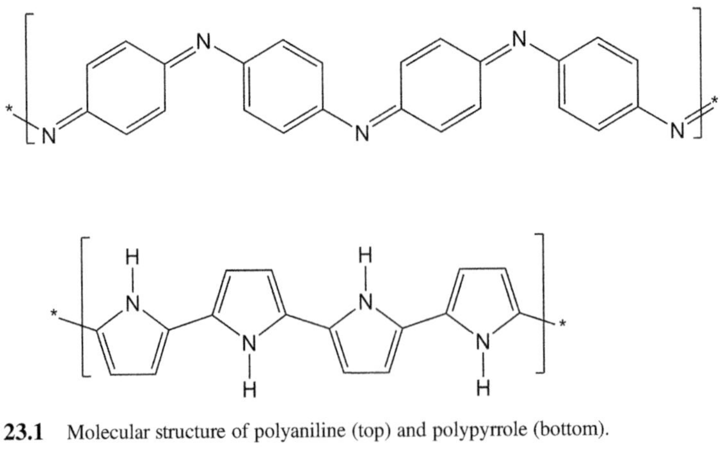

Figure 23.1 Molecular structure of polyaniline (top) and polypyrrole (bottom).

The highest reported stress that can be sustained by conducting polymers during actuator work cycles is in the range of 20 to 34 MPa for PPY films. Therefore, devices that require low strain actuation would be a suitable place to employ these actuators. With this in mind, we have developed a low powered pump suitable for operating fluid handling within a microfluidic chip. Figure 23.2 shows a fully functioning diaphragm pump based on a hybrid material of PPY and poly(dimethylsiloxane) (PDMS) (PPY-PDMS).[28] PPY actuation is due to electrochemical oxidation and reduction of the membrane that causes it to swell or contract. In this PPY-PDMS diaphragm design, a PPY membrane was grown on each side of poly(vinylidene fluoride) (PVDF) membrane. The two membranes on each side were connected to a potentiostat such that one worked as the working electrode and the other as the counter electrode. When electrical voltage is applied, one side swells while the other side contracts to produce maximum mechanical force during actuation. This micropump produced a maximum flow rate of 52 $\mu L\ min^{-1}$ and a nominal minimum flow rate of 18 $\mu L\ min^{-1}$ at ± 1.5 V supply voltage.[29]

It has been argued that the low actuating strain can be mechanically amplified (levers, bellows, hinges, etc.) to produce useful movements, but higher forces are needed to operate these amplifiers. To improve the mechanical performance of CPs, single-walled nanotubes (SWNTs) and multiwalled nanotubes (MWNTs) have been added to various CP matrices, and this has produced significant improvements in strength and stiffness. For example, it has been shown that the modulus of PANI can be increased by up to four times through the incorporation of relatively small amounts ($<2\%$) of nanotubes.

Even with their impressive physical changes, electrorheological materials have not solved the microactuator problem. This is due to the complex fabrication schemes required to incorporate these materials into microfluidic manifolds. Therefore,

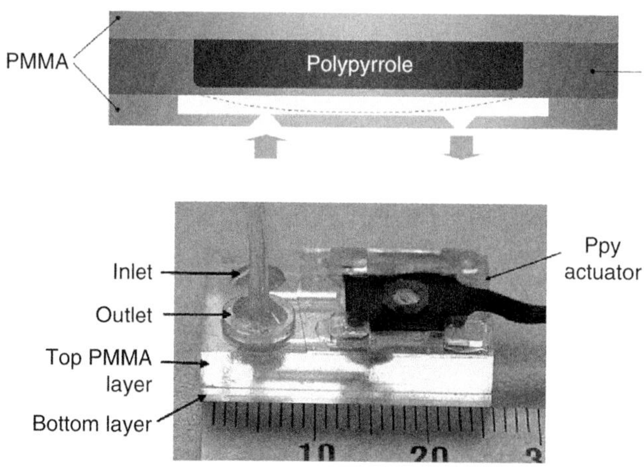

Figure 23.2 Schematic of PPY-PDMS diaphragm pump (above) and photograph of fabricated microfluidic pump (below). (See color insert.)

controlling physical properties with temperature,[1,2] photon irradiation,[11,12] or specific chemicals (pH, ionic strength)[13,14] would be of great benefit for rapid prototyping.

23.2.3 Photorheological Materials

Chemistry is fundamentally important for the emergence of the new molecular designs and synthetic schemes that will underpin the development of molecules that are better able to transduce external stimuli into a reversible change in viscosity, even to the point of gelation. Some of the most successful approaches in this field employ the use of photochemistry.

Wolff and Muller[30] were among the first to report the ability to switch between sol and gel states by the conformational changes that can be brought about by irradiation with light. Their studies concerned the selective production of the unstable 9-methylanthracene by preorganization in micellar solutions of cetyltrimethylammonium bromide and subsequent irradiation with light (Fig. 23.3).

As a side effect the rheological properties of the solutions also changed when irradiated, which led the authors to investigate further. The addition of 9-methylanthracene led to a strong enhancement of viscosity, upon photodimerization ($\lambda > 300$ nm), viscosity was reduced to approximately half the value. Subsequent irradiation with light of shorter wavelength ($\lambda < 300$ nm) led to an increase in the viscosity.

The *cis-trans* isomerization in azo compounds and stilbene has been used for switching viscosity.[31,32] Shinkai et al. developed a cholesterol-based gelator containing azobenzene; irradiation of the system at 330 to 380 nm led to isomerization of the *trans* azo linkage to its *cis* conformation and thereby disrupted gelation.[33] Irradiation of the sol at wavelengths longer than 460 nm resulted in reversal of the isomerization and re-formation of the gel. Pozzo and coworkers[34] employed the

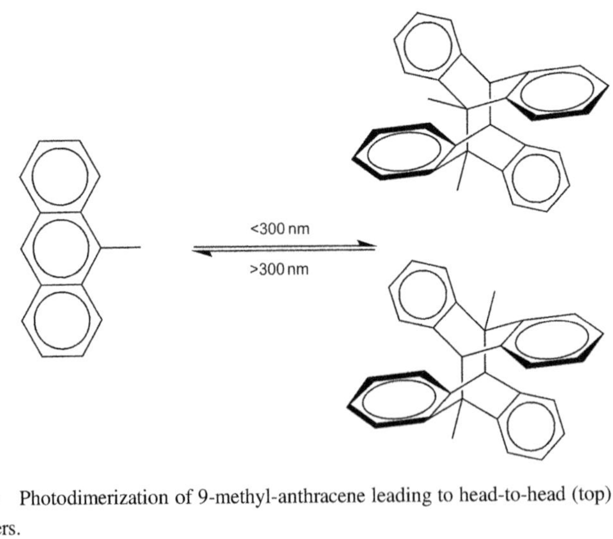

Figure 23.3 Photodimerization of 9-methyl-anthracene leading to head-to-head (top) and head-to-tail (bottom) dimers.

photochromic equilibrium of 3,3-diphenyl-*3H*-naphthopyran. Two forms of the naphthopyran can be distinguished: a colored "open" form and a colorless "closed" form. The naphthopyran unit in its closed form does not influence the stacking process of the carbamate and the molecule acts as an efficient gelator (Fig. 23.4). Photoisomerization of the acoplanar "closed" isomer to the planar "open" isomer induces a large conformational change, preventing the carbamate group from stacking and disrupting the gel. Heating converts the pyran unit back into the closed form, and on cooling the colorless gel is again obtained.

Room temperature ionic liquids (ILs) are composed solely of ions and show melting points below 100°C. The positive and negative charges carried by these ions produce a variety of electrostatic environments that should affect the behavior of electrons in ionic liquid systems. Because chemical reactions are, in a sense, relocation processes of valence electrons, the charge distribution in the ionic liquids should profoundly affect the chemical reactions proceeding within them. Due to these unique properties, ILs have been investigated for many applications such as synthesis,[35] catalysis,[36] polymerization,[37–39] separation,[40] electrochemistry,[41,42] and electrochemical devices.[7,43] Advocates of ILs claim they are a safer and greener alternative to conventional solvents for many synthetic processes. However, electrolyte solutions, including ILs, always need to be contained and there is the possibility of leakage due to physical stress. Therefore, it is inevitable that solid-state ion conductors will be developed, and in recent years, considerable attention has been focused on ILs in this regard.

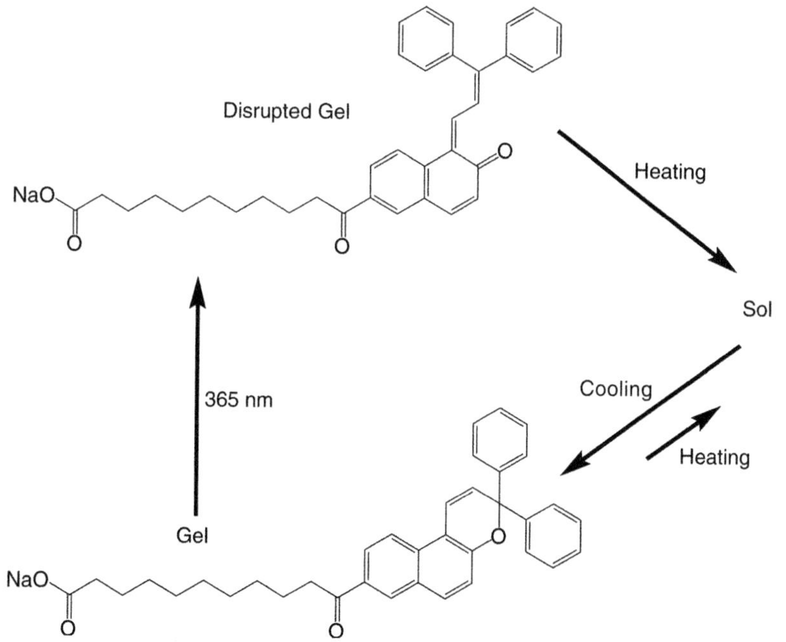

Figure 23.4 Light-induced switching between open (top) and closed (bottom) isomers of naphthopyrans.

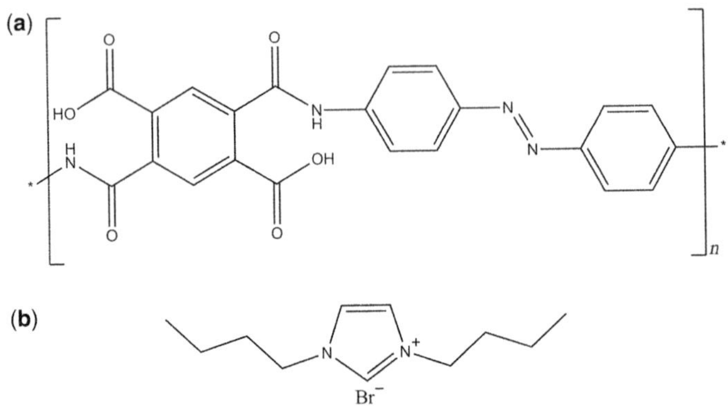

Figure 23.5 Structure of gel components (a) poly(amideacid) azobenzene, (b) 1,3-dibutylimidazolium bromide [bbim][Br].

On the other hand, Tamada et al.[44] have investigated stimulus-responsive gels utilizing the photochemical reaction of a polymeric azobenzene unit doped with IL (Fig. 23.5). Photoisomerization of the azobenzene group resulted in shrinkage of the irradiated site. It was also reported that ionic conductivity of the gel could be controlled by photoirradiation. The ionic conductivity of the gel decreased after UV light irradiation; this effect was coupled with an increase in viscosity, in turn suppressing diffusion of the component ions within the gel.

The chemical structure of the spirobenzopyran-modified gel and the effect of its photoisomerization are illustrated in Figure 23.6.[6] In an acidic aqueous system in the dark, spirobenzopyran is in a protonated open-ring form (Fig. 23.6a). When irradiated by blue light, the spirobenzopyran isomerizes immediately to a closed-ring form, dissociating protons and losing positive charges (Fig. 23.6b). When the light is turned off, the chromophore returns spontaneously to the protonated open-ring

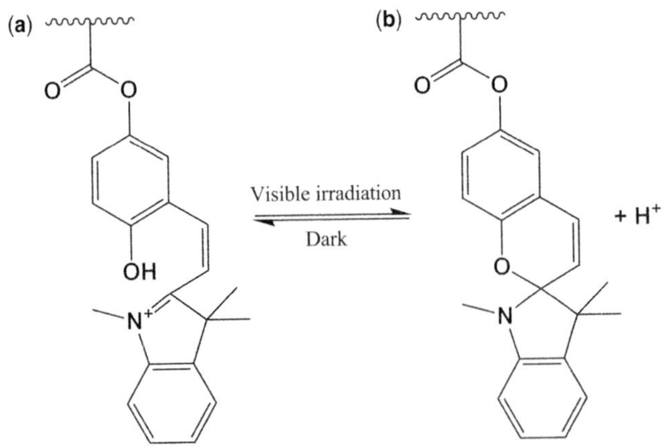

Figure 23.6 Chemical structure of spirobenzopyran in its protonated open-ring form (a) and its photoinduced closed-ring form (b).

form, which is more thermodynamically stable in the dark than the closed-ring form. When spirobenzopyran is incorporated into a hydrogel, the photoisomerization affects the hydration of the gel significantly. Under dark conditions, the open form protonated spirobenzopyran (hydrophilic) gel is yellow. When the gel is irradiated with blue light, the gel decolorizes, indicating the closed-ring form (hydrophobic) is present and consequently the gel reduces in size due to induced dehydration of water from the polymer chain.

More recently, Sugiura et al.[11] developed a fully functional microvalve based on this photoresponsive behavior, which was composed of poly(*N*-isopropylacrylamide) functionalized with the chromophore spirobenzopyran (pSPNIPAAm). The microvalve was fabricated in a polydimethylsiloxane (PDMS) microchannel by *in situ* photopolymerization. Blue light irradiation (18 to 30 s) to the gel induced photoisomerization of the spirobenzopyran chromophore which resulted in shrinkage due to dehydration of the gel, thus causing the microvalves to open, as seen in Figure 23.7. In this example, localized irradiation enabled independent control of three photoresponsive polymer gel microvalves, which had been fabricated on a single microchip.

23.2.4 Chemically Controlled Flow Systems Based on Responsive Hydrogels?

Chemically responsive hydrogels transduce chemical energy directly to mechanical energy without the need for external power sources, which makes them attractive for applications that cannot accommodate the cost or weight of electrical power. Eddington and colleagues pioneered the introduction of hybrid hydrogels into microfluidic channels.[14,45] These pH-responsive hydrogels composed of poly(hydroxyethylmethacrylate-acrylic acid) were applied as actuators or valves depending on microchannel fabrication, and were integrated into microfluidic manifolds to form so-called smart channels.

The latter can be defined as microchannels that allow fluid to flow under one set of conditions, but completely seals off flow under other conditions. This effect has been demonstrated in a number of ways; for example, by patterning an array of hydrogels within a microchannel or by patterning two strips of responsive hydrogels along the walls of the microchannels (Fig. 23.8).

When acidic solution flows through the channel, the hydrogel contracts, and when basic solution flows through, the hydrogel expands to occlude the channel by increasing the resistance of fluid flow.[46] The hybrid hydrogel-PDMS membrane valve operates by coupling the volume expansion of pH-responsive hydrogels to deform a flexible membrane to inhibit another channel. The stimulus to trigger the valve is isolated from the regulated stream by an impermeable PDMS membrane.

These types of concepts and ideas are of real importance for demonstrating the use of direct molecular interactions to control the way in which liquids move though microfluidic channels—no conventional valves are needed, and the material can be completely embedded within the manifold, with no need for external connections to circuitry or power sources.

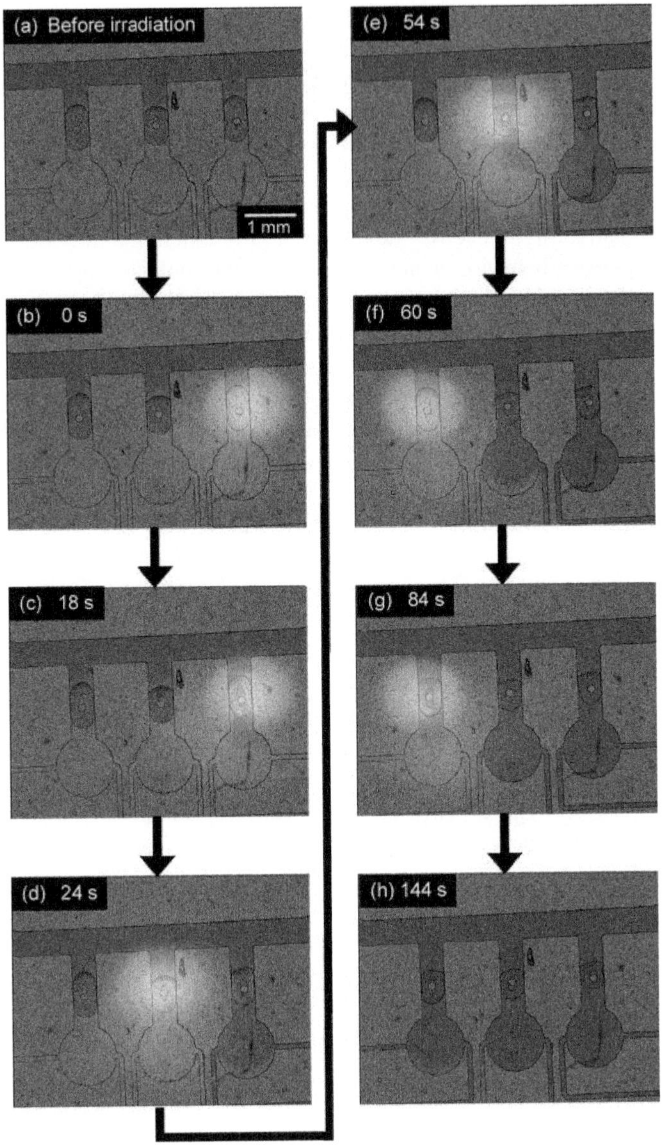

Figure 23.7 Independent control of multiple pSPNIPAAm gel microvalves by means of local light irradiation. (a) Blue dye introduced into the main microchannel. (b) Light locally irradiated to the gel microvalve on the right. (c) Right side microvalve opened after 18 s light irradiation. (d) Light locally irradiated to the center microvalve. (e) Center microvalve opened after 30 s light irradiation. (f) Light locally irradiated to left side microvalve. (g) Left side microvalve opened after 24 s blue light irradiation. (h) After light irradiation, the chambers were filled with the blue dye solution.[11] (Reprinted with permission from S. Sugiura et al., *Sens. Actuators A* **2007**, *140*, 176–184. Copyright 2007 Elsevier.) (See color insert.)

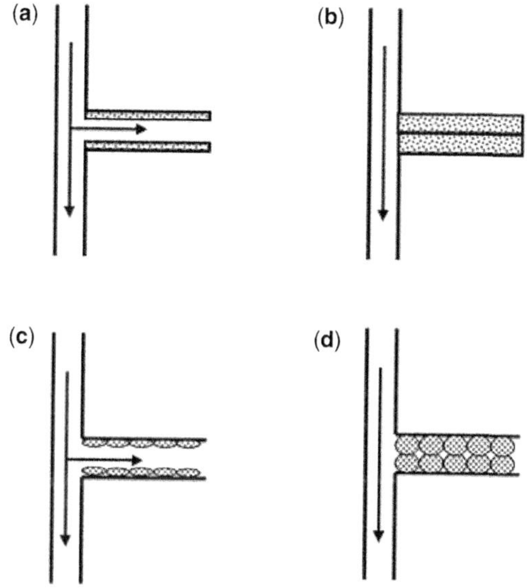

Figure 23.8 Schematic smart channel design with a strip (a) of hydrogels that swell to close the channel, (b) under a given stimulus, or multiple posts of hydrogels (c) that swell to close the channel (d). (Adapted from Reference 45.)

23.3 MOLECULAR RECOGNITION/TRANSDUCTION

The typical sensor design encompasses a molecular recognition element, signal transducer, support matrix, and CPU interface integrated physically to give an operational sensing system. The idea of using tailored materials that can perform some if not all functions of a sensor is very appealing and not beyond present day capabilities. Published examples of optical and electrochemical sensors have employed vast varieties of materials in the last decade, yet most sensors are still based on the pen-like probe, which requires after-treatment such as washing and addition of reagents for recalibration. It is expected that in future sensor platforms, preference will be given to systems that can self-calibrate, hence avoiding the current need for after-treatment of the sensing surface. Surprisingly, examples of this approach are scarcely found in the sensor literature. Detailed below are some examples of the application of the adaptive materials concept to chemical sensing systems.

23.3.1 Electrochemical Sensors

A major problem in practical applications of electrochemical sensor systems stems from the adsorption of surface-active materials onto the working electrode. This adsorption causes electrode fouling, which can have a very negative impact on sensor characteristics like baseline stability and response slope. Such surface fouling is probably the most serious barrier to the development of reliable continuous monitoring devices. Various protective coatings, such as Nafion or agarose gel, have been proposed for excluding surface-active macromolecules and minimizing surface-fouling effects. Wang's group has demonstrated an alternative approach to reduction of electrode fouling through the use of adaptive nanowires.[47] These consist of alkanethiol-coated gold nanowires, containing a short nickel (magnetic) segment. The nanowire-based adaptive protective system enables the user to control exposure of the surface during measurement mode (active) and protection of the surface in the protective mode (passive) between repetitive measurements (Fig. 23.9). This is accomplished by switching magnetically the surface orientation of the nanowires, between vertical and horizontal positions. This leads to opening and closing of the surface, respectively, to allow the measurement and protect the transducer between repetitive runs.

The optical images of Figure 23.9 shed useful insights into the protective action of the bisegment adaptive nanowires. The top view of the vertical nanowires (left) indicates that the island-like bundle structure of the vertically oriented nanowires exposes a major portion of the surface of the glassy-carbon disk (dark region). This vertical (active) state thus allows facile deposition and stripping of metals on the electrode surface and hence convenient electrochemical measurements. In contrast, the surface is fully covered by these nanowire bundles when the adaptive nanowires are reoriented in the horizontal position (right). Such a passive state completely blocks the surfactant access to the surface to offer the necessary protection between repetitive measurements.

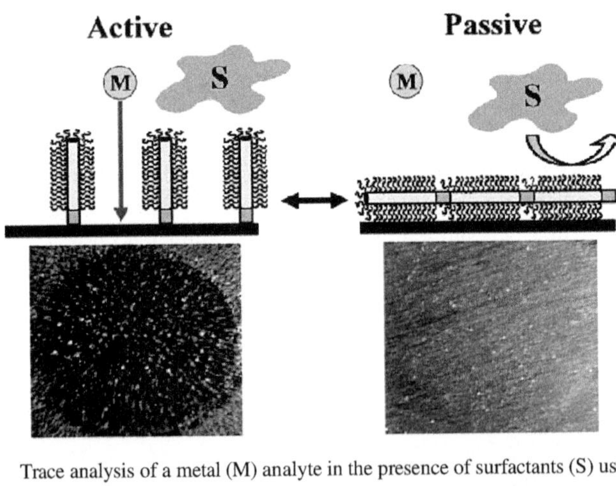

Figure 23.9 Trace analysis of a metal (M) analyte in the presence of surfactants (S) using the vertically "active" and horizontally "passive" aligned nanowires. Such adaptive operation leads to "opening" and "closing" of the surface to allow measurement and protection of the transducer between measurements. Also shown are the optical images (top view) of the glassy-carbon disk electrode covered with the vertically (left) and horizontally (right) aligned nanowires.[47] (Reprinted with permission from R. Laocharoensuk et al., *J. Am. Chem. Soc.* **2007**, *129*, 7774–7775. Copyright 2007 American Chemical Society.)

The ability to switch the operation of electrochemical metal sensors between active and passive modes on demand offers substantial improvements in their stability in the presence of common surfactants, as demonstrated in stripping-voltammetric signals obtained from cadmium in the presence of gelatin and Tween 80. Bare electrodes display a substantial diminution of the cadmium peak in the presence of both surfactants. In contrast, the adaptive-nanowire electrode system exhibits a highly stable response with a negligible change of the peak current over multiple measurements.

23.3.2 Optical Sensors

Typically an optical sensing platform consists of the molecular recognition and transduction agents immobilized within a membrane or on an active surface, and this is exposed to the sample. Such exposure of chemically active surfaces to real samples can lead to changes in the surface and bulk characteristics over time, due to various interactions with the sample. Active membrane components may leach out into the sample, binding sites may be blocked or changed in form, or surfaces may become fouled. Over the past several years, we have investigated the concept of adaptive optical sensing materials based on the following principles:

- The sensor surface should be in an inactive or passive state when a measurement is not being conducted.
- The surface is converted into an active state under an external stimulus (optical in this case).
- The active surface binds with the target species and generates a signal that enables the analytical measurement to be made.
- After the measurement is completed, the target species is expelled by an external stimulus (optical) and the surface returns to its inactive form.

In this way, it may be possible to maintain sensing surfaces in an inactive form that would remain relatively unchanged over time, potentially extending the sensor's useful lifetime by minimizing poisoning effects.

Within our research group we have explored the opportunities afforded by surface-based photoswitchable chemical sensors.[48,49] We focus on spiropyrans and related systems like spiroxazines, as they are a well-studied system that can be photonically switched between two states.[50–54] Spiropyran (SP) undergoes a heterocyclic ring cleavage at the C—O spiro bond that results in the formation of a planar, zwitterionic, and highly conjugated chromophore that absorbs strongly in the visible region, this being the merocyanine (MC) isomer, which exhibits metal ion-binding behavior resulting in a shift in the absorbance spectrum, and a corresponding color change, see Figure 23.10.[55,56] For certain divalent metal ions, solution-phase studies have shown that a 2:1 complex of the form MC_2-metal ion exists. Consequently, it was thought that covalent immobilization of the spiropyran molecules to a polymer backbone might inhibit or completely eliminate the ability of the MC-metal ion complex to form. The immobilization strategy therefore needed to allow enough flexibility for

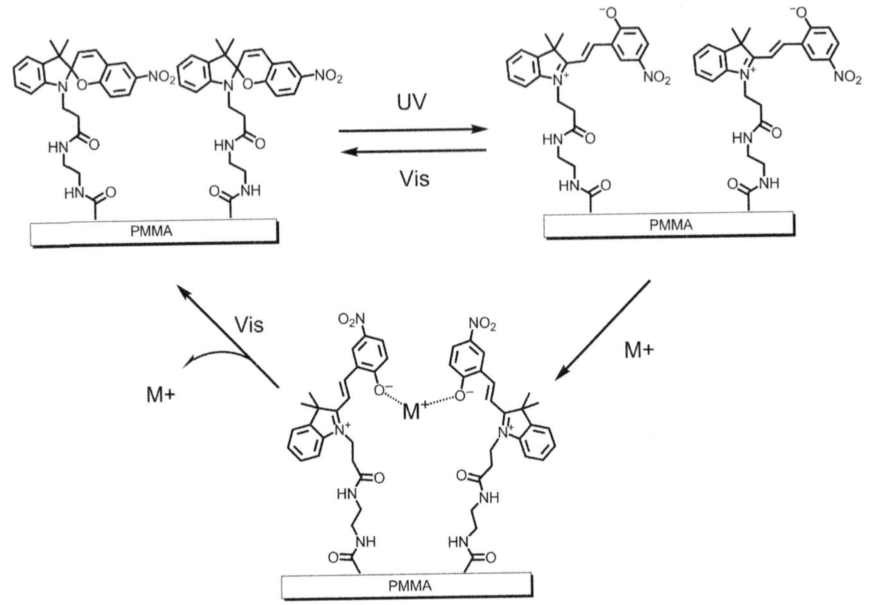

Figure 23.10 Representation of photocontrolled ion-binding at an SP-modified surface. (a) Colorless SP-immobilized surface. (b) On illumination with UV light, the surface becomes active and bright purple due to the photoisomerization of SP to MC. On illumination of this surface with visible light MC is switched back to SP. (c) Exposure of activated surface to an aqueous solution of divalent metal ions leads to formation of the complex MC-M^+ and further color change of the surface. Irradiation of this surface with green light leads to transformation of MC-M^+ to SP. The cycle is closed and the surface is returned to the passive, colorless state.

the spiropyran molecules not only to photoisomerize between the active and passive forms, but also to coordinate with the metal ions in the correct stoichiometric ratio. We therefore used a series of diamino alkyl linkers, ranging in length from two to eight carbons, as a method to covalently attach a carboxylate version of the spiropyran to poly(methyl methacrylic acid). The spiropyran-modified polymer substrates were treated to a series of experiments to determine whether the tether length influenced the efficiency of photoswitching, ion complexation, and photoinduced dissociation of the MC-metal ion complex, and regeneration of surface-bound spiropyran for further measurements. It was expected that formation of the 2:1 complex would be affected by molecular flexibility, and therefore tether length would be important. It was found that the ability of the SP to photoisomerize dramatically improves with increasing tether length, as evidenced by the increasing absorbance at 570 nm, the absorbance maximum for the covalently bound MC form. This was also the case for metal ion complexation; see Figure 23.11 for UV-vis spectra of MC complexation with cobalt(II) chloride. It was observed that metal ion complexation only occurred when there was a separation of at least eight carbon spacer atoms between the SP molecule and the polymer backbone. We also found it was possible to dissociate the MC-Co^{2+} complex with visible light. This was the first demonstration of the cycling

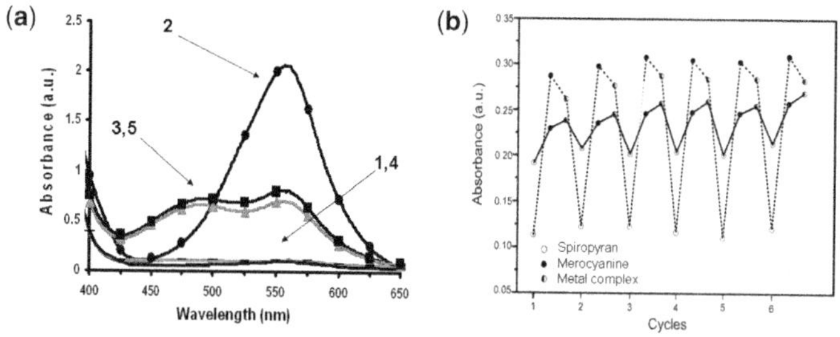

Figure 23.11 UV-vis analysis of photocontrolled ion-binding. (a) Regeneration of spiropyran and merocyanine–Co^{2+} complex in acetonitrile through the following steps: (1) spiropyran, (2) merocyanine, (3) complex, (4) spiropyran, (5) complex. (b) Cycling of spiropyran to merocyanine and generation of merocyanine-Cu^{2+} complex on SP-modified surface, 560 nm (dashed line) and 431 nm (full line).

of covalently immobilized spiropyran on a solid support to merocyanine for repeat cycling through uptake, detection, and release of metal ions.

These results demonstrate that a spiropyran-modified polymer can adapt its functionality through reversible molecular rearrangements triggered by external stimuli (photons). In principle, this means that the immobilized chemorecognition sites can be maintained in an inactive or passive form until a measurement is required. At this point, the surface is illuminated with UV photons, which triggers the molecular rearrangement to the active form. Furthermore, the polymer is self-indicating, as the presence of the active form is easily identified via the intense purple color, and therefore some degree of self-diagnostics can be easily incorporated into measurements. In the active form, binding with metal ions such as Co^{2+}, Cu^{2+}, and Cd^{2+} can occur,[55,56] and once again it is self-indicating, as complexation shifts the absorbance of the active site and the color changes to pink from purple. When the measurement has been completed, illumination with white light expels the guest ion and returns the polymer to the inactive (colorless) form. The implications of this research are that it may be possible, using this approach, to maintain a sensing surface in a passive mode that does not interact significantly with the external environment. When a measurement is required, the active form can be created and the population of active sites monitored via the intensity of the purple color. The presence of the target species (e.g., Co^{2+}) can then be measured by ratioing the absorbance at, for example, 430 nm and 570 nm. A decrease at 570 nm with an accompanying increase at 430 nm is indicative of the presence of Co^{2+}. This raises the prospect of having sensing surfaces that do not change characteristics in a significant manner over time, potentially extending the lifetime of the sensing surface. The self-indicating nature of spiropyran is another powerful feature, which provides a degree of self-diagnostics and internal referencing of analytical measurements. In addition, covalent attachment to the polymer substrate prevents leaching of active sites into the sample. Samanta and Locklin[57] have designed photochromic polymer brushes for photoswitchable surface wetting based on our polymer merocyanine-Co^{2+} complex. They reported a photoinduced enhancement of contact angle change with increasing linker length of spiropyran from the substrate surface. Perhaps more importantly, they also observed a significant enhancement of contact change (approximately 35 degrees) when the spiropyran-modified substrate was irradiated in the presence of cobalt(II) chloride. To our knowledge, this is the largest photoinduced change in contact angle that has been observed on a flat surface. More importantly, this photoinduced surface wetting was shown to be reproducible, up to five times, with no sign of degradation.[57] These results are direct confirmation through independent techniques of an interpretation of the switching and binding behavior of these molecular photoswitches.

Furthermore, we show the excellent potential of light-emitting diodes as light sources and detectors for photoswitching between the two isomeric states of spiropyran and measurement of ion complexation, see Figure 23.11b.[58,59] A simple, low-cost, low-power experimental setup provides spatial and temporal control of surface illumination and surface binding. This, coupled with low irradiance, is shown to generate significant improvement in fatigue resistance of SP-modified polymeric films, and may prove to be an important step towards more sophisticated materials capable of switching reversibly between active and passive forms, and simultaneously providing a number of transduction modes for gathering information about the molecular environment in the immediate vicinity of the binding site when in the active mode.

23.4 CONCLUSIONS

The key to many future disruptive technologies lies in the development of materials that exhibit stimulus-responsive behavior. The area has advanced rapidly in recent years, as the science and technology of molecular and nanoscale control and characterization of the surface and bulk structure of materials continues to develop. The range of materials that can be switched between dramatically different modes of behavior is expanding rapidly, and in this chapter, we have only been able to provide a high level introduction into some of the exciting possibilities that can arise from these developments—materials whose physical and chemical properties can be controlled using an external stimulus. These materials have the potential to be incorporated into a wide range of specialist and consumer products within the next five years that could dramatically impact on society. Furnishings and clothes that change color, textiles that can sense and communicate, chemical sensors whose surface binding activity can be turned on and off, and microfluidic systems with biomimetic components such as soft polymer pumps and valves that could revolutionize the way analytical measurements are made. All in all, it seems clear that there are exciting times ahead in sensor science aligned with adaptive or stimuli-responsive materials!

ACKNOWLEDGMENTS

We would like to acknowledge support from Science Foundation Ireland under the CLARITY grant (07/CE/I1147) and Light Activated Molecular Switches (07/RFP/MASF812) awards.

REFERENCES

1. S. Shenker, S. Ratnasamy, B. Karp, R. Govindan, D. Estrin, *ACM SIGCOMM Comput. Commun. Rev.* **2003**, *33*, 137–142.
2. D. Diamond, *Anal. Chem.* **2004**, *76*, 278A–286A.
3. D. Diamond, *NATO Security through Science, Ser. A: Chem. Biol.* **2006**, *2*, 121–146.
4. R. Byrne, D. Diamond, *Nature Mater.* **2006**, *5*, 421–424.
5. N. Reber, A. Kuchel, R. Spohr, A. Wolf, M. Yoshida, *J. Membr. Sci.* **2001**, *193*, 49–58.
6. A. Szilagyi, K. Sumaru, S. Sugiura, T. Takagi, T. Shinbo, M. Zrinyi, T. Kanamori, *Chem. Mater.* **2007**, *19*, 2730–2732.
7. W. Lu, A. G. Fadeev, B. Qi, E. Smela, B. R. Mattes, J. Ding, G. M. Spinks, J. Mazurkiewicz, D. Zhou, G. G. Wallace, D. R. MacFarlane, S. A. Forsyth, M. Forsyth, *Science* **2002**, *297*, 983–987.
8. G. M. Spinks, B. Xi, D. Zhou, V.-T. Truong, G. G. Wallace, *Synth. Meth.* **2004**, *140*, 273–280.
9. H. Hartshorne, C. J. Backhouse, W. E. Lee, *Sens. Actuators B* **2004**, *B99*, 592–600.
10. I. Willner, E. Katz, *Angew. Chem. Int. Ed.* **2003**, *42*, 4576–4588.
11. S. Sugiura, K. Sumaru, K. Ohi, K. Hiroki, T. Takagi, T. Kanamori, *Sens. Actuators A* **2007**, *140*, 176–184.
12. M. Kameda, K. Sumaru, T. Kanamori, T. Shinbo, *J. Appl. Polym. Sci.* **2003**, *88*, 2068–2072.
13. S. J. Kim, G. M. Spinks, S. Prosser, P. G. Whitten, G. G. Wallace, S. I. Kim, *Nature Mater.* **2006**, *5*, 48–51.
14. D. T. Eddington, R. H. Liu, J. S. Moore, D. J. Beebe, *Lab Chip* **2001**, *1*, 96–99.
15. F. M. Winnik, D. G. Whitten, *Langmuir* **2007**, *23*.
16. T. C. Halsey, J. E. Martin, D. Adolf, *Phys. Rev. Lett.* **1992**, *68*, 1519–1522.
17. K. D. Weiss, J. D. Carlson, *Int. J. Mod. Phys. B* **1992**, *6*, 2609–2623.
18. X. Niu, L. Liu, W. Wen, P. Sheng, *J. Intell. Mater. Syst. Struct.* **2007**, *18*, 1187–1190.
19. X. Niu, L. Liu, W. Wen, P. Sheng, *Phys. Rev. Lett.* **2006**, *97*, 044501–0445014.
20. T. Hao, *Adv. Mater.* **2001**, *13*, 1847–1857.
21. T. Hao, A. Kawai, F. Ikazaki, *Int. J. Mod. Phys. B* **1999**, *13*, 1758–1766.
22. T. Hao, A. Kawai, F. Ikazaki, *Langmuir* **1998**, *14*, 1256–1262.
23. W. Wen, X. Huang, S. Yang, K. Lu, P. Sheng, *Nature Mater.* **2003**, *2*, 727–730.
24. W. Wen, X. Huang, P. Sheng, *Soft Matter* **2008**, *4*, 200–210.
25. A. Gozdalik, H. Wycislik, J. Plocharski, *Synth. Meth.* **2000**, *109*, 147–150.
26. Y. T. Lim, J. H. Park, O. O. Park, *J. Colloid Interface Sci.* **2002**, *245*, 198–203.
27. J. H. Kim, K. T. Lau, R. Shepherd, Y. Wu, G. Wallace, D. Diamond, *Sens. Actuators A* **2008**, *A148*, 239–244.
28. J. Causley, S. Stitzel, S. Brady, D. Diamond, G. Wallace, *Synth. Meth.* **2005**, *151*, 60–64.
29. J. Kim, K. T. Lau, D. Diamond, *IEEE International Workshop on Embedded Processors, Sensors, and Actuators (EPSA2008)*, Taichung, Taiwan, Jun 11–13, **2008**.
30. T. Wolff, N. Muller, *J. Photochem.* **1983**, *23*, 131.
31. S. Miljanic, L. Frkanec, Z. Meic, M. Zinic, *Langmuir* **2005**, *21*, 2754–2760.
32. M. J. Earle, C. M. Gordon, N. V. Plechkova, K. R. Seddon, T. Welton, *Anal. Chem.* **2007**, *79*, 758–764.
33. K. Murata, M. Aoki, T. Suzuki, T. Harada, H. Kawabata, T. Komori, F. Ohseto, K. Ueda, S. Shinkai, *J. Am. Chem. Soc.* **1994**, *116*, 6664–6676.
34. S. A. Ahmed, X. Sallenave, F. Fages, G. Mieden-Gundert, W. M. Mueller, U. Mueller, F. Voegtle, J.-L. Pozzo, *Langmuir* **2002**, *18*, 7096–7101.
35. P. Wasserscheid, in *Organic Synthesis Highlights V* (Eds. H.-G. Schmalz, T. Wirth), Weinheim: Wiley-VCH, **2003**, 105–117.
36. G. A. Olah, T. Mathew, A. Goeppert, B. Toeroek, I. Bucsi, X.-Y. Li, Q. Wang, E. R. Marinez, P. Batamack, R. Aniszfeld, G. K. S. Prakash, *J. Am. Chem. Soc.* **2005**, *127*, 5964–5969.
37. C. Hardacre, J. D. Holbrey, S. P. Katdare, K. R. Seddon, *Green Chem.* **2002**, *4*, 143–146.
38. R. Vijayaraghavan, D. R. MacFarlane, *Chem. Commun.* **2004**, 700–701.
39. R. Vijayaraghavan, D. R. MacFarlane, *Aust. J. Chem.* **2004**, *57*, 129–133.

40. K. E. Gutowski, G. A. Broker, H. D. Willauer, J. G. Huddleston, R. P. Swatloski, J. D. Holbrey, R. D. Rogers, *J. Am. Chem. Soc.* **2003**, *125*, 6632–6633.
41. A. I. Bhatt, I. May, V. A. Volkovich, M. E. Hetherington, B. Lewin, R. C. Thied, N. Ertok, *J. Chem. Soc. Dalton Trans.* **2002**, 4532–4534.
42. J. M. Pringle, M. Forsyth, G. G. Wallace, D. R. MacFarlane, *Macromolecules* **2006**, *39*, 7193–7195.
43. P. C. Howlett, D. R. MacFarlane, A. F. Hollenkamp, *Electrochem. Solid State Lett.* **2004**, *7*, A97–A101.
44. M. Tamada, T. Watanabe, K. Horie, H. Ohno, *Chem. Commun.* **2007**, 4050–4052.
45. D. T. Eddington, D. J. Beebe, *Adv. Drug Deliv. Rev.* **2004**, *56*, 199–210.
46. Q. Yu, J. M. Bauer, J. S. Moore, D. J. Beebe, *Appl. Phys. Lett.* **2001**, *78*, 2589–2591.
47. R. Laocharoensuk, A. Bulbarello, S. B. Hocevar, S. Mannino, B. Ogorevc, J. Wang, *J. Am. Chem. Soc.* **2007**, *129*, 7774–7775.
48. R. J. Byrne, S. E. Stitzel, D. Diamond, *J. Mater. Chem.* **2006**, *16*, 1332–1337.
49. D. Diamond, R. Byrne, S. Stitzel (Invent DCU Ltd., Ire.). WO, **2007**.
50. J. C. Crano, T. Flood, D. Knowles, A. Kumar, B. Van Gemert, *Pure Appl. Chem.* **1996**, *68*, 1395–1398.
51. T. Yoshida, A. Morinaka, *J. Photochem. Photobiol. A* **1992**, *63*, 227–234.
52. I. Willner, S. Rubin, R. Shatzmiller, T. Zor, *J. Am. Chem. Soc.* **1993**, *115*, 8690–8694.
53. G. Berkovic, V. Krongauz, V. Weiss, *Chem. Rev.* **2000**, *100*, 1741–1753.
54. R. Byrne, K. J. Fraser, E. Izgorodina, D. R. MacFarlane, M. Forsyth, D. Diamond, *Phys. Chem. Chem. Phys.* **2008**, *10*, 5919–5924.
55. H. Görner, A. K. Chibisov, *J. Chem. Soc. Faraday Trans.* **1998**, *94*, 2557–2564.
56. A. K. Chibisov, H. Görner, *Chem. Phys.* **1998**, *237*, 425–442.
57. S. Samanta, J. Locklin, *Langmuir* **2008**, *24*, 9558–9565.
58. S. Stitzel, R. Byrne, D. Diamond, *J. Mater. Sci.* **2006**, *41*, 5841–5844.
59. A. Radu, S. Scarmagnani, R. Byrne, C. Slater, K. T. Lau, D. Diamond, *J. Phys. D* **2007**, *40*, 7238–7244.

Chapter 24

Hybrid Nanomaterials Research: Is It *Really* Interdisciplinary?

ISMAEL RAFOLS, MARTIN MEYER, AND JAE-HWAN PARK

24.1 INTRODUCTION

One of the central tenets of the current discourse on innovation is that the most important scientific and technological breakthroughs are the result of interdisciplinary endeavors. This idea has become particularly significant in those areas of science and technology that are perceived as being a result of technological convergence, such as nanotechnology or its cognate developments:

> Revolutionary advances at the interfaces between previously separate fields of science and technology are ready to create key *NBIC transforming tools (nano-, bio, info-, and*

The Supramolecular Chemistry of Organic–Inorganic Hybrid Materials. Edited by Knut Rurack and Ramón Martínez-Máñez

cognitive-based technologies), including scientific instruments, analytical methodologies, and radically new materials systems. The innovative momentum in these interdisciplinary areas must not be lost but harnessed to accelerate unification of the disciplines. (Roco and Bainbridge,[1] p. 3)

The field of hybrid organic–inorganic nanomaterials explored in this book fits nicely with this view of research as driven by applications whose development requires interdisciplinary effort. Along these lines, recent reviews have presented hybrid organic–inorganic nanomaterials as exploiting interdisciplinary interactions between supramolecular chemistry, materials science research, and nanotechnology.[2] However, interdisciplinarity is a rather nebulous and polysemous concept. A giant collaboration building a high energy collider is surely very different from a project on climate change in a social science institute, or a laboratory working on molecular motors. Although each of these endeavors may claim to be interdisciplinary, the underlying cognitive and social processes are markedly dissimilar.

In which sense is hybrid nanomaterials research interdisciplinary? Of the many existing perspectives on interdisciplinarity, this chapter presents an exploration of hybrid nanomaterials based on the knowledge sources of the field, using bibliometric data. Since this data is limited to the contents of the book, it should be stressed that the analysis and conclusions pertain to hybrid nanomaterials in the sense of this book, that is, in terms of supramolecular chemistry*. The main findings are that hybrid nanomaterials research is very fragmented among different materials-centered clusters [e.g. carbon nanotubes (CNTs) and quantum dots (QDs)], and applications-centered topics, with each of these topics drawing on disciplines from chemistry and materials sciences, and to a lesser extent the biological sciences.

The chapter is organized as follows. We begin with a "cautionary" review of the current interest in—and rhetoric on—interdisciplinarity in the context of the new discourses on science,[3] which stress the need for science to be legitimated by producing explicit social benefits. Second, we briefly review various approaches to the assessment of interdisciplinarity and present the conceptual framework and methodology of this investigation. Third, we show and discuss the empirical results and, finally, we summarize the findings.

24.2 THE CURRENT RISE OF INTERDISCIPLINARITY IN CONTEXT

Since the early 1990s there has been a boom in the (self-reported) adoption of interdisciplinarity by both scientists and policy makers.[4] Occurrences of the term *interdisciplinary* in scientific papers increased fivefold in 15 years, from 550 per year in 1993 to more than 3100 in 2007.† Policy reports have highlighted the importance of interdisciplinarity for strategic technologies such as nanotechnology.[5] There has been a

*Therefore, one should be cautious when extrapolating the conclusions from supramolecular chemistry to other fields related to hybrid nanomaterials, such as catalysis or separation research.

†This search was carried out using ISI's Science Citation Index database (i.e. not including the Social Sciences or Humanities) for Articles, Reviews, Notes and Letters only.

surge in programs fostering interdisciplinary collaboration and technological convergence, such as NEST (New and Emerging Science and Technology) in the 6th EU Framework Programme, and large investments have been made into interdisciplinary research centers, such as Cornell's Nanobiotechnology Center (NBTC).

While this enshrinement of interdisciplinarity as a positive scientific norm may today seem "natural," we should remember that it runs contrary to the beliefs of the early twentieth century:

> Only by strict specialization can the scientific worker become fully conscious, for once and perhaps never again in his lifetime, that he has achieved something that will endure. A really definitive and good accomplishment is today always a specialized accomplishment. And whoever lacks the capacity to put on blinders, so to speak, . . . may as well stay away from science. (Weber,[6] p. 135)

Hence, this surge in interdisciplinarity needs to be put into context. Although some interdisciplinary academic and research programs were developed in the mid-twentieth century,[7] interdisciplinarity first emerged as a social and policy issue at the time of student unrest over higher education reforms in the 1960s and 1970s,[8] and then regained prominence in the 1990s with the advent of a series of policy studies claiming that the science and technology system was undergoing major structural changes.[9] Among these studies, the most influential are Gibbons et al.'s contributions,[3,10] which depict a transition towards a new research mode (*Mode 2*) characterized by production of knowledge in the context of application, with more transient organizational settings, wider societal considerations for the evaluation of its worth, and which includes transdisciplinary research and heterogeneity of skills among its key attributes. Other models that have focused on the stronger interaction between academia and economic or government actors, such as Leydesdorff and Etzkowitz's Triple Helix,[11] also suggest indirectly a move towards greater interdisciplinarity in order to address societal or industrial demands.

However, the fact that this literature is prescriptive rather than descriptive suggests that "the discourse on interdisciplinarity is, in effect, a discourse on innovation in knowledge production," (Weingart,[8] p. 30) where "interdisciplinarity provides a means for steering and coordinating strategic investment in research across a range of partners," (Lowe and Phillipson,[12] p. 167). In other words, science is being pushed to be more interdisciplinary in order to better fulfill its new social contract, which entails a more direct interaction with societal actors via technology transfer or public engagement. In this context, the call for interdisciplinarity may become a convenient tool used to foster reform in scientific institutions (e.g., to justify the erosion of the tenure track system[13]), rather than a requirement of "organic" or "internal" developments in science (as we might argue was previously the case in areas such as biophysics).

In summary, the current discourse on science implicitly takes a prescriptive stance and believes that research *ought* to be more interdisciplinary in order to be more successful at dealing with societal problems and supporting innovation and competitiveness. However, since you can't get an *ought* from an *is*, when an emergent field such as hybrid nanomaterials is presented as interdisciplinary, one cannot help but wonder: is it *really*? Or is it rather that there is an expectation (or desire?) that it become interdisciplinary?

24.3 THE ASSESSMENT OF INTERDISCIPLINARITY

There is an increasing consensus (e.g., as shown in the report on interdisciplinarity of the U.S. National Academies[14]) that the key characteristic of interdisciplinarity is knowledge integration, that is, the combination and/or the fusion of concepts or theories, tools or techniques, and information or data from various bodies of specialized knowledge. However, there is no agreement in the literature about how to assess the degree of interdisciplinarity:[15] it has been measured in a variety of ways, including the affiliations of researchers involved in collaborations,[16] their educational background,[17] the citation flows among disciplines,[18] and the co-occurrence of disciplines in references/citations,[19] or in article headings.[20]

In our view, there are three questions that need to be raised with regard to the degree and type of interdisciplinarity of a field:[‡]

1. How diverse and how coherent is its cognitive structure? This question explores *which bodies of knowledge* the field is building on, and to what extent these fields are becoming integrated.
2. What are the drivers of its knowledge integration? This point aims to elucidate *why* the emergent field needs to reach out to various knowledge sources.
3. What are the strategies for knowledge acquisition? This examines *how* laboratories garner knowledge from various disciplinary bases.

The first question looks into interdisciplinarity proper. Since interdisciplinarity is an epistemic characteristic, it has to be assessed by looking into the contents of research (e.g., examining the concepts or techniques used, or analyzing their proxies, such as the structure of publications) rather than its social practices (e.g., collaborations). The second question (*why*) enquires into the motivations for pursuing interdisciplinary research and can be investigated by looking into the characteristics shared among the different bodies of knowledge that are brought together (e.g., instrumentation or application objectives). Schmidt[21] proposed that these drivers can be epistemological (sharing theories and concepts, e.g. in complex systems studies), methodological (e.g., the atomic force microscope), ontological (i.e. objects-centered, such as carbon-nanotubes) or problem-oriented (e.g., climate change research).

Finally, the third point (*how*) looks into the social mechanisms and processes that constitute the practice of interdisciplinarity (such as recruitment, interaction within joint facilities, sharing of research material or data, etc.). Here we should like to emphasize that interdisciplinary research "does not necessarily imply collaboration between researchers from different disciplines," (Bordons et al.,[15] p. 440) and even when it does, the term *collaboration* encompasses a variety of practices.[22,23] This is why in hyped fields such as nanotechnology, where the organizations and educational degrees are sometimes relabeled without much substantial transformation in order to cater to shifting funding demands, the number of collaborations among diverse

[‡]The questions apply equally to enquiry into interdisciplinarity of organizations, but for the purposes of this chapter, we will focus on research fields, themes, or topics.

disciplinary affiliations cannot always be assumed to be a reliable indicator of interdisciplinarity.[24]

Since the aim of this chapter is to provide a general outlook on interdisciplinarity in hybrid nanomaterials, we have focused our efforts on the knowledge structure and briefly touched upon the drivers for integration, both of which can be explored using bibliographic data. An exploration of the interdisciplinary practices (e.g., collaborations or recruitment) falls beyond the scope of this investigation.

24.4 DATA AND METHODS

If the boundaries between well-established research fields are always fuzzy and controversial, we have to recognize that the delineation of emergent areas of research is, inevitably, a highly arbitrary exercise. For hybrid nanomaterials, we have aimed to represent the field as characterized in this book, which takes a supramolecular chemistry perspective. To do so, we asked one of the editors, K. Rurack, to provide a representative review list of the field (which comprised 28 publications), and each of the contributors to select the five most relevant review publications in the 2002 to 2007 period for their chapter. This yielded a total of 117 unique review publications, with only 16 overlaps (9 by the editor) among the nominated publications. Four reviews were unused because they could not be localized in bibliometric databases. The list of 113 reviews is presented in Appendix 1.

This set of publications allowed us to conduct a bibliometric analysis of the field, following the methodology we recently developed.[25] Bibliographic records were downloaded using data from the ISI Web of Science and processed using the bibliometric program Bibexcel,[26] the statistical packet R,[27] and the network analysis software Pajek.[28] We then looked into the coherence of the network obtained by linking reviews via the normalized number of shared references (bibliographic coupling). The network was divided into research topics by means of cluster analysis (software SPSS 15.0 for Windows, using hierarchical clustering, with Ward's method and squared Euclidian distance). While allowing for the creation of between 5 to 15 clusters, 10 clusters were obtained, then simplified to 7 since 3 of them were tightly connected (these clusters were merged into the Hybrid Silica Materials cluster). Second, we looked at the distribution of disciplines cited in the references of reviews (where each reference is assigned to a discipline according to ISI Subject Category classification[§]). This distribution allowed us to assess the degree of interdisciplinarity for each cluster in terms of the number of ISI subject categories, Shannon entropy, and Rao–Stirling diversity.[¶] Figure 24.1 provides a schematic description of the methodology.

[§]The category "Multidisciplinary Sciences" was removed because it is perceived as misleading, given that its journals are multidisciplinary, but not necessarily the publications.

[¶]Let p_i be the proportion of references in discipline i, and s_{ij} the Salton's cosine similarity of disciplines i and j according to their citing patterns. Then Shannon entropy (or diversity) is defined as $H = -\sum_i p_i \ln p_i$ and Rao–Stirling diversity is defined as $\Delta = 1 - \sum_{i,j} s_{ij} p_i p_j$. See Rafols and Meyer[25] for details. Porter et al.[19] also introduced a formulation equivalent to Rao–Stirling's diversity.

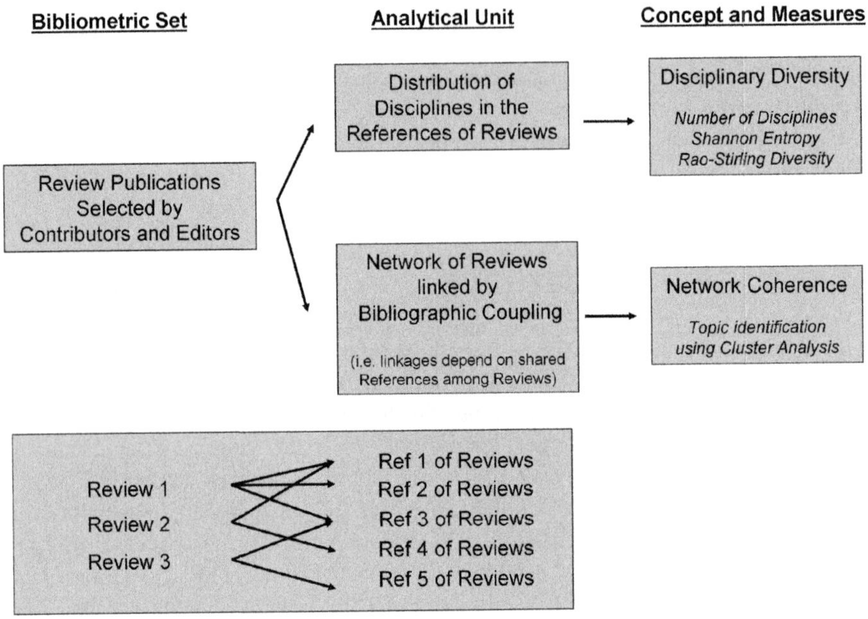

Figure 24.1 Schematic representation of the methodology employed to map the knowledge structure of hybrid nanomaterials research. See Rafols and Meyer[25] for details.

24.5 KNOWLEDGE STRUCTURE IN HYBRID NANOMATERIALS

24.5.1 Main Topics of Research Clusters

Figure 24.2 represents the knowledge structure of hybrid nanomaterials obtained from the similarity between publications in terms of shared references. For the sake of clarity, only the publications in clusters are displayed (66 out of 113, the remaining 47 publications being weakly linked), and only lines with values above 0.03 are drawn. The fact that the network is very spread, in spite of a very low similarity threshold (0.03, or about 3% shared references), indicates that hybrid nanomaterials is not (yet?) a coherent field, as one would expect from more mature areas. Cluster analysis allows us to identify groups of publications that share a certain focus as listed in Table 24.1. The label and description of each cluster highlights the main *specific* topic shared among the publications in the cluster. However, each of the reviews will generally also touch upon other topics (e.g., although all reviews on biosensing are related to gold nanoparticles, some of them also discuss QDs and CNTs for biosensing).

The map of hybrid nanomaterials in Figure 24.2 not only helps us view which are the distinct research topics but also how they are related to each other. It illustrates, for example, that the cluster on coordination polymers, crystal engineering, and metal-organic frameworks (top left) is very weakly related to the rest of the network.

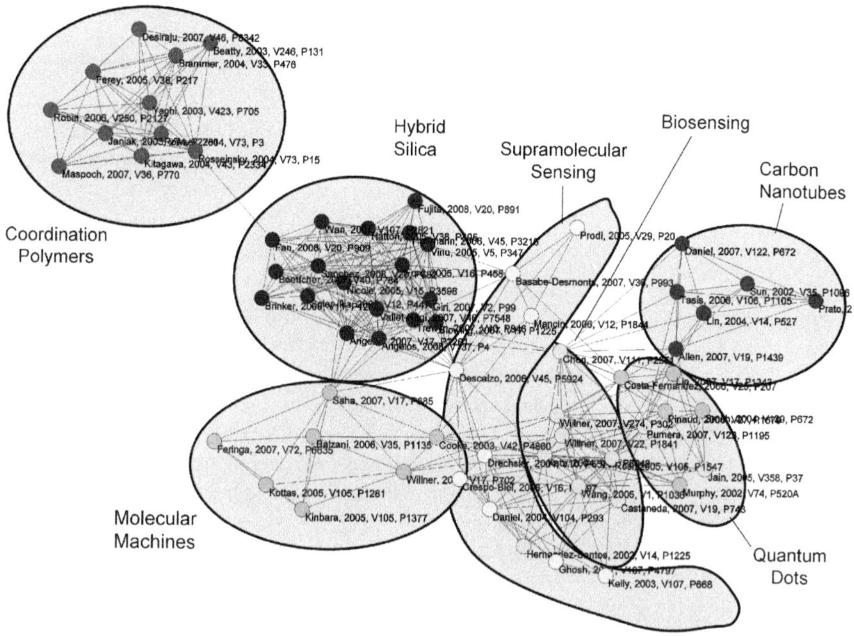

Figure 24.2 Knowledge structure of hybrid nanomaterials research. The nodes of the network represent review publications. The labels indicate first author and year of publication. The lines represent the similarity between nodes, as measured by the normalized number of shared references (bibliographic coupling).

It also shows that silica hybrid materials and molecular machines are quite close to each other, which is due to the use of porous silica sol-gel to provide a stable matrix for encapsulating molecular machines in a macroscopic solid. However, since Figure 24.2 is a two-dimensional projection of a much higher-dimensional space, one has to be cautious in the interpretation of apparent proximity: QD and CNT clusters appear close to each other only because they are both interconnected with the biosensing cluster, not because they are directly related to each other.

The main insight to draw from Figure 24.2 is that hybrid nanomaterials as a field is represented by a "periphery" of research on specific support materials (QDs, CNTs, hybrid silica materials, and coordination polymers), articulated around a "center" focused on their applications in sensing, biosensing, and, to a lesser extent, on molecular machines. This result is consistent with the general pattern that more applied research efforts (in this case in sensing) integrate more specialized areas (in this case, in inorganic support materials). It may be worth noting that the materials doing the functionalization (biomolecules, enzymes, or organic groups) do not form clusters, supporting the vision of the specific inorganic supports as generic platforms for a myriad of different organic hosts. The fact that one of the reviews contributed by the editors of this book[2] occupies a central position in the network is a reminder that this portrait of hybrid nanomaterials is unavoidably a partial view.

Table 24.1 Description of Central Topics in the Clusters of the Knowledge Structure of Hybrid Nanomaterials, According to Bibliographic Coupling Networks (see Fig. 24.2)

Topic Cluster	Description
Biosensing	Use of gold supports with biomolecules and enzymes as functionalization materials, for biosensing applications (electrochemical and optical).
Quantum Dots (QDs)	Use of quantum dots, luminescence as a detection method, with organic groups and biomolecules as functionalization materials, in sensing applications.
Carbon Nanotubes (CNTs)	Use of CNTs as inorganic support, focus on electrochemistry as detection method, organic groups and biomolecules as functionalization materials, and their application in sensing.
Supramolecular Chemistry and Sensing	Use of gold and silica supports with organic groups as functionalization materials, and investigation of the mechanisms involved in host–guest chemistry, for sensing applications.
Molecular Machines	Use, for molecular machine functions, of gold and silica supports and the chemical mechanisms involved in the function (host–guest chemistry).
Hybrid Silica Materials	Use of silica with periodically mesoporous organosilicas (PMOs) morphology, with organic groups as functionalization materials, for applications in sensing.
Coordination Polymers	Use of hybrid materials composed of metal ions and organic groups (metal–organic framework based on chemical engineering) and their preparation and/or fabrication.

The reviews that were not found to belong to a cluster tell us about areas within hybrid nanomaterials that may be either more peripheral or situated among stronger clusters. These include areas such as molecular imprinting, self-assembled monolayers, or biomimetic materials. A larger analysis would be necessary to be able to have significant evidence of the relative position of these other areas in Figure 24.2.

24.5.2 Disciplinary Diversity

Although Figure 24.2 told us about the local structure of knowledge in hybrid nanomaterials, it did not explain the position of the field in the larger map of science—in terms of disciplines. In order to do this, we looked into the references contained in the review publications for the whole field and for each of the clusters. The results are listed in Table 24.2 and presented over a map of science in Figure 24.3 (see Leydesdorff and Rafols[29] for details of the construction of the map). Overall, the references come from three research areas: first, from the core chemistry areas (multidisciplinary chem., analytical chem., organic chem., inorganic chem.) in the center of the

Table 24.2 Percentage of References in Each Discipline (ISI Subject Categories, Left Column) in the Publications of a Topic Cluster (Top Row)[a]

% References in Discipline Per Topic Cluster	Topic Cluster							
Main Disciplines (ISI Subject Category)	Hyb. Nano. (ALL)	Bio-sens.	QDs	CNTs	Sup. Sens.	Mol. Mac.	Hyb. Sil.	Coor. Poly.
Chemistry, Multidiscip.	23.5	22.6	20.7	20.6	26.0	39.8	18.4	43.9
Chemistry, Phys.	21.6	11.6	16.3	17.3	27.1	8.4	31.2	13.5
Materials Sci., Multidiscip.	13.9	9.0	13.5	19.4	11.5	5.4	24.1	9.9
Chemistry, Analytical	5.3	15.7	9.1	2.4	5.4	0.6	0.8	0.0
Physics, Applied	4.4	3.8	5.2	6.3	3.6	3.2	4.4	1.4
Nanosci. & Nanotech.	3.9	7.1	4.9	9.3	3.6	3.0	3.9	0.6
Phys., Condensed Matter	3.7	2.4	4.4	3.7	3.4	2.8	4.1	2.2
Chem., Inorganic & Nucl.	2.6	0.4	0.2	0.3	0.9	4.9	0.5	20.2
Polymer Science	2.4	0.2	0.0	5.1	1.4	0.8	0.9	0.3
Electrochemistry	2.3	7.7	0.5	1.7	0.9	0.4	0.4	0.0
Phys., At., Mol. & Chem.	2.0	0.9	2.5	4.5	3.6	3.8	0.9	0.6
Biochem. & Mol. Biology	1.7	2.8	3.0	1.1	1.0	4.4	0.5	0.1
Chemistry, Organic	1.4	0.5	0.5	0.7	1.2	10.2	0.4	1.0
Biotech. & App. Microbiol.	1.0	4.4	6.7	0.4	0.5	0.3	0.1	0.1
Biophysics	1.0	3.3	1.5	0.6	0.6	1.9	0.0	0.0
Physics, Multidiscip.	0.9	0.3	0.2	0.6	1.7	2.4	0.1	0.2
Chemistry, Applied	0.8	0.0	0.0	0.1	0.5	0.0	2.8	0.7
Biochem. Res. Methods	0.8	1.6	3.0	0.7	0.6	0.3	0.1	0.0
Crystallography	0.6	0.0	0.0	0.1	0.1	1.0	0.0	4.4
Engineering, Chem.	0.5	0.0	0.0	0.2	0.7	0.0	1.4	0.2
Materials Sci., Ceramics	0.4	0.1	0.2	0.0	0.1	0.0	1.0	0.0
Cell Biology	0.4	0.4	0.5	0.1	0.0	2.8	0.1	0.0

[a]The shaded values signal the topic cluster with the relative highest percentage of references.

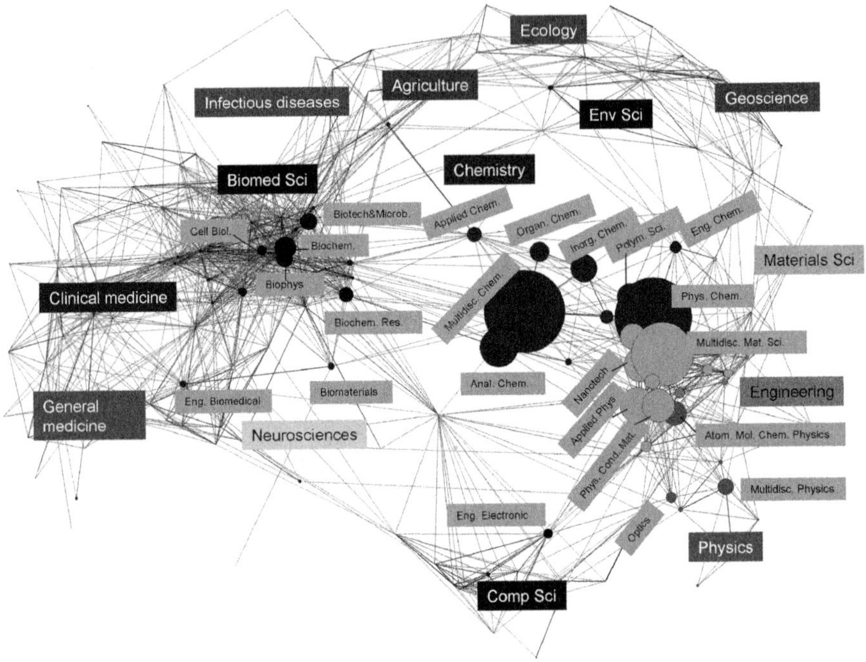

Figure 24.3 Disciplinary distribution in the map of science of the references in the review publications for hybrid nanomaterials. Each node represents one discipline (e.g., organic chemistry or cell biology). The node size is proportional to the number of references in a given discipline. The lines and distances portray the similarities between disciplines according to their citing patterns over all science—the more similar their citations to *all* other areas of science, the shorter and darker the lines between them. Light yellow labels display single disciplines. Colored labels are aggregate disciplinary areas (e.g., physics). Backbone map based on Leydesdorff and Rafols.[29] (See color insert.)

map; second from materials science areas (multidisciplinary materials sciences, applied physics, nanoscience and nanotechnology, and condensed matter physics**) to the right of the map, as well as the chemistry areas that sit between the chemistry and materials sciences (e.g., physical chemistry and polymer science). Finally, on the left of the map, there is a small but remarkable percentage of references (above 5%) from the biological sciences (biochemistry, biotechnology, biophysics, and cell biology).

A look into the disciplinary profile of each of the clusters helps to understand its knowledge sources and the differences and similarities between clusters. For example, biosensing has a high percentage of references from the biological sciences (arguably

**In spite of the labels, all four of these disciplines are closely interrelated via cross-citations. Factor analysis of cross-citations shows, for example, that the loading factor of applied physics is 0.83 in the materials science dimension and only 0.25 in the physics dimension—which is formed by more basic research subjects such as astronomy, nuclear physics, or fluids and plasmas.

due to applications), together with a relatively high proportion of analytical chemistry and electrochemistry (as one expects in sensing). At the opposite extreme, hybrid silica materials draws its references from materials science, and from chemistry areas close to materials science (such as physical chemistry) while it has very few references to biological sciences. Figure 24.4 illustrates these differences using as a background the map of science.

The diversity of disciplines of one cluster also offers us a perspective on its degree of interdisciplinarity. Figure 24.5 presents the Shannon and Rao–Stirling measures of diversity for each of the clusters. The results show that biosensing and QDs are the most interdisciplinary clusters in two respects: they both have the widest spread of references across disciplines (this is shown by the high value in Shannon's indicator), and the disciplines they cite are very varied, that is, they draw on disparate disciplines (as shown by the high value of the Rao–Stirling measure). This is illustrated in the case of biosensing in Figure 24.4, with publication activity in very distant nodes: in biological sciences as well as in chemistry and materials science. Overall, from Figure 24.5, we can see that those with the highest diversity are biosensing and QDs, whereas those with the lowest diversity are coordination polymers and hybrid silica materials, which are more narrowly focused on inorganic chemistry, first, and materials sciences, second.

The data presented in Figure 24.5 allows comparison among the diverse clusters but does not tell us whether the hybrid nanomaterials field as a whole has a high value of disciplinary diversity. To do this, we need to look at *external* benchmarks. Taking as a benchmark six review publications from the fields of multidisciplinary chemistry and physical chemistry, we obtained a Shannon diversity value of 1.5 ± 0.3 (Rao–Stirling, 0.27 ± 0.09), well below the average value in hybrid nanomaterials, 2.1 ± 0.2 (Rao–Stirling, 0.44 ± 0.09). This suggests that the field of hybrid nanomaterials is indeed more interdisciplinary than the main disciplines within which it is embedded.

24.5.3 Drivers of Knowledge Integration

What is the reason *why* hybrid nanomaterials research draws on the bodies of knowledge shown in Figures 24.2 and 24.3? Or, in Schmidt's framework[21] (see Section 24.3), what is the type of interdisciplinarity that is pursued in this field (epistemological, methodological, ontological, problem-oriented)? The previous mapping exercises help us explore these questions. First of all, the field is clearly not defined by shared theoretical frameworks or methods. Second, publication clustering around materials (inorganic supports) suggests object-centered interdisciplinarity in each of these clusters. This means that in the design, preparation, and use of a specific support material such as QDs, there is a need to integrate knowledge from studies of its physical properties (materials sciences area), its functionalization with biomolecules or organic groups (chemistry area), and the application of QDs (e.g., as markers in biological samples). Third, the fact that the core clusters of the network are focused on sensing and that this topic is also a recurrent theme in other materials-based clusters, seems to indicate that the field of hybrid nanomaterials as a whole is articulated around applications. The fact that the best known inorganic support, that is, gold nanoparticles,

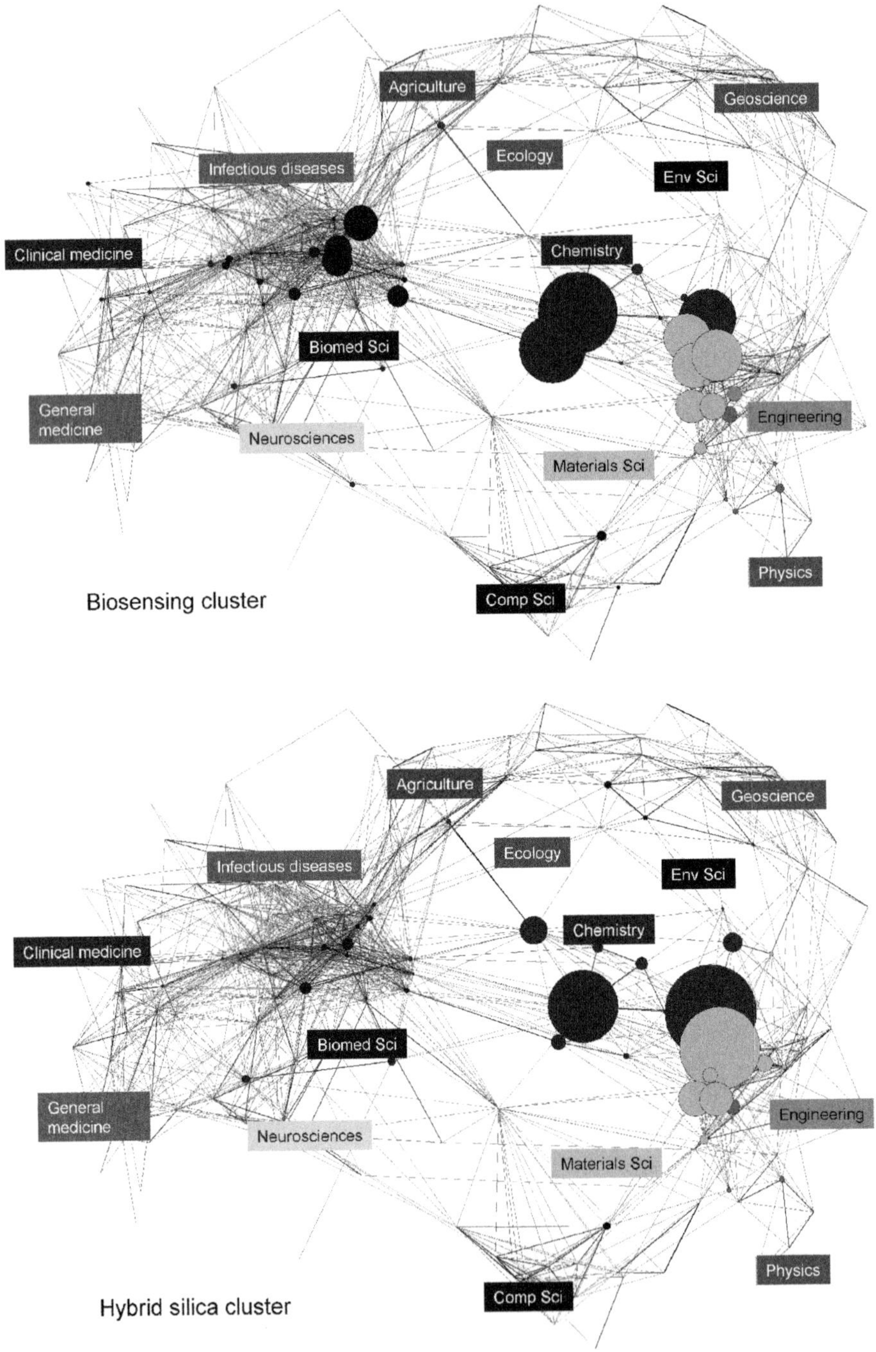

Figure 24.4 Disciplinary distribution in the map of science of the references in the review publications for two selected clusters. Top: Biosensing. Bottom: Hybrid silica materials. See text and Figure 24.3 for details. (See color insert.)

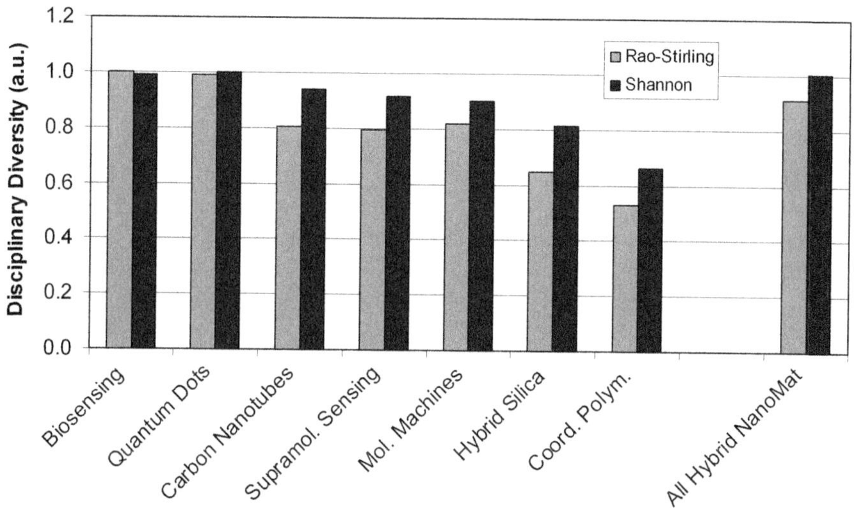

Figure 24.5 Measures of Shannon and Rao–Stirling diversity (in arbitrary units, a.u.) in relation to the publication disciplines of the references in the reviews. See Section 24.4, Data and Methods, for their definition. Both measures are normalized dividing by the largest value in the series.

falls mainly in the biosensing cluster (and partly in the supramolecular chemistry and sensing cluster) rather than making a cluster of its own, supports the view that the more consolidated research focuses on applications rather than on the materials themselves. In summary, following Schmidt's typology, the field becomes coherent by means of problem-oriented interdisciplinarity—yet the fact that it is still very segregated into materials-based topics as opposed to applications topics reveals that it is far from mature.

24.6 SUMMARY AND CONCLUSIONS

This chapter has explored whether and how the field of hybrid nanomaterials is interdisciplinary, with the aim of understanding the underlying knowledge structure of this book, hence it investigated hybrid nanomaterials only from the perspective of supramolecular chemistry. The results of a bibliometric analysis based on 113 review publications selected by the contributors suggest that integration of different disciplines is carried out first, around research on specific inorganic support materials (QDs, CNTs, etc.), and second, around research on applications in sensing and molecular machines, as shown in Figure 24.2. The centrality of sensing applications and their pervasive presence in other clusters suggests that they constitute the key integrative driver for interdisciplinarity in hybrid nanomaterials. However, the fact that the cognitive structure is very scattered and that many clusters are materials-centered may indicate that the field has not yet reached maturity in terms of applications.

Regarding disciplinary sources, hybrid nanomaterials are more diverse than the norm in the chemistry disciplines in which it is embedded. As shown in Figure 24.3, it draws mainly on disciplines from chemistry (organic, inorganic, physical) and materials science (multidisciplinary materials science, nanotechnology, condensed matter, applied physics), as well as from certain areas of biology (biophysics, biotechnology, biochemistry), which provide either biomimetic ideas or arenas of application.

The chapter started by locating the current discussion on interdisciplinarity in the context of a changing science-society contract. In this context, interdisciplinarity is being supported by the funding bodies both as a means to increase social relevance and as a justification for organizational reforms. Do these generic policy drivers for interdisciplinarity apply to the field of hybrid nanomaterials? First, regarding social relevance, the fact that sensing applications constitute the main driver for research in hybrid nanomaterials suggests that this field of research is indeed striving to make technological contributions. However, as mentioned, the field appears to be in its early phases, and more driven by science-push (creating technological opportunities) than by demand-pull dynamics (searching for solutions to specific problems), which would fully comply with the social relevance rationale.

Second, the question of organizational reforms (such as creation of interdisciplinary centers) unveils the biggest limitation of the current exploration: here we have focused on cognitive mapping at the expense of looking at the type of research practices (collaboration, recruitment, etc.) by which knowledge integration is achieved in hybrid nanomaterials.[††] Anecdotal evidence suggests that, as in other nanotechnology fields, such as lab-on-a-chip, deep interdisciplinary collaborations across organizations, or intense disciplinary mixing of researchers within one laboratory, are much less common that one would expect from the discourse.[30] Hence further research on this other aspect, the social and organizational issues in interdisciplinary practices, is definitely needed in order to better understand the development of hybrid nanomaterials and the adequacy of the current policy frameworks.

ACKNOWLEDGMENTS

This research was supported by an EU postdoctoral Marie-Curie Fellowship to IR. We are much indebted to K. Rurack for the collection of selected reviews from the contributors and his advice regarding the interpretation of the bibliometric analysis. We thank J. Gläser, G. Laudel, L. Leydesdorff, A. L. Porter, and SPRU colleagues W. E. Steinmueller and A. Stirling for fruitful discussions.

[††]Research on cognitive maps (based on easily accessible databases) seemed more appropriate than looking into research practices (accessible only via extensive interviews) given the time and resource limitations, and the exploratory nature of this chapter.

REFERENCES

1. M. C. Roco, W. S. Bainbridge, *Converging Technologies for Improving Human Performance. Nanotechnology, Biotechnology, Information Technology and Cognitive Science*, Arlington, VA: National Science Foundation, **2002**.
2. A. B. Descalzo, R. Martínez-Máñez, F. Sancenón, K. Hoffmann, K. Rurack, *Angew. Chem. Int. Ed.* **2006**, *45*, 5924–5948.
3. M. Gibbons, C. Limoges, H. Nowotny, S. Schwartzman, P. Scott, M. Trow, *The New Production of Knowledge: The Dynamics of Science and Research in Contemporary Societies*, London: Sage, **1994**.
4. T. Braun, A. Schubert, *Scientometrics* **2003**, *58*, 183–189.
5. I. Malsch, *IPTS Report* **1997**, *13*.
6. M. Weber, in *From Max Weber: Essays in Sociology* (Eds. H. H. Gerth, C. W. Mills), New York: Oxford University Press, **1946(1919)**, 129–156.
7. J. T. Klein, in *Practising Interdisciplinarity* (Eds. P. Weingart, N. Stehr), Toronto: University of Toronto Press, **2000**, 3–24.
8. P. Weingart, in *Practising Interdisciplinarity* (Eds. P. Weingart, N. Stehr), Toronto: University of Toronto Press, **2000**, 25–41.
9. L. K. Hessels, H. van Lente, *Res. Policy* **2008**, *37*, 740–760.
10. M. Gibbons, *Nature* **1999**, *402*, c81–c84.
11. L. Leydesdorff, E. Etzkowitz, *Sci. Public Policy* **1998**, *25*, 195–203.
12. P. Lowe, J. Phillipson, *J. Agric. Econ.* **2006**, *57*, 165–184.
13. S. Fuller, http://www.interdisciplines.org/interdisciplinarity/papers/3, **2003**.
14. *Facilitating Interdisciplinary Research*, Washington, D.C.: National Academies Press, **2005**.
15. M. Bordons, F. Morillo, I. Gómez, in *Handbook of Quantitative Science and Technology Research* (Eds. H. F. Moed, W. Glänzel, U. Schmoch), Dordrecht: Kluwer, **2004**, 437–456.
16. J. Schummer, *Scientometrics* **2004**, *59*, 425–465.
17. L. Sanz-Menéndez, B. M., M. A. Zulueta, *Res. Evaluat.* **2001**, *10*, 47–58.
18. A. L. Porter, D. E. Chubin, *Scientometrics* **1985**, *8*, 161–176.
19. A. L. Porter, A. S. Cohen, J. D. Roessner, M. Perreault, *Scientometrics* **2007**, *72*, 117–147.
20. R. J. W. Tijssen, *Res. Policy* **1992**, *21*, 27–44.
21. J. C. Schmidt, *Innov. Eur. J. Soc. Sci. Res.* **2007**, *20*, 313–328.
22. G. Laudel, *Int. J. Technol. Manage.* **2001**, *22*, 762–781.
23. G. Laudel, *Res. Evaluat.* **2002**, *11*, 3–15.
24. I. Rafols, M. Meyer, *Scientometrics* **2007**, *70*, 633–650.
25. I. Rafols, M. Meyer, *Scientometrics* **2009**, *81*, DOI 10.1007/s11192-009-0041-y.
26. O. Persson, *Bibexcel. A tool-box programme for bibliometric analysis.* http://www.umu.se/inforsk/Bibexcel/ **2008**.
27. R Development Core Team, *R: A language and environment for statistical computing*, URL http://www.R-project.org, R Foundation for Statistical Computing, Vienna, Austria, **2007**.
28. V. Batagelj, A. Mrvar, *Pajek. Program for Large Network Analysis.* http://vlado.fmf.uni-lj.si/pub/networks/pajek/, **2008**.
29. T. Rafols, L. Leydesdorff, *J. Am. Soc. Inf. Sci. Technol.* **2009**, *60*, 1823–1835.
30. I. Rafols, *Innov Eur. J. Soc. Sci. Res.* **2007**, *20*, 395–412.

Chapter 25

Supramolecular Chemistry Meets Hybrid (Nano)Materials: A Brief Look Ahead

KNUT RURACK AND RAMÓN MARTÍNEZ-MÁÑEZ

25.1 INTRODUCTION

The thematic outline of the book was conceived during winter 2006/2007. The contents of the book thus basically reflect the specialized areas of research at the intersection of *supramolecular chemistry* and *hybrid (nano)materials* that were established at that time. However, because of the vibrant drive of research in these two areas, it is not surprising that significant new developments are being conceived, realized, and published almost quarterly. Of course, once new findings are reported, especially on novel composite materials, it is difficult to forecast whether they will become important

The Supramolecular Chemistry of Organic–Inorganic Hybrid Materials. Edited by Knut Rurack and Ramón Martínez-Máñez

within the context of supramolecular chemistry and all its applications, will hide in a niche, or will even mature into a specialized field that may merit its own chapter in a future (edition of the) book, or even a book in itself. This last chapter will thus highlight a few selected examples of relatively young hybrid materials that we think will accumulate critical mass in the short term and which might have both a bright future and a strong impact on the supramolecular chemistry of hybrid materials.

25.2 (BIO)ORGANICALLY FUNCTIONALIZED NANODIAMOND

Diamond nanoparticles or ultradispersed diamond (UDD) or detonation diamond or nanodiamond (ND), although known for almost half a century,[1] sank virtually into oblivion until the late 1980s.[2,3] This fate was most likely connected to their discovery as a product of the detonation of carbon-based explosives at high temperatures and pressure under oxygen deficit during cold war times. Moreover, although the first successful findings were reported by U.S. scientists,[1] for a long time, intensive research in the field happened only in several research institutes in the USSR that worked on the development of explosive technologies and the synthesis of superhard materials,[4] under special security measures. The first report of the surface modification of ND dates back to 1991[5] and the first publication in the field from researchers outside the (former) USSR in recent years appeared in 1995 from a group of the Chinese Academy of Sciences at Lanzhou.[6] Functionalization with biomolecules through physisorption was accomplished by 2003,[7] which seems to be a method yielding sufficiently stable composites to allow transfection of NDs into a cell.[8] A mixed physicochemical immobilization strategy for biomolecules onto ND through electrostatic binding of a biomolecule-functionalized polylysine coating layer around the nanoparticles was published in 2004.[9] In 2006 and 2008, Krüger's group reported on the covalent attachment of peptides and biotin to ND (Fig. 25.1).[10,11]

The importance of these accomplishments becomes apparent when considering the properties of ND that render this material very promising for numerous potential applications in (bio)chemical analysis and biomedical therapy. Diamond nanoparticles are extremely hard, possess excellent chemical inertness yet are very biocompatible and show low cytoxicity.[12,13] Besides their optical transparency, a remarkable property that qualifies NDs as optical probes is the fact that they fluoresce in the 400 to 600 nm region upon excitation with UV light, or NIR light through multiphoton excitation processes. The intrinsic photoluminescence (PL) arises from structural defects and impurities (dopants).[14] It can be enhanced significantly by the formation of nitrogen-vacancy (NV) centers under proton or electron beam irradiation followed by annealing.[15] Since the photoluminescence features are defect-borne and different defect sites can be created, the PL features depend on the method of preparation.[16] Accordingly, intensely fluorescent NDs (FNDs) with red as well as green emission (gFNDs) have been obtained and utilized in bioimaging to date (Fig. 25.2).[12,17] Further advantageous characteristics of FNDs are their stability against photobleaching (imaging of a single FND particle in a cell can be performed for hours),[18] the absence of fluorescence intermittency (or blinking, as observed for quantum dots

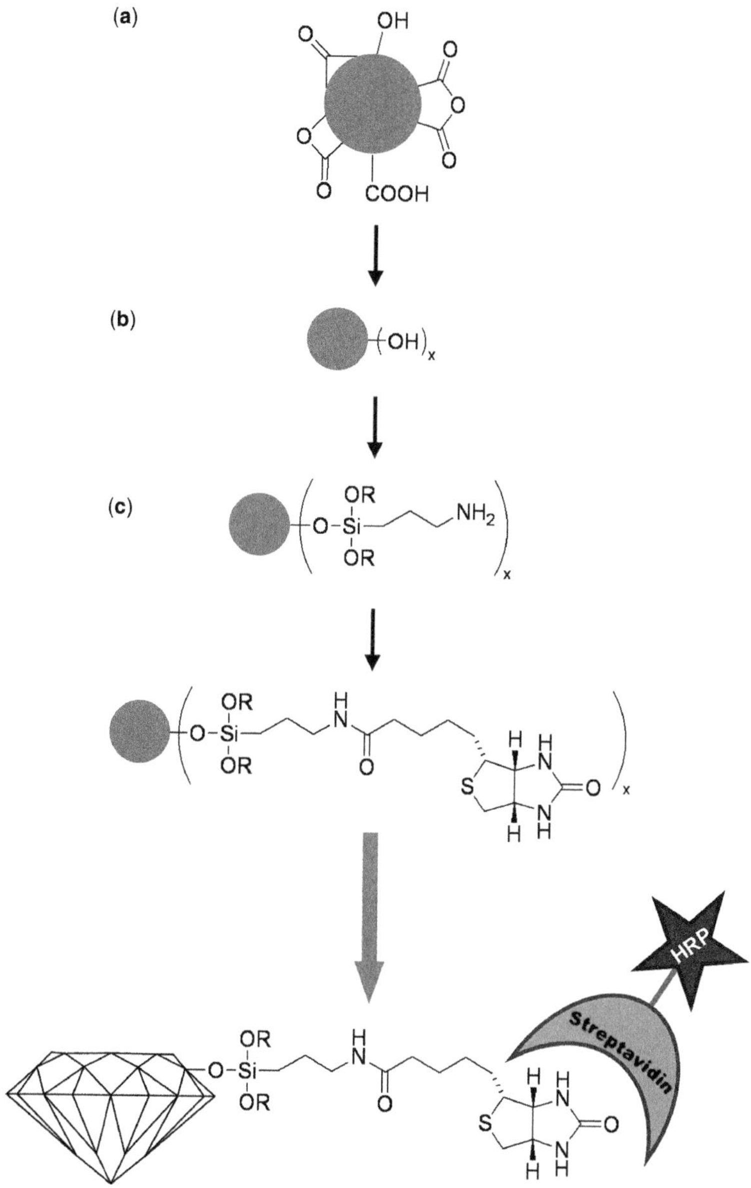

Figure 25.1 Biotinylation of nanodiamond and recognition protocol with horseradish peroxidase-labeled streptavidin. Conditions and reagents: (a) $BH_3 \cdot THF$, THF, reflux, 72 h; (b) APTMS, acetone, room temperature, 16 h; (c) biotin, EDC, DMAP, CH_2Cl_2, 0°C → room temperature, 65 h.[11]

for instance) and the comparatively long decay times of the NV emission (approximately 20 ns).[19] Of course, not only the intrinsic fluorescence of pristine NDs or processed FNDs can be utilized for signaling, but conjugation of other emitters to the particles is also possible as has been shown in a first report on the covalent attachment

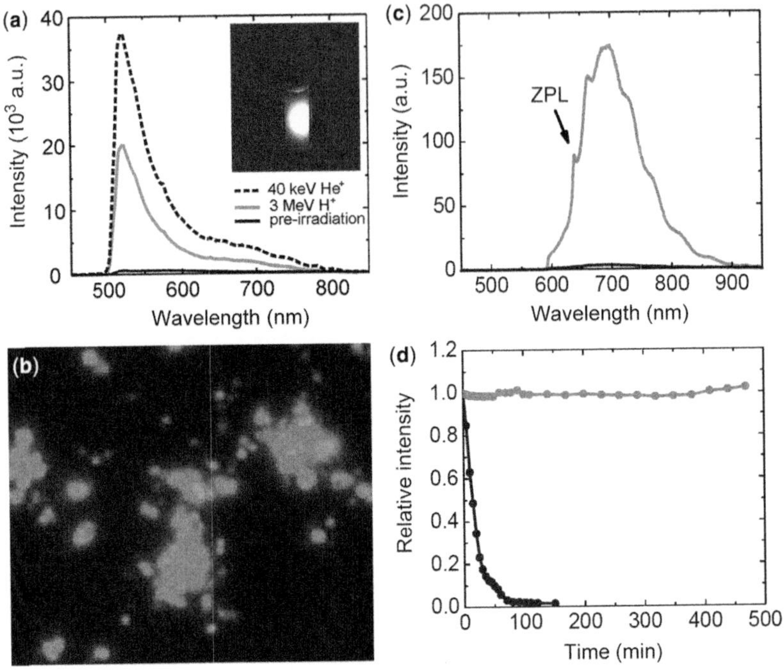

Figure 25.2 (a) Fluorescence spectra of 350 nm gFNDs prepared by either 40 keV He^+ or 3 MeV H^+ irradiation, followed by thermal annealing at 800°C and surface oxidation at 490°C. Both spectra were obtained for gFND suspensions with the same concentration (1 mg mL^{-1}) by laser excitation at λ_{ex} = 473 nm and emission collection at λ_{em} > 500 nm. Inset: a photo showing the fluorescence image of a cuvette containing 350 nm gFND suspension excited by a 473 nm laser. (b) Epifluorescence images of fluorescent ND (obtained with 40× objective). (c) Fluorescence spectra of annealed nanodiamonds with (grey) or without (black) proton beam irradiation. The excitation was made at 510 to 560 nm, and the emission was collected at a wavelength of >590 nm. (d) Photostability tests of fluorescent ND (grey) and fluorescent polystyrene nanospheres (black) excited under the same conditions. The fluorescence intensity was obtained by integrating over the wavelength range of 590 to 900 nm for each sample.[12,17] (Reprinted with permission from S.-J. Yu et al., *J. Am. Chem. Soc.* **2005**, *127*, 17604–17605. Copyright 2005 American Chemical Society; T.-L. Wee et al., *Diam. Relat. Mat.*, **2009**, *18*, 567–573. Copyright 2009 Elsevier.)

of a fluorescent dye, pyrene, through a short spacer to ND.[20] More insight into functionalized ND and FND materials can be found in several recent reviews.[14,21,22] Finally, besides all the favorable properties it is not unimportant for the potential commercial success of a hybrid material that methods to synthesize ND at industrial scale have been established.[23]

25.3 ORGANICALLY MODIFIED NANOSTRUCTURED CARBON MATERIALS

Whereas nanodiamond can be considered as an analog to solid nanoparticles, mesoporous carbon materials such as the CMK materials[24] introduced in Chapter 2 are the carbonic twins of mesoporous silica materials. Drawing from the rich knowledge on the functionalization of more traditional forms of carbon such as activated carbon, glassy carbon, or carbon nanotubes, a number of organically modified mesoporous carbon substrates have also been prepared in recent years.[25] The use and preparation of porous carbon became readily accessible when Ryoo et al. demonstrated in 1999 that mesoporous silica can be used as a mold for the preparation of porous carbon by simply filling the mesopores with a carbon precursor such as sucrose which could be processed to obtain, after further elimination of the silica scaffolding, the corresponding mesoporous carbon derivative.[26] Besides mesoporous silica, aluminosilica,[27] colloidal particles,[28,29] and crystals[30] have also been used as hard templates for the preparation of porous carbon. Recently, some methods involving the direct synthesis using surfactant and block-copolymer supramolecular templates have also been reported.[31,32] In analogy to mesoporous silica surfaces, porous carbon can be functionalized with different organic groups in order to fine-tune the interaction with different guest molecules.[33,34]

Because of their biocompatibility, chemical stability, high thermal and electrical conductivity, sorption ability, tuneable surfaces area, pore-size distribution and straightforward functionalization chemistry, porous carbons have found application in diverse topical areas such as sensors,[35] fuel cells,[36,37] hydrogen storage,[38] and sorption.[39–41] One particular property that distinguishes porous carbon from porous silica materials is the electrical conductivity of the former that has no counterpart in siliceous-based scaffoldings. This feature opens the route for certain applications

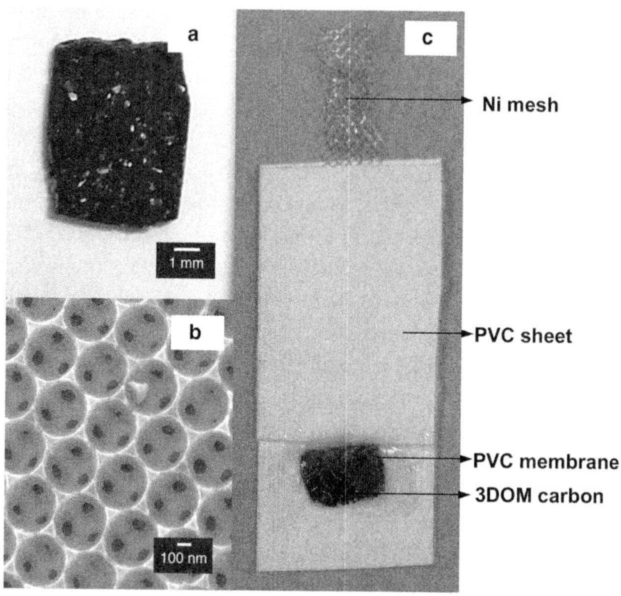

Figure 25.3 (a) Photograph of monolithic porous carbon. (b) SEM image of porous carbon. (c) Photograph of porous carbon-contacted ISE.[42] (Reprinted with permission from C.-Z. Lai et al., *Anal. Chem.* **2007**, *79*, 4621–4626. Copyright 2007 American Chemical Society.)

such as the preparation of ion-selective electrodes (ISEs) with high ionic and electric conductivity. For instance, three-dimensionally ordered porous carbon has been used as an intermediate layer between an ionophore-doped polymeric membrane and a metal contact in order to prepare new solid contact ISEs (SC-ISEs) that show excellent indication performance and a very low long term drift of the potential (Fig. 25.3).[42]

However, despite the growing use of this type of carbon material to enhance electrode performance, hydrogen storage, or other applications mentioned above, the development of hybrid mesoporous carbon materials with supramolecular functionalities has not seen much activity yet. Since this interesting and biocompatible support platform offers plenty of room for exploration in the context of this book, progress is soon expected here.

25.4 ORGANICALLY MODIFIED NANOSTRUCTURED SILICON CARBIDE

Leaving all-carbon for mixed carbon materials, the next section focuses on silicon carbide (SiC) and related materials. Like ND or mesoporous carbon, SiC is also characterized by favorable physical and mechanical properties such as low density, high strength, high thermal conductivity, stability at high temperature, low thermal expansion, high refractive index, wide (tunable) band gap, and chemical inertness.[43] In analogy to ND, SiC is biocompatible and nontoxic.[44,45] Since the discovery of SiC in the 1880s/1890s,[46] carbon and silicon building blocks have been assembled in almost limitless ways in the materials sciences, including more than 200 polymorphic forms or polytypes of SiC.[47] In recent years, especially advances in thin-film technology have fuelled the research on new forms of SiC that can be exploited in applications relating to nanotechnology. Accordingly, SiC materials of reduced dimension have been developed lately, including nanotubes,[48] nanowires,[49] and nanocrystals,[50] the latter of which can be considered as analogs of the well-known semiconductor quantum dots. Like for these relatives, SiC nanoparticles also show luminescence that obeys similar confinement effects. However, due to the different chemical and electronic nature, the photoluminescence is found between about 380 and 520 nm upon excitation in the UV.[51] Whereas these SiC nanocrystals are stoichiometric clusters, nonstoichiometric variants have also been reported, leading eventually to heterofullerenes.[43] Porous SiC materials have also been known for many years,[52] and have been used for instance as channels in artificial membranes for protein separation.[53]

In contrast to this extremely rich inorganic chemistry of SiC, approaches to modify its surface with organic or biomolecular groups have been very scarce.[54] Only in 2005, Cicero and Catellani embarked on a theoretical study to explore some of the basic possibilities of surface functionalization,[55] basically triggered by their previous study a year earlier in which they found that for SiC, irrespective of coverage, a silicon-terminated surface is highly hydrophilic whereas a carbon-terminated surface is highly hydrophobic.[56] The latter suggests that by controlling the terminating atomic layer on patterns or islands of SiC substrates, nanoscopically structured sensing surfaces with extreme yet well-defined properties might result. However, the first

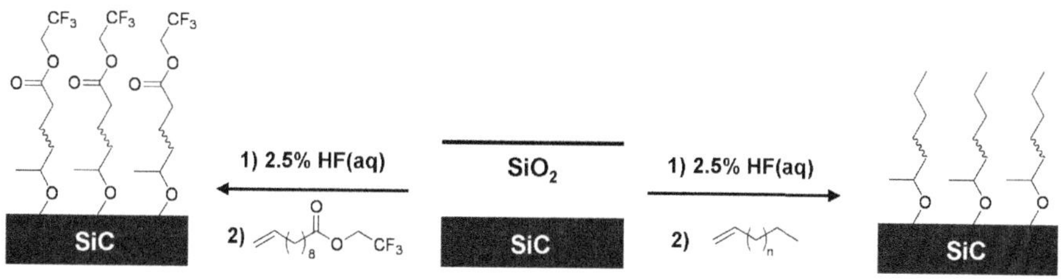

Figure 25.4 Functionalization of SiC with alkyl and ester groups.[57]

report on the covalent functionalization of SiC with organic groups, leading to alkyl- and ester-terminated SiC surfaces, has only been published in 2008 (Fig. 25.4).[57]

Combining aspects of carbon and silicon chemistry while at the same time expanding the tool box of the periodic table, recently, a first report on the templated synthesis of mesoporous silicon oxycarbide (SiOC) and silicon carbonitride (SiCN) as analogs of the well-known mesoporous silica materials discussed in many chapters of the book has appeared (Fig. 25.5),[58] opening an even wider horizon for the exploration of SiC-related nanomaterials in the fields covered in this book.

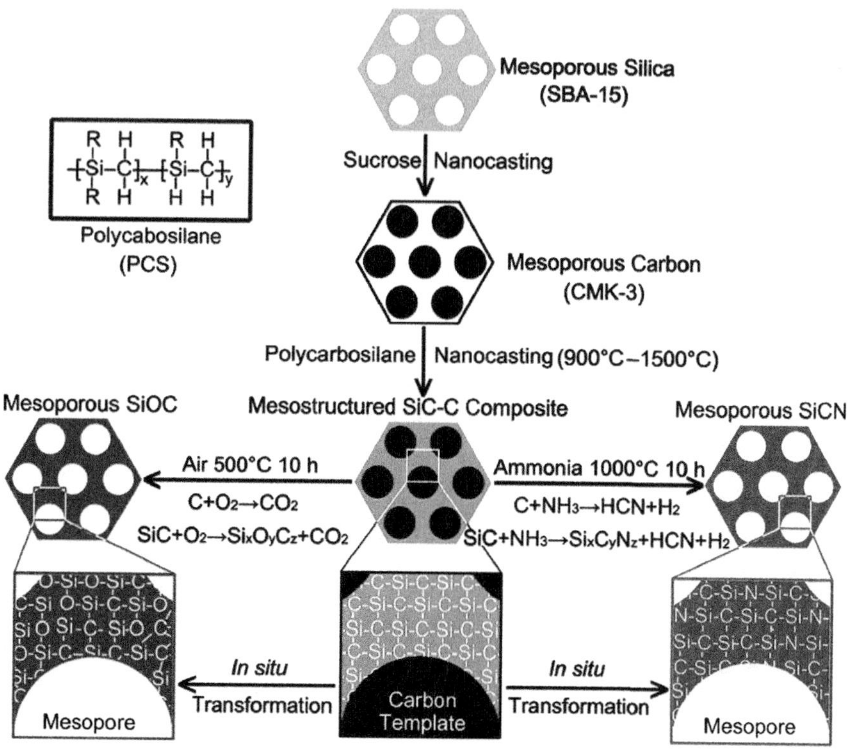

Figure 25.5 Synthetic approach to mesoporous SiC-based ceramics.[58] (Reprinted with permission from Y. Shi et al., *Chem. Mater.* **2007**, *19*, 1761–1771. Copyright 2007 American Chemical Society.)

25.5 INTRINSICALLY LUMINESCENT METAL ORGANIC FRAMEWORKS (MOFs)

As is discussed in Chapter 7 of this book, these highly porous hybrid materials play a significant role in separation, storage, and heterogeneous catalysis yet have been introduced to chemical sensing only to a minor degree. Recent advances in MOF design, however, lead to the assumption that progress lies ahead. One of the problems in utilizing MOFs for chemical sensing is signal generation. The coordination polymer network usually does not contain a species that undergoes a distinct change of a molecular or atomic property upon interaction with an analyte and can transduce this change to the macroscopic world through a signal measureable by a device or instrument. Since the surface modification of MOFs with organic dyes or even the implementation of such dyes into the polymer is not straightforward, another approach seems more promising and has indeed been realized lately: the implementation of atomic/ionic rare earth emitters. The trick here is to conceive a specific kind of interaction between the luminescent center and the target analyte so that the MOF stays intact yet the photophysical properties of the emitter centers are perturbed. At the same time, the organic component of the MOF should efficiently sensitize the rare earth luminescence. The first publications dealing with lanthanide-containing MOFs date only a decade back.[59] However, it was not until 2007 that attempts to use lanthanide-based MOFs for sensing have been reported, in this particular case on the detection of Zn^{2+}.[60] Here, the authors achieved the indication of the analyte not by monitoring the near infrared f-f luminescence of the Nd^{3+} metal ion employed in the framework, but they observed the changes in a transition at approximately 350 nm stemming from oxide-to-Nd^{3+} charge-transfer transitions that were modulated in the presence of Zn^{2+} and led to enhanced emission (Fig. 25.6).

In a second example, Chen et al. implemented pyridine groups into the coordination polymer in such a way that the pyridine nitrogen atoms did not participate in

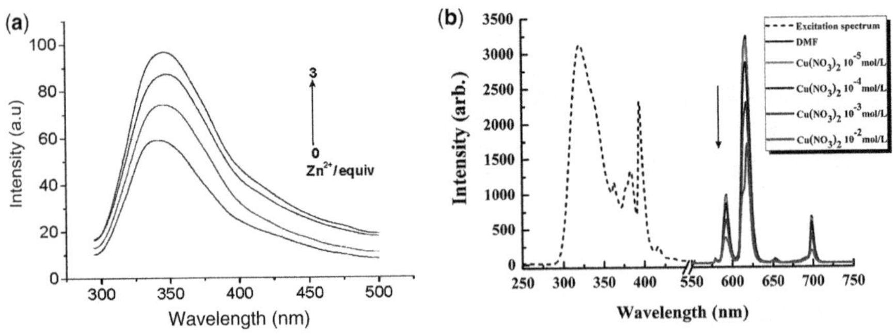

Figure 25.6 (a) Emission spectra of $\{[Nd_4(ox)_4(NO_3)_2(OH)_2(H_2O)_2]\cdot 5H_2O\}_n$ (ox = oxalate) in DMSO (10^{-3} M) at room temperature (excited at 285 nm) upon the addition of approximately 0 to 3 equiv of Zn^{2+} ions, respectively, as indicated by the arrow. (b) Excitation (- - - -) and PL spectra (-) of solid $[Eu(pdc)_{1.5}]\cdot$DMF activated in DMF solutions of $Cu(NO_3)_2$ at different concentrations (excited and monitored at 321 nm and 618 nm, respectively).[60,61] (Reprinted with permission from L.-Z. Zhang et al., *Inorg. Chem.* **2007**, *46*, 622–624. Copyright 2007 American Chemical Society; B. Chen et al., *Angew. Chem. Int. Ed.* **2009**, *48*, 500–503. Copyright Wiley-VCH Verlag GmbH & Co. KGaA.)

the network architecture, but pointed into the channels of the MOF.[61] They used pyridine-3,5-dicarboxylate (pdc) and the Eu^{3+} ions clearly favored to bind to the Lewis acidic carboxylate sites upon framework formation. After diffusion into the pores, interaction of Cu^{2+} and other transition metal ions that show a preference for the Lewis basic pyridine nitrogen atom then leads to an interaction with the pyridine moieties, entailing a quenching of luminescence. According to the authors, this quenching does not arise from the commonly found quenching interactions of open d shell metal ions with fluorophores through electron or energy transfer pathways,[62] but simply from a shift in the absorption spectrum of pyridine-3,5-dicarboxylate upon target metal coordination that leads to poorer sensitization of the Eu^{3+} luminescence.

25.6 NANOMATERIAL-CONFINED IONIC LIQUIDS

Whereas at least a few proofs of principle on using lanthanide-based MOFs for sensing applications have been reported (see the last section), Section 25.6 is even more speculative since it brings together two concepts, the combination of which has to the best of our knowledge not been achieved yet, nanomaterial-confined ionic liquids and sensing. Room temperature ionic liquids (ILs) have several unique properties that render them promising especially in sensing applications requiring separation steps, including high thermal stability, tuneable viscosity, and good miscibility with water and organic solvents.[63] These features have been exploited for the separation and extraction of various metal ions, in particular heavy and transition metal ions, and many organic compounds ranging from BTEX and PAHs to charged drugs and DNA. These applications have been realized mostly in either liquid chromatography (LC) or capillary electrophoresis (CE) systems, yet examples describing the use of ILs in fluorescence-based sensing have also been reported.[64] On the other hand, recalling the almost limitless functionalization chemistry of (especially) mesoporous silica supports, see for example, Chapters 3, 16, or 18 in this book, it is probably not surprising that the first steps toward the realization of the concept of supported ionic liquids, that is, ordered mesoporous silicas containing covalently linked ionic species typically used in ILs, were published some years ago (Fig. 25.7a).[65] Moreover, the use of

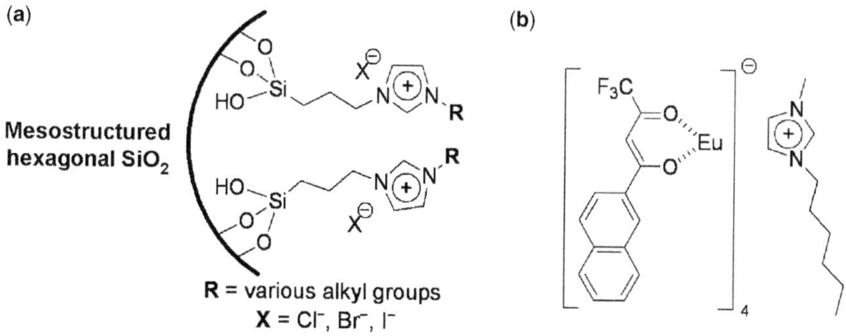

Figure 25.7 (a) First approach to supported ionic liquids.[65] (b) Components of the lanthanide complex-IL mixture employed for loading the sol–gel pores of brightly emissive ionogel monoliths.[68]

such functionalized silica materials in gated chemical sensing applications has also been recently accomplished (see also Chapter 19).[66,67] These materials can be seen as nanosized analogs of LC or CE columns yet still lack an active element for signaling. As the latter unit, however, again lanthanide ions might be a good starting option. Recently, studies by Lunstroot et al.[68] have shown that dissolving lanthanide(III) complexes in the ionic liquid 1-hexyl-3-methylimidazolium bis(trifluoromethylsulfonyl)-imide followed by confinement of the lanthanide-doped IL mixtures in the pores of a nanoporous silica network yielded highly emissive IL-silica ionogel hybrids.[68] The hybrids were obtained by a nonhydrolytic sol–gel route, employing tetramethoxy-silane (TMOS) and methyltrimethoxysilane (MTMOS), formic acid, and the pure ionic liquid. After monolith formation, the IL was removed by Soxhlet extraction. The luminescent complex was then introduced into the porous network by dipping the IL-extracted gel monoliths into a solution of the lanthanide complex in the same IL (Fig. 25.7b). The important findings were that the confinement into the hybrids did not change the luminescence properties of the rare earth emitters, that is, the luminescence lifetimes in the ILs and the ionogels were virtually identical, producing luminescent, ion-conductive organic–inorganic hybrid materials with facile color adaption.[69] The combination of the latter strategy with the covalent attachment of IL units to mesoporous supports and the application of such future materials in chemical sensing is now eagerly awaited.

25.7 CONCLUSIONS

Within the scope and the space limitations of the book, it was of course not possible to encompass the field of the supramolecular chemistry of organic–inorganic hybrid materials in its entirety. Thus, the choice of the five specialized research directions that in our opinion can gain importance in the field in the near future briefly touched here is also only a subjective one. Following other paths to elaborate, for instance fluorescent magnetic nanostructures,[70] particle-reinforced polymers based on metal-oxo clusters,[71] or novel silica morphologies such as silica nanohelices[72] deserves equally strong attention. Finally, during the process of preparing this book, it occurred to us that the partly very diverse strategies followed by most of the researchers engaged in this highly cross-disciplinary research area are perhaps not adequately described by the terms *supramolecular chemistry* and *hybrid (nano)materials* alone anymore, but that a suggestive term put forward by R. Backov lately might subsume all these different concepts, strategies and approaches even better: *integrative chemistry*.[73,74] The future will decide.

REFERENCES

1. P. S. DeCarli, J. C. Jamieson, *Science* **1961**, *133*, 1821–1822.
2. V. M. Titov, V. F. Anisichkin, I. Y Mal'kov, *Combust. Explos.* **1989**, *25*, 372–379.
3. N. R. Greiner, D. S. Phillips, J. D. Johnson, F. Volk, *Nature* **1988**, *333*, 440–442.
4. V. V. Danilenko, *Phys. Solid State* **2004**, *46*, 595–599.

5. V. F. Loktev, V. I. Makal'skii, I. V. Stoyanova, A. V. Kalinkin, V. A. Likholobov, V. N. Mit'kin, *Carbon* **1991**, *29*, 817–819.
6. T. Jiang, K. Xu, *Carbon* **1995**, *33*, 1663–1671.
7. G. U. Ostrovidova, A. V. Makeev, A. V. Biryukov, S. K. Gordeev, *Mater. Sci. Eng. C* **2003**, *23*, 377–381.
8. J. Opitz, A. Pohl, J. Schreiber, M. Mkandawire, U. Krause-Buchholz, G. Rodei, W. Pompe, T. Gubarevich, V. Lapina, *VDI Rep.* **2008**, *2027*, 105–110.
9. L.-C. L. Huang, H. C. Chang, *Langmuir* **2004**, *20*, 5879–5884.
10. A. Krüger, Y. Liang, G. Jarre, J. Stegk, *J. Mater. Chem.* **2006**, *16*, 2322–2328.
11. A. Krueger, J. Stegk, Y. Liang, L. Lu, G. Jarre, *Langmuir* **2008**, *24*, 4200–4204.
12. S.-J. Yu, M.-W. Kang, H.-C. Chang, K.-M. Chen, Y.-C. Yu, *J. Am. Chem. Soc.* **2005**, *127*, 17604–17605.
13. S. Vial, C. Mansuy, S. Sagan, T. Irinopoulou, F. Burlina, J.-P. Boudou, G. Chassaing, S. Lavielle, *ChemBioChem* **2008**, *9*, 2113–2119.
14. K. B. Holt, *Philos. Trans. R. Soc. A* **2007**, *365*, 2845–2861.
15. F. Treussart, V. Jacques, E. Wu, T. Gacoin, P. Grangier, J.-F. Roch, *Physica B* **2006**, *376–377*, 926–929.
16. H.-C. Chang, K. Chen, S. Kwok, *Astrophys. J.* **2006**, *639*, L63–L66.
17. T.-L. Wee, Y.-W. Mau, C.-Y. Fang, H.-L. Hsu, C.-C. Han, H.-C. Chang, *Diam. Relat. Mat.* **2009**, *18*, 567–573.
18. Y.-R. Chang, H.-Y. Lee, K. Chen, C.-C. Chang, D.-S. Tsai, C.-C. Fu, T.-S. Lim, Y.-K. Tzeng, C.-Y. Fang, C.-C. Han, H.-C. Chang, W. Fann, *Nature Nanotechnol.* **2008**, *3*, 284–288.
19. C.-C. Fu, H.-Y. Lee, K. Chen, T. S. Lim, H.-Y. Wu, P.-K. Lin, P.-K. Wie, P.-H. Tsao, H.-C. Chang, W. Fann, *Proc. Natl. Acad. Sci. U.S.A.* **2007**, *104*, 727–732.
20. W. S. Yeap, S. Chen, K. P. Loh, *Langmuir* **2009**, *25*, 185–191.
21. A. Krueger, *Chem. Eur. J.* **2008**, *14*, 1382–1390.
22. V. Vaijayanthimala, H. C. Chang, *Nanomedicine* **2009**, *4*, 47–55.
23. V. Y. Dolmatov, M. V. Veretennikova, V. A. Marchukov, V. G. Sushchev, *Phys. Solid State* **2004**, *46*, 611–615.
24. H. J. Shin, R. Ryoo, M. Kruk, M. Jaroniec, *Chem. Commun.* **2001**, 349–350.
25. A. Stein, Z. Wang, M. A. Fierke, *Adv. Mater.* **2009**, *21*, 265–293.
26. R. Ryoo, S. H. Joo, S. Jun, *J. Phys. Chem. B* **1999**, *103*, 7743–7746.
27. H. Yang, D. Zhao, *J. Mater. Chem.* **2005**, *15*, 1217–1231.
28. K. P. Gierszal, M. Jaroniec, *J. Am. Chem. Soc.* **2006**, *128*, 10026–10027.
29. W. M. Qiao, Y. Song, S. H. Hong, S. Y. Lim, S. H. Yoon, Y. Korai, I. Mochida, *Langmuir* **2006**, *22*, 3791–3797.
30. K. T. Lee, J. C. Lytle, N. S. Ergang, S. M. Oh, A. Stein, *Adv. Funct. Mater.* **2005**, *15*, 547–553.
31. Y. Meng, D. Gu, F. Zhang, Y. Shi, H. Yang, Z. Li, C. Yu, B. Tu, D. Zhao, *Angew. Chem. Int. Ed.* **2005**, *44*, 7053–7059.
32. Y. Wan, X. Qian, N. Jia, Z. Wang, H. Li, D. Zhao, *Chem. Mater.* **2008**, *20*, 1012–1018.
33. C. Liang, Z. Li, S. Dai, *Angew. Chem. Int. Ed.* **2008**, *47*, 3696–3717.
34. Y. Wan, Y. Shi, D. Zhao, *Chem. Mater.* **2008**, *20*, 932–945.
35. A. Walcarius, A. Kuhn, *Trends Anal. Chem.* **2008**, *27*, 593–603.
36. S. H. Joo, S. J. Choi, I. Oh, J. Kwak, Z. Lin, O. Terasaki, R. Ryoo, *Nature* **2001**, *412*, 169–172.
37. M.-L. Lin, C.-C. Huang, M.-Y. Lo, C.-Y. Mou, *J. Phys. Chem. C* **2008**, *112*, 867–873.
38. R. Gadiou, S.-E. Saadallah, T. Piquero, P. David, J. Parmentier, C. Vix-Guterl, *Microporous Mesoporous Mater.* **2005**, *79*, 121–128.
39. A. Vinu, C. Streb, V. Murugesan, M. Hartmann, *J. Phys. Chem. B* **2003**, *107*, 8297–8299.
40. A. Vinu, K. Z. Hossian, P. Srinivasu, M. Miyahara, S. Anandan, N. Gokulakrishnan, T. Mori, K. Ariga, V. V. Balasubramanian, *J. Mater. Chem.* **2007**, *17*, 1819–1825.
41. A. Vinu, M. Miyahara, V. Sivamurugan, T. Mori, K. Ariga, *J. Mater. Chem.* **2005**, *15*, 5122–5127.
42. C.-Z. Lai, M. A. Fierke, A. Stein, P. Bühlmann, *Anal. Chem.* **2007**, *79*, 4621–4626.
43. P. Mélinon, B. Masenelli, F. Tournus, A. Perez, *Nature Mater.* **2007**, *6*, 479–490.

44. A. Naji, M. F. Harmand, *Biomaterials* **1991**, *12*, 690–694.
45. S. Santavirta, M. Takagi, L. Nordsletten, A. Anttila, R. Lappalainen, Y. T. Konttinen, *Arch. Orthop. Trauma Surg.* **1998**, *118*, 89–91.
46. C. F. Mabery, *J. Am. Chem. Soc.* **1900**, *22*, 706–707.
47. G. C. Trigunayat, *Solid State Ion.* **1991**, *48*, 3–70.
48. W. Han, S. Fan, Q. Li, W. Liang, B. Gu, D. Yu, *Chem. Phys. Lett.* **1997**, *265*, 374–378.
49. J. Q. Hu, Q. K. Lu, K. B. Tang, B. Deng, R. R. Jiang, Y. T. Qian, W. C. Yu, G. E. Zhou, X. M. Liu, J. X. Wu, *J. Phys. Chem. B* **2000**, *104*, 5251–5254.
50. Y. Kamlag, A. Goossens, I. Colbeck, J. Schoonman, *Chem. Vapor Depos.* **2003**, *9*, 125–129.
51. J. Y. Fan, X. L. Wu, R. Kong, T. Qiu, G. S. Huang, *Appl. Phys. Lett.* **2005**, *86*, 171903.
52. J. Y. Fan, X. L. Wu, P. K. Chu, *Prog. Mater. Sci.* **2006**, *51*, 983–1031.
53. A. J. Rosenbloom, D. M. Sipe, Y. Shishkin, Y. Ke, R. P. Devaty, W. J. Choyke, *Biomed. Microdev.* **2004**, *6*, 261–267.
54. R. Yakimova, R. M. Petoral, Jr, G. R. Yazdi, C. Vahlberg, A. Lloyd Spetz, K. Uvdal, *J. Phys. D Appl. Phys.* **2007**, *40*, 6435–6442.
55. G. Cicero, A. Catellani, *J. Chem. Phys.* **2005**, *122*, 214716.
56. G. Cicero, A. Catellani, G. Galli, *Phys. Rev. Lett.* **2004**, *93*, 016102.
57. M. Rosso, A. Arafat, K. Schroën, M. Giesbers, C. S. Roper, R. Maboudian, H. Zuilhof, *Langmuir* **2008**, *24*, 4007–4012.
58. Y. Shi, Y. Wan, Y. Zhai, R. Liu, Y. Meng, B. Tu, D. Zhao, *Chem. Mater.* **2007**, *19*, 1761–1771.
59. T. M. Reineke, M. Eddaoudi, M. Fehr, D. Kelley, O. M. Yaghi, *J. Am. Chem. Soc.* **1999**, *121*, 1651–1657.
60. L.-Z. Zhang, W. Gu, B. Li, X. Liu, D.-Z. Liao, *Inorg. Chem.* **2007**, *46*, 622–624.
61. B. Chen, L. Wang, Y. Xiao, F. R. Fronczek, M. Xue, Y. Cui, G. Qian, *Angew. Chem. Int. Ed.* **2009**, *48*, 500–503.
62. K. Rurack, *Spectrochim. Acta Part A* **2001**, *57*, 2161–2195.
63. J. Liu, J. Å. Jönsson, G. Jiang, *Trends Anal. Chem.* **2005**, *24*, 20–27.
64. S. Pandey, *Anal. Chim. Acta* **2006**, *556*, 38–45.
65. B. Gadenne, P. Hesemann, J. J. E. Moreau, *Chem. Commun.* **2004**, 1768–1769.
66. C. Coll, R. Martínez-Máñez, M. D. Marcos, F. Sancenón, J. Soto, *Angew. Chem. Int. Ed.* **2007**, *46*, 1675–1678.
67. C. Coll, R. Casasús, E. Aznar, M. D. Marcos, R. Martínez-Máñez, F. Sancenón, J. Soto, P. Amorós, *Chem. Commun.* **2007**, 1957–1959.
68. K. Lunstroot, K. Driesen, P. Nockemann, C. Görller-Walrand, K. Binnemans, S. Bellayer, J. Le Bideau, A. Vioux, *Chem. Mater.* **2006**, *18*, 5711–5715.
69. K. Lunstroot, K. Driesen, P. Nockemann, K. Van Hecke, L. Van Meervelt, C. Görller-Walrand, K. Binnemans, S. Bellayer, L. Viau, J. Le Bideau, A. Vioux, *Dalton Trans.* **2009**, 298–306.
70. A. Quarta, R. Di Corato, L. Manna, A. Ragusa, T. Pellegrino, *IEEE Trans. Nanobiosci.* **2007**, *6*, 298–308.
71. F. Faccini, H. Fric, U. Schubert, E. Wendel, O. Tsetsgee, K. Müller, H. Bertagnolli, A. Venzo, S. Gross, *J. Mater. Chem.* **2007**, *17*, 3297–3307.
72. T. Delclos, C. Aimé, E. Pouget, A. Brizard, I. Huc, M.-H. Delville, R. Oda, *Nano Lett.* **2008**, *8*, 1929–1935.
73. R. Backov, *Soft Matter* **2006**, *2*, 452–464.
74. E. Prouzet, S. Ravaine, C. Sanchez, R. Backov, *New J. Chem.* **2008**, *32*, 1284–1299.

Appendix 1

Reviews found in each of the topic clusters shown in Figure 24.2.

BIOSENSING

Castaneda MT; Alegret S; Merkoci A, Electrochemical sensing of DNA using gold nanoparticles, *Electroanalysis*, **2007**, *19*, 743–753.

Chen D; Wang G; Li JH, Interfacial bioelectrochemistry: Fabrication, properties and applications of functional nanostructured biointerfaces, *J. Phys. Chem. C*, **2007**, *111*, 2351–2367.

Hernández-Santos D; González-García MB; Costa García A, Metal-nanoparticles based electroanalysis, *Electroanalysis*, **2002**, *14*, 1225–1235.

Jain KK, Nanotechnology in clinical laboratory diagnostics, *Clin. Chim. Acta*, **2005**, *358*, 37–54.

Katz E; Willner I, Integrated nanoparticle-biomolecule hybrid systems: Synthesis, properties, and applications, *Angew. Chem. Int. Ed.*, **2004**, *43*, 6042–6108.

Pumera M; Sanchez S; Ichinose I; Tang J, Electrochemical nanobiosensors, *Sens. Actuators B*, **2007**, *123*, 1195–1205.

Rosi NL; Mirkin CA, Nanostructures in biodiagnostics, *Chem. Rev.*, **2005**, *105*, 1547–1562.

Wang J, Nanomaterial-based amplified transduction of biomolecular interactions, *Small*, **2005**, *1*, 1036–1043.

Willner I; Baron R; Willner B, Integrated nanoparticle-biomolecule systems for biosensing and bioelectronics, *Biosens. Bioelectron.*, **2007**, *22*, 1841–1852.

Willner I; Basnar B; Willner B, Nanoparticle-enzyme hybrid systems for nanobiotechnology, *FEBS J.*, **2007**, *274*, 302–309.

QUANTUM DOTS

Costa-Fernández JM; Pereiro R; Sanz-Medel A, The use of luminescent quantum dots for optical sensing, *TrAC Trends Anal. Chem.*, **2006**, *25*, 207–218.

Lin CAJ; Liedl T; Sperling RA; Fernández-Argüelles MT; Costa-Fernández JM; Pereiro R; Sanz-Medel A; Chang WH; Parak WJ, Bioanalytics and biolabeling with semiconductor nanoparticles (quantum dots), *J. Mater. Chem.*, **2007**, *17*, 1343–1346.

Murphy CJ, Optical sensing with quantum dots, *Anal. Chem.*, **2002**, 520A–526A.

Pinaud F; Michalet X; Bentolila LA; Tsay JM; Doose S; Li JJ; Iyer G; Weiss S, Advances in fluorescence imaging with quantum dot bio-probes, *Biomaterials*, **2006**, *27*, 1679–1687.

The Supramolecular Chemistry of Organic–Inorganic Hybrid Materials. Edited by Knut Rurack and Ramón Martínez-Máñez

Smith AM; Nie SM, Chemical analysis and cellular imaging with quantum dots, *Analyst*, **2004**, *129*, 672–677.

CARBON NANOTUBES

Allen BL; Kichambare PD; Star A, Carbon nanotube field-effect-transistor-based biosensors, *Adv. Mater.*, **2007**, *19*, 1439–1451.

Daniel S; Rao TP; Rao KS; Rani SU; Naidu GRK; Lee HY; Kawai T, A review of DNA functionalized/grafted carbon nanotubes and their characterization, *Sens. Actuators B*, **2007**, *122*, 672–682.

Lin Y; Taylor S; Li HP; Fernando KAS; Qu LW; Wang W; Gu LR; Zhou B; Sun YP, Advances toward bioapplications of carbon nanotubes, *J. Mater. Chem.*, **2004**, *14*, 527–541.

Prato M; Kostarelos K; Bianco A, Functionalized carbon nanotubes in drug design and discovery, *Acc. Chem. Res.*, **2008**, *41*, 60–68.

Sun YP; Fu KF; Lin Y; Huang WJ, Functionalized carbon nanotubes: Properties and applications, *Acc. Chem. Res.*, **2002**, *35*, 1096–1104.

Tasis D; Tagmatarchis N; Bianco A; Prato M, Chemistry of carbon nanotubes, *Chem. Rev.*, **2006**, *106*, 1105–1136.

SUPRAMOLECULAR CHEMISTRY AND SENSING

Basabe-Desmonts L; Reinhoudt DN; Crego-Calama M, Design of fluorescent materials for chemical sensing, *Chem. Soc. Rev.*, **2007**, *36*, 993–1017.

Crespo-Biel O; Ravoo BJ; Reinhoudt DN; Huskens J, Noncovalent nanoarchitectures on surfaces: From 2D to 3D nanostructures, *J. Mater. Chem.*, **2006**, *16*, 3997–4021.

Daniel MC; Astruc D, Gold nanoparticles: Assembly supramolecular chemistry, quantum-size-related properties, and applications toward biology, catalysis, and nanotechnology, *Chem. Rev.*, **2004**, *104*, 293–346.

Descalzo AB; Martínez-Máñez R; Sancenón F; Hoffmann K; Rurack K, The supramolecular chemistry of organic-inorganic hybrid materials, *Angew. Chem. Int. Ed.*, **2006**, *45*, 5924–5948.

Drechsler U; Erdogan B; Rotello VM, Nanoparticles: Scaffolds for molecular recognition, *Chem.-Eur. J.*, **2004**, *10*, 5570–5579.

Ghosh SK; Pal T, Interparticle coupling effect on the surface plasmon resonance of gold nanoparticles: From theory to applications, *Chem. Rev.*, **2007**, *107*, 4797–4862.

Kelly KL; Coronado E; Zhao LL; Schatz GC, The optical properties of metal nanoparticles: The influence of size shape, and dielectric environment, *J. Phys. Chem. B*, **2003**, *107*, 668–677.

Mancin F; Rampazzo E; Tecilla P; Tonellato U, Self-assembled fluorescent chemosensors, *Chem.-Eur. J.*, **2006**, *12*, 1844–1854.

Prodi L, Luminescent chemosensors: From molecules to nanoparticles, *New J. Chem.*, **2005**, *29*, 20–31.

MOLECULAR MACHINES

Balzani V; Credi A; Silvi S; Venturi M, Artificial nanomachines based on interlocked molecular species: Recent advances, *Chem. Soc. Rev.*, **2006**, *35*, 1135–1149.

Cooke G, Electrochemical and photochemical control of host-guest complexation at surfaces, *Angew. Chem. Int. Ed.*, **2003**, *42*, 4860–4870.

Feringa BL, The art of building small: From molecular switches to molecular motors, *J. Org. Chem.*, **2007**, *72*, 6635–6652.

Kinbara K; Aida T, Toward intelligent molecular machines: Directed motions of biological and artificial molecules and assemblies, *Chem. Rev.*, **2005**, *105*, 1377–1400.

Kottas GS; Clarke LI; Horinek D; Michl J, Artificial molecular rotors, *Chem. Rev.*, **2005**, *105*, 1281–1376.

Saha S; Leung KCF; Nguyen TD; Stoddart JF; Zink JI, Nanovalves, *Adv. Funct. Mater.*, **2007**, *17*, 685–693.

Willner I; Basnar B; Willner B, From molecular machines to microscale motility of objects: Application as "smart materials", sensors, and nanodevices, *Adv. Funct. Mater.*, **2007**, *17*, 702–717.

HYBRID SILICA MATERIALS

Angelos S; Johansson E; Stoddart JF; Zink JI, Mesostructured silica supports for functional materials and molecular machines, *Adv. Funct. Mater.*, **2007**, *17*, 2261–2271.

Angelos S; Liong M; Choi E; Zink JI, Mesoporous silicate materials as substrates for molecular machines and drug delivery, *Chem. Eng. J.*, **2008**, *137*, 4–13.

Boettcher SW; Fan J; Tsung CK; Shi QH; Stucky GD, Harnessing the sol-gel process for the assembly of non-silicate mesostructured oxide materials, *Acc. Chem. Res.*, **2007**, *40*, 784–792.

Brinker CJ; Dunphy DR, Morphological control of surfactant-templated metal oxide films, *Curr. Opin. Colloid Interface Sci.*, **2006**, *11*, 126–132.

Fan J; Boettcher SW; Tsung CK; Shi Q; Schierhorn M; Stucky GD, Field-directed and confined molecular assembly of mesostructured materials: Basic principles and new opportunities, *Chem. Mater.*, **2008**, *20*, 909–921.

Ford DM; Simanek EE; Shantz DF, Engineering nanospaces: Ordered mesoporous silicas as model substrates for building complex hybrid materials, *Nanotechnology*, **2005**, *16*, S458–S475.

Fujita S; Inagaki S, Self-organization of organosilica solids with molecular-scale and mesoscale periodicities, *Chem. Mater.*, **2008**, *20*, 891–908.

Giri S; Trewyn BG; Lin VSY, Mesoporous silica nanomaterial-based biotechnological and biomedical delivery systems, *Nanomedicine*, **2007**, *2*, 99–111.

Hatton B; Landskron K; Whitnall W; Perovic D; Ozin GA, Past present, and future of periodic mesoporous organosilicas—The PMOs, *Acc. Chem. Res.*, **2005**, *38*, 305–312.

Hoffmann F; Cornelius M; Morell J; Fröba M, Silica-based mesoporous organic-inorganic hybrid materials, *Angew. Chem. Int. Ed.*, **2006**, *45*, 3216–3251.

Nicole L; Boissière C; Grosso D; Quach A; Sanchez C, Mesostructured hybrid organic-inorganic thin films, *J. Mater. Chem.*, **2005**, *15*, 3598–3627.

Sanchez C; Boissière C; Grosso D; Laberty C; Nicole L, Design, synthesis, and properties of inorganic and hybrid thin films having periodically organized nanoporosity, *Chem. Mater.*, **2008**, *20*, 682–737.

Slowing II; Trewyn BG; Giri S; Lin VSY, Mesoporous silica nanoparticles for drug delivery and biosensing applications, *Adv. Funct. Mater.*, **2007**, *17*, 1225–1236.

Soler-Illia GJAA; Innocenzi P, Mesoporous hybrid thin films: The physics and chemistry beneath, *Chem.-Eur. J.*, **2006**, *12*, 4478–4494.

Trewyn BG; Slowing II; Giri S; Chen HT; Lin VSY, Synthesis and functionalization of a mesoporous silica nanoparticle based on the sol-gel process and applications in controlled release, *Acc. Chem. Res.*, **2007**, *40*, 846–853.

Vallet-Regí M; Balas F; Arcos D, Mesoporous materials for drug delivery, *Angew. Chem. Int. Ed.*, **2007**, *46*, 7548–7558.

Vinu A; Hossain KZ; Ariga K, Recent advances in functionalization of mesoporous silica, *J. Nanosci. Nanotechnol.*, **2005**, *5*, 347–371.

Wan Y; Zhao DY, On the controllable soft-templating approach to mesoporous silicates, *Chem. Rev.*, **2007**, *107*, 2821–2860.

COORDINATION POLYMERS

Beatty AM, Open-framework coordination complexes from hydrogen-bonded networks: Toward host/guest complexes, *Coord. Chem. Rev.*, **2003**, *246*, 131–143.

BRAMMER L, Developments in inorganic crystal engineering, *Chem. Soc. Rev.*, **2004**, *33*, 476–489.

DESIRAJU GR, Crystal engineering: A holistic view, *Angew. Chem. Int. Ed.*, **2007**, *46*, 8342–8356.

FÉREY G; MELLOT-DRAZNIEKS C; SERRE C; MILLANGE F, Crystallized frameworks with giant pores: Are there limits to the possible?, *Acc. Chem. Res.*, **2005**, *38*, 217–225.

JANIAK C, Engineering coordination polymers towards applications, *Dalton Trans.*, **2003**, *74*, 2781–2804.

KITAGAWA S; KITAURA R; NORO S, Functional porous coordination polymers, *Angew. Chem. Int. Ed.*, **2004**, *43*, 2334–2375.

MASPOCH D; RUIZ-MOLINA D; VECIANA J, Old materials with new tricks: Multifunctional open-framework materials, *Chem. Soc. Rev.*, **2007**, *36*, 770–818.

ROBIN AY; FROMM KM, Coordination polymer networks with O- and N-donors: What they are, why and how they are made, *Coord. Chem. Rev.*, **2006**, *250*, 2127–2157.

ROSSEINSKY MJ, Recent developments in metal-organic framework chemistry: Design discovery, permanent porosity and flexibility, *Microporous Mesoporous Mater.*, **2004**, *73*, 15–30.

ROWSELL JLC; YAGHI OM, Metal-organic frameworks: A new class of porous materials, *Microporous Mesoporous Mater.*, **2004**, *73*, 3–14.

YAGHI OM; O'KEEFFE M; OCKWIG NW; CHAE HK; EDDAOUDI M; KIM J, Reticular synthesis and the design of new materials, *Nature*, **2003**, *423*, 705–714.

REVIEWS NOT CLUSTERED

ARIGA K; HILL JP; JI QM, Layer-by-layer assembly as a versatile bottom-up nanofabrication technique for exploratory research and realistic application, *Phys. Chem. Chem. Phys.*, **2007**, *9*, 2319–2340.

ARIGA K; NAKANISHI T; HILL JP, Self-assembled microstructures of functional molecules, *Curr. Opin. Colloid Interface Sci.*, **2007**, *12*, 106–120.

BAUGHMAN RH; ZAKHIDOV AA; DE HEER WA, Carbon nanotubes—the route toward applications, *Science*, **2002**, *297*, 787–792.

BRUNET E, Asymmetric induction under confinement, *Chirality*, **2002**, *14*, 135–143.

BUCK SM; KOO YEL; PARK E; XU H; PHILBERT MA; BRASUEL MA; KOPELMAN R, Optochemical nanosensor PEBBLEs: Photonic explorers for bioanalysis with biologically localized embedding, *Curr. Opin. Chem. Biol.*, **2004**, *8*, 540–546.

BURNS A; OW H; WIESNER U, Fluorescent core-shell silica nanoparticles: Towards "Lab on a Particle" architectures for nanobiotechnology, *Chem. Soc. Rev.*, **2006**, *35*, 1028–1042.

CALLIES M; QUÉRÉ D, On water repellency, *Soft Matter*, **2005**, *1*, 55–61.

COZZOLI PD; PELLEGRINO T; MANNA L, Synthesis, properties and perspectives of hybrid nanocrystal structures, *Chem. Soc. Rev.*, **2006**, *35*, 1195–1208.

DIAMOND D, Internet-scale sensing, *Anal. Chem.*, **2004**, *76*, 278A–286A.

DÍAZ-GARCÍA ME; BADÍA LAÍÑO R, Molecular imprinting in sol-gel materials: Recent developments and applications, *Microchim. Acta*, **2005**, *149*, 19–36.

EUSTIS S; EL-SAYED MA, Why gold nanoparticles are more precious than pretty gold: Noble metal surface plasmon resonance and its enhancement of the radiative and nonradiative properties of nanocrystals of different shapes, *Chem. Soc. Rev.*, **2006**, *35*, 209–217.

FEIZI T; FAZIO F; CHAI WG; WONG CH, Carbohydrate microarrays—a new set of technologies at the frontiers of glycomics, *Curr. Opin. Struct. Biol.*, **2003**, *13*, 637–645.

FENG W; PATEL SH; YOUNG MY; ZUNINO JL; XANTHOS M, Smart polymeric coatings—Recent advances, *Adv. Polym. Technol.*, **2007**, *26*, 1–13.

HANSEN DE, Recent developments in the molecular imprinting of proteins, *Biomaterials*, **2007**, *28*, 4178–4191.

HAO T, Electrorheological suspensions, *Adv. Colloid Interface Sci.*, **2002**, *97*, 1–35.

HE L; TOH CS, Recent advances in analytical chemistry—A material approach, *Anal. Chim. Acta*, **2006**, *556*, 1–15.

Holder E; Tessler N; Rogach AL, Hybrid nanocomposite materials with organic and inorganic components for opto-electronic devices, *J. Mater. Chem.*, **2008**, *18*, 1064–1078.

Holthoff EL; Bright FV, Molecularly imprinted xerogels as platforms for sensing, *Acc. Chem. Res.*, **2007**, *40*, 756–767.

Houston JE; Kim HI, Adhesion friction, and mechanical properties of functionalized alkanethiol self-assembled monolayers, *Acc. Chem. Res.*, **2002**, *35*, 547–553.

Huynh WU; Dittmer JJ; Teclemariam N; Milliron DJ; Alivisatos AP; Barnham KWJ, Charge transport in hybrid nanorod-polymer composite photovoltaic cells, *Phys. Rev. B*, **2003**, *67*, 115326.

Kendig MW; Buchheit RG, Corrosion inhibition of aluminum and aluminum alloys by soluble chromates, chromate coatings, and chromate-free coatings, *Corrosion*, **2003**, *59*, 379–400.

Lee SW; Chang WJ; Bashir R; Koo YM, "Bottom-up" approach for implementing nano/microstructure using biological and chemical interactions, *Biotechnol. Bioprocess Eng.*, **2007**, *12*, 185–199.

Lehn JM, Toward complex matter: Supramolecular chemistry and self-organization, *Proc. Natl. Acad. Sci. U. S. A.*, **2002**, *99*, 4763–4768.

Love JC; Estroff LA; Kriebel JK; Nuzzo RG; Whitesides GM, Self-assembled monolayers of thiolates on metals as a form of nanotechnology, *Chem. Rev.*, **2005**, *105*, 1103–1169.

Maenosono S; Okubo T; Yamaguchi Y, Overview of nanoparticle array formation by wet coating, *J. Nanopart. Res.*, **2003**, *5*, 5–15.

Martínez-Máñez R; Sancenón F, Chemodosimeters and 3D inorganic functionalised hosts for the fluoro-chromogenic sensing of anions, *Coord. Chem. Rev.*, **2006**, *250*, 3081–3093.

Matsushita Y, Creation of hierarchically ordered nanophase structures in block polymers having various competing interactions, *Macromolecules*, **2007**, *40*, 771–776.

Metrangolo P; Resnati G; Pilati T; Liantonio R; Meyer F, Engineering functional materials by halogen bonding, *J. Polym. Sci. Pol. Chem. Part A*, **2007**, *45*, 1–15.

Olson DA; Chen L; Hillmyer MA, Templating nanoporous polymers with ordered block copolymers, *Chem. Mater.*, **2008**, *20*, 869–890.

Onclin S; Ravoo BJ; Reinhoudt DN, Engineering silicon oxide surfaces using self-assembled monolayers, *Angew. Chem. Int. Ed.*, **2005**, *44*, 6282–6304.

Pagliaro M; Ciriminna R; Man MWC; Campestrini S, Better chemistry through ceramics: The physical bases of the outstanding chemistry of ORMOSIL, *J. Phys. Chem. B*, **2006**, *110*, 1976–1988.

Paulusse JMJ; Sijbesma RP, Molecule-based rheology switching, *Angew. Chem. Int. Ed.*, **2006**, *45*, 2334–2337.

Pirrung MC, How to make a DNA chip, *Angew. Chem. Int. Ed.*, **2002**, *41*, 1276–1289.

Pron A; Rannou P, Processible conjugated polymers: From organic semiconductors to organic metals and superconductors, *Prog. Polym. Sci.*, **2002**, *27*, 135–190.

Rogach AL; Eychmüller A; Hickey SG; Kershaw SV, Infrared-emitting colloidal nanocrystals: Synthesis, assembly, spectroscopy, and applications, *Small*, **2007**, *3*, 536–557.

Sarikaya M; Tamerler C; Jen AKY; Schulten K; Baneyx F, Molecular biomimetics: Nanotechnology through biology, *Nat. Mater.*, **2003**, *2*, 577–585.

Shan J; Tenhu H, Recent advances in polymer protected gold nanoparticles: Synthesis, properties and applications, *Chem. Commun.*, **2007**, 4580–4598.

Shchukin DG; Möhwald H, Self-repairing coatings containing active nanoreservoirs, *Small*, **2007**, *3*, 926–943.

Shimizu T; Masuda M; Minamikawa H, Supramolecular nanotube architectures based on amphiphilic molecules, *Chem. Rev.*, **2005**, *105*, 1401–1443.

Sullivan TP; Huck WTS, Reactions on monolayers: Organic synthesis in two dimensions, *Eur. J. Org. Chem.*, **2003**, 17–29.

Tamerler C; Sarikaya M, Molecular biomimetics: Utilizing nature's molecular ways in practical engineering, *Acta Biomater.*, **2007**, *3*, 289–299.

Tao F; Bernasek SL, Understanding odd-even effects in organic self-assembled monolayers, *Chem. Rev.*, **2007**, *107*, 1408–1453.

Umezawa Y; Aoki H, Ion channel sensors based on artificial receptors, *Anal. Chem.*, **2004**, *76*, 320A–326A.

Van Dijk MA; Lippitz M; Orrit M, Far-field optical microscopy of single metal manoparticies, *Acc. Chem. Res.*, **2005**, *38*, 594–601.

Verma A; Rotello VM, Surface recognition of biomacromolecules using nanoparticle receptors, *Chem. Commun.*, **2005**, 303–312.

Walcarius A; Mandler D; Cox JA; Collinson M; Lev O, Exciting new directions in the intersection of functionalized sol-gel materials with electrochemistry, *J. Mater. Chem.*, **2005**, *15*, 3663–3689.

Ye L; Mosbach K, Molecular imprinting: Synthetic materials as substitutes for biological antibodies and receptors, *Chem. Mater.*, **2008**, *20*, 859–868.

Index

Note: Abbreviations are not listed separately and directly in the index. Please, refer to the list of abbreviations for the full expressions first.

The Supramolecular Chemistry of Organic–Inorganic Hybrid Materials. Edited by Knut Rurack and Ramón Martínez-Máñez

Printed and bound by CPI Group (UK) Ltd, Croydon, CR0 4YY
16/09/2026
14963410-0002